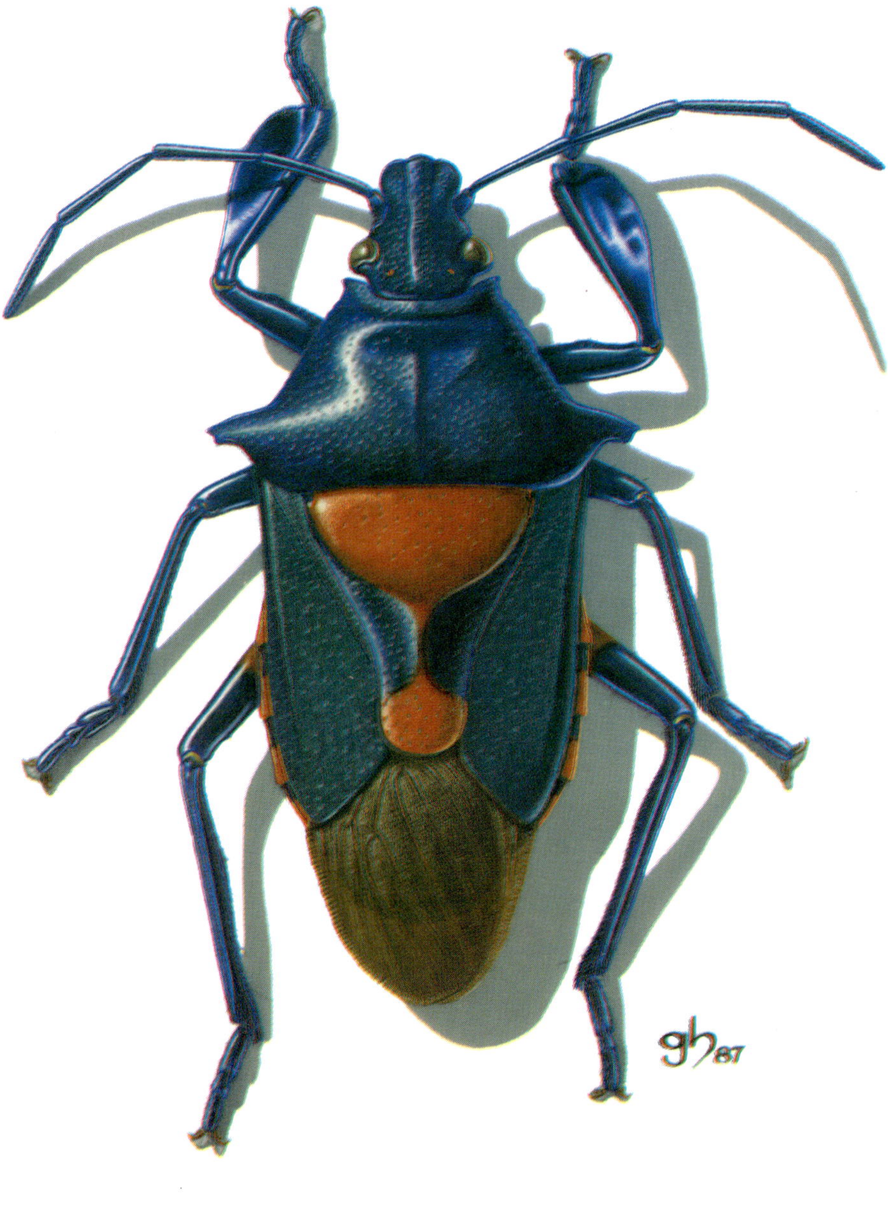

The male of *Euthyrhynchus floridanus* (L.) (Pentatomidae: Asopinae)

Catalog of the Heteroptera, or True Bugs,

of Canada and the Continental United States

Catalog of the Heteroptera, or True Bugs,

of Canada and the Continental United States

Edited By
THOMAS J. HENRY
Systematic Entomology Laboratory
Biosystematics and Beneficial Insects Institute
Agricultural Research Service
United States Department of Agriculture
C/O National Museum of Natural History
Washington, DC 20560

and

RICHARD C. FROESCHNER
Department of Entomology
National Museum of Natural History
Smithsonian Institution
Washington, DC 20560

E.J.BRILL
LEIDEN NEW YORK KØBENHAVN KÖLN

Acquisitions Editor: Ross H. Arnett, Jr.
Production Editor and Jacket Design: A. Lance Coalson

ISBN 0-916846-44-X; 90-04-08482-7

987654321

Library of Congress Cataloging-in-Publication Data

Catalog of the Heteroptera, or true bugs, of Canada and the continental United States / edited by Thomas J. Henry and Richard C. Froeschner.

P. cm.

Bibliography: p.

Includes index.

ISBN 0-916846-44-X (U.S.) : $58.50

1. Hemiptera--Canada. 2. Hemiptera--United States. I. Henry, Thomas J. II. Froeschner, Richard C.

QL522.1.C2C37 1988 87-38212

595.7'54'097--dc19 CIP

Typesetting by ***FAST TYPE, Inc.***

Manufactured in the United States of America

CONTENTS

INTRODUCTION

In science all modern works are the result of contributions by many persons of yesterday as well as of today. The present catalog is no exception. It has four predecessors that covered the same group of insects in the same general geographic area north of Mexico: Uhler's (1886) Check-list of the Hemiptera Heteroptera of North America (included 589 species north of Mexico); Banks' (1910) Catalogue of the Nearctic Hemiptera (1,268 species north of Mexico); Van Duzee's (1916) Check List of the Hemiptera of America North of Mexico (1,469 species); and Van Duzee's (1917) Catalogue of the Hemiptera of America North of Mexico (1,625 species). For comparison with those totals (Table l), the 3,834 species listed in this volume represent an increase of 136 percent over the 1917 enumeration.

Past colleagues who contributed to the preliminary plans and beginnings of this catalog were Jon L. Herring (1922-1985), Roland F. Hussey (1896-1967), and Robert L. Usinger (1912-1968), specialists in the Heteroptera who realized the need for such a venture. With sincere appreciation we dedicate this volume to these scholars. Unfortunately, events prevented their developing it far enough to establish a final form; the present form resulted from the experience and decisions of the coeditors.

The Early Workers

The study of the insect order Heteroptera in North America can be divided into several phases. The earliest, starting approximately in the mid-1700's, was one of itinerant American colonists and European travelers whose collections went to European scholars such as Linnaeus, Fabricius, and DeGeer. Among these collectors were the early American botanist John Bartram from Pennsylvania and the Swedish botanist Peter Kalm, after whom Stål named our common, widespread *Lygaeus kalmii*.

American entomology began with Thomas Say whose concise and precise descriptions of many forms of American organisms, including Heteroptera, were only a small part of his contributions to American natural history. The study of Heteroptera as a specialty was not taken up until 1862 when Philip R. Uhler, librarian and later Provost of the Peabody Institute in Baltimore, Maryland, began his forty years of contributions to heteropterology and homopterology, including numerous studies on insects collected by the United States government's early exploration of the western United States. Contemporary with Uhler were the Abbe Leon Provancher, a Canadian clergyman who prepared a synopsis of Canadian Heteroptera, and Otto Heidemann, a German-American who began his study of American Heteroptera after he was fifty years old and adapted his early skills as an engraver to beautifully illustrate many of the species he described. Included in the period from the later 1800's were the eminent American heteropterists Edwin P. Van Duzee, Jose de la Torre-Bueno, Harry Barber, and Nathan Banks. These men built the foundation of American heteropterology by adding to and adapting the work of such European greats as Carl Stål, William L. Distant, Odo M. Reuter, and Evald Bergroth. This period was summarized in Van Duzee's (1917) Catalog.

The post-Van Duzee Catalog period continued and increased in tempo when Van Duzee, Barber, and Torre Bueno were joined by numerous enthusiastic and productive newcomers like Carl J. Drake, Herbert B. Hungerford, Roland F. Hussey, Harry H. Knight, Waldo M. McAtee, and Willis S. Blatchley; later came numerous others.

COMPARISON TO EARLIER WORK

	VAN DUZEE		PRESENT LIST	
	Genera	Species	Genera	Species
Acanthosomatidae	2	4	2	4
Alydidae	11	19	11	30
Anthocoridae	13	35	23	89
Aradidae	9	61	10	123
Belostomatidae	3	20	3	21
Berytidae	6	8	7	12
Ceratocombidae	1	3	1	4
Cimicidae	3	4	8	15
Coreidae	30	82	33	88
Corixidae	6	62	17	125
Cydnidae	9	29	13	43
Dipsocoridae	0	0	1	2
Enicocephalidae	2	2	5	10
Gelastocoridae	2	5	2	7
Gerridae	7	20	8	46
Hebridae	2	6	3	15
Hydrometridae	1	2	1	9
Largidae	3	12	7	21
Leptopodidae	0	0	1	1
Lygaeidae	56	187	81	318
Macroveliidae	1	1	2	2
Mesoveliidae	1	1	1	3
Microphysidae	0	0	4	4
Miridae	135	496	223	1930
Nabidae	5	21	10	34
Naucoridae	2	13	5	21
Nepidae	3	8	3	13
Notonectidae	2	17	3	32
Ochteridae	1	3	1	6
Pentatomidae	52	161	60	222
Phymatidae	2	12	3	27
Piesmatidae	1	1	1	7
Pleidae	1	1	2	5
Polyctenidae	0	0	1	2
Pyrrhocoridae	2	10	1	9
Reduviidae	43	103	46	157
Rhopalidae	6	25	10	39
Saldidae	8	32	12	69
Schizopteridae	1	1	4	4
Scutelleridae	14	26	15	34
Tessaratomidae	0	0	1	1
Thaumastocoridae	0	0	1	1
Thyreocoridae	1	17	4	41
Tingidae	22	99	22	154
Veliidae	3	16	5	34
TOTALS	472	1625	677	3834

Table 1. Comparison of the numbers of Canadian and U.S. Heteroptera included in the Van Duzee (1917) catalog and the present list.

Some of the Comprehensive Works

Many authors have contributed to the vast base of knowledge currently available on Heteroptera, but only a few have drawn together and presented information across several families. The following brief outline, presented alphabetically by author or editor, gives some of the more comprehensive taxonomic works on Heteroptera that fit into this category and, because of our catalog format, may not always be repeatedly cited in the respective family introductions. Users are also directed to the separate family introductions and lower categories within the catalog body (subfamilies, tribes, and genera) for citations of numerous other important, but more restricted, literature, especially some of the excellent world catalogs (e.g. Carvalho, 1957-1960; Drake and Ruhoff, 1965; and Slater, 1964) and monographic treatments of particular families (e.g. Harris, 1928; Kelton, 1980; Knight, 1941, 1968; and Hungerford, 1948).

Blatchley's (1926) "Heteroptera of Eastern North America" is probably the single most useful reference to the bugs of the U.S. and Canada. Although more than 60 years have rendered certain parts out of date, its broad coverage with keys to families, genera, and species, accompanied by numerous valuable biological notes and illustrations, makes this volume a necessary point of reference for the North American fauna. Even though Blatchley covers only the eastern part of our region, his keys, notes, and introductory remarks provide a much broader perspective.

"The Hemiptera or Sucking Insects of Connecticut," edited by W. E. Britton (1923), is another regional work valuable to the heteropterist. Specialists of the respective families included such notables as H. G. Barber, H. H. Knight, H. M. Parshley, H. Osborn, J. R. de la Torre-Bueno, and E. P. Van Duzee. Although all family treatments in this work are useful, most important by far is Knight's monograph of the northeastern Miridae. It represents the first comprehensive treatment of that family for a significant part of North America and contains keys to all taxonomic categories of that region and many new species descriptions.

Brooks and Kelton's (1967) "Aquatic and Semiaquatic Heteroptera of Alberta, Saskatchewan, and Manitoba...." treats 12 of the aquatic families. Keys to genera and 95 species, accompanied by good illustrations and notes on many species, make this an important reference for the group in northcentral North America.

Froeschner's (1941-1962) "Hemiptera of Missouri" series is the only comprehensive treatment for the bugs of the midwestern United States. Its keys, numerous biological notes, host records, and many original illustrations still make it an important reference.

Hungerford's (1920) "Biology and Ecology of Aquatic and Semiaquatic Hemiptera" is the only treatment of its kind for the aquatic bugs in North America. The serious student of aquatic bugs should refer to its habitat keys, keys to genera and species, numerous illustrations, and hundreds of valuable notes on biology and ecology.

McPherson's (1982) "Pentatomoidea of Northeastern North America" is primarily a compilation of the literature for five pentatomoid families. Although limited in geographic coverage, this work's keys to families, genera, and species, and the almost exhaustive literature review for the Northeast, make it important for the study of these groups.

"The Semiaquatic and Aquatic Hemiptera of California," edited by A. S. Menke (1979), is a well-done update of R. L. Usinger's (1956) "Aquatic Hemiptera." Numerous illustrations and widely applicable keys to genera by well-chosen specialists make this recent treatise a good reference for the aquatic bugs.

Slater and Baranowski's (1978) "How to Know the True Bugs" is the latest attempt to cover all North American families of Heteroptera. Numerous illustrations and well-prepared keys to the families of nymphs and adults and the more common genera in each family (all genera for the Miridae) make it easy for specialists and nonspecialists to identify with confidence the more common bugs in our region.

Although now many years out of date, Torre-Bueno's (1939-1946) never completed "Synopsis of the Hemiptera of North America" is an attempt to provide keys to genera and species for a large number of the families found in North America. Some of the keys are based on published descriptions, so some weaknesses in them should be expected, but even so, this series is a useful starting point for many of the terrestrial families.

Users of this catalog may also be interested in "The Heteropterists' Newsletter," currently edited by Carl W. Schaefer. This informal publication, available upon request[1], appears irregularly and contains numerous notes and articles and a list of over 400 workers around the world interested in Heteroptera.

Hemiptera versus Heteroptera

The taxonomic level and name appropriate for the group of insects treated in this list still are not widely agreed upon. The confusion arises not from priority or from original definitions, but from the lack of agreement among modern workers. The oldest name accompanying this group was "Hemiptera" given by Linnaeus (1758), who based his classification of insect orders on wing development or modifications, Hemiptera being those characteristically with the wing leathery on the basal half and membranous on the apical half. Subsequently, Fabricius (1775), one of Linnaeus' students, proposed a new classification of insect orders based on mouthparts; he called this order "Ryngota," which he later (1803) emended to "Rhyngota." This latter name was further emended to "Rhynchota" by Burmeister (1835). Other names proposed for Linnaeus' group "Hemiptera" are universally placed in synonymy and do not figure in the current indecision.

The remainder of the modern problem involves terms proposed by Latreille (1810) when he recognized that "Hemiptera" contained two discrete "sections," which he called "Heteroptera" and "Homoptera."

The current disagreement involves at least three questions: First, is there a single order with two suborders (Latreille's "sections"), or two separate orders? The Linnaean term "Hemiptera" is used in the inclusive sense versus an equally insistent application of it to the more restricted group resulting from elevation of Latreille's "section" Heteroptera to full order. This is in contrast to another full order, the "Homoptera." A third point of confusion involves some authors' insistence to extend the rules of the *International Code of Zoological Nomenclature* (which do not deal with taxa above family-group names) directing that when a category is divided into two or more subgroups, one of these must take the same name as the group and thus become the "nominate" subgroup. In this usage, unless a modifier or suffix is added, the term "Hemiptera" would automatically have a double—and hence confusing—meaning.

Any decision offered as a solution to the dilemma surrounding the term "Hemiptera" will likely find objectors. However, general practice among students of the group is slowly forging a partial solution. Specialization by modern individual workers seldom crosses the line between "Homoptera" and the second group, regardless of what name is used for it. This practical isolation of the two sister units gives workers a sense of two widely separated groups, even though they share a number of derived (synapomorphic) characters. To avoid the confusion of inclusiveness often attributed to the term "Hemiptera," these workers have widely substituted the term "Heteroptera" (or Hemiptera-Heteroptera) and have seldom

[1] Copies now may be obtained by writing to Dr. Schaefer at: Dept. Ecol. & Evol. Biol., U-43, Univ. Connecticut, Storrs, CT 06268 [USA]

committed themselves to taking a position on the question of its taxonomic level. We note that homopterists do not qualify their names by using Hemiptera-Homoptera.

For the purpose of the present list the use of "Heteroptera" as an order is adopted. But the reader must be aware that this is an oversimplified solution. Some heteropterists and homopterists suggest that as many as four equal groups should be recognized (i.e. Heteroptera, auchenorrhynchus Homoptera, sternorrhynchus Homoptera, and Peloridiidae). With that suggestion are proposed new names and changes of some of the old ones.

Despite the uncomfortable feeling of uncertainty about the proper name to use and the group's level in the taxonomic hierarchy, one must recognize that effective communication is a matter of conventional acceptance and that a worker who developed in an area where "Hemiptera" implied an order with two suborders will tend to be at odds with a person who learned each of the groups as full orders. Our knowledge of insect groups is still incomplete so we may well expect series of changes before the final answer is reached. Fiat does not change biological truths--if it did we would still be where we were during the Dark Ages. Accepting that philosophical attitude, to communicate one needs only to use a current, widely used word with a precise definition. For the clearly recognizable, monophyletic group comprising the subject of this catalog, that term is "Heteroptera."

Classification Within the Heteroptera

Classification within the Heteroptera is also in a state of flux with conflicting categories, varying contents of categories, and some differences in terminology. Current research indicates more changes to come. See Schuh (1986, An. Rev. Ent., 31: 67-93) for a review and comments on the influence of cladistic methods in heteropteran classification.

The catalog follows a modified arrangement of the families of Heteroptera given by Štys and Kerzhner (1975, Acta Ent. Bohem., 72: 65-79) and Slater (1982, Synop. Classif. Hem., pp. 417-447) as outlined below. Because the former authors considered Heteroptera a suborder and the present catalog treats Heteroptera as an order distinct from Homoptera, we substitute the designation suborder for their infraorder, following Slater who used the order name Hemiptera. In addition to these changes, Rolston and McDonald's (1979, J. N.Y. Ent. Soc., 87: 189) recognition of the family Cyrtocoridae is followed, Andersen's (1982, Entomonograph, 3: 388-389) modifications are incorporated into the Gerromorpha, Štys' (1983, Acta Ent. Bohem., 80: 275) family Stemmocryptidae is added to the Dipsocoromorpha, and the phymatids are conservatively retained as a family. Families in brackets do not occur in our area; exclusively fossil families are not included.

Although the above list reflects the current classification of the Heteroptera, we have arranged the catalog alphabetically by family, not suborder and superfamily. Subgroups below family also are arranged alphabetically by subfamily, tribe, genus, species, and subspecies. This nonphylogenetic ordering, we feel, will best aid most users of this catalog, especially nonspecialists, in quick and easy reference to the respective groups.

Using the Catalog

A primary aim of this catalog is to offer an accounting for each species as originally proposed and for the first usage only of all its name combinations (including valid names, synonymies, and misspellings) that have been published for our area. We follow the policy outlined by the 1985 *Code of Zoological Nomenclature* for nomenclatorial purposes that a dissertation for an advanced educational degree is not published unless it satifies the criteria presented in Articles 8 and 9.

The deadline for literature coverage in this volume has been extended numerous times. Even minor revisions required months of additional preparation. Each extension

WORLD CLASSIFICATION:

Suborder Enicocephalomorpha
Superfamily Enicocephaloidea: Enicocephalidae

Suborder Dipsocoromorpha
Superfamily Dipsocoroidea: Ceratocombidae, Dipsocoridae, [Hypsipterygidae], Schizopteridae, and [Stemmocryptidae]

Suborder Gerromorpha (=Amphibiocorisae)
Superfamily Gerroidea: Gerridae, [Hermatobatidae], and Veliidae
Superfamily Hebroidea: Hebridae
Superfamily Hydrometroidea: Hydrometridae, Macroveliidae, and [Paraphrynoveliidae]
Superfamily Mesovelioidea: Mesoveliidae (including Madeoveliidae)

Suborder Leptopodomorpha
Superfamily Leptopodoidea: [Leotichidae], Leptopodidae, [Omaniidae], and Saldidae

Suborder Nepomorpha (=Cryptocerata =Hydrocorisae)
Superfamily Corixoidea: Corixidae
Superfamily Naucoroidea: Naucoridae (including Aphelocheiridae)
Superfamily Nepoidea: Belostomatidae and Nepidae
Superfamily Notonectoidea: [Helotrephidae], Notonectidae, and Pleidae
Superfamily: Gelastocoroidea (= Ochteroidea): Gelastocoridae and Ochteridae

Suborder Cimicomorpha
Superfamily Thaumastocoroidea: Thaumastocoridae
Superfamily Joppeicoidea: [Joppeicidae]
Superfamily Tingoidea: Tingidae and [Vianaididae]
Superfamily Miroidea: Microphysidae and Miridae
Superfamily Cimicoidea: Anthocoridae, Cimicidae, [Medocostidae], Nabidae, [Plokiophilidae], Polyctenidae, and [Velocipodidae]
Superfamily Reduvioidea: [Pachynomidae], Phymatidae, and Reduviidae

Suborder Pentatomomorpha
Superfamily Aradoidea: Aradidae and [Termitaphididae]
Superfamily Idiostoloidea: [Idiostolidae]
Superfamily Piesmatoidea: Piesmatidae
Superfamily Lygaeoidea: Berytidae, [Colobathristidae], Lygaeidae, and [Malcidae]
Superfamily Pyrrhocoroidea: Largidae and Pyrrhocoridae
Superfamily Coreoidea: Alydidae, Coreidae, [Hyocephalidae], Rhopalidae, and [Stenocephalidae]
Superfamily Pentatomoidea: Acanthosomatidae, [Aphylidae], [Canopidae], Cydnidae, [Cyrtocoridae], [Dinidoridae], [Eumenotidae], [Lestoniidae], [Megarididae], Pentatomidae, [Phloeidae], [Plataspidae], Scutelleridae, Tessaratomidae, [Thaumastellidae], Thyreocoridae, and [Urostylidae]

necessarily delayed completion and each delay allowed the coverage to become further outdated. Finally, for the sake of everyone involved, especially our contributing authors, who have patiently revised their chapters over the years, we set a cut-off date at June 30, 1986. Even with this determined "final" date, important works continued to appear. Alas! Users should be aware that authors have been given the option to incorporate major new taxonomic literature through December 31, 1986. We the editors accept responsibility for any confusion resulting from this decision (or indecision).

In ways not always clearly explainable, insects sometimes are reported for territories in which they are not truly known to occur. For the area under consideration here a number of such species have been reported. All such species deemed never to have been part of our established fauna will be given a direct comment of exclusion and a probable reason for that action. Such exclusions will be made under the appropriate family or generic heading if that taxon belongs to our fauna. Such names will be entered in the index of this volume so that they can be traced to their treatment.

Two families that have been recorded erroneously as members of our fauna are omitted here and not mentioned again in the catalog body. Rathvon (1869, Hist. Lancanster Co., Pa., p. 548) reported *Canopus globus* (Fabricius), family Canopidae Amyot and Serville, 1843, from Pennsylvania. Because this species has never again been reported for North America it and its family are excluded from the faunal list. Banks (1910, Cat. Nearc. Hem.-Het., p. 93) recorded *Cyrtocoris trigonus* (Germar), family Cyrtocoridae Distant, 1880, from California. Although this record was accepted by Horvath (1916, An. Mus. Nat. Hung., 4: 221), it has been omitted from our lists since that time and no additional reports have appeared.

Among those things excluded from this catalog is the presentation of taxonomic synonymies not mentioned in literature pertaining to our area or every published record reporting each species from our region. As a compromise, "Notes" are used to direct interested readers to the more extensive or detailed literature. The extent of this note option varies and was left to the discretion of contributing authors.

We also realize, more than anyone, the immensity of even the present task and the probability for error or omission. For these and other reasons this manuscript has been stored on word-processing diskettes with the hope that it can be updated upon the receipt of corrections and new literature so that in a "few" years a new, more perfect, catalog can be offered.

Each of the 45 families begins with an intentionally brief introduction that includes a short description of the group and their general habits, miscellaneous notes, and the major literature pertaining to it.

As a supplement to the family introductions, we have included habitus drawings that are representative of typical genera found in our area. We feel this aspect of the catalog will not only be more aesthetically appealing, but also taxonomically useful by illustrating some of the morphological diversity found in the various families.

The first proposal of a new and available name is indicated by presenting it without punctation between the taxon and author's name; citations for subsequent uses at different taxonomic levels, different combinations, misidentifications, and subsequent misspellings will be marked by a colon placed between the taxon and author's name. Misspellings are marked further with "[*sic*]."

Generic headings are centered. Information under the generic name includes the earliest proposal of each appropriate generic name followed by its type-species and the method of its designation, e.g., Monotypic or Original designation. If the first usage was preoccupied this is indicated after that entry. Names proposed as replacements for preoccupied names are entered and appropriately cited. When the hierarchical position of a generic name has changed, as from subgenus to genus, that action is documented by a citation.

Subgeneric names are arranged alphabetically as centered headings under their inclusive generic heading; the nominate subgenus, when in our fauna, is appropriately entered as such, but when not present its absence is commented upon in a "Note" under the "Genus" heading.

Under the generic category all included species are listed alphabetically; in a genus divided into subgenera the species are listed alphabetically under their appropriate subgeneric heading.

Species and subspecies are presented as left-margin headings followed by the author's name and year of proposal. Under each such category information is presented at three levels (that under the first two will be arranged chronologically): The original proposal and combination for each new name; the first of each subsequent usage of different name combinations; and a summary of published distribution information, with that for our area given by states and/or provinces. Any modification of the above pattern will be explained in a "Note" following the appropriate taxon.

For each original proposal of a species-group name from our area, a bracketed notation of the state or province of origin of the type specimen will follow the citation; for species described from localities outside our area, the type locality will be summarized by the name of the country or region of origin. If the latter case contains further distribution data establishing that form in our fauna, such information also will be included in the brackets (e.g., [holotype Mexico; also paratypes from U.S.]). If no holotype was designated and more than one locality is listed in the original description, all localities for the syntype series are given. Subsequent type-locality designations (i.e. localities for lectotypes and neotypes) are listed after the locality given in the original description.

Species not known to have subspecific divisions have all appropriate citations below binomen heading arranged chronologically. Species with subspecies have the appropriate subspecies presented as alphabetically arranged trinomial headings following the species binomen. Literature citations for each subspecies are arranged chronologically under that subspecies. A "Note" is provided when the nominate subspecies does not occur in our area.

Species-group taxa originally proposed before 1961 as "varieties" or "forms" are (according to the 1985 *International Code of Zoological Nomenclature*) to be interpreted as proposed in the sense of a subspecies. This interpretation for trinomials has been followed here unless a note explains otherwise; but in citing literature the original author's designation as a "variety" or "form" is copied.

The original spelling of species group names has been retained unless there is a compelling reason to change them. Where modifications have been made, an explanatory note is provided. The -i versus -ii ending for patronymic names has been particularly problematic. If certain names have been latinized (e.g. Fabricius), the proper termination is -ii; if the name has never been latinized, then the proper termination is -i. It was not possible to research the latinization or lack thereof for all of the names upon which -ii endings were founded; therefore, most of them remain despite objections by some contributors who were allowed to follow their convictions on this matter.

Abbreviations

Literature citations in the text are markedly abbreviated to save space but with attention to adequacy for the reader with experience to recognize the source. The reader not familiar with that literature need only consult the author-year entries in the terminal bibliography to find full titles of the books and serial publications.

The one exception to the abbreviated citations in the text occurs in Ashlock and Slater's Lygaeidae. The precatalog agreement to allow their inclusion of an all-inclusive bibliography to the North American Lygaeidae required us to add low-case letters to certain dates in the literature cited to identify publications given in their "Ref." section. Users of this catalog should note that these specially marked references may also be cited in other family treatments, in which case, the low-case letters are to be ignored.

Under distribution, all countries outside of Canada and the United States are spelled out. North America is abbreviated as N. Am.; the United States as U.S. Canadian and U.S. distributions in the catalog body are integrated alphabetically by the abbreviation as indicated in Table 2.

ABBREVIATIONS

CANADIAN PROVINCES

Alta. = Alberta	Ont. = Ontario
B.C. = British Columbia	P.Ed. = Prince Edward Island
Man. = Manitoba	Que. = Quebec
N.B. = New Brunswick	Sask. = Saskatchewan
N.T. = Northwest Territories	Yuk. = Yukon (formerly Yukon Territory)
Nfld. = Newfoundland (including Labrador)	

UNITED STATES

Ala. = Alabama	Mo. = Missouri
Alk. = Alaska	Mont. = Montana
Ariz. = Arizona	N.C. = North Carolina
Ark. = Arkansas	N.D. = North Dakota
Cal. = California	N.H. = New Hampshire
Col. = Colorado	N.J. = New Jersey
Conn. = Connecticut	N.M. = New Mexico
D.C. = District of Columbia (Washington, D.C.)	N.Y. = New York
	Neb. = Nebraska
Del. = Delaware	Nev. = Nevada
Fla. = Florida	Oh. = Ohio
Ga. = Georgia	Ok. = Oklahoma
Haw. = Hawaii	Ore. = Oregon
Ia. = Iowa	Pa. = Pennsylvania
Id. = Idaho	R.I. = Rhode Island
Ill. = Illinois	S.C. = South Carolina
Ind. = Indiana	S.D. = South Dakota
Ks. = Kansas	Tenn. = Tennessee
Ky. = Kentucky	Tex. = Texas
La. = Louisiana	Ut. = Utah
Mass. = Massachusetts	Va. = Viriginia
Md. = Maryland	Vt. = Vermont
Me. = Maine	W.Va. = West Virginia
Mich. = Michigan	Wash. = Washington
Minn. = Minnesota	Wis. = Wisconsin

Table 2. Abbreviations for Canadian provinces and states in the United States.

Contributing Authors

PETER D. ASHLOCK
Department of Entomology
University of Kansas
Lawrence, Kansas 66045

RICHARD C. FROESCHNER
Department of Entomology
National Museum of Natural History
Smithsonian Institution, NHB-127
Washington, D.C. 20560

THOMAS J. HENRY
Systematic Entomology Laboratory
Biosystematics and Biological Insects Institute
Agricultural Research Service, USDA, NHB-168
c/o National Museum of Natural History
Washington, D.C. 20560

JOHN D. LATTIN
Department of Entomology
Oregon State University
Corvallis, Oregon 97331

DAN A. POLHEMUS
3115 So. York
Englewood, Colorado 80110

JOHN T. POLHEMUS
3115 So. York
Englewood, Colorado 80110

ALEX SLATER
Department of Entomology
University of Kansas
Lawrence, Kansas 66045

CECIL L. SMITH
Department of Entomology
University of Georgia
Athens, Georgia 30602

ALFRED G. WHEELER, JR.
Bureau of Plant Industry
Pennsylvania Department of Agriculture
Harrisburg, Pennsylvania 17110

Acknowledgments

We are indebted to many people who have contributed to the completion of this catalog. We especially thank the contributing authors. We are sure there are many times when they became frustrated and wondered if they would ever see their family sections in print.

Numerous colleagues have reviewed parts of the manuscript. For their many useful suggestions we thank the following: L. A. Kelton (Miridae); I. M. Kerzhner (Nabidae); J. Maldonado (Reduviidae), A. S. Menke (Belostomatidae);J. T. & D. A. Polhemus (Gerridae, Hydrometridae, Mesoveliidae, and Veliidae); L. H. Rolston (Pentatomidae); C. W. Schaefer (Coreidae, Largidae, Pyrrhocoridae, and Rhopalidae); J. C. Schaffner (Miridae); R. T. Schuh (Miridae and Leptopodomorpha); M. D. Schwartz (Miridae); J. A. Slater (Lygaeidae; see also Miridae acknowledgments); Alex Slater (aquatic and semiaquatic families); G. M. Stonedahl (Miridae); P. Štys (Acanthosomatidae, Alydidae, Anthocoridae, Aradidae, and Belostomatidae, and review of introduction and format); F. C. Thompson (review of introduction and general format); and A. G. Wheeler, Jr. (Anthocoridae, Berytidae, Ceratocombidae, Dipsocoridae, Largidae, Microphysidae, Nabidae, Pyrrhocoridae, Rhopalidae, and review of general and family introductions).

Curtis Sabrosky and George Steyskal kindly shared their views on nomenclatural problems.

The typing of the manuscript for this publication involved months of tedious work. We gratefully acknowledge Vera Lee, Helen Proctor (both of Systematic Entomology Laboratory, BBII, Agricultural Research Service, USDA, at Washington, DC and Beltsville, Md., respectively), and Silver West (Department of Entomology, Smithsonian Institution) for their laborious efforts.

F. C. Thompson spent considerable time programing to correct several typing errors and to generate the comprehensive index. His efforts saved us hundreds of hours of manual labor, for which we are very grateful.

We are appreciative of the following people and organizations for permission to use their illustrations appearing in this catalog.--Original artwork: Frontispiece--Gustava Hormiga; figs. 13-16--Caroline Herbert; figs. 116-121 & 147--Elizabeth Myers; figs. 35, 123, 126, & 145--Young Sohn. Figures after Brooks and Kelton (1967) and Kelton (1980)--L. A. Kelton figs.; after Froeschner (1941-1962)-- Elsie Herbold Froeschner; fig. 102 after Truxal (1979)--F. S. Truxal and the University of California Press, Berkeley; and figs. after Usinger (1956)--University of California Press, Berkeley, California.

T. J. Henry

R. C. Froeschner

Family Acanthosomatidae Signoret, 1863

The Acanthosomatids

By Richard C. Froeschner

The majority of the New World genera within this family are Neotropical in distribution, but the two occurring in North America are shared with the Eurasian land mass and hence are Holarctic. Members of these two genera appear most abundant in Canada and the northern half of the United States. Scattered published observations indicate that these insects show decided preference for shrubs and trees rather than forbs. Hibernation is endured only by adults. At least one species in North America appears to undergo two generations per year. Literature reports from Europe and from North America credit members of this family with exhibiting considerable maternal instinct expressed by the adult female remaining astride the egg cluster and the newly emerged nymphs until the latter wander away in their quest for food. None of the North American species are considered to have any economic importance. Jones and McPherson (1980, J. Ga. Ent. Soc., 15: 286-289) presented interesting observations on the biologies and habits of three North American species.

This group has been variously treated as a full family, a subfamily, or a tribe of a more comprehensive version of the Pentatomidae; the current practice is to afford them family status. The most significant taxonomic treatments of this group in North America were by Van Duzee (1904, Trans. Am. Ent. Soc., 30: 73-75) and Torre-Bueno (1939, Ent. Am., 19: 244-245). A worldwide revision and classification, with keys to genera, was offered by Kumar (1974, Australian J. Zool., Suppl. Ser., 34: 1-60), and is adopted here. A key to the genera of the Western Hemisphere was given by Rolston and Kumar (1974, J. N.Y. Ent. Soc., 82: 271-273).

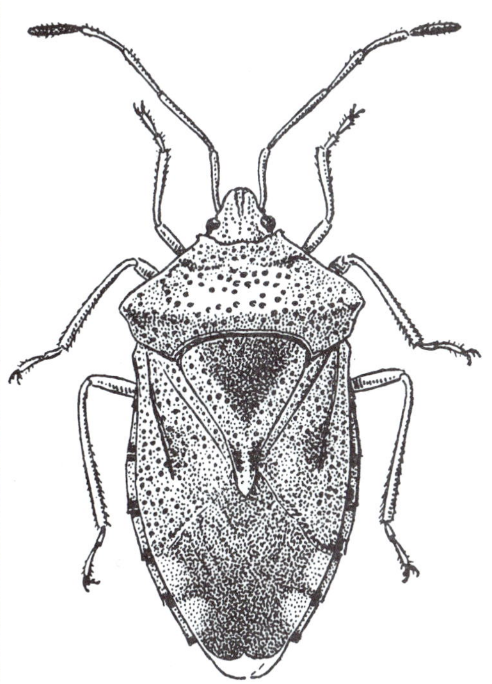

Fig. 1 *Elasmucha lateralis* [p. 2] (After Drake, 1922).

Subfamily Acanthosomatinae Signoret, 1863

Genus *Elasmosthethus* Fieber, 1861

1861 *Elasmostethus* Fieber, Europ. Hem., 78: 328. Type-species: *Cimex dentatus* De Geer, 1773, a junior synonym of *Cimex interstinctus* Linnaeus, 1758. Designated by Stål, 1864, An. Soc. Ent. France, ser. 4, 4: 54.

Elasmostethus atricornis (Van Duzee), 1904

1904 *Acanthosoma atricornis* Van Duzee, Trans. Am. Ent. Soc., 30: 75. [Ind., N.Y., Que.].

1907 *Elasmostethus atricornis*: Bergroth, Ent. News, 18: 48.

Distribution: Ill., Ind., Md., Mich., Mont., N.Y., Oh., Que., S.C.

Elasmostethus cruciatus (Say), 1831

1831 *Edessa cruciata* Say, Descrip. Het. Hem. N. Am., p. 2. ["U.S."].

1837 *Acanthosoma borealis* Westwood, Hope Cat., 1: 30. ["America Boreali"]. Synonymized by Distant, 1900, Proc. Zool. Soc. London, 54: 818.

1859 *Acanthosoma boreale*: Dohrn, Cat. Hem., p. 19.

1861 *Acanthosoma cruciata*: Uhler, Proc. Ent. Soc. Phila., 1: 23.

1907 *Elasmostethus cruciatus*: Bergroth, Ent. News, 18: 49.

Distribution: Alta., B.C., Cal., Col., Conn., Ga., Ill., Mass., Me., Mich., N.C., N.H., N.J., N.M., N.S., N.Y., Nev., Nfld., Ont., Ore., Que., S.C., Tex., Ut., Wash., Va., Vt.

Elasmostethus interstinctus (Linnaeus), 1758

1758 *Cimex interstinctus* Linnaeus, Syst. Nat., edit. 10, p. 445. [Europe].

1904 *Acanthosoma cruciata* var. *cooleyi* Van Duzee, Trans. Am. Ent. Soc., 30: 75. [Mont.]. Synonymized by Barber, 1932, Proc. Ent. Soc. Wash., 34: 65.

1907 *Elasmostethus cooleyi*: Bergroth, Ent. News, 18: 49.

1932 *Elasmostethus interstinctus*: Barber, Proc. Ent. Soc. Wash., 34: 65.

Distribution: Alk., Mont., "northwestern Canada" (Asia, Europe).

Note: Torre-Bueno (1939, Ent. Am., 19: 245) listed *cooleyi* as a "var." of *cruciata*, apparently following the original proposal and overlooking Barber's synonymizing of it.

Genus *Elasmucha* Stål, 1864

1864 *Elasmucha* Stål, An. Soc. Ent. France, ser. 4, 4: 54. Type-species: *Cimex ferrugator* Fabricius, 1787. Designated by Kirkaldy, 1909, Cat. Hem., 1: XXXII, 175.

1866 *Meadorus* Mulsant and Rey, Hist. Nat. Punaises France, Pentatomides, p. 315. Type-species: *Meadorus interstinctus*: Mulsant and Rey, 1866 (not *Cimex interstinctus* Linnaeus, 1758), a misidentification of *Cimex griseus* Linnaeus, 1758. Designated by Kirkaldy, 1909, Cat. Hem., 1: XXXII, 175. Synonymized by Kumar, 1974, Australian J. Zool., Suppl. Ser., 34: 48.

Note: Westwood's (1837, Hope Cat., p. 30) unlocalized species, *Acanthosoma picicolor*, was synonymized under the North American *lateralis* by Distant (1900, Proc. Zool. Soc. London, 54: 817) and this was followed by Van Duzee in several of his works. Kirkaldy (1909, Cat. Hem. Het., 1: 177) used Westwood's name for a Eurasian species and pointed out that North American records for it were "in error."

Elasmucha lateralis (Say), 1831 [Fig. 1]

1831 *Edessa lateralis* Say, Descrip. Hem. Het. N. Am., p. 3. ["Northwest Territory and Canada"].

1837 *Edessa nebulosa* Kirby, Fauna Bor. Am., 4: 277. ["New York to Cumberland-house and in Lat. 65," "Borealis-Americana"]. Synonymized by Uhler, 1878, Proc. Boston Soc. Nat. Hist., 19: 381.
1837 *Acanthosoma affinis* Westwood, Hope Cat., 1: 30. [America Boreali]. Synonymized by Distant, 1900, Proc. Zool. Soc. London, 54: 818.
1851 *Acanthosoma nebulosum*: Dallas, List Hem. Brit. Mus., 1: 307.
1872 *Acanthosoma lateralis*: Stål, K. Svens. Vet.-Akad. Handl., 10(4): 61.
1907 *Elasmucha lateralis*: Bergroth, Ent. News, 18: 49.
1908 *Clinocoris lateralis*: Van Duzee, Can. Ent., 40: 109.
1916 *Meadorus lateralis*: Van Duzee, Check List Hem., p. 8.
Distribution: Alk., Alta., B.C., Conn., Mass., Me., Mich., Minn., N.H., N.J., N.S., N.Y., Nfld., Oh., Ont., Pa., Que., R.I., S.C., Va., Vt.

Family Alydidae Amyot and Serville, 1843

The Broad-Headed Bugs

By Richard C. Froeschner

The few North American members of this relatively small family are usually found on foliage and flowers of various plants along roadsides and woods, and other places away from cultivated plants. The immature stages of these insects sometimes run on the surface of the ground where their slender form may allow a quick glance to assume they are large ants. Our species, being of no economic significance, are seldom noted. They are essentially sap-feeders but their small numbers keep them from conspicuously harming vegetation. Several members of the subfamily Alydinae have been reported as using their beak to probe the fluids on decomposing animal carcasses, but this practice is not an essential or even common part of their life cycle. Schaefer (1980, J. Ks. Ent. Soc., 53: 115-122) and Schaefer and Mitchell (1983, An. Ent. Soc. Am., 76: 591-615) summarized the feeding records for the family and concluded the subfamily Alydinae essentially feeds on leguminous plants while the subfamily Leptocorisinae and probably the Micrelytrinae generally feed on grasses. In the Orient the genera *Leptocorisa* Berthold and *Stenocoris* Burmeister (both subfamily Leptocorisinae) often damage rice crops. Schaefer and Pupedis (1981, J. Ks. Ent. Soc., 54: 143-152) described the presence of stridulatory mechanisms on some members

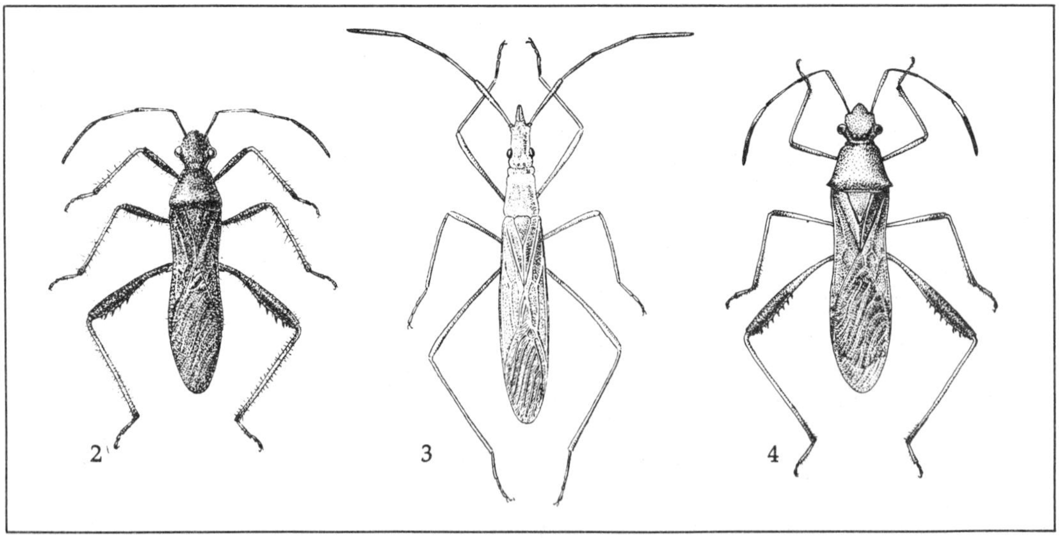

Figs. 2-4: 2, *Alydus eurinus* [p. 6]; 3, *Protenor belfragei* [p. 11]; 4, *Megalotomus quinquespinosus*[p. 8] (After Froeschner, 1942, except fig. 3, after Drake, 1922).

of the subfamily Alydinae and hypothesized these as "aggregating and/or premating isolating mechanisms."

Only thirty-two species are known from America north of Mexico, less than a third of them ranging as far north as Canada. Earlier, most authors treated this group as a subfamily of Coreidae, but more recent studies revealed sufficient reasons for according it family rank. Two authors have treated this family in its entirety for our region: Fracker (1918, An. Ent. Soc. Am., 11: 255-280, as a subfamily) and Torre-Bueno (1941, Ent. Am., 21: 78-88). The content of the family Alydidae was greatly changed by Bliven (1973, Occ. Ent., 1: 125) when he transferred the subfamilies Leptocorisinae and Micrelytrinae to the family Coreidae, and then incorporated into the Alydidae from the Largidae the three genera *Arhaphe* Herrich-Schaeffer, *Japetus* Distant, and *Jarhaphetus* Bliven for which he proposed the new subfamily Arhaphinae. These changes, made with a minimum discussion of critical and exact characters, have not been followed by other authors and are not accepted in the present catalog, which utilizes the traditional composition of Alydidae.

Subfamily Alydinae Amyot and Serville, 1843

Genus *Alydus* Fabricius, 1803

1796 *Coriscus* Schrank, Samm. Phys. Aufs., p. 121. Placed on *Official Index of Rejected and Invalid Names in Zoology* by the Int. Comm. on Zool. Nomen., 1950, Bull. Zool. Nomen., 4: 465.

1803 *Alydus* Fabricius, Syst. Rhyn., p. 248. Type-species: *Cimex calcaratus* Linnaeus, 1758. Designated by Int. Comm. Zool. Nomen., Bull. Zool. Nomen., 4: 464.

Note: Blatchley (1926, Het. E. N. Am., p. 266) considered the various color forms of *conspersus* and *eurinus* simply points on a continuous series of variation and doubted that they needed trinomials.

Alydus conspersus Montandon, 1893

1893 *Alydus conspersus* Montandon, Proc. U.S. Nat. Mus., 16: 49. ["Dak.," Col., Ia., Mass., Mich.].

1926 *Coriscus conspersus*: Blatchley, Het. E. N. Am., p. 266.

Distribution: Alta., Ariz., Col., "Dak.," Ia., Ill., Ind., Mass., Me., Mich., N.J., N.Y., Oh., Ont., Pa., Que., Ut., Wis.

Alydus conspersus conspersus Montandon, 1893

1893 *Alydus conspersus* Montandon, Proc. U.S. Nat. Mus., 16: 49.

1918 *Alydus conspersus conspersus*: Fracker, An. Ent. Soc. Am., 11: 271.

Distribution: Alta, Col., "Dak.," Ia., Ill., Ind., Mass., Me., Mich., N.J., N.Y., Oh., Ont., Pa., Que., Ut., Wis.

Alydus conspersus infuscatus Fracker, 1918

1918 *Alydus conspersus infuscatus* Fracker, An. Ent. Soc. Am., 11: 271. [Wis.].

Distribution: Alta., B.C., Col., Wis.

Alydus conspersus rufescens Barber, 1911

1911 *Alydus rufescens* Barber, J. N.Y. Ent. Soc., 19: 29. [Ariz.].

1918 *Alydus conspersus rufescens*: Fracker, An. Ent. Soc. Am., 11: 271.

Distribution: Ariz.

Note: Barber (1924, J. N.Y. Ent. Soc., 32: 134) restored this taxon to species level, but most subsequent authors followed Fracker (1918, above) and treated it as a trinomial.

Alydus eurinus (Say), 1825 [Fig. 2]

1825 *Lygaeus eurinus* Say, J. Acad. Nat. Sci. Phila., 4: 324. [Ark., Mo.].
1852 *Alydus ater* Dallas, List Hem. Brit. Mus., 2: 478. [N. Am.]. Synonymized by Uhler, 1872, Prelim. Rept. U.S. Geol. Surv. Mont., p. 401.
1852 *Alydus calcaratus*: Dallas, List Hem. Brit. Mus., 2: 478 (in part).
1870 *Alydus (Alydus) eurinus*: Stål, K. Svens. Vet.-Akad. Handl., 9(1): 213 (in part).
1870 *Alydus (Alydus) ater*: Stål, K. Svens. Vet.-Akad. Handl., 9(1): 213.
1872 *Alydus eurinus*: Uhler, Prelim. Rept. U.S. Geol. Surv. Mont., 5: 401.
1869 *Alydus curius* [*sic*]: Rathvon, Hist. Lancaster Co., Pa., p. 549.
1885 *Alydus pluto*: Provancher, Nat. Can., 3: 56.
1887 *Alydus vicarius*: Provancher, Nat. Can., 3: 175.
1910 *Alydus urinus* [*sic*]: Smith, Cat. Ins. N.J., p. 147.
1926 *Coriscus eurinus*: Blatchley, Het. E. N. Am., p. 265.
Distribution: Alta., Ariz., Ark., B.C., Cal., Col., Conn., "Dakota," D.C., Fla., Ga., Ia., Ill., Ind., Ks., Ky., Mass., Me., Mo., Mont., N.C., N.H., N.J., N.M., N.Y., Neb., Oh., Ont., Pa., Que., S.C., Tex., Ut., Va., Wis.

Alydus eurinus eurinus (Say), 1825

1825 *Lygaeus eurinus* Say, J. Acad. Nat. Sci. Phila., 4: 324.
1818 *Alydus eurinus eurinus* Fracker, An. Ent. Soc. Am., 11: 269.
Distribution: Same as for species.

Alydus eurinus obesus Fracker, 1918

1918 *Alydus eurinus obesus* Fracker, An. Ent. Soc. Am., 11: 270. [Ill.].
Distribution: Alta., Ill., Mo., Oh.

Alydus pilosulus Herrich-Schaeffer, 1848

1847 *Alydus pilosulus* Herrich-Schaeffer, Wanz. Ins., 8: 101. [Mexico].
1852 *Alydus pilosulus*: Dallas, List Hem Brit. Mus., 2: 478.
1869 *Alydus vittinasus* [*sic*]: Rathvon, Hist. Lancaster Co., p. 549.
1926 *Coriscus pilosulus*: Blatchley, Het. E. N. Am., p. 267.
Distribution: Cal., D.C., Del., Fla., Ia., Ill., Ind., Ks., La., Mass., Md., Me., Mo., N.C., N.J., N.Y., Neb., Ok., Pa., S.C., Tex., Ut., Va., Wis.
Note: The combination "*Alydus vittinosus*" was a Say manuscript name found on specimens in the Harris Collection according to Uhler (1878, Proc. Boston Soc. Nat. Hist., 19: 384).

Alydus pluto Uhler, 1872

1872 *Alydus pluto* Uhler, Prelim. Rept. U.S. Geol. Surv. Mont., 5: 401. [Col., Id., Ks., La.].
Distribution: Alta., Ariz., B.C., Cal., Col., Id., Ks., La., N.M., Ore., Tex., Ut., Wash.

Alydus scutellatus Van Duzee, 1903

1903 *Alydus scutellatus* Van Duzee, Trans. Am. Ent. Soc., 29: 108. [N.M.].
Distribution: Alta., B.C., Col., Ia., Mont., N.M.

Alydus tomentosus Fracker, 1918

1918 *Alydus tomentosus* Fracker, An. Ent. Soc. Am., 11: 267. [Col.].
Distribution: Col.

Genus *Burtinus* Stål, 1859

1859 *Burtinus* Stål, Öfv. K. Svens. Vet.-Akad. Förh., 16: 458, 489. Type-species: *Burtinus notatipennis* Stål, 1859. Monotypic.

Burtinus notatipennis Stål, 1859

1859 *Burtinus notatipennis* Stål, Öfv. K. Svens. Vet.-Akad. Förh., 16: 459. [Colombia, Mexico].

1910 *Burtinus notatipennis*: Barber, J. N.Y. Ent. Soc., 18: 37.

Distribution: Tex. (Mexico to Colombia).

Genus *Hyalymenus* Amyot and Serville, 1843

1843 *Hyalymenus* Amyot and Serville, Hist. Nat. Ins., Hem., p. 224. Type-species: *Alydus dentatus* Fabricius, 1803. Designated by Van Duzee, 1916, Check List Hem., p. 13.

1893 *Galeottus* Distant, Biol. Centr.-Am., Rhyn., 1: 459. Type-species: *Galeottus formicarius* Distant, 1893. Monotypic. Synonymized by Van Duzee, 1917, Univ. Cal. Publ. Ent., 2: 111.

Note: Banks' (1910, Cat. Nearc. Hem.-Het., p. 74) "Texas" record for the South American *Hyalymenus pulcher* (Stål) was accepted but queried by Van Duzee (1917, Univ. Cal. Publ. Ent., 2: 111); Torre-Bueno (1939, Bull. Brook. Ent. Soc., 34: 182) ignored that record and listed the species only for its type-locality "Honduras." The present tabulation follows Torre-Bueno in deleting it from our list. The nominate subgenus *Hyalymenus* does not occur in our region.

Subgenus *Tivarbus* Stål, 1859

1859 *Tivarbus* Stål, Öfv. K. Svens. Vet.-Akad. Förh., 16: 459. Type-species: *Cimex sinuatus* Fabricius, 1787. Designated by Van Duzee, 1917, Univ. Cal. Publ. Ent., 2: 111.

1870 *Hyalymenus* (*Tivarbus*): Stål, K. Svens. Vet.-Akad. Handl., 9(1): 211.

Note: Torre-Bueno (1939, Bull. Brook. Ent. Soc., 34: 177-197) reviewed the species of this subgenus and presented a key to them.

Hyalymenus longispinus Stål, 1870

1870 *Hyalymenus* (*Tivarbus*) *longispinus* Stål, K. Svens. Vet.-Akad. Handl., 9(1): 213 [Cuba].

1910 *Hyalymenus longispinus*: Banks Cat. Nearc. Hem., p. 74.

Distribution: Fla. (Greater Antilles).

Hyalymenus notatus Torre-Bueno, 1939

1933 *Hyalymenus longispinus*: Torre-Bueno, Bull. Brook. Ent. Soc., 28: 30.

1939 *Hyalymenus* (*Tivarbus*) *notatus* Torre-Bueno, Bull. Brook. Ent. Soc., 34: 181, 189. [Fla.].

Distribution: Fla.

Hyalymenus potens Torre-Bueno, 1939

1939 *Hyalymenus* (*Tivarbus*) *potens* Torre-Bueno, Bull. Brook. Ent. Soc., 34: 181, 187. [Fla.].

Distribution: Fla.

Hyalymenus subinermis Van Duzee, 1923

1923 *Hyalymenus subinermis* Van Duzee, Proc. Cal. Acad. Sci., ser. 4, 12: 134. [Mexico].

1939 *Hyalymenus (Tivarbus) subinermis*: Torre-Bueno, Bull. Brook. Ent. Soc., 34: 183.
Distribution: Ariz. (Mexico).

Hyalymenus tarsatus (Fabricius), 1803
1803 *Alydus tarsatus* Fabricius, Syst. Rhyn., p. 250. [Brazil].
1876 *Hyalymenus tarsatus*: Uhler, Bull. U.S. Geol. Geogr. Surv. Terr., 1: 294.
1917 *Galeottus formicarius* Distant, Biol. Centr.-Am., Rhyn., 1: 459. [Guatemala, Nicaragua]. Synonymized by Van Duzee, 1917, Univ. Cal. Publ. Ent., 2: 111.
1906 *Galeottus formicarius*: Barber, Mus. Brook. Inst. Arts Sci., Sci. Bull., 1: 269.
Distribution: Ariz., Cal., Tex. (Mexico to Brazil).

Genus *Megalotomus* Fieber, 1860

1860 *Megalotomus* Fieber, Europ. Hem., p. 58. Type-species: *Alydus limbatus* Herrich-Schaeffer, 1835, a junior synonym of *Cimex junceus* Scopoli, 1763. Designated by Reuter, 1888, Acta Soc. Sci. Fenn., 15: 763.

Megalotomus quinquespinosus (Say), 1825 [Fig. 4]
1825 *Lygaeus 5-spinosus* [*sic*] Say, J. Acad. Nat. Sci. Phila., 4: 323. [U.S.].
1846 *Alydus cruentus* Herrich-Schaeffer, Wanz. Ins., 8: 100. [N. Am.]. Synonymized by Uhler, 1861, Proc. Ent. Soc. Phila., 1: 23.
1870 *Alydus (Megalotomus) quinquespinosus*: Stål, K. Svens. Vet.-Akad. Handl., 9(1): 214.
1875 *Alydus quinquespinosus*: Uhler, Rept. U.S. Geol. Geogr. Surv. Terr., 5: 832.
1876 *Megalotomus quinquespinosus*: Uhler, Bull. U.S. Geol. Geogr. Surv. Terr., 1: 294.
Distribution: Alta., Ariz., B.C., Cal., Col., Del., D.C., Fla., Ia., Ill., Ind., Ks., Ky., Mass., Me., Mich., Mo., N.C., N.J., Oh., Ont., Pa., Que., S.C., Ut., Wash., Wis.

Megalotomus rufipes (Westwood), 1842
1842 *Alydus rufipes* Westwood, Hope Cat., 2: 19. ["America Aequinoct."].
1956 *Megalotomus rufipes*: Hussey, Fla. Ent., 39: 88.
Distribution: Fla. (Cuba).

Genus *Stachyocnemus* Stål, 1870

1870 *Stachyocnemus* Stål, K. Svens. Vet.-Akad. Handl., 9(1): 215. Type-species: *Alydus apicalis* Dallas, 1852. Monotypic.

Stachyocnemus apicalis (Dallas), 1852
1852 *Alydus apicalis* Dallas, List Hem. Brit. Mus., 2: 479. [Fla.].
1870 *Stachyocnemus apicalis*: Stål, K. Svens. Vet.-Akad. Handl., 9(1): 215.
1918 *Stachyocnemus apicalis apicalis*: Fracker, An. Ent. Soc. Am., 11: 276.
Distribution: Alta., Ariz., Cal., Col., D.C., "Dak.," Fla., Ind., Mont., N.C., N.J., N.M., N.Y., S.C., Tex. (Mexico).

Stachyocnemus cinereus Fracker, 1918
1918 *Stachyocnemus apicalis cinereus* Fracker, An. Ent. Soc. Am., 11: 276. [Col.].
1940 *Stachyocnemus cinereus*: Torre-Bueno, Bull. Brook. Ent. Soc., 35: 159.
Distribution: Ariz., Col., Ind., Mont.

Genus *Tollius* Stål, 1870

1870 *Alydus* (*Tollius*) Stål, K. Svens. Vet.-Akad. Handl., 9(1): 213. Type-species: *Alydus curtulus* Stål, 1859. Monotypic.
1873 *Tollius*: Stål, K. Svens. Vet-Akad. Handl., 11(2): 89.

Tollius curtulus (Stål), 1859
1859 *Alydus curtulus* Stål, Freg. Eug. Resa Jord., 3: 234. [Cal.].
1870 *Alydus* (*Tollius*) *curtulus*: Stål, K. Svens. Vet.-Akad. Handl., 9(1): 213.
1876 *Tollius curtulus*: Uhler, Bull. U.S. Geol. Geogr. Surv. Terr., 1: 294.
Distribution: Alta., B.C., Cal., Col., Ill., N.Y., Ore., Ut. (Mexico).

Tollius quadratus Van Duzee, 1921
1921 *Tollius quadratus* Van Duzee, Proc. Cal. Acad. Sci., ser. 4, 11: 113. [Cal.].
Distribution: Cal., S.D.

Tollius setosus Van Duzee, 1906
1906 *Tollius setosus* Van Duzee, Ent. News, 17: 386. [Ariz., Mont., Ut.].
Distribution: Ariz., B.C., Cal., Mont., N.Y., Ut.

Tollius vanduzeei Torre-Bueno, 1940
1913 *Tollius setosus*: Torre-Bueno, Ent. News, 24: 23.
1940 *Tollius vanduzeei* Torre-Bueno, Bull. Brook. Ent. Soc., 35: 159. [Cal.].
Distribution: Cal.

Subfamily Leptocorisinae Stål, 1872

Note: Ahmad (1965, Bull. Brit. Mus. (Nat. Hist.), Ent., Suppl., 5: 1-156) presented a worldwide revision of this subfamily.

Genus *Stenocoris* Burmeister, 1839

1839 *Stenocoris* Burmeister, Handb. Ent., 2: 1010. Type-species: *Cimex tipuloides* De Geer, 1773. Designated by Int. Comm. Zool. Nomen., Opinion 800, 1967, Bull. Zool. Nomen., 24: 10.
Note: The nominate subgenus *Stenocoris* does not occur in our region.

Subgenus *Oryzocoris* Ahmad, 1965

1965 *Stenocoris* (*Oryzocoris*) Ahmad, Bull. Brit. Mus. (Nat. Hist.), Ent., Suppl., 5: 11, 60. Type-species: *Cimex filiformis* Fabricius, 1775. Original designation.

Stenocoris filiformis (Fabricius), 1775
1775 *Cimex filiformis* Fabricius, Syst. Ent., p. 727. ["America"].
1794 *Gerris filiformis*: Fabricius, Ent. Syst., 4: 191.
1951 *Leptocorixa filiformis*: Hussey, Fla. Ent., 33: 150.
1965 *Stenocoris* (*Oryzocoris*) *filiformis*: Ahmad, Bull. Brit. Mus. (Nat. Hist.), Ent., Suppl., 5: 64.
Distribution: Fla. (Mexico to Brazil, West Indies).

Stenocoris furcifera (Westwood), 1842
1842 *Leptocorisa furcifera* Westwood, Hope Cat., 2: 18. [Brazil].

1965 *Stenocoris (Oryzocoris) furcifera*: Ahmad, Bull. Brit. Mus. (Nat. Hist.), Ent., Suppl., 5: 67.
Distribution: Fla. (Mexico to Brazil).

Subfamily Micrelytrinae Stål, 1867

Genus *Cydamus* Stål, 1860

1860 *Cydamus* Stål, K. Sven. Vet.-Akad. Handl., 2(7): 33. Type-species: *Cydamus adspersipes* Stål, 1860. Monotypic.

Cydamus abditus Van Duzee, 1925
1925 *Cydamus abditus* Van Duzee, Proc. Cal. Acad. Sci., ser. 4, 14: 394. [Ariz.].
Distribution: Ariz.

Cydamus borealis Distant, 1881
1881 *Cydamus borealis* Distant, Biol. Centr.-Am., Rhyn., 1: 159. [Guatemala].
1906 *Cydamus borealis*: Snow, Trans. Ks. Acad. Sci., 20: 151.
Distribution: Tex. (Guatemala).

Genus *Darmistus* Stål, 1859

1859 *Darmistus* Stål, Öfv. K. Svens. Vet.-Akad. Förh., 16: 469. Type-species: *Darmistus subvittatus* Stål, 1859. Monotypic.

Darmistus crassicornis Van Duzee, 1937
1937 *Darmistus crassicornis* Van Duzee, Pan-Pac. Ent., 13: 28. [Tex.].
Distribution: Tex.

Darmistus duncani Van Duzee, 1937
1937 *Darmistus duncani* Van Duzee, Pan-Pac. Ent., 13: 29. [Ariz.].
Distribution: Ariz.

Darmistus subvittatus Stål, 1859
1859 *Darmistus subvittatus* Stål, Öfv. K. Svens. Vet.-Akad. Förh., 16: 469. [Mexico].
1895 *Darmistus subvittatus*: Gillette and Baker, Col. Agr. Exp. Stn. Bull., 31: 19.
Distribution: Ariz., Cal., Col., Tex.

Genus *Esperanza* Barber, 1906

1906 *Esperanza* Barber, Mus. Brook. Inst. Arts Sci., Sci. Bull., 1: 269. Type-species: *Esperanza texana* Barber, 1906. Monotypic.

Esperanza texana Barber, 1906
1906 *Esperanza texana* Barber, Mus. Brook. Inst. Arts Sci., Sci. Bull., 1: 270. [Tex.].
Distribution: Fla., Ga., La., Miss., S.C., Tex. (Mexico).
Note: Wheeler and Henry (1984, Fla. Ent., 67: 525) presented an outline of the known biology, distribution, and hosts for this species.

Genus *Protenor* Stål, 1867

1867 *Protenor* Stål, Öfv. K. Svens. Vet.-Akad. Förh., 24: 543. Type-species: *Protenor belfragei* Haglund, 1868. First included species.

Protenor australis Hussey, 1925

1914 *Protenor belfragei*: Barber, Bull. Am. Mus. Nat. Hist., 33: 521.
1925 *Protenor australis* Hussey, J. N.Y. Ent. Soc., 33: 64. [Fla.].
Distribution: Fla.

Protenor belfragei Haglund, 1868 [Fig. 3]

1868 *Protenor belfragei* Haglund, Stett. Ent. Zeit., 29: 162. [Ill.].
1872 *Tetrarhinus quebecensis* Provancher, Nat. Can., 4: 76. [Que.]. Synonymized by Van Duzee, 1912, Can. Ent., 44: 319.
Distribution: Cal., Col., Ia., Ill., Ind., Mass., Me., Mich., N.J., N.Y., Oh., Ont., Que., Tex., Wis.

Genus *Rimadarmistus* Bliven, 1956

1956 *Rimadarmistus* Bliven, New Hem. W. St., p. 6. Type-species: *Rimadarmistus messor* Bliven, 1956. Original designation.

Rimadarmistus deprecator Bliven, 1956

1956 *Rimadarmistus deprecator* Bliven, New Hem. W. St., p. 7. [Cal.].
Distribution: Cal.

Rimadarmistus messor Bliven, 1956

1956 *Rimadarmistus messor* Bliven, New Hem. W. St., p. 7. [Cal.].
Distribution: Cal.

Family Anthocoridae Fieber, 1837

The Minute Pirate Bugs

By Thomas J. Henry

The family Anthocoridae, often called flower bugs or minute pirate bugs, is mostly predatory, feeding on aphids, mites, thrips, scales, and other arthropods and their eggs and larvae. A few, however, feed on pollen or other plant material. Some species like the insidious pirate bug, *Orius insidiosus* (Say), if inadvertently exposed to human skin, will probe and "bite," especially if perspiration is present. These bugs occupy a wide variety of habitats, including on and beneath loose or dead bark, in decaying vegetation, stored grain, insect galleries in fungi, bird nests, mammal burrows, guano in bat caves, or on various epiphytes like bromeliads and orchids. Anthocorids are often intercepted in interstate and international shipments of bulbs, cut flowers, nursery stock, and stored grain products. Many species are attracted to lights.

Members of this family are characterized by their small size (1.5-5.0 mm), flattened body, pointed head, ocelli, three-segmented rostrum, four-segmented antenna, distinct cuneus, wing membrane without closed cells, and three-segmented tarsus. The body may be dull or shiny, and pubescent or glabrous. Either fully winged (macropterous) or short-winged (brachypterous) forms may occur. Members of several genera undergo "traumatic insemination" in mating where the male punctures and fertilizes the female through the abdominal wall.

The predatory habits of these bugs make them potentially useful in controlling certain agricultural pests. The native North American *Orius insidiosus* and *O. tristicolor* (White) have been released in Hawaii to control certain lepidopterous larvae. Attempts also have been made to introduce exotic anthocorids into North America. *Montantoniola moraguezi* (Puton) has been introduced from the Philippines and Hawaii to control the Cuban laurel thrips, *Gynaikothrips ficorum* (Marchal), in California (pers. comm., A. R. Hardy, Calif. Dept. Food & Agric., Sacramento). *Tetraphleps abdulghani* Ghauri and *T. raoi* Ghauri from India and Pakistan have been intro-

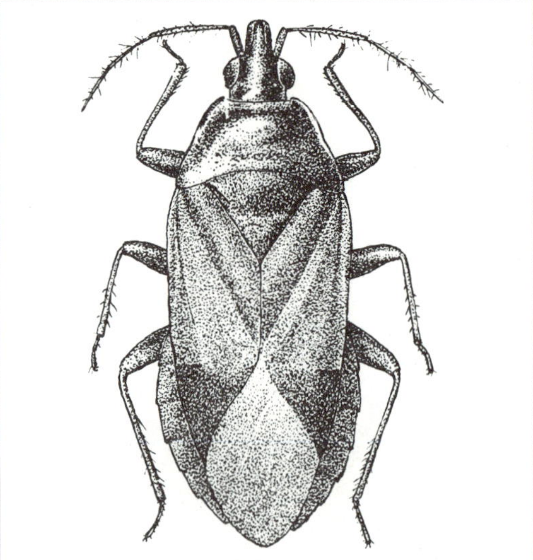

Fig. 5 *Lyctocoris campestris* [p. 24] (After Froeschner, 1949).

duced in British Columbia, New Brunswick, and Nova Scotia to control balsam woolly aphid, *Adelges piceae* (Ratzeburg) (pers. comm., L. A. Kelton, Biosyst. Res. Inst., Ottawa), but to date apparently neither has become established.

Pioneer work on the North American Anthocoridae was done by Reuter. Reuter's first paper treating the American forms (1871, Öfv. K. Svens. Vet.-Akad. Förh., 28(5): 557-568) was followed by a world monograph (1884, Acta Soc. Sci. Fenn., 14: 555-758). Poppius (1909, Acta. Soc. Sci. Fenn., 27: 1-43) added many Old World genera and species. Blatchley's (1926) "Heteroptera of Eastern North America" remains a slightly dated standard for the family in the eastern states. Carayon's (1972, An. Soc. Ent. France, 8: 309-349) "Caractères systématiques et classification des Anthocoridae" gave a revised classification of subfamilies and tribes. Carayon's scheme, with modifications suggested by Štys (1975, Acta Univ. Carol. Biol., 4: 159-162), is followed in this catalog. Péricart (1972, Hemiptères Anthocoridae...de l'Quest-Paléarctique, 402 pp.) gave a good review of the European fauna. Herring (1976, Fla. Ent., 59: 143-150) provided the only identification key to include all of the North American genera. Kelton's (1978) well-illustrated "Anthocoridae of Canada" is the single most important publication for identifying nearctic Anthocoridae.

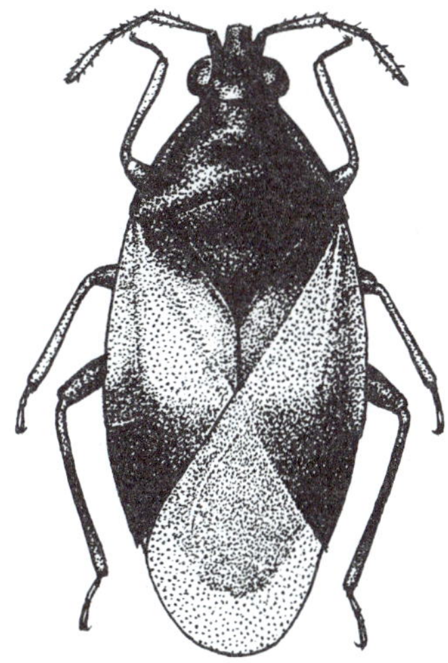

Fig. 6 *Orius insidiosus* [p. 18] (After Froeschner, 1949).

Subfamily Anthocorinae Reuter, 1884

Tribe Anthocorini Reuter, 1884

Genus *Acompocoris* Reuter, 1875

1875 *Acompocoris* Reuter, Bih. K. Svens. Vet.-Akad. Handl., 3: 63. Type-species: *Lygaeus pygmaeus* Fallén, 1807. Monotypic.

Acompocoris lepidus (Van Duzee), 1921

1921 *Tetraphleps lepidus* Van Duzee, Proc. Cal. Acad. Sci., (4)11: 142. [Cal.].
1962 *Acompocoris lepidus*: Kelton and Anderson, Can. Ent., 94: 1307.
Distribution: Alta., B.C., Cal., N.T.

Acompocoris pygmaeus (Fallén), 1807

1807 *Lygaeus pygmaeus* Fallén, Monogr. Cimic. Suec., p. 73. [Europe].
1977 *Acompocoris pygmaeus*: Kelton, Can. Ent., 109: 243.
Distribution: N.B., N.S., Ont., P.Ed. (Palearctic).

Genus *Anthocoris* Fallén, 1814

1814 *Anthocoris* Fallén, Spec. Nova. Hem. Disp. Meth., p. 9. Type-species: *Cimex nemorum* Linnaeus, 1758. Designated by Westwood, 1840, Intr. Mod. Class. Ins., 2 (Synopsis): 122.

Anthocoris albiger Reuter, 1884
1884 *Anthocoris albiger* Reuter, Acta Soc. Sci. Fenn., 14: 624. [Mexico].
1916 *Anthocoris albiger*: Van Duzee, Check List Hem., p. 34.
Distribution: Cal., N.M., Tex., Ut. (Mexico).

Anthocoris antevolens White, 1879
1879 *Anthocoris antevolens* White, Ent. Month. Mag., 16: 146. [Cal.].
Distribution: Alk., Alta, B.C., Cal., Col., Id., Mont., N.T., Nev., Nfld., Ont., Sask., Wyo.

Anthocoris bakeri Poppius, 1913
1913 *Anthocoris bakeri* Poppius, An. Ent. Soc. Belg., 57: 14. [Cal.].
1914 *Anthocoris ornatus* Van Duzee, Trans. San Diego Soc. Nat. Hist., 2: 14. [Cal.]. Synonymized by Hill, 1957, Pan-Pac. Ent., 33: 174.
1917 *Anthocoris bakeri* var. *ornatus*: Van Duzee, Univ. Cal. Publ. Ent., 2: 293.
Distribution: Cal.

Anthocoris confusus Reuter, 1884
1884 *Anthocoris confusus* Reuter, Acta Soc. Sci. Fenn., 14: 625. [Europe].
1946 *Anthocoris confusus*: Procter, Biol. Surv. Mt. Desert Reg., 7: 77.
Distribution: Me., N.S., Ont., P.Ed., Tenn.

Anthocoris dimorphicus Anderson and Kelton, 1963
1963 *Anthocoris dimorphicus* Anderson and Kelton, Can. Ent., 95: 440. [Ont.].
Distribution: Alta., Ont., N.T., Sask., Yuk.

Anthocoris fulvipennis Reuter, 1884
1884 *Anthocoris fulvipennis* Reuter, Acta Soc. Sci. Fenn., 14: 623. [Mexico].
1904 *Anthocoris fulvipennis*: Uhler, Proc. U.S. Nat. Mus., 27: 363.
Distribution: Cal., N.M. (Mexico).

Anthocoris musculus (Say), 1832
1832 *Reduvius musculus* Say, Descrip. Het. Hem. N. Am., p. 32. ["N.W. Territory"].
1852 *Anthocoris borealis* Dallas, List Hem. Brit. Mus., 2: 588. Synonymized by Blatchley, 1926, Het. E. N. Am., p. 635.
1876 *Anthocoris musculus*: Uhler, Bull. U.S. Geol. Geogr. Surv. Terr., 1: 321.
1918 *Anthocorus* [*sic*] *musculus*: Torre-Bueno, Can. Ent., 50: 25.
Distribution: Alk., Alta., B.C., Col., Ill., Ind., Ks., Man., Mo., N.B., N.C., N.S., N.Y., Nfld., Ont., Ore., P.Ed., Que., Sask.

Anthocoris nemoralis (Fabricius), 1794
1794 *Acanthia nemoralis* Fabricius, Ent. Syst., 4: 76. [Europe].
1963 *Anthocoris nemoralis*: Anderson and Kelton, Can. Ent., 95: 439.
Distribution: B.C., Ont. (Palearctic).

Anthocoris nigripes Reuter, 1884
1884 *Anthocoris nigripes* Reuter, Acta Soc. Sci. Fenn., 14: 623. [Mexico].
1904 *Anthocoris nigripes*: Uhler, Proc. U.S. Nat. Mus., 27: 363.
Distribution: N.M. (Mexico).

Anthocoris tomentosus Péricart, 1971
1884 *Anthocoris melanocerus* Reuter, Acta Soc. Sci. Fenn., 14: 634. [Col.]. Preoccupied.

1957 *Anthocoris melanoceros* [*sic*]: Hill, Pan-Pac. Ent., 33: 172.
1971 *Anthocoris tomentosus* Péricart, Bull. Soc. Linn. De Lyon, 40: 98. New name for *Anthocoris melanocerus* Reuter.
Distribution: Alk., Alta., B.C., Cal., Col., Id., Man., N.T., Nev., Yuk., Ut.

Anthocoris tristis Van Duzee, 1921
1921 *Anthocoris tristis* Van Duzee, Proc. Cal. Acad. Sci., (4)11: 138. [Cal.].
Distribution: Cal.

Anthocoris whitei Reuter, 1884
1884 *Anthocoris whitei* Reuter, Acta Soc. Sci. Fenn., 14: 628. [Cal.].
1927 *Anthocoris bakeri*: Downes, Proc. Ent. Soc. B.C., 23: 11.
Distribution: Cal., Id., B.C.
Note: Kelton (1978, Anthocorid. Can., p. 41) noted that Canadian records of *A. bakeri* should be referred to *A. whitei*.

Genus *Coccivora* McAtee and Malloch, 1925

1925 *Coccivora* McAtee and Malloch, Proc. Biol. Soc. Wash., 38: 146. Type-species: *Coccivora californica* McAtee and Malloch, 1925. Original designation.

Coccivora californica McAtee and Malloch, 1925
1925 *Coccivora californica* McAtee and Malloch, Proc. Biol. Soc. Wash., 38: 146. [Cal.].
Distribution: Cal.

Genus *Elatophilus* Reuter, 1884

1884 *Elatophilus* Reuter, Acta Soc. Sci. Fenn., 14: 616. Type-species: *Temnostethus nigrellus* Zetterstedt, 1838. Designated by Kirkaldy, 1906, Trans. Am. Ent. Soc., 32: 120.
1926 *Xenotracheliella* Drake and Harris, Proc. Biol. Soc. Wash., 39: 38. Type-species: *Xenotracheliella inimica* Drake and Harris, 1926. Original designation. Synonymized by Kelton and Anderson, 1962, Can. Ent., 94: 1306.
Note: Revision and key of North American spp. given by Kelton (1976, Can. Ent., 108: 631-634).

Subgenus *Elatophilus* Reuter, 1884

1884 *Elatophilus* Reuter, Acta Soc. Sci. Fenn., 14: 616. Type-species: *Temnostethus nigrellus* Zetterstedt, 1838. Designated by Kirkaldy, 1906, Trans. Am. Ent. Soc., 32: 120.

Elatophilus brimleyi Kelton, 1977 New Subgeneric Combination
1977 *Elatophilus brimleyi* Kelton, Can. Ent., 109: 1017. [Ont.].
Distribution: Ont.

Elatophilus dimidiatus (Van Duzee), 1921 New Subgeneric Combination
1921 *Anthocoris dimidiatus* Van Duzee, Proc. Cal. Acad. Sci., (4)11: 139. [Cal.].
1976 *Elatophilus dimidiatus*: Kelton, Can. Ent., 108: 632.
Distribution: Cal.

Elatophilus inimicus (Drake and Harris), 1926 New Subgeneric Combination
1926 *Xenotracheliella inimica* Drake and Harris, Proc. Biol. Soc. Wash., 39: 38. [N.Y.].
1926 *Xenotracheliella vicaria* Drake and Harris, Proc. Biol. Soc. Wash., 39: 39. [Mich.]. Synonymized by Kelton, 1976, Can. Ent., 108: 631.
1962 *Elatophilus inimicus*: Kelton and Anderson, Can. Ent., 94: 1306.

1962 *Elatophilus vicarius*: Kelton and Anderson, Can. Ent., 94: 1306.
Distribution: Conn., Man., Mich., N.C., N.Y., Ont., Que.

Elatophilus minutus Kelton, 1976 New Subgeneric Combination
1976 *Elatophilus minutus* Kelton, Can. Ent., 108: 632. [Que.].
Distribution: Alta., Man., Ont., Que., Sask.

Elatophilus oculatus (Drake and Harris), 1926 New Subgeneric Combination
1926 *Xenotracheliella oculata* Drake and Harris, Proc. Biol. Soc. Wash., 39: 40. [Ariz.].
1962 *Elatophilus oculatus*: Kelton and Anderson, Can. Ent., 94: 1306.
Distribution: Ariz.

Elatophilus pullus Kelton and Anderson, 1962 New Subgeneric Combination
1962 *Elatophilus pullus* Kelton and Anderson, Can. Ent., 94: 1306. [B.C.].
Distribution: B.C., Ore.

Subgenus *Euhadrocerus* Reuter, 1884

1884 *Euhadrocerus* Reuter, Acta Soc. Sci. Fenn., 14: 619. Type-species: *Elatophilus crassicornis* Reuter, 1875. Monotypic.

Elatophilus pinophilus Blatchley, 1928
1928 *Elatophilus* (*Euhadrocerus*) *pinophilus* Blatchley, Ent. News, 39: 87. [Fla.].
1962 *Elatophilus pinaphilus* [*sic*]: Kelton and Anderson, Can. Ent., 94: 1306.
Distribution: Fla.

Genus *Melanocoris* Champion, 1900

1900 *Melanocoris* Champion, Biol. Centr.-Am., Rhyn., 2: 329. Type-species: *Melanocoris obovatus* Champion, 1900. Monotypic.

Melanocoris longirostris Kelton, 1977
1977 *Melanocoris longirostris* Kelton, Can. Ent., 108: 246. [B.C.].
Distribution: Ariz., B.C., Col., N.M., Ut.

Melanocoris nigricornis Van Duzee, 1921
1921 *Melanocoris nigricornis* Van Duzee, Proc. Cal. Acad. Sci., (4)11: 143. [Cal.].
1926 *Tetraphleps novitus* Drake and Harris, Proc. Biol. Soc. Wash., 39: 41. [Col.]. Synonymized by Kelton and Anderson, 1962, Can. Ent., 94: 1307.
Distribution: B.C., Cal., Col.

Melanocoris pingreensis (Drake and Harris), 1926
1926 *Tetraphleps pingreensis* Drake and Harris, Proc. Biol. Soc. Wash., 39: 42. [Col.].
1962 *Melanocoris pingreensis*: Kelton and Anderson, Can. Ent., 94: 1307.
Distribution: Col.

Genus *Temnostethus* Fieber, 1860

1860 *Temnostethus* Fieber, Wien. Ent. Monat., 4: 263. Type-species: *Anthocoris pusillus* Herrich-Schaeffer, 1850. Designated by Kirkaldy, 1906, Trans. Am. Ent. Soc., 32: 120.

Temnostethus fastigiatus Drake and Harris, 1926
1926 *Temnostethus fastigiatus* Drake and Harris, Proc. Biol. Soc. Wash., 39: 40. [Cal.].
Distribution: Cal.

Temnostethus gracilis Horvath, 1907

1907 *Temnostethus pusillus* var. *gracilis* Horvath, An. Mus. Nat. Hung., 5: 310. [Europe].

1977 *Temnostethus gracilis*: Kelton, Can. Ent., 109: 243.

Distribution: N.S. (Palearctic).

Genus *Tetraphleps* Fieber, 1860

1860 *Tetraphleps* Fieber, Wien. Ent. Monat., 4: 262. Type-species: *Anthocoris vittatus* Fieber, 1860. Monotypic.

Note: Kelton (1966, Can. Ent., 98: 199-204) provided a key to the North American species.

Tetraphleps canadensis Provancher, 1886

1886 *Tetraphleps canadensis* Provancher, Pet. Faune Can. Ent., 3: 90. [Que.]. Lectotype designated by Kelton, 1968, Nat. Can., 95: 1072.

1912 *Lyctocoris canadensis*: Van Duzee, Can. Ent., 44: 320.

1920 *Tetraphleps americana* Parshley, Can. Ent., 52: 84. [Me.]. Synonymized by Drake and Harris, 1928, Can. Ent., 60: 50.

1922 *Tetraphleps osborni* Drake, Tech. Publ. 16, N.Y. St. Coll. For., 22: 67. [N.Y.]. Synonymized by Kelton and Anderson, 1962, Can. Ent., 94: 1307.

1926 *Acompocoris* (*Tetraphleps*) *osborni*: Drake and Harris, Proc. Biol. Soc. Wash., 39: 43.

1926 *Tetraphleps edacis* Drake and Harris, Proc. Biol. Soc. Wash., 39: 43. [N.Y.]. Synonymized by Kelton and Anderson, 1962, Can. Ent., 94: 1307.

Distribution: Alk., Alta., B.C., Man., Me., Mich., N.B., N.H., N.S., N.Y., Nfld., Ont., Que., Sask., Wis.

Tetraphleps feratis (Drake and Harris), 1926

1926 *Acompocoris feratis* Drake and Harris, Proc. Biol. Soc. Wash., 39: 41. [B.C.].

1962 *Tetraphleps feratis*: Kelton and Anderson, Can. Ent., 94: 1307.

Distribution: Alta., B.C.

Tetraphleps latipennis Van Duzee, 1921

1921 *Tetraphleps latipennis* Van Duzee, Proc. Cal. Acad. Sci., (4)11: 140. [Cal.].

1926 *Tetraphleps profugus* Drake and Harris, Proc. Biol. Soc. Wash., 39: 42. [B.C.]. Synonymized by Kelton and Anderson, 1962, Can. Ent., 94: 1307.

Distribution: Alta., B.C., Cal., Id., Man., N.B., N.S., Nfld., Ore., Sask., Wyo.

Tetraphleps pilosipes Kelton and Anderson, 1962

1962 *Tetraphleps pilosipes* Kelton and Anderson, Can. Ent., 94: 1307. [B.C.].

Distribution: Alk., B.C., Man., Nfdl, Ore., Yuk.

Tetraphleps uniformis Parshley, 1920

1906 *Tetraphleps* n. sp.: Slosson, Ent. News, 17: 326.

1920 *Tetraphleps uniformis* Parshley, Can. Ent., 52: 85. [N.H.].

1921 *Tetraphleps furvus* Van Duzee, Proc. Cal. Acad. Sci., (4)11: 141. [Col.]. Synonymized by Kelton, 1966, Can. Ent., 98: 202.

1922 *Tetraphleps concolor* Drake, Tech. Publ. 16, N.Y. St. Coll. For., 22: 68. *Nomen nudum*.

Distribution: Alta., B.C., Col., Man., Me., N.B., N.H., N.S., Nfld., Me., N.Y., Ont., Que., Sask., Yuk.

Tribe Oriini Carayon, 1955

Genus *Macrotracheliella* Champion, 1900

1900 *Macrotracheliella* Champion, Biol. Centr.-Am., Rhyn., 2: 322. Type-species: *Macrotracheliella laevis* Champion, 1900. Monotypic.

Macrotracheliella laevis Champion, 1900

1900 *Macrotracheliella* laevis Champion, Biol. Centr.-Am., Rhyn., 2: 322. [Mexico, Panama].

Distribution: Fla. (Mexico, Panama).

Note: The nominate subspecies *M. laevis laevis* is not known from the U.S. or Canada.

Macrotracheliella laevis floridana Drake and Harris, 1926

1926 *Macrotracheliella laevis floridana* Drake and Harris, Proc. Biol. Soc. Wash., 39: 37. [Fla.].

Distribution: Fla.

Macrotracheliella nigra Parshley, 1917

1917 *Macrotracheliella nigra* Parshley, Ent. News, 28: 38. [Mass.].

Distribution: Ark., B.C., Fla., Man., Mass., N.J., N.S., N.Y., Ont., Que., R.I.

Genus *Orius* Wolff, 1811

1811 *Orius* Wolff, Icon. Cimic. Descrip., 5: iv, fig. 161. Type-species: *Salda nigra* Wolff, 1811. Monotypic.

1860 *Triphleps* Fieber, Wien. Ent. Monat. 4: 266. Type-species: *Salda nigra* Wolff, 1811. Designated by Kirkaldy, 1906, Trans. Am. Ent. Soc., 32: 120. Synonymy by virtue of common type-species.

Note: Kelton (1963, Can. Ent., 95: 631-636) and Herring (1966, An. Ent. Soc. Am., 59: 1093-1109) reviewed the species of *Orius* in North America and in the Western Hemisphere, respectively. Kelton (1963, above) clarified that *Orius niger* (Wolff) does not occur in North America, even though Van Duzee (1917, Univ. Cal. Publ. Ent., 2: 294) listed it from California. Herring (1966, above) noted that the American species of *Orius* do not fit satisfactorily into the subgenera given by Wagner (192, Not. Ent., 32; 56-58).

Orius candiope Herring, 1966

1966 *Orius candiope* Herring, An. Ent. Soc. Am., 59: 1098. [Ia.].

Distribution: Ia.

Orius diespeter Herring, 1966

1966 *Orius diespeter* Herring, An. Ent. Soc. Am., 59: 1098. [B.C.].

Distribution: Alta., B.C.

Orius harpocrates Herring, 1966

1966 *Orius harpocrates* Herring, An. Ent. Soc. Am., 59: 1097. [Cal.].

Distribution: Cal.

Orius insidiosus (Say), 1832 [Fig. 6]

1832 *Reduvius insidiosus* Say, Descrip. Het. Hem. N. Am., p. 32. [U.S.].

1855 *Anthocoris pseudo-chinche* Fitch, Trans. N.Y. St. Agr. Soc., 15: 527. Synonymized by Stål, 1873, K. Svens. Vet.-Akad. Handl, 11(2): 102.

1870 *Anthocoris insidiosus*: Riley, Second An. Rept. Nox. Ins. Mo., p. 27.

1871 *Triphleps latulus* Reuter, Öfv. K. Svens. Vet.-Akad. Förh., 28(5): 565. [N.J., S.C.]. Synonymized by Reuter, 1884, Acta Soc. Sci. Fenn., 14: 651.
1871 *Triphleps rugicollis* Reuter, Öfv. K. Svens. Vet.-Akad. Förh., 28(5): 565. [Tex.]. Synonymized by Reuter, 1884, Acta Soc. Sci. Fenn., 14: 651.
1871 *Triphleps insidiosus*: Walker, Cat. Hem. Brit. Mus., 5: 157.
1917 *Triphleps insidiosus* var. *latulus*: Van Duzee, Univ. Cal. Publ. Ent., 2: 294.
1923 *Triphleps insidiosa*: Parshley, Conn. Geol. Nat. Hist. Surv. Bull., 34: 668.
1926 *Orius insidiosus*: Blatchley, Het. E. N. Am., p. 637.
1967 *Orus* [*sic*] *insidiosus*: Wray, Ins. N.C., 3rd. suppl., p. 26.
1979 *Oris* [*sic*] *insidiosis* [*sic*]: Scarbrough and Sraver, Proc. Ent. Soc. Wash., 81: 636.
Distribution: Ala., B.C.(?), Cal., Col., Conn., Fla., Ia., Ill., Ind., Ks., La., Man., Mass., Md., Mich., Miss., Mo., N.C., N.H., Neb., N.J., N.Y., Oh., Ont., Pa., Que., S.C., Tenn., Tex., Ut., Va., Wis. (Hawaii, Mexico to South America, West Indies).
Note: Kelton (1978, Anthocorid. Can., p. 49) noted that Lord's (1949, Can. Ent., 81:219) record from N.S. and Strickland's (1953, Can. Ent., 85:199) record from Alta. should refer to *O. tristicolor*. Tonks' (1953, Proc. Ent. Soc. B.C., 49: 28) B.C. record and other records from northwestern U.S. and western Canada may also refer to *O. tristicolor*. Ryerson and Stone (1979, Ent. Soc. Am. Bull., 25: 131-135) provided a useful selected bibliography for *O. insidiosus* and *O. tristicolor*.

Orius minutus (Linneaus), 1758
1758 *Cimex minutus* Linnaeus, Syst. Nat., p. 446. [Europe].
1953 *Orius minutus*: Tonks, Proc. Ent. Soc. B.C., 49: 27.
Distribution: B.C., Ore., Wash. (Palearctic).

Orius pumilio (Champion), 1900
1900 *Triphleps pumilio* Champion, Biol. Centr.-Am., Rhyn., 2: 327. [Guatemala].
1926 *Orius pumilio*: Blatchley, Het. E. N. Am., p. 638.
Distribution: Fla. (Central America, Mexico, West Indies).

Orius tristicolor (White), 1879
1879 *Triphleps tristicolor* White, Ent. Month. Mag., 16: 145. [Cal.].
1919 *Tetraphleps insidiosus* var. *tristicolor*: Parshley, Occas. Pap. Mus. Zool., Univ. Mich., 71: 28.
1926 *Orius insidiosus* var. *tristicolor*: Blatchley, Het. E. N. Am., p. 637.
1944 *Orius tristicolor*: Harris and Shull, Ia. St. Coll. J. Sci., 18: 207.
Distribution: Alk., Alta., Ariz., B.C., Cal., Col., Id., Man., Mass., Me., Mich., N.B., N.H., N.M., N.S., N.Y., Neb., Nfld., Ont., Ore., Que., Sask., Tex., Ut., Wash., Wyo., Yuk. (Mexico to South America, West Indies).
Note: Kelton (1963, Can. Ent., 95: 632) confirmed that *O. tristicolor* is a distinct species. Herring (1966, An. Ent. Soc. Am., 59: 1108) said "*O. tristicolor* is primarily a western species, but occurs across the northern U.S. and Canada, and through Mexico into Central and South America." Ryerson and Stone (1979, Bull. Ent. Soc. Am., 25: 131-135) provided a useful selected bibliography for *O. tristicolor* and *O. insidiosus*.

Genus *Paratriphleps* Champion, 1900

1900 *Paratriphleps* Champion, Biol. Centr.-Am., Rhyn., 2: 328. Type-species: *Paratriphleps laeviusculus* Champion, 1900. Monotypic.

Note: Possible synonymy of the two included species was pointed out by Barber (1939, Sci. Surv. Porto Rico Virgin Is., 14: 404) and Bachelor and Baranowski (1975, Fla. Ent., 58: 157.

Paratriphleps laeviusculus Champion, 1900

1900 *Paratriphleps laeviusculus* Champion, Biol. Centr.-Am., Rhyn., 2: 329. [Panama].
1975 *Paratriphleps laeviusculus*: Bacheler and Baranowski, Fla. Ent., 58: 157.
Distribution: Fla. (Puerto Rico, Mexico, Panama).
Note: Bacheler and Baranowski (1975, above) gave information on biology, described the egg and nymphs, and illustrated the five nymphal instars and adult.

Paratriphleps pallidus (Reuter), 1884

1884 *Brachysteles pallidus* Reuter, Acta Soc. Sci. Fenn., 14: 672. [St. Jean and St. Thomas Is.].
1886 *Brachysteles pallidus*: Uhler, Check-list Hem. Het., p. 21.
1939 *Paratriphleps pallidus*: Barber, Sci. Surv. Porto Rico Virgin Is., 14: 404.
Distribution: Fla., "Southern United States" (West Indies).

Subfamily Lasiochilinae Carayon, 1972

Tribe Lasiochilini Carayon, 1972

Genus *Lasiochilus* Reuter, 1871

1871 *Lasiochilus* Reuter, Öfv. K. Svens. Vet.-Akad. Förh., 28(5): 562. Type-species: *Lasiochilus pallidulus* Reuter, 1871. Monotypic.

Subgenus *Dilasia* Reuter, 1871

1871 *Dilasia* Reuter, Öfv. K. Svens. Vet.-Akad. Förh., 28(5): 563. Type-species: *Dilasia fuscula* Reuter, 1871. Monotypic.
1884 *Lasiochilus* (*Dilasia*) Reuter, Acta Soc. Sci. Fenn., 14: 574.

Lasiochilus fusculus (Reuter), 1871

1871 *Dilasia fuscula* Reuter, Öfv. K. Svens. Vet.-Akad. Förh., 28(5): 563. [S.C., Tex.].
1884 *Lasiochilus* (*Dilasia*) *fusculus*: Reuter, Acta Soc. Sci. Fenn., 14: 576.
Distribution: Fla., Ind., Ill., Miss., N.J., N.Y., Ont., Que., S.C., Tenn., Tex. (West Indies).

Subgenus *Lasiochilus* Reuter, 1884

1884 *Lasiochilus* (*Lasiochilus*) Reuter, Acta Soc. Sci. Fenn., 14: 569. Type-species: *Lasiochilus pallidulus* Reuter, 1871. Monotypic.

Lasiochilus comitalis Drake and Harris, 1926

1926 *Lasiochilus comitalis* Drake and Harris, Proc. Biol. Soc. Wash., 39: 34. [N.C.].
Distribution: N.C.

Lasiochilus divisus Champion, 1900

1900 *Lasiochilus divisus* Champion, Biol. Centr.-Am., Rhyn., 2: 310. [Mexico].
1926 *Lasiochilus divisus*: Drake and Harris, Proc. Biol. Soc. Wash., 39: 35.
Distribution: Fla. (Central America, Mexico, West Indies).

Lasiochilus gerhardi Blatchley, 1926
1926 *Lasiochilus gerhardi* Blatchley, Het. E. N. Am., p. 627. [Fla.].
Distribution: Fla., Mass.

Lasiochilus hirtellus Drake and Harris, 1926
1926 *Lasiochilus hirtellus* Drake and Harris, Proc. Biol. Soc. Wash., 39: 33. [Ala.].
Distribution: Ala., Fla., La., Tex. (Mexico).
Note: Blatchley (1928, J. N.Y. Ent. Soc., 36: 8) suggested that this species is probably a synonym of *Lasiochilus pallidulus* Reuter.

Lasiochilus mirificus Drake and Harris, 1926
1926 *Lasiochilus mirificus* Drake and Harris, Proc. Biol. Soc. Wash., 39: 35. [Tex.].
Distribution: Tex.

Lasiochilus pallidulus Reuter, 1871
1871 *Lasiochilus pallidulus* Reuter, Öfv. K. Svens. Vet.-Akad. Förh., 28(5): 562. [S.C.].
1884 *Lasiochilus (Lasiochilus) pallidulus*: Reuter, Acta Soc. Sci. Fenn., 14: 571.
Distribution: Fla., S.C., Tex. (Central America, Mexico, West Indies).

Genus *Plochiocoris* Champion, 1900

1900 *Plochiocoris* Champion, Biol. Centr.-Am., Rhyn., 2: 314. Type-species: *Plochiocoris longicornis* Champion, 1900. Monotypic.

Plochiocoris comptulus Drake and Harris, 1926
1926 *Plochiocoris comptulus* Drake and Harris, Proc. Biol. Soc. Wash., 39: 36. [Tex.].
Distribution: Tex.

Subfamily Lyctocorinae Reuter, 1884

Tribe Dufouriellini Van Duzee, 1916

Note: Some recent authors have used the tribal name Cardiastethini Carayon, 1972, but Štys (1975, Acta Univ. Carol., 4: 161) considered it a synonym of Dufouriellini. Kelton (1978, Anthocorid. Can., p. 55) used the name Cardiastethini without comment.

Genus *Alofa* Herring, 1976

1976 *Alofa* Herring, Fla. Ent., 59: 150. Type-species: *Cardiastethus sodalis* White, 1878. Original designation.

Alofa sodalis (White), 1878
1878 *Cardiastethus sodalis* White, An. Mag. Nat. Hist., 1: 372. [Haw.].
1966 *Buchananiella sodalis*: Herring, Proc. Ent. Soc. Wash., 68: 127.
1976 *Alofa sodalis*: Herring, Fla. Ent., 59: 150.
Distribution: "N. Am." [no definite locality given] (Africa, Central and South America, Pacific Islands, West Indies).

Genus *Amphiareus* Distant, 1904

1904 *Amphiareus* Distant, An. Mag. Nat. Hist., Ser. 7, 15: 220. Type-species: *Xylocoris fulvescens* Walker, 1872, a junior synonym of *Xylocoris constrictus* Stål, 1860. Monotypic.

1871 *Poronotus* Reuter, Öfv. K. Svens. Vet.-Akad. Förh., 28(5): 561. Preoccupied. Type-species: *Xylocoris constrictus* Stål, 1860. Designated by Kirkaldy, 1906, Trans. Am. Ent. Soc., 32: 120.

1904 *Poronotellus* Kirkaldy, Ent., 37: 280. New name for *Poronotus* Reuter. Synonymized by Herring, 1965, Proc. Ent. Soc. Wash., 67: 203.

Amphiareus constrictus (Stål), 1860

1860 *Xylocoris constrictus* Stål, K. Svens. Vet.-Akad. Handl., 2(7): 44. [Brazil].
1926 *Poronotus constrictus*: Blatchley, Het. E. N. Am., p. 644.
1960 *Amphiareus constrictus*: Hiura, Bull. Osaka Mus. Nat. Hist., 12: 46.
Distribution: Fla. (Brazil, Mexico; Circumtropical).

Genus *Cardiastethus* Fieber, 1860

1860 *Cardiastethus* Fieber, Wien. Ent. Monat., 4: 266. Type-species: *Cardiastethus luridellus* Fieber, 1860. Designated by Kirkaldy, 1906, Trans. Am. Ent. Soc., 32: 121.

1871 *Dasypterus* Reuter, Öfv. K. Svens. Vet.-Akad. Förh., 28(5): 564. Type-species: *Xylocoris limbatellus* Stål, 1860. Designated by Kirkaldy, 1906, Trans. Am. Ent. Soc., 32: 121. Synonymized by Reuter, 1884, Acta Soc. Sci. Fenn., 14: 692.

Cardiastethus assimilis (Reuter), 1871

1871 *Dasypterus assimilis* Reuter, Öfv. K. Svens. Vet.-Akad. Förh., 28(5): 564. [S.C., Tex.].
1884 *Cardiastethus assimilis*: Reuter, Acta Soc. Sci. Fenn., 14: 593.
Distribution: Fla., S.C., Tex., Va. (West Indies).

Cardiastethus borealis Kelton, 1977

1977 *Cardiastethus borealis* Kelton, Can. Ent., 109: 246. [B.C.]
Distribution: B.C., Man., N.S.

Cardiastethus cavicollis Blatchley, 1934

1934 *Cardiastethus cavicollis* Blatchley, Trans. Am. Ent. Soc., 60: 12. [Cal.].
Distribution: Cal.

Cardiastethus flaveolus Blatchley, 1928

1928 *Cardiastethus flaveolus* Blatchley, Ent. News, 39: 85. [Fla.].
Distribution: Fla.

Cardiastethus luridellus (Fieber), 1860

1860 *Cardiastethus luridellus* Fieber, Wien. Ent. Monat., 4: 271. [Pa.].
Distribution: Pa.

Cardiastethus pergandei Reuter, 1884

1884 *Cardiastethus pergaudei* [*sic*] Reuter, Acta Soc. Sci. Fenn., 14: 695. [D.C.].
1916 *Cardiastethus pergandii* [*sic*]: Van Duzee, Check List Hem., p. 35.
1930 *Cardiastethus pergandei*: Torre-Bueno, Bull. Brook. Ent. Soc., 25: 20.
Distribution: D.C.

Genus *Dufouriellus* Kirkaldy, 1906

1906 *Dufouriellus* Kirkaldy, Trans. Am. Ent. Soc., 32: 121. Type-species: *Xylocoris ater* Dufour, 1833. Original designation.

Dufouriellus ater (Dufour), 1833

1833 *Xylocoris ater* Dufour, An. Soc. Ent. France, 2: 106. [Europe].
1909 *Xylocoris ater*: Oshanin, Verz. Palearc. Hem., 3: 637.

1916 *Dufouriellus ater*: Van Duzee, Check List Hem., p. 35
Distribution: B.C., Cal., Id., Ky., N.C., N.Y., Ont.

Genus *Physopleurella* Reuter, 1884

1884 *Physopleurella* Reuter, Acta Soc. Sci. Fenn., 14: 678. Type-species: *Cardiastethus mundulus* White, 1877. Monotypic.

Physopleurella floridana Blatchley, 1925
1925 *Physopleurella floridana* Blatchley, Ent. News, 34: 47. [Fla.].
Distribution: Fla. (Jamaica).

Tribe Lyctocorini Reuter, 1884

Genus *Lyctocoris* Hahn, 1836

1836 *Lyctocoris* Hahn, Wanz. Ins., 3: 19. Type-species: *Cimex domesticus* Schilling, 1834, a junior synonym of *Acanthia campestris* Fabricius, 1794. Designated by Kirkaldy, 1906, Trans. Am. Ent. Soc., 32: 119.
Note: Kelton (1967, Can. Ent., 99: 807-814) reviewed and provided keys for species in North America.

Subgenus *Dolichomerium* Kirkaldy, 1900

1871 *Dolichomerus* Reuter, Öfv. K. Svens. Vet.-Akad. Förh., 28(5): 557. Preoccupied. Type-species: *Dolichomerus elongatus* Reuter, 1871. Designated by Kirkaldy, 1906, Trans. Am. Ent. Soc., 32: 119.
1884 *Lyctocoris* (*Dolichomerus*): Reuter, Acta Soc. Sci. Fenn., 14: 564.
1900 *Dolichomerium* Kirkaldy, Ent., 33: 242. New name for *Dolichomerus* Reuter.

Lyctocoris elongatus (Reuter), 1871
1871 *Dolichomerus elongatus* Reuter, Öfv. K. Svens. Vet.-Akad. Förh., 28: 558. [S.C.].
1884 *Lyctocoris elongatus*: Reuter, Acta Soc. Sci. Fenn., 14: 565.
Distribution: Ala., Fla., Ga., Id., N.C., N.J., S.C., Tex.

Lyctocoris stalii (Reuter), 1871
1871 *Dolichomerus stalii* Reuter, Öfv. K. Svens. Vet.-Akad. Förh., 28(5): 558. [S.C., Tex.].
1872 *Lyctocoris campestris*: Walker, Cat. Hem. Brit. Mus., 5: 154.
1879 *Dolichomerus reuteri* White, Ent. Month. Mag., 16: 146. [Ga., Mo.]. Synonymized by Reuter, 1884, Acta Soc. Sci. Fenn., 14: 564.
1926 *Lyctocoris elongatus*: Blatchley, Het. E. N. Am., p. 625 (in part).
Distribution: Ala., B.C., Cal., Fla., Ga., Ind., Man., Mo., N.C., N.Y., S.C., Tex.
Note: Blatchley (1928, J. N.Y. Ent. Soc., 36: 7) said his (1926, above) Ala. record refers to this species.

Subgenus *Lyctocoris* Hahn, 1836

1836 *Lyctocoris* Hahn, Wanz. Ins., 3: 19. Type-species: *Cimex domesticus* Schilling, 1834, a junior synonym of *Acanthia campestris* Fabricius, 1794. Designated by Kirkaldy, 1906, Trans. Am. Ent. Soc., 32: 119.

1884 *Lyctocoris* (*Lyctocoris*): Reuter, Acta Soc. Sci. Fenn., 14: 561.

Lyctocoris campestris (Fabricius), 1794 [Fig. 5]

1794 *Acanthia campestris* Fabricius, Ent. Syst., 4: 75. [Europe].
1834 *Cimex domesticus* Schilling, Isis, p. 738. [Europe]. Synonymized by Reuter, 1871, Öfv. K. Svens. Vet.-Akad. Förh., 28(3): 409.
1852 *Xylocoris americanus* Dallas, List Hem. Brit. Mus., p. 589. [North America]. Synonymized by Van Duzee, 1916, Check List Hem., p. 34.
1871 *Lyctocoris fitchi* Reuter, Öfv. K. Svens. Vet.-Akad. Förh., 28(5): 557. [N.Y.]. Synonymized by Reuter, 1884, Acta Soc. Sci. Fenn., 14: 561.
1871 *Lyctocoris domesticus*: Uhler, Proc. Boston Soc. Nat. Hist., 14: 106.
1884 *Lyctocoris campestris*: Reuter, Acta Soc. Sci. Fenn., 14: 561.
1886 *Lyctocoris americanus*: Uhler, Check-list Hem. Het., p. 21.
Distribution: Ala., B.C., Col., Fla., Id., Ill., Ind., Ks., Man., Md., Me., Mo., N.C., N.Y., Ont., Que., Tex., Ut., Wis. (Palearctic, Mexico, "cosmopolitan")

Lyctocoris canadensis Kelton, 1967

1967 *Lyctocoris canadensis* Kelton, Can. Ent., 99: 810. [Que.].
Distribution: Man., Que.

Lyctocoris doris Van Duzee, 1921

1921 *Lyctocoris doris* Van Duzee, Proc. Cal. Acad. Sci., (4)11: 139. [Cal.].
Distribution: Cal.

Lyctocoris okanaganus Kelton and Anderson, 1962

1962 *Lyctocoris okanaganus* Kelton and Anderson, Can. Ent., 94: 1303. [B.C.].
Distribution: B.C.

Lyctocoris rostratus Kelton and Anderson, 1962

1962 *Lyctocoris rostratus* Kelton and Anderson, Can. Ent., 94: 1304. [B.C.].
Distribution: B.C.

Lyctocoris tuberosus Kelton and Anderson, 1962

1962 *Lyctocoris tuberosus* Kelton and Anderson, Can. Ent., 94: 1303. [B.C.].
Distribution: B.C., Col., S.D.

Tribe Scolopini Carayon, 1954

Genus *Calliodis* Reuter, 1871

1871 *Calliodis* Reuter, Öfv. K. Svens. Vet.-Akad. Förh., 28(5): 558. Type-species: *Calliodis picturata* Reuter, 1871. Monotypic.
1884 *Asthenidea* Reuter, Acta Soc. Sci. Fenn., 14: 602. Type-species: *Asthenidea temnostethoides* Reuter, 1884. Designated by Kirkaldy, 1906, Trans. Am. Ent. Soc., 32: 120. Synonymized by Carayon, 1972, An. Soc. Ent. France, 8: 341.

Calliodis pallescens (Reuter), 1884

1884 *Asthenidea pallescens* Reuter, Acta Soc. Sci. Fenn., 14: 605. [Mexico].
1926 *Asthenidea pallescens*: Blatchley, Het. E. N. Am., p. 631.
1972 *Calliodis pallescens*: Carayon, An. Soc. Ent. France, 8: 347.
Distribution: Fla. (Mexico).

Calliodis semipicta (Blatchley), 1926

1926 *Asthenidea semipicta* Blatchley, Het. E. N. Am., p. 632. [Fla.].
1972 *Calliodis semipicta*: Carayon, An. Soc. Ent. France, 8: 347.
Distribution: Fla.

Calliodis temnostethoides (Reuter), 1884
1884 *Asthenidea temnostethoides* Reuter, Acta Soc. Sci. Fenn., 14: 605. [Ill.].
1972 *Calliodis temnostethoides*: Carayon, An. Soc. Ent. France, 8: 347.
Distribution: Conn., Fla., Ill., Mo., N.B., N.S., N.Y., Ont., P.Ed., Que., Tex.

Genus *Nidicola* Harris and Drake, 1941

1941 *Nidicola* Harris and Drake, Ia. St. Coll. J. Sci., 15: 343. Type-species: *Nidicola marginata* Harris and Drake, 1941. Original designation.

Note: Drake and Herring (1964, Proc. Biol. Soc. Wash., 77: 53-64) reviewed this genus, provided a key to species (p. 56-57), and described the four new species, *N. aglaia*, *N. engys*, *N. etes*, and *N. mitra*, all of which were intercepted in the U.S. on plant material coming from Guatemala or Mexico.

Nidicola jaegeri Peet, 1979
1979 *Nidicola jaegeri* Peet, An. Ent. Soc. Am., 72: 430. [Cal.].
Distribution: Cal.
Note: Peet (1979, above) provided a biological outline for this species.

Nidicola marginata Harris and Drake, 1941
1941 *Nidicola marginata* Harris and Drake, Ia. St. Coll. J. Sci., 15: 344. [Ariz.].
Distribution: Ariz., Cal., N.M.

Genus *Scoloposcelis* Fieber, 1864

1864 *Scolopscelis* Fieber, Wien. Ent. Monat., 7: 62. Type-species: *Xylocoris crassipes* Flor, 1860, a junior synonym of *Anthocoris pulchellus* Zetterstedt, 1838. Monotypic.

Scoloposcelis basilicus Drake and Harris, 1926
1926 *Scolopscelis basilicus* Drake and Harris, Proc. Biol. Soc. Wash., 39: 44. [N.M.].
Distribution: Ariz., N.M.

Scoloposcelis flavicornis Reuter, 1871
1871 *Scoloposcelis flavicornis* Reuter, Öfv. K. Svens. Vet.-Akad. Förh., 28(5): 561. [Tex.].
1926 *Scoloposcelis mississippensis* Drake and Harris, Proc. Biol. Soc. Wash., 39: 43. [Miss.]. Synonymized by Kelton, 1976, Can. Ent., 108: 196.
1926 *Scoloposcelis occidentalis* Drake and Harris, Proc. Biol. Soc. Wash., 39: 45. [Cal.]. Synonymized by Kelton, 1976, Can. Ent., 108: 196.
1976 *Scoloposcelis mississipensis* [*sic*]: Kelton, 1976, Can. Ent., 108: 196.
Distribution: Alta., B.C., Cal., Fla., Ind., Man., Miss., N.B., N.S., Ont., Pa., Que., Sask., Tex., Yuk. (Guatemala, Mexico).

Genus *Solenonotus* Reuter, 1871

1871 *Solenonotus* Reuter, Öfv. K. Svens. Vet.-Akad. Förh., 28(5): 559. Type-species: *Anthocoris sulcifer* Stål, 1860. Monotypic.

Solenonotus angustatus Poppius, 1913
1913 *Solenonotus angustatus* Poppius, An. Soc. Ent. Belg., 57: 13. [Cal.].
Distribution: Cal.

Tribe Xylocorini Carayon, 1972

Genus *Xylocoris* Dufour, 1831

1831 *Xylocoris* Dufour, An. Sci. Nat. Paris, 22: 423. Type-species: *Xylocoris rufipennis* Dufour, 1831, a junior synonym of *Lygaeus cursitans* Fallén, 1807. Monotypic.

1860 *Piezostethus* Fieber, Wien. Ent. Monat., 4: 265. Type-species: *Xylocoris rufipennis* Dufour. Designated by Kirkaldy, 1906, Trans. Am. Ent. Soc., 32: 119. Synonymy by virture of common type-species.

Note: Carayon (1972, An. Ent. Soc. France, 8: 579-606) reviewed the subgeneric classification of *Xylocoris*.

Subgenus *Arrostelus* Kirkaldy, 1906

1884 *Arrostus* Reuter, Acta Soc. Sci. Fenn., 14: 589. Preoccupied. Type-species: *Piezostethus flavipes* Reuter, 1875. Monotypic.

1906 *Arrostelus* Kirkaldy, Trans. Am. Ent. Soc., 32: 119. New name for *Arrostus* Reuter.

Xylocoris flavipes (Reuter), 1875

1875 *Piezostethus flavipes* Reuter, Bih. K. Svens. Vet.-Akad. Handl., 3(1): 65. [Europe].

1961 *Xylocoris flavipes*: Bibby, J. Econ. Ent., 54: 328.

Distribution: Ariz., D.C., Ga., Ks., Md., Tex. (Africa, Asia, Australia, Europe, South America).

Note: *Xylocoris flavipes* has been reported in numerous publications as an important predator of common stored grain pests. There are specimens in the USNM (D.C., Ks., Md., Tex.) collected in stored grain as early as 1933, but Bibby (1961, above) apparently is the first to publish a North American record. Arbogast et al. (1971, An. Ent. Soc. Am., 64: 1131-1134) studied the developmental stages; Arbogast (1975, Environ. Ent., 4: 825-831) gave information on the influence of temperature and humidity; Arbogast et al. (1977, J. Ga. Ent. Soc., 12: 58-64) studied longevity; Arbogast (1978, Proc. Second Int. Work Comm. Stored-Prod. Ent., pp. 91-105; 1979, Ent. Exp. Appl., 25: 128-135) documented cannibalism; Arbogast (1976, J. Ga. Ent. Soc., 11: 67-71), Jay et al. (1968, J. Ga. Ent. Soc., 3: 126-130), LeCato and Davis (1973, Fla. Ent., 56: 57-59), and Press et al. (1975, J. Ga. Ent. Soc., 10: 76-78) reported on the biology and predatory habits.

Subgenus *Proxylocoris* Carayon, 1972

1972 *Xylocoris* (*Proxylocoris*) Carayon, An. Ent. Soc. France (new ser.), 8: 594. Type-species: *Piezostethus afer* Reuter, 1884. Original designation.

Xylocoris galactinus (Fieber), 1837

1837 *Anthocoris galactinus* Fieber, Beitr. Gesch. Nat., 1: 107. [Europe].

1884 *Piezostethus galactinus*: Reuter, Acta Soc. Sci. Fenn., 14: 590.

1914 *Piezostethus flaccidus* Van Duzee, Trans. San Diego Soc. Nat. Hist., 2: 14. [Cal.]. Synonymized by Kelton, 1977, Can. Ent., 109: 1017.

1916 *Xylocoris galactinus*: Van Duzee, Check List Hem., p. 34.

1916 *Xylocoris flaccidus*: Van Duzee, Check List Hem., p. 34.

Distribution: Alta., B.C., Cal., Fla., Ga., Id., Ill., Man., Mo., N.J., N.Y., Ont., Que., Sask. (Palearctic).

Xylocoris sordidus (Reuter), 1871
1871 *Piezostethus sordidus* Reuter, Öfv. K. Svens. Vet.-Akad. Förh., 28(5): 560. [Tex., Brazil].
1871 *Piezostethus binotatus* Reuter, Öfv. K. Svens. Vet.-Akad. Förh., 28(5): 560. [S.C.]. Synonymized by Reuter, 1884, Acta Soc. Sci. Fenn., 14: 564.
1884 *Piezostethus sordidus* var. *binotatus*: Reuter, Acta Soc. Sci. Fenn., 14: 591.
1916 *Xylocoris sordidus*: Van Duzee, Check List Hem., p. 34.
1917 *Xylocoris sordidus* var. *binotatus*: Van Duzee, Univ. Cal. Publ. Ent., 2: 291.
Distribution: Ariz., Cal., Fla., Ga., Ks., Mass., Md., N.M., N.J., N.Y., Pa., S.C., Tenn., Tex. (Central America, Mexico, West Indies).
Note: Arbogast et al. (1983, Am. Midl. Nat., 109: 398-405) studied the biology and habits of this species in storage ecosystems in Georgia; Arbogast et al. (1985, Ent. News, 96: 53-58) studied the developmental stages.

Subgenus *Xylocoris* Dufour, 1831

1831 *Xylocoris* Dufour, An. Sci. Nat. Paris, 22: 423. Type-species: *Xylocoris rufipennis* Dufour, 1831, junior synonym of *Lygaeus cursitans* Fallén, 1807. Monotypic.

Xylocoris betulinus Drake and Harris, 1926
1926 *Xylocoris betulinus* Drake and Harris, Proc. Biol. Soc. Wash., 39: 37. [N.Y.].
Distribution: N.Y.

Xylocoris californicus (Reuter), 1884
1884 *Piezostethus californicus* Reuter, Acta Soc. Sci. Fenn., 14: 600.
1916 *Xylocoris californicus*: Van Duzee, Check List Hem., p. 34.
Distribution: Cal., N.M., Ut.

Xylocoris cursitans (Fallén), 1807
1807 *Lygaeus cursitans* Fallén, Monogr. Cimic. Suec., p. 74. [Europe].
1916 *Xylocoris cursitans*: Van Duzee, Check List Hem., p. 34
1962 *Xylocoris vicarius*: Anderson, Can. Ent., 94: 1326.
Distribution: Alta., B.C., Conn., Id., Ind., Mich., N.J., N.S., N.Y., Ont., Ore., Que. (Palearctic).

Xylocoris discalis (Van Duzee), 1914
1914 *Scoloposcelis discalis* Van Duzee, Trans. San Diego Soc. Nat. Hist., 2: 15. [Cal.]. Lectotype designated by Kelton, 1976, Can. Ent., 108: 193.
1921 *Xylocoris discalis*: Van Duzee, Proc. Cal. Acad. Sci., (4)11: 138.
Distribution: Ariz., Cal.
Note: Kelton (1976, Can. Ent., 108: 193) clarified the generic placement of this species.

Xylocoris hirtus Kelton, 1976
1976 *Xylocoris hirtus* Kelton, Can. Ent., 108: 193. [Sask.].
Distribution: N.Y., Ont., Que., Sask.

Xylocoris pilipes Kelton, 1976
1976 *Xylocoris pilipes* Kelton, Can. Ent., 108: 194. [N.Y.].
Distribution: N.Y.

Xylocoris punctatus Kelton, 1976
1976 *Xylocoris punctatus* Kelton, Can. Ent., 108: 196. [Ut.].
Distribution: Ut.

Xylocoris umbrinus Van Duzee, 1921

1921 *Xylocoris umbrinus* Van Duzee, Proc. Cal. Acad. Sci., (4)11: 137. [Cal.].

Distribution: Alta., B.C., Cal., Id., Man., Sask.

Xylocoris vicarius (Reuter), 1884

1884 *Piezostethus vicarius* Reuter, Acta Soc. Sci. Fenn., 14: 599. [Colombia, "Americae septemtrionalis"].

1916 *Xylocoris vicarius*: Van Duzee, Check List Hem., p. 34.

Distribution: Fla., Mass., N.J., N.Y. (South America).

Note: Blatchley (1926, Het. E. N. Am., p. 630) gave the first definite state record for this species.

Family Aradidae Spinola, 1837

(= Dysodiidae Reuter, 1912; Meziridae Oshanin, 1908)

The Flat Bugs

By Richard C. Froeschner

Close observation has shown that the generally accepted picture of flat bugs living and feeding on fungi under bark is not descriptive of all forms. As early as the turn of the century, Schwarz (1901, Proc. Ent. Wash., 4: 391) noted "... the family may be divided into two classes, according to food habits, the one appearing to feed upon a blackish mould under the bark, while others live outside the bark of dead trees, upon a whitish fungus."

Usinger (1936, An. Ent. Soc. Am., 29: 491, 507) reported four species of the genus *Aradus* Curtis as frequenting and feeding on exposed surfaces of "shelf fungi" (or "ball fungi"), *Polyporus* Fries, in California. Also for the western United States he noted several occurrences of *Mezira reducta* Van Duzee with termites, even to being present in undisturbed termite galleries. He hypothesized that both insect forms fed on fungi that may have been present, the termites chewing off the hyphae that grew into the tunnels and the flat bugs using their long threadlike stylets to feed on fungal parts embedded in the gallery walls.

The North American species *Aradus kormilevi* Heiss was reported (under the combination *Aradus cinnamomeus* Panzer, a European species) in association with the southern pine beetle, *Dendroctonus frontalis* Zimmerman by Overgaard (1968, J. Econ. Ent., 61: 1199) and by Moser et al. (1971, An. Ent. Soc. Am., 64: 73). The former author noted it as a "predator of the Southern Pine Beetle" but without details of the observation; the latter author cited its feeding habits from previous authors and characterized it as being "saprophagous on pines."

In Europe *Aradus cinnamomeus* attacks and causes the death of young shoots on pine trees. There, Tanada (1959, An. Rev. Ent., 4: 281) and Franz (1961, An. Rev. Ent., 6: 186) reported this flat bug was successfully controlled by field application of the parasitic fungus genus *Beauveria* Vuillemin (Fungi Imperfecti).

The limited information on the biology of North American flat bugs allows only an outline of their behavior. During the cold months the individuals are immobilized by the chilled air; most of our species hibernating under loose bark or other objects where they were feeding. As the weather warms the insects begin to move around and resume feeding. Mating takes place in the natural feeding areas, not in the flights which take them to new feeding grounds. Because the fungi that serve as food for the mycetophagous forms live on dead trees only at a certain stage in the tree's decay, the flat bugs must seek fresh food sources every year or so. Eggs are laid on or near the food that will be utilized for nourishment by the newly hatched and developing immatures.

For America north of Mexico the following list enumerates 123 species in 10 genera.

For study of flat bugs in the area of this

list four works are useful: the *Synopsis* of the family by Torre-Bueno (1939, Ent. Am., 19: 258-276), which provides keys to most North American species; Parshley's (1921, Trans. Am. Ent. Soc., 47: 1-106), scholarly *Essay*, which provides many biological details for members of the genus *Aradus*; the worldwide comprehensive *Classification of the Aradidae* by Usinger and Matsuda (1959); and Matsuda's (1977, Can. Dept. Agr., Publ. 1634) coverage of the Canadian species.

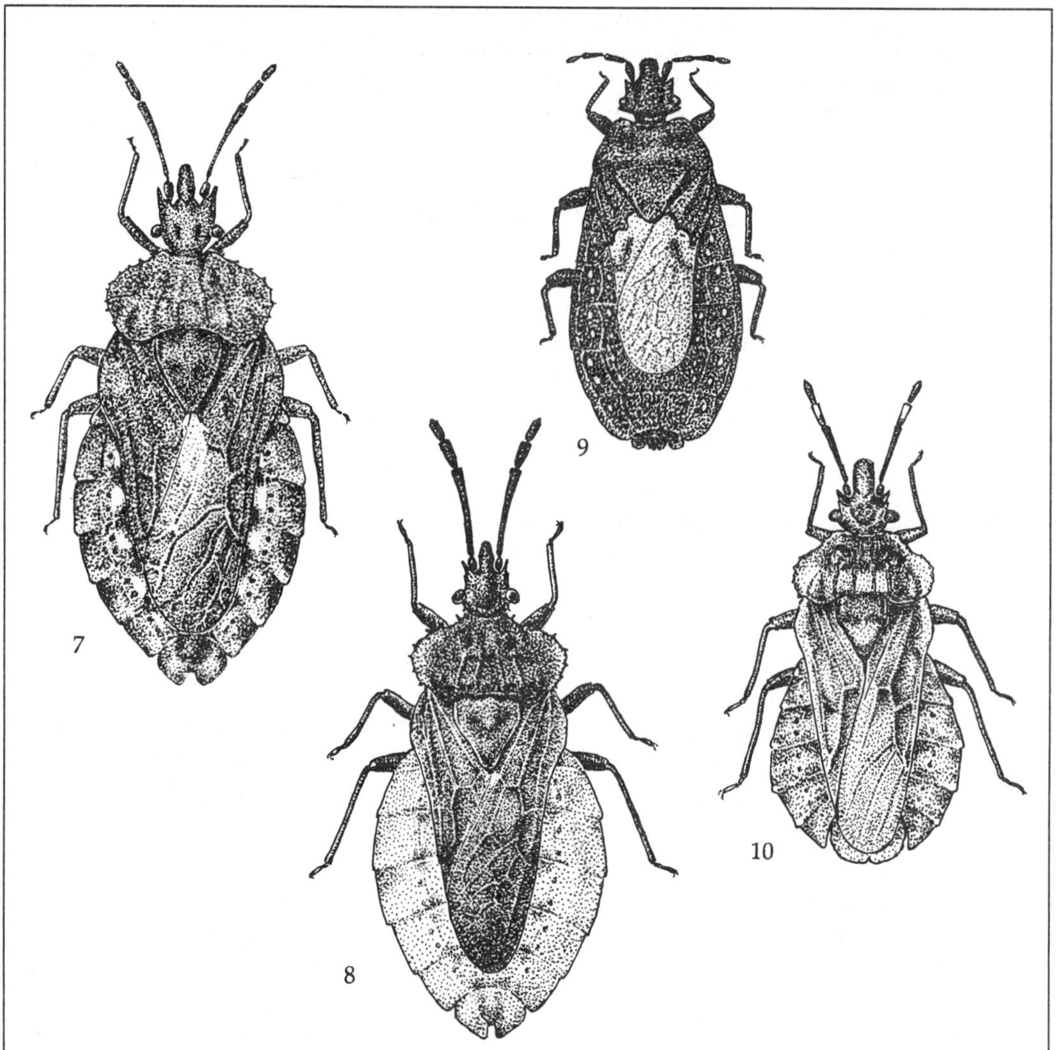

Figs. 7-10: 7, Aradus *acutus* [p.33]; 8, *Aradus inornatus* [p. 36]; 9, *Neuroctenus simplex* [p. 46]; 10, *Aradus cincticornis* [p. 34] (After Froeschner, 1942).

Subfamily Aneurinae Douglas and Scott, 1865

Genus *Aneurus* Curtis, 1825

1825 *Aneurus* Curtis, Brit. Ent., 7: pl. 86. Type-species: *Acanthia laevis* Fabricius, 1775. Monotypic.

Note: Four subgenera have been placed within this genus: *Aneurillus* Kormilev, *Aneurus* Curtis, *Aneurosoma* Champion, and *Iralunelus* Štys. All species in the area north of Mexico appear to belong to the nominate subgenus. A revision of thirty-two North and Central American and West Indian species was provided by Picchi (1977, Quaest. Ent., 13: 255-308, 267-270).

Subgenus *Aneurus* Curtis, 1825

1825 *Aneurus* Curtis, Brit. Ent., 7: pl. 86. Type-species: *Acanthia laevis* Fabricius, 1775. Monotypic.

1967 *Aneurus* (*Aneurus*): Kormilev, Opusc. Ent., 100:1

Aneurus arizonensis Picchi, 1977

1977 *Aneurus arizonensis* Picchi, Quaest. Ent., 13: 280. [Ariz.].

Distribution: Ariz.

Aneurus borealis Picchi, 1977

1977 *Aneurus borealis* Picchi, Quaest. Ent., 13: 282. [Alk.].

Distribution: Alk., B.C., Me., Ont., Que., Sask., Yuk.

Aneurus deborahae Picchi, 1977

1977 *Aneurus deborahae* Picchi, Quaest. Ent., 13: 274. [Cal.].

Distribution: Cal.

Aneurus fiskei Heidemann, 1904

1904 *Aneurus fiskei* Heidemann, Proc. Ent. Soc. Wash., 6: 164. [Ga., N.C., Pa., Va.].

1968 *Aneurus* (*Aneurus*) *fiskei*: Kormilev, Proc. U.S. Nat. Mus., 125 (3657): 3.

Distribution: Conn., D.C., Ga., Ill., Ind., Md., Mich., N.C., N.J., N.Y., Oh., Pa., Tenn., Tex., Va. (Mexico).

Aneurus inconstans Uhler, 1871

1871 *Aneurus inconstans* Uhler, Proc. Bost. Soc. Nat. Hist., 14: 105. [Mass.].

1968 *Aneurus* (*Aneurus*) *inconstans*: Kormilev, Proc. U.S. Nat. Mus., 125 (3657): 3.

Distribution: Alta., B.C., Conn., D.C., Del., Ind., Mass., Md., Me., N.C., N.H., N.J., N.S., N.Y., Ont., Pa., Que., R.I., S.D., Va., Vt.

Note: Under his description of *A. inconstans*, Uhler (1871, Proc. Bost. Soc. Nat. Hist., 14: 105) mentioned a Say manuscript name, "*Aradus sanguineus,*" on specimens in the T.W. Harris collection.

Aneurus leptocerus Hussey, 1957

1898 *Aneurus tenuicornis* Champion, Biol. Centr.-Am., Rhyn., 2: 116. [Guatemala, Panama]. Preoccupied.

1916 *Aneurus tenuicornis*: Van Duzee, Check List Hem., p. 17.

1957 *Aneurus leptocerus* Hussey, Fla. Ent., 40: 80. New name for *Aneurus tenuicornis* Champion.

Distribution: Ala., Fla., Ga. (Guatemala, Panama).

Aneurus minutus Bergroth, 1886

1886 *Aneurus minutus* Bergroth, Verh. Zool.-Bot. Gesell. Wien, 36: 58. [Tex.].

1968 *Aneurus* (*Aneurus*) *minutus*: Kormilev, Proc. U.S. Nat. Mus., 125 (3657): 3.

Distribution: Ariz., Fla., Ga., Oh.(?), Tex. (Mexico, West Indies).

Note: Blatchley (1926, Het. E. N. Am., p. 330) questioned the "Ohio" record.

Aneurus politus Say, 1832

1832 *Aneurus politus* Say, Descrip. Het. Hem. N. Am., p. 31. [Fla.]. Neotype designated by Picchi, 1977, Quaest. Ent., 13: 271.

1968 *Aneurus* (*Aneurus*) *politus*: Kormilev, Proc. U.S. Nat. Mus., 125 (3657): 3.

Distribution: Ala., Fla., Ga., Miss., N.Y.

Note: The "Cuba" records of *politus* by Uhler (1878, Proc. Bost. Soc. Nat. Hist., 19: 421) apparently belong to *A. patriciae* Picchi, the only species of *Aneurus* listed for Cuba in Picchi's revision.

Aneurus pygmaeus Kormilev, 1966

1966 *Aneurus pygmaeus* Kormilev, Proc. U.S. Nat. Mus., 119 (3548): 6. [Tex.].

1968 *Aneurus* (*Aneurus*) *pygmaeus*: Kormilev, Proc. U.S. Nat. Mus., 125 (3657): 3.

Distribution: Cal., Fla., Ga., Miss., Tex.

Aneurus roseae Picchi, 1977

1977 *Aneurus roseae* Picchi, Quaest. Ent., 13: 279. [Tex.].

Distribution: Tex.

Aneurus simplex Uhler, 1871

1871 *Aneurus simplex* Uhler, Proc. Bost. Soc. Nat. Hist., 14: 106. [Mass.]. Lectotype designated by Parshley, 1922, Ent. News, 33: 43.

1873 *Aneurus septentrionalis* Walker, Cat. Hem. Brit. Mus., 7: 30. [N.S., Ont.]. Synonymized by Picchi, 1977, Quaest. Ent., 13: 278.

1968 *Aneurus* (*Aneurus*) *simplex*: Kormilev, Proc. U.S. Nat. Mus., 125 (3657): 2.

1968 *Aneurus* (*Aneurus*) *septentrionalis*: Kormilev, Proc. U.S. Nat. Mus., 125 (3657): 2.

Distribution: Alk., Alta., B.C., Col., Id., Man., Mass., Me., Mont., N.C., N.H., N.J., N.S., N.T., N.Y., Ont., Ore., Que., Vt., Wash., Wyo., Yuk.

Subfamily Aradinae Amyot and Serville, 1843

Genus *Aradus* Fabricius, 1803

1803 *Aradus* Fabricius, Syst. Rhyn., p. 116. Type-species: *Cimex betulae* Linnaeus, 1758. Designated by Latreille, 1810, Consid. Gen. Crust. Ins., p. 433.

Note: Parshley (1921, Trans. Am. Ent. Soc., 47: 1-106, pls. 1-7) revised and gave a key (pp. 23-29) to 60 New World species. Rathvon (1869, Hist. Lancaster Co., Pa., p. 549) listed two species of *Aradus*, *sanguineus* (see *Anerus inconstans*) and *penultimus*, neither of which appears to have been validly published.

Subgenus *Aradus* Fabricius, 1803

1803 *Aradus* Fabricius, Syst. Rhyn., p. 116. Type-species: *Cimex betulae* Linnaeus, 1758. Designated by Latreille, 1810, Consid. Gen. Crust. Ins., p. 433.

1873 *Aradus* (*Aradus*): Stål, K. Svens. Vet.-Akad. Handl., 11(2): 136.

Aradus abbas Bergroth, 1889
1889 *Aradus abbas* Bergroth, Bull. Soc. Ent. Belg., 33: clxxx. [Que.].
1916 *Aradus* (*Aradus*) *abbas*: Van Duzee, Check List Hem., p. 16.
Distribution: Alk., Alta., B.C., Cal., Conn., D.C., Fla., Id., Ill., Man., Mass., Me., Mich., N.B., N.C., N.H., N.J., N.T., N.Y., Nfld., Ont., Pa., Que., Sask., Va., Vt., Wash., Wis.

Aradus acutus Say, 1832 [Fig. 7]
1832 *Aradus acutus* Say, Descrip. Het. Hem. N. Am., p. 28. [Fla., Ind.].
1847 *Aradus americanus* Herrich-Schaeffer, Wanz. Ins., 8: 115. [N. Am.]. Synonymized by Stål, 1873, K. Svens. Vet.-Akad. Handl., 11(2): 136.
1873 *Aradus* (*Aradus*) *acutus*: Stål, K. Svens. Vet.-Akad. Handl., 11(2): 136.
Distribution: Ala., Alk., Cal., Col., D.C., Del., Fla., Ga., Id., Ill., Ind., Ks., Man., Md., Me., Miss., Mo., Mont., N.C., N.H., Oh., Pa., S.C., Tex., Ut., Wash.

Aradus aequalis Say, 1832
1832 *Aradus aequalis* Say, Descrip. Het. Hem. N. Am., p. 29. [Ind.].
1873 *Aradus* (*Aradus*) *aequalis*: Stål, K. Svens. Vet.-Akad. Handl., 11(2): 136.
1903 *Aradus druryi* Obsborn, Oh. Nat., 4: 39. [Oh.]. Synonymized by Bergroth, 1913, Can. Ent., 45: 1.
Distribution: D.C., Ill., Ind., Md., Me., N.J., N.Y., Oh., Ok., Ont., Pa., Que., Tex., Va., Vt.

Aradus alaskanus Kormilev and Heiss, 1979
1979 *Aradus alaskanus* Kormilev and Heiss, Ber. Nat.-Med. Ver. Innsbruck, 66: 47. [Alk.].
Distribution: Alk.

Aradus ampliatus Uhler, 1876
1876 *Aradus ampliatus* Uhler, Bull. U.S. Geol. Geogr. Surv. Terr., 1: 321. [Cal.].
1916 *Aradus* (*Aradus*) *ampliatus*: Van Duzee, Check List Hem., p. 16.
Distribution: Cal., Ut.

Aradus antennalis Parshley, 1921
1921 *Aradus* (*Aradus*) *cinnamomeus* var. *antennalis* Parshley, Trans. Am. Ent. Soc., 47: 97. [B.C.].
1980 *Aradus antennalis*: Heiss, Ber. Nat.-Med. Ver. Innsbruck, 67: 112.
Distribution: B.C., Cal., Id., Neb., Wash.

Aradus apicalis Van Duzee, 1920
1917 *Aradus* (*Aradus*) *duzeei*: Van Duzee, Univ. Cal. Publ. Ent., 2: 130 (in part).
1920 *Aradus apicalis* Van Duzee, Proc. Cal. Acad. Sci., ser. 4, 9: 331. [Cal.].
1921 *Aradus* (*Aradus*) *apicalis*: Parshley, Trans. Am. Ent. Soc., 47: 46.
Distribution: Cal.

Aradus approximatus Parshley, 1921
1921 *Aradus* (*Aradus*) *approximatus* Parshley, Trans. Am. Ent. Soc., 47: 72. [Miss.].
Distribution: Ga., Ind., Me., Miss., N.J., N.Y., Que.

Aradus arizonicus Parshley, 1921
1921 *Aradus* (*Aradus*) *arizonicus* Parshley, Trans. Am. Ent. Soc., 47: 83. [Ariz.].
Distribution: Ariz.

Aradus barberi Kormilev, 1966
1966 *Aradus* (*Aradus*) *barberi* Kormilev, Proc. U.S. Nat. Mus., 119(3548): 4. [Col.].
Distribution: Col.

Aradus basalis Parshley, 1921
1921 *Aradus* (*Aradus*) *basalis* Parshley, Trans. Am. Ent. Soc., 47: 54. [N.H.].
Distribution: Me., N.H., N.Y.

Aradus behrensi Bergroth, 1886
1886 *Aradus behrensi* Bergroth, Wien. Ent. Zeit., 5: 97. [Cal.].
1916 *Aradus* (*Aradus*) *behrensi*: Van Duzee, Check List Hem., p. 16.
Distribution: B.C., Cal., Ore., Wash.

Aradus blaisdelli Van Duzee, 1920
1876 *Aradus inornatus*: Uhler, Bull. U.S. Geol. Geogr. Surv. Terr., 1: 323 (in part).
1920 *Aradus blaisdelli* Van Duzee, Proc. Cal. Acad. Sci., ser. 4, 9: 333. [Cal.].
1921 *Aradus* (*Aradus*) *blaisdelli*: Parshley, Trans. Am. Ent. Soc., 47: 70.
Distribution: B.C., Cal., Id., Mont., Nev., Ore., Sask., Wash.

Aradus borealis Heidemann, 1909
1909 *Aradus borealis* Heidemann, Proc. Ent. Soc. Wash., 11: 190. [Mich., N.H.].
1916 *Aradus* (*Aradus*) *borealis*: Van Duzee, Check List Hem., p. 16.
Distribution: Cal., Me., Mich., N.H., Ont., Que., Sask.

Aradus breviatus Bergroth, 1887
1887 *Aradus breviatus* Bergroth, Rev. d'Ent., 6: 245. [Fla.].
1916 *Aradus* (*Aradus*) *breviatus*: Van Duzee, Check List Hem., p. 16.
Distribution: D.C., Fla., N.J.

Aradus brevicornis Kormilev, 1980
1980 *Aradus* (*Aradus*) *brevicornis* Kormilev, Proc. Ent. Soc. Wash., 82: 105. [Cal.].
Distribution: Cal.

Aradus brunnicornis Blatchley, 1926
1926 *Aradus* (*Aradus*) *brunnicornis* Blatchley, Het. E. N. Am., p. 311. [Fla.]. Lectotype designated by Blatchley, 1930, Blatchleyana, p. 64.
Distribution: Fla., N.C.

Aradus carolinensis Kormilev, 1964
1964 *Aradus* (*Aradus*) *carolinensis* Kormilev, Arkiv Zool., ser. 2, 16: 476. [N.C.].
Distribution: N.C.

Aradus cincticornis Bergroth, 1906 [Fig. 10]
1906 *Aradus cincticornis* Bergroth, Can. Ent., 38: 198. [Ala.].
1916 *Aradus* (*Aradus*) *cincticornis*: Van Duzee, Check List Hem., p. 16.
Distribution: Ala., Mo.
Note: Bergroth (1913, Can. Ent., 45: 3) noted that specimens of this species he examined were identified by "the unpublished name *A*[*radus*] *nasutus* Uhl."

Aradus coarctatus Heidemann, 1907
1907 *Aradus coarctatus* Heidemann, Proc. Ent. Soc. Wash., 8: 69. [Cal.].
1916 *Aradus* (*Aradus*) *coarctatus*: Van Duzee, Check List Hem., p. 16.
Distribution: B.C., Cal.

Aradus coloradensis Kormilev, 1964
1964 *Aradus coloradensis* Kormilev, Arkiv Zool., ser. 2, 16: 476. [Col.].
Distribution: Col.

Aradus compressus Heidemann, 1907
1907 *Aradus compressus* Heidemann, Proc. Ent. Soc. Wash., 8: 70. [Wash.].
1916 *Aradus* (*Aradus*) *compressus*: Van Duzee, Check List Hem., p. 16.
Distribution: B.C., Cal., Ore., Wash.

Aradus concinnus Bergroth, 1892

1892 *Aradus* (*Piestosoma*) *concinnus* Bergroth, Proc. Ent. Soc. Wash., 2: 337. [Cal.].
1921 *Aradus* (*Aradus*) *concinnus*: Parshley, Trans. Am. Ent. Soc., 47: 49.
Distribution: Cal.
Note: Usinger (1936, An. Ent. Soc. Am., 2: 495) placed *Aradus depictus* Van Duzee as a junior synonym of this species, but later (1943, Pan-Pac. Ent., 19: 138) retracted the synonymy.

Aradus consors Parshley, 1921

1921 *Aradus* (*Aradus*) *consors* Parshley, Trans. Am. Ent. Soc., 47: 56. [Mass.].
Distribution: Mass.

Aradus crenatus Say, 1832

1832 *Aradus crenatus* Say, Descrip. Het. Hem. N. Am., p. 28. [U.S.].
1916 *Aradus* (*Aradus*) *crenatus*: Van Duzee, Check List Hem., p. 16.
Distribution: Ala., Conn., D.C., Ga., Ill., Ind., Md., Mich., Mo., N.C., N.Y., Oh., Ont., Pa., Que., Va. (Mexico).
Note: Heiss (1980, Ber. Nat.-Med. Ver. Innstruck, 67: 104) redefined this species.

Aradus curticollis Bergroth, 1913

1913 *Aradus curticollis* Bergroth, Can. Ent., 45: 2. [N.C.].
1916 *Aradus* (*Aradus*) *curticollis*: Van Duzee, Check List Hem., p. 16.
Distribution: Ga., N.C.

Aradus debilis Uhler, 1876

1876 *Aradus debilis* Uhler, Bull. U.S. Geol. Geogr. Surv. Terr., 1: 322. [B.C.].
1916 *Aradus* (*Aradus*) *debilis*: Van Duzee, Check List Hem., p. 16.
Distribution: B.C., Cal., Col., Id., Mass., Mont., N.Y., Ore., Wash.

Aradus depictus Van Duzee, 1917

1917 *Aradus depictus* Van Duzee, Proc. Cal. Acad. Sci., ser. 4, 7: 253. [Cal.].
1921 *Aradus* (*Aradus*) *depictus*: Parshley, Trans. Am. Ent. Soc., 47: 47.
Distribution: B.C., Cal., Ore.
Note: Usinger (1936, An. Ent. Soc. Am., 29: 495) placed this species as a junior synonym of *Aradus concinnus*, but Torre-Bueno (1939, Ent. Am., 19: 282) and Usinger (1943, Pan-Pac. Ent., 19: 138) treated it as a valid.

Aradus duzeei Bergroth, 1892

1892 *Aradus Duzeei* [*sic*] Bergroth, Proc. Ent. Soc. Wash., 2: 333. [Ont., Pa.].
1910 *Aradus vanduzeei* [*sic*]: Heidemann, Proc. Ent. Soc. Wash., 12: 47.
1916 *Aradus* (*Aradus*) *duzeei*: Van Duzee, Check List Hem., p. 16.
1917 *Aradus duzei* [*sic*]: Van Duzee, Proc. Cal. Acad. Sci., ser. 4, 7: 253.
Distribution: Ind., Mass., Md., Mo., N.J., N.Y., Oh., Ont., Pa., Que., Va.

Aradus evermanni Van Duzee, 1920

1920 *Aradus evermanni* Van Duzee, Proc. Cal. Acad. Sci., ser. 4, 9: 338. [Cal.].
1921 *Aradus* (*Aradus*) *evermanni*: Parshley, Trans. Am. Ent. Soc., 47: 91.
Distribution: Ariz., Cal., Tex.

Aradus falleni Stål, 1860

1860 *Aradus* (*Aradus*) *Falleni* [*sic*]: Stål, K. Svens. Vet.-Akad. Handl., 2(7): 68. [Brazil].
1918 *Aradus fallini* [*sic*]: Johnson and Ledig, J. Ent. Zool., 10: 4.
Distribution: Ariz., Ark., B.C., Cal., Conn., D.C., Fla., Ga., Ill., Ind., La., Md., Miss., Mo., N.C., N.J., N.M., N.Y., Ok., Pa., R.I., Tex., Va., W.Va. (Mexico to Brazil, West Indies).

Aradus funestus Bergroth, 1913
1876 *Aradus tuberculifer*: Uhler, Bull. U.S. Geol. Geogr. Surv. Terr., 1: 321 (in part).
1913 *Aradus funestus* Bergroth, Can. Ent., 45: 4. [Canada; U.S. including Col.].
1916 *Aradus* (*Aradus*) *funestus*: Van Duzee, Check List Hem., p. 16.
Distribution: Alk., Alta., Ariz., B.C., Cal., Col., D.C., Id., Mont., N.Y., Nev., Ont., Ore., Que., Ut., Wash., Wyo.

Aradus furnissi Usinger, 1936
1936 *Aradus* (*Aradus*) *furnissi* Usinger, An. Ent. Soc. Am., 29: 500. [Cal.].
Distribution: Cal., Id.

Aradus furvus Parshley, 1921
1921 *Aradus* (*Aradus*) *furvus* Parshley, Trans. Am. Ent. Soc., 47: 155. [Ariz.].
Distribution: Ariz.

Aradus fuscipennis Usinger, 1936
1936 *Aradus* (*Aradus*) *fuscipennis* Usinger, An. Ent. Soc. Am., 29: 504. [Wash.].
Distribution: Wash.

Aradus fuscomaculatus Stål, 1859
1859 *Aradus fuscomaculatus* Stål, K. Svens. Freg. Eug. Resa Jorden, 3: 260. [Cal.].
1873 *Aradus* (*Aradus*) *fuscoannulatus* [*sic*]: Stål, K. Svens. Vet.-Akad. Handl., 11(2): 136.
1916 *Aradus* (*Aradus*) *fuscomaculatus*: Van Duzee, Check List Hem., p. 16.
Distribution: B.C., Cal., Ore., Wash.

Aradus gracilicornis Stål, 1873
1873 *Aradus* (*Aradus*) *gracilicornis* Stål, K. Svens. Vet.-Akad. Handl., 11(2): 136. [Cuba].
1906 *Aradus gracilicornis*: Bergroth, Can. Ent., 38: 200.
1916 *Aradus* (*Aradus*) *gracilicornis*: Van Duzee, Check List Hem., p. 16.
Distribution: Ala., Ariz., Fla., Ga., Miss., N.C., N.M., Tex. (Cuba).

Aradus gracilis Parshley, 1929
1929 *Aradus* (*Aradus*) *gracilis* Parshley, Can. Ent., 61: 245. [Alta.].
Distribution: Alta.

Aradus hesperius Parshley, 1921
1921 *Aradus* (*Aradus*) *hesperius* Parshley, Trans. Am. Ent. Soc., 47: 71. [Ariz.].
Distribution: Ariz., Col.

Aradus implanus Parshley, 1921
1921 *Aradus* (*Aradus*) *implanus* Parshley, Trans. Am. Ent. Soc., 47: 45. [Ont.].
Distribution: D.C., Ill., Ind., Mich., Ont., Pa., Que.

Aradus inornatus Uhler, 1876 [Fig. 8]
1876 *Aradus inornatus* Uhler, Bull. U.S. Geol. Geogr. Surv. Terr., 1: 323. [Md.].
1916 *Aradus* (*Aradus*) *inornatus*: Van Duzee, Check List Hem., p. 16.
Distribution: D.C., Ga., Ill., Man., Mass., Md., Me., Mich., Mo., N.C., N.H., N.J., Neb., Ont., Pa., Que., S.C., S.D., Tenn., Tex., Va., W.Va., Wis.
Note: Earlier B.C. and Cal. records belong under *Aradus blaisdelli* Van Duzee.

Aradus insignitus Parshley, 1921
1921 *Aradus* (*Aradus*) *insignitus* Parshley, Trans. Am. Ent. Soc., 47: 75. [Mass.].
Distribution: Mass.

Aradus insolitus Van Duzee, 1916
1916 *Aradus insolitus* Van Duzee, Univ. Cal. Publ. Ent., 1: 233. [Cal.].
1917 *Aradus* (*Aradus*) *insoletis* [*sic*]: Van Duzee, Univ. Cal. Publ. Ent., 2: 135.

1921 *Aradus* (*Aradus*) *insolitus*: Parshley, Trans. Am. Ent. Soc., 47: 98.
Distribution: Alta., B.C., Cal., Id., Ore.

Aradus intectus Parshley, 1921
1921 *Aradus* (*Aradus*) *intectus* Parshley, Trans. Am. Ent. Soc., 47: 42. [Col.].
Distribution: B.C., Col., Mont., Wyo., Yuk.

Aradus intermedius Usinger, 1936
1936 *Aradus* (*Aradus*) *intermedius* Usinger, An. Ent. Soc. Am., 29: 498. [Cal.].
Distribution: Cal.

Aradus kormilevi Heiss, 1980
1873 *Aradus* (*Aradus*) *cinnamomeus*: Stål, K. Svens. Vet.-Akad. Handl., 11(2): 137.
1876 *Aradus cinnamomeus*: Uhler, Bull. U.S. Geol. Geogr. Surv. Terr., 1: 321 (in part).
1980 *Aradus* (*Aradus*) *kormilevi* Heiss, Ber. Nat.-Med. Ver. Innsbruck, 67: 113. [N.J.].
Distribution: Ala., Alta., B.C., Cal., Col., D.C., Fla., Ga., Ks., Man. Mass., Md., Miss., Mo., Mont., N.B., N.J., N.S., Neb., Ont., Pa., Sask., Tex., Va., W.Va., Wyo.
Note: The true *Aradus cinnamomeus* Panzer, according to Heiss (1980, above, p. 103), is a "westpalaearctic species" and does not occur in North America.

Aradus lawrencei Kormilev, 1966
1966 *Aradus lawrencei* Kormilev, Psyche, 73: 27. [N.H.].
Distribution: N.H.

Aradus leachi Van Duzee, 1929
1929 *Aradus leachi* Van Duzee, Pan-Pac. Ent., 5: 186. [Cal.].
1939 *Aradus* (*Aradus*) *leachi*: Torre-Bueno, Ent. Am., 19: 262.
Distribution: Cal.

Aradus linsleyi Usinger, 1936
1936 *Aradus linsleyi* Usinger, An. Ent. Soc. Am., 29: 493. [Cal.].
1939 *Aradus* (*Aradus*) *linsleyi*: Torre-Bueno, Ent. Am., 19: 267.
Distribution: Cal.

Aradus lugubris Fallén, 1807
1807 *Aradus lugubris* Fallén, Monogr. Cimic. Suec. p. 34.
1837 *Aradus affinis* Kirby, Fauna Bor.-Am., 4: 279. [Canada "lat. 65"]. Synonymized by Bergroth, 1886, Wien. Ent. Zeit., 5: 97.
1832 *Aradus rectus* Say, Descrip. Het. Hem. N. Am., p. 29. [Fla., Mo.]. Synonymized by Bergroth, 1886, Wien. Ent. Zeit., 5: 97.
1873 *Aradus fenestratus* Walker, Cat. Hem. Brit. Mus., 7: 36. [Ont.; "Rocky Mountains"; N.S.]. Synonymized by Bergroth, 1913, Can. Ent., 45: 5.
1916 *Aradus* (*Aradus*) *lugubris*: Van Duzee, Check List Hem., p. 16.
1926 *Aradus fenestralis* [*sic*]: Blatchley, Het. E. N. Am., p. 311.
Distribution: Alk., Alta., Ariz., B.C., Cal., Col., D.C., "Hudson B.T.," Id., Ill., Man., Mass., Me., Mich., Minn., Mo., Mont., N.C., N.D., N.H., N.M., N.S., N.T., N.Y., Neb., Nev., Nfld., Ont., Ore., Pa., Que., R.I., Sask., Ut., Wash., Wis., Yuk. (Palearctic).

Aradus lugubris lugubris Fallén, 1807
1807 *Aradus lugubris* Fallén, Monogr. Cimic. Suec., p. 34.
1900 *Aradus lugubris* var. *lugubris*: Reuter, Medd. Soc. Fauna Flora Fenn., 26: 134.
Distribution: Same as for species.

Aradus lugubris nigricornis Reuter, 1900
1900 *Aradus lugubris* var. *nigricornis* Reuter, Medd. Soc. Fauna Flora Fenn., 26: 134, 138. [Finland].

1921 *Aradus (Aradus) lugubris* var. *nigricornis*: Parshley, Trans. Am. Ent. Soc., 47: 82.
Distribution: Alk., Ariz., B.C., Cal., Col., Id., Mass., Me., N.M., N.T., Nev., Ont., Ore., Wash., Wis.

Aradus marginatus Uhler, 1893
1893 *Aradus marginatus* Uhler, Proc. Ent. Soc. Wash., 2: 381. [Ut.].
1916 *Aradus (Aradus) marginatus*: Van Duzee, Check List Hem., p. 16.
Distribution: Ut.

Aradus medioximus Parshley, 1921
1921 *Aradus (Aradus) medioximus* Parshley, Trans. Am. Ent. Soc., 47: 58. [Cal.].
Distribution: B.C., Cal., Ia., Ore.

Aradus montanus Bergroth, 1913
1913 *Aradus montanus* Bergroth, Can. Ent., 45: 1. [Col.].
1916 *Aradus (Aradus) montanus*: Van Duzee, Check List Hem., p. 16.
Distribution: Col., Mont., Que.

Aradus occidentalis Kormilev, 1980
1980 *Aradus occidentalis* Kormilev, Proc. Ent. Soc. Wash., 82: 106. [Wash.].
Distribution: Wash.

Aradus opertaneus Parshley, 1921
1921 *Aradus (Aradus) opertaneus* Parshley, Trans. Am. Ent. Soc., 47: 63. [Minn.].
Distribution: Minn.

Aradus orbiculus Van Duzee, 1920
1920 *Aradus orbiculus* Van Duzee, Proc. Cal. Acad. Sci., ser. 4, 9: 337. [Cal.].
1921 *Aradus (Aradus) orbiculus*: Parshley, Trans. Am. Ent. Soc., 47: 97.
1944 *Aradus orbiculatus* [*sic*]: Harris and Shull, Ia. St. Coll. J. Sci., 18: 203.
Distribution: Cal., Id., Ore.

Aradus oregonicus Kormilev, 1978
1978 *Aradus oregonicus* Kormilev, Proc. Ent. Soc. Wash., 80: 229. [Ore.].
Distribution: Ore.

Aradus ornatus Say, 1832
1832 *Aradus ornatus* Say, Descrip. Het. Hem. N. Am., p. 29. [Ind.].
1916 *Aradus (Aradus) ornatus*: Van Duzee, Check List Hem., p. 16.
Distribution: D.C., Ga., Ind., Md., N.Y., Oh., Pa., Va.

Aradus ovatus Kormilev, 1966
1966 *Aradus ovatus* Kormilev, Proc. U.S. Nat. Mus., 119(3548): 3. [Ks.].
Distribution: Ks.

Aradus oviventris Kormilev, 1966
1966 *Aradus oviventris* Kormilev, Psyche, 73: 26. [Ariz.].
Distribution: Ariz.

Aradus paganicus Parshley, 1929
1929 *Aradus (Aradus) paganicus* Parshley, Can. Ent., 61: 244. [B.C.].
Distribution: B.C., Ont.

Aradus pannosus Van Duzee, 1920
1920 *Aradus pannosus* Van Duzee, Proc. Cal. Acad. Sci., ser. 4, 9: 322. [Cal.].
1921 *Aradus (Aradus) pannosus*: Parshley, Trans. Am. Ent. Soc., 47: 39.
Distribution: Cal.

Aradus pannosus incomptus Parshley, 1921
1921 *Aradus (Aradus) pannosus* var. *incomptus* Parshley, Trans. Am. Ent. Soc., 47: 39. [Cal.].
Distribution: Cal.

Aradus pannosus pannosus Van Duzee, 1920
1920 *Aradus pannosus* Van Duzee, Proc. Cal. Acad. Sci., ser. 4, 9: 332. [Cal.].
1921 *Aradus (Aradus) pannosus*: Parshley, Trans. Am. Ent. Soc., 47: 39.
Distribution: Cal.

Aradus parshleyi Van Duzee, 1920
1920 *Aradus parshleyi* Van Duzee, Proc. Cal. Acad. Sci., ser. 4, 9: 336. [B.C.].
1921 *Aradus (Aradus) parshleyi*: Parshley, Trans. Am. Ent. Soc., 47: 78.
Distribution: Alta., B.C., Cal.

Aradus parvicornis Parshley, 1921
1921 *Aradus (Aradus) parvicornis* Parshley, Trans. Am. Ent. Soc., 47: 62. [N.M.].
1924 *Aradus parvicornis*: Downes, Proc. Ent. Soc. Wash., 21: 28.
Distribution: B.C., Cal., Id., N.M., Ore.

Aradus patibulus Van Duzee, 1927
1927 *Aradus patibulus* Van Duzee, Pan-Pac. Ent., 3: 140. [Cal.].
1939 *Aradus (Aradus) patibulus*: Torre-Bueno, Ent. Am., 19: 266.
Distribution: Cal.

Aradus persimilis Van Duzee, 1916
1916 *Aradus persimilis* Van Duzee, Univ. Cal. Publ. Ent., 1: 232. [Cal.].
1917 *Aradus (Aradus) persimilis*: Van Duzee, Univ. Cal. Publ. Ent., 2: 131.
Distribution: Alta., B.C., Cal., Col., Mont., Nfld., Wash.

Aradus proboscideus Walker, 1873
1873 *Aradus proboscideus* Walker, Cat. Hem. Brit. Mus., 7: 35. [Canada].
1903 *Artus* [*sic*] *luteolus* Fyles, Can. Ent., 35: 75. [Que.]. Synonymized by Parshley, 1921, Trans. Am. Ent. Soc., 47: 51.
1904 *Aradus hubbardi* Heidemann, Proc. Ent. Soc. Wash., 6: 232. [Ariz., B.C., Col., Ore., Wyo., Ut.]. Synonymized by Parshley, 1921, Trans. Am. Ent. Soc., 47: 51.
1916 *Aradus (Aradus) proboscideus*: Van Duzee, Check List Hem., p. 16.
Distribution: Alk., Alta., Ariz., B.C., Cal., Col., Id., Man., Mass., Me., Mont., N.B., N.H., N.M., N.S., N.Y., Nev., Ont., Ore., Que., Ut., Wash., Wyo.

Aradus quadrilineatus Say, 1825
1825 *Aradus quadrilineatus* Say, J. Acad. Nat. Sci. Phila., 4: 326. [Mo.].
1887 *Aradus robustus*: Provancher, Pet. Faune Ent. Can., 3: 165.
1916 *Aradus (Aradus) quadrilineatus*: Van Duzee, Check List Hem., p. 16.
1918 *Aradus lineatus* [*sic*]: Johnson and Ledig, J. Ent. Zool., 10: 4.
Distribution: Alta., B.C., Conn., D.C., Fla., Ga., Ill., Ind., Ia., Ks., La., Man., Mass., Md., Me., Mich., Minn., Mo., N.B., N.C., N.H., N.J., N.T., N.Y., Oh., Ont., Pa., Que., R.I., Sask., Va., Wis., Yuk. (Panama).

Aradus robustus Uhler, 1871
1871 *Aradus robustus* Uhler, Proc. Bost. Soc. Nat. Hist., 14: 104. [Mass.].
1873 *Aradus (Aradus) robustus*: Stål, K. Svens. Vet.-Akad. Handl., 11(2): 136.
1887 *Aradus quadrilineatus*: Provancher, Pet. Faune Ent. Can., 3: 166.
Distribution: Conn., D.C., Del., Fla., Ia., Ill., Ind., Ks., Mass., Md., Me., Mich., Minn.,

Miss., Mo., N.C., N.D., N.H., N.J., N.T., N.Y., Neb., Oh., Ok., Ont., Pa., Que., R.I., Tenn., Tex., Wis.

Aradus robustus insignis Parshley, 1921

1921 *Aradus* (*Aradus*) *robustus* var. *insignis* Parshley, Trans. Am. Ent. Soc., 47: 42. [Tex.].

Distribution: Mich., N.C., Neb., Tex.

Aradus robustus robustus Uhler, 1871

1871 *Aradus robustus* Uhler, Proc. Bost. Soc. Nat. Hist., 14: 104.

1921 *Aradus* (*Aradus*) *robustus* var. *robustus*: Parshley, Trans. Am. Ent. Soc., 47: 41.

Distribution: Same as for species.

Aradus saileri Kormilev, 1966

1966 *Aradus saileri* Kormilev, Proc. U.S. Nat. Mus., 119(3548): 2. [Alk.].

Distribution: Alk.

Aradus saskatchewanensis Matsuda, 1980

1980 *Aradus saskatchewanensis* Matsuda, Can. Ent., 112: 855. [Sask.].

Distribution: Sask.

Aradus serratus Usinger, 1936

1936 *Aradus* (*Aradus*) *serratus* Usinger, An. Ent. Soc. Am., 29: 496. [Alta.].

Distribution: Alta.

Aradus shermani Heidemann, 1907

1907 *Aradus shermani* Heidemann, Proc. Ent. Soc. Wash., 8: 68. [N.C.].

1916 *Aradus* (*Aradus*) *shermani*: Van Duzee, Check List Hem., p. 16.

Distribution: Ala., Fla., Ga., Me., N.C., N.J., Ont., Pa., Que., Sask.

Aradus signaticornis Sahlberg, 1848

1848 *Aradus signaticornis* Sahlberg, Monogr. Geocorisae Fenn., p. 141. [Finland].

1971 *Aradus martini* Matsuda, Can. Ent., 103: 1195. [Yuk.]. Synonymized by Heiss, 1980, Ber. Nat.-Med. Ver. Innsbruck, 67: 114.

1977 *Aradus* (*Aradus*) *martini*: Matsuda, Can. Dept. Agri. Publ., 1634: 29.

Distribution: Alk., Yuk. (Palearctic).

Aradus similis Say, 1832

1832 *Aradus similis* Say, Descrip. Het. Hem. N. Am., p. 28. [U.S.].

1873 *Aradus fascicornis* Walker, Cat. Hem. Brit. Mus., 7: 36. [N.S.]. Synonymized by Parshley, 1921, Trans. Am. Ent. Soc., 47: 64.

1873 *Aradus* (*Aradus*) *similis*: Stål, K. Svens. Vet.-Akad. Hand., 11(2): 136.

1887 *Aradus centriguttatus* Bergroth, Rev. d'Ent., 6: 246. Synonymized by Bergroth, 1892, Proc. Ent. Soc. Wash., 2: 335. See note below.

1917 *Aradus similis* var. *centriguttatus*: Van Duzee, Univ. Cal. Publ. Ent., 2: 131.

1917 *Aradus similis* var. *similis*: Van Duzee, Univ. Cal. Publ. Ent., 2: 131.

Distribution: Ala., Alk., Conn., D.C., Fla., Ga., Ia., Ill., Ind., Ks., Mass., Md., Me., Mich., Miss., Mo., N.C., N.H., N.J., N.S., N.Y., Oh., Ont., Pa., Que., S.C., Tex., Va., Wis.

Note: The species-group name *centriguttatus*, which was originally described as a distinct species, has been categorized differently by different authors. Bergroth (1892, Proc. Ent. Soc. Wash., 2: 335) and Matsuda (1977, Can. Dept. Agr. Publ., 1634: 64) placed it as a junior synonym of *similis*. Van Duzee (1917, Univ. Cal. Publ. Ent., 2: 131) and Parshley (1921, Trans. Am. Ent. Soc., 47: 66) treated it as a "variety" of *similis*. Torre-Bueno (1939, Ent. Am., vol. 19) made no entry of it. Kormilev (1964, Arkiv Zool., ser. 2, 16: 475) commented "we cannot consider it as a geographical subspecies" but did not suggest a placement. The present list recognizes the abundance of intermediate individuals and the lack of a convincing argument for retentions of an infra-specific rank for *centriguttatus* as sufficient reason to treat it as a junior synonym of *similis*.

Aradus snowi Van Duzee, 1920
1920 *Aradus snowi* Van Duzee, Proc. Cal. Acad. Sci., ser. 4, 9: 339. [Ariz.].
1921 *Aradus (Aradus) snowi*: Parshley, Trans. Am. Ent. Soc., 47: 94.
Distribution: Ariz., N.M., Tex. (Mexico).

Aradus subruficeps Hussey, 1953
1953 *Aradus (Aradus) subruficeps* Hussey, Occas. Pap. Mus. Zool., Univ. Mich., 550: 1. [Mich.].
Distribution: Mich.

Aradus taylori Van Duzee, 1920
1920 *Aradus taylori* Van Duzee, Proc. Cal. Acad. Sci., ser. 4, 9: 335. [B.C.].
1939 *Aradus (Aradus) taylori*: Torre-Bueno, Ent. Am., 19: 263.
Distribution: B.C., Cal., Id., Ore., Ut.
Note: Synonymized by Parshley (1921, Trans. Am. Ent. Soc., 47: 51) under *A. proboscideus*, but resurrected by Van Duzee (1927, Pan-Pac. Ent., 3: 140) and keyed as valid by Torre-Bueno (1939, Ent. Am., 19: 363).

Aradus tuberculifer Kirby, 1837
1837 *Aradus tuberculifer* Kirby, Fauna Bor.-Am., 4: 278. [Canada (Boreal America, "Lat. 65")].
1873 *Aradus caliginosus* Walker, Cat. Hem. Brit. Mus., 7: 36. ["Hudson's Bay"]. Synonymized by Bergroth, 1913, Can. Ent., 45: 4.
1916 *Aradus (Aradus) tuberculifer*: Van Duzee, Check List Hem., p. 16.
1926 *Aradus (Aradus) tuberculifera* [*sic*]: Blatchley, Het. E. N. Am., p. 309.
Distribution: Alk., Alta., B.C., Cal., Col., Man., Me., Mich., Minn., N.T., N.Y., Ont., Que., Vt.

Aradus uniannulatus Parshley, 1921
1921 *Aradus (Aradus) uniannulatus* Parshley, Trans. Am. Ent. Soc., 47: 90. [Alta.].
Distribution: Alta., Cal., Col., D.C., Man., Mich., N.T., N.Y., Pa., Que., Tex., Yuk.

Aradus uniformis Heidemann, 1904
1904 *Aradus uniformis* Heidemann, Proc. Ent. Soc. Wash., 6: 231. [Mass., Pa., Va.].
1916 *Aradus (Aradus) uniformis*: Van Duzee, Check List Hem., p. 16.
Distribution: Fla., Mass., Md., N.C., N.J., N.Y., Ont., Pa., Va.

Aradus vadosus Van Duzee, 1920
1920 *Aradus vadosus* Van Duzee, Proc. Cal. Acad. Sci., ser., 4: 9: 334. [B.C.].
1921 *Aradus (Aradus) vadosus*: Parshley, Trans. Am. Ent. Soc., 47: 58.
Distribution: B.C., Id., Mont.

Aradus vandykei Van Duzee, 1927
1927 *Aradus vandykei* Van Duzee, Pan-Pac. Ent., 3: 139. [Ore.].
1939 *Aradus (Aradus) vandykei*: Torre-Bueno, Ent. Am., 19: 261.
Distribution: Ore.

Subgenus *Quilnus* Stål, 1873

1873 *Aradus (Quilnus)* Stål, K. Svens. Vet.-Akad. Handl., 11(2): 137. Type-species: *Aradus parvicollis* Stål, 1873. Designated by Oshanin, 1912, Kat. Paläark. Hem., p. 47.

Aradus heidemanni Bergroth, 1906
1906 *Aradus* (*Quilnus*) *Heidemanni* [*sic*] Bergroth, Can. Ent., 38: 200. [B.C., Ore.].
Distribution: Alta., B.C., Cal., Col., N.M., Mont., Ore., Wash.

Aradus niger Stål, 1873
1873 *Aradus* (*Quilnus*) *niger* Stål, K. Svens. Vet.-Akad. Handl., 11(2): 137. [S.C.].
1895 *Aradus obliquus* Uhler, Bull. Col. Agr. Exp. Stn., 31: 58. *Nomen nudum*.
Distribution: Ala., Col., D.C., Fla., Mass., Me., Mo., N.B., N.C., N.H., N.J., N.S., N.Y., Ont., Que., S.C., Tex., Va., Wash. (Mexico).
Note: Bergroth (1906, Can. Ent., 38: 200) noted that the *A. obliquus* record above applied to this species.

Aradus nigrinus Parshley, 1921
1921 *Aradus* (*Quilnus*) *nigrinus* Parshley, Trans. Am. Ent. Soc., 47: 101. [Ariz.].
1929 *Aradus nigrinus*: Parshley, Can. Ent., 61: 246.
Distribution: Alta., Ariz.

Aradus nigrinus canadensis Parshley, 1929
1929 *Aradus nigrinus canadensis* Parshley, Can. Ent., 61: 246. [Alta.].
Distribution: Alta.

Aradus nigrinus nigrinus Parshley, 1921
1921 *Aradus* (*Quilnus*) *nigrinus* Parshley, Trans. Am. Ent. Soc., 47: 101.
1929 *Aradus nigrinus nigrinus*: Parshley, Can. Ent., 61: 246.
Distribution: Ariz.

Aradus usingeri Kormilev, 1978
1978 *Aradus* (*Quilnus*) *usingeri* Kormilev, Proc. Ent. Soc. Wash., 80: 230. [Ore.].
Distribution: Ore.

Subfamily Calisiinae Stål, 1873

Genus *Calisius* Stål, 1860

1860 *Calisius* Stål, K. Svens. Vet.-Akad. Handl., 2(7): 68. Type-Species: *Calisius pallipes* Stål, 1860. Monotypic.

Calisius anaemus Bergroth, 1913
1904 *Calisius pallipes*: Heidemann, Proc. Ent. Soc. Wash., 6: 229.
1913 *Calisius anaemus* Bergroth, Can. Ent., 45: 7. [Fla.].
Distribution: Fla.

Calisius contubernalis Bergroth, 1913
1913 *Calisius contubernalis* Bergroth, Can. Ent., 45: 6. [Fla.; Guadeloupe Island].
Distribution: Fla. (West Indies).

Calisius texasanus Kormilev, 1968
1968 *Calisius texasanus* Kormilev, Rev. Fac. Agron. Univ. Centr. Venezuela, 5: 45. [Tex.].
Distribution: Tex.

Subfamily Carventinae Usinger, 1950

Genus *Acaricoris* Harris and Drake, 1944

1944 *Acaricoris* Harris and Drake, Proc. Ent. Soc. Wash., 46: 128. Type-species: *Acaricoris ignotus* Harris and Drake, 1944. Monotypic.

Acaricoris floridus Drake, 1957
1957 *Acaricoris floridus* Drake, Proc. Biol. Soc. Wash., 70: 35. [Fla.].
Distribution: Fla.

Acaricoris ignotus Harris and Drake, 1944
1944 *Acaricoris ignotus* Harris and Drake, Proc. Ent. Soc. Wash., 46: 128. [La.].
Distribution: Ark., Ga., La., Miss.

Genus *Proxius* Stål, 1873

1873 *Proxius* Stål, K. Svens. Vet.-Akad. Handl., 11(2): 142. Type-species: *Proxius incrustatus* Stål, 1873. Monotypic.

Note: Key to three subgenera and six species provided by Usinger and Matsuda (1959, Classif. Aradidae, pp. 113-114).

Subgenus *Neoproxius* Usinger and Matsuda, 1959

1959 *Proxius* (*Neoproxius*) Usinger and Matsuda, Classif. Aradidae, p. 113. Type-species: None designated; included species: *gypsatus* Bergroth, *palliatus* Champion, *personatus* Champion, and *schwarzii* Heidemann.

Proxius gypsatus Bergroth, 1898
1898 *Proxius gypsatus* Bergroth, Ent. Month. Mag., 34: 100. [Venezuela].
1913 *Proxius gypsatus*: Bergroth, Can. Ent., 45: 8.
Distribution: Fla. (Panama, Venezuela).
Note: Bergroth (1913, Can. Ent., 45: 8) mentioned a Uhler manuscript combination "*Syrtidea diffrata*" under this species.

Proxius schwarzii Heidemann, 1904
1904 *Proxius schwarzii* Heidemann, Proc. Ent. Soc. Wash., 6: 230. [Fla.].
1959 *Proxius* (*Neoproxius*) *schwarzii*: Usinger and Matsuda, Classif. Aradidae, p. 113.
Distribution: Fla.

Subfamily Mezirinae Oshanin, 1908

Genus *Aphleboderrhis* Stål, 1860

1860 *Aphleboderrhis* Stål, K. Svens. Vet.-Akad. Handl., 2(7): 67. Type-species: *Aphleboderrhis pilosa* Stål, 1860. Monotypic.

Note: Champion (1898, Biol. Centr.-Am., Rhyn., 2: 78-79) gave a key to two species, including the North American one.

Aphleboderrhis pubescens (Walker), 1873
1873 *Aradus pubescens* Walker, Cat. Hem. Brit. Mus., 7: 38. [Brazil].
1914 *Aphleboderrhis pubescens*: Barber, J. N.Y. Ent. Soc., 22: 171.
Distribution: Tex. (Brazil, Colombia, Peru).

Genus *Mezira* Amyot and Serville, 1843

1843 *Mezira* Amyot and Serville, Hist. Nat. Ins. Hem., p. 305. Type-species: *Mezira granulata* Amyot and Serville, 1843. Preoccupied. Next available name is *Brachyrhynchus abdominalis* Stål, 1873. Monotypic.
Note: Key to seven species in America north of Mexico provided by Usinger (1936, An. Ent. Soc. Am., 29: 509). Key to eighty-nine New World species provided by Kormilev (1971, Proc. Ent. Soc. Wash., 73: 283-290).

Mezira emarginata (Say), 1832
1832 *Aradus emarginatus* Say, Descrip. Het. Hem. N. Am., p. 30. [Mexico].
1873 *Mezira modesta*: Walker, Cat. Hem. Brit. Mus., 7: 23.
1898 *Brachyrrhynchus* [*sic*] *emarginatus*: Champion, Biol. Centr.-Am., Rhyn., 2: 102.
1904 *Brachyrhynchus emarginatus*: Uhler, Proc. U.S. Nat. Mus., 27: 363.
1905 *Brachyrhynchus moestus* [*sic*]: Sherman, Ent. News, 16:8.
1914 *Mezira emarginata*: Barber, Bull. Am. Mus. Nat. Hist., 33: 517.
1916 *Mezira* (*Arictus*) *emarginata*: Van Duzee, Check List Hem., p. 17.
Distribution: Ark., Ariz., Cal., Fla., Miss., N.C., N.M., Nev., Tex. (Mexico).

Mezira granulata, (Say) 1832
1832 *Aradus granulatus* Say, Descrip. Het. Hem. N. Am., p. 30. [Fla., Ind.].
1853 *Dysodius parvulus* Herrich-Schaeffer, Wanz. Ins., 9: 139. [Md.]. Synonymized by Stål, 1873, K. Svens. Vet.-Akad. Handl., 11(2): 145.
1873 *Brachyrhynchus* (*Arictus*) *granulatus*: Stål, K. Svens. Vet.-Akad. Handl., 11(2): 145.
1887 *Brachyrhynchus lobatus*: Provancher, Pet. Faune Ent. Can., 3: 168.
1887 *Brachyrhynchus granulatus*: Provancher, Pet. Faune Ent. Can., 3: 168.
1909 *Mezira granulata*: Van Duzee, Bull. Buff. Soc. Nat Sci., 9: 175.
1916 *Mezira* (*Arictus*) *granulata*: Van Duzee, Check List Hem., p. 17.
Distribution: Ala., Ariz., D.C., Fla., Ga., Ind., Md., Mo., N.C., S.C., Tex. (Cuba, Mexico).

Mezira lobata (Say), 1832
1832 *Aradus lobatus* Say, Descrip. Het. Hem. N. Am., p. 30. [La.].
1873 *Brachyrhynchus* (*Arictus*) *lobatus*: Stål, K. Svens. Vet.-Akad. Handl., 11(2): 145.
1876 *Brachyrhynchus lobatus*: Uhler, Bull. U.S. Geol. Geogr. Surv. Terr., 1: 323.
1876 *Brachyrhynchus granulatus*: Uhler, Bull. U.S. Geol. Geogr. Surv. Terr., 1: 323.
1892 *Brachyrrhynchus* [*sic*] *lobatus*: Bergroth, Proc. Ent. Soc. Wash., 2: 336.
1914 *Mezira lobatus* [*sic*]: Barber, Bull. Am. Mus. Nat. Hist., 33: 517.
1916 *Mezira* (*Arictus*) *lobata*: Van Duzee, Check List Hem., p. 17.
1926 *Mezira lobata*: Blatchley, Het. E. N. Am., p. 322.
Distribution: Cal., "Canada," D.C., Fla., Ga., Ill., Ind., Md., Mich., Mo., N.C., N.Y., Oh., Pa., Tex. (Mexico).
Note: "Canada" was included in the distribution of this species as late as 1971 by Kormilev (Proc. Ent. Soc. Wash., 73: 290). Matsuda (1977, Can. Dept. Agr. Publ. 1634) included no species of *Mezira* in his synopsis of the Aradidae of that country, but because this species ranges as far north as N.Y. and Mich., there is a probability that it does occur in southern Canada (perhaps Ont. and Que.).

Mezira novella Blatchley, 1924

1924 *Mezira novella* Blatchley, Ent. News, 35: 88. [Fla.]. See note below for type selection.

Distribution: Fla.

Note: Blatchley (1930, Blatchleyana, p. 63) selected a lectotype from among the original series from Cape Sable, Florida. Kormilev (1982, J. Nat. Hist., 16: 775) wrote, "Blatchley's type is lost," and then selected a "neotype" from the Blatchley specimens from the same locality.

Mezira pacifica Usinger, 1936

1876 *Brachyrhynchus moestus*: Uhler, Bull. U.S. Geol. Geogr. Surv. Terr., 1: 323.

1916 *Mezira (Arictus) moesta*: Van Duzee, Check List Hem., p. 17.

1921 *Mezira moesta*: Parshley, Proc. Ent. Soc. B.C., 18: 15.

1936 *Mezira pacifica* Usinger, An. Ent. Soc. Am., 29: 506. [Cal.].

Distribution: Ariz., B.C., Cal., Id., Ore., Wash., Wyo.

Mezira reducta Van Duzee, 1927

1927 *Mezira reducta* Van Duzee, Pan-Pac. Ent., 3: 142. [Cal.].

Distribution: Cal.

Mezira sayi Kormilev, 1982

1982 *Mezira (Mezira) sayi* Kormilev, J. Nat. Hist., 16: 777. [Fla.].

Distribution: Fla., Ga., Ind., S.C.

Mezira smithi Kormilev, 1982

1982 *Mezira smithi* Kormilev, J. Ga. Ent. Soc., 17: 336. [Ga.].

Distribution: Ga.

Mezira vanduzeei Usinger, 1936

1927 *Mezira granulata*: Van Duzee, Pan-Pac. Ent., 3: 142.

1936 *Mezira vanduzeei* Usinger, An. Ent. Soc. Am., 29: 507. [Ariz.].

Distribution: Ariz.

Genus *Nannium* Bergroth, 1898

1898 *Nannium* Bergroth, Ent. Month. Mag., 34: 100. Type-Species: *Nannium parvum* Bergroth, 1898. Original designation.

Nannium pusio Heidemann, 1909

1909 *Nannium pusio* Heidemann, Proc. Ent. Soc. Wash., 11: 189. [Oh.].

Distribution: Oh.

Genus *Neuroctenus* Fieber, 1960

1860 *Neuroctenus* Fieber, Europ. Hem., p. 34. Type-species: *Neuroctenus brasiliensis* Mayr, 1866, a junior synonym of *Brachyrhynchus punctulatus* Burmeister, 1835. Designated by Van Duzee, 1916, Check List Hem., p. 17.

Note: Key to five species of North America provided by Kormilev (1982, Wasman J. Biol., 40: 8-9).

Neuroctenus arizonicus Kormilev, 1982

1982 *Neuroctenus arizonicus* Kormilev, Wasmann J. Biol., 40: 9. [Ariz.].

Distribution: Ariz.

Neuroctenus elongatus Osborn, 1903
1903 *Neuroctenus elongatus* Osborn, Oh. Nat., 4: 41. [Oh.].
Distribution: D.C., Ind., Oh., N.C., Pa.

Neuroctenus hopkinsi Heidemann, 1904
1904 *Neuroctenus hopkinsi* Heidemann, Proc. Ent. Soc. Wash., 6: 163. [N.C.].
Distribution: Ga., Md., N.C.

Neuroctenus pseudonymus Bergroth, 1898
1887 *Neuroctenus ovatus*: Bergroth, Öfv. F. Vet.-Soc. Förh., 29: 183 (in part).
1898 *Neuroctenus pseudonymus* Bergroth, Wien. Ent. Zeit., 17: 27. [N.C.].
1903 *Neuroctenus pseudonemus* [*sic*]: Heidemann, Proc. Ent. Soc. Wash., 5: 310.
1905 *Brachyrhynchus ovatus*: Sherman, Ent. News, 16: 8.
1964 *Neuroctenus pseudomymus* [*sic*]: Balduf, Proc. Ent. Soc. Wash., 66: 3.
Distribution: D.C., Ind., N.C., Oh., Tenn., Tex.

Neuroctenus simplex (Uhler), 1876 [Fig. 9]
1876 *Brachyrhynchus simplex* Uhler, Bull. U.S. Geol. Geogr. Surv. Terr., 1: 323. [Fla., Ill., "Indian Territory," Md., Mo., "New England," Pa., Tex., Cuba].
1887 *Neuroctenus simplex*: Bergroth, Öfv. F. Vet.-Soc. Förh., 29: 182.
Distribution: "Carolina," Conn., D.C., Fla., Ga., Ill., Ks., Mass., Md., Me., Mo., Mont., N.C., N.J., N.Y., Oh., Pa., S.C., Tex. (Cuba).

Genus *Notapictinus* Usinger and Matsuda, 1959

1959 *Notapictinus* Usinger and Matsuda, Classif. Aradidae, pp. 203, 361. Type-species: *Pictinus dominicus* Usinger, 1936. Original designation.
Note: Key to twenty-four species provided by Kormilev (1964, Arkiv Zool., ser. 2, 16: 469-471).

Notapictinus aurivillii (Bergroth), 1887
1887 *Pictinus aurivillii* Bergroth, Rev. d'Ent., 6: 247. [Ga.]. Lectotype designated by Kormilev, 1959, Rev. Soc. Uruguaya Ent., 3: 32.
1959 *Notapictinus aurivillii*: Usinger and Matsuda, Classif. Aradidae, p. 362.
Distribution: Fla., Ga., La.
Note: Bergroth (1886, Verh. Zool.-Bot. Ges. Wien, 36: 60) first reported and characterized Georgia specimens of this taxon as an unnamed "race" of *Pictinus cinctipes* Stål.

Family Belostomatidae Leach, 1815

The Giant Water Bugs or Electric Light Bugs

By John T. Polhemus, Dan A. Polhemus, and Thomas J. Henry

The giant water bugs are large, ovate, dorsoventrally flattened insects with powerful raptorial forelegs. Species live below the surface of lotic and lentic habitats, respiring via two straplike appendages at the tip of the abdomen which act as air siphons. Seven genera occur worldwide of which three, *Abedus* Stål, *Belostoma* Latreille, and *Lethocerus* Mayr, are found in North America, the latter two as far north as southern Canada.

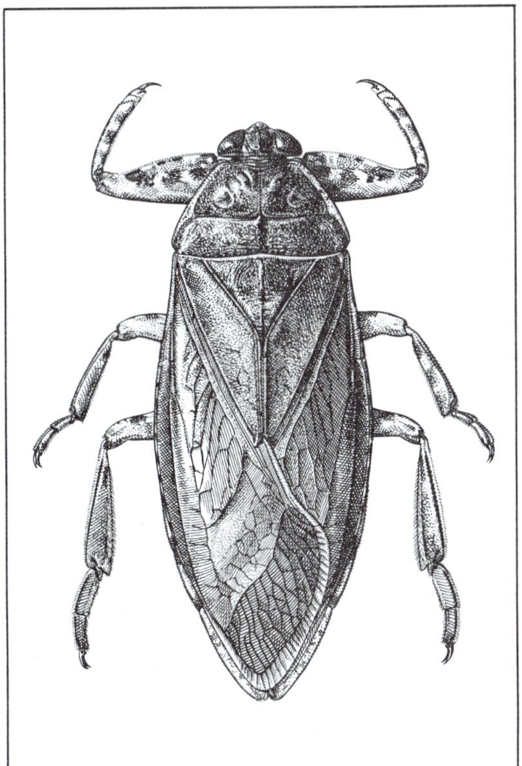

Fig. 11 *Lethocerus americanus* [p. 54] (After Usinger, 1956).

Belostomatids are voracious predators and will attack prey many times their size, including fish, frogs, and small birds, although their typical diet consists of smaller organisms such as tadpoles and insect larvae. Victims are subdued with powerful hydrolytic enzymes injected through the beak; the bite can produce a painful swelling in humans (Rees and Offord, 1969, Na-

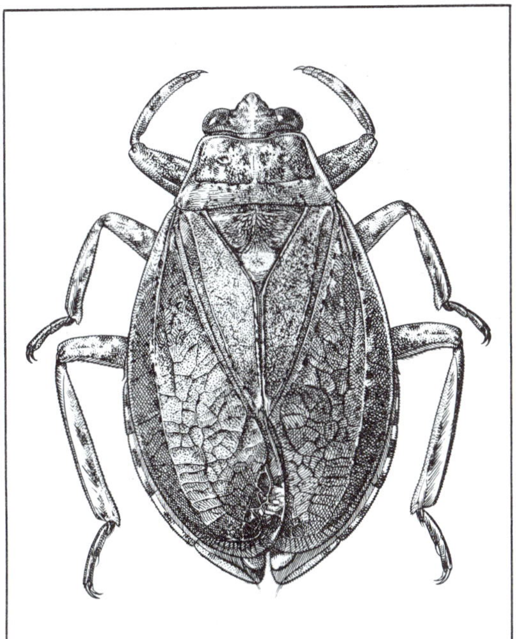

Fig. 12 *Abedus indentatus* [p. 50] (After Usinger, 1956).

ture, 221: 675-677). Eggs of *Lethocerus* are laid above water on vegetation and other protruding objects, but those of *Abedus* and *Belostoma* are laid on the backs of males, who carry them until they hatch, an unusual case of sex role reversal (Smith, 1976, An. Ent. Soc. Am., 69: 740-747; 1976, J. Ks. Ent. Soc., 49: 333-343). Kraus (1985, Pan-Pac. Ent., 61: 54-57) more recently reported that eggs of *Abedus indentatus* (Haldeman) may be laid on the backs of other females in the absence of sufficient space on available males. The eggs hatch in one to two weeks, with total developmental time ranging from one to two months (Hungerford, 1920, Univ. Ks. Sci. Bull., 11: 1-328; Smith, 1974, Psyche, 81:272-283). Adults often overwinter buried in mud, and will feign death if caught and handled (Severin and Severin, 1911, Behav. Monogr., 1: 1-44); in summer large numbers of *Lethocerus* adults may be attracted to outdoor lights, thus the colloquial name "electric light bugs."

The subfamily classification was established by Lauck and Menke (1961, An. Ent. Soc. Am., 54: 644-657). The major works for our fauna are cited under the respective genera.

The names *Belostoma marginata* and *B. reversapenne* listed by Rathvon (1869, Hist. Lancaster Co., Pa., p. 550) from Pennsylvania are *nomina nuda*. At this time, they cannot be associated with any described species.

Subfamily Belostomatinae Leach, 1815

Genus *Abedus* Stål, 1862

1862 *Abedus* Stål, Ent. Zeit, 23: 461. Type-species: *Abedus ovatus* Stål, 1862. Designated by Kirkaldy, 1906, Trans. Am. Ent. Soc., 32: 151.

1862 *Serphus* Stål, Ent. Zeit, 23: 462. Preoccupied. Type-species: *Belostoma dilatata* Say, 1832. Monotypic. Synonymized by Montandon, 1900, Bull. Soc. Sci. Buc.-Roum., 9: 272.

1863 *Pedinocoris* Mayr, Verh. Zool.-Bot. Ges. Wien, 13: 341. Type-species: *Pedinocoris macronyx* Mayr, 1863, a junior synonym of *Zaitha indentata* Haldeman, 1854. Designated by Kirkaldy, 1898, Ent., 31: 2. Synonymized by Montandon, 1900, Bull. Soc. Sci. Buc.-Roum., 9: 271.

1863 *Stenoscytus* Mayr, Verh. Zool.-Bot. Ver. Ges. Wien, 13: 343. Type-species: *Stenoscytus mexicanus* Mayr, 1863, a junior synonym of *Abedus ovatus* Stål, 1862. Synonymy by virtue of shared type-species.

1951 *Parabedus* De Carlo, Rev. Soc. Ent. Arg., 15: 71. Type-species: *Abedus breviceps*, Stål, 1862. Synonymized and designated by Menke, 1960, Univ. Cal. Publ. Ent., 16: 400.

Note: Menke (1960, Univ. Cal. Publ. Ent., 16: 393-440) reviewed the genus, proposed a subgeneric arrangement (followed here), and gave a key to species, and later (1979, Bull. Cal. Ins. Surv., 21: 84-85) presented an updated key to U.S. species.

Subgenus *Abedus* Stål, 1862

1862 *Abedus* Stål, Ent. Zeit, 23: 461. Type-species: *Abedus ovatus* Stål, 1862. Designated by Kirkaldy, 1906, Trans. Am. Ent. Soc., 32: 151.

1903 *Abedus* (*Abedus*): Montandon, Bull. Soc. Sci. Bucarest, 12: 111.

Abedus breviceps Stål, 1862

1862 *Abedus breviceps* Stål, Ent. Zeit., 23: 462. [Mexico].

1875 *Abedus ovatus*: Uhler, Rept. U.S. Geol. Geogr. Surv. Terr., 5: 840.

1877 *Abedus breviceps*: Uhler, An. Rept. Chief Eng., Append. NN, p. 1332.
1960 *Abedus (Abedus) breviceps*: Menke, Univ. Cal. Publ. Ent., 16: 414.
Distribution: Ariz., N.M., Tex. (Guatemala, Mexico).
Note: Menke (1960, above, 16: 413) noted that earlier records of *A. ovatus* Stål in the U.S. referred to either *A. breviceps* or *A. vicinus sonorensis* Menke [as *signoreti sonorensis*].

Abedus ovatus Stål, 1862
1862 *Abedus ovatus* Stål, Ent. Zeit., 23: 461. [Mexico].
1960 *Abedus (Abedus) ovatus*: Menke, Univ. Cal. Publ. Ent., 16: 411.
1977 *Abedus ovatus*: Menke, Southwest. Nat., 22: 118.
Distribution: Ariz. (Guatemala, Mexico).
Note: Menke (1960, Univ. Cal. Publ. Ent., 16: 413) considered early records of this species from the U.S. to be misidentifications of either *A. breviceps* Stål or *A. vicinus sonorensis* Menke [as *A. signoreti sonorensis*], but later (1977, above) documented an authentic specimen from Arizona.

Abedus parkeri Menke, 1966
1966 *Abedus (Abedus) parkeri* Menke, Contr. Sci. L. A. Co. Mus. Nat. Hist., 118: 1. [Mexico].
1977 *Abedus parkeri*: Menke, Southwest. Nat., 22: 118.
Distribution: Ariz. (Mexico).

Subgenus *Deinostoma* Kirkaldy, 1897

1862 *Serphus* Stål, Ent. Zeit., 23: 462. Preoccupied. Type-species: *Belostoma dilatata* Say, 1832. Monotypic.
1897 *Deinostoma* Kirkaldy, Ent., 30: 258. New name for *Serphus* Stål.
1903 *Abedus (Deinostoma)*: Montandon, Bull. Soc. Sci. Buc.-Roum., 12: 111.

Abedus herberti Hidalgo, 1935
1935 *Abedus herberti* Hidalgo, Univ. Ks. Sci. Bull., 22: 507. [Ariz.].
1938 *Abedus drakei* De Carlo, Rev. Soc. Ent. Arg., 10: 41. [Ariz.]. Synonymized by Menke, 1960, Univ. Cal. Publ. Ent., 16: 421.
1948 *Abedus stali* De Carlo, Comun. Mus. Arg. Cienc. Nat., 5: 21. [Ariz.]. Synonymized by Menke, 1960, Univ. Cal. Publ. Ent., 16: 421.
1960 *Abedus (Deinostoma) herberti*: Menke, Univ. Cal. Publ. Ent., 16: 421.
Distribution: Ariz., N.M., Ut. (Mexico).
Note: De Carlo (1963, An. Soc. Cien. Argent., 175: 78) resurrected his species *A. drakei* and *A. stali*, but Menke (1977, Southwest. Nat., 22: 122) argued to maintain them as junior synonyms. Smith (1974, Psyche, 81: 272-283, and 1975, Pan-Pac., 51: 259-267) presented details of biology, habits, etc. for this species in Arizona.

Abedus herberti herberti Hidalgo, 1935
1876 *Serphus dilatatus*: Uhler, Bull. U.S. Geol. Geogr. Surv. Terr., 1: 338.
1886 *Pedinocoris brachonyx*: Uhler, Check-list Hem. Het., p. 28 (in part).
1909 *Abedus macronyx*: Kirkaldy and Torre-Bueno, Proc. Ent. Soc. Wash., 10: 189 (in part).
1909 *Abedus dilatatus*: Kirkaldy and Torre-Bueno, Proc. Ent. Soc. Wash., 10: 189 (in part).
1909 *Abedus indentata*: Kirkaldy and Torre-Bueno, Proc. Ent. Soc. Wash., 10: 189 (in part).

1910 *Pedinocoris indentata*: Banks, Cat. Nearc. Hem.-Het., p. 8 (in part).
1917 *Abedus indentatus*: Van Duzee, Univ. Cal. Publ. Ent., 2: 471 (in part).
1935 *Abedus hungerfordi*: Hidalgo, Univ. Ks. Sci. Bull., 22: 505 (in part).
1935 *Abedus montandoni*: Hidalgo, Univ. Ks. Sci. Bull., 22: 504 (in part).
1935 *Abedus herberti* Hidalgo, Univ. Ks. Sci. Bull., 22: 507. [Ariz.].
1960 *Abedus (Deinostoma) herberti herberti*: Menke, Univ. Cal. Publ. Ent., 16: 421.
Distribution: Ariz., N.M. (Mexico).

Abedus herberti utahensis Menke, 1960
1960 *Abedus (Deinostoma) herberti utahensis* Menke, Univ. Cal. Publ. Ent., 16: 423. [Ut.].
1963 *Abedus utahensis*: De Carlo, An. Soc. Cien. Argent., 175: 72.
Distribution: Ariz., Ut.
Note: Known only from the Virgin River drainage of southwestern Utah and northeastern Arizona. De Carlo (1963, above) gave *A. herberti utahensis* species status, but Menke (1977, Southwest. Nat., 22: 122) maintained it as a subspecies.

Abedus indentatus (Haldeman), 1854 [Fig. 12]
1854 *Zaitha indentata* Haldeman, Proc. Acad. Nat. Sci. Phila., 6: 364. [Cal.]. Neotype designated by Menke, 1960, Univ. Cal. Publ. Ent., 16: 429.
1863 *Pedinocoris macronyx* Mayr, Verh. Zool.-Bot. Ges. Wien, 13: 350. [Cal.]. Synonymized by Menke, 1960, Univ. Cal. Publ. Ent., 16: 427.
1863 *Pedinocoris brachonyx* Mayr, Verh. Zool.-Bot. Ges. Wien., 13: 351. [Cal.]. Synonymized by Uhler, 1877, An. Rept. Chief Eng., Append. NN, p. 1331.
1873 *Belostoma brachonyx*: Walker, Cat. Hem. Brit. Mus., 8: 176.
1873 *Belostoma macronyx*: Walker, Cat. Hem. Brit. Mus., 8: 176.
1876 *Serphus dilatatus*: Uhler, Bull. U.S. Geol. Geogr. Surv. Terr., 1: 338.
1877 *Pedinocoris indentata*: Uhler, An. Rept. Chief Eng., Append. NN, p. 1331.
1909 *Abedus indentata*: Kirkaldy and Torre-Bueno, Proc. Ent. Soc. Wash., 10: 189 (in part).
1909 *Abedus macronyx*: Kirkaldy and Torre-Bueno, Proc. Ent. Soc. Wash., 10: 189 (in part).
1909 *Abedus dilatatus*: Kirkaldy and Torre-Bueno, Proc. Ent. Soc. Wash., 10: 189 (in part).
1917 *Abedus indentatus*: Van Duzee, Univ. Cal. Publ. Ent., 2: 471.
1932 *Abedus hungerfordi* De Carlo, Rev. Soc. Ent. Arg., 5: 123. [Cal.]. Synonymized by Menke, 1960, Univ. Cal. Publ. Ent., 16: 427.
1948 *Abedus mayri* De Carlo, Com. Mus. Arg. Cienc. Nat., 5: 13. [Cal.]. Synonymized by Usinger, 1956, Aquat. Ins. Cal., p. 205.
1960 *Abedus (Deinostoma) indentatus*: Menke, Univ. Cal. Publ. Ent., 16: 427.
Distribution: Cal. (Mexico).
Note: De Carlo (1963, An. Soc. Cien. Argent., 175: 77) refuted the above synonymy of his species *A. hungerfordi* and *A. mayri*, but Menke (1977, Southwest. Nat., 22: 122) maintained them as junior synonyms.

Subgenus *Microabedus* Hussey and Herring, 1950

1950 *Abedus (Microabedus)* Hussey and Herring, Fla. Ent., 33: 85. Type-species: *Abedus cantralli* Hussey and Herring, 1950, a junior synonym of *Belostoma immaculata* Say, 1832. Original designation.

Abedus immaculatus (Say), 1832

1832 *Belostoma fluminea* var. *immaculata* Say, Descrip. Het. Hem. N. Am., p. 37. [U.S.].

1950 *Abedus (Microabedus) cantralli* Hussey and Herring, Fla. Ent., 33: 84. [Fla.]. Synonymized by Hussey and Herring, 1950, Fla. Ent., 33: 155.

1950 *Abedus (Microabedus) immaculatus*: Hussey and Herring, Fla. Ent., 33: 154.

Distribution: Fla., Ga., Miss.

Subgenus *Pseudoabedus* De Carlo, 1951

1951 *Abedus* (*Pseudoabedus*) De Carlo, Rev. Soc. Ent. Arg., 15: 71. Type-species: *Abedus signoreti* Mayr, 1871. Designated by Menke, 1960, Univ. Cal. Publ. Ent., 16: 400.

Abedus vicinus Mayr, 1871

1871 *Abedus vicinus* Mayr, Verh. Zool.-Bot. Ges. Wien, 21: 405. [Mexico]. Lectotype designated by Menke, 1960, Univ. Cal. Publ. Ent., 16: 408.

1960 *Abedus* (*Pseudoabedus*) *signoreti vicinus*: Menke, Univ. Cal. Publ. Ent., 16: 408.

Distribution: Ariz. (Mexico).

Note: Menke (1977, Southwest. Nat., 22: 116) resurrected *A. vicinus* to species status. The nominate subspecies *A. vicinus vicinus* does not occur in our region.

Abedus vicinus sonorensis Menke, 1960

1960 *Abedus* (*Pseudoabedus*) *signoreti sonorensis* Menke, Univ. Cal. Publ. Ent., 16: 409. [Mexico].

1977 *Abedus vicinus sonorensis*: Menke, Southwest. Nat., 22: 117.

Distribution: Ariz. (Mexico).

Note: Most early records of *A. ovatus* Stål belong to this subspecies or the species *A. breviceps* (Menke, 1960, above, 16: 413).

Genus *Belostoma* Latreille, 1807

1807 *Belostoma* Latreille, Gen. Crust. Ins., 3: 144. Type-species: *Belostoma testaceopallidum* Latreille, 1807. Monotypic.

1843 *Zaitha* Amyot and Serville, Hist. Nat. Ins., Hem., p. 430. Type-species: *Zaitha stollii* Amyot and Serville, 1843. Designated by Kirkaldy, 1906, Trans. Am. Ent. Soc., 32: 151. Synonymized by Montandon, 1900, Bull. Soc. Sci. Buch.-Roum., 9: 9.

1847 *Perthostoma* Leidy, J. Acad. Nat. Sci. Phila., (2)1: 59. Type-species: *Perthostoma testaceum* Leidy, 1847. Designated by Kirkaldy, 1906, Trans. Am. Ent. Soc., 32: 151. Synonymized by Montandon, 1900, Bull. Soc. Sci. Buch.-Roum., 9: 267.

Note: Menke (1958, Bull. Cal. Acad. Sci., 57: 154-174) gave a key to the N. Am. species. Lauck (Bull. Chicago Acad. Sci., 11: 34-81 (1962); 82-101 (1963); and 102-154 (1964)) revised the genus in a 3-part series of papers.

Belostoma bakeri Montandon, 1913

1913 *Belostoma bakeri* Montandon, Bull. Soc. Sci. Buch.-Roum., 22: 123. ["Amerika bor."]. Lectotype designated by Lauck, 1964, Bull. Chicago Acad. Sci., 11: 146.

1959 *Belostoma confusum*: Lauck, Bull. Chicago Acad. Sci., 11: 4 (in part).

Distribution: Ariz., Cal., Nev., N.M., Ore., Tex., Ut., Wash. (Mexico).

Note: See Lauck (1964, Bull. Chicago. Acad. Sci., 11: 144-145) for a list of probable misidentifications of this species. Cal. records for *B. apache* Kirkaldy probably refer to *B. bakeri*.

Belostoma confusum Lauck, 1959

1959 *Belostoma confusum* Lauck, Bull. Chicago Acad. Sci., 11: 4. [holotype from Mexico; also from Cal. and Tex.].

Distribution: Ariz., Tex. (Mexico).

Note: Menke (1979, Bull. Cal. Ins. Surv., 21: 81) pointed out that Lauck's (1959, above) record of *B. confusum* for Cal. was a misidentification of *B. bakeri.*

Belostoma ellipticum Latreille, 1817

1817 *Belostoma ellipticum* Latreille, *In* Humboldt et Bonpland, Voy. Reg. Equin. Nouv. Cont., 2: 105. [No type or locality specified].

1901 *Belostoma ellipticum:* Kirkaldy and Torre-Bueno, Proc. Ent. Soc. Wash., 10: 191.

Distribution: Tex. (Central America, Mexico, West Indies).

Note: Lauck (1962, Bull. Chicago Acad. Sci., 11: 62) treated this name as a *nomen dubium,* since there are no known types or type locality, but deferred such action pending study of the original description. Menke (1979, Bull. Cal. Ins. Surv., 21: 81) examined the original description and concluded that Lauck probably misidentified *B. ellipticum,* so the occurrence of this species in North America is suspect.

Belostoma flumineum Say, 1832

1831 *Belostoma fluminea* Say, Descrip. N. Sp. N. Am. Ins. La., p. 12. [U.S.]. Neotype from Pa. designated by Lauck, 1964, Bull. Chicago Acad. Sci., 11: 138.

1847 *Perthostoma auranticum* Leidy, J. Acad. Nat. Sci. Phila., 1: 60. [Pa.]. Synonymized by Uhler, 1876, Bull. U.S. Geol. Geogr. Surv. Terr., 1: 338.

1847 *Perthostoma auranticum* var. *immaculatum* Leidy, J. Acad. Nat. Sci., 1: 60. [Pa.]. Synonymized by Menke, 1958, Bull. So. Cal. Acad. Sci., 57: 161.

1863 *Zaitha fluminea*: Dufour, An. Ent. Soc. France, (4)3: 388.

1891 *Zaitha fusciventris*: Townsend, Proc. Ent. Soc. Wash., 2: 55.

1905 *Belostoma flumineum*: Torre-Bueno, J. N.Y. Ent. Soc., 13: 44.

1952 *Belostoma bakeri*: Ellis, Am. Midl. Nat., 48: 327.

Distribution: Ala., Ariz., Ark., B.C., Cal., Col., Conn., D.C., Del., Fla., Ga., Ia., Ill., Ind., Ks., Ky., La., Man., Mass., Md., Me., Mich., Minn., Miss., Mo., N.B., N.C., N.D., N.H., N.J., N.M., N.S., N.Y., Neb., Nev., Oh., Ok., Ont., Ore., Pa., Que., S.C., Tenn., Tex., Va., Wis. (Mexico).

Belostoma fusciventre (Dufour), 1863

1863 *Zaitha fusciventris* Dufour, An. Soc. Ent. France, (4)3: 389. [Mexico].

1906 *Belostoma fusciventris*: Snow, Trans. Ks. Acad. Sci., 20: 180.

1938 *Belostoma fusciventre*: De Carlo, An. Mus. Arg. Cienc. Nat., 39: 222 (in part).

1959 *Belostoma thomasi* Lauck, Bull. Chicago Acad. Sci., 11: 2. [Mexico]. Synonymized by Menke, 1979, Bull. Cal. Ins. Surv., 21: 81.

Distribution: La., Texas (Central America, Mexico).

Note: Menke (1958, Bull. So. Cal. Acad. Sci., 57: 165-166) noted that Cal. records for this species "most certainly were result of misidentifications" and, later (1979, Bull. Cal. Ins. Surv., 21: 82), listed only the verified U.S. localities La. and Tex.

Belostoma lutarium (Stål), 1855

1855 *Zaitha lutaria* Stål, Öfv. K. Svens. Vet.-Akad. Förh., 12: 190. [North America].

1886 *Zaitha aurantiaca*: Uhler, Check-list Hem. Het., p. 28.

1907 *Belostoma aurantiacum*: Torre-Bueno and Brimley, Ent. News, 18: 435.

1910 *Belostoma lutarium*: Montandon, Bull. Soc. Sci. Buch.-Roum., 18: 187.

Distribution: Ala., Ark., Conn., Fla., Ga., Ill., Ind., Ks., La., Mass., Md., Mich., Miss.,

Mo., N.J., N.C., Oh., Ok., R.I., S.C., Tenn., Tex., Va.

Belostoma saratogae Menke, 1958

1958 *Belostoma saratogae* Menke, Bull. So. Cal. Acad. Sci., 57: 169. [Cal.].

Distribution: Cal.

Note: Known only from the thermal Saratoga Spring in Death Valley, Cal.

Belostoma subspinosum (Palisot), 1820

1820 *Nepa subspinosa* Palisot, Ins. Rec. Afr. Am., p. 236. [Dominican Republic]. Neotype from Dominican Republic designated by Lauck, 1962, Bull. Chicago Acad. Sci., 11: 67.

1909 *Belostoma subspinosum*: Kirkaldy and Torre-Bueno, Proc. Ent. Soc. Wash., 10: 192.

Distribution: Ariz., Cal., Tex. (Mexico to Panama, West Indies).

Note: The subspecies *B. subspinosum subspinosum* does not occur in the U.S.

Belostoma subspinosum bifoveatum (Haldeman), 1852

1852 *Zaitha bifoveata* Haldeman, Exp. Surv. Valley Great Salt Lake, Ut., Append. C., p. 370. [Tex.]. Synonymized with *cupreomicans* Stål by Mayr, 1871, Ver. Zool.-Bot. Ges. Wien, 21: 412; dates clarified and name resurrected by Menke, 1979, Bull. Cal. Ins. Surv., 21: 81.

1854 *Zaitha cupreomicans* Stål, Öfv. K. Svens. Vet.-Akad. Förh., 11(8): 240. [Mexico]. Synonymized by Menke, 1979, Bull. Cal. Ins. Surv., 21: 81.

1886 *Zaitha anurus*: Uhler, Check-list Hem. Het., p. 28 (in part).

1886 *Zaitha boscii*: Uhler, Check-list Hem. Het., p. 28.

1906 *Belostoma anurus*: Snow, Trans. Ks. Acad. Sci., 20: 180.

1909 *Belostoma boscii*: Kirkaldy and Torre-Bueno, Proc. Ent. Soc. Wash., 10: 190.

1916 *Belostoma bifoveata*: Van Duzee, Check List Hem., p. 53.

1917 *Belostoma bifoveatum*: Van Duzee, Univ. Cal. Publ. Ent., 2: 468.

1959 *Belostoma suspinosum cupreomicans*: Lauck, Bull. Chicago Acad. Sci., 11: 9.

1979 *Belostoma suspinosum bifoveatum*: Menke, Bull. Cal. Ins. Surv., 21: 81.

Distribution: Ariz., Cal., Tex. (Mexico to Panama).

Note: Lauck (1962, Bull. Chicago Acad. Sci., 11: 64-65) considered *B. boscii* Lepeletier and Serville, 1825, a *nomen dubium*, and applied the next available name *cupreomicans* for this subspecies. Menke (1979, above) pointed out that *B. bifoveatum* had priority. Records of *B. boscii* should be referred to this subspecies.

Belostoma testaceum (Leidy), 1847

1847 *Perthostoma testaceum* Leidy, J. Acad. Nat. Sci. Phila., 1: 66. [Pa.]. Neotype from Pa. designated by Lauck, 1964, Bull. Chicago Acad. Sci., 11: 125.

1852 *Zaitha reticulata* Haldeman, Exp. Surv. Val. Great Salt Lake, Ut., Append. C, p. 370. [Tex.]. Synonymized by Mayr, 1871, Verh. Zool.-Bot. Ges. Wien, 21: 417.

1863 *Zaitha testacea*: Mayr, Verh. Zool.-Bot. Ges. Wien, 13: 354.

1905 *Belostoma testaceum*: Torre-Bueno, J. N.Y. Ent. Soc., 13: 44.

Distribution: Ala., D.C., Fla., Ga., La., Md., Miss, N.C., N.J., N.Y., Mich., Pa., S.C., Tex., Va.

Subfamily Lethocerinae Lauck and Menke, 1961

Genus *Lethocerus* Mayr, 1853

1847 *Iliastus* Gistel, Handb. Naturges. Reiche, p. 490. Type-species: *Nepa grandis* Linnaeus, 1758. Monotypic. Suppressed by Int. Comm. Zool. Nomen., 1983, Opinion 1248, Bull. Zool. Nomen., 40: 81.

1853 *Lethocerus* Mayr, Verh. Zool.-Bot. Ver. Wien, 2: 15. Type-species: *Lethocerus cordofanus* Mayr, 1853, a junior synonym of *Belostoma fakir* Gistel, 1847. Monotypic.

1866 *Amorgius* Stål, Berl. Ent. Zeit., 10: 168. Type-species: *Belostoma collosicum* [as *collossicum*] Stål, 1855. Synonymized by Torre-Bueno, 1908, J. N.Y. Ent. Soc., 16: 237.

1901 *Belostoma* (*Montandonista*) Kirkaldy, Ent., 34: 6. Type-species: *Belostoma americanum* Leidy, 1847. Designated by Kirkaldy, 1906, Trans. Am. Ent. Soc., 32: 151.

Note: See Menke (1979, Bull. Zool. Nomen., 35: 236-238) for history and reasoning behind suppressing the generic name *Ilastes* Gistel. Menke (1963, An. Ent. Soc. Am., 56: 261-267) and De Carlo (1964, Physis, 24: 337-350) reviewed the genus and provided keys to species. Menke (1963) recognized two subgenera.

Subgenus *Benacus* Stål, 1861

1861 *Benacus* Stål, Öfv. K. Svens. Vet.-Akad. Förh., 18(4): 205. Type-species: *Belostoma haldemanum* Leidy, 1847, a junior synonym of *Belostoma grisea* Say, 1831. Monotypic.

1961 *Lethocerus* (*Benacus*): Lauck and Menke, An. Ent. Soc. Am., 54: 647.

Lethocerus griseus (Say), 1832

1831 *Belostoma grisea* Say, Descrip. N. Sp. N. Am. Ins. La., p. 37. [U.S.].

1847 *Belostoma haldemanum* Leidy, J. Acad. Nat. Sci., Phila., 1: 66. [U.S.]. Synonymized by Uhler, 1876, Bull. U.S. Geol. Geogr. Surv. Terr., 1: 237.

1854 *Belostoma harpax* Stål, Öfv. K. Svens. Vet.-Akad. Förh., 11(8): 240. ["America borealis"]. Synonymized by Stål, 1861, Öfv. K. Svens. Vet.-Akad. Förh., 18(4): 205.

1861 *Benacus haldemanus*: Stål, Öfv. K. Svens. Vet.-Akad. Förh., 18(4): 205.

1876 *Benacus griseus*: Uhler, Bull. U.S. Geol. Geogr. Surv. Terr., 1: 237.

1963 *Lethocerus* (*Benacus*) *griseus*: Menke, An. Ent. Soc. Am., 56: 267.

Distribution: Ala., Ark., D.C., Fla., Ga., Ia., Ill., Ind., Ks., Ky., La., Man., Mass., Md., Mich., Minn., Miss., Mo., N.C., N.J., N.Y., Neb., Oh., Ont., Pa., Que., S.C., Tenn., Tex., Va., W.Va., Wis. (Mexico to Guatemala, West Indies).

Subgenus *Lethocerus* Mayr, 1853

1853 *Lethocerus* Mayr, Verh. Zool.-Bot. Ver. Wien, 2: 15. Type-species: *Lethocerus cordofanus* Mayr, 1853, a junior synonym of *Belostoma fakir* Gistel, 1847. Monotypic.

1961 *Lethocerus* (*Lethocerus*): Lauck and Menke, An. Ent. Soc. Am., 54: 647.

Lethocerus americanus (Leidy), 1847 [Fig. 11]

1847 *Belostoma grandis* var. *americanum* Leidy, J. Acad. Nat. Sci. Phila., 1: 66. [U.S.].

1854 *Belostoma impressum* Haldeman, Proc. Acad. Nat. Sci. Phila., 6: 364. [Cal.]. Synonymized by Menke, 1963, An. Ent. Soc. Am., 56: 263.

1861 *Belostoma griseum*: Stål, Öfv. K. Svens. Vet.-Akad. Förh., 18(4): 206.

1863 *Belostoma litigiosum* Dufour, An. Soc. Ent. France, 32: 382. ["Nordamerica"]. Synonymized by Menke, 1963, An. Ent. Soc. Am., 56: 263.

1863 *Belostoma obscurum* Dufour, An. Soc. Ent. France, 32: 382. ["Nordamerica"]. Synonymized by Mayr, 1871, Verh. Zool.-Bot. Ges. Wien, 21: 427 (as synonym of *Belostoma grisea* Say); synonymized with *L. americanus* by Menke, 1963, An. Ent. Soc. Am., 56: 263.
1871 *Belostoma griseum*: Mayr, Verh. Zool.-Bot. Ges. Wien, 21: 427.
1906 *Belostoma americanum*: Snow: Trans. Ks. Acad. Sci., 20: 153.
1907 *Benacus griseus*: Howard, Ins. Book, pl. 29, fig. 36.
1908 *Lethocerus americanus*: Torre-Bueno, J. N.Y. Ent. Soc., 16: 237.
1908 *Lethocerus obscurus*: Torre-Bueno, J. N.Y. Ent. Soc., 16: 237.
1963 *Lethocerus (Lethocerus) americanus*: Menke, An. Ent. Soc. Am., 56: 263.
Distribution: Alta., B.C., Cal., Col., D.C., Del., Fla.(?), Ia., Id., Ill., Ind., Ks., Man., Mass., Md., Me., Mich., Minn., Miss., Mo., N.B., N.D., N.J., N.M., N.S., N.Y., Neb., Nev., Nfld., Oh., Ont., Ore., Que., Sask., Tex., Ut., Va., Wash., Wis., Wyo. (Mexico?).
Note: According to Menke (1963, above) records of this species in southeastern U.S. and Mexico need verification. Credit for the above synonymy is given to Menke (1963, above) who studied type material.

Lethocerus angustipes (Mayr), 1871
1871 *Belostoma angustipes* Mayr, 1871, Verh. Zool.-Bot. Ges. Wien, 21: 427. [Mexico]. Lectotype from Mexico designated by Menke, 1960, Pan-Pac. Ent., 36: 104.
1909 *Lethocerus angustipes*: Kirkaldy and Torre-Bueno, Proc. Ent. Soc. Wash., 10: 188.
1963 *Lethocerus (Lethocerus) angustipes*: Menke, An. Ent. Soc. Am., 56: 264.
Distribution: Cal., Nev. (Mexico).

Lethocerus annulipes (Herrich-Schaeffer), 1845
1845 *Belostoma annulipes* Herrich-Schaeffer, Wanz. Ins., 8: 28. [South America].
1933 *Lethocerus annulipes*: Cummings, Univ. Ks. Sci. Bull., 21: 203.
1963 *Lethocerus (Lethocerus) annulipes*: Menke, An. Ent. Soc. Am., 56: 264.
Distribution: Fla.(?) (South America, West Indies).
Note: This is a neotropical species. Cummings' (1933, above) Palm Beach, Fla. record may have been a chance introduction from the West Indies. Records of this species from Cal., Col., and Tex. are based on misidentifications.

Lethocerus medius (Guérin-Méneville), 1857
1857 *Belostoma medium* Guérin-Méneville, Hist. Is. Cuba, pt. 2, 7: 175. [Cuba].
1962 *Lethocerus medius*: Menke, Proc. Biol. Soc. Wash., 75: 62.
1963 *Lethocerus (Lethocerus) medius*: Menke, An. Ent. Soc. Wash., 56: 264.
Distribution: Ariz., N.M., Tex. (Mexico to Panama).
Note: Menke (1962, above) discussed the identity of this species, and later (1963, above) doubted the authenticity of a Nebraska record.

Lethocerus uhleri (Montandon), 1896
1896 *Belostoma uhleri* Montandon, 1896, An. Soc. Ent. Belg., 40: 513. [Fla., Ks., Pa.].
1907 *Amorgius (Montandonista) uhleri*: Torre-Bueno and Brimley, Ent. News, 18: 434.
1909 *Amorgius uhleri*: Van Duzee, Bull. Buffalo Soc. Nat. Sci., 9: 184.
1910 *Lethocerus uhleri*: Smith, Ins. N.J., p. 168.
1914 *Lethocerus (Belostoma) uhleri*: Barber, Bull. Am. Mus. Nat. Hist., 33: 498.
1963 *Lethocerus (Lethocerus) uhleri*: Menke, An. Ent. Soc. Am., 56: 263.
Distribution: Ala., Ark., Fla., Ga., Ill., Ind., Ks., La., Mass., Md., Miss., Mo., N.C., N.J., N.Y., Neb., Ok., Ont., Pa., S.C., Tenn., Tex., Wis. (Mexico).
Note: Menke (1963, above) doubted the authenticity of a Utah label on a specimen of this species.

Family Berytidae Fieber, 1851

(= Neididae Kirkaldy, 1902; Berytinidae Southwood and Leston, 1959)

The Stilt Bugs

By Richard C. Froeschner and Thomas J. Henry

The general appearance of the slender cylindrical body held high by the very long, thread-thin legs of most species is aptly reflected in the common name "stilt bugs."

Adults and nymphs wander over plant surfaces and generally feed upon the sap from the tender growth, but in several instances adults and nymphs have been reported attacking and feeding on insect eggs or soft-bodied insects. Apparently this animal food is not essential to development, as caged specimens with access only to growing plants reach adulthood and lay fertile eggs. Eggs are glued to various parts of the plants. Three or four generations may be passed during one growing season. In North America these insects are generally harmless to man's crops, but sporadically, for reasons not explained, the common *Jalysus wickhami* Van Duzee may damage certain cultivated plants (Wheeler and Henry, 1981, An. Ent. Soc. Am., 74: 606-615). Wheeler and Schaefer (1982, An. Ent. Soc. Am., 75: 498-506) summarized the known host plants for the Berytidae of the world.

The names Neididae, long used by American workers, and Berytinidae, more recently used by several European heteropterists, have been unnecessarily proposed for the family. The genus *Berytus* Fabricius is available for the stem of the earliest proposed group names, even though it is a synonym under *Neides* Latreille as a result of both genera having the same type-species.

The major works on this family in America north of Mexico are by McAtee (1919, J. N.Y. Ent. Soc., 27: 79-92); Harris (1941, Bull. Brook. Ent. Soc., 36: 105-109); and Torre-Bueno (1941, Ent. Am., 21: 101-107), the latter presenting keys through species.

Subfamily Berytinae Fieber, 1851

Genus *Berytinus* Kirkaldy, 1900

1900 *Berytinus* Kirkaldy, Ent., 33: 241. Type-species: *Cimex clavipes* Fabricius, 1775. Original designation.

Berytinus minor (Herrich-Schaeffer), 1835 [Fig. 13]

1835 *Berytus minor* Herrich-Schaeffer, Nomen. Ent., 1: 43. [Europe].
1935 *Berytus minor*: Walley, Can. Ent., 67: 160.
1941 *Berytinus minor*: Torre-Bueno, Ent. Am., 21: 103.

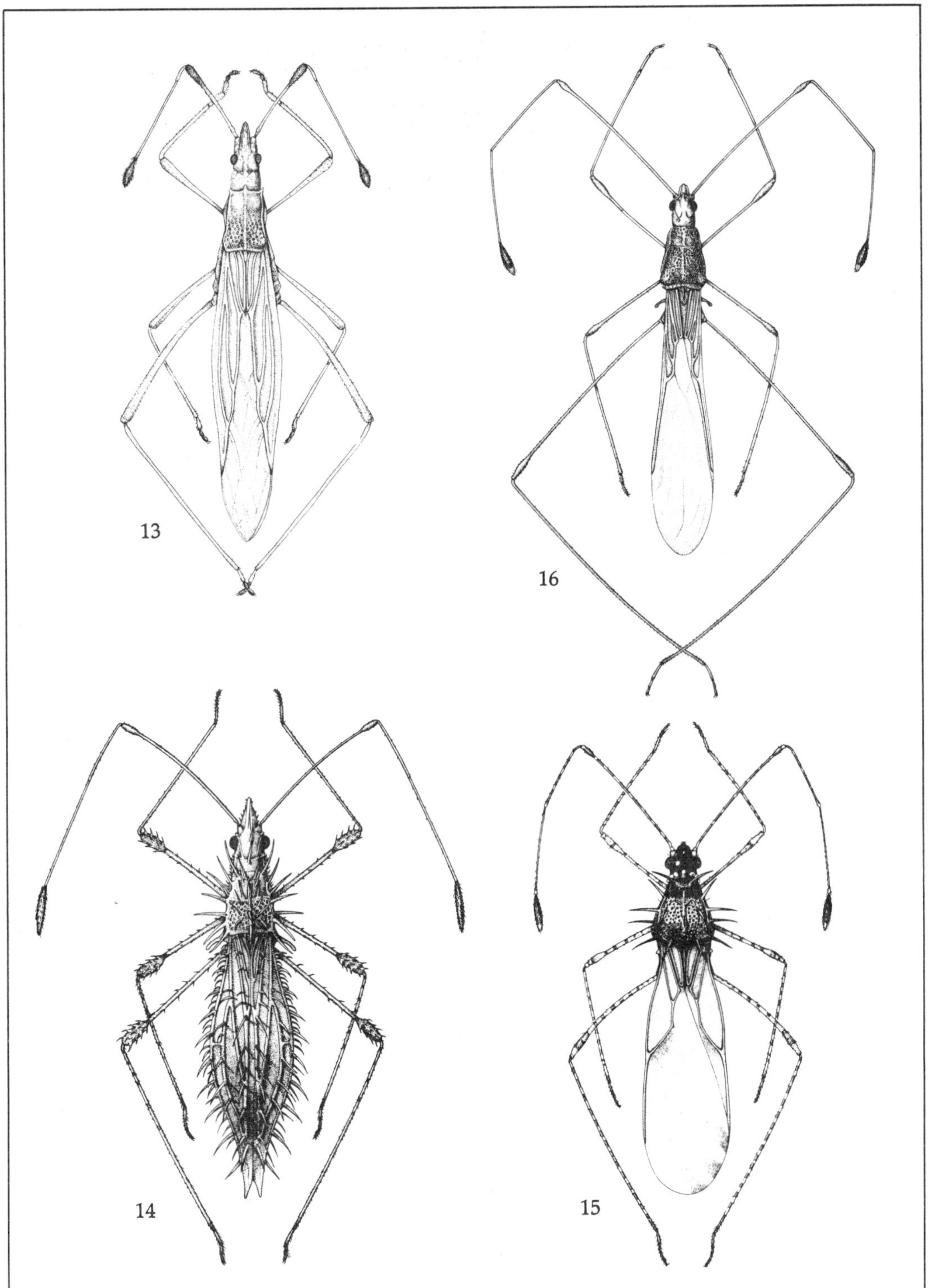

Figs. 13-16: 13, *Berytinus minor* [p. 56]; 14, *Acanthophysa echinata* [p. 58]; 15, *Pronotocantha annulata* [p. 60]; 16, *Metacanthus multispinus* [p. 60] (Originals).

Distribution: Conn., Mass., Me., Mich., N.H., N.J., N.Y., Oh., Ont., Pa., Que., W.Va. (Europe, North Africa).

Note: Wheeler (1970, Can. Ent., 102: 876 and 1971, Can. Ent., 103: 497), summarized the known literature and gave notes on hosts, distribution, and wing polymorphism for this introduced species.

Genus *Neides* Latreille, 1802.

1802 *Neides* Latreille, Hist. Nat. Crust. Ins., 3: 246. Type-species: *Cimex tipularis* Linnaeus, 1758. Designated by Latreille, 1810, Cons. Gen. Crust., Arach. Ins., p. 433.

1803 *Berytus* Fabricius, Syst. Rhyn., p. 264. Type-species: *Cimex tipularis* Linnaeus, 1758. Designated by Reuter, 1888, Acta Soc. Sci. Fenn., 15(2): 548 (separate, p. 176). Synonymy by virtue of shared type-species.

Neides muticus (Say), 1832

1832 *Berytus muticus* Say, Descrip. Het. Hem. N. Am., p. 13. ["North-west Territory"].

1859 *Neides gracilipes* Stål, Svens. Freg. Eug. Resa Jord., 3: 235. [Cal.]. Synonymized by Uhler, 1886, Check-list Hem. Het., p. 13.

1871 *Neides decurvatus* Uhler, Proc. Boston Soc. Nat. Hist., 14: 100. [N.H.]. Synonymized by Stål, 1874, K. Svens. Vet.-Akad. Handl., 12(1): 128.

1872 *Neides decurvatus* Uhler, Prelim. Rept. U.S. Geol. Surv. Mont., p. 402 (as new species). [Ariz., Col., "Washington Territory"]. Preoccupied.

1876 *Neides muticus*: Uhler, Bull. U.S. Geol. Geogr. Surv. Terr., 1: 299.

1885 *Capys muticus*: Provancher, Pet. Faune Ent. Can., 3: 58.

1894 *Jalysus muticus*: Lethierry and Severin, Gen. Cat. Hem., 2: 132.

Distribution: Every conterminous U.S. state and all Canadian Provinces bordering the U.S. (Mexico).

Note: Wheeler (1978, An. Ent. Soc. Am., 71: 733-736) studied life history and described and illustrated the fifth-instar nymph.

Subfamily Metacanthinae Douglas and Scott, 1865

Genus *Acanthophysa* Uhler, 1893

1893 *Acanthophysa* Uhler, N. Am. Fauna, 7: 261. Type-species: *Acanthophysa echinata* Uhler, 1893. Monotypic.

1919 *Saurocoris* McAtee, J. N.Y. Ent. Soc., 27: 89. Type-species: *Saurocoris instans* McAtee, 1919. Original designation. Synonymized by Van Duzee, 1929, Pan-Pac. Ent., 5: 166.

Acanthophysa echinata Uhler, 1893 [Fig. 14]

1893 *Acanthophysa echinata* Uhler, N. Am. Fauna, 7: 261. [Cal., N.M.].

1919 *Saurocoris instans* McAtee, J. N.Y. Ent. Soc., 27: 91. [Cal.]. Synonymized by Van Duzee, 1929, Pan-Pac. Ent., 5: 166.

1975 *Acanthophysa instans*: Benedict and Cothran, An. Ent. Soc. Am., 68: 898.

Distribution: Ariz., Cal., Id., N.M., Ore., Ut., Wash.

Acanthophysa idaho Harris, 1941

1941 *Acanthophysa idaho* Harris, Bull. Brook. Ent. Soc., 36: 108. [Id.].

Distribution: Cal., Id., Ore.

Genus *Jalysus* Stål, 1862

1862 *Jalysus* Stål, K. Svens. Vet.-Akad. Handl., 3(6): 59. Type-species: *Jalysus macer* Stål, 1859. Designated by Van Duzee, 1916, Check List Hem., p. 17.

Jalysus caducus (Distant), 1893

1893 *Neides caducus* Distant, Biol. Centr.-Am., Rhyn., 2: 460. [Mexico, Panama].

1911 *Jalysus elongatus* Barber, J. N.Y. Ent. Soc., 19: 23. [Ariz.]. Synonymized by Barber, 1948, Bull. Ent. Soc., 43: 21.

1948 *Jalysus caducus*: Barber, Bull. Brook. Ent. Soc., 43: 21.

Distribution: Ariz., Tex. (Mexico, Panama).

Jalysus spinosus (Say), 1824

1824 *Berytus spinosus* Say, Am. Ent., 1: 28. ["United States"].

1842 *Neides trispinus* [*sic*]: Westwood, Cat. Hem., p. 6 (in list).

1842 *Neides trispinosus* Westwood, Cat. Hem., p. 24. [Pa.]. Synonymized by Uhler, 1886, Check-list Hem. Het., p. 13.

1859 *Neides spinosus*: Dohrn, Cat. Hem., p. 29.

1874 *Jalysus spinosus*: Stål, K. Svens. Vet.-Akad. Handl., 12(1): 129.

Distribution: Ala., Ark., Conn., Del., Fla., Ga., Ia., Ill., Ind., Ks., Ky., La., Mass., Md., Me., Mich., Minn., Miss., Mo., N.C., N.D., N.H., N.J., N.Y., Neb., Oh., Ok., Pa., Que., R.I., S.C., S.D., Tenn., Tex., Vt., Va., W.Va., Wis.

Note: Wheeler and Henry (1981, An. Ent. Soc. Am., 74: 606-615) clarified the taxonomic confusion between this species and *J. wickhami*, reviewed hosts and distribution, and keyed and illustrated the fifth-instar nymphs. Wheeler (1986, Ent. News, 97: 63-65) gave additional biological information.

Jalysus wickhami Van Duzee, 1906

1906 *Jalysus wickhami* Van Duzee, Ent. News, 17: 387. [Ariz., Cal.].

1914 *Jalysus spinosus* var. *wickhami*: Van Duzee, Trans. San Diego Soc. Nat. Hist., 2: 5.

1919 *Jalysus spinosus* subspecies *wickhami*: McAtee, J. N.Y. Ent. Soc., 27: 86.

Distribution: Every continental state, except Alk., N.H., and Vt. (Mexico).

Note: Harris (1941, Bull. Brook. Ent. Soc., 36: 105) re-established this form at full species level. While some subsequent authors have used this unit in the trinomial, Wheeler and Henry (1981, An. Ent. Soc. Am., 74: 606-615) showed that *wickhami* is distinct from *spinosus*. The latter authors reviewed hosts and distribution and described, keyed, and illustrated the fifth-instar nymph.

Genus *Metacanthus* Costa, 1847

1847 *Metacanthus* Costa, Cimic. Reg. Neapol. Cent., p. 258. Type-species: *Berytus elegans* Costa, 1847, a junior synonym of *Berytus meridionalis* Costa, 1844. Monotypic.

1919 *Aknisus* McAtee, J. N.Y. Ent. Soc., 27: 81. Type-species: *Aknisus calvus* McAtee, 1919. Original designation. Synonymized by Froeschner, 1985, Smithson. Contr. Zool., 407: 10.

Metacanthus calvus (McAtee), 1919, New Combination

1919 *Aknisus calvus* McAtee, J. N.Y. Ent. Soc., 27: 85. [Cal.].

Distribution: Cal.

Metacanthus multispinus (Ashmead), 1887, New Combination [Fig. 16]

1887 *Hoplinus multispinus* Ashmead, Ent. Am., 3: 155. [Fla.].

1909 *Jalysus perclavatus* Van Duzee, Bull. Buff. Soc. Nat. Hist., 9: 163. [Fla.]. Synonymized by Barber, 1911, J. N.Y. Ent. Soc., 19: 24.

1908 *Jalysus multispinosus* [*sic*]: Barber, J. N.Y. Ent. Soc., 16: 248.

1911 *Jalysus* (*Hoplinus*) *multispinosus* [*sic*]: Barber, J. N.Y. Ent. Soc., 19: 24.

1912 *Julisus* [*sic*] *multispinosus* [*sic*]: Hunter, Proc. Ent. Soc. Wash., 14: 64.

1914 *Jalysus multispinus*: Barber, Bull. Am. Mus. Nat. Hist., 33: 517.

1919 *Aknisus multispinus*: McAtee, J. N.Y. Ent. Soc., 27: 82.

1957 *Aknisus multispinuus* [*sic*]: Glick, U.S. Dept. Agr. Tech. Bull., 1158: 6.

1961 *Aknisus multispinosus* [*sic*]: Bibby, J. Econ. Ent., 54: 328.

Distribution: Ala., Ariz., D.C., Fla., Ga., Ind., Ia., Ks., La., Miss., Mo., N.J., Ok., Ore., Tex. (Netherlands Antilles).

Note: Van Duzee (1929, Pan-Pac. Ent., 5: 166) expressed a desire to maintain his *perclavatus* as a distinct species but offered no separating features; all other authors have concurred with Barber's (1911, above) synonymy.

Metacanthus tenellus Stål, 1859

1859 *Metacanthus tenellus* Stål, Svens. Freg. Eug. Res. Jord., 4: 236. [Ecuador].

1941 *Jalysus tenellus*: Harris, Bull. Brook. Ent. Soc., 36: 107.

Distribution: Tex. (West Indies, Mexico to South America).

Note: Froeschner (1981, Smithson. Contr. Zool., 322: 17) and most authors subsequent to Harris (1941, above) have placed *tenellus* in the genus *Metacanthus*.

Genus *Pronotacantha* Uhler, 1893

1886 *Acantholaena* Uhler, Check-list Hem. Het., p. 13. *Nomen nudum*. Included species: *Acantholaena annulata* Uhler, 1886. *Nomen nudum*.

1893 *Pronotacantha* Uhler, N. Am. Fauna, 7: 260. Type-species: *Pronotacantha annulata* Uhler, 1893. Monotypic.

Pronotacantha annulata Uhler, 1893 [Fig. 15]

1886 *Acantholaena annulata* Uhler, Check-list Hem. Het., p. 13. *Nomen nudum*.

1893 *Pronotacantha annulata* Uhler, N. Am. Fauna, 7: 260. [Cal.].

Distribution: Ariz., Cal., N.M., Nev., Tex., Ut. (Mexico).

Genus *Protacanthus* Uhler, 1893

1893 *Protacanthus* Uhler, Proc. Zool. Soc. London, p. 707. Type-species: *Protacanthus decorus* Uhler, 1893. Monotypic.

Protacanthus decorus Uhler, 1893

1893 *Protacanthus decorus* Uhler, Proc. Zool. Soc. London, p. 708. [St. Vincent, West Indies].

1909 *Metacanthus decorus*: Van Duzee, Bull. Buff. Soc. Nat. Sci., 9: 164.

Distribution: Fla., Tex. (Colombia, Costa Rica, Venezuela, West Indies).

Family Ceratocombidae Fieber, 1861

The Ceratocombids

By Thomas J. Henry

Ceratocombids are a small family, until recently, combined with the Dipsocoridae. They are generally characterized by their long slender antennae, 2-segmented tarsi on all legs, bristlelike setae on the antennae, head, and tibiae, and by a short distinct fracture at the middle of the costa on the hemelytra. All species are less than 2 millimeters long. Only two genera and four species are recorded from the United States.

Ceratocombids probably are predators that feed on small co-existing arthropods. One species was reported biting man, undoubtedly a case of probing to feed after accidentally landing on the skin. These bugs have been collected in leaf litter and other ground debris, in rotting wood, and at the base of grass clumps.

The major literature pertaining to this family includes: Reuter's (1891, Acta Soc. Sci. Fenn., 19(6): 1-27) "Monographia Ceratocombidarum"; McAtee and Malloch's (1925, Proc. U.S. Nat. Mus., 67: 1-42) "Revision of the bugs of the family Cryptostemmatidae"; Emsley's (1969, Mem. Am. Ent. Soc., 25: 1-154) monograph of the Schizopteridae; and Štys' treatises on the families Dipsocoridae (1970, Acta Ent. Bohem., 67: 21-46) and Ceratocombidae (1982, Acta Ent. Bohem., 79: 354-376).

In this family, several new generic and subgeneric combinations are recognized because certain authors have created new generic-group name synonyms but did not transfer in print all species involved. The physical combinations listed herein technically must be considered new.

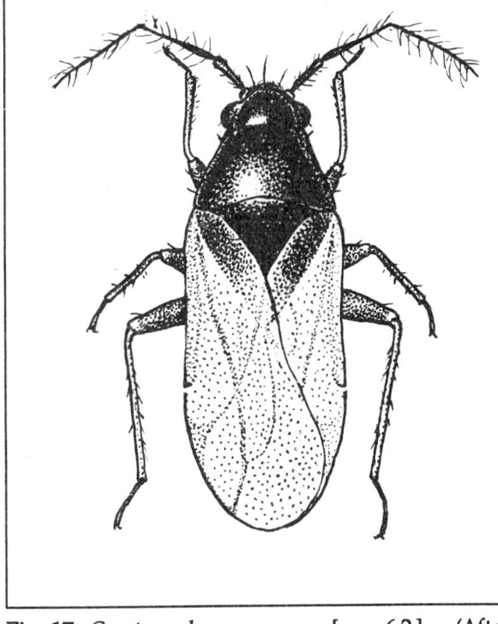

Fig. 17 *Ceratocombus vagans* [p. 62] (After Froeschner, 1949).

Subfamily Ceratocombinae Fieber, 1861

Tribe Ceratocombini Fieber, 1861

Genus *Ceratocombus* Signoret, 1852

1852 *Ceratocombus* Signoret, An. Soc. Ent. France, (2)10: 542. Type-species: *Astemma mulsanti* Signoret, 1852, a junior synonym of *Anthocoris coleoptratus* Zetterstedt, 1819. Monotypic.

Note: McAtee and Malloch (1925, Proc. U.S. Nat. Mus., 67: 5) provided a key to the New World species.

Subgenus *Ceratocombus* Signoret, 1852

1852 *Ceratocombus* Signoret, An. Soc. Ent. France, (2)10: 542. Type-species: *Astemma mulsanti* Signoret, 1852, a junior synonym of *Anthocoris coleoptratus* Zetterstedt, 1819. Monotypic.

1891 *Ceratocombus* (*Xylonannus*) Reuter, Acta Soc. Sci. Fenn., 19(6): 8. Type-species: *Ceratocombus corticalis* Reuter, 1891. Designated by Oshanin, 1912, Kat. Paläark. Hem., p. 85. Synonymized by Linnavouri, 1951, An. Ent. Fenn., 17: 97.

Ceratocombus hesperus McAtee and Malloch, 1925

1925 *Ceratocombus hesperus* McAtee and Malloch, Proc. U.S. Nat. Mus., 67: 6. [Cal.].

1925 *Ceratocombus* (*Ceratocombus*) *hesperus*: McAtee and Malloch, Proc. U.S. Nat. Mus., 67: 6.

1978 *Ceratocombus hesperius* [*sic*]: Slater and Baranowski, How To Know True Bugs, p. 206.

Distribution: Cal.

Ceratocombus niger Uhler, 1904, New Subgeneric Combination

1904 *Ceratocombus niger* Uhler, Proc. U.S. Nat. Mus., 27: 361. [N.M.].

Distribution: N.M.

Note: McAtee and Malloch (1925, Proc. U.S. Nat. Mus., 67: 8) suggested that their species, *C. vagans*, might be a synonym of this species, but that the "type" of *C. niger* is lost and Uhler's second specimen is damaged. This species is placed in the subgenus *Ceratocombus* because of the relationship to *C. vagans* inferred by McAtee and Malloch.

Ceratocombus vagans McAtee and Malloch, 1925, New Subgeneric Combination [Fig. 17]

1925 *Ceratocombus* (*Xylonannus*) *vagans* McAtee and Malloch, Proc. U.S. Nat. Mus., 67: 7. [Md.].

Distribution: Fla., Md., N.Y., Ont., Va. (Panama).

Note: Blatchley (1926, Het. E. N. Am., p. 649) stated that "The description of *C. niger* Uhler (1904, 361) from New Mexico agrees in all particulars with that of *vagans* M. & M...."

Genus *Leptonannus* Reuter, 1891

1891 *Ceratocombus* (*Leptonannus*) Reuter, Acta Soc. Sci. Fenn., 19(6): 5. Type-species: *Ceratocombus biguttulus*: Reuter, 1891. Monotypic.

1912 *Leptonannus*: Reuter, Öfv. F. Vet.-Soc. Förh., 54A(7): 65.

Note: Although this genus has been reduced to a subgenus since Reuter's (1912, above) treatment (e.g. McAtee and Malloch, 1925, Proc. U.S. Nat. Mus., 67: 4), Štys (1970, Acta Ent. Bohem., 67: 21) recognized generic status.

Leptonannus latipennis (Uhler), 1904, New Combination

1904 *Ceratocombus latipennis* Uhler, Proc. U.S. Nat. Mus., 27: 362. [N.M.].

1904 *Ceratocombus brasiliensis*: Uhler, Proc. U.S. Nat. Mus., 27: 361.

1925 *Ceratocombus* (*Leptonannus*) *latipennis*: McAtee and Malloch, Proc. U.S. Nat. Mus., 67: 9.

Distribution: N.M.

Note: McAtee and Malloch (1925, Proc. U.S. Nat. Mus., 67: 6) referred Uhler's (1904, above) record of *C. brasiliensis* from N.M. to this species. Slater and Baranowski (1978, How To Know True Bugs, p. 206) suggest that *C. latipennis* is a synonym of *C. vagans* McAtee and Malloch.

Family Cimicidae Latreille, 1802

The Bed Bugs

By Richard C. Froeschner

The infamous bed bug and related parasites, all of which feed on warm-blooded vertebrate animals, comprise the family Cimicidae. All are flightless with the wings greatly reduced to small pads. Creeping onto their sleeping or resting normal prey, they obtain a meal by painlessly withdrawing its blood. Their bite can be acutely painful to victims other than their normal host; thus, species that normally feed on birds nesting under the eaves of buildings can become conspicuously present if they move into the buildings and bite humans. Only occasionally do they cling to their hosts long enough to be carried to new places. Hypothetically, mankind first acquired his unwanted companion, the bed bug, *Cimex lectularius* Linnaeus, when he reached the caveman stage and shared those retreats with the host-bats of the bed bugs. The bugs, secreting themselves in crevices in man's possessions, were then carried throughout the world by man himself.

Members of the bed bug family have long been suspected of disease dissemination. Their habits of repeated feedings, each followed by a retreat from the host, logically place them in a position to transmit pathogens from one individual to another. Ryckman et al. (1981, Bull. Soc. Vector Ecol., 6: 98-99) gave a summary discussion of such a role and reported "Bed bugs have been indicted for transmission of many disease-producing pathogens to man but have never really been proven to be responsible for epidemics or serious outbreaks of disease." That summary further stated that the bed bug has been "suspected" of transmitting of 41 human disease organisms as well as contributing to certain kinds of human nutritional deficiencies and allergies.

Hayes et al. (1977, J. Med. Ent., 14: 257-262) and Rush et al. (1980, An. Ent. Soc. Am., 73: 315-322) studied the first known examples of "biologic virus transmission" by a member of this family. *Oeciacus vicarius*

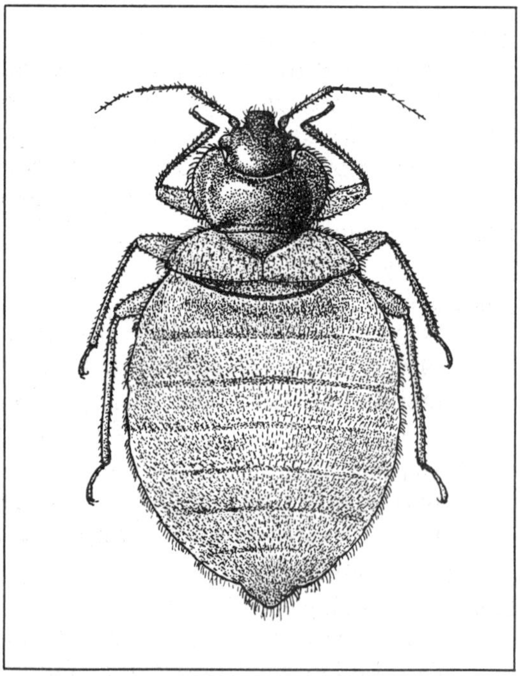

Fig. 18 *Cimex lectularius* [p. 66] (After Froeschner, 1949).

Horvath serves as both the vector and the overwintering host for the Fort Morgan virus, a virus related to western equine encephalitis but found only in a bird-bug-bird cycle in these insects, in the cliff swallows, *Petrochelidon pyrrhonota* (Viellot), whose mud nests they inhabit, and in house sparrows, *Passer domesticus* (Linnaeus), that sometimes use the empty cliff swallow nests.

The family now contains 91 species, of which 16 are known to occur in the Western Hemisphere north of Mexico. The starting point for the study of these insects is the *Monograph of Cimicidae* by Usinger (1966). It contains extensive and intensive studies of the 74 species of the world known at that time. Since then 19 new species have been described. The most recent summarizations can be found in Ryckman et al. (1981, Bull. Soc. Vector Ecol., 6: 93-142), which is a checklist of the species of the world and a bibliography including a comprehensive list of cimicid literature of the Americas, cross indexed by species and countries (by states within the U.S.).

Subfamily Cimicinae Latreille, 1802

Genus *Cimex* Linnaeus, 1785

1758 *Cimex* Linnaeus, Syst. Nat., 10th edit., p. 441. Type-species: *Cimex lectularius* Linnaeus, 1758. Designated by Opinion 81, Int. Comm. Zool. Nomen., 1924, Smithson. Misc. Coll., 73: 19.

1775 *Acanthia* Fabricius, Syst. Ent., p. 693. Type-species: *Cimex lectularius* Linnaeus, 1758. Designated by Opinion 81, Int. Comm. Zool. Nomen., 1924, Smithson. Misc. Coll., 73: 19. Synonymy by virtue of shared type-species.

1829 *Clinocoris* Fallén, Hem. Suec., p. 141. Type-species: *Cimex lectularius* Linnaeus, 1758. Designated by Opinion 81, Int. Comm. Zool. Nomen., 1924, Smithson. Misc. Coll., 73: 19. Synonymy by virtue of shared type-species.

1899 *Klinophilos* Kirkaldy, Ent., 32: 219. Type-species: *Cimex lectularius* Linnaeus, 1758. Designated by the Int. Comm. Zool. Nomen., 1924, Smithson. Misc. Coll., 73: 19. Synonymy by virtue of shared type-species.

1903 *Clinophilus* Blanford, Nature, 69: 200. Proposed as emendation for *Klinophilos* Kirkaldy, 1899.

Cimex adjunctus Barber, 1939

1912 *Cimex pilosellus*: Horvath, An. Mus. Nat. Hung., 10: 259 (in part).

1939 *Cimex adjunctus* Barber, Proc. Ent. Soc. Wash., 41: 244. [Pa.].

Distribution: Ala., Col., Del., Fla., Ga., Ia., Ind., Ky., Md., Me., N.C., N.H., N.J., N.Y., Neb., Oh., Pa., S.C., Tex., Va., W.Va.

Cimex antennatus Usinger and Ueshima, 1965

1963 *Cimex pilosellus* strain A: Ueshima, Chromosoma, 14: 512.

1965 *Cimex antennatus* Usinger and Ueshima, Pan-Pac. Ent., 41: 115. [Cal.].

Distribution: Cal., Nev.

Cimex brevis Usinger and Ueshima, 1965

1965 *Cimex brevis* Usinger and Ueshima, Pan-Pac. Ent., 41: 117. [Minn.].

Distribution: Ill., Mich., Minn., Que.

Cimex incrassatus Usinger and Ueshima, 1965
1961 *Cimex pilosellus*: Bradshaw and Ross, J. Ariz. Acad. Sci., 1: 110.
1963 *Cimex pilosellus* strain B: Ueshima, Chromosoma, 14: 512.
1965 *Cimex incrassatus* Usinger and Ueshima, Pan-Pac. Ent., 41: 115. [Ariz.].
Distribution: Ariz., Cal., Nev., Ut. (Guatemala, Mexico).

Cimex latipennis Usinger and Ueshima, 1965
1965 *Cimex latipennis* Usinger and Ueshima, Pan-Pac. Ent., 41: 114. [Ore.].
Distribution: Cal., Id., Mont., Ore.

Cimex lectularius Linnaeus, 1758 [Fig. 18]
1758 *Cimex lectularius* Linnaeus, Syst. Nat., 10th edit., p. 441. [Europe, England].
1871 *Acanthia lectularia*: Walker, Cat. Hem. Brit. Mus., 7: 43
1887 *Cimex lectularius*: Provancher, Nat. Can., 1887, p. 170.
1905 *Clinocoris lectularius*: Girault, Psyche, 12: 61.
Distribution: "Truly cosmopolitan" (Usinger, 1966, Monogr. Cimicidae, p. 315) from every state.

Cimex pilosellus (Horvath), 1910
1898 *Cimex pipistrelli*: Chittenden, U.S. Dept. Agr., Div. Ent., Tech. Bull., (new ser.), 18: 97.
1910 *Clinocoris pilosellus* Horvath, Ent. Month. Mag., Ser. 2, 21: 12. [B.C.].
1912 *Cimex pilosellus*: Horvath, An. Mus. Nat. Hist. Hung., 10: 259.
Distribution: Alta., Ariz., B.C., Cal., Col., Id., Mont., Neb., Nev., Ore., Wash.

Genus *Oeciacus* Stål, 1873

1873 *Oeciacus* Stål, K. Svens. Vet.-Akad. Handl., 11(2): 104. Type-species: *Cimex hirundinis* Lamarck, 1816. Monotypic.

Oeciacus vicarius Horvath, 1912
1870 *Cimex lunifrontis* Cooper, Ornith., 1: 105. [Cal.]. See note.
1890 *Acanthia papistrilla* [*sic*]: Gillette, Ent. News, 1: 26.
1892 *Acanthia hirundinis*: Osborn, Can. Ent., 24: 264.
1895 *Cimex hirundinis*: Gillette and Baker, Col. Agr. Exp. Stn. Bull., Tech. Ser., 1: 56.
1912 *Oeciacus vicarius* Horvath, An. Mus. Nat. Hung., 10: 161 (in part). [Cal.].
Distribution: Alta., B.C., Cal., Col., Ia., Me., N.C., N.H., N.S., N.Y., Neb., Ont., Ore., Ut.
Note: List (1925, Proc. Biol. Soc. Wash., 38: 108) transferred the Mexican record by Horvath (1912, above) to *Hesperocimex coloradensis* List. The International Commission on Zoological Nomenclature (1985, Opinion 1360, Bull. Zool. Nomen., 42: 347-348) has suppressed the combination *Cimex lunifrontis* Cooper, which was not used since its proposal, and conserved *Oeciacus vicarius*.

Subfamily Haematosiphoninae Jordan and Rothschild, 1912

Genus *Cimexopsis* List, 1925

1925 *Cimexopsis* List, Proc. Biol. Soc. Wash., 38: 106. Type-species: *Cimexopsis nyctalis* List, 1925. Monotypic.

Cimexopsis nyctalis List, 1925

1925 *Cimexopsis nyctalis* List, Proc. Biol. Soc. Wash., 38: 106. [D.C.].

1942 *Haematosiphon nyctalis*: Eichler, Mitt. Zool. Mus. Berlin, 25: 296.

Distribution: Ark., D.C., Fla., Ga., Ia., Ill., Ind., Me., Minn., Miss., N.Y., Neb., Oh., Pa., S.C., Va.

Genus *Haematosiphon* Champion, 1900

1900 *Haematosiphon* Champion, Biol. Centr.-Am., Rhyn., 2: 337. Type-species: *Acanthia inodora* Duges, 1892. Monotypic.

Haematosiphon inodorus (Duges), 1892

1892 *Acanthia inodora* Duges, La Nat., ser. 2, 2: 169. [Mexico].

1900 *Haematosiphon inodora*: Champion, Biol. Centr.-Am., Rhyn., 2: 337.

Distribution: Ariz., Cal., Ks., N.M., Ok., Tex. (Mexico).

Note: Usinger (1966, Monogr. Cimicidae, p. 478) wrote that the Florida record by Blatchley (1928, Fla. Ent., 12: 43) pertains to *Ornithocoris pallidus* Usinger.

Genus *Hesperocimex* List, 1925

1925 *Hesperocimex* List, Proc. Biol. Soc. Wash., 38: 104. Type-species: *Hesperocimex coloradensis* List, 1925. Monotypic.

Hesperocimex coloradensis List, 1925

1912 *Oeciacus vicarius*: Horvath, An. Mus. Nat. Hung., 10: 261 (in part).

1925 *Hesperocimex coloradensis* List, Proc. Biol. Soc. Wash., 38: 104. [Col.].

Distribution: B.C., Cal., Col., Neb., Ore. (Mexico).

Hesperocimex sonorensis Ryckman, 1958

1912 *Oeciacus vicarius*: Horvath, An. Mus. Nat. Hung., 10: 261 (in part).

1955 *Hesperocimex coloradensis*: Lee and Ryckman, Proc. Ent. Soc. Wash., 57: 164.

1958 *Hesperocimex sonorensis* Ryckman, An. Ent. Soc. Am., 51: 33. [Mexico].

Distribution: Ariz. (Mexico).

Genus *Ornithocoris* Pinto, 1927

1927 *Ornithocoris* Pinto, Rev. Biol. Hyg., 1: 17. Type-species: *Ornithocoris toledoi* Pinto, 1927. Monotypic.

Ornithocoris pallidus Usinger, 1959

1928 *Hesperocimex inodorus*: Blatchley, Fla. Ent., 12: 43.

1959 *Ornithocoris pallidus* Usinger, Ent., 92: 219. [Brazil].

Distribution: Fla., Ga. (Brazil).

Note: Usinger (1966, Monogr. Cimicidae, p. 466) concluded that this species had been introduced into the United States by unknown means.

Genus *Synxenoderus* List, 1925

1925 *Synxenoderus* List, Proc. Biol. Soc. Wash., 38: 108. Type-species: *Synxenoderus comosus* List. Monotypic.

Synxenoderus comosus List, 1925

1925 *Synxenoderus comosus* List, Proc. Biol. Soc. Wash., 38: 108. [Cal.].

Distribution: Cal., Nev.

Subfamily Primicimicinae Ferris and Usinger, 1955

Genus *Primicimex* Barber, 1941

1941 *Primicimex* Barber, J. Wash. Acad. Sci., 31: 315. Type-species: *Primicimex cavernis* Barber, 1914. Monotypic.

Primicimex cavernis Barber, 1941
1941 *Primicimex cavernis* Barber, J. Wash. Acad. Sci., 3: 315. [Tex.].
Distribution: Tex. (Guatemala, Mexico).

Family Coreidae Leach, 1815

The Coreid Bugs

By Richard C. Froeschner

The greatest number of species of Coreidae occurs in the tropical parts of the world and relatively few, about 87, are known from America north of Mexico. All members of this family are essentially phytophagous and generally concentrate their attacks on tender shoots and leaves. There are occasional reports of individuals imbibing fluids from decomposing animal carcasses but this is not a regular habit of the species involved (Adler and Wheeler, 1984, J. Ks. Ent. Soc., 57: 21-27).

Predictably those species feeding on cultivated plants have gained the designation of "pest," a category that includes relatively few species of the family. Among the pest species in North America are the squash bugs of the genus *Anasa* Amyot and Serville, especially *Anasa tristis* (De Geer), which concentrate their attentions on cucurbitaceous plants and cause wilting and death of parts or all of the plant victims by the direct effect of their feeding or by carrying on their body surfaces the *Bacillus* that causes a wilt disease of such plants. Another often reported pest species is one of the leaf-footed bugs, *Leptoglossus phyllopus* (Linnaeus), which in the southern states attacks a variety of field, garden and orchard crops such as cotton, tomatoes, asparagus, potatoes, oranges, peaches, and others. Recently, Bolkan et al. (1984, J. Econ. Ent., 77: 1163-1165) found that *Leptoglossus clypealis* (Heidmann) causes annual losses of 30% or more to the pistachio crop in California. A tabulation of the food plants of Coreidae was given by Schaefer and Mitchell (1983, An. Ent. Soc. Am., 76: 595-608).

Members of the genus *Chelinidea* Uhler concentrate their feeding on cacti. This food preference led to their introduction from the Americas into Australia here they contribute to the biological control of cacti, *Opuntia* spp., which escaped from cultivation and became serious pests on range land. This role for these insects was summarized by Herring (1980, Proc. Ent. Soc. Wash., 82: 237-251).

The biology of Coreidae is typically hemipterous. To assure that the immature stages hatch on or close to an appropriate food plant—many of which are annuals—the highly mobile adults hibernate and in spring seek the newly developing food plants on which the eggs are laid in exposed clusters. In one Old World species the eggs reportedly are stuck on the back of adult individuals of the same species—usually on males. In all species the young individuals pass through five stages, suck the plant's juices and in due time transform into adults. There may be one or more generations per year depending in part upon the species being considered and in part upon the region of the country in which they develop. Adults and nymphs are capable of exuding a strong-smelling fluid when disturbed.

The classification of the Coreidae has changed significantly during the last few

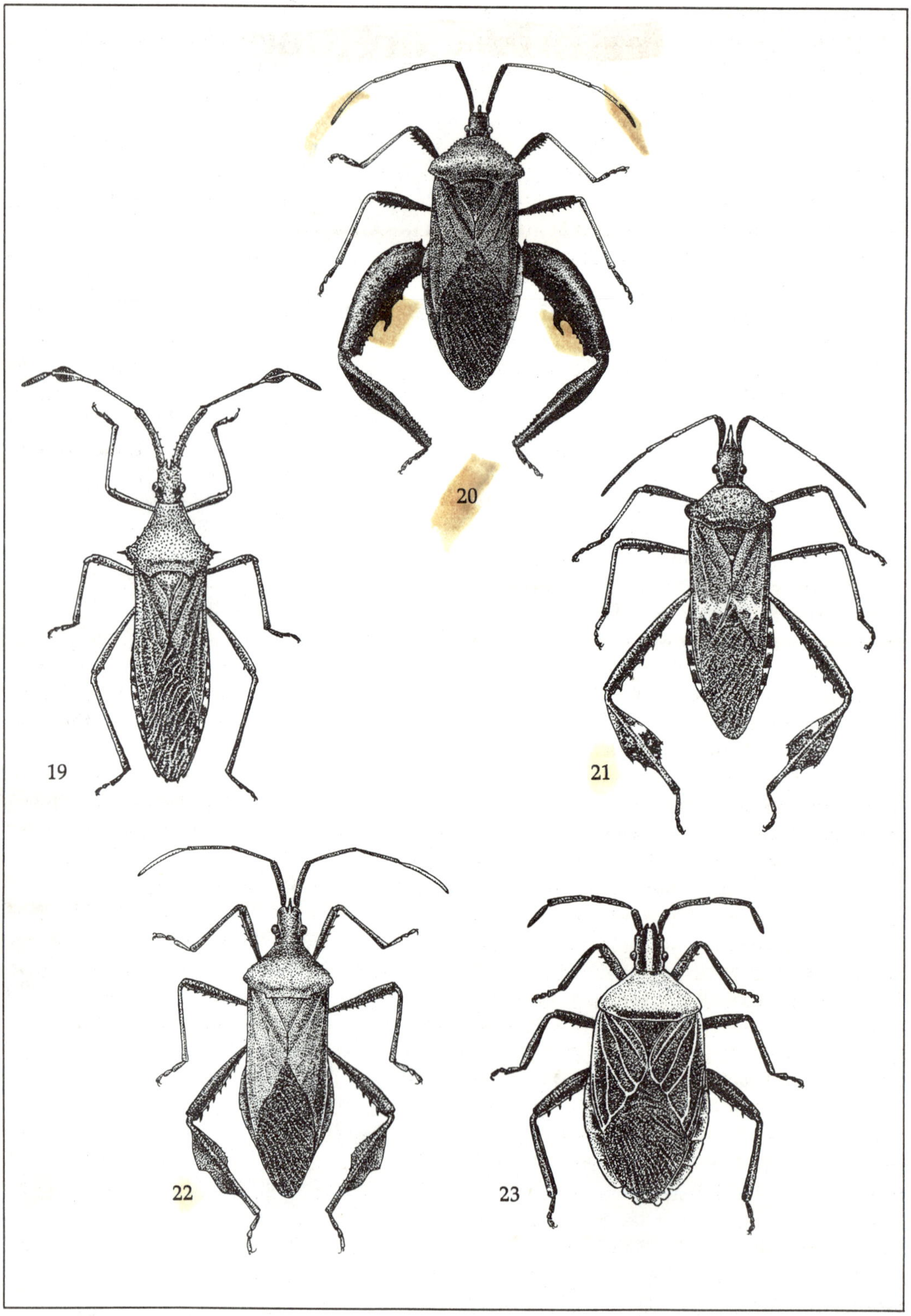

Figs. 19-23: 19, *Chariesterus antennator* [p. 79]; 20, *Acanthocephala femorata* [p. 73]; 21, *Leptoglossus clypealis* [p.76]; 22, *Acanthocephala terminalis* [p. 74]; 23, *Chelinidea vittiger* [p.80] (After Froeschner, 1942)

decades. Two groups formerly included as subfamilies are now accorded full family rank: the Alydidae and Rhopalidae (=Corizidae). The suprageneric-infrafamilial arrangement, which for nearly 100 years was mostly that of Stål (1867, Öfv. K. Svens. Vet.-Akad. Förh., 24: 534-551 and 1870, K. Svens. Vet.-Akad. Handl., 9(1): 125-228), underwent modifications as a result of intense studies by Schaefer and by O'Shea (see literature cited), separately and in coauthorship. Combining the present classification with the keys in Torre-Bueno's synopsis (1941, Ent. Am., 21: 44-77) forms the most useful way for identifying specimens in our fauna.

Literature contains North American records for six species of Coreidae that are excluded from this catalog: *Anisoscelis prominulus, Gonocerus obsoletus, Paryphes rufoscutellatus, Anasa uhleri, Sephina limbata,* and *Thasus acutangula*. The reasons for excluding the first three named species are given in the following paragraphs; for the last three species under the appropriate genus in the text.

The combination *"Anisoscelis prominulus"* was listed by Uhler (1871, Proc. Boston Soc. Nat. Hist., 14: 99) as a Say manuscript name found on a Pennsylvania specimen in the T. W. Harris collection; in this case Uhler presented the name in the discussion under his new species *Metapodius instabilis*, a name now considered a junior synonym of *Acanthocephala terminalis* (Dallas). All subsequent listings of the Say combination place it in synonymy as a Say manuscript name; but earlier Rathvon (1869, Hist. Lancaster Co., Pa., p. 549) had used it for a species in his list of Hemiptera in Pennsylvania but credited it to Harris— possibly Harris had Say-identified specimens before him. Apparently no one has associated a characterization, figure, or description with it; hence it is an invalid name and should be dropped from lists. For another Harris use of Say's combination see the lygaeid species *Eremocoris ferus* (Linnaeus).

The *"Gonocerus obsoletus,"* described from "Nordamerika" by Herrich-Schaeffer (1840, Wanz. Ins., 6:10, plate 83, figure 567), has had varying treatment in lists of North American forms. Uhler's (1886) Check-list omitted it; Lethierry and Severin (1894, Cat. Gen. Hem., 2: 66) placed it in the African genus *Cletus* Stål, a placement followed in the Banks (1910) Catalogue. Van Duzee, in his (1916, p. 13) Check List and (1917, p. 106) Catalogue, placed it in synonymy under *Zicca taeniola* Dallas; this allowed him to include *Z. taeniola* from our fauna and to comment that if this synonymy is accurate, Herrich - Schaeffer's name should be used as it has priority over Dallas'. Torre - Bueno (1941, Ent. Am.) omitted it from his Synopsis.

Herrich-Schaeffer's original description and its accompanying illustration for *Gonocerus obsoletus* present a species differing from all others in the region of this catalog. While the head shape, long first antennal segment, and the angularly projecting humeri are suggestive of *Zicca taeniola*, the absence of serrations on the side margins of the pronotum plus the slender unspined posterior femora prevent such placement. *"Gonocerus obsoletus"* is here omitted from the North American fauna with the conclusion that its country of origin was originally misstated.

The species *"Paryphes rufoscutellatus* (Gray)" has been variously listed for and excluded from the fauna of the United States, all such actions apparently stemming from the 1876 listing by Uhler (Bull. U.S. Geol. Geogr. Surv., 1: 293), who gave records for "California, Cape Saint Lucas (J. Xanthus), and Mexico." Because no subsequent United States findings have been reported, the present list follows Barber (1926, J. N.Y. Ent. Soc., 34: 215), who concluded that in the absence of specimens, the earlier "California" records all should be interpreted as "Lower California."

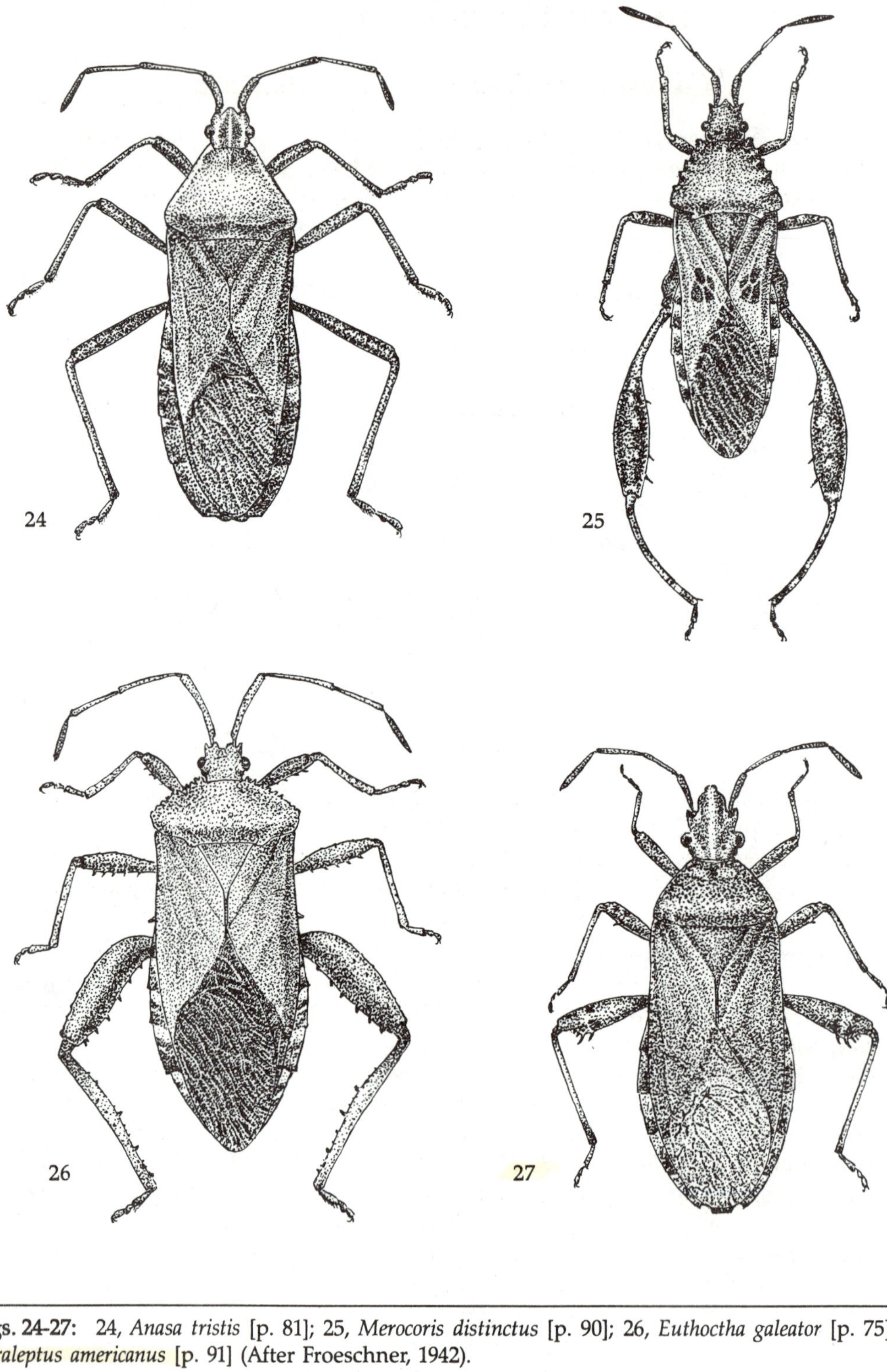

24 25 26 27

Figs. 24-27: 24, *Anasa tristis* [p. 81]; 25, *Merocoris distinctus* [p. 90]; 26, *Euthoctha galeator* [p. 75]; 27, *Ceraleptus americanus* [p. 91] (After Froeschner, 1942).

Subfamily Coreinae Leach, 1815

Tribe Acanthocephalini Stål, 1870

Genus *Acanthocephala* Laporte, 1833

1833 *Acanthocephala* Laporte, Mag. Zool., 2: 29. Type-species: *Lygaeus compressipes* Fabricius, 1794, a junior synonym of *Cimex latipes* Drury, 1782. Monotypic.

1842 *Metapodius* Westwood, Hope Cat., 2: 4. Unnecessary new name for *Acanthocephala* Laporte.

Subgenus *Acanthocephala* Laporte, 1833

1833 *Acanthocephala* Laporte, Mag. Zool., 2: 29. Type-species: *Lygaeus compressipes* Fabricius, 1794, a junior synonym of *Cimex latipes* Drury, 1782. Monotypic.

1870 *Acanthocephala* (*Acanthocephala*): Stål, K. Svens. Vet.-Akad. Handl., 9(1): 149.

Acanthocephala declivis (Say), 1832

1832 *Rhinuchus declivis* Say, Ins. La., p. 10. [Ga., La.].

1832 *Anisoscelis declivis*: Say, Descrip. Het. Hem. N. Am., p. 12.

1870 *Acanthocephala* (*Acanthocephala*) *declivis*: Stål, K. Svens. Vet.-Akad. Handl., 9(1): 150.

1876 *Acanthocephala declivis*: Uhler, Bull. U.S. Geol. Geogr. Surv. Terr., 5: 297.

1953 *Acanthocephala declevis* [*sic*]: Elkins, J. Ks. Ent. Soc., 26: 139.

Distribution: Ariz., Fla., Ga., La., Mo., N.C., N. M., S.C., Tex. (Greater Antilles, Guatemala, Mexico).

Subgenus *Metapodiessa* Kirkaldy, 1902

1902 *Acanthocephala* (*Metapodiessa*) Kirkaldy, 1902, Ent., 35: 137. Type-species: *Cimex femoratus* Fabricius, 1775. Designated by Van Duzee, 1917, Univ. Cal. Publ. Ent., 2: 85.

Note: The name *Metapodius* Westwood (1842, see genus above) used by Stål (1870, K. Svens. Vet.-Akad. Handl., 9(1): 150) for a subgenus of *Acanthocephala* was originally proposed as a new name for the supposedly preoccupied *Acanthocephala* Laporte and hence must take the same type-species as the nominate subgenus and must be associated there. Kirkaldy (1902, above) recognized this and substituted *Acanthocephala* (*Metapodiessa*) for the group of species diagnosed by Stål as *Acanthocephala* (*Metapodius*).

Acanthocephala confraterna (Uhler), 1871

1871 *Metapodius confraternus* Uhler, Proc. Boston Soc. Nat. Hist., 14: 99. [Fla.].

1916 *Acanthocephala* (*Metapodiessa*) *confraterna*: Van Duzee, Check List Hem., p. 10.

1918 *Acanthocephala confraterna*: Gibson and Holdridge, Can. Ent., 50: 239.

Distribution: Ala., Fla., Ga., S.C., Tex.

Acanthocephala femorata (Fabricius), 1775 [Fig. 20]

1775 *Cimex femoratus* Fabricius, Syst. Ent., p. 708. ["India" in error].

1832 *Rhinuchus nasulus* Say, Ins. La., p. 10. [Ga., Fla., La.]. Synonymized by Stål, 1870, K. Svens. Vet.-Akad. Handl., 9(1): 150.

1832 *Anisoscelis nasulus*: Say, Descrip. Het. Hem. N. Am., p. 13.

1842 *Metapodius obscurus* Westwood, Hope Cat., 2: 15. [N. Am.]. Synonymized by Stål, 1870, K. Svens. Vet.-Akad. Handl., 9(1): 150.

1852 *Metapodius femoratus*: Dallas, List Hem. Brit. Mus., 2: 430.
1870 *Acanthocephala (Metapodius) femorata*: Stål, K. Svens. Vet.-Akad. Handl., 9(1): 150.
1876 *Acanthocephala femorata*: Uhler, Bull. U.S. Geol. Geogr. Surv. Terr., 5: plate 20, figure 23 [caption reads thus; text, page 297, used combination *Metapodius femoratus*].
1916 *Acanthocephala* (*Metapodiessa*) *femorata*: Van Duzee, Check List Hem., p. 10.
Distribution: Fla., Ga., Ks., La., Miss, Mo., N.C., Ok., S.C., Tex. (Guatemala, Mexico).
Note: See comments under *Acanthocephala thomasi* (Uhler).

Acanthocephala terminalis (Dallas), 1852 [Fig. 22]
1852 *Metapodius terminalis* Dallas, List Hem. Brit. Mus., 2: 431. [N. Am.].
1869 *Anisosceolis* [*sic*] *prominulus*: Rathvon, Hist. Lancaster Co., Pa., p. 549.
1870 *Acanthocephala* (*Metapodius*) *terminalis*: Stål, K. Svens. Vet.-Akad. Handl., 9(1): 151.
1871 *Metapodius instabilis* Uhler, Proc. Boston Soc. Nat. Hist., 14: 98. [N.C., Pa.]. Synonymized by Gibson and Holdridge, 1918, Can. Ent., 50: 240.
1900 *Metapodius femoratus*: Smith, Ins. N.J., p. 122.
1908 *Acanthocephala terminalis*: Torre-Bueno, J. N.Y. Ent. Soc., 16: 227.
1916 *Acanthocephala* (*Metapodiessa*) *terminalis*: Van Duzee, Check List Hem., p. 10.
Distribution: Col., Conn., Fla., Ill., Ind., La., Mass., Md., Minn., Mo., N.C., N.J., N.Y., Ok., Pa., S.C., Tenn., Wis.
Note: Yonke and Medler (1963, An. Ent. Soc. Am., 62: 474-476) presented descriptions and illustrations of the immature stages.

Acanthocephala thomasi (Uhler), 1872
1872 *Metapodius thomasii* [*sic*] Uhler, U.S. Geol. Geogr. Surv. Terr. 5: 399. [Ariz.].
1875 *Metapodius granulosus*: Uhler, Wheeler's Surv. 100th Merid., 5: 831.
1881 *Acanthocephala granulosa*: Distant, Biol. Centr.-Am., Rhyn., 1: 120 (in part).
1916 *Acanthocephala* (*Metapodiessa*) *granulosa*: Van Duzee, Check List Hem., p. 10.
1926 *Acanthocephala thomasi*: Barber, J. N.Y. Ent. Soc., 34: 209.
Distribution: Ariz., Cal., Tex.
Note: No explained refutation of Barber's (1926, above) resurrection of this name from synonymy under *A. femorata* (Fabricius) was found. Apparently many of the later authors, including Torre-Bueno (1941, Ent. Am., 21: 47), overlooked Barber's action. Also, although *A. granulosa* Dallas, 1852, was considered a junior synonym of *A. femorata* (Fabricius) by Barber (1926, above), all records for it in the U.S. should be applied to *A. thomasi*.

Tribe Acanthocerini Bergroth, 1913

Note: O'Shea (1980, Studies Neotrop. Fauna Environ., 15: 57-80) revised this New World tribe and presented a key (p. 58) to its thirteen known genera. The results of that revision are followed here.

Genus *Acanthocerus* Palisot, 1818

1818 *Acanthocerus* Palisot, Ins. Rec. Afr. Am., pp. 201, 204. Type-species: *Acanthocerus crucifer* Palisot, 1818, a junior synonym of *Cimex cruciger* Tigny, 1813. Monotypic.
1833 *Hymeniphera* Laporte, Mag. Zool., 2: 43. Unnecessary new name for *Acanthocerus* Palisot.

Acanthocerus lobatus (Burmeister), 1835

1835 *Crinocerus lobatus* Burmeister, Handb. Ent., 2: 318. [Cuba].

1914 *Acanthocerus lobatus*: Barber, J. N.Y. Ent. Soc., 22: 171.

1916 *Hymenifera lobata*: Van Duzee, Check List Hem., p. 12.

Distribution: Fla., N.M.(?) (Cuba, Jamaica).

Note: O'Shea (1980, Studies Neotrop. Fauna Environ., 15: 61) questioned the New Mexico records.

Genus *Euthochtha* Mayr, 1865

1865 *Euthochtha* Mayr, Verh. Zool.-Bot. Ges. Wien, 15: 431. Type-species: *Coreus galeator* Fabricius, 1803. Monotypic.

Euthochtha galeator (Fabricius), 1803 [Fig. 26]

1803 *Coreus galeator* Fabricius, Syst. Rhyn., p. 191. ["Carolina"].

1852 *Crinocerus galeator*: Dallas, List Hem. Brit. Mus., 2: 408.

1865 *Euthochtha galeator*: Mayr, Verh. Zool.-Bot. Ges. Wien, 15: 431.

1868 *Euthochtha Galeator* [*sic*]: Stål, K. Svens. Vet.-Akad. Handl., 7(11): 49.

1878 *Euthoctha* [*sic*] *galeator*: Uhler, Proc. Bost. Soc. Nat. Hist., 19: 381.

1905 *Euthocta* [*sic*] *galeator*: Torre-Bueno, J. N.Y. Ent. Soc., 13: 36.

1908 *Acanthocerus galeator*: Torre-Bueno, J. N.Y. Ent. Soc., 16: 227.

Distribution: Conn., Fla., Ill., Ind., Ks., Mass., Mich., Mo., N.C., N.J., N.Y., Neb., Oh., Ont., Pa., S.C., Tex., Va., Wisc. (Greater Antilles).

Note: Yonke and Medler (1969, An. Ent. Soc. Am., 62: 469-473) presented descriptions and illustrations of the immatures stages.

Genus *Sagotylus* Mayr, 1865

1865 *Sagotylus* Mayr, Verh. Zool.-Bot. Ges. Wien, 15: 431. Type-species: *Crinocerus triguttatus* Herrich-Schaeffer, 1842, a junior synonym of *Coreus confluentus* Say, 1832. Monotypic.

Sagotylus confluens (Say), 1832

1832 *Coreus confluentus* Say, Descrip. Het. Hem. N. Am., p. 11. [Mexico].

1842 *Crinocerus triguttatus* Herrich-Schaeffer, Wanz. Ins., 6: 86. [Mexico]. Synonymized by Uhler, 1876, Bull. U.S. Geol. Geogr. Surv. Terr., 1: 297.

1876 *Sagotylus confluentus*: Uhler, Bull. U.S. Geol. Geogr. Surv. Terr. Terr., 1: 297.

1886 *Sagotylus diffusa*: Uhler, Check-list Hem. Het., p. 9.

1909 *Spartocera confluentus*: Van Duzee, Bull. Buff. Soc. Nat. Hist., 9: 159.

1910 *Sagotylus triguttatus*: Banks, Cat. Nearc. Hem.-Het., p. 80.

1916 *Corecoris confluentus*: Van Duzee, Check List Hem., p. 12.

1923 *Coreocoris* [*sic*] *fusca*: Barber, Bull. Univ. Ia. Studies Nat. Hist., 10: 18.

1980 *Sagotylus confluens*: O'Shea, Studies Neotrop. Fauna Environ., 15: 69.

Distribution: Ariz., Cal., Fla. (Mexico to Belize)

Tribe Anisoscelidini Amyot and Serville, 1843

Note: A key to the North American genera of this tribe was presented by Hussey (1953, Bull. Brook. Ent. Soc., 48: 33-34). A recent treatment of this tribe by E. Osuna (1985 [1984], Boletin Entomologia Venezolana, New Series, 3(5-8): 77-145), containing numerous generic changes, was unavailable for inclusion here.

Genus *Chondrocera* Laporte, 1832

1832 *Chondrocera* Laporte, Mag. Zool., 2: 44. Type-species: *Chondrocera laticornis* Laporte, 1832. Monotypic.

Chondrocera laticornis Laporte, 1831

1831 *Chondrocera laticornis* Laporte, Mag. Zool., 2: 45. [Cuba].
1910 *Chondrocera laticornis*: Banks, Cat. Nearc. Hem.-Het., p. 77.
Distribution: Fla. (West Indies).

Genus *Leptoglossus* Guérin-Méneville, 1831

1831 *Leptoglossus* Guérin-Méneville, Voyage Coquil. Zool., plate 12, figure 9. Type-species: *Leptoglossus dilaticollis* Guérin-Méneville, 1831. Monotypic.
1862 *Theognis* Stål, Stett. Ent. Zeit, 23: 294. Type-species: *Hypselonotus scriptus* Hahn, 1826, a junior synonym of *Cimex stigma* Herbst, 1784. Designated by Van Duzee, 1917, Univ. Cal. Publ. Ent., 2: 88.
Note: The present list follows Allen's (1969, Ent. Am., 45: 35-140) revision of *Leptoglossus* with *Theognis* as a junior synonym of it. If *Theognis* were given generic status as was done by Hussey (1953, Bull. Brook. Ent. Soc., 48: 32-34), all our species would belong to it and *Leptoglossus* would be left with but one species, a South American form. Hussey (*ibid*, pp. 30-32) gave a key to the North American species under the generic name *Theognis*.

Leptoglossus ashmeadi Heidemann, 1909

1909 *Leptoglossus ashmeadi* Heidemann, Bull. Buff. Soc. Nat. Hist., 9: 237. [Fla.].
1953 *Theognis ashmeadi*: Hussey, Bull. Brook. Ent. Soc., 48: 31.
Distribution: Ala., Fla., Miss.

Leptoglossus brevirostris Barber, 1918

1906 *Leptoglossus* "sp.?": Barber, Bull. Mus. Brook. Inst. Arts. Sci., Sci. Bull, 1: 266. [Tex.].
1910 *Leptoglossus stigma* variety *minor*: Heidemann, Proc. Ent. Soc. Wash., 12: 191.
1918 *Leptoglossus brevirostris* Barber, Bull. Brook. Ent. Soc., 13: 35. [Ariz.].
1953 *Theognis brevirostris*: Hussey, Bull. Brook. Ent. Soc., 48: 30.
Distribution: Ariz., Cal., Tex. (Mexico).

Leptoglossus clypealis Heidemann, 1910 [Fig. 21]

1910 *Leptoglossus clypealis* Heidemann, Proc. Ent. Soc. Wash., 12: 195. [Col.].
1953 *Theognis clypealis*: Hussey, Bull. Brook. Ent. Soc., 48: 30.
Distribution: Ariz., Cal., Col., Ia., Ks., N.M., Ore., Tex., Ut.

Leptoglossus concolor (Walker), 1871

1871 *Anisoscelis concolor* Walker, Cat. Hem. Brit. Mus., 4: 128. [Mexico].
1956 *Leptoglossus stigma*: Hussey, Fla. Ent., 39: 88.
1969 *Leptoglossus concolor*: Allen, Ent. Am., 45: 118.
Distribution: Fla. (Greater Antilles, Mexico to Panama).
Note: Hussey's (1956, above) Florida record of *L. stigma* was referred to *L. concolor* by Allen (1969, above, p. 122).

Leptoglossus corculus (Say), 1832

1832 *Anisoscelis corculus* Say, Descrip. Het. Hem. N. Am., p. 12. [Fla.].
1865 *Theognis excellens* Mayr, Verh. Zool.-Bot. Ges. Wien, 15: 434. [Ga.]. Synonymized by Stål, 1870, K. Svens. Vet.-Akad. Handl., 9(1): 165.

1870 *Leptoglossus corculus*: Stål, K. Svens. Vet.-Akad. Handl., 9(1): 165.
1953 *Theognis corculus*: Hussey, Bull, Brook. Ent. Soc., 48: 30
Distribution: Ala., Ark., D.C., Del., Fla., Ga., Md., Miss., N.C., N.J., N.Y., Oh., Pa., S.C., Tenn., Tex., Va.
Note: Allen (1969, Ent. Am., 45: 130-131) considered all records of *L. corculus* for Ariz., Cal., Col., and N.M. as belonging to *L. occidentalis*.

Leptoglossus fulvicornis (Westwood), 1842
1842 *Anisoscelis fulvicornis* Westwood, Hope Cat., 2: 17. [South America-?].
1870 *Leptoglossus fulvicornis*: Stål, K. Svens. Vet.-Akad. Handl., 9(1): 161.
1910 *Leptoglossus magnoliae* Heidemann, Proc. Ent. Soc. Wash., 12: 191. [D.C.]. Synonymized by and lectotype designated by Allen, 1969, Ent. Am., 45: 79.
1953 *Theognis fulvicornis*: Hussey, Bull. Brook. Ent. Soc., 48: 30.
Distribution: Ala., D.C., Fla., Ga., Mass., Md., N.C., N.Y., Pa., S.C., Va.
Note: Westwood (1842, above) himself questioned the original locality as South America. Stål (1870, above, p. 161) gave the type-locality as unknown. The species is reported only from the United States in subsequent literature.

Leptoglossus gonagra (Fabricius), 1775
1775 *Cimex gonagra* Fabricius, Syst. Ent., p. 708. [St. Thomas Island].
1910 *Leptoglossus gonager* [*sic*]: Heidemann, Proc. Ent. Soc. Wash., 12: 191.
1910 *Leptoglossus gonagra*: Banks, Cat. Nearc. Hem.-Het., p. 78.
1952 *Theognis gonagra*: Hussey, Fla. Ent., 35: 117.
Distribution: Fla., La., Mo., Tex. (Mexico to Argentina).

Leptoglossus occidentalis Heidemann, 1910
1910 *Leptoglossus occidentalis* Heidemann, Proc. Ent. Soc. Wash., 12: 196. [Ut.]. Lectotype designated by Allen, 1969, Ent. Am., 45: 131.
1953 *Theognis occidentalis*: Hussey, Bull. Brook. Ent. Soc., 48: 29.
Distribution: Ala., Ariz., B.C., Cal., Col., Ia., Id., Ks., Mont., N.M., Ore., Tex., Ut., Wash.

Leptoglossus oppositus (Say), 1832
1832 *Anisoscelis oppositus* Say, Descrip. Het. Hem. N. Am., p. 12. [Ind.].
1842 *Anisoscelis tibialis* Herrich-Schaeffer, Wanz. Ins., 7: 12. [N. Am.]. Synonymized by Stål, 1870, K. Svens. Vet.-Akad. Handl., 9(1): 164.
1870 *Leptoglossus oppositus*: Stål, K. Svens. Vet.-Akad. Handl., 9(1): 163.
1953 *Theognis oppositus*: Hussey, Bull. Brook. Ent. Soc., 48: 30.
Distribution: Ala., Ariz., Ark., D.C., Fla., Ga., Ia., Ill., Ind., Ky., La., Md., Minn., Miss., Mo., N.C., N.J., N.M., N.Y., Oh., Ok., Pa., S.C., Tex., Va., Wis. (Mexico).

Leptoglossus phyllopus (Linnaeus), 1767
1767 *Cimex phyllopus* Linnaeus, Syst. Nat., edit. 12, 1(2): 731.
1832 *Anisoscelis albicinctus* Say, Descrip. Het. Hem. N. Am., p. 12. [Fla.]. Synonymized by Stål, 1870, K. Svens. Vet.-Akad. Handl., 9(1): 161.
1835 *Anisoscelis phyllopus*: Burmeister, Handb. Ent., 2: 332.
1842 *Anisoscelis phyllopa* [*sic*]: Westwood, Hope Cat., 2: 16.
1852 *Anisoscelis confusa* Dallas, List Hem. Brit. Mus., 2: 453. [Fla., Brazil]. Synonymized by Stål, 1870, K. Svens. Vet.-Akad. Handl, 9(1): 161.
1870 *Leptoglossus phyllopus*: Stål, K. Svens. Vet.-Akad. Handl., 9(1): 161.
1871 *Anisoscelis albicincta*: Walker, Cat. Hem. Brit. Mus., 4: 124.
1952 *Theognis phyllopus*: Hussey, Fla. Ent., 35: 117.
Distribution: Ala., Ark., Cal., Fla., Ga., Ia., Ks., La., Miss., Mo., N.C., N.Y., Neb., Ok., S.C., Tex., Va. (Mexico to Brazil).

Leptoglossus zonatus (Dallas), 1852
1852 *Anisoscelis zonatus* Dallas, List Hem. Brit. Mus., 2: 452. [Mexico].
1876 *Leptoglossus zonatus*: Uhler, Bull. U.S. Geol. Geogr. Surv. Terr., 5: 298.
1953 *Theognis zonatus*: Hussey, Bull. Brook. Ent. Soc., 48: 32.
Distribution: Ariz., Cal., Ks., Tex. (Mexico to Brazil).

Genus *Narnia* Stål, 1862

1862 *Narnia* Stål, Stett. Ent. Zeit., 23: 294, 296. Type-species: *Narnia femorata* Stål, 1862. Monotypic.
Note: Brailovsky (1975, Rev. Soc. Mex. Hist. Nat., 36: 169-176) reviewed the species.

Narnia coachellea Bliven, 1956
1956 *Narnia coachellea* Bliven, New Hem. W. St., p. 6. [Cal.].
Distribution: Cal.

Narnia femorata Stål, 1862
1862 *Narnia femorata* Stål, Stett. Ent. Zeit., 23: 296. [Mexico].
1906 *Narnia (Narnia) femorata*: Van Duzee, Ent. News, 17: 384, 386.
Distribution: Ariz., Cal., N.M., Ok., Tex.

Narnia inornata Distant, 1893
1893 *Narnia inornata* Distant, Biol. Centr.-Am., Rhyn., 2: 361. [Mexico].
1910 *Narnia inornata*: Banks, Cat. Nearc. Hem.-Het., p. 80.
1916 *Narnia (Xerocoris) inornata*: Van Duzee, Check List Hem., p. 11.
Distribution: Ariz., Cal. (Mexico).
Note: Mann (1969, Bull. U.S. Nat. Mus., 256: 133) expressed the belief that many records of this species for Mexico and the United States actually belong to *Narnia pallidocornis*.

Narnia marquezi Brailovsky, 1975
1975 *Narnia marquezi* Brailovsky, Rev. Soc. Mex. Hist. Nat., 36: 169. [Ariz.].
Distribution: Ariz. (Mexico).

Narnia pallidicornis Stål, 1870
1870 *Narnia pallidicornis* Stål, K. Svens. Vet.-Akad. Handl., 9(1): 166. [Tex.].
1906 *Narnia (Narnia) pallidicornis*: Van Duzee, Ent. News, 17: 385.
Distribution: Ariz., Cal., N.M., Tex. (Mexico).
Note: See comment under *Narnia inornata*.

Narnia snowi Van Duzee, 1906
1906 *Narnia (Xerocoris) snowi* Van Duzee, Ent. News, 17: 384. [Tex.].
1910 *Narnia snowi*: Banks, Cat. Nearc. Hem.-Het., p. 78.
Distribution: Ariz., Cal., Col., N.M., Tex.

Narnia wilsoni Van Duzee, 1906
1906 *Narnia (Xerocoris) wilsoni* Van Duzee, Ent. News, 17: 385. [Cal.].
1910 *Narnia wilsoni*: Banks, Cat. Nearc. Hem.-Het., p. 78.
Distribution: Ariz., Cal.

Tribe Chariesterini Stål, 1867

Note: Yonke (1972, Proc. Ent. Soc. Wash., 74: 284) gave a key to the genera of Chariesterini.

Genus *Chariesterus* Laporte, 1832

1833 *Chariesterus* Laporte, Mag. Zool., 2: 44. Type-species: *Chariesterus gracilicornis* Laporte, 1833, a junior synonym of *Pendulinus armatus* Thunberg, 1825. Original designation.

Note: Ruckes (1955, Am. Mus. Novit., 1721: 1-16) presented a revision and key to the species of *Chariesterus*.

Chariesterus albiventris Burmeister, 1835

1835 *Chariesterus albiventris* Burmeister, Handb. Ent., 2: 317. [Mexico].
1906 *Chariesterus albiventris*: Snow, Trans. Ks. Acad. Sci., 20: 161.
Distribution: Ariz., Tex. (Guatemala, Mexico,).

Chariesterus antennator (Fabricius), 1803 [Fig. 19]

1803 *Coreus antennator* Fabricius, Syst. Rhyn., p. 198. ["Carolina"].
1832 *Gonocerus antennator*: Say, Descrip. Het. Hem. N. Am., p. 768.
1832 *Gonocerus dubius* Say, Descrip. Het. Hem. N. Am., p. 768. [Ind., Pa.]. Synonymized by Stål, 1870, K. Svens. Vet.-Akad. Handl., 9(1): 178.
1840 *Coreus* (*Chariesterus*) *antennator*: Blanchard, Hist. Nat. Ins., Hem., p. 120.
1842 *Chariesterus moestus*: Herrich-Schaeffer, Wanz. Ins., 7: 3.
1852 *Chariesterus antennator*: Dallas, List Hem. Brit. Mus., 2: 510.
Distribution: Col., Fla., Ga., Ill., Ind., Ks., Mich., Miss., N.C., N.J., N.Y., Ok., Pa., S.C., Tex., Va.

Chariesterus balli Fracker, 1919

1919 *Chariesterus balli* Fracker, An. Ent. Soc. Am., 12: 229. [Cal.].
Distribution: Cal. (Mexico)

Chariesterus cuspidatus Distant, 1892

1892 *Chariesterus cuspidatus* Distant, Biol. Centr.-Am., Rhyn., 1: 364.
1906 *Chariesterus cuspidatus*: Barber, Bull. Brook. Inst. Sci., 1: 266.
Distribution: Tex. (Mexico to Venezuela).

Tribe Chelinideini Blatchley, 1926

Genus *Chelinidea* Uhler, 1863

1863 *Chelinidea* Uhler, Proc. Ent. Soc. Phila., p. 365. Type-species: *Chelinidea vittiger* Uhler, 1863. Monotypic.

Note: Hamlin (1924, An. Ent. Soc. Am., 17: 193-208) presented considerable biological data for the species of *Chelinidea*. Herring (1980, Proc. Ent. Soc. Wash., 82: 237-251) revised and keyed the species.

Chelinidea canyona Hamlin, 1923

1923 *Chelinidea canyona* Hamlin, Proc. Royal Soc. Queensland, 35: 44. [Tex.].
1926 *Chelinidea canyoni* [*sic*]: Barber, J. N.Y. Ent. Soc., 34: 213.
Distribution: Ark., Tex. (Mexico; introduced into Australia).

Chelinidea hunteri Hamlin, 1923

1912 *Chelinidea* species: Hunter et al., U.S. Dept. Agr., Bur. Ent. Bull., 113: 20. [Ariz.].
1923 *Chelinidea hunteri* Hamlin, Proc. Royal Soc. Queensland, 35: 43. [Mexico].
Distribution: Ariz., Tex. (Mexico).

Chelinidea tabulata (Burmeister), 1835
1835 *Gonocerus tabulatus* Burmeister, Handb. Ent., 2: 311. [Mexico].
1906 *Chelinidea tabulata*: Barber, Brook. Inst. Sci. Bull., 1: 267.
Distribution: Ariz., Cal., Col., Tex., Ut. (Mexico to Venezuela).

Chelinidea vittiger Uhler, 1863 [Fig. 23]
1863 *Chelinidea vittiger* Uhler, Proc. Ent. Soc. Phil., 2: 366. [Mont.]. Lectotype designated by McAtee, 1919, Bull. Brook. Ent. Soc., 14: 11.
1919 *Chelinidea vittiger vittiger*: McAtee, Bull. Brook. Ent. Soc., 14: 31.
1919 *Chelinidea vittiger vittiger* var. *artuflava* McAtee, Bull. Brook. Ent. Soc., 14: 11. [Ariz.]. Synonymized by Herring, 1980, Proc. Ent. Soc. Wash., 82: 246.
1919 *Chelinidea vittiger aequoris* McAtee, Bull. Brook. Ent. Soc., 14: 12. [Tex.]. Synonymized by Herring, 1980, Proc. Ent. Soc. Wash., 82: 246.
1919 *Chelinidea vittiger aequoris* var. *artuatra* McAtee, Bull. Brook. Ent. Soc., 14: 12. [N.C.]. Synonymized by Herring, 1980, Proc. Ent. Soc. Wash., 82: 246.
1923 *Chelinidea vittigera* [*sic*] var. *texana* Hamlin, Proc. Royal Soc. Queensland, 35: 45. [Tex.]. Synonymized by Herring, 1980, Proc. Ent. Soc. Wash., 82: 246.
Distribution: Ala., Alta., Ariz., Cal., Col., Fla., Ga., Id., La., Mo., Mont., N.C., N.M., Neb., Ok., Ore., S.C., Tenn., Tex., Ut., Va., Wyo. (Mexico; introduced into Australia).

Tribe Coreini Leach, 1815

Genus *Althos* Kirkaldy, 1904

1852 *Margus* Dallas, List. Hem. Brit. Mus., 2: 523. Preoccupied. Type-species: *Margus pectoralis* Dallas, 1852. Designated by Van Duzee, 1916, Check List Hem., p. 12.
1904 *Althos* Kirkaldy, Ent., 37: 280. New name for *Margus* Dallas.
Note: Although Kirkaldy (1904, above) proposed the new name *Althos* for members of this genus, he did not provide the written combinations for all of the included species. Therefore, two of the combinations appearing below are technically new.

Althos inconspicuus (Herrich-Schaeffer), 1840, New Combination
1840 *Syromastes inconspicuus* Herrich-Schaeffer, Wanz. Ins., 6: 14. [Mexico].
1871 *Margus inconspicuus*: Uhler, Prelim. Rept. U.S. Geol. Surv. Wyo., p. 471.
1923 *Margus inconspicuous* [*sic*]: Fracker, An. Ent. Soc. Am., 16: 167.
Distribution: Ariz., Cal., Col., N.M., Tex. (Mexico).

Althos obscurator (Fabricius), 1803
1803 *Coreus obscurator* Fabricius, Syst. Rhyn., p. 20. [South America].
1909 *Althos obscurator*: Van Duzee, Bull. Buff. Soc. Nat. Sci., 9: 160.
1916 *Margus obscurator*: Van Duzee, Check List Hem., p. 12.
Distribution: Fla., S.C. (Mexico to Brazil).

Althos repletus (Van Duzee), 1925, New Combination
1925 *Margus repletus* Van Duzee, Proc. Cal. Acad. Sci., 14: 393. [Cal.].
Distribution: Cal.

Genus *Anasa* Amyot and Serville, 1843

1843 *Anasa* Amyot and Serville, Hist. Nat. Hem., p. 209. Type-species: *Anasa cornuta* Amyot and Serville, 1843. Monotypic.

Note: Brailovsky (1985, Monogr. Inst. Biol. Univ. Nal. Aut., Méx., 2: 1-266) revised and presented a key to the species of *Anasa*. Uhler's (1886, Check-list Hem. Het., p. 11) report of the Mexican species *Anasa uhleri* Stål from the western United States was followed by Van Duzee (1916, Check List Hem. Het., p. 13 and 1917, Univ. Cal. Publ. Ent., 2: 104), Torre-Bueno (1941, Ent. Am., 21: 74), and certain other authors; however, Barber (1926, J. N.Y. Ent. Soc., 34: 215) concluded that this occurrence was "undoubtedly cited in error" and probably belonged to "a large variety of *tristis* from Arizona which had been confused with *uhleri*." The latter author presented additional features for separating the two. Brailovsky (1985, above) accepted Barber's treatment.

Anasa andresii (Guérin-Méneville), 1857

1857 *Coreus* (*Gonocerus*) *andresii* Guérin-Méneville, Hist. Ins. Cuba, 7: 383. [Cuba].
1870 *Anasa* (*Oriterus*) *andresii*: Stål, K. Svens. Vet.-Akad. Handl., 9(1): 190.
1871 *Gonocerus affiliatus*: Walker, Cat. Hem. Brit. Mus., 4: 185.
1881 *Anasa andresii* Distant, Biol. Centr.-Am., Rhyn., 2: 141.
1917 *Anasa andrewsi* [*sic*]: Van Duzee, Check List Hem., p. 13.
Distribution: Ariz., Cal., Fla., La., N.M., Tex. (Mexico to Colombia).

Anasa armigera (Say), 1825

1825 *Coreus armigerus* Say, J. Acad. Nat. Sci. Phil., 4: 319. ["Missouri Territory"].
1868 *Anasa* (*Anasa*) *armigera*: Stål, K. Svens. Vet.-Akad. Handl., 7(11): 57.
1876 *Anasa armigera*: Uhler, Bull. U.S. Geol. Geogr. Surv. Terr., 5: 293.
Distribution: Conn., D.C., Del., Fla., Ga., Ia., Ill., Ind., Ks., La., Mass., Md., Mich., Miss., Mo., N.C., N.J., N.Y., Oh., Ok., Ont., Pa., S.C., Tex. Va.
Note: Parshley (1918, J. Econ. Ent., 11: 471-472) reported this species as feeding on the native star cucumber (*Sicyos angulatus* Linnaeus) and cultivated cucumbers.

Anasa maculipes Stål, 1862

1862 *Anasa maculipes* Stål, Stett. Ent. Zeit., 23: 299. [Mexico].
1985 *Anasa maculipes*: Brailovsky, Monogr. Inst. Biol. Univ. Nal. Aut. Méx., 2: 143.
Distribution: Ariz., Tex. (Mexico to Costa Rica).

Anasa repetita Heidemann, 1905

1905 *Anasa repetita* Heidemann, Proc. Ent. Soc. Wash., 7: 11. [D.C., Md.].
Distribution: Conn., D.C., Ind., Mass., Md., Mo., N.Y.

Anasa scorbutica (Fabricius), 1775

1775 *Cimex scorbuticus* Fabricius, Syst. Ent., p. 706. [N. Am.].
1818 *Acanthocerus nebulosus* Palisot, Ins. Rec. Afr. Am., p. 205. [U.S.]. Synonymized by Stål, 1870, K. Svens. Vet.-Akad. Handl., 9(1): 192.
1868 *Anasa* (*Anasa*) *scorbutica*: Stål, K. Svens. Vet.-Akad. Handl., 7(11): 56, 57.
1870 *Anasa* (*Acanthocerus*) *scorbutica*: Stål, K. Svens. Vet.-Akad., Handl., 9(1): 192.
1871 *Anasa scorbutica*: Walker, Cat. Hem. Brit. Mus., 4: 108.
Distribution: Fla., Ok., Tex. (Greater and Lesser Antilles, Mexico to Argentina, Galapagos Islands).

Anasa tristis (De Geer), 1775 [Fig. 24]

1773 *Cimex tristis* De Geer, Mem. Ins., 3: 340. ["Carolina"].
1790 *Cimex moestus* Gmelin, Syst. Nat., ed. 13, 1(4): 2168. [Pa.]. Synonymized by Stål, 1870, K. Svens. Vet.-Akad. Handl., 9(1): 189.
1803 *Coreus rugator* Fabricius, Syst. Rhyn., p. 192. ["Carolina"]. Synonymized by Stål, 1870, K. Svens. Vet.-Akad. Handl., 9(1): 189.

1825 *Coreus ordinatus* Say, J. Acad. Nat. Sci. Phila., 4: 318. [Fla., Md., Pa.]. Synonymized by Stål, 1870, K. Svens. Vet.-Akad. Handl., 9(1): 189.
1831 *Oriterus destructor* Hahn, Wanz. Ins., 1: 8. [N. Am.]. Synonymized by Stål, 1870, K. Svens. Vet.-Akad. Handl., 9(1): 189
1835 *Gonocerus rugator*: Burmeister, Handb. Ent., 2: 311.
1852 *Gonocerus tristis*: Dallas, List Hem. Brit. Mus., 2: 499.
1861 *Gonocerus obliquus* Uhler, Proc. Ent. Soc. Phila., 1: 23. [Cal.]. Synonymized by Brailovsky, 1985, Monogr. Inst. Biol. Univ. Nal. Aut. Méx., 2: 197.
1868 *Anasa (Oriterus) tristis*: Stål, K. Svens. Vet.-Akad. Handl., 7(11): 56.
1869 *Coreus tristis*: Rathvon, Hist. Lancaster Co., Pa., p. 549.
1870 *Anasa obliqua*: Stål, K. Svens. Vet.-Akad. Handl., 9(1): 197.
1878 *Anasa tristis*: Uhler, Proc. Boston Soc. Nat. Hist., 19: 385.
Distribution: Ala., Ariz., Ark., B.C., Cal., Col., Conn., Fla., Ga., Ia., Id., Ill., Ind., Ks., La., Mass., Md., Mich., Minn., Miss., Mo., N.C., N.J., N.M., N.Y., Neb., Oh., Ok., Ont., Ore., Pa., Que., S.C., Tenn., Tex., Ut., Va., Wash., Wis. (Mexico to Brazil)

Genus *Catorhintha* Stål, 1859

1859 *Catorhintha* Stål, Öfv. K. Vet.-Akad. Förh., 16: 470. Type-species: *Lygaeus guttula* Fabricius, 1794. Designated by Van Duzee, 1916, Check List Hem., p. 12.

Catorhintha divergens Barber, 1926
1926 *Catorhintha divergens* Barber, J. N.Y. Ent. Soc., 34: 214. [Fla.].
Distribution: Fla. (Cuba, Mexico).

Catorhintha guttula (Fabricius), 1794
1794 *Lygaeus guttula* Fabricius, Ent. Sys., 4: 162. [Insular America].
1871 *Catorhintha guttula*: Uhler, Prelim. Rept. U.S. Geol. Surv. Wyo., p. 471.
Distribution: Ariz., Col., Fla., N.M., Nev., Ok., S.C., Tex. (Mexico to Colombia).

Catorhintha mendica Stål, 1870
1870 *Catorhintha mendica* Stål, K. Svens. Vet.-Akad. Handl., 9(1): 187. [Tex., Mexico].
1965 *Catorintha* [*sic*] *mendica*: Schaefer, Misc. Publ., Ent. Soc. Am., 5: 30.
Distribution: Ariz., Col., Conn., Fla., Ia., Ill., Ind., Ks., Mich., Minn., Mo., N.Y., Neb., Oh., Ok., Pa., S.D., Tex., Wis. (Mexico, West Indies).
Note: Hoebeke and Wheeler (1982, Ent. News, 93(1): 29-31) summarized biological information for this species.

Catorhintha selector Stål, 1859
1859 *Catorhintha selector* Stål, Öfv. K. Vet.-Akad. Förh., 16: 471. [Mexico].
1875 *Catorhintha selector*: Uhler, Rept. U.S. Geol. Geogr. Surv. Terr., 5: 832.
Distribution: Ariz., N.M., Ok., Tex. (Mexico).

Catorhintha selector selector Stål, 1859
1859 *Catorhintha selector* Stål, Öfv. K. Vet.-Akad. Förh., 16: 471. [Mexico].
1875 *Catorhintha selector*: Uhler, Rept. U.S. Geol. Geogr. Surv. Terr., 5: 832.
1941 *Catorhintha selector selector*: Torre-Bueno, Ent. Am., 21: 72.
Distribution: Ariz., N.M., Tex. (Mexico).

Catorhintha selector texana Stål, 1870
1870 *Catorhintha texana* Stål, K. Svens. Vet.-Akad., Handl., 9(1): 188. [Tex.].
1941 *Catorhintha selector* var. *texana*: Torre-Bueno, Ent. Am., 21: 72.
Distribution: Ariz., N.M., Ok., Tex. (Mexico).

Catorhintha viridipes Blatchley, 1926
1926 *Catorhintha borinquensis* var. *viridipes* Blatchley, Het. E. N. Am., p. 247. [Fla.].
1941 *Catorhintha viridipes*: Torre-Bueno, Ent. Am., 21: 71.
Distribution: Fla.

Genus *Cimolus* Stål, 1862

1862 *Cimolus* Stål, Stett. Ent. Zeit., 23: 302. Type-species: *Cimolus vitticeps* Stål, 1862. Monotypic.

Cimolus obscurus Stål, 1870
1870 *Cimolus obscurus* Stål, K. Svens. Vet.-Akad. Handl., 9(1): 89. [S.C., Tex.].
Distribution: La., S.C., Tex.

Genus *Ficana* Stål, 1862

1862 *Catorhintha* (*Ficana*) Stål, Stett. Ent. Zeit., 23: 303. Type-species: *Gonocerus apicalis* Dallas, 1852. Monotypic.
1867 *Ficana*: Stål, Öfv. K. Vet.-Akad. Förh., 24: 548.

Ficana apicalis (Dallas), 1852
1852 *Gonocerus apicalis* Dallas, List Hem. Brit. Mus., 2: 499. [Mexico].
1872 *Ficana apicalis*: Uhler, Prelim. Rept. U.S. Geol. Surv. Mont., p. 401.
1923 *Catorhintha apicalis*: Fracker, An. Ent. Soc. Am., 16: 169.
Distribution: Ariz., Cal., Col. (Mexico).

Genus *Hypselonotus* Hahn, 1831

1831 *Hypselonotus* Hahn, Wanz. Ins., 1: 186. Type-species: *Hypselonotus interruptus* Hahn, 1831. Designated by Van Duzee, 1916, Check List Hem., p. 13.

Note: Whitehead (1974, J. Wash. Acad. Sci., 64: 231) reviewed and provided a key to species of *Hypselonotus*, and concluded "the only Nearctic species is *H. punctiventris*," apparently overlooking Torre-Bueno's (1931, Bull. Brook. Ent. Soc., 26: 137) material or record of *Hypselonotus lineatus* Stål from Texas, a species reported by Whitehead (above) with eastcoast Mexican records reasonably close to Texas.

Hypselonotus lineatus Stål, 1862
1862 *Hypselonotus lineatus* Stål, Stett. Ent. Zeit., 23: 297. [Mexico].
1931 *Hypselonotus lineatus*: Torre-Bueno, 1931, Bull. Brook. Ent. Soc., 26: 137.
Distribution: Tex. (Mexico to Peru).

Hypselonotus punctiventris Stål, 1862
1862 *Hypselonotus punctiventris* Stål, Stett. Ent. Zeit., 23: 297. [Mexico].
1871 *Hypselonotus fulvus*: Walker, Cat. Hem. Brit. Mus., 4: 137.
1906 *Hypselonotus punctiventris*: Barber, Brook. Inst. Arts Sci., Sci. Bull., 1: 268.
1916 *Hypselonotus fulvus* var. *venosus*: Van Duzee, Check List Hem., p. 13.
Distribution: Ariz., Tex. (Mexico to Guatemala).

Genus *Madura* Stål, 1860

1860 *Madura* Stål, K. Svens. Vet.-Akad. Handl., 2(7): 35. Type-species: *Madura fusco-clavata* [*sic*] Stål, 1868. Monotypic.

Madura perfida Stål, 1862
1862 *Madura perfida* Stål, Stett. Ent., Zeit., 23: 304. [Mexico].
1906 *Madura perfida*: Barber, Brook. Inst. Arts Sci., Sci. Bull., 1: 267.
Distribution: Tex. (Mexico to Panama, West Indies).

Genus *Namacus* Amyot and Serville, 1843

1843 *Namacus* Amyot and Serville, Hist. Nat. Hem., p. 242. Type-species: *Namacus transvirgatus* Amyot and Serville, 1843. Monotypic.

Namacus annulicornis Stål, 1870
1870 *Namacus annulicornis* Stål, K. Svens. Vet.-Akad. Handl., 9(1): 186. [Mexico].
1914 *Namacus annulicornis*: Barber, Bull. Am. Mus. Nat. Hist., 33: 520.
Distribution: Fla. (Mexico).

Genus *Nisoscolopocerus* Barber, 1928

1928 *Nisoscolopocerus* Barber, J. N.Y. Ent. Soc., 36: 25. Type-species: *Nisoscolopocerus apiculatus* Barber, 1928. Monotypic.

Nisoscolopocerus apiculatus Barber, 1928
1875 *Dasycoris humilis*: Uhler, Report U.S. Geol. Geogr. Surv. Terr., 5: 834.
1916 *Coriomeris humilis*: Van Duzee, Check List Hem., p. 14 (in part).
1928 *Nisoscolopocerus apiculatus* Barber, J. N.Y. Ent. Soc., 36: 26. [Col.].
1941 *Nissoscolopocerus* [*sic*] *apiculatus*: Torre-Bueno, Ent. Am., 21: 71.
Distribution: Alta., Col., N.M., Neb.

Genus *Scolopocerus* Uhler, 1875

1875 *Scolopocerus* Uhler, Report U.S. Geol. Geogr. Surv. Terr., 5: 832. Type-species: *Scolopocerus secundarius* Uhler, 1975. Monotypic.

Scolopocerus granulosus Barber, 1914
1914 *Scolopocerus granulosus* Barber, J. N.Y. Ent. Soc., 22: 166. [Tex.].
Distribution: Ariz., Tex.

Scolopocerus secundarius Uhler, 1875
1875 *Scolopocerus secundarius* Uhler, Wheeler's Surv. 100th Merid., 5: 833. [Ariz.].
Distribution: Ariz., Cal., Col., Nev., Tex. (Mexico).

Scolopocerus uhleri Distant, 1881
1881 *Scolopocerus uhleri* Distant, Biol. Centr.-Am., Rhyn., 1: 164. [Mexico].
1906 *Scolopocerus uhleri*: Barber, Brook. Inst., Arts Sci., Sci. Bull., 1: 269.
Distribution: Ariz., N.M. (Mexico).

Genus *Sethenira* Spinola, 1837

1837 *Sethenira* Spinola, Essai Hem., p. 196. Type-species: *Sethenira testacea* Spinola, 1837. Monotypic.

Sethenira ferruginea Stål, 1870
1870 *Sethenira ferruginea* Stål, K. Svens. Vet.-Akad. Handl., 9(1): 182. [Cuba].
1956 *Sethenira ferruginea*: Hussey, Fla. Ent., 39: 88.
Distribution: Fla. (Cuba).

Genus *Zicca* Amyot and Serville, 1843

1843 *Zicca* Amyot and Serville, Hist. Nat. Hem., p. 240. Type-species: *Zicca massulata* Amyot and Serville, 1843, a junior synonym of *Cimex nigropunctatus* De Geer, 1773. Monotypic.

Zicca taeniola (Dallas), 1852

1852 *Clavigralla taeniola* Dallas, List Hem. Brit. Mus., 2: 514. [Venezuela].
1916 *Zicca taeniola*: Van Duzee, Check List Hem., p. 13.
Distribution: Fla. (Greater Antilles, Guatemala, Costa Rica).
Note: Van Duzee (1916, above) listed this species from the United States with a question mark. Hussey (1956, Fla. Ent., 39: 88) reported specimens from Florida as the "first definite records."

Tribe Discogastrini Stål, 1867

Genus *Savius* Stål, 1862

1862 *Savius* Stål, K. Svens. Vet.-Akad. Handl, 3(6): 58. Type-species: *Paryphes suturellus* Stål, 1862, a junior synonym of *Homoeocerus diversicornis* Westwood, 1842. Monotypic.

Savius jurgiosus Stål, 1862

1862 *Savius jurgiosus* Stål, Stett. Ent. Zeit., 23: 296. [Mexico].
1926 *Savius jurgiosus*: McAtee, Ent. News, 37: 13.
Distribution: Tex. (Mexico).

Tribe Leptoscelidini Stål, 1867

Note: Schaefer (1968, Univ. Conn., Occas. Pap., Biol. Ser., 1: 153, 196) considered that *Phthia* appropriately should be removed from this tribe and possibly have its own tribe erected; but he deferred formal action until further studies are made on additional leptoscelidines.

Genus *Amblyomia* Stål, 1870

1870 *Amblyomia* Stål, K. Svens. Vet.-Akad. Handl., 9(1): 171. Type-species: *Amblyomia bifasciata* Stål, 1870. Monotypic.

Amblyomia bifasciata Stål, 1870

1870 *Amblyomia bifasciata* Stål, K. Svens. Vet.-Akad. Handl., 9(1): 172. [Mexico].
1910 *Amblyomia bifasciata*: Banks, Cat. Nearc. Hem.-Het., p. 77.
Distribution: "Western States" (Mexico).
Note: Van Duzee (1917, Univ. Cal. Publ. Ent., 2: 92) repeated "Western States" but with a question mark. Barber (1926, J. N.Y. Ent. Soc., 34: 210-211) decided this was based on a misidentification.

Genus *Phthia* Stål, 1862

1862 *Phthia* Stål, Stett. Ent. Zeit., 23: 294. Type-species: *Cimex lunatus* Fabricius, 1787. Designated by Van Duzee, 1916, Check List Hem., p. 11.

Phthia picta (Drury), 1770

1770 *Cimex pictus* Drury, Illust. Exotic Ent., 1: 107. [Antigua].

1803 *Lygaeus dispar* Fabricius, Sys. Rhyn., p. 214. [America meridionali]. Synonymized by Stål, K. Svens. Vet.-Akad. Handl., 7(11): 53.

1876 *Phthia picta*: Uhler, Bull. U.S. Geol. Geogr. Surv. Terr., 5: 299.

1919 *Phthia picta* var. *dispar*: McAtee, Bull. Brook. Ent. Soc., 14: 13.

Distribution: Cal., Fla., Tex. (Mexico to Brazil, West Indies).

Tribe Nematopodini Amyot and Serville, 1843

Note: O'Shea (1980, Studies Neotr. Fauna Environ., 15: 197-225) revised this New World tribe and presented a key (pp. 198-199) to its seventeen known genera. The results of that revision are followed here.

Genus *Mamurius* Stål, 1862

1862 *Mamurius* Stål, Stett. Ent. Zeit., 23: 278, 293. Type-species: *Mamurius mopsus* Stål, 1862. Monotypic.

Mamurius mopsus Stål, 1862

1862 *Mamurius mopsus* Stål, Stett. Ent. Zeit., 23: 293. [Mexico].

1910 *Mamurius mopsus*: Barber, J. N.Y. Ent. Soc., 18: 37.

1979 *Mammurius* [*sic*] *mopsus*: O'Shea, Ent. News, 90: 45.

Distribution: Ariz. (Mexico).

Genus *Mozena* Amyot and Serville, 1843

1843 *Mozena* Amyot and Serville, Hist. Nat. Hem., p. 192. Type-species: *Mozena spinicrus* Amyot and Serville, 1843, a junior synonym of *Archimerus brunnicornis* Herrich-Schaeffer, 1842. Monotypic.

Subgenus *Mozena* Amyot and Serville, 1843

1843 *Mozena* Amyot and Serville, Hist. Nat. Hem., p. 192. Type-species: *Mozena spinicrus* Amyot and Serville, 1843, a junior synonym of *Archimerus brunnicornis* Herrich-Schaeffer, 1842. Monotypic.

1870 *Mozena* (*Mozena*): Stål, K. Svens. Vet.-Akad. Handl., 9(1): 134.

Mozena affinis (Dallas), 1852

1852 *Archimerus affinis* Dallas, List Hem. Brit. Mus., 2: 417. [Mexico].

1941 *Mozena* (*Mozena*) *affinis*: Torre-Bueno, Ent. Am., 21: 56.

1942 *Mozena affinis*: Torre-Bueno, Bull. Brook. Ent. Soc., 37: 180.

Distribution: Ariz., Tex. (Mexico).

Mozena arizonensis Ruckes, 1955

1876 *Mozena lineolata*: Uhler, Bull. U.S. Geol. Geogr. Surv. Terr., 5: 296.

1916 *Mozena* (*Mozena*) *lineolata*: Van Duzee, Check List Hem., p. 11.

1955 *Mozena arizonensis* Ruckes, Am. Mus. Nov., 1702: 2. [Ariz.].

Distribution: Ariz. (Mexico).

Note: Ruckes (1955, above, p. 1) pointed out that the true *M. lineolata* (Herrich-Schaeffer) [using a misspelling "*lanceolata*"] does not occur in America north of Mexico.

Mozena brunnicornis (Herrich-Schaeffer), 1840
1840 *Archimerus brunnicornis* Herrich-Schaeffer, Wanz. Ins., 6: 26. [Mexico].
1910 *Mozena brunnicornis*: Banks, Cat. Nearc. Hem.-Het., p. 80.
1916 *Mozena (Mozena) brunnicornis*: Van Duzee, Check List Hem., p. 11.
Distribution: Ariz. (Mexico).

Mozena buenoi Hussey, 1958
1941 *Mozena nestor*: Torre-Bueno, Ent. Am., 21: 55.
1958 *Mozena buenoi* Hussey, Fla. Ent., 41: 142. [Tex.].
Distribution: Ariz., N.M., Tex. (Guatemala, Mexico).

Mozena lunata (Burmeister), 1835
1835 *Archimerus lunatus* Burmeister, Handb. Ent., 2: 322. [Mexico].
1876 *Mozena lunata*: Uhler, Bull. U.S. Geol. Geogr. Surv. Terr., 5: 295.
1916 *Mozena (Mozena) lunata*: Van Duzee, Check List. Hem., p. 11.
Distribution: Ariz., N.M., Tex. (Guatemala, Mexico).

Mozena lunata lunata (Burmeister), 1835
1835 *Archimerus lunatus* Burmeister, Handb. Ent., 2: 322. [Mexico].
1876 *Mozena lunata*: Uhler, Bull. U.S. Geol. Geogr. Surv. Terr., 5: 295.
1955 *Mozena lunata lunata*: Ruckes, Am. Mus. Nov., 1702: 7. [Tex.].
Distribution: Ariz., N.M., Tex.

Mozena lunata rufescens Ruckes, 1955
1955 *Mozena lunata rufescens* Ruckes, Am. Mus. Nov., 1702: 7. [Tex.].
Distribution: Tex.

Mozena lurida (Dallas), 1852
1852 *Archimerus luridus* Dallas, List Hem. Brit. Mus., 2: 417. [Honduras].
1906 *Mozena lurida*: Barber, Brook. Inst. Arts Sci., Sci. Bull., 1: 111
1916 *Mozena (Mozena) lurida*: Van Duzee, Check List Hem., p. 11.
Distribution: Tex. (Belize, Costa Rica, Mexico).
Note: Torre-Bueno (1941, Ent. Am., 21: 56) omitted this species from his key "because unidentifiable from the description and because of the absence of specimens."

Subgenus *Rhombogaster* Dallas, 1852

1852 *Rhombogaster* Dallas, List Hem. Brit. Mus., 2: 379, 415. Type-species: *Archimerus luteus* Herrich-Schaeffer, 1840. Monotypic.
1870 *Mozena (Rhombogaster)*: Stål, K. Svens. Vet.-Akad. Handl., 9(1): 135.

Mozena obesa Montandon, 1899
1899 *Mozena obesa* Montandon, Bull. Soc. Sci. Bucarest, 8: 190. [Fla.].
1916 *Mozena (Rhombogaster) obesa*: Van Duzee, Check List Hem., p. 11.
Distribution: Fla., Ks., Miss., Mo., Neb., S.C., S.D.

Genus *Piezogaster* Amyot and Serville, 1843

1843 *Piezogaster* Amyot and Serville, Hist. Nat. Hem., p. 197. Type-species: *Piezogaster albonotatus* Amyot and Serville, 1843, a junior synonym of *Coreus calcarator* Fabricius, 1803. Monotypic.
1862 *Capaneus* Stål, Stett. Ent. Zeit., 23: 277, 279. Type-species: *Capaneus rubronotatus* Stål, 1862. Monotypic. Synonymized by O'Shea, 1980, Studies Neotrop. Fauna Environ., 15: 212.

Note: O'Shea (1980, Stud. Neotrop. F. Environ., 15: 213) explained that Burmeister's (1835, Handb. Ent., 2: 321) proposal of *Archimerus* as a replacement name for *Pachymeria* Laporte, 1833, failed to include the latter's type-species and thus misled Stål (1876, Öfv. K. Vet.-Akad. Förh., 24(7): 538), whose definition was followed by American authors; and that Burmeister's "*Archimerus*" belongs in the Meropachydinae. Our species, formerly placed in *Archimerus*, now belong in the genus *Piezogaster*.

Piezogaster ashmeadi (Montandon), 1899

1899 *Archimerus ashmeadi* Montandon, Bull. Sci. Bucharest, 8: 194. [Fla.].

1980 *Piezogaster ashmeadi*: O'Shea, Studies Neotrop. Fauna Environ., 15: 214.

Distribution: Fla.

Piezogaster auriculatus (Stål), 1862

1862 *Capaneus auriculatus* Stål, Stett. Ent. Zeit., 23: 289. [Mexico].

1876 *Xuthus auriculatus*: Uhler, Bull. U.S. Geol. Geogr. Surv. Terr., 5: 296.

1881 *Capaneus auriculatus*: Distant, Biol. Centr.-Am., Rhyn., 1: 112.

1916 *Capaneus* (*Xuthus*) *auriculatus*: Van Duzee, Check List Hem., p. 11.

1980 *Piezogaster auriculatus*: O'Shea, Studies Neotrop. Fauna Environ., 15: 214.

Distribution: N.M., Tex. (Guatemala, Mexico).

Piezogaster calcarator (Fabricius), 1803

1803 *Coreus calcarator* Fabricius, Syst. Rhyn., p. 192. ["Carolina"].

1825 *Coreus alternatus* Say, J. Acad. Nat. Sci. Phila., 4: 317. ["Missouri Territory"]. Synonymized by O'Shea, 1980, Studies Neotrop. Fauna Environ., 15: 214.

1835 *Archimerus squalus* Klug, *In* Burmeister, Handb. Ent., 2: 321. [N. Am.]. Synonymized by Van Duzee, 1916, Check List Hem., p. 12.

1842 *Archimerus rubiginosus*: Herrich-Schaeffer, Wanz. Ins., 6: 83.

1843 *Piezogaster albonotatus* Amyot and Serville, Hist. Nat. Hem., p. 197. [N. Am.]. Synonymized by Van Duzee, 1916, Check List Hem., p. 12.

1868 *Archimerus calcarator*: Stål, K. Svens. Vet.-Akad. Handl., 7(11): 47.

1909 *Archimerus alternatus*: Van Duzee, Bull. Buff. Soc. Nat. Sci., 9: 159.

1980 *Piezogaster calcarator*: O'Shea, Studies Neotrop. Fauna. Environ., 15: 214.

Distribution: Col., Fla., Ill., Ks., Mich., Miss., N.C., N.J., Ok., S.C., Tenn., Wis.

Note: O'Shea (1980, above) decided to follow Lethierry and Severin (1894, Cat. Gen. Hem., 2: 17) in synonymizing *Coreus alternatus* Say under *P. calcarator* rather than accept the separating characters used by Van Duzee (1909, above) and other North American hemipterists. Yonke and Medler (1969, An. Ent. Soc. Am., 62: 477-480) presented descriptions and illustrations of the immature stages under the name *Archimerus alternatus*.

Piezogaster indecorus (Walker), 1871

1871 *Archimerus indecorus* Walker, Cat. Hem. Brit. Mus., 4: 64. [Mexico].

1909 *Archimerus indecorus*: Van Duzee, Bull. Buff. Soc. Nat. Hist., 9: 169.

Distribution: Ariz. (Mexico).

Note: O'Shea (1980, Studies Neotrop. Fauna Environ., 15: 215) did not include the Ariz. record in the distribution of this species when he transferred it to *Piezogaster*.

Genus *Thasus* Stål, 1865

1865 *Thasus* Stål, An. Ent. Soc. France, ser. 4, 5: 174. Type-species: *Pachylis gigas* Klug, 1835. Designated by Van Duzee, 1916, Check List Hem., p. 11.

Note: In literature there has been uncertainty over the identification of the species in this genus occurring north of Mexico. Uhler, on several occasions, reported it for the region as *gigas*, and in this was followed by Van Duzee in his Check List, his Catalogue, and for nymphs in his 1923 paper (Proc. Cal. Acad. Sci., ser. 4, 12: 132).
In 1940 (Bull. Brook. Ent. Soc., 35: 45) Torre-Bueno reported *T. acutangula* Stål from Arizona, but later in that year (1940, Bull. Brook. Ent. Soc., 35: 102), using notes from Barber, changed that identification to *gigas*; then in 1941 (Ent. Am., 21: 54) he listed *acutangula* for the region and rejected *gigas* "as probably a misidentification"; the following year (1942, Bull. Brook. Ent. Soc., 37: 184) he expressed doubts about Van Duzee's (1923, above) use of *gigas* and suggested *acutangulus*. Until all specimens concerned are reidentified by clearly stated characters, the present list places the records under the name used in the Van Duzee Catalog. O'Shea (1980, Studies Neotrop. Fauna Environ., 15: 219) listed both forms for the United States without clarifying comment.

Thasus gigas (Klug), 1835
1835 *Pachylis gigas* Klug, *In* Burmeister, Handb. Ent., 2: 338. [Mexico].
1875 *Pachylis gigas*: Uhler, Rept. Geol. Geogr. Surv. Terr., 5: 831.
1881 *Thasus gigas*: Distant, Biol. Centr.-Am., Rhyn., 1: 108.
1940 *Thasus acutangulus*: Torre-Bueno, Bull. Brook. Ent. Soc., 35: 45.
Distribution: Ariz., N.M. (Mexico).

Tribe Spartocerini Amyot and Serville, 1843

Genus *Sephina* Amyot and Serville, 1843

1843 *Sephina* Amyot and Serville, Hist. Nat. Hem., p. 185. Type-species: *Lygaeus pustulatus* Fabricius, 1803. Monotypic.
Note: Barber (1926, J. N.Y. Ent. Soc., 34: 212) presented reasons for dropping *Sephina limbata* Stål from the list of species occurring north of the Rio Grande River.

Sephina grayi Van Duzee, 1909
1909 *Sephina grayi* Van Duzee, Ent. News, 20: 232. [Fla.].
Distribution: Fla.
Note: Blatchley (1928, J. N.Y. Ent. Soc., 36: 3) wrote that this species may be only a variety of *Sephina gundlachi*.

Sephina gundlachi (Guérin-Méneville), 1857
1857 *Coreus (Sephina) gundlachi* Guérin-Méneville, Hist. Ins. Cuba., p. 377. [Cuba].
1910 *Sephina gundlachi*: Banks Cat. Nearc. Hem.-Het., p. 81.
Distribution: Fla. (West Indies).
Note: Blatchley (1928, J. N.Y. Ent. Soc., 36: 3) established this species as a member of the United States fauna with personally collected Florida specimens.

Genus *Spartocera* Laporte, 1833

1833 *Spartocera* Laporte, Mag. Zool., 2: 43. Type-species: *Spartocerus* [*sic*] *geniculatus* Burmeister, 1835. First included species.
Note: The present usage of species names follows Barber (1926, J. N.Y. Ent. Soc., 34: 212).

Spartocera diffusa (Say), 1832
1832 *Coreus diffusus* Say, Descrip. Het. Hem. N. Am., p. 770. [Ga.].

1870 *Spartocera diffusa*: Stål, K. Svens. Vet.-Akad. Handl., 9(1): 175.
1876 *Spartocera cinnamomea*: Uhler, Bull. U.S. Geol. Geogr. Surv. Terr., 5: 291.
1916 *Corecoris diffusus*: Van Duzee, Check List Hem., p. 12.
Distribution: Fla., Ga., N.C., N.M., Tex.

Subfamily Meropachydinae Stål, 1867

Tribe Merocorini Stål, 1870

Genus *Merocoris* Perty, 1833

1833 *Merocoris* Perty, Del. Anim. Artic., p. 170. Type-species: *Merocoris tristis* Perty, 1833. Monotypic.

Note: Literature shows an unusual lack of agreement among heteropterists as to the taxonomic level of the categories within this genus. Here each is given as a full species.

Merocoris curtatus McAtee, 1919
1919 *Merocoris typhaeus curtatus* McAtee, Bull. Brook. Ent. Soc., 14: 14. [Cal.].
1941 *Merocoris curtatus*: Torre-Bueno, Ent. Am., 21: 45.
Distribution: Ariz., Cal., Col., N.M., Tex.

Merocoris distinctus Dallas, 1852 [Fig. 25]
1852 *Merocoris distinctus* Dallas, List Hem. Brit. Mus., 2: 419. [Mo.].
1870 *Corynocoris distinctus*: Stål, K. Svens. Vet.-Akad. Handl., 9(1): 130.
1916 *Corynocoris typhaeus* var. *distinctus*: Van Duzee, Check List Hem., p. 10.
Distribution: Ala., Alta., Ark., Col., Ia., Ill., Ks., Mass., Md., Mo., N.J., N.M., N.Y., Oh., Ok., S.C., Va.

Merocoris typhaeus (Fabricius), 1798
1798 *Lygaeus typhaeus* Fabricius, Ent. Syst. Suppl, p. 537. ["Carolina"].
1803 *Coreus acridioides* Fabricius, Syst. Rhyn., p. 200. ["Carolina"]. Synonymized by Stål, 1870, K. Svens. Vet.-Akad. Handl., 9(1): 129.
1843 *Merocoris rugosus* Amyot and Serville, Hist. Nat. Hem., p. 244. ["Caroline"]. Synonymized by Stål, 1870, K. Svens. Vet.-Akad. Handl., 9(1): 129.
1852 *Merocoris typhaeus*: Dallas, List Hem. Brit. Mus., 2: 419.
1869 *Menocoris* [*sic*] *acridoides*: Rathvon, Hist. Lancaster Co., Pa., p. 549.
1870 *Corynocoris typhaeus*: Stål, K. Svens. Vet.-Akad. Handl., 9(1): 129.
1916 *Corynocoris typhaeus typhaeus*: Van Duzee, Check List Hem., p. 10.
Distribution: Ala., Ariz., Ark., Cal., Col., Fla., Ia., Ill., Ks., Md., Me., Mo., N.C., N.J., N.M., N.Y., Ok., Pa., S.C., Tex., Ut., Va., Wis.

Subfamily Pseudophloeinae Stål, 1867

Genus *Ceraleptus* Costa, 1847

1847 *Ceraleptus* Costa, Cimic. Reg. Neap. Cent., 2: 375. Type-species: *Coreus gracilicornis* Herrich-Schaeffer, 1835. Designated by Oshanin,

1912, Kat. Paläark. Hem., p. 23.

Note: Froeschner (1963, J. Ks. Ent. Soc., 36: 109-113) revised the Western Hemisphere species.

Ceraleptus americanus Stål, 1870 [Fig. 27]

1870 *Ceraleptus americanus* Stål, K. Svens. Vet.-Akad. Handl., 9(1): 219. [Tex.].

1948 *Coraleptus* [*sic*] *americanus*: Sherman, Ent. News, 59: 16.

Distribution: Ariz., Ark., Cal., D.C., Fla., Ind., La., Miss., Mo., N.C., N.Y., Tex., Ut.

Note: Uhler's (1876, Bull. U.S. Geol. Geogr. Surv. Terr., 1: 299) record for "Mexico" belongs under *Ceraleptus pacificus*.

Ceraleptus denticulatus Froeschner, 1963

1963 *Ceraleptus denticulatus* Froeschner, J. Ks. Ent. Soc., 36: 110. [Cal.].

Distribution: B.C., Cal., Wash.

Ceraleptus pacificus Barber, 1914

1876 *Ceraleptus americanus*: Uhler, Bull, U.S. Geol. Geogr. Surv. Terr., 1: 299 (in part).

1914 *Ceraleptus pacificus* Barber, J. N.Y. Ent. Soc., 22: 167. [Ore.]. Lectotype designated by Froeschner, 1963, J. Ks. Ent. Soc., 36: 112.

Distribution: Ariz., B.C., Cal., Id., Nev., Ore., Wash. (Mexico).

Ceraleptus probolus Froeschner, 1963

1963 *Ceraleptus probolus* Froeschner, J. Ks. Ent. Soc., 36: 112. [Tex.].

Distribution: Ks., Tex.

Genus *Coriomeris* Westwood, 1842

1834 *Merocoris* Hahn, Wanz. Ins., 2: 105. Preoccupied. Type-species: *Cimex denticulatus* Scopoli, 1764. Designated by Kirkaldy, 1900, Ent., 33: 240.

1842 *Coriomeris* Westwood, Hope Cat., 2: 6. New name for *Merocoris* Hahn.

Note: Dolling and Yonke (1976, An. Ent. Soc. Am., 69: 1147-1152) revised *Coriomeris* for the New World and gave a key (p. 1152) to the species.

Coriomeris humilis (Uhler), 1872

1871 *Dasycoris pilicornis*: Uhler, Prelim. Rept. U.S. Geol. Surv. Wyo., p. 471.

1872 *Dasycoris humilis* Uhler, Prelim. Rept. U.S. Geol. Surv. Mont., p. 403 (in part). [Col.]. Lectotype designated by Dolling and Yonke, 1976, An. Ent. Soc. Am., 69: 1147.

1894 *Coreus humilis*: Lethierry and Severin, Cat. Gen. Hem., 2: 96 (in part).

1895 *Dasycoris nigricornis*: Gillette and Baker, Bull. Col. Agr. Exp. Stn., 31: 19 (in part).

1916 *Coriomeris humilis*: Van Duzee, Check List Hem., p. 14 (in part).

Distribution: Alk., Alta., Ariz., B.C., Cal., Col., Conn., Ia., Id., Ill., Ks., Man., Mich., Minn., Mont., N.M., N.T., Neb., Oh., Ont., Ore., Que., S.D., Sask., Tex., Ut., Wash., Wis., Wyo., Yuk.

Note: Judging from Dolling and Yonke's (1976, An. Ent. Soc. Am., p. 69: 6) distribution for this species, Slater and Schaefer's (1963, Bull. Brook. Ent. Soc., 58: 114) querying of Florida records is supported and that record is here dropped as too far from the otherwise known range.

Coriomeris insularis Dolling and Yonke, 1976

1976 *Coriomeris insularis* Dolling and Yonke, An. Ent. Soc. Am., 69: 1149. [B.C.].

Distribution: B.C.

Coriomeris occidentalis Dolling and Yonke, 1976

1872 *Dasycoris humilis*: Uhler, Prelim. Rept. U.S. Geol. Surv. Mont., p. 403 (in part).

1894 *Coreus humilis*: Lethierry and Severin, Cat Gen. Hem., 2: 96 (in part).

1895 *Dasycoris nigricornis*: Gillette and Baker, Bull. Col. Agr. Exp. Stn., 31: 19 (in part).

1916 *Coriomeris nigricornis*: Van Duzee, Check List Hem., p. 14 (in part).

1976 *Coriomeris occidentalis* Dolling and Yonke, An. Ent. Soc. Am., 69: 1150. [Cal.].

Distribution: Cal., Ore.

Family Corixidae Leach, 1815

The Water Boatmen

By John T. Polhemus, Richard C. Froeschner, and Dan A. Polhemus

The water boatmen are small insects characterized by their elongate form, short unsegmented labium, one-segmented front tarsi, and short antennae [antenna shorter than head]. Species are commonly found in lakes, ponds, and slow sections of streams, where they cruise near the bottom, propelling themselves with their oar-like hind legs in a manner reminiscent of Notonectidae, but with the dorsal side up rather than down. While ordinarily thought of as fresh water bugs, many corixids also inhabit saline waters, and several species of *Trichocorixa* Kirkaldy apparently are physiologically adapted to high concentrations of salt (Scudder, 1976, pp. 263-269, Marine Insects).

Corixids primarily forage on bottom ooze, consuming algae, protozoa, and metazoa, but some prey on free-swimming mosquito larvae (Sailer and Lienk, 1954, Mos. News, 14: 14-16), brine shrimp, and other aquatic animals (Jansson and Scudder, 1976, J. Ent. Soc. B.C., 69: 44-45). Eggs are laid on various underwater objects and organisms, including crayfish (Griffith, 1945, Univ. Ks. Sci. Bull., 30: 241-365), and hatch in one to two weeks, with total developmental time averaging approximately two months (Peters and Spurgeon, 1971, Am. Midl. Nat., 86: 197-207; Dodson, 1975, Am. Midl. Nat., 94: 257-266). Adults commonly overwinter, and may be found swimming under ice or trapped in air pockets within it (Hussey, 1921, Bull. Brook. Ent. Soc., 16: 131-136). Dispersal flights are typical of many species and winds may carry them more than 5,000 feet into the air (Glick, 1939, U.S. Dept. Agri., Tech. Bull., 673: 25, 63-64); during those flights large numbers of individuals may be attracted to lights (Young, 1966, Ent. Month. Mag., 101: 217-229). Stridulation is used as a mating signal in many corixid genera, with songs being species and sex specific (Jansson, 1976, An. Zool. Fenn., 13: 48-62).

One manuscript name, "*Corixa annexa* Uhler," appeared a few times in our literature. Its first use by Walker (1873, Cat. Hem. Brit. Mus., 8: 199) for some Canadian records was the basis for subsequent listings. The name thus cannot be entered in our list.

Corixidae occur world wide and are found over the entire United States and Canada. The major work on the group for our region is that of Hungerford (1948, Univ. Ks. Sci. Bull., 32: 1-827), which treats the family for the Western Hemisphere and pro/vides keys to the majority of species. Brooks and Kelton (1967, Mem. Ent. Soc. Can., 51: 1-92) and Stonedahl and Lattin (1986, Ore. St. Univ. Agr. Exp. Stn. Bull., 150: 1-84) provided very useful faunal treatments for the Canadian Prairie Provinces and Oregon and Washington that include keys and many figures.

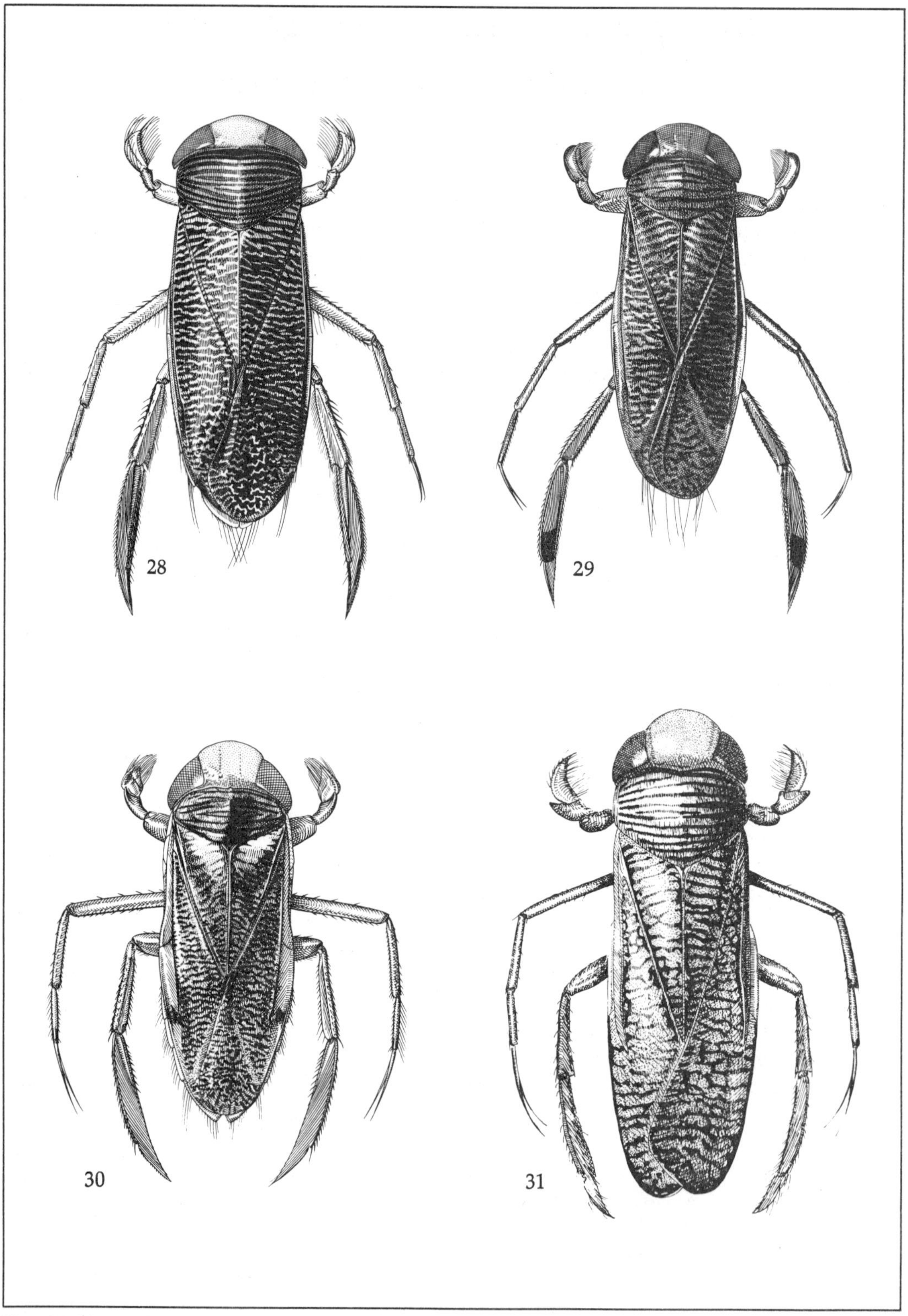

Figs. 28-31: 28, *Hesperocorixa vulgaris* [p. 103]; 29, *Callicorixa vulnerata* [p. 96]; 30, *Corisella decolor* [p. 99]; 31, *Trichocorixa reticulata* [p. 115] (After Usinger, 1956).

Subfamily Corixinae Enderlein, 1915

Tribe Corixini Enderlein, 1915

Genus *Arctocorisa* Wallengren, 1894

1894 *Arctocorisa* Wallengren, Ent. Tidskr., 15: 159. Type-species: *Corisa carinata* Sahlberg, 1819. Designated by Kirkaldy, 1906, Trans. Am. Ent. Soc., 32: 152.

Note: Many authors have used the spelling "*Arctocorixa*" for this genus.

Arctocorisa chanceae Hungerford, 1926

1926 *Arctocorixa* [*sic*] *chancei* [*sic*] Hungerford, An. Ent. Soc. Am., 19: 461. [Alk.].
1930 *Arctocorixa* [*sic*] *chanceae*: Walley, Can. Ent., 62: 281.
1940 *Corixa* (*Arctocorisa*) *chanceae*: Hutchinson, Trans. Conn. Acad. Arts. Sci., 33: 424.
1948 *Arctocorisa chanceae*: Hungerford, Univ. Ks. Sci. Bull., 32: 589.
Distribution: Alk., Man., Nfld., Que.

Arctocorisa convexa (Fieber), 1851

1851 *Corisa* [*sic*] *convexa* Fieber, Act. Reg. Bohem. Soc. Sci., Pragae, 1: 37. [Nfld.].
1909 *Arctocorisa convexa*: Kirkaldy and Torre-Bueno, Proc. Ent. Soc. Wash., 10: 194.
1916 *Arctocorixa* [*sic*] (*Arctocorixa* [*sic*]) *convexa*: Van Duzee, Check List Hem., p. 54.
1926 *Arctocorixa* [*sic*] *convexa*: Blatchley, Het. E. N. Am., p. 1072.
1931 *Sigara convexa*: Lundblad, Zool. Anz., 96: 87.
1940 *Corixa* (*Arctocorisa*) *convexa*: Hutchinson, Trans. Conn. Acad. Arts Sci., 33: 424.
Distribution: Alta., B.C., Man., N.T., Nfld., Que.
Note: Walley's (1931, Can. Ent., 63: 238) note that this species was described from Alaska is a *lapsus*.

Arctocorisa lawsoni Hungerford, 1948

1948 *Arctocorisa lawsoni* Hungerford, Univ. Ks. Sci. Bull., 32: 600. [Col.].
Distribution: Col., Wyo.

Arctocorisa planifrons (Kirby), 1837

1837 *Corixa planifrons* Kirby, Fauna Bor.-Am., 4: 284. [N.T.]. Lectotype designated by Hungerford, 1948, Univ. Ks. Sci. Bull., 32: 603.
1837 *Corixa carinata* Kirby, Fauna Bor.-Am., 4: 284. [N.T.]. Preoccupied. Synonymized by Hungerford, 1948, Univ. Ks. Sci. Bull., 32: 603.
1851 *Corixa kirbyi* Fieber, Act. Reg. Bohem. Soc. Sci., Pragae, 1: 4343. Unnecessary new name for *Corixa carinata* Kirby.
1859 *Corisa* [*sic*] *kirbyi*: Dohrn, Cat. Hem., p. 55.
1859 *Corisa* [*sic*] *planifrons*: Dohrn, Cat. Hem., p. 55.
1909 *Arctocorisa planifrons*: Kirkaldy and Torre-Bueno, Proc. Ent. Soc. Wash., 10: 196.
1916 *Arctocorixa* [*sic*] (*Arctocorixa* [*sic*]) *Kirbyi* [*sic*]: Van Duzee, Check List Hem., p. 54.
1916 *Arctocorixa* [*sic*] (*Arctocorixa* [*sic*]) *planifrons*: Van Duzee, Check List Hem., p. 54.
1926 *Arctocorixa* [*sic*] *planifrons*: Blatchley, Het. E. N. Am., p. 1073.
Distribution: Alk., Nfld., N.T.

Arctocorisa sutilis (Uhler), 1876

1876 *Corixa sutilis* Uhler, Bull. U.S. Geol. Geogr. Surv., 1: 339. [Col.]. Lectotype designated by Hungerford, 1948, Univ. Ks. Sci. Bull., 32: 594.

1877 *Corisa* [*sic*] *sutilis*: Uhler, An. Rept. Chief. Eng., Append. NN, p. 1332.
1900 *Corisa* [*sic*] *convexa*: Heidemann, Proc. Wash. Acad. Sci., 2: 506.
1909 *Arctocorisa sutilis*: Kirkaldy and Torre-Bueno, Proc. Ent. Soc. Wash., 10: 197.
1916 *Arctocorixa* [*sic*] (*Arctocorixa* [*sic*]) *sutilis*: Van Duzee, Check List Hem., p. 54.
Distribution: Alk., Alta., B.C., Col., Man., Mon., N.B., N.T., Ont., Que., Ut., Wyo., Yuk.

Genus *Callicorixa* White, 1873

1873 *Callicorixa* White, Ent. Month. Mag., 10: 62, 75. Type-species: *Corisa* [*sic*] *praeusta* Fieber, 1848. Designated by Kirkaldy, 1898, Ent., 31: 253.
1901 *Corisa* [*sic*] (*Callicorisa* [*sic*]): Kirkaldy, J. Quekett Microsc. Club, (2)48: 40.
Note: Kirkaldy (1906, Trans. Am. Ent. Soc., 32: 152) reelevated this name to generic status.

Callicorixa alaskensis Hungerford, 1926
1837 *Corixa striata* var. B Kirby, Fauna Bor.-Am., 4: 284.
1848 *Corisa* [*sic*] *praeusta*: Fieber, Bull. Soc. Nat. Moscou, 21: 521.
1909 *Callicorixa praeusta*: Kirkaldy and Torre-Bueno, Proc. Ent. Soc. Wash., 10: 194.
1926 *Callicorixa alaskensis* Hungerford, An. Ent. Soc. Am., 19: 462. [Alk.].
1974 *Callicorixa alaskaensis* [*sic*]: Bay, An. Rev. Ent., 19: 447.
Distribution: Alk., Alta., B.C., Man., Mich., Mont., N.B., N.H., N.S., N.T., N.Y., Nfld., Ont., Pa., Que., Sask., Ut., Wash., Wyo., Yuk.

Callicorixa audeni Hungerford, 1928
1928 *Callicorixa audeni* Hungerford, Can. Ent., 60: 229. [B.C.].
1930 *Callicorixa canadensis* Walley, Can. Ent., 62: 80. [Que.]. Synonymized by Hungerford, 1948, Univ. Ks. Sci. Bull., 32: 464.
Distribution: Alk., Alta., B.C., Cal., Col., Id., Me., Man., Mich., Minn., Mont., N.B., N.D., N.S., N.T., Nev., Nfld., Ont., Ore., P.Ed., Que., Sask., Ut., Wash., Wis., Wyo., Yuk.

Callicorixa producta (Reuter), 1880
1880 *Corisa* [*sic*] *praeusta producta* Reuter Medd. Soc. Fauna Flora Fenn., 5: 193. [Sweden].
Distribution: Alk., Man., Yuk. (Scandanavian Peninsula east to Germany; Siberia)
Note: The present generic combination was made by Lundblad (1927, Ent. Tidsk., 48: 85). The nominate subspecies does not occur in the New World.

Callicorixa producta noorvikensis Hungerford, 1926
1926 *Callicorixa noorvikensis* Hungerford, An. Ent. Soc. Am., 19: 462. [Alk.].
1948 *Callicorixa producta noorvikensis* Hungerford, 1948, Univ. Ks. Sci. Bull., 32: 477.
Distribution: Alk., Man., Yuk. (Siberia).

Callicorixa scudderi Jansson, 1979
1979 *Callicorixa scudderi* Jansson, Pan-Pac. Ent., 54: 263. [B.C.].
Distribution: B.C., Ore., Wash.

Callicorixa tetoni Hungerford, 1948
1948 *Callicorixa tetoni* Hungerford, Univ. Ks. Sci. Bull., 32: 469. [Wyo.].
Distribution: Wyo.

Callicorixa vulnerata (Uhler), 1861 [Fig. 29]
1861 *Corixa vulnerata* Uhler, Proc. Acad. Nat. Sci. Phila., 13: 284. [Wash.].
1878 *Corisa* [*sic*] *vulnerata*: Uhler, Bull. U.S. Geol. Geogr. Surv. Terr., 4: 509.
1909 *Arctocorisa vulnerata*: Kirkaldy and Torre-Bueno, Proc. Ent. Soc. Wash., 10: 197.
1916 *Arctocorixa* [*sic*] (*Arctocorixa* [*sic*]) *vulnerata*: Van Duzee, Check List Hem., p. 54.
1926 *Arctocorixa* [*sic*] *vulnerata*: Blatchley, Het. E. N. Am., p. 1081.

1948 *Callicorixa vulnerata*: Hungerford, Univ. Ks. Sci. Bull., 32: 481.
Distribution: Alk., Cal., B.C., Id., Mont., Ore., Ut., Wash., Wyo
Note: Hungerford (1948, above, 32: 483) could not verify Uhler's (1876, Bull. U.S. Geol. Geogr. Surv., 1: 340) disjunct distribution record for Illinois.

Genus *Cenocorixa* Hungerford, 1948

1948 *Cenocorixa* Hungerford, Univ. Ks. Sci. Bull., 32: 564. Type-species: *Arctocorixa* [*sic*] *wileyi* [*sic*] Hungerford, 1926. Original designation.
Note: Jansson (1972, An. Zool. Fenn., 9: 120-129; 1973, Behav., 46: 1-36) discussed stridulation in this genus and (1972, Can. Ent., 104: 449-450) presented a revised key to the species.

Cenocorixa andersoni Hungerford, 1948
1948 *Cenocorixa andersoni* Hungerford, Univ. Ks. Sci. Bull, 32: 573. [Wash.].
1956 *Cenocorixa malkini* Hungerford, J. Ks. Ent. Soc., 29: 40. [Wash.]. Synonymized by Jansson, 1972, Can. Ent., 104: 454.
1960 *Cenocorixa downesi* Lansbury, Proc. Ent. Soc. B.C., 57: 40. [B.C.]. Synonymized by Jannson, 1972, Can. Ent., 104: 454.
Distribution: B.C., Ore., Wash.

Cenocorixa bifida (Hungerford), 1926
1926 *Arctocorixa* [*sic*] *bifida* Hungerford, Can. Ent., 58: 268. [Alta.].
1948 *Cenocorixa bifida*: Hungerford, Univ. Ks. Sci. Bull., 32: 569.
Distribution: Alk., Alta., B.C., Col., Id., Man., Minn., Mont., N.D., R.I., Sask., Ut., Wyo.
Note: Lauck (1979, Bull. Cal. Ins. Surv., 21: 113) deleted the Cal. records by reidentifying the two Hungerford specimens as *Cenocorixa kuiterti* Hungerford and *Cenocorixa wileyae* (Hungerford). Scudder (1966, J. Ent. Soc. B.C., 63: 33-40) described the immature stages and gave a key to separate this species from *C. expleta* (Uhler).

Cenocorixa bifida bifida (Hungerford), 1926
1926 *Arctocorixa* [*sic*] *bifida* Hungerford, Can. Ent., 58: 268. [Alta.].
1948 *Cenocorixa bifida*: Hungerford, Univ. Ks. Sci. Bull., 32: 569.
1948 *Cenocorixa kuiterti*: Hungerford, Univ. Ks. Sci. Bull., 32: 571 (in part).
Distribution: Alta., Col., Id., Man., Minn., Mont., N.D., Sask., Ut.

Cenocorixa bifida hungerfordi Lansbury, 1960
1960 *Cenocorixa hungerfordi* Lansbury, Proc. Ent. Soc. B.C., 57: 36. [B.C.].
1972 *Cenocorixa bifida hungerfordi*: Jansson, Can. Ent., 104: 450.
Distribution: B.C., Wash.

Cenocorixa blaisdelli (Hungerford), 1930
1930 *Arctocorixa* [*sic*] *blaisdelli* Hungerford, Pan-Pac. Ent., 7: 26. [Cal.].
1931 *Sigara blaisdelli*: Jaczewski, Arch. Hydrobiol., 23: 511.
1948 *Cenocorixa blaisdelli*: Hungerford, Univ. Ks. Sci. Bull., 32: 574.
1953 *Coenocorixa* [*sic*] *blaisdelli*: Kellen, J. Econ. Ent., 46: 913.
1960 *Cenocorixa columbiensis* Lansbury, Proc. Ent. Soc. B.C., 57: 38. [B.C.]. Synonymized by Jansson, 1972, Can. Ent., 104: 457.
Distribution: B.C., Cal., Wash.

Cenocorixa dakotensis (Hungerford), 1928
1928 *Arctocorixa* [*sic*] *dakotensis* Hungerford, Can. Ent., 60: 229. [N.D.].
1948 *Cenocorixa dakotensis*: Hungerford, Univ. Ks. Sci. Bull., 32: 567.

Distribution: Alta., Ia., Ill., Man., Minn., N.D., N.T., S.D., Sask., Wis.

Cenocorixa expleta (Uhler), 1895

1895 *Corisa* [*sic*] *expleta* Uhler, Col. Agr. Exp. Stn. Bull., 31, Tech. ser. 1: 63. [Col.].
1909 *Arctocorisa expleta*: Kirkaldy and Torre-Bueno, Proc. Ent. Soc. Wash., 10: 195.
1916 *Arctocorixa* [*sic*] (*Arctocorixa* [*sic*]) *expleta*: Van Duzee, Check List Hem., p. 54.
1948 *Cenocorixa expleta*: Hungerford, Univ. Ks. Sci. Bull., 32: 576.
Distribution: Alta., B.C., Col., Man., N.D., S.D., Sask., Wash
Note: Blatchley (1926, Het. E. N. Am., p. 1073) deleted a N.J. record as "probably" a wrong determination. Scudder (1966, J. Ent. Soc. B.C., 63: 33-40) described the immature stages and provided a key to separate this species from *C. bifida* (Hungerford).

Cenocorixa kuiterti Hungerford, 1948

1948 *Cenocorixa kuiterti* Hungerford, Univ. Ks. Sci. Bull., 32: 571. [Cal.].
1948 *Cenocorixa bifida*: Hungerford, Univ. Ks. Sci. Bull., 32: 569 (in part).
1948 *Cenocorixa utahensis*: Hungerford, Univ. Ks. Sci. Bull., 32: 569 (in part).
Distribution: Cal.
Note: Jansson (1972, Can. Ent., 104: 450) transferred a Ut. record as "probably belongs to *C. bifida*."

Cenocorixa utahensis (Hungerford), 1925

1925 *Arctocorixa* [*sic*] *utahensis* Hungerford, Bull. Brook. Ent. Soc., 20: 22. [Ut.].
1948 *Cenocorixa utahensis*: Hungerford, Univ. Ks. Sci. Bull., 32: 580.
Distribution: Alta., Ariz., B.C., Col., Ia., Id., Ks., Man., N.D., N.M., Nev., Ont., Ore., S.D., Sask., Tex., Ut., Wash., Wis.
Note: Lauck (1979, Bull. Cal. Ins. Surv., 21: 116) transferred the Cal. records to *Cenocorixa kuiterti* Hungerford.

Cenocorixa wileyae (Hungerford), 1926

1926 *Arctocorixa* [*sic*] *wileyi* [*sic*] Hungerford, Can. Ent., 58: 271. [Ut.].
1948 *Cenocorixa wileyae*: Hungerford, Univ. Ks. Sci. Bull., 32: 578.
1948 *Cenocorixa bifida*: Hungerford, Univ. Ks. Sci. Bull., 32: 569 (in part).
Distribution: Ariz., Cal., Col., N.M., Nev., Ore., Ut., Wash.

Genus *Centrocorisa* Lundblad, 1928

1928 *Centrocorisa* Lundblad, Ent. Tidsk. 49: 68. Type-species: *Corisa kollari* Fieber, 1851. Monotypic.

Centrocorisa nigripennis (Fabricius), 1803

1803 *Sigara nigripennis* Fabricius, Syst. Rhyn., p. 105. [West Indies].
1843 *Corixa cubae* Guérin-Méneville, Icon. Reg. Anim., p. 345. [Cuba]. Synonymized by Hungerford, 1939, An. Ent. Soc. Am., 32: 588.
1894 *Corisa* [*sic*] *cubae*: Uhler, Proc. Royal Zool. Soc. London, p. 224.
1909 *Callicorixa kollari*: Kirkaldy and Torre-Bueno, Proc. Ent. Soc. Wash., 10: 194.
1948 *Centrocorisa nigripennis*: Hungerford, Univ. Ks. Sci. Bull., 32: 437.
Distribution: Tex. (Mexico, West Indies).
Note: The present usage accepts Hungerford's (1948, above) conclusions that *Centrocorisa kollari* (Fieber) is a neotropical species ranging no farther north than Mexico, and that *Centrocorisa nigripennis* is the proper name for this form.

Genus *Corisella* Lundblad, 1928

1928 *Corisella* Lundblad, Zool. Anz., 79: 158. Type-species: *Corixa mercenaria* Say, 1832. Monotypic.

Corisella decolor (Uhler), 1871 [Fig. 30]

1871 *Corixa decolor* Uhler, Am. J. Sci. Arts, ser. 3, 1: 106. [Cal.].
1875 *Corisa* [*sic*] *dispersa* Uhler, Rept. U.S. Geol. Geogr. Surv. Terr., 5: 841. [Cal.]. Synonymized by Hungerford, 1948, Univ. Ks. Sci. Bull., 32: 267.
1876 *Corisa* [*sic*] *decolor*: Uhler, Bull. U.S. Geol. Geogr. Surv. Terr., 1: 340.
1909 *Arctocorisa decolor*: Kirkaldy and Torre-Bueno, Proc. Ent. Soc. Wash., 10: 195.
1909 *Arctocorisa dispersa*: Kirkaldy and Torre-Bueno, Proc. Ent. Soc. Wash., 10: 195.
1916 *Arctocorixa* [*sic*] (*Arctocorixa* [*sic*]) *decolor*: Van Duzee, Check List Hem., p. 54.
1916 *Arctocorixa* [*sic*] (*Arctocorixa* [*sic*]) *dispersa*: Van Duzee, Check List Hem., p. 54.
1948 *Corisella decolor*: Hungerford, Univ. Ks. Sci. Bull., 53: 267.
Distribution: B.C., Cal., Col., Id., Mont., Nev., Ore., Tex., Ut., Wyo.

Corisella edulis (Champion), 1901

1901 *Corixa edulis* Champion, Biol. Centr.-Am., Rhyn., 2: 380. [Mexico].
1916 *Arctocorixa* [*sic*] (*Arctocorixa* [*sic*]) *edulis*: Van Duzee, Check List Hem., p. 54.
1925 *Arctocorixa* [*sic*] *edulis*: Hungerford, Bull. Brook. Ent. Soc., 20: 142.
1928 *Corisella edulis*: Lundblad, Zool. Anz., 79: 158.
Distribution: Ariz., D.C., Fla., Ga., Ia., Ks., Minn., Miss., N.M., Neb., Nev., Ok., Ore., Tenn., Tex., Ut., Wis. (Mexico).

Corisella inscripta (Uhler), 1894

1894 *Corixa inscripta* Uhler, Proc. Cal. Acad. Sci., ser. 2, 4: 294. [Mexico].
1909 *Arctocorisa inscripta*: Kirkaldy and Torre-Bueno, Proc. Ent. Soc. Wash., 10: 195.
1916 *Arctocorixa* [*sic*] (*Arctocorixa* [*sic*]) *inscripta*: Van Duzee, Check List Hem., p. 54.
1948 *Corisella inscripta*: Hungerford, Univ. Ks. Sci. Bull., 32: 279.
1969 *Corisela* [*sic*] *inscripta*: Smith and Enns, J. Ks. Ent. Soc., 10: 195.
Distribution: Ariz., B.C., Cal., Col., Id., Mo., N.M., Ore., Tex., Ut., Wash. (Mexico).

Corisella tarsalis (Fieber), 1851

1851 *Corisa* [*sic*] *tarsalis* Fieber, Act. Reg. Bohem. Soc. Sci., Pragae, 1: 19. [Pa.].
1877 *Corixa tumida* Uhler, Bull. U.S. Geol. Geogr. Surv. Terr., 3: 454. [Col.]. Synonymized by Hungerford, 1948, Univ. Ks. Sci. Bull., 32, 260.
1895 *Corisa* [*sic*] *tumida*: Uhler, Bull. Col. Agr. Exp. Stn., 31: 64.
1909 *Arctocorisa tarsalis*: Kirkaldy and Torre-Bueno, Proc. Ent. Soc. Wash., 10: 197.
1909 *Arctocorisa mercenaria*: Kirkaldy and Torre-Bueno, Proc. Ent. Soc. Wash., 10: 196 (in part).
1916 *Arctocorixa* [*sic*] (*Arctocorixa* [*sic*]) *tarsalis*: Van Duzee, Check List Hem., p. 54.
1916 *Arctocorixa* [*sic*] (*Arctocorixa* [*sic*]) *tumida*: Van Duzee, Check List Hem., p. 54.
1926 *Arctocorixa* [*sic*] *tarsalis*: Blatchley, Het. E. N. Am., p. 1075.
1931 *Sigara tarsalis*: Lundblad, Zool. Anz., 96: 86.
1936 *Corisella tumida*: Walley, Can. Ent., 68: 62.
1948 *Corisella tarsalis*: Hungerford, Univ. Ks. Sci. Bull., 32: 260.
Distribution: Alta., Ariz., Cal., Col., Ks., Man., Mont., N.M., Ok., Ont., Pa., S.D., Sask., Tex., Ut., Wis. (Mexico).
Note: Hungerford (1948, above, 32: 275) corrected the identification of the N.M. record of Kirkaldy and Torre-Bueno (1909, above) as this species and concluded "as yet there are no true records of *C. mercenaria* (Say) in the United States."

Genus *Hesperocorixa* Kirkaldy, 1908

1908 *Arctocorixa* (*Hesperocorixa*) Kirkaldy, Can. Ent., 40: 118. Type-species: *Arctocorisa brimleyi* Kirkaldy, 1908. Monotypic.

1926 *Hesperocorixa*: Blatchley, Het. E. N. Am., p. 1081.

Note: Dunn (1974, Proc. Acad. Nat. Sci. Phila., 131: 158-190) revised the genus and provided a phylogenetic study.

Hesperocorixa atopodonta (Hungerford), 1927

1878 *Corixa alternata*: Uhler, Proc. Bost. Soc. Nat. Hist. 19: 446 (in part).
1914 *Arctocorixa* [*sic*] *nitida*: Parshley, Psyche, 21: 140.
1916 *Arctocorisa dubia* Abbott, Ent. News, 27: 342. [Mass.]. Preoccupied.
1926 *Arctocorixa* [*sic*] *dubia*: Blatchley, Het. E. N. Am., p. 1072.
1927 *Arctocorixa* [*sic*] *atopodonta* Hungerford, Bull. Brook. Ent. Soc., 22: 35. New name for *Arctocorisa dubia* Abbott.
1936 *Sigara* (*Anticorixa*) *atopodonta*: Jaczewski, Proc. Royal Ent. Soc. London, ser. B., 5: 43.
1948 *Hesperocorixa atopodonta*: Hungerford, Univ. Ks. Sci. Bull., 32: 502.
Distribution: Alta., B.C., Col., Conn., Ks., Man., Mass., Me., Mich., Minn., Mont., N.B., N.D., N.H., N.J., N.Y., N.S., N.T., Nfld., Oh., Ont., Ore., P.Ed., Pa., Que., R.I., Sask., Wis., Wash.

Hesperocorixa brimleyi (Kirkaldy), 1908

1908 *Arctocorisa* (*Hesperocorixa*) *brimleyi* Kirkaldy, Can. Ent., 40: 120. [N.C.]. Lectotype designated by Hungerford, 1948, Univ. Ks. Sci. Bull., 32: 511.
1916 *Arctocorixa* [*sic*] (*Hesperocorixa*) *brimleyi*: Van Duzee, Check List Hem., p. 54.
1926 *Hesperocorixa brimleyi*: Blatchley, Het. E. N. Am., p. 1081.
1940 *Corixa* (*Hesperocorixa*) *brimleyi*: Hutchinson, Trans. Conn. Acad. Arts Sci., 33: 413.
Distribution: Ala., Fla., Ga., N.C., N.J.

Hesperocorixa escheri (Heer), 1853

1853 *Corisa* [*sic*] *escheri* Heer, Die Ins. Tiert. Oen., 3: 87. ["New Georgia"; Wash.].
1979 *Hesperocorixa escheri*: Dunn, Proc. Acad. Nat. Sci. Phila., 131: 183.
Distribution: Wash.
Note: Dunn (1979, above) considered the identity of this species uncertain. See note under *Hesperocorixa scabricula* Walley.

Hesperocorixa georgiensis (Egbert), 1946

1946 *Anticorixa georgiensis* Egbert, J. Ks. Ent. Soc., 19: 133. [Ga.].
1948 *Hesperocorixa georgiensis*: Hungerford, Univ. Ks. Sci. Bull., 32: 537.
Distribution: Ga.

Hesperocorixa harrisi (Uhler), 1878, New Emendation

1878 *Corisa* [*sic*] *harrisii* [*sic*] Uhler, Proc. Bost. Soc. Nat. Hist., 19: 444. [Mass.].
1909 *Arctocorisa harrisii* [*sic*]: Kirkaldy and Torre-Bueno, Proc. Ent. Soc. Wash., 10: 195.
1914 *Corixa harrisii* [*sic*]: Parshley, Psyche, 21: 140.
1916 *Arctocorixa* [*sic*] (*Arctocorixa* [*sic*]) *harrisii* [*sic*]: Van Duzee, Check List Hem., p. 54.
1930 *Arctocorixa* [*sic*] *harrissii* [*sic*]: Walley, Can. Ent., 62: 281.
1940 *Sigara harrisii* [*sic*]: Hungerford, J. Ks. Ent. Soc., 13: 9.
1948 *Hesperocorixa harrisii* [*sic*]: Hungerford, Univ. Ks. Sci. Bull., 32: 528.

Distribution: Ind., Mass., N.J., N.S., N.Y., Ont., P.Ed., Que., R.I., Vt.

Hesperocorixa interrupta (Say), 1825

1825 *Corixa interrupta* Say, J. Acad. Nat. Sci. Phila., 4: 328. [Mo.].

1851 *Corisa* [*sic*] *interrupta*: Fieber, Act. Reg. Bohem. Soc. Sci., Pragae, 1: 27.

1909 *Arctocorisa interrupta*: Kirkaldy and Torre-Bueno, Proc. Ent. Soc. Wash., 10: 195.

1916 *Arctocorixa* [*sic*] (*Arctocorixa* [*sic*]) *interrupta*: Van Duzee, Check List Hem., p. 54.

1929 *Sigara interrupta*: Lundblad, Ent. Tidsk., 50: 27.

1936 *Sigara* (*Anticorixa*) *interrupta*: Jaczewski, Proc. Royal Ent. Soc. Lond., ser. B, 5: 42.

1948 *Hesperocorixa interrupta*: Hungerford, Univ. Ks. Sci. Bull., 32: 529.

Distribution: Ala., Ark., Conn., D.C., Fla., Ga., Ill., Ind., La., Mass., Md., Me., Mich., Minn., Mo., N.C., N.H., N.J., N.Y., Neb., Oh., Ont., Pa., Que., R.I., S.C., Va., Vt., W.Va., Wis

Note: Bobb (1953, Va. J. Sci., 4: 111-115) provided notes on life history. Hungerford (1948, above, 32: 548) concluded that this is a species of the eastern United States and Canada and rejected records of the western United States and countries south of the United States.

Hesperocorixa kennicotti (Uhler), 1897, New Emendation

1897 *Corixa kennicottii* [*sic*] Uhler, Trans. Md. Acad. Sci., 1: 393. [Me.]. Lectotype designated by Hungerford, 1948, Univ. Ks. Sci. Bull., 32: 513.

1909 *Arctocorisa kennicottii* [*sic*]: Kirkaldy and Torre-Bueno, Proc. Ent. Soc. Wash., 10: 195.

1916 *Arctocorixa* [*sic*] (*Arctocorixa* [*sic*]) *kennicottii* [*sic*]: Van Duzee, Check List Hem., p. 54.

1928 *Arctocorixa* [*sic*] *kennecottii* [*sic*]: Hungerford, An. Ent. Soc. Am., 21: 140.

1936 *Sigara* (*Anticorixa*) *kennicottii* [*sic*]: Jaczewski, Proc. Royal Ent. Soc. London, ser. B, 5: 42.

1940 *Corixa* (*Hesperocorixa*) *kennicottii* [*sic*]: Hutchinson, Trans. Conn. Acad. Arts Sci., 33: 413.

1943 *Hesperocorixa kennicottii* [*sic*]: Walton, Trans. Soc. Brit. Ent., 32: 512.

Distribution: Conn., Ill., Mass., Md., Me., Mich., Minn., N.H., N.J., N.Y., Oh., Ont., Va., Wash., Wis.

Hesperocorixa laevigata (Uhler), 1893

1893 *Corisa* [*sic*] *laevigata* Uhler, Proc. Ent. Soc. Wash., 2: 384. [Cal.].

1909 *Arctocorisa laevigata*: Kirkaldy and Torre-Bueno, Proc. Ent. Soc. Wash., 10: 196.

1916 *Arctocorixa* [*sic*] (*Arctocorixa* [*sic*]) *laevigata*: Van Duzee, Check List Hem., p. 54.

1936 *Sigara* (*Anticorixa*) *laevigata*: Jaczewski, Proc. Royal Ent. Soc. London., ser. B, 5: 42.

1943 *Corixa laevigata*: Walton, Trans. Soc. Brit. Ent., 8: 161.

1948 *Hesperocorixa laevigata*: Hungerford, Univ. Ks. Sci. Bull., 32: 521.

Distribution: Alta., Ariz., B.C., Cal., Col., Id., Ill., Ks., Man., Md., Minn., Miss., Mont., N.C., N.D., N.M., N.Y., Neb., Nev., Oh., Ok., Ont., Ore., R.I., S.D., Sask., Tex., Ut., Wash., Wis., Wyo. (Mexico).

Hesperocorixa lobata (Hungerford), 1925

1925 *Arctocorixa lobata* Hungerford, Bull. Brook. Ent. Soc., 20: 143. [Minn.].

1936 *Sigara* (*Anticorixa*) *lobata*: Jaczewski, Proc. Royal Ent. Soc. Lond., ser. B, 5: 42.

1948 *Hesperocorixa lobata*: Hungerford, Univ. Ks. Sci. Bull., 32: 546.

Distribution: Ga., Mass., Md., Me., Mich., Minn., N.H., N.J., N.Y., R.I., Wis.

Hesperocorixa lucida (Abbott), 1916

1916 *Arctocorisa lucida* Abbott, Ent. News, 27: 341. [Conn.].

1917 *Arctocorixa* [*sic*] (*Arctocorixa* [*sic*]) *lucida*: Van Duzee, Univ. Cal. Publ. Ent., 2: 482.

1926 *Arctocorixa* [*sic*] *lucida*: Blatchley, Het. E. N. Am., p. 1072.

1936 *Sigara* (*Anticorixa*) *lucida*: Jaczewski, Proc. Royal Ent. Soc. Lond., ser. B, 5: 43.

1948 *Hesperocorixa lucida*: Hungerford, Univ. Ks. Sci. Bull., 32: 517.

Distribution: Ark., Conn., D.C., Fla., Ga., Ill., Mass., Md., Mich., N.J., N.Y., Ont., Pa., R.I., S.C., Tex., Va., Wis.

Hesperocorixa martini (Hungerford), 1928

1928 *Arctocorixa* [*sic*] *martini* Hungerford, Ent. News, 39: 157. [Ga.].

1936 *Sigara* (*Anticorixa*) *martini*: Jaczewski, Proc. Royal Ent. Soc. Lond., ser. B, 5: 42.

1948 *Hesperocorixa martini*: Hungerford, Univ. Ks. Sci. Bull., 32: 542.

Distribution: Fla., Ga.

Hesperocorixa michiganensis (Hungerford), 1926

1926 *Arctocorixa* [*sic*] *michiganensis* Hungerford, Bull. Brook. Ent. Soc., 21: 197. [Mich.].

1936 *Sigara* (*Anticorixa*) *michiganensis*: Jaczewski, Proc. Royal Ent. Soc. Lond., ser. B, 5: 43.

1948 *Hesperocorixa michiganensis*: Hungerford, Univ. Ks. Sci. Bull., 32: 506.

Distribution: Alta., B.C., Man., Mass., Mich., Minn., N.B., N.D., N.H., N.Y., Ont., Que., S.D., Sask., Wis.

Hesperocorixa minor (Abbott), 1913

1913 *Arctocorisa nitida* var. *minor* Abbott, Bull. Brook. Ent. Soc., 8: 82. [Ga.].

1916 *Arctocorixa* [*sic*] (*Arctocorixa* [*sic*]) *nitida minor*: Van Duzee, Check List Hem., p. 54.

1926 *Arctocorixa* [*sic*] *nitida minor*: Blatchley, Het. E. N. Am., p. 1070.

1939 *Arctocorixa* [*sic*] *minor*: Millspaugh, Field and Lab., 7: 85.

1948 *Hesperocorixa minor*: Hungerford, Univ. Ks. Sci. Bull., 32: 514.

Distribution: Ala., Ga., Mass., Md., Miss., N.C., N.J., N.J., Pa., R.I., S.C., Tex., Va., Wash.

Hesperocorixa minorella (Hungerford), 1926

1926 *Arctocorixa* [*sic*] *minorella* Hungerford, Bull. Brook. Ent. Soc., 21: 197. [Mich.].

1936 *Sigara* (*Anticorixa*) *minorella*: Jaczewski, Proc. Royal Ent. Soc. Lond., ser. B, 5: 43.

1948 *Hesperocorixa minorella*: Hungerford, Univ. Ks. Sci. Bull., 32: 499.

Distribution: Alta., Conn., Ks., Man., Me., Mich., Minn., N.B., N.H., N.S., N.T., Nfld., Ont., Que., Wis.

Hesperocorixa nitida (Fieber), 1851

1851 *Corisa* [*sic*] *nitida* Fieber, Act. Reg. Bohem. Soc. Sci., Pragae, 1: 28. [N.C.].

1909 *Arctocorisa nitida*: Kirkaldy and Torre-Bueno, Proc. Ent. Soc. Wash., 10: 196.

1916 *Arctocorixa* [*sic*] (*Arctocorixa* [*sic*]) *nitida*: Van Duzee, Check List Hem., p. 54.

1929 *Sigara nitida*: Lundblad, Arch. Hydrobiol., 20: 300.

1936 *Sigara* (*Anticorixa*) *nitida*: Jaczewski, Proc. Royal Ent. Soc. Lond., ser. B, 5: 42.

1948 *Hesperocorixa nitida*: Hungerford, Univ. Ks. Sci. Bull., 32: 539.

Distribution: Ala., Conn., Fla., Ga., Ia., Ill., Ind., Ks., Ky., La., Mass., Md., Mich., Minn., Miss., Mo., N.C., N.Y., Oh., Pa., R.I., S.C., Tenn., Tex., Va., Wash.

Note: Hungerford (1948, above, 32: 539) transferred Parshley's (1914, Psyche, 21: 140) record for this species from Maine to *Hesperocorixa atopodonta* (Hungerford).

Hesperocorixa obliqua (Hungerford), 1925
1925 *Arctocorixa* [*sic*] *obliqua* Hungerford, Bull. Brook. Ent. Soc., 20: 142. [Ks.].
1936 *Sigara* (*Anticorixa*) *obliqua*: Jaczewski, Proc. Royal Ent. Soc. Lond., ser. B, 5: 42.
1948 *Hesperocorixa obliqua*: Hungerford, Univ. Ks. Sci. Bull., 32: 543.
Distribution: Alk., Ark., Cal., Col., Ia., Ill., Ind., Ks., Mass., Mich., Minn., Miss., Mo., N.J., N.Y., Oh., Ok., Pa., Tenn., Tex., W.Va.

Hesperocorixa scabricula (Walley), 1936
1876 *Corixa escheri*: Uhler, Bull. U.S. Geol. Geogr. Surv. Terr., 5: 341.
1909 *Arctocorisa escheri*: Kirkaldy and Torre-Bueno, Proc. Ent. Soc. Wash., 10: 195.
1916 *Arctocorixa* [*sic*] (*Arctocorixa*) [*sic*] *escheri*: Van Duzee, Check List Hem., p. 54.
1936 *Arctocorixa* [*sic*] *scabricula* Walley, Can. Ent., 68: 56. [Ont.]. Synonymized with *H. escheri* (Heer) by Hungerford, 1948, Univ. Ks. Sci. Bull., 32: 519; resurrected by Dunn, 1979, Proc. Acad. Nat. Sci. Phila., 131: 183.
1948 *Hesperocorixa escheri*: Hungerford, Univ. Ks. Sci. Bull., 32: 519.
Distribution: Ill., Man., Mich., Minn., N.B., Ont., P.Ed., Que., Sask., Wis
Note: Hungerford (1948, above) considered *Corisa escheri* Heer (1853, Die Ins. Tiert. Oenin. Radob., 3: 87) to be a recognizable North American species of *Hesperocorixa* and placed *H. scabricula* as a junior synonym. Dunn (1979, above) investigated the nomenclatural history of *escheri* and concluded that it should be considered "*Incertae Sedis*." He thus established the identity of *Hesperocorixa scabricula* (Walley) but left that of *H. escheri* unresolved.

Hesperocorixa semilucida (Walley), 1930
1930 *Arctocorixa* [*sic*] *semilucida* Walley, Can. Ent., 62: 284. [Ont.].
1936 *Sigara* (*Anticorixa*) *semilucida*: Jaczweski, Proc. Royal Ent. Soc. Lond., ser. B, 5: 43.
1948 *Hesperocorixa semilucida*: Hungerford, Univ. Ks. Sci. Bull., 32: 508.
Distribution: Del., Fla., Ill., La., Mass., Mich., N.C., N.J., N.Y., Ont., Tenn., Wis.

Hesperocorixa vulgaris (Hungerford), 1925 [Fig. 28]
1925 *Arctocorixa* [*sic*] *vulgaris* Hungerford, Bull. Brook. Ent. Soc., 20: 143. [Ks.].
1936 *Sigara* (*Anticorixa*) *vulgaris*: Jaczewski, Proc. Royal Ent. Soc. Lond., ser. B, 5: 42.
1948 *Hesperocorixa vulgaris*: Hungerford, Univ. Ks. Sci. Bull., 32: 530.
Distribution: Alta., B.C., Cal., Conn., D.C., Ga., Ia., Ill., Ind., Ks., Man., Mass., Me., Mich., Minn., Miss., Mont., N.B., N.C., N.D., N.H., N.J., N.T., N.Y., Neb., Ont., Oh., Ore., Pa., Que., R.I., S.D., Sask., Tex., Va., Vt., Wash., Wis.

Genus *Morphocorixa* Jaczewski, 1931

1931 *Sigara* (*Morphocorixa*) Jaczewski, An. Mus. Zool. Pol., 9: 197. Type-species: *Arctocorixa* [*sic*] *compacta* Hungerford, 1925. Original designation.
1948 *Morphocorixa*: Hungerford, Univ. Ks. Sci. Bull., 32: 422.

Morphocorixa compacta (Hungerford), 1925
1925 *Arctocorixa* [*sic*] *compacta* Hungerford, Bull. Brook. Ent. Soc., 20: 22. [Tex.].
1948 *Morphocorixa compacta*: Hungerford, Univ. Ks. Sci. Bull., 32: 423.
Distribution: Tex. (Mexico).

Morphocorixa lundbladi (Jaczewski), 1931
1931 *Sigara* (*Morphocorixa*) *lundbladi* Jaczewski, An. Mus. Zool. Pol., 9: 197. [Mexico].
1948 *Morphocorixa lundbladi*: Hungerford, Univ. Ks. Sci. Bull., 32: 425.
Distribution: Ariz. (Mexico).

Genus *Palmacorixa* Abbott, 1912

1912 *Palmacorixa* Abbott, Ent. News, 23: 337. Type-species: *Palmacorixa gillettei* Abbott, 1912. Monotypic.

Palmacorixa buenoi Abbott, 1913

1913 *Palmacorixa buenoi* Abbott, Can. Ent., 45: 113. [N.Y.]. Lectotype designated by Hungerford, 1948, Univ. Ks. Sci. Bull., 32: 247.

Distribution: Ala., D.C., Del., Fla., Ga., Ia., Ind., Ks., La., Mass., Mich., Minn., N.C., N.J., N.Y., Ont., Pa., Que., S.C., S.D., Tex., Va., W.Va., Wis.

Palmacorixa gillettei Abbott, 1912

1912 *Palmacorixa gillettii* [*sic*] Abbott, Ent. News, 23: 337. [Col.]. Lectotype designated by Walley, 1930, Can. Ent., 62: 101.

1913 *Palmacorixa gillettei*: Abbott, Can. Ent., 45: 113.

Distribution: Col., Ia., Ks., Man., Mich., Minn., Ont., Que., Sask., Wis.

Note: As pointed out by Hungerford (1948, Univ. Ks. Sci. Bull., 32: 243), this species was named for Professor Gillette by Abbott who, after the original spelling ending in double-i, later always used the "ei" ending--apparently considering the original spelling a typographical error.

Palmacorixa gillettei confluens Walley, 1930

1930 *Palmacorixa gillettii* [*sic*] *confluens* Walley, Can. Ent., 62: 103. [Ont.].

1948 *Palmacorixa gillettei confluens*: Hungerford, Univ. Ks. Sci. Bull., 32: 245.

Distribution: Ont.

Palmacorixa gillettei gillettei Abbott, 1912

1912 *Palmacorixa gillettii* [*sic*] Abbott, Ent. News, 23: 337.

1930 *Palmacorixa gillettii* [*sic*] *gillettii* [*sic*]: Walley, Can. Ent., 62: 101.

1948 *Palmacorixa gillettei gillettei*: Hungerford, Univ. Ks. Sci. Bull., 32: 243.

Distribution: Col., Ia., Mich., Minn., Que.

Palmacorixa janeae Brooks, 1959

1959 *Palmacorixa janeae* Brooks, Proc. Ent. Soc. Wash., 61: 179. [Sask.].

Distribution: Man., Sask.

Palmacorixa nana Walley, 1930

1930 *Palmacorixa nana* Walley, Can. Ent., 62: 106. [Que.].

Distribution: Ks., Mich., Minn., N.C., N.Y., Oh., Que., Va., Wis.

Palmacorixa nana nana Walley, 1930

1930 *Palmacorixa nana* Walley, Can. Ent., 62: 106. [Que.].

1948 *Palmacorixa nana nana*: Hungerford, Univ. Ks. Sci. Bull., 32: 249.

Distribution: Mich., Minn., N.C., N.Y., Que.

Palmacorixa nana walleyi Hungerford, 1948

1948 *Palmacorixa nana walleyi* Hungerford, Univ. Ks. Sci. Bull., 32: 251. [Ks.].

Distribution: Ks., N.M., Tex.

Genus *Pseudocorixa* Jaczewski, 1931

1931 *Pseudocorixa* Jaczewski, An. Mus. Zool. Pol., 9: 220. Type-species: *Corixa guatemalensis* Champion, 1901. Monotypic.

Pseudocorixa beameri (Hungerford), 1928

1928 *Arctocorixa* [*sic*] *beameri* Hungerford, An. Ent. Soc. Am., 21: 142. [Ariz.].

1948 *Pseudocorixa beameri*: Hungerford, Univ. Ks. Sci. Bull., 32: 411.
Distribution: Ariz. (Mexico).

Genus *Ramphocorixa* Abbott, 1912

1912 *Ramphocorixa* Abbott, Can. Ent., 64: 120. Type-species: *Ramphocorixa balanodis* Abbott, 1912, a junior synonym of *Corixa acuminata* Uhler, 1897. Monotypic.

Ramphocorixa acuminata (Uhler), 1897

1897 *Corixa acuminata* Uhler, Trans. Md. Acad. Sci., 1: 392. [Tex.]. Lectotype designated by Hungerford, 1948, Univ. Ks. Sci. Bull., 32: 450.
1876 *Corixa alternata*: Forbes, Bull. Ill. Mus. Nat. Hist., 1: 4.
1905 *Corisa scutellata*: Crevecoeur, Trans. Ks. Acad. Sci., 19: 234.
1912 *Ramphocorixa balanodis* Abbott, Can. Ent., 64: 118. [Mo.]. Synonymized by Hungerford, 1917, J. N.Y. Ent. Soc., 25: 2.
1916 *Rhamphocorixa* [*sic*] *acuminata*: Torre-Bueno, An. Ent. Soc. Am., 9: 362.
1917 *Ramphocorixa acuminata*: Hungerford, J. N.Y. Ent. Soc., 25: 2.
1920 *Rhamphocorixa* [*sic*] *balanodis*: Torre-Bueno, Bull. Brook. Ent. Soc., 15: 88.
1960 *Rhaphocorixa* [*sic*] *acuminata*: Lansbury, Ent. News, 71: 244.
Distribution: Ala., Col., D.C., Ga., Ill., Ks., Minn., Miss., Mo., Oh., Ok., S.D., Tex., Wis. (Mexico).
Note: Griffith (1945, Univ. Ks. Sci. Bull., 30: 241-365) studied the biology and morphology of this species.

Ramphocorixa rotundocephala Hungerford, 1927

1927 *Ramphocorixa rotundocephala* Hungerford, Am. Mus. Nov., 278: 1. [Haiti].
1948 *Ramphocorixa rotundocephala*: Hungerford, Univ. Ks. Sci. Bull., 32: 452.
Distribution: Ariz. (Mexico, West Indies).

Genus *Sigara* Fabricius, 1775

1775 *Sigara* Fabricius, Syst. Ent., p. 691. Type-species: *Notonecta striata* Linneaus, 1758. Monotypic.

Notes:Three Old World species formerly listed in the American fauna are no longer included in this genus.

Pennsylvania records for *Sigara lateralis* (Leach) by Kirkaldy and Torre-Bueno (1909, Proc. Ent. Soc. Wash., 10: 196) and other authors, including Van Duzee (1917, Univ. Cal. Publ. Ent., 2: 481) and Blatchley (1926, Het. E. N. Am., p. 1079) were made in combinations of *lateralis* with *Arctocorisa*, *Arctocorixa*, and *Corixa*. Hungerford (1948, Univ. Ks. Sci. Bull., 32: 1-827) made no mention of it in his revision of the Corixidae of the Western Hemisphere. Without documentary specimens or recent literature reporting it for the continent, this palearctic species is here considered extralimital to this catalog.

The "*Corixa hieroglyphica* Dufour," for which Fieber (1849, Bull. Soc. Imp. Nat. Moscou, 21: 517) gave a Pennsylvania record, was treated as a junior synonym by the above-mentioned three earlier authors and was not mentioned by Hungerford. This name belongs to the Old World fauna and at this time must be excluded from our faunal list.

Sigara striata (Linnaeus): American entomologists, through the time of Van Duzee's (1917) catalog, often identified North American specimens of certain corixid species in combination with the generic names *Arctocorixa striata*, *Corisa striata*, and *Corixa striata*. But this is a palearctic species which later American entomologists, like Blatchley (1926, Het. E. N. Am.), Hungerford (1948, Univ. Ks. Sci. Bull., vol. 32), and Brooks and Kelton (1967, Mem. Ent. Soc. Can., 51: 1-92) dropped completely, even from synonymies. Old World entomologists like Wu (1935, Cat. Ins. Sinensium, p. 588) and Hoffmann (1941, Lingnan Sci. J., 20: 19), who derived information mainly from summarizing works such as Van Duzee's catalog, continued to report the species for Canada and the United States.

Subgenus *Allosigara* Hungerford, 1948

1948 *Sigara* (*Allosigara*) Hungerford, Univ. Ks. Sci. Bull., 32: 629. Type-species: *Arctocorisa decorata* Abbott, 1916. Monotypic.

Sigara decorata (Abbott), 1916

1916 *Arctocorisa decorata* Abbott, Ent. News, 27: 341. [Mass.].

1917 *Arctocorixa* [*sic*] (*Arctocorixa* [*sic*]) *decorata*: Van Duzee, Univ. Cal. Publ. Ent., 2: 479.

1926 *Arctocorixa* [*sic*] *decorata*: Blatchley, Het. E. N. Am., p. 1071.

1948 *Sigara* (*Allosigara*) *decorata*: Hungerford, Univ. Ks. Sci. Bull., 32: 629.

Distribution: Conn., Ill., Mass., Me., Minn., N.D., Wis.

Subgenus *Arctosigara* Hungerford, 1948

1948 *Sigara* (*Arctosigara*) Hungerford, Univ. Ks. Sci. Bull., 32: 614. Type-species: *Arctocorixa* [*sic*] *conocephala* Hungerford, 1926. Original designation.

Sigara bicoloripennis (Walley), 1936

1936 *Arctocorixa* [*sic*] *bicoloripennis* Walley, Can. Ent., 68: 55. [Ont.].

1948 *Sigara* (*Arctosigara*) *bicoloripennis*: Hungerford, Univ. Ks. Sci. Bull., 32: 623.

1960 *Sigara* (*Vermicorixa*) *bicoloripennis*: Lansbury, Proc. Ent. Soc. B.C., 57: 36.

1963 *Sigara bicoloripennis*: Judd, Can. Ent., 95: 1109.

Distribution: Alk., Alta., B.C., Man., Mich., Minn., N.B., N.D., Nfld., Ont., Que., S.D., Sask., Wis.

Sigara conocephala (Hungerford), 1926

1926 *Arctocorixa* [*sic*] *conocephala* Hungerford, Can. Ent., 58: 270. [Mich.].

1948 *Sigara* (*Arctosigara*) *conocephala*: Hungerford, Univ. Ks. Sci. Bull, 32: 614.

1953 *Sigara conocephala*: Strickland, Can. Ent., 85: 203.

Distribution: Alta., B.C., Man., Mich., Minn., N.D., N.S., N.T., Nfld., Ont., P.Ed., Que., S.D., Sask., Wis.

Sigara decoratella (Hungerford), 1926

1851 *Corisa* [*sic*] *limitata*: Fieber, Act. Reg. Bohem. Soc. Sci., Pragae, 1: 35 (in part).

1914 *Corixa alternata*: Parshley, Psyche, 21: 140 (in part).

1926 *Arctocorixa decoratella* Hungerford, Bull. Brook. Ent. Soc., 21: 195. [Mich.].

1948 *Sigara* (*Arctocorixa* [*sic*]) *decoratella*: Hungerford, Univ. Ks. Sci. Bull, 32: 616.

1953 *Sigara decoratella*: Strickland, Can. Ent., 85: 203.

Distribution: Alk., Alta., B.C., Conn., Ia., Man., Mass., Me., Mich., Minn., Mo., N.B., N.D., N.J., N.M., N.S., N.T., N.Y., Nfld., Ont., P.Ed., Pa., Que., R.I., S.D., Sask., Wash., Wis., Yuk.

Sigara penniensis (Hungerford), 1928

1928 *Arctocorixa* [*sic*] *penniensis* Hungerford, Can. Ent., 60: 228. [Mich.].

1948 *Sigara (Arctosigara) penniensis*: Hungerford, Univ. Ks. Sci. Bull., 32: 620.
Distribution: Alk., Alta., B.C., Man., Mich., Minn., N.B., N.J., N.T., N.Y., Nfld., Ont., P.Ed., Que., R.I., S.D., Sask., Wis., Yuk.

Subgenus *Lasiosigara* Hungerford, 1948

1948 *Sigara (Lasiosigara)* Hungerford, Univ. Ks. Sci. Bull., 32: 645. Type-species: *Notonecta lineata* Forster, 1771. Original designation.

Sigara lineata (Forster), 1771

1771 *Notonecta lineata* Forster, Nov. Spec. Ins., Cent., 1: 70. [N. Am.].
1790 *Notonecta noveboracensis* Gmelin, Syst. Nat., 13th ed., 4: 2119. Unnecessary new name for *N. lineata* Forster.
1872 *Corisa* [*sic*] *bilineata* Provancher, Nat. Can., 4: 108. [Que.]. Synonymized by Hungerford, 1939, An. Ent. Soc. Am., 32: 585.
1888 *Corisa* [*sic*] *bivittata* Provancher, Pet. Faune Ent. Can., 3: 203. Unnecessary new name for *C. bilineata* Provancher.
1900 *Corixa bilineata*: Osborn, Proc. Oh. Acad. Sci., 9: 37.
1909 *Arctocorisa bilineata*: Kirkaldy and Torre-Bueno, Proc. Ent. Soc. Wash., 10: 194.
1909 *Arctocorisa bivittata*: Kirkaldy and Torre-Bueno, Proc. Ent. Soc. Wash., 10: 194.
1909 *Arctocorisa lineata*: Kirkaldy and Torre-Bueno, Proc. Ent. Soc. Wash., 10: 196.
1909 *Arctocorisa noveboracensis*: Kirkaldy and Torre-Bueno, Proc. Ent. Soc. Wash., 10: 196.
1916 *Arctocorixa* [*sic*] (*Arctocorixa* [*sic*]) *bilineata*: Van Duzee, Check List Hem., p. 54.
1916 *Arctocorixa* [*sic*] (*Arctocorixa* [*sic*]) *lineata*: Van Duzee, Check List Hem., p. 54.
1926 *Arctocorixa* [*sic*] *bilineata*: Blatchley, Het. E. N. Am., p. 1077.
1926 *Arctocorixa* [*sic*] *lineata*: Blatchley, Het. E. N. Am., p. 1079.
1930 *Arctocorixa* [*sic*] *lineata*: Walley, Bull. Brook. Ent. Soc., 25: 203 (in part).
1948 *Sigara (Lasiosigara) lineata*: Hungerford, Univ. Ks. Sci. Bull., 32: 645.
Distribution: Ill., Man., Minn., Oh., Ont., Pa., Que., Sask., Wis.

Sigara trilineata (Provancher), 1872

1872 *Corisa* [*sic*] *trilineata* Provancher, Nat. Can., 4: 108. [Que.].
1888 *Corisa* [*sic*] *trivittata* Provancher, Pet. Faune Ent. Can., 3: 203. Unnecessary new name for *Corisa trilineata* Provancher.
1909 *Arctocorisa trilineata*: Kirkaldy and Torre-Bueno, Proc. Ent. Soc. Wash., 10: 197.
1909 *Arctocorisa trivittata*: Kirkaldy and Torre-Bueno, Proc. Ent. Soc. Wash., 10: 197.
1916 *Arctocorixa* [*sic*] (*Arctocorixa* [*sic*]) *trilineata*: Van Duzee, Check List Hem., p. 54.
1930 *Arctocorixa* [*sic*] *lineata*: Walley, Bull. Brook. Ent. Soc., 25: 203 (in part).
1948 *Sigara (Lasiosigara) trilineata*: Hungerford, Univ. Ks. Sci. Bull., 32: 649.
1953 *Sigara trilineata*: Strickland, Can. Ent., 85: 203.
Distribution: Alta., Man., Mich., Minn., N.B., N.T., Ont., Que., Sask., Wis., Wyo.
Note: Walley (1930, above) synonymized *trilineata* under *lineata* but Hungerford (1939, An. Ent. Soc. Am., 32: 585-586) again elevated it to species status.

Subgenus *Pediosigara* Hungerford, 1948

1948 *Sigara (Pediosigara)* Hungerford, Univ. Ks. Sci. Bull., 32: 638. Type-species: *Arctocorisa hydatotrephes* Kirkaldy, 1908. Original designation.

Sigara depressa Hungerford, 1948

1948 *Sigara (Pediosigara) depressa* Hungerford, Univ. Ks. Sci. Bull., 32: 640. [Va.].
Distribution: Va.

Sigara hydatotrephes (Kirkaldy), 1908

1908 *Arctocorisa hydatotrephes* Kirkaldy, Can. Ent., 40: 119. [N.C.].

1916 *Arctocorixa* [*sic*] (*Arctocorixa* [*sic*]) *hydatotrephes*: Van Duzee, Check List Hem., p. 54.

1948 *Sigara* (*Pediosigara*) *hydatotrephes*: Hungerford, Univ. Ks. Sci. Bull., 32: 638.

Distribution: Ala., Ga., N.C.

Sigara saileri Wilson, 1953

1953 *Sigara* (*Pediosigara*) *saileri* Wilson, Fla. Ent., 36: 67. [Miss.].

Distribution: Miss.

Note: Morphological and comparative notes were given by Jansson (1975, J. Ks. Ent. Soc., 48: 1-3).

Subgenus *Phaeosigara* Hungerford, 1948

1948 *Sigara* (*Phaeosigara*) Hungerford, Univ. Ks. Sci. Bull., 32: 725. Type-species: *Corisa* [*sic*] *signata* Fieber, 1851. Original designation.

Sigara berneri Hungerford and Hussey, 1957

1957 *Sigara berneri* Hungerford and Hussey, Quart. J. Fla. Acad. Sci., 20: 91. [Ga.].

Distribution: Ga.

Sigara bradleyi (Abbott), 1913

1913 *Arctocorisa bradleyi* Abbott, Bull. Brook. Ent. Soc., 8: 83. [Ga.].

1917 *Arctocorixa* [*sic*] (*Arctocorixa* [*sic*]) *bradleyi*: Van Duzee, Univ. Cal. Publ. Ent., 2: 479.

1926 *Arctocorixa* [*sic*] *abjecta* Blatchley, Het. E. N. Am., p. 1075. [Fla.]. Synonymized by Hungerford, 1948, Univ. Ks. Sci. Bull., 32: 725.

1926 *Arctocorixa* [*sic*] *bradleyi*: Blatchley, Het. E. N. Am., p. 1075.

1948 *Sigara* (*Phaeosigara*) *bradleyi*: Hungerford, Univ. Ks. Sci. Bull., 32: 725.

1951 *Sigara bradleyi*: Herring, Fla. Ent., 34: 26.

Distribution: Ala., Fla., Ga., Miss.

Sigara compressoidea (Hungerford), 1928

1928 *Arctocorixa* [*sic*] *compressoidea* Hungerford, Can. Ent., 60: 226. [Mich.].

1948 *Sigara* (*Phaeosigara*) *compressoidea*: Hungerford, Univ. Ks. Sci. Bull., 31: 735.

Distribution: Conn., D.C., Mass., Md., Me., Mich., Minn., N.B., N.C., N.J., N.S., N.Y., Nfld., Ont., Que., S.C., Wis.

Sigara dolabra Hungerford and Sailer, 1943

1943 *Sigara dolabra* Hungerford and Sailer, Bull. Brook. Ent. Soc., 37: 179. [Minn.].

1948 *Sigara* (*Phaeosigara*) *dolabra*: Hungerford, Univ. Ks. Sci. Bull., 32: 752.

Distribution: B.C., Mich., Minn., N.S., Nfld., Ont., Que., R.I., Wis.

Sigara mackinacensis (Hungerford), 1928

1928 *Arctocorixa* [*sic*] *mackinacensis* Hungerford, Can. Ent., 60: 228. [Mich.].

1948 *Sigara* (*Phaeosigara*) *mackinacensis*: Hungerford, Univ. Ks. Sci. Bull., 32: 740.

1963 *Sigara mackinacensis*: Judd, Can. Ent., 95: 1109.

Distribution: Conn., Mich., Minn., N.B., N.H., Ont., Pa., Que., Wis.

Sigara macrocepsoidea Hungerford, 1942

1942 *Sigara macrocepsoidea* Hungerford, Bull. Brook. Ent. Soc., 37: 128. [Ga.].

1948 *Sigara* (*Phaeosigara*) *macrocepsoidea*: Hungerford, Univ. Ks. Sci. Bull., 32: 751.

Distribution: Ga., Tex.

Sigara macropala (Hungerford), 1926
1926 *Arctocorixa* [*sic*] *macropala* Hungerford, Bull. Brook. Ent. Soc., 21: 196. [Mich.].
1927 *Arctocorisa maeropala* [*sic*]: Johnson, *In* Proctor, Biol. Surv. Mt. Desert Reg., 1: 40.
1948 *Sigara* (*Phaeosigara*) *macropala*: Hungerford, Univ. Ks. Sci. Bull., 32: 743.
Distribution: Fla., Mass., Md., Me., Mich., Minn., N.J., N.Y., R.I.

Sigara mississippiensis Hungerford, 1942
1942 *Sigara mississippiensis* Hungerford, Bull. Brook. Ent. Soc., 37: 129. [Miss.].
1948 *Sigara* (*Phaeosigara*) *mississippiensis*: Hungerford, Univ. Ks. Sci. Bull., 32: 742.
Distribution: Ala., D.C., Ga., Miss., S.C.

Sigara paludata Hungerford, 1942
1942 *Sigara paludata* Hungerford, Bull. Brook. Ent. Soc., 37: 127. [Miss.].
1948 *Sigara* (*Phaeosigara*) *paludata*: Hungerford, Univ. Ks. Sci. Bull., 32: 749.
Distribution: Ga., Miss.

Sigara quebecensis (Walley), 1930
1930 *Arctocorixa* [*sic*] *quebecensis* Walley, Can. Ent., 62: 281. [Que.].
1948 *Sigara* (*Phaeosigara*) *quebecensis*: Hungerford, Univ. Ks. Sci. Bull., 32: 738.
Distribution: Conn., N.H., Que.

Sigara sigmoidea (Abbott), 1913
1913 *Arctocorisa sigmoidea* Abbott, Bull. Brook. Ent. Soc., 8: 83. [Ga.].
1926 *Arctocorixa* [*sic*] *sigmoidea*: Blatchley, Het. E. N. Am., p. 1080.
1948 *Sigara* (*Phaeosigara*) *sigmoidea*: Hungerford, Univ. Ks. Sci. Bull., 32: 728.
1951 *Sigara sigmoidea*: Herring, Fla. Ent., 34: 27.
Distribution: Fla., Ga., N.C.

Sigara signata (Fieber), 1851
1851 *Corisa* [*sic*] *signata* Fieber, Act. Reg. Bohem. Soc. Sci., Pragae, 1: 21. [Pa.].
1909 *Arctocorisa signata*: Kirkaldy and Torre-Bueno, Proc. Ent. Soc. Wash., 10: 197.
1916 *Arctocorisa seriata* Abbott, Ent. News, 27: 342. [Mass.]. Synonymized by Hungerford, 1948, Univ. Ks. Sci. Bull., 32: 745.
1916 *Arctocorixa* [*sic*] (*Arctocorixa* [*sic*]) *signata*: Van Duzee, Check List Hem., p. 54.
1917 *Arctocorixa* [*sic*] (*Arctocorixa* [*sic*]) *seriata*: Van Duzee, Univ. Cal. Publ. Ent., 2: 483.
1917 *Arctocorixa* [*sic*] *scabra*: Parshley, Occas. Pap. Bost. Soc. Nat. Hist., 7: 119 (in part).
1926 *Arctocorixa* [*sic*] *signata*: Blatchley, Het. E. N. Am., p. 1074.
1929 *Sigara signata*: Lundblad, Arch. Hydrobiol., 20: 303.
1946 *Sigara seriata*: Proctor, Biol. Surv. Mt. Des. Reg., 7: 82.
1948 *Sigara* (*Phaeosigara*) *signata*: Hungerford, Univ. Ks. Sci. Bull., 32: 745.
Distribution: Conn., Fla., Ga., Ill., Ind., Man., Mass., Md., Me., Mich., Minn., N.B., N.H., N.J., N.Y., Nfld., Ont., P.Ed., Pa., Que., R.I., S.C., Wis.

Sigara variabilis (Hungerford), 1926
1926 *Arctocorixa* [*sic*] *variabolis* [*sic*] Hungerford, Bull. Brook. Ent. Soc., 21: 198. [Mich.].
1930 *Arctocorixa* [*sic*] *variabilis*: Walley, Can. Ent., 62: 281.
1948 *Sigara* (*Phaeosigara*) *variabilis*: Hungerford, Univ. Ks. Sci. Bull., 32: 733.
Distribution: Conn., D.C., Ill., Mich., N.B., N.J., Ont., P.Ed., Que., Va., Wis

Note: Hungerford (1948, above, 32: 733) noted the original spelling of the species name was a "typographical error."

Sigara zimmermanni (Fieber), 1851

1851 *Corisa* [*sic*] *zimmermanni* Fieber, Act. Reg. Behem. Soc. Sci., Pragae, 1: 21. ["Carolina," Pa.].

1909 *Arctocorisa zimmermanni*: Kirkaldy and Torre-Bueno, Proc. Ent. Soc. Wash., 10: 197.

1913 *Arctocorisa compressa*: Abbott, Bull. Brook. Ent. Soc., 8: 83. [Ga.]. Synonymized by Walley, 1936, Can. Ent., 68: 60.

1916 *Arctocorixa* [*sic*] (*Arctocorixa* [*sic*]) *zimmermanni*: Van Duzee, Check List Hem., p. 54.

1917 *Arctocorixa* [*sic*] (*Arctocorixa* [*sic*]) *compressa*: Van Duzee, Univ. Cal. Publ. Ent., 2: 479.

1929 *Sigara zimmermanni*: Lundblad, Arch. Hydrob., 20: 310.

1948 *Sigara* (*Phaeosigara*) *zimmermanni*: Hungerford, Univ. Ks. Sci. Bull., 32: 730.

Distribution: Ala., Conn., D.C., Fla., Ga., Mass., Me., Mich., Miss., N.C., N.Y., Pa., Tex., Va.

Subgenus *Pileosigara* Hungerford, 1948

1948 *Sigara* (*Pileosigara*) Hungerford, Univ. Ks. Sci. Bull., 32: 634. Type-species: *Arctocorixa* [*sic*] *douglasensis* Hungerford, 1926. Monotypic.

Sigara douglasensis (Hungerford), 1926

1926 *Arctocorixa* [*sic*] *douglasensis* Hungerford, Bull. Brook. Ent. Soc., 21: 196. [Mich.].

1948 *Sigara* (*Pileosigara*) *douglasensis*: Hungerford, Univ. Ks. Sci. Bull., 32: 634.

Distribution: Mass., Mich., Minn., N.B., N.J., Ont., Pa., Wis.

Subgenus *Subsigara* Stichel, 1935

1935 *Sigara* (*Subsigara*) Stichel, Illust. Best. Deut. Wanz., 11: 314. Type-species: *Corixa fossarum* Leach, 1818. Designated by China, 1938, Ent. Month. Mag., 74: 38.

Sigara fallenoidea (Hungerford), 1926

1926 *Arctocorixa* [*sic*] *fallenoidea* Hungerford, Can. Ent., 58: 270. [Man.].

1948 *Sigara* (*Subsigara*) *fallenoidea*: Hungerford, Univ. Ks. Sci. Bull., 32: 643.

Distribution: Alk., Alta., Man., N.T., Sask.

Subgenus *Vermicorixa* Walton, 1940

1940 *Corixa* (*Vermicorixa*) Walton, Trans. Conn. Acad. Arts Sci., 33: 343. Type-species: *Corixa lateralis* Leach, 1818. Original designation.

1948 *Sigara* (*Vermicorixa*): Hungerford, Univ. Ks. Sci. Bull., 32: 652.

Sigara alternata (Say), 1825

1825 *Corixa alternata* Say, J. Acad. Nat. Sci. Phila., 4: 329. [Mo.].

1851 *Corisa* [*sic*] *erichsonii* Fieber, Act. Freg. Bohem. Soc. Sci., Pragae, 1: 35. [Pa.]. Synonymized by Hungerford, 1948, Univ. Ks. Sci. Bull., 32: 654.

1851 *Corisa* [*sic*] *alternata*: Fieber, Act. Freg. Bohem. Soc. Sci., Pragae, 1: 43.

1876 *Corixa striata*: Uhler, Bull. U.S. Geol. Geogr. Surv. Terr., 1: 340 (in part).

1909 *Arctocorisa alternata*: Kirkaldy and Torre-Bueno, Proc. Ent. Soc. Wash., 10: 194.

1909 *Arctocorisa erichsonii*: Kirkaldy and Torre-Bueno, Proc. Ent. Soc. Wash., 10: 195.

1913 *Corixa alternata*: Abbott, Bull. Brook. Ent. Soc., 8: 87.

1916 *Arctocorisa parshleyi* Abbott, Ent. News, 27: 342. [R.I.]. Synonymized by Hungerford, 1948, Univ. Ks. Sci. Bull., 32: 654.

1916 *Arctocorixa*·[*sic*] (*Arctocorixa* [*sic*]) *erichsonii*: Van Duzee, Check List Hem., p. 54.
1917 *Arctocorixa* [*sic*] (*Arctocorixa* [*sic*]) *parshleyi*: Van Duzee, Univ. Cal. Publ. Ent., 2: 483.
1926 *Arctocorixa* [*sic*] *erichsonii*: Blatchley, Het. E. N. Am., p. 1078.
1929 *Sigara parshleyi*: Lundblad, Zool. Anz., 80: 193.
1948 *Sigara (Vermicorixa) alternata*: Hungerford, Univ. Ks. Sci. Bull., 32: 653.
Distribution: Alta., Ariz., Ark., B.C., Col., Conn., D.C., Ia., Ill., Ind., Ks., Man., Mass., Md., Mich., Minn., Mo., N.B., N.C., N.D., N.H., N.J., N.M., N.S., N.T., N.Y., Oh., Ok., Ont., Ore., P.Ed., Pa., Que., R.I., S.D., Sask., Tex., Va., W.Va., Wash., Wis., Wyo
Note: Blatchley (1926, above, p. 1076) rejected the earlier Florida record. California records were transferred to *Sigara vallis* by Lauck (1966, Pan-Pac. Ent., 42: 171).

Sigara defecta Hungerford, 1948
1948 *Sigara* (*Vermicorixa*) *defecta* Hungerford, Univ. Ks. Sci. Bull., 32: 699. [Minn.].
Distribution: Ill., Md., Mich., Minn., N.C., N.Y., Ont., Pa., Va., Wis.

Sigara gordita (Abbott), 1913
1913 *Corixa gordita* Abbott, Bull. Brook. Ent. Soc., 8: 84. [Ga.].
1926 *Arctocorixa* [*sic*] *gordita*: Blatchley, Het. E. N. Am., p. 1079.
1948 *Sigara* (*Vermicorixa*) *gordita*: Hungerford, Univ. Ks. Sci. Bull., 32: 712.
Distribution: Ga., Mass.

Sigara grossolineata Hungerford, 1948
1917 *Arctocorisa scabra*: Parshley, Occas. Pap. Bost. Soc. Nat. Hist., 7: 119 (in part).
1948 *Sigara* (*Vermicorixa*) *grossolineata* Hungerford, Univ. Ks. Sci. Bull., 32: 676. [Minn.].
1984 *Sigara grossolineata*: Stonedahl, J. N.Y. Ent. Soc., 92: 42.
Distribution: Alta., B.C., Cal., Col., Ia., Id., Ill., Ks., Me., Man., Mass., Mich., Minn., Mont., N.B., N.H., N.M., N.Y., Neb., Oh., Ok., Ont., Ore., Pa., Que., S.D., Sask, Ut., Wash., Wis., Wyo.

Sigara hubbelli (Hungerford), 1928
1928 *Arctocorixa* [*sic*] *hubbelli* Hungerford, Can. Ent., 60: 228. [Tenn.].
1948 *Sigara* (*Vermicorixa*) *hubbelli*: Hungerford, Univ. Ks. Sci. Bull., 32: 696.
1951 *Sigara hubbelli*: Herring, Fla. Ent., 34: 26.
Distribution: Ala., Ark., D.C., Fla., Ga., Ia., Ill., Ind., Ks., Ky., La., Md., Minn., Miss., Mo., N.H., N.C., Ont., Pa., Tenn., Va., W.Va.

Sigara johnstoni Hungerford, 1948
1948 *Sigara* (*Vermicorixa*) *johnstoni* Hungerford, Univ. Ks. Sci. Bull., 32: 693. [Minn.].
Distribution: Ill., Minn., N.B., Ont., Que., Wis.

Sigara knighti Hungerford, 1948
1948 *Sigara* (*Vermicorixa*) *knighti* Hungerford, Univ. Ks. Sci. Bull., 32: 695. [Minn.].
Distribution: Mich., Minn., N.B., N.S., Ont., Que., Wis.

Sigara krafti Stonedahl, 1984, New Subgeneric Combination
1984 *Sigara krafti* Stonedahl, J. N.Y. Ent. Soc., 92: 43. [Ore.].
Distribution: Ore., Wash.
Note: Stonedahl (1984, above) did not place this species in a subgenus, although he did infer placement by comparing *S. krafti* to other members of *Vermicorixa*.

Sigara mathesoni Hungerford, 1948
1917 *Arctocorixa* [*sic*] *scabra*: Parshley, Occas. Pap. Bost. Soc. Nat. Hist., 7: 119 (in part).

1948 *Sigara* (*Vermicorixa*) *mathesoni* Hungerford, Univ. Ks. Sci. Bull., 32: 683. [N.S.].
1960 *Sigara mathesoni*: Lansbury, Proc. Ent. Soc. B.C., 57: 42.
Distribution: Alta., Conn., Man., Mich., Minn., N.J., N.Y., N.S., Ont., Pa., Sask., Wis.

Sigara mckinstryi Hungerford, 1948
1948 *Sigara* (*Vermicorixa*) *mckinstryi* Hungerford, Univ. Ks. Sci. Bull., 32: 681. [Cal.].
1953 *Sigara mckinstryi*: Kellen, J. Econ. Ent., 46: 913.
Distribution: Cal., Ore., Wash.

Sigara modesta (Abbott), 1916
1916 *Arctocorisa modesta* Abbott, Ent. News, 27: 343. [D.C.]. Lectotype designated by Hungerford, 1948, Univ. Ks. Sci. Bull, 32: 667.
1917 *Arctocorixa* [*sic*] (*Arctocorixa* [*sic*]) *modesta*: Van Duzee, Univ. Cal. Publ. Ent., 2: 482.
1948 *Sigara* (*Vermicorixa*) *modesta*: Hungerford, Univ. Ks. Sci. Bull., 32: 666.
Distribution: Ark., B.C., Col., Conn., D.C., Del., Ill., Ind., Ks., La., Mass., Md., Mich., Minn., Miss., Mo., N.H., N.J., N.Y., Neb., Oh., Ok., Ont., Pa., Tenn., Tex., Va., W.Va. (West Indies).

Sigara mullettensis (Hungerford), 1928
1928 *Arctocorixa* [*sic*] *mullettensis* Hungerford, Can. Ent., 60: 230. [Mich.].
1930 *Arctocorixa* [*sic*] *impersonata* Walley, Bull. Brook. Ent. Soc., 25: 204. [Que.]. Synonymized by Walley, 1936, Can. Ent., 68: 60.
1948 *Sigara* (*Vermicorixa*) *mullettensis*: Hungerford, Univ. Ks. Sci. Bull., 32: 691.
Distribution: B.C., Conn., D.C., Man., Me., Mich., Minn., N.B., N.H., N.Y., Ont., Que., R.I., Sask., Va., Wis.

Sigara nevadensis (Walley), 1936
1936 *Arctocorixa* [*sic*] *nevadensis* Walley, Can. Ent., 68: 58. [Nev.].
1948 *Cenocorixa sorensoni* Hungerford, Univ. Ks. Sci. Bull., 32: 565. [Ut.]. Synonymized by Jansson, 1972, Can. Ent., 104: 458.
1948 *Sigara* (*Vermicorixa*) *nevadensis*: Hungerford, Univ. Ks. Sci. Bull., 32: 704.
Distribution: Nev., Ut., Wyo.

Sigara omani (Hungerford), 1930
1930 *Arctocorixa* [*sic*] *omani* Hungerford, Pan-Pac. Ent., 7: 25. [Nev.].
1931 *Sigara omani*: Jaczewski, Arch. Hydrobiol, 23: 513.
1948 *Sigara* (*Vermicorixa*) *omani*: Hungerford, Univ. Ks. Sci. Bull., 32: 701.
Distribution: Ariz., B.C., Cal., Id., Nev., Ore., Wash., Wyo.

Sigara pectenata (Abbott), 1913
1913 *Arctocorisa pectenata* Abbott, Bull. Brook. Ent. Soc., 8: 83. [Ga.].
1915 *Corixa pectenata*: Torre-Bueno, Ent. News, 26: 277.
1916 *Arctocorixa* [*sic*] (*Arctocorixa* [*sic*]) *pectenata*: Van Duzee, Check List Hem., p. 54.
1926 *Arctocorixa* [*sic*] *pectenata*: Blatchley, Het. E. N. Am., p. 1080.
1938 *Arctocorixa* [*sic*] *pectinata* [*sic*]: Brimley, Ins. N.C., p. 84.
1948 *Sigara* (*Vermicorixa*) *pectenata*: Hungerford, Univ. Ks. Sci. Bull., 32: 705.
Distribution: Ala., D.C., Ga., Ind., Md., Miss., Mo., N.C., N.J., Ok., Tenn., Tex., Va.

Sigara scabra (Abbott), 1913
1913 *Arctocorisa scabra* Abbott, Bull. Brook. Ent. Soc., 8: 83.
1917 *Arctocorixa* [*sic*] (*Arctocorixa* [*sic*]) *scabra*: Van Duzee, Univ. Cal. Publ. Ent., 2: 483.
1938 *Arctocorixa* [*sic*] *scabra*: Brimley, Ins. N.C., p. 84.
1948 *Sigara* (*Vermicorixa*) *scabra*: Hungerford, Univ. Ks. Sci. Bull., 32: 709.
Distribution: Ala., Ga., Miss., N.C.

Sigara solensis (Hungerford), 1926
1926 *Arctocorixa* [*sic*] *solensis* Hungerford, Bull. Brook. Ent. Soc., 21: 198. [Mich.].
1948 *Sigara* (*Vermicorixa*) *solensis*: Hungerford, Univ. Ks. Sci. Bull., 32: 687.
Distribution: Alk., Alta., B.C., Conn., Man., Mich., Minn., N.B., N.D., N.H., N.Y., Ont., P.Ed., Que., S.D., Sask., Wis.

Sigara stigmatica (Fieber), 1851
1851 *Corisa* [*sic*] *stigmatica* Fieber, Act. Reg. Bohem. Soc. Sci., Pragae, 1: 36. [N. Am.]. Lectotype from "Nord. Amer." designated by Hungerford, 1948, Univ. Ks. Sci. Bull., 32: 666.
1909 *Arctocorisa stigmatica*: Kirkaldy and Torre-Bueno, Proc. Ent. Soc. Wash., 10: 197.
1916 *Arctocorixa* [*sic*] (*Arctocorixa* [*sic*]) *stigmatica*: Van Duzee, Check List Hem., p. 54.
1948 *Sigara* (*Vermicorixa*) *stigmatica*: Hungerford, Univ. Ks. Sci. Bull., 32: 665.
Distribution: Va.
Note: Although described from "Nord Amer.," the only definite state record for this species is Va. (Bobb, 1974, Res. Div. Bull., Va. Polytech. Inst. St. Univ., 87: 167).

Sigara transfigurata (Walley), 1930
1930 *Arctocorixa* [*sic*] *transfigurata* Walley, Can. Ent., 62: 282. [Que.].
1948 *Sigara* (*Vermicorixa*) *transfigurata*: Hungerford, Univ. Ks. Sci. Bull., 32: 689.
Distribution: Mass., Mich., Ont., Que., Wis.

Sigara vallis Lauck, 1966
1948 *Sigara* (*Vermicorixa*) *alternata*: Hungerford, Univ. Ks. Sci. Bull., 32: 665 (in part).
1966 *Sigara vallis* Lauck, Pan-Pac. Ent., 42: 168. [Cal.].
1979 *Sigara* (*Vermicorixa*) *vallis*: Lauck, Bull. Cal. Ins. Surv., 21: 121.
Distribution: Cal.

Sigara vandykei Hungerford, 1948
1948 *Sigara* (*Vermicorixa*) *vandykei* Hungerford, Univ. Ks. Sci. Bull., 32: 685. [Cal.].
Distribution: Cal., Wash.

Sigara virginiensis Hungerford, 1948
1948 *Sigara* (*Vermicorixa*) *virginiensis* Hungerford, Univ. Ks. Sci. Bull., 32: 671. [Va.].
Distribution: D.C., Ga., Md., N.C., N.H., N.J., Oh., Pa., S.C., Tenn., Tex., Va., W.Va.

Sigara washingtonensis Hungerford, 1948
1948 *Sigara* (*Vermicorixa*) *washingtonensis* Hungerford, Univ. Ks. Sci. Bull., 32: 673. [Wash.].
Distribution: Alta., Ariz., B.C., Cal., Col., Id., Man., Mont., Nev., Ore., Sask., Ut., Wash., Wyo.

Subgenus *Xenosigara* Hungerford, 1948

1948 *Sigara* (*Xenosigara*): Hungerford, Univ. Ks. Sci. Bull., 32: 631. Type-species *Arctocorisa ornata* Abbott, 1916. Monotypic.

Sigara ornata (Abbott), 1916
1916 *Arctocorisa ornata* Abbott, Ent. News, 27: 341. [N.Y.].
1917 *Arctocorixa* [*sic*] (*Arctocorixa* [*sic*]) *ornata*: Van Duzee, Univ. Cal. Publ. Ent., 2: 483.
1926 *Arctocorixa* [*sic*] *ornata*: Blatchley, Het. E. N. Am., p. 1071.
1948 *Sigara* (*Xenosigara*) *ornata*: Hungerford, Univ. Ks. Sci. Bull., 32: 631.
Distribution: Conn., Mass., Me., N.B., N.J., N.Y., Ont., Que., R.I.

Note: Stonedahl (1984, J. N.Y. Ent. Soc., 92: 42) deleted the Oregon and Washington records for *S. ornata*.

Genus *Trichocorixa* Kirkaldy, 1908

1908 *Trichocorixa* Kirkaldy, Can. Ent., 40: 117. Type-species: *Corisa pygmaea* Fieber, 1851, a junior synonym of *Corisa verticalis* Fieber, 1851. Monotypic.

Note: Sailer (1948, Univ. Ks. Sci. Bull., 32: 289-407) presented a revision of *Trichocorixa* with a key to species; in a 1977 reprint of the work a corrected version of the key was given (pp. 302-304a).

Trichocorixa arizonensis Sailer, 1948

1948 *Trichocorixa arizonensis* Sailer, Univ. Ks. Sci. Bull., 32: 305. [Ariz.].

Distribution: Ariz.

Trichocorixa borealis Sailer, 1948

1948 *Trichocorixa borealis* Sailer, Univ. Ks. Sci. Bull., 32: 308. [Man.].

Distribution: Alta., Col., Ia., Man., Minn., N.D., N.T., Oh., Ont., Que., S.D., Sask., Wis.

Trichocorixa calva (Say), 1832

1832 *Corixa calva* Say, Descrip. Het. Hem. N. Am., p. 38. [U.S.]. Neotype from Ill. designated by Sailer, 1948, Univ. Ks. Sci. Bull., 32: 312.

1851 *Corisa* [*sic*] *burmeisterii* Fieber, Act. Reg. Bohem. Soc. Sci., Pragae, 1: 24. [N. Am.]. Synonymized by Sailer, 1948, Univ. Ks. Sci. Bull., 32: 310.

1931 *Trichocorixa burmeisterii*: Lundblad, Zool. Anz., 96: 85.

1948 *Trichocorixa calva*: Sailer, Univ. Ks. Sci. Bull., 32: 310.

Distribution: Ala., Ariz., Ark., Cal., Col., D.C., Fla., Ga., Ia., Ill., Ind., Ks., La., Mich., Minn., Miss., Mo., N.C., N.J., N.Y., Neb., Oh., Ok., Ont., Pa., S.C., S.D., Tenn., Tex., Va., Wis.

Note: The distribution of this species as mapped by Sailer (1948, above) rules out the earlier listings for Alaska.

Trichocorixa kanza Sailer, 1948

1943 *Tricocorixa* [*sic*] *verticalis*: Rau, Ent. News, 54: 258 (in part).

1948 *Trichocorixa kanza* Sailer, Univ. Ks. Sci. Bull., 32: 318. [Ks.].

Distribution: Ala., Ark., D.C., Del., Fla., Ga., Ia., Ill., Ks., La., Md., Miss., Mo., Neb., Ok., Pa., S.C., Tenn., Tex., Wis. (Mexico).

Trichocorixa louisianae Jaczewski, 1931

1914 *Arctocorisa* (*Corixa*) *reticulata*: Barber, Bull. Am. Mus. Nat. Hist., 33: 497.

1926 *Corixa reticulata*: Blatchley, Het. E. N. Am., p. 1084.

1931 *Trichocorixa louisianae* Jaczewski, Arch. Hydrobiol., 23: 516. [La.].

1974 *Trichocorixa louisiania* [*sic*]: Slater, Mem. Conn. Ent. Soc., p. 158.

Distribution: Ala., Conn., Fla., Ga., La., Mass., Miss., N.C., N.H., N.Y., Tex., Va. (Mexico, West Indies).

Trichocorixa macroceps (Kirkaldy), 1908

1908 *Arctocorisa macroceps* Kirkaldy, Can. Ent., 40: 119. [N.C.]. Neotype from Ga. designated by Sailer, 1948, Univ. Ks. Sci. Bull., 32: 329.

1913 *Corixa macroceps*: Abbott, Bull. Brook. Ent. Soc., 8: 86.

1926 *Trichocorixa micronectoides*: Blatchley, Het. E. N. Am., p. 1085.

1931 *Trichocorixa macroceps*: Lundblad, Zool. Anz., 96: 92.

Distribution: Ala., Ga., Ill., Mich., Miss., N.C., N.H., N.J., N.Y., S.C., Tex., Va.

Note: Blatchley (1926, above) listed the manuscript name "*Trichocorixa micronectoides*

Plunkett" in the discussion of this species. Sailer (1948, Univ. Ks. Sci. Bull., 32: 327) treated this *nomen nudum* as a synonym.

Trichocorixa minima (Abbott), 1913

1913 *Corixa minima* Abbott, Bull. Brook. Ent. Soc., 8: 86. [Ga.].

1926 *Corixa pulchra* Blatchley, Het. E. N. Am., p. 1085. [Fla.]. Synonymized by Sailer, 1948, Univ. Ks. Sci. Bull., 32: 332.

1948 *Trichocorixa minima*: Sailer, Univ. Ks. Sci. Bull., 32: 332.

Distribution: Fla., Ga. (West Indies).

Note: The definition and distribution of this species accepted by Sailer (1948, above) precludes acceptance of Blatchley's (1926, above) record for Mass.

Trichocorixa reticulata (Guérin-Méneville), 1857 [Fig. 31]

1857 *Corisa reticulata* Guérin-Méneville, Hist. de Cuba, 7: 423. [Cuba].

1859 *Corixa wallengreni* Stål, K. Svens. Freg. Eug. Resa Omkr. Jorden, 3: 268. [Cal.]. Synonymized by Sailer, 1946, Proc. Haw. Ent. Soc., 12: 619.

1909 *Arctocorisa wallengreni*: Kirkaldy and Torre-Bueno, Proc. Ent. Soc. Wash., 10: 197.

1916 *Arctocorixa* [*sic*] (*Arctocorixa* [*sic*]) *reticulata*: Van Duzee, Check List Hem., p. 54.

1916 *Arctocorixa* [*sic*] (*Arctocorixa* [*sic*]) *wallengreni*: Van Duzee, Check List Hem., p. 54.

1929 *Trichocorixa wallengreni*: Lundblad, Ent. Tidskr., 50: 24.

1946 *Trichocorixa reticulata*: Sailer, Proc. Haw. Ent. Soc., 12: 617.

Distribution: Cal., Fla., Ks., La., N.M., Nev., Tex. (Mexico, West Indies, South America, Pacific Islands).

Note: The above selected synonymy is abstracted from Sailer (1948, Univ. Ks. Sci. Bull., 32: 343-344). Sailer therein also reassigns some published records of "*Trichocorixa reticulata*" to *T. louisianae* and *T. naias* (now a junior synonym of *T. sexcincta*). Polhemus and Hendrickson (1974, Pan-Pac. Ent., 50: 52) discussed the occurrence of this species in salt water in the Gulf of California.

Trichocorixa sexcincta (Champion), 1901

1901 *Corixa sexlineata* Champion, Biol. Centr.-Am., Rhyn., 2: 379. [Mexico]. Preoccupied.

1901 *Corixa sexcincta* Champion, Biol. Centr.-Am., Rhyn., 2: xvi. New name for *Corixa sexlineata*.

1909 *Arctocorisa naias*: Kirkaldy and Torre-Bueno, Proc. Ent. Soc. Wash., 10: 196. Unnecessary new name for *Corixa sexlineata* Champion.

1927 *Trichocorixa championi* Jaczewski, An. Zool. Mus. Pol. Hist. Nat., 6: 367. Unnecessary new name for *Corixa sexlineata* Champion.

1943 *Tricocorixa* [*sic*] *verticalis*: Rau, Ent. News, 54: 258 (in part).

1948 *Trichocorixa naias*: Sailer, Univ. Ks. Sci. Bull., 32: 335.

1978 *Trichocorixa sexcincta*: Furth et al., Syst. Ent., 3: 147.

1982 *Trichocorixa sexcinta* [*sic*]: Sanderson, Aquat. Ins. N. S. Carolina, p. 6.27.

Distribution: Alta., Col., Conn., D.C., Fla., Ga., Ia., Ill., Ind., Ks., La., Man., Mass., Mich., Minn., Miss., Mo., N.S., N.Y., Ont., Pa., Que., S.D., Sask., Tex., Va., Wis. (Mexico, West Indies).

Note: Hussey (1950, Ent. News, 61: 13) discussed the nomenclatural problems of this species.

Trichocorixa uhleri Sailer, 1948

1948 *Trichocorixa uhleri* Sailer, Univ. Ks. Sci. Bull., 32: 348. [Col.].

Distribution: Ariz., Col., N.M., Tex.

Trichocorixa verticalis (Fieber), 1851
1851 *Corisa* [*sic*] *verticalis* Fieber, Act. Reg. Bohem. Soc. Sci., Pragae, 1: 24.
1913 *Corixa verticalis*: Abbott, Bull. Brook. Ent. Soc., 8: 87.
1929 *Trichocorixa verticalis*: Lundblad, Arch. Hydrobiol, 20: 312.
1943 *Tricocorixa* [*sic*] *verticalis*: Rau, Ent. News, 54: 258 (in part).
Distribution: Ala., Ariz., Cal., Col., Conn., Fla., Ga., Id., Ks., La., Man., Mass., Me., Minn., Miss., N.C., N.D., N.H., N.J., N.M., N.Y., Pa., Que., R.I., S.D., Sask., Tex., Ut., Va. (Mexico, West Indies).

Trichocorixa verticalis californica Sailer, 1948
1948 *Trichocorixa verticalis californica* Sailer, Univ. Ks. Sci. Bull., 32: 352. [Cal.].
Distribution: Cal.

Trichocorixa verticalis fenestrata Walley, 1930
1930 *Trichocorixa fenestrata* Walley, Can. Ent., 62: 81. [Que.].
1948 *Trichocorixa verticalis fenestrata*: Sailer, Univ. Ks. Sci. Bull., 32: 354.
Distribution: Que.

Trichocorixa verticalis interiores Sailer, 1948
1948 *Trichocorixa verticalis interiores* Sailer, Univ. Ks. Sci. Bull., 32: 354. [Tex.].
Distribution: Alta., Col., Id., Ks., Man., Minn., N.D., N.M., Neb., S.D., Sask., Tex., Ut.

Trichocorixa verticalis saltoni Sailer, 1948
1948 *Trichocorixa verticalis saltoni* Sailer, Univ. Ks. Sci. Bull., 32: 357. [Cal.].
Distribution: Ariz., Cal.

Trichocorixa verticalis sellaris (Abbott), 1913
1913 *Corixa sellaris* Abbott, Bull. Brook. Ent. Soc., 8: 85. [Ga.].
1931 *Trichocorixa sellaris*: Lundblad, Zool. Anz., 96: 91.
1948 *Trichocorixa verticalis sellaris*: Sailer, Univ. Ks. Sci. Bull., 32: 163.
Distribution: Conn., Ga., N.C., N.H., N.J., N.Y., Mass., Me., R.I.

Trichocorixa verticalis verticalis: (Fieber), 1851
1851 *Corisa* [*sic*] *verticalis* Fieber, Act. Reg. Bohem. Soc. Sci., Pragae, 1: 24.
1948 *Trichocorixa verticalis verticalis*: Sailer, Univ. Ks. Sci. Bull., 32: 358. [Cal.].
Distribution: Ala., Fla., La., Miss., N.C., N.J., N.Y., Tex., Va. (Mexico, West Indies).

Tribe Glaenocorisini Hungerford, 1948

Genus *Dasycorixa* Hungerford, 1948

1948 *Dasycorixa* Hungerford, Univ. Ks. Sci. Bull., 32: 142. Type-species: *Glaenocorisa hybrida* Hungerford, 1926. Original designation.

Dasycorixa hybrida (Hungerford), 1926
1926 *Glaenocorixa* [*sic*] *hybrida* Hungerford, Can. Ent., 58: 271. [Minn.].
1948 *Dasycorixa hybrida*: Hungerford, Univ. Ks. Sci. Bull., 32: 144.
Distribution: Alta., B.C., Minn.

Dasycorixa johanseni (Walley), 1931
1931 *Arctocorixa* [*sic*] *johanseni* Walley, Can. Ent., 63: 238. [Man.].
1948 *Dasycorixa johanseni*: Hungerford, Univ. Ks. Sci. Bull., 32: 142.
Distribution: Alta., B.C., Man., N.T., Que., Sask.

Dasycorixa rawsoni Hungerford, 1948
1948 *Dasycorixa rawsoni* Hungerford, Univ. Ks. Sci. Bull., 32: 145. [Sask.].
Distribution: Alta., B.C., Sask., Man., N.T.

Genus *Glaenocorisa* Thomson, 1869

1869 *Corisa* [*sic*] (*Glaenocorisa*) Thomson, Opusc. Ent., 1: 39. Type-species: *Corisa cavifrons* Thomson, 1869, a subspecies of *Corisa propinqua* Fieber, 1861. Monotypic.
1906 *Glaenocorisa*: Kirkaldy, Trans. Am. Ent. Soc., 32: 152.

Glaenocorisa propinqua (Fieber), 1861
1861 *Corisa* [*sic*] *propinqua* Fieber, Europ. Hem., p. 99. [Austria].
Distribution: Alk. (Europe).
Note: Kirkaldy (1906, Trans. Am. Ent. Soc., 32: 152) created this combination when he raised *Glaenocorisa* to generic status. Until China (1943, Gen. Names Brit. Ins., 8: 283) pointed out the seniority of *propinqua* over *cavifrons*, the latter was treated as the nominate form. The nominate subspecies does not occur in the Western Hemisphere.

Glaenocorisa propinqua cavifrons (Thomson), 1869
1869 *Corisa* [*sic*] (*Glaenocorisa*) *cavifrons* Thomson, Opusc. Ent., 1: 39. [Sweden].
1969 *Glaenocorisa cavifrons*: Lindroth and Ball, Kodiak I. Ref., p. 137.
Distribution: Alk. (Europe).

Tribe Graptocorixini Hungerford, 1948

Genus *Graptocorixa* Hungerford, 1930

1930 *Graptocorixa* Hungerford, Pan-Pac. Ent., 7: 22. Type-species: *Corixa abdominalis* Say, 1832. Original designation.

Graptocorixa abdominalis (Say), 1832
1832 *Corixa abdominalis* Say, Descrip. Het. Hem. N. Am., p. 38. [Mexico].
1894 *Corisa* [*sic*] *abdominalis*: Uhler, Proc. Cal. Acad. Sci., ser. 2, 4: 294.
1909 *Arctocorisa abdominalis*: Kirkaldy and Torre-Bueno, Proc. Ent. Soc. Wash., 10: 194.
1916 *Arctocorixa* [*sic*] (*Arctocorixa* [*sic*]) *abdominalis*: Van Duzee, Check List Hem., p. 52.
1925 *Arctocorixa* [*sic*] *abdominalis*: Hungerford, Bull. Brook. Ent. Soc., 20: 18.
1930 *Graptocorixa abdominalis*: Hungerford, Pan-Pac. Ent., 7: 22.
Distribution: Ariz., Cal., N.M., Nev., Ok., Tex., Ut. (Mexico).
Note: Banks' (1910, Cat. Nearct. Hem.-Het., p. 5) Florida record has not been accepted by any subsequent authors.

Graptocorixa californica (Hungerford), 1925
1925 *Arctocorixa* [*sic*] *californica* Hungerford, Bull. Brook. Ent. Soc., 20: 18. [Cal.].
1930 *Graptocorixa californica*: Hungerford, Pan-Pac. Ent., 7: 22.
Distribution: Cal., Ore.

Graptocorixa gerhardi (Hungerford), 1925
1925 *Arctocorixa* [*sic*] *gerhardi* Hungerford, Bull. Brook. Ent. Soc., 20: 21. [N.M.].
1930 *Graptocorixa gerhardi*: Hungerford, Pan-Pac. Ent., 7: 25.
Distribution: Ariz., N.M., Tex. (Mexico).

Graptocorixa serrulata (Uhler), 1897
1897 *Corixa serrulata* Uhler, Trans. Md. Acad. Sci., 1: 391. [Ariz.].
1909 *Arctocorisa serrulata*: Kirkaldy and Torre-Bueno, Proc. Wash. Ent. Soc., 10: 197.
1916 *Arctocorixa* [*sic*] (*Arctocorixa* [*sic*]) *serrulata*: Van Duzee, Check List Hem., p. 54.

1930 *Graptocorixa serrulata*: Hungerford, Pan-Pac. Ent., 7: 22.
Distribution: Ariz., Nev., N.M., Ore., Tex. (Mexico).
Note: Lauck (1979, Bull. Cal. Ins. Surv., 21: 93) did not accept Kirkaldy and Torre-Bueno's (1909, Proc. Ent. Soc. Wash., 10: 197) California record.

Graptocorixa uhleri (Hungerford), 1925
1925 *Arctocorixa* [*sic*] *uhleri* Hungerford, Bull. Brook. Ent. Soc., 20: 19. [Cal.].
1930 *Graptocorixa uhleri*: Hungerford, Bull. Brook. Ent. Soc., 7: 22.
Distribution: Cal., Nev.

Graptocorixa uhleroidea Hungerford, 1938
1938 *Graptocorixa uhleroidea* Hungerford, J. Ks. Ent. Soc., 11: 135.
Distribution: Cal.

Genus *Neocorixa* Hungerford, 1925

1925 *Neocorixa* Hungerford, Bull. Brook. Ent. Soc., 20: 19. Type-species: *Neocorixa snowi* Hungerford, 1925. Monotypic.

Neocorixa snowi Hungerford, 1925
1925 *Neocorixa snowi* Hungerford, Bull. Brook. Ent. Soc., 20: 20. [Ariz.].
Distribution: Ariz., N.M. (Mexico).

Subfamily Cymatiinae Hungerford, 1948

Genus *Cymatia* Flor, 1860

1860 *Cymatia* Flor, Arch. Naturk. Livlands, 3: 783. Type-species: *Sigara coleoptrata* Fabricius, 1777. Designated by Kirkaldy, 1898, Ent., 31: 252.

Cymatia americana Hussey, 1920
1920 *Cymatia americana* Hussey, Bull. Brook. Ent. Soc., 15: 80. [Minn.].
Distribution: Alk., Alta., B.C., Man., Mich., Minn., N.T., Ont., Que., S.D., Sask., Wis., Yuk.

Family Cydnidae Billberg, 1820

Burrowing Bugs

By Richard C. Froeschner

The burrowing habits of nearly all the Cydnidae hide most of their activities from human eyes. Generalizations drawn from scattered observations suggest there is one generation per year; the eggs are always laid on or in the soil; and nymphs of all species except those in the genus *Sehirus* Amyot and Serville enter the soil to feed on the roots of plants--the degree of their host specificity, if any, is not yet known. Nymphs of *Sehirus* are more like typical Pentatomidae in their habits and climb certain wild or domestic plants to feed upon succulent aerial plant parts. All North American species of this family are active above ground for part of their adult life and then, with the possible exception of the brachypterous species *Tominotus caecus* (Van Duzee), will come freely to lights. Some species show a strong positive reaction to certain chemicals; for instance *Pangaeus bilineatus* (Say) came frequently to traps baited with gerinol.

The present enumeration lists 43 species in 13 genera (including two species and 1 genus for which the records are questionable) in America north of Mexico.

The burrowing bugs are generally considered to be of no economic importance in North America. But for reasons not yet

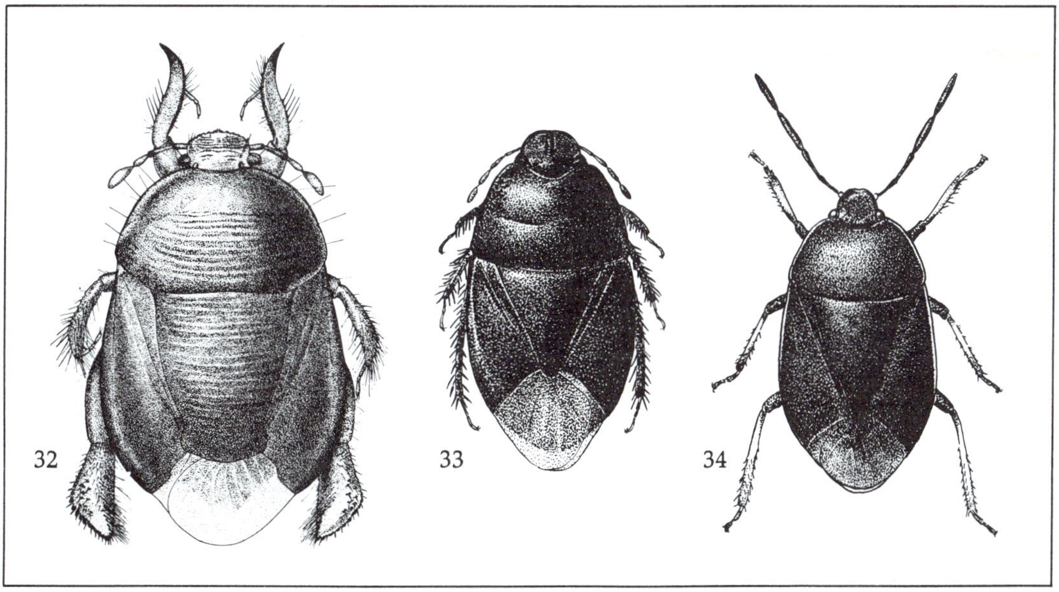

Figs. 32-34: 32, *Scaptocoris castaneus* [p. 128]; 33, *Pangaeus bilineatus* [p. 126]; 34, *Sehirus cinctus* [p. 129] (After Froeschner, 1941, except fig. 32, after Froeschner and Chapman, 1963).

know, in some years a species may become abundant and cause economic damage to a cultivated crop and then by the following year fall back to its regular ways and lesser numbers. *Pangaeus bilineatus* has been found thus damaging peanuts and spinach.

The soil-burrowing habit made these insects well suited for dispersal in the old sailing ships which often used soil ballast in their holds. To make space for legitimate cargoes, the soil ballast was unloaded by dumping on the countryside surrounding the port rather than into the water and blocking the channels. Frequent early records of distantly scattered specimens are probably explainable in this way; and North America's only burrowing bugs established from a distant land, *Aethus nigritus* (Fabricius) and *Scaptocoris castaneus* Perty, probably came in this manner (Hoebeke, 1978, U.S. Dept. Agr. Coop. Plant Pest Rept., 3: 376; Froeschner and Steiner, 1983, Ent. News, 94: 176).

Significant works, each covering all the known North American species of its time, were offered in Uhler's (1877) "Monograph"; Signoret's (1881-1884) "Revision"; Van Duzee's (1904) "Annotated List"; Torre-Bueno's (1939) "Synopsis"; and Froeschner's (1960) "monograph." Hoebeke and Wheeler (1984, Proc. Ent. Soc. Wash., 86: 741-743) presented modifications of the partial keys to North American genera as given by Slater and Baranowski (1878, How to Know True Bugs, pp. 33-37) and by McPherson (1982, Pentatomoidea Ne. N. Am., pp. 29-35) to accommodate the genus *Aethus* Dallas subsequently discovered in the United States. The suprageneric classification in Froeschner (1960) is followed here.

Subfamily Amnestinae Hart, 1919

Genus *Amnestus* Dallas, 1851

1851 *Amnestus* Dallas, List Hem. Brit. Mus., 1: 126. Type-species: *Cydnus spinifrons* Say, 1825. Original designation.

Amnestus basidentatus Froeschner, 1960

1960 *Amnestus basidentatus* Froeschner, Proc. U.S. Nat. Mus., 111: 634. [Tenn.].

Distribution: Ala., Ark., D.C., Fla., Ga., La., Miss., Mo., N.C., N.Y., Pa., S.C., Tenn., Tex., Va. (Cuba).

Amnestus pallidus Zimmer, 1910

1910 *Annectus* [*sic*] *pallidus* Zimmer, Can. Ent., 42: 166. [Neb.].

1916 *Amnestus pallidus*: Van Duzee, Check List Hem., p. 3.

Distribution: Ariz., Cal., Col., Ga., Ill., Ia., Ind., Ks., Ky., Md., Mass., Md., Mich., N.C., N.J., N.M., Neb., Ont., Ore., Pa., Que., Tenn., Tex., Va., Wash.

Amnestus pusillus Uhler, 1876

1876 *Amnestus pusillus* Uhler, Bull. U.S. Geol. Geogr. Surv. Terr., 1: 278. ["Indian Territory, Texas, Cuba, and generally the Eastern United States south of Cape Cod."].

Distribution: Ariz., Ark., Cal., Col., Ia., Ill., Ks., Ky., La., Mass., Md., Me., Mo., N.Y., Neb., Ont., Ore., Pa., Que., Tenn., Tex., Va., W.Va. (Guatemala, Mexico).

Amnestus pusio (Stål), 1860

1860 *Magoa pusio* Stål, K. Svens. Vet.-Akad. Handl., 2(7): 14. [Brazil].

1876 *Amnestus pusio*: Stål, K. Svens. Vet.-Akad. Handl., 14(4): 21.
Distribution: Tex. (Mexico to Brazil, West Indies).

Amnestus spinifrons (Say), 1825
1825 *Cydnus spinifrons* Say, J. Acad. Nat. Sci. Phila., 4: 316. [Mo., Pa.].
1851 *Amnestus spinifrons*: Dallas, List Hem. Brit. Mus., 1: 126.
Distribution: Ark., D.C., Fla., Ga., Ia., Ill., Ks., Mass., Md., Mich., Mo., N.C., N.J., N.Y., Oh., Ont., Pa., Que., Tenn., Tex.

Amnestus trimaculatus Froeschner, 1960
1960 *Amnestus trimaculatus* Froeschner, Proc. U.S. Nat. Mus., 111: 662. [Cuba].
1970 *Amnestus trimaculatus*: Froeschner and Baranowski, Fla. Ent., 53: 15.
Distribution: Fla. (Cuba).

Subfamily Cydninae Billberg, 1820

Genus *Aethus* Dallas, 1851

1851 *Aethus* Dallas, List Hem. Brit. Mus., 1: 112. Type-species: *Cimex indicus* Westwood, 1837. Designated by Van Duzee, 1914, Can. Ent., 46: 378.

Aethus nigritus (Fabricius), 1794
1794 *Cimex nigrita* Fabricius, Ent. Syst., 4: 123. [Germany].
1978 *Aethus nigritus*: Hoebeke, U.S. Dept. Agr. Coop. Plant Pest Rept., 3: 376.
Distribution: Conn., Del., N.J., N.Y., Pa. (Europe and North Africa east to Japan).
Note: Hoebeke and Wheeler (1984, Proc. Ent. Soc. Wash., 86: 738-744) depicted identification characteristics and summarized distribution and known natural history of this species.

Genus *Cydnus* Fabricius, 1803

1803 *Cydnus* Fabricius, Syst. Rhyn., p. 184. Type-species: *Cydnus tristis* Fabricius, 1844, a junior synonym of *Cimex aterrimus* Forster, 1771. Designated by Blanchard, Dict. Univ. Hist. Nat., 4: 505.

Cydnus aterrimus (Forster), 1771
1771 *Cimex aterrimus* Forster, Centuria, I: 71. ["Hispania ad fretum Gaditanum"].
1960 *Cydnus aterrimus*: Froeschner, Proc. U.S. Nat. Mus., 111: 408.
Distribution: Ala.(?) (Europe, North Africa, West Indies(?)).

Genus *Cyrtomenus* Amyot and Serville, 1843

1843 *Cyrtomenus* Amyot and Serville, Hist. Nat. Ins., Hem. p. 90. Type-species: *Cyrtomenus castaneus* Amyot and Serville, 1843, a junior synonym of *Pentatoma ciliata* Palisot, 1805. Designated by Kirkaldy, 1903, Ent., 36: 230.

Note: Signoret, (1881, An. Soc. Ent. France, ser. 6, 1: 25) listed *Cyrtomenus teter* Spinola for "Brazil, Costa Rica, San Francisco." Lethierry and Severin (1893, Gen. Cat. Hem., 1: 62) translated the last locality to "California," in which they were followed by Amyot and Serville, Hist. Nat. Ins., Hem., p. 90. Type-species: *Cyrtomenus castaneus* Amyot and Serville, 1843, a junior synonym of *Pentatoma ciliata* Palisot, 1805. Designated by Kirkaldy, 1903, Ent., 36: 230.

Subgenus *Cyrtomenus* Amyot and Serville, 1843

1843 *Cyrtomenus* Amyot and Serville, Hist. Nat. Ins., Hem., p. 90. Type-species: *Cyrtomenus castaneus* Amyot and Serville, 1843, a junior synonym of *Pentatoma ciliata* Palisot, 1805. Designated by Kirkaldy, 1903, Ent., 36: 230.

Cyrtomenus ciliatus (Palisot), 1805

1805 *Pentatoma ciliata* Palisot, Ins. Rec. Afr. Am., 11: 186. [U.S.].

1843 *Cyrtomenus castaneus* Amyot and Serville, Hist. Nat. Ins., Hem., p. 91. ["Amerique septentrionale"]. Synonymized by Froeschner, 1960, Proc. U.S. Nat. Mus., 111: 530.

1867 *Cyrtomenus mutabilis* [*sic*]: Walker, Cat. Hem. Brit. Mus., 1: 147.

1886 *Cyrtomenus mirabilis*: Uhler, Check-list Hem. Het., p. 3.

1960 *Cyrtomenus* (*Cyrtomenus*) *ciliatus*: Froeschner, Proc. U.S. Nat. Mus., 111: 530.

Distribution: Ala., Del., Fla, Ga., Ill., Ks., La., Miss., Mo., N.C., N.J., Ok., S.C., Tex., Va.

Note: The spelling "*mutabilis*" used by many authors was a misspelling on the caption for the plate accompanying the original description.

Cyrtomenus crassus Walker, 1867

1867 *Cyrtomenus crassus* Walker, Cat. Het. Brit. Mus., 1: 147. [Mexico].

1877 *Cyrtomenus obtusus* Uhler, Bull. U.S. Geol. Geogr. Surv. Terr., 3: 369. [Ariz., Tex.]. Synonymized by Froeschner, 1960, Proc. U.S. Nat. Mus., 111: 533.

1960 *Cyrtomenus* (*Cyrtomenus*) *crassus*: Froeschner, Proc. U. S. Nat. Mus., 111: 533.

Distribution: Ariz., N.M., Tex. (Mexico to Costa Rica, West Indies).

Subgenus *Syllobus* Signoret 1879

1879 *Syllobus* Signoret, Bull. Soc. Ent. France, ser. 5, 9 : clxxii. Type-species: *Cyrtomenus emarginatus* Stål, 1862. Monotypic.

1960 *Cyrtomenus* (*Syllobus*): Froeschner, Proc. U.S. Nat. Mus., 111: 517.

Cyrtomenus emarginatus Stål, 1862

1862 *Cyrtomenus emarginatus* Stål, Stett. Ent. Zeit., 23: 95. [Mexico].

1939 *Syllobus emarginatus*: Torre-Bueno, Bull. Brook. Ent. Soc., 34: 214.

1960 *Cyrtomenus* (*Syllobus*) *emarginatus*: Froeschner, Proc. U.S. Nat. Mus., 111: 518.

Distribution: Fla. (Mexico to Argentina).

Note: Froeschner (1960, above) suggested that the existence of a population in Florida recorded by Torre-Bueno (1939, above) needs verification.

Genus *Dallasiellus* Berg, 1901

1880 *Stenocoris* Signoret, Bull. Soc. Ent. France, ser. 5, 10: xliv. Preoccupied. Type-species: *Aethus longulus* Dallas, 1851. Monotypic.

1901 *Dallasiellus* Berg, Com. Mus. Nac. Buenos Aires, 1: 281. New name for *Stenocoris* Signoret.

Subgenus *Dallasiellus* Berg, 1901

1919 *Dallasiellus* Berg, Com. Mus. Nac. Buenos Aires, 1: 281. Type-species: *Aethus longulus* Dallas, 1851. Monotypic.

1960 *Dallasiellus* (*Dallasiellus*): Froeschner, Proc. U.S. Nat. Mus., 111: 573, 583.

Dallasiellus lugubris (Stål), 1860

1860 *Aethus lugubris* Stål, K. Svens. Vet.-Akad. Handl., 2(7): 13. [Brazil].

1960 *Dallasiellus* (*Dallasiellus*) *lugubris*: Froeschner, Proc. U.S. Nat. Mus., 111: 613.

Distribution: Ala., La., Miss., Tex. (Mexico to Argentina, West Indies).

Subgenus *Pseudopangaeus* Froeschner, 1960

1960 *Dallasiellus* (*Pseudopangaeus*) Froeschner, Proc. U.S. Nat. Mus., 111: 573. Type-species: *Pangaeus discrepans* Uhler, 1877. Original designation.

Dallasiellus californicus (Blatchley), 1929

1929 *Pangaeus californicus* Blatchley, Ent. News, 40: 74. [Cal.].

1960 *Dallasiellus* (*Pseudopangaeus*) *californicus*: Froeschner, Proc. U.S. Nat. Mus., 111: 575.

Distribution: Ariz., Cal., N.M. (Mexico).

Dallasiellus discrepans (Uhler), 1877

1877 *Pangaeus discrepans* Uhler, Bull. U.S. Geol. Geogr. Surv. Terr., 3: 386. ["Fort Cobb, Indian Territory"; Cal.].

1960 *Dallasiellus* (*Pseudopangaeus*) *discrepans*: Froeschner, Proc. U.S. Nat. Mus., 111: 577.

Distribution: Ariz., Cal., Col., Id., N.M., Ok., Ore., Pa.(?), Tex., Ut., Wash.

Dallasiellus puncticoria Froeschner, 1960

1960 *Dallasiellus* (*Pseudopangaeus*) *puncticoria* Froeschner, Proc. U.S. Nat. Mus., 111: 580. [Cal.].

Distribution: Cal. (Mexico).

Dallasiellus vanduzeei Froeschner, 1960

1960 *Dallasiellus* (*Pseudopangaeus*) *vanduzeei* Froeschner, Proc. U.S. Nat. Mus., 111: 575, 582. [Tex.].

Distribution: Cal., Tex.

Genus *Macroporus* Uhler, 1876

1876 *Macroporus* Uhler, Bull. U.S. Geol. Geogr. Surv. Terr., 1: 278. Type-species: *Macroporus repetitus* Uhler, 1877. Monotypic.

Macroporus repetitus Uhler, 1876

1876 *Macroporus repetitus* Uhler, Bull. U.S. Geol. Geogr. Surv. Terr., 1: 278. [Cal., Md.-latter in error].

Distribution: Cal.

Genus *Melanaethus* Uhler, 1876

1876 *Melanaethus* Uhler, Bull. U.S. Geol. Geogr. Surv. Terr., 1: 280. Type-species: *Melanaethus elongatus* Uhler, 1876, a junior synonym of *Aethus subglaber* Walker, 1867. Montypic.

1877 *Lobonotus* Uhler, Bull. U.S. Geol. Geogr. Surv., 3: 395. Preoccupied. Type-species: *Lobonotus anthracinus* Uhler, 1877. Monotypic. Synonymized by Froeschner, 1960, Proc. U.S. Nat. Mus., 111: 421.

1891 *Lobolophus* Bergroth, Rev. d'Ent., 10: 235. Unnecessary new name for *Lobonotus* Uhler.

Melanaethus anthracinus (Uhler), 1877

1877 *Lobonotus anthracinus* Uhler, Bull. U.S. Geol. Geogr. Surv. Terr., 3: 395. [Tex.].
1891 *Lobolophus anthracinus*: Bergroth, Rev. d'Ent., 10: 235.
1960 *Melanaethus anthracinus*: Froeschner, Proc. U.S. Nat. Mus., 111: 426.
Distribution: N.M., Tex. (Mexico).

Melanaethus cavicollis (Blatchley), 1924

1924 *Geotomus cavicollis* Blatchley, Ent. News, 35: 85. [Fla.].
1925 *Geocnethus cavicollis*: Hussey, J. N.Y. Ent. Soc., 33: 63.
1960 *Melanaethus cavicollis*: Froeschner, Proc. U.S. Nat. Mus., 111: 428.
Distribution: Ark., Fla., S.C.

Melanaethus crenatus (Signoret), 1883

1883 *Geotomus* (*Melanaethus*) *crenatus* Signoret, An. Soc. Ent. France, ser. 6, 3: 208. [Mexico].
1960 *Melanaethus crenatus*: Froeschner, Proc. U.S. Nat. Mus., 111: 430.
Distribution: Ariz., Tex. (Mexico).

Melanaethus noctivagus (Van Duzee), 1923

1923 *Geotomus noctivagus* Van Duzee, Proc. Cal. Acad. Sci., ser. 4, 12: 125. [Mexico].
1960 *Melanaethus noctivagus*: Froeschner, Proc. U.S. Nat. Mus., 111: 436.
Distribution: Ariz., Cal., Id., Tex. (Mexico).

Melanaethus pensylvanicus (Signoret), 1883

1876 *Cydnus* (*Melanaethus*) *picinus* Uhler, Bull. U.S. Geol. Geogr. Surv. Terr., 1: pl. 19, fig. 17 (figured and captioned, but not mentioned in text). Preoccupied.
1877 *Melanaethus picinus* Uhler, Bull. U.S. Geol. Geogr. Surv. Terr., 3: 391 (as new species). [Pa.]. Preoccupied.
1883 *Geotomus pensylvanicus* Signoret, An. Soc. Ent. France, ser. 6, 3: 207. New name for *Cydnus picinus* Uhler.
1960 *Melanaethus pensylvanicus*: Froeschner, Proc. U.S. Nat. Mus., 111: 441.
Distribution: Ala., Ark., Fla., Ga., Ill., Ks., La., Md., Miss., Mo., N.C., Neb., Ok., Pa., Tenn., Va.
Note: Froeschner (1960, above) explained the original one "n" French spelling of *M. pensylvanicus*.

Melanaethus planifrons Froeschner, 1960

1960 *Melanaethus planifrons* Froeschner, Proc. U.S. Nat. Mus., 111: 424, 443. [Cal.].
Distribution: Ariz., Cal. (Mexico).

Melanaethus punctatissimus (Signoret), 1883

1883 *Geotomus punctatissimus* Signoret, An. Soc. Ent. France, ser. 6, 3: 216. ["Sitka"].
1960 *Melanaethus punctatissimus*: Froeschner, Proc. U.S. Nat. Mus., 111: 445.
Distribution: Alk.

Melanaethus robustus Uhler, 1877

1877 *Melanaethus robustus* Uhler, Bull. U.S. Geol. Geogr. Surv. Terr., 3: 390. [Mass., Md.].
1883 *Geotomus* (*Melanaethus*) *robustus*: Signoret, An. Soc. Ent. France, ser. 6, 3: 59.

1893 *Geotomus robustus*: Lethierry and Severin, Gen. Cat. Hem., 1: 73.
Distribution: D.C., Fla., Ia., Ill., Ind., Ks., Mass., Md., Miss., Mo., N.J., Oh., Pa., Tex., Va.

Melanaethus subglaber (Walker), 1867
1867 *Aethus subglaber* Walker, Cat. Hem. Brit. Mus., 1: 150. ["North America"].
1876 *Melanaethus elongatus* Uhler, Bull. U.S. Geol. Geogr. Surv. Terr., 1: 280. [Cal.]. Preoccupied. Synonymized by Froeschner, 1960, Proc. U.S. Nat. Mus., 111: 438.
1883 *Geotomus parvulus* Signoret, An. Soc. Ent. France, ser. 6, 3: 208. Unecessary new name for *Melanaethus elongatus* Uhler.
1960 *Melanaethus subglaber*: Froeschner, Proc. U.S. Nat. Mus., 111: 438.
Distribution: Ariz., Cal., N.M., Nev., Tex., Ut. (Mexico, Galapagos Islands).

Melanaethus subpunctatus (Blatchley), 1926
1926 *Geotomus subpunctatus* Blatchley, Het. E. N. Am., p. 78. [Fla., Md.].
1960 *Melanaethus subpunctatus*: Froeschner, Proc. U.S. Nat. Mus., 111: 451.
Distribution: Ala., Ark., Fla., Ga., La., Md., N.C., Tex., Va.

Melanaethus uhleri (Signoret), 1883
1883 *Geotomus* (*Melanaethus*) *uhleri* Signoret, An. Soc. Ent. France, ser. 6, 3: 211. ["Amerique du Nord" - Ga. on type].
1886 *Melanaethus uhleri*: Uhler, Check-list Hem. Het., p. 3.
1893 *Geotomus uhleri*: Lethierry and Severin, Gen. Cat. Hem., 1: 74.
Distribution: Ala., Ark., Ga., Ks., Ok., Tenn., Tex.

Genus *Microporus* Uhler, 1872

1872 *Microporus* Uhler, Prelim. Rept. U. S. Geol. Surv. Mont., 5: 394. Type-species: *Microporus obliquus* Uhler, 1872. Monotypic.

Microporus obliquus Uhler, 1872
1872 *Microporus obliquus* Uhler, Prel. Rept. U.S. Geol. Surv. Mont., 5: 394. [Ut.].
1882 *Cydnus obliquus*: Signoret, An. Soc. Ent. France, ser. 6, 2: 161.
1916 *Aethus obliquus*: Van Duzee, Check List Hem., p. 3.
1939 *Aethus* (*Microporus*) *obliquus*: Torre-Bueno, Ent. Am., 19: 179.
Distribution: Ariz., Cal., Col., Ia., Id., Ill., Ind., Ks., La., N.M., Nev., Ok., Ore., S.C., S.D., Tex., Ut., Va., Wash. (Mexico).

Microporus testudinatus Uhler, 1876
1876 *Microporus testudinatus* Uhler, Bull. U.S. Geol. Geogr. Surv. Terr., 1: 276. [Cal.].
1881 *Aethus* (*Microporus*) *testudinatus*: Signoret, An. Soc. Ent. France, ser. 6, 1: 424.
1886 *Aethus testudinatus*: Uhler, Check-list Hem. Het., p. 3.
1893 *Cydnus testudinatus*: Lethierry and Severin, Gen. Cat. Hem., 1: 68.
Distribution: Cal. (Mexico).

Genus *Pangaeus* Stål, 1862

1862 *Pangaeus* Stål, Stett. Ent. Zeit., 23: 95. Type-species: *Aethus margo* Dallas, 1851, a junior synonym of *Cimex aethiops* Fabricius, 1787. Designated by Van Duzee, 1914, Can. Ent., 46: 378.

Subgenus *Homaloporus* Uhler, 1877

1877 *Homaloporus* Uhler, Bull. U.S. Geol. Geogr. Surv. Terr., 3: 376. Type-species: *Homaloporus congruus* Uhler, 1877. Monotypic.

1960 *Pangaeus* (*Homaloporus*): Froeschner, Proc. U.S. Nat. Mus., 111: 455.

Pangaeus bilineatus (Say), 1825 [Fig. 33]

1825 *Cydnus bilineatus* Say, J. Acad. Nat. Sci. Phila., 4: 315. [United States, Mo., Pa.].

1832 *Cydnus bilineatus* var. *picea* Say, Descrip. Het. Hem. N. Am., p. 10. [Ind.]. Synonymized by Froeschner, 1960, Proc. U.S. Nat. Mus., 111: 464.

1840 *Cydnus rugifrons* Herrich-Schaeffer, Wanz. Ins., 5: 97. [Ga.]. Synonymized by Froeschner, 1960, Proc. U.S. Nat. Mus., 111: 459.

1840 *Cydnus femoralis* Herrich-Schaeffer, Wanz. Ins., 5: 98. [Pa.]. Synonymized by Signoret, 1882, An. Soc. Ent. France, ser. 6, 2: 252.

1851 *Aethus bilineatus*: Dallas, List Hem. Brit. Mus., 1: 119.

1867 *Aethus femoralis*: Walker, Cat. Hem. Brit. Mus., 1: 150.

1867 *Aethus rugifrons*: Walker, Cat. Hem. Brit. Mus., 1: 150.

1869 *Sehirus bilineatus*: Rathvon, Hist. Lancaster, Pa., p. 548.

1876 *Pangaeus bilineatus*: Stål, K. Svens. Vet.-Akad. Handl., 14(4): 19.

1876 *Pangaeus femoralis*: Stål, K. Svens. Vet.-Akad. Handl., 14(4): 19.

1877 *Pangaeus rugifrons*: Uhler, Bull. U.S. Geol. Geogr. Surv. Terr., 3: 384.

1877 *Pangaeus piceatus*: Uhler, Bull. U.S. Geol. Geogr. Surv. Terr., 3: 388.

1880 *Pangaeus rufifrons* [*sic*]: Distant, Biol. Centr.-Am., Rhyn., 1: 7.

1882 *Pangoeus* [*sic*] *uhleri* Signoret, An. Soc. Ent. France, ser. 6, 2: 253. ["Caroline," Ga.]. Synonymized by Froeschner, 1960, Proc. U.S. Nat. Mus., 111: 459.

1882 *Pangoeus* [*sic*] *bilineatus*: Signoret, An. Soc. Ent. France, ser. 6, 2: 254.

1882 *Pangoeus* [*sic*] *spangbergi* Signoret, An. Soc. Ent. France, ser. , 2: 259. [Tex.]. Synonymized by Froeschner, 1960, Proc. U.S. Nat. Mus., 111: 460.

1886 *Pangaeus spangbergi*: Uhler, Check-list Hem. Het., p. 3.

1886 *Pangaeus uhleri*: Uhler, Check-list Hem. Het., p. 3.

1893 *Pangaeus margo*: Lethierry and Severin, Gen. Cat. Hem., 1: 70.

1904 *Pangaeus spahnbergi* [*sic*]: Van Duzee, Trans. Am. Ent. Soc., 30: 25.

1907 *Pangaelus* [*sic*] *uhleri*: Torre-Bueno and Brimley, Ent. News, 18: 441.

1926 *Pangaeus bilineatus* var. *picea*: Blatchley, Het. E. N. Am., p. 75.

1960 *Pangaeus* (*Homaloporus*) *bilineatus*: Froeschner, Proc. U.S. Nat. Mus., 111: 459.

Distribution: Ala., Ark., Ariz., Cal., Fla., Ga., Ill., Ind., Ks., La., Mass., Md., Miss., Mo., N.C., N.J., Neb., Ok., Pa., Que., S.C., S.D., Tenn., Tex., Va., W.Va. (Mexico, Guatemala, Bermuda).

Pangaeus congruus (Uhler), 1877

1877 *Homaloporus congruus* Uhler, Bull. U.S. Geol. Geogr. Surv. Terr., 3: 377. [Col., Tex.]

1881 *Homaloporus pangaeiformis* Signoret, An. Soc. Ent. France, ser. 6, 1: 331. [Mexico]. Synonymized by Froeschner, 1960, Proc. U.S. Nat. Mus., 111: 467.

1906 *Homaloporus pangaeiformis*: Snow, Trans. Ks. Acad. Sci., 20: 151.

1960 *Pangaeus* (*Homaloporus*) *congruus*: Froeschner, Proc. U.S. Nat. Mus., 111: 467.

Distribution: Ariz., Cal., Col., Ks., N.M., Tex., Ut. (Mexico).

Pangaeus punctilineus Froeschner, 1960

1960 *Pangaeus (Homaloporus) punctilinea* [*sic*] Froeschner, Proc. U.S. Nat. Mus., 111: 470. [Tex.].

Distribution: Tex.

Pangaeus setosus Froeschner, 1960

1960 *Pangaeus (Homaloporus) setosus* Froeschner, Proc. U.S. Nat. Mus., 111: 473. [Ariz.].

Distribution: Ariz., Tex. (Mexico).

Pangaeus tuberculipes Froeschner, 1960

1960 *Pangaeus (Homaloporus) tuberculipes* Froeschner, Proc. U.S. Nat. Mus., 111: 475. [Mexico].

Distribution: Tex. (Mexico).

Genus *Rhytidoporus* Uhler, 1877

1877 *Rhytidoporus* Uhler, Bull. U.S. Geol. Geogr. Surv. Terr., 3: 380. Type-species: *Rhytidoporus indentatus* Uhler, 1877. Monotypic.

Subgenus *Bergthora* Kirkaldy, 1904

1877 *Cryptoporus* Uhler, Bull. U.S. Geol. Geogr. Surv. Terr., 3: 381. Preoccupied. Type-species: *Cryptoporus compactus* Uhler, 1877. Monotypic.

1904 *Bergthora* Kirkaldy, Ent., 37: 280. New name for *Cryptoporus* Uhler.

Rhytidoporus compactus (Uhler), 1877

1877 *Cryptoporus compactus* Uhler, Bull. U.S. Geol. Geogr. Surv. Terr., 3: 382. [Tex.].

1882 *Aethus (Cryptoporus) compactus*: Signoret, An. Soc. Ent. France, ser. 6, 2: 41.

1886 *Aethus compactus*: Uhler, Check-list Hem. Het., p. 3.

1893 *Cydnus compactus*: Lethierry and Severin, Gen. Cat. Hem., 1: 65.

1904 *Bergthora compactus*: Kirkaldy, Ent., 37: 280.

1960 *Rhytidoporus (Bergthora) compactus*: Froeschner, Proc. U.S. Nat. Mus., 111: 390.

Distribution: Ariz., Cal., N.M., Tex. (Mexico).

Subgenus *Rhytidoporus* Uhler, 1877

1877 *Rhytidoporus* Uhler, Bull. U.S. Geol. Geogr. Surv. Terr., 3: 380. Type-species: *Rhytidoporus identatus* Uhler, 1877. Monotypic.

Rhytidoporus indentatus Uhler, 1877

1877 *Rhytidoporus indentatus* Uhler, Bull. U.S. Geol. Geogr. Surv. Terr., 3: 380. [Cuba, Fla.].

1886 *Aethus indentatus*: Uhler, Check-list Hem. Het., p.3

1910 *Cydnus indentatus*: Banks, Cat. Nearc. Hem.-Het., p. 99.

1960 *Rhytidoporus (Rhytidoporus) indentatus*: Froeschner, Proc. U.S. Nat. Mus., 111: 387.

Distribution: Fla. (West Indies).

Genus *Tominotus* Mulsant and Rey, 1866

1866 *Tominotus* Mulsant and Rey, An. Soc. Linn. Lyon, new ser., 13: 319. Type-species: *Cydnus* (*Tominotus*) *signoreti* Mulsant and Rey, 1866. Monotypic.
1876 *Trichocoris* Uhler, Bull. U.S. Geol. Geogr. Surv. Terr., 1: 277. Type-species: *Trichocoris conformis* Uhler, 1876. Monotypic. Synonymized by Froeschner, 1960, Proc. U.S. Nat. Mus., 111: 539.
1922 *Psectrocephalus* Van Duzee, Ent. News, 33: 270. Type-species: *Psectrocephalus caecus* Van Duzee, 1922. Monotypic. Synonymized by Froeschner, 1960, Proc. U.S. Nat. Mus., 111: 539.

Tominotus caecus (Van Duzee), 1922
1922 *Psectrocephalus caecus* Van Duzee, Ent. News, 33: 271. [Cal.].
1960 *Tominotus caecus*: Froeschner, Proc. U.S. Nat. Mus., 111: 549.
Distribution: Cal. (Mexico).

Tominotus communis (Uhler), 1877
1877 *Aethus communis* Uhler, Bull. U.S. Geol. Geogr. Surv. Terr., 3: 379. [Cuba, Fla.].
1882 *Aethus politus* Signoret, An. Soc. Ent. France, ser. 6, 2: 36. [Cal.]. Lectotype designated and synonymized by Froeschner, 1960., Proc. U.S. Nat. Mus., 111: 553.
1893 *Cydnus communis*: Lethierry and Severin, Gen Cat. Hem., 1: 65.
1893 *Cydnus politus*: Lethierry and Severin, Gen. Cat. Hem., 1: 67.
1960 *Tominotus communis*: Froeschner, Proc. U.S. Nat. Mus., 111: 551.
Distribution: Cal., Fla., Ga., Ill., Miss., N.C., S.C., Tenn., Tex. (Nicaragua, West Indies).

Tominotus conformis (Uhler), 1876
1876 *Trichocoris conformis* Uhler, Bull. U.S. Geol. Geogr. Surv. Terr., 1: 277. [Cal.].
1881 *Aethus conformis*: Signoret, An. Soc. Ent. France, ser. 6, 1: 425.
1893 *Cydnus conformis*: Lethierry and Severin, Gen. Cat. Hem., 1: 65.
1939 *Aethus* (*Trichocoris*) *conformis*: Torre-Bueno, Ent. Am., new ser., 19: 178.
Distribution: Ariz., Cal., Ut. (Mexico).

Tominotus unisetosus Froeschner, 1960
1960 *Tominotus unisetosus* Froeschner, Proc. U.S. Nat. Mus., 111: 543, 567. [Tex.].
Distribution: Tex. (Mexico to Costa Rica).

Subfamily Scaptocorinae Froeschner, 1960

Genus *Scaptocoris* Perty, 1830

1833 *Scaptocoris* Perty, Delect. Anim. Articul., p. 165. Type-species: *Scaptocoris castanea* Perty, 1833. Monotypic.
Note: Generic revision provided by Becker (1967, Arq. Zool., Sao Paulo, 15: 291-325).

Scaptocoris castaneus Perty, 1833 [Fig. 32]
1833 *Scaptocoris castanea* Perty, Delect. Anim. Articul., p. 166 [Brazil].
1963 *Scaptocoris castaneus*: Froeschner and Chapman, Ent. News, 74: 95.
Distribution: Ga., S.C. (South America).

Note: A South American species established near Charleston, S.C. and Little Cumberland Island, Ga. (Froeschner and Steiner, 1983, Ent. News, 94: 176).

Subfamily Sehirinae Amyot and Serville, 1843

Genus *Sehirus* Amyot and Serville, 1843

1843 *Sehirus* Amyot and Serville, Hist. Nat. Ins., Hemip., p. 96.
Type-species: *Cimex morio* Linnaeus, 1761. Designated by Reuter, 1888, Acta Soc. Sci. Fenn., 15: 768.

1866 *Canthophorus* Mulsant and Rey, Punaise France, Pentat., 2: 54. Type-species: *Cimex-dubius* Scopoli, 1763. Designated by Reuter, 1888, Acta Soc. Sci. Fenn., 15: 758. Synonymized by China, 1943, Gen. Names Brit. Ins., 8: 221.

Sehirus cinctus (Palisot), 1811 [Fig. 34]

1811 *Pentatoma cincta* Palisot, Ins. Rec. Afr. Am., p. 114. ["Africa" in error].
1832 *Cydnus lygatus* [*sic*] Descrip. Het. Hem. N. Am., p. 10. ["United States"]. Synonymized by Stål, 1876, K. Svens. Vet.-Akad. Handl., 14(4): 22.
1876 *Canthophorus cinctus*: Stål, K. Svens. Vet.-Akad. Handl., 14(4): 22.
1904 *Sehirus cinctus*: Van Duzee, Trans. Am. Ent. Soc., 30: 26.
1907 *Schirus* [*sic*] *cinctus*: Torre-Bueno and Brimley, Ent. News, 18: 441.
Distribution: Ala., Alta., Cal., Col., D.C., Fla., Ga., Ia., Ill., Ks., Ky., La., Man., Mass., Md., Me., Mich., Minn., Miss., Mo., Mont., N.C., N.D., N.H., N.M., N.Y., Neb., Nfld., Oh., Ok., Ont., Pa., Que., S.C., Tenn., Tex., Va., Wyo. (Mexico).

Sehirus cinctus albonotatus Dallas, 1851

1851 *Sehirus albonotatus* Dallas, List Hem. Brit. Mus., 1: 127. [Fla.; "N. America"]. Lectotype designated by Froeschner, 1960, Proc. U.S. Nat. Mus., 111: 360.
1960 *Sehirus cinctus albonotatus*: Froeschner, Proc. U.S. Nat. Mus., 111: 359.
Distribution: Alta., Cal., Col., Ia., Ill., Man., Mass., Me., Mich., Minn., Mont., N.D., N.H., N.Y., Nfld., Ont., Pa., Que., Vt., Wis., Wyo.
Note: Froeschner (1960, Proc. U.S. Nat. Mus., 111: 360) found Dallas' Fla. specimen to be *C. cinctus* and deleted that record for this subspecies.

Sehirus cinctus cinctus (Palisot), 1811

1811 *Pentatoma cincta* Palisot, Ins. Rec. Afr. Am., p. 114.
1832 *Cydnus lygatus* [*sic*] Say, Het. Hem. N. Am., p. 10. [U.S.]. Synonymized by Stål, 1865, Hem. Afr., p. 29.
1869 *Sehirus ligatus*: Rathvon, Hist. Lancaster Co., Pa., p. 548.
1960 *Sehirus cinctus cinctus*: Froeschner, Proc. U.S. Nat. Mus., 111: 359, 361.
Distribution: Ala., D.C., Fla., Ga., Ia., Ill., Ks., Ky., La., Mass., Md., Mich., Miss., Mo., N.C., N.M., N.Y., Neb., Oh., Ok., Pa., S.C., Tenn., Tex., Va., Wis. (Mexico).

Sehirus cinctus texensis Froeschner, 1960

1960 *Sehirus cinctus texensis* Froeschner, Proc. U.S. Nat. Mus., 111: 359, 363. [Tex.].
Distribution: Tex.

Family Dipsocoridae Dohrn, 1859

(= Cryptostemmatidae McAtee and Malloch, 1925)

The Dipsocorids

By Thomas J. Henry

This is a small family represented in the United States by only one genus and two species, both of which are less than two millimeters long. Distinctions of this family include 2-segmented tarsi, long, slender antennae with bristlelike setae, and a cuneuslike structure formed by a deep fracture on the apical third of each hemelytron.

Species of the family are cryptic in their habits, being found under stones and other objects, especially along streams. Little is known of their habits, but probably they prey on co-existing arthropods.

The primary literature on the family includes Reuter's (1891, Acta Soc. Sci. Fenn., 19(6): 1-27) "Monographia Ceratocombidarum"; McAtee and Malloch's (1925, Proc. U.S. Nat. Mus., 67: 1-42) "Revision of the bugs of the family Cryptostemmatidae"; Emsley's (1969, Mem. Am. Ent. Soc., 25: 1-154) monograph of the Schizopteridae; and Štys' (1970, Acta Ent. Bohem, 67: 21-46) treatise "On the morphology and classification of the family Dipsocoridae...."

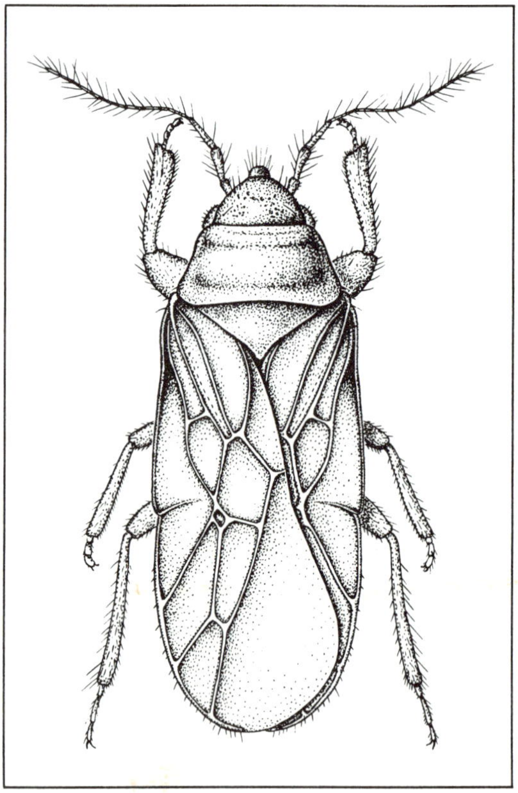

Fig. 35 *Cryptostemma uhleri* [p. 131] (Original).

Subfamily Dipsocorinae Dohrn, 1859

Tribe Dipsocorini Dohrn, 1859

Genus *Cryptostemma* Herrich-Schaeffer, 1835

1835 *Cryptostemma* Herrich-Schaeffer, Faunae Ins. German. Deut. Ins., p.11. Type-species: *Cryptostemma alienum* Herrich-Schaeffer, 1835. Monotypic.

Cryptostemma uhleri McAtee and Malloch, 1925 [Fig. 35]

1925 *Cryptostemma uhleri* McAtee and Malloch, Proc. U.S. Nat. Mus., 67: 10. [Mexico].

1945 *Cryptostemma uhleri:* Usinger, Ent. News, 56: 239.

Distribution: Ga. (Mexico, St. Vincent Island).

Note: Usinger (1945, above) noted that this species was common beneath stones along a stream in Ga.

Cryptostemma usingeri Wygodzinsky, 1955

1955 *Cryptostemma usingeri* Wygodzinsky, Pan-Pac. Ent., 31: 199. [Cal.]

Distribution: Cal.

Family Enicocephalidae Stål, 1860

(= Henicocephalidae Stål, 1865)

The Unique-Headed Bugs

By Richard C. Froeschner

The common name "unique-headed bugs" was coined as descriptive of the peculiar lobed appearance of the elongate head caused by a marked, transverse constriction: the eyes being situated laterally on the anterior lobe and the ocelli dorsally on the posterior lobe.

The species are generally relatively small, frail, winged insects that are sometimes carried to great heights by storm winds--some having been collected 5,000 feet in the air by traps on airplane wings. Their somber colors render them inconspicuous among particles of soil and fallen plant parts where many live; some species occur on foliage or wedged between the parts of the flowers. Kritsky (1977, Ent. News, 88: 105-106) described how these insects use the comblike structures on their tibiae as body-cleansing tools that pass the accumulated debris from the front legs to the middle legs to the hind legs, the latter cleaning each other. Most species fly and some perform a prenuptial flight-dance in patches of sunlight. Predaceous in food habits, they are generally considered to feed indiscriminately on any tiny invertebrates; one African species was found in the nests of an ant on which it feeds (Bergroth, 1915, Wien. Ent. Zeit., 34: 291-292).

Ten species are cataloged herein for America north of Mexico. A detailed revision for the family is in preparation, but for the present Kritsky's (1977, Ent. News, 88:

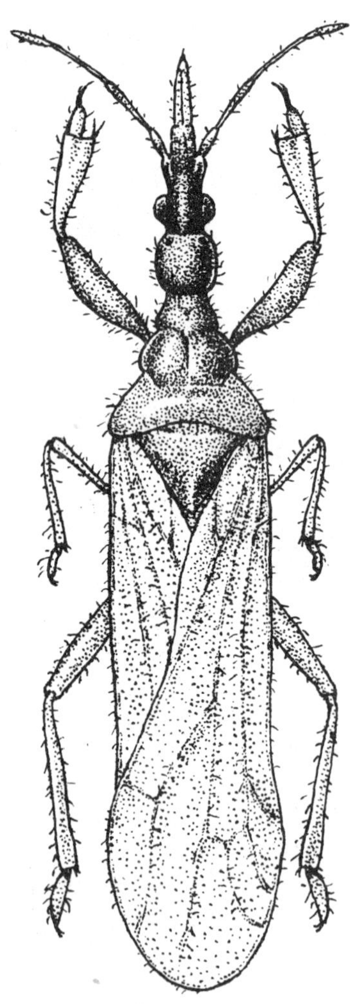

Fig. 36 *Systelloderes biceps* [p. 134] (After Froeschner, 1944).

168) key to the three subfamilies and nine genera in the Western Hemisphere is the most useful paper for identifying local specimens to genus. Concentrated and careful field studies on these insects should yield several interesting new forms and discover many details of their life activities. Special thanks are due George Steyskal (Retired, Systematic Entomology Laboratory, ARS, USDA) for his opinion on certain nomenclatorial matters.

Subfamily Aenictopechinae Usinger, 1932

Genus *Boreostolus* Wygodzinsky and Štys, 1970

1970 *Boreostolus* Wygodzinsky and Štys, Am. Mus. Nov., 2411: 2. Type-species: *Boreostolus americanus* Wygodzinsky and Štys, 1970. Original designation.

Boreostolus americanus Wygodzinsky and Štys, 1970
1970 *Boreostolus americanus* Wygodzinsky and Štys, Am. Mus. Nov., 2411: 7. [Col.].
Distribution: Cal., Col.

Subfamily Alienatinae Barber, 1953

Genus *Alienates* Barber, 1953

1953 *Alienates* Barber, Am. Mus. Nov., 1614: 2. Type-species: *Alienates insularis* Barber, 1953. Monotypic.

Alienates barberi Kritsky, 1981
1981 *Alienates barberi* Kritsky, Ent. News, 92: 130. [Ariz.].
Distribution: Ariz.

Subfamily Enicocephalinae Stål, 1860

Genus *Brevidorsus* Kritsky, 1977

1977 *Brevidorsus* Kritsky, Ent. News, 88: 164. Type-species: *Brevidorsus arizonensis* Kritsky, 1977. Monotypic.

Brevidorsus arizonensis Kritsky, 1977
1977 *Brevidosus arizonensis* Kritsky, Ent. News, 88: 164. [Ariz.].
Distribution: Ariz.

Genus *Hymenocoris* Uhler, 1892

1892 *Hymenocoris* Uhler, Trans. Md. Acad. Sci., 1: 181. Type-species: *Hymenocoris formicina* Uhler, 1892. Monotypic.
Note: Jeannel (1943, Bull. Soc. Ent. France, 48: 127) restored this taxon to generic rank. Kritsky (1978, Ent. News, 89: 76) presented a key to the two known species.

Hymenocoris formicina Uhler, 1892
1892 *Hymenocoris formicina* Uhler, Trans. Md. Acad. Sci., 1: 181. [Cal.].
1909 *Henicocephalus formicina*: Johannsen, Psyche, 16: 4.
1916 *Enicocephalus farmicinus* [*sic*]: Weiss, Ent. News, 27: 10.
1916 *Enicocephalus formacina* [*sic*]: Van Duzee, Check List Hem., p. 26.
1942 *Stenopirates* (*Hymenocoris*) *formicinus* [*sic*]: Jeannel, An. Soc. Ent. France, 110: 329.
1978 *Hymenocoris formicinia* [*sic*]: Kritsky, Ent. News, 89: 74.
Distribution: Cal. (Colombia).

Genus *Systelloderes* Blanchard, 1852

1852 *Systelloderes* Blanchard, Hist. Fis. Pol. Chile, Zool., 7: 224. Type-species: *Systelloderes moschatus* Blanchard, 1852. Monotypic.

1892 *Hymenodectes* Uhler, Trans. Am. Acad. Sci., 1: 180. Type-species: *Hymenodectes culicis* Uhler, 1892. Monotypic. Synonymized by Bergroth, 1915, Wien. Ent. Zeit., 34: 292.

Note: Kritsky (1978, Ent. News, 89: 66) provided a key to the species of North America and the Caribbean islands.

Systelloderes biceps (Say), 1832 [Fig. 36]
1832 *Reduvius biceps* Say, Descr. Het. Hem. N. Am., p. 32. [Pa.].
1892 *Hymenodectes culicis* Uhler, Trans. Md. Acad. Sci., 1: 181. [Ut.]. Synonymized by Bergroth, 1913, Ent. News, 24: 265.
1892 *Enicocephalus culicis*: Ashmead, Proc. Ent. Soc. Wash., 2: 329.
1898 *Henicocephalus culicis*: Champion, Biol. Centr.-Amer., Rhyn., 2: 162.
1913 *Hymenodectes biceps*: Bergroth, Ent. News, 24: 265.
1923 *Systelloderes biceps*: Parshley, Conn. Geol. Nat. Hist. Surv., Bull., 34: 694.
1927 *Systelloderus* [*sic*] *terrenus* Drake and Harris, Oh. J. Sci., 27: 103 [Ia., based on a nymph]. Synonymized by Kritsky, 1978, Ent. News, 89: 72.
1939 *Enicocephalus biceps*: Glick, U.S. Dept. Agr., Tech. Bull., 673: 24.
1943 *Systelloderus* [*sic*] *biceps*: McClure, Ent. News, 54: 10.
1945 *Systelloderes culicis*: Usinger, An. Ent. Soc. Am., 38: 340.
1978 *Systelloderes culicus* [*sic*]: Kritsky, Ent. News, 89: 70.
Distribution: Ariz., D.C., Fla., Ia., Ill., Ind., Ky., La., Md., Mo., N.C., N.Y., Pa., Que., R.I., Tenn., Ut., Va. (Cuba, Mexico to Panama)
Note: Ashmead (1892, see above) published his manuscript name "*schwarzii* Ashm. MS" in association with the combination *Enicocephalus culicis* and in a footnote wrote, "The specific name proposed by me must fall" [the proposal apparently had been made during the reading of his paper at an earlier meeting.].

Systelloderes crassatus Usinger, 1932
1932 *Systelloderus* [*sic*] *crassatus* Usinger, Pan-Pac. Ent., 8: 151. [Cal.].
1945 *Systelloderes crassatus*: Usinger, An. Ent. Soc. Am., 38: 340.
Distribution: Cal.

Systelloderes grandis Kritsky, 1978
1978 *Systelloderes grandes* [*sic*] Kritsky, Ent. News, 89: 70. [Ore.].
Distribution: Ore.
Note: The originally used spelling "*grandes*" is a misspelling which is here emended as above.

Systelloderes inusitatus Drake and Harris, 1927

1927 *Systelloderus* [*sic*] *inusitatus* Drake and Harris, Oh. J. Sci., 27: 102. [Miss.].

1945 *Systelloderes inusitatus*: Usinger, An. Ent. Soc. Am., 38: 340.

Distribution: Miss.

Systelloderes iowensis Drake and Harris, 1927

1927 *Systelloderus* [*sic*] *iowensis* Drake and Harris, Oh. J. Sci., 27: 102. [Ia.].

1945 *Systelloderes iowensis*: Usinger, An. Ent. Soc. Am., 38: 340.

Distribution: Ia

Note: Kritsky (1978, Ent. News, 89: 72) considered the status of this taxon as uncertain.

Systelloderes lateralis Kritsky, 1978

1978 *Systelloderes lateralus* [*sic*] Kritsky, Ent. News, 89: 68.

Distribution: Va.

Note: The originally used spelling "*lateralus*" is a misspelling which is here emended as above.

Family Gelastocoridae Kirkaldy, 1897

The Toad Bugs

By Dan A. Polhemus and John T. Polhemus

Toad bugs may be recognized by their short, broad form, protruding eyes, and raptorial forelegs. Their common name arises from their general bumpy appearance and hopping style of locomotion. Two genera, *Gelastocoris* Kirkaldy and *Nerthra* Say, occur in North America.

Species of *Gelastocoris* are usually found in damp open areas next to streams or ponds, especially on patches of wet sand. *Nerthra* species, by contrast, generally occur under stones on damp earth, although they have also been reported from rotting logs and leaf litter (Todd, 1955, Univ. Ks. Sci. Bull., 37: 277-475), under cow dung (Lauck and Wheatcroft, 1958, Ent. News, 69: 20), and in subaquatic habitats (La Rivers, 1953, Wasman J. Biol., 11: 83-96). The biology of this group is poorly studied; eggs are laid either in sand (*Gelastocoris*) or in mud under stones (*Nerthra*) in moist situations, and in the latter genus may be guarded by the female (Usinger, 1956, Aquatic Ins. Cal., 508 pp.). Development time ranges from 60 to 100 days, with both nymphs and adults preying on a variety of small insects. Ex-

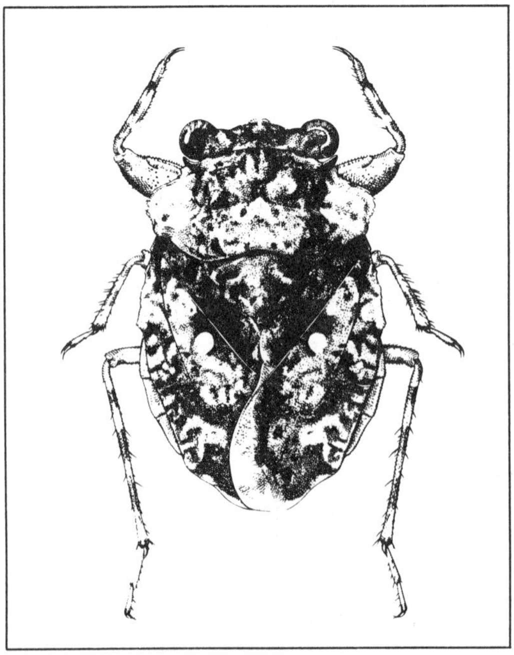

Fig. 37 *Gelastocoris oculatus* [p. 137] (After Usinger, 1956)

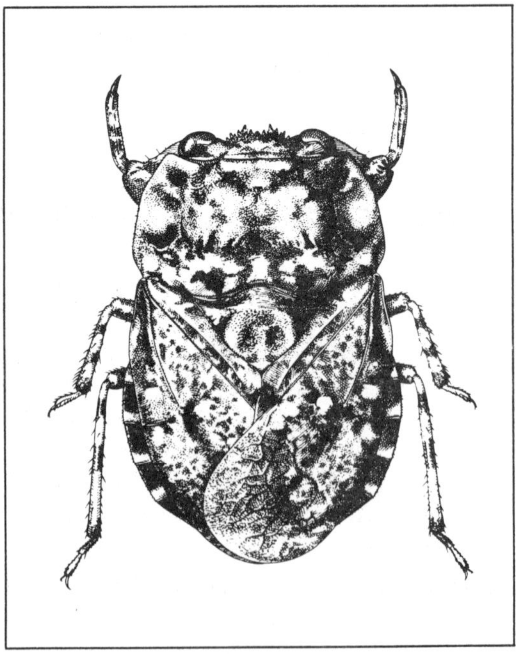

Fig. 38 *Nerthra martini* [p. 139] (After Usinger, 1956).

treme color polymorphisms are common within populations of *Gelastocoris* adults, causing them to resemble multicolored pebbles, and represent a likely case of frequency-dependent selection.

Gelastocoridae are a predominantly southern hemisphere group, with major centers of endemism in Australia, New Guinea, and South America. The genus *Gelastocoris* is restricted to the New World, whereas *Nerthra* is pantropical with a few extralimital temperate zone species. Our North American species are primarily southern and western in distribution, with the exception of *Gelastocoris oculatus* (Fabricius) which ranges over most of the continental United States and into southern Canada. Certain *Nerthra* species burrow into rotting logs and may be carried considerable distances over water via rafting (Todd, 1959, Nova Guinea, 10: 61-94), possibly accounting for odd distributions such as that of *Nerthra rugosa* (Desjardins) which occurs in Florida, Brazil, western Panama, and on the island of Mauritius in the Indian Ocean.

The major work on this family for North America is that of Todd (1955, *Op. cit.*), which provides keys including all our species. Keys to the California species and a review of what is known of the biology are found in Menke (1979, Bull. Cal. Ins. Surv., pp. 126-130); several earlier life history studies also exist for both *Gelastocoris* (Hungerford, 1922, Univ. Ks. Sci. Bull., 14: 145-171; Mackey, 1972, J. Tenn. Acad. Sci., 47: 153-155) and *Nerthra* (Usinger, 1956, *Op. cit.*). Parsons (1959, Bull. Mus. Comp. Zool., 122: 1-153 and 1960, Bull. Mus. Comp. Zool., 122: 299-357) published works on the internal and external anatomy, and Scudder (1890, Tert. Ins. N. Am., 734 pp.) reported the fossil genus *Necygonus* Scudder from the Florissant beds of Colorado.

The name Galgulidae is applied to this family in much of the earlier literature but, as stated by Menke (1979, *Op. cit.*), its usage by contemporary authors as the oldest family-group name is negated by Article 39 of the Code. The name Nerthridae (Kirkaldy and Torre-Bueno, 1909, Proc. Ent. Soc. Wash., 10: 173-215) is a junior synonym of Gelastocoridae. The name Mononychinae Fieber, 1851, has been applied to the subfamily Nerthrinae (China and Miller, 1955, An. Mag. Nat. Hist., 12: 257-267), but the latter name has gained acceptance among contemporary workers and is conserved under Article 40(a) of the 1985 ICZN.

Subfamily Gelastocorinae Champion, 1901

Genus *Gelastocoris* Kirkaldy, 1897

1802 *Galgulus* Latreille, Hist. Nat. Gen. Part. Crus. Ins., 3: 253. Preoccupied. Type-species: *Naucoris oculata* Fabricius 1798. Monotypic.

1897 *Gelastocoris* Kirkaldy, Ent., 30: 258. New name for *Galgulus* Latreille, 1802. Type-species: *Naucoris oculata* Fabricius, 1798.

Note: Todd (1955, Univ. Ks. Sci. Bull., 37: 277-475) provided a key to species including all North American taxa, accompanied by illustrations of pronotal structures and male genitalia.

Gelastocoris oculatus (Fabricius), 1798 [Fig. 37]

1798 *Naucoris oculata* Fabricius, Suppl. Ent. Syst., p. 525. [Carolina].

1802 *Galgulus oculatus*: Latreille, Hist. Nat. Gen. Part. Crus. Ins., 3: 253.

1869 *Galgulus variolatus* [*sic*]: Rathvon, Hist. Lancaster Co., Pa., p. 550.

1897 *Gelastocoris oculatus*: Kirkaldy, Ent., 30: 258.
1901 *Gelastocoris vicinus*: Champion, Biol. Centr.-Am., Rhyn., 2: 349 (in part).
1923 *Gelastocoris barberi* Torre-Bueno, Conn. Geol. Nat. Hist. Surv. Bull., 34: 393. [Ill.]. Synonymized by Martin, 1929, Univ. Ks. Sci. Bull., 18: 359.
1926 *Gelastocoris subsimilis* Blatchley, Het. E. N. Am., p. 1025. [Fla.]. Synonymized by Martin, 1929, Univ. Ks. Sci. Bull., 18: 359.
1929 *Gelastocoris californiensis* Melin, Zool. Bid. Fran. Upp., 12: 126. [Cal.]. Synonymized by Todd, 1955, Univ. Ks. Sci. Bull., 37: 300.
Distribution: Ala., Ariz., Ark., Cal., Col., D.C., Fla., Ga., Id., Ill., Ind., Ks., Ky., La., Md., Mich., Miss., Mo., N.J., N.M., N.C., Neb., Nev., Oh., Ok., Ont., Ore., Pa., S.C., Tenn., Tex., Ut., Wash., Wis. (Mexico)
Note: Hungerford (1922, Univ. Ks. Sci. Bull., 14: 145-171) and Mackey (1972, J. Tenn. Acad. Sci., 47: 153-155) provided notes on life history and biology. Deonier et al. (1976, Ent. News, 87: 257-264) supplied data on habitat preferences.

Gelastocoris oculatus oculatus (Fabricius), 1798
1798 *Naucoris oculata* Fabricius, Suppl. Ent. Syst., p. 525. [Carolina].
1955 *Gelastocoris oculatus oculatus*: Todd, Univ. Ks. Sci. Bull., 37: 298.
Distribution: Same as for species.

Gelastocoris oculatus variegatus (Guérin-Méneville), 1844
1844 *Galgulus variegatus* Guérin-Méneville, Icon. Reg. Anim. Cuv., 7: 352. [Mexico].
1876 *Galgulus variegatus*: Uhler, Bull. U.S. Geol. Geogr. Surv. Terr., 1: 336.
1955 *Gelastocoris oculatus variegatus*: Todd, Univ. Ks. Sci. Bull., 37: 311.
Distribution: Tex. (Mexico, Central America).

Gelastocoris rotundatus Champion, 1901
1901 *Gelastocoris rotundatus* Champion, Biol. Centr.-Am., Rhyn., 2: 347. [Mexico, Guatemala].
1929 *Gelastocoris rotundatus*: Martin, Univ. Ks. Sci. Bull., 18: 363.
Distribution: Ariz., Cal., Tex. (Mexico, Central America).

Subfamily Nerthrinae Kirkaldy, 1906

Genus *Nerthra* Say, 1832

1832 *Nerthra* Say, Descrip. Het. Hem. N. Am., p. 37. Type-species: *Nerthra stygica* Say, 1832. Monotypic.
1832 *Mononyx* Laporte, Mag. Zool., 2 (suppl.): 16. Type-species: *Naucoris raptoria* Fabricius, 1803. Monotypic. Synonymized by Todd, 1955, Univ. Ks. Sci. Bull., 37: 345.
1843 *Peltopterus* Guérin-Méneville, Rev. Zool., p. 113. Type-species: *Naucoris rugosa* Desjardins, 1837. Monotypic. Synonymized by Todd, 1955, Univ. Ks. Sci. Bull., 37: 345.
1925 *Glossoaspis* Blatchley, Ent. News, 36: 49. Type-species: *Glossoaspis brunnea* Blatchley, 1925. Monotypic. Synonymized by Todd, 1955, Univ. Ks. Sci. Bull., 37: 345.
Note: Confusion existed in the literature over the correct date for Laporte's description of *Mononyx*, but as discussed by Menke (1979, Bull. Cal. Ins. Surv., 21: 129), 1832 must stand as the correct date of citation, rather than 1833 as is often listed. Todd's (1955, above, pp. 297-298) key includes all North American species, accompanied by illustrations of the male and female genitalia.

Nerthra manni Todd, 1955

1955 *Nerthra manni* Todd, Univ. Ks. Sci. Bull., 37: 396. [Mexico].

1972 *Nerthra manni*: Polhemus, Proc. Ent. Soc. Wash., 74: 308.

Distribution: Ariz. (Mexico).

Note: Known in the U.S. only from Santa Cruz and Cochise Counties in southern Arizona.

Nerthra martini Todd, 1954 [Fig. 38]

1876 *Mononyx badius*: Uhler, Bull. U.S. Geol. Geogr. Surv. Terr., 1: 337.

1894 *Mononyx stygicus*: Uhler, Proc. Cal. Acad. Sci., (2)4: 290.

1909 *Mononyx fuscipes*: Kirkaldy and Torre-Bueno, Proc. Ent. Soc. Wash., 10: 181.

1954 *Nerthra martini* Todd, Pan-Pac. Ent., 30: 113. [Cal.].

Distribution: Ariz., Cal. (Mexico)

Note: Usinger (1956, Aquatic Ins. Cal., pp. 209-210) provided notes on life history.

Nerthra rugosa (Desjardins), 1837

1837 *Naucoris rugosa* Desjardins, An. Soc. Ent. France, 6: 239. [Mauritius].

1925 *Glossoaspis brunnea* Blatchley, Ent. News, 36: 49. [Fla.]. Synonymized by Todd, 1955, Univ. Ks. Sci. Bull., 37: 413.

1955 *Nerthra rugosa*: Todd, Univ. Ks. Sci. Bull., 37: 412.

Distribution: Fla. (Brazil, Mauritius, Panama).

Nerthra stygica Say, 1832

1832 *Nerthra stygica* Say, Descrip. Het. Hem. N. Am., p. 37. [Ga.].

1863 *Mononyx stigicus*: Stål, Berl. Ent. Zeit., 7: 406.

Distribution: Fla., Ga., N.C.

Nerthra usingeri Todd, 1954

1954 *Nerthra usingeri* Todd, Pan-Pac. Ent., 30: 116. [Cal.].

Distribution: Ariz., Cal. (Mexico).

Note: Known in the U.S. only from the Colorado River drainage in the low deserts of California and Arizona.

Family Gerridae Leach, 1815

The Water Striders

By Cecil L. Smith

The Gerridae comprise the second largest family of the aquatic and semiaquatic Heteroptera in numbers of genera and species. This is true for both the world and North American faunas. More than 50 species and subspecies assigned to seven genera have been recorded from America north of Mexico.

Gerrids, or water striders, are the most conspicuous inhabitants of the water's surface because of their relatively large size and lack of secretiveness. They occur on ponds, lakes, rivers, streams, temporary puddles and even the open oceans. In our area *Gerris* Fabricius is the most ubiquitous, inhabiting situations varying from rain pools to the largest of rivers and lakes. *Trepobates* Uhler and *Rheumatobates* Bergroth are usually restricted to stagnant or slowly flowing water, and *Limnogonus* Stål and *Neogerris* Matsumura to stagnant water. *Metrobates* Uhler is most often encountered in streams and rivers, while *Halobates* Eschscholtz is found strictly in marine and estuarine environments. Most species display some degree of gregariousness, but not to the extent found in some of the veliid genera.

Wing polymorphism is characteristic in the majority of *Gerris* species, in which apterous, micropterous, brachypterous, and fully alate individuals are occasionally found in a single population. With the exception of the strictly apterous *Halobates*, alary polymorphism is found in all other North American genera.

Except for a few isolated observations of phytophagy, all members of the family are predaceous or necrophagous. Because their diet is restricted largely to organisms inhabiting the water's surface or fallen insects trapped in the surface film, gerrids are of little economic importance. Cannibalism is common in many species, particularly on young, weak, or disabled individuals. Polhemus and Chapman (1979, Bull. Cal. Ins. Surv., 21: 58-60) gave a good review of the literature dealing with biology and habits.

The North American species of Gerridae are relatively well known taxonomically because of the contributions by Drake and Harris (1934, An. Carnegie Mus., 23: 179-241) on *Gerris, Limnogonus,* and *Neogerris*; Herring (1961, Pac. Ins., 3: 223-305) on *Halobates*; and Hungerford (1954, Univ. Ks. Sci. Bull., 36, Pt. 1: 529-588) on *Rheumatobates*. Current keys are not available at this time for *Metrobates* and *Trepobates*, although the latter has been revised by P. D. Kittle (1977, Unpubl. Ph.D. Disser., Univ. Ark.).

No comprehensive studies have been conducted on the immature stages of the family, although the following regional works on *Gerris* are available: Calabrese (1974, Mem., Conn. Ent. Soc., pp. 227-266)--Conn.; Scudder and Jamieson (1972, J. Ent. Soc. B.C., 69: 72-79)--B.C.; and Sprague (1967, An. Ent. Soc. Am., 60: 1038-1044)--New England.

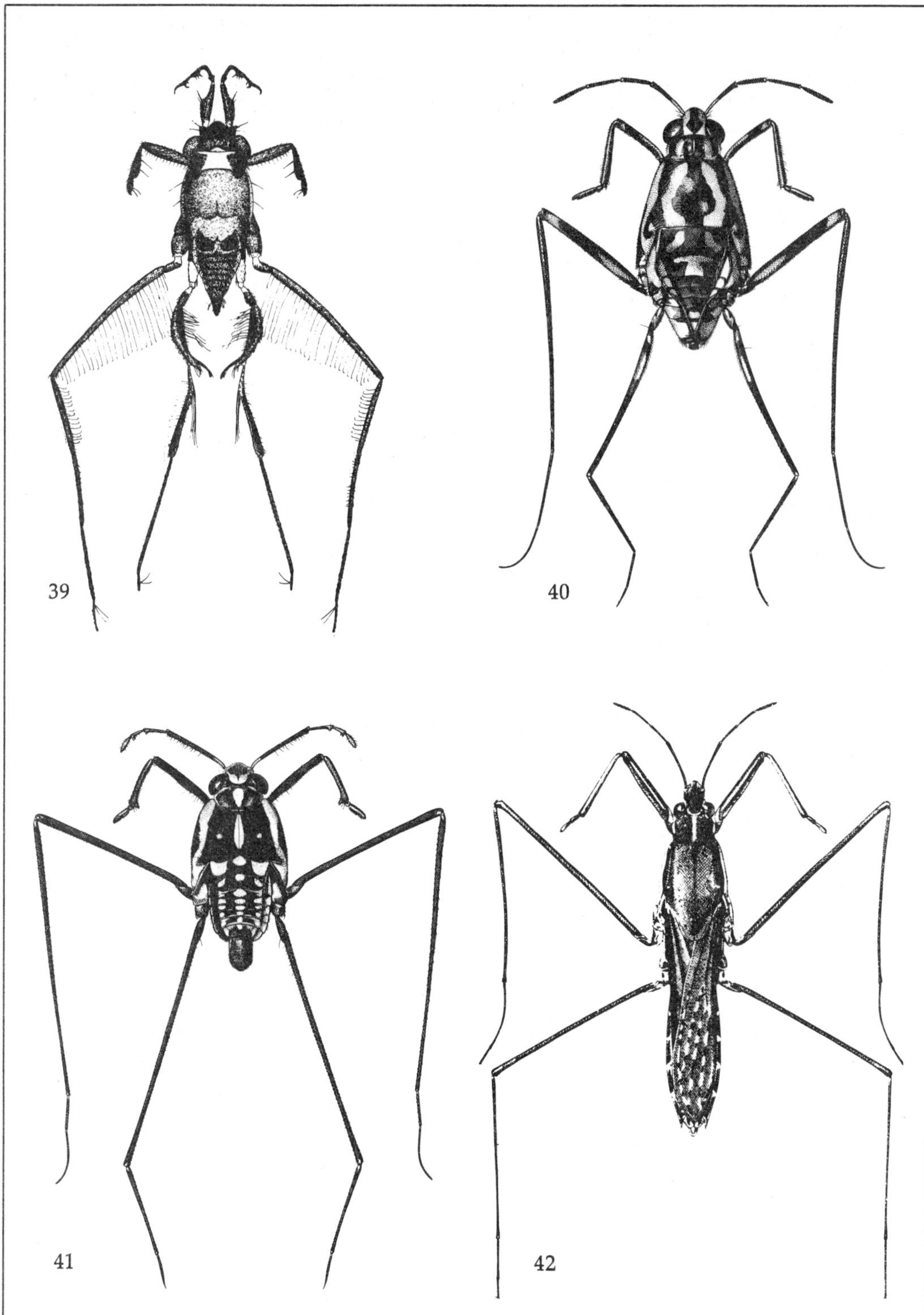

Figs. 39-42: 39, *Rheumatobates rileyi* [p. 148]; 40, *Trepobates becki* [p. 150]; 41, *Metrobates trux infuscatus* [p. 150]; 42, *Gerris remigis* [p. 143] (Fig. 39 after Brooks and Kelton, 1967; 40-42 after Usinger, 1956).

Matsuda (1960, Univ. Ks. Sci. Bull., 41: 25-632), Calabrese (1980, Misc. Publ. Ent. Soc. Am., 11: 1-119), and Andersen (1982, Entomonograph, 3: 183-239) have studied the higher classification. The latter paper should be consulted for updated keys to the subfamilies and genera of the world.

Subfamily Gerrinae Leach, 1815

Tribe Gerrini Leach, 1815

Genus *Gerris* Fabricius, 1794

1794 *Gerris* Fabricius, Ent. Syst., 4: 187. Type-species: *Cimex lacustris* Linnaeus, 1758. Designated by Latreille, 1810, Consid. Gén., pp. 259, 434.

1868 *Limnotrechus* Stål, Öfv. K. Svens. Vet.-Akad. Förh., 25: 395. Type-species: *Cimex lacustris* Linnaeus, 1758. Designated by Kirkaldy, 1906, Trans. Am. Ent. Soc., 32: 155. Synonymized by Torre-Bueno, 1905, J. N.Y. Ent. Soc., 13: 40.

1868 *Hygrotrechus* Stål, Öfv. K. Svens. Vet.-Akad. Förh., 25: 395. Type-species: *Cimex najas* DeGeer, 1773. Monotypic. Synonymized by Torre-Bueno, 1905, J. N.Y. Ent. Soc., 13: 40.

Note: Drake and Harris (1934, An. Carnegie Mus., 23: 179-240) reviewed the New World species. Walker (1873, Cat. Hem. Brit. Mus., 8: 165) listed "*Gerris aptera* Schum." from Ill. and Ind., but Schummel's (1832, Vers. Fam. Ruderwanzen, p. 34) species is a junior synonym of the European *Gerris najas* (DeGeer), which is not known to occur in North America.

Subgenus *Aquarius* Schellenberg, 1800

1800 *Aquarius* Schellenberg, Gesc. Land-u. Wasserwanz., p. 25. Type-species: *Aquarius paludum* Schellenberg 1800 [nec Fabricius, 1794], a junior synonym of *Cimex najas* De Geer, 1773. Designated by Kirkaldy, 1906, Trans. Am. Ent. Soc., 32: 155.

1906 *Gerris* (*Aquarius*): Kirkaldy, Trans. Am. Ent. Soc., 32: 155.

Note: *Gerris* (*Aquarius*) *uhleri* Drake and Hottes (1925, Proc. Biol. Soc. Wash., 38: 69), described from specimens labeled "Ariz.," was shown by Polhemus (1973, Great Basin Nat., 33: 114-115) to be from the Arize River in France and a junior synonym of the widespread European *Gerris paludum* Fabricius. Therefore, it should be eliminated from the list of North American species of Gerridae.

Gerris ampla Drake and Harris, 1938, New Subgeneric Combination

1938 *Gerris ampla* Drake and Harris, Pan-Pac. Ent., 14: 73. [Mexico].

1942 *Gerris ampla*: Kuitert, Univ. Ks. Sci. Bull., 28: 120.

Distribution: Ariz. (Mexico).

Note: The nominate subspecies *G. ampla ampla* Drake and Harris has not been reported from the U.S.

Gerris ampla arizonensis Kuitert, 1942, New Subgeneric Combination

1942 *Gerris ampla* var. *arizonensis* Kuitert, Univ. Ks. Sci. Bull., 28: 120. [Ariz.].

Distribution: Ariz.

Gerris conformis (Uhler), 1878
1878 *Hygrotrechus conformis* Uhler, Proc. Boston Soc. Nat. Hist., 19: 435. [Mass., Md.].
1905 *Gerris (Hygrotrechus) conformis*: Torre-Bueno, J. N.Y. Ent. Soc., 13: 40.
1908 *Gerris conformis*: Torre-Bueno, J. N.Y. Ent. Soc., 16: 233.
1911 *Gerris (Aquarius) conformis*: Torre-Bueno, Trans. Am. Ent. Soc., 37: 247.
Distribution: Conn., D.C., Del., Ga., La., Mass., Md., Me., Mich., Miss., N.C., N.H., N.J., N.S., N.Y., Oh., Pa., S.C., Tenn., Va., Wis. (Mexico).

Gerris nebularis Drake and Hottes, 1925
1925 *Gerris nebularis* Drake and Hottes, Proc. Biol. Soc. Wash., 38: 70. [Fla.].
1974 *Gerris (Aquarius) nebularis*: Calabrese, Mem., Conn. Ent. Soc., p. 228.
Distribution: Ala., Ark., Fla., Ga., Ill., Ind., Ia., Ks., La., Miss., Mo., N.C., N.J., N.Y., Neb., Oh., S.C., Tenn., Va.

Gerris nyctalis Drake and Hottes, 1925
1925 *Gerris nyctalis* Drake and Hottes, Oh. J. Sci., 25: 47. [Col.].
1974 *Gerris (Aquarius) nyctalis*: Calabrese, Mem., Conn. Ent. Soc., p. 228.
Distribution: Alta., B.C., Cal., Col., Id., Mass., Mont., Nfld., Que., Sask., Ut., Wash.
Note: Calabrese (1974, Mem., Conn. Ent. Soc., p. 228) wrote *G. nyctalis* "does not occur east of the Rocky Mountains in the United States" and that the Mass. record is probably in error. Polhemus and Chapman (1979, Bull. Cal. Ins. Surv., 21: 61) wrote they were "not convinced" that *G. nyctalis* is distinct from *G. remigis*.

Gerris remigis Say, 1832 [Fig. 42]
1832 *Gerris remigis* Say, Descrip. Het. Hem. N. Am., p. 35. [N.Y.].
1859 *Gerris orba* Stål, K. Svens. Freg. Eugen. Resa Jorden, p. 264. [Cal]. Synonymized by Drake and Harris, 1928, Oh. J. Sci., 28: 269.
1871 *Hygrotrechus robustus* Uhler, Am. J. Sci., Ser. 3, 1: 105. [Cal.]. Synonymized by Drake and Harris, 1928, Oh. J. Sci., 28: 269.
1872 *Hygrotrechus remigis*: Uhler, Prelim. Rept. U.S. Geol. Surv. Mont., p. 422.
1873 *Hydrometra orba*: Walker, Cat. Hem. Brit. Mus., 8: 152.
1873 *Gerris robustus*: Walker, Cat. Hem. Brit. Mus., 8: 165.
1893 *Hygrotrechus remiges* [*sic*]: Uhler, Proc. Ent. Soc. Wash., 2: 382.
1905 *Gerris (Hygrotrechus) remigis*: Torre-Bueno, J. N.Y. Ent. Soc., 13: 40.
1911 *Gerris (Aquarius) remigis*: Torre-Bueno, Trans. Am. Ent. Soc., 37: 246.
1914 *Hygrotrechus orba*: Van Duzee, Trans. San Diego Soc. Nat. Hist., 2: 32.
1916 *Gerris (Aquarius) robustus*: Van Duzee, Check List Hem., p. 49.
1974 *Gerris remigis caloregon* Calabrese, Ent. News, 85: 27. [Cal.]. Synonymized by Polhemus and Chapman, 1979, Bull. Cal. Ins. Surv., 21: 63.
1974 *Gerris remigis remigis*: Calabrese, Ent. News, 85: 27.
Distribution: Reported from all 48 contiguous states; B.C., Man., N.S., Nfld., Ont., Que. (Mexico to Guatemala).

Subgenus *Gerris* Fabricius, 1794

1794 *Gerris* Fabricius, Ent. Syst., 4: 187. Type-species: *Cimex lacustris* Linnaeus, 1758. Designated by Latreille, 1810, Consid. Gén., p. 434.
1975 *Gerris (Gerris)*: Andersen, Ent. Scand., 7 (Suppl.): 23.

Gerris alacris Hussey, 1921
1921 *Gerris alacris* Hussey, Psyche, 28: 11. [Mich.].

1974 *Gerris (Gerris) alacris*: Calabrese, Mem. Conn. Ent. Soc., p. 249.
Distribution: Conn., D.C., Ill., Ind., Ks., Me., Mich., Mo., N.C., N.J., Oh., S.C., Va.

Gerris argenticollis Parshley, 1916
1916 *Gerris argenticollis* Parshley, Ent. News, 27: 103. [Mass.].
1917 *Gerris (Gerris) argenticollis*: Van Duzee, Univ. Cal. Publ. Ent., 2: 428.
Distribution: Ark., Fla., Ga., Ill., Ind., La., Mass., Mich., Miss., Mo., N.C., N.J., N.Y., S.C., Va.

Gerris buenoi Kirkaldy, 1911
1905 *Gerris sulcatus*: Torre-Bueno, J. N.Y. Ent. Soc., 13: 40.
1911 *Gerris buenoi* Kirkaldy, Ent. News, 22: 246. ["British Columbia to Atlantic States"]. Lectotype from Col. designated by Menke and Polhemus, 1973, Pan-Pac. Ent., 49: 257.
1911 *Gerris (Gerris) buenoi*: Torre-Bueno, Trans. Am. Ent. soc., 37: 245.
Distribution: Alk., Alta., B.C., Col., Conn., Id., Man., Mass., Me., Mich., Minn., Mo., Mont., N.H., N.J., N.Y., Nfld., Ont., Que., R.I., S.C., S.D., Sask., Va., Wis., Wyo.

Gerris comatus Drake and Hottes, 1925
1925 *Gerris comatus* Drake and Hottes, Oh. J. Sci., 25: 48. [Col.].
1974 *Gerris (Gerris) comatus*: Calabrese, Mem. Conn. Ent. Soc., p. 250.
Distribution: Alta., Ariz., B.C., Col., Conn., Fla., Ia., Ill., Ind., Ks., Man., Md., Mich., Minn., Mo., Mont., N.H., N.J., N.M., N.Y., Neb., Oh., Ok., Ont., Pa., Que., S.C., S.D., Sask., Va., Wis., Wyo.

Gerris comatus comatus Drake and Hottes, 1925
1925 *Gerris comatus* Drake and Hottes, Oh. J. Sci., 25: 48. [Col.].
1925 *Gerris comatus* var. *comatus* Drake and Hottes, Proc. Biol. Soc. Wash., 38: 72.
Distribution: Same as for species.

Gerris comatus mickeli Drake and Hottes, 1925
1925 *Gerris comatus* var. *mickeli* Drake and Hottes, Proc. Biol. Soc. Wash., 38: 72. [Minn.].
Distribution: Col., Minn., Ore.

Gerris gillettei Lethierry and Severin, 1896
1895 *Limnotrechus productus* Uhler, Bull. Col. Agr. Exp. Stn., 31: 61. [Col.]. Preoccupied.
1895 *Hygrotrechus productus*: Heidemann, Proc. Ent. Soc. Wash., 3: 143.
1896 *Gerris Gillettei* [*sic*] Lethierry and Severin, Cat. Gén. Hém., 3: 60. New name for *Limnotrechus productus* Uhler.
1916 *Gerris (Gerris) gillettei*: Van Duzee, Check List Hem., p. 49.
1958 *Gerris gilletei* [*sic*]: Roback, Trans. Am. Ent. Soc., 84: 10.
Distribution: B.C., Cal., Col., Mont., Nev., Ore., Tex., Ut., Wash., Wyo.

Gerris incognitis Drake and Hottes, 1925
1925 *Gerris incognitus* Drake and Hottes, Proc. Biol. Soc. Wash., 38: 73. [B.C.].
1979 *Gerris (Gerris) incognitis*: Polhemus and Chapman, Bull. Cal. Ins. Surv., 21: 65.
Distribution: Alta., B.C., Cal., Col., Id., Man., Mont., Nev., Ont. Ore., Sask., Wash., Wyo.

Gerris incurvatus Drake and Hottes, 1925
1921 *Gerris marginatus*: Parshley, Proc. Ent. Soc. B.C., 18: 21.
1925 *Gerris incurvatus* Drake and Hottes, Proc. Biol. Soc. Wash., 38: 72. [Mont.].
1979 *Gerris (Gerris) incurvatus*: Polhemus and Chapman, Bull. Cal. Ins. Surv., 21: 65.

Distribution: B.C., Cal., Id., Ill., Mont., Nev., Ore., Tex., Wash., Wyo.
Note: Parshley's (1921, above) B.C. record was transferred to this species by Downes (1927, Proc. Ent. Soc. B.C., 23: 15).

Gerris insperatus Drake and Hottes, 1925
1925 *Gerris insperatus* Drake and Hottes, Proc. Biol. Soc. Wash., 38: 71. [Mexico].
1942 *Gerris inseperatus* [*sic*]: Kuitert, Univ. Ks. Sci. Bull., 28: 121.
1974 *Gerris* (*Gerris*) *insperatus*: Calabrese, Mem., Conn. Ent. Soc., p. 251.
Distribution: Ariz., Col., Fla., Ga., Ill., Ind., Minn., Miss., N.C., N.J., N.Y., Oh., Ont., Pa., Que., S.D., Tex., Va., W.Va. (Mexico).

Gerris marginatus Say, 1832
1832 *Gerris marginatus* Say, Descrip. Het. Hem. N. Am., p. 36. [U.S.].
1872 *Limnotrechus marginatus*: Uhler, Prelim. Rept. U.S. Geol. Surv. Mont., p. 423.
1905 *Gerris* (*Limnotrechus*) *marginatus*: Torre-Bueno, J. N.Y. Ent. Soc., 13: 40.
1911 *Gerris* (*Gerris*) *marginatus*: Torre-Bueno, Trans. Am. Ent. Soc., 37: 247.
1925 *Gerris marginata*: Frost, J. N.Y. Ent. Soc., 32: 180.
Distribution: "Every state in the United States" (Gonsoulin, 1974, Trans. Am. Ent. Soc., 100: 526), but Polhemus and Chapman (1979, Bull. Cal. Ins. Surv., 21: 66) treated California records as "misidentifications"; Man., N.S., Ont., Que. (Mexico to Brazil).

Gerris pingreensis Drake and Hottes, 1925
1925 *Gerris pingreensis* Drake and Hottes, Oh. J. Sci., 25: 49. [Col.].
Distribution: Alk., Alta., B.C., Col., Id., Man., Mont., N.T., Que., Sask., Yuk.

Genus *Limnogonus* Stål, 1868

1868 *Limnogonus* Stål, K. Svens. Vet.-Akad. Handl., 7: 133. Type-species: *Hydrometra hyalina* Fabricius, 1803. Designated by Kirkaldy, 1906, Trans. Am. Ent. Soc., 32: 155.
Note: Drake and Harris (1934, An. Carnegie Mus., 23: 179-241) reviewed the New World species.

Subgenus *Limnogonus* Stål, 1868

1868 *Limnogonus* Stål, K. Svens. Vet.-Akad. Handl., 7: 133. Type-species: *Hydrometra hyalina* Fabricius, 1803. Designated by Kirkaldy, 1906, Trans. Am. Ent. Soc., 32: 155.

Limnogonus franciscanus (Stål), 1859
1844 *Gerris marginatus* Guérin-Méneville, Icon. Règne Anim., 2: 351. [Cuba]. Preoccupied.
1859 *Gerris franciscana* Stål, K. Svens. Freg. Eugen. Resa Jorden, p. 265. [Cal.].
1873 *Hydrometra franciscana*: Walker, Cat. Hem. Brit. Mus., 8: 152.
1893 *Limnometra marginata*: Uhler, Proc. Zool. Soc. London, p. 706.
1896 *Gerris Guerini* [*sic*] Lethierry and Severin, Cat. Gén. Hém., 3: 61. [Cuba]. Synonymized by Drake and Harris, 1934, An. Carnegie Mus., 23: 206.
1898 *Limnogonus marginatus*: Champion, Biol. Centr.-Am., Rhyn., 2: 152.
1909 *Limnogonus guerini*: Kirkaldy and Torre-Bueno, Proc. Ent. Soc. Wash., 10: 210.
1910 *Limnogonus franciscana*: Banks, Cat. Nearc. Hem.-Het., p. 26.
1916 *Tenagogonus franciscanus*: Van Duzee, Check List Hem., p. 49.
1979 *Limnogonus* (*Limnogonus*) *franciscanus*: Polhemus and Chapman, Bull. Cal. Ins. Surv., 21: 66.

Distribution: Cal.(?), Fla., Tex. (Mexico to South America, West Indies).
Note: Polhemus and Chapman (1979, above, 21: 66) suggested that the locality label on the type of *G. franciscana* reading "California" is likely an error, and the specimen more plausibly originated in South America.

Genus *Limnoporus* Stål, 1868

1868 *Limnoporus* Stål, Öfv. K. Svens. Vet.-Akad. Förh., 25: 395. Type-species: *Gerris rufoscutellatus* Latreille, 1807. Monotypic.
Note: This genus was long considered a subgenus of *Gerris*, until Andersen (1975, Ent. Scand., 7(Suppl.): 23) elevated it to the genus level. Polhemus and Chapman (1979, Bull. Cal. Ins. Surv., 21: 65) noted that only the proportions of the antennal segments consistently separate *Limnoporus* from *Gerris*.

Limnoporus canaliculatus (Say), 1832
1832 *Gerris canaliculatus* Say, Descrip. Hem. Het. N. Am., p. 36. [Ga.].
1905 *Gerris (Limnotrechus) canaliculatus*: Torre-Bueno, J. N.Y. Ent. Soc., 13: 41.
1911 *Gerris (Gerris) canaliculatus*: Torre-Bueno, Trans. Am. Ent. Soc., 37: 248.
1979 *Limnoporus canaliculatus*: Polhemus and Chapman, Bull. Cal. Ins. Surv., 21: 65.
Distribution: Ala., Ark., Conn., Fla., Ga., Ia., Ill., La., Mass., Mich., Miss., Mo., N.C., N.J., N.Y., Oh., R.I., S.C., Tenn., Tex., Va.

Limnoporus dissortis (Drake and Harris), 1930
1930 *Gerris dissortis* Drake and Harris, Bull. Brook. Ent. Soc., 25: 145. [Oh.].
1967 *Gerris (Limnoporus) dissortis*: Sprague, An. Ent. Soc. Am., 60: 1043.
1979 *Limnoporus dissortis*: Polhemus and Chapman, Bull. Cal. Ins. Surv., 21: 66.
Distribution: Alta., B.C.(?), Conn., Del., Ia., Ind., Ill., Ks., Ky., Mass., Md., Me., Mich., Minn., Mo., N.J., N.Y., Neb., Nfld., Oh., Ont., Pa., R.I., W.Va., Wis.
Note: Scudder (1971, J. Ent. Soc. B.C., 68: 3) was unable to confirm occurence in B.C.

Limnoporus nearcticus (Kelton), 1961
1961 *Gerris nearcticus* Kelton, Can. Ent., 93: 663. [Yuk.].
1979 *Limnoporus nearcticus*: Polhemus and Chapman, Bull. Cal. Ins. Surv., 21: 66.
Distribution: Alk., Yuk.

Limnoporus notabilis (Drake and Hottes), 1925
1925 *Gerris notablis* [*sic*] Drake and Hottes, Oh. J. Sci., 25: 46. [Col.].
1925 *Gerris notabilis*: Drake and Hottes, Proc. Biol. Soc. Wash., 38: 73. [Emended spelling above].
1979 *Limnoporus notabilis*: Polhemus and Chapman, Bull. Cal. Ins. Surv., 21: 66.
Distribution: Alta., Ariz., B.C., Cal., Col., Ia., Id., Mont., N.M., Ore., S.D., Ut., Wash., Wyo.

Genus *Neogerris* Matsumura, 1913

1913 *Neogerris* Matsumura, Thous. Ins. Japan, Add., 1: 100. Type-species: *Neogerris boninensis* Matsumura, 1913. Monotypic.
1959 *Limnogonus (Limnogonellus)* Hungerford and Matsuda, J. Ks. Ent. Soc., 32: 40. Type-species: *Gerris parvula* Stål, 1868. Original designation. Synonymized by Hungerford and Matsuda, 1961, Ins. Matsum., 24: 112.
1961 *Limnogonus (Neogerris)*: Hungerford and Matsuda, Ins. Matsum., 24: 112.
Note: Drake and Harris (1934, An. Carnegie Mus., 23: 179-241) reviewed the New World

species. Andersen (1982, Semiaquat. Bugs, p. 238) is the most recent to give generic status to *Neogerris*.

Neogerris hesione (Kirkaldy), 1902

1902 *Gerris hesione* Kirkaldy, Ent., 35: 137. [Fla.].

1909 *Limnogonus hesione*: Kirkaldy and Torre-Bueno, Proc. Ent. Soc. Wash., 10: 210.

1916 *Tenagogonus hesione*: Van Duzee, Check List Hem., p. 49.

1921 *Gerris (Tenagogonus) hesione*: Torre-Bueno, Ent. News, 32: 274.

1925 *Gerris (Limnogonus) hesione*: Torre-Bueno, Can. Ent., 57: 31.

1959 *Limnogonus (Limnogonellus) hesione*: Hungerford and Matsuda, J. Ks. Ent. Soc., 32: 40.

1975 *Neogerris hesione*: Andersen, Ent. Scand., 7(Suppl.): 8.

Distribution: Ala., Alk., Ark., Fla., Ga., Ia., Ill., Ind., Ks., Ky., La., Mich., Miss., Mo., N.C., N.Y., Neb., Oh., Ok., Pa., S.C., Tenn., Tex., Va., W.Va. (Cuba, Panama).

Subfamily Halobatinae Bianchi, 1896

Genus *Halobates* Eschscholtz, 1822

1822 *Halobates* Eschscholtz, Entomogr., p. 106. Type-species: *Halobates micans* Eschscholtz, 1822. Designated by Laporte, 1833, Essai Classif. Syst. Hém., p. 24.

Note: Herring (1961, Pac. Ins., 3: 223-305) revised this genus.

Halobates micans Eschscholtz, 1822

1822 *Halobates micans* Eschscholtz, Entomogr., p. 107. [Southern Atlantic Ocean].

1894 *Halobates wüllerstorffi* [*sic*]: Wickham, Ent. News, 5: 34.

1901 *Halobates wuellerstorfi* [*sic*]: Slosson, Ent. News, 12: 11.

1909 *Halobates micans*: Kirkaldy and Torre-Bueno, Proc. Ent. Soc. Wash., 10: 212.

1920 *Halobates sericeus*: Hungerford, Ks. Univ. Sci. Bull., 11: 116.

Distribution: Coastal Tex. and Atlantic and Gulf coasts of Fla. (Circumtropical).

Halobates sericeus Eschscholtz, 1822

1822 *Halobates sericeus* Eschscholtz, Entomogr., p. 108. [Pacific Ocean near the Equator].

1909 *Halobates sericeus*: Kirkaldy and Torre-Bueno, Proc. Ent. Soc. Wash., 10: 213.

Distribution: Cal. coast (Widespread throughout Pacific Ocean).

Note: Except for Hungerford's record (1920, Ks. Univ. Sci. Bull., 11: 116) of *H. sericeus* from Florida, based on a misidentification of *H. micans* and Esaki's (1924, Psyche, 31: 117) notation as "widely distributed over the Pacific and Atlantic Oceans," this species has been reported only for Pacific Ocean areas.

Subfamily Rhagodotarsinae Lundblad, 1933

Genus *Rheumatobates* Bergroth, 1892

1892 *Rheumatobates* Bergroth, Ins. Life, 4: 321. Type-species: *Rheumatobates rileyi* Bergroth, 1892. Monotypic.

Note: Hungerford (1954, Univ. Ks. Sci. Bull., 36: 529-588) revised this genus. Spangler et al. (1985, Ent. News, 96: 196-200) provided a checklist and updated distributions.

Rheumatobates clanis Drake and Harris, 1932

1932 *Rheumatobates clanis* Drake and Harris, Pan-Pac. Ent., 8: 157. [British Honduras].

1958 *Rheumatobates clanis*: Herring, Pan-Pac. Ent., 34: 175.

Distribution: Fla. (British Honduras, Cuba, Guatemala, Jamaica).

Rheumatobates hungerfordi Wiley, 1923

1923 *Rheumatobates hungerfordi* Wiley, Can. Ent., 55: 202. [Tex.].

Distribution: Ariz., Ark., Ga., La., Mo., N.M., Ok., S.C., Tex., Ut. (Belize, Mexico).

Rheumatobates minutus Hungerford, 1936

1936 *Rheumatobates minutus* Hungerford, Carnegie Inst. Wash., Publ. No. 457, p. 147. [Mexico].

Rheumatobates minutus minutus Hungerford, 1936

1936 *Rheumatobates Minutus* Hungerford, Carnegie Inst. Wash., Publ. No. 457, p. 147. [Mexico].

1955 *Rheumatobates minutus*: Herring, Quart. J. Fla. Acad. Sci., 18: 122.

1985 *Rheumatobates minutus minutus*: Spangler, Froeschner, and Polhemus, Ent. News, 96: 198.

Distribution: Fla. (Costa Rica, Mexico, Panama, Peru, Puerto Rico, Trinidad).

Note: Herring (1958, Pan-Pac. Ent., 34: 175) suggested that populations of this species found in Florida were hurricane transported and temporary. A recent collection by D. A. Polhemus in the Everglades reveals that a breeding population is present.

Rheumatobates palosi Blatchley, 1926

1926 *Rheumatobates rileyi* var. *palosi* Blatchley, Het. E. N. Am., p. 984. [Ind.]. Lectotype designated by Blatchley, 1930, Blatchleyana, p. 66.

1974 *Rheumatobates palosi*: Bobb, Ins. Va., 87: 78.

Distribution: Ark., Fla., Ga., Ill., Ind., Ks., La., Mich., Minn., Miss., Mo., N.Y., Oh., Ok., Sask., Tenn., Tex., Va.

Rheumatobates rileyi Bergroth, 1892 [Fig. 39]

1892 *Rheumatobates rileyi* Bergroth, Ins. Life, 4: 321. [N.Y.].

1926 *Rheumatobates rileyi* var. *rileyi*: Blatchley, Het. E. N. Am., p. 982.

1929 *Rheumatobates ribleyi* [*sic*]: Torre-Bueno, Bull. Brook. Ent. Soc., 24: 93.

Distribution: D.C., Fla., Ga., Ia., Ill., Ind., Ks., La., Man., Mass., Md., Mich., Miss., N.C., N.J., N.Y., Oh., Que., S.C., Sask., Tenn., Va., Vt

Note: Subspecies status for this form was eliminated when Bobb (1974, Ins. Va., 87: 78) gave species status to the form formerly known as subspecies *palosi* (see that name).

Rheumatobates tenuipes Meinert, 1895

1895 *Rheumatobates tenuipes* Meinert, Ent. Medd., 5: 7. [D.C.].

Distribution: Ark., D.C., Fla., Ga., Ill., Ky., La., Md., Miss., Mo., N.C., N.J., N.Y., Ok., S.C., Tenn., Tex., Va. (Belize)

Rheumatobates trulliger Bergroth, 1915

1915 *Rheumatobates trulliger* Bergroth, Bull. Brook. Ent. Soc., 10: 63. [Ga.].

Distribution: Ark., Fla., Ga., Ks., Miss., Mo., Ok., Tenn., Tex.

Rheumatobates vegatus Drake and Harris, 1942

1942 *Rheumatobates vegatus* Drake and Harris, Rev. Brasil. Biol., 2: 401. [Cuba].

1949 *Rheumatobates crinitus* Herring, Fla. Ent., 32: 160. [Fla.]. Synonymized by Hungerford, 1954, Univ. Ks. Sci. Bull., 36: 544.
1951 *Rheumatobates vegatus crinitus*: Drake and Hottes, Proc. Biol. Soc. Wash., 64: 149.
Distribution: Fla. (Cuba, Puerto Rico).

Subfamily Trepobatinae Matsuda, 1960

Genus *Metrobates* Uhler, 1871

1871 *Metrobates* Uhler, Proc. Boston Soc. Nat. Hist., 14: 108. Type-species: *Metrobates hesperius* Uhler, 1871. Monotypic.
1898 *Trepobatopsis* Champion, Biol. Centr.-Am., Rhyn., 2: 158. Type-species: *Trepobatopsis denticornis* Champion, 1898. Monotypic. Synonymized by Anderson, 1932, Univ. Ks., Sci. Bull., 20: 300.
Note: Anderson (1932, Univ. Ks. Sci. Bull., 20: 297-311) revised the genus. Drake and Harris (1945, Rev. Brasil. Biol., 5:-180) gave a list of species.

Metrobates alacris Drake, 1955
1955 *Metrobates alacris* Drake, J. Ks. Ent. Soc., 28: 130. [La.].
Distribution: Ark., La., Tex.

Metrobates anomalus Hussey, 1948
1948 *Metrobates anomalus* Hussey, Fla. Ent., 31: 123. [Fla.].
Distribution: Fla.

Metrobates anomalus anomalus Hussey, 1948
1948 *Metrobates anomalus* Hussey, Fla. Ent., 31: 123. [Fla.].
1949 *Metrobates anomalus anomalus*: Hussey and Herring, Fla. Ent., 32: 168.
Distribution: Fla.

Metrobates anomalus comatipes Hussey and Herring, 1949
1949 *Metrobates anomalus comatipes* Hussey and Herring, Fla. Ent., 32: 169. [Fla.].
Distribution: Fla.

Metrobates artus Anderson, 1932
1932 *Metrobates artus* Anderson, J. Ks. Ent. Soc., 5: 56. [Tex.].
Distribution: Tex. (Mexico).

Metrobates denticornis (Champion), 1898
1898 *Trepobatopsis denticornis* Champion, Biol. Centr.-Am., Rhyn., 2: 158. [Mexico].
1928 *Trepobatopsis denticornis*: Drake and Harris, Oh. J. Sci. Bull., 28: 273.
1932 *Metrobates denticornis*: Anderson, Univ. Ks. Sci. Bull., 20: 300.
Distribution: Ariz., N.M., Tex. (Mexico, Central America).

Metrobates hesperius Uhler, 1871
1871 *Metrobates hesperius* Uhler, Proc. Boston Soc. Nat. Hist., 14: 109. [No locality given].
1897 *Halobatopsis beginii* Ashmead, Can. Ent., 29: 56. [Que.]. Synonymized by Torre-Bueno, 1911, Can. Ent., 43: 226.
Distribution: Ala., Ark., Conn., Del., Fla., Ga., Ia., Ill., Ind., Ks., La., Man., Mass., Md., Me., Mich., Minn., Miss., Mo., N.C., N.J., N.Y., Oh., Ont., Pa., Que., R.I., S.C., Tenn., Va. (Haiti).

Metrobates hesperius depilatus Hussey and Herring, 1949
1949 *Metrobates hesperius depilatus* Hussey and Herring, Fla. Ent., 32: 168. [Fla.]
Distribution: Fla., Ok.(?).

Metrobates hesperius hesperius Uhler, 1871
1871 *Metrobates hesperius* Uhler, Proc. Boston Soc. Nat. Hist., 14: 109.
1949 *Metrobates hesperius hesperius*: Hussey and Herring, Fla. Ent., 32: 167.
Distribution: Same as for species.

Metrobates hesperius ocalensis Hussey and Herring, 1949
1949 *Metrobates hesperius ocalensis* Hussey and Herring, Fla. Ent., 32: 167. [Fla.].
Distribution: Fla.

Metrobates trux (Torre-Bueno), 1921
1921 *Trepobatopsis trux* Torre-Bueno, Ent. News, 32: 274. [Col.].
1932 *Metrobates trux*: Anderson, Univ. Ks. Sci. Bull., 20: 305.
Distribution: Ariz., Cal., Col., Id., Ks., N.M., Ore., Tex., Wash., Wyo.

Metrobates trux infuscatus Usinger, 1953 [Fig. 41]
1953 *Metrobates trux infuscatus* Usinger, Pan-Pac. Ent., 29: 178. [Cal.].
Distribution: Ariz., Cal., Id., N.M., Ore., Wash., Wyo.

Metrobates trux trux (Torre-Bueno), 1921
1921 *Trepobatopsis trux* Torre-Bueno, Ent. News, 32: 274.
1953 *Metrobates trux*: Usinger, Pan-Pac. Ent., 29: 179.
Distribution: Ariz., Cal., Col., Id., Ks., N.M., Tex.

Genus *Trepobates* Uhler, 1883

1883 *Stephania* White, Rept. Voyage Challenger, 7: 79. Preoccupied. Type-species: *Halobates pictus* Herrich-Schaeffer, 1847. Original designation.
1894 *Trepobates* Uhler, Proc. Zool. Soc. London, p. 213. New name for *Stephania* White.
1899 *Kallistometra* Kirkaldy, Ent., 32: 28. Type-species: *Kallistometra taylori* Kirkaldy, 1899. Original designation. Synonymized by Drake and Harris, 1932, Psyche, 39: 112.
Note: Drake and Chapman (1953, Fla. Ent., 36: 109-112) reviewed this genus.

Trepobates becki Drake and Harris, 1932 [Fig. 40]
1932 *Trepobates becki* Drake and Harris, Bull. Brook. Ent. Soc., 27: 120. [Holotype Mexico; also Ariz. record].
Distribution: Ariz., Cal., N.M., Ut.(?) (Mexico).

Trepobates carri Kittle, 1982
1982 *Trepobates carri* Kittle, Proc. Ent. Soc. Wash., 84: 157. [Holotype Guatemala; also Tex. and other localities].
Distribution: Tex. (Cuba, Jamaica, Mexico to Honduras)

Trepobates citatus Drake and Chapman, 1953
1953 *Trepobates citatus* Drake and Chapman, Fla. Ent., 36: 111. [Fla.].
Distribution: Fla., Miss.

Trepobates floridensis Drake and Harris, 1928
1928 *Trepobates floridensis* Drake and Harris, Oh. J. Sci., 28: 273. [Fla.].
Distribution: Ala., Fla., Ga., Miss.

Trepobates inermis Esaki, 1926
1926 *Trepobates inermis* Esaki, An. Hist. Nat. Mus. Hung., 23: 140. [Md.].
Distribution: Conn., Fla., Ga., Ia., Ill., Ind., Ks., Ky., La., Mass., Md., Mich., Miss., N.J.,

N.Y., Oh., Ont., Pa., Tenn., Tex., Va., W.Va. (West Indies).

Trepobates knighti Drake and Harris, 1928

1928 *Trepobates knighti* Drake and Harris, Proc. Biol. Soc. Wash., 41: 28. [Mo.].

Distribution: Ark., Ia., Ill., Ind., Ks., La., Mich., Minn., N.D., Mo., Ok., S.D., Tex.

Trepobates pictus (Herrich-Schaeffer), 1847

1847 *Halobates pictus* Herrich-Schaeffer, Wanz. Ins., 8: 111. [N. Am.].

1883 *Stephania picta*: White, Rept. Voyage Challenger, 7: 79.

1894 *Trepobates pictus*: Uhler, Proc. Zool. Soc. London, 1894: 213.

Distribution: Ala., Ariz., Ark., Conn., D.C., Del., Fla., Ga., Ill., Ind., Ks., Ky., La., Mass., Md., Me., Mich., Minn., Miss., Mo., N.C., N.H., N.J., N.Y., Oh., Ont., Pa., R.I., S.C., Tenn., Tex., Va., W.Va. (South America, West Indies).

Trepobates subnitidus Esaki, 1926

1926 *Trepobates subnitidus* Esaki, An. Hist. Nat. Mus. Hung., 23: 141. [Ind.].

Distribution: Ala., Ark., Conn., D.C., Fla., Ga., Ia., Ill., Ind., Ks., Ky., La., Mass., Me., Md., Mich., Minn., Miss., Mo., N.C., N.H., N.J., N.M., N.Y., Neb., Oh., Ok., Ont., Pa., R.I., S.C., Tenn., Tex., Va., W.Va.

Trepobates taylori (Kirkaldy), 1899

1899 *Kallistometra taylori* Kirkaldy, Ent., 32: 28. [Jamaica].

1928 *Trepobates comitialis* Drake and Harris, Fla. Ent., 12: 7. [Grenada]. Synonymized by Drake and Harris, 1932, Psyche, 39: 112.

1932 *Trepobates taylori*: Drake and Harris, Bull. Brook. Ent. Soc., 27: 118.

Distribution: Ariz.(?), N.M.(?), Tex. (Mexico to South America, West Indies).

Trepobates trepidus Drake and Harris, 1928

1928 *Trepobates trepidus* Drake and Harris, Proc. Biol. Soc. Wash., 41: 27. [Mexico].

1952 *Trepobates trepidus*: Drake and Hottes, Great Basin Nat., 12: 38.

Distribution: Ariz.(?), N.M.(?), Tex. (Mexico to South America, West Indies).

Family Hebridae Amyot and Serville, 1843

The Velvet Water Bugs

By John T. Polhemus and Dan A. Polhemus

The velvet water bugs derive their common name from the thick, velvety hydrofuge pile that covers their surface. These insects are small and inconspicuous, resembling tiny veliids but distinguishable from the latter group by their apical claws, well-developed scutellum, and presence on the underside of the head of a groove in which the beak rests. Of the seven genera recognized worldwide, *Hebrus* Curtis, *Merragata* White, and *Lipogomphus* Berg occur in North America.

Hebrids are generally found in semiaquatic habitats, either in moist detritus near shore (*Hebrus*, *Lipogomphus*) or among floating aquatic plants (*Merragata*). Drake (1917, Oh. J. Sci., 17: 101-105) observed individuals of the latter genus in an aquarium, and noted that they were able to walk both on the water surface and along submerged portions of plants, their hydrofuge pile providing protection from wetting. Eggs are laid on moss or algae (Polhemus and Chapman, 1979, Bull. Cal. Ins. Surv., 21: 34-38), with total development time ranging from one to two months. Nymphs and adults are predaceous on Collembola and other small arthropods and have been reported to be cannibalistic (Williams, 1944, Proc. Haw. Ent. Soc., 12: 186-197). Wing polymorphism is common in hebrid species, but certain taxa are known only from either the winged or brachypterous morphs. Macropterous adults occasionally fly in great numbers, and have been taken at altitudes of up to 612 m. (2000 ft.) in aerial samples (Glick, 1939, U.S. Dept. Agr. Tech. Bull., 673: 1-150; Usinger, 1956, Aq. Ins. Cal., p. 220). Species are known to overwinter as adults in temperate regions.

Hebridae are a cosmopolitan group, ranging over essentially the entire United States and southern Canada. The major works on this family for the Americas are those of Drake and Harris (1943, Notas Mus.

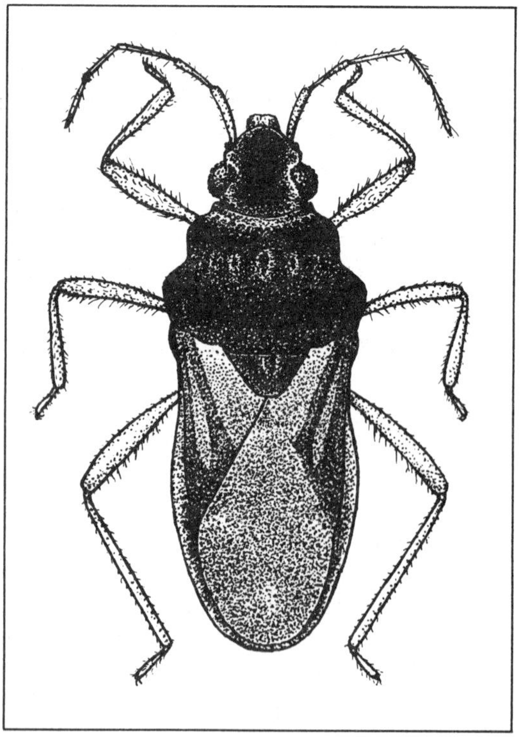

Fig. 43 *Hebrus concinnus* [p. 154] (After Froeschner, 1949).

La Plata, Zool., 8: 41-58), Drake and Chapman (1958, J. Wash. Acad. Sci., 48: 317-326), and Andersen (1981, Syst. Ent., 6: 377-412), the latter containing a key to genera. North American *Merragata* species were keyed by Drake (1917, Oh. J. Sci., 17: 101-105), but no complete key to *Hebrus* exists for our region. This family appears as Naeogeidae in some of the earlier literature, but this name is a junior synonym of Hebridae.

Subfamily Hebrinae Amyot and Serville, 1843

Genus *Hebrus* Curtis, 1833

1831 *Hebrus* Curtis, Guide Brit. Ins., col. 199. *Nomen nudum.*

1833 *Hebrus* Curtis, Ent. Mag., 1: 198. Type-species: *Lygaeus pusillus* Fallén, 1807. Monotypic.

1833 *Naeogeus* Laporte, Mag. Zool., 2: 34. Type-species: *Naeogeus erythrocephalus* Laporte, 1833. Monotypic. Synonymized by Harris, 1942, Pan-Pac. Ent., 18: 124.

Hebrus beameri Porter, 1952

1952 *Hebrus beameri* Porter, J. Ks. Ent. Soc., 25: 9. [Ks.].

1953 *Hebrus amnicus* Drake and Chapman, Great Basin Nat., 13: 9. [Ga.]. Synonymized by Polhemus and McKinnon, 1983, Proc. Ent. Soc. Wash., 85: 110.

Distribution: Ga., Ind., Ks., Oh.

Hebrus buenoi Drake and Harris, 1943

1943 *Hebrus buenoi* Drake and Harris, Notas Mus. La Plata, Zool., 8: 52. [N.Y.].

Distribution: Cal., Col., D.C., Fla., Ia., Id., Ill., Ks., La., Mass., Mich., Miss., Mo., N.J., N.Y., Neb., Oh., Ore., Pa., Va., Wis. (Mexico).

Note: Records of this species from the southern U.S. and Mexico refer to *H. buenoi furvus* Polhemus and Chapman.

Hebrus buenoi buenoi Drake and Harris, 1943

1943 *Hebrus buenoi* Drake and Harris, Notas Mus. La Plata, Zool., 8: 52. [N.Y.].

1970 *Hebrus buenoi buenoi*: Polhemus and Chapman, Proc. Ent. Soc. Wash., 72: 52.

Distribution: Cal., Col., D.C., Ia., Id., Ill., Ks., Mass., Mich., Mo., N.Y., Neb., Oh., Ore., Pa., Va., Wis.

Hebrus buenoi furvus Polhemus and Chapman, 1970

1926 *Naeogeus bilineatus*: Blatchley, Het. E. N. Am., p. 608.

1943 *Hebrus buenoi* Drake and Harris, Notas Mus. La Plata, Zool., 8: 52 (in part).

1970 *Hebrus buenoi furvus* Polhemus and Chapman, Proc. Ent. Soc. Wash., 72: 52. [La.].

Distribution: Fla., La., Miss. (Mexico).

Hebrus burmeisteri Lethierry and Severin, 1896

1835 *Hebrus pusillus* Burmeister, Handb. Ent., 2: 214. [Pa.]. Preoccupied.

1896 *Hebrus burmeisteri* Lethierry and Severin, Cat. Gen. Hem., 3: 51. New name for *Hebrus pusillus* Burmeister.

1909 *Naeogeus burmeisteri*: Kirkaldy and Torre-Bueno, Proc. Ent. Soc. Wash., 10: 214.

Distribution: D.C., Fla., Ga., Ia., Ill., Ks., Ky., Mass., Md., Mich., Mo., N.H., N.J., N.S., N.Y., Ont., Pa., Que., S.C., Sask., Va., Wis. (Mexico).

Hebrus comatus Drake and Harris, 1943
1943 *Hebrus comatus* Drake and Harris, Notas Mus. La Plata, Zool., 8: 53. [N.M.].
Distribution: Ks., N.M., Tex.

Hebrus concinnus Uhler, 1894 [Fig. 43]
1894 *Hebrus concinnus* Uhler, Proc. Zool. Soc. London, p. 221. [Grenada, British West Indies].
1909 *Naeogeus concinnus*: Kirkaldy and Torre-Bueno, Proc. Ent.Soc. Wash., 10: 214.
Distribution: Ala., Cal.(?), Col., Conn., Fla., Ill., La., Md., Miss., N.C., N.J., N.M., N.Y., Ok., Ont., Pa., Que., S.C., Tex., Wash.(?) (Mexico to South America, West Indies).
Note: Hungerford (1918, J. N.Y. Ent. Soc., 26: 12-18; 1920, Univ. Ks. Sci. Bull., 11: 1-328) provided notes on biology. Polhemus and Chapman (1979, Bull. Cal. Ins. Surv., 21: 36) suggested that the Cal. and Wash. records for this species, being very old, may have been based on misidentifications.

Hebrus consolidus Uhler, 1894
1894 *Hebrus consolidus* Uhler, Proc. Zool. Soc. London, p. 222. [Grenada].
1909 *Naeogeus consolidus*: Kirkaldy and Torre-Bueno, Proc. Ent. Soc. Wash., 10: 214.
Distribution: Fla., Ks., La., Miss. (Central America, West Indies).

Hebrus hubbardi Porter, 1952
1877 *Hebrus sobrinus*: Uhler, An. Rep. Chief Eng., Append. NN, p. 1330 (in part).
1952 *Hebrus hubbardi* Porter, J. Ks. Ent. Soc., 25: 10. [Cal.].
1952 *Hebrus piercei* Porter, J. Ks. Ent. Soc., 25: 147. [Ariz.]. Synonymized by Polhemus and Chapman, 1970, Proc. Ent. Soc. Wash., 72: 51.
Distribution: Ariz., Cal., Col., N.M., Nev., Tex.

Hebrus longivillus Polhemus and McKinnon, 1983
1983 *Hebrus longivillus* Polhemus and McKinnon, Proc. Ent. Soc. Wash., 85: 112. [Ariz.].
Distribution: Ariz.
Note: Some Arizona records of *Hebrus major* Champion may be misidentifications of this species.

Hebrus major Champion, 1898
1898 *Hebrus major* Champion, Biol. Centr.-Amer., Rhyn., 2: 118. [Mexico].
1958 *Hebrus major* Drake and Chapman, J. Wash. Acad. Sci., 48: 325.
Distribution: Ariz., Cal.(?), Tex. (Mexico).
Note: A record from California needs verification (Polhemus and Chapman, 1979, Bull. Cal. Ins. Surv., 21: 37).

Hebrus obscurus Polhemus and Chapman, 1966
1952 *Hebrus piercei*: Porter, J. Ks. Ent. Soc., 25: 147 (in part).
1966 *Hebrus obscura* [*sic*] Polhemus and Chapman, Proc. Ent. Soc. Wash., 68: 210. [Ariz.].
1979 *Hebrus obscurus*: Polhemus and Chapman, Bull. Cal. Ins. Surv., 21: 35.
Distribution: Ariz., Cal.

Hebrus sobrinus Uhler, 1877
1877 *Hebrus sobrinus* Uhler, Bull. U.S. Geol. Geogr. Surv. Terr., 3: 452. [Col.]. Neotype designated by J. Polhemus, 1977, Proc. Ent. Soc. Wash., 79: 237.
1877 *Hebrus sobrinus* Uhler, An. Rep. Chief Eng. Sec. War, 1877, appendix NN, p. 1330 (in part). [N.M.]. Preoccupied. Lectotype designated by J. Polhemus, 1975, Proc. Ent. Soc. Wash., 77: 128.

1893 *Hebrus pucellus* [*sic*]: Riley, N. Am. Fauna, 7: 250.

1909 *Naeogeus sobrinus*: Kirkaldy and Torre-Bueno, Proc. Ent. Soc. Wash., 10: 215.

Distribution: Ariz., Cal., Col., Ga., Ks., Mo., N.J., N.M., Nev., Tex., Va., Vt.

Note: Uhler described this species twice in the same year under the same name from localities in Col. and N.M., as discussed by J. Polhemus (1975, Proc. Ent. Soc. Wash., 77: 178; 1977, Proc. Ent. Soc. Wash., 79: 237).

Hebrus tuckahoanus Drake and Chapman, 1954

1954 *Hebrus tuckahoana* [*sic*] Drake and Chapman, Fla. Ent., 37: 152. [N.J.].

1958 *Hebrus tuckahoanus*: Drake and Chapman, J. Wash. Acad. Cal. Sci., 48: 322.

Distribution: Ill., N.J.

Genus *Merragata* White, 1877

1877 *Merragata* White, An. Mag. Nat. Hist., (4)20: 113. Type-species: *Merragata hebroides* White, 1877. Monotypic.

Note: Drake (1917, Oh. J. Sci., 17: 101-105) provided a key to North American species.

Merragata brunnea Drake, 1917

1917 *Merragata brunnea* Drake, Oh. J. Sci., 17: 105. [Oh.].

Distribution: "S. Canada," Fla., Ill., Ks., Mich., Minn., N.J., Neb., Oh., Tex.

Merragata hebroides White, 1877

1877 *Merragata hebroides* White, An. Mag. Nat. Hist., (4)20: 114. [Haw.].

1917 *Merragata foveata* Drake, Oh. J. Sci., 17: 103. [Oh.]. Synonymized by Blatchley, 1926, Het. E. N. Am., p. 610..

1921 *Merragata slossoni* Van Duzee, Proc. Cal. Acad. Sci., (4)11: 133. [Fla.]. Synonymized by Drake, 1952, Pan-Pac. Ent., 28: 194.

Distribution: Ala., Alta., Ariz., B.C., Cal., Col., Conn., Fla., Ia., Ill., Ind., Ks., Ky., La., Man., Mass., Mich., Minn., Mo., Miss., N.J., N.M., N.Y., Neb., Oh., Ont., Pa., Sask., Tex., Va., W.Va. (Hawaii, Mexico to Argentina, West Indies).

Genus *Lipogomphus* Berg, 1879

1879 *Lipogomphus* Berg, Hem. Argentina, p. 286. Type-species: *Lipogomphus lacuniferus* Berg, 1879. Monotypic.

Note: This genus was treated by Champion (1898, Biol. Centr.-Am., Rhyn., 2: 121) as a synonym of *Merragata*, but was resurrected by Andersen (1981, Syst. Ent., 6: 387).

Lipogomphus brevis (Champion), 1898

1898 *Merragata brevis* Champion, Biol. Centr.-Am., Rhyn., 2: 122. [Guatemala, Mexico, Panama].

1943 *Merragata brevis*: Drake and Harris, Notas Mus. La Plata, Zool., 8: 43.

1981 *Lipogomphus brevis*: Andersen, Syst. Ent., 6: 388.

Distribution: Cal.(?), Fla., Miss., Tex. (Mexico to Panama).

Note: Polhemus and Chapman (1979, Bull. Cal. Ins. Surv., 21: 38) remarked that the Cal. record needs verification.

Family Hydrometridae Billberg, 1820

The Marsh Treaders

By Cecil L. Smith

Although Hydrometridae are found in all zoogeographical regions, the majority of species are concentrated in tropical and subtropical regions. Worldwide, the family contains approximately 110 species in three subfamilies and seven genera (China and Usinger, 1949, Rev. Zool. Bot. Afr., 41: 314-319; Andersen, 1977, Ent. Scand., 8: 301-316). More than 100 of these, including the nine species known from North America, are in the cosmopolitan genus *Hydrometra* Latreille.

Hydrometrids are fragile, exceedingly slender bugs with long, threadlike legs and an elongated head. Their common names, water measurers or marsh treaders, stem from their usually slow, deliberate gait exhibited while traversing either a pleustonic or littoral substrate. In general, members of the family are found along margins or on emergent or floating vegetation of still or slowly moving water, rarely venturing out into the open water unless disturbed. Some primitive extralimital species are predominantly terrestrial, whereas others are commonly found on vertical rock surfaces above the water of streams or pools (Polhemus, 1982, Hem., Hydrometridae, pp. 313-314). Several records of occurrence for our species on brackish or saltwater have been reported (Herring, 1949, Fla. Ent., 31: 112-116; Polhemus and Chapman, 1979, Bull. Cal. Ins. Surv., 21: 43-45). The kinetics and dynamics of locomotion on the surface film have been studied for several species by Rudolph (1971, Forma Funct., 4: 454-464) and Andersen (1976, Viden. Medd. Natur. Foren., 139: 337-396)

The biology of the hydrometrids is, in most respects, similar to that noted for other families of the Gerromorpha - all species are apparently predators and/or scavengers, feeding mainly upon dead or disabled microarthropods that they can overcome on or

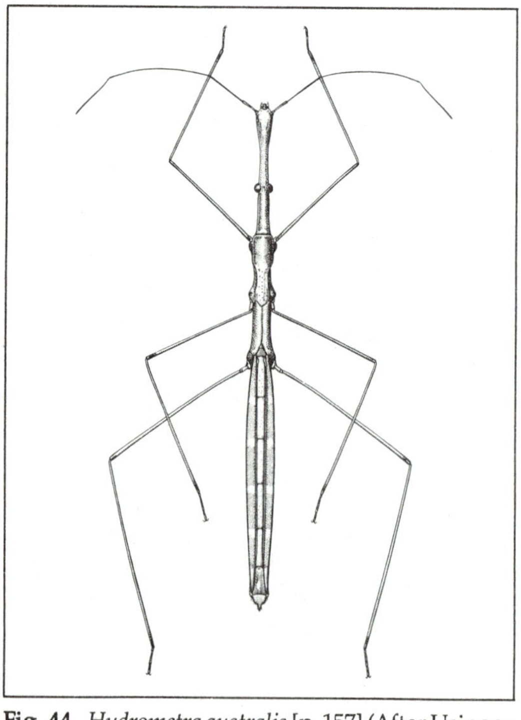

Fig. 44 *Hydrometra australis* [p. 157] (After Usinger, 1956).

within the surface film of the water or surrounding vegetation. Reported food sources include midges, collembolans, mayflies, mosquito larvae, and crustaceans. The life span of laboratory-reared individuals ranges from six to nine months, although the known extreme is 15 months.

Wing polymorphism occurs in most species, with the apterous or micropterous forms predominating. Macropterous individuals are relatively rare. Based on light trap records, flight is apparently a viable means of dispersal, but is rarely observed.

The taxonomy of the family is relatively stable although comprehensive keys are needed for the New World fauna. Andersen (1977, Ent. Scand., 8: 301-316) recently revised the family to the generic level and provided a cladistic analysis. The principal taxonomic works applicable to the North American species include Torre-Bueno's (1926, Ent. Am., 7: 83-128) key and descriptions for the Western Hemisphere species; Hungerford and Evans' (1934, An. Mus. Nat. Hung., 28: 31-112) key to world genera and species of *Hydrometra*; and Drake and Lauck's (1959, Great Basin Nat., 19: 43-52) checklist of New World species. Good reviews of hydrometrid literature are available in Polhemus and Chapman (1979, Bull. Cal. Ins. Surv., 21: 43-45) and Bennett and Cook (1981, Univ. Minn. Ag. Exp. Stn. Tech. Bull., 332: 3-58).

Subfamily Hydrometrinae Billberg, 1820

Genus *Hydrometra* Latreille, 1796

1796 *Hydrometra* Latreille, Précis Carac. Gén. Ins., p. 86. Type-species: *Cimex stagnorum* Linnaeus, 1758. Designated by Latreille, 1810, Consid. Gén., p. 434.

1835 *Limnobates* Burmeister, Handb. Ent., 2: 210. Type-species: *Cimex stagnorum* Linnaeus, 1758. Monotypic. Synonymy by virtue of common type-species.

Note: Hungerford and Evans (1934, An. Mus. Nat. Hung., 28: 31-112) published a key to the world species; Drake and Lauck (1959, Great Basin Nat., 19: 50-52) gave a list of the New World species.

Hydrometra aemula Drake, 1956

1956 *Hydrometra aemula* Drake, Proc. Biol. Soc. Wash., 69: 153. [Mexico].

1973 *Hydrometra aemula*: Polhemus, Great Basin Nat., 33: 114.

Distribution: Ariz. (Mexico).

Note: *H. ciliata* Mychajliw (1961, J. Ks. Ent. Soc., 34: 29), described from Mexico, was synonymized with *H. aemula* Drake by Polhemus (1973, Great Basin Nat., 33: 114).

Hydrometra australis Say, 1832 [Fig. 44]

1832 *Hydrometra lineata* var. *australis* Say, Descrip. Het. Hem. N. Am., p. 35. [La.].

1905 *Hydrometra australis*: Torre-Bueno, Can. Ent., 37: 264.

1926 *Hydrometra myrae* Torre-Bueno, Ent. Am., 7: 110. [Ga.]. Synonymized by Drake and Hottes, 1952, J. Ks. Ent. Soc., 25: 108.

1968 *Hydrometra martini*: Burdick, Pan-Pac. Ent., 44: 81 (in part).

Distribution: Cal., Fla., Ga., La., Miss., Tex., Va. (Mexico to Central America, West Indies).

Note: Lanciani (1971, An. Ent. Soc. Am., 64: 1254-1259) gave notes on biology [as *H. myrae*].

Hydrometra barei Hungerford, 1927

1927 *Hydrometra barei* Hungerford, An. Ent. Soc. Am., 20: 262. [Fla.].

Distribution: Fla.

Hydrometra beameri Mychajliw, 1961

1961 *Hydrometra beameri* Mychajliw, J. Ks. Ent. Soc., 34: 27. [Mexico].

Distribution: Tex. (Mexico).

Hydrometra consimilis Barber, 1923

1923 *Hydrometra consimilis* Barber, Am. Mus. Novit., 75: 9. [Puerto Rico].

1954 *Hydrometra consimilis*: Hungerford, J. Ks. Ent. Soc., 27: 80.

Distribution: Fla. (Mexico, West Indies).

Hydrometra hungerfordi Torre-Bueno, 1926

1923 *Hydrometra australis*: Hungerford, Can. Ent., 55: 55.

1926 *Hydrometra hungerfordi* Torre-Bueno, Ent. Am., 7: 107. [Ks.].

1958 *Hydrometra hungerfordii* [*sic*]: Roback, Trans. Am. Ent. Soc., 84: 11.

Distribution: Ark., D.C., Fla., Ga., Ks., La., Me., Miss., Mo., N.J., N.Y., S.C., Va. (Central America).

Hydrometra lillianis Torre-Bueno, 1926

1926 *Hydrometra lillianis* Torre-Bueno, Ent. Am., 7: 108. [Cal.].

Distribution: Cal.

Note: Polhemus and Chapman (1979, Bull. Cal. Ins. Surv., 21: 45) suggested that the only known specimens, including the types, may be labeled erroneously for Cal.

Hydrometra martini Kirkaldy, 1900

1832 *Hydrometra lineata* Say, Descrip. Hem. Het. N. Am., p. 35. [U.S.]. Preoccupied.

1886 *Limnobates lineata*: Uhler, Check-list Hem. Het., p. 26.

1900 *Hydrometra martini* Kirkaldy, Ent., 33: 175. New name for *Hydrometra lineata* Say.

Distribution: Ark., Ariz., B.C., Conn., D.C., Fla., Ga., Ia., Id., Ill., Ind., Ks., La., Md., Me., Man., Mass., Md., Mich., Minn., Miss., Mo., N.C., N.J., N.Y., Oh., Ont., Ore., Pa., Que., S.C., S.D., Tenn., Tex., Va.

Note: Polhemus and Chapman (1979, Bull. Cal. Ins. Surv., 21: 44) transferred a Cal. record under this name to *H. australis* and stated that this species may prove to be conspecific with *H. australis* Say. Martin (1900, Can. Ent., 32: 70-76 [as *H. lineata*]) and Sprague (1956, Univ. Ks. Sci. Bull., 38: 579-693) studied its biology and morphology.

Hydrometra wileyae Hungerford, 1923

1923 *Hydrometra wileyi* Hungerford, Can. Ent., 55: 54. [Tex.].

Distribution: Fla., Ks., La., Miss., Tex. (Mexico).

Note: Named after Mrs. Grace Wiley.

Family Largidae Amyot and Serville, 1843

The Largid Bugs

By Thomas J. Henry

The family Largidae is a small group of medium to large bugs. They lack ocelli, have a four-segmented rostrum, and the membrane of each hemelytron has seven to eight veins arising from two closed, basal cells. Members of the subfamily Arhaphinae Bliven are somewhat antlike, slender, with bulbous heads, constricted pronota, and, often, brachypterous black hemelytra, marked with large white or yellow spots. They usually are found running along sandy or gravelly soil littered with dried plant material and other debris. The Larginae Amyot and Serville are large, oval, often red and black, stout-bodied bugs. This subfamily usually is found on the foliage of plants. Immatures of several species, such as *Largus succinctus* (Linnaeus), are a beautiful steel blue, with a large, centrally located, orange to red spot on the dorsum.

This family, despite its small size, is in great need of revision. Most of the literature prior to 1970 is outdated and includes the Largidae with the family Pyrrhocoridae.

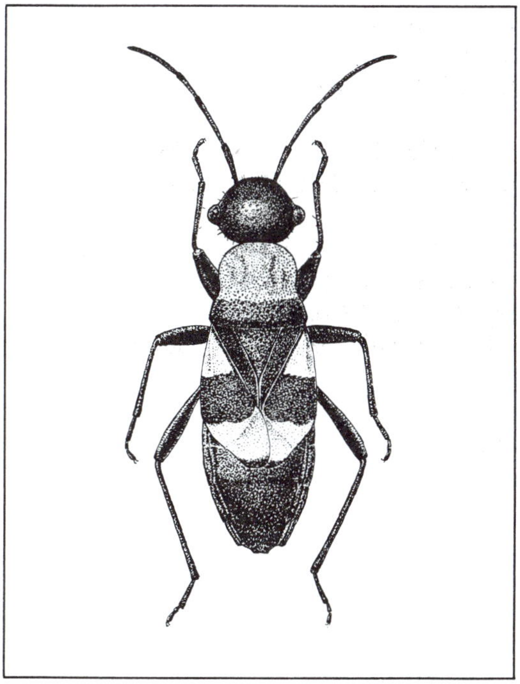

Fig. 45 *Araphe carolina* [p. 160] (After Froeschner, 1944).

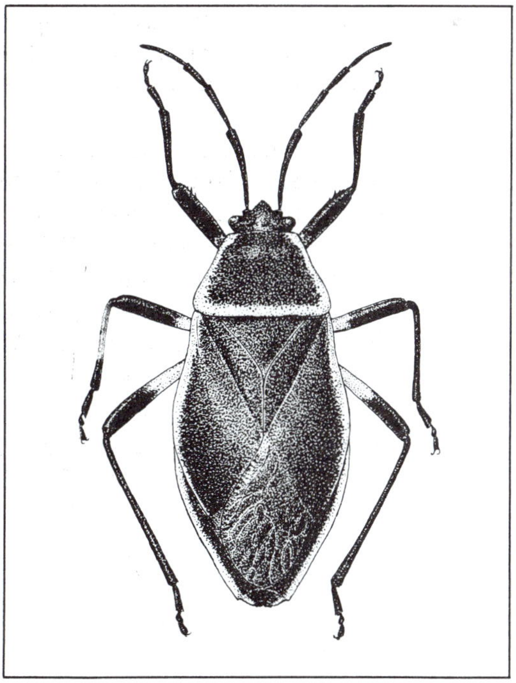

Fig. 46 *Largus succinctus* [p. 164] (After Froeschner, 1944).

China (1954, Ent. Month. Mag., 90: 188-189) clarified that Largidae is the proper name for the family. A summary of the early literature for the family was given in Hussey's (1929, Gen. Cat. Hem., 3: 1-144) catalog of the Pyrrhocoridae of the world. Torre-Bueno's (1941, Ent. Am., 21: 107-115) Synopsis, now badly outdated, provided keys to the North American genera and species included in the Hussey Catalog.

Most of the confusion in the North American genera and species resulted from a series of papers by two workers who disagreed on the definition of most of the taxa that they treated. Halstead provided keys to the U.S. species of *Arhaphe* Herrich-Schaeffer (1972, Pan-Pac. Ent., 48: 1-7) and *Largus* Hahn (1972, Pan-Pac. Ent., 48: 246-248), but a later paper by Bliven (1973, Occ. Ent., 1: 123-133) contradicted many of Halstead's species concepts and synonymies. A thorough revision of all genera and their species, with a careful study of types, will be necessary to sort out the confusion that now exists.

Subfamily Arhaphinae Bliven, 1973

Note: Halstead (1972, Pan-Pac. Ent., 48: 1-7) considered *Japetus* Distant and *Jarhaphetus* Bliven synonyms of *Arhaphe* Herrich-Schaeffer, but Bliven (1973, Occ. Ent., 1: 126-129) argued that all three genera are distinct. Until types of the included species are restudied, all three genera must be considered valid. A new name is provided for the preoccupied *Japetus* Distant under that generic listing. Bliven's (1973, above) rationale that this subfamily belongs in the family Alydidae is not followed in this catalog.

Genus *Arhaphe* Herrich-Schaeffer, 1850

1850 *Arhaphe* Herrich-Schaeffer, Wanz. Ins., 9: 175. Type-species: *Arhaphe carolina* Herrich-Schaeffer, 1850. Monotypic.

Arhaphe breviata Barber, 1924

1924 *Arhaphe breviata* Barber, Can. Ent., 56: 227. [Ks.].
Distribution: Ks.

Arhaphe carolina Herrich-Schaeffer, 1850 [Fig. 45]

1850 *Arhaphe carolina* Herrich-Schaeffer, Wanz. Ins., 9: 183. ["Carolina"].
1871 *Araphe* [*sic*] *carolina*: Uhler, Bull. U.S. Geol. Surv. Wyo., p. 471.
Distribution: Ala., Ariz.(?), Fla., Ga., La., Mo., N.C., N.M.(?), Tenn., Tex. (Mexico?).
Note: The western U.S. and Mexican records for this species need verification. Blatchley (1926, Het. E. N. Am., p. 440) gave biological notes.

Arhaphe cicindeloides Walker, 1873

1873 *Arhaphe cicindeloides* Walker, Cat. Hem. Brit. Mus., 6: 36. [Mexico].
1911 *Araphe* [*sic*] *cicindeloides*: Barber, J. N.Y. Ent. Soc., 19: 28.
1973 *Arhaphe snowi* Bliven, Occ. Ent., 1: 127. [See note below].
Distribution: Ariz., N.M. (Mexico).
Note: Uhler (1886, Check-list Hem. Het., p. 16) first synonymized this species with *A. carolina*, but all subsequent authors, except Bliven (1973, above), recognized it as distinct. Because no one has restudied Walker's type (s) [and in my

opinion, Distant's (1893, Biol. Centr.-Am., Rhyn., 1: plate 21, fig. 2) figure of *cicindeloides* agrees with current concepts of the species], it seems best to continue recognizing *A. cicindeloides* and consider Bliven's *snowi* a synonym. In addition, Halstead (1972, Pan-Pac. Ent., 48: 2) synonymized Bliven's (1956, New Hem. W. St., p. 11) *Jarhaphetus argutus* with *A. cicindeloides*, but Bliven (1973, above) argued to maintain the species and its generic placement. Again, until types are restudied, Bliven's species is recognized in the genus *Jarhaphetus* (which see). Torre-Bueno (1942, Bull. Brook. Ent. Soc., 37: 68-69) gave ecological notes on this species.

Genus *Jarhaphetus* Bliven, 1956

1956 *Jarhaphetus* Bliven, New Hem. W. St., p. 10. Type-species: *Jarhaphetus argutus* Bliven, 1956. Original designation.

Note: Halstead (1972, Pan-Pac. Ent., 48: 1) considered *Jarhaphetus* a synonym of *Arhaphe* Herrich-Schaeffer (along with *Japetus* Distant), but Bliven (1973, Occ. Ent., 1: 127) argued to maintain it as a separate genus.

Jarhaphetus argutus Bliven, 1956

1956 *Jarhaphetus argutus* Bliven, New Hem. W. St., p. 11. [Ariz.].

Distribution: Ariz.

Note: Halstead (1972, Pan-Pac. Ent., 48: 2) considered this species a junior synonym of *Arhaphe cicindeloides* Walker (which see). Bliven (1973, Occ. Ent., 1: 127) disagreed with Halstead and considered the species that Halstead and other American authors have called *A. cincindeloides* to be unnamed; he renamed this "misidentified" species *snowi*. Only the study of type material can clarify this confusion.

Genus *Pararhaphe*, New Name

1883 *Japetus* Distant, Biol. Centr.-Am., Rhyn., 1: 227. Preoccupied. [By *Japetus* Stål, 1863, Stett. Ent. Zeit., 24: 244]. Type-species: *Japetus sphaeroides* Distant, 1883. Monotypic.

Note: Barber (1911, J. N.Y. Ent. Soc., 19: 28) synonymized the genus *Japetus* with *Arhaphe* Herrich-Schaeffer but later reinstated it to generic status (Barber, 1924, Can. Ent., 56: 227). Halstead (1972, Pan-Pac. Ent., 48: 1), in a review of the genus, once more synonymized *Japetus* with *Arhaphe* (along with *Jarhaphetus*), but the latest treatment by Bliven (1973, Occ. Ent., 1: 128) again recognized it as a separate genus. *Pararhaphe sphaeroides* (Distant) is a new combination for the type-species.

Pararhaphe mimetica (Barber), 1911, New Combination

1911 *Araphe* [*sic*] *mimetica* Barber, J. N.Y. Ent. Soc., 19: 28. [Ariz.].

1913 *Arrhaphe* [*sic*] *mimetica*: Bergroth, Mem. Soc. Ent. Belg., 22: 166.

1916 *Arhaphe mimetica*: Van Duzee, Check List Hem., p. 24.

1924 *Japetus mimetica*: Barber, Can. Ent., 56: 227.

1929 *Japetus mimeticus*: Hussey, Gen. Cat. Hem., 3: 28.

Distribution: Ariz.

Note: Torre-Bueno (1942, Bull. Brook. Ent. Soc., 37: 68-69) gave ecological notes on this species.

Subfamily Larginae Amyot and Serville, 1843

Genus *Acinocoris* Hahn, 1834

1834 *Acinocoris* Hahn, Wanz. Ins., 2: 113. Type-species: *Acinocoris calidus* Hahn, 1834, a junior synonym of *Cimex lunaris* Gmelin, 1788. Monotypic.

Acinocoris lunaris (Gmelin), 1787

1787 *Cimex lunatus* Fabricius, Mant. Ins., 2: 302. ["Cajennae"]. Preoccupied.
1788 *Cimex lunaris* Gmelin, Sys. Nat., Ed. 13, 4: 2178. ["Cayenna"].
1876 *Acinocoris lunatas* [*sic*]: Uhler, Bull. U.S. Geol. Geogr. Surv. Terr., 1: 315.
1883 *Largus lunatus*: Distant, Biol. Centr.-Am., Rhyn., 1: 221.
1886 *Acinocoris lunatus*: Uhler, Check-list Hem. Het., p. 16.
1902 *Euryophthalmus lunaris*: Kirkaldy and Edwards, Wien. Ent. Zeit., 21: 163.
1918 *Euryophthalmus* (*Largus*) *lunatus*: Englehardt, Bull. Brook. Ent. Soc., 13: 40.
1929 *Acinocoris lunaris*: Hussey, Gen. Cat. Hem., 3: 11.
Distribution: Cal. (Mexico to South America, West Indies).

Genus *Largus* Hahn, 1831

1831 *Largus* Hahn, Wanz. Ins., 1: 13. Type-species: *Cimex humilis* Drury, 1782. Monotypic.

Largus bipustulatus Stål, 1861

1861 *Largus bipustulatus* Stål, Öfv. K. Svens. Vet.-Akad. Förh., 18(4): 196. [Mexico].
1912 *Largus bipustulatus*: Torre-Bueno, Ent. News, 23: 121.
1916 *Euryophthalmus bipustulatus*: Van Duzee, Check List Hem., p. 24.
Distribution: Tex. (Mexico, Central America).
Note: Halstead (1972, Pan-Pac. Ent., 48: 248) retracted his previous identification of this species (Halstead, 1970, Pan-Pac. Ent., 46: 46) from Brownsville, Tex., stating that he actually had *L. maculatus* Schmidt; however, he did not refute Torre-Bueno's (1912, above) record. Specimens from the United States determined as *L. bipustulatus* (including many in the U.S. Nat. Mus.) need verification.

Largus californicus (Van Duzee), 1923

1917 *Euryophthalmus convivus*: Van Duzee, Proc. Cal. Acad. Sci., (4)7: 256.
1923 *Euryophthalmus cinctus californicus* Van Duzee, Can. Ent., 55: 270. [Cal.].
1931 *Largus succinctus californicus*: Schmidt, Stett. Ent. Zeit., 92: 23.
1973 *Largus californicus*: Bliven, Occ. Ent., 1: 129.
Distribution: Cal., Col. (Mexico).
Note: Halstead (1972, Pan-Pac. Ent., 48: 246) considered this species, along with *L. semipletus* Bliven and *L. sculptilis* Bliven, a junior synonym of *L. cinctus* Herrich-Schaeffer. Bliven (1973, above), however, disagreed and reversed Halstead's action.

Largus cinctus Herrich-Schaeffer, 1842

1832 *Capsus succinctus* var. A Say, Descrip. Het. Hem. N. Am., p. 20. [Mexico].
1842 *Largus cinctus* Herrich-Schaeffer, Wanz. Ins., 7: 6. [Mexico].
1860 *Largus cinctus*: Uhler, Proc. Acad. Nat. Sci. Phila., 7: 230.
1916 *Euryophthalmus cinctus*: Van Duzee, Check List Hem., p. 24.

Distribution: Ariz., Cal., Col., N.M., Nev., Ore., Tex. (Mexico to South America.

Note: Bliven (1973, Occ. Ent., 1: 130) synonymized this species with *L. succinctus* (Linnaeus), basing most of his decision on tentative comments by previous authors; however, considering the confusion now existing in this genus, further study of types is necessary before such formal synonymy should be recognized. Slater and Baranowski (1978, How To Know True Bugs, p. 97) state: "There is need for further study to determine whether or not some of these "species" are actually distinct forms or merely geographic races."

Largus convivus Stål, 1861

1861 *Largus convivus* Stål, Öfv. K. Svens. Vet.-Akad. Förh, 18(4): 196. [Mexico].

1893 *Largus succinctus*: Osborn, Trans. Ia. Acad. Sci., 1: 120.

1906 *Largus convivus*: Barber, Mus. Brook Inst. Arts Sci., Sci. Bull., 1: 277.

1910 *Stiretrus ancherago* [*sic*]: Kirkaldy, Proc. Haw. Ent. Soc., 2: 124.

1916 *Euryophthalmus convivus*: Van Duzee, Check List Hem., p. 24.

Distribution: Cal., Col., Tex. (Mexico, Central America).

Note: Bliven (1973, Occ. Ent., 1: 130) synonymized this species with *L. succinctus*, but as stated under *L. cinctus*, because of the confusion existing in the genus, formal synonymy should await the study of types.

Largus davisi Barber, 1914

1876 *Largus succinctus*: Uhler, Bull. U.S. Geol. Geogr. Surv. Terr., 1: 315 (in part).

1914 *Largus davisi* Barber, Bull. Am. Mus. Nat. Hist., 33: 507. [Fla.].

1916 *Euryophthalmus davisi*: Van Duzee, Check List Hem., p. 24.

Distribution: Fla.

Note: Blatchley (1926, Het. E. N. Am., p. 439) gave notes on the biology and life stages of this species.

Largus maculatus Schmidt, 1931

1931 *Largus maculatus* Schmidt, Stett. Ent. Zeit., 92: 26. [Colombia, Costa Rica].

1970 *Largus bipustulatus*: Halstead, Pan-Pac. Ent., 46: 46.

1972 *Largus maculatus*: Halstead, Pan-Pac. Ent., 48: 248.

Distribution: Ariz.(?), Tex. (Colombia, Costa Rica).

Note: See note under *L. bipustulatus* Stål.

Largus pallidus Halstead, 1972

1972 *Largus pallidus* Halstead, Can. Ent., 104: 959. [Fla.].

Distribution: Fla.

Note: Bliven (1973, Occ. Ent., 1: 124) synonymized this species with *L. sellatus* without studying types; therefore, as with several other "controversial" species, formal synonymy should await the study of types during careful revisionary work.

Largus sculptilis Bliven, 1959

1959 *Largus sculptilis* Bliven, Occ. Ent., 1: 27. [Ariz.].

Distribution: Ariz.

Note: Halstead (1972, Pan-Pac. Ent., 48: 247) considered this species a synonym of *L. cinctus* Herrich-Schaeffer. Bliven (1973, Occ. Ent., 1: 131) argued to keep *L. sculptilis* as a distinct species and, in turn, synonymized Halstead's *L. semipunctatus* with *L. sculptilis*. These closely dated contradictions indicate that this group needs careful revision. Both of the above species are maintained as distinct from *L. cinctus* until types are studied.

Largus sellatus (Guérin-Méneville), 1857
1857 *Lygaeus* (*Largus*) *sellatus* Guérin-Méneville, Sagra's Hist. Cuba, 7: 401. [Cuba].
1914 *Largus sellatus*: Barber, Bull. Am. Mus. Nat. Hist., 33: 509.
1916 *Euryophthalmus sellatus*: Van Duzee, Check List Hem., p. 24.
Distribution: Fla. (Cuba).
Note: Myers (1927, An. Ent. Soc. Am., 20: 280) described some of the immature stages and gave biological notes.

Largus semipletus Bliven, 1959
1959 *Largus semipletus* Bliven, Occ. Ent., 1: 26. [Cal.].
Distribution: Cal.
Note: Halstead (1972, Pan-Pac. Ent., 46: 247) synonymized this species with *L. cinctus* Herrich-Schaeffer, but Bliven (1973, Occ. Ent., 1: 131) maintained it as distinct, commenting that "It is in many ways very close to *Wupatkius* [*semo* Bliven]." See notes under the genus *Wupatkius* Bliven.

Largus semipunctatus Halstead, 1970
1970 *Largus semipunctatus* Halstead, Pan-Pac. Ent., 46: 45. [Ariz.].
Distribution: Ariz.
Note: Bliven (1973, Occ. Ent., 1: 131) synonymized this species with his *L. sculptilis*, but as discussed under that species, it is maintained as distinct until types can be studied.

Largus succinctus (Linnaeus), 1763 [Fig. 46]
1763 *Cimex succinctus* Linnaeus, Cent. Ins. Rarior., p. 17. [Pa.].
1794 *Lygaeus succinctus*: Fabricius, Ent. Sys., 4: 170.
1835 *Largus succinctus*: Burmeister, Handb. Ent., 2: 283.
1969 *Lagrus* [*sic*] *cinctus*: Rathvon, Hist. Lancaster Co., Pa., p. 548.
1876 *Largus succinctus*: Uhler, Bull. U.S. Geol. Geogr. Surv. Terr., 1: 315 (in part).
1902 *Euryophthalmus succinctus*: Kirkaldy and Edwards, Wien. Ent. Zeit., 21: 162 (in part).
Distribution: Ariz.(?), Cal.(?), Col., Fla., Md., Minn., Miss., N.C., N.J., N.M.(?), N.Y., Ok., Pa., Tex., Va.
Note: Records of this species west of Texas may refer to *L. cinctus* if, indeed, the two are distinct species. See discussion under *L. cinctus*. Blatchley (1926, Het. E. N. Am., p. 438-439) gave biological notes on this species.

Genus *Stenomacra* Stål, 1870

1870 *Stenomacra* Stål, K. Svens. Vet.-Akad. Handl., 9(1): 90. Type-species: *Largus marginella* Herrich-Schaeffer, 1850. Designated by Van Duzee, 1916, Check List Hem., p. 24.

Stenomacra cliens (Stål), 1862
1862 *Theraneis cliens* Stål, Stett. Ent. Zeit., 23: 315. [Mexico].
1910 *Stenomacra marginella*: Barber, J. N.Y. Ent. Soc., 18: 37.
1914 *Stenomerca* [*sic*] *cliens*: Barber, J. N.Y. Ent. Soc., 22: 170.
1916 *Stenomacra cliens*: Van Duzee, Check List Hem., p. 24.
Distribution: Ariz., N.M. (Mexico, Costa Rica).

Stenomacra marginella (Herrich-Schaeffer), 1850
1850 *Largus marginella* Herrich-Schaeffer, Wanz. Ins., 9: 182. [Mexico].
1916 *Stenomacra marginella*: Van Duzee, Check List Hem., p. 24.

Distribution: Ariz., Cal. (Mexico to South America).

Genus *Wupatkius* Bliven, 1956

1956 *Wupatkius* Bliven, New Hem. W. St., p. 10. Type-species: *Wupatkius semo* Bliven, 1956. Original designation.

Note: Halstead (1972, Pan-Pac. Ent., 48: 248) synonymized this genus with *Largus*, considering the type-species a junior synonym of *L. convivus* Stål, but Bliven (1973, Occ. Ent., 1: 131) continued to recognize his monotypic genus. Until the "different" trichobothrial pattern noted by Bliven can be re-evaluated, this genus is recognized.

Wupatkius semo Bliven, 1956

1956 *Wupatkius semo* Bliven, New Hem. W. St., p. 10. [Ariz.].

Distribution: Ariz.

Note: This species and its generic placement need restudy, as noted above.

Family Leptopodidae Amyot and Serville, 1843

The Leptopodid Bugs

By Richard C. Froeschner

This family of slightly more than two dozen species is essentially an Old World group except for an amber fossil species in Mexico and a termite-nest frequenting species from Ecuador. The lone species in our region is *Patapius spinosus* Rossi, a native of the Mediterranean area in both Europe and North Africa. It was first reported from North America by Usinger (1941, Bull. Brook. Ent. Soc., 36: 164-165) on the basis of a specimen found under tree protectors in an orchard in California; it is now known to occur 300 miles inland in Nevada. These insects, despite their saldid-like aspect, are not restricted to areas along the margins of bodies of water but also occur in cultivated fields and under a variety of objects on very dry land.

The standard comprehensive works on this family are Horvath's (1911, An. Mus. Nat. Hung., 9: 358-370) "Revision" and Drake and Hoberlandt's (1951, Acta Ent. Mus. Nat. Pragae, 26: 1-5) "Check-List." Subsequent works by Schuh and Polhemus (1980, Syst. Ent., 29: 1-26; and 1980, Am. Mus. Nov., 2698: 1-5) modify the family definition and discuss phylogenetic matters concerning it.

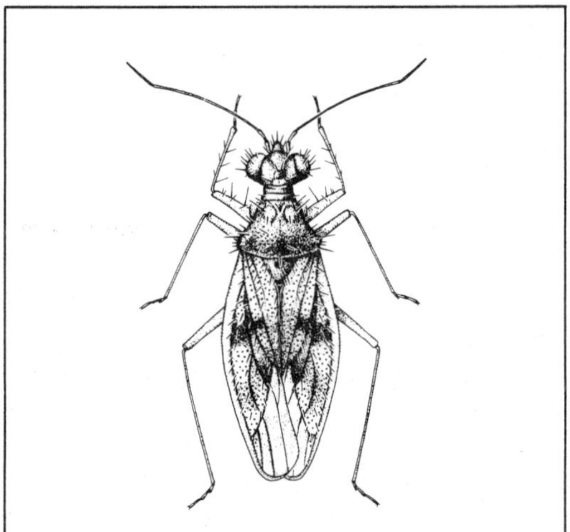

Fig. 47 *Patapius spinosus* [p. 166] (After Froeschner and Peña, 1985).

Subfamily Leptopodinae Amyot and Serville, 1843

Genus *Patapius* Horvath, 1912

1911 *Cryptoglena* Horvath, An. Mus. Nat. Hung., 9: 361, 368. Preoccupied. Type-species: *Acanthia spinosa* Rossi, 1790. Monotypic.

1912 *Patapius* Horvath, An. Mus. Nat. Hung., 10: 609. New name for *Cryptoglena* Horvath.

Patapius spinosus (Rossi), 1790 [Fig. 47]

1790 *Acanthia spinosa* Rossi, Fauna Etrusca, 2: 224. [Italy].

1941 *Patapius spinosus*: Usinger, Bull. Brook. Ent. Soc., 36: 164.

Distribution: Cal., Nev. (Canary Islands, Europe, North Africa; introduced into Chile).

Family Lygaeidae Schilling, 1829

(= Infericornes Amyot and Serville, 1843; Myodochidae Kirkaldy, 1899; Geocoridae Kirkaldy, 1902)

The Seed Bugs and Chinch Bugs

By Peter D. Ashlock and Alex Slater

With about 3,000 species in the world, the family Lygaeidae is second in numbers only to the Miridae in the Heteroptera. Here we list 322 species and 11 subspecies in 83 genera and 1 subgenus for America north of Mexico. J. A. Slater's (1964a) *Catalogue of the Lygaeidae of the World* listed 285 species and 21 subspecies in 73 genera and 1 subgenus for the same area. Not evident in these figures are the many synonymies and generic changes that have been made since 1964.

The most familiar Lygaeidae, large species of *Oncopeltus* Stål, *Lygaeus* Fabricius, and related genera (Lygaeinae), are usually bright red or orange marked with black, white, or occasionally yellow. Members of other subfamilies are more somber, ranging from pale straw yellow through various browns to black. Many species have black, brown, or white markings, and a few have tinges of red.

Lygaeids are morphologically diverse. They range in length in North America from about 20 mm (*Oncopeltus*) to less than 2 mm (*Plinthisus* Stephens, *Antillocoris* Kirkaldy, etc.). Some are long and narrow (*Ischnodemus* Fieber, *Paromius* Fieber); others are short and broad (*Geocoris* Fallén, *Isthmocoris* McAtee). Members can be recognized by the traditional characters. They have only about five veins in the membrane, with few if any crossveins, and the antennae are inserted fairly low on the head, below a line drawn between the center of the eye and the apex of the tylus.

The wings, while usually fully developed (macropterous), are often reduced in flightless morphs. Wing polymorphism is unrelated to sex in the Lygaeidae. The flightless form, when it is present, is the more common (*Blissus leucopterus* (Say) is an exception). Sweet (1964a) has shown that flightless morphs are usually found in species living in permanent habitats. Slater (1975a, 1977) has provided a convenient and generally useful classification of wing polymorphism in studies of the Australian and South African lygaeid faunas. Total aptery is unknown in the lygaeids, but microptery, in which the hemelytra become small pads with or without reduced membranes and so widely separated as to expose all of the abdominal segments, is common in the genus *Blissus* Burmeister. Staphylinoidy, with the clavus and corium short and fused, the claval commissure lengthened and straight, and the membrane absent, is seen in *Thylochromus* Barber, *Neosuris* Barber, and some *Plinthisus*. Coleoptery, with the corium and clavus fused and lengthened, the claval commissure straight and longer than normal, and the membrane reduced or absent, is found in *Ashlockaria* Harrington and *Cnemodus* Herrich-Schaeffer. More common is brachyptery, in which the clavus and corium may or may not be fused, but there is a distinct claval commissure, the corium is more or less shorter than in macropterous forms, and the membrane is obviously smaller than normal. The least obvious modification is termed submacroptery: the apparently normal hemelytron is actually shorter than those of macropterous individuals.

Many species, especially in the Rhyparochrominae and Pachygronthinae, have swollen fore femora armed with one to many spines; the fore femur in other groups usually has about the same structure as the mid and hind femora. Because of the enlarged fore femur, earlier workers suggested that many members of the Lygaeidae were predaceous. However, Puchkov (1956), Ashlock (1958), and notably Sweet (1960, 1964a, b) have all shown that the family consists primarily of insects that feed on dry mature seeds, and Sweet (1960) suggested that a suitable common name for the family Lygaeidae would be the "seed bugs." A few members of the family feed on the stems of grasses. The only predaceous lygaeids in the North American fauna are the Geocorinae, whose members also require seeds in their diet. Exclusively predaceous forms and species that feed on the blood of both vertebrates and insects (tribe Cleradini) are known outside of North America.

Oncopeltus fasciatus (Dallas), the large milkweed bug, is an important laboratory animal. Used to screen insecticides, it has been the subject of extensive physiological, morphological, and ecological studies. A comprehensive bibliography on *O. fasciatus* is in preparation by Alex Slater. Colonies are easy to maintain, for these bugs need only a supply of milkweed seeds and water. Sweet (1960, 1964a, b) demonstrated that most species of Rhyparochrominae are easy to rear on hulled sunflower seeds and water, though some diapausing species present problems. Potentially, many species of the family may be brought into the laboratory for intensive study.

Stridulation, when present, plays an important role in lygaeid mating behavior. Sweet (1964a, b) has provided most such data on the Lygaeidae, especially about mating and seed defense behavior. Ashlock and Lattin (1963) identified the North American genera with stridulating members.

Chromosomes of more than a tenth of the lygaeid species of the world have been studied (Ueshima and Ashlock, 1980); 74 of these are from America north of Mexico. Chromosome number ranges in the family from 10 to 30 (male diploids). Other characteristics of importance are chromosome size, the sex chromosome mechanism, presence or absence of the peculiar m chromosome (found in only a few families of Heteroptera and not at all outside the order), and positions of the m and sex chromosomes at first and second metaphase.

Several members of the Lygaeidae are pests in North America. The most outstanding offender over the years has been the chinch bug, *Blissus leucopterus*, a pest of corn (maize) and several small grains, including wheat. The number of overwintering adults, and the temperature and humidity (which, if high, encourage growth of a parasitic fungus) determine the magnitude of populations each year, and a better understanding of these factors, and better cultural practices, have lessened the importance of the species as a pest in recent decades. It nevertheless remains a major problem in the midwestern states. The hairy chinch bug, *Blissus leucopterus hirtus* Montandon, is a pest of lawn grasses in New England, and the southern chinch bug, *Blissus insularis* Barber, is a serious pest of St. Augustinegrass lawns in the southeastern and southcentral parts of the United States. The recently described *Blissus canadensis* Leonard (1970a) was reported to cause economic damage to barley in Alberta, Canada.

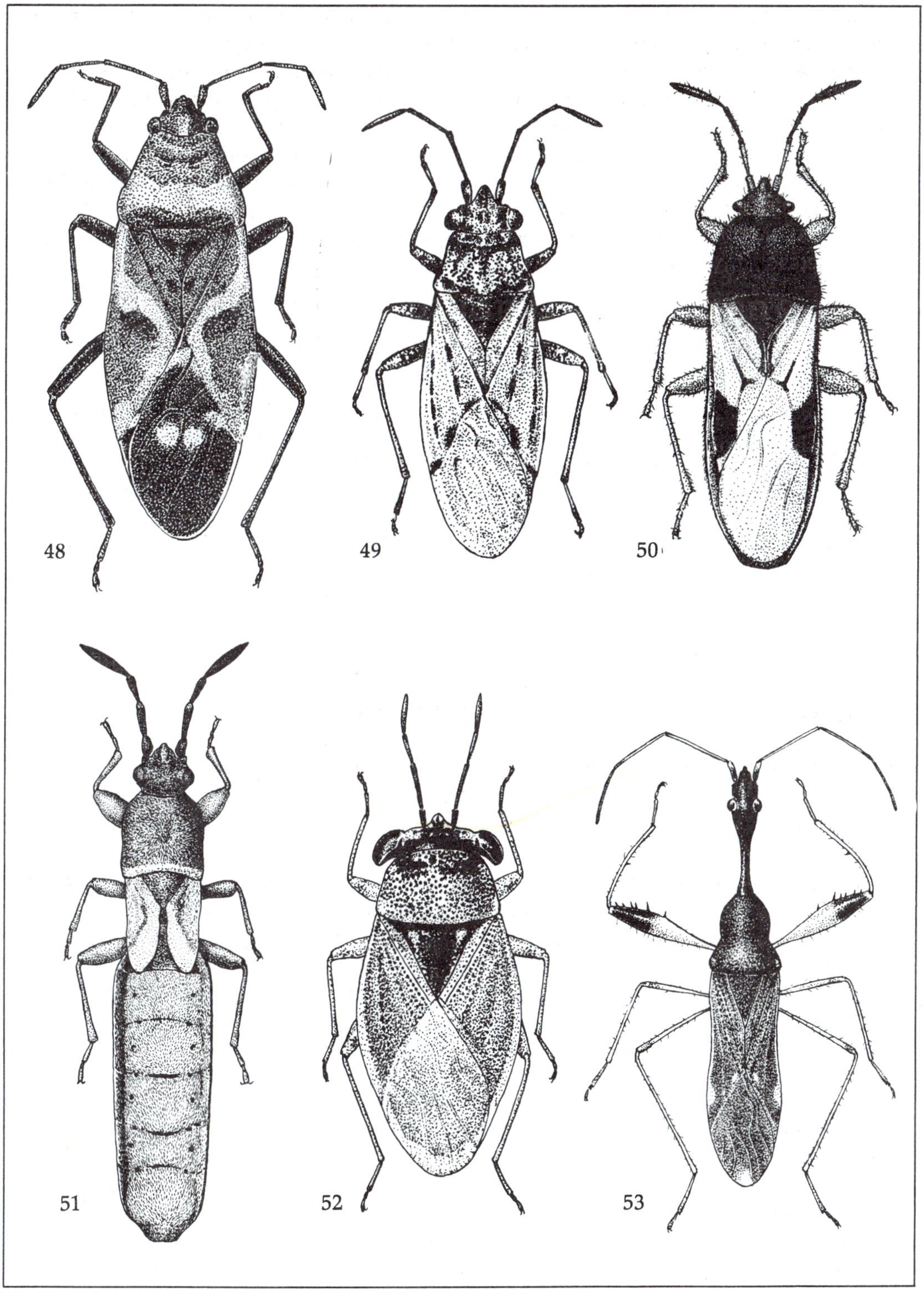

Figs. 48-53: 48, *Lygaeus kalmii kalmii* [p. 193]; 49, *Nysius niger* [p. 204]; 50, *Blissus leucopterus leucopterus* [p. 174]; 51, *Ischnodemus missouriensis* [p. 177]; 52, *Geocoris punctipes* [p. 185]; 53, *Myodocha serripes* [p. 230] (After Froeschner, 1944).

Several species of the orsilline genus *Nysius* Dallas are pests in various parts of the world. In North America, *Nysius raphanus* Howard, although feeding on seeds, also feeds on the vegetative parts of plants. It has been reported as a pest of alfalfa, sugar beets, several crucifers, and many other agricultural crops. The economic literature reports several members of the rhyparochromine tribe Myodochini as minor pests of strawberries and rice (Sweet, 1960).

Outside North America, various members of the Blissinae, as well as some members of the genus *Nysius*, are important pests. Several species of the genus *Oxycarenus* Fieber (Oxycareninae Stål) are important pests of cotton in the Old World and in South America, and *Elasmolomus sordidus* (Fabricius) is an important Old World pest of stored seed crops, especially peanuts (ground nuts) and wheat.

On the other hand, members of the genus *Geocoris* are beneficial insects. Prey of various species in the United States include economically important species of aphids, leafhoppers, mirids, *Nysius*, mealybugs, scales, weevils, hornworms and other lepidopterous pests, and mites (Tamaki and Weeks, 1972a). The potential of species of *Geocoris* as biological controls, then, is well founded, but may be more extensive than generally realized. Tamaki and Weeks (1972a) investigated the possibility of increasing populations of *Geocoris* in crop lands by increasing the supply of available seeds, with encouraging results. They used sunflower seeds; which species of seeds would best help exploit the potential of *Geocoris* spp. as native biological control agents remains to be seen.

Very little is known about the phylogenetic position of the Lygaeidae in the Heteroptera, but it is probably a relatively primitive group in the Pentatomomorpha. The family may be paraphyletic, for unpublished work suggests that the Coreidae and related families, as well as the Berytidae and the tropical family Colobathristidae, arose from within the Lygaeidae.

Keys and manuals for the identification of nearctic Lygaeidae are less than satisfactory. Recent world activity in the classification of the family has left most North American works outdated. The picture guide to the more common Hemiptera-Heteroptera by Slater and Baranowski (1978) helps to correct this situation, but about a third of the less common genera and most species are omitted. Barber in 1917 and 1918d provided a key to North American genera of Lygaeidae, and in 1923a wrote another for the genera of Lygaeidae found in Connecticut. Blatchley (1926a) covered the eastern United States, and Froeschner (1944) provided a key to the genera found in Missouri. All but Barber's works contain many excellent illustrations. The most comprehensive study of North American Lygaeidae is that of Torre-Bueno (1946a), but the work was based on published keys and where these were not available, on keys derived from descriptions of included species. Since Torre-Bueno's work was not based on actual specimens, his keys are only as reliable as the work upon which they were based. His new keys are generally unreliable. For immature lygaeids, there is Slater's (1951) key to the genera of midwestern nymphs and Sweet and Slater's (1961) key to fifth-instar nymphs of North America, which omits a few genera for which immatures were unavailable. A new comprehensive study is needed.

Keys to genera of several subfamilies and tribes of Lygaeidae and to species of many genera are available; these are cited in the body of the catalog and listed in the references.

This catalog of the Lygaeidae of America north of Mexico extends Slater's indispensable *Catalogue of the Lygaeidae of the World* (1964a) for America north of Mexico. Minor errors in the Slater catalog are corrected; we hope that the new errors we are bound to have introduced are not serious.

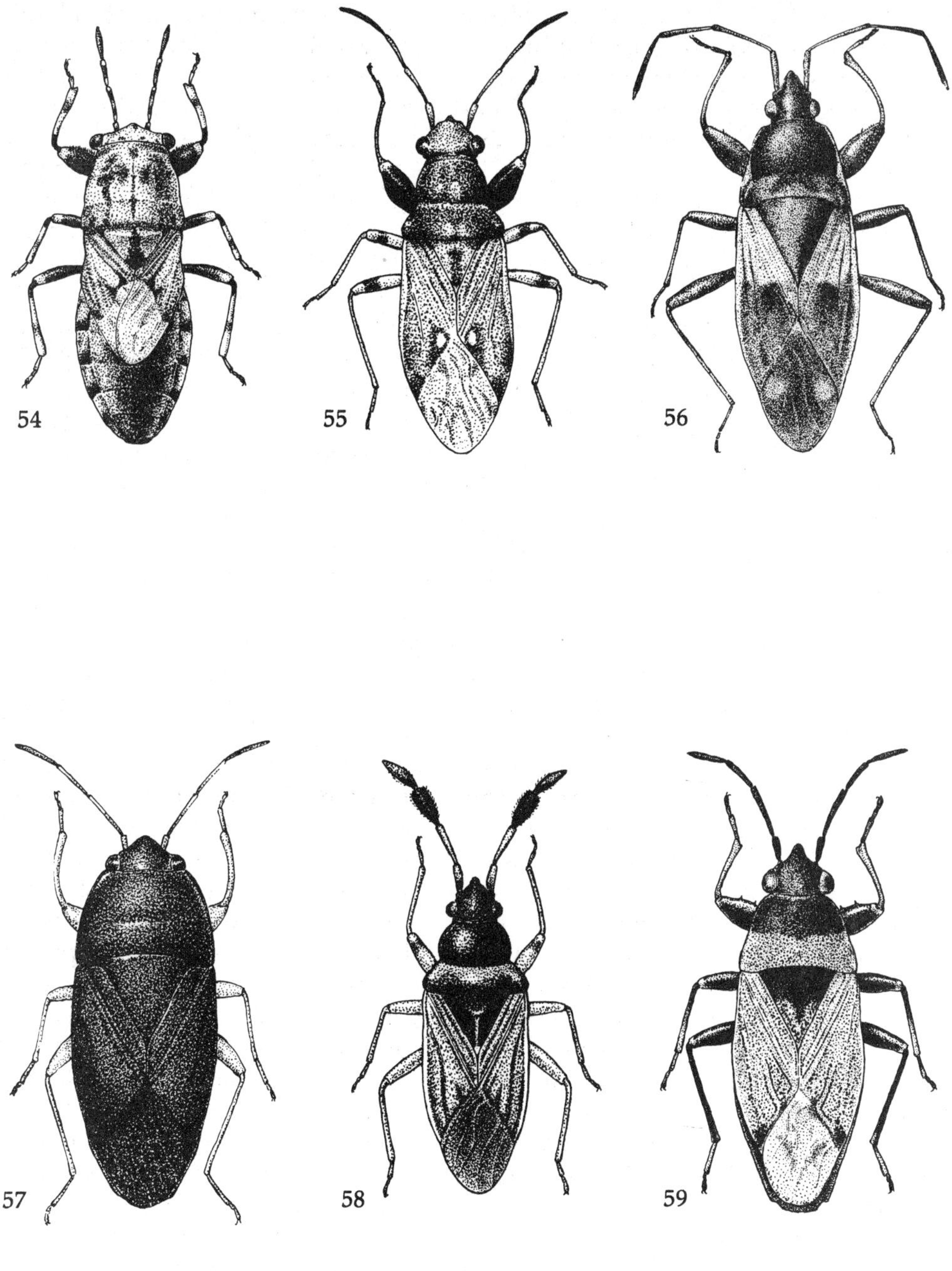

Figs. 54-59: 54, *Phlegyas abbreviatus* [p. 210]; 55, *Pseudopachybrachius basalis* [p. 235]; 56, *Eremocoris ferus* [p.214]; 57, *Delochilocoris illuminatus* [p. 219]; 58, *Ptochiomera nodosa* [p. 237]; 59, *Peritrechus fraternus* [p. 243] (After Froeschner, 1944).

The first citation in the references section [Ref.] for each species and higher taxon is a volume-and-page reference to that taxon in the Slater catalog. Other references locate keys to genera and species, and cite other information. The bibliography from 1981 to 1986 is limited to a Biosis survey.

Species that are incorrectly attributed to a particular region are difficult to account for in a catalog of that region. At least the following species have recently been identified as *not* occurring in our fauna. Barber (1918c) recorded *Acompus rufipes* (Wolff) from British Columbia, but Scudder (1961) showed that Barber's specimens were of a then undescribed genus and species from Victoria, Australia, rather than from Victoria, British Columbia. Barber (1932a) described *Exptochiomera nana* Barber from Massachusetts, but Sweet and Slater (1974) found that the description actually was based upon a mislabeled specimen of *Suffenus fusconervosus* (Motschulsky), an Asian and African species. *Oncopeltus varicolor* (Fabricius) has been dropped from the list of the lygaeid fauna of America north of Mexico on the advice of Flavia O'Rourke, whose revision of the New World species of the genus remains unpublished as of this writing. The only systematic change included is to transfer *Oncopeltus cayensis* Torre-Bueno from the subgenus *Erythrischius* Stål to the nominate subgenus.

The bibliography lists all references cited in the catalog itself, and all references known to us to the Lygaeidae of America north of Mexico from 1964 through 1986. A combination of Parshley's (1925) bibliography, Slater's (1964a) catalog, and the present bibliography should provide a complete listing of references to this date. It is too much to hope that we have not missed some papers mentioning North American Lygaeidae published since 1964. We would greatly appreciate information on such papers.

We acknowledge the late Jon L. Herring for his invitation to participate in the Heteroptera of North America catalog project and for his guidance in earlier stages of manuscript preparation. Thomas J. Henry and Richard C. Froeschner provided much assistance in later stages. Finally, we would like to thank James A. Slater for his review of a late version of the manuscript, and Kathleen Neeley and Judith Ende of the University of Kansas Libraries for performing a Biosis Previews computer search that aided in compilation of the bibliography.

Subfamily Blissinae Stål, 1862

Ref.: Slater, 1964a, 1: 435; Slater, 1979 (key genera world, cladogram of genera).

Genus *Blissus* Burmeister, 1835

1835 *Blissus* Burmeister, Handb. Ent., 2: 290. Type species: *Blissus hirtulus* Burmeister, 1835. Monotypic.

Note: The genus *Blissus* as treated here is based upon *B. leucopterus* and allies, a group confined to the New World. An appeal to the Int. Comm. Zool. Nomen. (Slater, 1961, Bull. Zool. Nomen., 18: 346-348) to set aside the palearctic *B. hirtulus* and designate *B. leucopterus* as the type species was rejected by the Commission (Opinion 705, China, 1964). Unless this decision is successfully reappealed, the generic name of this economically important genus will have to be changed.

Ref.: Slater, 1964a, 1: 437; Wagner, 1962 (comment on type-species); Leonard, 1968a (key spp. eastern N. Am.); Slater, Ashlock, and Wilcox, 1969 (comments on type-species); Slater, 1979. References mentioning genus only: Streu, 1973; Slater, 1978; Short and Koehler, 1979.

Blissus arenarius Barber, 1918

1918 *Blissus leucopterus* var. *arenarius* Barber, Bull. Brook. Ent. Soc., 13: 38. [N.J.].
1937 *Blissus arenarius*: Barber, Proc. Ent. Soc. Wash., 39: 81.
Distribution: Conn., Del., Fla., Ga., Md., N.B., N.C., N.J., N.S., N.Y., P.Ed., R.I., S.C., Va.
Ref.: Slater, 1964a, 1: 438.

Blissus arenarius arenarius Barber, 1918

1918 *Blissus leucopterus* var. *arenarius* Barber, Bull. Brook. Ent. Soc., 13: 38. [N.J.].
1966 *Blissus arenarius arenarius*: Leonard, Conn. Agr. Exp. Stn. Bull., 677: 16.
Distribution: Conn., Del., Md., N.B., N.J., N.S., N.Y., P.Ed., R.I., Va.
Ref.: Leonard, 1966.

Blissus arenarius maritimus Leonard, 1966

1966 *Blissus arenarius maritimus* Leonard, Conn. Agr. Exp. Stn. Bull., 677: 18. [Fla.].
Distribution: Fla., Ga., N.C., S.C.
Ref.: Leonard, 1966.

Blissus barberi Leonard, 1968

1968 *Blissus barberi* Leonard, An. Ent. Soc. Am., 61: 245. [Tex.].
Distribution: Tex.
Ref.: Leonard, 1968a.

Blissus breviusculus Barber, 1937

1937 *Blissus breviusculus* Barber, Proc. Ent. Soc. Wash., 39: 85. [Mass.].
Distribution: Conn., Mass., Me.
Ref.: Slater, 1964a, 1: 439; Leonard, 1968a.

Blissus canadensis Leonard, 1970

1970 *Blissus canadensis* Leonard, Can. Ent., 102: 1531. [Alta.].
Distribution: Alta., Mont., Sask.

Blissus insularis Barber, 1918

1888 *Blissus leucopterus*: Schwartz, Proc. Ent. Soc. Wash., 1: 105.
1918 *Blissus leucopterus* var. *insularis* Barber, Bull. Brook. Ent. Soc., 13: 38. [Fla.].
1937 *Blissus insularis*: Barber, Proc. Ent. Soc. Wash., 39: 82.
Distribution: Ala., Fla., Ga., La., Miss., N.C., Tex.
Note: West Indian records of this species are treated as three distinct species in Leonard (1968b, Proc. Ent. Soc. Wash., 70: 150-153).
Ref.: Slater, 1964a, 1: 466; Horn, 1962; Leonard, 1966, 1968a; Kerr, 1962, 1966; Stringfellow, 1967, 1968, 1969; Oliver and Komblas, 1968; Komarek, 1971; Miller [W.O.], 1971; Stroble, 1971; Reinert, 1972a, b, 1974, 1978a, b; Reinert and Kerr, 1973; Reinert and Dudeck, 1974; Simmands et al., 1976; Reinert et al., 1980, 1981; Newsom et al., 1976; Morgan et al., 1978; Bruton et al., 1979; Crocker and Simpson, 1981; Crocker et al., 1982; Reinert, 1982; Bruton et al., 1983; Reinert and Portier, 1983; Reinert et al., 1986.

Blissus iowensis Andre, 1937

1937 *Blissus iowensis* Andre, Ia. St. Coll. J. Sci., 11: 165. [Ia.].

Distribution: Ia., Ks.

Ref.: Slater, 1964a, 1: 448; Decker and Andre, 1938.

Blissus leucopterus (Say), 1832

1832 *Lygaeus leucopterus* Say, Descrip. Het. Hem. N. Am., p. 14. [Va.]. Neotype from Mo. designated by Leonard, 1968, An. Ent. Soc. Am., 61: 241.

1874 *Blissus leucopterus*: Stål, K. Svens. Vet.-Akad. Handl., 12(1): 133.

Distribution: Ala., Ark., Col., Conn., Del., Ga., Ill., Ia., Ind., Ks., Ky., La., Mass., Md., Me., Mich., Minn., Miss., Mo., N.B., N.C., N.H., N.J., N.S., N.Y., Neb., Nfld., Oh., Ok., Ont., Pa., Que., R.I., S.C., S.D., Tenn., Tex., Va., Vt., W.Va., Wis., Wyo.

Ref.: LeBaron, 1850; Fitch, 1855a, b, c; Smith [E.C.], 1855; Bird and Mitchner, 1953; Schread, 1963, 1970; Polivka and Irons, 1966; Labrador et al., 1967; Anon., 1969; Johnson [C.G.], 1969; Borror and White, 1970; Young, 1970; Komarek, 1971; Miller [W.O.], 1971; Randolph and Teetes, 1971; Little, 1972; Wood and Starks, 1972; Reis et al., 1976; Daly et al., 1978; Lamp and Holtzer, 1980; Jotwani, 1981; Gracen, 1986.

Blissus leucopterus hirtus Montandon, 1893

1857 *Micropus leucopterus*: Signoret, An. Soc. Ent. France, 3(5): 31.

1893 *Blissus hirtus* Montandon, An. Ent. Soc. Belg., 37: 405. [Pa.]. Neotype from Pa. designated by Leonard, 1968, An. Ent. Soc. Am., 61: 242.

1894 *Blissus leucopterus*: Van Duzee, Bull. Buff. Soc. Nat. Sci., 5: 173.

1918 *Blissus leucopterus* var. *hirtus*: Barber, Bull. Brook. Ent. Soc., 13: 38.

1966 *Blissus leucopterus hirtus*: Leonard, Conn. Agr. Exp. Stn. Bull., 677: 14.

Distribution: Conn., Del., Mass., Md., Me., Minn., N.B., N.H., N.J., N.S., N.Y., Nfld., Oh., Ont., Pa., Que., R.I., Va., Vt.

Ref.: Slater, 1964a, 1: 443 (as *B. hirtus*); Polivka, 1963; Leonard, 1966; Choban and Gupta, 1972 (embryology); Goble, 1972; Streu and Cruz, 1972; McEwen, 1973; Baker and Ratcliffe, 1977; Liu and McEwen, 1977, 1979; Mailloux and Streu, 1979, 1981 (biol.); Sears, 1980; Baker et al., 1981a, b; Mailloux and Streu, 1982 (biol.); Johnson-Cicalese et al., 1985.

Blissus leucopterus leucopterus (Say), 1832 [Fig. 50]

1832 *Lygaeus leucopterus* Say, Descrip. Het. Hem. N. Am., p. 14. [Va.].

1850 *Rhyparochromus devastator* LeBaron, Prairie Farmer, 10: 280. [Ill.]. Synonymized by Harris, 1852, Treat. Ins. Inj. Veg., p. 199.

1852 *Rhyparochromus leucopterus*: Harris, Treat. Ins. Inj. Veg., 2nd. ed., p. 198.

1852 *Blissus albipennis* Dallas, List Hem. Brit. Mus., 2: 583. [No locality given]. Synonymized by Horvath, 1899, Rev. d'Ent., 17: 279.

1855 *Micropus leucopterus*: Fitch, Cultivator, (3)3: 238.

1856 *Micropus leucopterus* var. *albivenosus* Fitch, 2nd Rept. Nox. Ben. Ins. N.Y., p. 291. [N.Y.]. Synonymized by Leonard, 1968, An. Ent. Soc. Am., 61: 241.

1856 *Micropus leucopterus* var. *apterus* Fitch, 2nd Rept. Nox. Ben. Ins. N.Y., p. 291. [N.Y.]. Synonymized by Leonard, 1968, An. Ent. Soc. Am., 61: 241.

1856 *Micropus leucopterus* var. *basalis* Fitch, 2nd Rept. Nox. Ben. Ins. N.Y., p. 291. [N.Y.]. Synonymized by Leonard, 1968, An. Ent. Soc. Am., 61: 241.

1856 *Micropus leucopterus* var. *dimidiatus* Fitch, 2nd Rept. Nox. Ben. Ins. N.Y., p. 291. [N.Y.]. Synonymized by Leonard, 1968, An. Ent. Soc. Am., 61: 241.

1856 *Micropus leucopterus* var. *femoratus* Fitch, 2nd Rept. Nox. Ben. Ins. N.Y., p. 291. [N.Y.]. Synonymized by Leonard, 1968, An. Ent. Soc. Am., 61: 241.

1856 *Micropus leucopterus* var. *fulvivenosus* Fitch, 2nd Rept. Nox. Ben. Ins. N.Y., p. 291. [N.Y.]. Synonymized by Leonard, 1968, An. Ent. Soc. Am., 61: 241.
1856 *Micropus leucopterus* var. *immarginatus* Fitch, 2nd Rept. Nox. Ben. Ins. N.Y., p. 291. [N.Y.]. Synonymized by Leonard, 1968, An. Ent. Soc. Am., 61: 241.
1856 *Micropus leucopterus* var. *nigricornis* Fitch, 2nd Rept. Nox. Ben. Ins. N.Y., p. 291. [N.Y.]. Synonymized by Leonard, 1968, An. Ent. Soc. Am., 61: 241.
1856 *Micropus leucopterus* var. *rufipedis* Fitch, 2nd Rept. Nox. Ben. Ins. N.Y., p. 291. [N.Y.]. Synonymized by Leonard, 1968, An. Ent. Soc. Am., 61: 241.
1872 *Ischnodemus leucopterus*: Walker, Cat. Hem. Brit. Mus., 5: 127.
1875 *Micropus leucopterus* var. *melanosus* Riley, 7th Rept. St. Ent. Mo., p. 22. [Mo.]. Synonymized by Leonard, 1968, An. Ent. Soc. Am., 61: 241.
1916 *Blissus leucopterus* var. *albivenosus*: Van Duzee, Check List Hem., p. 20.
1916 *Blissus leucopterus* var. *apterus*: Van Duzee, Check List Hem., p. 20.
1916 *Blissus leucopterus* var. *basalis*: Van Duzee, Check List Hem., p. 20.
1916 *Blissus leucopterus* var. *femoratus*: Van Duzee, Check List Hem., p. 20.
1916 *Blissus leucopterus* var. *fulvivenosus*: Van Duzee, Check List Hem., p. 20.
1916 *Blissus leucopterus* var. *immarginatus*: Van Duzee, Check List Hem., p. 20.
1916 *Blissus leucopterus* var. *nigricornis*: Van Duzee, Check List Hem., p. 20.
1916 *Blissus leucopterus* var. *rufipedis*: Van Duzee, Check List Hem., p. 20.
1926 *Blissus validus* Blatchley, Het. E. N. Am., p. 370. [Ind.]. Synonymized by Leonard, 1968, An. Ent. Soc. Am., 61: 241.
1966 *Blissus leucopterus leucopterus*: Leonard, Conn. Agr. Exp. Stn. Bull., 677: 11.
Distribution: Ala., Ark., Col., Ga., Ill., Ia., Ind., Ks., Ky., La., Mich., Minn., Miss., Mo., N.C., Neb., Oh., Ok., S.C., S.D., Tenn., Tex., Va., W.Va., Wis., Wyo.
Ref.: Slater, 1964a, 1: 448; Leonard, 1966, 1968a; Atkins et al., 1969; Parker and Randolph, 1972; Wilde and Morgan, 1978; Mize et al., 1980; Wilson and Burton, 1980; Smith [M.T.] et al., 1981; Peters, 1982; Starks et al., 1982; Merkle et al., 1983; Peters, 1983; Ahmad, Kindler, and Pruess, 1984; Ahmad, Pruess, and Kindler, 1984; Ramoska, 1984; Wilde, Kadoum, and Mize, 1984; Wilde and Mize, 1984; Wilde, Mize et al., 1984; Negron and Riley, 1985; Ramoska and Todd, 1985; Stuart et al., 1985; Mize and Wilde, 1986a, b, c; Wilde, Russ, and Mize, 1986; Wilde, Stuart, and Hatchett, 1986.

Blissus minutus (Blatchley), 1925
1925 *Ischnodemus pusillus* Blatchley, Ent. News, 36: 45. [Fla.]. Preoccupied.
1925 *Ischnodemus minutus* Blatchley, Ent. News, 36: 245. New name for *Ischnodemus pusillus* Blatchley.
1968 *Blissus minutus*: Leonard, An. Ent. Soc. Am., 61: 249.
Distribution: Fla.
Ref.: Slater, 1964a, 1: 482 (as *Ischnodemus minutus*); Leonard, 1968a.

Blissus mixtus Barber, 1937
1888 *Blissus leucopterus*: Riley, Ins. Life, 1: 126.
1937 *Blissus mixtus* Barber, Proc. Ent. Soc. Wash., 39: 85. [Cal.].
1951 *Blissus mixus* [*sic*]: Drake, J. Wash. Acad. Sci., 41: 321.
Distribution: Cal.
Ref.: Slater, 1964a, 1: 457; Prendergast, 1943 (biol.).

Blissus nanus Barber, 1937
1937 *Blissus nanus* Barber, Proc. Ent. Soc. Wash., 39: 82. [Ks.].

Distribution: Ks.
Ref.: Slater, 1964a, 1: 457.

Blissus occiduus Barber, 1918
1918 *Blissus occiduus* Barber, Bull. Brook. Ent. Soc, 13: 36. [Col.].
1919 *Blissus leucopterus*: Parshley, Occas. Pap. Mus. Zool., Univ. Mich., 71: 16.
1940 *Blissus occidus* [*sic*]: Drake, Notas Mus. Plata Inst. Mus., 5: 225.
Distribution: B.C., Col., N.M. (Mexico).
Ref.: Slater, 1964a, 1: 458.

Blissus omani Barber, 1937
1937 *Blissus omani* Barber, Proc. Ent. Soc. Wash., 39: 82. [Ariz.].
Distribution: Ariz.
Ref.: Slater, 1964a, 1: 458.

Blissus planarius Barber, 1937
1937 *Blissus planarius* Barber, Proc. Ent. Soc. Wash., 39: 83. [Wyo.].
Distribution: Col., Ks., Wyo.
Ref.: Slater, 1964a, 1: 459.

Blissus sweeti Leonard, 1968
1968 *Blissus sweeti* Leonard, An. Ent. Soc. Am., 61: 247. [Tex.].
Distribution: Tex. (Mexico).
Ref.: Leonard, 1968a.

Blissus villosus Barber, 1937
1937 *Blissus villosus* Barber, Proc. Ent. Soc. Wash., 39: 84. [Cal.].
Distribution: Cal.
Ref.: Slater, 1964a, 1: 461.

Blissus yumanus Drake, 1951
1951 *Blissus yumanus* Drake, J. Wash. Acad. Sci., 41: 321. [Ariz.].
Distribution: Ariz.
Ref.: Slater, 1964a, 1: 461.

Genus *Ischnodemus* Fieber, 1837

1837 *Ischnodemus* Fieber, Beitr. Ges. Natur. Heilwiss., p. 337. Type-species: *Ischnodemus quadratus* Fieber, 1837. Monotypic.
1837 *Micropus* Spinola, Essai Gen. Hem., p. 218. Preoccupied. Type-species: *Micropus genei* Spinola, 1837. Monotypic. Synonymized by Stål, 1859, Freg. Eug. Resa, p. 248.
Refs: Slater, 1964a, 1: 471, 1978, 1979 (key spp. world); Harrington, 1972.

Ischnodemus badius Van Duzee, 1909
1909 *Ischnodemus badius* Van Duzee, Bull. Buff. Soc. Nat. Sci., 9: 168. [Fla.].
Distribution: Fla., N.C., Tex., Va.
Ref.: Slater, 1964a, 1: 471; Davis and Gray, 1966; Harrington, 1972.

Ischnodemus brunnipennis (Germar), 1837
1837 *Pachymerus brunnipennis* Germar, Silb. Rev. Ent., 5: 140. [South Africa in error].
1972 *Ischnodemus brunnipennis*: Harrington, Univ. Conn. Occas. Pap., Biol. Sci. Ser., 2: 53.
Distribution: Cal., Fla., Ind., N.J., S.C., Tex.
Note: Harrington (1972, above) clarified the distribution of this species.

Ref.: Slater, 1964a, 1: 475; Harrington, 1972.

Ischnodemus conicus Van Duzee, 1909

1909 *Ischnodemus conicus* Van Duzee, Ent. News, 20: 234. [Tex.].

Distribution: Fla., La., Miss., N.C., S.C., Tex., Va.

Ref.: Slater, 1964a, 1: 476, Harrington, 1972.

Ischnodemus falicus (Say), 1832

1832 *Lygaeus falicus* Say, Descrip. Het. Hem. N. Am., p. 15. [Mo.].

1857 *Micropus falicus*: Signoret, An. Ent. Soc. France, (3)5: 27.

1874 *Ischnodemus falicus*: Stål, K. Svens. Vet.-Akad. Handl., 12(1): 131.

1916 *Ischnodemus fallicus* [*sic*]: Somes, J. Econ. Ent., 9: 44.

Distribution: Conn., Ia., Ill., Ind., Ks., La., Mass., Md., Me., Mo., N.C., N.D., N.J., N.Y., Oh., Ok., Ont., Pa., Que., S.D., Tex., Wis.

Ref.: Slater, 1964a, 1: 477; Harrington, 1972.

Ischnodemus fulvipes (De Geer), 1773

1773 *Cimex fulvipes* De Geer, Mem. Hist. Ins., 3: 555. [Surinam].

1859 *Micropus fulvipes*: Dohrn, Cat. Hem., p. 35.

1872 *Ischnodemus fulvipes*: Walker, Cat. Hem. Brit. Mus., 5: 127.

1972 *Ischnodemus fulvipes*: Harrington, Univ. Conn. Occas. Pap., Biol. Sci. Ser., 2: 55.

Distribution: Fla., Tex.(?) (Mexico to South America, West Indies).

Note: Snow (1906, Trans. Ks. Acad. Sci., 20: 152) questioned the Tex. record for this species.

Ref.: Slater, 1964a, 1: 478; Slater and Wilcox, 1969; Baranowski, 1979.

Ischnodemus hesperius Parshley, 1922

1922 *Ischnodemus brevicornis* Parshley, Tech. Bull. S.D. St. Coll., 2: 8. [S.D.]. Preoccupied.

1922 *Ischnodemus hesperius* Parshley, Bull. Brook. Ent. Soc., 17: 123. New name for *Ischnodemus brevicornis* Parshley.

1922 *Ischnodemus falicus*: Hussey, Occas. Pap. Mus. Zool., Univ. Mich., 115: 9.

Distribution: Ia., Ill., Ks., Man., Neb., S.D., Wis.

Ref.: Slater, 1964a, 1: 480; Harrington, 1972.

Ischnodemus lobatus Van Duzee, 1909

1909 *Ischnodemus lobatus* Van Duzee, Bull. Buff. Soc. Nat. Sci., 9: 169. [Fla.].

Distribution: Fla.

Ref.: Slater, 1964a, 1: 481.

Ischnodemus missouriensis Froeschner, 1944 [Fig. 51]

1944 *Ischnodemus missouriensis* Froeschner, Am. Midl. Nat., 31: 658. [Mo.].

Distribution: Mo.

Ref.: Slater, 1964a, 1: 482.

Ischnodemus praecultus Distant, 1882

1882 *Ischnodemus praecultus* Distant, Biol. Centr.-Am., Rhyn., 1: 196. [Guatemala].

1926 *Ischnodemus atramedius* Blatchley, Het. E. N. Am., p. 366. [Fla.]. Synonymized by Slater and Wilcox, 1969, Misc. Publ. Ent. Soc. Am., 6: 211.

1969 *Ischnodemus praecultus*: Slater and Wilcox, Misc. Publ. Ent. Soc. Am., 6: 211.

Distribution: Fla. (Mexico to South America).

Ref.: Slater, 1964a, 1: 473 (as *I. atramedius*); Harrington, 1972.

Ischnodemus robustus Blatchley, 1926

1926 *Ischnodemus robustus* Blatchley, Het. E. N. Am., p. 366. [Fla.].

Distribution: Fla.

Ref.: Slater, 1964a, 1: 485; Harrington, 1972.

Ischnodemus rufipes Van Duzee, 1909

1909 *Ischnodemus rufipes* Van Duzee, Bull. Buff. Soc. Nat. Sci., 9: 167. [Fla.]. Synonymized with *I. praecultus* Distant by Slater and Wilcox, 1969, Misc. Publ. Ent. Soc. Am., 6: 211; resurrected by Harrington, 1972, Univ. Conn. Occas. Pap., Biol. Sci. Ser., 2: 53.

1926 *Ischnodemus intermedius* Blatchley, 1926, Het. E. N. Am., p. 365. [Ind., Pa.]. Synonymized by Slater, 1979, Bull. Am. Mus. Nat. Hist., 165: 54.

Distribution: Ala., Fla., Ga., Ind., La., Miss., N.C., Pa., Tex., Va. (Mexico).

Ref.: Slater, 1964a, 1: 485; Harrington, 1972.

Ischnodemus slossonae Van Duzee, 1909

1909 *Ischnodemus slossoni* Van Duzee, Ent. News, 20: 233. [Fla., N.C.].

Distribution: Conn., Fla., Ks., Mo., N.C., Tex.

Note: Species named after Mrs. Annie Trumbull Slosson.

Ref.: Slater, 1964a, 1: 491; Harrington, 1972.

Genus *Toonglasa* Distant, 1893

1893 *Toonglasa* Distant, Biol. Centr.-Am., Rhyn., 1: 391. Type-species: *Toonglasa forficuloides* Distant, 1893. Monotypic.

1966 *Extarademus* Slater and Wilcox, An. Ent. Soc. Am., 59: 62. Type-species: *Micropus collaris* Signoret, 1857. Original designation. Synonymized by Slater and Brailovsky, 1983, An. Ent. Soc. Am., 76: 523.

Ref.: Slater and Wilcox, 1966 (key spp.)(as *Extarademus*); Slater, 1979 (key spp. world); Slater and Brailovsky, 1983a (key spp. world).

Toonglasa umbratus (Distant), 1893

1893 *Ischnodemus umbratus* Distant, Biol. Centr.-Am., Rhyn., 1: 391. [Guatemala].

1909 *Ischnodemus praecultus*: Van Duzee, Can. Ent., 41: 375.

1921 *Ischnodemus macer* Van Duzee, Proc. Cal. Acad. Sci., (4)11: 114. [Ariz.]. Synonymized by Slater and Brailovsky, 1983, An. Ent. Soc. Am., 76: 534.

1966 *Extarademus macer*: Slater and Wilcox, An. Ent. Soc. Am., 59: 68.

1983 *Toonglasa umbratus*: Slater and Brailovsky, An. Ent. Soc. Am., 76: 534.

Distribution: Ariz., Ia., Ks., Neb., N.M., Ore., Tex. (Mexico to Brazil).

Note: Distant (1893, Biol. Centr.-Am., Rhyn., 1: 391) described both *I. umbratus* and *I. cahabonensis* from Guatemala. Slater and Brailovsky (1983, An. Ent. Soc. Am., 76: 534) note that these two names refer to the same species and, as first revisors, select *umbratus* as the valid name of the species.

Ref.: Slater, 1964a, 1: 475 (as *I. cahabonensis*), 1: 481 (as *I. macer*), 1: 493 (as *I. umbratus*); Slater and Brailovsky, 1983a (descrip. fifth-instar nymph).

Subfamily Cyminae Baerensprung, 1860

Ref.: Slater, 1964a, 1: 383; Slater, 1963 (key genera nymphs U.S.); Hamid, 1975 (key tribes).

Tribe Cymini Baerensprung, 1860

Ref.: Slater, 1964a, 1: 385; Hamid, 1975 (key genera world).

Genus *Cymodema* Spinola, 1837

1837 *Cymodema* Spinola, Essai Hem., p. 213. Type-species: *Cymodema tabida* Spinola, 1837. Monotypic.

Ref.: Slater, 1964a, 1: 387; Hamid, 1975 (key spp. world).

Cymodema breviceps (Stål), 1874

1874 *Cymus breviceps* Stål, K. Svens. Vet.-Akad. Handl., 12(1): 127. [S.C.].
1876 *Cymodema tabida*: Uhler, Bull. U.S. Geol. Geogr. Surv. Terr., 1: 305.
1908 *Cymodema exiguum* Horvath, An. Hist.-Nat. Mus. Nat. Hung., 6: 559. [D.C.]. Synonymized by Van Duzee, 1909, Can. Ent., 41: 372.
1910 *Cymus exiguum*: Smith, Rept. N.J. St. Mus., 3rd ed., p. 142.
1924 *Cymus virescens*: Barber, Bull. Brook. Ent. Soc., 19: 89.
1975 *Cymodema breviceps*: Hamid, Occas. Publ. Ent. Soc. Nigeria, 14: 59.
Distribution: Cal., Col., D.C., Fla., Ill., Ind., La., Md., Mo., N.C., N.J., N.Y., S.C., Tex. (Mexico to South America, West Indies).
Ref.: Slater, 1964a, 1: 394 (as *Cymus breviceps*); Ashlock, 1960b (synonymy); Slater, 1963 (as *Cymus breviceps*).

Genus *Cymus* Hahn, 1832

1832 *Cymus* Hahn, Wanz. Ins., 1: 76. Type-species: *Lygaeus claviculus* Fallén, 1807. Designated by Distant, 1903, Fauna Brit. India, Rhyn., 2: 21.
1874 *Arphnus* Stål, K. Svens. Vet.-Akad. Handl., 12(1): 125. Type-species: *Oxycaraenus* [*sic*] *coriacipennis* Stål, 1859. Monotypic. Synonymized by Hamid, 1975, Occas. Publ. Ent. Soc. Nigeria, 14: 63.

Ref.: Slater, 1964a, 1: 389; Barber, 1924a (key spp.); Slater, 1952a, 1963 (key nymphs); Hamid, 1975 (key spp. world).

Cymus angustatus Stål, 1874

1874 *Cymus angustatus* Stål, K. Svens. Vet.-Akad. Handl., 12(1): 126. [N.J.].
1894 *Cymus augustatus* [*sic*]: Van Duzee, Bull. Buff. Soc. Nat. Sci., 5: 173.
1908 *Cymodema tabida*: Torre-Bueno, J. N.Y. Ent. Soc., 16: 229.
Distribution: Alta., Ariz., Conn., Col., D.C., Fla., Ga., Ia., Ill., Ind., Ks., La., Mass., Md., Me., Mich., Minn., Mo., N.C., N.H., N.J., N.Y., Neb., Oh., Ok., Ont., Pa., Que., R.I., S.C., Tex., Vt., Va. (Mexico).
Ref.: Slater, 1964a, 1: 392; Slater, 1952a; Hamid, 1971.

Cymus bellus Van Duzee, 1909

1909 *Cymus bellus* Van Duzee, Bull. Buff. Soc. Nat. Sci., 9: 167. [Fla.].
Distribution: Fla., Ga., La.
Ref.: Slater, 1964a, 1: 394.

Cymus californicus Hamid, 1975

1975 *Cymus californicus* Hamid, Occas. Publ. Ent. Soc. Nigeria, 14: 73. [Cal.].

Distribution: Cal., Col., Neb., Nev., Ore.

Cymus coriacipennis (Stål), 1859

1859 *Oxycaraenus* [*sic*] *coriacipennis* Stål, Freg. Eug. Resa, 3: 247. [Cal.].

1874 *Arphnus coriacipennis*: Stål, K. Svens. Vet.-Akad. Handl., 12(1): 126.

1914 *Arphanus* [*sic*] *coriacipennis*: Van Duzee, Trans. San Diego Soc. Nat. Hist., 2: 8.

1929 *Arphanus* [*sic*] *coriacpennis* [*sic*]: Van Duzee, Pan-Pac. Ent., 5: 189.

1929 *Arphnus profectus* Van Duzee, Pan-Pac. Ent., 5: 190. [Cal.]. Synonymized by Ashlock, 1961, Pan-Pac. Ent., 37: 18.

1929 *Arphnus tristis* Van Duzee, Pan-Pac. Ent., 5: 189. [Nev.]. Synonymized by Ashlock, 1961, Pan-Pac. Ent., 37: 18.

1975 *Cymus coriacipennis*: Hamid, Occas. Publ. Ent. Soc. Nigeria, 14: 78.

Distribution: B.C., Cal., Id., N.M., Nev., Ore., Ut., Wash.

Ref.: Slater, 1964a, 1: 386 (as *Arphnus coriacipennis*); Slater, 1952a (as *Arphnus profectus*); Ashlock, 1961 (as *Arphnus coriacipennis*).

Cymus discors Horvath, 1908

1871 *Cymus claviculus*: Provancher, Nat. Can., 3: 137.

1878 *Cymus clavulus* [*sic*]: Uhler, Proc. Boston Soc. Nat. Hist., 19: 395.

1886 *Cymus tabidus*: Provancher, Pet. Faune Ent. Can., 3: 72.

1893 *Cymodema tabida*: Osborn, Proc. Ia. Acad. Sci., 1: 104.

1908 *Cymus discors* Horvath, An. Hist.-Nat. Mus. Nat. Hung., 6: 559. [N.Y., Vt.].

Distribution: Col., Conn., D.C., Ga., Ia., Ill., Ind., Mass., Md., Me., Mich., Minn., N.C., N.H., N.J., N.S., N.Y., Oh., Ont., Que., R.I., Tenn., Vt., Wis.

Ref.: Slater, 1964a, 1: 401; Slater, 1952a; Hamid, 1971.

Cymus guatemalanus Distant, 1893

1893 *Cymus guatemalanus* Distant, Biol. Centr.-Am., Rhyn., 1: 390. [Guatemala].

1917 *Cymus guatemalanus*: Van Duzee, Univ. Cal. Publ. Ent., 2: 165.

1924 *Cymus reductus* Barber, Bull. Brook. Ent. Soc., 19: 88. [Ariz.]. Synonymized by Barber, 1949, Proc. Ent. Soc. Wash., 51: 276.

Distribution: Ariz., N.M. (Guatemala, Mexico).

Ref.: Slater, 1964a, 1: 408.

Cymus luridus Stål, 1874

1874 *Cymus luridus* Stål, K. Svens. Vet.-Akad. Handl., 12(1): 126. [N.J.].

Distribution: Alta., Ark., B.C., Cal., Col., Conn., Ga., Ia., Id., Ill., Ind., La., Man., Mass., Me., Mich., Minn., N.C., N.D., N.H., N.J., N.M., N.S., N.Y., Neb., Nev., Nfld., Ont., Ore., Pa., Que., R.I., S.D., Ut., Vt., Wash. (Mexico).

Ref.: Slater, 1964a, 1: 409; Slater, 1952a; Hamid, 1971.

Cymus nigrofemoralis Hamid, 1975

1975 *Cymus nigrofemoralis* Hamid, Occas. Publ. Ent. Soc. Nigeria, 14: 88. [Cal.].

Distribution: Cal.

Ref.: Hamid, 1975a.

Cymus robustus Barber, 1924

1924 *Cymus robustus* Barber, Bull. Brook. Ent. Soc., 19: 87. [N.Y.].

Distribution: Ia., Ill., Ind., Mich., N.Y., Oh., Ont., Wash.

Ref.: Slater, 1964a, 1: 415; Slater, 1952a.

Tribe Ninini Barber, 1956

Ref.: Slater, 1964a, 1: 422; Scudder, 1957b (key genera world).

Genus *Cymoninus* Breddin, 1907

1907 *Cymoninus* Breddin, Deut. Ent. Zeit., 1907: 38. Type-species: *Cymoninus subunicolor* Breddin, 1907, a junior synonym of *Ninus sechellensis* Bergroth, 1893. Original designation.

Ref.: Slater, 1964a, 1: 422; Scudder, 1957b (key spp. world).

Cymoninus notabilis (Distant), 1882

1882 *Ninus notabilis* Distant, Biol. Centr.-Am., Rhyn., 1: 191. [Guatemala].

1909 *Ninus notabilis*: Van Duzee, Bull. Buff. Soc. Nat. Sci., 9: 166.

1917 *Cymoninus notabilis*: Van Duzee, Univ. Cal. Publ. Ent., 2: 163.

Distribution: Fla., Ga., Miss., N.M., Tex. (Bermuda, Mexico to South America, West Indies).

Ref.: Slater, 1964a, 1: 422; Slater, 1963; Scudder, 1968.

Subfamily Geocorinae Baerensprung, 1860

Ref.: Slater, 1964a, 1: 520; Readio and Sweet, 1982 (key U.S. genera); Shelton et al., 1981.

Genus *Geocoris* Fallén, 1814

1814 *Geocoris* Fallén, Spec. Nov. Hem. Disp. Meth., p. 10. Type-species: *Cimex grylloides* Linnaeus, 1761. Designated by Oshanin, 1912, Kat. Paläark. Hem., p. 30.

1829 *Ophthalmicus* Schilling, Beitr. Ent., 1: 37, 62. Type-species: *Cimex grylloides* Linnaeus, 1761. Designated by Costa, 1843, An. Accad. Aspir. Nat. Napoli, 1: 293. Synonymized by Hahn, 1833, Wanz. Ins., 1: 233.

Ref.: Slater, 1964a, 1: 523; McAtee, 1914 (key spp., vars.); Barber, 1935a (partial key spp.); Readio and Sweet, 1982 (key spp. eastern U.S.). References mentioning *Geocoris* spp.: Bell and Whitcomb, 1962; Whitcomb and Bell, 1964; Wene et al., 1965; Leigh et al., 1966; Ridgway et al., 1967; Laster and Brazzel, 1968; Dinkins et al., 1970b; Dunbar, 1971; Fye, 1971, 1974; Rummel and Reeves, 1971; DeLoach and Peters, 1972; Fye and Carranza, 1972; Shanks and Finnigan, 1972; Shepard, Carner, and Turnipseed, 1974a, 1974b, 1977; Shepard, Sterling, and Walker, 1972; Taft et al., 1972; Watson [T.F.] et al., 1972; Barry, 1973; Pieters and Sterling, 1973a, b, 1974; Otman et al., 1974; Rakickas and Watson, 1974; Waddill, Shepard, Turnipseed, and Carner, 1974; Dietz et al., 1976a; Lopez and Teetes, 1976; Henneberry et al., 1977; Kinzer et al., 1977; Raney and Yeargan, 1977; Schultz et al., 1977; Byerly et al., 1978; Pieters, 1978; Semtner, 1979; Short and Koehler, 1979; Gutierrez et al., 1980; Irwin and Shepard, 1980; Shepard, 1980; Wilson and Gutierrez, 1980; Altieri et al., 1981; Morris and Smith, 1981; Roach and Hopkins, 1981; Ferguson and McPherson, 1982; Pitts and Pieters, 1982; Tamaki, Fox, and Featherston, 1982; Atim and Graham, 1983; Lentz et al., 1983; Scott et al., 1983; Ellington, Cardenas et al., 1984; Ellington, Kiser et al., 1984; Baker et al., 1985; Fleischer et al., 1985; Harris and Phillips, 1986.

Geocoris alboclavus Barber, 1949

1949 *Geocoris alboclavus* Barber, Pan-Pac. Ent., 24: 205. [Ariz.].

Distribution: Ariz.

Ref.: Slater, 1964a, 1: 527.

Geocoris atricolor Montandon, 1908

1908 *Geocoris atricolor* Montandon, Bull. Soc. Sci. Buch., 16: 228. [Col., Tex., Ut.].

Distribution: Alta., B.C., Cal., Col., Id., Nev., Ore., Tex., Ut., Wash., Wyo.

Ref.: Slater, 1964a, 1: 536; Knowlton, 1935a; Wene and Sheets, 1962; Dunbar and Bacon, 1972a; Benedict and Cothran, 1975a; Bisabri-Ershadi and Ehler, 1981; Moore et al., 1982.

Geocoris barberi Readio and Sweet, 1982

1982 *Geocoris barberi* Readio and Sweet, Misc. Publ. Ent. Soc. Am., 12(3): 58. [Ks.].

Distribution: Ks., S.D.

Geocoris beameri Barber, 1935

1935 *Geocoris beameri* Barber, J. N.Y. Ent. Soc., 43: 132. [Ariz.].

Distribution: Ariz.

Ref.: Slater, 1964a, 1: 537.

Geocoris bullatus (Say), 1832

1832 *Salda bullata* Say, Descrip. Het. Hem. N. Am., p. 18. [U.S.]. Neotype designated by Readio and Sweet, 1982, Misc. Publ. Ent. Soc. Am., 12(3): 31.

1852 *Ophthalmicus griseus* Dallas, List Hem. Brit. Mus., 2: 585. [N. Am.]. Synonymized by Uhler, 1876, Bull. U.S. Geol. Geogr. Surv. Terr., 1: 306.

1852 *Ophthalmicus borealis* Dallas, List Hem. Brit. Mus., 2: 585. [N. Am.]. Synonymized by Uhler, 1876, Bull. U.S. Geol. Geogr. Surv. Terr., 1: 306.

1859 *Geocoris griseus*: Dohrn, Cat. Hem., p. 35.

1874 *Geocoris bullatus*: Stål, K. Svens. Vet.-Akad. Handl., 12(1): 136.

1876 *Ophthalmicus bullatus*: Glover, Ms. Notes J. Hem., p. 55.

1876 *Ophthalmicus bullata*: Uhler, Bull. U.S. Geol. Geogr. Surv. Terr., 1: 306.

1914 *Geocoris bullatus* var. *borealis*: McAtee, Proc. Biol. Soc. Wash., 27: 132.

1922 *Geocoris bullatus bullatus*: Hussey, Occas. Pap. Mus. Zool., Univ. Mich., 118: 21.

1946 *Geocoris bullatus borealis*: Torre-Bueno, Ent. Am., 26: 47.

Distribution: Alk., Alta., B.C., Cal., Col., Conn., D.C., Ia., Id., Ill., Ind., Ks., Mass., Md., Me., Mich., N.H., N.J., N.S., N.Y., N.D., Neb., Nfld., Oh., Ont., Ore., Pa., Que., S.D., Sask., Ut., Vt., Va., Wash., Wis., Wyo., Yuk.

Ref.: Slater, 1964a, 1: 537; Clancy and Pierce, 1966; Dinkins et al., 1970; Johansen and Eves, 1972, 1973; Tamaki, 1972 (biol.); Tamaki and Weeks, 1972b (biol., economic importance); Tamaki and Weeks, 1973; Hagen and Hale, 1974; Walker [J.T.] et al., 1974; Whitcomb, 1974; Hagen et al., 1976; Walker and Turnipseed, 1976; Burgess, 1977; Tamaki and Olsen, 1977; Arnaud, 1978 (as tachinid host); Reinert, 1978b; Tamaki et al., 1978; Miller and Miller, 1979; Crocker and Whitcomb, 1980 (biol.); Richman et al., 1980; Froeschner and Halpin, 1981; Tamaki, 1981; Tamaki, Annis, and Weiss, 1981; Readio and Sweet, 1982; Tamaki, Weiss, and Long, 1983; Burgess et al., 1983; Chow et al., 1983 (behavior); Carroll and Hoyt, 1984; Kiman and Yeargan, 1985.

Geocoris bullatus obscuratus Montandon, 1908

1908 *Geocoris bullatus* var. *obscuratus* Montandon, Bull. Soc. Sci. Buch., 16: 220. [Col.].

Distribution: Col.

Note: Readio and Sweet (1982, Misc. Publ. Ent. Soc. Am., 12(3): 32-33) state that they were unable to locate the type-specimen of this form, and suggest that it may be a junior synonym of *Geocoris bullatus* (Say), a subspecies of *Geocoris pallens* Stål, or a distinct species.
Ref.: Slater, 1964a, 1: 540; Readio and Sweet, 1982.

Geocoris carinatus McAtee, 1914
1914 *Geocoris carinatus* McAtee, Proc. Biol. Soc. Wash., 27: 134. [Cal.].
Distribution: Cal.
Ref.: Slater, 1964a, 1: 541; Wene and Sheets, 1962.

Geocoris davisi Barber, 1935
1935 *Geocoris davisi* Barber, J. N.Y. Ent. Soc., 43: 133. [Nev.].
Distribution: Nev.
Ref.: Slater, 1964a, 1: 543.

Geocoris decoratus Uhler, 1877
1877 *Geocoris decoratus* Uhler, Bull. U.S. Geol. Geogr. Surv. Terr., 3: 410. [Col.].
1916 *Geocoris pallens decoratus*: Van Duzee, Check List Hem., p. 20.
Distribution: Col.
Note: When Barber (1949, Proc. Ent. Soc. Wash., 51: 274) elevated *G. pallens decoratus* to species, he evidently did not have Uhler's species before him. Readio and Sweet (1982, Misc. Publ. Ent. Soc. Am., 12(3): 43) located Uhler's type specimen in the U.S. National Museum of Natural History and stated that it is quite distinct from *G. pallens* and specimens identified by Barber and others as *G. decoratus*. They noted that the species is rare and all known specimens are from Colorado.
Ref.: Slater, 1964a, 1: 543.

Geocoris discopterus Stål, 1874
1874 *Geocoris discopterus* Stål, K. Svens. Vet.-Akad. Handl., 12(1): 136. [N.J.].
1914 *Geocoris bullatus discopterus*: McAtee, Proc. Biol. Soc. Wash., 27: 131.
Note: Barber (1949, Proc. Ent. Soc. Wash., 51: 274) re-elevated *G. bullatus discopterus* to species.
Distribution: Conn., Mass., Me., N.H., N.J., N.Y., Pa., Vt.
Ref.: Slater, 1964a, 1: 545; Readio and Sweet, 1982.

Geocoris duzeei Montandon, 1908
1908 *Geocoris duzeei* Montandon, Bull. Soc. Sci. Buch., 16: 231. [Col.].
Distribution: Col.
Ref.: Slater, 1964a, 1: 547.

Geocoris flavilineus Stål, 1874
1874 *Geocoris flavilineus* Stål, K. Svens. Vet.-Acad. Handl., 12(1): 135. [Colombia].
1982 *Geocoris flavilineus*: Readio and Sweet, Misc. Publ. Ent. Soc. Am., 12(3): 23.
Distribution: Tex. (Mexico to South America).
Ref.: Slater, 1964a, 1: 550; Readio and Sweet, 1982.

Geocoris floridanus Blatchley, 1926
1926 *Geocoris bullatus* var. *floridanus* Blatchley, Het. E. N. Am., p. 375. [Fla.].
1982 *Geocoris floridanus*: Readio and Sweet, Misc. Publ. Ent. Soc. Am., 12(3): 37.
Distribution: Ala., D.C., Fla., Ga., La., Md., N.C., S.C., Tex., Va.
Ref.: Slater, 1964a, 1: 540; Readio and Sweet, 1982.

Geocoris frisoni Barber, 1926

1926 *Geocoris frisoni* Barber, Bull. Brook. Ent. Soc., 21: 38. [Ill.].

Distribution: Ia., Ill., Ind., Ks., Mich., Mo., Neb., Tex., Wis.

Ref.: Slater, 1964a, 1: 550; Readio and Sweet, 1982.

Geocoris howardi Montandon, 1908

1908 *Geocoris howardi* Montandon, Bull. Soc. Sci. Buch., 16: 229. [Mich.].

1916 *Geocoris uliginosus* var. *howardi*: Van Duzee, Check List Hem., p. 120.

Note: Readio and Sweet (1982, Misc. Publ. Ent. Soc. Am., 12(3): 55) re-elevated *G. uliginosus howardi* to species.

Distribution: Alk., Mass., Me., Mich., Minn., N.D., N.H., N.J., N.Y., S.D., Vt., Wis., "South Central Canada."

Ref.: Slater, 1964a, 1: 589; Readio and Sweet, 1982.

Geocoris limbatus Stål, 1874

1874 *Geocoris limbatus* Stål, K. Svens. Vet.-Akad. Handl., 12(1): 136. [N.J.].

1908 *Geocoris uliginosus* var. *limbatus*: Montandon, Bull. Soc. Sci. Buch., 16: 227.

Distribution: Conn., Ia., Ill., Ind., Ks., Man., Mass., Md., Me., Mich., Minn., N.D., N.H., N.Y., Neb., Oh., Ont., Pa., Que., S.D., Sask., Vt., Wis.

Note: Readio and Sweet (1982, Misc. Publ. Ent. Soc. Am., 12(3): 52) re-elevated *G. uliginosus limbatus* to species.

Ref.: Slater, 1964a, 1: 590; Readio and Sweet, 1982.

Geocoris lividipennis Stål, 1862

1862 *Geocoris lividipennis* Stål, Stett. Ent. Zeit., 23: 311. [Mexico].

1946 *Geocoris lividipennis*: Torre-Bueno, Ent. Am., 26: 48.

Distribution: B.C.(?), Ks.(?), N.M., Tex. (Mexico to Central America, West Indies).

Ref.: Slater, 1964a, 1: 561; Readio and Sweet, 1982.

Geocoris nanus Barber, 1935

1914 *Geocoris bullatus* var. *bullatus*: McAtee, Proc. Biol. Soc. Wash., 27: 131 (in part).

1935 *Geocoris nanus* Barber, J. N.Y. Ent. Soc., 43: 134. [Ariz.].

Distribution: Ariz., Col.

Ref.: Slater, 1964a, 1: 569.

Geocoris omani Barber, 1935

1935 *Geocoris omani* Barber, J. N.Y. Ent. Soc., 43: 131. [Ariz.].

Distribution: Ariz.

Ref.: Slater, 1964a, 1: 572.

Geocoris pallens Stål, 1854

1854 *Geocoris pallens* Stål, Öfv. K. Svens. Vet.-Akad, Förh., 11(8): 236. [Cal.].

1895 *Ninyas pallens*: Gillette and Baker, Bull. Col. Agr. Exp. Stn., 31: 23.

1908 *Geocoris decoratus solutus* Montandon, Bull. Soc. Sci. Buch., 16: 223. [Col.]. Synonymized by Readio and Sweet, 1982, Misc. Publ. Ent. Soc. Am., 12(3): 40.

1914 *Geocoris bullatus solutus*: Montandon, Bull. Soc. Sci. Buch., 23: 240.

1917 *Geocoris pallens solutus*: Van Duzee, Univ. Cal. Publ. Ent., 2: 171.

1926 *Geocoris solutus*: Essig, Ins. W. N. Am., p. 305.

Distribution: Ariz., Ark., B.C., Cal., Col., Ia., Ill., Ind., Ks., Mo., Mont., N.M., Neb., Ok., Ore., S.D., Tex., Ut., Wyo. (Hawaii, Mexico to Central America).

Ref.: Slater, 1964a, 1: 573; McGregor and McDonough, 1917; York, 1944 (rearing); Wene and Sheets, 1962; Butler, 1966b; van den Bosch and Hagen, 1966; Falcon et al., 1968; Ridgway and Jones, 1968; Leigh et al., 1969; van den Bosch and Stern, 1969; van den Bosch et al., 1969, 1971; Hagen et al., 1971; Dunbar and Bacon, 1972a; Johansen, 1972; Johansen and Eves, 1972, 1973; Tamaki and Weeks, 1972a (biol., biocontrol potential); Ehler et al., 1973; Eveleens et al., 1973; Gordh and Coker, 1973; Tamaki and Weeks, 1973; Eveleens, 1974; Hagen and Hale, 1974; Irwin et al., 1974; Leigh et al., 1974; Benedict and Cothran, 1975a; Hussain, 1975; Jimenez, 1975, 1976; Van Steenwyk et al., 1975; Leigh and Gonzalez, 1976; Messenger et al., 1976; Saad and Bishop, 1976; Ehler, 1977; Gonzalez et al., 1977; Tamaki and Olsen, 1977; Ehler and Miller, 1978; Stoltz and Stern, 1978; Tamaki and Long, 1978; Yokoyama, 1978, 1980 (rearing); Bisabri-Ershadi and Ehler, 1981; Pape and Crowder, 1981; Tamaki, Annis, and Weiss, 1981; Gonzalez and Wilson, 1982; Readio and Sweet, 1982; Cohen, 1983; De Lima and Leigh, 1984; Yokoyama and Pritchard, 1984; Yokoyama et al., 1984; Henneberry and Clayton, 1985; Kiman and Yeargan, 1985; Stone and Fries, 1986; Trichilo and Leigh, 1986.

Geocoris paulus McAtee, 1914

1914 *Geocoris punctipes* var. *paulus* McAtee, Proc. Biol. Soc. Wash., 27: 130. [Cal.].

1935 *Geocoris paulus*: Barber, J. N.Y. Ent. Soc., 43: 136.

Distribution: Cal.

Ref.: Slater, 1964a, 1: 579.

Geocoris punctipes (Say), 1832 [Fig. 52]

1832 *Salda bullata* var. *punctipes* Say, Descrip. Het. Hem. N. Am., p. 19. [No locality given]. Neotype from D.C. designated by Readio and Sweet, 1982, Misc. Publ. Ent. Soc. Am., 12(3): 28.

1861 *Ophthalmicus luniger* Fieber, Wien. Ent. Monat., 5: 268. [Carolina]. Synonymized by Uhler, 1876, Bull. U.S. Geol. Geogr. Surv. Terr., 1: 306.

1876 *Ophthalmicus punctipes*: Glover, Ms. Notes J. Hem., p. 55.

1876 *Geocoris punctipes*: Uhler, Bull. U.S. Geol. Geogr. Surv. Terr., 1: 306.

1888 *Petritrechus luniger*: Riley, Ins. Life, 1: 130.

1908 *Geocoris pictipes* [*sic*]: Montandon, Bull. Soc. Sci. Buch., 16: 222.

1913 *Hypogeocoris punctipes*: Montandon, Bull. Acad. Roum. Buch., 2: 55.

1923 *Geocoris sonoraensis* Van Duzee, Proc. Cal. Acad. Sci., (4)12: 138. [Gulf of Cal. Isl.]. Synonymized by Readio and Sweet, 1982, Misc. Publ. Ent. Soc. Am., 12(3): 26.

1946 *Geocoris sonorensis* [*sic*]: Clancy, J. Econ. Ent., 39: 326.

Distribution: Ala., Ariz., Ark., Cal., Col., D.C., Del., Fla., Ga., Ia., Ill., Ind., Ks., La., Md., Me., Mich., Miss., Mo., N.C., N.J., N.M., N.Y., Nev., Ok., Pa., S.C., Tenn., Tex., Va. (Hawaii and Johnson Isl., Bahama Isl., Mexico to Colombia).

Ref.: Slater, 1964a, 1: 581; McGregor and McDonough, 1917; York, 1944 (rearing); Clancy, 1946; Dumas et al., 1962; Wene and Sheets, 1962; Butler, 1966a, b, 1967; van den Bosch and Hagen, 1966; Champlain and Sholdt, 1966, 1967a, b; Clancy and Pierce, 1966; Oatman and McMurty, 1966; Lingren and Ridgway, 1967; Oatman et al., 1967, 1981; Whitcomb, 1967, 1974; Lingren, Ridgway, Cowan et al., 1968; Lingren, Ridgway, and Jones, 1968; Dinkins et al., 1970a; Orphanides et al., 1971; Elsey, 1972, 1973; Dunbar, 1972; Dunbar and Bacon, 1972a, b; Oatman and Voth, 1972; Ridgway and Lingren, 1972; Bull, 1973; Bull et al., 1973; Gordh and Coker, 1973; Turnipseed, 1973; Greene et al., 1974; Irwin et al., 1974; Shepard et al., 1974; Waddill and Shepard, 1974; Walker et al., 1974; Crocker et al., 1975; Turnipseed et al., 1975; Van Steenwyk et al., 1975; Lopez et al., 1976; Menke and Greene, 1976; Schuster et al., 1976; Tejada and Blanc, 1976; Waive and Clower, 1976; Walker and Turnipseed, 1976; Gonzalez et al., 1977; Keever et al., 1977; Knausenberger and Allen, 1977; Ables et al., 1978; Arnaud, 1978 (as tachinid host); Livingston et al., 1978; Martin et al., 1978; Nadgauda and Pitre, 1978; Pitre et al., 1978; Stam et al., 1978; Whalon and Parker, 1978; Wilkinson, Biever, Ignoffo et al., 1978; Wilkinson, Biever, and Ignoffo, 1979; Lawrence and Watson, 1979; Marston and Thomas, 1979; McDaniel and Sterling, 1979; Morrison et al., 1979; Wyman et al., 1979; Chiravathanapong and Pitre, 1980; Crocker and Whitcomb, 1980 (biol.); Lynch and Garner, 1980; McCarty et al., 1980; Pyke et al., 1980; Richman et al., 1980; Roach, 1980; Bisabri-Ershadi and Ehler, 1981; Cohen, 1981 (rearing); Farlow and Pitre, 1981; Martin et al., 1981; McDaniel et al., 1981; Pape and Crowder, 1981; Thead et al., 1981; Agnew et al., 1982; Ali and Watson, 1982; Cohen, 1982; Deakle and Bradley, 1982; Hutchison and Pitre, 1982; McPherson, Smith, and Allen, 1982; Readio and Sweet, 1982; Cohen, 1983 (biol.); Cohen and Debolt, 1983 (rearing); Cosper et al., 1983; Farlow and Pitre, 1983; Gravena and Sterling, 1983; Huckaba et al., 1983; Hutchison and Pitre, 1983; Atim and Graham, 1984; Buschman et al., 1984; Cohen, 1984; Ferguson et al., 1984; Isenhour and Todd, 1984; Reed et al., 1984; Stone et al., 1984; Yokoyama and Pritchard, 1984; Yokoyama et al., 1984; Cohen, 1985a (biol.); 1985b (biol.); Gross et al., 1985; Henneberry and Clayton, 1985; Kiman and Yeargan, 1985; Naranjo and Stimac, 1985; Thead et al., 1985; Cohen and Urias, 1986; Herbert and Harper, 1986; Marti and Hamm, 1986; Rogers and Sullivan, 1986.

Geocoris scudderi Stål, 1874

1874 *Geocoris scudderi* Stål, K. Svens. Vet.-Akad. Handl., 12(1): 135. [Tex.].
Distribution: Tex.
Ref.: Slater, 1964a, 1: 583; Readio and Sweet, 1982.

Geocoris thoracicus (Fieber), 1861

1861 *Ophthalmicus thoracicus* Fieber, Wien. Ent. Monat., 5: 280. [Venezuela].
1982 *Geocoris thoracicus*: Readio and Sweet, Misc. Publ. Ent. Soc. Am., 12(3): 63.
Distribution: Tex. (Dutch West Indies, Mexico to Venezuela).
Ref.: Slater, 1964a, 1: 555, 586; Readio and Sweet, 1982.

Geocoris uliginosus (Say), 1832

1832 *Salda uliginosa* Say, Descrip. Het. Hem. N. Am., p. 19. [U.S.]. Neotype from N.Y. designated by Readio and Sweet, 1982, Misc. Publ. Ent. Soc. Am., 12(3): 50.
1852 *Ophthalmicus niger* Dallas, List Hem. Brit. Mus., 2: 586. [N. Am.]. Synonymized by Stål, 1874, K. Svens. Vet.-Akad. Handl., 12(1): 136.
1859 *Geocoris niger*: Dohrn, Cat. Hem., p. 35.
1861 *Ophthalmicus lateralis* Fieber, Wien. Ent. Monat., 5: 270. [Pa.]. Synonymized by Readio and Sweet, 1982, Misc. Publ. Ent. Soc. Am., 12(3): 48.
1869 *Opthalmicus* [*sic*] *uliginosus*: Rathvon, Hist. Lancaster Co., Pa., p. 549.
1874 *Geocoris uliginosus*: Stål, K. Svens. Vet.-Akad. Handl., 12(1): 136.
1886 *Geocoris fuliginosus* [*sic*]: Uhler, Check-list Hem. Het., p. 14.
1894 *Geocoris ater*: Lethierry and Severin, Cat. Gen. Hem., 2: 169.
1908 *Geocoris uliginosus* var. *speculator* Montandon, Bull. Soc. Sci. Buch., 16: 227.

[B.C., Mass., N.Y.]. Synonymized by Readio and Sweet, 1982, Misc. Publ. Ent. Soc. Am., 12(3): 48.
1914 *Geocoris uliginosus* var. *lateralis*: McAtee, Proc. Biol. Soc. Wash., 27: 135.
1947 *Geocoris uligiosus* [*sic*]: Dowdy, Ecol., 28: 428.
Distribution: Ala., Ark., Conn., Del., Fla., Ga., Ia., Ill., Ind., Ks., Ky., La., Md., Mass., Miss., Mo., N.C., N.J., N.M., N.Y., N.D., Neb., Oh., Ok., Pa., S.C., Tenn., Tex., Va., W.Va., Wis.
Ref.: Slater, 1964a, 1: 588; Bell and Whitcomb, 1964; Clancy and Pierce, 1966; Lingren, Ridgway, and Jones, 1968; Dinkins et al., 1970a; Greene et al., 1974; Walker et al., 1974; Whitcomb, 1974; Walker and Turnipseed, 1976; Knausenberger and Allen, 1977; Whitaker et al., 1977a; Arnaud, 1978 (as tachinid host); Reinert, 1978b; McDaniel and Sterling, 1979; Crocker and Whitcomb, 1980 (biol.); Pyke et al., 1980; Richman et al., 1980; Martin et al., 1981; Readio and Sweet, 1982; Gravena and Sterling, 1983; Ferguson and Allen, 1984; Braman et al., 1985.

Genus *Isthmocoris* McAtee, 1914

1914 *Isthmocoris* McAtee, Proc. Biol. Soc. Wash., 27: 127. Type-species: *Salda picea* Say, 1832. Original designation.
Note: Synonymized with *Hypogeocoris* Montandon by Barber (1923, Conn. Geol. Nat. Hist. Surv. Bull., 34: 719); resurrected by Readio and Sweet (1982, Misc. Publ. Ent. Soc. Am., 12(3): 9), who retained *Hypogeocoris* as a distinct genus; it is not found in America north of Mexico.
Ref.: Slater, 1964a, 1: 599 (as *Hypogeocoris*); McAtee, 1914 (key spp.); Readio and Sweet, 1982 (key spp.).

Isthmocoris imperialis (Distant), 1882
1882 *Geocoris imperialis* Distant, Biol. Centr.-Am., Rhyn., 1: 197. [Guatemala].
1914 *Isthmocoris imperialis*: McAtee, Proc. Biol. Soc. Wash., 27: 127.
1926 *Hypogeocoris imperialis*: Blatchley, Het. E. N. Am., p. 379.
Distribution: Ariz., Cal., Col., Fla., Ks., La., N.M., Ok., Tex., Ut. (Mexico to Central America).
Ref.: Slater, 1964a, 1: 599 (as *Hypogeocoris*); Readio and Sweet, 1982.

Isthmocoris piceus (Say), 1832
1832 *Salda picea* Say, Descrip. Het. Hem. N. Am., p. 18. [Mass.]. Neotype designated by Readio and Sweet, 1982, Misc. Publ. Ent. Soc. Am., 12(3): 12.
1876 *Ophthalmicus piceus*: Glover, Ms. Notes J. Hem., p. 55.
1876 *Geocoris piceus*: Uhler, Bull. U.S. Geol. Geogr. Surv. Terr., 1: 307.
1907 *Germalus piceus*: Montandon, An. Hist.-Nat. Mus. Nat. Hung., 5: 90.
1913 *Hypogeocoris piceus*: Montandon, Bull. Acad. Roum. Buch., 2: 55.
1914 *Isthmocoris piceus*: McAtee, Proc. Biol. Soc. Wash., 27: 127.
Distribution: Col., Conn., Fla., Ia., Ill., Ind., Ks., Mass., Mich., Md., Me., Miss., Mo., N.H., N.J., N.M., N.Y., Oh., Ont., Pa., R.I., Ut., Va., Vt., Wis.
Ref.: Slater, 1964a, 1: 600 (as *Hypogeocoris*); Readio and Sweet, 1982.

Isthmocoris slevini (Van Duzee), 1928
1928 *Hypogeocoris slevini* Van Duzee, Pan-Pac. Ent., 4: 190. [Cal.].
1982 *Isthmocoris slevini*: Readio and Sweet, Misc. Publ. Ent. Soc. Am., 12(3): 10.
Distribution: Cal. (Mexico).
Ref.: Slater, 1964a, 1: 601 (as *Hypogeocoris*).

Isthmocoris tristis (Stål), 1854
1854 *Geocoris tristis* Stål, Öfv. K. Svens. Vet.-Akad., Förh., 11(8): 236. [Cal.].
1872 *Ophthalmicus tristis*: Walker, Cat. Hem. Brit. Mus., 5: 137.
1914 *Isthmocoris tristis*: McAtee, Proc. Biol. Soc. Wash., 27: 127.
1946 *Hypogeocoris tristis*: Torre-Bueno, Ent. Am., 26: 43.
Distribution: Cal. (Mexico).
Ref.: Slater, 1964a, 1: 601 (as *Hypogeocoris*).

Subfamily Heterogastrinae Stål, 1872

Ref.: Slater, 1964a, 1: 739.

Genus *Heterogaster* Schilling, 1829

1829 *Heterogaster* Schilling, Beitr. Ent., 1: 37. Type-species: *Cimex urticae* Fabricius, 1775. Designated by Curtis, 1836, Brit. Ent. 2nd. ed., 13: 597.
1837 *Phygas* Fieber, Beitr. Ges. Nat.-Heilwiss., 1: 348. Preoccupied. Unnecessary new name for *Heterogaster* Schilling.
1851 *Phygadicus* Fieber, Abh. Konigl. Böhm. Ges. Wiss., (5)7: 461. Unecessary new name for *Heterogaster* Schilling.
Ref.: Slater, 1964a, 1: 746.

Heterogaster behrensii (Uhler), 1876
1876 *Phygadicus behrensii* Uhler, Bull. U.S. Geol. Geogr. Surv. Terr., 1: 312. [Cal.].
1894 *Heterogaster Behrensi* [*sic*]: Lethierry and Severin, Cat. Gen. Hem., 2: 176.
1916 *Heterogaster behrensii*: Van Duzee, Check List Hem., p. 20.
Distribution: B.C., Cal., Id., Ore., Ut. (Mexico).
Ref.: Slater, 1964a, 1: 753.

Heterogaster flavicosta Barber, 1939
1939 *Heterogaster flavicosta* Barber, Proc. Ent. Soc. Wash., 41: 173. [Tex.].
1964 *Heterogaster flavicostus* [*sic*]: Slater, Cat. Lygaeidae World, 1: 757.
Distribution: La., Tex.
Ref.: Slater, 1964a, 1: 757.

Subfamily Ischnorhynchinae Stål, 1872

Ref.: Slater, 1964a, 1: 351; Scudder, 1962a (key genera world).

Genus *Kleidocerys* Stephens, 1829

1829 *Kleidocerys* Stephens, Syst. Cat. Brit. Ins., 2: 342. Type-species: *Lygaeus resedae* Panzer, 1797. Designated by Barber, 1953, Proc. Ent. Soc. Wash., 55: 282.
1860 *Ischnorhynchus* Fieber, Europ. Hem., p. 51. Type-species: *Lygaeus resedae* Panzer, 1797. Designated by Oshanin, 1912, Kat. Paläark. Hem., p. 29. Synonymized by China, 1943, Gen. Names Brit. Ins., pt. 8, p. 237.

Ref.: Slater, 1964a, 1: 354; Barber, 1953b (key spp.); Scudder, 1962a (key spp. world); Brailovsky, 1976.

Kleidocerys dimidiatus Barber, 1953

1953 *Kleidocerys dimidiatus* Barber, Proc. Ent. Soc. Wash., 55: 280. [Ariz.].

Distribution: Ariz., Wash.

Ref.: Slater, 1964a, 1: 357.

Kleidocerys franciscanus (Stål), 1859

1859 *Cymus franciscanus* Stål, Freg. Eug. Resa, 3: 252. [Cal.]. Lectotype designated by Barber, 1956, Ent. News, 67: 205.

1914 *Ischnorhynchus franciscanus*: Van Duzee, Trans. San Diego Soc. Nat. Hist., 2: 8.

1916 *Ischnorrhynchus* [*sic*] *franciscanus*: Van Duzee, Check List Hem., p. 19.

1949 *Kleidocerys franciscanus*: Barber, Proc. Ent. Soc. Wash., 51: 273.

Distribution: Ariz., B.C., Cal., Id., Mont., Ore., Ut., Wash.

Ref.: Slater, 1964a, 1: 357.

Kleidocerys modestus Barber, 1953

1953 *Kleidocerys modestus* Barber, Proc. Ent. Soc. Wash., 55: 279. [Cal.].

Distribution: Ariz., Cal., Ore., Wash.

Ref.: Slater, 1964a, 1: 358.

Kleidocerys obovatus (Van Duzee), 1931

1931 *Ischnorrhynchus* [*sic*] *obovatus* Van Duzee, Pan-Pac. Ent., 7: 110. [Cal.].

1953 *Kleidocerys obovatus*: Barber, Proc. Ent. Soc. Wash., 55: 276.

Distribution: Cal.

Ref.: Slater, 1964a, 1: 358.

Kleidocerys ovalis Barber, 1953

1953 *Kleidocerys ovalis* Barber, Proc. Ent. Soc. Wash., 55: 278. [N.H.].

Distribution: Ariz., B.C., Cal., Col., Id., Mass., Me., Mich., Minn., N.D., N.H., N.Y., Ore., Ont., S.D., Ut., Wash., Wyo.

Ref.: Slater, 1964a, 1: 358.

Kleidocerys resedae (Panzer), 1797

1797 *Lygaeus resedae* Panzer, Faun. Ins. Germ., 40: 20. [Germany].

1819 *Lygaeus didymus* Zetterstedt, K. Svens. Vet.-Akad. Handl., p. 71. [Sweden]. Synonymized by Gistl, 1837, Faun. Germ., 1(2): 100.

1876 *Ischnorhynchus didymus*: Uhler, Bull. U.S. Geol. Geogr. Surv. Terr., 1: 305.

1876 *Ischnorynchus resedae*: Glover, Ms. Notes J. Hem., p. 34, 43, 104.

1883 *Cymus residae* [*sic*]: Brodie and White, Checklist Ins. Dom. Can., p. 60.

1908 *Kleidocerus* [*sic*] *resedae*: Torre-Bueno, J. N.Y. Ent. Soc., 16: 229.

1946 *Kleidocerys resedae*: Procter, Biol. Surv. Mt. Desert, 7: 71.

Distribution: Alk., Alta., B.C., Col., Id., Man., Mass., Me., Mich., N.H., N.Y., Nfld., Oh., Ont., Ore., Que., S.D., Wash. (Europe to Asia).

Ref.: Slater, 1964a, 1: 358; Jordan, 1933; Slater, 1952b; Scudder, 1963b; Kinzelbach, 1970; Wheeler, 1976a (life history); Melber et al., 1980 (egg predator).

Kleidocerys resedae fuscomaculatus Barber, 1953

1953 *Kleidocerys resedae fuscomaculatus* Barber, Proc. Ent. Soc. Wash., 55: 276. [Cal.].

Distribution: B.C., Cal., Ore., Wash.

Ref.: Slater, 1964a, 1: 367.

Kleidocerys resedae geminatus (Say), 1832

1832 *Lygaeus geminatus* Say, Descrip. Het. Hem. N. Am., p. 14. [Mo.].
1874 *Ischnorhynchus didymus*: Stål, K. Svens. Vet.-Akad. Handl., 12(1): 124 (in part).
1908 *Ischnorhynchus geminatus*: Horvath, An. Hist.-Nat. Mus. Nat. Hung., 6: 560.
1910 *Kleidocerys geminatus*: Torre-Bueno, J. N.Y. Ent. Soc., 18: 28.
1916 *Ischnorrhynchus* [*sic*] *geminatus*: Van Duzee, Check List Hem., p. 19.
1953 *Kleidocerys resedae geminatus*: Barber, Proc. Ent. Soc. Wash., 55: 276.
Distribution: Col., Conn., D.C., Fla., Ga., Ia., Ill., Ind., Ks., La., Mass., Md., Me., Mich., Minn., Mo., N.C., N.H., N.J., N.S., N.Y., Neb., Ok., Ont., Pa., Que., S.C., S.D., Tex. (Mexico to Central America).
Ref.: Slater, 1964a, 1: 367.

Kleidocerys resedae resedae (Panzer), 1797

1797 *Lygaeus resedae* Panzer, Fauna Ins. Ger., 40: 20. [Germany].
1953 *Kleidocerys resedae*: Barber, Proc. Ent. Soc. Wash., 55: 247.
Distribution: Alk., Alta., B.C., Col., Id., Man., Mass., Me., Mich., N.H., N.Y., Nfld., Oh., Ont., Ore., Que., S.D., Wash. (Asia, Europe).
Note: When Barber (1953, above) included *K. resedae fuscomaculatus* and *K. resedae geminatus* in *K. resedae*, he in effect also proposed the above trinomen.
Ref.: Slater, 1964a, 1: 358; Jordan, 1933; Slater, 1952a; Wheeler, 1976a (life history); Melber et al., 1980 (egg predator).

Kleidocerys virescens (Fabricius), 1794

1794 *Acanthia virescens* Fabricius, Ent. Syst., 4: 70. ["Americae meridionalis insulis Dom. Smidt"].
1882 *Ischnorhynchus championi* Distant, Biol. Centr.-Am., Rhyn., 1: 193. [Guatemala]. Synonymized by Ashlock, 1960, Proc. Biol. Soc. Wash., 73: 235.
1916 *Ischnorrhynchus* [*sic*] *championi*: Van Duzee, Check List Hem., p. 19.
1947 *Kleidocerys championi*: Barber, Mem. Soc. Cubana Hist. Nat., 19: 64.
1960 *Kleidocerys virescens*: Ashlock, Proc. Biol. Soc. Wash., 73: 235.
Distribution: Fla., Tex. (Mexico to Central America, West Indies).
Note: The present concept of this species was established by Ashlock (1960, above).
Ref.: Slater, 1964a, 1: 373.

Subfamily Lygaeinae Schilling, 1829

Ref.: Slater, 1964a, 1: 8; Barber, 1917 (key U.S. genera, as Lygaeini); Barber, 1921a (key U.S. genera, as subgenera of *Lygaeus*); Blatchley, 1926a (key eastern U.S. genera, Lygaeini); Torre-Bueno, 1946a (key N. Am. genera, as Lygaeini); Ashlock, 1975 (key U.S. genera); Brailovsky, 1978a (key Am. genera).

Genus *Craspeduchus* Stål, 1874

1874 *Lygaeus* (*Craspeduchus*) Stål, K. Svens. Vet.-Akad. Handl., 12(1): 105, 109. Type-species: *Lygaeus xanthostaurus* Herrich-Schaeffer, 1847. Designated by Van Duzee, 1916, Check List Hem., p. 18.
1874 *Lygaeus* (*Ochrostomus*) Stål, K. Svens. Vet.-Akad. Handl., 12(1): 105, 109. Type-species: *Lygaeus pulchellus* Fabricius, 1794. Designated by Van Duzee, 1916, Check List Hem., p. 18. Synonymized by Ashlock, 1975, J. Ks. Ent. Soc., 48: 27.

1964 *Craspeduchus*: Slater, Cat. Lygaeidae World, 1: 47.
1964 *Ochrostomus*: Slater, Cat. Lygaeidae World, 1: 153.
Ref.: Slater, 1964a, 1: 47, 153; Brailovsky, 1979b (revision, key spp.).

Craspeduchus pulchellus (Fabricius), 1794
1794 *Lygaeus pulchellus* Fabricius, Ent. Syst., 4: 159. [Saint Croix, West Indies].
1975 *Craspeduchus pulchellus*: Baranowski and Slater, Fla. Ent., 58: 297.
Distribution: Fla., Tex. (Central and South America, West Indies).
Note: This species was erroneously reported from Quebec [as *Lygaeus pulchellus*] by Provancher (1871, Nat. Can., 3(4): 137).
Ref.: Slater, 1964a, 1: 159 (as *Ochrostomus*); Baranowski and Slater, 1975 (life history).

Craspeduchus uhleri (Stål), 1874
1874 *Lygaeus (Craspeduchus) uhleri* Stål, K. Svens. Vet.-Akad. Handl., 12(1): 109. [Mexico].
1906 *Lygaeus uhleri*: Barber, Mus. Brook. Inst. Arts Sci., Sci. Bull., 1: 274.
1964 *Craspeduchus uhleri*: Slater, Cat. Lygaeidae World, 1: 47.
Distribution: Ariz., Cal., Tex. (Mexico to Central America).
Ref.: Slater, 1964a, 1: 47.

Genus *Lygaeospilus* Barber, 1921

1921 *Lygaeus (Lygaeospilus)* Barber, Proc. Ent. Soc. Wash., 23: 65. Type-species: *Aphanus tripunctatus* Dallas, 1852. Monotypic.
1948 *Lygaeospilus*: Barber, Oh. J. Sci., 48: 67.
Ref.: Slater, 1964a, 1: 73; Barber, 1921a (key spp., as *Lygaeus (Lygaeospilus)*); Torre-Bueno, 1946a (key U.S. spp., as *Lygaeus (Lygaeospilus)*); Scudder, 1981 (key spp.).

Lygaeospilus brevipilus Scudder, 1981
1981 *Lygaeospilus brevipilus* Scudder, Can. Ent., 113: 747. [B.C.].
Distribution: B.C.

Lygaeospilus fusconervosus Barber, 1948
1948 *Lygaeospilus fusconervosus* Barber, Oh. J. Sci., 48: 66. [Cal.].
Distribution: Cal.
Ref.: Slater, 1964a, 1: 74.

Lygaeospilus pusio (Stål), 1874
1874 *Melanocoryphus pusio* Stål, K. Svens. Vet.-Akad. Handl., 12(1): 112. [Tex.].
1893 *Lygaeus albulus* Distant, Biol. Centr.-Am., Rhyn., 1: 380. [Guatemala]. Synonymized by Barber, 1921, Proc. Ent. Soc. Wash., 23: 68.
1893 *Lygaeosoma solida* Uhler, N. Am. Fauna, 7: 262. [Cal.]. Synonymized with *Lygaeus albulus* Distant by Van Duzee, 1909, Bull. Buff. Soc. Nat. Sci., 9: 165.
1909 *Lygaeus albulus*: Van Duzee, Bull. Buff. Soc. Nat. Hist., 9: 165.
1912 *Lygaeus abulus* [*sic*]: Hunter et al., U.S. Dept. Agric. Ent. Bull., 113: 51.
1916 *Lygaeus (Melanopleurus) pusio*: Van Duzee, Check List Hem., p. 18.
1921 *Lygaeus (Lygaeospilus) pusio*: Barber, Proc. Ent. Soc. Wash., 23: 65, 68.
1948 *Lygaeospilus pusio*: Barber, Oh. J. Sci., 48: 66.
Distribution: Ariz., Cal., Col., Fla., Ok., Tex., Ut., Wyo. (Mexico to Guatemala).
Ref.: Slater, 1964a, 1: 74.

Lygaeospilus tripunctatus (Dallas), 1852

1852 *Aphanus tripunctatus* Dallas, List Hem. Brit. Mus., 2: 559. [Fla.].

1874 *Melanocoryphus (Melanocoryphus) obscuripennis* Stål, K. Svens. Vet.-Akad. Handl., 12(1): 112. [Tex.]. Synonymized by Barber, 1921, Proc. Ent. Soc. Wash., 23: 68.

1886 *Melanocoryphus obscuripennis*: Uhler, Check-list Hem. Het., p. 16.

1902 *Lygaeosoma parvula*: Montgomery, Ent. News, 13: 12. Uhler ms. name; *nomen nudum*.

1914 *Lygaeus albulus*: Barber, Bull. Am. Mus. Nat. Hist., 33: 509.

1916 *Lygaeus (Melanocoryphus) obscuripennis*: Van Duzee, Check List Hem., p. 18.

1916 *Rhyparochromus tripunctatus*: Van Duzee, Check List Hem., p. 22.

1921 *Lygaeus (Lygaeospilus) tripunctatus*: Barber, Proc. Ent. Soc. Wash., 23: 65.

1923 *Lygaeus tripunctatus*: Barber, Conn. Geol. Nat. Hist. Surv. Bull., 34: 713.

1948 *Lygaeospilus tripunctatus*: Barber, Oh. J. Sci., 48: 66.

Distribution: Ariz., Cal., Conn., Fla., Mass., Me., N.C., N.M., N.Y., Ok., Que., R.I., S.D., Tex., Ut. (Mexico).

Ref.: Slater, 1964a, 1: 74.

Genus *Lygaeus* Fabricius, 1794

1794 *Lygaeus* Fabricius, Ent. Syst., 4: 133. Type-species: *Cimex equestris* Linnaeus, 1758. Designated by Curtis, 1833, Brit. Ent., 10: 481.

1868 *Lygaeus (Graptolomus)* Stål, K. Svens. Vet.-Akad. Handl., 7(11): 73, 75. Type-species: *Lygaeus turcicus* Fabricius, 1803. Designated by China, 1943, Gen. Names Brit. Ins., B: 236. Synonymized by Reuter, 1885, Rev. d'Ent., 4: 201.

Ref.: Slater, 1964a, 1: 75; Barber, 1921a (key to U.S. spp.); Blatchley, 1926a (key spp. eastern U.S.); Torre-Bueno, 1946a (key U.S. spp.); Brailovsky, 1978b (key spp. Western Hemisphere).

Lygaeus formosus Blanchard, 1840

1840 *Lygaeus formosus* Blanchard, Hist. Nat. Ins., 3: 130. [Mexico].

1914 *Lygaeus formosus*: Barber, Bull. Am. Mus. Nat. Hist., 33: 509.

1916 *Lygaeus (Lygaeus) formosus*: Van Duzee, Check List Hem., p. 18.

Distribution: Fla. (Mexico to Venezuela, West Indies).

Ref.: Slater, 1964a, 1: 92; Brailovsky, 1978b (generic position).

Lygaeus kalmii Stål, 1874

1874 *Lygaeus (Graptolomus) kalmii* Stål, K. Svens. Vet.-Akad. Handl., 12(1): 107. [Cal., N.Y., Wis., Mexico].

1876 *Lygaeus kalmii*: Uhler, Bull. U.S. Geol. Geogr. Surv. Terr., 1: 302.

1876 *Lygaeus reclivatus*: Uhler, Bull. U.S. Geol. Geogr. Surv. Terr., 1: 302.

1876 *Lygaeus turcicus*: Uhler, Bull. U.S. Geol. Geogr. Surv. Terr., 1: 302.

1916 *Lygaeus (Lygaeus) kalmii*: Van Duzee, Check List Hem., p. 18.

Distribution: Alta., Ariz., B.C., Cal., Col., Conn., D.C., Fla., Ga., Ia., Ill., Ind., Id., Ks., Man., Mass., Md., Me., Mich., Minn., Miss., Mo., Mont., N.C. N.D., N.H., N.J., N.M., N.S., N.Y., Neb., Nev., Oh., Ok., Ont., Ore., Pa., Que., R.I., Sask., S.D., Tenn., Ut., Va., Vt., Wash., Wis., Wyo. (Mexico to Central America).

Ref.: Slater, 1964a, 1: 93; Jones [M.P.], 1935; Schechter and Brickley, 1959; Wadley, 1962; Kessel and Beams, 1963; Scudder, 1963b; Slifer and Sekhon, 1963; Caldwell, 1968, 1970, 1974; Caldwell and Hegmann, 1969; Slater and Knop, 1969 (re-evaluation of subspp. of *L. kalmii*); Borror and White, 1970; Duffey, 1970; Judd, 1970; Dingle and Caldwell, 1971; Duffey and Scudder, 1972; Dingle, 1974; Hogue, 1974; Price, 1975; Bowers et al., 1976; Dailey, 1977; Isman et al., 1977b; Nixon and McPherson, 1977; Wilson and Price, 1977; Arnaud, 1978 (as tachinid host); Dailey, 1977; Dailey et al., 1978; Lampman and Fashing, 1978; Aller and Caldwell, 1979; Aller et al., 1979; Price and Willson, 1979; Ignatowicz and Boczek, 1980; Milne and Milne, 1980; Trumbo and Fashing, 1980 (behavior); Chaplin and Chaplin, 1981; Duarte and Calabrese, 1982 (cytology); Burgess et al., 1983; Evans, 1983 (biol.); Root, 1986 (biol.).

Lygaeus kalmii angustomarginatus Parshley, 1919

1835 *Lygaeus turcicus*: Harris, Rept. Geol. Miner. Bot. Zool. Mass., p. 577.

1869 *Lygaeus reclivatus*: Rathvon, Hist. Lancaster Co., Pa., p. 549.

1874 *Lygaeus* (*Graptolomus*) *kalmii*: Stål, K. Svens. Vet.-Akad. Handl., 12(1): 107 (in part).

1919 *Lygaeus kalmii angustomarginatus* Parshley, Occas. Pap. Mus. Zool., Univ. Mich., 71: 14. [Conn.].

Distribution: Conn., D.C., Fla., Ga., Ia., Ill., Ind., Ks., Man., Mass., Md., Me., Mich., Minn., Miss., Mo., N.C., N.D., N.H., N.J., N.S., N.Y., Neb., Oh., Ok., Ont., Pa., Que., R.I., S.D., Tenn., Tex., Va., Vt., Wis.

Ref.: Slater, 1964a, 1: 95; Simanton and Andre, 1936 (biol.); Slater and Knop, 1969; Sauer and Feir, 1972 (predator); Hunt, 1979a, b (biol.); Wheeler, 1983a (food plant).

Lygaeus kalmii kalmii Stål, 1874 [Fig. 48]

1874 *Lygaeus* (*Graptolomus*) *kalmii* Stål, K. Svens. Vet.-Akad. Handl., 12(1): 107 (in part).

1893 *Lygaeus kalmii melanodermus* Montandon, An. Soc. Ent. Belg., 37: 400. [Mo.]. Synonymized by Parshley, 1923, Can. Ent., 55: 81.

1910 *Lygaeus kalmii* var. *melanodemus* [*sic*]: Banks, Cat. Nearc. Hem.-Het., p. 61.

1919 *Lygaeus kalmii kalmii*: Parshley, Occas. Pap. Mus. Zool., Univ. Mich., 71: 14.

Distribution: Alta., Ariz., B.C., Cal., Col., Id., Ks., Man., Mo., Mont., N.M., N.D., Neb., Nev., Ok., Ore., Sask., S.D., Tex., Ut., Wash., Wyo. (Mexico to Central America).

Note: All references to var. *melanodermus* Montandon since 1923 seem to have overlooked Parshley's (1923, above) synonymy, and no one has seen specimens since then. It seems best (J. A. Slater, pers. comm., agrees) to accept Parshley's synonymy and delete the form from our lists.

Ref.: Slater, 1964a, 1: 93; Slater and Knop, 1969.

Lygaeus nugatoria Kelso, 1937, *Nomen Nudum*

1937 *Lygaeus nugatoria* Kelso, *In* Girard, Univ. Wyo. Publ., 3(1): 37. [Wyo.]. *Nomen nudum; incerti sedis.*

Note: This name is included, along with *Lygaeus* sp., *Lygaeus* nymphs, and *Lygaeus lateralis*, in a list of the stomach contents of the sage grouse by Girard (1937, above). On page 2, Mr. L. H. Kelso is credited with examining the 33 stomachs used in the study and for the name which, to our knowledge, has never been published elsewhere.

Lygaeus reclivatus Say, 1825

1825 *Lygaeus reclivatus* Say, J. Nat. Sci. Phila., 4: 321. [Mo.].

1874 *Lygaeus* (*Graptolomus*) *reclivatus*: Stål, K. Svens. Vet.-Akad. Handl., 12(1): 107.

Distribution: Ariz., Cal., Col., Ia., Ks., Mo., N.M., Nev., Ok., Ore., Tex., Ut., Wyo. (Mexico to Central America).

Ref.: Ayala et al., 1975; Isman, 1979.

Lygaeus reclivatus enotus Say, 1832

1832 *Lygaeus reclivatus* var. *enotus* Say, Descrip. Het. Hem. N. Am., p. 14. [Mexico].

1843 *Lygaeus costalis* Herrich-Schaeffer, Wanz. Ins., 7: 22. [Mexico]. Synonymized by Barber, 1921, Proc. Ent. Soc. Wash., 23: 67.

1876 *Lygaeus costalis*: Uhler, Bull. U.S. Geol. Geogr. Surv. Terr., 1: 302.

1886 *Lygaeus* (*Graptolomus*) *costalis*: Distant, Biol. Centr.-Am., Rhyn., 2: 178.

1940 *Lygaeus trux*: Torre-Bueno, Bull. Brook. Ent. Soc., 35: 157.

1948 *Lygaeus reclivatus enotus*: Barber, Oh. J. Sci., 48: 66.

Distribution: Ariz., Cal., Tex. (Mexico to Costa Rica).

Ref.: Slater, 1964a, 1: 99.

Lygaeus reclivatus reclivatus Say, 1825

1825 *Lygaeus reclivatus* Say, J. Nat. Sci. Phila., 4: 321. [Mo.].

1874 *Lygaeus* (*Graptolomus*) *reclivatus*: Stål, K. Svens. Vet.-Akad. Handl., 12(1): 107.

1902 *Lygaeus turcicus reclivatus*: Heidemann, Proc. Ent. Soc. Wash., 5: 82.

Distribution: Ariz., Cal., Col., Ia., Mo., N.M., Ok., Ore., Tex. (Mexico to South America).

Note: The distribution given here is from Slater and Knop (1969, An. Ent. Soc. Am., 62: 1222-1232). Other localities cited by various authors are mostly based on misidentifications.

Ref.: Slater, 1964a, 1: 98, Slater and Knop, 1969.

Lygaeus truculentus Stål, 1862

1862 *Lygaeus truculentus* Stål, Stett. Ent. Zeit., 23: 308. [Mexico].

1867 *Lygaeus truculentus*: Uhler, Bull. U.S. Geol. Geogr. Surv. Terr., 1: 302.

1874 *Lygaeus* (*Graptolomus*) *truculentus*: Stål, K. Svens. Vet.-Akad. Handl., 12(1): 108.

1926 *Lygaeus trunculentus* [*sic*]: Essig, Ins. W. N. Am., p. 348.

Distribution: Ariz., Cal. (Mexico to South America).

Ref.: Slater, 1964a, 1: 101.

Lygaeus trux Stål, 1862

1862 *Lygaeus trux* Stål, Stett. Ent. Zeit., 23: 308. [Mexico].

1978 *Lygaeus trux*: Brailovsky, An. Inst. Biol. Univ. Nac. Aut. Méx., 49, Ser. Zool., 1: 140.

Distribution: Ariz., Tex. (Mexico).

Note: Ashlock (1977, Proc. Ent. Soc. Wash., 79: 575) found that the sole U.S. record (Ariz.), reported by Torre-Bueno (1940, Bull Brook. Ent. Soc., 35: 157; 1946a, Ent. Am., 26: 12), was based on a misidentified specimen of *L. reclivatus* Say. Subsequently, Brailovsky (1978, above) recorded the species from Arizona and Texas.

Ref.: Slater, 1964a, 1: 101.

Lygaeus turcicus Fabricius, 1803

1803 *Lygaeus turcicus* Fabricius, Syst. Rhyn., p. 218. [N.Y.].

1852 *Lygaeus trimaculatus* Dallas, List Hem. Brit. Mus., 2: 543. [N. Am.]. Synonymized by Barber, 1921, Proc. Ent. Soc. Wash., 23: 67.

1868 *Lygaeus* (*Graptolomus*) *turcicus*: Stål, K. Svens. Vet.-Akad. Handl., 7(11): 73.
Distribution: Ariz., Cal., Col., Conn., D.C., Ia., Ill., Ind., Ks., Mass., Md., Me., Mich., Minn., Mo., N.C., N.M., N.Y., Oh., Ok., Ont., Pa., Que., S.D., Va.
Ref.: Slater, 1964a, 1: 101; Slater, 1983 (biol.).

Genus *Melanopleurus* Stål, 1874

1874 *Lygaeus* (*Melanopleurus*) Stål, K. Svens. Vet.-Akad. Handl., 12(1): 105, 109. Type-species: *Lygaeus bicolor* Herrich-Schaeffer, 1850. Designated by Van Duzee, 1916, Check List Hem., p. 18.
1964 *Melanopleurus*: Slater, Cat. Lygaeidae World, 1: 146.
Ref.: Slater, 1964a, 1: 146; Barber, 1921a (key spp., as *Lygaeus* (*Melanopleurus*)); Torre-Bueno, 1946a (key spp., as *Lygaeus* (*Melanopleurus*)); Maldonado, 1974 (key spp.); Ashlock, 1975; Brailovsky, 1975a (key spp.); Brailovsky, 1979a (key spp.); Scudder, 1981 (key spp.).

Melanopleurus belfragei (Stål), 1874
1874 *Lygaeus* (*Melanopleurus*) *Belfragii* [*sic*] Stål, K. Svens. Vet.-Akad. Handl., 12(1): 109. [Tex.].
1876 *Melanopleurus Belfragii* [*sic*]: Uhler, Bull. U.S. Geol. Geogr. Surv. Terr., 1: 303.
1886 *Lygaeus Belfragii* [*sic*]: Uhler, Check-list Hem. Het., p. 16.
1916 *Lygaeus* (*Melanopleurus*) *belfragei*: Van Duzee, Check List Hem., p. 18.
1964 *Melanopleurus belfragei*: Slater, Cat. Lygaeidae World, 1:146.
Distribution: Ariz., Cal., N.M., Tex. (Mexico).
Ref.: Slater, 1964a, 1: 146; Stone and Fries, 1986.

Melanopleurus bicolor (Herrich-Schaeffer), 1850
1850 *Lygaeus bicolor* Herrich-Schaeffer, Wanz. Ins., 9: 195. [Brazil].
1904 *Lygaeus bicolor*: Snow, Univ. Ks. Sci. Bull., 2: 348.
1964 *Melanopleurus bicolor*: Slater, Cat. Lygaeidae World, 1: 146.
Distribution: Ariz., Cal., Tex. (Mexico to South America).
Ref.: Slater, 1964a, 1: 146.

Melanopleurus bistriangularis (Say), 1832
1832 *Lygaeus bistriangularis* Say, Descrip. Het. Hem. N. Am., p. 14. [Mexico].
1872 *Lygaeus bistriangularis*: Uhler, Prelim. Rept. U.S. Geol. Surv. Mont., p. 405 (in part).
1876 *Melanopleurus bistriangularis*: Uhler, Bull. U.S. Geol. Geogr. Surv. Terr., 1: 303.
Distribution: Ariz. (Mexico to South America).
Ref.: Slater, 1964a, 1: 147.

Melanopleurus bistriangularis bistriangularis (Say), 1832
1832 *Lygaeus bistriangularis* Say, Descrip. Het. Hem. N. Am., p. 14.
1974

Melanopleurus bistriangularis bistriangularis: Maldonado, Proc. Ent. Soc. Wash., 76: 26.
Distribution: Ariz. (Mexico to South America).
Note: Additional distributions [Cal., Col., N.M., S.D., Tex., Ut.] given by various authors probably refer to *Melanopleurus nubilus*, *M. tenor*, and *M. vasquezae*.

Melanopleurus bistriangularis marginellus (Dallas), 1852

1852 *Lygaeus marginellus* Dallas, List Hem. Brit. Mus., 2: 548. [Venezuela].

1916 *Lygaeus* (*Melanopleurus*) *bistriangularis* var. *marginellus*: Van Duzee, Check List Hem., p. 18.

1964 *Melanopleurus bistriangularis marginellus*: Slater, Cat. Lygaeidae World, 1: 148.

Distribution: Ariz. (Mexico to South America).

Ref.: Slater, 1964a, 1: 148.

Melanopleurus fuscosus Brailovsky, 1977

1977 *Melanopleurus fuscosus* Brailovsky, An. Inst. Biol. Univ. Nac. Aut. Méx. 48, Ser. Zool., 1: 129. [Cal.].

Distribution: Cal.

Melanopleurus nubilus Brailovsky, 1979

1979 *Melanopleurus nubilus* Brailovsky, An. Inst. Biol. Univ. Nac. Aut. Méx. 50, Ser. Zool., 1: 201. [Tex.].

Distribution: Tex.

Melanopleurus perplexus Scudder, 1981

1981 *Melanopleurus perplexus* Scudder, Can. Ent., 113: 751. [Sask.].

Distribution: Alta., Man., Sask.

Melanopleurus pyrrhopterus (Stål), 1874

1874 *Lygaeus* (*Ochrostomus*) *pyrrhopterus* Stål, K. Svens. Vet.-Akad. Handl., 12(1): 110. [Mexico].

1975 *Melanopleurus pyrrhopterus*: Ashlock, J. Kans. Ent. Soc., 48: 30.

Distribution: Alta., Ariz., Cal., Col., Id., Mont., Nev., Ore., Tex., Ut., Wyo. (Mexico to Central America).

Melanopleurus pyrrhopterus pyrrhopterus (Stål), 1874

1874 *Lygaeus* (*Ochrostomus*) *pyrrhopterus* Stål, K. Svens. Vet.-Akad. Handl., 12(1): 110. [Mexico].

1876 *Ochrostomus pyrrhopterus*: Uhler, Bull. U. S. Geol. Geogr. Surv. Terr., 1: 303.

1948 *Lygaeus pyrrhopterus*: Hayward, Ecol. Monogr., 18: 504.

1975 *Melanopleurus pyrrhopterus*: Ashlock, J. Ks. Ent. Soc., 48: 30.

Distribution: Ariz., Cal., Col., Id., Mont., Tex., Ut., Wyo. (Mexico to Central America).

Ref.: Slater, 1964a, 1: 160 (as *Ochrostomus*).

Melanopleurus pyrrhopterus melanopleurus (Uhler), 1893

1893 *Lygaeus melanopleurus* Uhler, N. Am. Fauna, 7: 262. [Cal., Col.].

1916 *Lygaeus* (*Ochrostomus*) *melanopleurus*: Van Duzee, Check List Hem., p. 18.

1946 *Lygaeus* (*Ochrostomus*) *pyrrhopterus* var. *melanopleurus*: Torre-Bueno, Ent. Am., 26: 16.

1964 *Ochrostomus pyrrhopterus melanopleurus*: Slater, Cat. Lygaeidae World, 1: 161.

1975 *Melanopleurus pyrrhopterus melanopleurus*: Ashlock, J. Ks. Ent. Soc., 48: 30.

Distribution: Ariz., Cal., Col., Nev.

Ref.: Slater, 1964a, 1: 161 (as *Ochrostomus*).

Melanopleurus tenor Brailovsky, 1979

1979 *Melanopleurus tenor* Brailovsky, An. Inst. Biol. Univ. Nac. Aut. Méx. 50, Ser. Zool., 1: 194. [Col.].

Distribution: Col.

Melanopleurus vasquezae Brailovsky, 1979

1979 *Melanopleurus vasquezae* Brailovsky, An. Inst. Biol. Univ. Nac. Aut. Méx. 50, Ser. Zool., 1: 197. [Ariz.].

Distribution: Ariz., Cal., Tex.

Genus *Neacoryphus* Scudder, 1965

1965 *Neacoryphus* Scudder, Proc. Ent. Soc. B.C., 62: 34. Type-species: *Lygaeus bicrucis* Say, 1825. Original designation.

Ref.: Slater, 1964a, 1: 119 (as *Lygaeus* (*Melanocoryphus*)); Blatchley, 1926a (key spp. eastern U.S.; as *Lygaeus* (*Melanocoryphus*)); Torre-Bueno, 1946a (key spp., as *Lygaeus* (*Melanocoryphus*)); Brailovsky, 1977b.

Neacoryphus admirabilis (Uhler), 1872

1871 *Lygaeus admirabilis* Uhler, U.S. Geol. Surv. Wyo., p. 471. *Nomen nudum.*

1872 *Lygaeus admirabilis* Uhler, Prelim. Rept. U.S. Geol. Surv. Mont., p. 405. [Col.].

1886 *Melanocoryphus admirabilis*: Uhler, Check-list Hem. Het., p. 16.

1916 *Lygaeus* (*Melanocoryphus*) *admirabilis*: Van Duzee, Check List Hem., p. 18.

1965 *Neacoryphus admirabilis*: Scudder, Proc. Ent. Soc. B.C., 62: 37.

Distribution: Ariz., Cal., Col., Fla., Id., Ks., Md., Nev., N.M., Tex., Ut., Wyo. (Mexico).

Ref.: Slater, 1964a, 1: 120 (as *Melanocoryphus*).

Neacoryphus bicrucis (Say), 1825

1825 *Lygaeus bicrucis* Say, J. Acad. Nat. Sci. Phila., 4: 322. [Ga., Mo.].

1859 *Lygaeus flavomarginellus* Stål, Freg. Eug. Resa, 3: 241. [Cal.]. Synonymized by Stål, 1874, K. Svens. Vet.-Akad. Handl., 12(1): 113.

1874 *Melanocoryphus* (*Melanocoryphus*) *bicrucis*: Stål, K. Svens. Vet.-Akad. Handl., 12(1): 113.

1876 *Melanopleurus bicrucis*: Uhler, Bull. U.S. Geol. Geogr. Surv. Terr., 1: 304.

1882 *Lygaeus* (*Melanocoryphus*) *bicrucis*: Distant, Biol. Centr.-Am., Rhyn., 1: 185.

1965 *Neacoryphus bicrucis*: Scudder, Proc. Ent. Soc. B.C., 62: 37.

Distribution: Ala., Ariz., B.C., Cal., Col., Conn., Fla., Ga., Ia., Id., Ill., Ind., Ks., La., Mass., Md., Me., Mich., Mo., N.C., N.J., N.M., N.Y., Nev., Oh., Ok., Ont., Pa., S.C., S.D., Tex., Ut., Wyo. (Mexico to Brazil).

Ref.: Slater, 1964a, 1: 126 (as *Melanocoryphus*); Lavigne, 1976; Brailovsky, 1977a; Solbreck, 1978, 1979; Solbreck and Pehrson, 1979; McLain, 1984a, b; McLain and Shure, 1985; McLain, 1986.

Neacoryphus circumlitus (Stål), 1862

1862 *Lygaeus circumlitus* Stål, Stett. Ent. Zeit., 23: 309. [Mexico].

1867 *Lygaeus circumcinctus* [*sic*]: Glover, Ms. Notes J. Hem., p. 45.

1906 *Lygaeus circumlitus*: Snow, Trans. Ks. Acad. Sci., 20: 178.

1916 *Lygaeus* (*Melanocoryphus*) *circumlitus*: Van Duzee, Check List Hem., p. 18.

1964 *Melanocoryphus circumlitus*: Slater, Cat. Lygaeidae World, 1: 129.

1965 *Neacoryphus circumlitus*: Scudder, Proc. Ent. Soc. B.C., 62: 137.

Distribution: Ariz., Col.(?), N.M., Tex. (Mexico to Panama).

Ref.: Slater, 1964a, 1: 129 (as *Melanocoryphus*).

Neacoryphus facetus (Say), 1832

1832 *Lygaeus facetus* Say, Descrip. Het. Hem. N. Am., p. 13. [Fla.].

1874 *Melanocoryphus* (*Melanocoryphus*) *facetus*: Stål, K. Svens. Vet.-Akad. Handl., 12(1): 113.

1874 *Melanocoryphus rubriger*: Stål, K. Svens. Vet.-Akad. Handl., 12(1): 113.
1902 *Melanocoryphus fascetus* [*sic*]: Heidemann, Proc. Ent. Soc. Wash., 5: 82.
1916 *Lygaeus* (*Melanocoryphus*) *facetus*: Van Duzee, Check List Hem., p. 18.
1964 *Melanocoryphus facetus*: Slater, Cat. Lygaeidae World, 1: 131.
1965 *Neacoryphus facetus*: Scudder, Proc. Ent. Soc. B.C., 62: 37.
Distribution: Fla., Ga., La., Md., N.J., N.C., Pa., S.C., Tex. (Mexico).
Note: Barber (1914, Bull. Am. Mus. Nat. Hist., 33: 509) and Blatchley (1926, Het. E. N. Am., p. 348) both state that western records of this species apply to *N. lateralis* (Dallas).
Ref.: Slater, 1964a, 1: 131 (as *Melanocoryphus*).

Neacoryphus lateralis (Dallas), 1852
1852 *Lygaeus lateralis* Dallas, List Hem. Brit. Mus., 2: 550. [Mexico].
1872 *Lygaeus facetus*: Uhler, Prelim. Rept. U.S. Geol. Surv. Mont., p. 405.
1874 *Melanocoryphus* (*Melanocoryphus*) *lateralis*: Stål, K. Svens. Vet.-Akad. Handl., 12(1): 113.
1876 *Melanocoryphus facetus*: Uhler, Rept. U.S. Geol. Geogr. Surv. Terr., 1: 304.
1882 *Lygaeus* (*Melanocoryphus*) *lateralis*: Distant, Biol. Centr.-Am., Rhyn., 1: 187.
1916 *Lygaeus* (*Melanocoryphus*) *facetus*: Van Duzee, Check List Hem., p. 18.
1965 *Neacoryphus lateralis*: Scudder, Proc. Ent. Soc. B.C., 62: 37.
Distribution: Ariz., B.C., Cal., Col., Ia., Id., Ks., Mont., N.M., Ok., S.D., Sask., Tex., Ut., Wyo. (Mexico).
Ref.: Slater, 1964a, 1: 134 (as *Melanocoryphus*); Brailovsky, 1977a; Stone and Fries, 1986.

Neacoryphus nigrinervis (Stål), 1874
1874 *Melanocoryphus* (*Melanocoryphus*) *nigrinervis* Stål, K. Svens. Vet.-Akad. Handl., 12(1): 112. [Mexico, Venezuela].
1886 *Melanocoryphus nigrinervis*: Uhler, Check-list Hem. Het., p. 16.
1916 *Lygaeus* (*Melanocoryphus*) *nigrinervis*: Van Duzee, Check List Hem., p. 18.
1948 *Lygaeus nigrinervis*: Barber, Oh. J. Sci., 48: 66.
1965 *Neacoryphus nigrinervis*: Scudder, Proc. Ent. Soc. B.C., 62: 37.
Distribution: Ariz., Col., Ut. (Mexico to South America, West Indies).
Ref.: Slater, 1964a, 1: 136 (as *Melanocoryphus*).

Neacoryphus rubicollis (Uhler), 1894
1894 *Melanocoryphus rubicollis* Uhler, Proc. Cal. Acad. Sci., 2nd ser., 4: 244. [Baja California].
1910 *Lygaeus rubricollis* [*sic*]: Banks, Cat. Nearc. Hem.-Het., p. 62.
1916 *Lygaeus* (*Melanocoryphus*) *rubicollis*: Van Duzee, Check List Hem., p. 18.
1965 *Neacoryphus rubicollis*: Scudder, Proc. Ent. Soc. B.C., 62: 37.
Distribution: Ariz., Cal., Ut. (Mexico).
Ref.: Slater, 1964a, 1: 136 (as *Melanocoryphus*).

Genus *Ochrimnus* Stål, 1874

1874 *Melanocoryphus* (*Ochrimnus*) Stål, K. Svens. Vet.-Akad. Handl., 12(1): 114.
Type-species: *Lygaeus collaris* Fabricius, 1803. Designated by Slater, 1964, Cat. Lygaeidae World, 1: 152.
1964 *Ochrimnus*: Slater, Cat. Lygaeidae World, 1: 152.

Ref.: Slater, 1964a, 1: 152, 153 (including most of *Ochrostomus*); Barber, 1921a (key spp., as *Lygaeus* (*Ochrostomus*)); Torre-Bueno, 1946a (key spp., as *Lygaeus* (*Ochrostomus*)); Brailovsky, 1982 (key spp.).

Ochrimnus barberi (Slater), 1964

1921 *Lygaeus* (*Ochrostomus*) *rubricatus* Barber, Proc. Ent. Soc. Wash., 12: 65. [Ariz.]. Preoccupied.

1960 *Lygaeus rubricatus*: Knowlton, Proc. Ut. Acad. Sci., 37: 53.

1964 *Ochrostomus barberi* Slater, Cat. Lygaeidae World, 1: 152. New name for *Lygaeus rubricatus* Barber.

1975 *Ochrimnus barberi*: Ashlock, J. Ks. Ent. Soc., 48: 30.

Distribution: Ariz., Cal., Ut. (Mexico).

Ref.: Slater, 1964a, 1: 154 (as *Ochrostomus*); Palmer, 1986.

Ochrimnus carnosulus (Van Duzee), 1914

1914 *Lygaeus carnosulus* Van Duzee, Trans. San Diego Soc. Nat. Hist., 2: 7. [Cal.].

1916 *Lygaeus* (*Ochrostomus*) *carnosulus*: Van Duzee, Check List Hem., p. 18.

1964 *Ochrostomus carnosulus*: Slater, Cat. Lygaeidae World, 1: 154.

1975 *Ochrimnus carnosulus*: Ashlock, J. Ks. Ent. Soc., 48: 30.

Distribution: Ariz., Cal., Tex. (Mexico and Central America).

Ref.: Slater, 1964a, 1: 154 (as *Ochrostomus*); Palmer, 1986.

Ochrimnus foederatus (Van Duzee), 1929

1929 *Lygaeus* (*Ochrostomus*) *foederatus* Van Duzee, Pan-Pac. Ent., 5: 187. [Ariz.].

1964 *Ochrostomus foederatus*: Slater, Cat. Lygaeidae World, 1: 156.

1982 *Ochrimnus foederatus*: Brailovsky, Folia Ent. Mex., 51: 80.

Distribution: Ariz., Nev.

Ref.: Slater, 1964a, 1: 154 (as *Ochrostomus*); Palmer, 1986.

Ochrimnus lineoloides (Slater), 1964

1852 *Lygaeus lineola* Dallas, List Hem. Brit. Mus., 2: 549. [Fla.]. Preoccupied.

1874 *Lygaeus* (*Ochrostomus*) *lineola*: Stål, K. Svens. Vet.-Akad. Handl., 12(1): 110.

1876 *Ochrostomus lineola*: Uhler, Bull. U.S. Geol. Geogr. Surv. Terr., 1: 303.

1964 *Ochrostomus lineoloides* Slater, Cat. Lygaeidae World, 1: 156. New name for *Lygaeus lineola* Dallas.

1975 *Ochrimnus lineoloides*: Ashlock, J. Ks. Ent. Soc., 48: 30.

Distribution: Fla., Ga., N.J., N.M., N.C., S.C., Tex., Va. (West Indies).

Ref.: Slater, 1964a, 1: 156 (as *Ochrostomus*); Palmer, 1986.

Ochrimnus mimulus (Stål), 1874

1874 *Melanocoryphus* (*Ochrimnus*) *mimulus* Stål, K. Svens. Vet.-Akad. Handl., 12(1): 113. [Tex.].

1886 *Melanocoryphus mimulus*: Uhler, Check-list Hem. Het., p. 16.

1909 *Lygaeus mimulus*: Van Duzee, Bull. Buff. Soc. Nat. Sci., 9: 164.

1916 *Lygaeus* (*Melanocoryphus*) *mimulus*: Van Duzee, Check List Hem., p. 18.

1926 *Lygaeus* (*Ochrostomus*) *mimulus*: Blatchley, Het. E. N. Am., p. 346.

1964 *Ochrimnus mimulus*: Slater, Cat. Lygaeidae World, 1: 153.

Distribution: Ala., Fla., La., Miss., N.C., S.C., Tex., Va.

Ref.: Slater, 1964a, 1: 153; Palmer, 1986.

Ochrimnus tripligatus (Barber), 1914

1914 *Lygaeus* (*Ochrostomus*) *tripligatus* Barber, Bull. Am. Mus. Nat. Hist., 33: 510. [Fla.].

1964 *Ochrostomus tripligatus*: Slater, Cat. Lygaeidae World, 1: 162.
1975 *Ochrimnus tripligatus*: Ashlock, J. Ks. Ent. Soc., 48: 30.
Distribution: Fla. (West Indies).
Ref.: Slater, 1964a, 1: 162 (as *Ochrostomus*).

Genus *Oncopeltus* Stål, 1868

1868 *Lygaeus* (*Oncopeltus*) Stål, K. Svens. Vet.-Akad. Handl., 7(11): 70, 75. Type-species: *Cimex famelicus* Fabricius, 1781. Designated by Distant, 1903, Fauna Brit. India, Rhyn., 2: 4.
1872 *Oncopeltus*: Stål, Öfv. K. Svens. Vet.-Akad. Förh., 29(7): 40.
Ref.: Slater, 1964a, 1: 162; Blatchley, 1926a (key spp. eastern U.S.); Torre-Bueno, 1946a (key spp.).

Subgenus *Oncopeltus* Stål, 1868

1868 *Lygaeus* (*Oncopeltus*) Stål, K. Svens. Vet.-Akad. Handl., 7(11): 70, 75. Type-species: *Cimex famelicus* Fabricius, 1781. Designated by Distant, 1903, Fauna Brit. India, Rhyn., 2: 4.

Oncopeltus cayensis Torre-Bueno, 1944, New Subgeneric Combination
1933 *Oncopeltus sexmaculatus*: Torre-Bueno, Bull. Brook. Ent. Soc., 28: 31.
1944 *Oncopeltus* (*Erythrischius*) *cayensis* Torre-Bueno, Bull. Brook. Ent. Soc., 39:135. [Fla.].
Distribution: Fla. (Bahamas).
Ref.: Slater, 1964a, 1: 176.

Oncopeltus guttaloides Slater, 1964
1843 *Lygaeus gutta* Herrich-Schaeffer, Wanz. Ins., 7: 20. [Mexico]. Preoccupied.
1876 *Oncopeltus guttas* [*sic*]: Uhler, Bull. U.S. Geol. Geogr. Surv. Terr., 1: 302.
1882 *Oncopeltus gutta*: Distant, Biol. Centr.-Am., Rhyn., 1: 174.
1916 *Oncopeltus* (*Oncopeltus*) *gutta*: Van Duzee, Check List Hem., p. 18.
1964 *Oncopeltus guttaloides* Slater, Cat. Lygaeidae World, 1: 167. New name for *Lygaeus gutta* Herrich-Schaeffer.
Distribution: Ariz., Cal. (Mexico to Central America, West Indies).
Ref.: Slater, 1964a, 1: 167.

Oncopeltus sanguineolentus Van Duzee, 1914
1914 *Oncopeltus sanguineolentus* Van Duzee, Trans. San Diego Soc. Nat. Hist., 2: 6. [Cal.].
1916 *Oncopeltus* (*Oncopeltus*) *sanguineolentus*: Van Duzee, Check List Hem., p. 18.
Distribution: Cal. (Mexico).
Ref.: Slater, 1964a, 1: 171.

Oncopeltus sexmaculatus Stål, 1874
1874 *Oncopeltus* (*Oncopeltus*) *sexmaculatus* Stål, K. Svens. Vet.-Akad. Handl., 12(1): 102. [Mexico].
1914 *Oncopeltus sex-maculatus* [*sic*]: Barber, Bull. Am. Mus. Nat. Hist., 33: 509.
1916 *Oncopeltus* (*Oncopeltus*) *sexmaculatus*: Van Duzee, Check List Hem., p. 18.
1916 *Oncopeltus* (*Erythrischius*) *sex-maculatus* [*sic*]: Van Duzee, Check List Hem., p. 18.
Distribution: Fla., Tex. (Mexico to Central America, West Indies).

Note: The Florida record for this species may apply to *O.* (*O.*) *cayensis* Torre-Bueno (J. A. Slater, pers. comm.).
Ref.: Slater, 1964a, 1: 171.

Subgenus *Erythrischius* Stål, 1874

1874 *Oncopeltus* (*Erythrischius*) Stål, K. Svens. Vet.-Akad. Handl., 12(1): 102. Type-species: *Lygaeus fasciatus* Dallas, 1852. Designated by Van Duzee, 1916, Check List Hem., p. 18.

Oncopeltus cingulifer Stål, 1874

1874 *Oncopeltus* (*Erythrischius*) *cingulifer* Stål, K. Svens. Vet.-Akad. Handl., 12(1): 103. [Colombia, Honduras, Mexico].
1978 *Oncopeltus cingulifer*: Dingle, Evol. Ins. Migr. Diap., p. 225.
1980 *Oncopeltus cingulifer antillensis* Dingle et al., Evol., 34: 357. *Nomen nudum.*
Distribution: Fla., Tex. (Mexico to South America).
Ref.: Slater, 1964a, 1: 176, 1972a; Chaplin, 1973; Ayers et al., 1974; Root and Chaplin, 1976; Dingle, 1978; Blakley and Dingle, 1978; Blakley and Goodner, 1978; Isman, 1979; Blakley, 1980, 1981; O'Rourke, 1980; Dingle, Alden, et al., 1980; Dingle, Blakley, and Miller, 1980; Griffith, 1980; Alden et al., 1983; Leslie and Dingle, 1983.

Oncopeltus fasciatus (Dallas), 1852

1842 *Lygaeus aulicus*: Herrich-Schaeffer, Wanz. Ins., 6: 76.
1852 *Lygaeus fasciatus* Dallas, List Hem. Brit. Mus., 2: 538. New name for *Lygaeus aulicus* Herrich-Schaeffer, not Fabricius. [Fla., Mexico, Colombia, Guyana, Brazil].
1874 *Oncopeltus* (*Erythrischius*) *fasciatus*: Stål, K. Svens. Vet.-Akad. Handl., 12(1): 103.
1876 *Erythrischius fasciatus*: Uhler, Bull. U.S. Geol. Geogr. Surv. Terr., 1: 303.
1882 *Oncopeltus fasciatus*: Distant, Biol. Centr.-Am., Rhyn., 1: 176.
Distribution: Ariz., Cal., Conn., Fla., Ia., Ill., Ind., Ks., La., Mass., Md., Me., Mich., Minn., Mo., N.C., N.J., N.Y., Oh., Ok., Ont., Pa., Que., S.D., Tenn., Tex., Va., Wis. (Mexico to South America, West Indies).
Ref.: Slater, 1964a, 1: 177; Andre, 1934; Bonhag and Wick, 1953; Caldwell, 1970, 1974; Dingle and Caldwell, 1971; Dingle, 1972; Arnaud, 1978 (as tachinid host); Dailey, 1977; Dailey et al., 1978; Aller et al., 1979; Aller and Caldwell, 1979; Isman, 1979; Trumbo and Fashing, 1980; Alden et al., 1983; Leslie and Dingle, 1983.

Oncopeltus sandarachatus (Say), 1832

1832 *Lygaeus sandarachatus* Say, Descrip. Het. Hem. N. Am., p. 13. [Mexico].
1876 *Erythrischius sandarachatus*: Uhler, Bull. U.S. Geol. Geogr. Surv. Terr., 1: 303.
1882 *Oncopeltus sandarachatus*: Distant, Biol. Centr.-Am., Rhyn., 2: 176.
Distribution: Cal., Fla. (Mexico to South America, West Indies).
Ref.: Slater, 1964a, 1: 183; McGhee, 1972; Alden et al., 1983; Leslie and Dingle, 1983.

Genus *Orsillacis* Barber, 1914

1886 *Orsillacis* Uhler, Check-list Hem. Het., p. 14. *Nomen nudum.*
1914 *Orsillacis* Barber, J. N.Y. Ent. Soc., 22: 169. Type-species: *Orsillacis producta* Barber, 1914. Monotypic.
Ref.: Slater, 1964a, 1: 185.

Orsillacis producta Barber, 1914

1886 *Orsillacis producta* Uhler, Check-list Hem. Het., p. 14. *Nomen nudum.*

1914 *Orsillacis producta* Barber, J. N.Y. Ent. Soc., 22: 169. [Ariz.].

Distribution: Ariz.

Ref.: Slater, 1964a, 1: 185.

Subfamily Orsillinae Stål, 1872

Ref.: Slater, 1964a, 1: 233; Barber, 1917 (key genera); Blatchley 1926a (key genera); Torre-Bueno, 1946a (key genera); Ashlock, 1967 (key tribes, genera).

Tribe Metrargini Kirkaldy, 1902

Ref.: Slater, 1964a, 1: 346; Ashlock, 1967 (key world genera).

Genus *Xyonysius* Ashlock and Lattin, 1963

1963 *Xyonysius* Ashlock and Lattin, An. Ent. Soc. Am., 56: 702. Type-species: *Nysius californicus* Stål, 1859. Original designation.

Ref.: Slater, 1964a, 1: 346 (as *Nysius*, in part), 2: 1626; Barber, 1947a (key spp., as *Nysius*, in part).

Xyonysius adjunctor (Barber), 1947

1947 *Nysius adjunctor* Barber, J. Wash. Acad. Sci., 37: 357. [Tex.].

1967 *Xyonysius adjunctor*: Ashlock, Univ. Cal. Publ. Ent., 48: 47.

Distribution: Ariz., Tex. (Mexico).

Ref.: Slater, 1964a, 1: 257 (as *Nysius*).

Xyonysius basalis (Dallas), 1852

1852 *Nysius basalis* Dallas, List Hem. Brit. Mus., 2: 553. [Jamaica, Brazil].

1894 *Nysius inaequalis* Uhler, Proc. Zool. Soc. London, 1894: 183. [Fla., Cuba, Grenada]. Synonymized by Barber, 1923, Am. Mus. Novit., 75: 12.

1946 *Nysius californicus* var. *inaequalis*: Torre-Bueno, Ent. Am., 26: 23.

1963 *Xyonysius basalis*: Ashlock and Lattin, An. Ent. Soc. Am., 56: 702.

Distribution: Fla. (Mexico to South America, West Indies).

Ref.: Slater, 1964a, 1: 260 (as *Nysius basalis*).

Xyonysius californicus (Stål), 1859

1859 *Nysius californicus* Stål, Freg. Eug. Resa, Ins., 242. [Cal.].

1963 *Xyonysius californicus*: Ashlock and Lattin, An. Ent. Soc. Am., 56: 702.

Distribution: Ala., Ariz., Cal., Col., Conn., Fla., Ga., Ia., Ill., Ks., La., Md., Me., Miss., Mo., N.C., N.D., N.M., N.Y., Nev., Ok., Ore., S.C., S.D., Tex., Ut., Va., Wash., Wyo. (Mexico to South America, West Indies).

Ref.: Slater, 1964a, 1: 262 (as *Nysius californicus*); Hantsbarger, 1957 (as *Nysius californicus*); Hurd and Linsley, 1975; Lavigne, 1976; Stone and Fries, 1986.

Xyonysius californicus alabamensis (Baker), 1906

1906 *Nysius californicus alabamensis* Baker, Invert. Pac., 1: 135. [Ala.].

1963 *Xyonysius californicus alabamensis*: Ashlock and Lattin, An. Ent. Soc. Am., 56: 702.

Distribution: Ala., Ariz., Fla., Ill., Md., N.C., Ont., S.C., Tex., Va. (Mexico, West Indies).
Note: This form is doubtfully distinct from *Xyonysius californicus* (Stål).
Ref.: Slater, 1964a, 1: 264 (as *Nysius californicus alabamensis*).

Tribe Nysiini Uhler, 1852

Ref.: Ashlock, 1967a (key world genera).

Genus *Nysius* Dallas, 1852

1852 *Nysius* Dallas, List Hem. Brit. Mus., 2: 551. Type-species: *Lygaeus thymi* Wolff, 1804. Designated by Oshanin, 1912, Kat. Paläark. Hem., p. 28, and by action of the Int. Comm. Zool. Nomen., 1955, Opinion 319.
Ref.: Slater, 1964a, 1: 253; Barber, 1947a (key spp.). References mentioning genus only: Krombein, 1969; Rogers and Lavigne, 1972; Pinto and Frommer, 1980; Stone and Fries, 1986.

Nysius angustatus Uhler, 1872
1871 *Nysius angustatus* Uhler, U.S. Geol. Surv. Wyo., p. 472. *Nomen nudum*.
1872 *Nysius angustatus* Uhler, Prelim. Rept. U.S. Geol. Surv. Mont., p. 406. [Col.].
1906 *Nysius coloradensis* Baker, Invert. Pac., 1: 135. [Col.]. Synonymized by Barber, 1947, J. Wash. Acad. Sci., 37: 358.
Distribution: Ala., Alta., Ariz., B.C., Cal., Col., D.C., Ia., Id., Ill., Ind., Ks., La., Md., Me., Man., Mass., Mich., Minn., Miss., N.D., N.H., N.J., N.M., N.Y., Neb., Nev., Oh., Ont., Ore., Pa., Que., S.D., Sask., Tex., Ut., Wash., Wyo. (Mexico).
Note: Slater (1964, below) notes that records of *N. angustatus* and *N. ericae* (Schilling) [= *N. niger* Baker] "are irrecoverably confused."
Ref.: Slater, 1964a, 1: 258; Aldrich, 1915; Arnaud, 1978 (as tachinid host); Burgess et al., 1983.

Nysius fuscovittatus Barber, 1958
1958 *Nysius fuscovittatus* Barber, Proc. Ent. Soc. Wash., 60: 70. [Alk.].
Distribution: Alk., Alta., Yuk.
Ref.: Slater, 1964a, 1: 280.

Nysius grandis Baker, 1906
1906 *Nysius coloradensis* var. *grandis* Baker, Invert. Pac., 1: 136. [Col.].
1947 *Nysius grandis*: Barber, J. Wash. Acad. Sci., 37: 359.
Distribution: Alta., Ariz., Col., Man. (Mexico to Central America).
Ref.: Slater, 1964a, 1: 280.

Nysius groenlandicus (Zetterstedt), 1838
1838 *Lygaeus groenlandicus* Zetterstedt, Ins. Lapponica Desc., p. 262. [Greenland].
1858 *Nysius thymi*: Stål, Stett. Ent. Zeit., 19: 179. [Alk.].
1876 *Nysius groenlandicus*: Uhler, Bull. U.S. Geol. Geogr. Surv. Terr., 1: 304.
1890 *Nysius ericae* var. *obscuratus* Horvath, Rev. d'Ent., 9: 188. [China, Siberia, Turkestan]. Synonymized by Lindberg, 1935, Exped. Karak. Zool. Hem., p. 415.
1931 *Nysius ericae* var. *obscuratus*: Lindroth, Zool. Bidr. Fran. Uppsala, 13: 150.
1934 *Nysius ericae groenlandicus*: Hutchinson, Mem. Conn. Acad. Arts Sci., 10: 123.

Distribution: Alk., Man., Nfld., Ont., P.Ed., Que. (China, Greenland, Iceland, Northern Europe to Siberia).

Ref.: Slater, 1964a, 1: 280; Böcher, 1972, 1975a, b, 1976, 1978.

Nysius insoletus Barber, 1947

1947 *Nysius insoletus* Barber, J. Wash. Acad. Sci., 37: 364. [Ut.].

Distribution: Col., Ut.

Ref.: Slater, 1964a, 1: 284.

Nysius monticola Distant, 1893

1893 *Nysius monticola* Distant, Biol. Centr.-Am., Rhyn., 1: 385. [Guatemala].

1947 *Nysius monticola*: Barber, J. Wash. Acad. Sci., 37: 361.

1964 *Nysius monticolus* [*sic*]: Slater, Cat. Lygaeidae World, 1: 288.

Distribution: Ariz., Col., Tex. (Mexico to Central America).

Ref.: Slater, 1964a, 1: 288.

Nysius niger Baker, 1906 [Fig. 49]

1906 *Nysius angustatus*: Baker, Invert. Pac., 1: 135.

1906 *Nysius angustatus* var. *niger* Baker, Invert. Pac., 1: 135. [Wash.].

1908 *Nysius ericae*: Horvath, An. Hist.-Nat. Mus. Nat. Hung., 6: 5.

1917 *Nysius ericae* var. *niger*: Van Duzee, Univ. Cal. Publ. Ent., 2: 160.

1977 *Nysius niger*: Ashlock, Proc. Ent. Soc. Wash., 79: 576.

Distribution: Alk., Alta., Ariz., B.C., Cal., Col., Conn., Fla., Ia., Id., Ill., Ind., Ks., La., Man., Mass., Me., Mich., Minn., Miss., Mo., Mont., N.B., N.C., N.D., N.H., N.J., N.M., N.Y., Neb., Nev., Nfld., Oh., Ok., Ont., Ore., Pa., Que., S.D., Tex., Ut., Va., Vt., Wash., Wyo. (Bermuda, Mexico to Central America, West Indies).

Note: *N. angustatus* var. *niger* Baker was synonymized with *N. ericae* (Schilling) by Barber (1947, J. Wash. Acad. Sci., 37: 360), but later resurrected by Ashlock (1977, Proc. Ent. Soc. Wash., 79: 576). Slater (1964, below) notes that records of this species are irrecoverably confused with those of *N. angustatus* Uhler; they are equally confused with those of *N. raphanus* Howard.

Ref.: Slater, 1964a, 1: 269 (as *N. ericae*, in part); Milliken and Wadley, 1913 (as *N. ericae*); Jones [M.P.], 1935 (as *N. ericae*); Knowlton, 1935a (as *N. ericae*); Little, 1972 (as *N. ericae*); Arnaud, 1978 (as tachinid host); Burgess et al., 1983 (vector, food plant, as *N. ericae*); Burgess and Weegar, 1986 (rearing, as *N. ericae*).

Nysius paludicola Barber, 1949

1949 *Nysius paludicola* Barber, Bull. Brook. Ent. Soc., 44: 144. [Wash.].

1964 *Nysius paludicolus* [*sic*]: Slater, Cat. Lygaeidae World, 1: 290.

Distribution: Alta., Wash.

Ref.: Slater, 1964a, 1: 290.

Nysius raphanus Howard, 1872

1872 *Nysius raphanus* Howard, Can. Ent., 4: 219. [Ks.].

1873 *Nysius destructor* Riley, 5th An. Rept. St. Board Agr. for 1872, 5: 113. [U.S.]. Invalid name proposed in synonymy; see Riley's (1873, above) footnote, p. 111.

1894 *Nysius strigosus* Uhler, Proc. Cal. Acad. Sci., 2nd ser., 4: 238. [Mexico]. Synonymized by Barber, 1947, J. Wash. Acad. Sci., 37: 360.

1895 *Nysius minutus* Uhler, Col. Agr. Exp. Stn. Bull., 31: 22. [Col.]. Synonymized by Barber, 1947, J. Wash. Acad. Sci., 37: 360.

1910 *Nysius strigosus*: Banks, Cat. Nearc. Hem.-Het., p. 62.

1915 *Nysius ericae minutus*: Cockerell, Can. Ent., 47: 281.

1917 *Nysius angustatus*: Watson, Fla. Bull., 135: 72.
1918 *Nysius ericae*: Milliken, J. Agr. Res., 13: 571.
Distribution: Alta., Ariz., B.C., Cal., Col., Fla., Ia., Id., Ks., La., Md., Mo., N.C., N.D., N.J., N.M., N.Y., Nev., Ok., Ont., Ore., S.D., Sask., Tex., Ut., Va. (Mexico, West Indies).
Ref.: Slater, 1964a, 1: 292; Milliken, 1918 (as *N. ericae*); Wene, 1953; Leigh, 1961, Leigh et al., 1966; Carillo, 1967b; Tappan, 1967, 1970; Barnes, 1970; Carillo and Cattagirone, 1970; Wood and Starks, 1972; Byers, 1973; Watts, 1973; Hurd and Linsley, 1975; Estrada, 1976; Arnaud, 1978 (as tachinid host).

Nysius scutellatus Dallas, 1852
1852 *Nysius scutellatus* Dallas, List Hem. Brit. Mus., 2: 553. [Jamaica].
1977 *Nysius scutellatus*: Ashlock, Proc. Ent. Soc. Wash., 79: 577.
Distribution: Fla., Ga., N.C., S.C., Va. (West Indies).
Ref.: Slater, 1964a, 1: 295.

Nysius tenellus Barber, 1947
1906 *Nysius senecionis*: Baker, Invert. Pac., 1: 136.
1908 *Nysius strigosus*: Horvath, An. Hist.-Nat. Mus. Nat. Hung., 6: 558.
1947 *Nysius tenellus* Barber, J. Wash. Acad. Sci., 37: 361. [Cal.].
Distribution: Ariz., B.C., Cal., Col., Fla., Id., N.M., Nev., Ore., Tex., Ut., Wash. (Mexico, Central America, West Indies).
Ref.: Slater, 1964a, 1: 296; Carillo, 1967a, b; Carillo and Cattagirone, 1970.

Nysius thymi (Wolff), 1804
1804 *Lygaeus thymi* Wolff, Icon. Cimic. Des. Illus., 4: 149. [Europe].
1878 *Nysius thymi*: Uhler, Proc. Boston Soc. Nat. Hist., 19: 395.
1886 *Nysius Groenlandicus* [*sic*]: Provancher, Pet. Faune Ent. Can., 3: 70.
1897 *Nysius thyus* [*sic*]: King, Can. Ent., 29: 103.
1905 *Nysius angustatus*: Van Duzee, N.Y. St. Mus. Bull., 97: 549.
Distribution: Alk., Alta., B.C., Col., Conn., Id., Ind., Man., Md., Me., Mass., N.D., N.H., N.M., N.Y., Oh., Ont., Que., Sask., Ut., Vt.
Note: Records of this species are confused with those of others in the genus.
Ref.: Slater, 1964a, 1: 297; Nokkala and Nokkala, 1985.

Tribe Orsillini Stål, 1872

Ref.: Slater, 1964a, 1: 234; Ashlock, 1967 (key to world genera).

Genus *Belonochilus* Uhler, 1871

1871 *Belonochilus* Uhler, Proc. Boston Soc. Nat. Hist., 14: 104. Type-species: *Lygaeus numenius* Say, 1832. Monotypic.
Ref.: Slater, 1964a, 1: 234.

Belonochilus numenius (Say), 1832
1832 *Lygaeus numenius* Say, Descrip. Het. Hem. N. Am., p. 15. [U.S.].
1871 *Belonochilus numenius*: Uhler, Proc. Boston Soc. Nat. Hist., 14: 104.
1876 *Belonocheilus* [*sic*] *numenius*: Glover, Ms. Notes J. Hem., p. 45.
1886 *Belonochitres* [*sic*] *numenius*: Provancher, Pet. Faune Ent. Can., 3: 86.
1894 *Orsillacis producta*: Osborn, Proc. Ia. Acad. Sci., 1(4): 120.

1905 *Orsillacus* [*sic*] *productus*: Crevecoeur, Trans. Ks. Acad. Sci., 19: 232.
1914 *Belanochilus* [*sic*] *numineus* [*sic*]: Van Duzee, Trans. San Diego Soc. Nat. Hist., 2: 7.
Note: *Belonochilus mexicanus* Distant (1893, Biol. Centr.-Am., Rhyn., 1: 386) was synonymized under *B. numenius* by Usinger (1941, Bull. Brook. Ent. Soc., 36: 130).
Distribution: Ariz., Cal., Conn., D.C., Ia., Ill., Ind., Ks., Md., Mass., Mo., N.C., N.H., N.J., N.Y., Oh., Ont., Pa. (Mexico).
Ref.: Slater, 1964a, 1: 235.; Wheeler, 1984b.

Genus *Neortholomus* Hamilton, 1983

1983 *Neortholomus* Hamilton, Univ. Ks. Sci. Bull, 52(7): 201. Type-species: *Lygaeus scolopax* Say, 1832. Original designation.
Ref.: Slater, 1964a, 1: 334 (as *Ortholomus*); Hamilton, 1983 (key spp., cladogram of spp.).

Neortholomus arphnoides (Baker), 1906
1906 *Ortholomus arphnoides* Baker, Invert. Pac., 1: 140. [Cal.]. Lectotype designated by Hamilton, 1983, Univ. Ks. Sci. Bull., 52(7): 205.
1910 *Nysius arphnoides*: Banks, Cat. Nearc. Hem.-Het., p. 62.
1914 *Ortholomus arphanoides* [*sic*]: Van Duzee, Trans. San Diego Soc. Nat. Hist., 2: 8.
1918 *Ortholomis* [*sic*] *arphnoides*: Johnson and Ledig, Pomona J. Ent. Zool., 10: 4.
1983 *Neortholomus arphnoides*: Hamilton, Univ. Ks. Sci. Bull., 52(7): 203.
Distribution: Cal.
Ref.: Slater, 1964a, 1: 335 (as *Ortholomus*); Hamilton, 1983.

Neortholomus jamaicensis (Dallas), 1852
1852 *Nysius jamaicensis* Dallas, List Hem. Brit. Mus., 2: 555. [Jamaica].
1893 *Nysius providus* Uhler, Proc. Zool. Soc. London, 1893: 705. *Nomen nudum*.
1894 *Nysius providus* Uhler, Proc. Zool. Soc. London, 1894: 182. [Fla., West Indies, Mexico to South America]. Synonymized by Distant, 1901a, An. Mag. Nat. Hist., (7)7: 539.
1907 *Ortholomus jamaicensis*: Van Duzee, Bull. Buff. Soc. Nat. Sci., 8: 16.
1983 *Neortholomus jamaicensis*: Hamilton, Univ. Ks. Sci. Bull., 52: 208.
Distribution: Ariz., Cal., Fla., Tex. (Mexico to South America, Tahiti, West Indies).
Ref.: Slater, 1964a, 1: 335 (as *Ortholomus*); Hamilton, 1983.

Neortholomus koreshanus (Van Duzee), 1909
1909 *Belonochilus koreshanus* Van Duzee, Bull. Buff. Soc. Nat. Sci., 9: 165. [Fla.]. Lectotype designated by Hamilton, 1983, Univ. Ks. Sci. Bull., 52: 217.
1947 *Ortholomus koreshanus*: Barber, Mem. Soc. Cubana Hist. Nat., 19: 62.
1983 *Neortholomus koreshanus*: Hamilton, Univ. Ks. Sci. Bull., 52: 215.
Distribution: Fla. (West Indies).
Ref.: Slater, 1964a, 1: 337 (as *Ortholomus*); Hamilton, 1983.

Neortholomus nevadensis (Baker), 1906
1906 *Ortholomus nevadensis* Baker, Invert. Pac., 1: 139. [Nev.]. Lectotype designated by Hamilton, 1983, Univ. Ks. Sci. Bull., 52: 217.
1983 *Neortholomus nevadensis*: Hamilton, Univ. Ks. Sci. Bull., 52: 217.
Distribution: Ariz., Cal., Id., Nev., Ore.
Ref.: Slater, 1964a, 1: 337 (as *Ortholomus*); Hamilton, 1983.

Neortholomus scolopax (Say), 1832

1832 *Lygaeus scolopax* Say, Descrip. Het. Hem. N. Am., p. 15. [Ind., Ia., Mo.]. Neotype from Ks. designated by Hamilton, 1983, Univ. Ks. Sci. Bull., 52: 224.

1872 *Nysius Saint-Cyri* [*sic*] Provancher, Nat. Can., 4: 77. [Que.]. Synonymized by Usinger, 1941, Bull. Brook. Ent. Soc., 36: 129.

1874 *Nysius* (*Ortholomus*) *longiceps* Stål, K. Svens. Vet.-Akad. Handl., 12(1): 120. ["Carolina," Ill., N.J., N.Y., Wis.]. Synonymized by Barber, 1923, Conn. Geol. Nat. Hist. Surv. Bull, 34: 714.

1874 *Belonochilus scolopax*: Stål, K. Svens. Vet.-Akad. Handl., 12(1): 122.

1876 *Nysius scolopax*: Glover, Ms. Notes J. Hem., p. 45 (also lists in *Lygaeus*).

1876 *Orsillus scolopax*: Uhler, Bull. U.S. Geol. Geogr. Surv. Terr., 1: 305.

1886 *Harmostes fraterculus*: Provancher, Pet. Faune Ent. Can., 3: 86.

1894 *Nysius providus*: Uhler, Proc. Zool. Soc. London, 1894: 183 (in part).

1901 *Nysius jamaicensis*: Distant, An. Mag. Nat. Hist., (7)7: 539.

1906 *Ortholomus longiceps*: Baker, Invert. Pac., 1: 138.

1906 *Ortholomus Uhleri* [*sic*] Baker, Invert. Pac., 1: 139. [Wis.]. Synonymized by Blatchley, 1926, Het. E. N. Am., p. 349.

1906 *Ortholomus longiceps* var. *cookii* Baker, Invert. Pac., 1: 139. [Cal.]. Synonymized by Hamilton, 1983, Univ. Ks. Sci. Bull., 52: 222.

1907 *Ortholomus jamaicensis*: Van Duzee, Bull. Buff. Soc. Nat. Sci., 8: 16.

1910 *Nysius uhleri*: Banks, Cat. Nearc. Hem.-Het., p. 62.

1917 *Ortholomus scolopax*: Van Duzee, Univ. Cal. Publ. Ent., 2: 157.

1918 *Ortholomis* [*sic*] *longiceps*: Johnson and Ledig, Pomona J. Ent. Zool., 10: 4 (also lists *Ortholomis* [*sic*] *langiceps* [*sic*]).

1918 *Ortholomis* [*sic*] *cookii*: Johnson and Ledig, Pomona J. Ent., 10: 4.

1925 *Ortholomus ocolopa* [*sic*]: Weiss and West, Psyche, 32: 239.

1964 *Ortholomus scolopax cookii*: Slater, Cat. Lygaeidae World, 1: 343.

1983 *Neortholomus scolopax*: Hamilton, Univ. Ks. Sci. Bull., 52: 222.

Distribution: B.C., Cal., Col., Conn., Fla., Ia., Id., Ill., Ind., Ks., La., Man., Md., Me., Mass., Mich., Mo., N.C., N.D., N.H., N.J., N.S., N.Y., Neb., Oh., Ok., Ont., Pa., Que., S.C., S.D., Sask., Tex., Wis. (Mexico,Guatemala).

Ref.: Slater, 1964a, 1: 341, 343 (as *Ortholomus*); Hamilton, 1983.

Subfamily Oxycareninae Stål, 1862

Ref.: Slater, 1964a, 1: 622; Barber, 1917 (key U.S. genera); Torre-Bueno, 1946a (key U.S. genera).

Genus *Crophius* Stål, 1874

1874 *Crophius* Stål, K. Svens. Vet.-Akad. Handl., 12(1): 141. Type-species: *Lygaeus disconotus* Say, 1832. Designated by Van Duzee, 1910, Bull. Buff. Soc. Nat. Sci., 9: 391.

1893 *Mayana* Distant, Biol. Centr.-Am., Rhyn., 1: 378. Type-species: *Mayana costata* Distant, 1893. Designated by Van Duzee, 1916, Check List Hem., p. 21. Synonymized by Van Duzee, 1910, Bull. Buff. Soc. Nat. Sci., 9: 389.

Ref.: Slater, 1964a, 1: 635; Van Duzee, 1910a (key spp.); Barber, 1938b (key spp.); Torre-Bueno, 1946a (key spp.).

Crophius albidus Barber, 1938
1938 *Crophius albidus* Barber, J. N.Y. Ent. Soc., 46: 316. [Ut.].
Distribution: Ut.
Ref.: Slater, 1964a, 1: 635.

Crophius angustatus Van Duzee, 1910
1910 *Crophius angustatus* Van Duzee, Bull Buff. Soc. Nat. Sci., 9: 395. [Cal., Ut.].
Distribution: B.C., Cal., Col., Ore., Ut.
Ref.: Slater, 1964a, 1: 635.

Crophius bohemani (Stål), 1859
1859 *Cymus bohemani* Stål, Freg. Eug. Resa, 3: 251. [Cal.].
1874 *Crophius bohemani*: Stål, K. Svens. Vet.-Akad. Handl., 12(1): 142.
Distribution: Ariz., B.C., Cal., Col., Id., Ore., Ut., Wyo.
Ref.: Slater, 1964a, 1: 636.

Crophius disconotus (Say), 1832
1832 *Lygaeus disconotus* Say, Descrip. Het. Hem. N. Am., p. 14. [Mo.].
1874 *Crophius disconotus*: Stål, K. Svens. Vet.-Akad. Handl., 12(1): 142.
1886 *Oxycarenus disconotus*: Provancher, Pet. Faune Ent. Can., 3: 75.
1922 *Crophius disconatus* [*sic*]: Osborn and Drake, N.Y. St. Coll. For., Tech. Publ. 16, 22(5): 21.
Distribution: Ala., B.C., Cal., Col., Conn., D.C., Mass., Me., Mo., N.H., N.J., N.Y., Pa., Ont., Que., Sask., Ut., Wyo. (Mexico).
Ref.: Slater, 1964a, 1: 637; Lavigne, 1976; Brailovsky and Barrera, 1979.

Crophius heidemanni Van Duzee, 1910
1910 *Crophius heidemanni* Van Duzee, Bull Buff. Soc. Nat. Sci., 9: 393. [Ariz.].
Distribution: Ariz., Cal. (Mexico).
Ref.: Slater, 1964a, 1: 638; Brailovsky and Barrera, 1979.

Crophius impressus Van Duzee, 1910
1910 *Crophius impressus* Van Duzee, Bull. Buff. Soc. Nat. Sci., 9: 396. [Cal.].
Distribution: Cal., Ut.
Ref.: Slater, 1964a, 1: 638.

Crophius ramosus Barber, 1938
1938 *Crophius ramosus* Barber, J. N.Y. Ent. Soc., 46: 315. [Id.].
Distribution: B.C., Id., Sask., Ut.
Ref.: Slater, 1964a, 1: 638.

Crophius scabrosus (Uhler), 1904
1904 *Oxycarenus scabrosus* Uhler, Proc. U.S. Nat. Mus., 27: 352. [Ut.].
1910 *Crophius scabrosus*: Van Duzee, Bull. Buff. Soc. Nat. Sci., 9: 397.
1916 *Crophius* (*Mayana*) *scabrosus*: Van Duzee, Check List Hem., p. 21.
Distribution: Ariz., Cal., Col., Id., N.M., Neb., Ut. (Mexico).
Ref.: Slater, 1964a, 1: 639; Brailovsky and Barrera, 1979; Pinto and Frommer, 1980 (as near *scabrosus*).

Crophius schwarzi Van Duzee, 1910
1910 *Crophius schwarzi* Van Duzee, Bull. Buff. Soc. Nat. Sci., 9: 389, 392. [Ariz.].
Distribution: Ariz.
Ref.: Slater, 1964a, 1: 639.

Genus *Dycoderus* Uhler, 1901

1901 *Dycoderus* Uhler, Proc. Ent. Soc. Wash., 4: 507. Type species: *Dycoderus picturatus* Uhler, 1901. Monotypic.
Ref.: Slater, 1964a, 1: 639.

Dycoderus picturatus Uhler, 1901
1901 *Dycoderus picturatus* Uhler, Proc. Ent. Soc. Wash., 4: 508. [Ariz., Col.].
Distribution: Ariz., Col.
Ref.: Slater, 1964a, 1: 639.

Subfamily Pachygronthinae Stål, 1865

Ref.: Slater, 1964a, 1: 709; Slater, 1955 (key tribes, genera world).

Tribe Pachygronthini Stål, 1865

Ref.: Slater, 1964a, 1: 710; Slater, 1955 (key genera world).

Genus *Oedancala* Amyot and Serville, 1843

1843 *Oedancala* Amyot and Serville, Hist. Nat. Ins. Hem., p. 258. Type-species: *Oedancala dorsilinea* Amyot and Serville, 1843, a junior synonym of *Lygaeus crassimanus* Fabricius, 1803. Monotypic.
Ref.: Slater, 1964a, 1: 711; Slater, 1955 (key spp.), 1966 (key spp., fifth-instar nymphs).

Oedancala bimaculata (Distant), 1893
1893 *Pachygrontha bimaculata* Distant, Biol. Centr.-Am., Rhyn., 1: 393. [Panama].
1910 *Pachygrontha bimaculata*: Banks, Cat. Nearc. Hem.-Het., p. 60.
1955 *Oedancala bimaculata*: Slater, Philip. J. Sci., 84: 99.
Distribution: Tex. (Mexico to Brazil and Paraguay, West Indies).
Ref.: Slater, 1964a, 1: 711.

Oedancala crassimana (Fabricius), 1803
1803 *Lygaeus crassimanus* Fabricius, Syst. Rhyn., p. 233. [Carolina].
1843 *Oedancala dorsilinea* Amyot and Serville, Hist. Nat. Ins. Hem., p. 258. [Am.]. Synonymized by Barber, 1949, Proc. Ent. Soc. Wash., 51: 273.
1850 *Pachymerus crassimanus*: Herrich-Schaeffer, Wanz. Ins., 9: 147.
1874 *Oedancala crassimana*: Stål, K. Svens. Vet.-Akad. Handl., 12(1): 139 (in part).
1876 *Oedancala cubana*: Uhler, Bull. U.S. Geol. Geogr. Surv. Terr., 1: 307.
1910 *Oedancala dorsalis*: Banks, Cat. Nearc. Hem.-Het., p. 60.
Distribution: Ala., Fla., Ga., La., Md., N.C., N.J., N.Y., Ok., S.C., Tex. (South America, West Indies).
Ref: Slater, 1964a, 1: 712; Slater, 1952c.

Oedancala cubana Stål, 1874
1874 *Oedancala cubana* Stål, K. Svens. Vet.-Akad. Handl., 12(1): 139. [Cuba].
1975 *Oedancala cubana*: Slater, Fla. Ent., 58: 70.
Distribution: Fla. (West Indies).
Ref.: Slater, 1964a, 1: 713.

Oedancala dorsalis (Say), 1832

1832 *Pamera dorsalis* Say, Descrip. Het. Hem. N. Am., p. 17. [Ind.].

1872 *Oedancala dorsilinea*: Walker, Cat. Hem. Brit. Mus., 5: 145.

1873 *Oedancala dorsalis*: Walker, Cat. Hem. Brit. Mus. Suppl., p. 52.

1874 *Oedancala crassimana*: Stål, K. Svens. Vet.-Akad. Handl., 12(1): 139 (in part).

1876 *Plociomerus dorsalis*: Glover, Ms. Notes J. Hem., p. 56.

1912 *Ordancala* [*sic*] *dorsalis*: Banks, Ent. News, 23: 105.

Distribution: Col., Conn., D.C., Fla., Ia., Ill., Ind., Ks., Mass., Md., Me., Mich., Mo., N.C., N.H., N.J., N.Y., Neb., Oh., Ok., Ont., Pa., Que., R.I., S.C., S.D., Tenn., Tex., Va., Vt., Wis.

Ref.: Slater, 1964a, 1: 714; Slater, 1952c.

Genus *Pachygrontha* Germar, 1837

1837 *Pachygrontha* Germar, Silb. Rev. Ent., 5: 152. Type-species: *Pachygrontha lineata* Germar, 1837. Monotypic.

Ref.: Slater, 1964a, 1: 717, 1955 (key spp. world).

Pachygrontha compacta Distant, 1893

1893 *Pachygrontha compacta* Distant, Biol. Centr.-Am., Rhyn., 1: 393. [Guatemala, Panama].

1955 *Pachygrontha compacta*: Slater, Philip. J. Sci., 84: 39.

Distribution: Tex. (Central America to Brazil, West Indies).

Ref.: Slater, 1964a, 1: 721.

Tribe Teracriini Stål, 1872

Ref.: Slater, 1964a, 1: 728; Barber, 1917 (key genera); Torre-Bueno, 1946a (key genera); Slater, 1955 (key genera world).

Genus *Phlegyas* Stål, 1865

1865 *Phlegyas* Stål, Hem. Afr., 2: 145. Type-species: *Phlegyas annulicrus* Stål, 1869. First included species, Stål, 1869, Berl. Ent. Zeit., 13: 230.

1876 *Helonotus* Uhler, Bull. U.S. Geol. Geogr. Surv. Terr., 1: 312. Preoccupied. Type-species: *Helonotus abbreviatus* Uhler. Monotypic. Synonymized by Lethierry and Severin, 1894, Cat. Gen. Hem., 2: 179.

1886 *Peliopelta* Uhler, Check-list Hem. Het., p. 15. New name for *Helonotus* Uhler. Synonymized by Berg, 1892, Nac. Soc. Cient. Argent., 33: 160.

1888 *Helonotocoris* Lethierry, An. Mus. Civ. Stor. Nat. Hist. Gen., (2)6: 463. New name for *Helonotus* Uhler. Synonymized by Lethierry and Severin, 1894, Cat. Gen. Hem., 2: 179.

Ref.: Slater, 1964a, 1: 733; Slater, 1955 (key spp.).

Phlegyas abbreviatus (Uhler), 1876 [Fig. 54]

1876 *Helonotus abbreviatus* Uhler, Bull. U.S. Geol. Geogr. Surv. Terr., 1: 313. [Ill., Md., D.C., Mass., Mich., Mo., N.C., N.J., N.Y., Neb., Pa.].

1876 *Pelionotus abbreviatus*: Glover, Ms. Notes J. Hem., pl. 2.

1883 *Henestris* [*sic*] *abbreviatus*: Brodie and White, Checklist Ins. Dom. Can., p. 60.

1886 *Peliopelta abbreviata*: Uhler, Check-list Hem. Het., p. 15.
1905 *Phlegyas annulicrus*: Wirtner, An. Carnegie Mus., 3: 193.
1905 *Phlegyas abbreviatus*: Van Duzee, Bull. N.Y. St. Mus., 97: 549.
1910 *Phlegyas annulicornis* [*sic*]: Banks, Cat. Nearc. Hem.-Het., p. 60 (in part).
Distribution: B.C., Cal., Col., Conn., Fla., Ga., Ia., Ill., Ind., Ks., Ky., Mass., Md., Me., Mich., Minn., Miss., Mo., N.C., N.H., N.J., N.Y., Neb., Oh., Ok., Ont., Pa., Que., R.I., S.C., Tenn., Tex., Va., Vt., Wyo.
Ref.: Slater, 1964a, 1: 734; Slater, 1952c.

Phlegyas annulicrus Stål, 1869
1869 *Phlegyas annulicrus* Stål, Berl. Ent. Zeit., 13: 230. [Carolina, N.J., Tex.]. Lectotype designated by Slater, 1955, Philip. J. Sci., 84: 145.
1893 *Peliopelta tropicalis* Distant, Biol. Centr.-Am., Rhyn., 1: 411. [Guatemala, Mexico]. Synonymized by Barber, 1949, Proc. Ent. Soc. Wash., 51: 276.
1910 *Phlegyas annulicornis* [*sic*]: Banks, Cat. Nearc. Hem.-Het., p. 60 (in part).
1916 *Phlegyas tropicalis*: Van Duzee, Check List Hem., p. 20.
Distribution: Ariz., B.C., Cal., Fla., Ga., Ks., La., Miss., N.C., N.J., Ok., S.C., Tex., Ut. (Mexico).
Ref.: Slater, 1964a, 1: 735.

Subfamily Rhyparochrominae Amyot and Serville, 1843

Ref.: Slater, 1964a, 2: 779; Slater and Woodward, 1982 (cladogram of tribes); Slater, 1986 (world zoogeogr.).

Tribe Antillocorini Ashlock, 1964

Ref.: Slater, 1964a, 2: 843; Slater, 1980 (key genera New World).

Genus *Antillocoris* Kirkaldy, 1904

1894 *Pygaeus* Uhler, Proc. Zool. Soc. London, 1894: 187. Preoccupied. Type-species: *Pygaeus pallidus* Uhler, 1894. Monotypic.
1904 *Antillocoris* Kirkaldy, Ent., 37: 280. New name for *Pygaeus* Uhler.
Ref.: Slater, 1964a, 2: 843; Barber, 1952a (key spp.).

Antillocoris discretus Barber, 1952
1914 *Pygaeus pallidus*: Barber, Bull. Am. Mus. Nat. Hist., 33: 516.
1952 *Antillocoris discretus* Barber, Bull. Brook. Ent. Soc., 47: 86. [Fla.].
Distribution: Fla., Ga., N.J., Tex.
Ref.: Slater, 1964a, 2: 843; Scudder, 1968.

Antillocoris minutus (Bergroth), 1895
1892 *Rhyparochromus minutus* Osborn, Proc. Ia. Acad. Sci., 1: 122. *Nomen nudum.*
1894 *Pygaeus pallidus*: Uhler, Proc. Zool. Soc. London, 1894: 187 (in part).
1894 *Salacia pilosulus*: Van Duzee, Bull. Buff. Soc. Nat. Sci., 5: 174.
1895 *Cligenes minutus* Bergroth, Rev. d'Ent., 14: 143. [Mass.].
1905 *Phygaeus* [*sic*] *pallidus*: Wirtner, An. Carnegie Mus., 3: 194.

1910 *Cligenes pallidus*: Banks, Cat. Nearc. Hem.-Het., p. 63.
1911 *Antillocoris pallidus*: Gibson, 41st An. Rept. Ent. Soc. Ont., p. 119.
1952 *Antillocoris minutus*: Barber, Bull. Brook. Ent. Soc., 47: 85.
Distribution: Conn., Fla.(?), Ia., Ill., Ind., La., Mass., Md., Me., Mo., N.C., N.H., N.J., N.Y., Neb., Oh., Pa., Ont., Que., Tex.
Ref.: Slater, 1964a, 2: 843; Sweet, 1964b.

Antillocoris pallidus (Uhler), 1894
1894 *Pygaeus pallidus* Uhler, Proc. Zool. Soc. London, 1894: 187. [Grenada]. Lectotype designated by Scudder, 1967, Bull. Brit. Mus. (Nat. Hist.), Ent., 20: 276.
1952 *Antillocoris pallidus*: Barber, Bull. Brook. Ent. Soc., 47: 85. [Fla.(?)].
Distribution: Fla.(?). (Central and South America, West Indies).
Ref.: Slater, 1964a, 2: 845.

Antillocoris pilosulus (Stål), 1874
1874 *Salacia pilosula* Stål, K. Svens. Vet.-Akad. Handl., 12(1): 158. [N.J., Tex.]. Lectotype designated by Scudder, 1977, Ent. Scand., 8: 33.
1895 *Cligenes pilosulus*: Bergroth, Rev. d'Ent., 14: 143.
1909 *Pygaeus pilosulus*: Van Duzee, Can. Ent., 41: 374.
1912 *Antillocoris (Cligenes) pilosulus*: Torre-Bueno, Can. Ent., 44: 212.
1916 *Antillocoris pilosulus*: Van Duzee, Check List Hem., p. 22.
Distribution: Ala., Ark., Conn., Fla., Ia., Ill., Ind., La., Mass., Mo., N.C., N.J., N.Y., Oh., Ok., Ont., Que., Tenn., Tex.
Ref.: Slater, 1964a, 2: 845; Sweet, 1964b.

Genus *Botocudo* Kirkaldy, 1904

1874 *Salacia* Stål, K. Svens. Vet.-Akad. Handl., 12(1): 154. Preoccupied. Type-species: *Aphanus diluticornis* Stål, 1858. Designated by Scudder, 1962, Can. Ent., 94: 764.
1904 *Botocudo* Kirkaldy, Ent., 37: 280. New name for *Salacia* Stål.
Ref.: Slater, 1964a, 2: 846.

Botocudo delineatus (Distant), 1893
1893 *Salacia delineata* Distant, Biol. Centr.-Am., Rhyn., 2: 406. [Panama].
1946 *Cligenes delineatus*: Torre-Bueno, Ent. Am., 26: 110.
1964 *Botocudo delineatus*: Slater, Cat. Lygaeidae World, 2: 847.
Distribution: Cal., Tex. (Mexico, Panama).
Ref.: Slater, 1964a, 2: 847.

Botocudo modestus (Barber), 1948
1918 *Cligenes delineata*: Barber, J. N.Y. Ent. Soc., 26: 59.
1948 *Cligenes modesta* Barber, Proc. Ent. Soc. Wash., 50: 157. [Ark.].
1964 *Botocudo modestus*: Slater, Cat. Lygaeidae World, 2: 847.
Distribution: Ark., Cal., Id., Mo., Ore., Tex., Ut.
Ref.: Slater, 1964a, 2: 847.

Genus *Cligenes* Distant, 1893

1893 *Cligenes* Distant, Biol. Centr.-Am., Rhyn., 1: 405. Type-species: *Cligenes distinctus* Distant, 1893, Monotypic.
1893 *Tomopelta* Uhler, Proc. Zool. Soc. London, p. 708. Type-species: *Tomopelta munda* Uhler, 1893. Monotypic. Synonymized by Scudder, 1962, Can. Ent., 94: 764.
Ref.: Slater, 1964a, 2: 848.

Cligenes distinctus Distant, 1893

1893 *Cligenes distinctus* Distant, Biol. Centr.-Am., Rhyn., 1: 405. [Panama]. Lectotype designated by Scudder, 1967, Bull. Brit. Mus. (Nat. Hist.), Ent., 20: 264.

1893 *Tomopelta munda* Uhler, Proc. Zool. Soc. London, p. 709. [St. Vincent Island]. Synonymized by Scudder, 1962, Can. Ent., 94: 764; lectotype designated by Scudder, 1967, Bull. Brit. Mus. (Nat. Hist.) Ent., 20: 273.

1926 *Tomopelta munda*: Blatchley, Het. E. N. Am., p. 405.

Distribution: Fla. (West Indies, Panama).

Ref.: Slater, 1964a, 2: 849.

Genus *Paurocoris* Slater, 1980

1980 *Paurocoris* Slater, Syst. Ent., 5: 220. Type-species: *Paurocoris wygodzinskyi* Slater, 1980. Original designation.

Ref.: Slater, 1980 (key spp.).

Paurocoris punctatus (Distant), 1893

1893 *Salacia punctata* Distant, Biol. Centr.-Am., Rhyn., 1: 406. [Panama].

1910 *Cligenes punctatus*: Banks, Cat. Nearc. Hem.-Het., p. 63.

1916 *Antillocoris punctatus*: Van Duzee, Check List Hem., p. 22.

1967 *Stygnocoris punctatus*: Scudder, Bull. Brit. Mus. (Nat. Hist.), Ent., 20: 278.

1977 *Bathydema punctatus*: Ashlock, Proc. Ent. Soc. Wash., 79: 578.

1980 *Paurocoris punctatus*: Slater, Syst. Ent., 5: 222.

Distribution: Ga., Tex. (Mexico, Panama).

Note: Ashlock (1977, Proc. Ent. Soc. Wash., 79: 578) questioned Banks' (1910, Cat. Nearc. Hem.-Het., p. 63) record from Tex. M. H. Sweet (pers. comm.), however, has collected the species in Tex.

Ref.: Slater, 1964a, 2: 846 (as *Antillocoris punctatus*).

Tribe Drymini Stål, 1872

Ref.: Slater, 1964a, 2: 870-871; Barber, 1918a (key genera, as Lethaeini); Torre-Bueno, 1946a (as Lethaeini); Ashlock, 1979 (key Am. genera).

Genus *Drymus* Fieber, 1860

1860 *Drymus* Fieber, Europ. Hem., pp. 46, 178. Type-species: *Drymus pilipes* Fieber, 1861. Designated by Distant, 1904, Fauna Brit. India, Rhyn., 2: 92.

Ref.: Slater, 1964a, 2: 874; Slater, 1986 (zoogeogr.).

Subgenus *Drymus* Fieber, 1860

1860 *Drymus* Fieber, Europ. Hem., p. 46, 178. Type-species: *Drymus pilipes* Fieber, 1861.

Note: Le Quisne (1956, Ent. Month. Mag., 92: 337) divided the British *Drymus* into two subgenera; Sweet (1964b, Ent. Am., 44: 1-201) placed the North American species.

Drymus crassus Van Duzee, 1910

1910 *Drymus crassus* Van Duzee, Trans. Am. Ent. Soc., 36: 76. [N.C., N.H., N.J.].

1964 *Drymus* (*Drymus*) *crassus*: Sweet, Ent. Am., (n.s.), 44: 101.

Distribution: Conn., Ia., Ill., Ind., Ks., Mass., Me., N.C., N.H., N.J., N.Y., Ok., Pa., Que., S.C., S.D., Tex.

Ref.: Slater, 1964a, 2: 876; Sweet, 1964b.

Drymus unus (Say), 1832
1832 *Pamera una* Say, Descrip. Het. Hem. N. Am., p. 16. [Ind.].
1876 *Megalonotus una*: Glover, Ms. Notes. J. Hem., p. 56.
1882 *Rhyparochromus una*: Distant, Biol. Centr.-Am., Rhyn., 1: 215.
1886 *Megatonotus* [*sic*] *unus*: Provancher, Pet. Faune Ent. Can., 3: 88.
1886 *Megalonotus unus*: Uhler, Check-list Hem. Het., p. 15.
1908 *Drymus unus*: Horvath, Ann. Hist.-Nat. Mus. Nat. Hung., 6: 564.
1964 *Drymus* (*Drymus*) *unus*: Sweet, Ent. Am., (n.s.), 44: 95.
Distribution: Conn., Ia., Ill., Ind., Mass., Md., Me., Mich., Mo., N.B., N.C., N.H., N.J., N.S., N.T., N.Y., Ont., Pa., Que.,Tex.
Ref.: Slater, 1964a, 2: 882; Sweet, 1964b.

Genus *Eremocoris* Fieber, 1860

1860 *Eremocoris* Fieber, Europ. Hem., pp. 49, 187. Type-species: *Lygaeus erraticus* Fabricius, 1794, a junior synonym of *Cimex abietis* Linnaeus, 1758. Designated by Distant, 1910, Fauna Brit. India, Rhyn., 5: 84.
Ref.: Slater, 1964a, 2: 896; Barber, 1928a (key spp. eastern U.S.); Sweet, 1977a (key spp. east of 100th meridian, N. Am.); Slater, 1986 (zoogeogr.).

Eremocoris borealis (Dallas), 1852
1852 *Rhyparochromus borealis* Dallas, List Hem. Brit. Mus., 2: 565. [Hudson's Bay]. Lectotype designated by Scudder, 1967, Bull. Brit. Mus. (Nat. Hist.), Ent., 20: 259.
1872 *Rhyparochromus ferus*: Walker, Cat. Hem. Brit. Mus., 5: 89.
1977 *Eremocoris borealis*: Sweet, Ent. News, 88: 174.
Distribution: Alk., Alta., B.C., Conn., Mich., Mont., N.C., N.H., Nfld., Ore., Tenn., Wis. [See note under *E. ferus*].
Note: This species was synonymized with *E. ferus* (Say) by Uhler (1871, Proc. Boston Soc. Nat. Hist., 14: 103); resurrected by Sweet (1977, Ent. News, 88: 174).
Ref.: Slater, 1964a, 2: 904 (as *E. ferus*, in part); Sweet, 1977a.

Eremocoris canadensis Walley, 1929
1929 *Eremocoris canadensis* Walley, Can. Ent., 61: 41. [B.C.].
Distribution: B.C., Id.
Ref.: Slater, 1964a, 2: 902.

Eremocoris cupressicola Ashlock, 1979
1979 *Eremocoris cupressicola* Ashlock, Pan-Pac. Ent., 55: 149. [Cal.].
Distribution: Cal.

Eremocoris depressus Barber, 1928
1928 *Eremocoris depressus* Barber, Proc. Ent. Soc. Wash., 30: 59. [N.C.].
Distribution: La., Miss., Mo., N.C., N.J., S.C., Va.
Ref.: Slater, 1964a, 2: 902.

Eremocoris dimidiatus Van Duzee, 1921
1921 *Eremocoris dimidiatus* Van Duzee, Proc. Cal. Acad. Sci., (4), 11: 116. [Col.].
Distribution: Col.
Ref.: Slater, 1964a, 2: 902.

Eremocoris ferus (Say), 1832 [Fig. 56]
1832 *Pamera fera* Say, Descrip. Het. Hem. N. Am., p. 16. [U.S.]. Neotype from Va. designated by Sweet, Ent. News, 88: 173.

1871 *Eremocoris ferus*: Uhler, Proc. Boston Soc. Nat. Hist., 14: 103.
Distribution: Conn., Ia., Ill., Ind., Ks., La., Mass., Mo., N.C., N.H., N.J., N.Y., Oh., Pa., S.C., Tex., Va., Vt.
Note: Sweet (1977a, Ent. News, 88: 174-175) indicated that old northern records refer to *E. borealis*, and records west of the 100th meridian refer to a complex of as yet undescribed species.
Ref.: Slater, 1964a, 2: 904; Sweet, 1964b, 1977a.

Eremocoris inquilinus Van Duzee, 1914
1914 *Eremocoris inquilinus* Van Duzee, Trans. San Diego Soc. Nat. Hist., 2: 10. [Cal.].
Distribution: Cal.
Ref.: Slater, 1964a, 2: 906.

Eremocoris melanotus Walley, 1929
1929 *Eremocoris melanotus* Walley, Can. Ent., 61: 42. [B.C.].
Distribution: B.C., Id.
Ref.: Slater, 1964a, 2: 907.

Eremocoris obscurus Van Duzee, 1906
1906 *Eremocoris obscurus* Van Duzee, Ent. News, 17: 388. [B.C.].
Distribution: B.C., Cal., Id.
Ref.: Slater, 1964a, 2: 908.

Eremocoris opacus Van Duzee, 1921
1921 *Eremocoris opacus* Van Duzee, Proc. Cal. Acad. Sci., (4)11: 117. [Cal.].
Distribution: Cal.
Ref.: Slater, 1964a, 2: 908; Ashlock and O'Brien, 1964; Arnaud, 1978 (as tachinid host).

Eremocoris semicinctus Van Duzee, 1921
1921 *Eremocoris semicinctus* Van Duzee, Proc. Cal. Acad. Sci., (4)11: 115. [Cal.].
Distribution: Cal., Id.
Ref.: Slater, 1964a, 2: 915.

Eremocoris setosus Blatchley, 1926
1926 *Eremocoris setosus* Blatchley, Het. E. N. Am., p. 433 (as Barber ms.). [Ind., N.J., N.Y., Va.]. Lectotype from Ind. designated by Blatchley, 1930, Blatchleyana, 1: 64.
1928 *Eremocoris plebejus* var. *setosus* Barber, Proc. Ent. Soc. Wash., 30: 60. [Va.].
Distribution: D.C., Fla., Ga., Ind., Mass., N.C., N.J., N.Y., Oh., Que., S.C., Va.
Note: McAtee (1927, Bull. Brook. Ent. Soc., 22: 276) established Blatchley as the author of this species; specific rank reestablished by Torre-Bueno (1946a, Ent. Am., 26: 130).
Ref.: Slater, 1964a, 1: 915; Sweet, 1964b.

Genus *Gastrodes* Westwood, 1840

1829 *Platygaster* Schilling, Beitr. Ent., 1: 37, 82. Preoccupied. Type-species: *Cimex abietus*: Schilling, 1829 (not Linnaeus, 1758).
1840 *Gastrodes* Westwood, Introd. Mod. Class. Ins., 2 (Synop.): 122. Type-species: *Gastrodes abietum* Bergroth, 1914. Designated by action of Int. Comm. Zool. Nomen., 1950, Bull. Zool. Nomen., 4: 473. See also Int. Comm. Zool. Nomen., 1954, Opinion 246.
Ref.: Slater, 1964a, 2: 916; Usinger, 1938a (key spp.); Slater, 1986 (zoogeogr.).

Gastrodes arizonensis Usinger, 1938
1938 *Gastrodes arizonensis* Usinger, Proc. Cal. Acad. Sci., (4)23: 292, 297. [Ariz.].
Distribution: Ariz.
Ref.: Slater, 1964a, 2: 923.

Gastrodes conicola Usinger, 1933
1933 *Gastrodes conicola*: Usinger, Pan-Pac. Ent., 9: 127. [Cal.].
1964 *Gastrodes conicolus* [*sic*]: Slater, Cat. Lygaeidae World, 2: 923.
Distribution: Cal.
Ref.: Slater, 1964a, 2: 923.

Gastrodes intermedius Usinger, 1938
1938 *Gastrodes intermedius* Usinger, Proc. Cal. Acad. Sci., (4)23: 292, 297. [B.C.].
Distribution: B.C.
Ref.: Slater, 1964a, 2: 930.

Gastrodes pacificus (Provancher), 1886
1886 *Platygaster pacificus* Provancher, Pet. Faune Ent. Can., 3: 205. [B.C.].
1894 *Gastrodes pacificus*: Lethierry and Severin, Cat. Gen. Hem., 2: 231.
Distribution: B.C., Cal., Col., Neb., Nev., Ore., Ut., Wash.
Ref.: Slater, 1964a, 2: 931.

Gastrodes walleyi Usinger, 1938
1938 *Gastrodes ferrugineus*: Banks, Cat. Nearc. Hem.-Het., p. 64. [N. states].
1938 *Gastrodes walleyi* Usinger, Proc. Cal. Acad. Sci., (4)23: 300. [Ont.].
Distribution: Ont.
Ref.: Slater, 1964a, 2: 931.

Genus *Scolopostethus* Fieber, 1860

1860 *Scolopostethus* Fieber, Europ. Hem., pp. 49, 188. Type-species: *Scolopostethus cognatus* Fieber, 1961. Designated by Distant, 1903, Fauna Brit. India, Rhyn., 2: 92.
Ref.: Slater, 1964a, 2: 952; Barber, 1917 (key spp. N.Y.).

Scolopostethus atlanticus Horvath, 1893
1893 *Scolopostethus atlanticus* Horvath, Rev. d'Ent., 12: 239. [Mass., N.J.]. Lectotype from N.Y. designated by Scudder, 1970, An. Hist.-Nat. Mus. Nat. Hung., 62: 198. See note below.
Distribution: B.C., Col., Conn., Mass., Man., Me., N.H., N.J., N.M., N.Y., Nfld., Ont., Que., R.I., Wash.
Note: Scudder (1970, above) selected a specimen from New York as a lectotype for this species, but New York was not one of the original localities. However, correspondence between Froeschner and Vasarhelyi confirms that this is a Scudder *lapsus* for N.J., and we accept the New Jersey lectotype.
Ref.: Slater, 1964a, 2: 959; Sweet, 1964b.

Scolopostethus diffidens Horvath, 1893
1886 *Plociomerus nodosus*: Provancher, Pet. Faune Ent. Can., 3: 77.
1892 *Scolopostethus affinis*: Harrington, Ottawa Nat., 6: 27.
1893 *Scolopostethus diffidens* Horvath, Rev. d'Ent., 12: 240. [Mass.]. Lectotype designated by Scudder, 1970, An. Hist.-Nat. Mus. Nat. Hung., 62: 200.
Distribution: B.C., Cal., Conn., Id., Mass., Me., N.J., N.S., N.Y., Ont., Que., Wash.
Ref.: Slater, 1964a, 2: 965; Sweet, 1964b.

Scolopostethus pacificus Barber, 1918
1918 *Scolopostethus pacificus* Barber, J. N.Y. Ent. Soc., 26: 64. [Cal.].
Distribution: B.C., Cal., Id., Ore.
Ref.: Slater, 1964a, 2: 969.

Scolopostethus thomsoni Reuter, 1874
1870 *Lygaeus (Scolopostethus) decoratus*: Thomson, Opusc. Ent., 12: 201.
1874 *Scolopostethus thomsoni* Reuter, An. Soc. Ent. Fr., (5)4: 562. [Sweden]. New name for *Lygaeus (Scolopostethus) decoratus* Thomson, not Hahn.
1894 *Scolopostethus thomsoni*: Van Duzee, Ent. News, 5: 108.
1922 *Scolopostethus thompsoni* [*sic*]: Walley, Can. Ent., 67: 152.
1966 *Scolopistethus* [*sic*] *thomsoni*: Pielou, Can. Ent., 98: 1235.
Distribution: Alk., Ariz., B.C., Cal., Col., Conn., Ia., Id., Ill., Ind., Mass., Me., N.H., N.J., N.M., N.S., N.Y., Nfld., Oh., Ont., Que., S.D., Ut. (Europe, North Africa).
Ref.: Slater, 1964a, 2: 976; Butler [E.A.], 1923; Eyles, 1964; Sweet, 1964b; Woodroffe, 1967; Slater (zoogeogr.).

Scolopostethus tropicus (Distant), 1882
1882 *Eremocoris tropicus* Distant, Biol. Centr.-Am., Rhyn., 1: 218. [Guatemala]. Lectotype designated by Scudder, 1967, Bull. Brit. Mus. (Nat. Hist.), Ent., 20: 282.
1910 *Eremocoris tropicus*: Banks, Cat. Nearc. Hem.-Het., p. 64. [W. states].
1914 *Scolopostethus tropicus*: Van Duzee, Trans. San Diego Soc. Nat. Hist., 2: 10.
Distribution: B.C. Cal., Id., Ore. (Guatemala).
Ref.: Slater, 1964a, 2: 980.

Genus *Thylochromus* Barber, 1928

1928 *Thylochromus* Barber, Bull. Brook. Ent. Soc., 23: 264. Type-species: *Thylochromus nitidulus* Barber, 1928. Monotypic.
Ref.: Slater, 1964a, 2: 994.

Thylochromus nitidulus Barber, 1928
1928 *Thylochromus nitidulus* Barber, Bull. Brook. Ent. Soc., 23: 265. [Cal.].
Distribution: Cal., Ore.
Ref.: Slater, 1964a, 2: 994; Ashlock and O'Brien, 1964; Arnaud, 1978 (as tachinid host); Slater, 1986 (zoogeogr.).

Genus *Togodolentus* Barber, 1918

1918 *Togodolentus* Barber, J. N.Y. Ent. Soc., 26: 64. Type-species: *Togodolentus genuinus* Barber, 1918, a junior synonym of *Eremocoris wrighti* Van Duzee, 1914. Monotypic.
Ref.: Slater, 1964a, 2: 994; Slater, 1986 (zoogeogr.).

Togodolentus wrighti (Van Duzee), 1914
1914 *Eremocoris wrighti* Van Duzee, Trans. San Diego Soc. Nat. Hist., 2: 9. [Cal.].
1918 *Togodolentus genuinus* Barber, J. N.Y. Ent. Soc., 26: 64. [Cal.]. Synonymized by Blatchley, 1934, Trans. Am. Ent. Soc., 60: 9.
1934 *Togodolentus wrighti*: Blatchley, Trans. Am. Ent. Soc., 60: 9.
Distribution: Cal.
Ref.: Slater, 1964a, 2: 995.

Tribe Gonianotini Stål, 1872

Ref.: Slater, 1964a, 2: 1397; Slater and Ashlock, 1966 (key Am. genera); Kiritshenko and Scudder, 1973 (key genera world); Scudder, 1984 (addition to Kiritshenko and Scudder 1973 key).

Genus *Atrazonotus* Slater and Ashlock, 1966

1966 *Atrazonotus* Slater and Ashlock, Proc. Ent. Soc. Wash., 68: 154. Type-species: *Dorachosa illuminatus* var. *umbrosus* Distant, 1893. Monotypic.

Ref.: Slater and Ashlock, 1966.

Atrazonotus umbrosus (Distant), 1893

1885 *Microtoma carbonaria*: Popenoe, Trans. Ks. Acad. Sci., 9: 63.

1893 *Dorachosa illuminatus* var. *umbrosus* Distant, Biol. Centr.-Am., Rhyn., 1: 409. [Tex., Guatemala, Mexico, Panama]. Lectotype from Tex. designated by Slater and Ashlock, 1966, Proc. Ent. Soc. Wash., 68: 153.

1893 *Delochilocoris umbrosa*: Bergroth, Rev. d'Ent., 12: 154.

1894 *Microtoma atrata*: Van Duzee, Bull. Buff. Soc. Nat. Sci., 5: 175.

1894 *Delochilocoris illuminatus umbrosus*: Lethierry and Severin, Gen. Cat. Hem., 2: 227.

1895 *Rhyparochromus* (*Dorachosa*) *illuminatus*: Gillette and Baker, Bull. Col. Agr. Exp. Stn., 31: 26.

1905 *Delochilocoris illuminata*: Wirtner, An. Carnegie Mus., 3: 194.

1908 *Aphanus umbrosus*: Horvath, An. Hist.-Nat. Mus. Nat. Hung., 6: 561.

1910 *Arphanus* [*sic*] *umbrosus*: Smith Rept. N.J. St. Mus., 1909: 144.

1920 *Aphanus plenus*: Britton, Conn. Geol. Nat. Hist. Surv. Bull., 31: 79.

1952 *Alphanus* [*sic*] *umbrosus*: Slater, Proc. Ia. Acad. Sci., 59: 536.

1957 *Rhyparochromus umbrosus*: Slater and Hurlbutt, Proc. Ent. Soc. Wash., 59: 74, 78.

1960 *Delochilocoris umbrosus*: Ashlock, Proc. Biol. Soc. Wash., 73: 237.

1966 *Atrazonotus umbrosus*: Slater and Ashlock, Proc. Ent. Soc. Wash., 68: 155.

Distribution: Ariz., Cal., Col., Conn., D.C., Fla., Ia., Ill., Ind., Ks., Ky., La., Mass., Mo., N.D., N.J., N.Y., Oh., Ok., Ont., Pa., R.I., S.C., S.D., Tex. (Mexico to Central America).

Ref.: Slater, 1964a, 2: 1419 (as *Delochilocoris*); Sweet, 1964b (as *Delochilocoris*); Slater, 1986 (zoogeogr.).

Genus *Delochilocoris* Bergroth, 1893

1893 *Dorachosa* Distant, Biol. Centr.-Am., Rhyn., 1: 409. Preoccupied. Type-species: *Dorachosa illuminatus* Distant, 1893. Monotypic.

1893 *Delochilocoris* Bergroth, Rev. d'Ent., 12: 154. New name for *Dorachosa* Distant.

Ref.: Slater, 1964a, 2: 1418.

Delochilocoris caliginosus (Distant), 1882

1882 *Trapezonotus caliginosus* Distant, Biol. Centr.-Am., Rhyn., 1: 216. [Guatemala, Panama]. Lectotype from Guatemala designated by Scudder, 1967, Bull. Brit. Mus. (Nat. Hist.), Ent., 20: 259.

1918 *Trapezonotus caliginosus*: Barber, J. N.Y. Ent. Soc., 26: 57. [Ariz.].

1966 *Delochilocoris caliginosus*: Slater and Ashlock, Proc. Ent. Soc. Wash., 68: 153.

Distribution: Ariz., Cal. (Central America).
Ref.: Slater, 1964a, 2: 1484 (as *Trapezonotus*); Slater, 1986 (zoogeogr.).

Delochilocoris illuminatus (Distant), 1893 [Fig. 57]
1893 *Dorachosa illuminatus* Distant, Biol. Centr.-Am., Rhyn., 1: 409. [Guatemala, Mexico]. Lectotype from Guatemala designated by Slater and Ashlock, 1966, Proc. Ent. Soc. Wash., 68: 152.
1918 *Aphanus illuminatus*: Barber, J. N.Y. Ent. Soc., 26: 61.
1960 *Delochilocoris illuminatus*: Ashlock, Proc. Biol. Soc. Wash., 73: 237.
Distribution: Fla., Mo. (Mexico to Central America).
Ref.: Slater, 1964a, 2: 1419; Slater and Ashlock, 1966.

Genus *Emblethis* Fieber, 1860

1860 *Emblethis* Fieber, Europ. Hem., pp. 51, 197. Type-species: *Emblethis platychilus* Fieber, 1861, a junior synonym of *Lygaeus verbasci* Fabricius, 1803. Designated by Van Duzee, 1917, Univ. Cal. Publ. Ent., 2: 195.
Ref.: Slater, 1964a, 2: 1422; Slater, 1986 (zoogeogr.).

Emblethis vicarius Horvath, 1908
1872 *Emblethis arenarius*: Uhler, Prelim. Rept. U.S. Geol. Surv. Mont., p. 407.
1878 *Gonianotus marginepunctatus*: Uhler, Proc. Boston Soc. Nat. Hist., 19: 392.
1900 *Emblethis griseus*: Forbes, Rept. Ill. Ent., 21: 93.
1908 *Emblethis vicarius* Horvath, Ann. Hist.-Nat. Mus. Nat. Hung., 6: 563. [Col.]. Lectotype designated by Scudder, 1970, An. Hist.-Nat. Mus. Nat. Hung., 62: 205.
Distribution: Alta., Ariz., B.C., Cal., Col., Conn., D.C., Fla., Ga., Ia., Id., Ill., Ind., Ks., Man., Mass., Minn., Mo., Mont., N.D., N.H., N.J., N.M., N.Y., Neb., Nev., Oh., Ok., Ore., Ont., Que., R.I., S.D., Sask., Tex., Ut., Wash., Wyo. (Mexico to South America).
Ref.: Slater, 1964a, 2: 1439; Cockerell, 1893; Sweet, 1964b; Arnaud, 1978 (as tachinid host).

Genus *Malezonotus* Barber, 1918

1918 *Malezonotus* Barber, J. N.Y. Ent. Soc., 2: 54. Type-species: *Trapezonotus rufipes* Stål, 1874. Original designation.
Ref.: Slater, 1964a, 2: 1455; Ashlock, 1958 (key spp.), 1963 (key spp., suppl.); Slater, 1986 (zoogeogr.).

Malezonotus angustatus (Van Duzee), 1910
1910 *Rhyparochromus angustatus* Van Duzee, Trans. Am. Ent. Soc., 36: 74. [B.C., Wash.].
1918 *Malezonotus angustatus*: Barber, J. N.Y. Ent. Soc., 26: 55.
Distribution: B.C., Cal., Id., Nev., Ore., Ut., Wash.
Ref.: Slater, 1964a, 2: 1455.

Malezonotus arcuatus Ashlock, 1958
1958 *Malezonotus arcuatus* Ashlock, An. Ent. Soc. Am., 51: 206. [Wash.].
Distribution: B.C., Wash.
Ref.: Slater, 1964a, 2: 1456.

Malezonotus barberi Ashlock, 1958
1958 *Malezonotus barberi* Ashlock, An. Ent. Soc. Am., 51: 206. [Cal.].
Distribution: Cal., Id., Nev., Ore., Ut.
Ref.: Slater, 1964a, 2: 1456.

Malezonotus fuscosus Barber, 1918
1890 *Trapezonotus rufipes*: Smith, Geol. Surv. N.J., 2(2): 423.
1910 *Sphragisticus rufipes*: Smith, Rept. N.J. St. Mus., p. 144.
1918 *Malezonotus fuscosus* Barber, J. N.Y. Ent. Soc., 26: 56. [N.J., N.Y.].
1946 *Malezonotus rufipes*: Torre-Bueno, Ent. Am., 26: 115 (in part).
Distribution: Conn., N.J., N.Y.
Ref.: Slater, 1964a, 2: 1456; Sweet, 1964b.

Malezonotus grossus Van Duzee, 1935
1935 *Malezonotus grossus* Van Duzee, Pan-Pac. Ent., 11: 28. [Cal.].
Distribution: B.C., Cal., Ore.
Ref.: Slater, 1964a, 2: 1456.

Malezonotus obrieni Ashlock, 1963
1963 *Malezonotus obrieni* Ashlock, Pan-Pac. Ent., 39: 264. [Cal.].
Distribution: Cal., Ore.
Ref.: Slater, 1964a, 2: 1627.

Malezonotus rufipes (Stål), 1874
1874 *Trapezonotus rufipes* Stål, K. Svens. Vet.-Akad. Handl., 12(1): 159. [Tex.]. Lectotype designated by Ashlock, 1958, An. Ent. Soc. Am., 51: 202.
1876 *Megalonotus sodalicius*: Uhler, Rept. U.S. Geol. Geogr. Surv. Terr., 5: 835 (in part).
1906 *Trapezonatus* [*sic*] *rufipes*: Slosson, Ent. News, 17: 326.
1914 *Sphagisticus* [*sic*] *rufipes*: Torre-Bueno, Ent. News, 25: 230.
1917 *Trapezonotus sodalicus* [*sic*]: Van Duzee, Univ. Cal. Publ. Ent., 2: 190.
1918 *Malezonotus rufipes*: Barber, J. N.Y. Ent. Soc., 26: 54.
1918 *Malezonotus soldalicius* [*sic*]: Barber, J. N.Y. Ent. Soc., 26: 55 (in part).
1926 *Malezonotus sodalicius*: Blatchley, Het. E. N. Am., p. 425.
Distribution: Ariz., Ark., Ks., La., Miss., Mo., Ok., Tex., Va.
Ref.: Slater, 1964a, 2: 1456.

Malezonotus sodalicius (Uhler), 1876
1876 *Megalonotus sodalicius* Uhler, Rept. U.S. Geol. Geogr. Surv. Terr., 5: 835 (in part). [Nev.].
1910 *Rhyparochromus sodalicius*: Van Duzee, Trans. Am. Ent. Soc., 36: 75.
1914 *Rhyparochromus sodalicus* [*sic*]: Van Duzee, Trans. San Diego Soc. Nat. Hist., 2: 9.
1918 *Malezonotus soldalicius* [*sic*]: Barber, J. N.Y. Ent. Soc., 26: 55, 57 (in part).
1919 *Malezonotus sodalicius*: Parshley, Occas. Pap. Mus. Zool. Univ. Mich., 71: 17.
Distribution: B.C., Cal., Col., Id., Nev., Ore., Ut., Wash., Wyo.
Ref.: Slater, 1964a, 2: 1457.

Genus *Spinigernotus* Scudder, 1984

1984 *Spinigernotus* Scudder, Can. Ent., 116: 1293. Type-species: *Sphragisticus simulatus* Barber, 1918. Original designation.

Spinigernotus simulatus (Barber), 1918
1918 *Sphragisticus simulatus* Barber, J. N.Y. Ent. Soc., 26: 58. [N.M.].
1984 *Spinigernotus simulatus*: Scudder, Can. Ent., 116: 1295.
Distribution: N.M., Tex. (Mexico).
Ref.: Slater, 1964a, 2: 1396 (as *Sphragisticus*).

Genus *Trapezonotus* Fieber, 1860

1860 *Trapezonotus* Fieber, Europ. Hem., p. 50, 190. Type-species: *Lygaeus agrestis* Fallén, 1807, a junior synonym of *Cimex arenarius* Linnaeus, 1758. Designated by Van Duzee, 1917, Univ. Cal. Publ. Ent., 2: 192.

Ref.: Slater, 1964a, 2: 1477.

Trapezonotus arenarius (Linnaeus), 1758

1758 *Cimex arenarius* Linnaeus, Syst. Nat., 10th ed., p. 448. [Europe].

1807 *Lygaeus agrestis* Fallén, Monogr. Cimic. Suec., p. 66. [Sweden]. Synonymized by Horvath, 1899, Rev. d'Ent., 17: 276.

1913 *Trapezonotus agrestis*: Gibson, Rept. Ent. Soc. Ont., 1912: 136.

1916 *Trapezonotus arenarius*: Van Duzee, Check List Hem., p. 23.

Distribution: B.C., Man., Mass., Me., N.H., N.Y., N.T., Ont., Que., Sask.

Ref.: Slater, 1964a, 2: 1478; Sweet, 1964b; Remold, 1962; Coulianos and Kugelberg, 1973; Slater, 1986 (zoogeogr.).

Trapezonotus derivatus Barber, 1918

1918 *Trapezonotus derivatus* Barber, J. N.Y. Ent. Soc., 26: 57. [Ariz.].

Distribution: Ariz.

Ref.: Slater, 1964a, 2: 1484.

Trapezonotus diversus Barber, 1918

1918 *Trapezonotus diversus* Barber, J. N.Y. Ent. Soc., 26: 57. [Cal.].

Distribution: Cal.

Ref.: Slater, 1964a, 2: 1487.

Trapezonotus vandykei Van Duzee, 1937

1937 *Trapezonotus vandykei* Van Duzee, Pan-Pac. Ent., 13: 30. [Col.].

Distribution: Col.

Ref.: Slater, 1964a, 2: 1489.

Tribe Lethaeini Stål, 1872

Ref.: Slater, 1964a, 2: 806; Barber, 1917 (key to genera); Blatchley, 1926a (key eastern U.S. genera); Torre-Bueno, 1946a (key genera).

Genus *Cistalia* Stål, 1874

1874 *Cistalia* Stål, K. Svens. Vet.-Akad. Handl., 12(1): 164. Type-species: *Lygaeus signoretii* Guérin-Méneville, 1857.

Note: Slater (1964, Cat. Lygaeidae World, 2: 813), in a footnote, explained that Van Duzee selected *L. signoretii* Guérin-Méneville as the type of *Cistalia* because it was one of the two species Stål originally included in his genus. Barber (1938, Proc. Ent. Soc. Wash., 40: 88) showed that Stål did not have *signoretii* Guérin-Méneville, but *explanata* Barber. Thus, this is a case of a misidentifed type-species and technically requires action by the Int. Comm. Zool. Nomen.

Ref.: Slater, 1964a, 2: 812; Slater and Baranowski, 1973 (key spp.).

Cistalia explanata Barber, 1938

1874 *Cistalia Signoretii* [*sic*]: Stål, K. Svens. Vet.-Akad. Handl., 12(1): 175.

1894 *Cistalia signoreti* [*sic*]: Lethierry and Severin, Cat. Gen. Hem., 2: 224 (in part).

1938 *Cistalia explanata* Barber, Proc. Ent. Soc. Wash., 40: 88. [Tex.].

1964 *Cistalia signoretii*: Slater, Cat. Lygaeidae World, 2: 815 (in part).
Distribution: Cal., Ks., La., Nev., Tex.
Ref.: Slater, 1964a, 2: 813.

Cistalia signoretii (Guérin-Méneville), 1857
1857 *Lygaeus* (*Platygaster*) *signoretii* Guérin-Méneville, Hist. Cuba Ins., p. 396. [Cuba].
1973 *Cistalia signoreti* [*sic*]: Slater and Baranowski, Fla. Ent., 56: 264.
Distribution: Fla. (South America, West Indies).
Note: Barber (1938, Proc. Ent. Soc. Wash., 40: 87) in effect transferred earlier North American records of *C. signoretii* to his new species *C. explanata* when he wrote that *C. signoretii* is not known to occur outside of Cuba.
Ref.: Slater, 1964a, 2: 813; Slater and Baranowski, 1973.

Genus *Cryphula* Stål, 1874

1874 *Cryphula* Stål, K. Svens. Vet.-Akad. Handl., 12(1): 164. Type-species: *Cryphula parallelogramma* Stål, 1874. Monotypic.
1882 *Trapezus* Distant, Biol. Centr.-Am., Rhyn., 1: 217. Type-species: None selected; three included species. Synonymized by Barber, 1918, Psyche, 25: 88.
Ref.: Slater, 1964a, 2: 814; Barber, 1955 (key spp.); Scudder, 1962b (key spp.)

Cryphula abortiva Barber, 1918
1918 *Cryphula abortiva* Barber, J. N.Y. Ent. Soc., 26: 63. [Ariz.].
Distribution: Ariz.
Ref.: Slater, 1964a, 2: 815.

Cryphula apicata (Distant), 1882
1882 *Trapezus apicatus* Distant, Biol. Centr.-Am., Rhyn., 1: 217. [Mexico, Guatemala].
1946 *Cryphula apicatus*: Torre-Bueno, Ent. Am., 26: 138.
1964 *Cryphula apicata*: Slater, Cat. Lygaeidae World, 2: 815.
Distribution: Ariz.(?) (Mexico, Central America).
Note: Barber's (1955, J. N.Y. Ent. Soc., 63: 135-137) revision of the genus does not list any locality north of Mexico for this species, and Torre-Bueno's (1946, above) record from Ariz. may be in error.
Ref.: Slater, 1964a, 2: 815.

Cryphula nitens Barber, 1955
1918 *Cryphula apicatus*: Barber, J. N.Y. Ent. Soc., 26: 63.
1955 *Cryphula nitens* Barber, J. N.Y. Ent. Soc., 63: 135. [Cal.].
Distribution: Ariz., Cal., Id., Tex., Ut.
Ref.: Slater, 1964a, 2: 816.

Cryphula parallelogramma Stål, 1874
1874 *Cryphula parallelogramma* Stål, K. Svens. Vet.-Akad. Handl., 12(1): 165. [Tex.]. Lectotype designated by Scudder, 1977, Ent. Scand., 8: 33.
Distribution: Tex.
Note: Uhler's (1904, Proc. U.S. Nat. Mus., 27: 354) N.M. record under this name combination may belong here or to *C. subunicolor* Barber.
Ref.: Slater, 1964a, 2: 816.

Cryphula subunicolor Barber, 1955
1955 *Cryphula subunicolor* Barber, J. N.Y. Ent. Soc., 63: 136. [Tex.]. Synonymized with *C. parallelogramma* by Scudder, 1962, Can. Ent., 94: 766; resurrected by Scudder, 1970, Can. Ent., 102: 99.

Distribution: Tex.
Note: See note under *Cryphula parallelogramma* Stål.
Ref.: Slater, 1964a, 2: 816 (as synonym of *C. parallelogramma*).

Cryphula trimaculata (Distant), 1882
1882 *Trapezus trimaculata* Distant, Biol. Centr.-Am., Rhyn., 1: 217. [Guatemala]. Lectotype designated by Scudder, 1967, Bull. Brit. Mus. (Nat. Hist.), Ent., 20: 282..
1910 *Cryphula parallelogramma*: Smith, Rept. N.J. St. Mus., p. 145.
1913 *Cryphula parallelograma* [*sic*]: Torre-Bueno, Can. Ent., 45: 59.
1946 *Cryphula paralellograma* [*sic*]: Torre-Bueno, Ent. Am., 26: 138.
1962 *Cryphula trimaculata*: Scudder, Can. Ent., 94: 766.
Distribution: Col., Conn., Fla., Ia., Ill., Ind., Ks., Mass., Mo., N.C., N.J., N.Y., S.C. (Mexico, Guatemala).
Ref.: Slater, 1964a, 2: 816; Sweet, 1964b.

Genus *Valtissius* Barber, 1918

1918 *Valtissius* Barber, J. N.Y. Ent. Soc., 26: 62. Type-species: *Petissius diversus* Distant, 1893. Monotypic.
Ref.: Slater, 1964a, 2: 841.

Valtissius diversus (Distant), 1893
1893 *Petissius diversus* Distant, Biol. Centr.-Am., Rhyn., 1: 407. [Guatemala, Panama]. Lectotype designated by Scudder, 1967, Bull. Brit. Mus. (Nat. Hist.), Ent., 20: 264.
1918 *Valtissius diversus*: Barber, J. N.Y. Ent. Soc., 26: 62.
Distribution: Fla., La., Mich., Tex. (Central America).
Ref.: Slater, 1964a, 2: 841.

Genus *Xestocoris* Van Duzee, 1906

1906 *Xestocoris* Van Duzee, Ent. News, 17: 389. Type-species: *Xestocoris nitens* Van Duzee, 1906. Monotypic.
Ref.: Slater, 1964a, 2: 842.

Xestocoris nitens Van Duzee, 1906
1906 *Xestocoris nitens* Van Duzee, Ent. News, 17: 390. [N.Y.].
Distribution: Conn., Ia, Ill., Mass., Me., N.H., N.Y., Ont., Que.
Ref.: Slater, 1964a, 2: 842; Sweet, 1964b.

Tribe Megalonotini Slater, 1957

Ref.: Slater, 1964a, 2: 1338.

Genus *Lamprodema* Fieber, 1860

1860 *Lamprodema* Fieber, Europ. Hem., pp. 48, 184. Type-species: *Lygaeus maurus* Fabricius, 1803. Designated by Distant, 1903, Fauna Brit. India, Rhyn., 2: 71.
Ref.: Slater, 1964a, 2: 1344.

Lamprodema maura (Fabricius), 1803
1803 *Lygaeus maurus* Fabricius, Syst. Rhyn., p. 238. [Austria].

1977 *Lamprodema maura*: Ashlock, Proc. Ent. Soc. Wash., 74: 579.
Distribution: Cal., N.S. (Europe to Siberia and India, North Africa).
Ref.: Slater, 1964a, 2: 1345 (as *Lamprodema maurum*).

Genus *Megalonotus* Fieber, 1860

1860 *Megalonotus* Fieber, Europ. Hem., p. 47. Type-species: *Lygaeus chiragra* Fabricius, 1794. Designated by China, 1941, Proc. Roy. Ent. Soc. London, (B) 10: 130.
Ref.: Slater, 1964a, 2: 1353.

Megalonotus sabulicola (Thomson), 1870
1870 *Lygaeus (Megalonotus) sabulicola* Thomson, Opusc. Ent., 2: 190. [Sweden].
1928 *Rhyparochromus chiragra californicus* Van Duzee, Pan-Pac. Ent., 5: 47. [Cal.]. Synonymized with *Megalonotus chiragra* Fabricius by Slater and Sweet (1958, Bull. Brook. Ent. Soc., 53: 103).
1947 *Rhyparochromus californicus*: Williams, Proc. Haw. Ent. Soc., 12: 648.
1958 *Megalonotus chiragra*: Slater and Sweet, Bull. Brook. Ent. Soc., 53: 102.
1961 *Megalonotus chiragra sabulicola*: Scudder, Proc. Ent. Soc. B.C., 58: 27.
1964 *Megalonotus chiragrus sabulicolus*: Slater, Cat. Lygaeidae World, 2: 1365.
1964 *Megalonotus sabulicolus*: Sweet, Ent. Am., 44: 133.
Distribution: B.C., Cal., Conn., Id., Mass., Md., N.D., Ore., Ut., Wash. (Europe, North Africa, Asia Minor).
Note: Sweet (1964, Ent. Am., 44: 133-134) reviewed the involved history of this taxon in North America and concluded that it is a distinct species.
Ref.: Slater, 1964a, 2: 1358 (as *M. chiragrus*, in part), 2: 1365 (as *M. chiragrus sabulicolus*); Slater and Sweet, 1958 (as *M. chiragra*); Sweet, 1964b; Ashlock, 1977; Slater, 1986 (zoogeogr.).

Genus *Sphragisticus* Stål, 1872

1872 *Sphragisticus* Stål, Öfv. Svens. Vet.-Akad. Förh., 29: 55. Type-species: *Lygaeus nebulosus* Fallén, 1807. Designated by Oshanin, 1912, Kat. Paläark. Hem., p. 37.
Ref.: Slater, 1964a, 2: 1391.

Sphragisticus nebulosus (Fallén), 1807
1807 *Lygaeus nebulosus* Fallén, Monogr. Cimic. Suec., p. 65. [Sweden].
1832 *Pamera fallax* Say, Descrip. Het. Hem. N. Am., p. 17. ["N.W. Territory"]. Synonymized by Stål, 1874, K. Svens. Vet.-Akad. Handl., 12(1): 159.
1872 *Rhyparochromus fallax*: Uhler, Prelim. Rept. U.S. Geol. Surv. Mont., p. 407.
1874 *Trapezonotus (Sphragisticus) nebulosus*: Stål, K. Svens. Vet.-Akad. Handl., 12(1): 159.
1876 *Trapezonotus nebulosus*: Uhler, Bull. U.S. Geol. Geogr. Surv. Terr., 1: 311.
1885 *Trapezonotus fallax*: Popenoe, Trans. Ks. Acad. Sci., 9: 63.
1894 *Sphragisticus nebulosus*: Lethierry and Severin, Cat. Gen. Hem., 2: 213.
Distribution: Alta., B.C., Cal., Col., Conn., Ia., Id., Ill., Ind., Ks., Man., Mass., Me., Mich., Minn., Mo., Mont., N.D., N.H., N.J., N.M., N.Y., Neb., Oh., Ok., Ont., Ore., Pa., Que., S.D., Sask., Tex., Ut., Wyo. (Europe to Siberia, North Africa).
Ref.: Slater, 1964a, 2: 1391; Sweet, 1964b; Pemberton and Hoover, 1980; Slater, 1986 (zoogeogr.).

Tribe Myodochini Boitard, 1827

Ref.: Slater, 1964a, 2: 1065; Barber, 1918d (key genera); Blatchley, 1926a (key genera eastern U.S.); Torre-Bueno, 1946a (key genera); Harrington, 1980 (key genera world, cladogram of genera).

Genus *Ashlockaria* Harrington, 1980

1980 *Ashlockaria* Harrington, Bull. Am. Mus. Nat. Hist., 167: 104. Type-species: *Cnemodus sobrius* Uhler, 1894. Monotypic.

Ashlockaria sobria (Uhler), 1894

1894 *Cnemodus sobrius* Uhler, Proc. Cal. Acad. Sci., (2)4: 241. [Cal., Mexico].
1921 *Ligyrocoris (Neoligyrocoris) sobrius*: Barber, J. N.Y. Ent. Soc., 29: 102, 105.
1946 *Ligyrocoris sobrius*: Torre-Bueno, Ent. Am., 26: 73.
1964 *Pseudopamera sobria*: Slater, Cat. Lygaeidae World, 2: 1166.
1980 *Ashlockaria sobrius*: Harrington, Bull. Am. Mus. Nat. Hist., 167: 104.
Distribution: Cal. (Mexico).
Ref.: Slater, 1964a, 2: 1166 (as *Pseudopamera*).

Genus *Caenopamera* Barber, 1918

1918 *Caenopamera* Barber, J. N.Y. Ent. Soc., 26: 44. Type-species: *Pseudopamera forreri* Distant, 1893. Monotypic.
Ref.: Slater, 1964a, 2: 1067.

Caenopamera forreri (Distant), 1893

1893 *Pseudopamera forreri* Distant, Biol. Centr.-Am., Rhyn., 1: 399. [Mexico]. Lectotype designated by Scudder, 1967, Bull. Brit. Mus. (Nat. Hist.), Ent., 20: 266.
1910 *Pseudopamera ferrerei* [*sic*]: Banks, Cat. Nearc. Hem.-Het., p. 67. [Ariz.].
1918 *Caenopamera forreri*: Barber, J. N.Y. Ent. Soc., 36: 45.
Distribution: Ariz. (Mexico).
Ref.: Slater, 1964a, 2: 1068.

Genus *Carpilis* Stål, 1874

1874 *Carpilis* Stål, K. Svens. Vet.-Akad. Handl., 12(1): 144. Type-species: *Carpilis ferruginea* Stål, 1874. Monotypic.
Ref.: Slater, 1964a, 2: 1069; Barber, 1928b (key spp.).

Carpilis barberi (Blatchley), 1924

1924 *Ptochiomera (Carpilis) barberi* Blatchley, Ent. News, 35: 89. [Fla.].
1926 *Ptochiomera barberi* Blatchley, Het. E. N. Am., p. 410.
1928 *Carpilis barberi*: Barber, J. N.Y. Ent. Soc., 36: 176.
Distribution: Fla., N.C.
Ref.: Slater, 1964a, 2: 1069.

Carpilis consimilis Barber, 1949

1886 *Carpilis ferruginea*: Provancher, Pet. Faune Ent. Can., 3: 78. [Que.].
1918 *Ptochiomera (Carpilis) ferruginea*: Barber, J. N.Y. Ent. Soc., 26: 46.
1922 *Ptochiomera ferruginea*: Barber, J. N.Y. Ent. Soc., 30: 111.
1949 *Carpilis consimilis* Barber, Proc. Ent. Soc. Wash., 51: 275. [N.Y.].

Distribution: Conn., N.J., N.Y., Mass., Me., Que.
Ref.: Slater, 1964a, 2: 1070; Sweet, 1964b.

Carpilis ferruginea Stål, 1874
1874 *Carpilis ferruginea* Stål, K. Svens. Vet.-Akad. Handl., 12(1): 153. [Tex.]. Lectotype designated by Scudder, 1977, Ent. Scand., 8: 31.
1918 *Ptochiomera ferruginea*: Barber, Psyche, 25: 77.
Distribution: N.M., Tex. (Mexico).
Ref.: Slater, 1964a, 2: 1070.

Genus *Cnemodus* Herrich-Schaeffer, 1850

1850 *Cnemodus* Herrich-Schaeffer, Wanz. Ins., 9: 184. Type-species: *Cnemodus brevipennis* Herrich-Schaeffer, 1850, a junior synonym of *Astemma mavortia* Say, 1832. Monotypic.
Ref.: Slater, 1964a, 2: 1071; Brailovsky and Barrera, 1984 (discussion).

Cnemodus hirtipes Blatchley, 1924
1914 *Cnemodus mavortius*: Barber, Bull. Am. Mus. Nat. Hist., 33: 515.
1924 *Cnemodus hirtipes* Blatchley, Ent. News, 35: 90. [Fla.].
Distribution: Fla., Ga.
Ref.: Slater, 1964a, 2: 1072.

Cnemodus inflatus Van Duzee, 1915
1915 *Cnemodus inflatus* Van Duzee, Pomona J. Ent. Zool., 7: 109. [N.C.].
1926 *Cnemodus mavortius inflatus*: Blatchley, Het. E. N. Am., p. 412.
1944 *Cnemodus mavortius*: Froeschner, Am. Midl. Nat., 31: 664 (in part).
Distribution: Ind., Mo., N.C.
Ref.: Slater, 1964a, 2: 1972.

Cnemodus mavortius (Say), 1832.
1832 *Astemma mavortia* Say, Descrip. Het. Hem. N. Am., p. 19. [Fla., Ind., Mo., Pa.].
1835 *Lygaeus mavortia*: Harris, Rept. Geol. Min. Bot. Zool. Mass., 2nd. ed., p. 578.
1850 *Cnemodus brevipennis* Herrich-Schaeffer, Wanz. Ins., 9: 184. [N. Am.]. Synonymized by Stål, 1867, Berlin. Ent. Zeit., 10: 161.
1872 *Thaumastopus alacris* Walker, Cat. Hem. Brit. Mus., 5: 147. [N. Am.]. Synonymized by Distant, 1901, An. Mag. Nat. Hist., (7)8: 483.
1876 *Cnemodus mavortius*: Uhler, Bull. U.S. Geol. Geogr. Surv. Terr., 1: 308.
1912 *Cnemodus mavortus* [*sic*]: Davis and Leng, J. N.Y. Ent. Soc., 20: 121.
Distribution: Ark., Col., Conn., Fla., Ga., Ia., Ill., Ind., Ks., Mass., Md., Me., Mo., N.C., N.J., N.Y., Ok., Pa., S.C., Tex., Va.
Ref.: Slater, 1964a, 2: 1072; Jones [M.P.], 1935; Sweet, 1964b.

Genus *Ereminellus* Harrington, 1980

1980 *Ereminellus* Harrington, Bull. Am. Mus. Nat. Hist., 167: 85. Type-species: *Exptochiomera arizonensis* Barber, 1932. Monotypic.
Ref.: Brailovsky and Barrera, 1984 (key spp.).

Ereminellus arizonensis (Barber), 1932
1932 *Exptochiomera arizonensis* Barber, J. N.Y. Ent. Soc., 40: 359. [Ariz.].
1980 *Ereminellus arizonensis*: Harrington, Bull. Am. Mus. Nat. Hist., 167: 85.
Distribution: Ariz., Cal., N.M., Nev. (Mexico).
Ref.: Slater, 1964a, 2: 1077 (as *Exptochiomera*).

Ereminellus vitabundus Brailovsky and Barrera, 1984
1984 *Ereminellus vitabundus* Brailovsky and Barrera, An. Inst Biol. Univ. Nac. Aut. Méx., 54, Ser. Zool. (1): 54, 60. [Cal.].
Distribution: Cal., Nev.

Genus *Froeschneria* Harrington, 1980

1980 *Froeschneria* Harrington, Bull. Am. Mus. Nat. Hist., 167: 101. Type-species: *Ligyrocoris multispinus* Stål, 1874. Original designation.
Ref.: Barber, 1921b (key spp. as *Ligyrocoris*).

Froeschneria multispinus (Stål), 1874
1874 *Ligyrocoris multispinus* Stål, K. Svens. Vet.-Akad. Handl., 12(1): 145. [Mexico]. Lectotype designated by Scudder, 1977, Ent. Scand., 8: 33.
1909 *Ligyrocoris multispinus*: Van Duzee, Bull. Buff. Soc. Nat. Sci., 9: 171.
1914 *Ligyrocoris confraternus* Barber, Bull. Am. Mus. Nat. Hist., 33: 512. [Fla.]. Synonymized by Barber, 1921, J. N.Y. Ent. Soc., 29: 111.
1980 *Froeschneria multispinus*: Harrington, Bull. Am. Mus. Nat. Hist., 167: 101.
Distribution: Fla. (Mexico to South America, West Indies).
Ref.: Slater, 1964a, 2: 1091 (as *Ligyrocoris*).

Froeschneria piligera (Stål), 1862
1857 *Lygaeus* (*Beosus*) *abdominalis* Guérin-Méneville, Hist. Phys., Polit., Nat. Cuba, 7: 397-398. [Cuba]. Preoccupied.
1862 *Plociomera piligera* Stål, Stett. Ent. Zeit., 23: 312. [Mexico]. Lectotype designated by Scudder, 1977, Ent. Scand., 8: 33. Synonymized with *Lygaeus abdominalis* Guérin-Méneville by Stål, 1874, K. Svens. Vet.-Akad. Handl., 12(1): 146; resurrected by Slater, 1964, Cat. Lygaeidae World, 2: 1092.
1906 *Ligyrocoris constrictus*: Barber, Mus. Brook. Inst. Arts, Sci., Sci. Bull., 1: 275.
1914 *Ligyrocoris abdominalis*: Barber, Bull. Am. Mus. Nat. Hist., 33: 512.
1939 *Perigenes constrictus*: Glick, U.S. Dept. Agr. Tech. Bull., 673: 23.
1964 *Ligyrocoris piligerus*: Slater, Cat. Lygaeidae World, 2: 1092.
1973 *Ligyrocoris piliger*: Steyskal, Proc. Ent. Soc. Wash., 75: 278.
1980 *Froeschneria piligera*: Harrington, Bull. Am. Mus. Nat. Hist., 167: 101.
Distribution: Fla., Ill., La., Mass., Ont., Tex. (Mexico to South America, West Indies).
Ref.: Slater, 1964a, 2: 1092 (as *Ligyrocoris*).

Genus *Heraeus* Stål, 1862

1862 *Heraeus* Stål, Stett. Ent. Zeit., 23: 314. Type-species: *Lygaeus triguttatus* Guérin-Méneville, 1857. Monotypic.
Ref.: Slater, 1964a, 2: 1081.

Heraeus cinnamomeus Barber, 1948
1914 *Heraeus coquilletti*: Barber, J. N.Y. Ent. Soc., 22: 165 (in part).
1948 *Heraeus cinnamomeus* Barber, Oh. J. Sci., 48: 67. [Tex.].
Distribution: Tex.
Ref.: Slater, 1964a, 2: 1082.

Heraeus coquilletti Barber, 1914
1914 *Heraeus coquilletti* Barber, J. N.Y. Ent. Soc., 22: 165 (in part). [Cal.].
1914 *Heraeus nitens* Van Duzee, Trans. San Diego Soc. Nat. Hist., 2: 8. [Cal.]. Synonymized by Van Duzee, 1916, Check List Hem., p. 21.

Distribution: Cal.
Ref.: Slater, 1964a, 2: 1082.

Heraeus eximius Distant, 1882

1882 *Heraeus eximius* Distant, Biol. Centr.-Am., Rhyn., 1: 204. [Guatemala]. Lectotype designated by Scudder, 1967, Bull. Brit. Mus. (Nat. Hist.), Ent., 20: 265.
1916 *Heraeus eximius*: Van Duzee, Check List Hem., p. 21.
Distribution: Ariz. (Guatemala).
Ref.: Slater, 1964a, 2: 1082.

Heraeus plebejus Stål, 1874

1874 *Heraeus plebejus* Stål, K. Svens. Vet.-Akad. Handl., 12(1): 147. [N.J., Tex.]. Lectotype from Tex. designated by Scudder, 1977, Ent. Scand., 8: 34.
Distribution: Ariz., Conn., D.C., Fla., Ia., Ill., Ind., Ks., La., Mass., Mich., Mo., N.C., N.J., N.Y., Oh., Ok., Ont., Pa., Que., S.C., Tex. (Mexico, West Indies).
Ref.: Slater, 1964a, 2: 1083; Sweet, 1964a.

Heraeus triguttatus (Guérin-Méneville), 1857

1857 *Lygaeus (Plociomerus) triguttatus* Guérin-Méneville, Hist. Phys., Polit., Nat. Cuba, 7: 400. [Cuba].
1916 *Heraeus triguttatus*: Van Duzee, Check List Hem., p. 21.
Distribution: Ariz., Fla. (West Indies).
Ref.: Slater, 1964a, 2: 1084.

Genus *Kolenetrus* Barber, 1918

1918 *Kolenetrus* Barber, J. N.Y. Ent. Soc., 26: 49. Type-species: *Rhyparochromus plenus* Distant, 1882. Monotypic.
Ref.: Slater, 1964a, 2: 1085.

Kolenetrus plenus (Distant), 1882

1882 *Rhyparochromus plenus* Distant, Biol. Centr.-Am., Rhyn., 1: 216. [Guatemala]. Lectotype designated by Scudder, 1967, Bull. Brit. Mus. (Nat. Hist.), Ent., 20: 278.
1917 *Rhyparochromus plenus*: Van Duzee, Univ. Cal. Publ. Ent., 2: 813.
1918 *Kolenetrus plenus*: Barber, J. N.Y. Ent. Soc., 26: 50.
1922 *Kolenetrus planus* [*sic*]: Barber, J. N.Y. Ent. Soc., 30: 111.
Distribution: Ariz., B.C., Conn., Mass., Me., N.H., N.Y., Ont., Que. (Guatemala).
Ref.: Slater, 1964a, 2: 1086; Sweet, 1964b.

Genus *Ligyrocoris* Stål, 1872

1872 *Ligyrocoris* Stål, Öfv. Svens. Vet.-Akad. Förh., 29: 51. Type-species: *Cimex sylvestris* Linnaeus, 1758. Monotypic.
Ref.: Slater, 1964a, 2: 1086; Barber, 1921a (key spp.); Sweet, 1963 (key spp. northeastern U.S.); Sweet, 1986 (key spp. eastern U.S.).

Ligyrocoris balteatus Stål, 1874

1874 *Ligyrocoris balteatus* Stål, K. Svens. Vet.-Akad. Handl., 12(1): 145. [Mexico]. Lectotype designated by Scudder, 1977, Ent. Scand., 8: 30.
1903 *Ligyrocoris balteatus*: Van Duzee, Trans. Am. Ent. Soc., 29: 109.
1921 *Ligyrocoris halteatus* [*sic*]: Barber, J. N.Y. Ent. Soc., 29: 114.

Distribution: Cal., N.M. (Mexico to South America).
Ref.: Slater, 1964a, 2: 1087.

Ligyrocoris barberi Sweet, 1986
1986 *Ligyrocoris barberi* Sweet, J. N.Y. Ent. Soc., 94: 282. [Tex.].
Distribution: Ala., Ks., Miss., Mo., N.C., Ok., Tex.
Ref.: Sweet, 1986 (biol.).

Ligyrocoris caricis Sweet, 1963
1963 *Ligyrocoris caricis* Sweet, Psyche, 70: 17. [Conn.].
Distribution: Conn., Me., Que.
Ref.: Slater, 1964a, 2: 1087; Sweet, 1963, 1964a.

Ligyrocoris delitus Distant, 1882
1882 *Ligyrocoris delitus* Distant, Biol. Centr.-Am., Rhyn., 1: 201. [Guatemala]. Lectotype designated by Scudder, 1967, Bull. Brit. Mus. (Nat. Hist.), Ent., 20: 263.
1921 *Ligyrocoris delitus*: Barber, J. N.Y. Ent. Soc., 29: 103, 112.
Distribution: Ariz., Cal. (Mexico to Central America).
Ref.: Slater, 1964a, 2: 1088.

Ligyrocoris depictus Barber, 1921
1921 *Ligyrocoris depictus* Barber, J. N.Y. Ent. Soc., 29: 102, 109. [N.J.].
Distribution: Conn., Mass., Me., N.C., N.H., N.J., N.Y., Que.
Ref.: Slater, 1964a, 2: 1088; Sweet, 1964a.

Ligyrocoris diffusus (Uhler), 1871
1871 *Plociomerus diffusus* Uhler, Proc. Boston Soc. Nat. Hist., 14: 101. [N.H.].
1871 *Cymbogaster diffusus*: Uhler, U.S. Geol. Surv. Wyo., p. 472.
1874 *Ligyrocoris sylvestris*: Stål, K. Svens. Vet.-Akad. Handl., 12(1):145.
1890 *Pamera vicina*: Smith, Geol. Surv. N.J., 2(2): 423.
1910 *Ligyrocoris contractus*: Smith, Rept. N.J. St. Mus., p. 143.
1910 *Ligyrocoris diffusus*: Banks, Cat. Nearc. Hem.-Het., p. 65.
Distribution: Alta., Ark., Ariz., B.C., Cal., Col., Conn., D.C., Ia., Id., Ill., Ind., Ks., Mass., Md., Me., Mich., Mo., N.C., N.B., N.D., N.H., N.J., N.M., N.S., N.Y., Nfld., Ont., Pa., Que., R.I., S.D., Sask., Tex., Ut., Vt., Va., Wis. (Mexico).
Ref.: Slater, 1964a, 2: 1088; Sweet, 1964a; Thorpe and Harrington, 1979 (biol., parasite); Thorpe and Harrington, 1981 (behavior).

Ligyrocoris latimarginatus Barber, 1921
1921 *Ligyrocoris latimarginatus* Barber, J. N.Y. Ent. Soc., 29: 102, 107. [Cal.].
Distribution: B.C., Cal., Id., Ore., Wash.
Ref.: Slater, 1964a, 2: 1090.

Ligyrocoris litigiosus (Stål), 1862
1862 *Plociomera litigiosa* Stål, Stett. Ent. Zeit., 23: 313. [Mexico]. Lectotype designated by Scudder, 1977, Ent. Scand., 8: 32.
1921 *Ligyrocoris litigiosus*: Barber, J. N.Y. Ent. Soc., 29: 102, 109.
Distribution: Ariz., Fla. (Mexico to South America, West Indies).
Ref.: Slater, 1964a, 2: 1091.

Ligyrocoris obscurus Barber, 1921
1921 *Ligyrocoris obscurus* Barber, J. N.Y. Ent. Soc., 29: 102, 108. [Md.].
Distribution: Ill., Ind., Ks., Md., Mo.(?).
Ref.: Slater, 1964a, 2: 1092.

Ligyrocoris occultus (Barber), 1952
1952 *Pachybrachius occultus* Barber, J. N.Y. Ent. Soc., 60: 218. [Id.].
1980 *Ligyrocoris occultus*: Harrington, Bull. Am. Mus. Nat. Hist., 167: 100.
Distribution: Col., Id., Mont., Ut.
Ref.: Slater, 1964a, 2: 1135 (as *Pachybrachius*).

Ligyrocoris slossonae Barber, 1914
1914 *Ligyrocoris slossoni* Barber, Bull. Am. Mus. Nat. Hist., 33: 513. [Fla.].
Distribution: Fla., Ill.
Note: Species named after Mrs. Annie Trumbull Slosson.
Ref.: Slater, 1964a, 2: 1094.

Ligyrocoris sylvestris (Linnaeus), 1758
1758 *Cimex sylvestris* Linnaeus, Syst. Nat., 10th ed., p. 449. [Europe].
1832 *Pamera contracta* Say, Descrip. Het. Hem. N. Am., p. 16. ["N.W. Territory"]. Synonymized by Stål, 1874, K. Svens. Vet.-Akad. Handl., 12(1): 145.
1852 *Rhyparochromus vicinus* Dallas, List Hem. Brit. Mus., 2: 576. [N. Am.]. Synonymized by Barber, 1949, Proc. Ent. Soc. Wash., 51: 276. Lectotype from "N. Am." designated by Scudder, 1967, Bull Brit. Mus. (Nat. Hist.), Ent., 20: 284.
1874 *Pamera vicina*: Stål, K. Svens. Vet.-Akad. Handl., 12(1): 152.
1886 *Pamera bilobata*: Provancher, Pet. Faune Ent. Can., 3: 82.
1894 *Ligyrocoris contractus*: Lethierry and Severin, Cat. Gen. Hem., 2: 190.
1908 *Ligyrocoris diffusus*: Van Duzee, Can. Ent., 40: 110.
1926 *Orthaea vicina*: Blatchley, Het. E. N. Am., p. 398.
1927 *Ligyrocoris silvestris* [*sic*]: Johnson, Biol. Surv. Mt. Desert, 1: 46.
1946 *Orthoea* [*sic*] *vicina*: Torre-Bueno, Ent. Am., 26: 81.
Distribution: Alk., Alta., B.C., Col., Conn., Id., Ill., Ind., Ks., Mass., Me., Mich., Mo., N.C., N.D., N.H., N.J., N.S., N.Y., Nfld. (Lab.), Oh., Ont., Pa., Que., R.I., S.D., Sask., Tex., Ut., Vt., Wis. (Europe, North Africa).
Ref.: Slater, 1964a, 2: 1094; Sweet, 1964a; Froeschner and Halpin, 1981; Slater, 1986 (zoogeogr.).

Genus *Myodocha* Latreille, 1807

1807 *Myodocha* Latreille, Gen. Crust. Ins., 3: 126. Type-species: *Myodochus serripes* Olivier, 1811. Designated by Int. Comm. Zool. Nomen., 1963, Opinion 669.
1837 *Chiroleptes* Kirby, Fauna Boreal Am., 4: 280. Type-species: *Chiroleptes raptor* Kirby, 1837. Synonymized by Stål, 1874, K. Svens. Vet.-Akad. Handl., 12(1): 146.
Ref.: Slater, 1964a, 2: 1099; Blatchley, 1926a (key spp.); Torre-Bueno, 1946a (key spp.).

Myodocha annulicornis Blatchley, 1926
1926 *Myodocha annulicornis* Blatchley, Het. E. N. Am., p. 388. [Fla.].
Distribution: Fla.
Ref.: Slater, 1964a, 2: 1100.

Myodocha serripes Olivier, 1811 [Fig. 53]
1810 *Myodocha serripes* Latreille, Consid. Gen. Crust., Arach., Ins., 29: 433. *Nomen nudum*.
1811 *Myodochus* [*sic*] *serripes* Olivier, Hist. Nat. Ins., 8: 105. [No locality given].
1832 *Myodocha petiolata* Say, Descrip. Het. Hem. N. Am., p. 19. [U.S.]. Synonymized by Stål, 1874, K. Svens. Vet.-Akad. Handl., 12(1): 147.
1837 *Chiroleptes raptor* Kirby, Fauna Boreal Am., 4: 281. [N.Y.]. Synonymized by Stål,

1874, K. Svens. Vet.-Akad. Handl., 12(1): 147.
1859 *Myodocha opetilata* [*sic*]: LeConte, Complete Writ. T. Say, 1: 337.
1907 *Chiroleptes serripes*: Tucker, Ks. Univ. Sci. Bull., 4: 56.
Distribution: Col., Conn., Fla., Ia., Ill., Ind., Ks., La., Mass., Md., Me., Mich., Minn., Mo., N.C., N.H., N.J., N.M., N.Y., Neb., Oh., Ok., Ont., Pa., Que., S.C., Tenn., Tex., Vt., Va.
Ref.: Slater, 1964a, 2: 1101; Sweet, 1964a; Payne et al., 1968; Milne and Milne, 1980.

Genus *Neopamera* Harrington, 1980

1980 *Neopamera* Harrington, Bull. Am. Mus. Nat. Hist., 167: 107. Type-species: *Pamera bilobata* Say, 1831. Original designation.
Ref.: Barber, 1952b (key spp. as *Pachybrachius*).

Neopamera albocincta (Barber), 1953
1916 *Orthaea servillei*: Van Duzee, Check List Hem., p. 22.
1931 *Orthoea* [*sic*] *servillei*: Torre-Bueno, Bull. Brook. Ent. Soc., 26: 135.
1952 *Pachybrachius servillei*: Slater, Proc. Ia. Acad. Sci., 59: 534.
1953 *Pachybrachius albocinctus* Barber, J. N.Y. Ent. Soc., 60: 216. [Tex.].
1980 *Neopamera albocinctus*: Harrington, Bull. Am. Mus. Nat. Hist., 167: 107.
Distribution: Ala., D.C., Fla., Ga., Ill., La., Md., Mich., Miss., Mo., N.C., N.Y., Oh., Ont., S.C., Tex., Va. (Mexico to South America, West Indies).
Ref.: Slater, 1964a, 2: 1110 (as *Pachybrachius*); Sweet, 1964b (as *Pachybrachius*).

Neopamera bilobata (Say), 1832
1832 *Pamera bilobata* Say, Descrip. Het. Hem. N. Am., p. 17. [La., Mexico].
1980 *Neopamera bilobata*: Harrington, Bull. Am. Mus. Nat. Hist., 167: 107.
Distribution: Ala., Cal., Conn., Fla., Ga., Ill., Ia., La., Md., Miss., Mo., N.C., N.J., Oh., Ok., Ont., Pa., S.C., Tex. (Mexico to South America, West Indies).

Neopamera bilobata bilobata (Say), 1832
1832 *Pamera bilobata* Say, Descrip. Het. Hem. N. Am., p. 17.
1852 *Rhyparochromus lineatus*: Dallas, List Hem. Brit. Mus., 2: 575. See note below. Lectotype [no locality given] designated by Scudder, 1967, Bull. Brit. Mus. (Nat. Hist.), Ent., 20: 271.
1869 *Pachymera bilobata*: Rathvon, Hist. Lancaster Co., Pa., p. 549.
1876 *Plochiomera bilobata*: Glover, Ms. Notes J. Hem., p. 56.
1882 *Pamera dallasi* Distant, Biol. Centr.-Am., Rhyn., 1: 208. [N. Am., Mexico, Guatemala]. Synonymized by Distant, 1893, Biol. Centr.-Am. Rhyn., 1: 398. Lectotype designated herein; see note below.
1897 *Pamera vincta*: Quaintance, Bull. Agr. Exp. Stn. Fla., 42: 564.
1910 *Pamera lineata*: Banks, Cat. Nearc. Hem.-Het., p. 66.
1916 *Orthaea bilobata*: Van Duzee, Check List Hem., p. 22.
1916 *Orthaea dallasi*: Van Duzee, Check List Hem., p. 22.
1939 *Pachybrachius bilobatus*: Barber, Sci. Surv. Porto Rico and Virgin Is., 14: 352.
1946 *Orthoea* [*sic*] *dallasi*: Torre-Bueno, Ent. Am., 26: 81.
1952 *Pachybrachius bilobata*: Barber, J. N.Y. Ent. Soc., 60: 214.
1980 *Neopamera bilobata*: Harrington, Bull. Am. Mus. Nat. Hist., 167: 107.
Distribution: Cal., Conn., Fla., Ga., Ill., Ia., La., Md., Miss., Mo., N.C., N.J., Oh., Ok., Ont., Pa., S.C., Tex. (Bermuda, Mexico to South America, West Indies).
Note: Dallas (see above) based his combination "*Rhyparochromus lineatus*" on Fabri-

cius' original description of *Lygaeus lineatus* and other earlier literature but without descriptive comments. Stål (1868, Svens. Vet.-Akad. Handl., 11(17): 86) studied Fabricius' material and placed the Fabrician species in the Family "Capsidae," now the Miridae. Because Dallas credited the species to another author and his specimen stands in the British Museum (Natural History) simply as a misidentified specimen. Thus, when Distant (see above) proposed the combination *Pamera dallasi* with descriptive comments, he actually pro posed a new species based, in part, on Dallas' misidentified material. Scudder (1967, Bull. Brit. Mus. (Nat. Hist.), Ent., 20: 271) designated a lectotype for Dallas' combination "*Rhyparochromus lineatus*," but because that combination was originally a misidentification without descriptive comments, there was no such species and hence there could be no lectotype. If this lectotype is to be accepted, it would be more appropriate to use it to establish the name *Pamera dallasi* Distant. We therefore select as lectotype of *Pamera dallasi* Distant, 1882, the specimen in the British Museum (Natural History) selected by Scudder, 1967, as the lectotype of "*Rhyparochromus lineatus*" Dallas (not Fabricius), 1882.

Ref.: Slater, 1964a, 2: 1113 (as *Pachybrachius*); Quaintance, 1897 (as *Pamera vincta*); Sweet, 1964b (as *Pachybrachius*); Buschman et al., 1977 (egg predator, as *Pachybrachius*); Altieri and Whitcomb, 1980.

Neopamera bilobata scutellata (Dallas), 1852, New combination

1852 *Rhyparochromus scutellatus* Dallas, List Hem. Brit. Mus., 2: 575-576. [N. Am.]. Lectotype from "North Amer." designated by Scudder, 1967, Bull. Brit. Mus. (Nat. Hist.), Ent., 20: 279.

1909 *Pamera bilobata scutellata*: Van Duzee, Bull. Buff. Soc. Nat. Sci., 9: 171.

1914 *Pamera servillei*: Barber, Bull. Am. Mus. Nat. Hist., 33: 514.

1952 *Pachybrachius bilobata scutellatus*: Barber, J. N.Y. Ent. Soc., 60: 215.

1964 *Pachybrachius bilobatus scutellatus*: Slater, Cat. Lygaeidae World, 2: 1116.

Distribution: Ala., Fla., Ga. (Mexico, West Indies).

Note: This form is doubtfully distinct from *Neopamera bilobata* (Say).

Ref.: Slater, 1964a, 2: 1116 (as *Pachybrachius*).

Neopamera neotropicalis (Kirkaldy), 1909

1803 *Lygaeus serripes* Fabricius, Syst. Rhyn., p. 236. [Am. Merid.]. Preoccupied.

1909 *Orthaea neotropicalis* Kirkaldy, Can. Ent., 41: 31. New name for *Lygaeus serripes* Fabricius.

1960 *Pachybrachius neotropicalis*: Hussey, Fla. Ent., 43: 93.

1980 *Neopamera neotropicalis*: Harrington, Bull. Am. Mus. Nat. Hist., 167: 107.

Distribution: Fla. (Central to South America, West Indies).

Ref.: Slater, 1964a, 2: 1132 (as *Pachybrachius*).

Genus *Pachybrachius* Hahn, 1826

1826 *Pachybrachius* Hahn, Icon. Monogr. Cimic., p. 18. Type-species: *Pachybrachius luridus* Hahn, 1826. Monotypic.

Ref.: Slater, 1964a, 2: 1107; Barber, 1952b (key spp.).

Pachybrachius fracticollis (Schilling), 1829

1829 *Pachymerus fracticollis* Schilling, Beitr. Ent., 1: 82. [Germany].

1923 *Orthaea fracticollis*: Parshley, Ent. News, 34: 22.

1938 *Pachybrachius fracticollis*: Stichel, Ill. Bst. Deut. Wanz., 12: 416.

Distribution: Que. (Europe).
Note: Doubtfully established in North America.
Ref.: Slater, 1964a, 2: 1121; Walley, 1934; Pfaler, 1936.

Pachybrachius luridus Hahn, 1826
1826 *Pachybrachius luridus* Hahn, Icon. Monogr. Cimic., p. 18. [Germany].
1923 *Orthaea lurida*: Parshley, Ent. News, 34: 22.
Distribution: Que. (Europe to Japan).
Note: Doubtfully established in North America.
Ref.: Slater, 1964a, 2: 1128; Jordan, 1935.

Genus *Paromius* Fieber, 1860

1839 *Stenocoris* Rambur, Faune Ent. Andalousie, 2: 139. Preoccupied. Type-species: *Stenocoris gracilis* Rambur, 1839. Monotypic.
1860 *Paromius* Fieber, Europ. Hem., pp. 45, 170. New name for *Stenocoris* Rambur.
Ref.: Slater, 1964a, 2: 1148.

Paromius longulus (Dallas), 1852
1852 *Rhyparochromus longulus* Dallas, List Hem. Brit. Mus., 2: 578. [Type-locality unknown]. Lectotype [no locality given] designated by Scudder, 1967, Bull. Brit. Mus. (Nat. Hist.), Ent., 20: 271.
1874 *Pamera (Paromius) longula*: Stål, K. Svens. Vet.-Akad. Handl., 12(1): 148.
1876 *Pamera longula*: Uhler, Bull. U.S. Geol. Geogr. Surv. Terr., 1: 310.
1910 *Paromius longulus*: Banks, Cat. Nearc. Hem.-Het., p. 66.
1914 *Panonius* [*sic*] *longulus*: Barber, J. N.Y. Ent. Soc., 22: 269.
1926 *Orthaea longulus*: Blatchley, Het. E. N. Am., p. 399.
1933 *Orthoea* [*sic*] *(Paromius) longula*: Torre-Bueno, Bull. Brook. Ent. Soc., 28: 31.
1946 *Orthoea* [*sic*] *longulus*: Torre-Bueno, Ent. Am., 26: 79.
Distribution: Fla., Ga., La., Miss., Mo., N.C., S.C., Tenn., Tex. (Mexico to South America, West Indies).
Ref.: Slater, 1964a, 2: 1155.

Genus *Perigenes* Distant, 1893

1893 *Perigenes* Distant, Biol. Centr.-Am., Rhyn., 1: 396. Type-species: *Perigenes dispositus* Distant, 1893. Monotypic.
Ref.: Slater, 1964a, 2: 1159; Torre-Bueno, 1946a (key spp.).

Perigenes constrictus (Say), 1832
1832 *Pamera constricta* Say, Descrip. Het. Hem. N. Am., p. 15. [U.S.].
1835 *Lygaeus (Pachymera) constricta*: Harris, Rept. Geol. Min. Bot. Zool. Mass., p. 577.
1869 *Pachymera constricta*: Rathvon, Hist. Lancaster Co., Pa., p. 549.
1871 *Plociomerus constrictus*: Uhler, Proc. Boston Soc. Nat. Hist., 14: 101.
1874 *Ligyrocoris constrictus*: Stål, K. Svens. Vet.-Akad. Handl., 12(1): 146.
1876 *Plociomerus constricta*: Glover, Ms. Notes J. Hem., pp. 56, 58.
1903 *Perigenes fallax* Heidemann, Proc. Ent. Soc. Wash., 5: 156. [D.C.]. Synonymized by Van Duzee, 1909, Can. Ent. 41: 372.
1909 *Perigenes constrictus*: Van Duzee, Can. Ent., 41: 372.
1917 *Perigenes costalis*: Parshley, Occas. Pap. Boston Soc. Nat. Hist., 7: 49.
1947 *Peregenes* [*sic*] *constrictus*: Dowdy, Ecology, 28: 428.

Distribution: Alk., Cal., Col., Conn., D.C., Ia., Ill., Ind., Ks., Mass., Mich., Mo., N.C., N.D., N.H., N.J., N.Y., Ok., Oh., Pa., S.D., Tex., N.S., Ont., Que., Wis. (Mexico).
Ref.: Slater, 1964a, 2: 1160; Sweet, 1964b.

Perigenes similis Barber, 1906
1906 *Perigines* [*sic*] *similis* Barber, Mus. Brook. Inst. Arts Sci., Sci. Bull., 1: 276. [Tex.].
1916 *Perigenes similis*: Van Duzee, Check List Hem., p. 21.
1926 *Perigenes constrictus*: Blatchley, Het. E. N. Am., p. 404.
Distribution: Fla., Ga., Ia., Ill., Mo., Ok., S.D., Tex.
Ref.: Slater, 1964a, 2: 1161.

Genus *Prytanes* Distant, 1893

1893 *Prytanes* Distant, Biol. Centr.-Am., Rhyn., 1: 401. Type-species: *Prytanes globosus* Distant, 1893. Designated by Slater, 1964, Cat. Lygaeidae World, 2: 1163.
1928 *Exptochiomera* Barber, J. N.Y. Ent. Soc., 36: 175. Type-species: *Lygaeus* (*Beosus*) *minimus* Guérin-Méneville, 1857. Original designation. Synonymized by Harrington, 1980, Bull. Am. Mus. Nat. Hist., 167: 87.
Ref.: Slater, 1964a, 2: 1076 (as *Exptochiomera*); Barber, 1953a (key spp., as *Exptochiomera*).

Prytanes confusus (Barber), 1953
1876 *Ptochiomera minima*: Uhler, Bull. U.S. Geol. Geogr. Surv. Terr., 1: 308.
1886 *Plochiomera minima*: Provancher, Pet. Faune Ent. Can., 3: 87.
1928 *Exptochiomera minima*: Barber, J. N.Y. Ent. Soc., 36: 176.
1948 *Extochiomera* [*sic*] *albomaculatus*: Barber, Oh. J. Sci., 48: 68.
1953 *Exptochiomera confusa* Barber, Proc. Ent. Soc. Wash., 55: 21. [Tex.].
1980 *Prytanes confusa*: Harrington, Bull. Am. Mus. Nat. Hist., 167: 87.
Distribution: Fla., La., Tex. (Mexico to South America, West Indies).
Ref.: Slater, 1964a, 2: 1077 (as *Exptochiomera*).

Prytanes dissimilis (Barber), 1953
1953 *Exptochiomera dissimilis* Barber, Proc. Ent. Soc. Wash., 55: 22. [Fla.].
1980 *Prytanes dissimilis*: Harrington, Bull. Am. Mus. Nat. Hist., 167: 87.
Distribution: Fla. (West Indies).
Ref.: Slater, 1964a, 2: 1078 (as *Exptochiomera*).

Prytanes formosus (Distant), 1882
1882 *Plociomera formosa* Distant, Biol. Centr.-Am., Rhyn., 1: 210. [Guatemala]. Lectotype designated by Scudder, 1967, Bull. Brit. Mus. (Nat. Hist.), Ent., 20: 266.
1894 *Plociomera formosa*: Lethierry and Severin, Cat. Gén. Hem., 2: 194.
1910 *Ptochiomera formosa*: Banks, Cat. Nearc. Hem.-Het., p. 67.
1928 *Exptochiomera formosa*: Barber, J. N.Y. Ent. Soc., 36: 176.
1980 *Prytanes formosa*: Harrington, Bull. Am. Mus. Nat. Hist., 167: 87.
Distribution: Ariz., Tex. (Mexico to Central and South America).
Ref.: Slater, 1964a, 2: 1078 (as *Exptochiomera*).

Prytanes fuscicornis (Stål), 1874
1874 *Plociomera fuscicornis* Stål, K. Svens. Vet.-Akad. Handl., 12(1): 152. [Tex.]. Lectotype designated by Scudder, 1977, Ent. Scand., 8: 31.
1876 *Ptochiomera fuscicornis*: Uhler, Bull. U.S. Geol. Geogr. Surv. Terr., 1: 308.
1928 *Exptochiomera fuscicornis*: Barber, J. N.Y. Ent. Soc., 36: 176.
1980 *Prytanes fusicornis* [*sic*]: Harrington, Bull. Am. Mus. Nat. Hist., 167: 87.
Distribution: Ariz., Ark., Tex. (Mexico to South America).

Ref.: Slater, 1964a, 2: 1079 (as *Exptochiomera*).

Prytanes intercisus (Barber), 1932

1932 *Exptochiomera intercisa* Barber, J. N.Y. Ent. Soc., 40: 357. [Cuba, Fla.].

1946 *Exptochiomera intercissa* [*sic*]: Torre-Bueno, Ent. Am., 26: 87.

1980 *Prytanes intercisa*: Harrington, Bull. Am. Mus. Nat. Hist., 167: 87.

Distribution: Ark., Fla., La. (Cuba).

Ref.: Slater, 1964a, 2: 1079 (as *Exptochiomera*).

Prytanes oblongus (Stål), 1862

1862 *Plociomera oblonga* Stål, Stett. Ent. Zeit., 23: 313. [Mexico]. Lectotype designated by Scudder, 1977, Ent. Scand., 8: 33.

1916 *Ptochiomera oblonga*: Van Duzee, Check List Hem., p. 22.

1932 *Exptochiomera oblonga*: Barber, J. N.Y. Ent. Soc., 40: 363.

1980 *Prytanes oblonga*: Harrington, Bull. Am. Mus. Nat. Hist., 167: 87.

Distribution: Ariz., Ks., Ok., Tex. (Mexico to Central and South America, West Indies).

Ref.: Slater, 1964a, 2: 1080 (as *Exptochiomera*).

Genus *Pseudocnemodus* Barber, 1911

1911 *Pseudocnemodus* Barber, J. N.Y. Ent. Soc., 19: 25. Type-species: *Pseudocnemodus bruneri* Barber, 1911, a junior synonym of *Pterotmetus canadensis* Provancher, 1886. Monotypic.

Ref.: Slater, 1964a, 2: 1163; Brailovsky and Barrera, 1984 (discussion).

Pseudocnemodus canadensis (Provancher), 1886

1886 *Pterotmetus canadensis* Provancher, Pet. Faune Ent. Can., 3: 84. [Que.].

1911 *Pseudocnemodus bruneri* Barber, J. N.Y. Ent. Soc., 19: 26. [N.C., N.J., Neb.]. Synonymized by Van Duzee, 1912, Can. Ent., 44: 320.

1912 *Pseudocnemodus canadensis*: Van Duzee, Can. Ent., 44: 320.

Distribution: B.C., Conn., Ia., Ill., Ind., Ks., Mass., N.C., N.J., N.Y., Neb., Ont., Que, S.D.

Ref.: Slater, 1964a, 2: 1164; Sweet, 1964b; Brailovsky, 1981.

Genus *Pseudopachybrachius* Malipatil, 1978

1978 *Pseudopachybrachius* Malipatil, Aust. J. Zool., Suppl. Ser. No. 56: 63. Type-species: *Rhyparochromus gutta* Dallas, 1852. Original designation.

Ref.: Barber, 1952b (key spp. as *Pachybrachius*); Harrington, 1980; Zheng and Slater, 1984 (key spp. world, phylogeny).

Pseudopachybrachius basalis (Dallas), 1852 [Fig. 55]

1835 *Lygaeus* (*Pachymera*) *bilobata*: Harris, Rept. Geol. Min. Bot. Zool. Mass., p. 577.

1852 *Rhyparochromus basalis* Dallas, List Hem. Brit. Mus., 2: 575. [N.Am.]. Lectotype designated by Scudder, 1967, Bull. Brit. Mus. (Nat. Hist.), Ent., 20: 258.

1874 *Pamera basalis*: Stål, K. Svens. Vet.-Akad. Handl., 12(1): 152.

1874 *Pamera curvipes* Stål, K. Svens. Vet.-Akad. Handl., 12(1): 148. [S.C.]. Synonymized by Barber, 1952, J. N.Y. Ent. Soc. 60: 213. Lectotype designated by Scudder, 1977, Ent. Scand., 8: 31.

1907 *Orthaea* (*Pamera*) *basalis*: Tucker, Ks. Univ. Sci. Bull., 4: 56.

1908 *Orthoea* [*sic*] *basalis*: Torre-Bueno, J. N.Y. Ent. Soc., 16: 231.

1916 *Orthaea basalis*: Van Duzee, Check List Hem., p. 22.

1916 *Orthaea curvipes*: Van Duzee, Check List Hem., p. 22.

1946 *Pachybrachius basalis*: Procter, Biol. Surv. Mt. Desert Reg., 7: 73.
1980 *Pseudopachybrachius basalis*: Harrington, Bull. Am. Mus. Nat. Hist., 167: 95.
Distribution: Cal., Conn., D.C., Fla., Ga., Ia., Ill., Ind., Ks., La., Mass., Md., Me., Mich., Minn., Miss., Mo., N.C., N.J., N.M., N.Y., Neb., Oh., Ok., Pa., Que., S.C., Tex. (Mexico to Panama, West Indies).
Ref.: Slater, 1964a, 2: 1111 (as *Pachybrachius*); Sweet, 1964b (as *Pachybrachius*).

Pseudopachybrachius vinctus (Say), 1832
1832 *Pamera vincta* Say, Descrip. Het. Hem. N. Am., p. 16. [Fla.].
1852 *Rhyparochromus parvulus* Dallas, List. Hem. Brit. Mus., 2: 576. [N. Am.]. Synonymized by Uhler, 1894, Proc. Zool. Soc. London, 1894: 186. Lectotype from "North Amer. E.D." designated by Scudder, 1967, Bull. Brit. Mus. (Nat. Hist.), Ent., 20: 277.
1874 *Pamera parvula*: Stål, K. Svens. Vet.-Akad. Handl., 12(1): 148.
1876 *Plociomerus vincta*: Glover, Ms. Notes J. Hem., pp. 56, 58.
1916 *Orthaea vincta*: Van Duzee, Check List Hem., p. 22.
1939 *Pachybrachius vinctus*: Barber, Sci. Surv. Porto Rico and Virgin Is., 14: 352.
1946 *Orthoea* [*sic*] (*Orthoea*) [*sic*] *vincta*: Torre-Bueno, Ent. Am., 26: 81.
1953 *Pachybrachius vincta*: Barber, J. N.Y. Ent. Soc., 60: 213.
1978 *Pseudopachybrachius vinctus*: Malipatil, Aust. J. Zool., Suppl. Ser. No. 56: 63.
Distribution: Cal., Fla., Ga., La., Mo., N.C., Pa., S.C., Tex. (Hawaii, Mexico to South America, West Indies).
Ref.: Slater, 1964a, 2: 1144 (as *Pachybrachius*), 1972a (as *Pachybrachius*); Scudder, 1968 (as *Pachybrachius*).

Genus *Pseudopamera* Distant, 1882

1882 *Pseudopamera* Distant, Biol. Centr.-Am., Rhyn., 1: 209. Type-species: *Pseudopamera aurivilliana* Distant, 1882. Monotypic.
1921 *Ligyrocoris* (*Neoligyrocoris*) Barber, J. N.Y. Ent. Soc., 29: 104. Type-species: *Pseudopamera aurivilliana* Distant, 1882. Original designation. Synonymized by Slater, 1957, Bull. Brook. Ent. Soc., 52: 36.
Ref.: Slater, 1964a, 2: 1164; Barber, 1921b (key spp. as *Ligyrocoris* (*Neoligyrocoris*)), Torre-Bueno, 1946a (key spp., as *Ligyrocoris* (*Neoligyrocoris*)).

Pseudopamera aurivilliana Distant, 1882
1882 *Pseudopamera aurivilliana* Distant, Biol. Centr.-Am., Rhyn., 1: 209. [Mexico].
1906 *Ligyrocoris pseudoheraeus* Barber, Bull. Brook. Mus. Inst. Arts Sci., 1: 275. [Tex.]. Synonymized by Barber, 1921, J. N.Y. Ent. Soc., 29: 104.
1921 *Ligyrocoris* (*Neoligyrocoris*) *aurivillianus*: Barber, J. N.Y. Ent. Soc., 29: 101, 104.
1923 *Ligyrocoris aurivillianus*: Van Duzee, Proc. Cal. Acad. Sci., 12(ll): 139.
Distribution: Ariz., Tex. (Mexico).
Ref.: Slater, 1964a, 2: 1165.

Pseudopamera coloradensis (Barber), 1921
1921 *Ligyrocoris* (*Neoligyrocoris*) *coloradensis* Barber, J. N.Y. Ent. Soc., 29: 102, 105. [Col.].
1952 *Ligyrocoris coloradensis*: Slater, Proc. Ia. Acad. Sci., 59: 533.
1964 *Pseudopamera coloradensis*: Slater, Cat. Lygaeidae World, 2: 1165.
Distribution: Col., Ia.
Ref.: Slater, 1964a, 2: 1165.

Pseudopamera nitidicollis (Stål), 1874

1874 *Pamera nitidicollis* Stål, K. Svens. Vet.-Akad. Handl., 12(1): 150. [Tex.]. Lectotype designated by Scudder, 1977, Ent. Scand., 8: 33.
1914 *Ligyrocoris nitidicollis*: Barber, Bull. Am. Mus. Nat. Hist., 33: 514.
1916 *Orthaea nitidicollis*: Van Duzee, Check List Hem., p. 22.
1921 *Ligyrocoris (Neoligyrocoris) nitidicollis*: Barber, J. N.Y. Ent. Soc., 29: 101, 104.
1964 *Pseudopamera nitidicollis*: Slater, Cat. Lygaeidae World, 2: 1165.
Distribution: Ariz., N.M., Ok., Tex. (Mexico).
Ref.: Slater, 1964a, 2: 1165.

Pseudopamera nitidula (Uhler), 1893

1893 *Pamera nitidula* Uhler, N. Am. Fauna, 7(2): 262. [Cal., N.M., Tex.].
1910 *Ligyrocoris nitidula*: Banks, Cat. Nearc. Hem.-Het., p. 65.
1914 *Ligyrocoris nitidulus*: Van Duzee, Trans. San Diego Soc. Nat. Hist., 2: 9.
1921 *Ligyrocoris (Neoligyrocoris) nitidulus*: Barber, J. N.Y. Ent. Soc., 29: 102, 105.
1964 *Pseudopamera nitidula*: Slater, Cat. Lygaeidae World, 2: 1166.
Distribution: Ariz., Cal., Col., N.M., Tex., Ut. (Mexico).
Ref.: Slater, 1964a, 2: 1166.

Pseudopamera rubricata (Barber), 1921

1921 *Ligyrocoris (Neoligyrocoris) rubricatus* Barber, J. N.Y. Ent. Soc., 29: 102, 105. [Ariz.].
1960 *Ligyrocoris rubricatus*: Knowlton, Proc. Ut. Acad. Sci., 37: 53.
1964 *Pseudopamera rubricata*: Slater, Cat. Lygaeidae World, 2: 1166.
Distribution: Ariz., Cal., Ut.
Ref.: Slater, 1964a, 2: 1166.

Pseudopamera setosa (Stål), 1874

1874 *Pamera setosa* Stål, K. Svens. Vet.-Akad. Handl., 12(1): 150. [Tex.]. Lectotype designated by Scudder, 1977, Ent. Scand., 8: 34.
1882 *Heraeus percultus* Distant, Biol. Centr.-Am., Rhyn., 1: 205. [Guatemala]. Synonymized by Barber, 1914, Bull. Am. Mus. Nat. Hist., 33: 514. Lecotype designated by Scudder, 1967, Bull. Brit. Mus. (Nat. Hist.), Ent., 20: 277.
1914 *Ligyrocoris setosa*: Barber, Bull. Am. Mus. Nat. Hist., 33: 514.
1914 *Ligyrocoris percultus*: Van Duzee, Trans. San Diego Soc. Nat. Hist., 2: 9.
1916 *Ligyrocoris setosus*: Van Duzee, Check List Hem., p. 21.
1980 *Pseudopamera setosa*: Harrington, Bull. Am. Mus. Nat. Hist., 167: 102.
Distribution: Ariz., Cal., Tex. (Mexico to Central America).
Ref.: Slater, 1964a, 2: 1093 (as *Ligyrocoris*).

Genus *Ptochiomera* Say, 1832

1832 *Ptochiomera* Say, Descrip. Het. Hem. N. Am., p. 18. Type-species: *Ptochiomera nodosa* Say, 1832. Monotypic.
Ref.: Slater, 1964a, 2: 1167.

Ptochiomera nodosa Say, 1832 [Fig. 58]

1832 *Ptochiomera nodosa* Say, Descrip. Het. Hem. N. Am., p. 18. [U.S.].
1852 *Aphanus clavatus* Dallas, List Hem. Brit. Mus., 2: 560. [La.]. Synonymized by Stål, 1867, Berlin. Ent. Zeit., 10: 161. Lectotype designated by Scudder, 1967, Bull. Brit. Mus. (Nat. Hist.), Ent., 20: 261.
1867 *Plociomera (Pamera) nodosa*: Stål, Berlin. Ent. Zeit., 10: 161.
1869 *Plociomera nodosum*: Rathvon, Hist. Lancaster Co., Pa., p. 549.

1872 *Aphanus nodosus*: Walker, Cat. Hem. Brit. Mus., 5: 77.
1874 *Plociomera nodosa*: Stål, K. Svens. Vet.-Akad. Handl., 12(1): 153.
1912 *Plociomerus nodosus*: Van Duzee, Can. Ent., 44: 319.
Distribution: Ala., Conn., Fla., Ga., Ia., Ill., Ind., Ks., La., Mass., Md., Mo., N.C., N.J., N.Y., Neb., Oh., Ok., Pa., S.C., Tex. (Mexico).
Ref.: Slater, 1964a, 2: 1168; Sweet, 1964b.

Genus *Sisamnes* Distant, 1893

1893 *Sisamnes* Distant, Biol. Centr.-Am., Rhyn., 1: 402. Type-species: *Sisamnes contractus* Distant, 1893. Monotypic.
Ref.: Slater, 1964a, 2: 1177; Barber, 1953b (key spp.).

Sisamnes claviger (Uhler), 1895
1871 *Cymbogaster diffusus*: Uhler, U.S. Geol. Surv. Wyo., p. 472 (in part).
1872 *Plociomerus diffusus*: Uhler, Prelim. Rept. U.S. Geol. Surv. Mont., p. 407 (in part).
1895 *Ptochiomera clavigera* Uhler, Bull. Col. Agr. Exp. Stn., 31: 24. [Col., N.Y., Tex.].
1928 *Sisamnes clavigera*: Barber, J. N.Y. Ent. Soc., 36: 177.
1964 *Sisamnes clavigerus*: Slater, Cat. Lygaeidae World, 2: 1177.
1973 *Sisamnes claviger*: Steyskal, Proc. Ent. Soc. Wash., 75: 278.
Distribution: B.C., Col., Conn., Ia., Id., Ks., Mass., Mo., N.C., N.J., N.M., N.Y., Neb., Ore., Tex., Ut.
Ref.: Slater, 1964a, 2: 1177; Sweet, 1964b.

Sisamnes contractus Distant, 1893
1893 *Sisamnes contractus* Distant, Biol. Centr.-Am., Rhyn., 1: 402. [Guatemala]. Lectotype designated by Scudder, 1967, Bull. Brit. Mus. (Nat. Hist.), Ent., 20: 262.
1909 *Ptochiomera antennata* Van Duzee, Bull. Buff. Soc. Nat. Sci., 9: 172. [Ariz., Fla.]. Synonymized by Barber, 1953, Proc. Ent. Soc. Wash., 55: 26.
1928 *Sisamnes puberula*: Barber, J. N.Y. Ent. Soc., 36: 177.
1944 *Sisamnes antennata*: Froeschner, Am. Midl. Nat., 31: 664.
Distribution: Ariz., Cal., Fla., Mo., Tex. (Mexico to Central America).
Ref.: Slater, 1964a, 2: 1178.

Genus *Slaterobius* Harrington, 1980

1980 *Slaterobius* Harrington, Bull. Am. Mus. Nat. Hist., 167: 100. Type-species: *Heraeus insignis* Uhler, 1872. Original designation.
Ref.: Slater, 1964a, 2: 1178 (as *Sphaerobius*); Blatchley, 1926a (key spp. as *Sphaerobius*); Torre-Bueno, 1946a (key spp. as *Sphaerobius*).

Slaterobius insignis (Uhler), 1872
1872 *Heraeus insignis* Uhler, Prelim. Rept. U.S. Geol. Surv. Mont., p. 407. [Col., Minn., Ut.].
1893 *Sphaerobius insignis*: Uhler, Proc. Zool. Soc. London, 1893: 711.
1916 *Sphaerobius quadristriata*: Van Duzee, Check List Hem., p. 21.
1950 *Spherobius* [*sic*] *insignis*: Moore, Nat. Can., 77: 238.
1980 *Slaterobius insignis*: Harrington, Bull. Am. Mus. Nat. Hist., 167: 100.
Distribution: Alta., B.C., Cal., Col., Conn., Ia., Id., Ill., Man., Me., Minn., N.D., N.H., N.Y., Neb., Nfld., Ont., Que., S.D., Sask., Ut., Wis.
Ref.: Slater, 1964a, 2: 1179 (as *Sphaerobius*); Sweet, 1964b (as *Sphaerobius*); Kurczewski, 1972 (as *Sphaerobius*).

Slaterobius quadristriatus (Barber), 1911
1890 *Heraeus insignis*: Smith, Geol. Surv. N.J., 2(2): 423.
1910 *Heraeus orbicollis* Smith, Rep. N.J. St. Mus., p. 143. *Nomen nudum.*
1911 *Sphaerobius quadristriata* Barber, J. N.Y. Ent. Soc., 19: 24. [N.J.].
1964 *Sphaerobius quadristriatus*: Slater, Cat. Lygaeidae World, 2: 1180.
1980 *Slaterobius quadristriatus*: Harrington, Bull. Am. Mus. Nat. Hist., 167: 100.
Distribution: N.J.
Ref.: Slater, 1964a, 2: 1180 (as *Sphaerobius*).

Genus *Valonetus* Barber, 1918

1918 *Valonetus* Barber, J. N.Y. Ent. Soc., 26: 50. Type-species: *Valonetus pilosus* Barber, 1918, a junior synonym of *Plociomera puberula* Stål, 1874. Monotypic.
Ref.: Slater, 1964a, 2: 1183.

Valonetus puberulus (Stål), 1874
1874 *Plociomera puberula* Stål, K. Svens. Vet.-Akad. Handl., 12(1): 153. [Tex.]. Type lost according to Scudder, 1977, Ent. Scand., 8: 34.
1876 *Ptochiomera puberula*: Uhler, Bull. U.S. Geogr. Geol. Surv. Terr., 1: 308.
1918 *Valonetus pilosus* Barber, J. N.Y. Ent. Soc., 26: 50. [Tex.]. Synonymized by Barber, 1949, Proc. Ent. Soc. Wash., 51: 275.
1928 *Sisamnes puberula*: Barber, J. N.Y. Ent. Soc., 36: 176.
1949 *Valonetus puberula*: Barber, Proc. Ent. Soc. Wash., 51: 275.
Distribution: Ariz., Col., Mo., Tex.
Ref.: Slater, 1964a, 2: 1183.

Genus *Zeridoneus* Barber, 1981

1918 *Zeridoneus* Barber, J. N.Y. Ent. Soc., 26: 45. Type-species: *Perigenes costalis* Van Duzee, 1909. Monotypic.
Ref.: Slater, 1964a, 2: 1184.

Zeridoneus costalis (Van Duzee), 1909
1905 *Pamera constricta*: Van Duzee, Bull. N.Y. St. Mus., 97: 549.
1909 *Perigenes costalis* Van Duzee, Can. Ent., 41: 373. [N.Y., Oh.].
1910 *Ligyrocoris costalis*: Banks, Cat. Nearc. Hem.-Het., p. 65.
1918 *Zeridoneus costalis*: Barber, J. N.Y. Ent. Soc., 26: 45.
Distribution: Alta., Conn., Ia., Ill., Ind., Mass., Man., Md., Mo., N.C., N.D., N.Y., Oh., Ont., Que., S.D., Sask., Wis.
Ref.: Slater, 1964a, 2: 1184; Sweet, 1964b.

Zeridoneus knulli Barber, 1948
1948 *Zeridoneus knulli* Barber, Oh. J. Sci., 48: 69. [Tex.].
Distribution: Ks., Tex.
Ref.: Slater, 1964a, 2: 1185.

Zeridoneus petersoni Reichart, 1966
1966 *Zeridoneus petersoni* Reichart, Oh. J. Sci., 66: 347. [Ut.].
Distribution: Ut.

Genus *Zeropamera* Barber, 1948

1948 *Zeropamera* Barber, Pan-Pac. Ent., 24: 201. Type-species: *Zeropamera nigra* Barber, 1948. Monotypic.

Ref.: Slater, 1964a, 2: 1185.

Zeropamera nigra Barber, 1948

1948 *Zeropamera nigra* Barber, Pan-Pac. Ent., 24: 201. [Cal.].

Distribution: Cal., Ut.

Ref.: Slater, 1964a, 2: 1185; Ashlock, 1977.

Tribe Ozophorini Sweet, 1964

Ref.: Slater, 1964a, 2: 1040; Ashlock and Slater, 1982 (key Western Hemisphere genera).

Genus *Balboa* Distant, 1893

1893 *Balboa* Distant, Biol. Centr.-Am., Rhyn., 1: 408. Type-species: *Balboa variabilis* Distant, 1893. Monotypic.

Note: Synonymized with *Ozophora* Uhler by Barber (1918, Psyche, 25: 80); resurrected by Ashlock (1960, Proc. Biol. Soc. Wash., 73: 237).

Ref.: Slater, 1964a, 2: 1040.

Balboa ampliata (Barber), 1918

1918 *Ozophora ampliatus* Barber, J. N.Y. Ent. Soc., 26: 52. [Ariz.].

1946 *Dieuches occidentalis* Torre-Bueno, Bull. Brook. Ent. Soc., 41: 126. [Ariz.]. Synonymized by Eyles, 1969, New Zealand J. Sci., 12: 728.

1960 *Balboa ampliata*: Ashlock, Proc. Biol. Soc. Wash., 73: 237.

Distribution: Ariz., N.M., Tex.

Ref.: Slater, 1964a, 2: 1040.

Genus *Ozophora* Uhler, 1871

1871 *Ozophora* Uhler, Proc. Boston Soc. Nat. Hist., 14: 102. Type-species: *Ozophora picturata* Uhler, 1871. Monotypic.

1893 *Davila* Distant, Biol. Centr.-Am., Rhyn., 1: 394. Preoccupied. Type-species: None selected; *D. concavus* Distant and *D. pallescens* Distant included species. Synonymized by Uhler, 1894, Proc. Zool. Soc. London, 1894: 186.

1904 *Peggichisme* Kirkaldy, Ent., 37: 280. New name for *Davila* Distant.

Ref.: Slater, 1964a, 2: 1048; Slater and Baranowski, 1983 (key Fla. spp.); Slater, 1986 (zoogeogr.).

Ozophora angustata Barber, 1949

1949 *Ozophora angustata* Barber, Pan-Pac. Ent., 24: 202. [Tex.].

Distribution: Tex.

Ref.: Slater, 1964a, 2: 1048.

Ozophora burmeisteri (Guérin-Méneville), 1857

1857 *Lygaeus* (*Beosus*) *burmeisteri* Guérin-Méneville, Hist. Phys., Polit., Nat. Cuba, 7: 397. [Cuba].

1865 *Plociomerus burmeisteri*: Stål, An. Soc. Ent. Fr., (4)5: 187.

1906 *Ozophora burmeisteri*: Barber, Sci. Bull. Brook. Inst. Arts Sci., 1: 276.

Distribution: Fla., Tex. (Mexico, West Indies).

Ref.: Slater, 1964a, 2: 1049; Slater and Baranowski, 1983 (nymphs, eggs).

Ozophora caroli Slater and Baranowski, 1983
1983 *Ozophora caroli* Slater and Baranowski, Fla. Ent., 66: 422. [Fla.].
Distribution: Fla. (Mexico?).

Ozophora concava (Distant), 1893
1893 *Davila concava* Distant, Biol. Centr.-Am., Rhyn., 1: 394. [Mexico, Guatemala, Panama]. Lectotype from Panama designated by Scudder, 1967, Bull. Brit. Mus. (Nat. Hist.), Ent., 20: 261.
1916 *Ozophora concava*: Van Duzee, Check List Hem., p. 22.
1917 *Ozophora concavus*: Van Duzee, Univ. Cal. Publ. Ent., 2: 189.
Distribution: Ariz.(?) (Mexico, Central America).
Note: Slater (pers. comm.) stated that the Ariz. record for this species might be erroneous. Slater and Baranowski (1983, Fla. Ent., 66: 417) removed this species from the Fla. list. Van Duzee's (1916, above) distribution was a categorical group of states.
Ref.: Slater, 1964a, 2: 1050.

Ozophora depicturata Barber, 1928
1928 *Ozophora depicturata* Barber, Bull. Brook. Ent. Soc., 23: 266. [Cal.].
Distribution: Ariz., Cal.
Ref.: Slater, 1964a, 2: 1050.

Ozophora divaricata Barber, 1954
1954 *Ozophora divaricata* Barber, Am. Mus. Nov., 1682: 6. [Bimini, West Indies].
1926 *Ozophora pallescens*: Blatchley, Het. E. N. Am., p. 418 (in part).
1983 *Ozophora divaricata*: Slater and Baranowski, Fla. Ent., 66: 424.
Distribution: Fla. (West Indies).
Ref.: Baranowski and Slater, 1983 (key *Ozophora pallescens* group).

Ozophora gilva Slater and Baranowski, 1983
1926 *Ozophora pallescens*: Blatchley, Het. E. N. Am., p. 418 (in part).
1983 *Ozophora gilva* Slater and Baranowski, Fla. Ent., 66: 428. [Fla.].
Distribution: Fla.
Ref.: Slater and Baranowski, 1983 (nymphs).

Ozophora laticephala Slater and O'Donnell, 1979
1979 *Ozophora laticephala* Slater and O'Donnell, J. Ks. Ent. Soc., 52: 161. [Holotype from Jamaica; also from Fla.].
Distribution: Fla. (West Indies).
Ref.: Slater, 1972a (as *Ozophora* sp.); Slater and O'Donnell, 1979 (nymphs).

Ozophora levis Slater and Baranowski, 1983
1983 *Ozophora levis* Slater and Baranowski, Fla. Ent., 66: 433. [Fla.].
Distribution: Fla.

Ozophora maculata Slater and O'Donnell, 1979
1979 *Ozophora maculata* Slater and O'Donnell, J. Ks. Ent. Soc., 52: 167. [Holotype from Mexico; also from Ariz.].
Distribution: Ariz. (El Salvador, Guatemala, Mexico).

Ozophora picturata Uhler, 1871
1871 *Ozophora picturata* Uhler, Proc. Boston Soc. Nat. Hist., 14: 102. [Mass.].
Distribution: Ariz., Cal., Conn., Fla., Ga., Ia., Ill., Ind., Md., Mass., Mo., N.C., N.H., N.J., N.Y., Ok., Ont., Pa., S.C., Tex. (Mexico, West Indies).

Ref.: Slater, 1964a, 2: 1052; Sweet, 1964b; Payne et al., 1968; Slater and Baranowski, 1983 (nymphs).

Ozophora reperta Blatchley, 1926

1926 *Ozophora reperta* Blatchley, Het. E. N. Am., p. 416. [Fla.]. Lectotype designated by Blatchley, 1930, Blatchleyana, p. 63.

Distribution: Fla. (West Indies).

Ref.: Slater, 1964a, 2: 1053.

Ozophora trinotata Barber, 1914

1914 *Ozophora trinotatus* Barber, Bull. Am. Mus. Nat. Hist., 33: 515. [Fla.].

1931 *Ozophora binotata* [*sic*]: Torre-Bueno, Bull. Brook. Ent. Soc., 26: 135.

1964 *Ozophora trinotata*: Slater, Cat. Lygaeidae World, 2: 1053.

Distribution: Fla.

Ref.: Slater, 1964a, 2: 1053; Slater and Baranowski, 1983 (nymphs, eggs).

Ozophora unicolor Uhler, 1894

1894 *Ozophora unicolor* Uhler, Proc. Cal. Acad. Sci., (2)4: 242. [Baja Cal.].

1910 *Ozophora unicolor*: Banks, Cat. Nearc. Hem.-Het., p. 66.

Distribution: Cal.(?) (Mexico).

Note: Cal. may refer to Uhler's record above.

Ref.: Slater, 1964a, 2: 1054.

Tribe Plinthisini Slater and Sweet, 1961

Ref.: Slater, 1964a, 2: 781.

Genus *Plinthisus* Stephens, 1829

1829 *Plinthisus* Stephens, Nomencl. Brit. Ins., p. 65. Type-species: *Lygaeus brevipennis* Latreille 1807. Monotypic.

Ref.: Slater, 1964a, 2: 781; Barber, 1918a (key spp.); Slater, 1986 (zoogeogr.).

Plinthisus americanus Van Duzee, 1910

1910 *Plinthisus americanus* Van Duzee, Trans. Am. Ent. Soc., 36: 75. [Mass.].

1918 *Plinthisus compactus*: Barber, Proc. Ent. Soc. Wash., 20: 109 (in part).

1928 *Plinthisus contractus* [*sic*]: Barber, Ent. News, 39: 194.

Distribution: Conn., Mass., Me., N.H., N.Y., Ont., Que., S.D.(?), Sask.(?).

Note: This species was synonymized with *P. compactus* (Uhler) by Barber (1918, Proc. Ent. Soc. Wash., 20: 109); resurrected by Sweet (1964, Ent. Am., 44: 59).

Ref.: Slater, 1964a, 2: 791 (as syn. of *P. compactus*); Sweet, 1964b.

Plinthisus compactus (Uhler), 1904

1904 *Rhyparochromus compactus* Uhler, Proc. U.S. Nat. Mus., 27: 354. [N.M.].

1916 *Aphanus compactus*: Van Duzee, Check List Hem., p. 23.

1918 *Plinthisus compactus*: Barber, Proc. Ent. Soc. Wash., 20: 109 (in part).

Distribution: Ariz., Cal., N.M.

Ref.: Slater, 1964a, 2: 791.

Plinthisus indentatus Barber, 1918

1918 *Plinthisus indentatus* Barber, Proc. Ent. Soc. Wash., 20: 109. [Mont.].

Distribution: Mont., Sask.

Ref.: Slater, 1964a, 2: 793.

Plinthisus longisetosus Barber, 1918

1918 *Plinthisus longisetosus* Barber, Proc. Ent. Soc. Wash., 20: 110. [Cal.].

Distribution: Cal.

Ref.: Slater, 1964a, 2: 795.

Plinthisus martini Van Duzee, 1921

1921 *Plinthisus martini* Van Duzee, Proc. Cal. Acad. Sci., (4)11: 114. [Cal.].

Distribution: Cal.

Note: Synonymized with *P. longisetosus* Barber by Torre-Bueno (1944, Bull. Brook. Ent. Soc., 39: 170); resurrected by Ashlock (1977, Proc. Ent. Soc. Wash., 79: 758).

Ref.: Slater, 1964a, 2: 795 (as syn. of *P. longisetosus*).

Plinthisus pallidus Barber, 1918

1918 *Plinthisus pallidus* Barber, Proc. Ent. Soc. Wash., 20: 111. [Cal.].

Distribution: Cal.

Ref.: Slater, 1964a, 2: 797.

Tribe Rhyparochromini Amyot and Serville, 1843

Ref.: Slater, 1964a, 2: 1186; Scudder, 1984 (key to N. Am. genera).

Genus *Cordillonotus* Scudder, 1984

1984 *Cordillonotus* Scudder, Can. Ent., 116: 1297. Type-species: *Cordillonotus stellatus* Scudder, 1984. Original designation.

Note: This genus was tentatively placed in the tribe Rhyparochromini on the assumption that the immature stage has a Y-suture. If it does not, then *Cordillonotus* must be transferred to the Megalonotini.

Cordillonotus stellatus Scudder, 1984

1984 *Cordillonotus stellatus* Scudder, Can. Ent., 116: 1300. [Ore.].

Distribution: B.C., Cal., Ore., Wash.

Genus *Peritrechus* Fieber, 1860

1860 *Peritrechus* Fieber, Europ. Hem., p. 48, 183. Type-species: *Beosus angusticollis* Sahlberg, 1848. Designated by Distant, 1903, Fauna Brit. India, Rhyn., 2: 75.

Ref.: Slater, 1964a, 2: 1248.

Peritrechus distinguendus (Flor), 1860

1860 *Pachymerus distinguendus* Flor, Rhyn. Livl., p. 266. [Livlands].

1958 *Trapezonotus distinguendus*: Lindberg, Acta Zool. Fennica, 96: 12.

Distribution: Nfld. (Europe, Siberia).

Note: Scudder (1957, Ent. Month. Mag., 93: 244) first placed this species in *Peritrechus*).

Ref.: Slater, 1964a, 2: 1252; Pfaler, 1936 (as *Trapezonotus*).

Peritrechus fraternus Uhler, 1871 [Fig. 59]

1871 *Peritrechus fraternus* Uhler, Proc. Boston Soc. Nat. Hist., 14: 103. [Mass.].

1872 *Rhyparochromus punctatus* Provancher, Nat. Can., 4: 76. [Que.]. Preoccupied. See note below.

1886 *Ischnodemus falicus*: Provancher, Pet. Faune Ent. Can., 3: 75.

Distribution: B.C., Cal., Col., Ia., Id., Ill., Ind., Ks., Mass., Mo., N.C., N.D., N.H., N.J.,

N.M., N.Y., Ok., Ont., S.D., Ut. (Mexico).

Note: Provancher (1886, Pet. Faune Ent. Can., 3: 76) synonymized his *Rhyparochromus punctatus* under *Ischnodemus falicus*, but Van Duzee (1912, Can. Ent., 44: 319), after examining Provancher's specimens, wrote that those labeled *Ischnodemus falicus* actually were *Peritrechus fraternus*.

Ref.: Slater, 1964a, 2: 1254; Sweet, 1964b; Kelton, 1968a.

Peritrechus paludemaris Barber, 1914

1914 *Peritrechus paludemaris* Barber, Bull. Am. Mus. Nat. Hist., 33: 516. [Fla., Md., N.Y.].

Distribution: Fla., Mass., Md., N.Y.

Ref.: Slater, 1964a, 2: 1267; Sweet, 1964b.

Peritrechus saskatchewanensis Barber, 1918

1918 *Peritrechus saskatchewanensis* Barber, J. N.Y. Ent. Soc., 26: 60. [Cal., Sask.].

1922 *Peritrechus fraternus*: Hussey, Occas. Pap. Mus. Zool. Univ. Mich., 115: 10.

Distribution: Alta., B.C., Cal., Id., Man., N.D., Ont., Que., S.D., Sask., Ut.

Ref.: Slater, 1964a, 2: 1268.

Peritrechus tristis Van Duzee, 1906

1906 *Peritrechus tristis* Van Duzee, Ent. News, 17: 388. [B.C.].

Distribution: B.C., Cal., Id., Ore., Ut., Wash.

Ref.: Slater, 1964a, 2: 1268.

Genus *Uhleriola* Horvath, 1908

1908 *Uhleriola* Horvath, Ann. Hist.-Nat. Mus. Nat. Hung., 6: 562. Type-species: *Rhyparochromus floralis* Uhler, 1895. Monotypic.

Ref.: Slater, 1964a, 2: 1337.

Uhleriola floralis (Uhler), 1895

1895 *Rhyparochromus floralis* Uhler, Bull. Col. Agr. Exp. Stn., 31: 26. [Cal., Col., Mont.].

1908 *Uhleriola floralis*: Horvath, Ann. Hist.-Nat. Mus. Nat. Hung., 6: 563.

Distribution: Alta., Ariz., Cal., Col., Ia., Ill., Ks., Mont., N.D., Neb., S.D., Ut., Wyo.

Ref.: Slater, 1964a, 2: 1337; Slater, 1948; Slater, 1986 (zoogeogr.).

Tribe Stygnocorini Gulde, 1936

Ref.: Slater, 1964a, 2: 996.

Genus *Stygnocoris* Douglas and Scott, 1865

1860 *Stygnus* Fieber, Europ. Hem., pp. 49, 186. Preoccupied. Type-species: *Lygaeus rusticus* Fallén, 1897. Designated by Oshanin, 1912, Kat. Paläark. Hem., p. 35.

1865 *Stygnocoris* Douglas and Scott, Brit. Hem.-Het., p. 213. New name for *Stygnus* Fieber.

Ref.: Slater, 1964a, 2: 1011; Southwood and Leston, 1959 (key spp.).

Stygnocoris rusticus (Fallén), 1807

1807 *Lygaeus rusticus* Fallén, Monogr. Cimic. Suec., p. 70. [Sweden].

1908 *Stygnocoris rusticus*: Horvath, An. Hist.-Nat. Mus. Nat. Hung., 6: 559.

1916 *Stignocoris* [*sic*] *rusticus*: Van Duzee, Check List Hem., p. 23. [Canada].

Distribution: B.C., Conn., Ill., Me., Mich., N.S., N.Y., Ont., P.Ed., Pa., Que., Vt., W.Va., Wash., Wis. (Europe, North Africa).

Ref.: Slater, 1964a, 2: 1020; Butler [E.A.], 1923; Sweet, 1964b; Wheeler, 1983b (distribution, hosts); Slater, 1986 (zoogeogr.).

Stygnocoris sabulosus (Schilling), 1829
1807 *Lygaeus pedestris* Fallén, Monogr. Cimic. Suec., p. 70. [Sweden]. Preoccupied.
1829 *Pachymerus sabulosus* Schilling, Beitr. Ent., 1: 81. [Germany].
1918 *Stygnocoris pedestris*: Barber, J. N.Y. Ent. Soc., 24: 104.
1964 *Stygnocoris sabulosus*: Slater, Cat. Lygaeidae World, 2: 1025.
Distribution: B.C., Me., N.S., N.Y., Nfld., Ore., Que. (Europe, North Africa).
Ref.: Slater, 1964a, 2: 1025; Butler [E.A.], 1923 (as *S. pedestris*); Sweet, 1964b (as *S. pedestris*); Slater, 1986 (zoogeogr.).

Tribe Udeocorini Sweet, 1967

Ref.: Slater, 1964a, 2: 1057; Sweet, 1977b (zoogeography).

Genus *Neosuris* Barber, 1924

1924 *Neosuris* Barber, J. N.Y. Ent. Soc., 32: 133. Type-species: *Esuris castanea* Barber, 1911. Monotypic.
Ref.: Slater, 1964a, 2: 1106; Sweet, 1977b; Brailovsky, 1978c (key spp.).

Neosuris castanea (Barber), 1911
1911 *Esuris castanea* Barber, J. N.Y. Ent. Soc., 19: 27. [Ariz.].
1924 *Neosuris castanea*: Barber, J. N.Y. Ent. Soc., 32: 133.
1948 *Neosuris castanea fraterna* Barber, Psyche, 55: 86. [Id.]. Synonymized by Ashlock, 1977, Proc. Ent. Soc. Wash., 79: 580.
Distribution: Ariz., Cal., Col., Id., Tex., Ut. (Mexico).
Ref.: Slater, 1964a, 2: 1106; Sweet, 1977b.

Neosuris fulgida (Barber), 1918
1918 *Esuris fulgidus* Barber, J. N.Y. Ent. Soc., 26: 51. [Ariz.].
1946 *Neosuris fulgidus*: Torre-Bueno, Ent. Am., 26: 90.
1973 *Neosuris fulgida*: Steyskal, Proc. Ent. Soc. Wash., 75: 278.
Distribution: Ariz., Cal.
Ref.: Slater, 1964a, 2: 1107.

Genus *Tempyra* Stål, 1874

1874 *Tempyra* Stål, K. Svens. Vet.-Akad. Handl., 12(1): 154. Type-species: *Tempyra biguttula* Stål, 1874. Monotypic.
Ref.: Slater, 1964a, 2: 1063.

Tempyra biguttula Stål, 1874
1874 *Tempyra biguttula* Stål, K. Svens. Vet.-Akad. Handl., 12(1): 157. [Tex.]. Lectotype designated by Scudder, 1977, Ent. Scand., 8: 30.
1944 *Tempyra biguttata* [*sic*]: Froeschner, Am. Midl. Nat., 31: 664.
Distribution: Ks., Md., Mo., Tex. (Hawaii).
Ref.: Slater, 1964a, 2: 1064.

Tempyra testacea Barber, 1949
1949 *Tempyra testacea* Barber, Pan-Pac. Ent., 24: 203. [Ariz.].
Distribution: Ariz., Cal.
Ref.: Slater, 1964a, 2: 1064.

Family Macroveliidae McKinstry, 1942

Macroveliid Water Bugs

By Richard C. Froeschner

The family Macroveliidae, erected for the genus *Macrovelia* Uhler originally described in the family Veliidae, differs from the true water striders in having the tarsal claws inserted apically, rather than subapically, on the last tarsal segment. With few exceptions the family status has been accepted since its proposal.

Members frequent damp habitats, usually along moving waters. Although capable of walking on the surface of open water, they apparently seldom venture from protective vegetation. They are carnivores or scavengers feeding on the body fluids of other arthropods, mostly insects. Polhemus and Chapman (1979, Bull. Cal. Ins. Surv., 21: 46) reported that in aquaria eggs were glued to moss on exposed rocks.

Only two of the three known species of this New World family occur in our region. Polhemus and Chapman (1979, 21: 47-48) keyed and illustrated the U.S. species. Andersen (1982, Entomonograph, 3: 96-106) discussed the morphology and classification of this family.

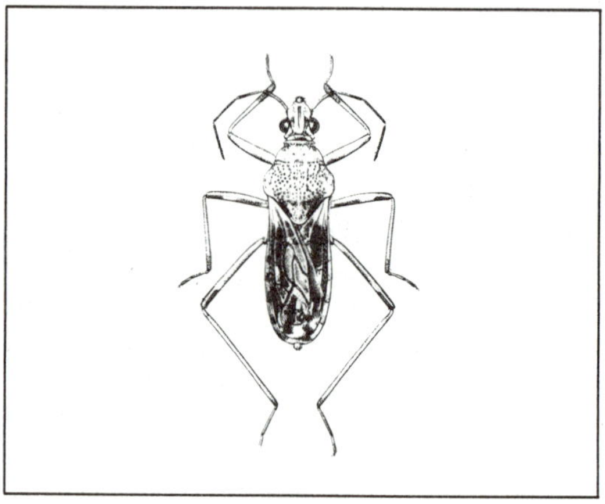

Fig. 60 *Macrovelia hornii* [p. 246] (After Usinger, 1956).

Genus *Macrovelia* Uhler, 1872

1872 *Macrovelia* Uhler, Prelim. Rept. U.S. Geol. Surv. Mont., p. 422. Type-species: *Macrovelia hornii* Uhler, 1872. Monotypic.

Macrovelia hornii Uhler, 1872 [Fig. 60]

1872 *Macrovelia Hornii* [*sic*] Uhler, Prelim. Rept. U.S. Geol. Surv. Mont., p. 422. [Ariz., Cal., N.M.]. Lectotype from N.M. designated by Drake and Chapman, 1963, Proc. Biol. Soc. Wash., 76: 234.

1909 *Macrovelia horni* [*sic*]: Kirkaldy and Torre-Bueno, Proc. Ent. Soc. Wash., 10: 207.

Distribution: Ariz., Cal., Col., N.D., N.M., Neb., Ore., S.D., Ut. (Mexico).

Genus *Oravelia* Drake and Chapman, 1963

1963 *Oravelia* Drake and Chapman, Proc. Biol. Soc. Wash., 76: 229. Type-species: *Oravelia pege* Drake and Chapman, 1963. Monotypic.

Oravelia pege Drake and Chapman, 1963

1963 *Oravelia pege* Drake and Chapman, Proc. Biol. Soc. Wash., 76: 231. [Cal.].

Distribution: Cal.

Family Mesoveliidae Douglas and Scott, 1867

The Water Treaders

By Cecil L. Smith

The Mesoveliidae (water treaders or pondweed bugs) are a family of small to medium-sized (3-6 mm), usually green or brownish bugs. Worldwide, there are only 9 genera containing 32 species, of which 22, including the 3 North American species, are in the cosmopolitan genus *Mesovelia* Mulsant and Rey. Two of our species are widely distributed; *Mesovelia amoena* Uhler and *M. mulsanti* White have been recorded from northern North America to the southern Neotropical region.

Considering the relative morphological homogeneity of the family, the habitat preferences of mesoveliids are surprisingly diverse. They are found on the open water of still or slowly flowing ponds, lakes, etc., covered with emergent or floating vegetation (preferred by many *Mesovelia* species) to the leaf-litter layer of forest floors far removed from bodies of water (Štys, 1976, Act. Ent. Bohem., 7: 388-403). A few species are hygropetric, living on watersoaked moss or seeping rock faces, while other cavernicolous species are found on the moist slime fungus covering the walls of lava tubes (Gagné and Howarth, 1975, Pac. Ins., 16: 399-413) or moist walls of coastal caves (Yuasa, 1929, An. Mag. Nat. Hist., 10: 346-349). Polhemus (1975, Pan-Pac. Ent., 51: 243-347) recorded the Mexican species *Speovelia mexicana* Polhemus from protected pools in the intertidal area at low tide.

Mesoveliids are predators and scavengers feeding on a variety of microarthropods and microcrustaceans at the water surface or on floating vegetation. Hungerford (1917, Psyche, 24: 73-84) records ostracods being attacked, while Brooks and Kelton (1967, Mem. Ent. Soc. Can., 51: 1-92) report observations of mosquito larvae and pupae as food sources.

Macropterous specimens, although considerably rarer than apterous individuals, have been recorded for all North American species with the exception of *M. cryptophila* Hungerford. The absence of wings is accompanied by reduction or loss of the ocelli, and simplification of the thoracic dorsum (Polhemus and Chapman, 1979, Bull. Cal. Ins. Surv., 21: 39-42). Thoracic polymorphism was studied by Galbreath (1975, Univ. Ks. Sci. Bull., 50: 459-482).

The majority of species and genera have been described subsequent to Horvath's (1915, An. Mus. Nat. Hung., 8: 535-556; 1929, Gen. Cat. Hem., 15 pp.) treatment of the world fauna. Particularly relevant to the nearctic

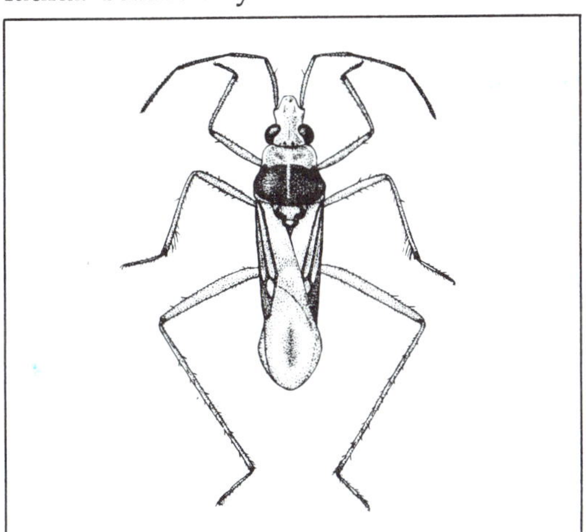

Fig. 61 *Mesovelia mulsanti* [p. 248] (After Froeschner, 1949).

species are Jaczewski's (1930, An. Mus. Zool. Pol., 9: 3-12) key to the American species, and Drake's (1948, Bol. Ent. Venez., 7: 145-147) checklist of American species. The latest major taxonomic work and review of the family with a world checklist of taxa, synonymies, etc. is by Andersen and Polhemus (1980, Ent. Scand., 11: 369-392). Excellent reviews of the literature pertaining to the biology, morphology and taxonomy of the North American species are provided by Polhemus and Chapman (1979, Bull. Cal. Ins. Surv., 21: 39-42) and Bennett and Cook (1981, Univ. Minn. Agr. Exp. Stn. Tech. Bull., 332: 3-58).

Subfamily Mesoveliinae Douglas and Scott, 1867

Genus *Mesovelia* Mulsant and Rey, 1852

1852 *Mesovelia* Mulsant and Rey, An. Soc. Linn. Lyon, p. 138. Type-species: *Mesovelia furcata* Mulsant and Rey, 1852. Monotypic.

Note: Andersen and Polhemus (1980, Ent. Scand., 11: 369-392) provided a checklist of the world's species.

Mesovelia amoena Uhler, 1894

1894 *Mesovelia amoena* Uhler, Proc. Zool. Soc. Lond., p. 218. [Grenada].

1924 *Mesovelia douglasensis* Hungerford, Can. Ent., 56: 142. [Mich.]. Tentatively synonymized by Jaczewski, 1930, An. Mus. Zool. Pol., 9: 4; synonymized by Andersen and Polhemus, 1980, Ent. Scand., 11: 389.

1926 *Mesovelia amoena*: Blatchley, Het. E. N. Am., p. 615.

Distribution: Cal., Fla., Ga., Ks., La., Mich., Miss., N.J., Nev., Tex. (Mexico to South America, West Indies, Hawaiian Is.).

Note: Hoffmann (1932, Can. Ent., 64: 118-130) studied the biology and described the immature stages.

Mesovelia cryptophila Hungerford, 1924

1924 *Mesovelia cryptophila* Hungerford, An. Ent. Soc. Am., 17: 454. [Mich.].

Distribution: Fla., Ga., Ia., Mich., Miss., N.J., Ok.

Note: Hoffmann (1932, Can. Ent., 64: 115-118) studied the biology and described the immature stage.

Mesovelia mulsanti White, 1879 [Fig. 61]

1879 *Mesovelia mulsanti* White, Trans. Ent. Soc. London, p. 268. [Brazil].

1884 *Mesovelia bisignata* Uhler, Stand. Nat. Hist., 2: 273. [Eastern U.S., Cuba]. Synonymized by Champion, 1898, Biol. Centr.-Am., Rhyn., 2: 123.

1898 *Mesovelia mulsanti*: Champion, Biol. Centr.-Am., Rhyn., 2: 123.

1905 *Mesoretia* [*sic*] *bisignata*: Crevecoeur, Trans. Ks. Acad. Sci., 19: 234.

1930 *Mesovelia mulsanti bisignata*: Jaczewski, An. Mus. Zool. Pol., 9: 7. Synonymized by Neering, 1954, Univ. Ks. Sci. Bull., 36: 144.

Distribution: B.C., Cal., Conn., Fla., Ga., Ia., Ill., Ind., Ks., La., Man., Mass., Md., Mich., Miss., Mo., N.C., N.J., N.Y., Oh., S.C., Tex., Va. (Mexico to Argentina, West Indies, Hawaiian Is.).

Note: Various aspects of the biology and morphology of *Mesovelia mulsanti* were studied by Hungerford (1917, Psyche, 24: 73-84), Hoffmann (1932, Can. Ent., 64: 90-115), Neering (1954, Univ. Ks. Sci. Bull., 36: 125-148) and Galbreath (1973, J. Ks. Ent. Soc., 46: 224-233; 1975, Univ. Ks. Sci. Bull., 50: 459-482; 1976, J. Ks. Ent. Soc., 49: 27-31; 1977, Trans. Ill. Acad. Sci., 69: 91-99).

Family Microphysidae Dohrn, 1859

The Microphysids

By Thomas J. Henry

Only four genera and species of this small family are known from Canada and the United States. Microphysids are characterized by the the presence of ocelli, a distinct cuneus, a single closed cell on the hemelytral membrane, and two-segmented tarsi. Brachyptery is common in some genera.

Microphysids generally are found in association with mosses and lichens growing on various trees, shrubs, or low-growing plants like mats of heather, *Calluna vulgaris* (L.), intermixed with moss, clumps of grass and other vegetation. Most are thought to prey on small arthropods such as mites, psocids, psyllids, springtails, thrips, and fly larvae. Eggs are laid in moss, lichen, or crevices of bark, overwinter, and hatch in the spring.

Péricart's (1972, Faune L'Europ. Bassin Medit., 7: 313-362) revision of the European Microphysidae contains keys, descriptions of adults and immatures and numerous notes on biology, ecology, and hosts. Adults and nymphs from North America can be identified to family using Slater and Baranowski's (1978) key in "How To Know The True Bugs." Kelton (1981, Can. Ent., 113: 1125) gave a key to the two known Canadian species.

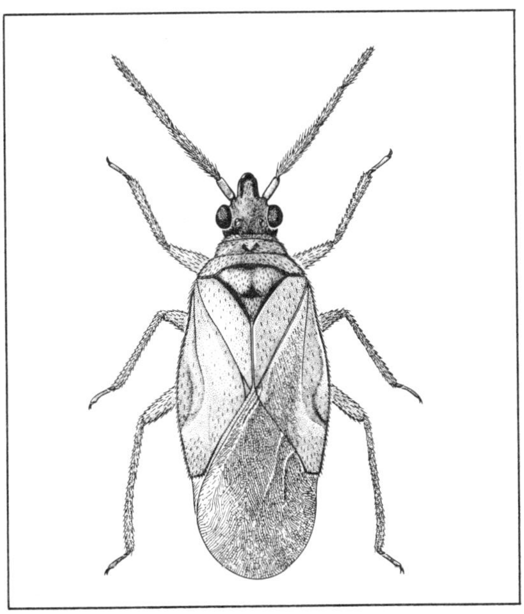

Figs. 62 *Loricula pselaphiformis* [p. 250] male (After Kelton, 1980).

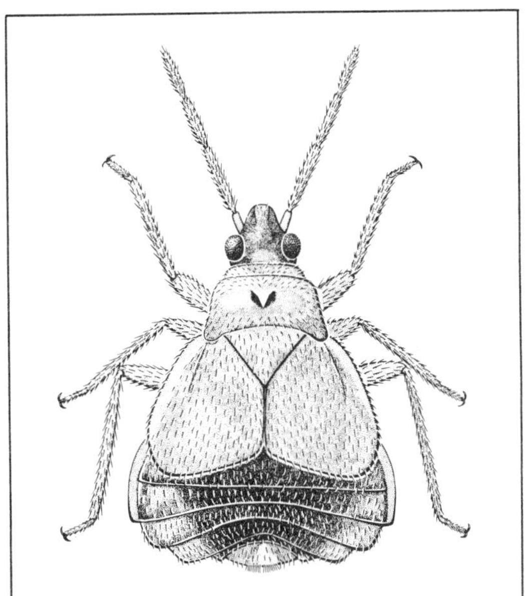

Fig. 63. *Loricula pselaphiformis* [p. 250] female (After Kelton, 1980).

Genus *Chinaola* Blatchley, 1928

1928 *Chinaola* Blatchley, Ent. News, 39: 87. Type-species: *Chinaola quercicola* Blatchley, 1928. Original designation.

Chinaola quercicola Blatchley, 1928

1928 *Chinaola quercicola* Blatchley, Ent. News, 39: 88. [Fla.].

Distribution: Fla.

Genus *Loricula* Curtis, 1833

1833 *Loricula* Curtis, Ent. Mag., 1: 197. Type-species: *Loricula pselaphiformis* Curtis, 1833. Monotypic.

Loricula pselaphiformis Curtis, 1833 [Figs. 62-63]

1833 *Loricula pselaphiformis* Curtis, Ent. Mag., 2: 198. [Britain].

1980 *Loricula pselaphiformis*: Kelton, Can. Ent., 112: 1085.

Distribution: N.S. (Asia, Europe, northern Africa).

Note: Péricart (1972, Faune L'Europ. Bassin Medit., 7: 319-321) illustrated the adult male and female and instars III to V and gave biological notes. Cobben (1968, Evol. Trends Het., p. 138) discussed a case of possible "haemocoelic fertilization" of eggs in this species. Kelton (1980, above) illustrated a macropterous male and brachypterous female and provided notes on hosts, habits, and general life history.

Genus *Mallochiola* Bergroth, 1925

1925 *Mallochiola* Bergroth, Bull. Brook. Ent. Soc., 20: 160. Type-species: *Idiotropus gagates* McAtee and Malloch, 1924. Monotypic.

Mallochiola gagates (McAtee and Malloch), 1924

1924 *Idiotropus gagates* McAtee and Malloch, Bull. Brook. Ent. Soc., 19: 71. [D.C., Md., Mexico].

1925 *Mallochiola gagates*: Bergroth, Bull. Brook. Ent. Soc., 20: 160.

1928 *Mallochiola* (*Idiotropis* [*sic*]) *gagates*: Blatchley, Ent. News, 39: 87.

Distribution: D.C., Md. (Mexico).

Note: The distribution of D.C., Md. and Tampico, Mexico suggests that this apparently rare species is much more widespread than records indicate or, perhaps, Mexico is a labeling error.

Genus *Myrmedobia* Baerensprung, 1857

1857 *Myrmedobia* Baerensprung, Berl. Ent. Zeit., 1: 161. Type-species: *Salda coleoptrata* Fallén, 1807. Designated by Kirkaldy, 1906, Trans. Am. Ent. Soc., 32: 122.

Myrmedobia exilis (Fallén), 1807

1807 *Lygaeus exilis* Fallén, Mon. Cimic. Suec., p. 73. [Europe].

1981 *Myrmedobia exilis*: Kelton, Can. Ent., 113: 1125.

Distribution: Nfld. (Asia, Europe)

Note: Kelton (1981, above) illustrated a macropterous adult male and brachypterous female and gave a summary of the life history and hosts. Seidenstücker (1950, Seidenberg. Biol., 31: 292) figured the fifth instar.

Family Miridae Hahn, 1833

(= Capsidae Burmeister, 1835)

The Plant Bugs

By Thomas J. Henry and A. G. Wheeler, Jr.

The family Miridae, commonly called plant bugs, is the largest of all heteropteran families, containing well over a third of the known species. It is estimated that the world fauna will approach 20,000 once the neotropical species are fully studied. In this catalog, we treat about 223 genera and 1,930 species from Canada and the continental United States. Owing to the great species diversity, wide range of feeding habits, and damage inflicted on fruits, food and field crops, and ornamental plants, the mirids are one of the most important of all insect groups.

Plant bugs range in size from about 1.5 mm for certain species of the spider web-inhabiting genus *Ranzovius* Distant or brachypterous forms of the garden fleahopper, *Halticus bractatus* (Say), to more than 15 mm in neotropical species of *Platytylus* Fieber. Many species are brightly colored as is expressed in the yellow and black fourlined plant bug, *Poecilocapsus lineatus* (Fabricius), the bright-red ash plant bug *Tropidosteptes cardinalis* Uhler, and the orange *Ceratocapsus aurantiacus* Henry, but most are less spectacular and, frequently, cryptically colored gray, black, or brown, blending in remarkably well with their surrounding habitat. Body surfaces may be dull, highly polished, smooth or heavily punctate. Vestiture or pubescence may be almost nonexistent as in species of the genera *Metriorrhynchomiris* Kirkaldy, *Proba* Distant, and *Tropidosteptes* Uhler or thick and woolly in others. Setae may be simple, silky or sericeous, or flattened and scalelike. Species of *Clivinema* Reuter have a thick covering of silvery sericeous pubescence and those of *Brooksetta* Kelton have numerous black, flattened, scalelike setae.

Mirids generally are fragile, delicate bugs, characterized by 4-segmented antennae, 4-segmented rostrum, 2 or 3-segmented tarsi, a single dorsal-abdominal scent gland, lack of ocelli (except in the subfamily Isometopinae), asymmetrical male parameres, hemelytral membrane with two closed cells, and a distinct cuneus at the apex of the corium in macropterous forms. Color, types of pubescence, and lengths, ratios, and proportions of antennal and rostral segments in relation to other body structures are important characters in diagnosing species. Male genitalia are necessary to identify many genera and species, especially those in the subfamilies Orthotylinae and Mirinae. Parameres, often complex and bizarre in shape, are most easily studied, but internal structures also reveal many specific distinctions (Kelton, 1959, Can. Ent., 91 (Suppl. 11): 1-72). Female structures, though less used than male genitalia, offer many characters that are useful in classifica-

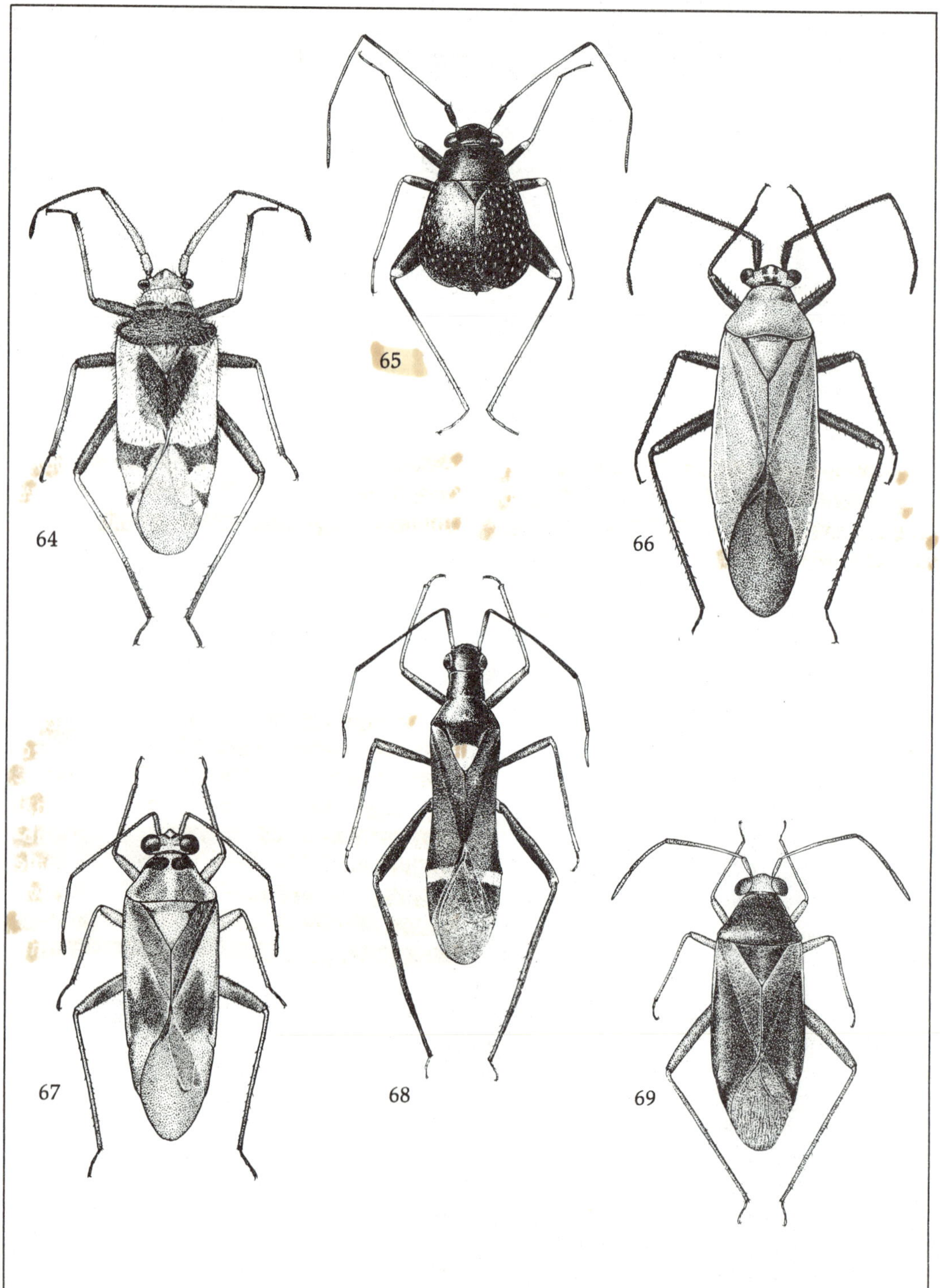

Figs. 64-69: 64, *Semium hirtum* [p.499]; 65, *Halticus bractatus* [p. 401]; 66, *Lopidea media* [p. 422]; 67, *Orthotylus ornatus* [p. 434]; 68, *Pseudoxenetus regalis* [p. 445]; 69, *Ceratocapsus modestus* [p. 396] (After Froeschner, 1949).

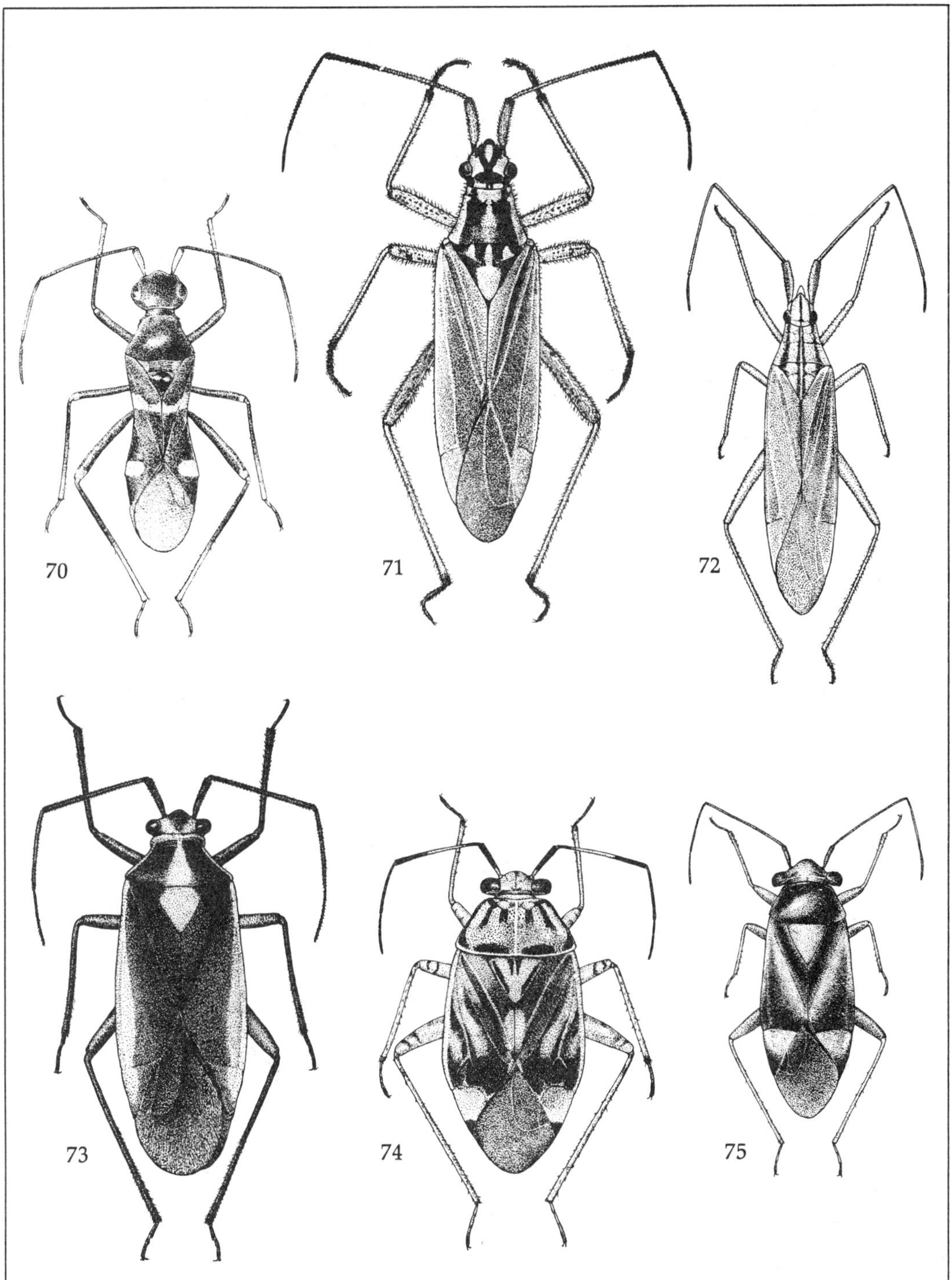

Figs. 70-75: 70, *Cyrtopeltocoris illini* [p. 455]; 71, *Leptopterna dolabrata* [p. 384]; 72, *Trigonotylus pulcher* [p. 390]; 73, *Prepops circumcinctus* [p.377]; 74, *Lygus lineolaris* [p. 323]; 75, *Bolteria luteifrons* [p. 297] (After Froeschner, 1949).

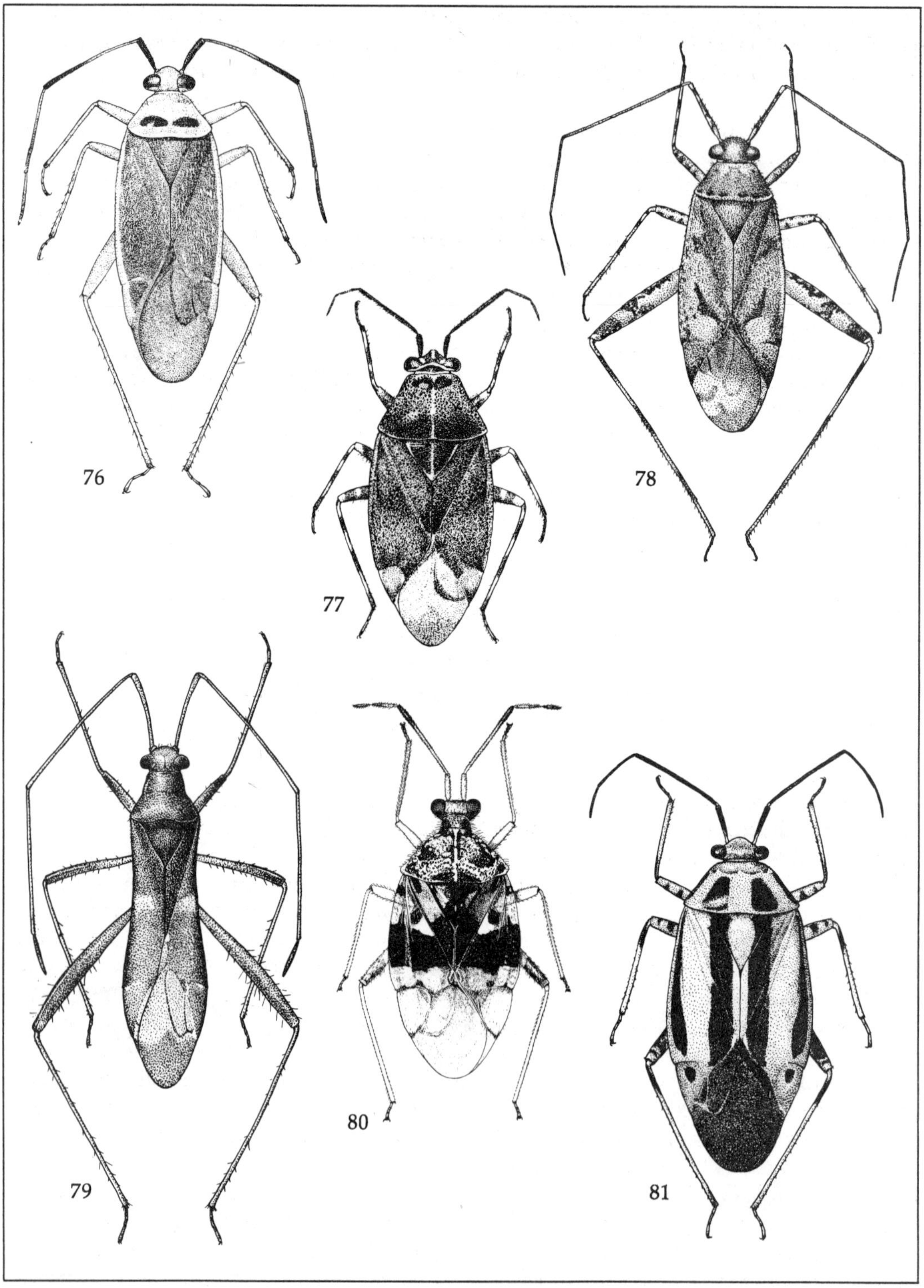

Figs. 76-81: 76, *Adelphocoris rapidus* [p. 295]; 77, *Deraeocoris poecilus* [p. 283]; 78, *Phytocoris canadensis* [p. 336]; 79, *Paraxenetus guttulatus* [p. 295]; 80, *Stethoconus japonicus* [p. 289]; 81, *Poecilocapsus lineatus* [p. 354] (All after Froeschner, 1949, except fig. 80, after Henry et al., 1986).

tion schemes, frequently even at the species level (Slater, 1950, Ia. St. Coll. J. Sci., 25: 1-81). Pretarsal structures aid in recognizing higher taxa (Schuh, 1976, Am. Mus. Nov., 2585: 1-26). The parempodia, arising between the claws, may be fleshy and apically convergent in the Orthotylinae, fleshy and apically divergent in the Mirinae, and simple and hairlike in most remaining subfamilies. Trends in number and pattern of femoral trichobothria, or sensory hairs, give additional information for interpreting relationships in higher categories (Schuh, 1976, Am. Mus. Nov., 2601: 1-39).

A large percent of the Miridae are phytophagous, attacking flowers, fruits, leaves, or stems. Feeding causes a wide array of damage symptoms. Leaf and stem feeders remove chlorophyll and plant juices, causing heavy mottling, withering and, frequently, lesions, all of which may significantly reduce plant vigor. Heavy infestations of certain species, like the honeylocust plant bug, *Diaphnocoris chlorionis* (Say), may cause defoliation of entire trees. The phlox plant bug, *Lopidea davisi* Knight, produces extensive chlorosis and death of foliage on phlox in the midwestern states. Species of the onion plant bug genus *Lindbergocapsus* Wagner have destroyed hundreds of acres of commercial onions. The garden fleahopper and *Spanagonicus albofasciatus* (Reuter) are serious pests of legumes and garden crops. The ash plant bugs, all belonging to the genus *Tropidosteptes*, heavily discolor and often crinkle ash foliage.

Species preferring reproductive structures may produce premature withering and petal drop on flowers or cause abortion of fruiting parts. *Lygus elisus* Van Duzee and *Lygus hesperus* Knight seriously hinder alfalfa seed production in the western states. The cotton fleahopper, *Pseudatomoscelis seriatus* (Reuter), one of the most important cotton pests, causes early abortion of cotton squares. *Leptopterna dolabrata* (Linnaeus) and *Amblytylus nasutus* (Kirschbaum) attack florets of Kentucky bluegrass and other grasses, severely reducing seed yield. *Neurocolpus nubilus* (Say) produces brown stippling and premature withering of the stamens on tulip tree flowers.

Many plant bugs are important predators of various arthropods like aphids, mealybugs, mites, phylloxerans, and scales. *Corticoris signatus* (Heidemann) and *Myiomma cixiiforme* (Uhler) may significantly reduce obscure scale populations on oak trees. *Ceratocapsus modestus* (Uhler) may be important in control of leaf-gall forming grape phylloxera. Some species, like the introduced *Campyloneura virgula* (Herrich-Schaeffer) in the Pacific Northwest or the native *Deraeocoris nebulosus* (Uhler) in eastern North America, are general predators attacking aphids, lace bugs, mites, and other arthropod pests. Most species of *Phytocoris* Fallén and *Deraeocoris* Kirschbaum are predatory, often stalking prey on the trunks and large branches of trees.

Scavenging habits are widespread in the family, with even the chiefly phytophagous species occasionally obtaining nutrients from other sources. Nearly all species will feed on dead or dying and parasitized individuals. The ubiquitous tarnished plant bug, *Lygus lineolarus* (Palisot), is recorded feeding on mummified or parasitized aphids. Most other species will prey on molting arthropods, including those of their own species. Species of *Dicyphus* Fieber living on glandular-hairy plants will feed on tiny insects entrapped on the sticky stems and leaves. The spider commensal *Ranzovius clavicornis* (Knight) feeds almost exclusively on cadavers of prey caught in the webs of the theridiid *Anelosimus studiosus* (Keyserling). Other mirids may take honeydew, carrion, excrement, and even vertebrate blood.

An overview of the major literature for the family must start with the works of Harry H. Knight, who has influenced the study of North American Miridae more than any other individual. His "Miridae of Connecticut" (1923), "Miridae of Illinois"

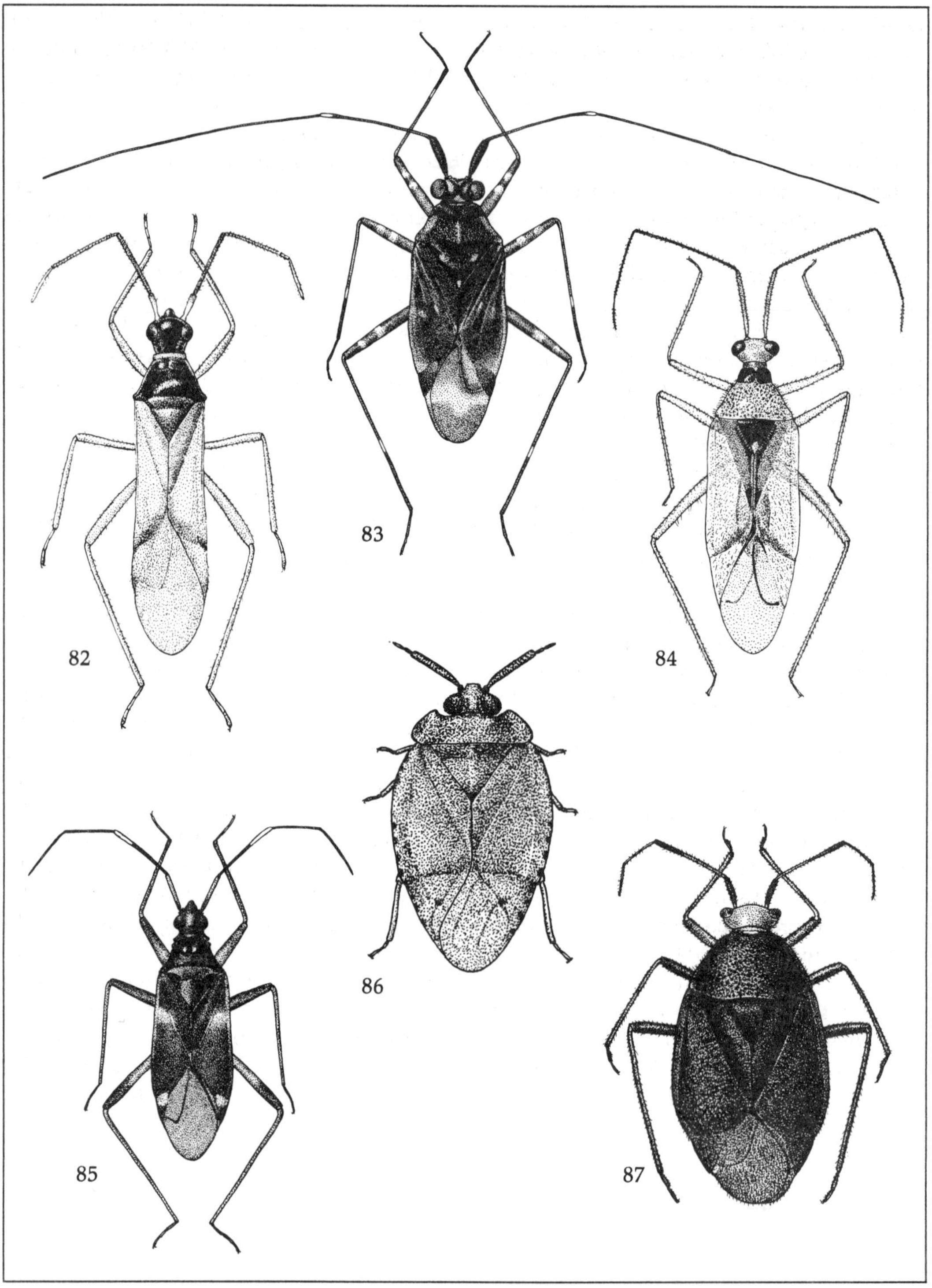

Figs. 82-87: 82, *Dicyphus agilis* [p. 262]; 83, *Cylapus tenuicornis* [p. 271]; 84, *Hyaliodes harti* [p. 289]; 85, *Fulvius imbecilis* [p. 272]; 86, *Diphleps unica* [p. 291]; 87, *Bothynotus modestus* [p. 273] (After Froeschner, 1949).

(1941), and the "Miridae of the Nevada Test Site and the Western United States" (1968) are outstanding regional taxonomic treatments necessary to anyone seriously interested in the family. Blatchley's "Heteroptera of Eastern North America" (1926) remains a useful standard for work on mirids (pp. 660-964), as well as for other bugs in the East. Kelton's "Plant Bugs of the Prairie Provinces" (1980) is the first comprehensive treatment of the mirids of central Canada and northcentral United States. Original keys to genera and species, habitus and genitalic illustrations for many taxa, and host-plant information make this a required reference. Slater and Baranowski's "How to Know the True Bugs" (1978) provides the only recent key to all the known North American genera. Their easy-to-use keys and numerous illustrations provide a good starting point for studying North American Miridae. Ultimately, any mirid worker needs to have available Carvalho's "Catalogue of the Miridae of the World" (1957-1960). This masterful work has done more than any other to promote the study of Miridae in the world today. Although the literature coverage stopped more than 30 years ago (1955), it remains an essential compendium of world biological and taxonomic literature.

In addition to the major taxonomic publications, there are several extensive state or regional mirid lists that provide useful distributional and host information. Some of these are: Henry and Smith (1979, J. Ga. Ent. Soc., 14: 212-220) for Georgia; Snodgrass et al. (1984, Proc. Ent. Soc. Wash., 86: 845-860) for the Mississippi Delta Region; and Wheeler et al. (1983, Trans. Am. Ent. Soc., 109: 127-159) for West Virginia. Braimah et al. (1982, Nat. Can., 109: 153-180) and Kelton (1983, Agr. Can. Monogr., 24: 1-201) reviewed the mirid fauna of apple and fruit crops in Canada.

Acknowledgments.-- Many individuals provided encouragement throughout the compilation of this catalog. We thank the late J. L. Herring (retired, Systematic Entomology Laboratory, IIBIII, Agricultural Research Service, USDA, c/o U. S. National Museum of Natural History, Washington, D.C.), who first invited us to coauthor the mirid fascicle in 1977. Unfortunately, his retirement and move from the Washington area in 1979 prevented his continued participation in the project [Dr. Herring died 25 Feb. 1985; see Henry and Froeschner, 1987, Proc. Ent. Soc. Wash., 89: 384-388)]. J. A. Slater (Department of Ecology and Evolutionary Biology, University of Connecticut, Storrs) generously allowed us to consult his unpublished checklist of the Miridae (for years through 1977). R. C. Froeschner (Department of Entomology, U. S. National Museum of Natural History, Washington, D.C.) provided much advice on nomenclatural problems, , leads to finding or citing difficult literature, and spent considerable time reviewing this fascicle. L. A. Kelton (Biosystematics Research Institute, Agriculture Canada, Ottawa), J. C. Schaffner (Department of Entomology, Texas A & M. University, College Station), R. T. Schuh (Department of Entomology, American Museum of Natural History, New York, N.Y.), M. D. Schwartz (AMNH), and G. L. Stonedahl (AMNH) also kindly read the manuscript and offered suggestions for its improvement.

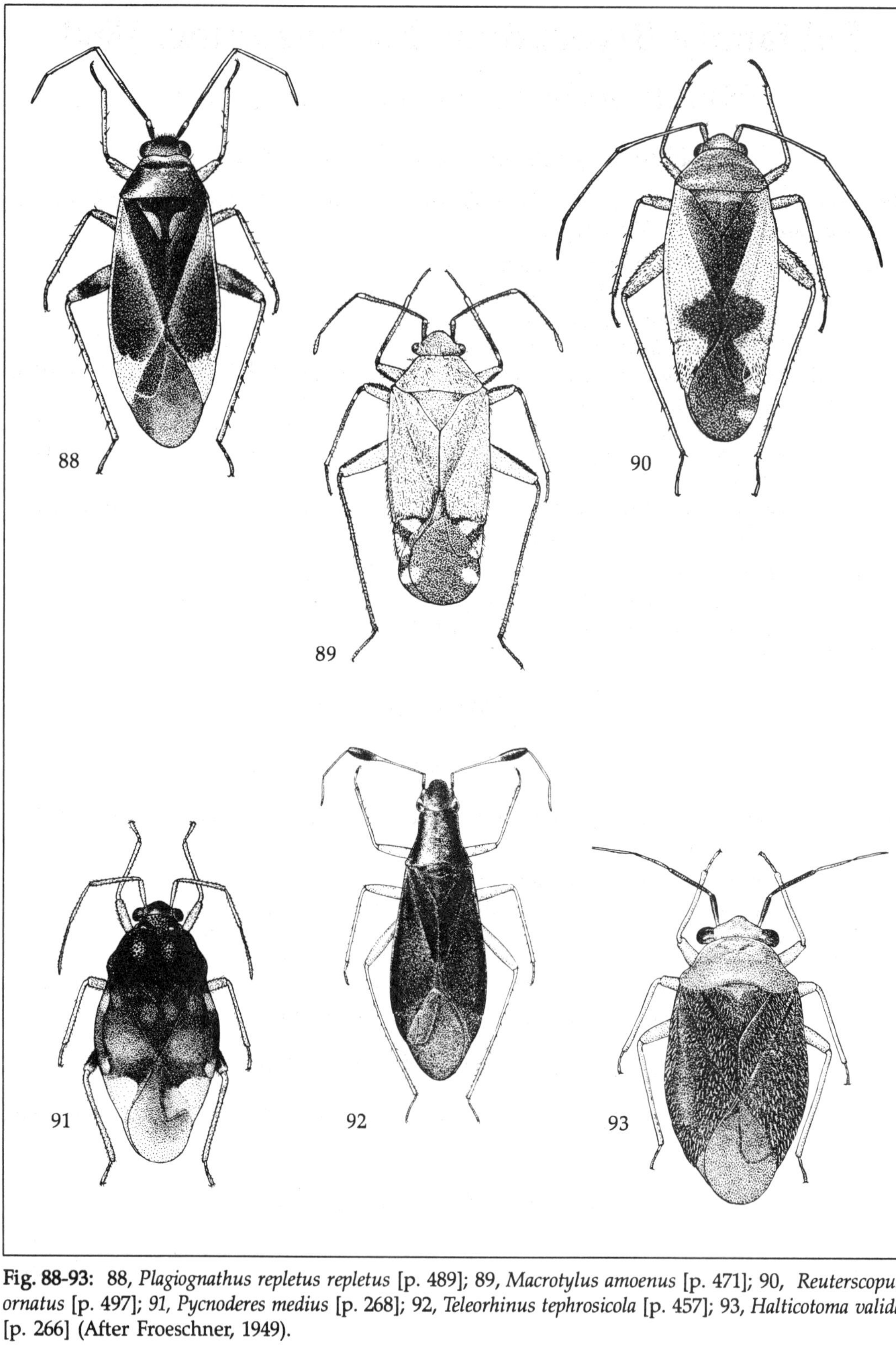

Fig. 88-93: 88, *Plagiognathus repletus repletus* [p. 489]; 89, *Macrotylus amoenus* [p. 471]; 90, *Reuterscopus ornatus* [p. 497]; 91, *Pycnoderes medius* [p. 268]; 92, *Teleorhinus tephrosicola* [p. 457]; 93, *Halticotoma valida* [p. 266] (After Froeschner, 1949).

Subfamily Bryocorinae Baerensprung, 1860

Tribe Bryocorini Baerensprung, 1860

Genus *Monalocoris* Dahlbom 1851

1851 *Monalocoris* Dahlbom, Öfv. K. Svens Vet.-Akad. Förh., p. 209. Type-species: *Cimex filicis* Linnaeus, 1758. Monotypic.

Monalocoris americanus Wagner and Slater, 1952

1876 *Monalocoris filicis*: Uhler, Bull. U.S. Geol. Geogr. Surv. Terr., 1: 315.
1907 *Monolocoris* [*sic*] *filicis*: Moore, Can. Ent., 39: 163.
1949 *Monalocoris filisis* [*sic*]: Hoffmann et al., Ecol. Monogr., 19: 19.
1952 *Monalocoris americanus* Wagner and Slater, Proc. Ent. Soc. Wash., 54: 279. [Minn.].
1984 *Monacoloris* [*sic*] *americanus*: Larochelle, Fabreries, Suppl. 3, p. 315.
Distribution: Alk., Alta., Col., Conn., Fla., Ia., Ill., Ind., Man., Mass., Md., Me., Minn., N.C., N.H., N.J., N.S., N.Y., Neb., Pa., Que., Sask., Tenn., Tex., Vt., W.Va., Wis. (Cuba).
Note: Wagner and Slater (1952, above) credit all North American records of palearctic *M. filicis* to *M. americanus*.

Monalocoris eminulus (Distant), 1893

1893 *Carmelus eminulus* Distant, Biol. Centr.-Am., Rhyn., 1: 445. [Mexico].
1952 *Monalocoris eminulus*: Carvalho, Bol. Mus. Nac., 118: 6.
Distribution: Fla. (Mexico).

Tribe Dicyphini Reuter, 1883

Genus *Campyloneura* Fieber, 1860

1860 *Campyloneura* Fieber, Europ. Hem., p. 67. Type-species: *Capsus virgula* Herrich-Schaeffer, 1836. Monotypic.

Campyloneura virgula (Herrich-Schaeffer), 1836

1836 *Capsus virgula* Herrich-Schaeffer, Wanz. Ins., 3: 51. [Europe].
1957 *Campyloneura virgula*: Downes, Proc. Ent. Soc. B.C., 54: 11.
Distribution: B.C., Cal., Ore., Wash. (Europe, Asia Minor, Turkestan).
Note: Lattin and Stonedahl (1984, Pan-Pac. Ent., 60: 4-7) gave additional host and distributional records and summarized the literature on this species.

Genus *Cyrtopeltis* Fieber, 1860

1860 *Cyrtopeltis* Fieber, Europ. Hem., p. 76. Type-species: *Cyrtopeltis geniculatus* Fieber, 1860. Monotypic.

Subgenus *Engytatus* Reuter, 1876

1876 *Engytatus* Reuter, Öfv. K. Svens. Vet.-Akad. Förh., 32(9): 82. Type-species: *Engytatus geniculatus* Reuter, 1876. Preoccupied. Next available name *Neosilia modesta* Distant, 1893. Monotypic.

Cyrtopeltis modesta (Distant), 1893

1876 *Engytatus geniculatus* Reuter, Öfv. K. Svens. Vet.-Akad. Förh., 32(9): 83. [Tex.]. Preoccupied.

1893 *Neosilia modesta* Distant, Biol. Centr.-Am., Rhyn., 1: 447. [Guatemala].

1909 *Cyrtopeltis varians*: Reuter, Acta Soc. Sci. Fenn., 36(2): 62.

1910 *Engytatus varians*: Reuter, Acta Soc. Sci. Fenn., 37(3): 151.

1954 *Cyrtopeltis (Engytatus) modesta*: Carvalho and Hussey, Occas. Pap. Mus. Zool., Univ. Mich., 552: 6.

1961 *Cyrtopeltis modestus* [*sic*]: Bibby, J. Econ. Ent., 54: 329.

Distribution: Ariz., Cal., Fla., Ga., Miss., Mo., N.C., S.C., Tex. (Hawaii, Mexico to South America, West Indies).

Note: Tanada and Holdaway (1954, Univ. Haw. Agr. Exp. Stn. Tech. Bull., 24: 1-40) studied *C. modesta* as a pest of tomato, *Lycopersicon esculentum* Mill., and summarized the economic literature.

Subgenus *Nesidiocoris* Kirkaldy, 1902

1902 *Nesidiocoris* Kirkaldy, Trans. Ent. Soc. London, p. 247. Type-species: *Nesidiocoris volucer* Kirkaldy, 1902, a junior synonym of *Cyrtopeltis tenuis* Reuter, 1895. Original designation.

Note: The International Commission on Zoological Nomenclature (1972, Bull. Zool. Nomen., 28: 32-33, Opinion 958) placed the economically important *C. tenuis* on the *Official List of Names in Zoology* to conserve it over senior synonym *Dicyphus tamaricis* Puton, 1886. Although *Nesidiocoris* generally has been kept as a subgenus of *Cyrtopeltis* by taxonomists, applied entomologists have recently given it generic status. To avoid confusion in the genus, we retain *Nesidiocoris* as a subgenus until a formal generic analysis can be made.

Cyrtopeltis tenuis Reuter, 1895

1895 *Cyrtopeltis tenuis* Reuter, Rev. d'Ent., 14: 139. [Madeira].

1909 *Cyrtopeltis tenuis*: Van Duzee, Bull. Buff. Soc. Nat. Sci., 9: 182.

1916 *Engytatus tenuis*: Van Duzee, Check List Hem., p. 43.

1952 *Cyrtopeltis (Nesidiocoris) tenuis*: China and Carvalho, An. Mag. Nat. Hist., (12)5: 160.

Distribution: Fla. (West Indies; Circumtropical).

Subgenus *Tupiocoris* China and Carvalho, 1952

1952 *Tupiocoris* China and Carvalho, An. Mag. Nat. Hist., (12)5: 162. Type-species: *Neoproba notata* Distant, 1893. Original designation.

Cyrtopeltis bakeri Knight, 1943

1943 *Cyrtopeltis bakeri* Knight, Pan-Pac. Ent., 19: 58. [Wash.].

1952 *Cyrtopeltis (Tupiocoris) bakeri*: China and Carvalho, An. Mag. Nat. Hist., (12)5: 165.

1952 *Cyrtopeltis (Tupiocoris) notatus*: China and Carvalho, An. Mag. Nat. Hist., (12)5: 162.

Distribution: B.C., Ore., Wash.

Cyrtopeltis melanocephala Reuter, 1909

1909 *Cyrtopeltis melanocephalus* Reuter, Acta Soc. Sci. Fenn., 36(2): 63. [Tex.].

1916 *Engytatus melanocephalus*: Van Duzee, Check List Hem., p. 43.

1952 *Cyrtopeltis (Tupiocoris) melanocephalus*: China and Carvalho, An. Mag. Nat. Hist., (12)5: 166.

1973 *Cyrtopeltis (Tupiocoris) melanocephala*: Steyskal, Stud. Ent., 16: 207.

Distribution: Ok., Tex.

Cyrtopeltis notata (Distant), 1893

1893 *Neoproba notata* Distant, Biol. Centr.-Am., Rhyn., 1: 432. [Panama].

1893 *Dicyphus secundus* Riley, N. Am. Fauna, 7: 250. *Nomen nudum.*

1898 *Dicyphus minimus* Quaintance, Fla. Agr. Exp. Stn. Bull., 48: 167 (validated Uhler ms. name). [Fla.]. Synonymized by Costa Lima, 1936, Terc. Cat. Ins. Brasil, p. 117.

1899 *Dicyphus minimus* Uhler, Ent. News, 10: 59. [Fla.]. Synonymized by Costa Lima 1936, Terc. Cat. Ins. Brasil, p. 117.

1914 *Dicyphus minutus*: Van Duzee, Trans. San Diego Soc. Nat. Hist., 2: 25. *Lapsus.*

1973 *Cyrtopeltis (Tupiocoris) notata*: Steyskal, Stud. Ent., 16: 207.

Distribution: Ariz., Cal., Col., D.C., Fla., Miss., N.M., Tex. (Mexico to South America, West Indies).

Note: Johnston (1930, J. Econ. Ent., 23: 642) discussed this species (as *D. minimus*) as a pest of tomato, *Lycopersicon esculentum* Mill.

Subgenus *Usingerella* China and Carvalho, 1952

1952 *Usingerella* China and Carvalho, An. Mag. Nat. Hist., (12)5: 159. Type-species: *Cyrtopeltis simplex* Reuter, 1909. Original designation.

Cyrtopeltis simplex Reuter, 1909

1909 *Cyrtopeltis simplex* Reuter, Acta Soc. Sci. Fenn., 36(2): 63. [Cal.].

1915 *Engytatus simplex*: LaFollett, J. Ent. Zool., 7: 128.

1952 *Cyrtopeltis (Usingerella) simplex*: China and Carvalho, An. Mag. Nat. Hist., (12)5: 165.

Distribution: Ariz., Cal.

Genus *Dicyphus* Fieber, 1858

1858 *Dicyphus* Fieber, Wien. Ent. Monat., 2: 326. Type-species: *Capsus pallidus* Herrich-Schaeffer, 1835. Designated by Kirkaldy, 1906, Trans. Am. Ent. Soc., 32: 129.

1865 *Idolocoris* Douglas and Scott, Brit. Hem., p. 374. Type-species: *Brachyceraea pallicornis* Fieber, 1861. Designated by Kirkaldy, 1906, Trans. Am. Ent. Soc., 32: 129. Synonymized by Poppius, 1914, Acta Soc. Sci. Fenn., 44(3): 141.

Note: McGavin (1982, Ent. Month. Mag., 118: 79-86) described the new genus *Neodicyphus* to accommodate *Dicyphus rhododendri* Dolling (type-species) and the species *D. californicus* (Stål) and *D. cucurbitaceus* (Spinola), but because he did not adequately evaluate the New World fauna of this generic complex, we are maintaining all species in *Dicyphus* until a comprehesive revision and placement of all the species is made.

Subgenus *Dicyphus* Fieber, 1858

1858 *Dicyphus* Fieber, Wien. Ent. Monat., 2: 326. Type-species: *Capsus pallidus* Herrich-Schaeffer, 1836. Designated by Kirkaldy, 1906, Trans. Am. Ent. Soc., 32: 129.

1951 *Dicyphus* (*Dicyphus*): Wagner, Soc. Sci. Fenn. Comm. Biol. 12(6): 8.

Dicyphus agilis (Uhler), 1877 [Fig. 82]

1877 *Idolocoris agilis* Uhler, Bull. U.S. Geol. Geogr. Surv. Terr., 3: 425. [Col.]. Lectotype designated by Kelton, 1980, Can. Ent., 112: 388.

1895 *Dicyphus californicus* var. *agilis*: Uhler, Bull. Col. Agr. Exp. Stn., 31: 46.

1898 *Dicyphus agilis*: Slosson, Ent. News, 9: 251.

1947 *Dicyphus cucurbitaceus*: Carvalho, Bol. Mus. Nac., 77: 20.

Distribution: Col.

Note: Kelton (1980, Can. Ent., 112: 387) clarified the status of *D. agilis* and indicated that North American records of *D. cucurbitaceus* (Spinola) are in error. All earlier records for this species in North America need reevaluation.

Dicyphus brachypterus Knight, 1943

1943 *Dicyphus brachypterus* Knight, Pan-Pac. Ent., 19: 53. [Wash.].

Distribution: Wash.

Dicyphus californicus (Stål), 1859

1859 *Capsus californicus* Stål, K. Svens. Freg. Eug. Resa Jorden, p. 259. [Cal.].

1873 *Cyllocoris californicus*: Walker, Cat. Hem. Brit. Mus., 6: 66.

1876 *Dicyphus californicus*: Reuter, Öfv. K. Svens. Vet.-Akad Förh., 32(9): 82.

1982 *Neodicyphus californicus*: McGavin, Ent. Month. Mag., 118: 79.

Distribution: Cal., Col. (Mexico).

Note: Wheeler and Henry (1977, Great Lakes Ent., 10: 150) indicated that eastern U.S. records (Ks., N.H., N.Y., Ont., and Pa.) for this species are in error.

Dicyphus confusus Kelton, 1980

1980 *Dicyphus confusus* Kelton, Can. Ent., 112: 389. [B.C.].

Distribution: Alta., B.C., Col., Id., Yuk.

Dicyphus crudus Van Duzee, 1916

1916 *Dicyphus crudus* Van Duzee, Univ. Cal. Publ. Ent., 1: 240. [Cal.].

Distribution: Cal.

Dicyphus diplaci Knight, 1968

1968 *Dicyphus diplaci* Knight, Brig. Young Univ. Sci. Bull., 9(3): 71. [Cal.].

Distribution: Cal.

Dicyphus discrepans Knight, 1923

1923 *Dicyphus discrepans* Knight, Conn. Geol. Nat. Hist. Surv. Bull., 34: 477. [N.Y.].

Distribution: Alta., B.C., Ind., Mich., Minn., N.B., N.D., N.H., N.Y., Ont., Ore., Que., Sask., Wash., Wis.

Dicyphus elongatus Van Duzee, 1917

1917 *Dicyphus elongatus* Van Duzee, Proc. Cal. Acad. Sci., (4)7: 269. [Cal.].

Distribution: B.C., Cal., Wash.

Dicyphus famelicus (Uhler), 1878

1869 *Phytocoris famelicus* Rathvon, Hist. Lancaster Co., Pa., p. 549. *Nomen nudum.*

1878 *Idolocoris famelicus* Uhler, Proc. Boston Soc. Nat. Hist., 19: 413. [N.H.].

1890 *Dicyphus famelicus*: Atkinson, J. Asiatic Soc. Bengal, 58(2): 128.

1918 *Dicyphus fameliculus* [*sic*]: Torre-Bueno, Can. Ent., 50: 25.
1928 *Dicyphus familicus* [*sic*]: Watson, Oh. Biol. Surv. Bull., 16: 37.
Distribution: Ill., Ind., Mass., Me., Mich., N.B., N.C., N.H., N.J., N.S., N.Y., Oh., Ont., Pa., Que., W.Va., Wis.

Dicyphus gracilentus Parshley, 1923
1923 *Dicyphus gracilentus* Parshley, Ent. News, 34: 21. [Ill.].
1926 *Dicyphus vestitus*: Blatchley, Het. E. N. Am., p. 910.
Distribution: Ill., Ind., Oh.

Dicyphus hesperus Knight, 1943
1943 *Dicyphus hesperus* Knight, Pan-Pac. Ent., 19: 56. [Id.].
Distribution: Alta., B.C., Cal., Col., Id., Ill., Man., Mich., Minn., Mont., N.B., N.D., Nev., Ont., Ore., Que., Sask., Ut., Wash.

Dicyphus paddocki Knight, 1968
1968 *Dicyphus paddocki* Knight, Brig. Young Univ. Sci. Bull., 9(3): 73. [Cal.].
Distribution: Cal.

Dicyphus pallicornis (Fieber), 1861
1861 *Brachyceraea pallicornis* Fieber, Europ. Hem., 2: 324. [Europe].
1957 *Dicyphus pallidicornis* [*sic*]: Downes, Proc. Ent. Soc. B.C., 54: 11.
1968 *Dicyphus pallicornis*: Knight, Brig. Young Univ. Sci. Bull., 9(3): 70.
Distribution: B.C., Cal., Ore., Wash. (Africa, Europe).
Note: Downes (1957, above) indicated that this species had been reported from Oregon, but we have been unable to find that published record.

Dicyphus phaceliae Knight, 1968
1968 *Dicyphus phaceliae* Knight, Brig. Young Univ. Sci. Bull., 9(3): 71. [Cal.].
Distribution: Cal.

Dicyphus rhododendri Dolling, 1972
1972 *Dicyphus rhododendri* Dolling, Ent. Month. Mag., 107: 244. [England].
1976 *Dicyphus rhododendri*: Henry and Wheeler, Proc. Ent. Soc. Wash., 78: 108.
1982 *Neodicyphus rhododendri*: McGavin, Ent. Month. Mag., 118: 79.
Distribution: Oh., Pa., Tenn., W.Va. (England).

Dicyphus ribesi Knight, 1968
1968 *Dicyphus ribesi* Knight, Brig. Young Univ. Sci. Bull., 9(3): 70. [Nev.].
Distribution: Nev.

Dicyphus rivalis Knight, 1943
1943 *Dicyphus rivalis* Knight, Pan-Pac. Ent., 19: 54. [Ore.].
Distribution: B.C., Cal., Ore.

Dicyphus rubi Knight, 1968
1923 *Dicyphus agilis*: Knight, Conn. Geol. Nat. Hist. Surv. Bull., 34: 477.
1968 *Dicyphus rubi* Knight, Brig. Young Univ. Sci. Bull., 9(3): 72. [N.Y.].
Distribution: B.C., Man., N.Y., Ont., Que., N.S., Sask., Wis.

Dicyphus rufescens Van Duzee, 1917
1917 *Dicyphus rufescens* Van Duzee, Proc. Cal. Acad. Sci., (4)7: 268. [Cal.].
Distribution: Cal.

Dicyphus similis Kelton, 1980
1980 *Dicyphus similis* Kelton, Can. Ent., 112: 389. [Alta.].

Distribution: Alta., B.C., N.B., N.H., N.S., N.Y., Nfld., Ont., Que., Sask.

Dicyphus stitti Knight, 1968

1968 *Dicyphus stitti* Knight, Brig. Young Univ. Sci. Bull., 9(3): 74. [Ariz.].
Distribution: Ariz.

Dicyphus tibialis Kelton, 1980

1980 *Dicyphus tibialis* Kelton, Can. Ent., 112: 390. [Alta.].
Distribution: Alta., B.C., Col., Mont.

Dicyphus tinctus Knight, 1943

1943 *Dicyphus tinctus* Knight, Pan-Pac. Ent., 19: 55. [Wash.].
Distribution: Wash.

Dicyphus usingeri Knight, 1943

1943 *Dicyphus usingeri* Knight, Pan-Pac. Ent., 19: 53. [Cal.].
Distribution: Cal.

Dicyphus vestitus Uhler, 1895

1895 *Dicyphus vestitus* Uhler, Bull. Col. Agr. Exp. Stn., 31: 46. [Col.].
1922 *Dicyphus notatus* Parshley, S.D. St. Coll. Tech. Bull., 2: 16. [S.D.]. Synonymized by Knight, 1927, Bull. Brook. Ent. Soc., 22: 104.
Distribution: B.C., Cal., Col., Id., Ill., Ind., Minn., N.H., N.M., N.Y., Oh., S.D.

Genus *Macrolophus* Fieber, 1858

1858 *Macrolophus* Fieber, Wien. Ent. Monat., 2: 326. Type-species: *Capsus nubilus* Herrich-Schaeffer, 1836. Designated by Kirkaldy, 1906, Trans. Am. Ent. Soc., 32: 129. See note below.

1923 *Tylocapsus* Van Duzee, Proc. Cal. Acad. Sci., (4)12: 151. Type-species: *Tylocapsus lopezi* Van Duzee, 1923. Original designation. Synonymized by Carvalho, 1955, Rev. Chil. Ent., 4: 224.

Notes: Knight (1926, Ent. News, 37: 313-316) provided a key to the North American species.

Apparently no one has recognized that *Capsus nubilus* Herrich-Schaeffer (1835, Faunae Ins. Germ. Deut., 135, 9), the type-species of *Macrolophus*, is a primary junior homonym of *Capsus nubilus* Say, 1832 [now in Nearctic genus *Neurocolpus* Reuter]. This widely distributed Palearctic species should take the next available name, which is *Phytocoris pygmaeus* Rambur (1839, Faune Ent. Andal., 2: 163), previously synonymized under *C. nubilus* by Reuter (1883, Hem. Gym. Europ., 3: 477). *Macrolophus pygmaeus* (Rambur), Revised Status and New Combination.

Macrolophus brevicornis Knight, 1926

1926 *Macrolophus brevicornis* Knight, Ent. News, 37: 315. [N.J.].
Distribution: Ia., Ill., Ks., Md., Mo., N.J.

Macrolophus lopezi (Van Duzee), 1923

1923 *Tylocapsus lopezi* Van Duzee, Proc. Cal. Acad. Sci., (4)12: 152. [Mexico].
1968 *Macrolophus lopezi*: Knight, Brig. Young Univ. Sci. Bull., 9(3): 75.
Distribution: Cal. (Mexico).

Macrolophus mimuli Knight, 1968

1968 *Macrolophus mimuli* Knight, Brig. Young Univ. Sci. Bull., 9(3): 75. [Ariz.].
Distribution: Ariz.

Macrolophus separatus (Uhler), 1894

1887 *Idolocoris famelicus*: Provancher, Pet. Faune Ent. Can., 3: 141.

1889 *Macrolophus seperatus* [*sic*]: Van Duzee, Can. Ent., 21: 4.

1894 *Dicyphus separatus* Uhler, Proc. Zool. Soc. London, p. 194. [Grenada].

1898 *Macrolophus separatus*: Osborn, Proc. Ia. Acad. Sci., 5: 233.

Distribution: Ala., Cal., Col., Conn., D.C., Fla., Ia., Ill., Ind., Mass., Md., Me., Mo., N.C., N.M., N.Y., Oh., Ont., Pa., Tex. (West Indies).

Macrolophus tenuicornis Blatchley, 1926

1926 *Macrolophus tenuicornis* Blatchley, Het. E. N. Am., p. 913. [Ind.]. Lectotype designated by Blatchley, 1930, Blatchleyana, pp. 65, 68.

1926 *Macrolophus longicornis* Knight, Ent. News, 37: 314. [N.Y.]. Synonymized by Blatchley, 1928, J. N.Y. Ent. Soc., 36: 14.

Distribution: Conn., Ill., Ind., Mo., N.S., N.Y., Neb., Ont., Pa., Que., W.Va.

Note: Wheeler et al. (1979, Melsheimer Ent. Ser., 27: 11-17) reported the seasonal history and habits on hayscentedfern, *Dennstaedtia punctilobula* (Michx.) Moore, and described the fifth-instar nymph.

Tribe Eccritotarsini Berg, 1884

Genus *Caulotops* Bergroth, 1898

1898 *Caulotops* Bergroth, Wien. Ent. Zeit., 17: 33. Type-species: *Caulotops puncticollis* Bergroth, 1898, a junior synonym of *Eccritotarsus platensis* Berg, 1883. Monotypic.

Note: Henry (1985, Fla. Ent., 68: 322) provided a key to the U.S. species.

Caulotops agavis Reuter, 1909

1909 *Caulotops agavis* Reuter, Acta Soc. Sci. Fenn., 36(2): 1. [Ariz.].

Distribution: Ariz.

Caulotops barberi Knight, 1926

1926 *Caulotops barberi* Knight, Bull. Brook. Ent. Soc., 21: 101. [Ariz.].

Distribution: Ariz.

Caulotops distanti (Reuter), 1905

1905 *Eurycipitia distanti* Reuter, Öfv. F. Vet.-Soc. Förh., 47(19): 4. [Venezuela].

1985 *Caulotops distanti*: Henry, Fla. Ent., 68: 320.

Distribution: Fla. (Costa Rica, Venezuela).

Note: This species, discovered on *Yucca* nursery stock in Fla., probably occurs in this country only in ornamental situations (Henry, 1985, above).

Genus *Cyrtocapsus* Reuter, 1876

1876 *Cyrtocapsus* Reuter, Öfv. K. Svens. Vet.-Akad. Förh., 32(9): 78. Type-species: *Capsus caligineus* Stål, 1859. Monotypic.

Note: Carvalho (1954, Bull. Brook. Ent. Soc., 49: 12-17) provided a key to species.

Cyrtocapsus caligineus (Stål), 1859

1859 *Capsus caligineus* Stål, K. Svens. Freg. Eug. Resa Jorden, p. 258. [Cal.].

1886 *Cyrtocapsus caligineus*: Uhler, Check-list Hem. Het., p. 19.

Distribution: Cal., Fla. (Central America, Mexico, Puerto Rico).

Cyrtocapsus caligineus aureopubescens Knight, 1926
1926 *Cyrtocapsus caligineus* var. *aureopubescens* Knight, Bull. Brook. Ent. Soc., 21: 102. [Fla.].
Distribution: Fla.

Cyrtocapsus caligineus caligineus (Stål), 1859
1859 *Capsus caligineus* Stål, K. Svens. Freg. Eug. Resa Jorden, p. 258. [Cal.].
Distribution: Cal.

Genus *Halticotoma* Townsend, 1892

1892 *Halticotoma* Townsend, Can. Ent., 24: 194. Type-species: *Halticotoma valida* Townsend, 1892. Monotypic.
1913 *Halticotoma* Reuter, An. Soc. Ent. Belg., 57: 278 (as Uhler ms. name). Preoccupied.
Note: Authorship of this genus was clarified by Wheeler (1976, Fla. Ent., 59: 71-76).

Halticotoma andrei Knight, 1968
1968 *Halticotoma andrei* Knight, Brig. Young Univ. Sci. Bull., 9(3): 78. [Ariz.].
Distribution: Ariz.

Halticotoma brunnea Knight, 1968
1968 *Halticotoma brunnea* Knight, Brig. Young Univ. Sci. Bull., 9(3): 78. [Wash.].
Distribution: Wash.

Halticotoma cornifer Knight, 1928
1928 *Halticotoma cornifer* Knight, Bull. Brook. Ent. Soc., 23: 242. [Ariz.].
Distribution: Ariz.

Halticotoma nicholi Knight, 1928
1928 *Halticotoma nicholi* Knight, Bull. Brook. Ent. Soc., 23: 241. [Ariz.].
Distribution: Ariz.

Halticotoma nicholi fulvicollis Knight, 1928
1928 *Halticotoma nicholi fulvicollis* Knight, Bull. Brook. Ent. Soc., 23: 241. [Ariz.].
Distribution: Ariz.

Halticotoma nicholi nicholi Knight, 1928
1928 *Halticotoma nicholi*: Knight, Bull. Brook. Ent. Soc., 23: 241. [Ariz.].
Distribution: Ariz.

Halticotoma valida Townsend, 1892 [Fig. 93]
1892 *Halticotoma valida* Townsend, Can. Ent., 24: 194. [N.M.].
1912 *Sixeonotus luteiceps*: Hunter et al., U.S. Dept. Agr., Bur. Ent. Bull., 113: 36.
1913 *Halticotoma valida* Reuter, An. Soc. Ent. Belg., 57: 279. [Ariz., Tex.]. Preoccupied.
1982 *Halticatoma* [*sic*] *valida*: Condit and Kate, Entomophaga, 27: 204.
Distribution: Ala., Ariz., Cal., Col., D.C., Del., Fla., Ga., Ia., Ind., Ks., Ky., La., Md., Miss., Mo., N.C., N.M., Neb., Nev., Oh., Ok., S.C., Tenn., Tex., Ut., Va.
Note: Wheeler (1976, Fla. Ent., 59: 71-76) studied life history, hosts, damage, and distribution, and clarified Townsend's unintentional validation of Uhler's manuscript name.

Genus *Hemisphaerodella* Reuter, 1908

1908 *Hemisphaerodella* Reuter, Wien. Ent. Zeit., 27: 297. Type-species: *Hemisphaerodella mira-*

bilis Reuter, 1908. Monotypic.

Hemisphaerodella mirabilis Reuter, 1908

1908 *Hemisphaerodella mirabilis* Reuter, Wien. Ent. Zeit., 27: 298. [West Indies]. Lectotype from Santo Domingo [now Dominican Republic] designated by Carvalho, 1955, Rev. Chilena Ent., 4: 223.

1928 *Hemisphaerodella mirabilis*: Blatchley, J. N.Y. Ent. Soc., 36: 13.

Distribution: Fla. (West Indies, South America).

Genus *Hesperolabops* Kirkaldy, 1902

1902 *Hesperolabops* Kirkaldy, Trans. Ent. Soc. London, p. 249. Type-species: *Hesperolabops gelastops* Kirkaldy, 1902. Monotypic.

1912 *Stylopidea* Hunter, Pratt, and Mitchell, U.S. Dept. Agr., Bur. Ent. Bull., 113:22. Type-species: *Stylopidea picta* Hunter, Pratt, and Mitchell, 1912. Monotypic. Synonymized by Knight, 1928, Proc. Ent. Soc. Wash., 30: 68.

Note: Froeschner (1967, Proc. U.S. Nat. Mus., 123(3614): 1-11) revised this genus and provided a key to the known species.

Hesperolabops gelastops Kirkaldy, 1902

1896 *Lobos* [*sic*] *hesperius*: Marlatt, Proc. Ent. Soc. Wash., 4: 45.

1902 *Hesperolabops gelastops* Kirkaldy, Trans. Ent. Soc. London, p. 250. [Mexico].

1912 *Stylopidea picta* Hunter, Pratt, and Mitchell, U.S. Dept. Agr., Bur. Ent. Bull., 113: 22. [Tex.]. Synonymized by Knight, 1928, Proc. Ent. Soc. Wash., 30: 68.

1916 *Hesperolabops gelastops*: Van Duzee, Check List Hem., p. 42

1928 *Hesperolabops picta*: Knight, Proc. Ent. Soc. Wash., 30: 68.

1973 *Hesperolabops pictus*: Steyskal, Stud. Ent., 16: 206.

1982 *Hesperalabops* [*sic*] *gelastops*: Condit and Kate, Entomophaga, 27: 204.

Distribution: Tex. (Mexico).

Note: Marlatt's (1896, above) record of "*Labos*" *hesperius* found on cacti undoubtedly refers to *H. gelastops*, not the true grass-feeding *Labops hesperius* Uhler.

Genus *Pycnoderes* Guérin-Méneville, 1857

1857 *Pycnoderes* Guérin-Méneville, Sagra's Hist. Cuba, 7: 404. Type-species: *Pycnoderes quadrimaculatus* Guérin-Méneville, 1857. Monotypic.

Pycnoderes angustatus Reuter, 1907

1907 *Pycnoderes angustatus* Reuter, Öfv. F. Vet.-Soc. Förh., 49(5): 2. [Jamaica].

1952 *Pycnoderes angustatus*: Hussey, Fla. Ent., 35: 118.

Distribution: Fla. (Jamaica).

Pycnoderes atratus (Distant), 1884

1884 *Eccritotarsus atratus* Distant, Biol. Cent.-Am., Rhyn., 1: 285. [Panama].

1894 *Eccritotarsus atratus*: Uhler, Proc. Zool. Soc. London, p. 193.

1908 *Pycnoderes atratus*: Reuter, An. Nat. Hofmus., 22: 160.

Distribution: Ariz., Cal., Tex. (Central America, West Indies).

Pycnoderes balli Knight, 1926

1926 *Pycnoderes balli* Knight, Bull. Brook. Ent. Soc., 21: 104. [Fla.].

Distribution: Fla., Tex.

Pycnoderes convexicollis Blatchley, 1926

1926 *Pycnoderes convexicollis* Blatchley, Ent. News, 37: 166. [Ind.]. Lectotype designated by Blatchley, 1930, Blatchleyana, pp. 64, 68.

Distribution: Ill., Ind.

Pycnoderes dilatatus Reuter, 1909

1909 *Pycnoderes dilatatus* Reuter, Acta Soc. Sci. Fenn., 36(2): 3. [D.C., Md.].

Distribution: D.C., Md., N.C., N.J.

Pycnoderes drakei Knight, 1926

1926 *Pycnoderes drakei* Knight, Bull. Brook. Ent. Soc., 21: 106. [Miss.].

Distribution: Miss.

Pycnoderes infuscatus Knight, 1926

1926 *Pycnoderes infuscatus* Knight, Bull. Brook. Ent. Soc., 21: 106. [N.C.].

Distribution: N.C.

Pycnoderes medius Knight, 1926 [Fig. 91]

1926 *Pycnoderes medius* Knight, Bull. Brook. Ent. Soc., 21: 105. [Mo.].

Distribution: Ill., Miss., Mo., Tenn.

Pycnoderes obscuratus Knight, 1926

1926 *Pycnoderes balli* var. *obscuratus* Knight, Bull. Brook. Ent. Soc., 21: 105. [Pa.].

1957 *Pycnoderes obscuratus*: Carvalho, Arq. Mus. Nac., 44: 123.

Distribution: Pa.

Pycnoderes quadrimaculatus Guérin-Méneville, 1857

1857 *Pycnoderes quadrimaculatus* Guérin-Méneville, Sagra's Hist. Cuba, 7: 404. [Cuba].

1894 *Pycnoderes quadrimaculatus*: Uhler, Proc. Cal. Acad. Sci., (2)4: 267.

1900 *Eccritotarsus quadrimaculatus*: Smith, Ins. N.J., p. 131.

Distribution: Ariz., Cal., Conn.(?), Fla., Mass.(?), Md.(?), N.H.(?), Tex. (Hawaii, Central America, Brazil, West Indies).

Note: Northern records for this species are suspect and need verification.

Genus *Sixeonotopis* Carvalho and Schaffner, 1974

1974 *Sixeonotopis* Carvalho and Schaffner, Rev. Brasil. Biol., 33 (supl.): 14. Type-species: *Sixeonotopis crassicornis* Carvalho and Schaffner, 1974. Original designation.

Sixeonotopis crassicornis Carvalho and Schaffner, 1974

1974 *Sixeonotopsis crassicornis* Carvalho and Schaffner, Rev. Brasil. Biol., 33 (supl.): 15. [Tex.].

Distribution: Tex.

Genus *Sixeonotus* Reuter, 1876

1876 *Sixeonotus* Reuter, Öfv. K. Vet.-Akad. Förh., 32(9): 77. Type-species: *Sixeonotus insignis* Reuter, 1876. Monotypic.

Sixeonotus albicornis Blatchley, 1926

1926 *Sixenotus* [*sic*] *albicornis* Blatchley, Ent. News, 37: 167. [Fla.]. Lectotype designated by Blatchley, 1930, Blatchleyana, pp. 64, 68.

1929 *Sixeonotus albicornis*: Knight, Bull. Brook. Ent. Soc., 24: 152.

Distribution: Ala., Fla., La., Miss.

Note: Knight (1929, Bull. Brook. Ent. Soc., 24: 152) considered *S. albicornis* a junior synonym of *S. insignis* Reuter. Blatchley (1930, above) and Carvalho (1957, Arq. Mus. Nac., 44: 125), however, listed *S. albicornis* as a valid species.

Sixeonotus albohirtus Knight, 1926
1926 *Sixeonotus albohirtus* Knight, Bull. Brook. Ent. Soc., 21: 107. [Fla.].
1926 *Sixenotus* [*sic*] *albohirtus*: Blatchley, Het. E. N. Am., p. 872, 874.
Distribution: Fla., Miss.

Sixeonotus areolatus Knight, 1928
1928 *Sixeonotus areolatus* Knight, Bull. Brook. Ent. Soc., 23: 243. [Miss.].
Distribution: Ala., Ark., Miss., Tex.

Sixeonotus basicornis Knight, 1928
1928 *Sixeonotus basicornis* Knight, Bull. Brook. Ent. Soc., 23: 248. [N.C.].
Distribution: N.C.

Sixeonotus bebbiae Knight, 1968
1968 *Sixeonotus bebbiae* Knight, Brig. Young Univ. Sci. Bull., 9(3): 76. [Ariz.].
Distribution: Ariz.

Sixeonotus brevirostris Knight, 1928
1928 *Sixeonotus brevirostris* Knight, Bull. Brook. Ent. Soc., 23: 235. [Col.].
Distribution: Col., Ks.

Sixeonotus brevis Knight, 1926
1926 *Sixeonotus brevis* Knight, Bull. Brook. Ent. Soc., 21: 107. [Miss.].
1926 *Sixenotus* [*sic*] *brevis*: Blatchley, Het. E. N. Am., p. 872-873.
Distribution: Miss.

Sixeonotus deflatus Knight, 1928
1928 *Sixeonotus deflatus* Knight, Bull. Brook. Ent. Soc., 23: 246. [N.Y.].
Distribution: N.Y.

Sixeonotus dextratus Knight, 1928
1928 *Sixeonotus dextratus* Knight, Bull. Brook. Ent. Soc., 23: 244. [Ariz.].
Distribution: Ariz.

Sixeonotus gracilis Blatchley, 1926
1926 *Sixenotus* [*sic*] *gracilis* Blatchley, Ent. News, 37: 168. [Fla.].
1957 *Sixeonotus gracilis*: Carvalho, Arq. Mus. Nac., 44: 125.
Distribution: Fla.

Sixeonotus insignis Reuter, 1876
1876 *Sixeonotus insignis* Reuter, Öfv. K. Svens. Vet.-Akad. Förh., 32(9): 78. [N.J., N.Y., Tex.].
1890 *Pycnoderes insignis*: Atkinson, J. Asiatic Soc. Bengal, 58(2): 41.
1900 *Eccritotarsus insignis*: Smith, Cat. Ins. N.J., p. 131.
1926 *Sixenotus* [*sic*] *insignis*: Blatchley, Ent. News, 37: 168.
Distribution: Col., Conn., Del., Fla., Ill., Mass., Md., Miss., Mo., N.H., N.J., N.Y., Oh., Pa., Que., Tex., Va.
Note: See notes under *S. albicornis* Blatchley and *S. recurvatus* Knight.

Sixeonotus nicholi Knight, 1928
1928 *Sixeonotus nicholi* Knight, Bull. Brook. Ent. Soc., 23: 242. [Ariz.].
Distribution: Ariz.

Sixeonotus pusillus Knight, 1928
1928 *Sixeonotus pusillus* Knight, Bull. Brook. Ent. Soc., 23: 248. [Tex.].
Distribution: Tex.

Sixeonotus recurvatus Knight, 1929
1929 *Sixeonotus recurvatus* Knight, Bull. Brook. Ent. Soc., 24: 153. [D.C.].
Distribution: D.C., Md., N.C.
Note: Knight (1929, above, pp. 152-153) considered all northern records for *S. insignis* to belong to *S. recurvatus*, but did not sort out previous records. In this catalog we are listing the distribution for both species as they appear in the literature.

Sixeonotus rostratus Knight, 1928
1928 *Sixeonotus rostratus* Knight, Bull. Brook. Ent. Soc., 23: 245. [Col.].
Distribution: Alta., Col., Sask.

Sixeonotus scabrosus (Uhler), 1895
1895 *Eccritotarsus scabrosus* Uhler, Bull. Col. Agr. Exp. Stn., p. 40. [Col.].
1909 *Sixeonotus luteiceps* Reuter, Acta Soc. Sci. Fenn., 36(2): 4. [Tex.]. Synonymized by Carvalho 1957, Arq. Mus. Nac., 44: 126.
1917 *Sixeonotus scabrosus*: Van Duzee, Univ. Cal. Publ. Ent., 2: 360.
Distribution: Col., Tex.

Sixeonotus tenebrosus (Distant), 1893
1893 *Eccritotarsus tenebrosus* Distant, Biol. Centr.-Am., Rhyn., 1: 441. [Guatemala].
1909 *Sixeonotus tenebrosus*: Van Duzee, Bull. Buff. Soc. Nat. Sci., 9: 178.
1926 *Sixenotus* [*sic*] *tenebrosus*: Blatchley, Ent. News, 37: 169.
Distribution: D.C., Fla. Ks., Md., W.Va. (Guatemala).

Sixeonotus unicolor Knight, 1928
1928 *Sixeonotus unicolor* Knight, Bull. Brook. Ent. Soc., 23: 247. [Miss.].
Distribution: Ill., Miss.

Genus *Sysinas* Distant, 1883

1883 *Sysinas* Distant, Biol. Centr.-Am., Rhyn., 1: 248. Type-species: *Sysinas linearis* Distant, 1883. Designated by Kirkaldy, 1906, Trans. Am. Ent. Soc., 32: 146.

Sysinas linearis Distant, 1883
1883 *Sysinas linearis* Distant, Biol. Centr.-Am., Rhyn., 1: 248. [Mexico].
1910 *Sysinas linearis*: Banks, Cat. Nearc. Hem.-Het., p. 38.
Distribution: N.Y. (Mexico).
Note: Blatchley (1926, Het. E. N. Am., p. 876) suggested that the Sea Cliff, N.Y., record reported by Banks (1910, above) represented an adventive specimen. We agree and believe the record should be removed from future U.S. lists.

Subfamily Cylapinae Kirkaldy, 1903

Note: Kelton (1985, Can. Ent., 117: 1071-1073) modified his subfamily key to the "Miridae of the Prairie Provinces" to include the Cylapinae.

Tribe Cylapini Kirkaldy, 1903

Genus *Cylapus* Say, 1832

1832 *Cylapus* Say, Descrip. Het. Hem. N. Am., p. 26. Type-species: *Capsus* (*Cylapus*) *tenuicornis* Say, 1832. Designated by Poppius, 1909, Acta Soc. Sci. Fenn., 37(4): 10.

Cylapus tenuicornis (Say), 1832 [Fig. 83]

1832 *Capsus* (*Cylapus*) *tenuicornis* Say, Descrip. Het. Hem. N. Am., p. 26. [Ind.].

1891 *Cylapus tenuicornis*: Heidemann, Proc. Ent. Soc. Wash., 2: 68.

1900 *Cylapus tenucornis* [*sic*]: Smith, Ins. N.J., p. 131.

Distribution: Conn., D.C., Ill., Ind., Md., N.J., N.Y., Ont., Pa., R.I., W.Va.

Genus *Peritropis* Uhler, 1891

1891 *Peritropis* Uhler, Proc. Ent. Soc. Wash., 2: 122. Type-species: *Peritropis saldaeformis* Uhler, 1891. Monotypic.

Peritropis husseyi Knight, 1923

1923 *Peritropis husseyi* Knight, Ent. News, 34: 50. [Mich.].

Distribution: Ala., Ill., Ind., Mich.

Peritropis saldaeformis Uhler, 1891

1891 *Peritropis saldaeformis* Uhler, Proc. Ent. Soc. Wash., 2: 122. [D.C., Ill.].

1920 *Peritropis saldiformis* [*sic*]: Bergroth, An. Soc. Ent. Belg., 60: 74. Unnecessary emendation.

Distribution: D.C., Fla., Ia., Ill., Md., Ok., Pa., Tex. (Brazil, Paraguay).

Note: Records from South America need verification.

Tribe Fulviini Uhler, 1886

Genus *Fulvius* Stål, 1862

1862 *Fulvius* Stål, Stett. Ent. Zeit., 23(7-9): 322. Type-species: *Fulvius anthocorides* Stål, 1862. Monotypic.

1875 *Teratodella* Reuter, Bih. K. Vet.-Akad. Handl., 3: 7. Type-species: *Teratodella anthocoroides* Reuter, 1875. Monotypic. Synonymized by Reuter, 1895, Ent. Tidskr., 16: 131.

1877 *Pamerocoris* Uhler, Bull. U.S. Geol. Geogr. Surv. Terr., 3: 424. Type-species: *Pamerocoris anthocoroides* Uhler, 1877. Monotypic. Synonymized by Reuter 1895, Ent. Tidskr., 16: 131.

Note: Kelton (1985, Can. Ent., 117: 1071) gave a key to the Canadian species.

Fulvius anthocoroides (Reuter), 1895

1875 *Teratodella anthocoroides* Reuter, Bih. K. Vet.-Akad. Handl., 3: 8. [Senegal (specimen intercepted in France)].

1894 *Fulvius atratus*: Uhler, Proc. Zool. Soc. London, p. 192 (in part).

1895 *Fulvius brevicornis* Reuter, Ent. Tidskr., 16: 138. Unnecessary new name for *Teratodella anthocoroides* Reuter.

1914 *Fulvius atratus*: Barber, Bull. Am. Mus. Nat. Hist., 33: 500.

1954 *Fulvius brevicornis*: Hussey, Fla. Ent., 37: 21.

Distribution: Fla. (Africa, Asia, South and Central America).

Note: Wheeler (1977, Proc. Ent. Soc. Wash., 79: 589-590) outlined the confusing history of the names *T. "anthocoroides"* Reuter (1875, above) and *F. "anthocorides"* Stål (1862, Stett, Ent. Zeit., 23(7-9): 322), and showed that these names apply to different species and do not conflict in homonymy, despite Walker's (1873, Cat. Hem. het. Brit. Mus., 6: 160) incorrect emendation [or misspelling] of Stål's *anthocorides* to *anthocoroides*.

Fulvius imbecilis (Say), 1832 [Fig. 85]

1832 *Capsus imbecilis* Say, Descrip. Het. Hem. N. Am., p. 25. [Ind.].

1886 *Poecilocytus imbecilis*: Uhler, Check-list Hem. Het., p. 19.

1895 *Fulvius Heinemanni* [*sic*] Reuter, Ent. Tidskr., 16: 142. [Md.]. Synonymized by Van Duzee, 1917, Univ. Cal. Publ. Ent., 2: 365.

1904 *Fulvius heidemanni*: Wirtner, An. Carnegie Mus., 3: 195.

1916 *Fulvius imbecilis*: Van Duzee, Check List Hem., p. 42.

Distribution: Ala., Alta., D.C., Del., Fla., Ill., Ind., La., Man., Md., Mich., Minn., Miss., Mo., N.C., N.J., Pa., Ont., Tenn., Va., W.Va.

Fulvius slateri Wheeler, 1977

1877 *Pamerocoris anthocoroides* Uhler, Bull. U.S. Geol. Geogr. Surv. Terr., 3: 424. [Col.]. Preoccupied.

1886 *Lygus brunneus*: Uhler, Check-list Hem. Het., p. 18.

1889 *Fulvius anthocoroides*: Van Duzee, Can. Ent., 21: 4.

1895 *Fulvius brunneus*: Reuter, Ent. Tidskr., 16: 140.

1895 *Fulvius Uhleri* [*sic*] Reuter, Ent. Tidskr., 16: 148. [See note below].

1977 *Fulvius slateri* Wheeler, Proc. Ent. Soc. Wash., 79: 588. New name for *Pamerocoris anthocoroides* Uhler.

Distribution: Cal., Col., Conn., D.C., Fla., Ia., Ill., Ind., Ks., Mass., Md., Me., N.H., N.J., N.Y., Ont., Pa., Que., R.I., Va., Wis. (Mexico).

Note: Reuter (1895, Ent. Tidskr., 16: 148) used the name *F. uhleri*, gave descriptive characters for it, and compared it to his new species *F. dubius* but did not give accompanying locality data or mention this species elsewhere in the paper. Later, he (1895, Ent. Tidskr., 16: 254) explained that *F. uhleri* was originally intended to be a replacement name for *F. anthocoroides* Uhler, until he discovered the latter was also a junior synonym of *F. brunneus* (Provancher) and, thus, *F. uhleri* was a *lapsus* and inadvertently left in the manuscript on page 148 [but not on pages 136, 140, and 141]. Because this name was unintentionally published and for the purpose of this catalog, we interpret it as unavailable and recognize Wheeler's (1977, above) replacement name, *F. slateri*. George Steyskal (Systematic Entomology Laboratory, retired, Agricultural Research Service, U.S. Department of Agriculture, Washington, DC), recommends that the Commission on Zoological Nomenclature be petitioned to place *F. uhleri* on the Official Index of Rejected and Invalid Names. See Wheeler (1977, above) for further clarification of this complicated nomenclatural history. See also *Plagiognathus obscurus* Uhler, the senior synonym of *Lygus brunneus* Provancher.

Subfamily Deraeocorinae Douglas and Scott, 1865

Tribe Clivinematini Reuter, 1876

Genus *Bothynotus* Fieber, 1864

1864 *Bothynotus* Fieber, Wien. Ent. Monat., 8: 76. Type-species: *Bothynotus minki* Fieber, 1864, a junior synonym of *Phytocoris pilosus* Boheman, 1852. Monotypic.

1917 *Neobothynotus* Wirtner, Ent. News, 28: 33. Type-species: *Neobothynotus modestus* Wirtner, 1917. Monotypic. Synonymized by Knight, 1917, Can. Ent., 49: 251.

Note: Henry (1979, Fla. Ent., 62: 232-244) reviewed the genus and provided a key to species.

Bothynotus albus Carvalho, 1953
1953 *Bothynotus albus* Carvalho, Fla. Ent., 36: 161. [Fla.].
Distribution: Fla.

Bothynotus barberi Knight, 1933
1933 *Bothynotus barberi* Knight, Ent. News, 44: 133. [Ariz.].
Distribution: Ariz.

Bothynotus floridanus Henry, 1979
1979 *Bothynotus floridanus* Henry, Fla. Ent., 62: 235. [Fla.].
Distribution: Fla.

Bothynotus johnstoni Knight, 1933
1933 *Bothynotus johnstoni* Knight, Ent. News, 44: 135. [Miss.]
1979 *Bothynotus johnsoni* [*sic*]: Henry and Smith, J. Ga. Ent. Soc., 14: 215.
Distribution: Fla., Ga., Miss.

Bothynotus modestus (Wirtner), 1917 [Fig. 87]
1917 *Neobothynotus modestus* Wirtner, Ent. News, 28: 34. [Pa.].
1917 *Bothynotus pilosus*: Knight, Can. Ent., 49: 251.
1920 *Bothynotus modestus*: Bergroth, An. Soc. Ent. Belg., 60: 69.
Distribution: Ia., Ill., Ks., Md., Mich., Miss.(?), Oh., Pa.
Note: Blatchley's (1926, Het. E. N. Am., p. 882) record of this species from Fla. probably refers to *B. floridanus* Henry (Henry, 1979, Fla. Ent., 62, 240); Henry also questioned the Miss. record by Khalaf (1971, Fla. Ent., 54: 340).

Genus *Clivinema* Reuter, 1876

1876 *Clivinema* Reuter, Öfv. K. Svens. Vet.-Akad. Förh., 32(9): 63. Type-species: *Clivinema villosa* Reuter, 1876. Monotypic.

Clivinema bonita Knight, 1928
1928 *Clivinema bonita* Knight, Proc. Biol. Soc. Wash., 41: 35. [Ariz.].
Distribution: Ariz.

Clivinema coalinga Bliven, 1966
1966 *Clivinema coalinga* Bliven, Occ. Ent., 1(9): 120. [Cal.].
Distribution: Cal.

Clivinema detecta Knight, 1928
1928 *Clivinema detecta* Knight, Proc. Biol. Soc. Wash., 41: 34. [N.M.].
Distribution: Ariz., Col., N.M.

Clivinema fusca Downes, 1924
1924 *Clivinema fusca* Downes, Proc. Ent. Soc. B.C., 21: 29. [B.C.].
Distribution: B.C.

Clivinema fuscinervis Knight, 1928
1928 *Clivinema fuscinervis* Knight, Proc. Biol. Soc. Wash., 41: 33. [Mont.].
Distribution: Mont.

Clivinema medialis Knight, 1928
1893 *Clivinema villosa*: Uhler, Proc. Ent. Soc. Wash., 2: 372.
1928 *Clivinema medialis* Knight, Proc. Biol. Soc. Wash., 41: 31. [Ut.].
Distribution: Col., Nev., Ut.

Clivinema regalis Knight, 1917
1917 *Clivinema regalis* Knight, Ent. News, 28: 5. [Tex.].
1951 *Ambracius harrisi* Carvalho, Arq. Mus. Nac., 42: 157. [Ariz.]. Synonymized by Carvalho, 1955, Rev. Chil. Ent., 4: 222.
Distribution: Ariz., Tex.

Clivinema serica Knight, 1928
1928 *Clivinema serica* Knight, Proc. Biol. Soc. Wash., 41: 34. [N.M.].
Distribution: N.M., Ut.

Clivinema sinuata Knight, 1928
1928 *Clivinema sinuata* Knight, Proc. Biol. Soc. Wash., 41: 32. [Cal.].
Distribution: Cal.

Clivinema sulcata Knight, 1928
1928 *Clivinema sulcata* Knight, Proc. Biol. Soc. Wash., 41: 32. [Cal.].
Distribution: Cal.

Clivinema villosa Reuter, 1876
1876 *Clivinema villosa* Reuter, Öfv. K. Svens. Vet.-Akad. Förh., 32(9): 63. [Tex.].
Distribution: Alta., Ariz., Cal., Col., Fla., Ok., Mass., Mont., Tex., Ut.

Genus *Largidea* Van Duzee, 1912

1912 *Largidea* Van Duzee, Bull. Buff. Soc. Nat. Sci., 10: 480. Type-species: *Largidea marginata* Van Duzee, 1912, a junior synonym of *Clivinema rubida* Uhler, 1904. Monotypic.

Largidea arizonae Knight, 1968
1968 *Largidea arizonae* Knight, Brig. Young Univ. Sci. Bull., 9(3): 85. [Ariz.].
Distribution: Ariz.

Largidea balli Knight, 1968
1968 *Largidea balli* Knight, Brig. Young Univ. Sci. Bull., 9(3): 86. [Ariz.].
Distribution: Ariz.

Largidea davisi Knight, 1917
1917 *Largidea davisi* Knight, Ent. News, 28: 7. [N.Y.].
Distribution: Mass., N.Y.

Largidea gerhardi Knight, 1968
1968 *Largidea gerhardi* Knight, Brig. Young Univ. Sci. Bull., 9(3): 86. [Col.].
Distribution: Col.

Largidea grossa Van Duzee, 1916
1916 *Largidea grossa* Van Duzee, Univ. Cal. Publ. Ent., 1: 238. [Cal.].
Distribution: Ariz., Cal., Ill., Ore., Wash.

Largidea nevadensis Knight, 1968
1968 *Largidea nevadensis* Knight, Brig. Young Univ. Sci. Bull., 9(3): 84. [Nev.].
Distribution: Nev.

Largidea pudica Van Duzee, 1925
1925 *Largidea pudica* Van Duzee, Proc. Cal. Acad. Sci., (4)14: 397. [Ore.].
Distribution: Cal., Ore.

Largidea rubida (Uhler), 1904
1904 *Clivinema rubida* Uhler, Proc. U.S. Nat. Mus., 27: 355. [N.M.].
1912 *Largidea marginata* Van Duzee, Bull. Buff. Soc. Nat. Sci., 10: 481. [Col.]. Synonymized by Knight, 1917, Ent. News, 28: 8.
1916 *Largidea rubida*: Van Duzee, Check List Hem., p. 42.
Distribution: Ariz., Col., N.M.

Largidea shoshonea Knight, 1968
1968 *Largidea shoshonea* Knight, Brig. Young Univ. Sci. Bull., 9(3): 87. [Col.].
Distribution: Alta., Col., Man., Mont., Sask., Wyo.

Largidea stitti Knight, 1968
1968 *Largidea stitti* Knight, Brig. Young Univ. Sci. Bull., 9(3): 86. [Ariz.].
Distribution: Ariz.

Tribe Deraeocorini Douglas and Scott, 1865

Genus *Deraeocapsus* Knight, 1921

1921 *Deraeocapsus* Knight, Minn. St. Ent. Rept., 18: 198. Type-species: *Deraeocoris ingens* Van Duzee, 1916. Original designation.

Deraeocapsus fraternus (Van Duzee), 1916
1916 *Deraeocoris fraternus* Van Duzee, Univ. Cal. Publ. Ent., 1: 238. [Cal.].
1921 *Deraeocapsus fraternus*: Knight, Minn. St. Ent. Rept., 18: 198.
Distribution: B.C., Cal.

Deraeocapsus ingens (Van Duzee), 1916
1916 *Deraeocoris ingens* Van Duzee, Univ. Cal. Publ. Ent., 1: 237. [Cal.].
1921 *Deraeocapsus ingens*: Knight, Minn. St. Ent. Rept., 18: 198.
Distribution: Cal.

Genus *Deraeocoris* Kirschbaum, 1856

1856 *Deraeocoris* Kirschbaum, Jahrb. Ver. Nat. Nassau, 10: 208. Type-species: *Cimex olivaceus* Fabricius, 1776. Designated by Distant, 1904, Fauna Brit. India, Rhyn., 2: 466.

1858 *Camptobrochis* Fieber, Wien. Ent. Monat., 2: 304. Type-species: *Lygaeus punctulatus* Fallén, 1829. Designated by Distant, 1904, Fauna Brit. India, Rhyn., 2: 460. Synonymized by Poppius, 1912, Acta Soc. Sci. Fenn., 41(3): 119.

1876 *Callicapsus* Reuter, Öfv. K. Svens. Vet.-Akad. Förh., 32(9): 75. Type-species: *Callicapsus histrio* Reuter, 1876. Monotypic. Synonymized by Reuter, 1909, Acta Soc. Sci. Fenn., 36(2): 52.

1876 *Euarmosus* Reuter, Öfv. K. Svens. Vet.-Akad. Förh., 32(9): 76. Type-species: *Euarmosus sayi* Reuter, 1876. Monotypic. Synonymized by Reuter, 1909, Acta. Soc. Sci. Fenn., 36(2): 52.

1884 *Cimatlan* Distant, Biol. Centr.-Am., Rhyn., 1: 281. Type-species: *Cimatlan delicatum* Distant, 1884. Monotypic. Synonymized by Carvalho, 1952, An. Acad. Brasil. Ci., 24: 53.

1904 *Mycterocoris* Uhler, Proc. U.S. Nat. Mus., 27: 358. Type-species: *Deraeocoris cerachates* Uhler, 1894. Monotypic. Synonymized by Reuter, 1909, Acta Soc. Sci. Fenn., 36(2): 52.

Note: Knight's revision (1921, Minn. St. Ent. Rept., 18: 76-210), with a key to species, remains the single most useful reference to this genus.

Deraeocoris albigulus Knight, 1921
1921 *Deraeocoris albigulus* Knight, Minn. St. Ent. Rept., 18: 71. [N.Y.].
Distribution: Alta., Ia., Ind., Man., Mich., Minn., N.Y., Pa., Sask., W.Va., Wis.

Deraeocoris alnicola Knight, 1921
1921 *Deraeocoris alnicola* Knight, Minn. St. Ent. Rept., 18: 132. [N.Y.].
1980 *Dereacoris* [*sic*] *alnicola*: Loan, Nat. Can., 107: 91.
Distribution: Conn., N.Y., Ont., Que.

Deraeocoris apache Knight, 1921
1921 *Deraeocoris apache* Knight, Minn. St. Ent. Rept., 18: 179. [Ariz.].
Distribution: Ariz.

Deraeocoris aphidiphagus Knight, 1921
1921 *Deraeocoris aphidiphagus* Knight, Minn. St. Ent. Rept., 18: 134. [N.Y.].
1979 *Deraeocoris aphidaphagus* [*sic*]: Henry and Smith, J. Ga. Ent. Soc., 14: 215.
Distribution: Alta., Ark., Col., Conn., D.C., Ga., Ill., Man., Md., Me., Mich., Minn., Mo., N.B., N.D., N.H., N.S., N.Y., Oh., Ont., P.Ed., Pa., Que., Sask., Va., W.Va., Wis.

Deraeocoris appalachianus Knight, 1921
1921 *Deraeocoris appalachianus* Knight, Minn. St. Ent. Rept., 18: 160. [N.C.].
Distribution: N.C.

Deraeocoris atriventris Knight, 1921
1921 *Deraeocoris atriventris* Knight, Minn. St. Ent. Rept., 18: 112. [Ariz.].
1933 *Deraecoris* [*sic*] *atriventris*: Knowlton, Ent. News, 44: 263.
Distribution: Ariz.. Ut.

Deraeocoris bakeri Knight, 1921
1921 *Deraeocoris bakeri* Knight, Minn. St. Ent. Rept., 18: 102. [Nev.].
1927 *Deraeocoris* (*Camptobrochis*) *bakeri*: Knight, Can. Ent., 59: 37.
1933 *Deraecoris* [*sic*] *bakeri*: Knowlton, Ent. News, 44: 263.

Distribution: Ariz., B.C., Cal., Col., Id., Nev., Ut.

Deraeocoris balli Knight, 1927
1927 *Deraeocoris balli* Knight, Bull. Brook. Ent. Soc., 22: 137. [Col.].
Distribution: Col.

Deraeocoris barberi Knight, 1921
1921 *Deraeocoris barberi* Knight, Minn. St. Ent. Rept., 18: 157. [Neb.].
Distribution: Ariz., Col. Mont., N.M., Neb.

Deraeocoris barberi barberi Knight, 1921
1921 *Deraeocoris barberi*: Knight, Minn. St. Ent. Rept., 18: 157.
Distribution: Same as for species.

Deraeocoris barberi lignipes Knight, 1921
1921 *Deraeocoris barberi* var. *lignipes* Knight, Minn. St. Ent. Rept., 18: 159. [Ariz.].
Distribution: Ariz.

Deraeocoris betulae Knight, 1921
1921 *Deraeocoris betulae* Knight, Minn. St. Ent. Rept., 18: 129. [N.Y.].
Distribution: Conn., Mass., Me., N.H., N.J., N.S., N.Y., Pa., Que.

Deraeocoris borealis (Van Duzee), 1920
1920 *Camptobrochys* [*sic*] *borealis* Van Duzee, Proc. Cal. Acad. Sci., (4)9: 354. [Me.].
1921 *Deraeocoris borealis*: Knight, Minn. St. Ent. Rept., 18: 119.
Distribution: B.C., Conn., Mass., Me., Mich., Minn., N.J., N.S., N.Y., Oh., Ont., Que., Wis.

Deraeocoris brevis (Uhler), 1904
1904 *Camptobrochis brevis* Uhler, Proc. U.S. Nat. Mus., 27: 359. [N.M.].
1916 *Camptobrochis* (*Camptobrochis*) *brevis*: Van Duzee, Check List Hem., p. 41.
1917 *Camptobrochys* [*sic*] (*Camptobrochys* [*sic*]) *brevis*: Van Duzee, Univ. Cal. Publ. Ent., 2: 354.
1921 *Deraeocoris* (*Camptobrochis*) *brevis*: Knight, Minn. St. Ent. Rept., 18: 103.
1957 *Deraeocoris brevis*: Carvalho, Arq. Mus. Nac., 44: 61.
Distribution: Alta., Ariz., B.C., Cal., Col., Id., Man., Mont., Nev., N.M., Sask., Ut., Wash.

Deraeocoris brevis brevis (Uhler), 1904
1904 *Camptobrochis brevis* Uhler, Proc. U.S. Nat. Mus., 27: 359.
1921 *Deraeocoris* (*Camptobrochis*) *brevis* : Knight, Minn. St. Ent. Rept., 18: 103.
Distribution: Same as for species.

Deraeocoris brevis piceatus Knight, 1921
1921 *Deraeocoris* (*Camptobrochis*) *brevis* var. *piceatus* Knight, Minn. St. Ent. Rept., 18: 105. [Col.].
1967 *Deraeocoris brevis piceatus*: McMullen and Jong, J. Ent. Soc. B.C., 64: 36.
Distribution: B.C., Cal., Col., Id., Mont., Nev., Ore., Wash.
Note: Westigard (1973, Can. Ent., 105: 1105-1111) studied the biology of this subspecies.

Deraeocoris bullatus Knight, 1921
1921 *Deraeocoris bullatus* Knight, Minn. St. Ent. Rept., 18: 147. [Ariz.].
Distribution: Ariz., Nev.

Deraeocoris cerachates Uhler, 1894

1894 *Deraeocoris cerachates* Uhler, Proc. Cal. Acad. Sci., (2)4: 265. [Cal.].

1904 *Mycterocoris cerachates*: Uhler, Proc. U.S. Nat. Mus., 27: 358.

1909 *Camptobrochis cerachates*: Reuter, Acta Soc. Sci. Fenn., 36(2): 53.

1916 *Camptobrochis (Mycterocoris) cerachates*: Van Duzee, Check List Hem., p. 41.

1917 *Camptobrochys* [*sic*] (*Mycterocoris*) *cerachates*: Van Duzee, Univ. Cal. Publ. Ent., 2: 355.

1921 *Deraeocoris californicus* Knight, Minn. St. Ent. Rept., 18: 185. [Cal.]. Synonymized by Razafimahatratra and Lattin, 1982, Pan-Pac. Ent., 58: 363.

1921 *Deraeocoris californicus* var. *bradleyi* Knight, Minn. St. Ent. Rept., 18: 187. [Cal.]. Synonymized by Razafimahatratra and Lattin, 1982, Pan-Pac. Ent., 58: 363.

1921 *Deraeocoris californicus* var. *desiccatus* Knight, Minn. St. Ent. Rept., 18: 187. [Cal.]. Synonymized by Razafimahatratra and Lattin, 1982. Pan-Pac. Ent., 58: 363.

1921 *Deraeocoris californicus* var. *rufocuneatus* Knight, Minn. St. Ent. Rept., 18: 187. [Cal.]. Synonymized by Razafimahatratra and Lattin, 1982, Pan-Pac. Ent., 58: 363.

Distribution: Cal., Col., N.M., Ore. (Mexico).

Deraeocoris cochise Razafimahatratra and Lattin, 1982

1982 *Deraeocoris cochise* Razafimahatratra and Lattin, Pan-Pac. Ent., 58: 358. [Ariz.].

Distribution: Ariz.

Deraeocoris comanche Knight, 1921

1921 *Deraeocoris comanche* Knight, Minn. St. Ent. Rept., 18: 177. [Ariz.].

Distribution: Ariz., N.M.

Deraeocoris convexulus Knight, 1921

1921 *Deraeocoris convexulus* Knight, Minn. St. Ent. Rept., 18: 148. [Cal.].

Distribution: Cal.

Deraeocoris davisi Knight, 1921

1921 *Deraeocoris davisi* Knight, Minn. St. Ent. Rept., 18: 140. [N.Y.].

Distribution: Ala., Ill., Ind., Mo., N.C., N.Y., Tex., W.Va.

Deraeocoris delicatus (Distant), 1884

1884 *Cimatlan delicatum* Distant, Biol. Centr.-Am., Rhyn., 1: 281. [Guatemala].

1916 *Cimatlan delicatum*: Van Duzee, Check List Hem., p. 41.

1957 *Deraeocoris delicatum*: Carvalho, Arq. Mus. Nac., 44: 63.

1973 *Deraeocoris delicatus*: Steyskal, Stud. Ent., 16: 206.

Distribution: Ariz. (Mexico to Central America).

Deraeocoris diveni Knight, 1921

1921 *Deraeocoris diveni* Knight, Minn. St. Ent. Rept., 18: 117. [Wyo.].

Distribution: Alta., Ore., Wyo.

Deraeocoris fasciolus Knight, 1921

1921 *Deraeocoris fasciolus* Knight, Minn. St. Ent. Rept., 18: 123. [N.Y.].

1966 *Deraecoris* [*sic*] *fasicolus* [*sic*]: Sanford and Herbert, Can. Ent., 98: 997.

Distribution: Alta., B.C., Ill., Ind., Man., Mass., Me., Mich., Minn., N.B., N.D., N.S., N.Y., Ont., Ore., Que., Sask., W.Va., Wis.

Deraeocoris fasciolus castus Knight, 1921

1921 *Deraeocoris fasciolus* var. *castus* Knight, Minn. St. Ent. Rept., 18: 125. [N.Y.].

Distribution: Mass., Mich., N.J., N.Y., Oh., Que.

Deraeocoris fasciolus fasciolus Knight, 1921
1921 *Deraeocoris fasciolus* Knight, Minn. St. Ent. Rept., 18: 123.
Distribution: Same as for species.

Deraeocoris fenestratus (Van Duzee), 1917
1917 *Camptobrochis fenestratus* Van Duzee, Proc. Cal. Acad. Sci., (4)7: 266. [Cal.].
1921 *Deraeocoris fenestratus*: Knight, Minn. St. Ent. Rept., 18: 174.
Distribution: Cal.

Deraeocoris franserensis Razafimahatratra and Lattin, 1982
1982 *Deraeocoris franserensis* Razafimahatratra and Lattin, Pan-Pac. Ent., 58: 352. [B.C.].
Distribution: B.C.

Deraeocoris fulgidus (Van Duzee), 1914
1914 *Camptobrochis fulgidus* Van Duzee, Trans. San Diego Soc. Nat. Hist., 2: 21. [Cal.].
1916 *Camptobrochis* (*Mycterocoris*) *fulgidus*: Van Duzee, Check List Hem., p. 41.
1917 *Camptobrochys* [*sic*] (*Mycterocoris*) *fulgidus*: Van Duzee, Univ. Cal. Publ. Ent., 2: 355.
1921 *Deraeocoris fulgidus*: Knight, Minn. St. Ent. Rept., 18: 149.
Distribution: Cal., Col., Nev., Wyo.

Deraeocoris fulvescens (Reuter), 1909
1909 *Camptobrochis* (*Euarmosus*) *fulvescens* Reuter, Acta Soc. Sci. Fenn., 36(2): 53. [Col.].
1914 *Camptobrochis fulvescens*: Van Duzee, Trans. San Diego Soc. Nat. Hist., 2: 21.
1916 *Camptobrochis* (*Euarmosus*) *fulvescens*: Van Duzee, Check List Hem., p. 41.
1917 *Camptobrochys* [*sic*] (*Euarmosus*) *fulvescens*: Van Duzee, Univ. Cal. Publ. Ent., 2: 355.
1921 *Deraeocoris fulvescens*: Knight, Minn. St. Ent. Rept., 18: 153.
1921 *Deraeocoris pilosus* Knight, Minn. St. Ent. Rept., 18: 169. [Ariz.]. Synonymized by Razafimahatratra and Lattin, 1982, Pan-Pac. Ent., 58: 363.
1973 *Deraeocoris pilosulus*: Steyskal, Stud. Ent., 16: 206.
Distribution: Ariz., Cal., Col., N.M.

Deraeocoris fulvus Knight, 1921
1921 *Deraeocoris fulvus* Knight, Minn. St. Ent. Rept., 18: 144. [Ariz.].
Distribution: Ariz.

Deraeocoris fusifrons Knight, 1921
1921 *Deraeocoris fusifrons* Knight, Minn. St. Ent. Rept., 18: 180. [Cal.].
Distribution: Cal.

Deraeocoris fusifrons fusifrons Knight, 1921
1921 *Deraeocoris fusifrons* Knight, Minn. St. Ent. Rept., 18: 180.
Distribution: Cal.

Deraeocoris fusifrons deletans Knight, 1921
1921 *Deraeocoris fusifrons* var. *deletans* Knight, Minn. St. Ent. Rept., 18: 181. [Cal.].
Distribution: Cal.

Deraeocoris gilensis Razafimahatratra and Lattin, 1982
1982 *Deraeocoris gilensis* Razafimahatratra and Lattin, Pan-Pac. Ent., 58: 360. [Ariz.].
Distribution: Ariz.

Deraeocoris grandis (Uhler), 1887
1887 *Camptobrochis grandis* Uhler, Ent. Am., 2: 230. [Ont.].
1909 *Camptobrochis (Euarmosus) grandis*: Reuter, Acta Soc. Sci. Fenn., 36(2): 53.
1917 *Camptobrochys* [*sic*] (*Euarmosus*) *grandis*: Van Duzee, Univ. Cal. Publ. Ent., 2: 355.
1921 *Deraeocoris grandis*: Knight, Minn. St. Ent. Rept., 18: 126.
Distribution: B.C., Cal., Col., D.C., Ia., Ill., Ind., Ks., Mass., Md., Me., Mich., Mo., N.J., N.M., N.Y., Oh., Ont., Pa., Que., Va.
Note: As noted by Knight (1921, above), Uhler's *D. grandis* was a composite of at least four species; some of the records above for this species may be based on misidentifications.

Deraeocoris hesperus Knight, 1921
1921 *Deraeocoris bakeri hesperus* Knight, Minn. St. Ent. Rept., 18: 159. [Cal.].
1957 *Deraeocoris hesperus*: Carvalho, Arq. Mus. Nac., 44: 66.
Distribution: Cal., Id.

Deraeocoris histrio (Reuter), 1876
1876 *Callicapsus histrio* Reuter, Öfv. K. Svens. Vet.-Akad. Förh., 32 (9): 75. [Tex., "Carolina"].
1905 *Calicapsus* [*sic*] *histrio*: Crevecoeur, Trans. Ks. Acad. Sci., 19: 233.
1909 *Camptobrochis (Callicapsus) histrio*: Reuter, Acta Soc. Sci. Fenn., 36(2): 54.
1917 *Camptobrochys* [*sic*] (*Callicapsus*) *histrio*: Van Duzee, Univ. Cal. Publ. Ent., 2: 355.
1921 *Deraeocoris (Camptobrochis) histrio*: Knight, Minn. St. Ent. Rept., 18: 100.
1926 *Deraeocoris* (*Comptobrochis* [*sic*]) *histrio*: Holmquist, An. Ent. Soc. Am., 19: 43.
1949 *Deraeocoris histrio*: Froeschner, Am. Midl. Nat., 42: 137.
1958 *Deraecoris* [*sic*] *historio* [*sic*]: Stearns, J. Econ. Ent., 51: 81.
Distribution: Alta., Cal., Col., D.C., Del., Ia., Ill., Ind., Ks., Man., Mich., Minn., Mo., Mont., N.C., N.Y., Neb., Pa., S.D., Sask., Tex., Wis.

Deraeocoris incertus Knight, 1921
1921 *Deraeocoris incertus* Knight, Minn. St. Ent. Rept., 18: 114. [Ore.].
1921 *Deraeocoris rufusculus* Knight, Minn. St. Ent. Rept., 18: 116. [Ore.]. Synonymized by Razafimahatratra and Lattin, 1982, Pan-Pac. Ent., 58: 363.
Distribution: Ariz., B.C., Cal., Col., Ore.

Deraeocoris kennicotti Knight, 1921
1921 *Deraeocoris kennicotti* Knight, Minn. St. Ent. Rept., 18: 166. [N.T.].
Distribution: Alta., Man., Me., N.T., Sask.

Deraeocoris knightonius Razafimahatratra and Lattin, 1982
1982 *Deraeocoris knightonius* Razafimahatratra and Lattin, Pan-Pac. Ent., 58: 361. [Cal.].
Distribution: Cal., Ore.

Deraeocoris laricicola Knight, 1921
1921 *Deraeocoris laricicola* Knight, Minn. St. Ent. Rept., 18: 164. [N.Y.].
Distribution: Ill., Ind., Man., Mass., Minn., N.Y., Sask.

Deraeocoris luridipes Knight, 1921
1921 *Deraeocoris (Camptobrochis) luridipes* Knight, Minn. St. Ent. Rept., 18: 110. [Cal.].
1957 *Deraeocoris luridipes*: Carvalho, Arq. Mus. Nac., 44: 68.
Distribution: Cal.

Deraeocoris madisonensis Akingbohungbe, 1972
1972 *Deraeocoris madisonensis* Akingbohungbe, An. Ent. Soc. Am., 65: 840. [Wis.].
Distribution: Wis.

Deraeocoris manitou (Van Duzee), 1920
1920 *Camptobrochys* [*sic*] *manitou* Van Duzee, Proc. Cal. Acad. Sci., (4)9: 355. [Col.].
1921 *Deraeocoris manitou*: Knight, Minn. St. Ent. Rept., 18: 153.
Distribution: Ariz., Col., Mo., N.M.

Deraeocoris manitou atratus Knight, 1921.
1921 *Deraeocoris manitou* var. *atratus* Knight, Minn. St. Ent. Rept., 18: 155. [N.M.].
Distribution: Ariz., N.M.

Deraeocoris manitou intermedius Knight, 1921
1921 *Deraeocoris manitou* var. *intermedius* Knight, Minn. St. Ent. Rept., 18: 154. [N.M.].
Distribution: N.M.

Deraeocoris manitou manitou (Van Duzee), 1920
1920 *Camptobrochys* [*sic*] *manitou* Van Duzee, Proc. Cal. Acad. Sci., (4)9: 355. [Col.].
1921 *Deraeocoris manitou*: Knight, Minn. St. Ent. Rept., 18: 153.
Distribution: Ariz., Col., Mo., N.M.

Deraeocoris mutatus Knight, 1921
1921 *Deraeocoris mutatus* Knight, Minn. St. Ent. Rept., 18: 161. [Cal.].
Distribution: Cal.

Deraeocoris navajo Knight, 1921
1921 *Deraeocoris navajo* Knight, Minn. St. Ent. Rept., 18: 155. [Ariz.].
Distribution: Ariz.

Deraeocoris nebulosus (Uhler), 1872
1872 *Camptobrochis nebulosus* Uhler, Prelim. Rept. U.S. Geol. Surv. Mont., p. 417. [Col.].
1903 *Camptobrochis nebulosis* [*sic*]: MacGillivray and Houghton, Ent. News, 14: 262.
1916 *Camptobrochis* (*Camptobrochis*) *nebulosus*: Van Duzee, Check List Hem., p. 41.
1917 *Camptobrochys* [*sic*] (*Camptobrochys* [*sic*]) *nebulosus*: Van Duzee, Univ. Cal., Publ. Ent., 2: 354.
1921 *Deraeocoris* (*Camptobrochis*) *nebulosus*: Knight, Minn. St. Ent. Rept., 18: 91.
1949 *Deraeocoris nebulosus*: Froeschner, Am. Midl. Nat., 42: 164.
1973 *Deraeocoris nubulosus* [*sic*]: Tugwell et al., Ark. Agr. Exp. Stn. Rept., 214: 15.
Distribution: Ala., Ariz., Ark., Cal., Col., Conn., D.C., Del., Fla., Ga., Ia., Ill., Ind., Ks., La., Mass., Md., Me., Minn., Mo., N.B., N.C., N.D., N.H., N.J., N.M., N.S., N.Y., Neb., Oh., Ont., Pa., Que., R.I., Tex., Va., W.Va., Wis.
Note: Wheeler et al. (1975, An. Ent. Soc. Am., 68: 1063-1068) studied biology, hosts, prey, and immature stages.

Deraeocoris nigrifrons Knight, 1921
1921 *Deraeocoris* (*Camptobrochis*) *nigrifrons* Knight, Minn. St. Ent. Rept., 18: 108. [Col.].
1957 *Deraeocoris nigrifrons*: Carvalho, Arq. Mus. Nac., 44: 70.
Distribution: Cal., Col., Wyo.

Deraeocoris nigritulus Knight, 1921
1909 *Camptobrochis* (*Euarmosus*) *nigrita* Reuter, Acta Soc. Sci. Fenn., 36(2): 55. [D.C.]. Preoccupied.
1917 *Camptobrochys* [*sic*] (*Euarmosus*) *nigritus*: Van Duzee, Univ. Cal. Publ. Ent., 2: 355.

1921 *Deraeocoris nigritulus* Knight, Minn. St. Ent. Rept., 18: 170. New name for *Camptobrochis nigrita* Reuter.
Distribution: D.C., Ga., Md., Oh., Va., W.Va.

Deraeocoris nitenatus Knight, 1921
1892 *Camptobrochis grandis*: Heidemann, Proc. Ent. Soc. Wash., 2: 220.
1909 *Camptobrochis* (*Euarmosus*) *nitens* Reuter, Acta Soc. Sci. Fenn., 36(2): 53. [D.C.]. Preoccupied.
1910 *Camptobrochis nitens*: Banks, Cat. Nearc. Hem.-Het., p. 43.
1916 *Camptobrochis* (*Euarmosus*) *grandis nitens*: Van Duzee, Check List Hem., p. 41.
1917 *Camptobrochys* [*sic*] (*Camptobrochys* [*sic*]) *grandis nitens*: Van Duzee, Univ. Cal. Publ. Ent., 2: 356.
1921 *Deraeocoris nitenatus* Knight, Minn. St. Ent. Rept., 18: 141. New name for *Camptobrochis nitens* Reuter.
1938 *Deraeocoris nitennatus* [*sic*]: Brimley, Ins. N.C., p. 80.
Distribution: Col., Conn., D.C., Ia., Ill., Man., Mass., Md., Me., Minn., N.C., N.D., N.J., N.Y., Ont., Pa., Que., Sask., Va., W.Va., Wis.

Deraeocoris nubilus Knight, 1921
1921 *Deraeocoris* (*Camptobrochis*) *nubilus* Knight, Minn. St. Ent. Rept., 18: 106. [N.Y.].
1926 *Camptobrochis nubilus*: Blatchley, Het. E. N. Am., p. 889.
1957 *Deraeocoris nubilus*: Carvalho, Arq. Mus. Nac., 44: 71.
1975 *Deraecoris* [*sic*] *nubilus*: Stinner, Proc. Pa. Acad. Sci., 49: 101.
Distribution: Conn., D.C., Id., Ill., Mass., Minn., Mont., N.C., N.H., N.S., N.Y., Neb., Nfld., Oh., Pa., Va., W.Va., Wis.

Deraeocoris nubilus nubilus Knight, 1921
1921 *Deraeocoris* (*Camptobrochis*) *nubilus* Knight, Minn. St. Ent. Rept., 18: 106. [N.Y.].
1957 *Deraeocoris nubilus* var. *nubilus*: Carvalho, Arq. Mus. Nac., 44: 71.
Distribution: Same as for species.

Deraeocoris nubilus obscuripes Knight, 1921
1921 *Deraeocoris* (*Camptobrochis*) *nubilus* var. *obscuripes* Knight, Minn. St. Ent. Rept., 18: 107. [Mont.].
1957 *Deraeocoris nubilus* var. *obscuripes*: Carvalho, Arq. Mus. Nac., 44: 71.
Distribution: Mont.

Deraeocoris ornatus Knight, 1921
1921 *Deraeocoris* (*Camptobrochis*) *ornatus* Knight, Minn. St. Ent. Rept., 18: 99. [S.D.].
1926 *Camptobrochis ornatus*: Blatchley, Het. E. N. Am., p. 890.
1949 *Deraeocoris ornatus*: Froeschner, Am. Midl. Nat., 42: 138.
Distribution: Alta., Ia., Ill., Man., Mo., Neb., S.D., Sask.

Deraeocoris piceicola Knight, 1927
1927 *Deraeocoris piceicola* Knight, Bull. Brook. Ent. Soc., 22: 136. [Col.].
Distribution: Alta., B.C., Col.

Deraeocoris picipes Knight, 1921
1921 *Deraeocoris incertus* var. *picipes* Knight, Minn. St. Ent. Rept., 18: 116. [Ariz.].
1921 *Deraeocoris incertus* var. *carneolus* Knight, Minn. St. Ent. Rept., 18: 116. [Ariz.]. Synonymized by Razafimahatratra and Lattin, 1982, Pan-Pac. Ent., 58: 353.
1982 *Deraeocoris picipes*: Razafimahatratra and Lattin, Pan-Pac. Ent., 58: 353.
Distribution: Ariz., Col., N.M., Ut.

Deraeocoris pinicola Knight, 1921

1921 *Deraeocoris pinicola* Knight, Minn. St. Ent. Rept., 18: 162. [N.Y.].

Distribution: Conn., D.C., Ia., Mass., Minn., N.H., N.Y., W.Va.

Deraeocoris poecilus McAtee, 1919 [Fig. 77]

1909 *Camptobrochis validus* var. *cunealis* Reuter, Acta Soc. Sci. Fenn., 36(2): 59. [D.C., Nev.]. Preoccupied.

1916 *Camptobrochis (Camptobrochis) validus cunealis*: Van Duzee, Check List Hem., p. 41.

1917 *Camptobrochys* [*sic*] (*Camptobrochys* [*sic*]) *validus cunealis*: Van Duzee, Univ. Cal. Publ. Ent., 2: 354.

1919 *Deraeocoris (Camptobrochis) poecilus* McAtee, Ent. News, 30: 246. New name for *Camptobrochis cunealis* Reuter.

1921 *Deraeocoris (Camptobrochis) cuneatus* Knight, Minn. St. Ent. Rept., 18: 96. Unnecessary new name for *Camptobrochis cunealis* Reuter.

1926 *Camptobrochis poecilus*: Blatchley, Het. E. N. Am., p. 890.

1949 *Deraeocoris poecilus*: Froeschner, Am. Midl. Nat., 42: 138.

Distribution: Conn., D.C., Ia., Ill., Ind., Mass., Md., Minn., Mo., N.C., N.H., N.J., N.Y., Nev.(?), Pa., S.C., W.Va., Wis.

Note: Reuter's (1909, above) description of *D. validus* var. *cunealis* contained syntypes from D.C., Ia., and Nev. D.C. and Ia. are well within the range of *poecilis*; Nev. is not, suggesting that the Ormsby, Nev., specimen represents a different species.

Deraeocoris quercicola Knight, 1921

1921 *Deraeocoris quercicola* Knight, Minn. St. Ent. Rept., 18: 138. [N.Y.].

1928 *Deraeocoris quericola* [*sic*]: Watson, Oh. Biol. Surv. Bull, 16: 36.

Distribution: Col., Conn., Del., Ga., Ia., Ill., Ind., Ks., Man., Mass., Mich., Minn., Mo., N.C., N.M., N.Y., Oh., Ont., Pa., Que., Sask., W.Va., Wis.

Deraeocoris quercicola pallens Knight, 1921

1921 *Deraeocoris quercicola* var. *pallens* Knight, Minn. St. Ent. Rept., 18: 140. [N.Y.].

1928 *Deraeocoris quericola* [*sic*] var. *pallens*: Watson, Oh. Biol. Surv. Bull, 16: 36.

Distribution: Col., Ill., Minn., N.M., N.Y., S.D.

Deraeocoris quercicola quercicola Knight, 1921

1921 *Deraeocoris quercicola* var. *quercicola*: Knight, Minn. St. Ent. Rept., 18: 138. [N.Y.].

Distribution: Same as for species.

Deraeocoris ruber (Linnaeus), 1758

1758 *Cimex ruber* Linnaeus, Syst. Nat., ed. 10, p. 446. [Europe].

1766 *Cimex segusinus* Müller, Manip. Ins. Taur., 3(7): 191. [Europe]. Synonymized by Horvath, 1899, Rev. d'Ent., 17: 277.

1775 *Cimex capillaris* Fabricius, Sys. Ent., p. 725. [Europe]. Synonymized by Reuter, 1879, Ent. Month. Mag., 15: 67.

1794 *Lygaeus danicus* Fabricius, Ent. Sys., 4: 181. [Europe]. Listed as variety of *sequsinus* by Reuter, 1896, Acta Soc. Sci. Fenn., 33(5): 34.

1878 *Capsus capillaris*: Uhler, Proc. Boston Soc. Nat. Hist., 19: 408.

1890 *Deraeocoris segusinus*: Atkinson, J. Asiatic Soc. Bengal, 58(2): 100.

1896 *Deraeocoris segusinus* var. *concolor* Reuter, Acta Soc. Sci. Fenn., 33(5): 34. [No specific locality given; Europe, Am. boreali for species].

1905 *Deraeocoris segusinus* var. *capilaris* [*sic*]: Heidemann, J. N.Y. Ent. Soc., 13: 48.

1917 *Deraeocoris ruber*: Van Duzee, Univ. Cal. Publ. Ent., 2: 356.
1917 *Deraeocoris ruber* var. *concolor*: Van Duzee, Univ. Cal. Publ. Ent., 2: 358.
1917 *Deraeocoris ruber* var. *danicus*: Van Duzee, Univ. Cal. Publ. Ent., 2: 357.
1917 *Deraeocoris ruber* var. *segusinus*: Van Duzee, Univ. Cal. Publ. Ent., 2: 358.
1921 *Deraeocoris ruber* var. *bicolor* Knight, Minn. St. Ent. Rept., 18: 193. [Conn.]. New Synonymy.
1921 *Deraeocoris ruber concolor*: Knight, Minn. St. Ent. Rept., 18: 194.
1921 *Deraeocoris ruber danicus*: Knight, Minn. St. Ent. Rept., 18: 193.
1921 *Deraeocoris ruber ruber*: Knight, Minn. St. Ent. Rept., 18: 191.
1921 *Deraeocoris ruber segusinus*: Knight, Minn. St. Ent. Rept., 18: 193.
Distribution: Conn., Mass., N.J., N.Y., Pa. (Europe, South America).
Note: Knight (1921, Minn. St. Ent. Rept., 18: 194) indicated that all varieties of *ruber* are "merely melanic color forms." We concur with this statement and, therefore, consider all named varieties of *ruber* synonymous with the nominate species/subspecies.

Deraeocoris rubripes Kelton, 1980
1980 *Deraeocoris rubripes* Kelton, Can. Ent., 112: 289. [Man.].
Distribution: Man.

Deraeocoris rubroclarus Knight, 1921
1921 *Deraeocoris rubroclarus* Knight, Minn. St. Ent. Rept., 18: 156. [Ore.].
Distribution: B.C., Ore., Wash.

Deraeocoris rufiventris Knight, 1921
1921 *Deraeocoris rufiventris* Knight, Minn. St. Ent. Rept., 18: 184. [Cal.].
Distribution: Cal., Ore.

Deraeocoris sayi (Reuter), 1876
1876 *Euarmosus sayi* Reuter, Öfv. K. Svens. Vet.-Akad. Förh., 32(9): 76. [Tex.].
1909 *Camptobrochis* (*Euarmosus*) *sayi*: Reuter, Acta Soc. Sci. Fenn., 36(2): 52.
1917 *Camptobrochys* [*sic*] (*Euarmosus*) *sayi*: Van Duzee, Univ. Cal. Publ. Ent., 2: 355.
1921 *Deraeocoris sayi*: Knight, Minn. St. Ent. Rept., 18: 174.
1921 *Deraeocoris sayi* var. *marginatus* Knight, Minn. St. Ent. Rept., 18: 176. [Ill.]. New Synonymy.
1921 *Deraeocoris sayi* var. *costalis* Knight, Minn. St. Ent. Rept., 18: 177. [N.Y.]. New Synonymy.
1921 *Deraeocoris sayi* var. *femoralis* Knight, Minn. St. Ent. Rept., 18: 177. [Minn.]. New Synonymy.
1921 *Deraeocoris sayi* var. *frontalis* Knight, Minn. St. Ent. Rept., 18: 177. [Ill.]. New Synonymy.
1921 *Deraeocoris sayi* var. *unicolor* Knight, Minn. St. Ent. Rept., 18: 177. [Ill.]. New Synonymy.
1921 *Deraeocoris sayi sayi*: Knight, Minn. St. Ent. Rept., 18: 175.
1941 *Deraeocoris* (*Euarmosus*) *sayi*: Knight, Ill. Nat. Hist. Surv. Bull., 22: 68.
Distribution: Ala., Cal., Col., Fla., Ga., Ia., Ill., Ind., La., Man., Mass., Mich., Minn., Miss., Mo., N.C., N.Y., Pa., Tex., Wis.
Note: As with the species *Deraeocoris ruber*, the varietal names given to this species are mere color forms without geographic significance and should not be given subspecific status. We consider them synonyms of the nominate species/subspecies.

Deraeocoris schwarzii (Uhler), 1893
1893 *Camptobrochis schwarzii* Uhler, Proc. Ent. Soc. Wash., 2: 375. [Ut.].
1916 *Camptobrochis* (*Camptobrochis*) *schwarzii*: Van Duzee, Check List Hem., p. 41.
1917 *Camptobrochys* [*sic*] (*Mycterocoris*) *schwarzii*: Van Duzee, Univ. Cal. Publ. Ent., 2: 355.
1921 *Deraeocoris schwarzii*: Knight, Minn. St. Ent. Rept., 18: 146.
Distribution: B.C., Cal., Col., Id., N.M., Ut., Wash., Wyo.

Deraeocoris shastan Knight, 1921
1921 *Deraeocoris shastan* Knight, Minn. St. Ent. Rept., 18: 133. [Cal.].
Distribution: Cal.

Deraeocoris schuhi Razafimahatratra and Lattin, 1982
1982 *Deraeocoris schuhi* Razafimahatratra and Lattin, Pan-Pac. Ent., 58: 359. [Cal.].
Distribution: Cal.

Deraeocoris tinctus Knight, 1921
1921 *Deraeocoris* (*Camptobrochis*) *tinctus* Knight, Minn. St. Ent. Rept., 18: 95. [Col.].
1957 *Deraeocoris tinctus*: Carvalho, Arq. Mus. Nac., 44: 81.
Distribution: Col.

Deraeocoris triannulipes Knight, 1921
1921 *Deraeocoris triannulipes* Knight, Minn. St. Ent. Rept., 18: 137. [Col.].
Distribution: Alta., Col., Man., Mich., Minn., Sask.

Deraeocoris triannulipes flavisignatus Knight, 1927
1927 *Deraeocoris triannulipes flavisignatus* Knight, Bull. Brook. Ent. Soc., 22: 138. [Minn.].
Distribution: Mich., Minn.

Deraeocoris triannulipes triannulipes Knight, 1921
1921 *Deraeocoris triannulipes* Knight, Minn. St. Ent. Rept., 18: 137.
Distribution: Alta., Col., Man., Mich., Minn., Sask.

Deraeocoris tsugae Bliven, 1956
1956 *Deraeocoris tsugae* Bliven, New Hem. W. St., p. 11. [Cal.].
Distribution: Cal.

Deraeocoris validus (Reuter), 1909
1909 *Camptobrochis validus* Reuter, Acta. Soc. Sci. Fenn., 36(2): 58. [Cal., Nev.].
1916 *Camptobrochis* (*Camptobrochis*) *validus*: Van Duzee, Check List Hem., p. 41.
1917 *Camptobrochys* [*sic*] (*Camptobrochys* [*sic*]) *validus*: Van Duzee, Univ. Cal. Publ. Ent., 2: 354.
1921 *Deraeocoris* (*Camptobrochis*) *validus*: Knight, Minn. St. Ent. Rept., 18: 108.
1957 *Deraeocoris validus*: Carvalho, Arq. Mus. Nac., 44: 82.
Distribution: B.C., Cal., N.M., Nev., Ore.
Note: Early eastern records of this species listed by Van Duzee (1917, above) from D.C., Ia., Mass., and Pa. refer to the eastern *D. poecilus* (McAtee).

Deraeocoris vanduzeei Knight, 1921
1921 *Deraeocoris vanduzeei* Knight, Minn. St. Ent. Rept., 18: 183. [Cal.].
Distribution: Cal.

Genus *Diplozona* Van Duzee, 1915

1915 *Diplozona* Van Duzee, Pomona J. Ent. Zool., 7: 112. Type-species: *Diplozona collaris* Van Duzee, 1915. Monotypic.

Diplozona collaris Van Duzee, 1915
1915 *Diplozona collaris* Van Duzee, Pomona J. Ent. Zool., 7: 114. [Fla.].
Distribution: Fla. (Puerto Rico).

Genus *Eurychilopterella* Reuter, 1909

1909 *Eurychilopterella* Reuter, Acta Soc. Sci. Fenn., 36(2): 59. Type-species: *Eurychilopterella luridula* Reuter, 1909. Monotypic.

Eurychilopterella barberi Knight, 1927
1927 *Eurychilopterella barberi* Knight, Bull. Brook. Ent. Soc., 22: 140. [Ariz.].
Distribution: Ariz.

Eurychilopterella brunneata Knight, 1927
1927 *Eurychilopterella brunneata* Knight, Bull. Brook. Ent. Soc., 22: 141. [Ill.].
Distribution: Ill.

Eurychilopterella luridula Reuter, 1909
1909 *Eurychilopterella luridula* Reuter, Acta Soc. Sci. Fenn., 36(2): 60. [D.C.].
Distribution: D.C., Ia., Ill., Minn., Mo., N.S., N.Y., Ont., P.Ed., S.C., Va., W.Va.

Genus *Eustictus* Reuter, 1909

1909 *Eustictus* Reuter, Acta Soc. Sci. Fenn., 36(2): 35. Type-species: *Megacoelum grossum* Uhler, 1887. Original designation.

Eustictus ainsliei Knight, 1926
1926 *Eustictus ainsliei* Knight, Psyche, 33: 121. [Fla.].
Distribution: Fla.

Eustictus albocuneatus Knight, 1927
1927 *Eustictus albocuneatus* Knight, Bull. Brook. Ent. Soc., 22: 139. [Ariz.].
Distribution: Ariz.

Eustictus albomaculatus Johnston, 1939
1939 *Eustictus albomaculatus* Johnston, Bull. Brook. Ent. Soc., 34: 129. [Tex.].
Distribution: Tex.

Eustictus catulus (Uhler), 1894
1894 *Megacoelum catulum* Uhler, Proc. Cal. Acad. Sci., (2)4: 257. [Pa., Tex., Mexico].
1909 *Eustictus catulus*: Reuter, Acta Soc. Sci. Fenn., 36(2): 86.
1916 *Cimatlan catulum*: Van Duzee, Check List Hem., p. 41.
Distribution: Conn., Ks., La., N.Y., Pa., Tex. (Mexico).

Eustictus claripennis Knight, 1925
1925 *Eustictus claripennis* Knight, Bull. Brook. Ent. Soc., 20: 40. [Tex.].
Distribution: Tex.

Eustictus clarus Knight, 1925
1925 *Eustictus claripennis* var. *clarus* Knight, Bull. Brook. Ent. Soc., 20: 40. [Ariz.].

1957 *Eustictus clarus*: Carvalho, Arq. Mus. Nac., 44: 85.
Distribution: Ariz., Tex.

Eustictus grossus (Uhler), 1887
1873 *Capsus filicornis* Walker, Cat. Hem. Brit. Mus., 6: 96. [Fla.]. Preoccupied.
1886 *Megacoelum filicorne*: Uhler, Check-list Hem. Het., p. 18.
1887 *Megacoelum grossum* Uhler, Ent. Am., 3: 70. [Md.].
1904 *Creontiades filicornis*: Distant, An. Mag. Nat. Hist., (7)13: 106.
1909 *Eustictus grossum*: Reuter, Acta Soc. Sci. Fenn., 36(2): 35.
1916 *Cimatlan grossum*: Van Duzee, Check List Hem., p. 41.
1923 *Eustictus grossus*: Knight, Conn. Geol. Nat. Hist. Surv. Bull., 34: 481.
1927 *Eustictus filicornis*: Knight, Bull. Brook. Ent. Soc., 22: 101.
Distribution: D.C., Fla., Mass., Md., Mo., N.C., N.J., N.Y., W.Va.

Eustictus hirsutipes Knight, 1925
1925 *Eustictus hirsutipes* Knight, Bull. Brook. Ent. Soc., 20: 39. [Ariz.].
Distribution: Ariz.

Eustictus knighti Johnston, 1930
1930 *Eustictus knighti* Johnston, Bull. Brook. Ent. Soc., 25: 297. [Tex.].
Distribution: Tex.

Eustictus minimus Knight, 1925
1925 *Eustictus minimus* Knight, Bull. Brook. Ent. Soc., 20: 37. [Tex.].
Distribution: Tex.

Eustictus morrisoni Knight, 1925
1925 *Eustictus morrisoni* Knight, Bull. Brook. Ent. Soc., 20: 38. [Ariz.].
Distribution: Ariz.

Eustictus mundus (Uhler), 1887
1887 *Megacoelum mundum* Uhler, Ent. Am., 3: 71. [Ga.].
1909 *Eustictus mundum*: Van Duzee, Bull. Buff. Soc. Nat. Sci., 9: 179.
1916 *Cimatlan mundum*: Van Duzee, Check List Hem., p. 41.
1926 *Eustictus mundus*: Blatchley, Het. E. N. Am., p. 885.
Distribution: Ala., Fla., Ga., N.C.

Eustictus necopinus Knight, 1923
1923 *Eustictus necopinus* Knight, Conn. Geol. Nat. Hist. Surv. Bull., 34: 481. [N.Y.].
Distribution: B.C., Conn., D.C., Man., Mass., Miss., Mo., N.Y., Ont.

Eustictus necopinus discretus Knight, 1923
1923 *Eustictus necopinus* var. *discretus* Knight, Conn. Geol. Nat. Hist. Surv. Bull., 34: 482. [Conn.].
Distribution: Conn., Mo.

Eustictus necopinus necopinus Knight, 1923
1923 *Eustictus necopinus* Knight, Conn. Geol. Nat. Hist. Surv. Bull., 34: 481. [N.Y.].
Distribution: Same as for species.

Eustictus obscurus Knight, 1925
1925 *Eustictus obscurus* Knight, Bull. Brook. Ent. Soc., 20: 37. [Ariz.].
Distribution: Ariz., N.M.

Eustictus pilipes Knight, 1926
1926 *Eustictus pilipes* Knight, Psyche, 33: 122. [Fla.].
Distribution: Fla.

Eustictus productus Knight, 1925
1925 *Eustictus productus* Knight, Bull. Brook. Ent. Soc., 20: 38. [Ariz.].
Distribution: Ariz.

Eustictus pubescens Knight, 1926
1926 *Eustictus pubescens* Knight, Psyche, 33: 123. [Fla.].
Distribution: Fla.

Eustictus pusillus (Uhler), 1887
1887 *Megacoelum pusillum* Uhler, Ent. Am., 3: 71. [Ariz.].
1906 *Megacoelum pusillus*: Snow, Trans. Kans. Acad. Sci., 20: 152.
1909 *Eustictus pusillus*: Reuter, Acta Soc. Sci. Fenn., 36(2): 86.
1916 *Cimatlan pusillum*: Van Duzee, Check List Hem., p. 41.
Distribution: Ariz., Cal., Col., Tex.

Eustictus salicicola Knight, 1923
1923 *Eustictus salicicola* Knight, Conn. Geol. Nat. Hist. Surv. Bull., 34: 482. [Minn.].
Distribution: Ia., Ill., Ks., Minn., Miss., Mo., N.D., Neb., Ok., S.D., Tex., Ut.

Eustictus spinipes Knight, 1926
1926 *Eustictus spinipes* Knight, Psyche, 33: 122. [Fla.].
Distribution: Fla.

Eustictus tibialis Knight, 1927
1927 *Eustictus tibialis* Knight, Bull. Brook. Ent. Soc., 22: 139. [Tex.].
Distribution: Tex.

Eustictus venatorius Van Duzee, 1912
1912 *Eustictus venatorius* Van Duzee, Bull. Buff. Soc. Nat. Sci., 10: 479. [Fla., N.Y.].
1916 *Cimatlan venatorius*: Van Duzee, Check List Hem., p. 41.
Distribution: Fla., N.Y.

Genus *Klopicoris* Van Duzee, 1915

1915 *Klopicoris* Van Duzee, Pomona J. Ent. Zool., 7: 115. Type-species: *Camptobrochis phorodendronae* Van Duzee, 1914. Original designation.

Klopicoris phorodendronae (Van Duzee), 1914
1914 *Camptobrochis phorodendronae* Van Duzee, Trans. San Diego Soc. Nat. Hist., 2: 22. [Cal.].
1915 *Klopicoris phorodendronae*: Van Duzee, Pomona J. Ent. Zool., 7: 116.
Distribution: Cal.

Genus *Strobilocapsus* Bliven, 1956

1956 *Strobilocapsus* Bliven, New Hem. W. St., p. 12. Type-species: *Strobilocapsus annulatus* Bliven, 1956. Original designation.

Strobilocapsus annulatus Bliven, 1956
1956 *Strobilocapsus annulatus* Bliven, New Hem. W. St., p. 13. [Cal.].
Distribution: Cal.

Tribe Hyaliodini Carvalho and Drake, 1943

Genus *Hyaliodes* Reuter, 1876

1876 *Hyaliodes* Reuter, Öfv. K. Svens. Vet.-Akad. Förh., 32(9): 83. Type-species: *Capsus vitripennis* Say, 1832. Monotypic.

Hyaliodes brevis Knight, 1941
- 1941 *Hyaliodes brevis* Knight, Ill. Nat. Hist. Surv. Bull., 22: 58. [Ill.].
- Distribution: Ia., Ill., Minn., Wis.

Hyaliodes harti Knight, 1941 [Fig. 84]
- 1941 *Hyaliodes harti* Knight, Ill. Nat. Hist. Surv. Bull., 22: 57. [Ill.].
- 1949 *Hyloides* [*sic*] *harti*: Hoffmann et al., Ecol. Monogr., 19: 19.
- Distribution: Alta., B.C., Ga., Ia., Ill., Ind., Man., Me., Mo., N.C., N.D., N.S., N.Y., Ont., Pa., Que., Sask., W.Va., Wis.

Hyaliodes vitripennis (Say), 1832
- 1832 *Capsus vitripennis* Say, Descrip. Het. Hem. N. Am., p. 24. [Ind., Pa.].
- 1870 *Hyaliodes vitripennis*: Riley, An. Rept. Nox. Ins. Mo., p. 137.
- 1875 *Campyloneura* (*Capsus*) *vitripennis*: Glover, Rept. Dept. Agr. U.S., p. 125.
- 1883 *Campyloneura vitripennis*: Saunders, Ins. Inj. Fruit, p. 228.
- 1909 *Hyaliodes vitripennis* var. *discoidalis* Reuter, Acta Soc. Sci. Fenn., 36(2): 6. [Pa.]. Synonymized by Larochelle, 1984, Fabreries, Suppl. 3, p. 315.
- 1916 *Hyaliodes vitripennis discoidalis*: Van Duzee, Check List Hem., p. 43.
- 1950 *Hyaloides* [*sic*] *vitripennis*: Moore, Nat. Can., 77: 247.
- 1950 *Hyaloides* [*sic*] *vitripennis discoidalis*: Moore, Nat. Can., 77: 247.
- 1980 *Hyloides* [*sic*] *vitripennis*: Bouchard et al., An. Soc. Ent. Que., 27: 84.
- Distribution: Ark., B.C., Conn., Del., Ia., Ill., Ind., Ks., Mass., Me., Mich., Mo., N.C., N.H., N.J., N.S., N.Y., Oh., Ont., Pa., Que., R.I., Tenn., Va., Vt., W.Va., Wis.

Genus *Paracarnus* Distant

1884 *Paracarnus* Distant, Biol. Centr.-Am., Rhyn., 1: 289. Type-species: *Paracarnus elongatus* Distant, 1884. Monotypic.

Paracarnus cubanus Bruner, 1934
- 1934 *Paracarnus cubanus* Bruner, Mem. Soc. Poey, 8: 36. [Cuba].
- 1982 *Paracarnus cubanus*: Henry and Wheeler, Fla. Ent., 65: 237.
- Distribution: Fla. (Cuba, Puerto Rico).

Genus *Stethoconus* Flor, 1861

1861 *Stethoconus* Flor, Rhyn. Livl., 2: 216. Type-species: *Capsus cyrtopeltis* Flor, 1860. Monotypic.

Note: Henry et al. (1986, Proc. Ent. Soc. Wash., 88: 728) transferred this genus to Hyaliodini.

Stethoconus japonicus Schumacher, 1917 [Fig. 80]
- 1917 *Stethoconus japonicus* Schumacher, Sitz. Gesell. Naturf. Freunde Berlin, 9: 345. [Japan].
- 1986 *Stethoconus japonicus*: Henry et al., Proc. Ent. Soc. Wash., 88: 722.
- Distribution: Md. (Japan).

Tribe Termatophylini Reuter, 1884

Genus *Conocephalocoris* Knight, 1927

1927 *Conocephalocoris* Knight, Bull. Brook. Ent. Soc., 22: 141. Type-species: *Conocephalocoris nasicus* Knight, 1927. Original designation.

Conocephalocoris nasicus Knight, 1927

1927 *Conocephalocoris nasicus* Knight, Bull. Brook. Ent. Soc., 22: 142. [Ariz.].
Distribution: Ariz.

Genus *Hesperophylum* Reuter and Poppius, 1912

1912 *Hesperophylum* Reuter and Poppius, Öfv. F. Vet.-Soc. Förh., 54A(1): 16. Type-species: *Hesperophylum heidemanni* Reuter and Poppius, 1912. Monotypic.

Hesperophylum arizonae Knight, 1968

1968 *Hesperophylum arizonae* Knight, Brig. Young Univ. Sci. Bull., 9(3): 80. [Ariz.].
Distribution: Ariz.

Hesperophylum heidemanni Reuter and Poppius, 1912

1912 *Hesperophylum heidemanni* Reuter and Poppius, Öfv. F. Vet.-Soc. Förh., 54A(1): 16. [N.H.].
Distribution: D.C., Ia., N.H.
Note: The Arizona record for *H. heidemanni* by Barber (1914, J. N.Y. Ent. Soc., 22: 170) is here referred to the species *H. arizonae*. Authorship for *Hesperophylum heidemanni* has been credited to Reuter and Poppius or R euter only, but in the introduction of their paper (Reuter and Poppius, 1912, above) Reuter acknowledges Poppius for his assistance in constructing the descriptions of species he previously recognized as new; thus, co-authorship is interpreted.

Subfamily Isometopinae Fieber, 1860

Note: Henry (1980, Proc. Ent. Soc. Wash., 82: 179-181) provided a key to the genera of the Western Hemisphere. Wheeler and Henry (1978, An. Ent. Soc. Am., 71: 610) gave a key to last-instar nymphs of five eastern U.S. species.

Although Akingbohungbe and Henry (1984, XVII Internat. Congr. Ent., Abst. Vol., P.13) proposed the name Myommaria [sic] and suggested according it full tribal status, we conserve the following two-tribe scheme of the subfamily, until further revisionary work is completed.

Tribe Diphlebini Bergroth, 1924

Genus *Diphleps* Bergroth, 1924

1924 *Diphleps* Bergroth, Not. Ent., 4: 4. Type-species: *Diphleps unica* Bergroth, 1924. Monotypic.

1924 *Teratodia* Bergroth, Not. Ent., 4: 7. Type-species: *Teratodia emoritura* Bergroth, 1924. Monotypic. Synonymized by Henry, 1977, Fla. Ent., 60: 201.

Note: Henry (1977, Fla. Ent., 60: 209) provided a key to the three known species.

Diphleps unica Bergroth, 1924 [Fig. 86]

1924 *Diphleps unica* Bergroth, Not. Ent., 4: 7. [Oh.].

1924 *Teratodia emoritura* Bergroth, Not. Ent., 4: 8. [Va.]. Synonymized by Henry, 1977, Fla. Ent., 60: 203.

Distribution: Fla., Ill., Mo., N.C., Oh., Pa., Va., W.Va.

Note: Wheeler and Henry (1978, An. Ent. Soc. Am., 71: 608-612) gave information on biology and described and illustrated the fifth instar.

Tribe Isometopini Fieber, 1860

Genus *Corticoris* McAtee and Malloch, 1922

1922 *Corticoris* McAtee and Malloch, Proc. Biol. Soc. Wash., 35: 95. Type-species: *Isometopus pulchellus* Heidemann, 1908. Original designation.

1924 *Dendroscirtus* Bergroth, Not. Ent., 4: 4. Unnecessary new name for *Corticoris* McAtee and Malloch.

Note: Henry and Herring (1979, Proc. Ent. Soc. Wash., 81: 84) provided a key to species; Henry (1984, Proc. Ent. Soc. Wash., 86: 341) gave a revised key.

Corticoris libertus (Gibson), 1917

1917 *Isometopus libertus* Gibson, Bull. Brook. Ent. Soc., 12: 76. [N.M.].

1924 *Dendroscirtus libertus*: Bergroth, Not. Ent., 4: 8.

1924 *Isometopus liberatus* [*sic*]: McAtee and Malloch, Bull. Brook. Ent. Soc., 19: 79.

1924 *Corticoris libertus*: McAtee and Malloch, Bull. Brook. Ent. Soc., 19: 79.

Distribution: Ariz., N.M.

Note: McAtee and Malloch (1924, above) suggested that *C. libertus* was a color variety of *Corticoris unicolor* (Heidemann), but Henry and Herring (1979, Proc. Ent. Soc. Wash., 81: 86) showed that the two species are distinct.

Corticoris pulchellus (Heidemann), 1908

1908 *Isometopus pulchellus* Heidemann, Proc. Ent. Soc. Wash., 9: 128. [N.Y., Va., W.Va.].

1922 *Corticoris pulchellus*: McAtee and Malloch, Proc. Biol. Soc. Wash., 35: 95.

1924 *Dendroscirtus pulchellus*: Bergroth, Not. Ent., 4: 8.

Note: Wheeler and Henry (1978, An. Ent. Soc. Am., 71: 608-610) gave information on biology and described and illustrated the fifth instar.

Distribution: Md., N.Y., Pa., Va., W.Va.

Corticoris signatus (Heidemann), 1908

1908 *Isometopus signatus* Heidemann, Proc. Ent. Soc. Wash., 9: 129. [Tex.].

1924 *Dendroscirtus signatus*: Bergroth, Not. Ent., 4: 8.

1924 *Corticoris signatus*: McAtee and Malloch, Bull. Brook. Ent. Soc. 19: 79.

Distribution: Ala., D.C., Fla., N.C., Ont., Pa., Tex., W.Va. (Mexico).

Note: Wheeler and Henry (1978, An. Ent. Soc. Am., 71: 608-610) gave information on feeding habits and biology and described and illustrated the fifth instar.

Corticoris unicolor (Heidemann), 1908

1908 *Isometopus unicolor* Heidemann, Proc. Ent. Soc. Wash., 9: 130. [Ariz.].

1917 *Myiomma media* Gibson, Bull. Brook. Ent. Soc., 12: 75. [Ariz.]. Synonymized by McAtee and Malloch, 1924, Bull. Brook. Ent. Soc., 19: 79.

1924 *Dendroscirtus unicolor*: Bergroth, Not. Ent., 4: 8.

1924 *Corticoris unicolor*: McAtee and Malloch, Bull. Brook. Ent. Soc., 19: 79.

Distribution: Ariz. (Mexico).

Genus *Lidopus* Gibson, 1917

1917 *Lidopus* Gibson, Bull. Brook. Ent. Soc., 12: 74. Type-species: *Lidopus heidemanni* Gibson, 1917. Monotypic.

Note: Henry (1980, Proc. Ent. Soc. Wash., 82: 186) provided a key to species.

Lidopus heidemanni Gibson, 1917

1917 *Lidopus heidemanni* Gibson, Bull. Brook. Ent. Soc., 12: 74. [Tex.].

Distribution: Ala., Fla., Ill., N.C., Tenn., Tex. (Mexico).

Note: Wheeler and Henry (1978, An. Ent. Soc. Am., 71: 610) included the fifth instar in a key to nymphs of the eastern U.S. isometopines.

Genus *Myiomma* Puton, 1872

1872 *Myiomma* Puton, Pet. Nouv. Ent., 44: 177. Type-species: *Myiomma fieberi* Puton, 1872. Monotypic.

1891 *Heidemannia* Uhler, Proc. Ent. Soc. Wash., 2: 119. Type-species: *Heidemannia cixiiformis* Uhler, 1891. Monotypic. Synonymized by Henry, 1979, Proc. Ent. Soc. Wash., 81: 553.

Note: Henry (1979, Proc. Ent. Soc. Wash., 81: 554, 556) provided a key to the New World species.

Myiomma cixiiforme (Uhler), 1891

1891 *Heidemannia cixiiformis* Uhler, Proc. Ent. Soc. Wash., 2: 121. [D.C., Md.]

1912 *Myiomma cixiiformis*: Reuter, Öfv. F. Vet.-Soc. Förh., 54A: 27.

1979 *Myiomma cixiiforme*: Henry, Proc. Ent. Soc. Wash., 81: 559.

Distribution: D.C., Del., Fla., Md., Ont., Pa., Que., Tex., Va., W.Va. (Mexico).

Note: Wheeler and Henry (1978, An. Ent. Soc. Am., 71: 608-612) gave information on feeding habits and biology and described and illustrated the fifth instar.

Genus *Wetmorea* McAtee and Malloch, 1924

1924 *Wetmorea* McAtee and Malloch, Bull. Brook. Ent. Soc., 19: 80. Type-species: *Wetmorea notabilis* McAtee and Malloch, 1924. Monotypic.

Wetmorea notabilis McAtee and Malloch, 1924

1924 *Wetmorea notabilis* McAtee and Malloch, Bull. Brook. Ent. Soc., 19: 80. [Ariz.]

Distribution: Ariz. (Mexico).

Note: Henry (1980, Proc. Ent. Soc. Wash., 82: 193) redescribed this species and gave new distribution records; Henry (1984, Proc. Ent. Soc. Wash., 86: 344) gave an additional distribution record.

Subfamily Mirinae Hahn, 1833

Tribe Herdoniini Distant, 1904

Genus *Barberiella* Poppius, 1914

1914 *Barberiella* Poppius, An. Soc. Ent. Belg., 58: 255. Type-species: *Barberiella formicoides* Poppius, 1914. Monotypic.

Barberiella formicoides Poppius, 1914
- 1914 *Barberiella formicoides* Poppius, An. Soc. Ent. Belg., 58: 256. [Tex.].
- 1918 *Zoshippus* sp.: Knight, J. N.Y. Ent. Soc., 26: 44.
- 1923 *Barberiella apicalis* Knight, Conn. Geol. Nat. Hist. Surv. Bull., 34: 657. [N.Y.]. Synonymized by Carvalho and Schaffner, 1975, Rev. Brasil. Biol., 35: 356.
- 1926 *Pilophorus brimleyi* Blatchley, Ent. News, 37: 165. [N.C.]. Synonymized by Knight, 1927, Bull. Brook. Ent. Soc., 22: 102.

Distribution: Ark., Ga., Ill., Miss., Mo., N.C., N.Y., Pa., Tex. (Mexico).

Note: Wheeler and Henry (1980, Proc. Ent. Soc. Wash., 82: 269-275) studied seasonal history and host plants and redescribed and illustrated the fifth-instar nymph of this myrmecomorphic species.

Genus *Closterocoris* Uhler, 1890

1890 *Closterocoris* Uhler, Trans. Md. Acad. Sci., 1: 76. Type-species: *Closterocoris ornata* Uhler, 1890, a junior synonym of *Pycnopterna amoena* Provancher, 1887. Monotypic.

Note: Kelton (1959, Can. Ent., 91: 47) and Schuh (1974, Ent. Am., 47: 316) agreed that this genus should be placed in the Mirinae rather than Phylinae.

Closterocoris amoenus (Provancher), 1887
- 1887 *Pycnopterna amoena* Provancher, Pet. Faune Ent. Can., 3: 114. [Can.]. Lectotype designated by Kelton, 1968, Nat. Can., 95: 1077.
- 1890 *Closterocoris ornata* Uhler, Trans. Md. Acad. Sci., 1: 77. [Cal.]. Synonymized by Van Duzee, 1912, Can. Ent., 44: 321.
- 1912 *Closterocoris amoenus*: Van Duzee, Can. Ent., 44: 321.

Distribution: Ariz., Cal., Col., Ks. (Mexico).

Note: Provancher (1887, above) described *amoena* from "Ottawa"; however, Kelton (1968, Nat. Can., 95: 1078) pointed out that this species is a western North American species and it is likely that his material originated from that area.

Genus *Cyphopelta* Van Duzee, 1910

1910 *Cyphopelta* Van Duzee, Trans. Am. Ent. Soc., 36: 81. Type-species: *Cyphopelta modesta* Van Duzee, 1910. Original designation.

Note: Kelton (1959, Can. Ent., 91: 47) and Schuh (1974, Ent. Am., 47: 316) agreed that this genus should be placed in the subfamily Mirinae, not Orthotylinae.

Cyphopelta modesta Van Duzee, 1910
- 1910 *Cyphopelta modesta* Van Duzee, Trans. Am. Ent. Soc., 36: 81. [Cal.].

Distribution: Cal.

Genus *Dacerla* Signoret, 1881

1881 *Dacerla* Signoret, Bull. Séan. Soc. Ent. France, ser. 6, p. clvii. Type-species: *Dacerla mediospinosa* Signoret, 1881. Monotypic.
1894 *Myrmecopsis* Uhler, Proc. Cal. Acad. Sci., (2)4: 276. Type-species: *Myrmecopsis inflatus* Uhler, 1894. Monotypic. Preoccupied.
Note: Carvalho and Usinger (1957, Wasmann J. Biol., 15: 4) provided a key to species.

Dacerla alata Carvalho and Usinger, 1957
1957 *Dacerla alata* Carvalho and Usinger, Wasmann J. Biol., 15: 7. [Cal.].
Distribution: Cal.

Dacerla mediospinosa Signoret, 1881
1881 *Dacerla mediospinosa* Signoret, Bull. Séan. Soc. Ent. France, ser. 6, p. clvii. [Cal.].
1894 *Myrmecopsis inflatus* Uhler, Proc. Cal. Acad. Sci., (2)4: 277. [Cal.]. Synonymized by Bergroth, 1897, Ent. News, 8: 95.
1897 *Dacerla inflata*: Bergroth, Ent. News, 8: 95.
1909 *Dacerla inflata* var. *rufuscula* Reuter, Acta Soc. Sci. Fenn., 36(2): 8. [Cal.].
Distribution: Cal.
Note: Carvalho and Usinger (1957, Wasmann J. Biol., 15: 1-13) discussed the identity of this species and gave notes on its color variation, habits, and relationship to certain species of ants.

Genus *Heidemanniella* Poppius, 1914

1914 *Heidemanniella* Poppius, An. Soc. Ent. Belg., 58: 258. Type-species: *Heidemanniella scutellaris* Poppius, 1914. Monotypic.
Note: Schuh (1974, Ent. Am., 47: 316) noted that this genus was closely related to *Closterocoris* and *Cyphopelta* and probably should be transferred from the Hallodapini to the Herdoniini; more recently he (Schuh, *in litt.*) has confirmed this placement.

Heidemanniella scutellaris Poppius, 1914
1914 *Heidemanniella scutellaris* Poppius, An. Soc. Ent. Belg., 58: 259. [Ariz.].
Distribution: Ariz.

Genus *Paradacerla* Carvalho and Usinger, 1957

1957 *Paradacerla* Carvalho and Usinger, Wasmann J. Biol., 15: 8. Type-species: *Dacerla formicina* Parshley, 1921. Original designation.
Note: Carvalho and Usinger (1957, above) provided a key to species; Kelton and Knight (1959, Can. Ent., 91: 122) gave a revised key.

Paradacerla downesi (Knight), 1927
1927 *Dacerla downesi* Knight, Ent. News, 38: 314. [Ore.].
1957 *Paradacerla downesi*: Carvalho and Usinger, Wasmann J. Biol., 15: 13.
Distribution: B.C., Id., Mont., Ore., Ut., Wash.

Paradacerla formicina (Parshley), 1921
1921 *Dacerla formicina* Parshley, Proc. Ent. Soc. B.C., 18: 18. [B.C.].
1957 *Paradacerla formicina*: Carvalho and Usinger, Wasmann J. Biol., 15: 12.
Distribution: B.C., Cal., Id., Ore.

Genus *Paraxenetus* Reuter, 1907

1907 *Paraxenetus* Reuter, An. Nat. Hofmus. Wien, 22: 46. Type-species: *Xenetus bracteatus* Distant, 1883. Designated by Van Duzee, 1916, Check List Hem., p. 37.

Paraxenetus guttulatus (Uhler), 1887 [Fig. 79]
1887 *Eucerocoris guttulatus* Uhler, Ent. Am., 3: 150. [Ill., Md., Tex.].
1907 *Paraxenetus guttulatus*: Reuter, An. Nat. Hofmus. Wien, 22: 47.
Distribution: D.C., Fla., Ia., Ill., Ind., Ks., Md., Miss., Mo., N.J., N.Y., Oh., Pa., Tex., Va., W.Va.

Tribe Mirini Hahn, 1833

Genus *Adelphocoris* Reuter, 1896

1896 *Adelphocoris* Reuter, Öfv. F. Vet.-Soc. Förh., 38: 168. Type-species: *Cimex seticornis* Fabricius, 1775. Monotypic.

Adelphocoris idahoensis Bliven, 1959
1959 *Adelphocoris idahoensis* Bliven, Occ. Ent., 1(3): 30. [Id.].
Distribution: Id.

Adelphocoris lineolatus (Goeze), 1778
1778 *Cimex lineolatus* Goeze, Ent. Beitr., 2: 267. [Europe].
1922 *Adelphocoris lineolatus*: Knight, Can. Ent., 53: 287.
Distribution: Alta., Conn., Ia., Ill., Ks., Man., Minn., Mo., Mont., Neb., N.S., N.Y., Oh., Ont., Pa., S.D., Sask., Vt., W.Va., Wis. (Palearctic).
Note: Craig (1963, Can. Ent., 95: 6-13) studied the biology in Sask.

Adelphocoris rapidus (Say), 1832 [Fig. 76]
1832 *Capsus rapidus* Say, Descrip. Het. Hem. N. Am., p. 20. [Ind.].
1845 *Capsus multicolor* Herrich-Schaeffer, Wanz. Ins., 8: 19. [N.Am.]. Synonymized by Reuter, 1909, Acta. Soc. Sci. Fenn., 36(2): 36.
1869 *Phytocoris rapidus*: Rathvon, Hist. Lancaster Co., Pa., p. 549.
1872 *Calocoris rapidus*: Uhler, Prelim. Rept. U.S. Geol. Surv. Mont., p. 410.
1876 *Calocoris multicolor*: Reuter, Öfv. K. Svens. Vet.-Akad. Förh., 32(9): 70.
1878 *Deraeocoris rapidus*: Uhler, Proc. Boston Soc. Nat. Hist., 19: 401.
1886 *Draeocoris* [*sic*] *rapidus*: Webster, An. Rept. U.S. Dept. Agr., p. 317.
1897 *Colocaris* [*sic*] *rapidus*: Seiss, Ent. News, 8: 67.
1907 *Calacoris* [*sic*] *rapidus*: Moore, Can. Ent., 39: 163.
1908 *Adelphocoris rapidus*: Van Duzee, Can. Ent., 40: 113.
1910 *Adelphacoris* [*sic*] *rapidus*: Smith, Ins. N.J., p. 163.
1950 *Adelphocaris* [*sic*] *rapidus*: Pielou, Can. Ent., 82: 150.
1978 *Adlephocoris* [*sic*] *rapidus*: Messina, J. N.Y. Ent. Soc., 86: 139.
Distribution: Alta., Ariz., Ark., B.C., Cal., Col., Conn., D.C., Del., Fla., Ga., Ia., Ill., Ind., Ks., Ky., La., Man., Mass., Me., Miss., Mo., N.C., N.H., N.J., N.M., N.Y., Neb., Oh., Ont., Pa., Que., R.I., S.D., Sask., Tenn., Tex., Ut., Va., Vt., W.Va., Wis., Wyo.
Note: Webster and Stoner (1914, J. N.Y. Ent. Soc., 22: 229-234) described immature stages and gave notes on seasonal history.

Adelphocoris superbus (Uhler), 1875

1875 *Calocoris superbus* Uhler, Wheeler's Rept. 100th Merid., 5: 838. [Cal.].

1909 *Adelphocoris superbus*: Reuter, Acta. Soc. Sci. Fenn., 36(2): 36.

1917 *Adelphocoris superbus* var. *borealis* Van Duzee, Proc. Cal. Acad. Sci., (4)7: 263. [Wash.]. New Synonymy.

Distribution: Alta., Ariz., B.C., Cal., Col., Ia., Id., Ks., Nev., N.D., N.M., S.D., Sask., Tex., Ut., Wash. (Mexico).

Note: No workers subsequent to Van Duzee (1917, above) have recognized the color variety *borealis*; it is here considered a synonym of the nominate species/subspecies. Slater and Baranowski (1978, How to Know True Bugs, p. 170) noted that the species status of *A. superbus* and its relationship to *A. rapidus* need further study.

Genus *Agnocoris* Reuter, 1875

1875 *Cyphodema* (*Agnocoris*) Reuter, Rev. Crit. Caps., 1: 82. Type-species: *Lygaeus rubicundus* Fallén, 1807. Monotypic.

1955 *Agnocoris*: Moore, Proc. Ent. Soc. Wash., 57: 175.

Agnocoris pulverulentus (Uhler), 1892

1892 *Hadronema pulverulentus* Uhler, Trans. Md. Acad. Sci., 1: 183. [D.C.]. Lectotype designated by Wagner and Slater, 1952, Proc. Ent. Soc. Wash., 54: 276.

1905 *Hadronema pulverulenta*: Crevecoeur, Trans. Ks. Acad. Sci., 19: 233.

1906 *Hadronema pulverulentum*: Snow, Trans. Ks. Acad. Sci., 20: 152.

1907 *Lygus* (*Hadronema*) *pulverulenta*: Tucker, Ks. Univ. Sci. Bull., 4: 60.

1909 *Lygus rubicundus*: Reuter, Acta. Soc. Sci. Fenn., 36(2): 45.

1916 *Lygus* (*Agnocoris*) *rubicundus*: Van Duzee, Check List Hem., p. 40 (in part).

1952 *Lygus* (*Agnocoris*) *pulverulentus*: Wagner and Slater, Proc. Ent. Soc. Wash., 54: 276.

1955 *Agnocoris pulverulenta*: Moore, Proc. Ent. Soc. Wash., 57: 176.

1955 *Agnocoris pulvurulentus* [*sic*]: Kelton, Can. Ent., 87: 282.

1959 *Agnocoris pulverulentus*: Carvalho, Arq. Mus. Nac., 48: 23.

Distribution: Alk., Alta., Cal., Col., Conn., D.C., Ga., Ia., Id., Ill., Ind., Ks., Man., Mass., Md., Mo., N.C., N.J., N.M., N.Y., Oh., Ont., Pa., Tex., Ut., W.Va., Wis.

Agnocoris rossi Moore, 1955

1941 *Lygus rubicundus*: Knight, Ill. Nat. Hist. Surv. Bull., 22: 153 (in part).

1955 *Agnocoris rossi* Moore, Proc. Ent. Soc. Wash., 57: 176. [Ill.].

Distribution: Ill., La., Mo., Tenn., Tex., Ut. (Mexico).

Agnocoris rubicundus (Fallén), 1807

1807 *Lygaeus rubicundus* Fallén, Monogr. Cimic. Suec., p. 84. [Europe].

1917 *Lygus rubicundus*: Knight, Cornell Univ. Agr. Exp. Stn. Bull., 391: 589.

1917 *Lygus rubicundus* var. *winnipegensis* Knight, Cornell Univ. Agr. Exp. Stn. Bull., 391: 591. [Man.]. Synonymized by Wagner and Slater, Proc. Ent. Soc. Wash., 54: 278.

1916 *Lygus* (*Agnocoris*) *rubicundus*: Van Duzee, Check List Hem., p. 40 (in part).

1955 *Agnocoris rubicundus*: Kelton, Can. Ent., 87: 282.

1956 *Agnocoris rubicunda*: Moore, J. Ks. Ent. Soc., 29: 37.

Distribution: Alta., Id., Man., Ont., Sask.

Note: Nearctic records for *A. rubicundus* have been confused with *A. pulverulentus* (Uhler). Moore (1956, above) and Kelton (1980, Agr. Can. Res. Publ. 1703, p. 80) clarified the N. Am. records, giving the distribution listed above.

Agnocoris utahensis Moore, 1955

1955 *Agnocoris utahensis* Moore, Proc. Ent. Soc. Wash., 57: 178. [Ut.].

Distribution: Alta., Cal., Id., Man., Ont., Ore., Sask., Ut., Wyo.

Genus *Allorhinocoris* Reuter, 1876

1876 *Allorhinocoris* Reuter, Pet. Nouv. Ent., 2: 33. Type-species: *Allorhinocoris prasinus* Reuter, 1876, a junior synonym of *Allorhinocoris flavus* Sahlberg, 1878. Monotypic.

Allorhinocoris flavus Sahlberg, 1878

1878 *Allorhinocoris flavus* Sahlberg, K. Svens. Vet.-Akad. Handl., 16(4): 24. [USSR].

1916 *Allorhinocoris flavus*: Van Duzee, Check List Hem., p. 38.

Distribution: Cal., Ore. (Europe, USSR).

Allorhinocoris speciosus Bliven, 1960

1960 *Allorhinocoris speciosus* Bliven, Occ. Ent., 1(4): 37. [Cal.].

Distribution: Cal.

Genus *Bolteria* Uhler, 1887

1887 *Bolteria* Uhler, Ent. Am., 3: 33. Type-species: *Bolteria amicta* Uhler, 1887. Monotypic.

Note: Knight (1971, Ia. St. J. Sci., 46: 87-88) provided a key to species.

Bolteria amicta Uhler, 1887

1887 *Bolteria amicta* Uhler, Ent. Am., 3: 33. [N.M.]. Lectotype designated by Knight, 1919, Bull. Brook. Ent. Soc., 14: 126.

Distribution: Ariz., N.M.

Bolteria arizonae Knight, 1971

1971 *Bolteria arizonae* Knight, Ia. St. J. Sci., 46: 89. [Ariz.].

Distribution: Ariz.

Bolteria atricornis Kelton, 1972

1972 *Bolteria atricornis* Kelton, Can. Ent., 104: 634. [N.M.].

Distribution: N.M.

Bolteria balli Knight, 1928

1928 *Bolteria balli* Knight, Bull. Brook. Ent. Soc., 23: 130. [Ut.].

Distribution: Ut.

Bolteria dakotae Knight, 1971

1971 *Bolteria dakotae* Knight, Ia. St. J. Sci., 46: 90. [S.D.].

Distribution: S.D.

Bolteria juniperi Knight, 1968

1968 *Bolteria juniperi* Knight, Brig. Young Univ. Sci. Bull., 9(3): 202. [Ut.].

Distribution: Ariz., Nev., Ut.

Bolteria luteifrons Knight, 1921 [Fig. 75]

1921 *Bolteria luteifrons* Knight, Bull. Brook. Ent. Soc., 16: 73. [N.C.].

Distribution: Ark., Col., Ia., Md., Me., Mo., N.C., N.H., Oh., Ont., Pa., Que., S.C., S.D., Va., W.Va.

Note: Wheeler and Henry (1977, Trans. Am. Ent. Soc., 103: 628-629) described the fifth-instar nymph and summarized the seasonal history on ornamental junipers, *Juniperus* spp.

Bolteria nevadensis Knight, 1971
1971 *Bolteria nevadensis* Knight, Ia. St. J. Sci., 46: 90. [Nev.].
Distribution: Nev.

Bolteria nicholi Knight, 1928
1928 *Bolteria nicholi* Knight, Bull. Brook. Ent. Soc., 23: 129. [Ariz.].
Distribution: Ariz., Nev.

Bolteria omani Knight, 1971
1971 *Bolteria omani* Knight, Ia. St. J. Sci., 46: 89. [Cal.].
Distribution: Cal.

Bolteria rubropallida (Knight), 1918
1918 *Dichrooscytus speciosus* var. *rubropallidus* Knight, Bull. Brook. Ent. Soc., 13: 115. [N.M.].
1919 *Dichrooscytus speciosus nigropallidus*: Knight, Bull. Brook. Ent. Soc., 14: 127. *Lapsus* for *D. rubropallidus*.
1928 *Bolteria rubropallida*: Knight, Bull. Brook. Ent. Soc., 23: 131.
Distribution: N.M.

Bolteria schaffneri Knight, 1971
1971 *Bolteria schaffneri* Knight, Ia. St. J. Sci., 46: 91. [Col.].
Distribution: Col.

Bolteria scutata Kelton, 1972
1972 *Bolteria scutata* Kelton, Can. Ent., 104: 632. [Cal.].
Distribution: Cal.

Bolteria scutellata Kelton, 1972
1972 *Bolteria scutellata* Kelton, Can. Ent., 104: 632. [Cal.].
Distribution: Cal.

Bolteria siouxan Knight, 1971
1971 *Bolteria siouxan* Knight, Ia. St. J. Sci., 46: 91. [Wyo.].
Distribution: Wyo.

Bolteria speciosa (Van Duzee), 1916
1916 *Dichrooscytus speciosus* Van Duzee, Univ. Cal. Publ. Ent., 1: 236. [Cal.].
1920 *Bolteria speciosa*: Knight, Bull. Brook. Ent. Soc., 14: 127.
Distribution: Cal.

Genus *Calocoris* Fieber, 1858

1858 *Calocoris* Fieber, Wien. Ent. Monat., 2: 305. Type-species: *Capsus affinis* Herrich-Schaeffer, 1835. Designated by Kirkaldy, 1906, Trans. Am. Ent. Soc., 32: 137.

Calocoris barberi Henry and Wheeler, New Name
1906 *Poecilocapsus sexmaculatus* Barber, Mus. Brook. Inst. Arts Sci., Sci. Bull., 1: 280. [Tex.]. Preoccupied.
1909 *Horcias sexmaculatus*: Reuter, Acta. Soc. Sci. Fenn., 36(2): 41.
1974 *Calocoris sexmaculatus*: Carvalho and Jurberg, Rev. Brasil. Biol., 34: 51.

Distribution: Col., Ok., Tex.
Note: Carvalho and Jurberg (1974, above) placed *sexmaculatus* Barber in *Calocoris* with a question, creating a secondary homonym with the European *Cimex sexmaculatus* Müller, 1776, a junior synonym of *Calocoris sexguttatus* (Fabricius), 1776.

Calocoris fasciativentris Stål, 1862
1862 *Calocoris fasciativentris* Stål, Stett. Ent. Zeit., 23(7-9): 320. [Mexico].
1872 *Calocoris palmeri* Uhler, Prelim. Rept. U.S. Geol. Surv. Mont., p. 410. [Ariz.]. Synonymized by Knight, 1942, Ent. News, 53: 158.
1875 *Calocoris palmerii* [*sic*]: Uhler, Wheeler's Rept. 100th Merid., 5: 838.
1916 *Adelphocoris palmeri*: Van Duzee, Check List Hem., p. 38.
1925 *Horcias fasciativentris*: Knight, Bull. Brook. Ent. Soc., 20: 50.
1927 *Calocoris fasciativentris*: Knight, Can. Ent., 59: 43.
Distribution: Ariz., Col. (Mexico).

Calocoris fasciativentris fasciativentris Stål, 1862
1862 *Calocoris fasciativentris* Stål, Stett. Ent. Zeit., 23(7-9): 320. [Mexico].
1925 *Horcias fasciativentris*: Knight, Bull. Brook. Ent. Soc., 20: 51.
Distribution: Ariz., Col. (Mexico).

Calocoris fasciativentris imitator (Knight), 1925
1925 *Horcias fasciativentris imitator* Knight, Bull. Brook. Ent. Soc., 20: 51. [Ariz.].
Distribution: Ariz.

Calocoris fulvomaculatus (De Geer), 1773
1773 *Cimex fulvomaculatus* De Geer, Mem. Hist. Ins., 3: 294. [Europe].
1886 *Calocoris fulvomaculatus*: Uhler, Check-list Hem. Het., p. 18.
Distribution: Alk., Ariz., B.C., Man., N.T., Yuk. (Asia Minor, Europe, northern Africa, USSR).

Calocoris norvegicus (Gmelin), 1788
1779 *Cimex bipunctatus* Fabricius, Reis. Norw. Bem. Nat. Oek., p. 346. [Europe]. Preoccupied.
1788 *Cimex norvegicus* Gmelin, Syst. Nat., ed. 13, 1(4): 2176. [Norway].
1873 *Capsus contiguus* Walker, Cat. Het. Brit. Mus., 6: 85. [N.Y.]. Synonymized by Distant, 1904, An. Mag. Nat. Hist., (7)13: 110.
1873 *Capsus stramineus* Walker, Cat. Het. Brit. Mus., 6: 96. [N.S.]. Synonymized by Distant, 1904, An. Mag. Nat. Hist., (7)13: 110.
1878 *Deraeocoris bipunctatus*: Uhler, Proc. Boston Soc. Nat. Hist., 19: 400.
1887 *Calocoris bipunctatus*: Provancher, Pet. Faune Nat. Can., p. 114.
1890 *Calocoris norvegicus*: Atkinson, J. Asiatic Soc. Bengal, 58(2): 75.
Distribution: B.C., Conn., Mass., Me., N.B., N.J., N.Y., N.S., Nfld., Que., Tex.(?), Wis. (Asia Minor, Europe, northern Africa).
Note: Reuter (1876, Öfv. K. Svens. Vet.-Akad. Förh., 32(9): 71) reported *C. norvegicus* [as *C. bipunctatus* (Fabricius)] from Tex., but Blatchley (1926, Het. E. N. Am., p. 747) questioned this record and Knight (1941, Ill. Nat. Hist. Surv. Bull., 22: 138) omitted Tex. from his list of localities.

Calocoris texanus Knight, 1942
1942 *Calocoris texanus* Knight, Ent. News, 53: 158. [Tex.].
Distribution: Tex.

Genus *Camptozygum* Reuter, 1896

1896 *Camptozygum* Reuter, Öfv. F. Vet.-Soc. Förh., 38: 160. Type-species: *Lygaeus pinastri* Fallén, 1807, a junior synonym of *Cimex aequalis* Villers, 1789. Monotypic.

Camptozygum aequale (Villers), 1789

1789 *Cimex aequalis* Villers, Car. Linn. Ent., 1: 529. [Europe].

1807 *Lygaeus pinastri* Fallén, Monogr. Cimic. Suec, p. 95. [Europe]. Synonymized by Stichel, 1958, Illus. Bestim. Wanz., 2(24): 749. New Synonymy [for variety/subspecies].

1852 *Capsus maculicollis* Mulsant and Rey, An. Soc. Linn. Lyon, p. 140. Synonymized by Reuter, 1888, Rev. Syn. Het., 15(2): 644. [Europe]. New Synonymy [for variety/subspecies].

1930 *Camptozygum pinastri* var. *fieberi* Stichel, Illus. Bestim. Deutsch. Wanz., 6/7: 189. [Europe]. New Synonymy.

1973 *Camptozygum aequale*: Wheeler and Henry, Proc. Ent. Soc. Wash., 75: 240.

1973 *Camptozygum aequale* var. *aequale*: Wheeler and Henry, Proc. Ent. Soc. Wash., 75: 241.

1973 *Camptozygum aequale* var. *fieberi*: Wheeler and Henry, Proc. Ent. Soc. Wash., 75: 241.

1973 *Camptozygum aequale* var. *maculicollis*: Wheeler and Henry, Proc. Ent. Soc. Wash., 75: 241.

1973 *Camptozygum aequale var. pinastri*: Wheeler and Henry, Proc. Ent. Soc. Wash., 75: 241.

Distribution: N.Y., Ont., Pa. (Palearctic).

Note: The varieties of *Camptozygum aequale* listed by Wheeler and Henry (1973, Proc. Ent. Soc. Wash., 75: 241) are only color forms found in nearly every population and should not be given subspecies status. We, therefore, consider them synonyms of the nominate species/ subspecies.

Genus *Capsus* Fabricius, 1803

1803 *Capsus* Fabricius, Syst. Rhyn., p. 241. Type-species: *Cimex ater* Linnaeus, 1758. Designated by Int. Comm. Zool. Nomen., 1954, Bull. Zool. Nomen., 8(15): 199.

1858 *Rhopalotomus* Fieber, Wien. Ent. Monat., 2: 307. Type-species: *Cimex ater* Linnaeus, 1758. Monotypic. Synonymized by Reuter, 1875, Bih. K. Vet.-Akad. Handl., 3(1): 21.

Capsus ater (Linnaeus), 1758

1758 *Cimex ater* Linnaeus, Syst. Nat., ed. 10, p. 447. [Europe].

1767 *Cimex semiflavus* Linnaeus, Sys. Ent., ed. 12, p. 725. [Europe]. Synonymized by Fallén, 1807, Monogr. Cimic. Suec., p. 98.

1781 *Cimex tyrannus* Fabricius, Spec. Ins., 2: 371. [Europe]. Synonymized by Kolenati, 1845, Melet. Ent., 2: 127.

1869 *Phytocoris lugubris* Rathvon, Hist. Lancaster Co., Pa., p. 549. *Nomen nudum.*

1873 *Capsus ater*: Walker, Cat. Het. Brit. Mus., 6: 77.

1878 *Rhopalotomus ater*: Uhler, Proc. Boston Soc. Nat. Hist., 19: 411.

1896 *Capsus ater* var. *tyrannus*: Reuter, Acta Soc. Sci. Fenn., 33(5): 15.

1896 *Capsus ater* var. *semiflavus*: Reuter, Acta. Soc. Sci. Fenn., 33(5): 15. Synonymized as subspecies by Larochelle, 1984, Fabreries, Suppl. 3, p. 287.

1917 *Capsus ater tyrannus*: Parshley, Occas. Pap. Boston Soc. Nat. Hist., 7: 81.
1917 *Capsus ater semiflavus*: Parshley, Occas. Pap. Boston Soc. Nat. Hist., 7: 82.
Distribution: Alk., Alta., B.C., Cal., Conn., D.C., Fla., Ia., Ill., Ind., Mass., Me., Mich., Mo., N.C., N.H., N.J., N.S., N.Y., Oh., Ont., Pa., Que., R.I., Vt., W.Va., Wash., Wis. (Europe, North Africa, USSR).
Note: *Capsus semiflavus* Linnaeus (1767, Syst. Nat., ed. 12, p. 725) was synonymized by Fallén (1807, Monogr. Cimic. Suec., p. 98; *Lygaeus tyrannus* Fabricius (1794, Ent. Syst., 4: 177) was synonymized by Kolenati (1845, Melet. Ent., 2: 127). *Cimex ater* Linnaeus has been placed on the *Official List of Specific Names in Zoology* (1985, Bull. Zool. Nomen., 42: 188), along with *Capsus ater* Jakovlev, 1899 [now in *Deraeocoris*] to prevent the latter from being rejected as a secondary homonym because of both temporarily being placed in the genus *Capsus*. Van Duzee (1917, Univ. Cal. Publ. Ent., 2: 338) incorrectly listed *Capsus flavipes* Provancher as a synonym of *C. ater*; Kelton (1968, Nat. Can., 95: 1075) clarified this confusion and transferred *C. flavipes* to the phyline genus *Microphylellus*.

Capsus cinctus (Kolenati), 1845
1845 *Heterotoma cinctus* Kolenati, Meletem. Ent., 2: 128. [Siberia].
1858 *Deraeocoris simulans* Stål, Stett. Ent. Zeit., 19: 186. [USSR]. Synonymized by Vinokurov, 1977, Ent. Obozr., 56: 108.
1926 *Capsus simulans*: Knight, Can. Ent., 58: 59.
1926 *Capsus simulans* var. *fulvipes* Knight, Can. Ent., 58: 59. [S.D.]. Synonymized by Vinokurov, 1977, Ent. Obozr., 56: 108.
1977 *Capsus cinctus*: Vinokurov, Ent. Obozr., 56: 108.
Distribution: Alk., Alta., Ia., Man., Minn., Mont., S.D., Sask., Wis., Wyo. (Siberia).

Genus *Coccobaphes* Uhler, 1878

1878 *Coccobaphes* Uhler, Proc. Boston Soc. Nat. Hist., 19: 401. Type-species: *Coccobaphes sanguinareus* Uhler, 1878, a junior synonym of *Capsus frontifer* Walker, 1873. Monotypic.

Coccobaphes frontifer (Walker), 1873
1873 *Capsus frontifer* Walker, Cat. Hem. Brit. Mus., 6: 94. ["North America"].
1878 *Coccobaphes sanguinareus* [*sic*] Uhler, Proc. Boston Soc. Nat. Hist., 19: 401. [N.C., N.H., "Canada"]. Synonymized by Henry, 1985, Proc. Ent. Soc. Wash., 87: 679.
1878 *Coccobaphes sanguinarius*: Uhler, Proc. Boston Soc. Nat. Hist., 19: 402.
1890 *Caccobaphes* [*sic*] *sanguinarius*: Smith, Ins. N.J., p. 126.
1983 *Coccobaphes sanguinarus* [*sic*]: Wheeler et al., Trans. Am. Ent. Soc., 109: 146.
1985 *Coccobaphes frontifer*: Henry, Proc. Ent. Soc. Wash., 87: 679.
Distribution: Conn., Fla., Ia., Ill., Ind., Mass., Me., Mich., Minn., Mo., N.C., N.H., N.J., N.Y., Oh., Ont., Pa., Tenn., Vt., W.Va., Wis.
Note: The spelling of *sanguinarius* in the original description with an "eus" ending is incorrect for this specific epithet and, before the synonymy of *sanguinarius* with *C. frontifer*, normally would have to remain as the "correct" spelling; however, Uhler (1878, above) in the same paper and in all subsequent papers spells the name with an "ius" ending, indicating that the original "eus" was a *lapsus* on Uhler's part or a printer's error. Wheeler (1982, Proc. Ent. Soc. Wash., 84: 177-183) studied seasonal history on red maple, *Acer rubrum* Linnaeus.

Genus *Creontiades* Distant, 1883

1883 *Creontiades* Distant, Biol. Centr.-Am., Rhyn., 1: 237. Type-species: *Megacoelum rubrinerve* Stål, 1862. Monotypic.

Creontiades debilis Van Duzee, 1915
- 1909 *Creontiades* sp.: Van Duzee, Bull. Buff. Soc. Nat. Sci., 9: 180.
- 1915 *Creontiades debilis* Van Duzee, Pomona J. Ent. Zool., 7: 111. [Ga.].
- Distribution: Ga., Fla., Miss., Tex. (West Indies).

Creontiades rubrinervis (Stål), 1862
- 1862 *Megacoelum rubrinerve* Stål, Stett. Ent. Zeit., 23: 321. [Mexico].
- 1894 *Creontiades rubrinerve*: Uhler, Proc. Cal. Acad. Sci., (2)4: 255.
- 1914 *Creontiades rubrinervis*: Barber, Bull. Am. Mus. Nat. Hist., 33: 501.
- 1914 *Creontiades femoralis* Van Duzee, Trans. San Diego Soc. Nat. Hist., 2: 19. [Cal.]. Synonymized by Carvalho, 1959, Arq. Mus. Nac., 48: 77.
- Distribution: Ariz., Cal., Fla., Ga., Tex. (Mexico to South America).

Creontiades signatus (Distant), 1884
- 1884 *Megacoelum signatum* Distant, Biol. Centr.-Am., Rhyn., 1: 269. [Mexico].
- 1906 *Creontiades signatus*: Snow, Trans. Ks. Acad. Sci., 20: 152.
- Distribution: Tex. (Mexico).

Genus *Dagbertus* Distant 1904

1904 *Dagbertus* Distant, An. Mag. Nat. Hist., (7)13: 203. Type-species: *Capsus darwini* Butler, 1877. Designated by Kirkaldy, 1906, Trans. Am. Ent. Soc., 36: 123.

Note: Carvalho and Fontes (1983, Rev. Brasil. Biol., 43: 157-176) revised this genus and provided a key to species. Henry (1985, J. N.Y. Ent. Soc., 93: 1122-1124) further clarified the status of the species mentioned in the U.S. literature.

Dagbertus fasciatus (Reuter), 1876
- 1876 *Lygus* (*Lygus*) *fasciatus* Reuter, Öfv. K. Svens. Vet.-Akad. Förh., 32(9): 72. [S.C.]. Lectotype designated by Henry, 1985, J. N.Y. Ent. Soc., 93: 1122.
- 1886 *Lygus fasciatus*: Uhler, Check-list Hem. Het., p. 18.
- 1955 *Dagbertus fasciatus*: Kelton, Can. Ent., 87: 284.
- Distribution: Conn., Fla., Ind., La., Mass., Me., Miss., N.C., N.J., N.Y., S.C., Tex.
- Note: Henry (1985, above, p. 1122) clarified confusion between this species and *D. hospitus*.

Dagbertus irroratus (Blatchley), 1926
- 1926 *Lygus* (*Neolygus*) *irroratus* Blatchley, Het. E. N. Am., p. 775. [Fla.]. Unneccessary lectotype designated by Blatchley, 1930, Blatchleyana, pp. 65, 68; holotype recognized for single specimen by Henry, 1985, J. N.Y. Ent. Soc., 93: 1123.
- 1959 *Lygus irroratus*: Carvalho, Arq. Mus. Nac., 48: 122.
- 1974 *Lygocoris irroratus*: Kelton, Can. Ent., 106: 379.
- 1985 *Dagbertus irroratus*: Henry, J. N.Y. Ent. Soc., 93: 1123.
- Distribution: Fla.

Dagbertus olivaceus (Reuter), 1907
- 1907 *Lygus olivaceus* Reuter, Öfv. F. Vet.-Soc. Förh., 49(5): 6. [Jamaica]. Lectotype designated by Henry, 1985, J. N.Y. Ent. Soc., 93: 1123.

1917 *Lygus olivaceus*: Knight, Cornell Univ. Agr. Exp. Stn. Bull., 391: 599.
1955 *Dagbertus olivaceus*: Kelton, Can. Ent., 87: 285.
1963 *Dagbertus olivaceous* [*sic*]: Wolfenbarger, Bull. Univ. Fla. Agr. Exp. Stn., 605A: 29.
Distribution: Fla. (West Indies).
Note: Henry (1985, above, p. 1122) clarified early distribution records and confusion between this species and *D. fasciatus*.

Dagbertus semipictus (Blatchley), 1926
1926 *Bolteria semipicta* Blatchley, Het. E. N. Am., p. 743. [Fla.].
1985 *Dagbertus semipictus*: Henry, J. N.Y. Ent. Soc., 93: 1123.
Distribution: Fla. (West Indies).
Note: Henry (1985, above) synonymized *Dagbertus parafasciatus* Maldonado, 1969, known from Puerto Rico, with *D. semipictus*. Leston (1979, Fla. Ent., 62: 377) previously had listed *D. parafasciatus* as a synonym of *Dagbertus fasciatus*. Carvalho and Fontes (1983, Rev. Brasil. Biol., 43: 75-176), apparently overlooking Leston's paper, listed *parafasciatus* as a valid species.

Genus *Derophthalma* Berg, 1883

1883 *Derophthalma* Berg, An. Soc. Cient. Arg., 16: 22. Type-species: *Derophthalma reuteri* Berg, 1883. Monotypic.
1905 *Cyrtocapsidea* Reuter, Öfv. F. Vet.-Soc. Förh., 47(19): 25. Type-species: *Cyrtocapsidea nebulosa* Reuter, 1905. Monotypic. Synonymized by Carvalho, 1952, An. Acad. Brasil. Ciênc., 24: 87.
Note: Carvalho and Gomes (1980, Experientiae, 26: 93-146) revised this genus and provided a key to species.

Derophthalma emissitia (Distant), 1893
1893 *Monalocorisca emissitia* Distant, Biol. Centr.-Am., Rhyn., 1: 443. [Hondurus, Mexico].
1980 *Derophthalma emissitia*: Carvalho and Gomes, Experientiae, 26: 108.
Distribution: Tex. (Mexico).
Note: Carvalho (1952, Bol. Mus. Nac. Zool., 11: 6) previously considered this species a junior synonym of *Derophthalma laterata* (Distant), 1893.

Derophthalma variegata (Blatchley), 1926
1926 *Cyrtocapsidea variegata* Blatchley, Het. E. N. Am., p. 785. [Fla.].
1959 *Derophthalma variegata*: Carvalho, Arq. Mus. Nac., 48: 81.
Distribution: Fla.

Genus *Dichrooscytus* Fieber, 1858

1858 *Dichrooscytus* Fieber, Wien. Ent. Monat., 2: 309. Type-species: *Lygaeus rufipennis* Fallén, 1807. Designated by Reuter, 1888, Acta. Soc. Sci. Fenn., 15(2): 760.
Note: Knight (1968, Brig. Young Univ. Sci. Bull., 9(3): 192-194) provided a key to species of the western U.S.; Kelton (1972, Can. Ent., 104: 1034-1035) gave a key to the Canadian species.

Dichrooscytus abietis Bliven, 1956
1956 *Dichrooscytus abietis* Bliven, New Hem. W. St., p. 13. [Cal.].
Distribution: Cal.

Dichrooscytus adamsi Knight, 1968
1968 *Dichrooscytus adamsi* Knight, Brig. Young Univ. Sci. Bull., 9(3): 200. [Wash.].
Distribution: Wash.

Dichrooscytus alpinus Kelton, 1972
1972 *Dichrooscytus alpinus* Kelton, Can. Ent., 104: 1463. [Col.].
Distribution: Alta., Col.

Dichrooscytus angustifrons Knight, 1968
1968 *Dichrooscytus angustifrons* Knight, Brig. Young Univ. Sci. Bull., 9(3): 197. [Col.].
Distribution: Col.

Dichrooscytus apicalis Knight, 1968
1968 *Dichrooscytus apicalis* Knight, Brig. Young Univ. Sci. Bull., 9(3): 201. [Nev.].
Distribution: Nev.

Dichrooscytus barberi Knight, 1925
1925 *Dichrooscytus barberi* Knight, Can. Ent., 57: 95. [Ariz.].
Distribution: Ariz.

Dichrooscytus brevirostris Kelton, 1972
1972 *Dichrooscytus brevirostris* Kelton, Can. Ent., 104: 1462. [Ariz.].
Distribution: Ariz.

Dichrooscytus convexifrons Knight, 1968
1968 *Dichrooscytus convexifrons* Knight, Brig. Young Univ. Sci. Bull., 9(3): 198. [Wyo.].
Distribution: Alta., B.C., Wyo.

Dichrooscytus cuneatus Knight, 1968
1968 *Dichrooscytus cuneatus* Knight, Brig. Young Univ. Sci. Bull., 9(3): 197. [Col.].
Distribution: Col.

Dichrooscytus deleticus Knight, 1968
1968 *Dichrooscytus deleticus* Knight, Brig. Young Univ. Sci. Bull., 9(3): 200. [Wyo.].
Distribution: Wyo.

Dichrooscytus dentatus Kelton, 1972
1972 *Dichrooscytus dentatus* Kelton, Can. Ent., 104: 1463. [Ariz.].
Distribution: Ariz.

Dichrooscytus elegans Heidemann, 1892
1892 *Dichroscytus* [*sic*] *elegans* Heidemann, Proc. Ent. Soc. Wash., 2: 225. [D.C.]. Lectotype designated by Wheeler and Henry, 1975, Trans. Am. Ent. Soc., 101: 363.
1904 *Dichrooscytus elegans*: Uhler, Proc. U.S. Nat. Mus., 27: 356 (in part).
1927 *Dichrooscytus tinctipennis* Knight, Proc. Biol. Soc. Wash., 40: 15. [N.Y.]. Synonymized by Wheeler and Henry, 1975, Trans. Am. Ent. Soc., 101: 362.
Distribution: D.C., Ga., Ia., Ill., Ks., Mass., Md., Mich., Minn., Miss., Mo., N.B., N.C., N.J., N.S., N.Y., Neb., Oh., Ont., Pa., P.Ed., Sask., Tex., Va., W.Va., Wis.
Note: Wheeler and Henry (1975, Trans. Am. Ent. Soc., 101: 362) explained Heidemann's (1892, above) unintentional validation of this species. Wheeler and Henry (1977, Trans. Am. Ent. Soc., 103: 631-633) described the fifth-instar nymph and summarized the seasonal history on ornamental junipers, *Juniperus* spp.

Dichrooscytus flagellatus Kelton, 1972
1972 *Dichrooscytus flagellatus* Kelton, Can. Ent., 104: 1039. [B.C.].
Distribution: B.C.

Dichrooscytus flagitiosus Kelton and Schaffner, 1972
1972 *Dichrooscytus flagitiosus* Kelton and Schaffner, Can. Ent., 104: 1442. [N.M.].
Distribution: N.M.

Dichrooscytus flavescens Knight, 1968
1968 *Dichrooscytus flavescens* Knight, Brig. Young Univ. Sci. Bull., 9(3): 196. [Wyo.].
Distribution: B.C., N.D., S.D., Wyo.

Dichrooscytus flavivenosus Knight, 1968
1968 *Dichrooscytus flavivenosus* Knight, Brig. Young Univ. Sci. Bull., 9(3): 197. [Ariz.].
Distribution: Ariz., Col., Nev.

Dichrooscytus fuscosignatus Knight, 1968
1968 *Dichrooscytus fuscosignatus* Knight, Brig. Young Univ. Sci. Bull., 9(3): 196. [Col.].
Distribution: Col.

Dichrooscytus irroratus Van Duzee, 1912
1912 *Dichrooscytus irroratus* Van Duzee, Bull. Buff. Soc. Nat. Sci., 10: 482. [Col.].
Distribution: B.C., Col., Id., N.D., N.M., Nev., S.D., Tex., Ut., Wyo.

Dichrooscytus junipericola Knight, 1968
1968 *Dichrooscytus junipericola* Knight, Brig. Young Univ. Sci. Bull., 9(3): 201. [Nev.].
Distribution: Nev.

Dichrooscytus lagopinus Bliven, 1956
1956 *Dichrooscytus lagopinus* Bliven, New Hem. W. St., p. 15. [Cal.].
Distribution: Cal.

Dichrooscytus latifrons Knight, 1968
1968 *Dichrooscytus latifrons* Knight, Brig. Young Univ. Sci. Bull., 9(3): 199. [Col.].
Distribution: Alta., Ariz., B.C., Col., Man., Mich., N.B., N.S., Nfld., Ont., P.Ed., Que., Sask., Yuk.

Dichrooscytus longirostris Kelton, 1972
1972 *Dichrooscytus longirostris* Kelton, Can. Ent., 104: 1461. [Cal.].
Distribution: Cal.

Dichrooscytus maculatus Van Duzee, 1912
1912 *Dichrooscytus maculatus* Van Duzee, Bull. Buff. Soc. Nat. Sci., 10: 483. [Fla.].
Distribution: Fla., Ga., N.C.

Dichrooscytus minimus Knight, 1968
1968 *Dichrooscytus minimus* Knight, Brig. Young Univ. Sci. Bull., 9(3): 202. [Ariz.].
Distribution: Ariz.

Dichrooscytus minutus Kelton and Schaffner, 1972
1972 *Dichrooscytus minutus* Kelton and Schaffner, Can. Ent., 104: 1441. [N.M.].
Distribution: N.M.

Dichrooscytus nitidus Knight, 1968
1968 *Dichrooscytus nitidus* Knight, Brig. Young Univ. Sci. Bull., 9(3): 195. [Col.].
Distribution: Col.

Dichrooscytus ochreus Kelton, 1972
1972 *Dichrooscytus ochreus* Kelton, Can. Ent., 104: 1461. [Cal.].
Distribution: Cal.

Dichrooscytus pinicola Knight, 1968
1968 *Dichrooscytus pinicola* Knight, Brig. Young Univ. Sci. Bull., 9(3): 200. [Nev.].
Distribution: Nev.

Dichrooscytus rainieri Knight, 1968
1968 *Dichrooscytus rainieri* Knight, Brig. Young Univ. Sci. Bull., 9(3): 199. [Wash.].
Distribution: B.C., Wash.

Dichrooscytus repletus (Heidemann), 1892
1892 *Lygus repletus* Heidemann, Proc. Ent. Soc. Wash., 2: 225. [W.Va.]. Lectotype designated by Wheeler and Henry, 1975, Trans. Am. Ent. Soc., 101: 361.
1918 *Dichrooscytus elegans* var. *viridicans* Knight, Bull. Brook. Ent. Soc., 13: 114. [N.Y.]. Synonymized by Wheeler and Henry, 1975, Trans. Am. Ent. Soc., 101: 361.
1923 *Dichrooscytus viridicans*: Knight, Conn. Geol. Nat. Hist. Surv. Bull., 34: 597.
1929 *Dichrooscytus repletus*: Knight and McAtee, Proc. U.S. Nat. Mus., 75: 19.
Distribution: D.C., Ia., Ill., Ind., Ks., Mass., Md., Minn., Miss., Mo., N.B., N.C., N.H., N.J., N.S., N.Y., Neb., Oh., Ont., P.Ed., Pa., Que., W.Va., Wis.
Note: Wheeler and Henry (1975, Trans. Am. Ent. Soc., 101: 361) explained Heidemann's (1892, above) unintentional validation of this species. Wheeler and Henry (1977, Trans. Am. Ent. Soc., 103: 633-637) described the fifth-instar nymph and summarized seasonal history on ornamental junipers, *Juniperus* spp.

Dichrooscytus rostratus Kelton, 1972
1972 *Dichrooscytus rostratus* Kelton, Can. Ent., 104: 1041. [Alta.].
Distribution: Alta., Ariz., B.C., Cal., Col., Wyo.

Dichrooscytus ruberellus Knight, 1968
1968 *Dichrooscytus ruberellus* Knight, Brig. Young Univ. Sci. Bull., 9(3): 199. [Col.].
Distribution: Alta., B.C., Col., Man., N.D., S.D., Sask., Wyo.

Dichrooscytus rubidus Kelton, 1972
1972 *Dichrooscytus rubidus* Kelton, Can. Ent., 104: 1042. [Alta.].
Distribution: Alta., B.C., Wash.

Dichrooscytus rubromaculatus Kelton, 1972
1972 *Dichrooscytus rubromaculatus* Kelton, Can. Ent., 104: 1464. [Cal.].
Distribution: Cal.

Dichrooscytus rufipennis (Fallén), 1807
1807 *Lygaeus rufipennis* Fallén, Monogr. Cimic. Suec., p. 84. [Europe].
1928 *Dichrooscytus rufipennis*: Knight, N.Y. Agr. Exp. Stn. Mem., 101: 129.
Distribution: Conn., N.Y., Ont. (Europe).

Dichrooscytus rufivenosus Knight, 1968
1968 *Dichrooscytus rufivenosus* Knight, Brig. Young Univ. Sci. Bull., 9(3): 196. [Col.].
Distribution: Ariz., Col., Ut.

Dichrooscytus rufusculus Kelton, 1972
1972 *Dichrooscytus rufusculus* Kelton, Can. Ent., 104: 1043. [B.C.].
Distribution: B.C.

Dichrooscytus rugosus Knight, 1968
1968 *Dichrooscytus rugosus* Knight, Brig. Young Univ. Sci. Bull., 9(3): 194. [Ut.].
Distribution: N.M., Ut.

Dichrooscytus sequoiae Bliven, 1954
1954 *Dichrooscytus sequoiae* Bliven, Bull. Brook. Ent. Soc., 49: 109. [Cal.].

Distribution: Cal.

Dichrooscytus suspectus Reuter, 1909
1886 *Dichrooscytus rufipennis*: Uhler, Check-list Hem. Het., p. 18.
1909 *Dichrooscytus suspectus* Reuter, Acta. Soc. Sci. Fenn., 36(2): 37. [D.C., Mass.].
Distribution: B.C., Col., Conn., D.C., Ind., Man., Mass., Md., Me., Mich., N.B., N.J., N.S., N.Y., Ont., Que., R.I., Sask., W.Va.

Dichrooscytus taosensis Kelton and Schaffner, 1972
1972 *Dichrooscytus taosensis* Kelton and Schaffner, Can. Ent., 104: 1443. [N.M.].
Distribution: N.M.

Dichrooscytus tescorum Bliven, 1956
1956 *Dichrooscytus tescorum* Bliven, New Hem. W. St., p. 14. [Cal.].
Distribution: Cal.

Dichrooscytus texanus Kelton and Schaffner, 1972
1972 *Dichrooscytus texanus* Kelton and Schaffner, Can. Ent., 104: 1443. [Tex.].
Distribution: Tex.

Dichrooscytus uhleri Wheeler and Henry, 1975
1904 *Dichrooscytus elegans* Uhler, Proc. U.S. Nat. Mus., 27: 356. [N.M.]. Preoccupied.
1975 *Dichrooscytus uhleri* Wheeler and Henry, Trans. Am. Ent. Soc., 101: 363. New name for *Dichrooscytus elegans* Uhler.
Distribution: Ariz., Cal., N.M., Ut.

Dichrooscytus utahensis Knight, 1968
1968 *Dichrooscytus utahensis* Knight, Brig. Young Univ. Sci. Bull., 9(3): 201. [Ut.].
Distribution: Ut.

Dichrooscytus visendus Bliven, 1956
1956 *Dichrooscytus visendus* Bliven, New Hem. W. St., p. 14. [Ore.].
Distribution: Ore.

Dichrooscytus vittatipennis Knight, 1968
1968 *Dichrooscytus vittatipennis* Knight, Brig. Young Univ. Sci. Bull., 9(3): 198. [Cal.].
Distribution: Cal.

Dichrooscytus vittatus Van Duzee, 1921
1916 *Dichrooscytus irroratus*: Van Duzee, Univ. Cal. Publ. Ent., 1: 236.
1921 *Dichrooscytus vittatus* Van Duzee, Proc. Cal. Acad. Sci., (4)11: 122. [Cal.].
Distribution: Cal.

Genus *Dolicholygus* Bliven, 1957

1957 *Dolicholygus* Bliven, Occ. Ent., 1(1): 1. Type-species: *Phytocoris scrophulariae* Bliven, 1956. Original designation.

Dolicholygus scrophulariae (Bliven), 1956
1956 *Phytocoris scrophulariae* Bliven, New Hem. W. St., p. 15. [Cal.].
1957 *Dolicholygus scrophulariae*: Bliven, Occ. Ent., 1(1): 1.
Distribution: Cal.

Genus *Ecertobia* Reuter, 1909

1909 *Ecertobia* Reuter, Acta. Soc. Sci. Fenn., 36(2): 10. Type-species: *Ecertobia decora* Reuter, 1909. Monotypic.

Ecertobia decora Reuter, 1909
1909 *Ecertobia decora* Reuter, Acta Soc. Sci. Fenn., 36(2): 36. [Col.].
Distribution: Col., S.D.

Genus *Ectopiocerus* Uhler, 1890

1890 *Ectopiocerus* Uhler, Trans. Md. Acad. Sci., 1: 73. Type-species: *Ectopiocerus anthracinus* Uhler, 1890. Monotypic.

Ectopiocerus anthracinus Uhler, 1890
1890 *Ectopiocerus anthracinus* Uhler, Trans. Md. Acad. Sci., 1: 74. [Cal.].
Distribution: Cal., Ore.
Note: Van Duzee's (1917, Univ. Cal. Publ. Ent., 2: 324) record for N.J. is undoubtedly based on a misidentification of the superficially similar *Teleorhinus tephrosicola* Knight.

Genus *Ganocapsus* Van Duzee, 1912

1912 *Ganocapsus* Van Duzee, Bull. Buff. Soc. Nat. Sci., 10: 481. Type-species: *Ganocapsus filiformis* Van Duzee, 1912. Original designation.

Ganocapsus filiformis Van Duzee, 1912
1912 *Ganocapsus filiformis* Van Duzee, Bull. Buff. Soc. Nat. Sci., 10: 481. [Ariz.].
Distribution: Ariz.

Genus *Garganus* Stål, 1862

1862 *Garganus* Stål, Stett. Ent. Zeit., 23(7-9): 321. Type-species: *Garganus albidivittis* Stål, 1862. Designated by Kirkaldy, 1906, Trans. Am. Ent. Soc., 32: 137.

Garganus albidivittis Stål, 1862
1862 *Garganus albidivittis* Stål, Stett. Ent. Zeit., 23(7-9): 322. [Mexico].
1886 *Garganus albidivittis*: Uhler, Check-list Hem. Het., p. 20.
Distribution: Tex. (Mexico, Central America).

Garganus fusiformis (Say), 1832
1832 *Capsus fusiformis* Say, Descrip. Het. Hem. N. Am., p. 24. [U.S.].
1845 *Capsus croceipes* Herrich-Schaeffer, Wanz. Ins., 8: 16. [Pa.]. Synonymized by Van Duzee, 1917, Univ. Cal. Publ. Ent., 2: 324.
1862 *Garganus fusiformis*: Stål, Stett. Ent. Zeit., 23(7-9): 322.
1873 *Cyllocoris fusiformis*: Walker, Cat. Hem. Brit. Mus., 6: 66.
1973 *Garganus fusciformis* [*sic*]: Pimentel and Wheeler, Environ. Ent., 2: 661.
Distribution: Ala., Conn., D.C., Del., Fla., Ga., Ill., Ind., Ks., La., Mass., Md., Mich., Miss., Mo., N.C., N.J., N.Y., Oh., Ont., Pa., S.C., Tenn., Va.

Garganus splendidus Distant, 1893
1893 *Garganus splendidus* Distant, Biol. Centr.-Am., Rhyn., 1: 429. [Mexico].
1945 *Garganus splendidus*: Carvalho, Bol. Mus. Nac. Zool., 45: 6.
Distribution: Ariz. (Mexico).

Genus *Irbisia* Reuter, 1875

1875 *Irbisia* Reuter, Pet. Nouv. Ent., 137: 548. Type-species: *Leptomerocoris sericans* Stål, 1858. Monotypic.

1894 *Thyrillus* Uhler, Proc. Cal. Acad. Sci., (2)4: 266. Type-species: *Rhopalotomus pacificus* Uhler, 1872. Synonymized by Reuter, 1896, Acta. Soc. Sci. Fenn., 33(5): 11.

Note: Schwartz (1984, J. N.Y. Ent. Soc., 92: 193-306) revised the genus and provided a key to species.

Irbisia bliveni Schwartz, 1984

1984 *Irbisia bliveni* Schwartz, J. N.Y. Ent. Soc., 92: 225. [Cal.].

Distribution: Cal., Ore., Wash.

Irbisia brachycera (Uhler), 1872

1872 *Rhopalotomus brachycerus* Uhler, Prelim. Rept. U.S. Geol. Surv. Mont., 1871, p. 416. [Col.]. Lectotype designated by Schwartz, 1984, J. N.Y. Ent. Soc., 92: 229.

1890 *Capsus brachycerus*: Atkinson, J. Asiatic Soc. Bengal, 54: 105.

1893 *Capsus brachycorus* [*sic*]: Cockerell, Trans. Am. Ent. Soc., 20: 363.

1894 *Thryillus brachycerus*: Uhler, Proc. Cal. Acad. Sci., (2)4: 267.

1921 *Irbisia arcuata* Van Duzee, Proc. Cal. Acad. Sci., (4)11: 148. [Col.]. Synonymized by Knight, 1941, Bull. Brook. Ent. Soc., 36: 79.

1921 *Irbisia paeta* Van Duzee, Proc. Cal. Acad. Sci., (4)11: 150. [Cal.]. Synonymized by Schwartz, 1984, J. N.Y. Ent. Soc., 92: 228.

1941 *Irbisia brachycera*: Knight, Bull. Brook. Ent. Soc., 36: 79.

1961 *Irbisia vestifera* Bliven, Occ. Ent., 1(5): 46. [Cal.]. Synonymized by Schwartz, 1984, J. N.Y. Ent. Soc., 92: 228.

1961 *Irbisia gorgoniensis* Bliven, Occ. Ent., 1(5): 47. [Cal.]. Synonymized by Schwartz, 1984, J. N.Y. Ent. Soc., 92: 228.

1961 *Irbisia tejonica* Bliven, Occ. Ent., 1(5): 48. [Cal.]. Synonymized by Schwartz, 1984, J. N.Y. Ent. Soc., 92: 228.

Distribution: Alta., Ariz., B.C., Cal., Col., Ia.(?), Id., Mont., N.D., N.M., Neb., Ore., S.D., Sask., Ut., Wash., Wyo. (Mexico).

Note: The Ia. record for this species is suspect (pers. comm., M. D. Schwartz, Am. Mus. Nat. Hist., N.Y.).

Irbisia californica Van Duzee, 1921

1914 *Irbisia sericans*: Childs, Month. Bull. Cal. St. Comm. Hort., 3: 220.

1915 *Irbisia brachycerus*: Essig, Month. Bull. Cal. St. Comm. Hort., Suppl., 4: 213.

1921 *Irbisia californica* Van Duzee, Proc. Cal. Acad. Sci., (4)11: 146. [Cal.].

1961 *Irbisia eurekae* Bliven, Occ. Ent., 1(5): 45. [Cal.]. Synonymized by Schwartz, 1984, J. N.Y. Ent. Soc., 92: 232.

1961 *Irbisia paenulata* Bliven, Occ. Ent., 1(5): 46. [Cal.]. Synonymized by Schwartz, 1984, J. N.Y. Ent. Soc., 92: 232.

1963 *Irbisia umbratica* Bliven, 1(7): 69. [Cal.]. Synonymized by Schwartz, 1984, J. N.Y. Ent. Soc., 92: 232.

Distribution: Cal., Ore. (Mexico).

Irbisia cascadia Schwartz, 1984

1984 *Irbisia cascadia* Schwartz, J. N.Y. Ent. Soc., 92: 235. [Ore.].

Distribution: Cal., Ore., Wash.

Irbisia castanipes Van Duzee, 1921

1921 *Irbisia castanipes* Van Duzee, Proc. Cal. Acad. Sci., (4)11: 145. [Cal.].

Distribution: Cal.

Irbisia cuneomaculata Blatchley, 1934
1934 *Irbisia cuneo-maculata* [*sic*] Blatchley, Trans. Am. Ent. Soc., 60: 13. [Cal.].
1959 *Irbisia cuneomaculata*: Carvalho, Arq. Mus. Nac., 48: 105.
Distribution: Cal.

Irbisia elongata Knight, 1941
1941 *Irbisia elongata* Knight, Bull. Brook. Ent. Soc., 36: 77. [Wyo.].
1963 *Irbisia retrusa* Bliven, Occ. Ent., 1(7): 71. [Cal.]. Synonymized by Schwartz, 1984, J. N.Y. Ent. Soc., 92: 241.
Distribution: Alta., B.C., Cal., Col., Id., Mont., Nev., Ore., Ut., Wash., Wyo.

Irbisia fuscipubescens Knight, 1941
1941 *Irbisia fuscipubescens* Knight, Bull. Brook. Ent. Soc., 36: 76. [Wash.].
Distribution: Alta., B.C., Id., Mont., Ore., Wash., Wyo.

Irbisia incomperta Bliven, 1963
1963 *Irbisia incomperta* Bliven, Occ. Ent., 1(7): 73. [Cal.].
Distribution: Cal.

Irbisia knighti Schwartz and Lattin, 1984
1984 *Irbisia knighti* Schwartz and Lattin, J. N.Y. Ent. Soc., 91: 413. [Wash.].
Distribution: B.C., Cal., Ore., Wash.

Irbisia limata Bliven, 1963
1963 *Irbisia limata* Bliven, Occ. Ent., 1(7): 78. [Cal.].
Distribution: Cal.

Irbisia mollipes Van Duzee, 1917
1917 *Irbisia sericans* var. *mollipes* Van Duzee, Proc. Cal. Acad. Sci., 7: 264. [Cal.].
1921 *Irbisia mollipes* Van Duzee, Proc. Cal. Acad. Sci., (4)11: 147.
1963 *Irbisia umbratica*: Bliven, Occ. Ent., 1(7): 70 (in part).
1963 *Irbisia upupa* Bliven, Occ. Ent., 1(7): 70. [Cal.]. Synonymized by Schwartz, 1984, J. N.Y. Ent. Soc., 92: 250.
Distribution: Cal.

Irbisia nigripes Knight, 1925
1925 *Irbisia nigripes* Knight, Can. Ent., 57: 94. [Id.].
Distribution: Alta., B.C., Id., Mont., Ore., Wash.

Irbisia oreas Bliven, 1963
1963 *Irbisia oreas* Bliven, Occ. Ent., 1(7): 72. [Ariz.].
Distribution: Ariz., Cal., Ut. (Mexico).

Irbisia pacifica (Uhler), 1872
1871 *Rhopalotomus pacificus* Uhler, Rept. U.S. Geol. Surv. Wyo., p. 471. *Nomen nudum*.
1872 *Rhopalotomus pacificus* Uhler, Prelim. Rept. U.S. Geol. Surv. Mont., p. 415. [Cal., Id., Mont.]. Lectotype from Id. designated by Schwartz, 1984, J. N.Y. Ent. Soc., 92: 257.
1890 *Capsus pacificus*: Atkinson, J. Asiatic Soc. Bengal, 58(2): 106.
1894 *Thyrillus pacificus*: Uhler, Proc. Cal. Acad. Sci., (2)4: 267.
1914 *Irbisia pacificus*: Van Duzee, Trans. San Diego Soc. Nat. Hist., 2: 24.
1968 *Irbisia pacifica*: Knight, Brig. Young Univ. Sci. Bull., 9(3): 185.
Distribution: B.C., Cal., Col., Id., Mont., Nev., Ore., Ut., Wash., Wyo. (Mexico).

Irbisia panda Bliven, 1963
1963 *Irbisia panda* Bliven, Occ. Ent., 1(7): 77. [Cal.].
Distribution: Cal.

Irbisia sericans (Stål), 1858

1858 *Leptomerocoris sericans* Stål, Stett. Ent. Zeit., 19: 188. [Alk.].

1879 *Irbisia sericans*: Reuter, Öfv. F. Vet.-Soc. Förh., 21: 58

1886 *Orthocephalus saltator*: Uhler, Check-list Hem. Het., p. 20.

1899 *Orthocephalus saltator*: Schwarz, Fur Seals and Fur-Seal Isl. N. Pac., 3: 5.

1900 *Irbisia (Leptomerocoris) sericans*: Heidemann, Proc. Wash. Acad. Sci., 2: 504.

Distribution: Alk., B.C., Cal., Ore., Wash. (USSR).

Note: Henry and Kelton (1986, J. N.Y. Ent. Soc., 94: 51-55) clarified the records of *O. saltator* above.

Irbisia serrata Bliven, 1963

1910 *Capsus solani*: Heidemann, Proc. Ent. Soc. Wash., 12: 200 (in part).

1910 *Irbisia brachycerus*: Cockerell, Can. Ent., 42: 370.

1913 *Irbisia brachycerus*: Vosler, Month. Bull. Cal. St. Comm. Hort., 2: 553 (in part).

1914 *Irbisia brachycerus*: Van Duzee, Trans. San Diego Soc. Nat. Hist., 2: 24 (in part).

1921 *Irbisia mollipes*: Van Duzee, Proc. Cal. Acad. Sci., (4)11: 147 (in part).

1921 *Irbisia brachycerus*: Van Duzee, Proc. Cal. Acad. Sci., (4)11: 149, 152.

1963 *Irbisia inurbana*: Bliven, Occ. Ent., 1(7): 81 (in part).

1963 *Irbisia serrata* Bliven, Occ. Ent., 1(7): 82. [Cal.].

Distribution: Ariz., B.C., Cal., Col., Id., N.M., Nev., Ore., Ut., Wash., Wyo.

Irbisia setosa Van Duzee, 1921

1921 *Irbisia mollipes*: Van Duzee, Proc. Cal. Acad. Sci., (4)11: 147 (in part).

1921 *Irbisia setosa* Van Duzee, Proc. Cal. Acad. Sci., (4)11: 149. [Cal.].

1963 *Irbisia ustricula* Bliven, Occ. Ent., 1(7): 79. [Cal.]. Synonymized by Schwartz, 1984, J. N.Y. Ent. Soc., 92: 269.

Distribution: Cal.

Irbisia shulli Knight, 1941

1941 *Irbisia shulli* Knight, Bull. Brook. Ent. Soc., 36: 75. [Id.].

Distribution: B.C., Col., Id., Mont., Ore., Ut., Wash., Wyo.

Irbisia silvosa Bliven, 1961

1921 *Irbisia sita*: Van Duzee, Proc. Cal. Acad. Sci., (4)11: 150 (in part).

1961 *Irbisia silvosa* Bliven, Occ. Ent., 1(5): 48. [Cal.].

1963 *Irbisia upupa*: Bliven, Occ. Ent., 1(7): 70 (in part).

1963 *Irbisia paulula* Bliven, Occ. Ent., 1(7): 75. [Cal.]. Synonymized by Schwartz, 1984, J. N.Y. Ent. Soc., 92: 273.

1963 *Irbisia inurbana*: Bliven, Occ. Ent., 1(7): 81 (in part).

Distribution: Cal., Ore. (Mexico).

Irbisia sita Van Duzee, 1921

1921 *Irbisia sita* Van Duzee, Proc. Cal. Acad. Sci., (4)11: 150. [Cal.].

1963 *Irbisia neptis* Bliven, Occ. Ent., 1(7): 74. [Cal.]. Synonymized by Schwartz, 1984, J. N.Y. Ent. Soc., 92: 278.

Distribution: Cal.

Irbisia solani (Heidemann), 1910

1910 *Capsus solani* Heidemann, Proc. Ent. Soc. Wash., 12: 200. [Cal.]. Lectotype designated by Schwartz, 1984, J. N.Y. Ent. Soc., 92: 281.

1913 *Irbisia brachycerus*: Vosler, Month. Bull. Cal. St. Comm. Hort., 2: 551.

1914 *Irbesia* [*sic*] *brachycerus*: Childs, Month. Bull. Cal. St. Comm. Hort., 3: 220.

1914 *Irbisia brachycerus* var. *solani*: Van Duzee, Trans. San Diego Soc. Nat. Hist., 2: 24 (in part).

1915 *Irbisia sericans*: Essig, Month. Bull. Cal. St. Comm. Hort., Suppl., 4: 213.
1916 *Irbisia solani*: Van Duzee, Check List Hem., p. 38.
1917 *Lygus brachycerus*: Van Duzee, Proc. Cal. Acad. Sci., (4)7: 265 (in part).
1917 *Lygus solani*: Van Duzee, Proc. Cal. Acad. Sci., (4)7: 266.
1921 *Irbisia setosa*: Van Duzee, Proc. Cal. Acad. Sci., (4)11: 149 (in part).
1931 *Irbisia brachycera* var. *solani*: Knowlton, Ent. News, 42: 68.
1963 *Irbisia incomperta*: Bliven, Occ. Ent., 1(7): 73 (in part).
1963 *Irbisia paulula*: Bliven, Occ. Ent., 1(7): 75 (in part).
1963 *Irbisia lactertosa* Bliven, Occ. Ent., 1(7): 76. [Cal.]. Synonymized by Schwartz, 1984, J. N.Y. Ent. Soc., 92: 280.
1963 *Irbisia inurbana* Bliven, Occ. Ent., 1(7): 81. [Cal.]. Synonymized by Schwartz, 1984, J. N.Y. Ent. Soc., 92: 280.
1963 *Irbisia serrata*: Bliven, Occ. Ent., 1(7): 82 (in part).
Distribution: Cal., Ore., Ut., Wash.

Genus *Knightomiris* Kelton, 1973

1973 *Knightomiris* Kelton, Can. Ent., 105: 1417. Type-species: *Lygus distinctus* Knight, 1917. Original designation.

Knightomiris distinctus (Knight), 1917
1917 *Lygus distinctus* Knight, Cornell Univ. Agr. Exp. Stn. Bull., 391: 594. [Ariz.].
1973 *Knightomiris distinctus*: Kelton, Can. Ent., 105: 1417.
Distribution: Ariz. (Mexico).

Genus *Lampethusa* Distant, 1884

1884 *Lampethusa* Distant, Biol. Centr.-Am., Rhyn., 1: 303. Type-species: *Lampethusa anatina* Distant, 1884. Monotypic.

Lampethusa anatina Distant, 1884
1884 *Lampethusa anatina* Distant, Biol. Centr.-Am., Rhyn., 1: 303. [Guatemala].
1906 *Lampethusa anatina*: Snow, Trans. Ks. Acad. Sci., 20: 179.
Distribution: Ariz., Tex. (Mexico to South America).

Lampethusa nicholi Knight, 1933
1933 *Lampethusa nicholi* Knight, Pan-Pac. Ent., 9: 71. [Ariz.].
Distribution: Ariz., Tex.

Genus *Lygidea* Reuter, 1879

1875 *Lygidea* Reuter, Pet. Nouv. Ent., 1: 547. *Nomen nudum*.
1879 *Lygidea* Reuter, Öfv. F. Vet.-Soc. Förh., 21: 54. Type-species: *Deraeocoris illotus* Stål, 1858. Monotypic.

Lygidea annexa (Uhler), 1872
1872 *Lygus annexus* Uhler, Prelim. Rept. U.S. Geol. Surv. Mont., p. 413. [Col.].
1917 *Lygidea annexus*: Knight, Cornell Univ. Agr. Exp. Stn. Bull., 391: 639.
1959 *Lygidea annexa*: Carvalho, Arq. Mus. Nac., 48: 112.
Distribution: Cal., Col., Minn.

Lygidea essigi Van Duzee, 1925
1925 *Lygidea essigi* Van Duzee, Proc. Cal. Acad. Sci., (4)14: 394. [Cal.].
Distribution: Cal.

Lygidea mendax Reuter, 1909
1909 *Lygidea mendax* Reuter, Acta Soc. Sci. Fenn., 36(2): 47. [N.Y.].
Distribution: Conn., Ia., Ind., Ky., Mass., Me., Mich., N.B., N.H., N.J., N.S., N.Y., Oh., Ont., Pa., Que., W.Va.
Note: Schaefer (1974, Mem., Conn. Ent. Soc., pp. 101-116) outlined the economic history of this former apple pest.

Lygidea morio Reuter, 1909
1909 *Lygidea morio* Reuter, Acta Soc. Sci. Fenn., 36(2): 47. [Cal.].
Distribution: Cal.

Lygidea obscura Reuter, 1909
1909 *Lygidea rubecula* var. *obscura* Reuter, Acta Soc. Sci. Fenn., 36(2): 46. [N.Y.].
1923 *Lygidea obscura*: Knight, Conn. Geol. Nat. Hist. Surv. Bull., 34: 570.
Distribution: B.C., Ill., Mich., Mo., N.S., N.Y., Oh., Ont., Pa., W.Va.

Lygidea rosacea Reuter, 1909
1909 *Lygidea rubecula* var. *rosacea* Reuter, Acta Soc. Sci. Fenn., 36(2): 46. [Ill.].
1923 *Lygidea rosacea*: Knight, Conn. Geol. Nat. Hist. Surv. Bull., 34: 570.
Distribution: Alta., Col., Ia., Ill., Man., Minn., Mo., Oh., Ont., S.D., Sask., Wis.

Lygidea rubecula (Uhler), 1895
1895 *Neoborus rubeculus* Uhler, Bull. Col. Agr. Exp. Stn., 31: 37. [Col., Ill., Mich.].
1909 *Lygidea rubecula*: Reuter, Acta Soc. Sci. Fenn., 36(2): 45.
Distribution: Cal., Col., Ill., Man., Me., Mich., N.C., N.H., N.Y., Ont., Vt.

Lygidea rubecula infuscata Reuter, 1909
1909 *Lygidea rubecula* var. *infuscata* Reuter, Acta Soc. Sci. Fenn., 36(2): 46. [Col.].
Distribution: Col.

Lygidea rubecula rubecula (Uhler), 1895
1895 *Neoborus rubeculus* Uhler, Bull. Col. Agr. Exp. Stn., 31: 37. [Col., Ill., Mich.].
1909 *Lygidea rubecula*: Reuter, Acta Soc. Sci. Fenn., 36(2): 45.
Distribution: Same as for species.

Lygidea salicis Knight, 1939
1939 *Lygidea salicis* Knight, Bull. Brook. Ent. Soc., 34: 22. [Minn.].
Distribution: Alta., Ill., Man., Mich., Minn., N.Y., Ont., Que., Sask., W.Va.

Lygidea viburni Knight, 1923
1923 *Lygidea viburni* Knight, Conn. Geol. Nat. Hist. Surv. Bull., 34: 569. [N.Y.].
Distribution: Mass., N.H., N.Y., W.Va.

Genus *Lygocoris* Reuter, 1875

1875 *Lygus* (*Lygocoris*) Reuter, Rev. Crit. Caps., 1(2): 81. Type-species: *Cimex pabulinus* Linnaeus, 1761. Designated by Kirkaldy, 1906, Trans. Am. Ent. Soc., 32: 139.
1959 *Lygocoris*: Carvalho, Arq. Mus. Nac., 48: 132.
Note: Kelton (1971, Mem. Ent. Soc. Can., 83: 1-87) reviewed the genus and provided a key to the Canadian species. The type-species of *Lygocoris* was formally established as *Cimex pabulinus* Linnaeus by the Int. Comm. Zool. Nomen. (1963, Opinion 667, Bull. Zool. Nomen., 20: 270-271).

Subgenus *Apolygus* China, 1941

1941 *Lygus* (*Apolygus*) China, Proc. R. Ent. Soc. London., Ser. B, 10: 60. Type-species: *Lygaeus limbatus* Fallén, 1807. Designated by Int. Comm. Zool. Nomen., 1963, Opinion 667, Bull. Zool. Nomen., 20: 270-271.

Lygocoris lucorum (Meyer-Dür), 1843

1843 *Capsus lucorum* Meyer-Dür, Verz. Schw. Rhyn., p. 46. [Europe].
1886 *Lygus lucorum*: Uhler, Check-list Hem. Het., p. 18.
1955 *Lygus* (*Apolygus*) *lucorum*: Kelton, Can. Ent., 87: 278.
1959 *Lygocoris* (*Apolygus*) *lucorum*: Carvalho, Arq. Mus. Nac., 48: 138.
1974 *Apolygus lucorum*: Slater, Mem. Conn. Ent. Soc., p. 156.
Distribution: Me., Que. (Europe).
Note: Kelton (1971, Mem. Ent. Soc. Can., 83: 8) clarified past confusion related to this species.

Subgenus *Lygocoris* Reuter, 1875

1875 *Lygus* (*Lygocoris*) Reuter: Rev. Crit. Caps., 1(2): 81. Type-species: *Cimex pabulinus* Linnaeus, 1761. Designated by Kirkaldy, 1906, Trans. Am. Ent. Soc., 32: 139.

Lygocoris pabulinus (Linnaeus), 1761

1761 *Cimex pabulinus* Linnaeus, Faun. Suec., p. 253. [Europe].
1886 *Lygus pabulinus*: Uhler, Check-list Hem. Het., p. 18.
1887 *Lygus contaminatus*: Provancher, Pet. Faune Ent. Can., 3: 121.
1903 *Lygus chagnoni* Stevenson, Can. Ent., 35: 214. [Que.]. Synonymized by Van Duzee, 1912, Bull. Buff. Soc. Nat. Sci., 10: 512.
1909 *Lygus pabulinus* var. *signifer* Reuter, Acta Soc. Sci. Fenn., 36(2): 42. [Md.]. New Synonymy.
1910 *Lygus pabulinus signifer*: Banks, Cat. Nearc. Hem.-Het., p. 46
1916 *Lygus* (*Lygocoris*) *pabulinus*: Van Duzee, Check List Hem., p. 39.
1916 *Lygus* (*Lygocoris*) *pabulinus* var. *signifer*: Van Duzee, Check List Hist., p. 39.
1955 *Lygus* (*Lygus*) *pabulinus*: Kelton, Can. Ent., 87: 278.
1959 *Lygocoris* (*Lygocoris*) *pabulinus*: Carvalho, Arq. Mus. Nac., 48: 134.
1959 *Lygocoris* (*Lygocoris*) *pabulinus* var. *signifer*: Carvalho, Arq. Mus. Nac., 48: 134.
1967 *Lygus* (*Neolygus*) *pabulinus*: Wray, Ins. N.C., 3rd. Suppl., p. 26.
Distribution: Alk., Alta, B.C., Col., Conn., D.C., Del., Ga., Ia., Ill., Ind., Man., Mass., Md., Me., Mich., Mont., N.B., N.C., N.H., N.J., N.M., N.S., N.Y., Nfld., Oh., Ont., Ore., Pa., Que., R.I., S.D., Sask., Tenn., Va., Vt., W.Va., Wash., Wis., Wyo., Yuk. (Europe, Asia).
Note: The variety *signifer* Reuter (1909, above), although listed by Banks (1910, above), Van Duzee (1916, above; 1917, Univ. Cal. Publ. Ent., 2: 341), and Parshley (1920, Can. Ent., 52: 86), was not recognized by Kelton in his revision of this genus. We have studied a male in the USNM collection from Plummer's Island, Md., labeled by Heidemann as *Lygus pabulinus* var. *signifer*. This specimen, typical of the nominate subspecies/species, is here considered a synonym.

Subgenus *Neolygus* Knight, 1917

1917 *Lygus* (*Neolygus*) Knight, Cornell Univ. Agr. Exp. Stn. Bull., 391: 561. Type-species: *Lygus communis* Knight, 1916. Original designation.

Lygocoris aesculi (Knight), 1953
1953 *Neolygus aesculi* Knight, Ia. St. Coll. J. Sci., 27: 511. [Mo.].
1959 *Lygocoris (Neolygus) aesculi*: Carvalho, Arq. Mus. Nac., 48: 140.
Distribution: Mo.

Lygocoris alni (Knight), 1917
1909 *Lygocoris viridis* Reuter, Acta Soc. Sci. Fenn., 36(2): 42 (in part).
1917 *Lygus (Neolygus) alni* Knight, Cornell Univ. Agr. Exp. Stn. Bull., 391: 607. [N.Y.].
1917 *Lygus alni*: Parshley, Occas. Pap. Boston Soc. Nat. Hist., 7: 87.
1941 *Neolygus alni*: Knight, Ill. Nat. Hist. Surv. Bull., 22: 154.
1959 *Lygocoris (Neolygus) alni*: Carvalho, Arq. Mus. Nac., 48: 140.
Distribution: B.C., Ill., Minn., N.H., N.S., N.Y., Ont., Que., W.Va.

Lygocoris atricallus Kelton, 1971
1971 *Lygocoris (Neolygus) atricallus* Kelton, Can. Ent., 103: 1109. [Ont.].
Distribution: Ont., Que.

Lygocoris atrinotatus (Knight), 1917
1917 *Lygus (Neolygus) atrinotatus* Knight, Cornell Univ. Agr. Exp. Stn. Bull., 391: 617. [Pa.].
1941 *Neolygus atrinotatus*: Knight, Ill. Nat. Hist. Surv. Bull., 22: 162.
1959 *Lygocoris (Neolygus) atrinotatus*: Carvalho, Arq. Mus. Nac., 48: 140.
1980 *Lygocoris knighti*: Wheeler, Ent. News, 91: 25 (in part).
Distribution: D.C., Ga., N.C., Oh., Pa., S.C., Tenn., W.Va.
Note: Wheeler and Henry (1983, Proc. Ent. Soc. Wash., 85: 26-31) studied the seasonal history and host plants and described the fifth-instar nymph.

Lygocoris atritylus (Knight), 1917
1917 *Lygus (Neolygus) atritylus* Knight, Cornell Univ. Agr. Exp. Stn. Bull., 391: 606. [Vt.].
1917 *Lygus atritylus*: Parshley, Occas. Pap. Boston Soc. Nat. Hist., 7: 87.
1925 *Neolygus atritylus*: Knight, Can. Ent., 57: 182.
1959 *Lygocoris (Neolygus) atritylus*: Carvalho, Arq. Mus. Nac., 48: 140.
Distribution: Alta., B.C., Col., Man., Minn., N.B., N.H., N.S., N.Y., Ont., P.Ed., Sask., Vt.

Lygocoris belfragii (Reuter), 1876
1876 *Lygus (Lygus) Belfragii* [*sic*] Reuter, Öfv. K. Svens. Vet.-Akad. Förh., 32(9): 71. [N.Y.].
1886 *Lygocoris belfragii*: Provancher, Pet. Faune Ent. Can., 3: 20.
1910 *Lygocoris belfragei* [*sic*]: Banks, Cat. Nearc. Hem.-Het., p. 45.
1917 *Lygus (Neolygus) belfragii*: Knight, Cornell Univ. Agr. Exp. Stn. Bull., 391: 630.
1917 *Lygus belfragei* [*sic*]: Parshley, Occas. Pap. Boston Soc. Nat. Hist., 7: 90.
1941 *Neolygus belfragii*: Knight, Ill. Nat. Hist. Surv. Bull., 22: 162.
1959 *Lygocoris (Neolygus) belfragii*: Carvalho, Arq. Mus. Nac., 48: 141.
Distribution: Conn., Ga., Ill., Man., Mass., Me., Minn., N.B., N.C., N.H., N.S., N.Y., Ont., Pa., P.Ed., Que., R.I., Tenn., Vt., W.Va., Wis.

Lygocoris betulae (Knight), 1953
1953 *Neolygus betulae* Knight, Ia. St. Coll. J. Sci., 27: 512. [Mo.].
1959 *Lygocoris (Neolygus) betulae*: Carvalho, Arq. Mus. Nac., 48: 141.
Distribution: Ia., Mo.

Lygocoris canadensis (Knight), 1917
1917 *Lygus (Neolygus) canadensis* Knight, Cornell Univ. Agr. Exp. Stn. Bull., 391: 634. [Ont.].

1917 *Lygus (Neolygus) canadensis* var. *binotatus* Knight, Cornell Univ. Agr. Exp. Stn. Bull., 391: 635. [N.J.]. Synonymized by Kelton, 1971, Mem. Ent. Soc. Can., 83: 39.
1941 *Neolygus canadensis*: Knight, Ill. Nat. Hist. Surv. Bull., 22: 164.
1941 *Neolygus canadensis* var. *binotatus*: Knight, Ill. Nat. Hist. Surv. Bull., 22: 164.
1959 *Lygocoris (Neolygus) canadensis*: Carvalho, Arq. Mus. Nac., 48: 141.
1959 *Lygocoris (Neolygus) canadensis* var. *binotatus* Carvalho, Arq. Mus. Nac., 48: 141.
Distribution: Ill., Man., Minn., N.J., N.Y., Nfld., Oh., Ont., Que., Wis.

Lygocoris carpini (Knight), 1939
1939 *Neolygus carpini* Knight, Bull. Brook. Ent. Soc., 34: 21. [Minn.].
1959 *Lygocoris (Neolygus) carpini*: Carvalho, Arq. Mus. Nac., 48: 141.
Distribution: Ia., Ill., Minn.

Lygocoris caryae (Knight), 1917
1917 *Lygus (Neolygus) caryae* Knight, Cornell Univ. Agr. Exp. Stn. Bull., 391: 615. [N.Y.].
1917 *Lygus (Neolygus) caryae* var. *subfuscous* Knight, Cornell Univ. Agr. Exp. Stn. Bull., 391: 616. [N.Y.]. Synonymized by Kelton, 1971, Mem. Ent. Soc. Can., 83: 28.
1917 *Lygus caryae*: Parshley, Occas. Pap. Boston Soc. Nat. Hist., 7: 88.
1941 *Neolygus caryae*: Knight, Ill. Nat. Hist. Surv. Bull., 22: 161.
1958 *Neolygus caryae caryae*: Rings, J. Econ. Ent., 51: 31.
1958 *Neolygus caryae subfuscous*: Rings, J. Econ. Ent., 51: 31.
1959 *Lygocoris (Neolygus) caryae*: Carvalho, Arq. Mus. Nac., 48: 141.
Distribution: Conn., Ga., Ia., Ill., Ind., Mass., Miss., Me., Mich., Minn., Mo., N.C., N.H., N.Y., Oh., Ont., Pa., Que., Tenn., Tex., Vt., W.Va., Wis.

Lygocoris clavigenitalis (Knight), 1917
1917 *Lygus (Neolygus) clavigenitalis* Knight, Cornell Univ. Agr. Exp. Stn. Bull., 391: 632. [Mass.].
1917 *Lygus clavigenitalis*: Parshley, Occas. Pap. Boston Soc. Nat. Hist., 7: 90.
1941 *Neolygus clavigenitalis*: Knight, Ill. Nat. Hist. Surv. Bull., 22: 163.
1959 *Lygocoris (Neolygus) clavigenitalis*: Carvalho, Arq. Mus. Nac., 48: 141.
Distribution: Conn., Ga., Mass., Md., Me., N.C., N.S., Oh., Ont., Que.

Lygocoris communis (Knight), 1916
1878 *Lygus invitus*: Uhler, Bull. U.S. Geol. Geogr. Surv. Terr., 4: 506.
1916 *Lygus communis* Knight, Can. Ent., 48: 346. [N.Y.].
1916 *Lygus communis* var. *novascotiensis* Knight, Can. Ent., 48: 349. [N.S.]. Synonymized by Kelton, 1971, Mem. Ent. Soc. Can., 83: 9.
1917 *Lygus (Neolygus) communis*: Knight, Cornell Univ. Agr. Exp. Stn. Bull., 391: 620.
1941 *Neolygus communis*: Knight, Ill. Nat. Hist. Surv. Bull., 22: 159.
1959 *Lygocoris (Neolygus) communis*: Carvalho, Arq. Mus. Nac., 48: 141.
Distribution: Alk., Alta., B.C., Cal., Col., Conn., Ia., Id., Ill., Ind., Man., Mass., Me., Mich., Minn., Mont., N.B., N.C., N.H., N.J., N.S., N.T., N.Y., Nfld., Oh., Ont., Pa., P.Ed., Que., Sask., Va., Vt., W.Va., Wis., Wyo., Yuk.
Note: Boivin (1983, Bibliogr. Ent. Soc. Am., 2: 1-9) provided a bibliography of economic literature for this species, the pear plant bug.

Lygocoris contaminatus (Fallén), 1807
1807 *Lygaeus contaminatus* Fallén, Monogr. Cimic. Suec., p. 76. [Europe].
1886 *Lygus contaminatus*: Uhler, Check-list Hem. Het., p. 18.
1917 *Lygus (Neolygus) confusus* Knight, Cornell Univ. Agr. Exp. Stn. Bull., 391: 606.

[Me.]. Synonymized by Kelton, 1971, Mem. Ent. Soc. Can., 83: 11.
1917 *Lygus confusus*: Parshley, Occas. Pap. Boston Soc. Nat. Hist., 7: 87.
1924 *Neolygus confusus*: Downes, Proc. Ent. Soc. B.C., 21: 29.
1955 *Lygus (Neolygus) contaminatus*: Kelton, Can. Ent., 87: 280.
1959 *Lygocoris (Apolygus) contaminatus*: Carvalho, Arq. Mus. Nac., 48: 137.
1971 *Lygocoris (Neolygus) contaminatus*: Kelton, Mem. Ent. Soc. Can., 83: 11.
1981 *Lygocoris contaminatus*: Froeschner and Halpin, Proc. Biol. Soc. Wash., 94: 425.
Distribution: Alk., Alta., B.C., Cal., Me., N.B., N.H., N.S., Nfld., Ont., P.Ed., Que., Yuk.
Note: Kelton (1971, above, p. 11) clarified past confusion involving this species.

Lygocoris deraeocorides (Knight), 1925
1925 *Lygus (Neolygus) deraeocorides* Knight, Bull. Brook. Ent. Soc., 20: 50. [Ariz.].
1927 *Neolygus deraeocorides*: Knight, Can. Ent., 59: 43.
1959 *Lygocoris (Neolygus) deraeocoroides* [*sic*]: Carvalho, Arq. Mus. Nac., 48: 142.
1971 *Lygocoris deraeocoroides* [*sic*]: Kelton, Mem. Ent. Soc. Can., 83: 49.
Distribution: Ariz.

Lygocoris fagi (Knight), 1917
1917 *Lygus (Neolygus) fagi* Knight, Cornell Univ. Agr. Exp. Stn. Bull., 391: 603. [N.Y.].
1917 *Lygus fagi*: Parshley, Occas. Pap. Boston Soc. Nat. Hist., 7: 86.
1941 *Neolygus fagi*: Knight, Ill. Nat. Hist. Surv. Bull., 22: 161.
1959 *Lygocoris (Neolygus) fagi*: Carvalho, Arq. Mus. Nac., 48: 142.
Distribution: Ind., Mass., N.C., N.H., N.S., N.Y., Oh., Ont., Pa., Que., Tenn., Vt., W.Va., Wyo.

Lygocoris geminus (Knight), 1941
1941 *Neolygus geminus* Knight, Ill. Nat. Hist. Surv. Bull., 22: 163. [Ill.].
1959 *Lygocoris (Neolygus) geminus*: Carvalho, Arq. Mus. Nac., 48: 142.
Distribution: Ill.

Lygocoris geneseensis (Knight), 1917
1917 *Lygus (Neolygus) geneseensis* Knight, Cornell Univ. Agr. Exp. Stn. Bull., 391: 609. [N.Y.].
1917 *Lygus geneseensis*: Parshley, Occas. Pap. Boston Soc. Nat. Hist., 7: 87.
1941 *Neolygus geneseensis*: Knight, Ill. Nat. Hist. Surv. Bull., 22: 159.
1959 *Lygocoris (Neolygus) geneseensis*: Carvalho, Arq. Mus. Nac., 48: 142.
Distribution: Ga., Ia., Ill., Mass., Md., Mich., Miss., Mo., N.Y., Oh., Ont., Pa., Va., W.Va., Wis.

Lygocoris hirticulus (Van Duzee), 1916
1904 *Lygus* species nova?: Wirtner, An. Carnegie Mus., 3: 196.
1912 *Lygus tenellus* Van Duzee, Bull. Buff. Soc. Nat. Sci., 10: 484. [N.Y.]. Preoccupied.
1916 *Lygus (Lygocoris) hirticulus* Van Duzee, Check List Hem., p. 40. New name for *Lygus tenellus* Van Duzee.
1917 *Lygus (Neolygus) hirticulus*: Knight, Cornell Univ. Agr. Exp. Stn. Bull., 391: 633.
1917 *Lygus hirticulus*: Parshley, Occas. Pap. Boston Soc. Nat. Hist., 7: 90.
1941 *Neolygus hirticulus*: Knight, Ill. Nat. Hist. Surv. Bull., 22: 163.
1959 *Lygocoris (Neolygus) hirticulus*: Carvalho, Arq. Mus. Nac., 48: 142.
Distribution: Conn., Fla., Ia., Ill., Ind., Mass., Md., Me., Minn., N.B., N.C., N.H., N.J., N.S., N.Y., Oh., Ont., P.Ed., Pa., Que., Va., Vt., W.Va., Wis.
Note: Wheeler and Henry (1977, Great Lakes Ent., 10: 148) clarified the Wirtner (1904, above) record.

Lygocoris inconspicuus (Knight), 1917
1917 *Lygus* (*Neolygus*) *inconspicuus* Knight, Cornell Univ. Agr. Exp. Stn. Bull., 391: 612. [N.Y.].
1918 *Lygus inconspicuus*: Knight, Bull. Brook. Ent. Soc., 13: 44.
1941 *Neolygus inconspicuus*: Knight, Ill. Nat. Hist. Surv. Bull., 22: 161.
1959 *Lygocoris* (*Neolygus*) *inconspicuus*: Carvalho, Arq. Mus. Nac., 48: 143.
Distribution: Conn., Ga., Ia., Ill., Ind., Md., Mich., Minn., N.C., N.Y., Ok., Ont., Que., Va., Wis.

Lygocoris invitus (Say), 1832
1832 *Capsus invitus* Say, Descrip. Het. Hem. N. Am., p. 24. [Ind.].
1854 *Phytocoris lineolatus* Emmons, Nat. Hist. N.Y. Insects, 5: pl. 30, fig. 7. [N.Y.]. Synonymized by Van Duzee, 1916, Check List Hem., p. 40.
1877 *Lygus invitus*: Uhler, Wheeler's Ann. Rept., p. 1328.
1917 *Lygus* (*Neolygus*) *invitus*: Knight, Cornell Univ. Agr. Exp. Stn. Bull., 391: 604.
1941 *Neolygus invitus*: Knight, Ill. Nat. Hist. Surv. Bull., 22: 157.
1959 *Lygocoris* (*Neolygus*) *invitus*: Carvalho, Arq. Mus. Nac., 48: 143.
Distribution: Conn., Ia., Ill., Ind., Ks., Mass., Md., Me., Mich., Minn., Miss., Mo., N.C., N.H., N.J., N.S., N.Y., Oh., Ont., Pa., Que., Va., Vt., Wis.
Note: Kelton (1971, Mem. Ent. Soc. Can., 83: 20) noted that several early records for this species referred to *L. communis* and *L. inconspicuus*.

Lygocoris johnsoni (Knight), 1917
1917 *Lygus* (*Neolygus*) *johnsoni* Knight, Cornell Univ. Agr. Exp. Stn. Bull., 391: 629. [Conn.].
1917 *Lygus johnsoni*: Parshley, Occas. Pap. Boston Soc. Nat. Hist., 7: 86.
1941 *Neolygus johnsoni*: Knight, Ill. Nat. Hist. Surv. Bull., 22: 162.
1959 *Lygocoris* (*Neolygus*) *johnsoni*: Carvalho, Arq. Mus. Nac., 48: 144.
Distribution: Conn., Ind., N.Y., Oh., Ont., Tenn., Va., Vt.

Lygocoris knighti Kelton, 1971
1971 *Lygocoris* (*Neolygus*) *knighti* Kelton, Can. Ent., 103: 1107. [Ont.].
1980 *Lygocoris knighti*: Wheeler, Ent. News, 91: 25 (in part).
Distribution: Man., N.Y., Ont., Pa.
Note: Wheeler (1980, above, 91: 25-26) gave hosts and distribution for this species; Wheeler and Henry (1983, Proc. Ent. Soc. Wash., 85: 26-31) transferred the N.C. and W.Va. records to *L. atrinotatus* and clarified confusion between hosts of *L. knighti* and *L. atrinotatus* (Knight).

Lygocoris laureae (Knight), 1917
1917 *Lygus* (*Neolygus*) *laureae* Knight, Cornell Univ. Agr. Exp. Stn. Bull., 391: 636. [N.Y.].
1938 *Lygus laurae* [*sic*]: Brimley, Ins. N.C., p. 77.
1959 *Lygocoris* (*Neolygus*) *laureae*: Carvalho, Arq. Mus. Nac., 48: 144.
Distribution: Ga., N.C., N.Y., Ont., Pa., Que., S.C., Tenn., W.Va.

Lygocoris neglectus (Knight), 1917
1917 *Lygus* (*Neolygus*) *neglectus* Knight, Cornell Univ. Agr. Exp. Stn. Bull., 391: 619. [Mass.].
1917 *Lygus neglectus*: Parshley, Occas. Pap. Boston Soc. Nat. Hist., 7: 88.
1941 *Neolygus neglectus*: Knight, Ill. Nat. Hist. Surv. Bull., 22: 162.
1959 *Lygocoris* (*Neolygus*) *neglectus*: Carvalho, Arq. Mus. Nac., 48: 144.
Distribution: Ala., La., Mass., Me., Miss., N.C., Oh.

Lygocoris nyssae (Knight), 1918
1918 *Lygus (Neolygus) nyssae* Knight, Bull. Brook. Ent. Soc., 13: 43. [Ala.].
1941 *Neolygus nyssae*: Knight, Ill. Nat. Hist. Surv. Bull., 22: 164.
1959 *Lygocoris (Neolygus) nyssae*: Carvalho, Arq. Mus. Nac., 48: 144.
Distribution: Ala., Conn., Ill., Miss., Oh., Pa., W.Va.

Lygocoris omnivagus (Knight), 1917
1917 *Lygus (Neolygus) omnivagus* Knight, Cornell Univ. Agr. Exp. Stn. Bull., 391: 627. [N.Y.].
1917 *Lygus omnivagus*: Parshley, Occas. Pap. Boston Soc. Nat. Hist., 7: 89.
1941 *Neolygus omivagus*: Knight, Ill. Nat. Hist. Surv. Bull., 22: 163.
1959 *Lygocoris (Neolygus) omnivagus*: Carvalho, Arq. Mus. Nac., 48: 144.
Distribution: Conn., Fla., Ga., Ia., Ill., Man., Mass., Me., Mich., Minn., N.B., N.C., N.H., N.S., N.Y., Ont., Pa., Que., R.I., Tenn., Va., Vt., W.Va., Wis.

Lygocoris ostryae (Knight), 1917
1917 *Lygus (Neolygus) ostryae* Knight, Cornell Univ. Agr. Exp. Stn. Bull., 391: 635. [N.Y.].
1917 *Lygus ostryae*: Parshley, Occas. Pap. Boston Soc. Nat. Hist., 7: 90.
1941 *Neolygus ostryae*: Knight, Ill. Nat. Hist. Surv. Bull., 22: 164.
1959 *Lygocoris (Neolygus) ostryae*: Carvalho, Arq. Mus. Nac., 48: 144.
Distribution: Ill., Mass., N.Y., Ont., Pa., Que., Vt.

Lygocoris parrotti (Knight), 1919
1919 *Lygus (Neolygus) parrotti* Knight, Bull. Brook. Ent. Soc., 14: 21. [N.Y.].
1959 *Lygocoris (Neolygus) parrotti*: Carvalho, Arq. Mus. Nac., 48: 144.
Distribution: N.Y., Ont., Que.

Lygocoris parshleyi (Knight), 1917
1917 *Lygus (Neolygus) parshleyi* Knight, Cornell Univ. Agr. Exp. Stn. Bull., 391: 611. [N.H.].
1917 *Lygus (Neolygus) parshleyi* var. *shermani* Knight, Cornell Univ. Agr. Exp. Stn. Bull., 391: 612. [N.C.]. Synonymized by Kelton, 1971, Mem. Ent. Soc. Can., 83: 49.
1917 *Lygus parshleyi*: Parshley, Occas. Pap. Boston Soc. Nat. Hist., 7: 87.
1938 *Lygus parshleyi* var. *shermani*: Brimley, Ins. N.C., p. 77.
1959 *Lygocoris (Neolygus) parshleyi*: Carvalho, Arq. Mus. Nac., 48: 145
1959 *Lygocoris (Neolygus) parshleyi* var. *shermani*: Carvalho, Arq. Mus. Nac., 48: 145.
Distribution: Me., N.C., N.H., N.Y.

Lygocoris piceicola Kelton, 1971
1971 *Lygocoris (Neolygus) piceicola* Kelton, Can. Ent., 103: 1107. [N.B.].
Distribution: Man., N.B., Ont., Que.

Lygocoris quercalbae (Knight), 1917
1917 *Lygus (Neolygus) quercalbae* Knight, Cornell Univ. Agr. Exp. Stn. Bull., 391: 624. [N.Y.].
1917 *Lygus quercialbae* [*sic*]: Parshley, Occas. Pap. Boston Soc. Nat. Hist., 7: 89.
1941 *Neolygus quercalbae*: Knight, Ill. Nat. Hist. Surv. Bull., 22: 160.
1959 *Lygocoris (Neolygus) quercalbae*: Carvalho, Arq. Mus. Nac., 48: 145.
Distribution: Conn., Ia., Ill., Ind., Man., Mass., Md., Mich., Minn., Mo., N.S., N.Y., Ont., Pa., Que., Va., W.Va., Wis.

Lygocoris semivittatus (Knight), 1917
1917 *Lygus (Neolygus) semivittatus* Knight, Cornell Univ. Agr. Exp. Stn. Bull., 391: 626. [N.Y.].

1918 *Lygus semivittatus*: Knight, Bull. Brook. Ent. Soc., 13: 44.
1927 *Neolygus semivittatus*: Knight, Can. Ent., 59: 43
1959 *Lygocoris* (*Neolygus*) *semivittatus*: Carvalho, Arq. Mus. Nac., 48: 145.
Distribution: Ala., Fla., Minn., Miss., Mo., N.Y., Ont., Pa., Que., Tenn., Tex., Va., W.Va., Wis.

Lygocoris tiliae (Knight), 1917
1917 *Lygus* (*Neolygus*) *tiliae* Knight, Cornell Univ. Agr. Exp. Stn. Bull., 391: 613. [N.Y.].
1917 *Lygus tiliae*: Parshley, Occas. Pap. Boston Soc. Nat. Hist., 7: 87.
1941 *Neolygus tiliae*: Knight, Ill. Nat. Hist. Surv. Bull., 22: 161.
1959 *Lygocoris* (*Neolygus*) *tiliae*: Carvalho, Arq. Mus. Nac., 48: 145.
Distribution: Conn., Ga., Ia., Ill., Mass., Minn., Miss., Mo., N.H., N.Y., Oh., Ont., Pa., Que., Vt., W.Va., Wis.

Lygocoris tiliae heterophyllus (Knight), 1918
1918 *Lygus* (*Neolygus*) *tiliae* var. *heterophyllus* Knight, Bull. Brook. Ent. Soc., 13: 44. [Ga.].
1959 *Lygocoris* (*Neolygus*) *tiliae* var. *heterophyllus*: Carvalho, Arq. Mus. Nac., 48: 145.
Distribution: Ga., Miss.
Note: The holotype data in the original description of this subspecies was listed as "Fla., Georgia." This is a typographical error for "Ila [Madison Co.], Georgia."

Lygocoris tiliae tiliae (Knight), 1917
1917 *Lygus* (*Neolygus*) *tiliae* Knight, Cornell Univ. Agr. Exp. Stn. Bull., 391: 613. [N.Y.].
Distribution: Same as for species.

Lygocoris tinctus (Knight), 1941
1941 *Neolygus tinctus* Knight, Ill. Nat. Hist. Surv. Bull., 22: 157. [Ind.].
1959 *Lygocoris* (*Neolygus*) *tinctus*: Carvalho, Arq. Mus. Nac., 48: 145.
1976 *Lygocoris tinctus*: Wheeler and Henry, An. Ent. Soc. Am., 69: 1095.
Distribution: Ill., Ind., N.Y., Pa., W.Va.
Note: Wheeler and Henry (1976, above) described the egg and fifth-instar nymph and gave information about its habits on honeylocust, *Gleditsia triacanthos* L.

Lygocoris univittatus (Knight), 1917
1917 *Lygus* (*Neolygus*) *univittatus* Knight, Cornell Univ. Agr. Exp. Stn. Bull., 391: 623. [N.Y.].
1920 *Lygus univittatus*: Wellhouse, J. Econ. Ent., 13: 389.
1941 *Neolygus univittatus*: Knight, Ill. Nat. Hist. Surv. Bull., 22: 160.
1959 *Lygocoris* (*Neolygus*) *univittatus*: Carvalho, Arq. Mus. Nac., 48: 145.
Distribution: N.Y., Ont., Que.

Lygocoris viburni (Knight), 1917
1917 *Lygus* (*Neolygus*) *viburni* Knight, Cornell Univ. Agr. Exp. Stn. Bull., 391: 609. [N.Y.].
1917 *Lygus viburni*: Parshley, Occas. Pap. Boston Soc. Nat. Hist., 7: 87.
1941 *Neolygus viburni*: Knight, Ill. Nat. Hist. Surv. Bull., 22: 159.
1959 *Lygocoris* (*Neolygus*) *viburni*: Carvalho, Arq. Mus. Nac., 48: 145.
Distribution: Conn., Ill., Mich., Minn., N.B., N.S., N.Y., Oh., Ont., P.Ed., Pa., Que., Vt., W.Va.

Lygocoris vitticollis (Reuter), 1876
1876 *Lygus* (*Lygus*) *vitticollis* Reuter, Öfv. K. Svens. Vet.-Akad. Förh., 32: 71. [Tex.].
1885 *Lygus invitus*: Forbes, Fourteenth Rept. St. Ent. Ill., p. 11.
1886 *Lygus monachus* Uhler, Can. Ent., 18: 208. [Ill., Mass., Mo., N.H., Que.]. Syn-

onymized by Reuter, 1909, Acta Soc. Sci. Fenn., 36(2): 43.

1887 *Lygus monachus* Uhler, U.S. Dept. Agr., Div. Ent. Bull., 13: 63. [Ill., Mass. Mo., N.H., Que.]. Preoccupied by and a junior synonym of *L. monachus* Uhler, 1886.

1917 *Lygus (Neolygus) vitticollis*: Knight, Cornell Univ. Agr. Exp. Stn. Bull., 391: 618.

1917 *Lygus vitticollis*: Parshley, Occas. Pap. Boston Soc. Nat. Hist., 7: 88.

1941 *Neolygus vitticollis*: Knight, Ill. Nat. Hist. Surv. Bull., 22: 162.

1959 *Lygocoris (Neolygus) vitticollis*: Carvalho, Arq. Mus. Nac., 48: 145.

Distribution: Conn., Ia., Ill., Ind., Ks., Mass., Me., Mich., Minn., Miss., Mo., N.B., N.C., N.H., N.J., N.S., N.Y., Oh., Ont., Pa., Que., R.I., Tex., Va., Vt., W.Va., Wis.

Note: Wheeler (1982, Proc. Ent. Soc. Wash., 84: 177-183) studied the seasonal history of *L. vitticollis* on red maple, *Acer rubrum* L.

Lygocoris walleyi Kelton, 1971

1971 *Lygocoris (Neolygus) walleyi* Kelton, Can. Ent., 103: 1110. [Ont.].

Distribution: Ont.

Genus *Lygus* Hahn, 1833

1833 *Lygus* Hahn, Wanz. Ins., 1: 147. Type-species: *Cimex pratensis* Linnaeus, 1758. Proposed by Carvalho et al., 1961, Bull. Zool. Nomen., 18: 281-284, and designated by Int. Comm. Zool. Nomen., 1963, Opinion 667, Bull. Zool. Nomen., 20: 270-271.

Note: Kelton (1975, Mem. Ent. Soc. Can., 95: 1-101) revised the genus and provided a key to the North American species. He noted (p. 5) that records of *L. rugulipennis* Poppius from Alaska (and we add Utah) "probably should refer to *L. perplexus* Stanger or some other species" and that most North American literature dealing with *L. pratensis* (Linnaeus) should refer to *L. lineolaris* (Palisot). Graham et al. (1984, U.S. Dept. Agr., Bibliogr. Lit. Agr., 30: 1-205) provided an extensive world listing of literature for *Lygus* species and "related" genera.

Lygus abroniae Van Duzee, 1918

1918 *Lygus abroniae* Van Duzee, Proc. Cal. Acad. Sci., (4)8: 289. [Cal.].

1925 *Camptobrochis slevini* Van Duzee, Proc. Cal. Acad. Sci., (4)14: 395. [Cal.]. Synonymized by Carvalho, 1959, Arq. Mus. Nac., 48: 114.

Distribution: Cal.

Lygus aeratus Knight, 1917

1917 *Lygus (Lygus) aeratus* Knight, Cornell Univ. Agr. Exp. Stn. Bull., 391: 580. [Cal.].

1917 *Lygus aeriginosus*: Van Duzee, Proc. Cal. Acad. Sci., (4)7: 266.

Distribution: Cal., Nev., Ore.

Note: Van Duzee's *aeriginosus* is an error for Knight's *aeratus*, according to Carvalho (1959, Arq. Mus. Nac., 48: 147).

Lygus atriflavus Knight, 1917

1917 *Lygus (Lygus) atriflavus* Knight, Cornell Univ. Agr. Exp. Stn. Bull., 391: 572. [N.M.].

1955 *Liocoris atriflavus*: Kelton, Can. Ent., 87: 550.

Distribution: Ariz., B.C., Cal., Col., Id., Mont., N.M., Ore., Ut., Wash.

Lygus atritibialis Knight, 1941

1941 *Lygus (Lygus) atritibialis* Knight, Ill. Nat. Hist. Surv. Bull., 22: 152. [Ill.].

Distribution: B.C., Ill., Ks., Man., Mich., Minn., Neb., S.D., Sask., Wis.

Lygus borealis (Kelton), 1955

1955 *Liocoris borealis* Kelton, Can. Ent., 87: 488. [Sask.].

1959 *Lygus borealis*: Carvalho, Arq. Mus. Nac., 48: 148.
Distribution: Alk., Alta., B.C., Col., Ia., Id., Man., Mich., Minn., Mont., N.D., N.T., Nfld., Neb., Ont., Que., S.D., Sask., Wash., Wyo., Yuk.

Lygus bradleyi Knight, 1917
1917 *Lygus (Lygus) bradleyi* Knight, Cornell Univ. Agr. Exp. Stn. Bull., 391: 581. [Cal.].
Distribution: B.C., Cal., Ore.

Lygus ceanothi Knight, 1941
1941 *Lygus ceanothi* Knight, Ia. St. Coll. J. Sci., 15: 269. [Col.].
1942 *Lygus regulus* Stanger, Univ. Cal. Publ. Ent., 7: 162. [Cal.]. Synonymized by Carvalho, 1959, Arq. Mus. Nac., 48: 148.
1942 *Lygus maculosus* Stanger, Univ. Cal. Publ. Ent., 7: 165. [Cal.]. Synonymized by Kelton, 1975, Mem. Ent. Soc. Can., 95: 13.
1944 *Lygus ceanothi* var. *rufus* Knight, Ia. St. Coll. J. Sci., 18: 477. [Wash.]. New Synonymy.
Distribution: B.C., Cal., Col., Id., Mont., Nev., Ore., S.D., Wash., Wyo.
Note: Kelton (1975, above) apparently overlooked the variety *L. ceanothi rufus*, but his treatment of other color forms of *Lygus* prompts us to list this "var." as a synonym of *L. ceanothi*.

Lygus columbiensis Knight, 1917
1909 *Lygus pratensis*: Adams, Mich. Biol. Surv., p. 261.
1917 *Lygus (Lygus) columbiensis* Knight, Cornell Univ. Agr. Exp. Stn. Bull., 391: 571. [B.C.].
1917 *Lygus superiorensis* Knight, Cornell Univ. Agr. Exp. Stn. Bull., 391: 572. [Mich.]. Synonymized by Kelton, 1975, Mem. Ent. Soc. Can., 95: 53.
1927 *Lygus punctatus*: Knight, Can. Ent., 59: 42.
1955 *Liocoris columbiensis*: Kelton, Can. Ent., 87: 554.
Distribution: Alk., Alta., B.C., Col., Id., Man., Mich., Minn., Mont., N.T., Nfld., Ont., Ore., Que., S.D., Sask., Ut., Wash., Wyo.
Note: The *L. punctatus* record by Knight (1927, above) is a misidentification, according to Kelton (1975, Mem. Ent. Soc. Can., 95: 53).

Lygus convexicollis Reuter, 1876
1876 *Lygus (Lygus) convexicollis* Reuter, Öfv. K. Svens. Vet.-Akad. Förh., 32(9): 72. [Cal.].
1917 *Lygus (Lygus) convexicollis* var. *coloratus* Knight, Cornell Univ. Agr. Exp. Stn. Bull., 391: 569. [Cal.]. Synonymized by Kelton, 1975, Mem. Ent. Soc. Can., 95: 11.
1942 *Lygus dolichorhynchus* Stanger, Univ. Cal. Publ. Ent., (4)7: 166. [Cal.]. Synonymized by Kelton, 1975, Mem. Ent. Soc. Can., 95: 11.
Distribution: Alta., B.C., Cal., Col., Id., Nev., Ore., S.D., Sask.

Lygus desertus Knight, 1944
1944 *Lygus desertus* Knight, Ia. St. Coll. J. Sci., 18: 471. [Ariz.].
1955 *Liocoris desertus*: Kelton, Can. Ent., 87: 548.
1968 *Lygus desertinus* Knight, Brigham Young Univ. Sci. Bull., 9(3): 189. New name for *Lygus desertus* Knight. See note below.
Distribution: Alta., Ariz., B.C., Cal., Col., Ia., Id., Man., Minn., Mont., N.D., N.M., N.T., Ore., S.D., Sask., Ut., Wash. (Mexico).
Note: Muminov (1986, Izv. Akad. Nauk Tadzhik S.S.R., Otd. Biol. Nauk, 101: 41) has shown that *Capsus desertus* Becker, 1864 [treated as a *nomen oblitum*] is a syn-

onym of *Deraeocoris* (*Camptobrochis*) *serenus* Douglas and Scott, 1868; thus, Knight's original name *L. desertus* is no longer a secondary junior homonym.

Lygus elisus Van Duzee, 1914

1914 *Lygus pratensis* var. *elisus* Van Duzee, Trans. San Diego Soc. Nat. Hist., 2: 20. [Cal.].

1916 *Lygus elisus*: Van Duzee, Check List Hem., p. 40.

1917 *Lygus* (*Lygus*) *elisus* var. *viridiscutatus* Knight, Cornell Univ. Agr. Exp. Stn. Bull., 391: 575. [Cal.]. Synonymized by Kelton, 1975, Mem. Ent. Soc. Can., 95: 36.

1931 *Lygus elisius* [*sic*]: Shull and Wakeland, J. Econ. Ent., 24: 326.

1941 *Lygus nigrosignatus* Knight, Ia. St. Coll. J. Sci., 15: 270. [Id.]. Synonymized by Kelton, 1975, Mem. Ent. Soc. Can., 95: 37.

1955 *Liocoris elisus*: Kelton, Can. Ent., 87: 548.

1955 *Liocoris nigrosignatus*: Kelton, Can. Ent., 87: 552.

Distribution: Alk., Alta., Ariz., B.C., Cal., Col., Ia., Id., Ill., Ind., Ks., Man., Minn., Mo., Mont., N.D., N.M., N.T., Neb., Nev., Ok., S.D., Sask., Tex., Ut., Wash., Wyo. (Hawaii, Mexico).

Note: Shull (1933, Id. Agr. Exp. Stn. Res. Bull., 11: 1-42) described the nymphal stages.

Lygus hesperus Knight, 1917

1917 *Lygus* (*Lygus*) *elisus* var. *hesperus* Knight, Cornell Univ. Agr. Exp. Stn. Bull., 391: 575. [Cal.].

1917 *Lygus elisus* var. *hesperius* [*sic*]: Van Duzee, Univ. Cal. Publ. Ent., 2: 821.

1933 *Lygus hesperus*: Shull, J. Econ. Ent., 26: 1076.

1955 *Liocoris hesperus*: Kelton, Can. Ent., 87: 550.

1976 *Lygus hesperius* [*sic*]: Kumar et al., Univ. Wyo. Agr. Exp. Stn. Sci. Monogr., 32: 38.

1976 *Lygus herperus* [*sic*]: Moffett et al., J. Ariz. Acad. Sci., 11: 48.

Distribution: B.C., Cal., Id., Mont., Neb., Nev., Ore., S.D., Tex., Ut., Wyo.

Note: Kelton (1975, Mem. Ent. Soc. Can., 95: 35) suggested that the records for *hesperus* from Ia., Ill., Mich., and Minn. may refer to other species of *Lygus*. The nymphal stages were described by Shull (1933, Id. Agr. Exp. Stn. Res. Bull., 11: 1-42). Strong et al. (1970, Hilgardia, 40: 105-147) studied the reproductive biology.

Lygus humeralis Knight, 1917

1917 *Lygus* (*Lygus*) *humeralis* Knight, Cornell Univ. Agr. Exp. Stn. Bull., 391: 570. [B.C.].

1941 *Lygus ceanothi* var. *delecticus* Knight, Ia. St. Coll. J. Sci., 15: 270. [Id.]. Synonymized by Kelton, 1975, Mem. Ent. Soc. Can., 95: 14 [spelled *deleticus*].

1955 *Liocoris humeralis*: Kelton, Can. Ent., 87: 547.

Distribution: Alta., B.C., Cal., Col., Id., Mont., N.M., Ore., Ut., Wash., Wyo.

Lygus lineolaris (Palisot), 1818 [Fig. 74]

1818 *Capsus lineolaris* Palisot, Ins. Rec. Afr. Am., p. 187. [E. U.S.].

1818 *Capsus linearis* [*sic*]: Palisot, Ins. Rec. Afr. Am., p. 187, pl. 11, no. 7.

1832 *Capsus oblineatus* Say, Descrip. Het. Hem. N. Am., p. 21. [Ind., Mo., N.T.]. Synonymized by Uhler, 1872, Prelim. Rept. U.S. Geol. Surv. Mont., p. 413.

1841 *Phytocoris lineolaris*: Harris, Rept. Inj. Ins. Mass., p. 161.

1864 *Phytocoris linearis* [*sic*]: Walsh, Proc. Boston Soc. Nat. Hist., 9: 313.

1872 *Lygus lineolaris*: Uhler, Prelim. Rept. U.S. Geol. Surv. Mont., p. 413.

1872 *Capsus flavonotatus* Provancher, Nat. Can., 4: 103. [Que.]. Synonymized by Van Duzee, 1912, Can. Ent., 44: 321.

1873 *Capsus strigulatus* Walker, Cat. Het. Brit. Mus., 6: 94. [Can.]. Synonymized by Knight, 1921, Rept. St. Ent. Minn., 18: 197.
1886 *Deraeocoris flavonotatus*: Uhler, Check-list Hem. Het., p. 19.
1886 *Lygus pratensis*: Uhler, Check-list Hem. Het., p. 18.
1887 *Lygus flavonotatus*: Provancher, Pet. Faune Ent. Can., 3: 120.
1887 *Lygus flavonomaculatus* [*sic*]: Van Duzee, Can. Ent., 19: 71.
1900 *Lygus oblineatus*: Smith, Ins. N.J., 2: 124.
1904 *Camptobrochys* [*sic*] *strigulatus*: Distant, An. Mag. Nat. Hist., (7)13: 111.
1916 *Camptobrochis* (*Camptobrochis*) *strigulatus*: Van Duzee, Check List Hem., p. 41.
1917 *Camptobrochys* [*sic*] (*Camptobrochys*) *strigulatus*: Van Duzee, Univ. Cal. Publ. Ent., 2: 354.
1917 *Lygus pratensis* (*Lygus*) var. *oblineatus*: Knight, Cornell Univ. Agr. Exp. Stn. Bull., 391: 564.
1917 *Lygus pratensis* (*Lygus*) var. *rubidus* Knight, Cornell Univ. Agr. Exp. Stn. Bull., 391: 565. [Me.]. Listed as synonym by Kelton, 1975, Mem. Ent. Soc. Can., 95: 43.
1938 *Lygus strigulatus*: Brimley, Ins. N.C., p. 78.
1953 *Lygus lineolarius* [*sic*]: Strickland, Can. Ent., 85: 201.
1954 *Lygus linearis* [*sic*]: Glick and Lattimore, J. Econ. Ent., 47: 683.
1955 *Liocoris lineolaris*: Kelton, Can. Ent., 87: 552.
1961 *Lygus* (*Exolygus*) *linearis* [*sic*]: Davis, An. Ent. Soc. Am., 54: 345.
1963 *Lygus lineclaris* [*sic*]: Menhinick, U.S. Atomic Energy Comm., TID-19136: 15.
1979 *Lygus lincolaris* [*sic*]: Henry and Smith, J. Ga. Ent. Soc., 14: 217.

Distribution: Known from every state and province, south to southern Mexico. Some Mexican records may be confused with those for *L. mexicanus* Kelton (which see).

Note: Graham et al. (1984, U.S. Dept. Agr., Bibliogr. Lit. Agr., 30: 1-205) provided an extensive listing of literature treating the tarnished plant bug, *Lygus lineolaris*, other *Lygus* species, and "related" genera.

Lygus mexicanus Kelton, 1973

1884 *Lygus pratensis*: Distant, Biol. Centr.-Am., Rhyn., 1: 272.
1973 *Lygus mexicanus* Kelton, Can. Ent., 105: 1546. [Holotype from Mexico; also from Tex.].

Distribution: Tex. (Mexico).

Lygus nigropallidus Knight, 1917

1917 *Lygus* (*Lygus*) *nigropallidus* Knight, Cornell Univ. Agr. Exp. Stn. Bull., 391: 579. [N.M.].
1955 *Liocoris nigropallidus*: Kelton, Can. Ent., 547.

Distribution: Alta., Ariz., B.C., Cal., Col., Id., Mont., N.M., Ore., S.D., Sask., Ut., Wash., Wyo.

Lygus nubilatus Knight, 1917

1917 *Lygus* (*Lygus*) *nubilatus* Knight, Cornell Univ. Agr. Exp. Stn. Bull., 391: 584. [Cal.].
1917 *Lygus* (*Lygus*) *nubilosus* Knight, Cornell Univ. Agr. Exp. Stn. Bull., 391: 585. [N.M.]. Synonymized by Kelton, 1975, Mem. Ent. Soc. Can., 95: 29.
1942 *Lygus usingeri* Stanger, Univ. Cal. Publ. Ent., (4)7: 162. [Cal.]. Synonymized by Kelton, 1975, Mem. Ent. Soc. Can., 95: 29.
1955 *Liocoris nubilatus*: Kelton, Can. Ent., 87: 285.

Distribution: Ariz., B.C., Cal., Col., Id., N.M., Ore., Wash.

Lygus nubilus Van Duzee, 1914

1914 *Lygus distinguendus* var. *nubilus* Van Duzee, Trans. San Diego Soc. Nat. Hist., 2: 20. [Cal.].

1917 *Lygus nubilus* Van Duzee, Univ. Cal. Publ. Ent., 2: 350.

1917 *Lygus* (*Lygus*) *nubilus*: Knight, Cornell Univ. Agr. Exp. Stn. Bull., 391: 582.

1917 *Lygus* (*Lygus*) *ultranubilus* Knight, Cornell Univ. Agr. Exp. Stn. Bull., 391: 583. [N.M.]. Synonymized by Kelton, 1975, Mem. Ent. Soc. Can., 95: 28.

1954 *Lygus epelys* Hussey, Proc. Ent. Soc. Wash., 56: 196. [Conn.]. Synonymized by Kelton, 1975, Mem. Ent. Soc. Can., 95: 28.

1955 *Liocoris ultranubilus*: Kelton, Can. Ent., 87: 285.

Distribution: Alta., B.C., Cal., Col., Conn., Id., Mich., Mont., N.M., Neb., Ont., Ore., Que., S.D., Ut., W.Va., Wash.

Lygus oregonae Knight, 1944

1944 *Lygus oregonae* Knight, Ia. St. Coll. J. Sci., 18: 476. [Ore.].

Distribution: Ore.

Lygus perplexus Stanger, 1942

1942 *Lygus perplexus* Stanger, Univ. Cal. Publ. Ent., 7: 163. [Cal.].

Distribution: Alk., Alta., B.C., Cal., Col., Id., Mont., Ore., Wash., Wyo., Yuk.

Lygus plagiatus Uhler, 1895

1895 *Lygus plagiatus* Uhler, Bull. Col. Agr. Exp. Stn., 31: 35. [Col.].

1917 *Lygus* (*Lygus*) *plagiatus*: Knight, Cornell Univ. Agr. Exp. Stn. Bull., 391: 576.

1955 *Liocoris plagiatus*: Kelton, Can. Ent., 87: 552.

Distribution: Alta., Cal., Col., Conn., Ia., Ill., Ind., Ks., Man., Mass., Md., Me., Mich., Minn., Mo., N.C., N.D., N.J., N.M., N.Y., Neb., Ont., Pa., Que., R.I., Sask., W.Va., Wis., Wyo.

Lygus potentillae Kelton, 1973

1953 *Lygus varius*: Strickland, Can. Ent., 85: 202 (in part).

1955 *Liocoris varius*: Kelton, Can. Ent., 87: 553 (in part).

1973 *Lygus potentillae* Kelton, Can. Ent., 105: 1545. [Sask.].

Distribution: Alk., Alta., B.C., Col., Id., Man., Minn., Nfld., Ont., Que., S.D., Sask., Wyo., Yuk.

Lygus ravus Stanger, 1942

1942 *Lygus ravus* Stanger, Univ. Cal. Publ. Ent., 7: 164. [Cal.].

1944 *Lygus nigritus* Knight, Ia. St. Coll. J. Sci., 18: 472. [Wash.]. Synonymized by Kelton, 1975, Mem. Ent. Soc. Can., 95: 52.

Distribution: Alk., Alta., B.C., Cal., Col., Id., Man., N.B., N.T., Nfld., Ore., Que., Sask., Wash., Yuk.

Lygus robustus (Uhler), 1895

1895 *Camptobrochis robustus* Uhler, Bull. Col. Agr. Exp. Stn., 31: 39. [Col.].

1917 *Lygus* (*Lygus*) *robustus*: Knight, Cornell Univ. Agr. Exp. Stn. Bull., 391: 588.

1944 *Lygus fultoni* Knight, Ia. St. Coll. J. Sci., 18: 474. [Col.]. Synonymized by Kelton, 1975, Mem. Ent. Soc. Can., 95: 60.

1944 *Lygus brindleyi* Knight, Ia. St. Coll. J. Sci., 18: 475. [Id.]. Synonymized by Kelton, 1975, Mem. Ent. Soc. Can., 95: 60.

1955 *Liocoris robustus*: Kelton, Can. Ent., 87: 554.

Distribution: Ariz., B.C., Cal., Col., Id., Mont., N.M., Nev., Ore., Ut., Wyo.

Lygus rolfsi Knight, 1941

1941 *Lygus rolfsi* Knight, Ia. St. Coll. J. Sci., 15: 271. [Wash.].

Distribution: Cal., Ok., Ore., Wash.

Lygus rubroclarus Knight, 1917
1917 *Lygus* (*Lygus*) *vanduzeei* var. *rubroclarus* Knight, Cornell Univ. Agr. Exp. Stn. Bull., 391: 567. [N.S.].
1941 *Lygus frisoni* Knight, Ill. Nat. Hist. Surv. Bull., 22: 148. [Ill.]. Synonymized by Kelton, 1975, Mem. Ent. Soc. Can., 95: 58.
1953 *Lygus rubroclarus* Knight, Ia. St. Coll. J. Sci., 27: 518.
1955 *Liocoris rubroclarus*: Kelton, Can. Ent., 87: 553.
Distribution: Alta., B.C., Ia., Ill., Man., Mass., Me., Mich., Minn., N.B., N.C., N.H., N.S., N.Y., Nfld., Ont., Que., P.Ed., Sask., Vt., W.Va., Wis.

Lygus rubrosignatus Knight, 1923
1923 *Lygus pratensis* var. *rubrosignatus* Knight, Conn. Geol. Nat. Hist. Surv. Bull., 34: 576. [Mass.].
1953 *Lygus rubrosignatus*: Knight, Ia. St. Coll. J. Sci., 27: 518.
1955 *Liocoris rubrosignatus*: Kelton, Can. Ent., 87: 550.
Distribution: Alta., B.C., Id., Man., Mass., N.T., N.Y., Nfld., Sask., Yuk.

Lygus rufidorsus (Kelton), 1955
1955 *Liocoris rufidorsus* Kelton, Can. Ent., 87: 484. [Sask.].
1959 *Lygus rufidorsus*: Carvalho, Arq. Mus. Nac., 48: 155.
Distribution: Alta., B.C., Col., Man., Mich., N.B., N.H., N.S., N.T. Ont., P.Ed., S.D., Sask., Ut.

Lygus shulli Knight, 1941
1941 *Lygus shulli* Knight, Ia. St. Coll. J. Sci., 15: 272. [Id.].
1955 *Liocoris shulli*: Kelton, Can. Ent., 87: 552.
Distribution: Alta., Ariz., B.C., Cal., Col., Id., Man., N.M., Ore., S.D., Sask., Ut., Wash., Wyo.

Lygus solidaginis (Kelton), 1955
1955 *Liocoris solidaginis* Kelton, Can. Ent., 87: 489. [Sask.].
1959 *Lygus solidaginis*: Carvalho, Arq. Mus. Nac., 48: 156.
Distribution: Alta., B.C., Ia., Id., Man., Minn., Mont., N.D., Neb., S.D., Sask., Wyo.

Lygus striatus Knight, 1917
1917 *Lygus* (*Lygus*) *striatus* Knight, Cornell Univ. Agr. Exp. Stn. Bull., 391: 578. [Cal.].
1925 *Lygus striatus*: Knight, Can. Ent., 57: 182.
Distribution: Alk., B.C., Cal., Ore., Wash., Yuk.

Lygus unctuosus (Kelton), 1955
1955 *Liocoris unctuosus* Kelton, Can. Ent., 87: 486. [Sask.].
1959 *Lygus unctuosus*: Carvalho, Arq. Mus. Nac., 48: 156
1960 *Liocoris unctuosis* [*sic*]: Arrand, Proc. Ent. Soc. B.C., 57: 60.
Distribution: Alk., Alta., B.C., Col., Man., Mich., Minn., N.B., N.T., Ont., Que., S.D., Sask.

Lygus vanduzeei Knight, 1917
1909 *Lygus convexicollis*: Reuter, Acta. Soc. Sci. Fenn., 36(2): 43.
1917 *Lygus* (*Lygus*) *vanduzeei* Knight, Cornell Univ. Agr. Exp. Stn. Bull., 391: 565. [N.Y.].
1918 *Lygus vanduzeii* [*sic*]: Blackman, N.Y. St. Coll. For., Syracuse Tech. Publ. 10, 28: 136.
1955 *Liocoris vanduzeei*: Kelton, Can. Ent., 87: 553.
1979 *Lygus vanduzeeis* [*sic*]: Henry and Smith, J. Ga. Ent. Soc., 14: 217.
Distribution: Conn., Del., Ga., Ia., Ill., Man., Mass., Md., Me., Mich., Minn., N.B., N.C.,

N.D., N.H., N.S., N.Y., Oh., Ont., P.Ed., Pa., S.D., Sask., Va., Vt., W.Va., Wis., Wyo.

Lygus varius Knight, 1944
1944 *Lygus varius* Knight, Ia. St. Coll. J. Sci., 18: 473. [Col.].
1955 *Liocoris varius*: Kelton, Can. Ent., 87: 553.
Distribution: Alk., Alta., B.C., Cal., Col., Id., Man., Mont., N.T., Nev., Nfld., Ont., Ore., Que., S.D., Sask., Ut., Wash., Wyo., Yuk.

Genus *Metriorrhynchomiris* Kirkaldy, 1904

1876 *Poecilocapsus* (*Metriorhynchus*) Reuter, Öfv. K. Svens. Vet.-Akad. Förh., 32(9): 74. Preoccupied. Type-species: *Poecilocapsus* (*Metriorhynchus*) *affinis* Reuter, 1876, a junior synonym of *Capsus dislocatus* Say, 1832. Designated by Kirkaldy, 1904, Ent., 37: 280.
1904 *Metriorrhynchomiris* Kirkaldy, Ent., 37: 280. New name for *Metriorhynchus* Reuter.

Metriorrhynchomiris dislocatus (Say), 1832
1832 *Capsus dislocatus* Say, Descrip. Het. Hem. N. Am., p. 21. [Pa.].
1832 *Capsus goniphorus* Say, Descrip. Het. Hem. N. Am., p. 21. [U.S.]. As variety by Van Duzee, 1912, Bull. Buff. Soc. Nat. Sci., 10: 484. New Synonymy.
1845 *Capsus melaxanthus* Herrich-Schaeffer, Wanz. Ins., 8: 18, pl. 254, fig. 794. Synonymized by Van Duzee, 1916, Check List Hem., p. 39 [previously synonymized with "*Deraocoris*" *rapidus* (Say) by Uhler, 1878, Proc. Boston Soc. Nat. Hist., 19: 401].
1854 *Phytocoris coccineus* Emmons, Nat. Hist. N.Y. Agr., 5: plate 30, fig. 2. [N.Y.]. New Synonymy.
1869 *Capsis* [*sic*] *dislocatus*: Rathvon, Hist. Lancaster Co., Pa., p. 549.
1873 *Capsus limbatellus* Walker, Cat. Hem. Brit. Mus., 6: 93. [N.Y.]. Synonymized by Van Duzee, 1917, Univ. Cal. Publ. Ent., 2: 334.
1873 *Capsus marginatus* Walker, Cat. Hem. Brit. Mus., 6: 96. [N.Y.]. Synonymized by Van Duzee, 1917, Univ. Cal. Publ. Ent., 2: 334.
1876 *Poecilocapsus* (*Metriorhynchus*) *affinis* Reuter, Öfv. K. Svens. Vet.-Akad. Förh., 32(9): 74. [N.J.]. As variety by Van Duzee, 1912, Bull. Buff. Soc. Nat. Sci., 10: 484. New Synonymy.
1876 *Poecilocapsus* (*Metriorhynchus*) *marginalis* Reuter, Öfv. K. Svens. Vet.-Akad. Förh., 32(9): 75. [N.Y.]. As variety by Van Duzee, 1912, Bull. Buff. Soc. Nat. Sci., 10: 484. New Synonymy.
1878 *Lygus dislocatus*: Uhler, Bull. U.S. Geol. Geogr. Surv. Terr., 4: 406.
1886 *Poecilocapsus dislocatus*: Uhler, Check-list Hem. Het., p. 19.
1909 *Horcias dislocatus* var. *nigritus* Reuter, Acta Soc. Sci. Fenn., 36(2): 41. [Pa.]. New Synonymy.
1912 *Horcias dislocatus* var. *dislocatus*: Van Duzee, Bull. Buff. Soc. Nat. Sci., 10: 483
1912 *Horcias dislocatus* var. *goniphorus*: Van Duzee, Bull. Buff. Soc. Nat. Sci., 10: 484.
1912 *Horcias dislocatus* var. *affinis*: Van Duzee, Bull. Buff. Soc. Nat. Sci., 10: 484.
1912 *Horcias dislocatus* var. *marginalis*: Van Duzee, Buff. Buff. Soc. Nat. Sci., 10: 484.
1912 *Horcias dislocatus* var. *nigrita*: Van Duzee, Bull. Buff. Soc. Nat. Sci., 10: 484.
1912 *Horcias dislocatus* var. *pallipes* Van Duzee, Bull. Buff. Soc. Nat. Sci., 10: 484. [N.Y.]. New Synonymy.
1912 *Horcias dislocatus* var. *scutellatus* Van Duzee, Bull. Buff. Soc. Nat. Sci., 10: 484. [Conn., Me.]. Synonymized with var. *coccineus* Emmons by Van Duzee, 1917, Univ. Cal. Publ. Ent., 2: 334.
1912 *Horcias dislocatus* var. *thoracius* Van Duzee, Bull. Buff. Soc. Nat. Sci., 10: 484. [Ind.]. New Synonymy.

1916 *Horcias dislocatus* var. *coccineus*: Van Duzee, Check List Hem., p. 39.
1916 *Horcias dislocatus* var. *residuus* Van Duzee, Check List Hem., p. 39. Unnecessary name for *Horcias dislocatus thoracius* Van Duzee.
1923 *Horcias dislocatus* var. *rubellus* Knight, Conn. Geol. Nat. Hist. Surv. Bull., 34: 608. [Minn.]. New Synonymy.
1923 *Horcias dislocatus* var. *gradus* Knight, Conn. Geol. Nat. Hist. Surv. Bull., 34: 609. [Minn.]. New Synonymy.
1923 *Horcias dislocatus* var. *flavidus* Knight, Conn. Geol. Nat. Hist. Surv. Bull., 34: 609. [N.Y.]. New Synonymy.
1923 *Horcias dislocatus* var. *scutatus* Knight, Conn. Geol. Nat. Hist. Surv. Bull., 34: 609. [N.Y.]. New Synonymy.
1923 *Horcias dislocatus var. nigriclavus* Knight, Conn. Geol. Nat. Hist. Surv. Bull., 34: 609. [N.Y.]. New Synonymy.
1963 *Morcias* [*sic*] *dislocatus*: Hardee et al., J. Econ. Ent., 56: 556.
1974 *Metriorhychomiris* [*sic*] *dislocatus*: Carvalho and Jurberg, Rev. Brasil. Biol., 34: 50.
1974 *Metriorhynchomiris* [*sic*] *dislocatus* var. *affinis*: Carvalho and Jurberg, Rev. Brasil. Biol., 34: 50.
1974 *Metriorhynchomiris* [*sic*] *dislocatus* var. *coccineus*: Carvalho and Jurberg, Rev. Brasil. Biol., 34: 50.
1974 *Metriorhynchomiris* [*sic*] *dislocatus* var. *goniphorus*: Carvalho and Jurberg, Rev. Brasil. Biol., 34: 50.
1974 *Metriorhynchomiris* [*sic*] *dislocatus* var. *gradus*: Carvalho and Jurberg, Rev. Brasil. Biol., 34: 50.
1974 *Metriorhynchomiris* [*sic*] *dislocatus* var. *limbatellus*: Carvalho and Jurberg, Rev. Brasil., Biol., 34: 50.
1974 *Metriorhynchomiris* [*sic*] *dislocatus* var. *marginalis*: Carvalho and Jurberg, Rev. Brasil. Biol., 34: 50.
1980 *Metriorrhynchomiris dislocatus*: Kelton, Agr. Can., Res. Publ., 1703: 162.
Distribution: Alta., Col., Conn., D.C., Ga., Ia., Ill., Ind., Ks., Man., Mass., Mich., Me., Minn., Mo., N.H., N.J., N.Y., Oh., Ont., Pa., Que., Sask., Tex., Vt., W.Va., Wis.
Note: We consider all varieties of *M. dislocatus* merely color forms, not subspecies; therefore, they are here treated as synonyms of the nominate species/subspecies.

Metriorrhynchomiris fallax (Reuter), 1909
1909 *Horcias fallax* Reuter, Acta Soc. Sci. Fenn., 36(2): 42. [Pa.].
1974 *Metriorhynchomiris* [*sic*] *fallax*: Carvalho and Jurberg, Rev. Brasil. Biol., 34: 50.
Distribution: Ia., Ill., Ind., Mo., Oh., Pa., W.Va., Wis.

Metriorrhynchomiris illini (Knight), 1941
1941 *Horcias illini* Knight, Ill. Nat. Hist. Surv. Bull., 22: 172. [Ill.].
1974 *Metriorhynchomiris* [*sic*] *illini*: Carvalho and Jurberg, Rev. Brasil. Biol., 34: 50.
Distribution: Ill., Mo., Ont.

Genus *Monalocorisca* Distant, 1884

1884 *Monalocorisca* Distant, Biol. Centr.-Am., Rhyn., 1: 286. Type-species: *Monalocorisca granulata* Distant, 1884. Designated by Kirkaldy, 1906, Trans. Am. Ent. Soc., 32: 147.

Monalocorisca maculatus (Johnston), 1939
1939 *Neoborus maculatus* Johnston, Bull. Brook. Ent. Soc., 34: 132. [Tex.].

1959 *Tropidosteptes maculatus*: Carvalho, Arq. Mus. Nac., 48: 272.
1976 *Monalocoris* [*lapsus*] *maculatus*: Carvalho, Rev. Brasil. Biol., 36: 53.
Distribution: Tex.

Monalocorisca rostratus (Johnston), 1939
1939 *Neoborus rostratus* Johnston, Bull. Brook. Ent. Soc., 34: 130. [Tex.].
1959 *Tropidosteptes rostratus*: Carvalho, Arq. Mus. Nac., 48: 272.
1976 *Monalocorisca rostratus*: Carvalho, Rev. Brasil. Biol., 36: 55.
Distribution: Tex.

Genus *Neoborella* Knight, 1925

1925 *Neoborella* Knight, Bull. Brook. Ent. Soc., 20: 48. Type-species: *Neoborella tumida* Knight, 1925. Monotypic.

Note: Kelton and Herring (1978, Can. Ent., 110: 780) provided a key to species.

Neoborella canadensis Kelton and Herring, 1978
1978 *Neoborella canadensis* Kelton and Herring, Can. Ent., 110: 779. [Alta.].
Distribution: Alta., Sask.

Neoborella pseudotsugae Kelton and Herring, 1978
1978 *Neoborella pseudotsugae* Kelton and Herring, Can. Ent., 110: 779. [Ariz.].
Distribution: Ariz., N.M.

Neoborella tumida Knight, 1925
1925 *Neoborella tumida* Knight, Bull. Brook. Ent. Soc., 20: 48. [Ariz.].
Distribution: Ariz., Col.

Neoborella xanthenes Herring, 1972
1972 *Neoborella xanthenes* Herring, Proc. Ent. Soc. Wash., 74: 9. [Col.].
Distribution: Col.

Genus *Neoborops* Uhler, 1895

1895 *Neoborops* Uhler, Bull. Col. Agr. Exp. Stn., 31: 36. Type-species: *Neoborops vigilax* Uhler, 1895. Monotypic.

Neoborops vigilax Uhler, 1895
1895 *Neoborops vigilax* Uhler, Bull. Col. Exp. Stn., 31: 36. [Col.].
Distribution: Ariz., Col.
Note: A Pa. record listed by Van Duzee (1917, Univ. Cal. Publ. Ent., 2: 352) and others was based on the list of Pa. mirids given by Wirtner (1904, An. Carnegie Mus., 3: 197). Wheeler and Henry (1977, Great Lakes Ent., 10: 149) referred the Wirtner record to *Tropidosteptes amoenus* Reuter.

Genus *Neocapsus* Distant, 1884

1884 *Neocapsus* Distant, Biol. Centr.-Am., Rhyn., 1: 277. Type-species: *Neocapsus mexicanus* Distant, 1884. Monotypic.

Neocapsus cuneatus Distant, 1893
1893 *Neocapsus cuneatus* Distant, Biol. Centr.-Am., Rhyn., 1: 438. [Mexico].
1910 *Neocapsus mexicanus*: Banks, Cat. Nearc. Hem.-Het., p. 46.
1912 *Tropidosteptes imperialis* Van Duzee, Bull. Buff. Soc. Nat. Sci., 10: 487. [Tex.]. Synonymized by Carvalho, 1959, Arq. Mus. Nac., 48: 170.

1959 *Neocapsus cuneatus*: Carvalho, Arq. Mus. Nac., 48: 170.

Distribution: Ariz.(?), Tex.(?) (Mexico).

Note: Carvalho (1955, Rev. Chil. Ent., 4: 226) clarified taxonomic confusion between this species and *Rhasis leviscutatus* (Knight). Records for *N. cuneatus* from the U.S. probably are in error and need verification. The W.Va. record of this species by Wheeler et al. (1983, Trans. Am. Ent. Soc., 109: 148) is referred to *R. leviscutatus*.

Genus *Neurocolpus* Reuter, 1876

1876 *Neurocolpus* Reuter, Öfv. K. Svens. Vet.-Akad. Förh., 32(9): 69. Type-species: *Capsus nubilus* Say, 1832. Monotypic.

Note: Henry and Kim (1984, Trans. Am. Ent. Soc., 110: 1-75) revised this genus and provided a key to species. Although *N. mexicanus* Distant has been recorded from the U.S. (Carvalho, 1959, Arq. Mus. Nac., 48: 171), Henry and Kim (1984, above) indicated that these records probably are in error and refer to *N. knighti* Henry or *N. montanus* Knight.

Neurocolpus arizonae Knight, 1934

1892 *Neurocolpus nubilus*: Townsend, Can. Ent., 24: 193.

1934 *Neurocolpus arizonae* Knight, Bull. Brook. Ent. Soc., 29: 165. [Ariz.].

Distribution: Ariz., Col., N.M., Tex., Ut. (Mexico).

Neurocolpus brevicornis Henry, 1984

1984 *Neurocolpus brevicornis* Henry, *In* Henry and Kim, Trans. Am. Ent. Soc., 110: 15. [Ill.].

Distribution: Ill., Pa.

Neurocolpus flavescens Blatchley, 1928

1928 *Neurocolpus nubilus* var. *flavescens* Blatchley, J. N.Y. Ent. Soc., 36: 10. [Fla.]. Lectotype designated by Blatchley, 1930, Blatchleyana, p. 66-67; holotype recognized by Henry and Kim, 1984, Trans. Am. Ent. Soc., 110: 21.

1964 *Neurocolpus nubilus*: Frost, Fla. Ent., 47: 136.

1984 *Neurocolpus flavescens*: Henry and Kim, Trans. Am. Ent. Soc., 110: 20.

Distribution: Fla., Ga., Miss.

Neurocolpus jessiae Knight, 1934

1904 *Neurocolpus nubilus*: Uhler, Proc. U.S. Nat. Mus., 27: 356 (in part).

1904 *Neurocolpus nubilus*: Wirtner, An. Carnegie Mus., 3: 198 (in part).

1934 *Neurocolpus jessiae* Knight, Bull. Brook. Ent. Soc., 29: 163. [Mo.].

Distribution: Ark., D.C., Ia., Ill., La., Mass., Md., Mich., Miss., Mo., N.C. N.Y., Neb., Oh., Ok., Ont., Pa., Tex., Va., W.Va., Wis.

Note: Wheeler and Henry (1977, Great Lakes Ent., 10: 149) clarified Wirtner's (1904, above) record, which applied in part to *N. jessiae*, *N. nubilus*, and *N. tiliae*.

Neurocolpus johnstoni Knight, 1934

1934 *Neurocolpus johnstoni* Knight, Bull. Brook. Ent. Soc., 29: 166. [Tex.].

Distribution: Tex.

Neurocolpus knighti Henry, 1984

1984 *Neurocolpus knighti* Henry, *In* Henry and Kim, Trans. Am. Ent. Soc., 110: 51. [Tex.].

Distribution: Tex. (Mexico).

Neurocolpus longirostris Knight, 1968
1968 *Neurocolpus longirostris* Knight, Brig. Young Univ. Sci. Bull., 9(3): 210. [Wash.].
Distribution: Cal., Id., Mont., Ore., Wash., Wyo.

Neurocolpus montanus Knight, 1968
1968 *Neurocolpus montanus* Knight, Brig. Young Univ. Sci. Bull., 9(3): 208. [Ariz.].
1968 *Neurocolpus chiricahuae* Knight, Brig. Young Univ. Sci. Bull., 9(3): 209. [Ariz.]. Synonymized by Henry and Kim, 1984, Trans. Am. Ent. Soc., 110: 57.
1968 *Neurocolpus obsoletus* Knight, Brig. Young Univ. Sci. Bull., 9(3): 210. [Ariz.]. Synonymized by Henry and Kim, 1984, Trans. Am. Ent. Soc., 110: 57.
1968 *Neurocolpus stitti* Knight, Brig. Young Univ. Sci. Bull, 9(3): 208. [Ariz.]. Synonymized by Henry and Kim, 1984, Trans. Am. Ent. Soc., 110: 57.
Distribution: Ariz., N.M., Tex. (Mexico).

Neurocolpus nicholi Knight, 1968
1968 *Neurocolpus nicholi* Knight, Brig. Young Univ. Sci. Bull., 9(3): 209. [Ariz.].
Distribution: Ariz.

Neurocolpus nubilus (Say), 1832
1832 *Capsus nubilus* Say, Descrip. Het. Hem. N. Am., p. 22. [Ind.]. Neotype designated by Henry and Kim, 1984, Trans. Am. Ent. Soc., 110: 38.
1873 *Capsus hirsutulus* Walker, Cat. Hem. Brit. Mus., 6: 95. [Lake Huron]. Synonymized by Distant, 1904, An. Mag. Nat. Hist., (7)13: 110.
1876 *Neurocolpus nubilus*: Reuter, Öfv. K. Svens. Vet.-Akad. Förh., 32(9): 70.
1876 *Phytocoris nubilus*: Uhler, Bull. U.S. Geol. Geogr. Surv. Terr., 1: 317.
1900 *Neurocolpus nubilis* [*sic*]: Smith, Ins. N.J., p. 129.
1904 *Neurocolpus nubilis* [*sic*]: Wirtner, An. Carnegie Mus., 3: 198 (in part).
1904 *Neurocolpus hirsutulus*: Distant, An. Mag. Nat. Hist., 7(13): 205.
1920 *Neurocoplus* [*sic*] *nubilus*: Parshley, Can. Ent., 52: 86.
1934 *Neurocolpus rubidus* Knight, Bull. Brook. Ent. Soc., 29: 164. [N.Y.]. Synonymized by Henry and Kim, 1984, Trans. Am. Ent. Soc., 110: 37.
Distribution: Ala., Ark., B.C., Col., Conn., Fla., Ga., Ia., Ill., Ind., Ks., La., Man., Mass., Md., Me., Mich., Mo., N.C., N.H., N.J., N.Y., Neb., Oh., Ok., Ont., Que., R.I., Tenn., Tex., Va., Vt., Wash., W.Va., Wis.
Note: Lipsey (1970, Ent. News, 81: 213-219) reviewed host plants and (1970, Ent. News, 81: 257-262) reported on the life history of *N. nubilus*. Earlier records of *nubilus* from Mexico to Panama should be assigned in part to *N. mexicanus* and in part to other species described by Henry (1984, *In* Henry and Kim, above).

Neurocolpus pumilus Henry, 1984
1983 *Neurocolpus* n. sp.: Wheeler et al., Trans. Am. Ent. Soc., 109: 148.
1984 *Neurocolpus pumilus* Henry, *In* Henry and Kim, Trans. Am. Ent. Soc., 110: 43. [W.Va.].
Distribution: Conn., Mass., Md., N.C., N.J., N.Y., Pa., W.Va.

Neurocolpus simplex Van Duzee, 1918
1918 *Neurocolpus simplex* Van Duzee, Proc. Cal. Acad. Sci., (4)8: 281. [Cal.].
Distribution: Ariz., Cal.

Neurocolpus tiliae Knight, 1934
1904 *Neurocolpus nubilus*: Wirtner, An. Carnegie Mus., 3: 198 (in part).

1934 *Neurocolpus tiliae* Knight, Bull. Brook. Ent. Soc., 29: 162. [Minn.].
Distribution: D.C., Ia., Ill., Ind., Md., Mich., Minn., N.Y., Pa., Wis.

Genus *Notholopisca* Carvalho, 1975

1975 *Notholopus* (*Notholopisca*) Carvalho, Rev. Brasil. Biol., 35: 369. Type-species: *Calocorisca californica* Knight, 1933. Original designation.
1981 *Notholopisca*: Fontes, Arq. Mus. Nac., 56: 153.
Note: Carvalho (1975, Rev. Brasil. Biol., 35: 369-378) revised *Notholopus* Bergroth and provided a key to species, including *Notholopisca californica*.

Notholopisca californica (Knight), 1933
1933 *Calocorisca californica* Knight, Pan-Pac. Ent., 9: 69. [Cal.].
1955 *Notholopus californicus*: Carvalho, Rev. Chil. Ent., 4: 224.
1975 *Notholopus* (*Notholopisca*) *californica* [*sic*]: Carvalho, Rev. Brasil. Biol., 35: 370.
1981 *Notholopisca californica*: Fontes, Arq. Mus. Nac., 56: 153.
Distribution: Ariz., Cal.
Note: Fontes (1981, above) described and illustrated female genitalia.

Genus *Orthops* Fieber, 1858

1858 *Orthops* Fieber, Wien. Ent. Monat., 2: 311. Type-species: *Cimex kalmii* Linnaeus, 1758. Designated by Kirkaldy, 1906, Trans. Am. Ent. Soc., 32: 139.

Orthops scutellatus Uhler, 1877
1877 *Orthops scutellatus* Uhler, Bull. U.S. Geol. Geogr. Surv. Terr., 3: 420. [Col.].
1887 *Orthops pastinaceae*: Van Duzee, Can. Ent., 19: 71.
1890 *Lygus scutellatus*: Atkinson, J. Asiatic Soc. Bengal, 58(2): 92.
1900 *Lygus* (*Orthops*) *scutellatus*: Heidemann, Proc. Wash. Acad. Sci., 2: 505.
1908 *Lygus Pastinacae* [*sic*]: Horvath, An. Hist.-Nat. Mus. Nat. Hung., 6: 5.
1909 *Lygus campestris*: Reuter, Acta Soc. Sci. Fenn., 36(2): 44.
1917 *Lygus* (*Orthops*) *campestris*: Van Duzee, Univ. Cal. Publ. Ent., 2: 348 (in part).
Note: Wagner and Slater (1952, Proc. Ent. Soc. Wash., 54: 273) considered *O. scutellatus* distinct from the palearctic *O. campestris* (Linnaeus), 1758. Brittain (1919, Proc. N.S. Ent. Soc., 4: 76-81) described and illustrated the immatures (as *Lygus campestris*). Whitcomb (1953, Univ. Mass. Agr. Exp. Stn. Bull., 473: 1-15) studied biology.
Distribution: Alk., Alta., Ariz., B.C., Cal., Col., Conn., D.C., Id., Man., Me., N.H., N.S., N.Y., Nfld., Ont., Pa., Que., Sask., Tex., W.Va., Wyo.

Genus *Pachypeltocoris* Knight, 1953

1953 *Pachypeltocoris* Knight, Ia. St. Coll. J. Sci., 27: 514. Type-species: *Pachypeltocoris conspersus* Knight, 1953. Monotypic.

Pachypeltocoris conspersus Knight, 1953
1953 *Pachypeltocoris conspersus* Knight, Ia. St. Coll. J. Sci., 27: 514. [Mo.].
Distribution: Mo.

Genus *Pallacocoris* Reuter, 1876

1876 *Pallacocoris* Reuter, Öfv. K. Svens. Vet.-Akad. Förh., 32(9): 62. Type-species: *Pallacocoris suavis* Reuter, 1876. Monotypic.

Pallacocoris suavis Reuter, 1876
1876 *Pallacocoris suavis* Reuter, Öfv. K. Svens. Vet.-Akad. Förh., 32(9): 62. [Tex.].
1910 *Palaeocoris* [*sic*] *suavis*: Reuter, Acta Soc. Sci. Fenn., 37(3): 159.
Distribution: Ia., Tex.

Genus *Phytocoris* Fallén, 1814

1814 *Phytocoris* Fallén, Spec. Nova Hem., p. 10. Type-species: *Cimex populi* Linnaeus, 1758. Designated by Westwood, 1840, Intr. Mod. Classif. Ins., 2(Synop.): 122.

1876 *Compsocerocoris* Reuter, Öfv. K. Svens. Vet.-Akad. Förh., 32(9): 70. Type-species: *Compsocerocoris annulicornis* Reuter, 1876. Monotypic. Synonymized by Reuter, 1909, Acta Soc. Sci. Fenn., 36(2): 14.

1895 *Callodemus* Uhler, Bull. Col. Agr. Exp. Stn., 31: 33. Type-species: *Callodemus laevis* Uhler, 1895. Monotypic. Synonymized by Reuter, 1909, Acta Soc. Sci. Fenn., 36(2): 14.

Note: *Phytocoris* is the largest known genus in the Miridae. Knight (1923, Conn. Geol. Nat. Hist. Bull., 34: 615-617, 631-632, 644-645; 1941, Ill. Nat. Hist. Surv. Bull., 22: 184-185, 191-193, 199-202) gave keys to many of the eastern U.S. species; Knight (1968, Brig. Young Univ. Sci. Bull., 9(3): 211-214, 218-223, 236-237, 255-256, 247-249) keyed many western U.S. species. "*Phytocoris marmoratus* Fitch, MSS.," listed by Walker (1873, Cat. Hem., 6: 61) is a *nomen nudum* and not included in the following list.

Phytocoris abiesi Knight, 1974
1974 *Phytocoris abiesi* Knight, Ia. St. J. Res., 49: 124. [Cal.].
Distribution: Cal.

Phytocoris acaciae Knight, 1925
1925 *Phytocoris acaciae* Knight, Bull. Brook. Ent. Soc., 20: 53. [Ariz.].
Distribution: Ariz., N.M., Tex.

Phytocoris adenostomae Stonedahl, 1985
1985 *Phytocoris adenostomae* Stonedahl, J. N.Y. Ent. Soc., 93: 1271. [Cal.].
Distribution: Cal.

Phytocoris albellus Knight, 1934
1934 *Phytocoris albellus* Knight, Bull. Brook. Ent. Soc., 29: 14. [Ariz].
Distribution: Ariz., Cal.

Phytocoris albertae Knight, 1974
1923 *Phytocoris junceus*: Knight, Conn. Geol. Nat. Hist. Surv. Bull., 34: 621 (in part).
1974 *Phytocoris albertae* Knight, Ia. St. J. Res., 49: 131. [Alta.].
Distribution: Alta.

Phytocoris albiceps Knight, 1968
1968 *Phytocoris albiceps* Knight, Brig. Young Univ. Sci. Bull., 9(3): 234. [Cal.].
Distribution: Ariz., Cal.

Phytocoris albiclavus Knight, 1974
1974 *Phytocoris albiclavus* Knight, Ia. St. J. Res., 49: 127. [Mont.].
Distribution: Mont.

Phytocoris albidopictus Knight, 1961
1961 *Phytocoris albidopictus* Knight, Ia. St. J. Sci., 35: 476. [Ariz.].
Distribution: Ariz., Cal., N.M., Tex.

Phytocoris albidosquamus Knight, 1968
1968 *Phytocoris albidosquamus* Knight, Brig. Young Univ. Sci. Bull., 9(3): 232. [Nev.].

Distribution: Nev.

Phytocoris albifacies Knight, 1926
1926 *Phytocoris albifacies* Knight, Bull. Brook. Ent. Soc., 21: 159. [Miss.].
Distribution: Ill., Miss., Wis.

Phytocoris albifrons Knight, 1968
1968 *Phytocoris albifrons* Knight, Brig. Young Univ. Sci. Bull., 9(3): 241. [Ariz.].
Distribution: Ariz.

Phytocoris albitylus Knight, 1926
1926 *Phytocoris albitylus* Knight, Bull. Brook. Ent. Soc., 21: 162. [Fla.].
1926 *Phytocoris albitylus* Blatchley, Het. E. N.Am., p. 718 (as Knight ms. name). Preoccupied.
Distribution: Fla.

Phytocoris alboscutellatus Knight, 1968
1968 *Phytocoris alboscutellatus* Knight, Brig. Young Univ. Sci. Bull., 9(3): 242. [Ariz.].
Distribution: Ariz., Nev.

Phytocoris alpinus Kelton, 1979
1979 *Phytocoris alpinus* Kelton, Can. Ent., 111: 689. [Alta.].
Distribution: Alta.

Phytocoris americanus Carvalho, 1959
1909 *Phytocoris angustulus* Reuter, Acta Soc. Sci. Fenn., 36(2): 29. [W.Va.]. Preoccupied. Lectotype designated by Henry and Stonedahl, 1984, J. N.Y. Ent. Soc., 91: 444.
1959 *Phytocoris americanus* Carvalho, Arq. Mus. Nac., 48: 190. New name for *Phytocoris angustulus* Reuter.
Distribution: N.C., N.S., N.Y., Pa., Que., Vt., W.Va.

Phytocoris angustatus Knight, 1961
1961 *Phytocoris angustatus* Knight, Ia. St. J. Sci., 35: 483. [Ariz.].
Distribution: Ariz.

Phytocoris angusticollis Knight, 1925
1925 *Phytocoris angusticollis* Knight, Bull. Brook. Ent. Soc., 20: 57. [Ariz.].
Distribution: Ariz.

Phytocoris angustifrons Knight, 1926
1926 *Phytocoris angustifrons* Knight, Bull. Brook. Ent. Soc., 21: 164. [Fla.].
1926 *Phytocoris megalopsis* Blatchley, Het. E. N. Am., p. 713. [Fla.]. Synonymized by Knight, 1927, Bull. Brook. Ent. Soc., 22: 100.
Distribution: Fla., La., Miss.

Phytocoris annulicornis (Reuter), 1876
1876 *Compsocerocoris annulicornis* Reuter, Öfv. K. Svens. Vet.-Akad. Förh., 32(9): 70. [Tex.]. Lectotype designated by Henry and Stonedahl, 1984, J. N.Y. Ent. Soc., 91: 444.
1909 *Phytocoris annulicornis*: Reuter, Acta Soc. Sci. Fenn., 36(2): 33.
1909 *Phytocoris antennalis*: Van Duzee, Bull. Buff. Soc. Nat. Sci., 9: 180.
1910 *Phytocoris bipunctatus* Van Duzee, Trans. Am. Ent. Soc., 36: 77. [Fla.]. Synonymized and lectotype designated by Henry and Stonedahl, 1984, J. N.Y. Ent. Soc., 91: 444.
1959 *Phytocoris annulicornis bipunctatus*: Carvalho, Arq. Mus. Nac., 48: 190.
Distribution: Cal.(?), Col.(?), Fla., Me., N.J., N.M.(?), Oh., Pa. (Guatemala(?), Mexico, Panama(?)).

Note: Henry and Stonedahl (1984, above) questioned the far western (Cal., Col., N.M.) and southern (Guatemala and Panama) records.

Phytocoris antennalis Reuter, 1909

1909 *Phytocoris antennalis* Reuter, Acta Soc. Sci. Fenn., 36(2): 32. [D.C.]. Lectotype designated by Henry and Stonedahl, 1984, J. N.Y. Ent. Soc., 91: 445.

Distribution: D.C., Fla., Ia., Ill., Ind., Mass., Miss., N.C., N.J., N.Y., Ok., Pa., Va.

Phytocoris apache Knight, 1928

1928 *Phytocoris apache* Knight, Bull. Brook. Ent. Soc., 23: 41. [Ariz.].

Distribution: Ariz., N.M.

Phytocoris arundinicola Knight, 1941

1941 *Phytocoris arundinicola* Knight, Ill. Nat. Hist. Surv. Bull., 22: 198. [Ill.].

Distribution: Ill.

Phytocoris aurora Van Duzee, 1920

1920 *Phytocoris aurora* Van Duzee, Proc. Cal. Acad. Sci., (4)9: 340. [Cal.].

Distribution: Cal.

Phytocoris bakeri Reuter, 1909

1909 *Phytocoris bakeri* Reuter, Acta Soc. Sci. Fenn., 36(2): 28. [Cal.]. Lectotype designated by Henry and Stonedahl, 1984, J. N.Y. Ent. Soc., 91: 446.

Distribution: Cal.

Phytocoris balli Knight, 1926

1926 *Phytocoris balli* Knight, Bull. Brook. Ent. Soc., 21: 167. [Fla.].

Distribution: Fla.

Phytocoris becki Knight, 1968

1968 *Phytocoris becki* Knight, Brig. Young Univ. Sci. Bull., 9(3): 214. [Nev.].

Distribution: Nev.

Phytocoris blackwelli Bliven, 1966

1966 *Phytocoris blackwelli* Bliven, Occ. Ent., 1(9): 119. [Cal.].

Distribution: Cal.

Phytocoris borealis Knight, 1926

1926 *Phytocoris borealis* Knight, Bull. Brook. Ent. Soc., 21: 158. [Ont.].

Distribution: Man., N.Y., Ont.

Phytocoris breviatus Knight, 1968

1968 *Phytocoris breviatus* Knight, Brig. Young Univ. Sci. Bull., 9(3): 226. [Nev.].

Distribution: Nev.

Phytocoris brevicornis Knight, 1968

1968 *Phytocoris brevicornis* Knight, Brig. Young Univ. Sci. Bull., 9(3): 237. [Ariz.].

Distribution: Ariz.

Phytocoris brevifurcatus Knight, 1920

1920 *Phytocoris brevifurcatus* Knight, Bull. Brook. Ent. Soc., 15: 53. [N.Y.].

Distribution: Ill., N.Y., Ont., Que.

Phytocoris breviusculus Reuter, 1876

1876 *Phytocoris breviusculus* Reuter, Öfv. K. Svens. Vet.-Akad. Förh. 32(9): 68. [Tex.].

Distribution: Ala., Col., D.C., Ill., Ind., Ks., Miss., Mo., N.J., Oh., Pa., Tex., W.Va.

Note: Wheeler and Henry (1977, Trans. Am. Ent. Soc., 103: 639-643) studied the hosts and biology and redescribed and illustrated the adult and 5th-instar nymph.

Phytocoris brimleyi Knight, 1974
1974 *Phytocoris brimleyi* Knight, Ia. St. J. Res., 49: 130. [Ont.].
Distribution: Man., Ont.

Phytocoris brooksi Kelton, 1979
1979 *Phytocoris brooksi* Kelton, Can. Ent., 111: 689. [Man.].
Distribution: Man., Sask.

Phytocoris broweri Knight, 1974
1974 *Phytocoris broweri* Knight, Ia. St. J. Res., 49: 130. [Me.].
Distribution: Me.

Phytocoris buenoi Knight, 1920
1920 *Phytocoris buenoi* Knight, Bull. Brook. Ent. Soc., 15: 57. [N.Y.].
Distribution: Conn., Mass., N.Y., Ont., Que., W.Va.

Phytocoris californicus Knight, 1968
1968 *Phytocoris californicus* Knight, Brig. Young Univ. Sci. Bull., 9(3): 244. [Cal.].
Distribution: Cal.

Phytocoris calli Knight, 1934
1934 *Phytocoris calli* Knight, Bull. Brook. Ent. Soc., 29: 11: [Ut.].
Distribution: Ore., Ut.

Phytocoris calvus Van Duzee, 1920
1920 *Phytocoris calvus* Van Duzee, Proc. Cal. Acad. Sci., (4)9: 343. [Cal.].
Distribution: Cal.

Phytocoris canadensis Van Duzee, 1920 [Fig. 78]
1878 *Phytocoris inops* Uhler, Proc. Boston Soc. Nat. Hist., 19: 402. [N.H.]. Preoccupied. Lectotype designated by Henry and Stonedahl, 1984, J. N.Y. Ent. Soc., 91: 447.
1900 *Paracalocoris inops*: Smith, Ins. N.J., p. 129.
1920 *Phytocoris canadensis* Van Duzee, Proc. Cal. Acad. Sci., (4)9: 346. [Ont].
Distribution: Conn., Ga., Ia., Ill., Ind., Ks., Mass., Me., Mich., Minn., Mo., N.C., N.H., N.J., N.S., N.Y., Oh., Ont., P.Ed., Pa., Que., S.D., W.Va., Wis.

Phytocoris candidus (Van Duzee), 1918
1918 *Pallacocoris candidus* Van Duzee, Proc. Cal. Acad. Sci., (4)8: 288. [Cal.].
1968 *Phytocoris candidus*: Knight, Brig. Young Univ. Sci. Bull., 9(3): 215.
Distribution: Cal.

Phytocoris canescens Reuter, 1909
1909 *Phytocoris canescens* Reuter, Acta Soc. Sci. Fenn., 36(2): 30. [Cal.]. Lectotype designated by Henry and Stonedahl, 1984, J. N.Y. Ent. Soc., 91: 447.
1915 *Phytocoris cunescens* [*sic*]: LaFollette, J. Ent. Zool., 7: 129.
Distribution: Cal., Ut.

Phytocoris carnosulus Van Duzee, 1920
1920 *Phytocoris carnosulus* Van Duzee, Proc. Cal. Acad. Sci., (4)9: 347. [Ariz.].
Distribution: Ariz., Nev., Tex.

Phytocoris caryae Knight, 1923
1923 *Phytocoris caryae* Knight, Conn. Geol. Nat. Hist. Surv. Bull., 34: 652. [N.Y.].
Distribution: Ill., N.Y.

Phytocoris cercocarpi Knight, 1928
1928 *Phytocoris cercocarpi* Knight, Bull. Brook. Ent. Soc., 23: 39. [Col.].
Distribution: Col.

Phytocoris chiricahuae Knight, 1968
1968 *Phytocoris chiricahuae* Knight, Brig. Young Univ. Sci. Bull., 9(3): 239. [Ariz.].
Distribution: Ariz.

Phytocoris commissuralis Van Duzee, 1920
1920 *Phytocoris commissuralis* Van Duzee, Proc. Cal. Acad. Sci., (4)9: 351. [Cal.].
Distribution: B.C., Cal.

Phytocoris comulus Knight, 1928
1928 *Phytocoris comulus* Knight, Bull. Brook. Ent. Soc., 23: 38. [Col.].
Distribution: Ariz., Col., Neb., N.M.

Phytocoris confluens Reuter, 1909
1909 *Phytocoris puella* var. *confluens* Reuter, Acta Soc. Sci. Fenn., 36(2): 20. [D.C.]. Lectotype designated by Henry and Stonedahl, 1984, J. N.Y. Ent. Soc., 91: 448.
1923 *Phytocoris confluens*: Knight, Conn. Geol. Nat. Hist. Surv. Bull., 34: 650.
Distribution: Conn., D.C., Ill., Ks., Mass., Md., Miss., Mo., N.C., N.J., N.Y., Oh., Pa., S.C., W.Va., Wis.

Phytocoris consors Van Duzee, 1918
1918 *Phytocoris consors* Van Duzee, Proc. Cal. Acad. Sci., (4)8: 287. [Cal.].
Distribution: Cal., Nev.

Phytocoris conspersipes Reuter, 1909
1892 *Phytocoris eximus* [*sic*]: Heidemann, Proc. Ent. Soc. Wash., 2: 225.
1905 *Phytocoris breviusculus*: Van Duzee, Rept. N.Y. St. Ent., N.Y. St. Mus. Bull., 97: 551.
1909 *Phytocoris conspersipes* Reuter, Acta Soc. Sci. Fenn., 36(2): 22. [D.C.]. Lectotype designated by Henry and Stonedahl, 1984, J. N.Y. Ent. Soc., 91: 448.
Distribution: Alta., Conn., D.C., Man., Mass., Md., N.C., N.J., N.Y., Ont., S.C., Sask., Va., W.Va.

Phytocoris conspicuus Johnston, 1930
1930 *Phytocoris conspicuus* Johnston, Bull. Brook. Ent. Soc., 25: 295. [Tex.].
Distribution: Col., Tex.

Phytocoris conspurcatus Knight, 1920
1920 *Phytocoris conspurcatus* Knight, Bull. Brook. Ent. Soc., 15: 61. [N.Y.].
1944 *Phytocoris conspurgatus* [*sic*]: Milne and Milne, Ent. Am., 24: 34.
1960 *Phytocoris conspercatus* [*sic*]: LeRoux, An. Soc. Ent. Que., 6: 96.
Distribution: Alta., B.C., Conn., D.C., Ia., Ill., Ks., Man., Mass., Md., Mich., Minn., Miss., Mo., N.H., N.J., N.S., N.Y., Oh., Ont., Pa., Que., S.D., Sask., Va., W.Va., Wis.

Phytocoris contrastus Knight, 1968
1968 *Phytocoris contrastus* Knight, Brig. Young Univ. Sci. Bull., 9(3): 259. [Nev.].
Distribution: Nev.

Phytocoris corticevivens Knight, 1920
1920 *Phytocoris corticevivens* Knight, Bull. Brook. Ent. Soc., 15: 63. [N.Y.].
1984 *Phytocoris corticevirens* [*sic*]: Larochelle, Fabreries, Suppl. 3, p. 263.
Distribution: Conn., Ia., Ill., Md., Me., Minn., Mo., N.J., N.S., N.Y., Oh., Ont., Que., Tex., Wis.

Phytocoris cortitectus Knight, 1920
1920 *Phytocoris cortitectus* Knight, Bull. Brook. Ent. Soc., 15: 55. [N.Y.].
Distribution: Ill., N.H., N.J., N.S., N.Y., Ont., Pa.

Phytocoris crawfordi Knight, 1974
1974 *Phytocoris crawfordi* Knight, Ia. St. J. Res., 49: 129. [Ont.].
Distribution: Ont.

Phytocoris cunealis Van Duzee, 1914
1914 *Phytocoris cunealis* Van Duzee, Trans. San Diego Soc. Nat. Hist., 2: 16. [Cal.]. Lectotype designated by Henry and Stonedahl, 1984, J. N.Y. Ent. Soc. 91: 449.
Distribution: Cal.

Phytocoris cuneotinctus Knight, 1925
1925 *Phytocoris cuneotinctus* Knight, Bull. Brook. Ent. Soc., 20: 55. [N.M.].
Distribution: N.M., Nev., Ut.

Phytocoris davisi Knight, 1923
1923 *Phytocoris davisi* Knight, Conn. Geol. Nat. Hist. Surv. Bull., 34: 624. [N.J.].
Distribution: Miss., N.J., N.Y.

Phytocoris decurvatus Knight, 1968
1968 *Phytocoris decurvatus* Knight, Brig. Young Univ. Sci. Bull., 9(3): 226. [Nev.].
Distribution: Nev.

Phytocoris dentatus Knight, 1974
1974 *Phytocoris dentatus* Knight, Ia. St. J. Res., 49: 125. [B.C.].
Distribution: B.C.

Phytocoris depictus Knight, 1923
1923 *Phytocoris depictus* Knight, Conn. Geol. Nat. Hist. Surv. Bull., 34: 654. [N.Y.].
Distribution: D.C., Ill., Minn., Mo., N.Y., Oh., Wis.

Phytocoris deserticola Knight, 1968
1968 *Phytocoris deserticola* Knight, Brig. Young Univ. Sci. Bull., 9(3): 251. [Nev.].
Distribution: Nev.

Phytocoris difficilis Knight, 1927
1927 *Phytocoris difficilis* Knight, Proc. Biol. Soc. Wash., 40: 17. [Md.].
Distribution: Md., N.J.

Phytocoris difformis Knight, 1934
1934 *Phytocoris difformis* Knight, Bull. Brook. Ent. Soc., 29: 8. [Ariz.].
Distribution: Ariz.

Phytocoris dimidiatus Kirschbaum, 1856
1856 *Phytocoris dimidiatus* Kirschbaum, Jahrb. Nat. Herz. Nass, 10: 182, 199. [Europe].
1923 *Phytocoris dimidiatus*: Knight, Conn. Geol. Nat. Hist. Surv. Bull., 34: 630.
Distribution: B.C., N.S., Ore., Wash. (Palearctic).
Note: Stonedahl (1983, Proc. Ent. Soc. Wash., 85: 469-470) summarized the distribution and literature and gave additional records for the Pacific Northwest.

Phytocoris diversus Knight, 1920
1920 *Phytocoris conspersipes diversus* Knight, Bull. Brook. Ent. Soc., 15: 60. [N.Y.].
1923 *Phytocoris diversus*: Knight, Conn. Geol. Nat. Hist. Surv. Bull., 34: 641.
Distribution: Ill., Mass., Me., N.H., N.Y., Minn., Pa., Que., W.Va.

Phytocoris dreisbachi Knight, 1974
1974 *Phytocoris dreisbachi* Knight, Ia. St. J. Res., 49: 125. [Mich.].
1974 *Phytocoris discoidalis* Henry, Ent. News, 85: 187. [Pa.]. Synonymized by Henry, 1982, Proc. Ent. Soc. Wash., 84: 337.
Distribution: Man., Mich., Pa., W.Va., Wis.

Phytocoris elongatus Knight, 1974
1974 *Phytocoris elongatus* Knight, Ia. St. J. Res., 49: 131. [Ut.].
Distribution: Col., Ut.

Phytocoris empirensis Knight, 1968
1968 *Phytocoris empirensis* Knight, Brig. Young Univ. Sci. Bull., 9(3): 246. [Ariz.].
Distribution: Ariz.

Phytocoris ephedrae Knight, 1961
1961 *Phytocoris ephedrae* Knight, Ia. St. J. Sci., 35: 478. [Ariz.].
Distribution: Ariz., Nev., Tex.

Phytocoris erectus Van Duzee, 1920
1920 *Phytocoris erectus* Van Duzee, Proc. Cal. Acad. Sci., (4)9: 345. [N.Y.].
Distribution: Ala., D.C., Ia., Ill., Ind., La., Mass., Md., Me., Mich., Minn., Miss., Mo., N.C., N.H., N.S., N.Y., Oh., Ont., Pa., Que., Sask., Ut., W.Va.

Phytocoris eurekae Bliven, 1966
1966 *Phytocoris eurekae* Bliven, Occ. Ent., 1(9): 116. [Cal.].
Distribution: Cal.

Phytocoris exemplus Knight, 1926
1926 *Phytocoris exemplus* Knight, Bull. Brook. Ent. Soc., 21: 163. [La.].
Distribution: La.

Phytocoris eximius Reuter, 1876
1876 *Phytocoris eximius* Reuter, Öfv. K. Svens. Vet.-Akad. Förh., 32(9): 67. [Tex.]. Lectotype designated by Henry and Stonedahl, 1984, J. N.Y. Ent. Soc., 91: 449.
1920 *Phytocoris penepectus* Knight, Bull. Brook. Ent. Soc., 15: 58. [Conn.]. Synonymized by Henry and Stonedahl, 1984, J. N.Y. Ent. Soc., 91: 449.
1923 *Phytocoris penepecten* [*sic*]: Knight, Conn. Geol. Nat. Hist. Surv. Bull., 34: 640.
Distribution: Col., Conn., D.C., Fla., Ga., Ill., Ind., Mass., Md., Me., Mo., N.C., N.J., N.Y., Ont., Pa., Que., Tex.
Note: Henry and Stonedahl (1984, above) clarified the identity of *P. eximius*, a species misidentified subsequent to Reuter's original description.

Phytocoris fenestratus Reuter, 1909
1909 *Phytocoris fenestratus* Reuter, Acta Soc. Sci. Fenn., 36(2): 24. [D.C.]. Lectotype designated by Henry and Stonedahl, 1984, J. N.Y. Ent. Soc., 91: 450.
Distribution: Conn., D.C., Ga., N.C., Pa., W.Va.

Phytocoris flavellus Knight, 1968
1968 *Phytocoris flavellus* Knight, Brig. Young Univ. Sci. Bull., 9(3): 234. [Nev.].
Distribution: Nev.

Phytocoris flaviatus Knight, 1968
1968 *Phytocoris flaviatus* Knight, Brig. Young Univ. Sci. Bull., 9(3): 241. [Ariz.].
Distribution: Ariz.

Phytocoris formosus Van Duzee, 1914
1914 *Phytocoris reuteri* Van Duzee, Trans. San Diego Soc. Nat. Hist., 2: 18. [Cal.]. Preoccupied. Lectotype designated by Henry and Stonedahl, 1984, J. N.Y. Ent. Soc., 91: 450.
1916 *Phytocoris formosus* Van Duzee, Check List Hem., p. 37. New name for *Phytocoris reuteri* Van Duzee.
Distribution: Cal.

Phytocoris fraterculus Van Duzee, 1918
1918 *Phytocoris fraterculus* Van Duzee, Proc. Cal. Acad. Sci., (4)8: 283. [Cal.].
Distribution: Ariz., Cal.

Phytocoris fulvipennis Knight, 1928
1928 *Phytocoris fulvipennis* Knight, Bull. Brook. Ent. Soc., 23: 31. [Fla.].
Distribution: Fla., Ga.

Phytocoris fulvus Knight, 1920
1920 *Phytocoris fulvus* Knight, Bull. Brook. Ent. Soc., 15: 59. [N.Y.].
Distribution: Me., N.Y., Ont., Pa., Que., W.Va.

Phytocoris fumatus Reuter, 1909
1909 *Phytocoris fumatus* Reuter, Acta Soc. Sci. Fenn., 36(2): 25. [D.C.]. Lectotype designated by Henry and Stonedahl, 1984, J. N.Y. Ent. Soc., 91: 451.
1909 *Phytocoris subnitidulus* Reuter, Acta Soc. Sci. Fenn., 36(2): 26. [Md.]. Synonymized by Knight, 1920, Bull. Brook. Ent. Soc., 15: 63. Lectotype designated by Henry and Stonedahl, 1984, J. N.Y. Ent. Soc., 91: 451.
Distribution: D.C., Ga., Ill., Mass., Md., Mo., N.D., N.J., N.Y., Pa.

Phytocoris fuscipennis Knight, 1934
1934 *Phytocoris fuscipennis* Knight, Bull. Brook. Ent. Soc., 29: 5. [Ariz.].
Distribution: Ariz.

Phytocoris fuscosignatus Knight, 1928
1928 *Phytocoris fuscosignatus* Knight, Bull. Brook. Ent. Soc., 23: 45. [Ore.].
Distribution: Ore.

Phytocoris geniculatus Van Duzee, 1918
1918 *Phytocoris geniculatus* Van Duzee, Proc. Cal. Acad. Sci., (4)8: 286. [Cal.].
Distribution: Cal., Nev. (Mexico).

Phytocoris gracillatus Knight, 1968
1968 *Phytocoris gracillatus* Knight, Brig. Young Univ. Sci. Bull., 9(3): 229. [Nev.].
Distribution: B.C., Nev., Ut., Wash.

Phytocoris heidemanni Reuter, 1909
1909 *Phytocoris heidemanni* Reuter, Acta Soc. Sci. Fenn., 36(2): 27. [N.M.]. Lectotype designated by Henry and Stonedahl, 1984, J. N.Y. Ent. Soc., 91: 451.
Distribution: Ariz., Col., N.M., Nev.

Phytocoris hettenshawi Bliven, 1956
1956 *Phytocoris hettenshawi* Bliven, New Hem. W. St., p. 18. [Cal.].
Distribution: Cal.

Phytocoris hirsuticus Knight, 1968
1968 *Phytocoris hirsuticus* Knight, Brig. Young Univ. Sci. Bull., 9(3): 223. [Nev.].
Distribution: Nev.

Phytocoris hirtus Van Duzee, 1918
1918 *Phytocoris hirtus* Van Duzee, Proc. Cal. Acad. Sci., (4)8: 284. [Cal.].
Distribution: Cal.

Phytocoris histriculus Van Duzee, 1920
1920 *Phytocoris histriculus* Van Duzee, Proc. Cal. Acad. Sci., (4)9: 346. [Cal.].
Distribution: Cal.

Phytocoris hopi Knight, 1928
1928 *Phytocoris hopi* Knight, Bull. Brook. Ent. Soc., 23: 42. [Col.].
Distribution: Ariz., Col., N.M.

Phytocoris husseyi Knight, 1923
1923 *Phytocoris husseyi* Knight, Conn. Geol. Nat. Hist. Surv. Bull., 34: 639. [Oh.].
Distribution: Minn., N.S., Oh., Pa., Que., W.Va.

Phytocoris hyampom Bliven, 1966
1966 *Phytocoris hyampom* Bliven, Occ. Ent., 1(9): 115. [Cal.].
Distribution: Cal.

Phytocoris infuscatus Reuter, 1909
1909 *Phytocoris puella* var. *infuscatus* Reuter, Acta Soc. Sci. Fenn., 36(2): 20. [D.C.]. Lectotype designated by Henry and Stonedahl, 1984, J. N.Y. Ent. Soc., 91: 452.
1914 *Phytocoris infuscatus*: Van Duzee, Trans. San Diego Soc. Nat. Hist., 2: 17.
Distribution: Conn., D.C., Del., Ga., Ia., Ill., Ind., Mass., Mich., Miss., Mo., N.C., N.Y., Oh., Ont., Pa., Que., W.Va.

Phytocoris ingens Van Duzee, 1920
1920 *Phytocoris ingens* Van Duzee, Proc. Cal. Acad. Sci., (4)9: 340. [Cal.].
Distribution: Cal.

Phytocoris inops Uhler, 1877
1877 *Phytocoris inops* Uhler, Bull. U.S. Geol. Geogr. Surv. Terr., 3: 413. [Col.]. Lectotype designated by Henry and Stonedahl, 1984, J. N.Y. Ent. Soc., 91: 452.
1890 *Neurocolpus inops*: Atkinson, J. Asiatic Soc. Bengal, 58(2): 68.
1909 *Phytocoris vittatus* Reuter, Acta Soc. Sci. Fenn., 36(2): 28. [N.Y.]. Synonymized and lectotype designated by Henry and Stonedahl, 1984, J. N.Y. Ent. Soc., 91: 453.
1909 *Phytocoris palmeri* Reuter, Acta Soc. Sci. Fenn., 36(2): 32. [Que.]. Synonymized and lectotype designated by Henry and Stonedahl, 1984, J. N.Y. Ent. Soc., 91: 453.
1928 *Phytocoris hesperius* Knight, Bull. Brook. Ent. Soc., 23: 44. [Col.]. Synonymized by Henry and Stonedahl, 1984, J. N.Y. Ent. Soc., 91: 453.
1968 *Phytocoris hesperellus* Knight, Brig. Young Univ. Sci. Bull., 9(3): 232. [Ut.]. Synonymized by Henry and Stonedahl, 1984, J. N.Y. Ent. Soc., 91: 453.
Distribution: Ariz., Cal., Col., Mass., Md., N.H., N.J., N.M., N.Y., Nev., Ore., Pa., Que., Tex., Ut., Wyo.
Note: Henry and Stonedahl (1984, above) clarified the identity of this much confused species.

Phytocoris interspersus Uhler, 1895
1895 *Phytocoris interspersus* Uhler, Bull. Col. Agr. Exp. Stn., 31: 32. [Col.]. Neotype designated by Henry and Stonedahl, 1984, J. N.Y. Ent. Soc., 91: 454.
1909 *Phytocoris interspersus* var. *signifer* Reuter, Acta Soc. Sci. Fenn., 36(2): 19. [No locality given]. New Synonymy.
Distribution: Ariz., B.C., Cal., Col., Id., N.M., Ut.
Note: The variety *signifer* has not been recognized since the original description and is here considered a synonym.

Phytocoris jucundus Van Duzee, 1914
1914 *Phytocoris jucundus* Van Duzee, Trans. San Diego Soc. Nat. Hist., 2: 17. [Cal.]. Lectotype designated by Henry and Stonedahl, 1984, J. N.Y. Ent. Soc., 91: 454.
Distribution: Cal., Id., Ore., Wash.

Phytocoris junceus Knight, 1923
1923 *Phytocoris junceus* Knight, Conn. Geol. Nat. Hist. Surv. Bull., 34: 621. [N.H.].
Distribution: Alta., Mont., N.H., W.Va.

Phytocoris junipericola Knight, 1927
1927 *Phytocoris junipericola* Knight, Proc. Biol. Soc. Wash., 40: 16. [D.C.].
1928 *Phytocoris radicola* Blatchley, J. N.Y. Ent. Soc., 36: 2. Synonymized by Carvalho, 1959, Arq. Mus. Nac., 48: 203.
Distribution: D.C., Ind., Md., N.C., Pa., W.Va.
Note: Wheeler and Henry (1977, Trans. Am. Ent. Soc., 103: 642-644) gave notes on seasonal history and described and illustrated the fifth-instar nymph.

Phytocoris juniperanus Knight, 1968
1968 *Phytocoris juniperanus* Knight, Brig. Young Univ. Sci. Bull., 9(3): 238. [Nev.].
Distribution: Nev.

Phytocoris kahtahbi Bliven, 1966
1966 *Phytocoris kahtahbi* Bliven, Occ. Ent., 1(9): 117. [Cal.].
Distribution: Cal.

Phytocoris ketinelbi Bliven 1966
1966 *Phytocoris ketinelbi* Bliven, Occ. Ent., 1(9): 118. [Cal.].
Distribution: Cal.

Phytocoris knowltoni Knight, 1974
1974 *Phytocoris knowltoni* Knight, Ia. St. J. Res., 49: 126. [Ut.].
Distribution: Ut.

Phytocoris lacunosus Knight, 1920
1920 *Phytocoris lacunosus* Knight, Bull. Brook. Ent. Soc., 15: 56. [N.Y.].
Distribution: Minn., N.C., N.D., N.Y., Que.

Phytocoris laevis (Uhler), 1895
1895 *Callodemus laevis* Uhler, Bull. Col. Agr. Exp. Stn., 31: 33. [Col.]. Lectotype designated by Henry and Stonedahl, 1984, J. N.Y. Ent. Soc., 91: 455.
1909 *Phytocoris laevis*: Reuter, Acta Soc. Sci. Fenn., 36(2): 14.
Distribution: Alta., Ariz., Col., N.M., S.D., Sask., Ut.

Phytocoris lasiomerus Reuter, 1909
1887 *Phytocoris scrupeus*: Provancher, Pet. Faune Ent. Can., 3: 108.
1887 *Phytocoris pallidicornis*: Van Duzee, Can. Ent., 19: 70.
1892 *Phytocoris annulicornis*: Osborn, Proc. Ia. Acad. Sci., 1: 123.
1909 *Phytocoris lasiomerus* Reuter, Acta Soc. Sci. Fenn., 36(2): 34. [N.Y.]. Lectotype designated by Henry and Stonedahl, 1984, J. N.Y. Ent. Soc., 91: 455.
Distribution: Alta., Col., Ia., B.C., Man., Mass., Me., N.D., N.H., N.S., N.Y., Ont., Que., Sask., Vt., Wash., Wis., Wyo.

Phytocoris laticeps Knight, 1968
1968 *Phytocoris laticeps* Knight, Brig. Young Univ. Sci. Bull., 9(3): 243. [Ut.].
Distribution: Ut.

Phytocoris lineatellus Knight, 1968
1968 *Phytocoris lineatellus* Knight, Brig. Young Univ. Sci. Bull., 9(3): 250. [Nev.].
Distribution: Nev.

Phytocoris lineatus Reuter, 1909
1909 *Phytocoris lineatus* Reuter, Acta Soc. Sci. Fenn., 36(2): 30. [Col.]. Lectotype designated by Henry and Stonedahl, 1984, J. N.Y. Ent. Soc., 91: 455.
Distribution: Col.

Phytocoris listi Knight, 1928
1928 *Phytocoris listi* Knight, Bull. Brook. Ent. Soc., 23: 30. [Col.].

Distribution: Col., S.D., Sask.

Phytocoris longihirtus Knight, 1968
1968 *Phytocoris longihirtus* Knight, Brig. Young Univ. Sci. Bull., 9(3): 218. [Nev.].
Distribution: Nev.

Phytocoris longirostris Knight, 1934
1934 *Phytocoris longirostris* Knight, Bull. Brook. Ent. Soc., 29: 6. [Ariz.].
Distribution: Ariz.

Phytocoris luteolus Knight, 1923
1923 *Phytocoris luteolus* Knight, Conn. Geol. Nat. Hist. Surv. Bull., 34: 649. [Conn.].
Distribution: Ala., Conn., La.

Phytocoris maritimus Van Duzee, 1920
1920 *Phytocoris maritimus* Van Duzee, Proc. Cal. Acad. Sci., (4)9: 349. [Cal.].
Distribution: Cal.

Phytocoris mellarius Knight, 1925
1925 *Phytocoris mellarius* Knight, Bull. Brook. Ent. Soc., 20: 56. [Ariz.].
Distribution: Ariz., Nev.

Phytocoris merinoi Knight, 1968
1968 *Phytocoris merinoi* Knight, Brig. Young Univ. Sci. Bull., 9(3): 227. [Nev.].
Distribution: Nev.

Phytocoris mesillae Knight, 1968
1968 *Phytocoris mesillae* Knight, Brig. Young Univ. Sci. Bull., 9(3): 258. [N.M.].
Distribution: N.M.

Phytocoris michiganae Knight, 1974
1974 *Phytocoris michiganae* Knight, Ia. St. J. Res., 49: 128. [Mich.].
Distribution: Mich., Wis.

Phytocoris miniatus Knight, 1961
1961 *Phytocoris miniatus* Knight, Ia. St. J. Sci., 35: 480. [Ariz.].
Distribution: Ariz., Ut.

Phytocoris minituberculatus Knight, 1968
1968 *Phytocoris minituberculatus* Knight, Brig. Young Univ. Sci. Bull., 9(3): 252. [Nev.].
Distribution: Nev.

Phytocoris minuendus Knight, 1968
1968 *Phytocoris minuendus* Knight, Brig. Young Univ. Sci. Bull., 9(3): 243. [Ariz.].
Distribution: Ariz.

Phytocoris minutulus Reuter, 1909
1909 *Phytocoris minutulus* Reuter, Acta Soc. Sci. Fenn., 36(2): 24. [Md.]. Neotype designated by Henry and Stonedahl, 1984, J. N.Y. Ent. Soc., 91: 456.
1938 *Phytocoris minutilus* [*sic*]: Brimley, Ins. N.C., p. 76.
Distribution: Md., Mass., N.C., N.H., N.Y., Pa., Que., Va., W.Va.

Phytocoris mirus Knight, 1928
1928 *Phytocoris mirus* Knight, Bull. Brook. Ent. Soc., 23: 35. [Col.].
Distribution: Ariz., Col.

Phytocoris montanae Knight, 1974
1974 *Phytocoris montanae* Knight, Ia. St. J. Res., 49: 128. [Mont.].
Distribution: Mont.

Phytocoris mundus Reuter, 1909
1892 *Phytocoris mundus* Heidemann, Proc. Ent. Soc. Wash., 2: 225. *Nomen nudum.*
1909 *Phytocoris mundus* Reuter, Acta Soc. Sci. Fenn., 36(2): 18. [D.C.]. Lectotype designated by Henry and Stonedahl, 1984, J. N.Y. Ent. Soc., 91: 456.
Distribution: D.C., Md., Me., N.C., N.J., N.Y., Pa., Va., W.Va.

Phytocoris neglectus Knight, 1920
1919 *Phytocoris eximius*: Parshley, Occas. Pap. Mus. Zool., Univ. Mich., 71: 31.
1920 *Phytocoris neglectus* Knight, Bull. Brook. Ent. Soc., 15: 54. [N.Y.].
Distribution: Alta., B.C., D.C., Ia., Ill., Man., Mass., Me., Mich., Minn., Miss., Mo., N.H., N.S., N.Y., Ont., Pa., Que., S.C., S.D., Sask., W.Va., Wis.

Phytocoris nicholi Knight, 1928
1928 *Phytocoris nicholi* Knight, Bull. Brook. Ent. Soc., 23: 29. [Ariz.].
Distribution: Ariz.

Phytocoris nigricollis Knight, 1923
1923 *Phytocoris nigricollis* Knight, Conn. Geol. Nat. Hist. Surv. Bull., 34: 636. [N.H.].
Distribution: N.C., N.H., N.S.

Phytocoris nigrifrons Van Duzee, 1920
1920 *Phytocoris nigrifrons* Van Duzee, Proc. Cal. Acad. Sci., (4)9: 352. [Cal.].
Distribution: Cal.

Phytocoris nigrisignatus Knight, 1934
1934 *Phytocoris nigrisignatus* Knight, Bull. Brook. Ent. Soc., 29: 13. [Tex.].
Distribution: Tex.

Phytocoris nigrolineatus Knight, 1968
1968 *Phytocoris nigrolineatus* Knight, Brig. Young Univ. Sci. Bull., 9(3): 224. [Nev.].
Distribution: Nev.

Phytocoris nobilis Stonedahl, 1984
1984 *Phytocoris nobilis* Stonedahl, Pan-Pac. Ent., 60: 47. [Ore.].
Distribution: Ore.

Phytocoris obtectus Knight, 1920
1920 *Phytocoris obtectus* Knight, Bull. Brook. Ent. Soc., 15: 58. [N.Y.].
Distribution: N.Y., Oh., W.Va.

Phytocoris occidentalis Stonedahl, 1984
1984 *Phytocoris occidentalis* Stonedahl, Pan-Pac. Ent., 60: 50. [Ore.].
Distribution: B.C., Cal., Ore.

Phytocoris olseni Knight, 1923
1923 *Phytocoris olseni* Knight, Conn. Geol. Nat. Hist. Surv. Bull., 34: 647. [N.J.].
Distribution: Fla., Miss., N.J., N.Y., Va.

Phytocoris onustus Van Duzee, 1920
1920 *Phytocoris eximius*: Parshley, Psyche, 27: 142.
1920 *Phytocoris onustus* Van Duzee, Proc. Cal. Acad. Sci., (4)9: 344. [N.Y.].
Distribution: Ga., Ill., Man., Mass., Me., N.H., N.S., N.Y., Ont., Pa., Que., Vt., W.Va.

Phytocoris oppositus Knight, 1926
1926 *Phytocoris oppositus* Knight, Bull. Brook. Ent. Soc., 21: 160. [Miss.].
Distribution: Miss.

Phytocoris osage Knight, 1953
1953 *Phytocoris osage* Knight, Ia. St. Coll. J. Sci., 27: 517. [Mo.].

Distribution: Mo.

Phytocoris osborni Knight, 1928
1928 *Phytocoris osborni* Knight, Bull. Brook. Ent. Soc., 23: 28. [Ia.].
Distribution: Ia., Ks., Neb.

Phytocoris pallidicornis Reuter, 1876
1876 *Phytocoris pallidicornis* Reuter, Öfv. K. Svens. Vet.-Akad. Förh., 32(9): 69. [Wis.].
1909 *Phytocoris pallicornis* [*sic*]: Reuter, Acta Soc. Sci. Fenn., 36(2): 33.
1920 *Phytocoris lasiomerus*: Parshley, Psyche, 27: 142.
Distribution: Alta., B.C., Col., Man., Mass., Me., Mich., Minn., Mont., N.D., N.H., N.J., N.Y., Ont., Sask., Wis.

Phytocoris pectinatus Knight, 1920
1920 *Phytocoris pectinatus* Knight, Bull. Brook. Ent. Soc., 15: 58. [Fla.].
Distribution: Fla.

Phytocoris piceicola Knight, 1928
1928 *Phytocoris piceicola* Knight, Bull. Brook. Ent. Soc., 23: 32. [Col.].
Distribution: Ariz., Col.

Phytocoris pinicola Knight, 1920
1920 *Phytocoris pinicola* Knight, Bull. Brook. Ent. Soc., 15: 59. [N.Y.].
Distribution: Conn., Man., Mass., Minn., Miss., Mo., N.J., N.Y.

Phytocoris plenus Van Duzee, 1918
1918 *Phytocoris plenus* Van Duzee, Proc. Cal. Acad. Sci., (4)8: 282. [Cal.].
Distribution: Cal., Nev.

Phytocoris pleuroimos Henry, 1984
1979 *Phytocoris intermedius* Henry, Melsheimer Ent. Ser., 27: 6. [Ga.]. Preoccupied.
1985 *Phytocoris pleuroimos* Henry, J. N.Y. Ent. Soc., 93: 1128. New name for *Phytocoris intermedius* Henry.
Distribution: Ga., N.C.

Phytocoris politus Reuter, 1909
1909 *Phytocoris politus* Reuter, Acta Soc. Sci. Fenn., 36(2): 21. [Nev.]. Lectotype designated by Henry and Stonedahl, 1984, J. N.Y. Ent. Soc., 91: 457.
1920 *Phytocoris rusticus* Van Duzee, Proc. Cal. Acad. Sci., (4)9: 348. [Cal.]. Synonymized by Henry and Stonedahl, 1984, J. N.Y. Ent. Soc., 91: 458.
Distribution: Ariz., Cal., N.M., Nev.

Phytocoris populi (Linnaeus), 1758
1758 *Cimex populi* Linnaeus, Syst. Nat., ed. 10., p. 449. [Europe].
1983 *Phytocoris populi*: Stonedahl, Proc. Ent. Soc. Wash., 85: 463.
Distribution: B.C., Wash. (Europe, North Africa).

Phytocoris proctori Knight, 1974
1974 *Phytocoris proctori* Knight, Ia. St. J. Res., 49: 129. [Me.].
Distribution: Me.

Phytocoris pseudonymus Hussey, 1957
1926 *Phytocoris angustifrons* Blatchley, Het. E. N. Am., p. 727 (as Knight ms. name). [Fla.]. Preoccupied.
1957 *Phytocoris pseudonymus* Hussey, Fla. Ent., 40: 80. New name for *Phytocoris angustifrons* Blatchley.

Distribution: Fla.

Phytocoris puella Reuter, 1876

1876 *Phytocoris puella* Reuter, Öfv. K. Svens. Vet.-Akad. Förh., 32(9): 69. [N.Y.]. Lectotype designated by Henry and Stonedahl, 1984, J. N.Y. Ent. Soc., 91: 458.

1904 *Phytocorus* [*sic*] *puella*: Wirtner, An. Carnegie Mus., 3: 199.

Distribution: Conn., D.C., Fla., Ill., Ind., Ks., Man., Mass., Md., Mich., Mo., N.C., N.J., N.Y., Oh., Ont., Pa., Que., R.I., W.Va., Wis.

Phytocoris pulchellus Knight, 1934

1934 *Phytocoris pulchellus* Knight, Bull. Brook. Ent. Soc., 29: 15. [Ariz.].

Distribution: Ariz.

Phytocoris pulchricollis Van Duzee, 1923

1923 *Phytocoris pulchricollis* Van Duzee, Proc. Cal. Acad. Sci., (4)12: 148. [Mexico].

1968 *Phytocoris pulchricollis*: Knight, Brig. Young Univ. Sci. Bull., 9(3): 255.

Distribution: Ariz., Nev. (Mexico).

Phytocoris purvus Knight, 1927

1927 *Phytocoris purvus* Knight, Proc. Biol. Soc. Wash., 40: 17. [Md.].

Distribution: D.C., Ia., Ill., Md., N.C., S.C., W.Va.

Phytocoris quadriannulipes Knight, 1968

1968 *Phytocoris quadriannulipes* Knight, Brig. Young Univ. Sci. Bull., 9(3): 228. [Ut.].

Distribution: Ut.

Phytocoris quadricinctus Knight, 1968

1968 *Phytocoris quadricinctus* Knight, Brig. Young Univ. Sci. Bull., 9(3): 256. [Ariz.].

Distribution: Ariz., Tex.

Phytocoris quercicola Knight, 1920

1920 *Phytocoris quercicola* Knight, Bull. Brook. Ent. Soc., 15: 60. [N.Y.].

Distribution: Fla., Ia., Ill., Mass., Md., Minn., N.C., N.Y., Ont., Va.

Phytocoris rainieri Knight, 1974

1974 *Phytocoris rainieri* Knight, Ia. St. J. Res., 49: 126. [Wash.].

Distribution: Wash.

Phytocoris ramosus Uhler, 1894

1894 *Phytocoris ramosus* Uhler, Proc. Cal. Acad. Sci., (2)4: 252. [Cal.]. Lectotype designated by Henry and Stonedahl, 1984, J. N.Y. Ent. Soc., 91: 458.

1925 *Phytocoris covilleae* Knight, Bull. Brook. Ent. Soc., 20: 54. [Ariz.]. Synonymized by Carvalho, 1959, Arq. Mus. Nac., 48: 214.

Distribution: Ariz., Cal., Nev., Ut.

Phytocoris relativus Knight, 1968

1968 *Phytocoris relativus* Knight, Brig. Young Univ. Sci. Bull., 9(3): 240. [Ariz.].

Distribution: Ariz., Ut.

Phytocoris reticulatus Knight, 1968

1968 *Phytocoris reticulatus* Knight, Brig. Young Univ. Sci. Bull., 9(3): 217. [Nev.].

Distribution: Nev.

Phytocoris rinconae Knight, 1968

1968 *Phytocoris rinconae* Knight, Brig. Young Univ. Sci. Bull., 9(3): 246. [Ariz.].

Distribution: Ariz.

Phytocoris rolfsi Knight, 1934
1934 *Phytocoris rolfsi* Knight, Bull. Brook. Ent. Soc., 29: 1. [Wash.].
Distribution: Ore., Wash.

Phytocoris roseipennis Knight, 1934
1934 *Phytocoris roseipennis* Knight, Bull. Brook. Ent. Soc., 29: 3. [Ariz.].
Distribution: Ariz.

Phytocoris roseotinctus Knight, 1925
1925 *Phytocoris roseotinctus* Knight, Bull. Brook. Ent. Soc., 20: 52. [Ariz.].
Distribution: Ariz., N.M.

Phytocoris roseus (Uhler), 1894
1894 *Compsocerocoris roseus* Uhler, Proc. Cal. Acad. Sci., (2)4: 253. [Cal.]. Lectotype designated by Henry and Stonedahl, 1984, J. N.Y. Ent. Soc., 91: 459.
1909 *Phytocoris roseus*: Reuter, Acta Soc. Sci. Fenn., 36(2): 27.
1920 *Phytocoris barbatus* Van Duzee, Proc. Cal. Acad. Sci., (4)9: 353. [Cal.]. Synonymized by Carvalho 1959, Arq. Mus. Nac., 48: 214.
Distribution: Cal. (Mexico).

Phytocoris rostratus Knight, 1968
1968 *Phytocoris rostratus* Knight, Brig. Young Univ. Sci. Bull., 9(3): 253. [Nev.].
Distribution: Nev.

Phytocoris rubellus Knight, 1926
1926 *Phytocoris rubellus* Knight, Bull. Brook. Ent. Soc., 21: 166. [S.D.].
1926 *Phytocoris rubellus* Blatchley, Het. E. N. Am., p. 730 (as Knight ms. name). [Ind., "Ind. west to Ia. and Ks."]. Preoccupied.
Distribution: Ia., Ill., Ind., Ks., Mo., S.D.

Phytocoris rubroornatus Knight, 1961
1961 *Phytocoris rubroornatus* Knight, Ia. St. J. Sci., 35: 482. [Ariz.].
Distribution: Ariz.

Phytocoris rubropictus Knight, 1923
1923 *Phytocoris rubropictus* Knight, Conn. Geol. Nat. Hist. Surv. Bull., 34: 619. [N.Y.].
Distribution: Me., N.Y.

Phytocoris rufoscriptus Van Duzee, 1914
1914 *Phytocoris rufoscriptus* Van Duzee, Trans. San Diego Soc. Nat. Hist., 2: 15. [Cal.]. Lectotype designated by Henry and Stonedahl, 1984, J. N.Y. Ent. Soc., 91: 459.
Distribution: Cal.

Phytocoris rufus Van Duzee, 1912
1912 *Phytocoris rufus* Van Duzee, Bull. Buff. Soc. Nat. Sci., 10: 477. [Fla.]. Lectotype designated by Henry and Stonedahl, 1984, J. N.Y. Ent. Soc., 91: 460.
Distribution: Fla., La., Miss.

Phytocoris sagax Van Duzee, 1920
1920 *Phytocoris sagax* Van Duzee, Proc. Cal. Acad. Sci., (4)9: 352. [Cal.].
Distribution: Cal.

Phytocoris salicis Knight, 1920
1920 *Phytocoris salicis* Knight, Bull. Brook. Ent. Soc., 15: 56. [N.Y.].
1958 *Phytocoris salacis* [*sic*]: Rings, J. Econ. Ent., 51: 31.
1976 *Phytocoris salicus* [*sic*]: Reid et al., Can. Ent., 108: 563.

Distribution: Conn., D.C., Fla., Ia., Ill., Ind., Man., Mass., Md., Me., Mich., Minn., Miss., N.C., N.D., N.H., N.J., N.Y., Oh., Ont., Pa., Que., S.D., Sask., W.Va., Wis.

Phytocoris santaritae Knight, 1968
1968 *Phytocoris santaritae* Knight, Brig. Young Univ. Sci. Bull., 9(3): 245. [Ariz.].
Distribution: Ariz.

Phytocoris schotti Knight, 1926
1926 *Phytocoris schotti* Knight, Bull. Brook. Ent. Soc., 21: 162. [N.J.].
Distribution: Ill., N.J.

Phytocoris schuylkillensis Henry, 1974
1974 *Phytocoris schuylkillensis* Henry, Ent. News, 85: 190. [Pa.].
Distribution: Pa.

Phytocoris seminotatus Knight, 1934
1934 *Phytocoris seminotatus* Knight, Bull. Brook. Ent. Soc., 29: 7. [Ariz.].
Distribution: Ariz.

Phytocoris sequoiae Bliven, 1954
1954 *Phytocoris sequoiae* Bliven, Bull. Brook. Ent. Soc., 49: 112. [Cal.].
Distribution: Cal.

Phytocoris sewardi Bliven, 1966
1966 *Phytocoris sewardi* Bliven, Occ. Ent., 1(9): 116. [Cal.].
Distribution: Cal.

Phytocoris signatipes Knight, 1926
1926 *Phytocoris signatipes* Knight, Bull. Brook. Ent. Soc., 21: 164. [Fla.].
Distribution: Fla.

Phytocoris simulatus Knight, 1928
1928 *Phytocoris simulatus* Knight, Bull. Brook. Ent. Soc., 23: 34. [Col.].
Distribution: Col., N.M.

Phytocoris sonorensis Van Duzee, 1920
1920 *Phytocoris sonorensis* Van Duzee, Proc. Cal. Acad. Sci., (4)9: 342. [Cal.].
Distribution: Cal.

Phytocoris spicatus Knight, 1920
1920 *Phytocoris spicatus* Knight, Bull. Brook. Ent. Soc., 15: 55. [N.Y.].
Distribution: Ia., Ill., Mass., Md., Me., Minn., N.C., N.S., N.Y., Que., W.Va.

Phytocoris squamosus Knight, 1934
1934 *Phytocoris squamosus* Knight, Bull. Brook. Ent. Soc., 29: 11. [Ariz.].
Distribution: Ariz., Cal., Nev.

Phytocoris stellatus Van Duzee, 1920
1920 *Phytocoris stellatus* Van Duzee, Proc. Cal. Acad. Sci., (4)9: 350. [Cal.].
Distribution: Alta., B.C., Cal., Col.

Phytocoris stitti Knight, 1961
1961 *Phytocoris stitti* Knight, Ia. St. J. Sci., 35: 474. [Ariz.].
Distribution: Ariz.

Phytocoris strigosus Knight, 1925
1925 *Phytocoris strigosus* Knight, Bull. Brook. Ent. Soc., 20: 51. [Ariz.].
Distribution: Ariz., N.M.

Phytocoris subcinctus Knight, 1968
1968 *Phytocoris subcinctus* Knight, Brig. Young Univ. Sci. Bull., 9(3): 254. [Ut.].
Distribution: Ut.

Phytocoris sublineatus Knight, 1968
1968 *Phytocoris sublineatus* Knight, Brig. Young Univ. Sci. Bull., 9(3): 254. [Ut.].
Distribution: Ut.

Phytocoris sulcatus Knight, 1920
1920 *Phytocoris sulcatus* Knight, Bull. Brook. Ent. Soc., 15: 64. [N.Y.].
Distribution: Conn., D.C., Ia., Ill., Ks., Mass., Mich., Minn., N.J., N.Y., Ont., Pa., Que., S.D., Va., Wis.

Phytocoris tanneri Knight, 1968
1968 *Phytocoris tanneri* Knight, Brig. Young Univ. Sci. Bull., 9(3): 257. [Ut.].
Distribution: N.M., Ut.

Phytocoris taos Knight, 1974
1974 *Phytocoris taos* Knight, Ia. St. J. Res., 49: 127. [N.M.].
Distribution: N.M.

Phytocoris taxodii Knight, 1926
1926 *Phytocoris taxodii* Knight, Bull. Brook. Ent. Soc., 21: 165. [La.].
Distribution: Ga., Ill., La., Miss.

Phytocoris tehamae Bliven, 1956
1956 *Phytocoris tehamae* Bliven, New Hem. W. St., p. 16. [Cal.].
Distribution: Cal.

Phytocoris tenuis Van Duzee, 1920
1920 *Phytocoris tenuis* Van Duzee, Proc. Cal. Acad. Sci., (4)9: 341. [Cal.].
Distribution: Cal.

Phytocoris texanus Knight, 1961
1961 *Phytocoris texanus* Knight, Ia. St. J. Sci., 35: 481. [Tex.].
Distribution: Tex.

Phytocoris tibialis Reuter, 1876
1876 *Phytocoris tibialis* Reuter, Öfv. K. Svens. Vet.-Akad. Förh., 32(9): 68. [Tex.]. Lectotype designated by Henry and Stonedahl, 1984, J. N.Y. Ent. Soc., 91: 460.
1887 *Compsocerocoris pallidicornis*: Provancher, Pet. Faune Ent. Can., 3: 108.
1904 *Phytocorus* [*sic*] *tibialis*: Wirtner, An. Carnegie Mus., 3: 199.
1928 *Phytocoris tibiolus* [*sic*]: Watson, Oh. Biol. Surv. Bull., 16: 31.
Distribution: Ala., Conn., D.C., Fla., Ia., Ind., Mass., Md., Me., Mich., Minn., Miss., Mo., N.C., N.J., N.Y., Oh., Ont., Pa., R.I., Tex., Va., Vt., W.Va., Wis. (Jamaica, Mexico, Central America).

Phytocoris tiliae (Fabricius), 1777
1777 *Cimex tiliae* Fabricius, Gen. Ins., p. 301. [Europe].
1873 *Phytocoris tiliae*: Walker, Cat. Hem. Brit. Mus., 6: 59.
Distribution: B.C., N.S., N.Y., Ore., Wash. (Europe, North Africa).
Note: Wheeler and Henry (1976, Ent. News, 87: 25) first reported this species in the U.S.; Stonedahl (1983, Proc. Ent. Soc. Wash., 85: 467-469) gave additional records for the Pacific Northwest.

Phytocoris tillandsiae Johnston, 1935
1935 *Phytocoris tillandsiae* Johnston, Bull. Brook. Ent. Soc., 30: 18. [Tex.].
Distribution: Miss., Tex.

Phytocoris tinctus Knight, 1928
1928 *Phytocoris tinctus* Knight, Bull. Brook. Ent. Soc., 33: 36. [Col.].
Distribution: Ariz., Col.

Phytocoris tricinctipes Knight, 1968
1968 *Phytocoris tricinctipes* Knight, Brig. Young Univ. Sci. Bull., 9(3): 230. [Nev.].
Distribution: Nev.

Phytocoris tricinctus Knight, 1968
1968 *Phytocoris tricinctus* Knight, Brig. Young Univ. Sci. Bull., 9(3): 256. [Ariz.].
Distribution: Ariz.

Phytocoris tuberculatus Knight, 1920
1920 *Phytocoris tuberculatus* Knight, Bull. Brook. Ent. Soc., 15: 64. [N.Y.].
1979 *Phytocoris tuberculata* [*sic*]: Henry and Smith, J. Ga. Ent. Soc., 14: 217.
Distribution: Ga., Ill., Ind., Mich., Mo., N.C., N.Y., Ok., Tex., Wis.

Phytocoris tucki Knight, 1953
1953 *Phytocoris tucki* Knight, Ia. St. Coll. J. Sci., 27: 515. [Mo.].
1959 *Phytocoris tuki* [*sic*]: Carvalho, Arq. Mus. Nac., 48: 219.
Distribution: Mo.

Phytocoris ulmi (Linnaeus), 1758
1758 *Cimex ulmi* Linnaeus, Syst. Nat., Ed. 10, p. 449. [Europe].
1923 *Phytocoris ulmi*: Knight, Conn. Geol. Nat. Hist. Surv. Bull., 34: 620.
Distribution: N.S. (Europe, North Africa, USSR).

Phytocoris umbrosus Knight, 1928
1928 *Phytocoris umbrosus* Knight, Bull. Brook. Ent. Soc., 23: 37. [Col.].
Distribution: Ariz., Col., N.M.

Phytocoris uniformis Knight, 1923
1923 *Phytocoris uniformis* Knight, Conn. Geol. Nat. Hist. Surv. Bull., 34: 643. [N.Y.].
Distribution: Mass., Md., Miss., N.C., N.Y., Pa., Va.

Phytocoris utahensis Knight, 1961
1961 *Phytocoris utahensis* Knight, Ia. St. J. Sci., 35: 473. [Ut.].
Distribution: Ut.

Phytocoris validus Reuter, 1909
1909 *Phytocoris validus* Reuter, Acta Soc. Sci. Fenn., 36(2): 31. [Col.]. Lectotype designated by Henry and Stonedahl, 1984, J. N.Y. Ent. Soc., 91: 461.
Distribution: Alta., Col., Man., S.D., Sask.

Phytocoris vanduzeei Reuter, 1912
1894 *Lygus vividus* Uhler, Proc. Cal. Acad. Sci., (2)4: 260. [Mexico]. Preoccupied. Lectotype designated by Henry and Stonedahl, 1984, J. N.Y. Ent. Soc., 91: 461.
1910 *Dichrooscytus marmoratus* Van Duzee, 1910, Trans. Am. Ent. Soc., 36: 78. [N.M.]. Preoccupied. Lectotype designated by Henry and Stonedahl, 1984, J. N.Y. Ent. Soc., 91: 461.
1912 *Phytocoris vanduzeei* Reuter, Öfv. F. Vet.-Soc. Förh., 54A(7): 30. New name for *Dichrooscytus marmoratus* Van Duzee.

1917 *Phytocoris vividus*: Knight, Cornell Univ. Agr. Exp. Stn. Bull., 391: 640.
1925 *Phytocoris nigripubescens* Knight, Bull. Brook. Ent. Soc., 20: 55. [Ariz.]. Synonymized by Henry and Stonedahl, 1984, J. N.Y. Ent. Soc., 91: 462.
Distribution: Ariz., Cal., N.M., Nev., Ut. (Mexico).
Note: Henry and Stonedahl (1984, above) discussed and clarified the complex nomenclatural history of this species.

Phytocoris varipes Boheman, 1852
1852 *Phytocoris varipes* Boheman, Öfv. K. Svens. Vet.-Akad. Förh., p. 107. [Europe].
1983 *Phytocoris varipes*: Stonedahl, Proc. Ent. Soc. Wash., 85: 465.
Distribution: Ore. (Europe, North Africa).

Phytocoris varius Knight, 1934
1934 *Phytocoris varius* Knight, Bull. Brook. Ent. Soc., 29: 9. [Ariz.].
Distribution: Ariz., Col.

Phytocoris vau Van Duzee, 1912
1912 *Phytocoris vau* Van Duzee, Bull. Buff. Soc. Nat. Sci., 10: 478. [Cal.].
Distribution: Cal.

Phytocoris ventralis Van Duzee, 1918
1918 *Phytocoris ventralis* Van Duzee, Proc. Cal. Acad. Sci., (4)8: 287. [Cal.].
Distribution: Cal., Nev.

Phytocoris venustus Knight, 1923
1923 *Phytocoris venustus* Knight, Conn. Geol. Nat. Hist. Surv. Bull., 34: 651. [Conn.].
Distribution: Ala., Conn., D.C., Ill., Md., N.Y., Pa., W.Va.

Phytocoris vinaceus Van Duzee, 1917
1917 *Phytocoris vinaceus* Van Duzee, Proc. Cal. Acad. Sci., (4)7: 263. [Cal.].
Distribution: Cal.

Phytocoris viridescens Knight, 1961
1961 *Phytocoris viridescens* Knight, Ia. St. J. Sci., 35: 483. [Col.].
Distribution: Col.

Phytocoris westwoodi Bliven, 1966
1966 *Phytocoris westwoodi* Bliven, Occ. Ent., 1(9): 110. [Cal.].
Distribution: Cal.

Phytocoris yollabollae Bliven, 1956
1956 *Phytocoris yollabollae* Bliven, New Hem. W. St., p. 17. [Cal.].
Distribution: Cal.

Phytocoris yuma Knight, 1961
1961 *Phytocoris yuma* Knight, Ia. St. J. Sci., 35: 479. [Ariz.].
Distribution: Ariz.

Phytocoris yuroki Bliven, 1954
1954 *Phytocoris yuroki* Bliven, Bull. Brook. Ent. Soc., 49: 110. [Cal.].
Distribution: Cal.

Genus *Pinalitus* Kelton, 1955

1955 *Pinalitus* Kelton, Can. Ent., 87: 282. Type-species: *Deraeocoris approximatus* Stål, 1858. Original designation.

Note: Kelton (1977, Can. Ent., 109: 1549-1554) provided a key to species.

Pinalitus approximatus (Stål), 1858

1858 *Deraeocoris approximatus* Stål, Stett. Ent. Zeit., 19: 185. [Alk.].

1879 *Lygus approximatus*: Reuter, Öfv. F. Vet.-Soc. Förh., 21: 53.

1955 *Pinalitus approximatus*: Kelton, Can. Ent., 87: 282.

Distribution: Alk., Alta., B.C., Col., Man., Me., N.B., N.C., N.H., N.S., N.T., N.Y., Nfld., Ont., P.Ed., Que., Sask., W.Va., Wyo. (Siberia).

Pinalitus rostratus Kelton, 1977

1977 *Pinalitus rostratus* Kelton, Can. Ent., 109: 1552. [N.S.].

Distribution: Alta., B.C., Col., N.B., N.M., N.S., P.Ed., Que., Yuk.

Pinalitus rubricatus (Fallén), 1807

1807 *Lygaeus rubricatus* Fallén, Monogr. Cimic. Suec., p. 91. [Europe].

1974 *Orthops rubricatus*: Henry and Wheeler, Proc. Ent. Soc. Wash., 76: 221.

1977 *Pinalitus rubricatus*: Kelton, Can. Ent., 109: 1549.

Distribution: N.S., N.Y., Pa. (Europe, Northern Africa, Northern Asia).

Note: Henry and Wheeler (1974, above) summarized the European literature and gave hosts.

Pinalitus rubrotinctus Knight, 1968

1968 *Pinalitus rubrotinctus* Knight, Brig. Young Univ. Sci. Bull., 9(3): 189. [Ariz.].

Distribution: Ariz., Col., N.M.

Pinalitus solivagus (Van Duzee), 1921

1921 *Lygidea solivaga* Van Duzee, Proc. Cal. Acad. Sci., (4)11: 119. [Cal.].

1968 *Pinalitus brevirostris* Knight, Brig. Young Univ. Sci. Bull., 9(3): 187. [Col.]. Synonymized by Kelton, 1977, Can. Ent., 109: 1550.

1968 *Pinalitus utahensis* Knight, Brig. Young Univ. Sci. Bull., 9(3): 188. [Ariz.]. Synonymized by Kelton, 1977, Can. Ent., 109: 1550.

1968 *Pinalitus solivagus*: Knight, Brig. Young Univ. Sci. Bull., 9(3): 188.

Distribution: Ariz., Cal., Col., Ore., Ut.

Genus *Platylygus* Van Duzee, 1915

1915 *Platylygus* Van Duzee, Pomona J. Ent. Zool., 7: 111. Type-species: *Lygidea rubecula* var. *lurida* Reuter, 1909. Monotypic.

Note: Kelton and Knight (1970, Can. Ent., 102: 1429-1460) revised the genus and provided a key to species.

Platylygus andrei Knight, 1970

1970 *Platylygus andrei* Knight, *In* Kelton and Knight, Can. Ent., 102: 1439. [Ariz.].

Distribution: Ariz.

Platylygus balli Knight, 1970

1970 *Platylygus balli* Knight, *In* Kelton and Knight, Can. Ent., 102: 1453. [Ariz.].

Distribution: Ariz., Col., N.M.

Platylygus contortae Kelton, 1970

1970 *Platylygus contortae* Kelton, *In* Kelton and Knight, Can. Ent., 102: 1443. [Cal.].

Distribution: Cal.

Platylygus crinitus Kelton, 1970

1970 *Platylygus crinitus* Kelton, *In* Kelton and Knight, Can. Ent., 102: 1435. [Cal.].

Distribution: Cal.

Platylygus fuliginosus Knight, 1918
1918 *Platylygus fuliginosus* Knight, Bull. Brook. Ent. Soc., 13: 17. [Ariz.].
Distribution: Ariz.

Platylygus grandis Knight, 1918
1918 *Platylygus grandis* Knight, Bull. Brook. Ent. Soc., 13: 17. [Ariz.].
Distribution: Ariz., B.C., Cal., Col., Mont., S.D.

Platylygus hirtus Knight, 1970
1970 *Platylygus hirtus* Knight, *In* Kelton and Knight, Can. Ent., 102: 1437. [Ariz.].
Distribution: Ariz.

Platylygus intermedius Knight, 1918
1918 *Platylygus intermedius* Knight, Bull. Brook. Ent. Soc., 13: 16. [Ariz.].
Distribution: Ariz., Col., N.M.

Platylygus knighti Kelton, 1970
1970 *Platylygus knighti* Kelton, *In* Kelton and Knight, Can. Ent., 102: 1455. [Ariz.].
Distribution: Ariz., Col.

Platylygus luridus (Reuter), 1909
1895 *Calocoris tinctus* Uhler, Bull. Col. Agr. Exp. Stn., 31: 34. [Col.]. Preoccupied.
1909 *Lygidea rubecula* var. *lurida* Reuter, Acta Soc. Sci. Fenn., 36(2): 46. [N.Y.].
1912 *Calocoris uhleri* Van Duzee, Bull. Buff. Soc. Nat. Sci., 10: 490. New name for *Calocoris tinctus* Uhler. Synonymized by Kelton and Knight, 1970, Can. Ent., 102: 1429, 1458.
1915 *Platylygus luridus*: Van Duzee, Pomona J. Ent. Zool., 7: 111.
1959 *Platylygus uhleri*: Carvalho, Arq. Mus. Nac., 48: 226.
Distribution: Alta., Ariz., B.C., Cal., Col., Ill., Man., N.B., N.H., N.M., N.S., N.Y., Ont., Que., Sask., W.Va., Wis., Wyo., Yuk.
Note: Rauf et al. (1984, Can. Ent., 116: 1213-1218 and 116: 1219-1225) discussed conelet abortion in pines from the feeding of *P. luridus* and reported on bionomics in Wisconsin.

Platylygus magnus Kelton, 1970
1970 *Platylygus magnus* Kelton, *In* Kelton and Knight, Can. Ent., 102: 1451. [Cal.].
Distribution: Cal.

Platylygus piceicola Kelton, 1970
1970 *Platylygus piceicola* Kelton, *In* Kelton and Knight, Can. Ent., 102: 1436. [Yuk.].
Distribution: Alta., Ariz., B.C., Col., Yuk.

Platylygus pilosipes Kelton, 1970
1970 *Platylygus pilosipes* Kelton, *In* Kelton and Knight, Can. Ent., 102: 1441. [Holotype from Mexico; also from Ariz.].
Distribution: Ariz. (Mexico).

Platylygus pseudotsugae Kelton, 1970
1970 *Platylygus pseudotsugae* Kelton, *In* Kelton and Knight, Can. Ent., 102: 1454. [B.C.].
Distribution: B.C., Ore.

Platylygus rolfsi Knight, 1970
1970 *Platylygus rolfsi* Knight, *In* Kelton and Knight, Can. Ent., 102: 1450. [Wash.].

Distribution: B.C., Cal., Ore., Wash.

Platylygus rubripes Knight, 1970

1970 *Platylygus rubripes* Knight, *In* Kelton and Knight, Can. Ent., 102: 1457. [Cal.].
Distribution: Alta., B.C., Cal., Col., Ore., Wash., Wyo.

Platylygus scutellatus Kelton, 1970

1970 *Platylygus scutellatus* Kelton, *In* Kelton and Knight, Can. Ent., 102: 1443. [Holotype from Mexico; also from U.S.].
Distribution: Ariz., N.M. (Mexico).

Platylygus usingeri Knight, 1970

1970 *Platylygus usingeri* Knight, *In* Kelton and Knight, Can. Ent., 102: 1442. [Cal.].
Distribution: Cal.

Platylygus vanduzeei Usinger, 1931

1931 *Platylygus vanduzeei* Usinger, Pan-Pac. Ent., 7: 129. [Ariz.].
Distribution: Ariz., Cal., Col., N.M., Nev., Ut.

Genus *Plesiocoris* Fieber, 1861

1861 *Plesiocoris* Fieber, Europ. Hem., p. 272. Type-species: *Lygaeus rugicollis* Fallén, 1807. Designated by Kirkaldy, 1906, Trans. Am. Ent. Soc., 32: 139.

Plesiocoris rugicollis (Fallén), 1807

1807 *Lygaeus rugicollis* Fallén, Monogr. Cimic. Suec., p. 76. [Europe].
1921 *Plesiocoris rugicollis*: Knight, Oh. J. Sci., 21: 109.
1953 *Pleciocoris* [*sic*] *rugicollis*: Strickland, Can. Ent., 85: 202.
Distribution: Alk., Alta., B.C. (Europe, Northern Asia).

Genus *Poecilocapsus* Reuter, 1876

1876 *Poecilocapsus* Reuter, Öfv. K. Svens. Vet.-Akad. Förh., 32(9): 73. Type-species: *Lygaeus lineatus* Fabricius, 1798. Designated by Kirkaldy, 1906, Trans. Am. Ent. Soc., 32: 140.

Poecilocapsus lineatus (Fabricius), 1798 [Fig. 81]

1798 *Lygaeus lineatus* Fabricius, Ent. Syst. Suppl., p. 541. ["America boreali"].
1832 *Capsus quadrivittatus* Say, Descrip. Het. Hem. N. Am., p. 20. Synonymized by Reuter, 1876, Öfv. K. Svens. Vet.-Akad. Förh., 32(9): 74.
1854 *Phytocoris bellus* Emmons, Nat. Hist. N.Y. Agr., 5: plate 30, fig. 1. Synonymized by Van Duzee, 1916, Check List Hem., p. 39.
1869 *Phytocoris vittatus*: Rathvon, Hist. Lancaster Co., Pa., p. 549.
1870 *Phytocoris lineatus*: Fitch, Trans. N.Y. Agr. Soc., 29: 513.
1875 *Lygus lineatus*: Glover, Rept. Dept. Agr. U.S., p. 125.
1876 *Poecilocapsus lineatus*: Reuter, Öfv. K. Vet.-Akad. Förh., 32(9): 74.
1884 *Poecilacapsus* [*sic*] *vittatus*: Uhler, Stand. Nat. Hist., 2: 286. *Lapsus*.
1915 *Poscilocapsus* [*sic*] *lineatus*: LaFollette, J. Ent. Zool., 7: 128.
Distribution: B.C., Conn., D.C., Fla., Ia., Ill., Ind., Man., Mass., Me., Mich., Mo., Ks., N.B., N.D., N.H., N.J., N.M., N.S., N.Y., Neb., Oh., Ont., Pa., Que., R.I., S.D., Sask., Tex., Wis. (Mexico).
Note: Wheeler and Miller (1981, Great Lakes Ent., 14: 23-35) reviewed seasonal history and host plants and discussed economic importance.

Poecilocapsus nigriger (Stål), 1862
1862 *Brachycoleus nigriger* Stål, Stett. Ent. Zeit., 23(7-9): 319. [Mexico].
1909 *Poecilocapsus nigriger*: Van Duzee, Bull. Buff. Soc. Nat. Sci., 9: 178.
1926 *Poecilocapsus lineatus nigriger*: Blatchley, Het. E. N. Am., p. 751.
Distribution: Cal., Fla. (Mexico).
Note: Blatchley (1926, above, p. 751) considered this species a southern variety or race of *Poecilocapsus lineatus*, but Carvalho (1954, Arq. Mus. Nac., 48: 230) listed it as a separate species.

Genus *Polymerus* Hahn, 1831

1831 *Polymerus* Hahn, Wanz. Ins., 1: 27. Type-species: *Polymerus holosericeus* Hahn, 1831. Monotypic.
1858 *Poeciloscytus* Fieber, Wien. Ent. Monat., 2: 311. Type-species: *Lygaeus unifasciatus* Fabricius, 1794. Designated by Distant, 1904, Faun. Brit. India, Rhyn., 2: 458. Synonymized by Knight, 1923, Conn. Geol. Nat. Hist. Surv. Bull., 34: 598.
1865 *Systratiotus* Douglas and Scott, Brit. Hem., p. 443. Type-species: *Lygaeus nigrita* Fallén, 1807. Monotypic. Unnecessary new name for *Polymerus* Hahn.
Note: The name "*Poeciloscytus pusillus* Reuter" listed by Popenoe (1885, Trans. Ks. Acad. Sci., 9: 63) apparently is a manuscript name that was never published by Reuter; it is omitted from the present list.

Polymerus americanus (Reuter), 1876
1876 *Systratiotus americanus* Reuter, Öfv. K. Svens. Vet.-Akad. Förh., 32(9): 73. [Tex.].
1890 *Poeciloscytus americanus*: Atkinson, J. Asiatic Soc. Bengal, 58(2): 94.
1907 *Polymerus americanus*: Tucker, Ks. Univ. Sci. Bull., 4: 60.
Distribution: Col., Fla., Ia., Ks., N.J., N.M., N.Y., Ont., Pa., Tex., Ut.
Note: This species was synonymized under *P. venaticus* (Uhler) by Van Duzee (1917, Univ. Cal. Publ. Ent., 2: 332), but Knight (1925, Can. Ent., 57: 253) considered both species distinct.

Polymerus balli Knight, 1925
1925 *Polymerus balli* Knight, Can. Ent., 57: 250. [Neb.].
Distribution: Alta., Col., Man., Neb., Sask.

Polymerus basalis (Reuter), 1876
1876 *Poeciloscytus basalis* Reuter, Öfv. K. Svens. Vet.-Akad. Förh., 32(9): 73. [Tex.].
1877 *Poeciloscytus sericeus* Uhler, Bull. U.S. Geol. Geogr. Surv. Terr., 3: 422. [Que. to Fla., w. to Tex. and N.M.]. Synonymized by Barber, 1906, Mus. Brook. Inst. Arts Sci., Sci. Bull., 1: 280.
1900 *Poecyloscytus* [*sic*] *basalis*: Osborn, Oh. Nat., 1: 12.
1923 *Polymerus basalis*: Knight, Conn. Geol. Nat. Hist. Surv. Bull., 34: 599.
Distribution: Ariz., Ark., Cal., Col., Conn., D.C., Fla., Ga., Ia., Ind., Ks., Mass., Md., Me., Minn., Miss., Mo., N.H., N.J., N.M., N.Y., Oh., Ok., Ont., Pa., Que., R.I., S.D., Tex., Va., Vt., W.Va., Wis. (Mexico).
Note: Although Knight (1926, Can. Ent., 58: 167) treated *P. sericeus* as a valid species, Carvalho (1959, Arq. Mus. Nac., 48: 239) listed it as a synonym of *P. basalis*.

Polymerus basalis basalis (Reuter), 1876.
1876 *Poeciloscytus basalis* Reuter, Öfv. K. Svens. Vet.-Akad. Förh, 32(9): 73. [Tex.].

1926 *Polymerus basalis*: Knight, Can. Ent., 58: 167.
Distribution: Same as for species.

Polymerus basalis fuscatus Knight, 1926
1926 *Polymerus basalis* var. *fuscatus* Knight, Can. Ent., 58: 167. [Fla.].
Distribution: Fla., Ind., Minn., N.Y.

Polymerus basivittis (Reuter), 1909
1909 *Poeciloscytus basivittis* Reuter, Acta Soc. Sci. Fenn., 36(2): 61. [Col.].
1925 *Polymerus basivittis*: Knight, Can. Ent., 57: 95.
Distribution: Alta., Col., Man., Mont., Sask., Wyo.

Polymerus basivittis basivittis (Reuter), 1909
1909 *Poeciloscytus basivittis* Reuter, Acta Soc. Sci. Fenn., 36(2): 61. [Col.].
1925 *Polymerus basivittis*: Knight, Can. Ent., 57: 95.
Distribution: Same as for species.

Polymerus basivittis pallidulus Knight, 1925
1925 *Polymerus basivittis* var. *pallidulus* Knight, Can. Ent., 57: 95. [Wyo.].
1959 *Polymerus basalis* var. *pallidulus*: Carvalho, Arq. Mus. Nac., 48: 233 (incorrect citation for *P. basivittis* var. *pallidulus*).
Distribution: Alta., Col., Mont., Wyo.

Polymerus brevirostris Knight, 1925
1925 *Polymerus brevirostris* Knight, Can. Ent., 57: 246. [Minn.].
Distribution: Alta., Ill., Man., Mass., Mich., Minn., N.D., N.Y., S.D., Sask., Wis.

Polymerus brevis Knight, 1925
1925 *Polymerus brevis* Knight, Can. Ent., 57: 249. [S.D.].
Distribution: S.D.

Polymerus chrysopsis Knight, 1925
1925 *Polymerus chrysopsis* Knight, Can. Ent., 57: 245. [Minn.].
Distribution: Ia., Ill., Man., Mass., Minn., N.D., S.D., Sask.

Polymerus costalis Knight, 1943
1943 *Polymerus costalis* Knight, Can. Ent., 75: 179. [Wash.].
Distribution: Wash.

Polymerus delongi Knight, 1925
1925 *Polymerus delongi* Knight, Can. Ent., 57: 252. [Fla.].
Distribution: Fla.

Polymerus diffusus (Uhler), 1872
1872 *Poeciloscytus diffusus* Uhler, Prelim. Rept. U.S. Geol. Surv. Mont., p. 415. [Col.].
1914 *Polymerus divergens* Parshley, Psyche, 21: 141. *Nomen nudum*.
1926 *Polymerus diffusus*: Knight, Can. Ent., 58: 165.
Distribution: Alta., Ariz., B.C., Cal., Col., Id., Nev., N.M., Ore., Sask., Ut., Wash.
Note: The Va. record listed by Carvalho (1959, Arq. Mus. Nac., 48: 235) undoubtedly is based on a misidentification and is dropped from the present list.

Polymerus elegans (Reuter), 1909
1909 *Poeciloscytus elegans* Reuter, Acta Soc. Sci. Fenn., 36(2): 60. [Cal.].
1959 *Polymerus elegans*: Carvalho, Arq. Mus. Nac., 48: 235.
Distribution: Cal.

Polymerus fasciolus Knight, 1943
1943 *Polymerus fasciolus* Knight, Can. Ent., 75: 181. [Ariz.].
Distribution: Ariz.

Polymerus flaviloris Knight, 1925
1925 *Polymerus flaviloris* Knight, Can. Ent., 57: 251. [Mont.].
Distribution: Col., Mont.

Polymerus flavocostatus Knight, 1926
1926 *Polymerus flavocostatus* Knight, Can. Ent., 58: 165. [Ia.].
Distribution: Ia., Ill., Mo., N.D., Neb.

Polymerus froeschneri Knight, 1953
1953 *Polymerus froeschneri* Knight, Ia. St. Coll. J. Sci., 27: 513. [Mo.].
Distribution: Mo.

Polymerus fulvipes Knight, 1923
1904 *Systratiotus americanus*: Wirtner, An. Carnegie Mus., 3: 196.
1923 *Polymerus fulvipes* Knight, Conn. Geol. Nat. Hist. Surv. Bull., 34: 603. [N.Y.].
Distribution: Conn., Mass., N.C., N.Y., Pa., S.D., W.Va.
Note: Wheeler and Henry (1977, Great Lakes Ent., 10: 148) clarified the Wirtner (1904, above) record.

Polymerus gerhardi Knight, 1923
1923 *Polymerus gerhardi* Knight, Conn. Geol. Nat. Hist. Surv. Bull., 34: 606. [Ind.].
Distribution: Ill., Ind., Miss., Ok., Tex.

Polymerus gracilentus Knight, 1925
1925 *Polymerus gracilentus* Knight, Can. Ent., 57: 249. [Cal.].
Distribution: Cal.

Polymerus hirtus Knight, 1943
1943 *Polymerus hirtus* Knight, Can. Ent., 75: 180. [Id.].
Distribution: Alta., Id., Sask.

Polymerus illini Knight, 1941
1941 *Polymerus illini* Knight, Ill. Nat. Hist. Surv. Bull., 22: 168. [Ill.].
Distribution: Ill.

Polymerus nigrigulis Knight, 1926
1926 *Polymerus nigrigulis* Knight, Can. Ent., 58: 166. [Mont.].
Distribution: Mont.

Polymerus nigropallidus Knight, 1923
1923 *Polymerus nigropallidus* Knight, Conn. Geol. Nat. Hist. Surv. Bull., 34: 599. [N.J.].
Distribution: N.J.
Note: Henry (1978, Proc. Ent. Soc. Wash., 80: 546) gave the host and new N.J. records for this poorly known species.

Polymerus nubilipes Knight, 1925
1925 *Polymerus nubilipes* Knight, Can. Ent., 57: 248. [Minn.].
Distribution: Minn., Wis.

Polymerus opacus Knight, 1923
1923 *Polymerus opacus* Knight, Conn. Geol. Nat. Hist. Surv. Bull., 34: 604. [N.Y.].
Distribution: Me., N.Y., Ont., Vt.

Polymerus pallescens (Walker), 1873
1873 *Capsus pallescens* Walker, Cat. Hem. Brit. Mus., 6: 94. ["St. Martins Falls, Albany River, Hudson's Bay"; Man.].
1959 *Polymerus pallescens*: Carvalho, Arq. Mus. Nac., 48: 238.
Distribution: Man.

Polymerus proximus Knight, 1923
1923 *Polymerus proximus* Knight, Conn. Geol. Nat. Hist. Surv. Bull., 34: 601. [Pa.].
Distribution: Ia., Ill., Ks., Minn., Mo., Neb., Oh., Ont., Pa., W.Va., Wis.

Polymerus punctipes Knight, 1923
1923 *Polymerus punctipes* Knight, Conn. Geol. Nat. Hist. Surv. Bull., 34: 602. [N.Y.].
Distribution: Fla., Ga., Ia., Ill., Ind., Md., Me., Minn., Mo., N.C., N.Y., Oh., Ont., Que., W.Va.

Polymerus relativus Knight, 1926
1926 *Polymerus relativus* Knight, Can. Ent., 58: 165. [Col.].
Distribution: Col., N.M., Nev., Ut.

Polymerus robustus Knight, 1925
1925 *Polymerus robustus* Knight, Can. Ent., 57: 251. [Cal.].
Distribution: Cal.

Polymerus rostratus Henry, 1978
1978 *Polymerus rostratus* Henry, Proc. Ent. Soc. Wash., 80: 543. [N.J.].
Distribution: N.J.

Polymerus rubrocuneatus Knight, 1925
1925 *Polymerus rubrocuneatus* Knight, Can. Ent., 57: 247. [N.D.].
Distribution: Alta., Man., Mont., N.D., S.D., Sask.

Polymerus rufipes Knight, 1926
1926 *Polymerus basalis* var. *rufipes* Knight, Can. Ent., 58: 167. [Col.].
1980 *Polymerus rufipes*: Kelton, Agr. Can., Res. Bull., 1703: 88.
Distribution: Alta., Col., Wyo.

Polymerus sculleni Knight, 1943
1943 *Polymerus sculleni* Knight, Can. Ent., 75: 180. [Ore.].
Distribution: Ore., Sask.

Polymerus severini Knight, 1925
1925 *Polymerus severini* Knight, Can. Ent., 57: 247. [S.D.].
Distribution: Alta., Minn., S.D.

Polymerus shawi Knight, 1943
1943 *Polymerus shawi* Knight, Can. Ent., 75: 179. [N.H.].
Distribution: N.H.

Polymerus standishi Knight, 1943
1943 *Polymerus standishi* Knight, Can. Ent., 75: 179. [Ok.].
Distribution: Ok.

Polymerus testaceipes (Stål), 1860
1860 *Deraeocoris testaceipes* Stål, Öfv. K. Svens. Vet.-Akad. Förh., 2(7): 50. [Brazil].
1910 *Poeciloscytus cuneatus*: Banks, Cat. Nearc. Hem.-Het., p. 49.
1926 *Polymerus flavocuneatus*: Knight, Can. Ent., 58: 186.
1926 *Polymerus clandestinus* Blatchley, Ent. News, 37: 164. [Fla.]. Synonymized by

Blatchley, 1926, Het. E. N. Am., p. 737.
1959 *Polymerus testaceipes*: Carvalho, Arq. Mus. Nac., 48: 240.
Distribution: Fla., Tex. (Mexico to South America, West Indies).

Polymerus tinctipes Knight, 1923
1923 *Polymerus tinctipes* Knight, Conn. Geol. Nat. Hist. Surv. Bull., 34: 600. [Md.].
Distribution: Md.

Polymerus tumidifrons Knight, 1925
1925 *Polymerus tumidifrons* Knight, Can. Ent., 57: 248. [Mont.].
Distribution: Alta., Mont., Ut.

Polymerus uhleri (Van Duzee), 1914
1894 *Poeciloscytus intermedius* Uhler, Proc. Cal. Acad. Sci., (2)4: 261. [Mexico]. Preoccupied.
1914 *Poeciloscytus uhleri* Van Duzee, Trans. San Diego Soc. Nat. Hist., 2: 23. New name for *Poeciloscytus intermedius* Uhler.
1959 *Polymerus uhleri*: Carvalho, Arq. Mus. Nac., 48: 240.
Distribution: Ariz., Cal., Col. (Mexico).

Polymerus unifasciatus (Fabricius), 1794
1794 *Lygaeus unifasciatus* Fabricius, Ent. Syst., 4: 178. [Europe].
1878 *Poeciloscytus unifasciatus*: Uhler, Bull. U.S. Geol. Geogr. Surv. Terr., 4: 507.
1925 *Polymerus unifasciatus lateralis*: Knight, Can. Ent., 57: 182.
1941 *Polymerus unifasciatus*: Knight, Ill. Nat. Hist. Surv. Bull., 22: 167.
Distribution: Alk., Alta., B.C., Cal., Col., Conn., Ia., Ill., Man., Me., Minn., N.D., N.Y., Ont., Que., Sask., Ut., Vt., (Asia, Europe).

Polymerus venaticus (Uhler), 1872
1872 *Poeciloscytus venaticus* Uhler, Prelim. Rept. U.S. Geol. Surv. Mont., p. 414. [Col.].
1872 *Rhopalotomus rubronotatus* Provancher, Nat. Can., 4: 105. Synonymized by Provancher, 1887, Pet. Faune Ent. Can., p. 125. Lectotype designated by Kelton, 1968, Nat. Can., 95: 1077.
1886 *Systratiotus venaticus*: Uhler, Check-list Hem. Het., p. 19.
1890 *Capsus rubronotatus*: Atkinson, J. Asiatic Soc. Bengal, 58(2): 106.
1907 *Polymerus venaticus*: Tucker, Ks. Univ. Sci. Bull., 4: 60.
1923 *Poeciloscylus* [*sic*] *venaticus*: Parshley, Proc. Acadian Ent. Soc., 8: 106.
Distribution: Alta., B.C., Cal., Col., Conn., D.C., Ia., Id., Ill., Ind., Ks., Man., Mass., Md., Me., Mich., Minn., Mo., N.C., N.D., N.H., N.J., N.M., N.S., N.Y., Neb., Oh., Ont., Pa., Que., Sask., Tex., Ut., Vt., W.Va., Wis.

Polymerus venustus Knight, 1923
1923 *Polymerus venustus* Knight, Conn. Geol. Nat. Hist. Surv. Bull., 34: 605. [Mich.].
Distribution: Fla., Ill., Mich., Miss., Oh., N.C., S.C., Va.

Polymerus vittatipennis Knight, 1943
1943 *Polymerus vittatipennis* Knight, Can. Ent., 75: 181. [Ariz.].
Distribution: Ariz., N.M., Ut.

Polymerus wheeleri Henry, 1979
1979 *Polymerus wheeleri* Henry, Melsheimer Ent. Ser., 27: 5. [W.Va.].
Distribution: W.Va.

Genus *Proba* Distant, 1884

1884 *Proba* Distant, Biol. Centr.-Am., Rhyn., 1: 269. Type-species: *Proba gracilis* Distant, 1884. Monotypic.

Proba californica (Knight), 1968

1968 *Pinalitus californica* Knight, Brig. Young Univ. Sci. Bull., 9(3): 189. [Cal.].
1977 *Proba californica*: Kelton, Can. Ent., 109: 1553.
Distribution: Cal.

Proba distanti (Atkinson), 1890

1884 *Lygus scutellatus* Distant, Biol. Centr.-Am., Rhyn., 1: 274. [Mexico, Guatemala]. Preoccupied.
1890 *Lygus distanti* Atkinson, J. Asiatic Soc. Bengal, 58(2): 92. New name for *Lygus scutellatus* Distant.
1906 *Lygus scutellatus*: Barber, Bull. Mus. Brook. Inst. Arts Sci., Sci. Bull., 1: 279.
1907 *Lygus distantii* [*sic*]: Tucker, Ks. Univ. Sci. Bull., 4: 60.
1959 *Proba distanti*: Carvalho, Arq. Mus. Nac., 48: 244.
1969 *Proba scutellatus*: Maldonado, Univ. Puerto Rico Agr. Exp. Stn. Tech. Pap., 45: 50.
Distribution: Ariz., Col., Ks., Tex. (Mexico to Panama).
Note: *Lygus scutellatus* Distant became a secondary homonym when *Orthops scutellatus* Uhler (which see) was temporarily placed in *Lygus*.

Proba hyalina Maldonado, 1969

1969 *Proba hyalina* Maldonado, Univ. Puerto Rico Agr. Exp. Stn. Tech. Pap., 45: 50. [Puerto Rico].
1982 *Proba hyalina*: Henry and Wheeler, Fla. Ent., 65: 238.
Distribution: Fla. (Puerto Rico).

Proba sallei (Stål), 1862

1862 *Lygus sallei* Stål, Stett. Ent. Zeit., 23(7-9): 321. [Mexico].
1952 *Proba sallei*: Carvalho, Bol. Mus. Nac. Zool., 118: 14.
Distribution: Ariz., Cal., Col., Id., N.M., Tex. (Mexico, Central America).

Proba vittiscutis (Stål), 1860

1860 *Deraeocoris vittiscutis* Stål, Öfv. K. Svens. Vet.-Akad. Förh., 2(7): 48. [Brazil].
1884 *Lygus cristatus* Distant, Biol. Centr.-Am., Rhyn., 1: 254. [Guatemala, Mexico, Panama]. Synonymized by Carvalho, 1952, Bol. Mus. Nac., 118: 5.
1927 *Lygus cristatus*: Knight, Can. Ent., 59: 43.
1952 *Proba vittiscutis*: Carvalho, Bol. Mus. Nac., 118: 5.
Distribution: Tex. (Mexico to South America).

Genus *Pycnocoris* Van Duzee, 1914

1914 *Pycnocoris* Van Duzee, Trans. San Diego Soc. Nat. Hist., 2: 23. Type-species: *Pycnocoris ursinus* Van Duzee, 1914. Monotypic.

Pycnocoris ursinus Van Duzee, 1914

1914 *Pycnocoris ursinus* Van Duzee, Trans. San Diego Soc. Nat. Hist., 2: 24. [Cal.].
Distribution: Cal.

Genus *Rhasis* Distant, 1893

1893 *Rhasis* Distant, Biol. Centr.-Am., Rhyn., 1: 436. Type-species: *Rhasis amplificatus* Distant, 1893. Monotypic.

Rhasis leviscutatus (Knight), 1925

1925 *Neocapsus cuneatus* var. *leviscutatus* Knight, Ent. News, 36: 79. [Miss.].

1941 *Neocapsus cuneatus*: Knight, Ill. Nat. Hist. Surv. Bull., 22: 147 (in part).

1955 *Rhasis laeviscutatus* [*sic*]: Carvalho, Rev. Chil. Ent., 4: 226.

1983 *Neocapsus cuneatus*: Wheeler et al., Trans. Am. Ent. Soc., 109: 148.

Distribution: Ill., Miss., Mo., N.C., Ok., Tex., W.Va.

Note: Carvalho (1955, above) clarified confusion between this species and *Neocapsus cuneatus* Distant (which see).

Genus *Salignus* Kelton, 1955

1955 *Salignus* Kelton, Can. Ent., 87: 283. Type-species: *Lygus distinguendus* Reuter, 1875. Original designation.

Salignus distinguendus (Reuter), 1875

1875 *Lygus distinguendus* Reuter, Pet. Nouv. Ent., 1(136): 544. [Siberia].

1909 *Lygus distinguendus*: Reuter, Acta Soc. Sci. Fenn., 36(2): 45.

1924 *Lygus distinguendas* [*sic*]: Downes, Proc. Ent. Soc. B.C., 21: 29.

1955 *Salignus distinguendus*: Kelton, Can. Ent., 87: 283.

Distribution: Alk., Alta., B.C., Cal., Col., N.M., Ut., Yuk. (China, Siberia).

Salignus distinguendus distinguendus (Reuter), 1875

1875 *Lygus distinguendus* Reuter, Pet. Nouv. Ent., 1(136): 544. [Siberia].

1917 *Lygus* (*Lygus*) *distinguendus*: Knight, Cornell Agr. Exp. Stn. Bull., 391: 587.

1955 *Salignus distinguendus*: Kelton, Can. Ent., 87: 283.

Distribution: Same as for species.

Salignus distinguendus tahoensis (Knight), 1917

1917 *Lygus* (*Lygus*) *distinguendus* var. *tahoensis* Knight, Cornell Univ. Agr. Exp. Stn. Bull., 391: 587. [Cal.].

1955 *Salignus distinguendus* var. *tahoensis*: Kelton, Can. Ent., 87: 283.

Distribution: Cal.

Genus *Stenotus* Jakovlev, 1877

1877 *Stenotus* Jakovlev, Bull. Soc. Nat. Mosc., 52(1): 288. Type-species: *Stenotus sareptanus* Jakovlev, 1877, a junior synonym of *Lygaeus binotatus* Fabricius, 1794. Designated by Kirkaldy, 1906, Trans. Am. Ent. Soc., 32: 139.

1858 *Oncognathus* Fieber, Wien. Ent. Monat., 2: 303. Type-species: *Lygaeus binotatus* Fabricius, 1794. Monotypic. Synonymized by Reuter, 1896, Acta Soc. Sci. Fenn., 33(2): 122.

Stenotus binotatus (Fabricius), 1794

1794 *Lygaeus binotatus* Fabricius, Ent. Syst., 4: 172. [Europe].

1886 *Oncognathus binotatus*: Uhler, Check-list Hem. Het., p. 18.

1896 *Stenotus binotatus*: Reuter, Acta Soc. Sci. Fenn. 33(2): 123.

Distribution: Ark., B.C., Conn., Ga., Ia., Ill., Ind., Ks., Man., Mass., Md., Mich., Minn., Mo., N.C., N.H., N.J., N.S., N.Y., Oh., Ont., Ore., Pa., Que., Tenn., W.Va., Wash., Wis., Wyo. (Africa, Asia, Europe, New Zealand).

Genus *Stittocapsus* Knight, 1942

1942 *Stittocapsus* Knight, Ent. News, 53: 156. Type-species: *Stittocapsus franseriae* Knight, 1942. Monotypic.

Stittocapsus franseriae Knight, 1942
1942 *Stittocapsus franseriae* Knight, Ent. News, 53: 156. [Ariz.].
Distribution: Ariz., Cal., Ut.

Genus *Taedia* Distant, 1883

1883 *Taedia* Distant, Biol. Centr.-Am., Rhyn., 1: 262. Type-species: *Taedia bimaculata* Distant, 1883. Monotypic.

1883 *Paracalocoris* Distant, Biol. Centr.-Am., Rhyn., 1: 263. Type-species: *Calocoris jurgiosus* Stål, 1862. Designated by Distant, 1904, Fauna Brit. India, Rhyn., 2: 449. Synonymized by Carvalho, 1952, An. Acad. Brasil. Ci., 24: 93.

Note: Species in this genus are extremely variable and Latin names have been assigned to many of the different color forms. As in *Metriorrhycomiris dislocatus* (Say), these color morphs have no geographic significance; therefore, we consider them synonyms. The genus is in need of revision. Knight (1930, An. Ent. Soc. Am., 23: 825-827) gave a key to the North American species.

Taedia adusta (McAtee), 1917
1917 *Paracalocoris adustus* McAtee, An. Ent. Soc. Am., 9: 377. [N.J.].
1959 *Taedia adustus*: Carvalho, Arq. Mus. Nac., 48: 258.
1973 *Taedia adusta*: Steyskal, Stud. Ent., 16: 208.
Distribution: Mich., N.J.

Taedia albifacies (Knight), 1930
1930 *Paracalocoris albifacies* Knight, An. Ent. Soc. Am., 23: 814. [Ariz.].
1959 *Taedia albifacies*: Carvalho, Arq. Mus. Nac., 48: 258.
Distribution: Ariz.

Taedia breviata (Knight), 1926
1926 *Paracalocoris breviatus* Knight, An. Ent. Soc. Am., 19: 372. [Ala.].
1926 *Paracalocoris heidemanni*: Blatchley, Het. E. N. Am., p. 697.
1959 *Taedia breviatus*: Carvalho, Arq. Mus. Nac., 48: 258.
1973 *Taedia breviata*: Steyskal, Stud. Ent., 16: 208.
Distribution: Ala.

Taedia casta (McAtee), 1917
1917 *Paracalocoris colon* var. *castus* McAtee, An. Ent. Soc. Am., 9: 382. [Md.].
1917 *Paracalocoris colon* var. *amiculus* McAtee, An. Ent. Soc. Am., 9: 384. [Md.]. New Synonymy.
1917 *Paracalocoris colon* var. *colonus* McAtee, An. Ent. Soc. Am., 9: 383. [Va.]. New Synonymy.
1930 *Paracalocoris castus*: Knight, An. Ent. Soc. Am., 23: 823.
1930 *Paracalocoris castus* var. *amiculus*: Knight, An. Ent. Soc. Am., 23: 824.

1930 *Paracalocoris castus* var. *colonus*: Knight, An. Ent. Soc. Am., 23: 823.
1959 *Taedia castus*: Carvalho, Arq. Mus. Nac., 48: 259.
1959 *Taedia castus* var. *amiculus*: Carvalho, Arq. Mus. Nac., 48: 259.
1959 *Taedia castus* var. *colonus*: Carvalho, Arq. Mus. Nac., 48: 259.
1973 *Taedia casta*: Steyskal, Stud. Ent., 16: 208.
Distribution: Col., D.C., Fla., Ill., Ind., Ky., Mass., Md., Mich., Mo., N.J., N.Y., Ont., Pa., Va., Vt., W.Va.

Taedia celtidis (Knight), 1930
1930 *Paracalocoris celtidis* Knight, An. Ent. Soc. Am., 23: 810. [Ia.].
1959 *Taedia celtidis*: Carvalho, Arq. Mus. Nac., 48: 259.
Distribution: Ia., Ill., Miss.

Taedia colon (Say), 1832
1832 *Capsus colon* Say, Descrip. Het. Hem. N. Am., p. 25. [Ind.].
1886 *Phytocoris colon*: Uhler, Check-list Hem. Het., p. 18.
1904 *Phytocorus* [*sic*] *colon*: Wirtner, An. Carnegie Mus., 3: 199.
1909 *Paracalocoris colon*: Reuter, Acta. Soc. Sci. Fenn., 36(2): 39.
1959 *Taedia colon*: Carvalho, Arq. Mus. Nac., 48: 259.
Distribution: Col., D.C., Ill., Ind., Ky., Mass., Md., Mich., Mo., N.J., N.Y., Ont., Pa., Va., Vt., W.Va., Wis.

Taedia deletica (Reuter), 1909
1909 *Paracalocoris deleticus* Reuter, Acta. Soc. Sci. Fenn., 36(2): 39. [Col.].
1959 *Taedia deleticus*: Carvalho, Arq. Mus. Nac., 48: 259.
1973 *Taedia deletica*: Steyskal, Stud. Ent., 16: 208.
Distribution: Col., N.M., Tex.

Taedia evonymi (Knight), 1930
1930 *Paracalocoris evonymi* Knight, An. Ent. Soc. Am., 23: 812. [Ia.].
1959 *Taedia evonymi*: Carvalho, Arq. Mus. Nac., 48: 259.
Distribution: Ill., Ia., Mo., N.Y., Oh.

Taedia externa (Herrich-Schaeffer), 1845
1845 *Capsus externus* Herrich-Schaeffer, Wanz. Ins., 8: 16. [Fla.].
1873 *Capsus incisus* Walker, Cat. Hem. Brit. Mus., 6: 92. [Fla.]. Synonymized by Knight, 1926, An. Ent. Soc. Am., 19: 376.
1904 *Resthenia incisa*: Distant, An. Mag. Nat. Hist., (7)13: 108.
1917 *Platytylellus incisus*: Van Duzee, Univ. Cal. Publ. Ent., 2: 312.
1926 *Paracalocoris externus*: Knight, An. Ent. Soc. Am., 19: 376.
1926 *Paracalocoris externus* var. *notatus* Knight, Ent. News, 37: 262. [Fla.]. New Synonymy.
1926 *Paracalocoris novellus* Blatchley, Ent. News, 37: 261. [Fla.]. Synonymized by Blatchley, 1926, Het. E. N. Am., p. 695.
1926 *Paracalocoris externus* var. *scissus* Knight, Ent. News, 37: 261. [Fla.]. New Synonymy.
1926 *Paracalocoris externus* var. *solutus* Knight, Ent. News, 37: 261. [Fla.]. New Synonymy.
1926 *Paracalocoris externus* var. *totus* Knight, Ent. News, 37: 262. [Fla.]. New Synonymy.
1959 *Taedia externus*: Carvalho, Arq. Mus. Nac., 48: 259
1959 *Taedia externus* var. *notatus*: Carvalho, Arq. Mus. Nac., 48: 260.

1959 *Taedia externus* var. *scissus*: Carvalho, Arq. Mus. Nac., 48: 260.
1959 *Taedia externus* var. *solutus*: Carvalho, Arq. Mus. Nac., 48: 260.
1959 *Taedia externus* var. *totus*: Carvalho, Arq. Mus. Nac., 48: 260.
1973 *Taedia externa*: Steyskal, Stud. Ent., 16: 208.
Distribution: Fla., Ga.

Taedia fasciola (Knight), 1930
1930 *Paracalocoris fasciolus* Knight, An. Ent. Soc. Am., 23: 819. [Ariz.].
1959 *Taedia fasciolus*: Carvalho, Arq. Mus. Nac., 48: 260.
1973 *Taedia fasciola*: Steyskal, Stud. Ent., 16: 208.
Distribution: Ariz.

Taedia floridana (Knight), 1926
1926 *Paracalocoris floridanus* Knight, An. Ent. Soc. Am., 19: 373. [Fla.].
1959 *Taedia floridanus*: Carvalho, Arq. Mus. Nac., 48: 260.
1973 *Taedia floridana*: Steyskal, Stud. Ent., 16: 208.
Distribution: Fla.

Taedia gleditsiae (Knight), 1926
1926 *Paracalocoris gleditsiae* Knight, An. Ent. Soc. Am., 19: 370. [Ia.].
1959 *Taedia gleditsiae*: Carvalho, Arq. Mus. Nac., 48: 260.
Distribution: Ia., Ill., Ind., Mo., Oh., Ont., Pa., W.Va., Wis.
Note: Wheeler and Henry (1976, An. Ent. Soc. Am., 69: 1096) described the fifth-instar nymph.

Taedia hawleyi (Knight), 1917
1917 *Paracalocoris hawleyi* Knight, *In* McAtee, An. Ent. Soc. Am., 9: 377. [N.Y.].
1917 *Paracalocoris hawleyi* var. *ancora* Knight, *In* McAtee, An. Ent. Soc. Am., 9: 378. [N.Y.]. New Synonymy.
1917 *Paracalocoris hawleyi* var. *fissus* McAtee, An. Ent. Soc. Am., 9: 379. [Md.]. New Synonymy.
1959 *Taedia hawleyi*: Carvalho, Arq. Mus. Nac., 48: 260
1959 *Taedia hawleyi* var. *ancora*: Carvalho, Arq. Mus. Nac., 48: 260.
1959 *Taedia hawleyi* var. *fissus*: Carvalho, Arq. Mus. Nac., 48: 260.
Distribution: D.C., Ind., Mass., Md., Me., N.Y., Oh.
Note: Hawley (1917, J. Econ. Ent., 10: 545-552) described and illustrated the immature stages and reported on its injury to hops, *Humulus lupulus* L.

Taedia heidemanni (Reuter), 1909
1909 *Paracalocoris heidemanni* Reuter, Acta Soc. Sci. Fenn., 36(2): 40. [W.Va.].
1917 *Paracalocoris heidemanni* var. *ablutus* McAtee, An. Ent. Soc. Am., 9: 386. [Md.]. New Synonymy.
1959 *Taedia heidemanni* Carvalho, Arq. Mus. Nac., 48: 260.
1959 *Taedia heidemanni* var. *ablutus*: Carvalho, Arq. Mus. Nac., 48: 260.
Distribution: Md., Va., W.Va.

Taedia johnstoni (Knight), 1930
1930 *Paracalocoris johnstoni* Knight, An. Ent. Soc. Am., 23: 815. [Tex.].
1959 *Taedia johnstoni*: Carvalho, Arq. Mus. Nac., 48: 260.
Distribution: Miss., Tex.

Taedia limba (McAtee), 1917
1917 *Paracalocoris limbus* McAtee, An. Ent. Soc. Am., 9: 380. [Ga.].

1930 *Paracalocoris limbus* var. *otiosus* Knight, An. Ent. Soc. Am., 23: 822. [Mass.]. New Synonymy.
1930 *Paracalocoris limbus* var. *suffusus* Knight, An. Ent. Soc. Am., 23: 822. [Md.]. New Synonymy.
1959 *Taedia limbus*: Carvalho, Arq. Mus. Nac., 48: 261.
1959 *Taedia limbus* var. *otiosus*: Carvalho, Arq. Mus. Nac., 48: 261.
1959 *Taedia limbus* var. *suffusus*: Carvalho, Arq. Mus. Nac., 48: 261.
Distribution: Ala., Ga., Mass., Md., N.Y., Oh., Pa., Va.

Taedia maculosa (Knight), 1930
1910 *Paracalocoris jurgiosus*: Banks, Cat. Nearc. Hem.-Het., p. 47.
1930 *Paracalocoris maculosus* Knight, An. Ent. Soc. Am., 23: 816. [Ariz.].
1959 *Taedia jurgiosus*: Carvalho, Arq. Mus. Nac., 48: 261 (in part).
1959 *Taedia maculosus*: Carvalho, Arq. Mus. Nac., 48: 261.
1973 *Taedia maculosa*: Steyskal, Stud. Ent., 16: 208.
Distribution: Ariz., Tex.
Note: We follow Knight (1930, above, pp. 810, 817), who considered all U.S. records of *Taedia jurgiosus* (Stål), 1862, misidentifications of *T. maculosa*. Carvalho (1959, Arq. Mus. Nac., 48: 261) listed *T. jurgiosa* from Tex., apparently overlooking Knight's discussion; the Tex. record is here transferred to *T. maculosa* until this confusion is clarified.

Taedia marmorata (Uhler), 1894
1894 *Paracalocoris marmoratus* Uhler, Proc. Cal. Acad. Sci., 4: 263. [Mexico].
1917 *Horcias marmoratus*: Van Duzee, Univ. Cal. Publ. Ent., 2: 335.
1917 *Paracalocoris acceptus* McAtee, An. Ent. Soc. Am., 9: 389. [N.M.]. Synonymized by Knight, 1926, An. Ent. Soc. Am., 19: 374.
1959 *Taedia marmoratus*: Carvalho, Arq. Mus. Nac., 48: 261.
1973 *Taedia marmorata*: Steyskal, Stud. Ent., 16: 208.
Distribution: Ariz., N.M., Tex. (Mexico).

Taedia multisignata (Reuter), 1909
1909 *Paracalocoris multisignatus* Reuter, Acta Soc. Sci. Fenn., 36(2): 40. [D.C.].
1959 *Taedia multisignatus*: Carvalho, Arq. Mus. Nac., 48: 261.
1973 *Taedia multisignata*: Steyskal, Stud. Ent., 16: 208.
Distribution: D.C., Ga., Md., Miss., Mo., N.C., N.Y., Oh., Ok., Tex., Va., W.Va.

Taedia nicholi (Knight), 1926
1926 *Paracalocoris nicholi* Knight, An. Ent. Soc. Am., 19: 372. [Ariz.].
1959 *Taedia nicholi*: Carvalho, Arq. Mus. Nac., 48: 262.
Distribution: Ariz.

Taedia pallidula (McAtee), 1917
1917 *Paracalocoris hawleyi* var. *pallidulus* McAtee, An. Ent. Soc. Am., 9: 380. [N.Y.].
1930 *Paracalocoris pallidulus*: Knight, An. Ent. Soc. Am., 23: 822.
1930 *Paracalocoris pallidulus* var. *albigulus* Knight, An. Ent. Soc. Am., 23: 823. [Minn.]. New Synonymy.
1959 *Taedia pallidulus*: Carvalho, Arq. Mus. Nac., 48: 262.
1959 *Taedia pallidulus* var. *albigulus*: Carvalho, Arq. Mus. Nac., 48: 262.
1973 *Taedia pallidula*: Steyskal, Stud. Ent., 16: 208.
Distribution: Man., Minn., N.D., N.Y., Oh., Ont., Que., Sask., Wis.

Taedia parenthesis (Knight), 1930

1930 *Paracalocoris parenthesis* Knight, An. Ent. Soc. Am., 23: 817. [Ariz.].

1959 *Taedia parenthesis*: Carvalho, Arq. Mus. Nac., 48: 262.

Distribution: Ariz.

Taedia salicis (Knight), 1926

1922 *Paracalocoris adustus*: Hussey, Occas. Pap. Mus. Zool., Univ. Mich., 118: 28.

1926 *Paracalocoris salicis* Knight, An. Ent. Soc. Am., 19: 367. [Minn.].

1959 *Taedia salicis*: Carvalho, Arq. Mus. Nac., 48: 262.

Distribution: Col., Ia., Ill., Ind., Mich., Minn., Mo., Neb., Pa., S.D.. Wis.

Taedia scrupea (Say), 1832

1832 *Capsus scrupeus* Say, Descrip. Het. Hem. N. Am., p. 23. [U.S.].

1850 *Capsus tetrastigma* Herrich-Schaeffer, Wanz. Ins., 9: 166. [Mexico]. Synonymized by Carvalho, 1959, Arq. Mus. Nac., 48: 262. See note below.

1876 *Phytocoris scrupeus*: Uhler, Bull. U.S. Geol. Geogr. Surv. Terr., 1: 317.

1909 *Paracalocoris scrupeus*: Reuter, Acta. Soc. Sci. Fenn., 36(2): 39.

1917 *Paracalocoris scrupeus* var. *ardens* McAtee, An. Ent. Soc. Am., 9: 375. [Ill.]. New Synonymy.

1917 *Paracalocoris scrupeus* var. *bicolor* McAtee, An. Ent. Soc. Am., 9: 376. [Tex.]. New Synonymy.

1917 *Paracalocoris scrupeus* var. *bidens* McAtee, An. Ent. Soc. Am., 9: 374. [Md.]. Synonymized by Larochelle, 1984, Fabreries, Suppl. 3, p. 270.

1917 *Paracalocoris scrupeus* var. *compar* McAtee, An. Ent. Soc. Am., 9: 373. [Neb.]. New Synonymy.

1917 *Paracalocoris scrupeus* var. *cunealis* McAtee, An. Ent. Soc. Am., 9: 373. [N.Y.]. Synonymized by Larochelle, Fabreries, Suppl. 3, p. 270.

1917 *Paracalocoris scrupeus* var. *delta* McAtee, An. Ent. Soc. Am., 9: 376. [Ill.]. Synonymized by Larochelle, 1984, Fabreries, Suppl. 3, p. 270.

1917 *Paracalocoris scrupeus* var. *diops* McAtee, An. Ent. Soc. Am., 9: 372. [N.J.]. New Synonymy.

1917 *Paracalocoris scrupeus* var. *lucidus* McAtee, An. Ent. Soc. Am., 9: 373. [N.Y.]. New Synonymy.

1917 *Paracalocoris scrupeus* var. *nubilus* McAtee, An. Ent. Soc. Am., 9: 375. [N.Y.]. New Synonymy.

1917 *Paracalocoris scrupeus* var. *par* McAtee, An. Ent. Soc. Am., 9: 373. [N.Y.]. New Synonymy.

1917 *Paracalocoris scrupeus* var. *percursus* McAtee, An. Ent. Soc. Am., 9: 372. [Md.]. New Synonymy.

1917 *Paracalocoris scrupeus* var. *rubidus* McAtee, An. Ent. Soc. Am., 9: 371. [N.Y.]. New Synonymy.

1917 *Paracalocoris scrupeus* var. *sordidus* McAtee, An. Ent. Soc. Am., 9: 374. [N.Y.]. New Synonymy.

1917 *Paracalocoris scrupeus* var. *triops* McAtee, An. Ent. Soc. Am., 9: 371. [N.Y.]. New Synonymy.

1917 *Paracalocoris scrupeus* var. *varius* McAtee, An. Ent. Soc. Am., 9: 376. [N.Y.]. New Synonymy.

1929 *Phytocoris schupeus* [*sic*]: Torre-Bueno, Bull. Brook. Ent. Soc., 24: 336.

1930 *Paracalocoris scrupens* [*sic*]: Stewart, J. N.Y. Ent. Soc., 38: 44.

1952 *Taedia scrupeus*: Carvalho, Bol. Mus. Nac., 118: 15.

1959 *Taedia scrupeus* var. *ardens*: Carvalho, Arq. Mus. Nac., 48: 263.
1959 *Taedia scrupeus* var. *bicolor*: Carvalho, Arq. Mus. Nac., 48: 263.
1959 *Taedia scrupeus* var. *bidens*: Carvalho, Arq. Mus. Nac., 48: 263.
1959 *Taedia scrupeus* var. *compar*: Carvalho, Arq. Mus. Nac., 48: 263.
1959 *Taedia scrupeus* var. *cunealis*: Carvalho, Arq. Mus. Nac., 48: 263.
1959 *Taedia scrupeus* var. *delta*: Carvalho, Arq. Mus. Nac., 48: 263.
1959 *Taedia scrupeus* var. *diops*: Carvalho, Arq. Mus. Nac., 48: 263.
1959 *Taedia scrupeus* var. *lucidus*: Carvalho, Arq. Mus. Nac., 48: 263.
1959 *Taedia scrupeus* var. *nubilus*: Carvalho, Arq. Mus. Nac., 48: 263.
1959 *Taedia scrupeus* var. *par*: Carvalho, Arq. Mus. Nac., 48: 263.
1959 *Taedia scrupeus* var. *percursus*: Carvalho, Arq. Mus. Nac., 48: 263.
1959 *Taedia scrupeus* var. *rubidus*: Carvalho, Arq. Mus. Nac., 48: 263.
1959 *Taedia scrupeus* var. *sordidus*: Carvalho, Arq. Mus. Nac., 48: 263.
1959 *Taedia scrupeus* var. *triops*: Carvalho, Arq. Mus. Nac., 48: 263.
1959 *Taedia scrupeus* var. *varius*: Carvalho, Arq. Mus. Nac., 48: 263.
1973 *Taedia scrupea*: Steyskal, Stud. Ent., 16: 208.
Distribution: Col., Conn., D.C., Ga., Ill., Ind., Ks., Ky., Mass., Md., Mo., N.J., N.M., N.Y., Neb., Oh., Ont., Pa., Que., R.I., Tex., Va., Vt., W.Va., Wis. (Mexico).
Note: Van Duzee (1917, Univ. Cal. Publ. Ent., 2: 326) first noted that the description [or figure] of *Capsus tetrastigmus* in the original description was reversed with *Prepops* [as *Capsus*] *divisus* (Herrich-Schaeffer) (which see). We follow Carvalho's (1959, Arq. Mus. Nac., 48: 262) interpretation of this confusion.

Taedia severini (Knight), 1926
1926 *Paracalocoris severini* Knight, An. Ent. Soc. Am., 19: 369. [S.D.].
1926 *Paracalocoris severini* var. *modestus* Knight, An. Ent. Soc. Am., 19: 369. [S.D.]. New Synonymy.
1926 *Paracalocoris severini* var. *nigriclavus* Knight, An. Ent. Soc. Am., 19: 370. [S.D.]. New Synonymy.
1959 *Taedia severini*: Carvalho, Arq. Mus. Nac., 48: 263.
1959 *Taedia severini* var. *modestus*: Carvalho, Arq. Mus. Nac., 48: 263.
1959 *Taedia severini* var. *nigriclavus*: Carvalho, Arq. Mus. Nac., 48: 263.
Distribution: S.D.

Taedia trivitta (Knight), 1930
1926 *Paracalocoris trivittatus* Knight, An. Ent. Soc. Am., 19: 371. [Miss.]. Preoccupied.
1930 *Paracalocoris trivittis* Knight, An. Ent. Soc. Am., 23: 812. New name for *P. trivittatus* Knight.
1959 *Taedia trivittis*: Carvalho, Arq. Mus. Nac., 48: 263.
1973 *Taedia trivitta*: Steyskal, Stud. Ent., 16: 208.
Distribution: Miss.

Taedia virgulata (Knight), 1930
1930 *Paracalocoris virgulatus* Knight, An. Ent. Soc. Am., 23: 820. [Ariz.].
1930 *Paracalocoris virgulatus* var. *stigmosus* Knight, An. Ent. Soc. Am., 23: 821. [Tex.]. New Synonymy.
1959 *Taedia virgulatus*: Carvalho, Arq. Mus. Nac., 48: 263
1959 *Taedia virgulatus* var. *stigmosus*: Carvalho, Arq. Mus. Nac., 48: 263.
1973 *Taedia virgulata*: Steyskal, Stud. Ent., 16: 208.
Distribution: Ariz., Tex.

Genus *Taylorilygus* Leston, 1952

1952 *Lygus* (*Taylorilygus*) Leston, Ent. Gaz., 3: 219. Type-species: *Lygus simonyi* Reuter, 1903. Original designation.

1955 *Taylorilygus*: Kelton, Can. Ent., 87: 281.

Taylorilygus pallidulus (Blanchard), 1852

1852 *Phytocoris pallidulus* Blanchard, *In* Gay's Hist. Fis. Pol. Chile, 7: 183. [Chile].

1861 *Lygus apicalis* Fieber, Europ. Hem., p. 275. [Spain]. Synonymized by Carvalho, 1959, Arq. Mus. Nac., 48: 265.

1876 *Lygus Carolinae* [*sic*] Reuter, Öfv. K. Svens. Vet.-Akad. Förh., 32(9): 71. ["Carolina"]. Synonymized by Carvalho, 1959, Arq. Mus. Nac., 48: 265.

1876 *Lygus* (*Lygus*) *prasinus* Reuter, Öfv. K. Svens. Vet.-Akad. Förh, 32(9): 72. [Tex.]. Synonymized by Reuter, 1909, Acta Soc. Sci. Fenn., 36(2): 43.

1909 *Lygus apicalis*: Reuter, Acta Soc. Sci. Fenn., 36(2): 43.

1909 *Lygus apicalis* var. *prasinus*: Reuter, Acta Soc. Sci. Fenn., 32(9): 72.

1914 *Lygus contaminatus*: Van Duzee, Trans. S. Diego Soc. Nat. Hist., 2: 20.

1917 *Lygus olivaceus* var. *viridiusculus* Knight, Cornell Univ. Agr. Exp. Stn. Bull., 391: 600. [Mass.]. Synonymized by Henry, 1985, J. N.Y. Ent. Soc., 93: 1131.

1917 *Lygus* (*Neolygus*) *carolinae*: Knight, Cornell Univ. Agr. Exp. Stn. Bull., 391: 638.

1952 *Lygus pallidulus*: Carvalho, Rev. Française d'Ent., 19: 185.

1952 *Lygus* (*Taylorilygus*) *apicalis*: Leston, Ent. Gaz., 3: 214.

1955 *Taylorilygus apicalis*: Kelton, Can. Ent., 87: 281.

1959 *Taylorilygus pallidulus*: Carvalho, Arq. Mus. Nac., 48: 265.

Distribution: Ala., Ark., Cal., Conn., Fla., Ga., La., Mass., Md., Me., Miss., Mo., N.C., Ok., Tex. (Africa, Asia, Europe, Mexico to South America, West Indies).

Note: Carvalho (1952, above, 19: 185) suggested that *Lygus apicalis* Fieber, 1861, was a junior synonym of *Taylorilygus pallidulus*, but did not make this synonymy formal until 1959 (above, 48: 265). Snodgrass et al. (1984, Proc. Ent. Soc. Wash., 86: 857-859) gave an extensive host list for Miss. Early records of the European species *Lygocoris contaminatus* (Fallén)(which see) and *Lygocoris lucorum* (Meyer-Dür)(which see) were referred to *Taylorilygus pallidulus* [as *Lygus apicalis*] by Knight (1917, above) and Blatchley (1926, Het. E. N. Am., p. 763); confirmed records for these two holarctic species were given by Kelton (1971, Mem. Ent. Soc. Can., 83: 8, 11).

Genus *Tropidosteptes* Uhler, 1878

1878 *Tropidosteptes* Uhler, Proc. Boston Soc. Nat. Hist., 19: 404. Type-species: *Tropidosteptes cardinalis* Uhler, 1878. Monotypic.

1884 *Neoborus* Distant, Biol. Centr.-Am., Rhyn., 1: 276. Type-species: *Neoborus saxeus* Distant, 1884. Designated by Kirkaldy, 1906, Trans. Am. Ent. Soc., 32: 140. Synonymized by Carvalho, 1954, An. Acad. Brasil. Ci., 26: 425.

1908 *Neoborus* (*Xenoborus*) Reuter, *In* Van Duzee, Can. Ent., 40: 112. Type-species: *Neoborus* (*Xenoborus*) *commissuralis* Reuter, 1887. Monotypic. Synonymized by Akingbohungbe et al., 1972, Univ. Wis. Res. Bull., R2396: 3.

1916 *Xenoborus*: Van Duzee, Univ. Cal. Publ. Ent., 1: 208.

Note: Akingbohungbe et al. (1972, Univ. Wis. Res. Bull., R2396: 2-3) treated the genus *Xenoborus* as a synonym without explaining their decision; Kelton (1978, Can. Ent.,

110: 471-473) formally synonymized *Xenoborus* and gave reasons for his action.

Tropidosteptes adeliae (Knight), 1929

1929 *Neoborus adeliae* Knight, Bull. Brook. Ent. Soc., 24: 3. [Tex.].

1929 *Neoborus adelia* [*sic*]: Johnston, Bull. Brook. Ent. Soc., 24: 218.

1959 *Tropidosteptes adeliae*: Carvalho, Arq. Mus. Nac., 48: 270.

Distribution: Tex.

Tropidosteptes adustus (Knight), 1929

1929 *Neoborus adustus* Knight, Bull. Brook. Ent. Soc., 24: 3. [Mo.].

1959 *Tropidosteptes adustus*: Carvalho, Arq. Mus. Nac., 48: 270.

Distribution: Ind., Mo., Pa., Tex.

Tropidosteptes amoenus Reuter, 1909

1887 *Orthops scutellatus*: Provancher, Pet Faune Ent. Can., 3: 124.

1894 *Neoborus saxeus*: Uhler, Proc. Cal. Acad. Sci., (2)4: 164.

1904 *Neoborops vigilax*: Wirtner, An. Carnegie Mus., 3: 197.

1909 *Tropidosteptes amoenus* Reuter, Acta Soc. Sci. Fenn., 36(2): 48. [D.C., Md., N.Y.].

1910 *Tropidosteptes saxeus*: Smith, Ins. N.J., p. 164.

1916 *Neoborus amoenus*: Dickerson and Weiss, J. N.Y. Ent. Soc., 24: 302.

Distribution: Alta., B.C., Conn., D.C., Fla., Ga., Ill., Ind., Ks., Man., Mass., Md., Me., Mo., N.D., N.J., N.S., N.Y., Oh., Ok., Ont., Pa., Que., S.D., Sask., Tex., Vt., W.Va., Wis.

Note: Dickerson and Weiss (1916, J. N.Y. Ent. Soc., 24: 302-306) described and illustrated immature stages. Wheeler and Henry (1977, Great Lakes Ent., 10: 149) clarified the Wirtner (1904, above) record of *Neoborops vigilax* Uhler (which see).

Tropidosteptes amoenus amoenus Reuter, 1909

1909 *Tropidosteptes amoenus* var. *amoena* [*sic*] Reuter, Acta. Soc. Sci. Fenn., 36(2): 49. [D.C., Md., N.Y.].

1973 *Tropidosteptes amoenus amoenus*: Akingbohungbe et al., Univ. Wis. Res. Bull., R2561: 8.

Distribution: Same as for species.

Tropidosteptes amoenus atriscutis (Knight), 1929

1929 *Neoborus amoenus* var. *atriscutis* Knight, Bull. Brook. Ent. Soc., 24: 10. [Ont.].

1959 *Tropidosteptes amoenus* var. *atriscutis*: Carvalho, Arq. Mus. Nac., 48: 270.

Distribution: Ont.

Tropidosteptes amoenus floridanus (Knight), 1929

1929 *Neoborus amoenus floridanus*: Knight, Bull. Brook. Ent. Soc., 24: 9. [Fla.].

1959 *Tropidosteptes amoenus floridanus*, Carvalho, Arq. Mus. Nac., 48: 270.

Distribution: Fla.

Tropidosteptes amoenus plagiatus Reuter, 1909

1909 *Tropidosteptes amoenus* var. *plagiata* Reuter, Acta. Soc. Sci. Fenn., 36(2): 49. [Locality not given.].

Distribution: No locality given.

Tropidosteptes amoenus scutellaris Reuter, 1909

1909 *Tropidosteptes amoenus* var. *scutellaris* Reuter, Acta. Soc. Sci. Fenn., 36(2): 49. [Ks.].

1923 *Neoborus amoenus scutellaris*: Knight, Conn. Geol. Nat. Hist. Surv. Bull., 34: 563.

Distribution: Ill., Ind., Ks., Mass., Md., N.H., N.Y., S.D., Vt.

Tropidosteptes amoenus signatus Reuter, 1909

1909 *Tropidosteptes amoenus* var. *signata* Reuter, Acta. Soc. Sci. Fenn., 36(2): 49. [D.C.].

1923 *Neoborus amoenus* var. *signatus*: Knight, Conn. Geol. Nat. Hist. Surv. Bull., 34: 562.

1973 *Tropidosteptes amoenus signatus*: Akingbohungbe et al., Univ. Wis. Res. Bull. R2561: 8.

Distribution: D.C., Ill., Md., Minn., N.Y., Que., Wis.

Tropidosteptes atratus (Knight), 1929

1929 *Neoborus atratus* Knight, Bull. Brook. Ent. Soc., 24: 2. [N.M.].

1959 *Tropidosteptes atratus*: Carvalho, Arq. Mus. Nac., 48: 217.

Distribution: N.M.

Tropidosteptes brooksi Kelton, 1978

1978 *Tropidosteptes brooksi* Kelton, Can. Ent., 110: 471. [Sask.].

Distribution: Man., Ont., Que., Sask.

Tropidosteptes canadensis Van Duzee, 1912

1912 *Tropidosteptes canadensis* Van Duzee, Bull. Buff. Soc. Nat. Sci., 10: 486. [Ont.].

1916 *Neoborus canadensis*: Van Duzee, Univ. Cal. Publ. Ent., 1: 237.

Distribution: Conn., Ia., Ill., Ind., Man., Mass., Md., Mich., Minn., Miss., N.Y., Oh., Ont., Pa., Que., S.D., Tex., Wis.

Note: The Cal. record for this species, questioned by Blatchley (1926, Het. E. N. Am., p. 790), is dropped from the present list.

Tropidosteptes cardinalis Uhler, 1878

1869 *Capsis* [*sic*] *cardinalis* Rathvon, Hist. Lancaster Co., Pa., p. 549. *Nomen nudum.*

1878 *Tropidosteptes cardinalis* Uhler, Proc. Boston Soc. Nat. Hist., 19: 404. [Mass.].

Distribution: Conn., Fla., Ia., Ill., Ind., Mass., Mich., Miss., Mo., N.H., N.J., N.Y., Oh., Ont., Pa., Tex., Vt., W.Va.

Note: Leonard (1916, Psyche, 23: 1-3) described and illustrated the nymphal stages.

Tropidosteptes chionanthi (Knight), 1927

1927 *Xenoborus chionanthi* Knight, Proc. Biol. Soc. Wash., 40: 14. [Md.].

1978 *Tropidosteptes chionanthi*: Kelton, Can. Ent., 110: 471.

Distribution: Md., Va.

Tropidosteptes chionanthi chionanthi (Knight), 1927

1927 *Xenoborus chionanthi* var. *chionanthi* Knight, Proc. Biol. Soc. Wash., 40: 14. [Md.].

1978 *Tropidosteptes chionanthi* var. *chionanthi*: Kelton, Can. Ent., 110: 471.

Distribution: Md., Va.

Tropidosteptes chionanthi nigrellus (Knight), 1927

1927 *Xenoborus chionanthi* var. *nigrellus* Knight, Proc. Biol. Soc. Wash., 40: 15. [Md.].

1978 *Tropidosteptes chionanthi* var. *nigrellus*: Kelton, Can. Ent., 110: 471.

Distribution: Md.

Tropidosteptes commissuralis (Reuter), 1908

1887 *Sthenarops chloris*: Provancher, Pet. Faune Ent. Can., 3: 134.

1908 *Neoborus* (*Xenoborus*) *commissuralis* Reuter, *In* Van Duzee, Can. Ent., 40: 112. [N.Y., Ont., Que.].

1909 *Tropidosteptes commissuralis*: Reuter, Acta. Soc. Sci. Fenn., 36(2): 51.

1917 *Xenoborus commissuralis*: Knight, Bull. Brook. Ent. Soc., 12: 82.

1950 *Zenoborus* [*sic*] *commissuralis*: Pielou, Can. Ent., 82: 150, 152.

Distribution: B.C., Ia., Ill., Ind., Man., Minn., N.S., N.Y., Ont., Que.

Tropidosteptes fasciolus (Knight), 1929
1929 *Neoborus fasciolus* Knight, Bull. Brook. Ent. Soc., 24: 8. [Cal.].
1959 *Tropidosteptes fasciolus*: Carvalho, Arq. Mus. Nac., 48: 271.
Distribution: Cal.

Tropidosteptes flaviceps (Knight), 1929
1929 *Neoborus flaviceps* Knight, Bull. Brook. Ent. Soc., 24: 9. [Ariz.].
1959 *Tropidosteptes flaviceps*: Carvalho, Arq. Mus. Nac., 48: 271.
Distribution: Ariz.

Tropidosteptes geminus (Say), 1832
1832 *Capsus geminus* Say, Descrip. Het. Hem. N. Am., p. 24. [Ind.].
1890 *Poecilocapsus geminus*: Atkinson, J. Asiatic Soc. Bengal, 58(2): 94.
1912 *Tropidosteptes geminus*: Van Duzee, Bull. Buff. Soc. Nat. Sci., 10: 487.
1917 *Neoborus geminus*: Van Duzee, Univ. Cal. Publ. Ent., 2: 351.
Distribution: Conn., Ill., Ind., La., Mass., N.Y., Oh., Ont., Que., W.Va., Wis.

Tropidosteptes glaber (Knight), 1923
1923 *Neoborus glaber* Knight, Conn. Geol. Nat. Hist. Surv. Bull., 34: 563. [N.Y.].
1959 *Tropidosteptes glaber*: Carvalho, Arq. Mus. Nac., 48: 271.
Distribution: Ga., Ia., Ill., Ind., Mich., Minn., Mo., N.D., N.H., N.Y., Oh., Ont., Que., S.D., Tex., W.Va., Wis.

Tropidosteptes illitus (Van Duzee), 1921
1921 *Neoborus illitus* Van Duzee, Proc. Cal. Acad. Sci., (4)11: 120. [Cal.].
1959 *Tropidosteptes illitus*: Carvalho, Arq. Mus. Nac., 48: 272.
Distribution: Cal.
Note: Usinger (1945, J. Econ. Ent., 38: 585-591) studied the biology of this *Fraxinus* pest.

Tropidosteptes imbellis Bliven, 1973
1973 *Tropidosteptes imbellis* Bliven, Occ. Ent., 1(10): 137. [Cal.].
Distribution: Cal.

Tropidosteptes neglectus (Knight), 1917
1917 *Xenoborus neglectus* Knight, Bull. Brook. Ent. Soc., 12: 82. [N.Y.].
1928 *Xenobrus* [*sic*] *neglectus*: Watson, Oh. Biol. Surv. Bull., 16: 36.
1959 *Tropidosteptes neglectus*: Carvalho, Arq. Mus. Nac., 48: 274.
Distribution: Ill., Mich., Miss., N.Y., Oh., Ont., Va., Wis.

Tropidosteptes osmanthicola (Johnston), 1935
1935 *Neoborus osmanthicola* Johnston, Bull. Brook. Ent. Soc., 30: 17. [Miss.].
1959 *Tropidosteptes osmanthicola*: Carvalho, Arq. Mus. Nac., 48: 272.
Distribution: Miss.

Tropidosteptes pacificus (Van Duzee), 1921
1921 *Neoborus pacificus* Van Duzee, Proc. Cal. Acad. Sci., (4)11: 121. [Ore.].
1959 *Tropidosteptes pacificus*: Carvalho, Arq. Mus. Nac., 48: 272.
Distribution: Ariz., Cal., Ore., Pa., Ut., Wash.
Note: Wheeler and Henry (1974, Coop. Econ. Ins. Rept., 24: 588-589) reported this species damaging ash, *Fraxinus* spp., in Pa., where it had been introduced with nursery stock from the Pacific Northwest.

Tropidosteptes palmeri (Reuter), 1908
1887 *Orthops scutellatus*: Provancher, Pet. Faune Ent. Can., 3: 124.
1908 *Neoborus amoenus* var. *palmeri* Reuter, *In* Van Duzee, Can. Ent., 40: 112. [Que.].
1909 *Tropidosteptes amoenus* var. *palmeri*: Reuter, Acta. Soc. Sci. Fenn., 36(2): 49.
1917 *Neoborus palmeri*: Knight, Bull. Brook. Ent. Soc., 12: 81.
1959 *Tropidosteptes palmeri*: Carvalho, Arq. Mus. Nac., 48: 272.
Distribution: Conn., Del., Ill., Ind., Man., Mich., Minn., Mo., N.Y., Pa., Que., S.D., Sask., W.Va.

Tropidosteptes pettiti Reuter, 1909
1887 *Trichia punctulata*: Provancher, Pet. Faune Ent. Can., 3: 133.
1892 *Neoborus pettitii* [*sic*]: Heidemann, Proc. Ent. Soc. Wash., 2: 226.
1905 *Neoborus petitii* [*sic*]: Crevecoeur, Trans. Ks. Acad. Sci., 19: 233.
1909 *Tropidosteptes pettiti* Reuter, Acta. Soc. Sci. Fenn., 36(2): 50. [Pa.].
1917 *Xenoborus pettiti*: Knight, Bull. Brook. Ent. Soc., 12: 82.
1928 *Xenobrus* [*sic*] *pettiti*: Watson, Oh. Biol. Surv. Bull., 16: 36.
Distribution: Conn., D.C., Ia., Ks., La., Man., Mass., Me., Minn., Miss., Mo., N.H., N.J., N.Y., Oh., Ont., Pa., Que.

Tropidosteptes plagifer Reuter, 1909
1909 *Tropidosteptes plagifer* Reuter, Acta Soc. Sci. Fenn., 36(2): 51. [Ont.].
1917 *Xenoborus plagifer*: Knight, Bull. Brook. Ent. Soc., 12: 82.
Distribution: Ill., Man., Minn., N.Y., Ont., Que., Wis.

Tropidosteptes populi (Knight), 1929
1929 *Neoborus populi* Knight, Bull. Brook. Ent. Soc., 24: 4. [Ill.].
1959 *Tropidosteptes populi*: Carvalho, Arq. Mus. Nac., 48: 272.
Distribution: Ill.

Tropidosteptes pubescens (Knight), 1917
1917 *Neoborus pubescens* Knight, Bull. Brook. Ent. Soc., 12: 81. [N.Y.].
1959 *Tropidosteptes pubescens*: Carvalho, Arq. Mus. Nac., 48: 272.
Distribution: Ill., Ind., Man., Mass., Mich., N.H., N.Y., Oh., Ont., Pa., Que., W.Va.

Tropidosteptes quercicola (Johnston), 1939
1939 *Neoborus quercicola* Johnston, Bull. Brook. Ent. Soc., 34: 130. [Tex.].
1959 *Tropidosteptes quercicola*: Carvalho, Arq. Mus. Nac., 48: 272.
Distribution: Tex.

Tropidosteptes rufivenosus (Knight), 1929
1929 *Neoborus rufivenosus* Knight, Bull. Brook. Ent. Soc., 24: 7. [N.M.].
1959 *Tropidosteptes rufivenosus*: Carvalho, Arq. Mus. Nac., 48: 272.
Distribution: N.M.
Note: Through a *lapsus*, Knight in the original description listed the type locality of this species as "Jemez Springs, Arizona"; the holotype (USNM) is clearly labeled as "Jemez Springs, N.M."

Tropidosteptes rufusculus (Knight), 1923
1923 *Neoborus rufusculus* Knight, Conn. Geol. Nat. Hist. Surv. Bull., 34: 564. [N.Y.].
1959 *Tropidosteptes rufusculus*: Carvalho, Arq. Mus. Nac., 48: 273.
Distribution: Ill., Minn., Miss., N.Y., Pa., Tex.

Tropidosteptes saxeus (Distant), 1884
1884 *Neoborus saxeus* Distant, Biol. Centr.-Am., Rhyn., 1: 276. [Mexico].

1959 *Tropidosteptes saxeus*: Carvalho, Arq. Mus. Nac., 48: 273.
Distribution: N.M. (Mexico to Panama).

Tropidosteptes selectus (Knight), 1929
1929 *Xenoborus selectus* Knight, Bull. Brook. Ent. Soc., 24: 10. [Mo.].
1978 *Tropidosteptes selectus*: Kelton, Can. Ent., 110: 471.
Distribution: Mo.

Tropidosteptes setiger Bliven, 1973
1973 *Tropidosteptes setiger* Bliven, Occ. Ent., 1(10): 140. [Cal.].
Distribution: Cal.

Tropidosteptes torosus Bliven, 1973
1973 *Tropidosteptes torosus* Bliven, Occ. Ent., 1(10): 138. [Cal.].
Distribution: Cal.

Tropidosteptes tricolor Van Duzee, 1912
1912 *Tropidosteptes tricolor* Van Duzee, Bull. Buff. Soc. Nat. Sci., 10: 487. [N.J.].
1917 *Neoborus tricolor*: Van Duzee, Univ. Cal. Publ. Ent., 2: 351.
Distribution: Ill., Ind., Mass., Miss., Mo., N.J.

Tropidosteptes turgidulus Bliven, 1973
1973 *Tropidosteptes turgidulus* Bliven, Occ. Ent., 1(10): 139. [Cal.].
Distribution: Cal.

Tropidosteptes viscicolus (Van Duzee), 1921
1921 *Neoborus viscicolus* Van Duzee, Proc. Cal. Acad. Sci., (4)11: 121. [Cal.].
1959 *Tropidosteptes viscicolus*: Carvalho, Arq. Mus. Nac., 48: 273.
Distribution: Cal.

Tropidosteptes vittifrons (Knight), 1929
1929 *Neoborus vittifrons* Knight, Bull. Brook. Ent. Soc., 24: 5. [Ariz.].
1959 *Tropidosteptes vittifrons*: Carvalho, Arq. Mus. Nac., 48: 273.
Distribution: Ariz., Nev., Ut.

Tropidosteptes vittifrons umbratus (Knight), 1929
1929 *Neoborus vittifrons umbratus* Knight, Bull. Brook. Ent. Soc., 24: 6. [Ariz.].
1959 *Tropidosteptes vittifrons umbratus*: Carvalho, Arq. Mus. Nac., 48: 273.
Distribution: Ariz.

Tropidosteptes vittifrons vittifrons (Knight), 1929
1929 *Neoborus vittifrons* Knight, Bull. Brook. Ent. Soc., 24: 5. [Ariz.].
1959 *Tropidosteptes vittifrons*: Carvalho, Arq. Mus. Nac., 48: 273.
Distribution: Same as for species.

Tropidosteptes vittiscutis (Knight), 1923
1923 *Neoborus vittiscutis* Knight, Conn. Geol. Nat. Hist. Surv. Bull., 34: 566. [Va.].
1959 *Tropidosteptes vittiscutis*: Carvalho, Arq. Mus. Nac., 48: 273.
Distribution: D.C., Ill., Md., Miss., Mo., Va.

Tropidosteptes wileyae (Knight), 1929
1929 *Neoborus wileyae* Knight, Bull. Brook. Ent. Soc., 24: 1. [Tex.].
1959 *Tropidosteptes wileyae*: Carvalho, Arq. Mus. Nac., 48: 273.
Distribution: Tex.

Tropidosteptes ygdrasilis Bliven, 1973
1973 *Tropidosteptes ygdrasilis* Bliven, Occ. Ent., 1(10): 142. [Cal.].

Distribution: Cal.

Genus *Xerolygus* Bliven, 1957

1957 *Xerolygus* Bliven, Occ. Ent., 1(1): 2. Type-species: *Xerolygus orocopiae* Bliven, 1957. Original designation.

Xerolygus orocopiae Bliven, 1957
1957 *Xerolygus orocopiae* Bliven, Occ. Ent., 1(1): 2. [Cal.].
Distribution: Cal.

Tribe Pithanini Douglas and Scott, 1865

Genus *Mimoceps* Uhler, 1890

1890 *Mimoceps* Uhler, Trans. Md. Acad. Sci., 1: 83. Type-species: *Mimoceps insignis* Uhler, 1890. Monotypic.

Mimoceps insignis Uhler, 1890
1887 *Globiceps flavomaculatus*: Provancher, Pet. Faune Ent. Can., 3: 147.
1890 *Mimoceps insignis* Uhler, Trans. Md. Acad. Sci., 1: 84. [Ill.].
1890 *Mimoceps gracilis* Uhler, Trans. Md. Acad. Sci., 1: 85. [N.Y., Ont., Wis.]. Synonymized by Kelton, 1980, Agr. Can., Res. Publ., 1703: 20.
1927 *Mimoceps insignis* var. *gracilis*: Knight, Can. Ent., 59: 41.
Distribution: Alta., Col., Ia., Id., Ill., Man., Minn., Mo., Mont., N.D., N.M., N.Y., Oh., Ont., Que., Sask., Ut., Wis.

Genus *Pithanus* Fieber, 1858

1858 *Pithanus* Fieber, Wien. Ent. Monat., 2: 303. Type-species: *Capsus maerkelii* Herrich-Schaeffer, 1838. Monotypic.

Pithanus maerkelii (Herrich-Schaeffer), 1838
1838 *Capsus maerkelii* Herrich-Schaeffer, Wanz. Ins., 4: 78. [Switzerland].
1915 *Pithanus maerkeli* [*sic*]: Olsen, Bull. Brook. Ent. Soc., 10: 35.
1916 *Pithanus maerkelii*: Van Duzee, Check List Hem., p. 37.
Distribution: Alta., B.C., Conn., Man., Mass., Me., N.B., N.S., N.Y., Nfld., Ont., Ore., P.Ed., Que., Sask., Wash. (Europe, USSR).
Note: Kelton (1966, Can. Ent., 98: 1305-1307) reviewed the North American distribution of this species and gave notes on its habits.

Tribe Resthenini Reuter, 1905

Genus *Oncerometopus* Reuter, 1876

1876 *Oncerometopus* Reuter, Öfv. K. Svens. Vet.-Akad. Förh., 32(9): 65. Type-species: *Oncerometopus nigriclavus* Reuter, 1876. Designated by Kirkaldy, 1906, Trans. Am. Ent. Soc., 32: 136.

Note: Knight (1928, J. N.Y. Ent. Soc., 36: 194) provided a key to species.

Oncerometopus atriscutis Knight, 1928
1928 *Oncerometopus atriscutis* Knight, J. N.Y. Ent. Soc., 36: 191. [Ariz.].
Distribution: Ariz., N.M.

Oncerometopus californicus Van Duzee, 1918
1918 *Oncerometopus californicus* Van Duzee, Proc. Cal. Acad. Sci., (4)8: 280. [Cal.].
Distribution: Cal.

Oncerometopus impictus Knight, 1928
1928 *Oncerometopus impictus* Knight, J. N.Y. Ent. Soc., 36: 190. [Col.].
Distribution: Ariz., Col., Wyo.

Oncerometopus nasutus Knight, 1928
1928 *Oncerometopus nasutus* Knight, J. N.Y. Ent. Soc., 36: 193. [Col.].
Distribution: Col.

Oncerometopus nicholi Knight, 1928
1928 *Oncerometopus nicholi* Knight, J. N.Y. Ent. Soc., 36: 190. [Ariz.].
Distribution: Ariz., Col., Ut.

Oncerometopus nigriclavus Reuter, 1876
1876 *Oncerometopus nigriclavus* Reuter, Öfv. K. Svens. Vet.-Akad. Förh., 32(9): 66. [Tex.].
Distribution: Ariz., Cal., Col., Ks., N.M., Nev., Ok., Tex. (Mexico).
Note: Smith's (1900, Ins. N.J., p. 128) New Jersey record undoubtedly is based on a misidentification, possibly of *O. nitens* Knight, which is not recorded from that state but may occur there.

Oncerometopus nitens Knight, 1928
1928 *Oncerometopus nitens* Knight, J. N.Y. Ent. Soc., 36: 192. [Miss.].
Distribution: Ala., Ga., Md., Miss., N.C., Tenn., W.Va.

Oncerometopus ruber Reuter, 1876
1876 *Oncerometopus ruber* Reuter, Öfv. K. Svens. Vet.-Akad. Förh., 32(9): 66. [Tex.].
Distribution: Col., Ks., Tex.

Genus *Opistheurista* Carvalho, 1959

1959 *Opistheurista* Carvalho, Arq. Mus. Nac., 48: 347. Type-species: *Opistheuria clandestina* Van Duzee, 1915. Monotypic.

Opistheurista clandestina (Van Duzee), 1915
1909 *Resthenia* sp.: Van Duzee, Bull. Buff. Soc. Nat. Sci., 9: 181.
1915 *Opistheuria clandestina* Van Duzee, Pomona J. Ent. Zool., 7: 110. [Fla.].
1918 *Opistheuria clandestina* var. *dorsalis* Knight, Bull. Brook. Ent. Soc., 13: 115. [Oh.]. Synonymized by Blatchley, 1926, Het. E. N. Am., p. 689.
1918 *Opistheuria clandestina* var. *ventralis* Knight, Bull. Brook. Ent. Soc., 13: 115. [Wis.]. Synonymized by Blatchley, 1926, Het. E. N. Am., p. 689.
1959 *Opistheurista clandestina*: Carvalho, Arq. Mus. Nac., 48: 347.
Distribution: Fla., Ill., Ind., La., Man., Minn., Mo., N.Y., Oh., Ok., Ont., Wis.
Note: Although Knight (1941, Ill. Nat. Hist. Surv. Bull., 22: 131) continued to use the varietal names *dorsalis* and *ventralis*, Blatchley (1926, Het. E. N. Am., p. 689) stated: "As the color, like the color of a horse, varies greatly, these names are

superfluous." We concur that these names represent mere color forms and should not be given subspecies status.

Genus *Platytylus* Fieber, 1858

1858 *Platytylus* Fieber, Wien. Ent. Monat., 2: 308. Type-species: *Capsus bicolor* Lepeletier and Serville, 1825. Monotypic.

Note: Carvalho and Schaffner (1976, Rev. Brasil. Biol., 35: 705-736) revised this genus and the closely related genus *Callichila* Reuter, and provided keys to species.

Platytylus bicinctus (Walker), 1873

1873 *Capsus bicinctus* Walker, Cat. Hem. Brit. Mus., 6: 100. [Mexico].

1906 *Resthenia ornaticollis*: Barber, Mus. Brook. Inst. Arts Sci., Sci. Bull., 1: 278.

1916 *Platytylellus ornaticollis*: Van Duzee, Check List Hem., p. 36.

1959 *Prepops ornaticollis*: Carvalho, Arq. Mus. Nac., 48: 340.

1976 *Platytylus bicinctus*: Carvalho and Schaffner, Rev. Brasil. Biol., 35: 721.

Distribution: Tex. (Mexico, Panama).

Note: Carvalho and Schaffner (1976, above) clarified confusion between this species and *Resthenia ornaticollis* Stål and transferred the latter to *Platytylus*.

Genus *Prepops* Reuter, 1905

1905 *Resthenia* (*Prepops*) Reuter, Öfv. F. Vet.-Soc. Förh., 47(19): 15. Type-species: *Resthenia frontalis* Reuter, 1905. Monotypic.

1907 *Platytylellus* Reuter, An. Nat. Hofmus. Wien, 22: 71. Type-species: *Resthenia nigripennis* Stål, 1860. Original designation. Synonymized by Carvalho, 1954, An. Acad. Brasil. Ci., 26: 426.

1908 *Opistheuria* Reuter, An. Nat. Hofmus. Wien, 22: 170. Type-species: *Resthenia latipennis* Stål, 1862. Monotypic. Synonymized by Carvalho, 1952, An. Acad. Brasil. Ci., 24: 98.

1954 *Prepops*: Carvalho, An. Acad. Brasil Ci., 26: 426.

Note: This genus contains many color forms named as varieties and now listed as subspecies that have no apparent geographic significance. Most of these "subspecies" should probably be reduced to synonyms of the nominate species in future lists, as we have done in the genus *Taedia*.

Prepops atripennis (Reuter), 1876

1876 *Resthenia atripennis* Reuter, Öfv. K. Svens. Vet.-Akad. Förh., 32(9): 65. [Tex.].

1917 *Platytylellus atripennis*: Van Duzee, Univ. Cal. Publ. Ent., 2: 310.

1959 *Prepops atripennis*: Carvalho, Arq. Mus. Nac., 48: 332.

Distribution: Ariz., Col., Fla., Nev., Tenn., Tex., Ut.

Prepops bivittis (Stål), 1862

1862 *Resthenia bivittis* Stål, Stett. Ent. Zeit., 23: 318. [Mexico].

1893 *Resthenia intercidenda* Distant, Biol. Centr.-Am., Rhyn., 1: 426. [Mexico]. Synonymized by Carvalho, 1952, Bol. Mus. Nac., 118: 8.

1895 *Resthenia bivittis*: Gillette and Baker, Bull. Col. Agr. Exp. Stn., 31: 29.

1914 *Platytylellus basivittis* [*sic*]: Van Duzee, Trans. San Diego Soc. Nat. Hist., 2: 25.

1914 *Resthenia intercidenda*: Barber, Bull. Am. Mus. Nat. Hist., 33: 501.

1916 *Platytylellus bivittis*: Van Duzee, Check List Hem., p. 36.

1938 *Platytylellus intercidendus*: Brimley, Ins. N.C., p. 76.

1959 *Prepops bivittis*: Carvalho, Arq. Mus. Nac., 48: 332.

Distribution: Alta., B.C., Cal., Col., Fla., Id., Man., N.C., Sask., Ut. (Mexico).

Prepops bivittis bivittis (Stål), 1862
1862 *Resthenia bivittis* Stål, Stett. Ent. Zeit., 23: 318. [Mexico].
1929 *Platytylellus bivittis*: Knight, Ent. News, 40: 192.
1959 *Prepops bivittis*: Carvalho, Arq. Mus. Nac., 48: 332.
Distribution: Same as for species.

Prepops bivittis evittatus (Knight), 1929
1929 *Platytylellus bivittis* var. *evittatus* Knight, Ent. News, 40: 192. [Ut.].
1959 *Prepops bivittis* var. *evittatus*: Carvalho, Arq. Mus. Nac., 48: 332.
Distribution: Ut.

Prepops borealis (Knight), 1923
1923 *Platytylellus borealis* Knight, Conn. Geol. Nat. Hist. Surv. Bull., 34: 551. [N.Y.].
1950 *Platytyllelus* [*sic*] *borealis*: Moore, Nat. Can., 77: 245.
1959 *Prepops borealis*: Carvalho, Arq. Mus. Nac., 48: 332.
Distribution: Alta., Conn., Man., Me., Mich., Minn., N.D., N.Y., Ont., Pa., Que., Sask., Wis.

Prepops borealis borealis (Knight), 1923
1923 *Platytylellus borealis* Knight, Conn. Geol. Nat. Hist. Surv. Bull., 34: 551. [N.Y.].
1929 *Platytylellus borealis*: Knight, Ent. News, 40: 192.
1959 *Prepops borealis*: Carvalho, Arq. Mus. Nac., 48: 332.
Distribution: Same as for species.

Prepops borealis notatus (Knight), 1929
1929 *Platytylellus borealis* var. *notatus* Knight, Ent. News, 40: 192. [N.D.].
1959 *Prepops borealis* var. *notatus*: Carvalho, Arq. Mus. Nac., 48: 333.
Distribution: N.D.

Prepops circumcinctus (Say), 1832 [Fig. 73]
1832 *Capsus circumcinctus* Say, Descrip. Het. Hem. N. Am., p. 23. [Ind.].
1886 *Resthenia circumcincta*: Uhler, Check-list Hem. Het., p. 17.
1893 *Lopidea circumcincta*: Uhler, Proc. Ent. Soc. Wash., 2: 372.
1916 *Platytylellus circumcinctus*: Van Duzee, Check List Hem., p. 36.
1950 *Platytyllelus* [*sic*] *circumcinctus*: Moore, Nat. Can., 77: 245.
1959 *Prepops circumcinctus*: Carvalho, Arq. Mus. Nac., 48: 333.
Distribution: D.C., Fla., Ill., Ind., La., Mass., Me., Miss., Mo., N.C., N.D., N.H., N.J., N.Y., Pa., Ut., Va. (Mexico).

Prepops concisus (Knight), 1929
1929 *Platytylellus concisus* Knight, Ent. News, 40: 190. [S.D.].
1959 *Prepops concisus*: Carvalho, Arq. Mus. Nac., 48: 334.
Distribution: S.D.

Prepops confraternus (Uhler), 1872
1872 *Resthenia confraterna* Uhler, Prelim. Rept. U.S. Geol. Surv. Mont., p. 411. [Col., Ill., Md., Pa., Wis.]
1875 *Resthenia* (*Capsus*) *confraterna*: Glover, Rept. Dept. Agr., p. 125.
1876 *Resthenia maculicollis* Reuter, Öfv. K. Svens. Vet.-Akad. Förh., 32(9): 65. [Tex.]. Synonymized by Banks, 1910, Cat. Nearc. Hem.-Het., p. 40.
1916 *Platytylellus confraternus*: Van Duzee, Check List Hem., p. 36.
1959 *Prepops confraternus*: Carvalho, Arq. Mus. Nac., 48: 334.

Distribution: Col., Conn., Fla., Ga., Ill., Mass., Md., Mich., Miss., N.H., N.J., N.M., N.Y., Pa., Tex., Wis.

Note: Knight (1926, Can. Ent., 58: 255) suggested that *P. confraternus* does not occur east of the 100th meridian even though Uhler's syntype series included eastern U.S. [and Knight (1926, Can. Ent., 58: 255) described the "variety" *collaris* from Fla.]. Because no lectotypes have been designated, we must interpret the more inclusive range. For additional discussion see Blatchley (1926, Het. E. N. Am., p. 683-684) and Knight (1927, Bull. Brook. Ent. Soc., 22: 99).

Prepops confraternus collaris (Knight), 1926

1926 *Platytylellus confraternus* var. *collaris* Knight, Can. Ent., 58: 255. [Fla.].

1959 *Prepops confraternus* var. *collaris*: Carvalho, Arq. Mus. Nac., 48: 334.

Distribution: Fla.

Prepops confraternus confraternus (Uhler), 1872

1872 *Resthenia confraternus* Uhler, Prelim. Rept. U.S. Geol. Surv. Mont., p. 411. [Col., Ill., Md., Pa., Wis.].

1926 *Platytylellus confraternus*: Knight, Can. Ent., 58: 255.

1959 *Prepops confraternus*: Carvalho, Arq. Mus. Nac., 48: 334.

Distribution: Same as for species.

Prepops diminutus Carvalho and Fontes, 1973

1973 *Prepops diminutus* Carvalho and Fontes, Rev. Brasil. Biol., 33: 539. [N.M.].

Distribution: N.M.

Prepops divisus (Herrich-Schaeffer), 1850

1850 *Capsus divisus* Herrich-Schaeffer, Wanz. Ins., 9: 167. [No locality given].

1873 *Capsus divisus*: Walker, Cat. Hem. Brit. Mus., 6: 91.

1906 *Resthenia divisus*: Barber, Mus. Brook. Inst. Arts Sci., Sci. Bull., 1: 278.

1916 *Platytylellus divisus*: Van Duzee, Check List Hem., p. 36.

1959 *Prepops divisus*: Carvalho, Arq. Mus. Nac., 48: 335.

Distribution: Fla., Tex. (Mexico).

Note: Van Duzee (1917, Univ. Cal. Publ. Ent., 2: 310) first noted that the description or figure of this species in the original description was reversed with that of *Capsus tetrastigmus* [which see under *Taedia scrupeus* (Say)]. We follow Carvalho's (1959, above, 48: 335) interpretation of this confusion.

Prepops eremicola (Knight), 1929

1871 *Resthenia eremicola* Uhler, Prelim. Rept. U.S. Geol. Surv. Wyo., p. 471. *Nomen nudum.*

1917 *Platytylellus eremicola*: Van Duzee, Univ. Cal. Publ. Ent., 2: 312. *Nomen nudum.*

1929 *Platytylellus eremicola* Knight, Ent. News, 40: 189. [Wyo.].

1959 *Prepops eremicola*: Carvalho, Arq. Mus. Nac., 48: 335.

Distribution: Alta., Col., Id., Man., Mont., N.M., Sask., Wyo.

Prepops fraterculus (Knight), 1923

1923 *Platytylellus insignis* var. *fraterculus* Knight, Conn. Geol. Nat. Hist. Surv. Bull., 34: 554. [Mich.].

1941 *Platytylellus fraterculus*: Knight, Ill. Nat. Hist. Surv. Bull., 22: 136.

1959 *Prepops fraterculus*: Carvalho, Arq. Mus. Nac., 48: 335.

Distribution: Ill., Ind., Md., Mich., Minn., Miss., N.C., Ont., Pa., Va., W.Va., Wis.

Prepops fraternus (Knight), 1923

1904 *Resthenia maculicollis*: Wirtner, An. Carnegie Mus., 3: 199.

1923 *Platytylellus fraternus* Knight, Conn. Geol. Nat. Hist. Surv. Bull., 34: 557. [W.Va.].

1950 *Platytyllelus* [*sic*] *fraternus*: Moore, Nat. Can., 77: 245.

1959 *Prepops fraternus*: Carvalho, Arq. Mus. Nac., 48: 335.

Distribution: Conn., D.C., Fla., Ga., Ill., Ind., Man., Md., Mich., Minn., Miss., Mo., N.C., N.H., N.Y., Pa., S.C., Tenn., Va., W.Va., Wis.

Note: Wheeler and Henry (1977, Great Lakes Ent., 10: 150) clarified the Wirtner (1904, above) record.

Prepops fraternus discifer (Knight), 1923

1923 *Platytylellus fraternus* var. *discifer* Knight, Conn. Geol. Nat. Hist. Surv. Bull., 34: 559. [Fla.].

1959 *Prepops fraternus* var. *discifer*: Carvalho, Arq. Mus. Nac., 48: 335.

Distribution: Fla., Md.

Prepops fraternus fraternus (Knight), 1923

1923 *Platytylellus fraternus* Knight, Conn. Geol. Nat. Hist. Surv. Bull., 34: 557. [W.Va.].

1959 *Prepops fraternus*: Carvalho, Arq. Mus. Nac., 48: 335.

Distribution: Same as for species.

Prepops fraternus regalis (Knight), 1923

1923 *Platytylellus fraternus* var. *regalis* Knight, Conn. Geol. Nat. Hist. Surv. Bull., 34: 559. [Fla.].

1959 *Prepops fraternus* var. *regalis*: Carvalho, Arq. Mus. Nac., 46: 335.

Distribution: D.C., Fla., N.C., N.J., Va.

Prepops fraternus rubromarginatus (Knight), 1923

1923 *Platytylellus fraternus* var. *rubromarginatus* Knight, Conn. Geol. Nat. Hist. Surv. Bull., 34: 558. [Mich.].

1959 *Prepops fraternus* var. *rubromarginatus*: Carvalho, Arq. Mus. Nac., 48: 335.

Distribution: Ill., Mich., N.H., N.Y., Pa.

Prepops insignis (Say), 1832

1832 *Capsus insignis* Say, Descrip. Het. Hem. N. Am., p. 32. [Ga., Pa.].

1872 *Resthenia insignis*: Uhler, Prelim. Rept. U.S. Geol. Surv. Mont., p. 411.

1916 *Platytylellus insignis*: Van Duzee, Check List Hem., p. 36.

1950 *Platytyllelus* [*sic*] *insignis*: Moore, Nat. Can., 77: 245.

1959 *Prepops insignis*: Carvalho, Arq. Mus. Nac., 48: 336.

Distribution: Col., Conn., Fla., Ga., Ia., Ill., Ind., Mass., Md., Me., Mich., Minn., N.D., N.H., N.J., N.Y., Ont., Pa., Que., R.I., Tex., Va., Vt., W.Va.

Prepops insitivus (Say), 1832

1832 *Capsus insitivus* Say, Descrip. Het. Hem. N. Am., p. 21. [Ind.].

1869 *Capsis* [*sic*] *insitivus*: Rathvon, Hist. Lancaster Co., Pa., p. 549.

1873 *Capsus xanthomelas* Walker, Cat. Hem. Brit. Mus., 6: 92. [Fla.]. Synonymized by Uhler, 1886, Check-list Hem. Het., p. 17.

1878 *Resthenia insitiva*: Uhler, Proc. Boston Soc. Nat. Hist., 19: 399.

1916 *Platytylellus insitivus*: Van Duzee, Check List Hem., p. 36.

1959 *Prepops insitivus*: Carvalho, Arq. Mus. Nac., 48: 337.

Distribution: Col., Conn., Fla., Ga., Ia., Ill., Ind., Mass., Md., Me., Mo., N.C., N.H., N.J., N.Y., Ont., Pa., R.I., Vt., W.Va.

Prepops insitivus insitivus (Say), 1832
1832 *Capsus insitivus* Say, Descrip. Het. Hem. N. Am., p. 21. [Ind.].
1923 *Platytylellus insitivus*: Knight, Conn. Geol. Nat. Hist. Surv. Bull., 34: 556.
1959 *Prepops insitivus*: Carvalho, Arq. Mus. Nac., 48: 337.
Distribution: Same as for species.

Prepops insitivus angusticollis (Knight), 1923
1923 *Platytylellus insitivus* var. *angusticollis* Knight, Conn. Geol. Nat. Hist. Surv. Bull., 34: 556. [Me.].
1959 *Prepops insitivus* var. *angusticollis*: Carvalho, Arq. Mus. Nac., 48: 337.
Distribution: Me.

Prepops nigricollis (Reuter), 1876
1876 *Resthenia nigricollis* Reuter, Öfv. K. Svens. Vet.-Akad. Förh., 32(9): 65. [N.J.].
1916 *Platytylellus nigricollis*: Van Duzee, Check List Hem., p. 36.
1950 *Platytyllelus* [*sic*] *nigricollis*: Moore, Nat. Can., 77: 245.
1959 *Prepops nigricollis*: Carvalho, Arq. Mus. Nac., 48: 339.
1976 *Prepops nigrocollis* [*sic*]: Reid et al., Can. Ent., 108: 563.
Distribution: Ia., Ill., Ind., Mass., Me., Mo., N.D., N.H., N.J., Ok., Ont., Pa., Tex., Va., W.Va.

Prepops nigripilus (Knight), 1929
1929 *Platytylellus nigripilus* Knight, Ent. News, 40: 189. [N.Y.].
1959 *Prepops nigripilus*: Carvalho, Arq. Mus. Nac., 48: 339.
Distribution: Alta., N.H., N.Y.

Prepops nigroscutellatus (Knight), 1923
1923 *Platytylellus nigroscutellatus* Knight, Conn. Geol. Nat. Hist. Surv. Bull., 34: 557. [N.Y.].
1959 *Prepops nigroscutellatus*: Carvalho, Arq. Mus. Nac., 48: 339.
Distribution: Ill., Mich., Mo., N.C., N.Y., Pa.

Prepops persignandus (Distant), 1883
1883 *Resthenia persignanda* Distant, Biol. Centr.-Am., Rhyn., 1: 257. [Mexico].
1906 *Resthenia persignanda*: Snow, Trans. Ks. Acad. Sci., 20: 179.
1916 *Platytylellus persignandus*: Van Duzee, Check List Hem., p. 36.
1959 *Prepops persignandus*: Carvalho, Arq. Mus. Nac., 48: 340.
Distribution: Ariz. (Mexico, Guatemala).

Prepops robustus (Reuter), 1913
1913 *Platytylellus robustus* Reuter, Öfv. F. Vet.-Soc. Förh., 55A(18): 32. [Ga.].
1959 *Prepops robustus*: Carvalho, Arq. Mus. Nac., 48: 341.
Distribution: Ga.

Prepops rubellicollis (Knight), 1923
1923 *Platytylellus rubellicollis* Knight, Conn. Geol. Nat. Hist. Surv. Bull., 34: 555. [Minn.].
1950 *Platytyllelus* [*sic*] *rubellicollis*: Moore, Nat. Can., 77: 245.
1959 *Prepops rubellicollis*: Carvalho, Arq. Mus. Nac., 48: 341.
Distribution: Alta., B.C., Man., Me., Mich., Minn., Neb., Ont., Que., Sask., Wis.

Prepops rubellicollis confluens (Knight), 1923
1923 *Platytylellus rubellicollis* var. *confluens* Knight, Conn. Geol. Nat. Hist. Surv. Bull., 34: 556. [Me.].

1950 *Platytyllelus* [*sic*] *rubellicollis confluens*: Moore, Nat. Can., 77: 245.
1959 *Prepops rubellicollis* var. *confluens*: Carvalho, Arq. Mus. Nac., 48: 341.

Prepops rubellicollis rubellicollis (Knight), 1923
1923 *Platytylellus rubellicollis* Knight, Conn. Geol. Nat. Hist. Surv. Bull., 34: 555. [Minn.].
1959 *Prepops rubellicollis*: Carvalho, Arq. Mus. Nac., 48: 341.
Distribution: Same as for species.

Prepops rubellicollis vittiscutis (Knight), 1923
1923 *Platytylellus rubellicollis* var. *vittiscutis* Knight, Conn. Geol. Nat. Hist. Surv. Bull., 34: 556. [Minn.].
1950 *Platytyllelus* [*sic*] *rubellicollis vittiscutis*: Moore, Nat. Can., 77: 245.
1959 *Prepops rubellicollis* var. *vittiscutis*: Carvalho, Arq. Mus. Nac., 48: 341.
Distribution: Me., Minn.

Prepops rubroscutellatus (Knight), 1929
1929 *Platytylellus rubroscutellatus* Knight, Ent. News, 40: 191. [N.M.].
1959 *Prepops rubroscutellatus*: Carvalho, Arq. Mus. Nac., 48: 341.
Distribution: Ariz., Col., N.M.

Prepops rubroscutellatus nigriscutis (Knight), 1929
1929 *Platytylellus rubroscutellatus* var. *nigriscutis* Knight, Ent. News, 40: 192. [N.M.].
1959 *Prepops rubroscutellatus* var. *nigriscutis*: Carvalho, Arq. Mus. Nac., 48: 341.
Distribution: N.M.

Prepops rubroscutellatus rubroscutellatus (Knight), 1929
1929 *Platytylellus rubroscutellatus* Knight, Ent. News, 40: 191. [N.M.].
1959 *Prepops rubroscutellatus*: Carvalho, Arq. Mus. Nac., 48: 341.
Distribution: Ariz., Col., N.M.

Prepops rubrovittatus (Stål), 1862
1862 *Resthenia rubrovittata* Stål, Stett. Ent. Zeit., 23(7-9): 318. [N.Am.].
1873 *Capsus rubrovittatus*: Walker, Cat. Hem. Brit. Mus., 6: 92.
1916 *Platytylellus rubrovittatus*: Van Duzee, Check List Hem., p. 36.
1959 *Prepops rubrovittatus*: Carvalho, Arq. Mus. Nac., 48: 341.
Distribution: Col., Fla., Ga., Ind., La., Mass., Me., Miss., Mo., N.J., N.M., N.S., N.Y., Tex., Wis. (Mexico).

Prepops zonatus (Knight), 1926
1926 *Platytylellus zonatus* Knight, Can. Ent., 58: 254. [Minn.].
1959 *Prepops zonatus*: Carvalho, Arq. Mus. Nac., 48: 343.
Distribution: Alta., Ill., Man., Mich., Minn., N.D., Sask., Wis.

Tribe Stenodemini China, 1943

Note: Knight (1968, Brig. Young Univ. Sci. Bull., 9(3): 177-178) provided a key containing most of the North American stenodemine genera. Carvalho (1975, Rev. Brasil. Biol., 35: 121-140) gave a key to the neotropical genera, and Eyles and Carvalho (1975, J. Nat. Hist., 9: 266-268) gave a key to the world genera.

Genus *Acetropis* Fieber, 1858

1858 *Acetropis* Fieber, Wien. Ent. Monat., 2: 302. Type-species: *Lopus carinatus* Herrich-Schaeffer, 1841. Monotypic.

Acetropis americana Knight, 1927
1927 *Acetropis americana* Knight, Ent. News, 38: 206. [Ore.].
Distribution: Ore.

Genus *Actitocoris* Reuter, 1880

1878 *Actinocoris* [*sic*] Reuter, Medd. Soc. Fauna Flora Fenn., 2: 194. *Lapsus*. Type-species: *Actinocoris signatus* Reuter, 1878. Monotypic.
1880 *Actitocoris* Reuter, Medd. Soc. Fauna Flora Fenn., 5: 167. Emendation to correct misspelling.

Actitocoris signatus Reuter, 1878
1878 *Actinocoris* [*sic*] *signatus* Reuter, Medd. Soc. Fauna Flora Fenn., 2: 194. [Finland, Europe].
1966 *Actitocoris signatus*: Kelton, Can. Ent., 98: 1307.
Distribution: Alta., N.T., Sask. (Europe, USSR).

Genus *Chaetofoveolocoris* Knight, 1968

1968 *Chaetofoveolocoris* Knight, Brig. Young. Univ. Sci. Bull., 9(3): 179. Type-species: *Megaloceroea hirsuta* Knight, 1928. Original designation.

Chaetofoveolocoris hirsutus (Knight), 1928
1928 *Megaloceroea hirsuta* Knight, Ent. News, 39: 248. [Ariz.].
1968 *Chaetofoveolocoris hirsuta*: Knight, Brig. Young Univ. Sci. Bull., 9(3): 180.
Distribution: Ariz., Tex.

Genus *Chaetomiris* Bliven, 1973

1973 *Chaetomiris* Bliven, Occ. Ent., 1(10): 134. Type-species: *Chaetomiris sequoiarum* Bliven, 1973. Monotypic.

Chaetomiris sequoiarum Bliven, 1973
1973 *Chaetomiris sequoiarum* Bliven, Occ. Ent., 1(10): 134. [Cal.].
Distribution: Cal.

Genus *Collaria* Provancher, 1872

1872 *Collaria* Provancher, Nat. Can., 4: 79. Type-species: *Collaria meilleurii* Provancher, 1872. Monotypic.
1876 *Trachelomiris* Reuter, Öfv. K. Svens. Vet.-Akad. Förh., 32(9): 61. Type-species: *Trachelomiris oculatus* Reuter, 1876. Monotypic. Synonymized by Reuter, 1905, Festsch. Palmén, 1: 47.
1878 *Nabidea* Uhler, Proc. Boston Soc. Nat. Hist., 19: 397. Type-species: *Nabidea coracina* Uhler, 1878. Monotypic. Synonymized by Uhler, 1887, Ent. Am., 2: 230.

Collaria meilleurii Provancher, 1872

1872 *Collaria meilleurii* Provancher, Nat. Can., 4: 79. [Que.]. Lectotype designated by Kelton, 1968, Nat. Can., 95: 1073.

1878 *Nabidea coracina* Uhler, Proc. Boston Soc. Nat. Hist., 19: 398. [N.H.]. Synonymized by Uhler, 1887, Ent. Am., 2: 230.

1887 *Trachelomiris meilleurii*: Van Duzee, Can. Ent., 19: 70.

1904 *Collaria oculata*: Wirtner, An. Carnegie, 3: 195 (in part).

1907 *Collaria meuilleuri* [*sic*]: Moore, Can. Ent., 39: 163.

1922 *Collaria meilleuri* [*sic*]: Hussey, Occas. Pap. Mus. Zool., Univ. Mich., 118: 26.

1952 *Collaria coracina*: Carvalho, An. Acad. Brasil. Ci., 24: 85.

1958 *Collaris* [*sic*] *meilleurii*: Messina, J. N.Y. Ent. Soc., 86: 139.

Distribution: Alta., Conn., Ill., Ind., Man., Mass., Me., Mich., Minn., N.B., N.C., N.H., N.S., N.Y., Oh., Ont., Pa., Que., Sask., Vt., W.Va., Wis.

Collaria oculata (Reuter), 1876

1876 *Trachelomiris oculata* Reuter, Öfv. K. Svens. Vet.-Akad. Förh., 32(9): 61. [N.Y., Tex.]

1894 *Collaria oculata*: Van Duzee, Bull. Buff. Soc. Nat. Sci., 5: 176.

1904 *Collaria oculata*: Wirtner, An. Carnegie Mus., 3: 195 (in part).

1905 *Collaris* [*sic*] *oculata*: Crevecoeur, Trans. Ks. Acad. Sci., 19: 233.

Distribution: Alta., Conn., D.C., Fla., Ga., Ia., Ill., Ind., Ks., Mass., Me., Mich., Minn., N.B., N.D., N.H., N.J., N.Y., Oh., Ont., Pa., Que., Tex., Va., Vt., W.Va., Wis. (Mexico).

Collaria oleosa (Distant), 1883

1883 *Trachelomiris oleosa* Distant, Biol. Centr.-Am., Rhyn., 1: 238. [Mexico].

1887 *Collaria explicata* Uhler, Ent. Am., 2: 230. [Cuba, "San Domingo"]. Synonymized by Carvalho, 1959, Arq. Mus. Nac., 48: 285.

1907 *Collaria oleosa*: Reuter, An. Nat. Hofmus. Wien, 22: 64.

1914 *Collaria explicata*: Barber, Bull. Am. Mus. Nat. Hist., 33: 500.

Distribution: Cal., Fla., La., Miss. (Mexico to South America, West Indies).

Genus *Dolichomiris* Reuter, 1882

1882 *Dolichomiris* Reuter, Öfv. F. Vet.-Soc. Förh., 25: 29. Type-species: *Dolichomiris linearis* Reuter, 1882. Monotypic.

1893 *Eioneus* Distant, Biol. Centr.-Am., Rhyn., 1: 416. Type-species: *Eioneus bilineatus* Distant, 1893. Monotypic. Synonymized by Reuter, 1909, Öfv. F. Vet.-Soc. Förh., 51A(13): 5.

Note: Eyles and Carvalho (1975, J. Nat. Hist., 9: 257-269) reviewed the genus and provided a key to species.

Dolichomiris linearis Reuter, 1882

1882 *Dolichomiris linearis* Reuter, Öfv. F. Vet.-Soc. Förh., 25: 29. [Africa].

1926 *Eioneus gutticornis* Blatchley, Het. E. N. Am., p. 674. [Fla.]. Synonymized by Knight, 1927, Bull. Brook. Ent. Soc., 22: 98.

1927 *Dolichomiris linearis*: Knight, Bull. Brook. Ent. Soc., 22: 98.

Distribution: Fla., Tex. (Africa, Southern Europe, Mexico to South America, West Indies).

Genus *Leptopterna* Fieber

1858 *Leptopterna* Fieber, Wien. Ent. Monat., 2: 302. Type-species: *Cimex dolabratus* Linnaeus, 1758. Monotypic. Placed on the Official List of Generic Names in Zoology by Int. Comm. Zool. Nomen., 1970, Bull. Zool. Nomen., 26: 203, Opinion 898.

1865 *Lopomorphus* Douglas and Scott, Brit. Hem., p. 293. Type-species: *Miris ferrugatus* Fallén, 1807. Designated by Kirkaldy, 1906, Trans. Am. Ent. Soc., 2: 144. Synonymized by Reuter, 1910, Acta Soc. Sci. Fenn., 37(3): 167.

Leptopterna dolabrata (Linnaeus), 1758 [Fig. 71]

1758 *Cimex dolabratus* Linnaeus, Syst. Nat., ed. 10, p. 449. [Europe].
1872 *Miris belangeri* Provancher, Nat. Can., 4: 78. [Que.]. Synonymized by Provancher, 1886, Pet. Faune Ent. Can., 3: 104.
1875 *Leptopterna dolabrata* van *aurantiaca* Reuter, Rev. Crit. Caps., 1(2): 16.[Scandinavia]. New Synonymy.
1878 *Lopomorphus dolabratus*: Uhler, Proc. Boston Soc. Nat. Hist., 19: 397.
1886 *Leptopterna dolabrata*: Provancher, Pet. Faune Ent. Can., 3: 104.
1903 *Leptopterna dolobrata* [*sic*]: Van Duzee, Trans. Am. Ent. Soc., 29: 109.
1907 *Leptoterna* [*sic*] *dolabrata*: Moore, Can. Ent., 39: 163.
1908 *Miris dolabratus*: Van Duzee, Can. Ent., 40: 111.
1916 *Miris dolobrata* [*sic*]: Leonard, Ent. News, 27: 236.
1941 *Miris dolabratus* var. *aurantiacus*: Knight, Ill. Nat. Hist. Surv. Bull., 22: 127.
1950 *Miris delabratus* [*sic*]: Moore, Nat. Can., 77: 245.
1964 *Leptopterna dolabratus* [*sic*]: Thompson, J. Econ. Ent., 57: 961.
1972 *Leptoterna* [*sic*] *dolabratus*: Beirne, Mem. Ent. Soc. Can., 85: 15.

Distribution: B.C., Col., Conn., D.C., Ga., Ia., Ill., Ind., Ky., Man., Mass., Md., Me., Minn., Mo., N.C., N.H., N.J., N.M., N.S., N.Y., Oh., Ont., Ore., Pa., Que., R.I., Ut., Va., W.Va., Wash., Wis., Wyo. (Europe, USSR).

Note: Osborn (1918, J. Agr. Res., 15: 175-200) described and illustrated the immature stages. Jewett and Townsend (1947, Ky. Agr. Exp. Stn. Bull., 508: 3-8) discussed seasonal history and habits in Kentucky. Reuter's (1875, above) "*aurantiaca*" is a mere color form of *L. dolabrata*.

Leptopterna ferrugata (Fallén), 1807

1807 *Miris ferrugatus* Fallén, Monogr. Cimic. Suec., p. 107. [Europe].
1872 *Leptopterna amoena* Uhler, Prelim. Rept. U.S. Geol. Surv. Mont., p. 409. [Col.]. Synonymized by Carvalho, 1959, Arq. Mus. Nac., 48: 292.
1876 *Leptopterna amoenus*: Uhler, Bull. U.S. Geol. Geogr. Surv. Terr., 1: 316.
1895 *Leptoterna* [*sic*] *amoena*: Gillette and Baker, Bull. Col. Agr. Exp. Stn., 31: 29.
1900 *Leptopterna ferrugata*: Heidemann, Proc. Wash. Acad. Sci., 2: 504.
1907 *Miris amoenus*: Tucker, Ks. Univ. Sci. Bull., 4: 61.
1908 *Miris ferrugatus*: Horvath, An. Mus. Nat. Hung., 6: 5.
1981 *Leptotnerna* [*sic*] *ferrugata*: Thomas and Werner, Univ. Ariz. Exp. Stn. Tech. Bull., 243: 10.

Distribution: Alk., Alta., Col., Conn., Ia., Id., Ky., Minn., Mo., Nev., Que., Ut., Wash., Wyo. (Europe, USSR).

Leptopterna silacea Bliven, 1973

1973 *Leptopterna silacea* Bliven, Occ. Ent., 1(10): 136. [Cal.].

Distribution: Cal.

Genus *Litomiris* Slater, 1956

1956 *Litomiris* Slater, Proc. Ent. Soc. Wash., 58: 118. Type-species: *Megaloceroea debilis* Uhler, 1872. Original designation.

Note: Knight (1968, Brig. Young Univ. Sci. Bull., 9(3): 178-179) provided a key to species.

Litomiris curtus (Knight), 1928
1928 *Megaloceroea curta* Knight, Ent. News, 39: 247. [Wyo.].
1956 *Litomiris curta*: Slater, Proc. Ent. Soc. Wash., 58: 120.
1968 *Litomiris curtus*: Knight, Brig. Young Univ. Sci. Bull., 9(3): 179.
Distribution: Id., Mont., S.D., Wyo.

Litomiris debilis (Uhler), 1872
1872 *Megaloceroea debilis* Uhler, Prelim. Rept. U.S. Geol. Surv. Mont., p. 408. [Col.].
1900 *Megalocoerea* [*sic*] *debilis*: Osborn, Oh. Nat., 1: 12.
1909 *Megaloceraea* [*sic*] *debilis*: Reuter, Acta Soc. Sci. Fenn., 36(2): 5.
1956 *Litomiris debilis*: Slater, Proc. Ent. Soc. Wash., 58: 119.
Distribution: Alta., B.C., Col., Ia., Id., Man., Mont., Nev., Oh.(?), Sask., Wash., Wis., Wyo.

Litomiris gracilis (Van Duzee), 1914
1914 *Stenodema gracilis* Van Duzee, Trans. San Diego Soc. Nat. Hist., 2: 25. [Cal.].
1916 *Megaloceroea gracilis*: Van Duzee, Check List Hem., p. 36.
1956 *Litomiris gracilis*: Slater, Proc. Ent. Soc. Wash., 58: 120.
Distribution: B.C., Cal.

Litomiris punctatus (Knight), 1928
1928 *Megaloceroea punctatus* Knight, Ent. News, 39: 249. [Ariz.].
1956 *Litomiris punctata*: Slater, Proc. Ent. Soc. Wash., 58: 120.
1968 *Litomiris punctatus*: Knight, Brig. Young Univ. Sci. Bull., 9(3): 179.
Distribution: Ariz.

Litomiris rubicundus (Uhler), 1872
1872 *Megaloceroea rubicunda* Uhler, Prelim. Rept. U.S. Geol. Surv. Mont., p. 409. [Col.].
1956 *Litomiris rubicunda*: Slater, Proc. Ent. Soc. Wash., 58: 120.
1968 *Litomiris rubicundus*: Knight, Brig. Young Univ. Sci. Bull., 9(3): 179.
Distribution: Ariz., Col., Ia., N.M., Tex.

Litomiris tritavus Bliven, 1973
1973 *Litomiris tritavus* Bliven, Occ. Ent., 1(10): 135. [Cal.].
Distribution: Cal.

Genus *Megaloceroea* Fieber, 1858

1858 *Megaloceroea* Fieber, Wien. Ent. Monat., 2: 301. Type-species: *Megaloceroea longicornis* Fallén, 1807, a junior synonym of *Cimex recticornis* Geoffroy, 1785. Monotypic.

1861 *Megaloceraea* Fieber, Europ. Hem., p. 243. Unnecessary emendation for *Megaloceroea* Fieber.

Megaloceroea koebelei Van Duzee, 1921
1921 *Megaloceroea koebelei* Van Duzee, Proc. Cal. Acad. Sci., (4)11: 118. [Ariz.].
1956 *Megaloceraea* [*sic*] *koebelei*: Slater, Proc. Ent. Soc. Wash., 58: 120.
Distribution: Ariz.

Note: Slater (1956, above) suggested that this species is more nearly related to species of the genus *Stenodema* Laporte.

Megaloceroea letcheri Knight, 1928

1928 *Megaloceroea letcheri* Knight, Ent. News, 39: 250. [Ariz.].

1956 *Megaloceraea* [*sic*] *letcheri*: Slater, Proc. Ent. Soc. Wash., 58: 120.

Distribution: Ariz.

Note: Slater (1956, above) suggested that this species, like *M. koebelei*, is more nearly related to species belonging to the genus *Stenodema* Laporte.

Megaloceroea recticornis (Geoffroy), 1785

1775 *Cimex linearis* Fuessly, Verz. Schw. Ins., p. 26. Preoccupied.

1785 *Cimex recticornis* Geoffroy, *In* Fourcroy, Ent. Paris, p. 209. [Europe].

1922 *Megaloceraea* [*sic*] *recticornis*: Knight, Can. Ent., 53: 286.

1927 *Megaloceroea recticornis*: Knight, Can. Ent., 59: 41.

Distribution: B.C., Conn., Ia., Id., Ill., Me., N.C., N.H., N.Y., Ont., Ore., Que., Va., W.Va., Wash., Wis., Wyo. (Europe, U.S.S.R.).

Note: Slater (1956, above) described the second-through fifth-instar nymphs.

Genus *Porpomiris* Berg, 1883

1883 *Porpomiris* Berg, An. Soc. Cient. Arg., 16: 8. Type-species: *Porpomiris picturatus* Berg, 1883. Monotypic.

1909 *Mesomiris* Reuter, Acta Soc. Sci. Fenn., 36(2): 4. Type-species: *Mesomiris curtulus* Reuter, 1909. Monotypic. Synonymized by Carvalho, 1952, An. Acad. Brasil. Ci., 24: 84.

Porpomiris albescens (Van Duzee), 1925

1925 *Mesomiris albescens* Van Duzee, Pan-Pac. Ent., 2: 35. [Ariz.].

1952 *Porpomiris albescens*: Carvalho, An. Acad. Brasil. Ci., 24: 84.

Distribution: Ariz.

Porpomiris curtulus (Reuter), 1909

1909 *Mesomiris curtulus* Reuter, Acta Soc. Sci. Fenn., 36(2): 5. [Md.].

1952 *Porpomiris curtulus*: Carvalho, An. Acad. Brasil. Ci., 24: 84.

Distribution: Col.(?), Conn., La., Md., Mass., N.J., N.Y., Neb.(?), Oh.(?), Pa.(?), Ut.(?).

Note: Records of this coastal species from inland localities undoubtedly are in error.

Genus *Stenodema* Laporte, 1833

1833 *Stenodema* Laporte, Mag. Zool. Suppl, 2: 40. Type-species: *Cimex virens* Linnaeus, 1767. Monotypic. Placed on the Official List of Generic Names in Zoology by the Int. Comm. Zool. Nomen., 1970, Bull. Zool. Nomen., 26: 203, Opinion 898.

Note: Kelton (1961, Can. Ent., 93: 452) provided a key to the North American species, and noted that records of *S. virens* (Linnaeus) from North America are based on misidentifications but did not refer the records to particular species.

Subgenus *Brachystira* Fieber 1858

1858 *Brachystira* Fieber, Wien. Ent. Monat., 2: 301. Type-species: *Miris calcarata* Fallén, 1807. Monotypic. Placed on the Official List of Generic Names in Zoology by the Int. Comm. Zool. Nomen., 1970, Bull. Zool. Nomen., 26: 203, Opinion 898.

1858 *Brachytropis* Fieber, Wien. Ent. Monat., 2: 343. Type-species: *Miris calcarata* Fallén, 1807. Unnecessary new name for *Brachystira* Fieber.
1906 *Stenodema* (*Brachystira*): Kirkaldy, Trans. Am. Ent. Soc., 32: 144.

Stenodema trispinosa Reuter, 1904
1873 *Miris calcaratus*: Walker, Cat. Hem. Brit. Mus., 6: 48.
1876 *Brachystira calcaratus*: Uhler, Bull. U.S. Geol. Geogr. Surv. Terr., 1: 316.
1886 *Brachytropis calcarata*: Uhler, Check-list Hem. Het., p. 17.
1904 *Stenodema trispinosum* Reuter, Öfv. F. Vet.-Soc. Förh., 46(15): 4. [Europe].
1904 *Miris instabilis*: Wirtner, An. Carnegie Mus., 3: 195 (in part).
1904 *Brachytropis calcarata*: Wirtner, An. Carnegie Mus., 3: 195 (in part).
1908 *Stenodema trispinosa*: Van Duzee, Can. Ent., 40: 111.
1917 *Stenodema* (*Brachystira*) *trispinosum*: Van Duzee, Univ. Cal. Publ. Ent., 2: 303.
Distribution: Alk., Alta., B.C., Cal., Col., Conn., D.C., Ga., Ill., Ind., Man., Mass., Md., Me., Mich., Minn., Mo., Mont., N.B., N.C., N.J., N.M., N.S., N.T., N.Y., Nfld., Oh., Ont., Ore., Pa., Que., Sask., Tenn., Tex., Wash., W.Va., Wis., Wyo., Yuk. (Asia, Europe).
Note: Wheeler and Henry (1977, Great Lakes Ent., 10: 148) clarified the Wirtner (1904, above) records.

Subgenus *Stenodema* Laporte, 1833

1833 *Stenodema* Laporte, Mag. Zool. Suppl., 2: 40. Type-species: *Cimex virens* Linnaeus, 1767. Monotypic.

Stenodema dorsalis (Say), 1832
1832 *Miris dorsalis* Say, Descrip. Het. Hem. N. Am., p. 26. [U.S.].
1917 *Stenodema dorsalis*: Van Duzee, Univ. Cal. Publ. Ent., 2: 304.
Distribution: U.S.
Note: Kelton (1961, Can. Ent., 93: 452) proposed that *Miris dorsalis*, mentioned only in catalogs since the original description, be considered a *nomen oblitum*. We are listing this species here because the 50-year rule used to determine *nomina oblita* (1961, ICZN, Article 23b) required action by the Commission, and this term is not used in the 1985 Code. Further, we have restudied Say's brief, but very diagnostic, original description and believe *M. dorsalis* should be associated with a stenodemine species that is common throughout much of eastern North America. Because a clarification of this problem requires more space than this catalog will allow, we are deferring action to a future paper.

Stenodema falki Bliven, 1958
1958 *Stenodema falki* Bliven, Occ. Ent., 1(2): 13. [Cal.].
Distribution: Cal.

Stenodema imperii Bliven, 1958
1958 *Stenodema imperii* Bliven, Occ. Ent., 1(2): 14. [Cal.].
Distribution: Cal.

Stenodema pilosipes Kelton, 1961
1961 *Stenodema pilosipes* Kelton, Can. Ent., 93: 453. [Alta.].
Distribution: Alta., Ariz., B.C., Cal., Col., Man., Mont., N.M., N.T., Ore., Sask., Wash., Yuk.

Stenodema sequoiae Bliven, 1955

1955 *Stenodema sequoiae* Bliven, Stud. Ins. Redwood Empire, p. 8. [Cal.].
Distribution: Cal.

Stenodema vicina (Provancher), 1872

1869 *Miris bivittatus* Rathvon, Hist. Lancaster Co., Pa., p. 549. *Nomen nudum.*
1872 *Miris vicinus* Provancher, Nat. Can., 4: 77. [Que.]. Lectotype designated by Kelton, 1968, Nat. Can., 95: 1072.
1875 *Miris instabilis* Uhler, Wheeler's Rept. 100th Mer., 5: 836. [Col.]. Synonymized with *S. affinis* by Van Duzee, 1908, Can. Ent., 40: 111; synonymized with *S. vicina* by Van Duzee, 1916, Check List Hem., p. 36.
1876 *Miris affinis* Reuter, Öfv. K. Svens. Vet.-Akad. Förh., 32(9): 59. [N.J., Pa., Wis.]. Synonymized by Van Duzee, 1916, Check List Hem., p. 36.
1898 *Stenodema instabilis*: Bergroth, Wien. Ent. Zeit., 17: 33.
1904 *Miris instabilis*: Wirtner, An. Carnegie Mus., 3: 195 (in part).
1904 *Brachytropis calcarata*: Wirtner, An. Carnegie Mus., 3: 195 (in part).
1905 *Miris rubellus* Van Duzee, N.Y. St. Ent., p. 550. *Nomen nudum.*
1908 *Stenodema affinis*: Van Duzee, Can. Ent., 40: 111.
1916 *Stenodema vicinum*: Van Duzee, Check List Hem., p. 36.
1973 *Stenodema vicina*: Steyskal, Stud. Ent., 16: 208.
Distribution: Alta., B.C., Cal., Col., Conn., Del., Ga., Ill., Ind., Ks., Man. Mass., Me., Mich., Minn., N.C., N.H., N.J., N.M., N.S., N.Y., Nfld., Oh., Ont., Ore., Pa., Que., Sask., Va., Vt., W.Va., Wis., Wyo.
Note: Wheeler and Henry (1977, Great Lakes Ent., 10: 148) clarified the Wirtner (1904, above) records.

Genus *Teratocoris* Fieber, 1858

1858 *Teratocoris* Fieber, Wien. Ent. Monat., 2: 302. Type-species: *Capsus antennatus* Boheman, 1852. Monotypic.

Teratocoris borealis Kelton, 1966

1966 *Teratocoris borealis* Kelton, Can. Ent., 98: 1269. [N.T.].
Distribution: N.T.

Teratocoris caricis Kirkaldy, 1909

1895 *Teratocoris longicornis* Uhler, Bull. Col. Agr. Exp. Stn., 31: 29. [Col.]. Preoccupied.
1909 *Teratocoris caricis* Kirkaldy, Can. Ent., 41: 390. New name for *Teratocoris longicornis* Uhler.
Distribution: Alk., Alta., B.C., Cal., Col., Man., N.T., Nfld., Que., Sask., Wyo., Yuk.
Note: Kelton (1966, Can. Ent., 98: 1269) clarified the nomenclatural problems between this species and *Teratocoris saundersi* Douglas and Scott.

Teratocoris discolor Uhler, 1887

1887 *Teratocoris discolor* Uhler, Ent. Am., 3: 68. [Col., Mass., Mo.].
Distribution: Alta., B.C., Col., Ia., Ill., Ind., Man., Mass., Mich., Mo., N.Y., Oh., Ont., Que., S.D., Sask., Ut., Wis.

Teratocoris paludum Sahlberg, 1870

1870 *Teratocoris paludum* Sahlberg, Not. Förh. Sällsk. Fauna Flora Fenn., 11: 291. [Europe].
1922 *Teratocoris paludum*: Knight, Can. Ent., 53: 286.

Distribution: Alk., Alta., B.C., Cal., Ia., Ill., Man., Minn., N.Y., Neb. Nfld., Ont., Que., S.D., Sask., Ut., Yuk. (Europe, Northern Asia).

Teratocoris saundersi Douglas and Scott, 1869

1869 *Teratocoris saundersi* Douglas and Scott, Ent. Month. Mag., 5: 260. [England].

1887 *Teratocoris herbaticus* Uhler, Ent. Am., 3: 67. [Nfld.]. Synonymized by Kelton, 1966, Can. Ent., 98: 1266.

1921 *Teratocoris saundersi*: Knight, Oh. J. Sci., 21: 111.

Distribution: Alk., B.C., Col., Conn., Man., N.T., Nfdl., Que., Sask., Wyo., Yuk. (Northern Asia, Europe).

Teratocoris viridis Douglas and Scott, 1867

1867 *Teratocoris viridis* Douglas and Scott, Ent. Month. Mag., 4: 46. [England].

1966 *Teratocoris viridis*: Kelton, Can. Ent., 98: 1268.

Distribution: Alk., N.T., Yuk. (Europe, Northern Asia).

Genus *Trigonotylus* Fieber, 1858

1858 *Trigonotylus* Fieber, Wien. Ent. Monat., 2: 302. Type-species: *Miris ruficornis* Fallén, 1807, a junior synonym of *Cimex ruficornis* Geoffroy, 1785. Monotypic.

1876 *Callimiris* Reuter, Öfv. K. Svens. Vet.-Akad. Förh., 32(9): 60. Type-species: *Callimiris uhleri* Reuter, 1876. Designated by Kirkaldy, 1906, Trans. Am. Ent. Soc., 32: 144. Synonymized by Reuter, 1909, Acta Soc. Sci. Fenn., 36(2): 5.

Note: Carvalho and Wagner (1957, Arq. Mus. Nac., 43: 121-155) and Kelton (1971, Can. Ent., 103: 685-705) provided revisions and keys to species.

Trigonotylus americanus Carvalho, 1957

1957 *Trigonotylus americanus* Carvalho, *In* Carvalho and Wagner, Arq. Mus. Nac., 43: 126. [Ut.].

Distribution: Alta., Ariz., B.C., Col., Ia., Id., Man., N.M., Nev., S.D., Sask., Ut., Wyo.

Note: Wheeler and Henry (1985, Proc. Ent. Soc. Wash., 87: 700) noted that the New York record for *T. americanus*, a western species, given by Hardee et al. (1963, J. Econ. Ent., 56: 555-559) probably refers to *T. coelestialium* (Kirkaldy) or [here added] *T. saileri* Carvalho.

Trigonotylus antennatus Kelton, 1970

1970 *Trigonotylus antennatus* Kelton, Can. Ent., 102: 337. [B.C.].

Distribution: B.C., Cal., Col., Man., Ore., Sask., Ut.

Trigonotylus brooksi Kelton, 1970

1970 *Trigonotylus brooksi* Kelton, Can. Ent., 102: 334. [Sask.].

Distribution: B.C., Sask.

Trigonotylus canadensis Kelton, 1970

1970 *Trigonotylus canadensis* Kelton, Can. Ent., 102: 336. [Sask.].

Distribution: Alta., Man., Sask.

Trigonotylus coelestialium (Kirkaldy), 1902

1902 *Megaloceroea coelestialium* Kirkaldy, Trans. Ent. Soc. London, p. 266. [China].

1904 *Megaloceroea ruficornis*: Wirtner, An. Carnegie Mus., 3: 195.

1956 *Trigonotylus coelestialium*: Wagner, Schrift. Naturwiss. Ver. Schl.-Holst., 28: 71.

1976 *Trigonotylus colestialium* [*sic*]: Reid et al., Can. Ent., 108: 563.

Distribution: Alk., Alta., Ark., Ga., Ia., Ill., Ky., La., Man., Mich., Minn., Miss., N.B.,

N.C., N.H., N.S., N.Y., Oh., Ont., Pa., P.Ed., Que., S.D., Sask., Tenn., W.Va., Wyo. (Asia, Europe).

Note: According to Carvalho and Wagner (1957, Arq. Mus. Nac., 43: 130) and Kelton (1971, Can. Ent., 103: 701), many early records for *Trigonotylus ruficornis* (Geoffroy) in N. Am. refer to *Trigonotylus coelestialium*. Wheeler and Henry (1985, Proc. Ent. Soc. Wash., 87: 699-713) studied the life history, host plants, and damage to small grains and illustrated the adult and nymphal stages.

Trigonotylus confusus Reuter, 1909

1909 *Trigonotylus confusus* Reuter, Acta Soc. Sci. Fenn., 36(2): 6. [Md.].

1926 *Trigonotylus longicornis* Blatchley, Het. E. N. Am., p. 677. [Fla.]. Synonymized by Carvalho and Wagner, 1957, Arq. Mus. Nac., 43: 131.

Distribution: Fla., La., Md., Miss., Tex., Va.

Trigonotylus doddi (Distant), 1904

1904 *Miris doddi* Distant, An. Mag. Nat. Hist., (7)13: 269. [Queensland].

1904 *Miris dohertyi* Distant, Fauna Brit. India, Rhyn., 2: 425. [India]. Synonymized by Eyles, 1975, J. Nat. Hist., 9: 162.

1907 *Trigonotylus tenuis*: Van Duzee, Bull. Buff. Soc. Nat. Sci., 8: 30.

1926 *Trigonotylus brevipes*: Blatchley, Het. E. N. Am., p. 676.

1957 *Trigonotylus californicus*: Carvalho, *In* Carvalho and Wagner, Arq. Mus. Nac, 43: 128. Synonymized (in part) by Kelton, 1971, Can. Ent., 103: 697.

1957 *Trigonotylus dohertyi*: Carvalho and Wagner, Arq. Mus. Nac., 43: 132.

1970 *Trigonotylus doherti* [*sic*]: Kelton, Can. Ent., 102: 334.

1975 *Trigonotylus doddi*: Eyles, 1975, J. Nat. Hist., 9: 162.

Distribution: Alta., Ala., Ariz., Ark., Cal., Col., Del., Fla, Ga., La., Man., Md., N.C., N.J., Nev., S.C., Sask., Tenn., Tex., Ut., Va. (Africa, Asia, Australia, Europe, South Pacific Islands, Mexico to South America, West Indies).

Note: Northern records of this mainly southern species, e.g., New York (Pimentel and Wheeler, 1973, Environ. Ent., 2: 661), are based on misidentifications.

Trigonotylus flavicornis Kelton, 1970

1970 *Trigonotylus flavicornis* Kelton, Can. Ent., 102: 335. [Sask.].

Distribution: Man., Sask.

Trigonotylus longipes Slater and Wagner, 1955

1955 *Trigonotylus longipes* Slater and Wagner, Deut. Ent. Zeit., 2: 101. [Ut.].

1957 *Trigonotylus californicus* Carvalho, *In* Carvalho and Wagner, Arq. Mus. Nac., 43: 128. [Cal.]. Synonymized (in part) by Kelton, 1971, Can. Ent., 103: 694.

Distribution: B.C., Cal., Col., Ks., N.M., Neb., Nev., Ore., Tex., Ut., Wash. (Mexico).

Trigonotylus pulcher Reuter, 1876 [Fig. 72]

1876 *Trigonotylus pulcher* Reuter, Öfv. K. Svens. Vet.-Akad. Förh., 32(9): 59. [Tex.].

Distribution: Ariz., Cal., Col., Ks., La., Miss., Mo., N.M., Ok., Tex. (Mexico).

Note: The N.Y. (Drake, 1922, Tech. Publ. N.Y. St. Coll. For., Syracuse, 16: 71) and N.J., Mass., and Me. (Van Duzee, 1917, Univ. Cal. Publ. Ent., 2: 307) records for this species probably are based on misidentifications.

Trigonotylus ruficornis (Geoffroy), 1785

1785 *Cimex ruficornis* Geoffroy, *In* Fourcroy, Ent. Paris, p. 209. [Europe].

1872 *Trigonotylus ruficornis*: Uhler, Prelim. Rept. U.S. Geol. Surv. Mont., p. 409.

1872 *Miris viridis* Provancher, Nat. Can., 4: 78. [Que.]. Lectotype designated by Kelton, 1968, Nat. Can., 95: 1073. Synonymized by Kelton, 1971, Can. Ent., 103: 702.

1900 *Megalocraea* [*sic*] (*Trigonotylus*) *ruficornis*: Heidemann, Proc. Wash. Acad. Sci., 2: 503.

1952 *Trigonostylus* [*sic*] *ruficornis*: Frost, J. N.Y. Ent. Soc., 60: 238.

1957 *Trigonotylus montanus* Carvalho, *In* Carvalho and Wagner, Arq. Mus. Nac., 43: 137. [Col.]. Synonymized by Kelton, 1971, Can. Ent., 103: 702.

1957 *Trigonotylus viridis*: Carvalho and Wagner, Arq. Mus. Nac., 43: 152.

1972 *Trogonotylus* [*sic*] *ruficornis*: Beirne, Mem. Ent. Soc. Can., 85: 35.

Distribution: Alk., Alta., B.C., Cal., Col., Man., Mass., Mich., N.B., N.H., N.J., N.S., N.T., N.Y., Nfld., Ont., P.Ed., Que., Sask., Ut., Wash., Yuk. (Asia, Europe).

Note: *Trigonotylus ruficornis* has been confused with *T. coelestialium*, so many of the records in the literature are incorrect; Kelton's (1971, above) confirmed distribution is followed here. Records from Ariz., Ia., Ill., Ind., Ks., Minn., Neb., Oh., Pa., and Va. need verification.

Trigonotylus saileri Carvalho, 1957

1957 *Trigonotylus saileri* Carvalho, *In* Carvalho and Wagner, Arq. Mus. Nac., 43: 145. [Md.].

Distribution: Ariz., Fla., La., Md., Mass., Mo., N.B., N.C., N.J., N.S., N.Y., P.Ed., S.C., Tex., Va.

Trigonotylus slateri Carvalho, 1957

1957 *Trigonotylus slateri* Carvalho, *In* Carvalho and Wagner, Arq. Mus. Nac., 43: 147. [Tex.].

Distribution: Ariz., N.M., Tex. (Mexico).

Trigonotylus tarsalis (Reuter), 1876

1876 *Callimiris tarsalis* Reuter, Öfv. K. Svens. Vet.-Akad. Förh., 32(9): 60. [Tex., Wis.].

1909 *Trigonotylus tarsalis*: Reuter, Acta Soc. Sci. Fenn., 36(2): 6.

Distribution: Alta., Col., Conn., Ia., Ill., Ks., Man., Mass., Minn., Mo., N.B., N.D., N.H., N.M., N.S., N.Y., Neb., Nfld., Ont., P.Ed., Que., S.D., Sask., Tex., Ut., Wis.

Trigonotylus uhleri (Reuter), 1876

1876 *Callimiris uhleri* Reuter, Öfv. K. Svens. Vet.-Akad. Förh., 32(9): 60. [Boreal America].

1905 *Calimiris* [*sic*] *uhleri*: Crevecoeur, Trans. Ks. Acad. Sci., 19: 233.

1910 *Trigonotylus uhleri*: Smith, Ins. N.J., p. 163.

Distribution: Col.(?), Conn., Del., Ks.(?), Mass., Md., Miss., N.B., N.H., N.J., N.Y., P.Ed., Que., Va.

Note: Kelton (1971, Can. Ent., 103: 690) questioned Col. and Ks. records for this coastal species. We agree and suggest that these states be removed from future lists.

Subfamily Orthotylinae Van Duzee, 1916

Tribe Ceratocapsini Van Duzee, 1916

Genus *Ceratocapsus* Reuter, 1876

1876 *Ceratocapsus* Reuter, Öfv. K. Svens. Vet.-Akad. Förh., 32(9): 87. Type-species: *Ceratocapsus lutescens* Reuter, 1876. Designated by Kirkaldy, 1906, Trans. Am. Ent. Soc., 32: 127.

1876 *Trichia* Reuter, Öfv. K. Svens. Vet.-Akad. Förh., 32(9): 81. Preoccupied. Type-species: *Trichia punctulata* Reuter, 1876. Monotypic.

1887 *Melinna* Uhler, Ent. Am., 3: 68. Preoccupied. Type-species: *Melinna modesta* Uhler, 1887. Monotypic. Synonymized by Reuter, 1905, Öfv. F. Vet.-Soc. Förh., 47(19): 34.

1903 *Tiryus* Kirkaldy, Wien. Ent. Zeit., 22: 14. New name for *Trichia* Reuter. Synonymized by Reuter, 1908, An. Nat. Hofmus., 22: 174.

1903 *Hypereides* Kirkaldy, Wien. Ent. Zeit., 22: 14. New name for *Melinna* Uhler. Synonymized by Reuter, 1910, Acta Soc. Sci. Fenn., 37(3): 166.

Note: Knight (1941, Ill. Nat. Hist. Surv. Bull., 22: 108-109) and Henry (1979, Proc. Ent. Soc. Wash., 81: 403-407) provided partial keys to the eastern U.S. species of *Ceratocapsus*. Carvalho et al. (1983, U.S. Dept. Agr. Tech. Bull., 1676: 1-58) studied the South American species and summarized the predaceous habits of this New World genus.

Ceratocapsus advenus Blatchley, 1926

1926 *Ceratocapsus advenus* Blatchley, Het. E. N. Am., p. 823. [Fla.]. Neotype designated by Henry, 1979, Proc. Ent. Soc. Wash., 81: 407.

Distribution: Fla.

Ceratocapsus apicalis Knight, 1925

1925 *Ceratocapsus apicalis* Knight, Bull. Brook. Ent. Soc., 20: 46. [Tex.].

Distribution: Ariz., Col., Mo., S.D., Tex.

Ceratocapsus apicatus Van Duzee, 1921

1921 *Ceratocapsus apicatus* Van Duzee, Proc. Cal. Acad. Sci., (4)11: 128. [Cal.].

Distribution: B.C., Cal.

Ceratocapsus aurantiacus Henry, 1978

1978 *Ceratocapsus aurantiacus* Henry, Proc. Ent. Soc. Wash., 80: 385. [Fla.].

Distribution: Fla.

Ceratocapsus balli Knight, 1927

1927 *Ceratocapsus balli* Knight, Oh. J. Sci., 27: 146. [Fla.].

Distribution: Fla.

Ceratocapsus barbatus Knight, 1927

1887 *Melinna modesta*: Uhler, Ent. Am., 3: 69 (in part).

1892 *Melinna modesta*: Heidemann, Proc. Ent. Soc. Wash., 2: 225.

1927 *Ceratocapsus barbatus* Knight, Oh. J. Sci., 27: 150. [Va.].

Distribution: D.C., Md., Pa., Va., W.Va.

Note: Wheeler and Henry (1975, Trans. Am. Ent. Soc., 101: 360) discussed the confusion between this species and *Ceratocapsus modestus* (Uhler).

Ceratocapsus barberi Knight, 1930
1930 *Ceratocapsus barberi* Knight, Bull. Brook. Ent. Soc., 25: 190. [Ariz.].
Distribution: Ariz.

Ceratocapsus biformis Knight, 1927
1927 *Ceratocapsus biformis* Knight, Oh. J. Sci., 27: 152. [Col.].
Distribution: Ariz., Col., Mont.

Ceratocapsus bifurcus Knight, 1927
1927 *Ceratocapsus bifurcus* Knight, Oh. J. Sci., 27: 152. [Fla.].
Distribution: Fla.
Note: Van Duzee's (1909, Bull. Buff. Soc. Nat. Hist., 9: 182) "*Ceratocapsus* sp. nov?" probably refers to this species.

Ceratocapsus blatchleyi Henry, 1979
1979 *Ceratocapsus blatchleyi* Henry and Smith, J. Ga. Ent. Soc., 14: 214. *Nomen nudum.*
1979 *Ceratocapsus blatchleyi* Henry, Proc. Ent. Soc. Wash., 81: 408. [Ga.].
Distribution: Ga., Miss., N.J.

Ceratocapsus camelus Knight, 1930
1930 *Ceratocapsus camelus* Knight, Bull. Brook. Ent. Soc., 25: 187. [Ill.].
Distribution: Ill., Mo., Wis.

Ceratocapsus cecilsmithi Henry, 1979
1979 *Ceratocapsus cecilsmithi* Henry, Proc. Ent. Soc. Wash., 81: 409. [Ga.].
Distribution: Ga., N.J.

Ceratocapsus clavicornis Knight, 1925
1925 *Ceratocapsus clavicornis* Knight, Bull. Brook. Ent. Soc., 20: 47. [Ariz.].
Distribution: Ariz.

Ceratocapsus complicatus Knight, 1927
1927 *Ceratocapsus complicatus* Knight, Oh. J. Sci., 27: 148. [Mo.].
Distribution: Fla., Ill., Md., Miss., Mo., N.C., Tex., Va., W.Va.

Ceratocapsus cunealis Henry, 1985
1985 *Ceratocapsus cunealis* Henry, Proc. Ent. Soc. Wash., 87: 387. [B.C.].
Distribution: B.C., Id.

Ceratocapsus decurvatus Knight, 1930
1930 *Ceratocapsus decurvatus* Knight, Bull. Brook. Ent. Soc., 25: 194. [N.Y.].
Distribution: Ill., Md., N.Y., Pa.

Ceratocapsus denticulatus Knight, 1925
1925 *Ceratocapsus denticulatus* Knight, Bull. Brook. Ent. Soc., 20: 46. [Ariz.].
Distribution: Ariz.

Ceratocapsus digitulus Knight, 1923
1923 *Ceratocapsus digitulus* Knight, Conn. Geol. Nat. Hist. Surv. Bull., 34: 533. [N.Y.].
1980 *Ceratocapsus digitalus* [*sic*]: Kelton, Agr. Can., Res. Publ., 1703: 196, Map 61.
Distribution: Ill., Man., Mass., Md., Mo., N.C., N.Y., Ont., Que., Va.

Ceratocapsus divaricatus Knight, 1927
1927 *Ceratocapsus divaricatus* Knight, Oh. J. Sci., 27: 145. [Fla.].
Distribution: Fla.

Ceratocapsus downesi Knight, 1927
1919 *Ceratocapsus modestus*: Parshley, Occas. Pap. Univ. Mich., Mus. Zool., 71: 32.

1924 *Ceratocapsus fusiformis*: Downes, Proc. Ent. Soc. B.C., 21: 30.
1927 *Ceratocapsus downesi* Knight, Oh. J. Sci., 27: 151. [B.C.].
Distribution: B.C.

Ceratocapsus drakei Knight, 1923
1923 *Ceratocapsus drakei* Knight, Conn. Geol. Nat. Hist. Surv. Bull., 34: 533. [N.Y.].
Distribution: Alta., N.Y., Que., Sask.

Ceratocapsus elongatus (Uhler), 1894
1894 *Melinna elongatus* Uhler, Proc. Cal. Acad. Sci., (2)4: 257. [Mexico, Cal., Col., Fla., Tex.].
1910 *Ceratocapsus elongata*: Banks, Cat. Nearc. Hem.-Het., p. 32.
1914 *Ceratocapsus elongatus*: Van Duzee, Trans. San Diego Soc. Nat. Hist., 2: 27.
1916 *Tiryas* [*sic*] *elongatus*: Van Duzee, Check List Hem., p. 44.
1917 *Tiryus elongatus*: Van Duzee, Univ. Cal. Publ. Ent., 2: 381.
Distribution: Ariz., Cal., Col., Fla., Tex. (Mexico).

Ceratocapsus fanseriae Knight, 1930
1930 *Ceratocapsus fanseriae* Knight, Bull. Brook. Ent. Soc., 25: 191. [Ariz.].
Distribution: Ariz., N.M.

Ceratocapsus fasciatus (Uhler), 1877
1877 *Megacoelum fasciatus* Uhler, Bull. U.S. Geol. Geogr. Surv. Terr., 3: 421. [Col., Ill., Md., Mo., Pa., Tex.].
1887 *Melinna fasciata*: Uhler, Ent. Am., 3: 69.
1907 *Ceratocapsus fasciatus*: Tucker, Ks. Univ. Sci. Bull., 4: 59.
1910 *Ceratocapsus fasciatum*: Banks, Cat. Nearc. Hem.-Het., p. 32.
1910 *Ceratoscopus* [*sic*] *fasciatus*: Smith, Cat. Ins. N.J., p. 161.
Distribution: Cal., Col., D.C., Del., Ga., Ia., Ill., Ind., Ks., Mass., Md., Mich., Miss., Mo., N.C., N.J., N.Y., Oh., Ont., Pa., Tex., Va., W.Va., Wis.

Ceratocapsus fascipennis Knight, 1930
1930 *Ceratocapsus fascipennis* Knight, Bull. Brook. Ent. Soc., 25: 189. [Ariz.].
Distribution: Ariz.

Ceratocapsus fulvipennis Knight, 1927
1927 *Ceratocapsus fulvipennis* Knight, Oh. J. Sci., 27: 153. [Col].
Distribution: Col.

Ceratocapsus fuscinus Knight, 1923
1923 *Ceratocapsus fuscinus* Knight, Conn. Geol. Nat. Hist. Surv. Bull., 34: 531. [N.Y.].
Distribution: D.C., Fla., Ga., Ill., Ind., La., Md., Minn., Miss., Mo., N.C., N.Y., Oh., Ont., Pa., Que., W.Va.

Ceratocapsus fuscosignatus Knight, 1927
1927 *Ceratocapsus fuscosignatus* Knight, Oh. J. Sci., 27: 149. [Fla.].
Distribution: Ariz., Ark., Cal., Fla., Ia., N.M., Tex.

Ceratocapsus fusiformis Van Duzee, 1917
1917 *Ceratocapsus fusiformis* Van Duzee, Proc. Cal. Acad. Sci., (4)7: 270. [Cal.].
Distribution: Cal., Col., Nev.
Note: Downes (1935, Proc. Ent. Soc. B.C., 31: 47) noted that earlier records of *C. fusiformis* from B.C. actually belonged under *C. downesi*.

Ceratocapsus geminatus Knight, 1930
1930 *Ceratocapsus geminatus* Knight, Bull. Brook. Ent. Soc., 25: 192. [Col.].
Distribution: Col., Man., Sask.

Ceratocapsus hirsutus Henry, 1979
1979 *Ceratocapsus hirsutus* Henry, Proc. Ent. Soc. Wash., 81: 410. [Ga.].
Distribution: Ga.

Ceratocapsus husseyi Knight, 1930
1923 *Ceratocapsus sericus*: Knight, Conn. Geol. Nat. Hist. Surv. Bull., 34: 532 (in part).
1930 *Ceratocapsus husseyi* Knight, Bull. Brook. Ent. Soc., 25: 196. [Mich.].
Distribution: Ill., Mich., Pa.
Note: Knight (1930, above) pointed out that the allotype of *C. sericus* was a different species and used it as the holotype of *C. husseyi.*

Ceratocapsus incisus Knight, 1923
1923 *Ceratocapsus incisus* Knight, Conn. Geol. Nat. Hist. Surv. Bull., 34: 532. [N.Y.].
Distribution: Ia., Ill., Ind., Mo., N.Y., Oh., Ont., Pa., W.Va., Wis.

Ceratocapsus insperatus Blatchley, 1928
1928 *Ceratocapsus insperatus* Blatchley, J. N.Y. Ent. Soc., 36: 12. [Fla.].
Distribution: Fla., Ga., Miss.

Ceratocapsus juglandis Knight, 1930
1925 *Ceratocapsus denticulatus*: Knight, Bull. Brook. Ent. Soc., 20: 46 (in part).
1930 *Ceratocapsus juglandis* Knight, Bull. Brook. Ent. Soc., 25: 193. [Ariz.].
Distribution: Ariz.
Note: Knight (1930, above) pointed out that the allotype female of *C. denticulatus* is a female of *juglandis.*

Ceratocapsus keltoni Henry, 1985
1985 *Ceratocapsus keltoni* Henry, Proc. Ent. Soc. Wash., 87: 389. [Oh.].
Distribution: Oh., Ont.

Ceratocapsus knighti Henry, 1979
1979 *Ceratocapsus knighti* Henry, Proc. Ent. Soc. Wash., 81: 413. [Ga.].
Distribution: Ga.

Ceratocapsus lutescens Reuter, 1876
1876 *Ceratocapsus lutescens* Reuter, Öfv. K. Svens. Vet.-Akad. Förh., 32(9): 87. [Tex.].
Distribution: Fla., Ks., Mo., N.Y., Tex.
Note: Henry (1979, Proc. Ent. Soc. Wash., 81: 414) noted that all specimens he examined from the East [of Texas] were either *C. bifurcus* Knight or *C. rubricornis* Knight, suggesting that all records except those from Tex. need verification.

Ceratocapsus luteus Knight, 1923
1923 *Ceratocapsus luteus* Knight, Conn. Geol. Nat. Hist. Surv. Bull., 34: 527. [N.Y.].
Distribution: Ill., N.Y., Oh., W.Va.

Ceratocapsus mcateei Knight, 1927
1927 *Ceratocapsus mcateei* Knight, Oh. J. Sci., 27: 151. [Md.].
Distribution: Md.

Ceratocapsus minutus (Uhler), 1893

1893 *Melinna minuta* Uhler, Proc. Zool. Soc. London, p. 713. [St. Vincent Island].

1916 *Ceratocapsus minutus*: Van Duzee, Check List Hem., p. 44.

Distribution: Fla., Ks. (West Indies).

Ceratocapsus modestus (Uhler), 1887 [Fig. 69]

1887 *Melinna modesta* Uhler, Ent. Am., 3: 69 (in part). [Ill., Mass., Md., N.Y., Ont., Pa.]. Lectotype designated [no state locality given] by Henry, 1979, Proc. Ent. Soc. Wash., 81: 415.

1906 *Hypereides modesta*: Kirkaldy, Trans. Am. Ent. Soc., 32: 138.

1910 *Ceratocapsus modesta*: Banks, Cat. Nearc. Hem.-Het., p. 33.

1910 *Ceratoscopus* [*sic*] *modestus*: Smith, Cat. Ins. N.J., p. 161.

1916 *Ceratocapsus modestus*: Van Duzee, Check List Hem., p. 44.

Distribution: Col., Conn., D.C., Del., Fla., Ga., Ia., Ill., Ind., Ks., Man., Mass., Md., Mich., Miss., N.C., N.J., N.M., N.Y., Oh., Ont., Pa., Que., S.D., Sask., W.Va., Wis.

Note: Downes (1935, Proc. Ent. Soc. B.C., 31: 47) reported that previous records for this species (which he had emended to *C. fusiformis* in 1927, Proc. Ent. Soc. B.C., 23: 22) in B.C. belonged to *C. downesi*. Uhler's (1894, Proc. Zool. Soc. London, p. 191) record from Grenada is clearly an error and is dropped from the present distribution. Wheeler and Henry (1975, Trans. Am. Ent. Soc., 101: 360) discussed the confusion between this species and *C. barbatus* Knight. Wheeler and Henry (1978, Melsheimer Ent. Ser., 25: 6-10) reported on seasonal history and predaceous habits and described the fifth-instar nymph.

Ceratocapsus neoboroides Knight, 1930

1930 *Ceratocapsus neoboroides* Knight, Bull. Brook. Ent. Soc., 25: 197. [Ariz.].

Distribution: Ariz.

Ceratocapsus nevadensis Knight, 1968

1968 *Ceratocapsus nevadensis* Knight, Brig. Young Univ. Sci. Bull., 9(3): 156. [Nev.].

Distribution: Ariz., Nev.

Ceratocapsus nigellus Knight, 1923

1923 *Ceratocapsus nigellus* Knight, Conn. Geol. Nat. Hist. Surv. Bull., 34: 528. [N.Y.].

1958 *Ceratocapsis* [*sic*] *nigellus*: Rings, J. Econ. Ent., 51: 31.

1977 *Ceratocapsus* sp.: Wheeler and Henry, Great Lakes Ent., 10: 155.

Distribution: Ga., Ia., Ill., Ind., Mass., Md., Minn., Mo., N.C., N.J., N.Y., Oh., Pa., Va., W.Va., Wis.

Note: Henry (1979, Proc. Ent. Soc. Wash., 81: 415) clarified the Wheeler and Henry (1977, above) record.

Ceratocapsus nigrocephalus Knight, 1923

1923 *Ceratocapsus nigrocephalus* Knight, Conn. Geol. Nat. Hist. Surv. Bull., 34: 534. [N.H.].

Distribution: Ia., Man., Mich., Minn., Miss., N.H., Ont., Que., S.D., Sask.

Ceratocapsus nigrocuneatus Knight, 1968

1968 *Ceratocapsus nigrocuneatus* Knight, Brig. Young Univ. Sci. Bull., 9(3): 157. [Nev.].

Distribution: Nev.

Ceratocapsus nigropiceus Reuter, 1907
1907 *Ceratocapsus nigropiceus* Reuter, Öfv. F. Vet.-Soc. Förh., 49(5): 13. [Jamaica].
1982 *Ceratocapsus nigropiceus*: Henry and Wheeler, Fla. Ent., 65: 235.
Distribution: Fla. (Jamaica).

Ceratocapsus oculatus Knight, 1930
1930 *Ceratocapsus oculatus* Knight, Bull. Brook. Ent. Soc., 25: 190. [Tex.].
Distribution: Tex.

Ceratocapsus piceatus Henry, 1979
1979 *Ceratocapsus piceatus* Henry, Proc. Ent. Soc. Wash., 81: 415. [Pa.].
Distribution: D.C., Ga., Mass., N.Y., Pa.

Ceratocapsus pilosulus Knight, 1930
1923 *Ceratocapsus pilosus* Knight, Conn. Geol. Nat. Hist. Surv. Bull., 34: 526. [Mass.]. Preoccupied.
1930 *Ceratocapsus pilosulus* Knight, Bull. Brook. Ent. Soc., 25: 198. New name for *Ceratocapsus pilosus* Knight.
Distribution: Ia., Ill., Ind., Man., Mass., Mich., Minn., N.Y., Ont., Wis.

Ceratocapsus proximus Blatchley, 1934
1934 *Ceratocapsus proximus* Blatchley, Trans. Am. Ent. Soc., 60: 14. [Cal.].
Distribution: Cal.

Ceratocapsus pubescens Henry, 1979
1979 *Ceratocapsus pubescens* Henry, Proc. Ent. Soc. Wash., 81: 417. [Ga.].
Distribution: Ga.

Ceratocapsus pumilus (Uhler), 1887
1887 *Melinna pumila* Uhler, Ent. Am., 3: 69. [Mass. to Ill., Tex.].
1909 *Ceratocapsus pumilus*: Van Duzee, Bull. Buff. Soc. Nat. Sci., 9: 182.
1910 *Ceratocapsus pumila*: Banks, Cat. Nearc. Hem.-Het., p. 33.
1910 *Ceratoscopus [sic] pumilus*: Smith, Cat. Ins. N.J., p. 161.
Distribution: Ark., Col., Conn., D.C., Fla., Ill., Ind., Ks., La., Mass., Md., Me., Mich. Mo., N.D., N.J., N.S., N.Y., Oh., Ont., P.Ed., Pa., Que., Tex., W.Va.

Ceratocapsus punctulatus (Reuter), 1876
1876 *Trichia punctulata* Reuter, Öfv. K. Vet.-Akad. Förh., 32(9): 82. [Tex.].
1903 *Tiryus punctulatus*: Kirkaldy, Wien. Ent. Zeit., 22: 14.
1909 *Tiryas* [*sic*] *punctulatus*: Reuter, Acta. Soc. Sci. Fenn., 36(2): 71.
1910 *Ceratocapsus punctulatus*: Bergroth, An. Soc. Ent. Belg., 54: 68.
1910 *Tiryas* [*sic*] *punctulata*: Banks, Cat. Nearc. Hem.-Het., p. 35.
1984 *Ceratocapsus punctatus* [*sic*]: Snodgrass et al., Proc. Ent. Soc. Wash., 86: 852.
Distribution: Fla., Ga., Miss., N.C., Tex. (Mexico, West Indies).
Note: Downes' (1924, Proc. Ent. Soc. B.C., 21: 30) record for this species [in genus *Tiryus*] from B.C. is certainly based on a misidentification and therefore dropped from the present list.

Ceratocapsus quadrispiculus Knight, 1927
1927 *Ceratocapsus quadrispiculus* Knight, Oh. J. Sci., 27: 148. [La.].
Distribution: Ill., La., Md., Miss., Tex., W.Va.

Ceratocapsus rubricornis Knight, 1927
1927 *Ceratocapsus rubricornis* Knight, Oh. J. Sci., 27: 145. [Miss.].
Distribution: D.C., Del., Ill., Miss., Pa., W.Va.

Ceratocapsus rufistigmus Blatchley, 1926
1926 *Ceratocapsus rufistigmus* Blatchley, Het. E. N. Am., p. 829 (as Knight ms. name). [Fla.].
Distribution: Fla.

Ceratocapsus sericus Knight, 1923
1923 *Ceratocapsus sericus* Knight, Conn. Geol. Nat. Hist. Surv. Bull., 34: 530 (in part). [N.J.].
Distribution: Ill., Mich., N.J., N.Y., Pa., Wis.
Note: See note under *C. husseyi.*

Ceratocapsus seticornis Knight, 1953
1953 *Ceratocapsus seticornis* Knight, Ia. St. Coll. J. Sci., 27: 510. [Mo.].
Distribution: La., Miss., Mo.

Ceratocapsus setosus Reuter, 1909
1886 *Bryocoris pteridis*: Uhler, Check-list Hem. Het., p. 19.
1904 *Melinna* species nova?: Wirtner, An. Carnegie Mus., 3: 198.
1909 *Ceratocapsus setosus* Reuter, Acta Soc. Sci. Fenn., 36(2): 70. [D.C., Pa.].
Distribution: Cal.(?), D.C., Fla., Ga., Ill., Ind., Ky., Md., Miss., N.C., N.J., N.Y., Oh., Pa., Va., W.Va.
Note: The Cal. record is suspect and probably should be dropped from future lists. Wheeler and Henry (1977, Great Lakes Ent., 10: 149) clarified the Wirtner (1904, above) record.

Ceratocapsus spinosus Henry, 1978
1978 *Ceratocapsus spinosus* Henry, Proc. Ent. Soc. Wash., 80: 383. [Pa.].
Distribution: Miss., Pa.

Ceratocapsus taxodii Knight, 1927
1927 *Ceratocapsus taxodii* Knight, Oh. J. Sci., 27: 143. [Fla.].
Distribution: Ill., Fla., La., Miss., Mo., Tenn.

Ceratocapsus tricolor Knight, 1927
1927 *Ceratocapsus tricolor* Knight, Oh. J. Sci., 27: 153. [Col.].
Distribution: Col.

Ceratocapsus truncatus Knight, 1930
1930 *Ceratocapsus truncatus* Knight, Bull. Brook. Ent. Soc., 25: 195. [Fla.].
Distribution: Fla.

Ceratocapsus uniformis Knight, 1927
1927 *Ceratocapsus uniformis* Knight, Oh. J. Sci., 27: 147. [Mo.].
Distribution: D.C., Ill., Ind., Md., Miss., Mo., Oh., Pa., Va., W.Va.

Ceratocapsus vicinus Knight, 1923
1923 *Ceratocapsus vicinus* Knight, Conn. Geol. Nat. Hist. Surv. Bull., 34: 529. [N.Y.].
Distribution: Ill., Mo., N.J., N.Y., Pa., W.Va.

Ceratocapsus wheeleri Henry, 1979
1979 *Ceratocapsus wheeleri* Henry, Proc. Ent. Soc. Wash., 81: 421. [N.C.].
Distribution: La., N.C.

Genus *Pamillia* Uhler, 1887

1887 *Pamillia* Uhler, Ent. Am., 3: 31. Type-species: *Pamillia behrensii* Uhler, 1887. Monotypic.

Pamillia affinis Knight, 1925
1925 *Pamillia affinis* Knight, Bull. Brook. Ent. Soc., 20: 45. [N.M.].
1958 *Pamilia* [*sic*] *affinis*: Carvalho, Arq. Mus. Nac., 47: 120.
Distribution: N.M.

Pamillia behrensii Uhler, 1887
1887 *Pamillia behrensii* Uhler, Ent. Am., 3: 31. [Cal.].
1958 *Pamilia* [*sic*] *behrensii*: Carvalho, Arq. Mus. Nac., 47: 120.
1974 *Pamilia* [*sic*] *behrensi* [*sic*]: Schuh, Ent. Am., 47: 280.
Distribution: Cal.

Pamillia nyctalis Knight, 1925
1925 *Pamillia nyctalis* Knight, Bull. Brook. Ent. Soc., 20: 44. [Ariz.].
1958 *Pamilia* [*sic*] *nyctalis*: Carvalho, Arq. Mus. Nac., 47: 120.
Distribution: Ariz.

Pamillia pilosella Knight, 1925
1925 *Pamillia pilosella* Knight, Bull. Brook. Ent. Soc., 20: 45. [Ariz.].
1958 *Pamilia* [*sic*] *pilosella*: Carvalho, Arq. Mus. Nac., 47: 121.
Distribution: Ariz.

Genus *Pilophoropsis* Poppius, 1914

1914 *Pilophoropsis* Poppius, An. Soc. Ent. Belg., 58: 249. Type-species: *Pilophoropsis brachypterus* Poppius, 1914. Monotypic.

1927 *Renodaella* Knight, Ent. News, 38: 306. Type-species: *Renodaella nicholi* Knight, 1927. Original designation. Synonymized by Carvalho, 1955, Rev. Chil. Ent., 4: 227.

Note: Knight (1968, Brig. Young Univ. Sci. Bull, 9(3): 158) provided a key to species.

Pilophoropsis brachyptera Poppius, 1914
1914 *Pilophoropsis brachypterus* Poppius, An. Soc. Ent. Belg., 58: 249. [Ariz.].
1968 *Pilophoropsis balli* Knight, Brig. Young Univ. Sci. Bull., 9(3): 158. [Ariz.]. Synonymized by Polhemus and Polhemus, 1985, Pan-Pac. Ent., 61: 29.
1973 *Pilophoropsis brachyptera*: Steyskal, Stud. Ent., 16: 207.
Distribution: Ariz.

Pilophoropsis nicholi (Knight), 1927
1927 *Renodaella nicholi* Knight, Ent. News, 38: 306. [Ariz.].
1952 *Pilophoropsis nicholi*: Carvalho, An. Acad. Brasil. Ci., 24: 83.
Distribution: Ariz.

Genus *Schaffneria* Knight, 1966

1966 *Schaffneria* Knight, Ia. St. Coll. J. Sci., 41: 1. Type-species: *Schaffneria schaffneri* Knight,

1966. Original designation.

Note: Knight (1966, above) provided a key to species.

Schaffneria bureni Knight, 1966

1966 *Schaffneria bureni* Knight, Ia. St. Coll. J. Sci., 41: 5. [La.].

Distribution: La.

Schaffneria davisi (Knight), 1923

1923 *Pamillia davisi* Knight, Conn. Geol. Nat. Hist. Surv. Bull., 34: 535. [N.J.].

1958 *Pamilia* [*sic*] *davisisi* [*sic*]: Carvalho, Arq. Mus. Nac., 47: 121.

1966 *Schaffneria davisi*: Knight, Ia. St. Coll. J. Sci., 41: 1.

Distribution: Man., N.J.

Schaffneria hungerfordi Knight, 1966

1966 *Schaffneria hungerfordi* Knight, Ia. St. Coll. J. Sci., 41: 3. [Mich.].

Distribution: Mich.

Schaffneria pilophoroides (Knight), 1930

1930 *Ceratocapsus pilophoroides* Knight, Bull. Brook. Ent. Soc., 25: 197. [Ks.].

1966 *Schaffneria pilophoroides*: Knight, Ia. St. Coll. J. Sci., 41: 1.

Distribution: Ks.

Schaffneria schaffneri Knight, 1966

1966 *Schaffneria schaffneri* Knight, Ia. St. Coll. J. Sci., 41: 2. [Tex.].

Distribution: Ariz., Sask., Tex.

Tribe Halticini Kirkaldy, 1906

Genus *Anapus* Stål, 1858

1858 *Anapus* Stål, Stett. Ent. Zeit., 19: 188. Type-species: *Anapus kirschbaumi* Stål, 1858. Monotypic.

Note: Schuh and Lattin (1980, Am. Mus. Novit., 2697: 1-11) discussed the relationship of this genus to *Myrmecophyes* Fieber and other Halticini.

Anapus americanus Knight, 1959

1959 *Anapus americanus* Knight, Ia. St. Coll. J. Sci., 33: 421. [Wash.].

Distribution: Ut., Wash.

Genus *Halticus* Hahn, 1832

1832 *Halticus* Hahn, Wanz. Ins., 1: 113. Type-species: *Acanthia pallicornis* Fabricius, 1794, a junior synonym of *Cicada aptera* Linnaeus, 1758. Monotypic.

1865 *Halticocoris* Douglas and Scott, Brit. Hem., 1: 478. Unnecessary new name for *Halticus* Hahn, 1832.

Note: Henry (1983, Proc. Ent. Soc. Wash., 85: 610-611) gave a key to New World species.

Halticus apterus (Linnaeus), 1758

1758 *Cicada aptera* Linnaeus, Syst. Nat., ed. 10, 1: 438. [Europe].

1794 *Acanthia pallicornis* Fabricius, Ent. Syst., 4: 69. [Europe]. Synonymized by Thomson, 1871, Opusc. Ent., 4: 441.

1878 *Halticocoris pallicornis*: Uhler, Proc. Boston Soc. Nat. Hist., 19: 411.

1886 *Halticus apterus*: Uhler, Check-list Hem. Het., p. 20.

1983 *Halticus apterous* [*sic*]: Henry, Proc. Ent. Soc. Wash., 85: 610.

Distribution: Me., N.S., Ont., Que., Wis.(?). (Europe, northern Africa, USSR).

Note: Knight (1917, Can. Ent., 49: 249) questioned early U.S. records of this species; Henry (1983, above) summarized the confirmed distribution.

Halticus bractatus (Say), 1832 [Fig. 65]

1832 *Capsus (Cylapus) bractatus* Say, Descrip. Het. Hem. N. Am., p. 26. [Ind.].

1886 *Halticus bractatus*: Uhler, Check-list Hem. Het., p. 20.

1887 *Rhinocloa* [*sic*] *citri* Ashmead, Ent. Am., 3: 155. [Fla.]. Synonymized by Reuter, 1909, Acta Soc. Sci. Fenn., 36(2): 71.

1890 *Rhinacloa citri*: Atkinson, J. Asiatic Soc. Bengal, 58(2): 170.

1890 *Halticus minutus* Uhler, *In* Popenoe, Second Rept. Ks. Exp. Stn. p. 212. [Ks.]. Preoccupied. Synonymized by Reuter, 1909, Acta Soc. Sci. Fenn., 36(2): 71.

1890 *Agalliastes bractatus*: Uhler, *In* Popenoe, Second Rept. Ks. Exp. Stn., p. 212.

1892 *Halticus Uhleri* [*sic*] Giard, Comp. Rend. Soc. Biol., 9(4): 81. New name for *Halticus minutus* Uhler. Synonymized by Parshley, 1915, Psyche, 22: 23.

1890 *Halticus bracteatus* [*sic*]: Atkinson, J. Asiatic Soc. Bengal, 58(2): 119.

1909 *Haltica* [*sic*] *Uhleri* [*sic*]: Van Duzee, Bull. Buff. Soc. Nat. Hist., 9: 181.

1922 *Halticus citri*: Hussey, Occas. Pap. Mus. Zool., Univ. Mich., 118: 31.

Distribution: Col., Conn., Fla., Ga., Ia., Ind., Ks., Md., Mich., Mo., N.C., N.J., N.Y., Oh., Ont., Pa., Que., S.C., Ut., W.Va., Wis. (Hawaii, Mexico to South America, West Indies).

Note: Beyer (1921, U.S. Dept. Agr. Bull., 964: 1-27) discussed biology and injury to crop plants, and described and illustrated nymphal stages.

Halticus intermedius Uhler, 1904

1904 *Halticus intermedius* Uhler, Proc. U.S. Nat. Mus., 27: 360. [N.M.].

1926 *Halticus apterus*: Sibley et al., Bull. Lloyd Libr., 27(Ent. Ser. 5): 95.

Distribution: Ariz., Col., Conn., Man., Mich., Miss., N.C., N.D., N.M., N.Y., Oh., Ont., Pa., Que., Sask., W.Va.

Note: The specimen "taken in July at McLean Bogs, N.Y." (Sibley et al., 1926, above) was listed as *H. intermedius* by Knight (1928, Mem. N.Y. Agr. Exp. Stn., 101: 118).

Genus *Labops* Burmeister, 1835

1835 *Labops* Burmeister, Handb. Ent., 2: 279. Type-species: *Labops diopsis* Burmeister, 1835, a junior synonym of *Capsus sahlbergi* Fallén, 1829. Monotypic.

Note: Slater (1954, Bull. Brook. Ent. Soc., 49: 57-65, 89-94) reviewed the genus and provided a key to species.

Labops brooksi Slater, 1954

1954 *Labops brooksi* Slater, Bull. Brook. Ent. Soc., 49: 64. [Alta.].

Distribution: Alta., B.C., Man., Sask.

Labops burmeisteri Stål, 1858

1858 *Labops burmeisteri* Stål, Stett. Ent. Zeit., 19: 189. [USSR].

1926 *Labops burmeisteri*: Knight, Can. Ent., 58: 60.

Distribution: Alk., B.C., N.Y., Ont., Que., Wis., Yuk. (USSR).

Note: Knight (1926, above) listed this species from Ont., but questioned earlier records for Alk. and N.Y.

Labops chelifer Slater, 1954
1954 *Labops chelifer* Slater, Bull. Brook. Ent. Soc., 49: 90. [N.T.].
Distribution: N.T.

Labops hesperius Uhler, 1872
1872 *Labops hesperius* Uhler, Prelim. Rept. U.S. Geol. Surv. Mont., p. 416. [Col., Man., Mont., N.T.].
1889 *Labops hesperia*: Van Duzee, Can. Ent., 21: 4.
Distribution: Alta., Ariz., B.C., Cal., Col., Id., Man., Mont., N.D., N.H.(?), N.M., N.T., N.Y.(?), Neb., Ont., Ore., Que., S.D., Sask., Ut., Wash., Wyo.
Note: Northeastern U.S. records for this species probably apply to *L. hirtus* Knight.

Labops hirtus Knight, 1922
1922 *Labops hirtus* Knight, Can. Ent., 54: 258. [N.Y.].
1926 *Labops hesperius*: Blatchley, Het. E. N. Am., p. 797 (in part).
Distribution: Alta., B.C., Cal., Col., Id., Mass., Me., Mont., N.H., N.J., N.M., N.Y., Neb., Ont., Ore., Que., Sask., Ut., Vt., Wash., Wyo.

Labops tumidifrons Knight, 1922
1922 *Labops tumidifrons* Knight, Can. Ent., 54: 259. [B.C.].
Distribution: Alta., B.C., S.D., Sask., Ut.

Labops utahensis Slater, 1954
1945 *Labops tumidifrons*: Knowlton, J. Econ. Ent., 38: 707.
1954 *Labops utahensis* Slater, Bull. Brook. Ent. Soc., 49: 89. [Ut.].
Distribution: Col., Ut.
Note: Although no mention was made, specimens reported as *L. tumidfrons* by Knowlton (1945, above) were used in the original description of *L. utahensis*.

Labops verae Knight, 1929
1929 *Labops verae* Knight, Can. Ent., 61: 214. [Wash.].
Distribution: Alta., B.C., Man., N.T., Wash.

Genus *Labopella* Knight, 1929

1929 *Labopella* Knight, Can. Ent., 61: 215. Type-species: *Labopella claripennis* Knight, 1929. Monotypic.

Labopella claripennis Knight, 1929
1929 *Labopella claripennis* Knight, Can. Ent., 61: 215. [N.M.].
Distribution: N.M., Tex.

Genus *Myrmecophyes* Fieber, 1870

1870 *Myrmecophyes* Fieber, Verb. Zool.-Bot. Gesell. Wien, 20: 253. Type-species: *Myrmecophyes oshanini* Fieber, 1870, a junior synonym of *Diplacus alboornatus* Stål, 1855. Monotypic.

Myrmecophyes oregonensis Schuh and Lattin, 1980
1980 *Myrmecophyes oregonensis* Schuh and Lattin, Am. Mus. Novit., 2697: 4. [Ore.].
Distribution: Ore.

Genus *Orthocephalus* Fieber, 1858

1858 *Orthocephalus* Fieber, Wien. Ent. Monat., 2: 316. Type-species: *Lygaeus brevis* Panzer, 1798. Designated by Kirkaldy, 1906, Trans. Am. Ent. Soc., 32: 131.

Orthocephalus coriaceus (Fabricius), 1777

1777 *Acanthia coriacea* Fabricius, Gen. Ins., p. 299. [Europe].

1807 *Capsus mutabilis* Fallén, Monogr. Cimic. Suec., p. 98. [Sweden]. Synonymized by Thomson, 1871, Opusc. Ent., 4: 432.

1917 *Orthocephalus mutabilis*: Knight, Can. Ent., 49: 249.

1955 *Orthocephalus coriaceus*: Carvalho, Beitr. Ent., 5: 336.

Distribution: Me., N.Y., Oh., Pa., Va., W.Va. (Europe, North Africa).

Note: Wheeler (1985, Proc. Ent. Soc. Wash., 87: 85-93) reported on seasonal history and host plants and described nymphal stages.

Orthocephalus saltator (Hahn), 1835

1835 *Capsus saltator* Hahn, Wanz. Ins., 3: 11. [Europe].

1986 *Orthocephalus saltator*: Henry and Kelton, J. N.Y. Ent. Soc., 94: 51.

Distribution: N.B., N.S., N.Y., Ont., Pa., Va. (Europe, Asia).

Note: Henry and Kelton (1985, above) clarified previous misidentifications of this species and listed chicory, *Cichorium intybus* L., as the primary host in eastern U.S.

Tribe Orthotylini Van Duzee, 1916

Genus *Acaciacoris* Schaffner, 1977

1977 *Acaciacoris* Schaffner, Folia Ent. Mex., 38: 7. Type-species: *Heterocordylus xerophilus* Schaffner, 1967. Original designation.

Note: Schaffner (1977, Folia Ent. Mex., 38: 11) provided a key to the three known species, one of which occurs in the U.S.

Acaciacoris acaciae (Knight), 1918

1918 *Heterocordylus acaciae* Knight, Bull. Brook. Ent. Soc., 13: 111. [Tex.].

1977 *Acaciacoris acaciae*: Schaffner, Folia Ent. Mex., 38: 7.

Distribution: Tex. (Mexico).

Genus *Apachemiris* Carvalho and Schaffner, 1974

1974 *Apachemiris* Carvalho and Schaffner, Rev. Brasil. Biol., 33: 65. Type-species: *Apachemiris areolatus* Carvalho and Schaffner, 1974. Original designation.

Apachemiris areolatus Carvalho and Schaffner, 1974

1974 *Apachemiris areolatus* Carvalho and Schaffner, Rev. Brasil. Biol., 33: 66. [Tex.].

Distribution: N.M., Tex.

Genus *Argyrocoris* Van Duzee, 1912

1912 *Argyrocoris* Van Duzee, Bull. Buff. Soc. Nat. Sci., 10: 478. Type-species: *Argyrocoris scurrilis* Van Duzee, 1912. Monotypic.

Argyrocoris scurrilis Van Duzee, 1912

1912 *Argyrocoris scurrilis* Van Duzee, Bull. Buff. Soc. Nat. Hist., 10: 479. [Ariz.].

1963 *Argyrocoris scurrilus* [*sic*]: Watts, An. Ent. Soc. Am., 56: 377.

Distribution: Ariz., Col., N.M., Tex., Ut.

Genus *Ballella* Knight, 1959

1959 *Ballella* Knight, Ia. St. Coll. J. Sci., 33: 422. Type-species: *Ballella basicornis* Knight, 1959. Monotypic.

Ballella basicornis Knight, 1959

1959 *Ballella basicornis* Knight, Ia. St. Coll. J. Sci., 33: 422. [Ariz.].

Distribution: Ariz.

Genus *Blepharidopterus* Kolenati, 1845

1845 *Blepharidopterus* Kolenati, Melet. Ent., 2: 107. Type-species: *Lygaeus angulatus* Fallén, 1829. Designated by Kirkaldy, 1906, Trans. Am. Ent. Soc., 32: 128.

Blepharidopterus angulatus (Fallén), 1807

1807 *Lygaeus angulatus* Fallén, Monogr. Cimic. Suec., p. 81. [Sweden].

1922 *Blepharidopterus angulatus*: Knight, Can. Ent., 53: 285.

Distribution: Alta., B.C., N.S., P.Ed. (Europe, northern Africa, USSR).

Genus *Brachynotocoris* Reuter, 1880

1880 *Brachynotocoris* Reuter, Öfv. F. Vet.-Soc. Förh., 22: 22. Type-species: *Brachynotocoris puncticornis* Reuter, 1880. Monotypic.

Brachynotocoris puncticornis Reuter, 1880

1880 *Brachynotocoris puncticornis* Reuter, Öfv. F. Vet.-Soc. Förh., 22: 23. [Europe].

1892 *Orthotylus delicatus* Heidemann, Proc. Ent. Soc. Wash., 2: 226. [D.C.]. Preoccupied.

1917 *Orthotylus* (*Diommatus*) *delicatus*: Van Duzee, Univ. Cal. Publ. Ent., 2: 396.

1927 *Diaphnidia heidemanni* Knight, Proc. Biol. Soc. Wash., 40: 13. [D.C.]. Synonymized by Wheeler and Henry, 1980, Proc. Ent. Soc. Wash., 82: 568.

1961 *Brachynotocoris heidemanni*: Kelton, Can. Ent., 93: 568.

1968 *Labopidea utahensis* Knight, Brig. Young Univ. Sci. Bull., 9(3): 97. [Ut.]. Synonymized by Kelton, 1979, Can. Ent., 111: 757.

1980 *Brachynotocoris puncticornis*: Wheeler and Henry, Proc. Ent. Soc. Wash., 82: 568.

Distribution: D.C., Md., N.Y., Ut. (Europe).

Note: The record of this introduced species from Ut. [as *Labopidea utahensis*] indicates a separate introduction on its host, *Fraxinus excelsior* L. Wheeler and Henry (1980, above) summarized literature and gave biological notes.

Genus *Brooksetta* Kelton, 1979

1979 *Brooksella* Kelton, Can. Ent., 111: 949. Preoccupied. Type-species: *Asciodema inconspicua* Uhler, 1893. Original designation.

1979 *Brooksetta* Kelton, Can. Ent., 111: 1423. New name for *Brooksella* Kelton.

Note: Kelton (1979, Can. Ent., 111: 954) provided a key to the species of *Brooksetta* [as *Brooksella*]. Although Kelton (1979, above, 111: 1423) provided a replacement name for *Brooksella*, he did not formally make the new generic combinations for all species; therefore, because of this technicality, we present several new combinations.

Brooksetta althaeae (Hussey), 1924, New Combination

1891 *Orthotylus (Psallus) delicatus* Cook, Bull. Mich. Agr. Coll. Exp. Stn., 76: 10. [Mich.]. Preoccupied.
1922 *Orthotylus delicatus*: Hussey, Psyche, 29: 230.
1924 *Orthotylus althaeae* Hussey, Bull. Brook. Ent. Soc., 19: 165. New name for *Orthotylus delicatus* Cook.
1927 *Melanotrichus althaeae*: Knight, Can. Ent., 59: 142.
1958 *Orthotylus (Melanotrichus) althaeae*: Carvalho, Arq. Mus. Nac., 47: 113.
1979 *Brooksella althaeae*: Kelton, Can. Ent., 111: 953.
Distribution: Col., Ia., Ill., Ind., Mich., Minn., Mo., Wyo.

Brooksetta azteci (Knight), 1968, New Combination

1968 *Melanotrichus azteci* Knight, Brig. Young Univ. Sci. Bull., 9(3): 122. [Ariz.].
1979 *Brooksella azteci*: Kelton, Can. Ent., 111: 953.
Distribution: Ariz.

Brooksetta chelifer (Knight), 1927

1927 *Melanotrichus chelifer* Knight, Can. Ent., 59: 144. [Ariz.].
1927 *Melanotrichus brevirostris* Knight, Can. Ent., 59: 144. [Ut.]. Synonymized by Kelton, 1979, Can. Ent., 111: 952.
1958 *Orthotylus (Melanotrichus) chelifer*: Carvalho, Arq. Mus. Nac., 47: 113.
1958 *Orthotylus (Melanotrichus) brevirostris*: Carvalho, Arq. Mus. Nac., 47: 113.
1968 *Melanotrichus custeri* Knight, Brig. Young Univ. Sci. Bull., 9(3): 122. [S.D.]. Synonymized by Kelton, 1979, Can. Ent., 111: 953.
1979 *Brooksella chelifer*: Kelton, Can. Ent., 111: 952.
1980 *Brooksetta chelifer*: Kelton, Agr. Can. Res. Publ., 1703: 223.
Distribution: Alta. Ariz., Cal., Col., Man., N.M., Nev., S.D., Sask., Ut., Wyo.

Brooksetta ferox (Van Duzee), 1916, New Combination

1916 *Orthotylus ferox* Van Duzee, Proc. Cal. Acad. Sci., (4)6: 94. [Cal.].
1917 *Orthotylus (Orthotylus) ferox*: Van Duzee, Univ. Cal. Publ. Ent., 2: 391.
1927 *Melanotrichus ferox*: Knight, Can. Ent., 59: 142.
1958 *Orthotylus (Melanotrichus) ferox*: Carvalho, Arq. Mus. Nac., 47: 114.
1979 *Brooksella ferox*: Kelton, Can. Ent., 111: 950.
Distribution: Cal.

Brooksetta inconspicua (Uhler), 1893

1893 *Asciodema inconspicua* Uhler, Proc. Ent. Soc. Wash., 2: 376. [Ut.].
1914 *Orthotylus inconspicuus*: Van Duzee, Trans. San Diego Soc. Nat. Hist., 2: 27.
1917 *Orthotylus (Diommatus) inconspicuus*: Van Duzee, Univ. Cal. Publ. Ent., 2: 396.
1927 *Melanotrichus inconspicuus*: Knight, Can. Ent., 59: 142.
1927 *Melanotrichus atricornis* Knight, Can. Ent., 59: 145. [B.C.]. Synonymized by Kelton, 1979, Can. Ent., 111: 951.
1958 *Orthotylus (Melanotrichus) inconspicuus*: Carvalho, Arq. Mus. Nac., 47: 115.
1958 *Orthotylus (Melanotrichus) atricornis*: Carvalho, Arq. Mus. Nac., 47: 115.
1979 *Brooksella inconspicua*: Kelton, Can. Ent., 111: 950.
1980 *Brooksetta inconspicua*: Kelton, Agr. Can. Res. Publ., 1703: 222.

Distribution: Alta., B.C., Cal., Col., Man., Sask., Ut.

Brooksetta incurva (Knight), 1927

1927 *Melanotrichus incurvus* Knight, Can. Ent., 59: 143. [Col.].
1958 *Orthotylus (Melanotrichus) incurvus*: Carvalho, Arq. Mus. Nac., 47: 116.
1979 *Brooksella incurva*: Kelton, Can. Ent., 111: 953.
1980 *Brooksetta incurva*: Kelton, Agr. Can. Res. Publ., 1703: 222.
Distribution: Alta., Ariz., Col., Ia., Man., N.M., Sask., Wyo.

Brooksetta malvastri (Knight), 1968, New Combination

1968 *Melanotrichus malvastri* Knight, Brig. Young Univ. Sci. Bull., 9(3): 120. [Cal.].
1979 *Brooksella malvastri*: Kelton, Can. Ent., 111: 952.
Distribution: Ariz., Cal.

Brooksetta nevadensis (Knight), 1968, New Combination

1968 *Melanotrichus nevadensis* Knight, Brig. Young Univ. Sci. Bull., 9(3): 122. [Nev.].
1979 *Brooksella nevadensis*: Kelton, Can. Ent., 111: 950.
Distribution: Cal., Nev.

Brooksetta nicholi (Knight), 1927, New Combination

1927 *Melanotrichus nicholi* Knight, Can. Ent., 59: 145. [Ariz.].
1958 *Orthotylus (Melanotrichus) nicholi*: Carvalho, Arq. Mus. Nac., 47: 116.
1979 *Brooksella nicholi*: Kelton, Can. Ent., 111: 952.
Distribution: Ariz.

Brooksetta shoshonea (Knight), 1968, New Combination

1968 *Melanotrichus shoshonea* Knight, Brig. Young Univ. Sci. Bull., 9(3): 124. [Wyo.].
1979 *Brooksella shoshonea*: Kelton, Can. Ent., 111: 950.
Distribution: Id., Wyo.

Brooksetta tibialis (Van Duzee), 1916, New Combination

1916 *Orthotylus (Orthotylus) tibialis* Van Duzee, Proc. Cal. Acad. Sci., (4)6: 93. [Cal.].
1927 *Melanotrichus tibialis*: Knight, Can. Ent., 59: 142.
1958 *Labopidea tibialis*: Carvalho, Arq. Mus. Nac., 47: 82.
1979 *Brooksella tibialis*: Kelton, Can. Ent., 111: 950.
Distribution: B.C., Cal., Nev.

Brooksetta viridicata (Uhler), 1895

1895 *Orthotylus viridicatus* Uhler, Bull. Col. Agr. Exp. Stn., 31: 48. [Col.].
1916 *Orthotylus (Orthotylus) viridicatus*: Van Duzee, Check List Hem., p. 45.
1927 *Melanotrichus viridicatus*: Knight, Can. Ent., 59: 143.
1958 *Orthotylus (Melanotrichus) viridicatus*: Carvalho, Arq. Mus. Nac., 47: 119.
1979 *Brooksella viridicata*: Kelton, Can. Ent., 111: 952.
1980 *Brooksetta viridicata*: Kelton, Agr. Can. Res. Publ., 1703: 223.
Distribution: Alta., Cal., Col., Id., Man., N.M., Nev., Sask., Ut., Wyo.

Genus *Ceratopidea* Knight, 1968

1968 *Ceratopidea* Knight, Brig. Young Univ. Sci. Bull., 9(3): 100. Type-species: *Ceratopidea daleae* Knight, 1968. Original designation.

Ceratopidea daleae Knight, 1968

1968 *Ceratopidea daleae* Knight, Brig. Young Univ. Sci. Bull., 9(3): 100. [Nev.].
Distribution: Nev.

Genus *Cyrtorhinus* Fieber, 1858

1858 *Cyrtorhinus* Fieber, Wien. Ent. Monat., 2: 313. Type-species: *Capsus elegantulus* Meyer-Dür, a junior synonym of *Capsus caricis* Fallén, 1807. Monotypic.

Cyrtorhinus caricis (Fallén), 1807

1807 *Capsus caricis* Fallén, Monogr. Cimic. Suec., p. 102. [Sweden].
1955 *Cyrtorhinus caricis*: Carvalho and Southwood, Bol. Mus. Goeldi, 11: 42.
Distribution: Alk., Alta., Col., Man., Sask. (Europe).
Note: Carvalho and Southwood (1955, above) referred most previous records of *C. caricis* in N. Am. to the phyline *Tytthus vagus* (Knight). Kelton (1980, Agr. Can. Res. Publ. 1703, p. 262) gave additional records.

Genus *Daleapidea* Knight, 1968

1968 *Daleapidea* Knight, Brig. Young Univ. Sci. Bull., 9(3): 101. Type-species: *Daleapidea daleae* Knight, 1968. Original designation.

Daleapidea albescens (Van Duzee), 1918

1918 *Hadronema albescens* Van Duzee, Proc. Cal. Acad. Sci., (4)8: 297. [Cal.].
1928 *Hadronema* (*Aoplonema*) *albescens*: Knight, Can. Ent., 60: 181.
1968 *Daleapidea albescens*: Knight, Brig. Young Univ. Sci. Bull., 9(3): 102.
Distribution: Ariz., Cal., Nev.

Daleapidea daleae Knight, 1968

1968 *Daleapidea daleae* Knight, Brig. Young Univ. Sci. Bull., 9(3): 101. [Nev.].
Distribution: Cal., Nev.

Genus *Diaphnidia* Uhler, 1895

1895 *Diaphnidia* Uhler, Bull. Col. Agr. Exp. Stn., 31: 43. Type-species: *Diaphnidia debilis* Uhler, 1895. Designated by Kirkaldy, 1906, Trans. Am. Ent. Soc., 32: 128.

Note: Kelton (1965, Can. Ent., 97: 1028, 1030) provided a key to species.

Diaphnidia debilis Uhler, 1895

1895 *Diaphnidia debilis* Uhler, Bull. Col. Agr. Exp. Stn., 31: 43. [Col.].
Distribution: B.C., Cal., Col.
Note: Kelton (1965, Can. Ent., 97: 1027) noted that records for *D. debilis* in eastern N. Am. are based on misidentifications.

Diaphnidia richardsi Kelton, 1965

1965 *Diaphnidia richardsi* Kelton, Can. Ent., 97: 1027. [Cal.].
Distribution: Cal., Tex.

Genus *Diaphnocoris* Kelton, 1961

1961 *Diaphnocoris* Kelton, Can. Ent., 93: 566. Type-species: *Diaphnidea pellucida* Uhler, 1895, a junior synonym of *Malacocoris provancheri* Burque, 1887. Original designation.

Note: Kelton (1965, Can. Ent., 97: 1028, 1030) provided a key to species.

Diaphnocoris chlorionis (Say), 1832

1832 *Capsus chlorionis* Say, Descrip. Het. Hem. N. Am., p. 25. [Ind.].

1886 *Lygus chlorionis*: Uhler, Check-list Hem. Het., p. 18.
1905 *Orthotylus chlorionis*: Van Duzee, N.Y. St. Mus. Bull., 97: 552.
1965 *Diaphnocoris chlorionis*: Kelton, Can. Ent., 97: 1025.
1967 *Labopidea ainsliei*: Wray, Ins. N.C., 3rd. suppl., p. 27.
Distribution: Cal., Conn., D.C., Del., Ia., Ill., Ind., Mich., Miss., Mo. N.C., N.Y., Oh., Ont., Pa., Que., S.C., Tex., Va., Vt., W.Va., Wis.
Note: Wheeler and Henry (1976, An. Ent. Soc. Am., 69: 1095-1104) studied the hosts and biology of *D. chlorionis* and other species found on honeylocust, *Gleditsia triacanthos* L., described the egg and fifth-instar nymph, and clarified misidentifications. A B.C. record for this species by Downes (1927, Proc. Ent. Soc. B.C., 23: 14) needs verification.

Diaphnocoris provancheri (Burque), 1887
1887 *Malacocoris provancheri* Burque, *In* Provancher, Pet. Faune Ent. Can., 3: 144. [Canada]. Lectotype designated by Kelton, 1968, Nat. Can., 95: 1076.
1895 *Diaphnidia pellucida* Uhler, Bull. Col. Agr. Exp. Stn., 31: 44. [Col.]. Synonymized by Kelton, 1980, Can. Ent., 112: 343.
1904 *Diaphnidia* species nova?: Wirtner, An. Carnegie Mus., 3: 200.
1907 *Orthotylus translucens* Tucker, Ks. Univ. Sci. Bull., 4: 58. [Col.]. Synonymized by Kelton, 1965, Can. Ent., 97: 1025.
1912 *Diaphnidia provancheri*: Van Duzee, Can. Ent., 44: 322.
1915 *Diaphnidia pellicucuida* [*sic*]: LaFollette, J. Ent. Zool., 7: 128.
1950 *Diaphnida* [*sic*] *pellucida*: Moore, Nat. Can., 77: 246.
1950 *Diaphnida* [*sic*] *provancheri*: Moore, Nat. Can., 77: 246.
1961 *Diaphnocoris pellucida*: Kelton, Can. Ent., 93: 566.
1961 *Diaphnocoris provancheri*: Kelton, Can. Ent., 93: 566.
1977 *Diaphnidea* [*sic*] *pellucida*: Nixon and McPherson, Great Lakes Ent., 10: 215.
Distribution: Alta., Ark., B.C., Cal., Col., Conn., D.C., Ga., Ky., Ia., Ill., Ind., Man., Mass., Me., Mich., Minn., Mo., N.B., N.H., N.J., N.S., N.Y., Oh., Ont., Ore., P.Ed., Pa., Que., Sask., Tenn., Va., W.Va., Wash.
Note: Wheeler and Henry (1977, Great Lakes Ent., 10: 150) clarified the Wirtner (1904, above) record; Henry and Wheeler (1979, Proc. Ent. Soc. Wash., 81: 60-63) clarified records and hosts of *O. translucens* Tucker.

Diaphnocoris ulmi (Knight), 1927
1927 *Orthotylus ulmi* Knight, Can. Ent., 59: 179. [Minn.].
1965 *Diaphnocoris ulmi*: Kelton, Can. Ent., 97: 1025.
Distribution: Alta., B.C., Cal., Col., Ga., Ia., Man., Minn., Mo., N.H., N.Y., Ont., Que., Sask., Tex.

Genus *Dichaetocoris* Knight, 1968

1968 *Dichaetocoris* Knight, Brig. Young Univ. Sci. Bull., 9(3): 109. Type-species: *Dichaetocoris pinicola* Knight, 1968. Original designation.

Note: Knight (1968, above) provided a key to species.

Dichaetocoris anasazi Polhemus, 1984
1984 *Dichaetocoris anasazi* Polhemus, Pan-Pac. Ent., 60: 33. [Ut.].
Distribution: Col., Ut.

Dichaetocoris coloradensis Knight, 1968
1968 *Dichaetocoris coloradensis* Knight, Brig. Young Univ. Sci. Bull., 9(3): 116. [Col.].
Distribution: Ariz., Col., Nev.

Dichaetocoris geronimo Polhemus, 1985
1985 *Dichaetocoris geronimo* Polhemus, Pan-Pac. Ent., 61: 146. [Ariz.].
Distribution: Ariz.

Dichaetocoris juniperi Knight, 1968
1968 *Dichaetocoris juniperi* Knight, Brig. Young Univ. Sci. Bull., 9(3): 113. [Nev.].
Distribution: Nev.

Dichaetocoris merinoi Knight, 1968
1968 *Dichaetocoris merinoi* Knight, Brig. Young Univ. Sci. Bull., 9(3): 111. [Nev.].
Distribution: Cal., Nev.

Dichaetocoris minimus Knight, 1968
1968 *Dichaetocoris minimus* Knight, Brig. Young Univ. Sci. Bull., 9(3): 116. [Ut.].
Distribution: Ut.

Dichaetocoris mojave Polhemus, 1985
1985 *Dichaetocoris mojave* Polhemus, Pan-Pac. Ent., 61: 147. [Nev.].
Distribution: Nev.

Dichaetocoris nevadensis Knight, 1968
1968 *Dichaetocoris nevadensis* Knight, Brig. Young Univ. Sci. Bull., 9(3): 113. [Nev.].
Distribution: Cal., Col., Nev., Ut.

Dichaetocoris piceicola (Knight), 1927
1927 *Orthotylus piceicola* Knight, Can. Ent., 59: 180. [Col.].
1958 *Orthotylus* (*Orthotylus*) *piceicola*: Carvalho, Arq. Mus. Nac., 47: 107.
1985 *Dichaetocoris piceicola*: Polhemus, Pan-Pac. Ent., 61: 150.
Distribution: Col.

Dichaetocoris pinicola Knight, 1968
1968 *Dichaetocoris pinicola* Knight, Brig. Young Univ. Sci. Bull., 9(3): 110. [Nev.].
Distribution: Cal., Nev., Ut.

Dichaetocoris spinosus (Knight), 1925
1925 *Orthotylus spinosus* Knight, Bull. Brook. Ent. Soc., 20: 43. [Ariz.].
1968 *Dichaetocoris spinosus*: Knight, Brig. Young Univ. Sci. Bull., 9(3): 116.
Distribution: Ariz., Col.

Dichaetocoris utahensis Knight, 1968
1968 *Dichaetocoris utahensis* Knight, Brig. Young Univ. Sci. Bull., 9(3): 114. [Ut.].
Distribution: Ut.

Genus *Ephedrodoma* Polhemus and Polhemus, 1984

1984 *Ephedrodoma* Polhemus and Polhemus, Proc. Ent. Soc. Wash., 86: 550. Type-species: *Ephedrodoma multilineata* Polhemus and Polhemus, 1984. Original designation.

Ephedrodoma multilineata Polhemus and Polhemus, 1984
1984 *Ephedrodoma multilineata* Polhemus and Polhemus, Proc. Ent. Soc. Wash., 86: 551. [Nev.].

Distribution: Ariz., Cal., N.M., Nev., Tex., Ut.

Genus *Fieberocapsus* Carvalho and Southwood, 1955

1955 *Fieberocapsus* Carvalho and Southwood, Bol. Mus. Goeldi, 11: 33. Type-species: *Tytthus flaveolus* Reuter, 1871. Original designation.

Fieberocapsus flaveolus (Reuter), 1871
1871 *Tytthus flaveolus* Reuter, Not. Förh. Sällsk. Fauna Fl. Fenn., 11: 323. [Europe].
1980 *Fieberocapsus flaveolus*: Kelton, Can. Ent., 112: 342.
Distribution: Alta., Man., Ont., Sask., Yuk. (Europe).

Genus *Hadronema* Uhler, 1872

1872 *Hadronema* Uhler, Prelim. Rept. U.S. Geol. Surv. Mont, p. 412. Type-species: *Hadronema militaris* Uhler, 1872. Monotypic.

Note: Knight (1928, Can. Ent., 60: 180-181) provided a key to species.

Subgenus *Aoplonema* Knight, 1928

1928 *Hadronema* (*Aoplonema*) Knight, Can. Ent., 60: 177. Type-species: *Hadronema princeps* Uhler, 1894. Original designation.

Hadronema decoratum Uhler, 1894
1894 *Hadronema decorata* Uhler, Proc. Cal. Acad. Sci., (2)4: 250. [Mexico].
1928 *Hadronema* (*Aoplonema*) *decorata*: Knight, Can. Ent., 60: 181.
1973 *Hadronema decoratum*: Steyskal, Stud. Ent., 16: 207.
Distribution: Ariz., Cal. (Mexico).

Hadronema echinatum Gruetzmacher and Schaffner, 1977
1977 *Hadronema* (*Aoplonema*) *echinata* Gruetzmacher and Schaffner, Southwest. Ent., 2: 53. [Tex.].
Distribution: Tex.

Hadronema festivum Van Duzee, 1910
1910 *Hadronema festiva* Van Duzee, Trans. Am. Ent. Soc., 36: 80. [N.M.].
1928 *Hadronema* (*Aoplonema*) *festiva*: Knight, Can. Ent., 60: 180.
1973 *Hadronema festivum*: Steyskal, Stud. Ent., 16: 207.
Distribution: Ariz., N.M.

Hadronema princeps Uhler, 1894
1894 *Hadronema princeps* Uhler, Proc. Cal. Acad. Sci., (2)4: 251. [Cal., Ore., Wash.].
1928 *Hadronema* (*Aoplonema*) *princeps*: Knight, Can. Ent., 60: 181.
Distribution: Alta., B.C., Cal., Col., Id., Mont., Ore., Sask., Wash., Wyo.

Hadronema uhleri Van Duzee, 1928
1914 *Hadronema robusta*: Van Duzee, Trans. San Diego Soc. Nat. Hist., 2: 28.
1928 *Hadronema uhleri* Van Duzee, Pan-Pac. Ent., 4: 182. [Cal.].
1928 *Hadronema* (*Aoplonema*) *uhleri*: Knight, Can. Ent., 60: 181.
Distribution: Alta., Ariz., Cal., Col., Id., Nev., Ut., Wash.

Hadronema uniforme Knight, 1928
1928 *Hadronema (Aoplonema) uniformis* Knight, Can. Ent., 60: 181. [Ore.].
1973 *Hadronema uniforme*: Steyskal, Stud. Ent., 16: 207.
Distribution: Id., Nev., Ore.

Subgenus *Hadronema* Uhler, 1872

1872 *Hadronema* Uhler, Prelim. Rept. U.S. Geol. Surv. Mont., p. 412. Type-species: *Hadronema militaris* Uhler, 1872. Monotypic.

Hadronema bispinosum Knight, 1928
1928 *Hadronema (Hadronema) bispinosa* Knight, Can. Ent., 60: 179. [Wyo.].
1973 *Hadronema bispinosum*: Steyskal, Stud. Ent., 16: 207.
Distribution: Alta., Col., S.D., Sask., Wyo.

Hadronema breviatum Knight, 1928
1928 *Hadronema (Hadronema) breviata* Knight, Can. Ent., 60: 177. [Wyo.].
1973 *Hadronema (Hadronema) breviatum*: Steyskal, Stud. Ent., 16: 207.
Distribution: Ariz., Ut., Wyo.

Hadronema militare Uhler, 1872
1872 *Hadronema militaris* Uhler, Prelim. Rept. U.S. Geol. Surv. Mont., p. 412. [Cal., Col., Ut., Wyo.].
1872 *Oncotylus militaris*: Uhler, Prelim. Rept. U.S. Geol. Surv. Wyo., p. 471.
1928 *Hadronema (Hadronema) militaris*: Knight, Can. Ent., 60: 177.
1941 *Hadronema militare*: Knight, Ill. Nat. Hist. Surv. Bull., 22: 84.
Distribution: Alta., Ariz., B.C., Cal., Col., Ia., Id., Ind., Ks., Man., Mich., Mo., Mont., N.M., N.Y., S.D., Sask., Ut., Wis., Wyo. (Mexico).

Hadronema pictum Uhler, 1895
1895 *Hadronema picta* Uhler, Bull. Col. Agr. Exp. Stn., 31: 31. [Col., "Dak."].
1928 *Hadronema (Hadronema) picta*: Knight, Can. Ent., 60: 180.
1973 *Hadronema pictum*: Steyskal, Stud. Ent., 16: 207.
Distribution: Alta., Ariz., Col., Ks., N.D., N.M., Neb., Nev., S.D.

Hadronema simplex Knight, 1928
1928 *Hadronema (Hadronema) simplex* Knight, Can. Ent., 60: 178. [Wyo.].
Distribution: Alta., Col., Id., Man., Sask., Wyo.

Hadronema sinuatum Knight, 1928
1928 *Hadronema (Hadronema) sinuata* Knight, Can. Ent., 60: 179. [Ariz.].
1973 *Hadronema sinuatum*: Steyskal, Stud. Ent., 16: 207.
Distribution: Ariz., N.M., Ut.

Hadronema splendidum Gibson, 1918
1918 *Hadronema splendida* Gibson, Can. Ent., 50: 84. [N.M.].
1928 *Hadronema (Hadronema) splendida*: Knight, Can. Ent., 60: 181.
1973 *Hadronema splendidum*: Steyskal, Stud. Ent., 16: 207.
Distribution: Ariz., N.M., Ut.

Genus *Heterocordylus* Fieber, 1858

1858 *Heterocordylus* Fieber, Wien. Ent. Monat., 2: 316. Type-species: *Capsus tumidicornis* Kirschbaum, 1855, a junior synonym and homonym of *Capsus tumidicornis* Herrich-Schaeffer, 1835. Designated by Kirkaldy, 1906, Trans. Am. Ent. Soc., 32: 127.

Note: Schaffner (1967, J. Ks. Ent. Soc., 40: 579) discussed the relationships between *Heterocordylus* and *Acaciacoris* Schaffner.

Heterocordylus malinus Slingerland, 1909

1909 *Heterocordylus malinus* Slingerland, Proc. 54th An. Meet. W. N.Y. Hort. Soc., p. 90. [N.Y.].

1909 *Heterocordylus malinus* Reuter, Acta Soc. Sci. Fenn., 36(2): 71. [Ill., N.Y.]. Preoccupied.

1949 *Heterocordylus malina*: Froeschner, Am. Midl. Nat., 42: 142.

Distribution: Col., Ga., Ia., Ill., Ind., Ky., La., Mich., Minn., Miss., Mo., N.C., N.H., N.Y., Oh., Ont., Pa., Que., Tex., W.Va., Wis.

Note: Wheeler (1983, J. Wash. Acad. Sci., 73: 60-64) clarified the authorship of this species. Schaefer (1974, Mem., Conn. Ent. Soc., pp. 101-116) outined its history as an apple pest.

Genus *Heterotoma* Lepeletier and Serville, 1825

1825 *Heterotoma* Lepeletier and Serville, Encyclo. Meth., 10: 326. Type-species: *Cimex spissicornis* Fabricius, 1776, a junior synonym of *Cimex meriopterus* Scopoli, 1763. Monotypic.

Heterotoma merioptera (Scopoli), 1763

1763 *Cimex meriopterus* Scopoli, Ent. Carn., p. 131. [Europe].

1917 *Heterotoma merioptera*: Knight, Can. Ent., 49: 250.

1957 *Heterotoma meriopterum*: Downes, Proc. Ent. Soc. B.C., 54: 11.

1974 *Heterotoma meriopterus*: Slater, Mem., Conn. Ent. Soc., p. 156.

Distribution: B.C., N.S., N.Y., (Europe).

Note: Kelton (1982, Agr. Can. Monogr., 24: 99-100) summarized the N. Am. hosts and distribution and provided an illustration of the adult.

Genus *Hyalochloria* Reuter, 1907

1907 *Hyalochloria* Reuter, Öfv. F. Vet.-Soc. Förh., 49(5): 18. Type-species: *Hyalochloria caviceps* Reuter, 1907. Designated by Van Duzee, 1916, Univ. Cal. Publ. Ent., 1: 218.

Note: Henry (1978, Trans. Am. Ent. Soc., 104: 70-71) provided a key to the thirteen known species of *Hyalochloria*.

Hyalochloria caviceps Reuter, 1907

1907 *Hyalochloria caviceps* Reuter, Öfv. F. Vet.-Soc. Förh., 49(5): 20. [Jamaica].

1916 *Hyalochloria caviceps*: Van Duzee, Univ. Cal. Publ. Ent., 1: 218.

Distribution: Fla. (West Indies).

Note: Henry (1978, Trans. Am. Ent. Soc., 104: 73-75) described the male and illustrated the adult.

Genus *Ilnacora* Reuter, 1876

1876 *Ilnacora* Reuter, Öfv. K. Svens. Vet.-Akad. Förh., 32(9): 85. Type-species: *Ilnacora divisa* Reuter, 1876. Monotypic.

1876 *Corinala* Reuter, Öfv. K. Svens. Vet.-Akad. Förh., 32(9): 86. Type-species: *Ilnacora stalii* Reuter, 1876. Monotypic. Synonymized by Carvalho, 1958, Arq. Mus. Nac., 47: 77.

1877 *Sthenarops* Uhler, Bull. U.S. Geol. Geogr. Surv. Terr., 3: 418. Type-species: *Sthenarops chloris* Uhler, 1877. Designated by Kirkaldy, 1906, Trans. Am. Ent. Soc., 32: 129. Synonymized by Uhler, 1894, Proc. Cal. Acad. Sci., (2)4: 268.

Note: Knight (1963, Ia. St. J. Sci., 38: 176-178) and Knight and Schaffner (1976, Ia. St. J. Res., 50: 404-405) provided keys to species.

Ilnacora albifrons Knight, 1963

1963 *Ilnacora albifrons* Knight, Ia. St. J. Sci., 38: 165. [S.D.].

Distribution: Alta., Col., Ia., Ks., Man., Ok., Ont., Ore., S.D., Sask., Wash., Wis.(?), Wyo.

Note: Akingbohungbe et al. (1972, Univ. Wis. Res. Bull., R2396: 10) recorded Wis. with a question.

Ilnacora arizonae Knight, 1963

1963 *Ilnacora arizonae* Knight, Ia. St. J. Sci., 38: 171. [Ariz.].

Distribution: Ariz.

Ilnacora chloris (Uhler), 1877

1877 *Sthenarops chloris* Uhler, Bull. U.S. Geol. Geogr. Surv. Terr., 3: 419. [Col.].

1894 *Ilnacora chloris*: Uhler, Proc. Cal. Acad. Sci., (2)4: 268.

1916 *Ilnacora (Ilnacora) chloris*: Van Duzee, Check List Hem., p. 45.

Distribution: Col., Ks., N.M., Que., S.D., Wyo. (Mexico).

Ilnacora divisa Reuter, 1876

1876 *Ilnacora (Ilnacora) divisa* Reuter, Öfv. K. Svens. Vet.-Akad. Förh., 32(9): 86. [Tex.].

Distribution: Col., Ia., Ill., Ks., Minn., Mo., N.J., S.D., Tex., Wis.

Ilnacora furcata Knight, 1963

1963 *Ilnacora furcata* Knight, Ia. St. J. Sci., 38: 168. [N.M.].

Distribution: Ariz., N.M.

Ilnacora illini Knight, 1941

1941 *Ilnacora illini* Knight, Ill. Nat. Hist. Surv. Bull, 22: 83. [Ill.].

Distribution: Ia., Ill., Mo., Ok.

Ilnacora malina (Uhler), 1877

1877 *Sthenarops malina* Uhler, Bull. U.S. Geol. Geogr. Surv. Terr., 3: 419. [E. U.S., Ill. to Tex.].

1900 *Ilnacora malina*: Slosson, Ent. News, 11: 323.

1916 *Ilnacora (Corinala) malina*: Van Duzee, Check List Hem., p. 45.

Distribution: Conn., Ia., Ill., Ind., Ks., Mass., Md., Me., Mich., Mo., N.C., N.H., N.J., N.Y., Neb., Oh., Ok., Ont., Pa., Que., Tenn., Tex., W.Va., Wis.

Ilnacora nicholi Knight, 1963

1963 *Ilnacora nicholi* Knight, Ia. St. J. Sci., 38: 169. [Ariz.].

Distribution: Ariz.

Ilnacora nigrinasi (Van Duzee), 1916
1916 *Orthotylus* (*Orthotylus*) *nigrinasi* Van Duzee, Proc. Cal. Acad. Sci., (4)6: 104. [Tex.].
1958 *Ilnacora nigrinasi*: Carvalho, Arq. Mus. Nac., 47: 78.
Distribution: Tex.

Ilnacora santacatalinae Knight, 1963
1963 *Ilnacora santacatalinae* Knight, Ia. St. J. Sci., 38: 176. [Ariz.].
Distribution: Ariz.

Ilnacora spicata Knight, 1963
1963 *Ilnacora spicata* Knight, Ia. St. J. Sci., 38: 167. [Ariz.].
Distribution: Ariz.

Ilnacora stalii Reuter, 1876
1876 *Ilnacora* (*Corinala*) *stalii* Reuter, Öfv. K. Svens. Vet.-Akad. Förh., 32(9): 86. [N.Y., Tex.].
1887 *Parthenicus psalliodes*: Provancher, Pet. Faune Ent. Can., 3: 146.
1887 *Ilnacora stalii*: Van Duzee, Can. Ent., 19: 72.
1897 *Ilnacora stalli* [*sic*]: Seiss, Ent. News, 8: 67.
1904 *Ilnacora stali* [*sic*]: Wirtner, An. Carnegie Mus., 3: 200.
1979 *Ilnocora* [*sic*] *stalii*: Henry and Smith, J. Ga. Ent. Soc., 14: 214.
Distribution: Alta., Ariz., Ark., Col., D.C., Del., Ga., Ia., Ill., Ind., Ks., Man., Mich., Miss., Mo., Mont., N.C., N.J., N.M., N.Y., Oh., Ont., Pa., Sask., S.D., Tex., Va., W.Va., Wis., Wyo.

Ilnacora texana Knight and Schaffner, 1976
1976 *Ilnacora texana* Knight and Schaffner, Ia. St. J. Res., 50: 402. [Tex.].
Distribution: Tex.

Ilnacora vittifrons Knight, 1963
1963 *Ilnacora vittifrons* Knight, Ia. St. J. Sci., 38: 166. [S.D.].
Distribution: Ia., Ks., Man., N.D., S.D., Sask., Wis.(?), Wyo.

Genus *Ilnacorella* Knight, 1925

1925 *Ilnacorella* Knight, Can. Ent., 57: 91. Type-species: *Ilnacorella nigrisquamosa* Knight, 1925. Original designation.

Ilnacorella argentata Knight, 1925
1925 *Ilnacorella argentata* Knight, Can. Ent., 57: 94. [Mont.].
Distribution: Id., Mont., Ore., Ut.

Ilnacorella insignis (Van Duzee), 1916
1916 *Orthotylus* (*Orthotylus*) *insignis* Van Duzee, Proc. Cal. Acad. Sci., (4)6: 92. [Cal.].
1955 *Labopidea insignis*: Carvalho, Rev. Chil. Ent., 4: 224.
1979 *Ilnacorella insignis*: Kelton, Can. Ent., 111: 756.
Distribution: B.C., Cal.

Ilnacorella nigrisquamosa Knight, 1925
1925 *Illnacorella nigrisquamosa* Knight, Can. Ent., 57: 92. [Col.].
Distribution: Col., Wyo.

Ilnacorella sulcata Knight, 1925
1925 *Ilnacorella sulcata* Knight, Can. Ent., 57: 93. [Alta.].
Distribution: Alta., Col., Id., Wyo.

Ilnacorella viridis (Uhler), 1895
1895 *Ilnacora viridis* Uhler, Bull. Col. Agr. Exp. Stn., 31: 41. [Col.].
1916 *Ilnacora* (*Ilnacora*) *viridis*: Van Duzee, Check List Hem., p. 45.
1958 *Ilnacorella viridis*: Carvalho, Arq. Mus. Nac., 47: 79.
Distribution: Col., N.M.

Genus *Jobertus* Distant, 1893

1893 *Jobertus* Distant, Biol. Centr.-Am., Rhyn., 1: 421. Type-species: *Jobertus chrysolectrus* Distant, 1893. Monotypic.

Note: Maldonado (1980, J. Agr. Univ. Puerto Rico, 64: 308) gave a key to the six known species.

Jobertus chrysolectrus Distant, 1893
1893 *Jobertus chrysolectrus* Distant, Biol. Centr.-Am., Rhyn., 1: 421. [Mexico].
1982 *Jobertus chrysolectrus*: Henry and Wheeler, Fla. Ent., 65: 235.
Distribution: Fla. (Mexico, Central America, West Indies).

Genus *Labopidea* Uhler, 1877

1877 *Labopidea* Uhler, Bull. U.S. Geol. Geogr. Surv. Terr., 3: 415. Type-species: *Labopidea chloriza* Uhler, 1877, a junior synonym of *Tinicephalus simplex* Uhler, 1872. Monotypic.

Note: Kelton (1979, Can. Ent., 111: 753-758) revised this genus and provided a key to species.

Labopidea arizonae Knight, 1928
1928 *Labopidea arizonae* Knight, Can. Ent., 60: 233. [Ariz.].
Distribution: Ariz., Cal.

Labopidea brooksi Kelton, 1979
1979 *Labopidea brooksi* Kelton, Can. Ent., 111: 754. [Alta].
Distribution: Alta., Sask.

Labopidea discolor (Sahlberg), 1878
1878 *Orthotylus discolor* Sahlberg, K. Svens. Vet.-Akad. Handl., 16(4): 29. [Europe].
1979 *Labopidea discolor*: Kelton, Can. Ent., 111: 754.
Distribution: Man. (Palearctic).

Labopidea nigripes (Reuter), 1909
1909 *Orthotylus nigripes* Reuter, Acta. Soc. Sci. Fenn., 36(2): 68. [Nev.].
1916 *Labopidea nigripes*: Van Duzee, Univ. Cal. Publ. Ent., 1: 219.
Distribution: B.C., Cal., Col., Nev., Ore.

Labopidea nigrisetosa Knight, 1925
1925 *Labopidea nigrisetosa* Knight, Can. Ent., 57: 94. [Wyo.].
Distribution: Alta., B.C., Id., Ore., Wash., Wyo.

Labopidea pallida Knight, 1928
1928 *Labopidea pallida* Knight, Can. Ent., 60: 233. [Wash.].
Distribution: Alta., Sask., Wash.

Labopidea simplex (Uhler), 1872
1872 *Tinicephalus simplex* Uhler, Prelim. Rept. U.S. Geol. Surv. Mont., p. 417. [Col.].
1877 *Labopidea chloriza* Uhler, Bull. U.S. Geol. Geogr. Surv. Terr., 3: 416. [Ut.]. Syn-

onymized by Kelton, 1979, Can. Ent., 111: 753.
1909 *Hyoidea grisea* Reuter, Acta. Soc. Sci. Fenn., 36(2): 73. [Cal.]. Synonymized by Van Duzee, 1916, Univ. Cal. Publ. Ent., 1: 220.
1916 *Labopidea simplex*: Van Duzee, Univ. Cal. Publ. Ent., 1: 220.
1916 *Labopidea simplex* var. *simplex*: Van Duzee, Univ. Cal. Publ. Ent., 1: 220.
Distribution: Alta., Ariz., B.C., Cal., Col., Id., Man., Mont., N.M., Sask., Ut., Wyo.

Genus *Lindbergocapsus* Wagner, 1960

1960 *Lindbergocapsus* Wagner, Not. Ent., 40: 112. Type-species: *Lindbergocapsus lenaensis* Wagner, 1960, a junior synonym of *Orthotylus lenensis* Lindberg, 1928. Original designation.
1979 *Labopidicola* Kelton, Can. Ent., 111: 757. Type-species: *Labopidea idahoensis* Knight, 1968. Original designation. Synonymized by Henry, 1985, J. N.Y. Ent. Soc., 93: 1124.
Note: Henry (1982, Proc. Ent. Soc. Wash., 84: 1-15) reviewed this genus (as *Labopidicola*) and provided a key to the N. Am. species.

Lindbergocapsus ainsliei (Knight), 1928
1928 *Labopidea ainsliei* Knight, Can. Ent., 60: 235. [Ia.].
1979 *Labopidicola ainsliei*: Kelton, Can. Ent., 111: 757.
1985 *Lindbergocapsus ainsliei*: Henry, J. N.Y. Ent. Soc., 93: 1125.
Distribution: Ia., Ill., Mich., Pa., Tenn.

Lindbergocapsus allii (Knight), 1923
1923 *Orthotylus translucens*: Glenn, J. Econ. Ent., 16: 79.
1923 *Labopidea allii* Knight, Bull. Brook. Ent. Soc., 18: 31. [Mo.].
1979 *Labopidicola allii*: Kelton, Can. Ent., 111: 757.
1985 *Lindbergocapsus allii*: Henry, J. N.Y. Ent. Soc., 93: 1125.
Distribution: Ark., Ia., Ill., Ind., Ks., Ky., Mich., Miss., Mo., N.C., Ok., Pa., Ut., W.Va.
Note: Henry (1982, Proc. Ent. Soc. Wash., 84: 1-15) reviewed the literature on *L. allii* as a pest of onions, *Allium* spp.

Lindbergocapsus cepulus (Henry), 1982
1982 *Labopidicola cepula* Henry, Proc. Ent. Soc. Wash., 84: 8. [Tex.].
1985 *Lindbergocapsus cepulus*: Henry, J. N.Y. Ent. Soc., 93: 1125.
Distribution: Tex.

Lindbergocapsus geminatus (Johnston), 1930
1930 *Labopidea geminata* Johnston, Bull. Brook. Ent. Soc., 25: 298. [Tex.].
1979 *Labopidicola geminata*: Kelton, Can. Ent., 111: 758.
1982 *Labopidicola geminatus*: Henry, Proc. Ent. Soc. Wash., 84: 9.
1985 *Lindbergocapsus geminatus*: Henry, J. N.Y. Ent. Soc., 93: 1125.
Distribution: Ark., Tex.

Lindbergocapsus idahoensis (Knight), 1968
1968 *Labopidea idahoensis* Knight, Brig. Young Univ. Sci. Bull., 9(3): 97. [Id.].
1979 *Labopidicola idahoensis*: Kelton, Can. Ent., 111: 757.
1985 *Lindbergocapsus idahoensis*: Henry, J. N.Y. Ent. Soc., 93: 1125.
Distribution: Alta., B.C., Col., Id., Man., Ore., Sask., Ut.

Lindbergocapsus planifrons (Knight), 1928
1928 *Labopidea planifrons* Knight, Can. Ent., 60: 234. [S.D.].

1979 *Labopidicola planifrons*: Kelton, Can. Ent., 111: 757.
1985 *Lindbergocapsus planifrons*: Henry, J. N.Y. Ent. Soc., 93: 1125.
Distribution: Ia., Man., S.D., Sask.

Genus *Lopidea* Uhler, 1872

1872 *Lopidea* Uhler, Prelim. Rept. U.S. Geol. Surv. Mont., p. 411. Type-species: *Capsus medius* Say, 1832. Monotypic.

1876 *Lomatopleura* Reuter, Öfv. K. Svens. Vet.-Akad. Förh., 32(9): 67. Type-species: *Lomatopleura caesar* Reuter, 1876. Monotypic. Synonymized by Knight, 1917, Ent. News, 28: 455.

Note: Two names, *L. fuscicornis* and *L. strigata*, used by Osborn (1898, Proc. Ia. Acad. Sci., 5: 233) apparently are Uhler manuscript names and, hence, are not included as valid species in the following list.

Lopidea aculeata Van Duzee, 1917
1917 *Lopidea aculeata* Van Duzee, Proc. Cal. Acad. Sci., (4)7: 271. [Wash.].
Distribution: B.C., Cal., Ore., Wash.

Lopidea amorphae Knight, 1923
1923 *Lopidea amorphae* Knight, Ent. News, 34: 65. [Minn.].
Distribution: Ia., Ill., Ks., Minn., Neb., S.D.

Lopidea ampla Van Duzee, 1917
1917 *Lopidea ampla* Van Duzee, Proc. Cal. Acad. Sci., (4)7: 272. [Cal.].
Distribution: B.C., Cal., Wash.

Lopidea angustata Knight, 1965
1965 *Lopidea angustata* Knight, Ia. St. J. Sci., 40: 12. [Cal.].
Distribution: Cal.

Lopidea anisacanthi Knight, 1962
1962 *Lopidea anisacanthi* Knight, Ia. St. J. Sci., 37: 31. [Ariz.].
Distribution: Ariz.

Lopidea apache Knight, 1918
1918 *Lopidea apache* Knight, Ent. News, 29: 173. [Ariz.].
Distribution: Ariz.

Lopidea arizonae Knight, 1918
1918 *Lopidea arizonae* Knight, Ent. News, 29: 172. [Ariz.].
Distribution: Ariz., N.M.

Lopidea arkansae Knight, 1965
1965 *Lopidea arkansae* Knight, Ia. St. J. Sci., 40: 4. [Ark.].
Distribution: Ark.

Lopidea audeni Knight, 1965
1965 *Lopidea audeni* Knight, Ia. St. J. Sci., 40: 9. [B.C.].
Distribution: B.C.

Lopidea balli Knight, 1923
1923 *Lopidea balli* Knight, Ent. News, 34: 66. [Col.].
Distribution: Alta., Col., Neb., Sask.

Lopidea barberi Knight, 1962
1962 *Lopidea barberi* Knight, Ia. St. J. Sci., 37: 33. [Ariz.].
Distribution: Ariz.

Lopidea becki Knight, 1968
1968 *Lopidea becki* Knight, Brig. Young Univ. Sci. Bull., 9(3): 98. [Nev.].
Distribution: Nev.

Lopidea bifurca Van Duzee, 1921
1921 *Lopidea bifurca* Van Duzee, Proc. Cal. Acad. Sci., (4)11: 126. [Ore.].
Distribution: Ore.

Lopidea bisselli Knight, 1965
1965 *Lopidea bisselli* Knight, Ia. St. J. Sci., 40: 4. [Ga.].
Distribution: Ga.

Lopidea bullata Knight, 1923
1923 *Lopidea bullata* Knight, Ent. News, 34: 71. [Cal.].
Distribution: Cal.

Lopidea bullata bullata Knight, 1923
1923 *Lopidea bullata* Knight, Ent. News, 34: 71. [Cal.].
Distribution: Cal.

Lopidea bullata fusca Knight, 1923
1923 *Lopidea bullata* var. *fusca* Knight, 1923, Ent. News, 34: 71. [Cal.].
Distribution: Cal.

Lopidea caesar (Reuter), 1876
1876 *Lomatopleura caesar* Reuter, Öfv. K. Svens. Vet.-Akad. Förh., 32(9): 67. [Pa.].
1917 *Lopidea caesar*: Knight, Ent. News, 28: 455.
Distribution: Cal.(?), Col., Conn., Ga., Ia., Mich., N.C., N.J., N.Y., Pa., Tex. (Mexico?).
Note: Records from Cal. and Mexico are very likely based on misidentifications.

Lopidea calcaria Knight, 1965
1965 *Lopidea calcaria* Knight, Ia. St. J. Sci., 40: 11. [Ore.].
Distribution: Ore.

Lopidea calli Knight, 1962
1962 *Lopidea calli* Knight, Ia. St. J. Sci., 37: 29. [Ut.].
Distribution: Col., Ut.

Lopidea chamberlini Knight, 1965
1965 *Lopidea chamberlini* Knight, Ia. St. J. Sci., 40: 12. [Ore.].
Distribution: Ore.

Lopidea chandleri Moore, 1956
1956 *Lopidea chandleri* Moore, Ent. News, 67: 40. [Ill.].
Distribution: Ill.

Lopidea chelifer Knight, 1923
1923 *Lopidea chelifer* Knight, Ent. News, 34: 67. [N.M.].
Distribution: Col., N.M.

Lopidea confluenta (Say), 1832
1832 *Capsus confluentus* Say, Descrip. Het. Hem. N. Am., p. 23. [Mo.].
1876 *Lopidea confluens*: Reuter, Öfv. K. Svens. Vet.-Akad. Förh., 32(9): 66.

1894 *Lopidea confluenta*: Van Duzee, Bull. Buff. Soc. Nat. Sci., 5: 176.
Distribution: Ala., Col., Ga., Ia., Ill., Ind., Ks., Mass., Md., Me., Mich., Miss., Mo., Mont., N.C., N.H., N.J., N.Y., Oh., Ont., Pa., Que., Tex., W.Va.

Lopidea confraterna (Gibson), 1918
1918 *Hadronema confraterna* Gibson, Can. Ent., 50: 83. [N.M.].
1918 *Lopidea occidentalis* Van Duzee, Proc. Cal. Acad. Sci., (4)8: 296. [Cal.]. Synonymized by Carvalho, 1958, Arq. Mus. Nac., 47: 86.
1918 *Lopidea lepidii* Knight, Ent. News, 29: 175. [Ariz.]. Synonymized by Henry, 1985, J. N.Y. Ent. Soc., 93: 1125.
1928 *Lopidea confraterna*: Knight, Can. Ent., 60: 177.
Distribution: Alta., Ariz., Cal., Col., Ks., Mont., N.D., N.M.

Lopidea cuneata Van Duzee, 1910
1910 *Lopidea cuneata* Van Duzee, Trans. Am. Ent. Soc., 36: 79. [N.Y.].
Distribution: Ia., Ill., Minn., N.Y., W.Va.
Note: The Cal. specimen mentioned in the original description undoubtedly refers to a similar-appearing species.

Lopidea dakota Knight, 1923
1895 *Lomatopleura caesar*: Uhler, Bull. Col. Agr. Exp. Stn., 31: 31.
1923 *Lopidea dakota* Knight, Ent. News, 34: 67. [N.D.].
Distribution: Alta., B.C., Col., Man., Minn., Mont., N.D., Neb., S.D., Sask., Ut., Wyo.

Lopidea davisi Knight, 1917
1917 *Lopidea davisi* Knight, Ent. News, 28: 458. [Md.].
Distribution: Ark., D.C., Ill., Ind., Md., Minn., Mo., N.Y., Oh., S.D., Tenn., Va., W.Va.
Note: Cory and McConnell (1927, Univ. Md. Agr. Exp. Stn. Bull., 292: 15-22) discussed seasonal history and injury to phlox, *Phlox paniculata* L., and described the nymphal stages.

Lopidea dawsoni Knight, 1965
1965 *Lopidea dawsoni* Knight, Ia. St. J. Sci., 40: 7. [Neb.].
Distribution: Neb.

Lopidea denmani Knight and Schaffner, 1972
1972 *Lopidea denmani* Knight and Schaffner, Ia. St. J. Res., 47: 107. [Col.].
Distribution: Col., N.M.

Lopidea deserta Knight, 1968
1968 *Lopidea deserta* Knight, Brig. Young Univ. Sci. Bull., 9(3): 99. [Nev.].
Distribution: Nev.

Lopidea discreta Van Duzee, 1921
1921 *Lopidea discreta* Van Duzee, Proc. Cal. Acad. Sci., (4)11: 127. [Cal.].
Distribution: Cal.

Lopidea drakei Knight, 1962
1962 *Lopidea drakei* Knight, Ia. St. J. Sci., 37: 30. [Ore.].
Distribution: Ore., Wash.

Lopidea eremita Van Duzee, 1923
1923 *Lopidea eremita* Van Duzee, Proc. Cal. Acad. Sci., (4)12: 154. [Mexico].
1933 *Lopidea eremita*: Usinger, Pan-Pac. Ent., 9: 172.
Distribution: Ariz., Cal. (Mexico).

Lopidea eriogoni Knight, 1965
1965 *Lopidea eriogoni* Knight, Ia. St. J. Sci., 40: 10. [Ore.].
Distribution: Ore.

Lopidea falcata Knight, 1923
1923 *Lopidea falcata* Knight, Ent. News, 34: 72. [N.M.].
Distribution: N.M. (Mexico).

Lopidea falcicula Knight, 1923
1923 *Lopidea falcicula* Knight, Ent. News, 34: 68. [Col.].
Distribution: Col.

Lopidea fallax Knight, 1923
1923 *Lopidea fallax* Knight, Ent. News, 34: 69. [Cal.].
Distribution: Cal.

Lopidea flavicostata Knight and Schaffner, 1968
1968 *Lopidea flavicostata* Knight and Schaffner, Ia. St. J. Sci., 43: 75. [Cal.].
Distribution: Cal.

Lopidea fuscina Knight, 1923
1923 *Lopidea fuscina* Knight, Ent. News, 34: 68. [Cal.].
Distribution: Cal.

Lopidea fuscosa Knight, 1968
1968 *Lopidea fuscosa* Knight, Brig. Young Univ. Sci. Bull., 9(3): 100. [Nev.].
Distribution: Nev.

Lopidea gainesi Knight, 1962
1962 *Lopidea gainesi* Knight, Ia. St. J. Sci., 37: 32. [Tex.].
Distribution: Tex.

Lopidea garryae Knight, 1918
1918 *Lopidea garryae* Knight, Ent. News, 29: 175. [Ariz.].
Distribution: Ariz.

Lopidea heidemanni Knight, 1917
1904 *Lopidea* species nova?: Wirtner, An. Carnegie Mus., 3: 199.
1917 *Lopidea heidemanni* Knight, Ent. News, 28: 456. [N.Y.].
Distribution: Ark., Conn., D.C., Ga., Ia., Ill., Ind., Mass., Md., Minn., Miss., Mo., N.C., N.J., N.Y., Oh., Pa., Va., Vt., W.Va.
Note: Wheeler and Henry (1977, Great Lakes Ent., 10: 150) clarified the Wirtner (1904, above) record.

Lopidea hesperus (Kirkaldy), 1902
1873 *Capsus coccineus* Walker, Cat. Het. Brit. Mus., 6: 93. [Fla.]. Preoccupied.
1902 *Lomatopleura hesperus* Kirkaldy, Trans. Ent. Soc. London, p. 252. New name for *Capsus coccineus* Walker.
1904 *Lomatopleura coccineus*: Distant, An. Mag. Nat. Hist., (7)13: 109.
1916 *Lomatopleura hesperius* [*sic*]: Van Duzee, Check List Hem., p. 44.
1917 *Lomatopleura hesperia* [*sic*]: Van Duzee, Univ. Cal. Publ. Ent., 2: 383.
1917 *Lopidea hesperus*: Knight, Ent. News, 28: 455.
1917 *Lopidea reuteri* Knight, Ent. News, 28: 459. [Mo.]. Synonymized by Henry, 1985, J. N.Y. Ent. Soc., 93: 1126.
1926 *Lopidea hesperia* [*sic*]: Blatchley, Het. E. N. Am., p. 837.

Distribution: Conn., Fla., Ga., Ill., Mass., Mich., Minn., Miss., Mo., N.C., N.J., N.Y., Pa., Va., W.Va.

Lopidea incurva Knight, 1918
1918 *Lopidea incurva* Knight, Ent. News, 29: 214. [Mo.].
Distribution: D.C., Ia., Ill., Ind., Ks., Mo., Neb., Oh., Pa., Vt., W.Va., Wis.
Note: Wheeler and Henry (1976, An. Ent. Soc. Am., 69: 1101) studied the biology and habits of this species on honeylocust, *Gleditsia triacanthos* L., and described the egg and fifth-instar nymph.

Lopidea instabilis (Reuter), 1909
1909 *Lomatopleura instabile* Reuter, Acta Soc. Sci. Fenn., 36(2): 72. [D.C., Md., N.Y.].
1917 *Lopidea instabilis*: Knight, Ent. News, 28: 455.
1918 *Lopidea instabile*: Blackman, N.Y. St. Coll. For., Tech. Publ. 10, 28: 136.
Distribution: Conn., D.C., Fla., Ga., Md., Miss., Mo., Neb., N.C., N.D., N.Y., S.D., W.Va.

Lopidea intermedia Knight, 1918
1918 *Lopidea intermedia* Knight, Ent. News, 29: 210. [Tex.].
Distribution: Tex.

Lopidea johnstoni Knight, 1965
1965 *Lopidea johnstoni* Knight, Ia. St. J. Sci., 40: 3. [Miss.].
Distribution: Miss.

Lopidea knowltoni Knight, 1965
1965 *Lopidea knowltoni* Knight, Ia. St. J. Sci., 40: 6. [Ut.].
Distribution: Nev., Ut.

Lopidea lateralis Knight, 1918
1918 *Lopidea lateralis* Knight, Ent. News, 29: 174. [Ariz.].
Distribution: Ariz.

Lopidea lathyri Knight, 1923
1923 *Lopidea lathyrae* Knight, Ent. News, 34: 66. [Minn.].
1941 *Lopidea lathyri*: Knight, Ill. Nat. Hist. Surv. Bull., 22: 91.
Distribution: Alta., Ill., Man., Minn., N.D., Ok., Sask.

Lopidea major Knight, 1918
1918 *Lopidea major* Knight, Ent. News, 29: 215. [Tex.].
Distribution: Tex.

Lopidea malvastri Knight, 1962
1962 *Lopidea malvastri* Knight, Ia. St. J. Sci., 37: 35. [Wyo.].
Distribution: Wyo.

Lopidea marginalis (Reuter), 1909
1909 *Lomatopleura instabile* var. *marginalis* Reuter, Acta Soc. Sci. Fenn., 36(2): 72. [D.C., Md., N.Y.].
1923 *Lopidea marginalis*: Knight, Conn. Geol. Nat. Hist. Surv. Bull., 34: 508.
Distribution: Conn., D.C., Ind., Md., Minn., N.Y., Ont., W.Va.

Lopidea marginata Uhler, 1894
1894 *Lopidea marginata* Uhler, Proc. Cal. Acad. Sci., (2)4: 249. [Ariz., Cal., Col., Mexico].
Distribution: Ariz., Cal., Col. (Mexico).

Lopidea media (Say), 1832 [Fig. 66]

1832 *Capsus medius* Say, Descrip. Het. Hem. N. Am., p. 22. [Ind.].

1872 *Lopidea media*: Uhler, Prelim. Rept. U.S. Geol. Surv. Mont., p. 412.

1873 *Capsus floridanus* Walker, Cat. Het. Brit. Mus., 6: 97. [Fla.]. Synonymized by Knight, 1962, Ia. St. J. Sci., 37: 37; confirmed by Henry, 1985, J. N.Y. Ent. Soc., 93: 1127.

1902 *Lopidea marginata*: Slosson, Ent. News, 13: 8.

1904 *Lopidea floridana*: Distant, An. Mag. Nat. Hist., (7)13: 108.

Distribution: Alta., Cal., Col., Conn., D.C., Fla., Ga., Ia., Ind., Ks., Man., Mass., Md., Me., Mo., Mont., N.B., N.C., N.D., N.H., N.J., N.Y., Ok., Ont., Que., S.D., Sask., Ut., Va., W.Va., Wyo. (Mexico).

Lopidea medleri Akingbohungbe, 1972

1972 *Lopidea medleri* Akingbohungbe, An. Ent. Soc. Am., 65: 840. [Wis.].

Distribution: Wis.

Lopidea minima Knight, 1918

1918 *Lopidea minima* Knight, Ent. News, 29: 176. [Ariz.].

1918 *Hadronema infans* Van Duzee, Proc. Cal. Acad. Sci., (4)8: 296. [Cal.]. Synonymized by Carvalho 1958, Arq. Mus. Nac., 47: 86.

Distribution: Ariz., Cal. (Mexico).

Note: Knight (1928, Can. Ent., 60: 177) commented on *H. infans* before Carvalho (1958, above): "...I am unable to distinguish between this species and *Lopidea minima* Kngt."

Lopidea minor Knight, 1918

1918 *Lopidea minor* Knight, Ent. News, 29: 213. [Col.].

Distribution: Alta., Col., Ia., Ill., Ks., Man., Miss., N.C., N.D., N.Y., S.D., Sask., Tex., Wyo.

Lopidea mohave Knight, 1923

1923 *Lopidea mohave* Knight, Ent. News, 34: 70. [Cal.].

Distribution: B.C., Cal.

Lopidea navajo Knight, 1918

1918 *Lopidea navajo* Knight, Ent. News, 29: 173. [Ariz.].

Distribution: Ariz.

Lopidea nevadensis Knight, 1962

1962 *Lopidea nevadensis* Knight, Ia. St. J. Sci., 37: 35. [Nev.].

Distribution: Nev.

Lopidea nicholella Knight, 1965

1965 *Lopidea nicholella* Knight, Ia. St. J. Sci., 40: 2. [Ariz.].

Distribution: Ariz.

Lopidea nicholi Knight, 1923

1923 *Lopidea nicholi* Knight, Ent. News, 34: 70. [Wash.].

Distribution: Wash.

Lopidea nigridea Uhler, 1895

1895 *Lopidea nigridea* Uhler, Bull. Col. Agr. Exp. Stn., 31: 30. [Col.].

1898 *Lopidea nigrida* [*sic*]: Osborn, Proc. Ia. Acad. Sci., 5: 233.

Distribution: Cal., Col., Ia.

Lopidea nigridea hirta Van Duzee, 1921
1921 *Lopidea nigridea* var. *hirta* Van Duzee, Proc. Cal. Acad. Sci., (4)11: 128. [Cal.].
Distribution: Cal.

Lopidea nigridea nigridea Uhler, 1895
1895 *Lopidea nigridea* Uhler, Bull. Col. Agr. Exp. Stn., 31: 30. [Col.].
1921 *Lopidea nigridea*: Van Duzee, Proc. Cal. Acad. Sci., (4)11: 128.
Distribution: Cal., Col., Ia.

Lopidea oregona Hsiao, 1942
1942 *Lopidea oregona* Hsiao, Pan-Pac. Ent., 18: 160. [Ore.].
Distribution: Ore.

Lopidea paddocki Knight, 1962
1962 *Lopidea paddocki* Knight, Ia. St. J. Sci., 37: 34. [Cal.].
Distribution: Cal.

Lopidea petalostemi Knight, 1965
1941 *Lopidea minor*: Knight, Ill. Nat. Hist. Surv. Bull., 22: 88 (in part).
1965 *Lopidea petalostemi* Knight, Ia. St. J. Sci., 40: 3. [Ia.].
Distribution: Ia., Ill., N.Y.

Lopidea phlogis Knight, 1965
1950 *Lopidea minor*: Wray, Ins. N.C., Suppl. 2, p. 12.
1965 *Lopidea phlogis* Knight, Ia. St. J. Sci., 40: 3. [N.C.].
Distribution: N.C., W.Va.

Lopidea picta Knight, 1918
1918 *Lopidea picta* Knight, Ent. News, 29: 214. [Col.].
Distribution: Col.

Lopidea polingorum Knight, 1965
1965 *Lopidea polingorum* Knight, Ia. St. J. Sci., 40: 13. [Tex.].
Distribution: Tex.

Lopidea pteleae Knight and Schaffner, 1968
1968 *Lopidea pteleae* Knight and Schaffner, Ia. St. J. Sci., 43: 75. [Tex.].
Distribution: Tex.

Lopidea puella Van Duzee, 1921
1921 *Lopidea puella* Van Duzee, Proc. Cal. Acad. Sci., (4)11: 126. [Cal.].
Distribution: Cal.

Lopidea rainieri Knight, 1965
1965 *Lopidea rainieri* Knight, Ia. St. J. Sci., 40: 8. [Wash.].
Distribution: Wash.

Lopidea robiniae (Uhler), 1861
1861 *Capsus robiniae* Uhler, Proc. Ent. Soc. Phila., 1: 24. [Md.].
1872 *Lopidea media* var. *robiniae*: Uhler, Prelim. Rept. U.S. Geol. Surv. Mont., p. 412.
1916 *Lopidea robiniae*: Van Duzee, Check List Hem., p. 44.
Distribution: Conn., D.C., Ga., Ind., La., Mass., Md., Miss., Mo., N.C., N.Y., Oh., Ont, Pa.
Note: Leonard (1916, Ent. News, 27: 49-54) described and illustrated the immature stages.

Lopidea robusta (Uhler), 1894
1894 *Hadronema robusta* Uhler, Proc. Cal. Acad. Sci., (2)4: 250. [Mexico].
1914 *Hadronema robusta*: Van Duzee, Trans. San Diego Soc. Nat. Hist., 2: 28.
1928 *Lopidea robusta*: Van Duzee, Pan-Pac. Ent., 4: 182.
Distribution: B.C., Cal., Col., Ks., Tex. (Mexico).

Lopidea rolfsi Knight, 1965
1965 *Lopidea rolfsi* Knight, Ia. St. J. Sci., 40: 9. [Wash.].
Distribution: Wash.

Lopidea rubella Knight, 1965
1962 *Lopidea hesperus*: Knight, Ia. St. J. Sci., 37: 36.
1965 *Lopidea rubella* Knight, Ia. St. J. Sci., 40: 13. [Fla.].
Distribution: Fla.

Lopidea rubrofusca Knight, 1965
1965 *Lopidea rubrofusca* Knight, Ia. St. J. Sci., 40: 13. [Ut.].
Distribution: Ut.

Lopidea salicis Knight, 1917
1917 *Lopidea salicis* Knight, Ent. News, 28: 457. [N.Y.].
Distribution: Ia., Ill., Ks., Minn., Mo., N.Y., W.Va.

Lopidea sayi Knight, 1918
1918 *Lopidea sayi* Knight, Ent. News, 29: 212. [S.C.].
Distribution: Md., Oh., S.C., Va.

Lopidea sculleni Knight, 1965
1965 *Lopidea sculleni* Knight, Ia. St. J. Sci., 40: 9. [Ore.].
Distribution: Ore.

Lopidea scutata Knight, 1962
1962 *Lopidea scutata* Knight, Ia. St. J. Sci., 37: 34. [Ariz.].
Distribution: Ariz., Nev., Ut.

Lopidea serica Knight, 1923
1923 *Lopidea serica* Knight, Ent. News, 34: 69. [Col.].
Distribution: Alta., Col., Man., Sask.

Lopidea staphyleae Knight, 1917
1917 *Lopidea staphyleae* Knight, Ent. News, 28: 460. [N.Y.].
Distribution: Conn., Ia., Ill., Ind., Ks., Mass., Mich., Minn., Mo., N.Y., Ok., S.C., Va.

Lopidea staphyleae staphyleae Knight, 1917
1917 *Lopidea staphyleae* Knight, Ent. News, 28: 460. [N.Y.].
Distribution: Same as for species.

Lopidea staphyleae sanguinea Knight, 1917
1917 *Lopidea staphyleae* var. *sanguinea* Knight, Ent. News, 28: 461. [Mass.].
Distribution: Conn., Mass., W.Va.

Lopidea stitti Knight, 1962
1962 *Lopidea stitti* Knight, Ia. St. J. Sci., 37: 31. [Ariz.].
Distribution: Ariz.

Lopidea taurina Van Duzee, 1921
1921 *Lopidea taurina* Van Duzee, Proc. Cal. Acad. Sci., (4)11: 125. [Ore.].
Distribution: B.C., Ore.

Lopidea taurula Knight, 1923
1923 *Lopidea taurula* Knight, Ent. News, 34: 68. [Ore.].
Distribution: Id., Ore.

Lopidea teton Knight, 1923
1923 *Lopidea teton* Knight, Ent. News, 34: 70. [Minn.].
Distribution: Alta., Col., Ia., Ks., Man., Minn., Mont., N.D., Nev., S.D., Sask., Tex.

Lopidea texana Knight, 1918
1918 *Lopidea texana* Knight, Ent. News, 29: 215. [Tex.].
Distribution: Tex.

Lopidea trispicata Knight, 1965
1965 *Lopidea trispicata* Knight, Ia. St. J. Sci., 40: 14. [Fla.].
Distribution: Fla.

Lopidea usingeri Van Duzee, 1933
1933 *Lopidea usingeri* Van Duzee, Pan-Pac. Ent., 9: 96. [Cal.].
Distribution: Cal.

Lopidea utahensis Knight, 1965
1965 *Lopidea utahensis* Knight, Ia. St. J. Sci., 40: 6. [Ut.].
Distribution: Ut.

Lopidea ute Knight, 1923
1923 *Lopidea ute* Knight, Ent. News, 34: 70. [Col.].
Distribution: Col.

Lopidea wilcoxi Knight, 1965
1965 *Lopidea wilcoxi* Knight, Ia. St. J. Sci., 40: 11. [Wash.].
Distribution: Wash.

Lopidea wileyae Knight, 1923
1923 *Lopidea wileyi* Knight, Ent. News, 34: 71. [Tex.].
1968 *Lopidea wileyae*: Knight and Schaffner, Ia. St. J. Sci., 43: 76.
Distribution: Tex. (Mexico).

Lopidea wisteriae Moore, 1956
1956 *Lopidea wisteriae* Moore, Ent. News, 67: 39. [Ill.].
Distribution: Ill.

Lopidea yakima Knight, 1923
1923 *Lopidea yakima* Knight, Ent. News, 34: 69. [Wash.].
Distribution: B.C., Wash.

Lopidea yampae Knight, 1965
1965 *Lopidea yampae* Knight, Ia. St. J. Sci., 40: 7. [Col.].
Distribution: Alta., Col.

Genus *Lopidella* Knight, 1925

1925 *Lopidella* Knight, Bull. Brook. Ent. Soc., 20: 41. Type-species: *Lopidella flavoscuta* Knight, 1925. Original designation.

Lopidella flavoscuta Knight, 1925
1925 *Lopidella flavoscuta* Knight, Bull. Brook. Ent. Soc., 20: 41. [Ariz.].
Distribution: Ariz.

Genus *Macrotyloides* Van Duzee, 1916

1916 *Macrotyloides* Van Duzee, Univ. Cal. Publ. Ent., 1: 222. Type-species: *Macrotylus vestitus* Uhler, 1890. Original designation.

Macrotyloides symmetricus Knight, 1928

1928 *Macrotyloides symmetricus* Knight, Can. Ent., 60: 235. [Col.].

Distribution: Col.

Macrotyloides vestitus (Uhler), 1890

1890 *Macrotylus vestitus* Uhler, Trans. Md. Acad. Sci., 1: 88. [Cal.].

1916 *Macrotyloides vestitus*: Van Duzee, Univ. Cal. Publ. Ent., 1: 222.

1916 *Macrotyloides apicalis* Van Duzee, Univ. Cal. Publ. Ent., 1: 223. [Cal.]. Synonymized by Carvalho, 1958, Arq. Mus. Nac., 47: 89.

Distribution: Ariz., Cal., Col., Id.

Note: The Pa. record by Van Duzee (1917, Univ. Cal. Publ. Ent., 2: 400) was based on a record given by Wirtner (1904, An. Carnegie Mus., 3: 201). Wheeler and Henry (1977, Great Lakes Ent., 10: 150) referred Wirtner's record to *Icodema nigrolineatum* (Knight).

Genus *Mecomma* Fieber, 1858

1858 *Mecomma* Fieber, Wien. Ent. Monat., 2: 313. Type-species: *Capsus ambulans* Fallén, 1807. Monotypic.

Mecomma angustatum (Uhler), 1895

1895 *Globiceps angustatus* Uhler, Bull. Col. Agr. Exp. Stn., 31: 42. [Col.].

1922 *Globiceps dispar*: Knight, Can. Ent., 53: 285.

1955 *Mecomma mimetica* Carvalho and Southwood, Bol. Mus. Goeldi, 11: 57. [B.C.]. Synonymized by Kelton and Knight, 1962, Can. Ent., 94: 1300.

1962 *Mecomma angustata*: Kelton and Knight, 1962, Can. Ent., 94: 1300.

1973 *Mecomma mimeticum*: Steyskal, Stud. Ent., 16: 207.

1980 *Mecomma angustatum*: Kelton, Agr. Can. Publ., 1703: 261.

Distribution: Alk., Alta., B.C., Col., Man., N.H., N.T., Ont., Que., Sask., Wash., Wyo., Yuk.

Mecomma antennatum Van Duzee, 1917

1917 *Mecomma antennata* Van Duzee, Proc. Cal. Acad. Sci., (4)7: 275. [Cal.].

1973 *Mecomma antennatum*: Steyskal, Stud. Ent., 16: 207.

Distribution: Cal., Ore., Wash.

Mecomma bradora Kelton, 1960

1932 *Mecomma ambulans*: Walley, Can. Ent., 64: 152.

1960 *Mecomma bradora* Kelton, Can. Ent., 92: 572. [Que.].

Distribution: Que.

Mecomma gilvipes (Stål), 1858

1858 *Leptomerocoris giloipes* [*sic*] Stål, Stett. Ent. Zeit., 19: 187. [Alk.].

1873 *Leptomerocoris gilvipes*: Walker, Cat. Hem. Brit. Mus., 6: 144.

1879 *Chlamydatus gilvipes*: Reuter, Öfv. F. Vet.-Soc. Förh, 21: 57.

1883 *Mecomma gilvipes*: Reuter, Acta Soc. Sci. Fenn., 13(3): 386.

1887 *Chlamydatus luctuosus* Provancher, Pet. Faune Ent. Can., 3: 137. [Que.]. Synonymized by Kelton and Knight, 1962, Can. Ent., 94: 1298; lectotype designated by Kelton, 1968, Nat. Can., 95: 1076.

1900 *Mecomma (Leptomerocoris) gilvipes*: Heidemann, Proc. Wash. Acad. Sci., 2: 504.

1955 *Mecomma luctuosa*: Carvalho and Southwood, Bol. Mus. Goeldi, 11: 59.

1955 *Mecomma luctuosa pacifica* Carvalho and Southwood, Bol. Mus. Goeldi, 11: 61. [Wash.]. Synonymized by Kelton and Knight, 1962, Can. Ent., 94: 1298.

1973 *Mecomma luctuosum*: Steyskal, Stud. Ent., 16: 207.

1973 *Mecomma luctuosum pacificum*: Steyskal, Stud. Ent., 16: 207.

Distribution: Alk., Alta., B.C., Cal., Fla., Ill., Man., Me., Mich., N.H., N.J., N.S., N.T., N.Y., Nfld., Ont., Ore., Que., S.C., Sask., W.Va., Wash.

Genus *Melanotrichus* Reuter, 1875

1875 *Melanotrichus* Reuter, Rev. Crit. Caps., 1: 92. Type-species: *Phytocoris flavosparsus* Sahlberg, 1842. Designated by Kirkaldy, 1906, Trans. Am. Ent. Soc., 32: 127.

Note: In most Old World literature, this genus is treated as a subgenus of *Orthotylus* Fieber, but most American workers have followed Knight and others in recognizing *Melanotrichus* as a genus, as we do in this catalog.

Melanotrichus albocostatus (Van Duzee), 1918

1918 *Orthotylus albocostatus* Van Duzee, Proc. Cal. Acad. Sci., (4)8: 299. [Cal.].

1927 *Melanotrichus albocostatus*: Knight, Can. Ent., 59: 147.

1958 *Orthotylus (Melanotrichus) albocostatus*: Carvalho, Arq. Mus. Nac., 47: 112.

Distribution: Alta., Ariz., B.C., Cal., Col., Id., Nev., Sask., Ut., Wash.

Melanotrichus atriplicis Knight, 1968

1968 *Melanotrichus atriplicis* Knight, Brig. Young Univ. Sci. Bull., 9(3): 125. [Nev.].

Distribution: Alta., Nev.

Melanotrichus brindleyi Knight, 1968

1968 *Melanotrichus brindleyi* Knight, Brig. Young Univ. Sci. Bull., 9(3): 127. [Wyo.].

Distribution: Alta., Id., Man., Minn., Sask., Wyo.

Melanotrichus catulus (Van Duzee), 1916

1904 *Orthotylus* species nova?: Wirtner, An. Carnegie Mus., 3: 200.

1916 *Orthotylus catulus* Van Duzee, Proc. Cal. Acad. Sci., (4)6: 106. [N.Y.].

1917 *Orthotylus (Orthotylus) catulus*: Van Duzee, Univ. Cal. Publ. Ent., 2: 395.

1941 *Melanotrichus catulus*: Knight, Ill. Nat. Hist. Surv. Bull., 22: 97.

1958 *Orthotylus (Melanotrichus) catulus*: Carvalho, Arq. Mus. Nac., 47: 113.

Distribution: Conn., D.C., Ia., Ill., Me., Minn., Mo., N.Y., Ont., Pa., Que., W.Va.

Note: Wheeler and Henry (1977, Great Lakes Ent., 10: 150) clarified the Wirtner (1904, above) Pa. record.

Melanotrichus coagulatus (Uhler), 1877

1877 *Macrocoleus coagulatus* Uhler, Bull. U.S. Geol. Geogr. Surv. Terr., 3: 417. [Col.].

1907 *Megalocoleus (Macrocoleus) coagulatus*: Tucker, Ks. Univ. Sci. Bull., 4: 58.

1909 *Orthotylus coagulatus*: Reuter, Acta. Soc. Sci. Fenn., 36(2): 68.

1916 *Orthotylus (Orthotylus) coagulatus*: Van Duzee, Check List Hem., p. 45.

1927 *Melanotrichus coagulatus*: Knight, Can. Ent., 59: 142.

1958 *Orthotylus (Melanotrichus) coagulatus*: Carvalho, Arq. Mus. Nac., 47: 113.

Distribution: Alta., Ariz., Cal., Col., Ia., Id., Man., Sask., Tex., Ut., Wyo. (Mexico).

Note: Van Duzee (1917, Univ. Cal. Publ. Ent., 2: 392) noted that eastern records of this species probably refer to *M. flavosparsus*. Blatchley (1926, Het. E. N. Am.,

p. 854) also omitted this species from his eastern U.S. treatment; therefore, records from Me., N.H., N.J., N.Y., and Pa. are dropped from the present list. Wheeler and Henry (1977, Great Lakes Ent., 10: 150) referred Wirtner's (1904, An. Carnegie Mus., 3: 201) record of *Macrocoleus coagulatus* to *Melanotrichus flavosparsus*.

Melanotrichus concolor (Kirschbaum), 1856

1856 *Capsus concolor* Kirschbaum, Jahrb. Ver. Naturk. Nassau, 10: 249. [Europe].
1922 *Orthotylus concolor*: Knight, Can. Ent., 53: 284.
1927 *Melanotrichus concolor*: Knight, Can. Ent., 59: 142.
1958 *Orthotylus* (*Orthotylus*) *concolor*: Carvalho, Arq. Mus. Nac., 47: 98.
Distribution: B.C., Cal., Mass., Que., Wis.(?) (Europe, North Africa).
Note: Waloff (1966, J. Appl. Ecol., 3: 294) summarized the Pacific Northwest distribution and relative abundance of this introduced species that feeds on broom, *Cytisus scoparius* (L.) Wimm.

Melanotrichus elongatus Kelton, 1980

1980 *Melanotrichus elongatus* Kelton, Can. Ent., 112: 338. [Alta.].
Distribution: Alta., Sask.

Melanotrichus eurotiae Knight, 1968

1968 *Melanotrichus eurotiae* Knight, Brig. Young Univ. Sci. Bull., 9(3): 128. [Nev.].
Distribution: Nev.

Melanotrichus flavosparsus (Sahlberg), 1842

1842 *Phytocoris flavosparsus* Sahlberg, Acta Soc. Sci. Fenn., 1(2): 411. [Europe].
1872 *Lygus unicolor:* Provancher, Nat. Can., 4: 105 (in part). [Type-locality not given; Que?]. Synonymized by Van Duzee, 1912, Can. Ent., 44: 322; further clarified by Kelton, 1968, Nat. Can., 95: 1074.
1875 *Orthotylus* (*Melanotrichus*) *flavosparsus*: Reuter, Bih. Svens. Vet.-Akad. Handl., 3(1): 35.
1877 *Poeciloscytus sericeus*: Provancher, Pet. Faune Ent. Can., 3: 127.
1877 *Oncotylus pulchellus*: Provancher, Pet. Faune Ent. Can., 3: 148.
1900 *Macrocoleus chlorionis*: Forbes, 21st Rept. St. Ent. Ill., p. 90.
1904 *Macrocoleus coagulatus*: Wirtner, An. Carnegie Mus., 3: 201.
1907 *Orthotylus flavosparus*: Tucker, Trans. Ks. Acad. Sci., 20: 191.
1908 *Orthotylus chlorionis*: Horvath, An. Hist.-Nat. Mus. Nat. Hung., 6: 5.
1917 *Orthotylus flavisparus* [*sic*]: Adkins, Oh. J. Sci., 18: 61.
1941 *Melanotrichus flavosparsus*: Knight, Ill. Nat. Hist. Surv. Bull., 22: 96.
Distribution: Alta., B.C., Cal., Conn., D.C., Fla., Ga., Ill., Ind., Ks., Man., Me., Mich., Mo., N.J., N.S., N.Y., Oh., Ont., Pa., R.I., Que., Sask., Va., W.Va., Wis. (Europe, North Africa, South America).

Melanotrichus knighti Polhemus, 1985

1968 *Dichaetocoris brevirostris* Knight, Brig. Young Univ. Sci. Bull., 9(3): 115. [Nev.]. Preoccupied.
1985 *Melanotrichus knighti* Polhemus, Pan-Pac. Ent., 61: 149. New name for *Dichaetocoris brevirostris* Knight.
Distribution: Nev.

Melanotrichus leviculus Knight, 1927

1927 *Melanotrichus leviculus* Knight, Can. Ent., 59: 146. [N.Y.].
1958 *Orthotylus* (*Melanotrichus*) *leviculus*: Carvalho, Arq. Mus. Nac., 47: 116.

Distribution: Alta., Man., N.Y., Sask., Tex.

Melanotrichus mimus Knight, 1927
1927 *Melanotrichus mimus* Knight, Can. Ent., 59: 147. [Ariz.].
1958 *Orthotylus* (*Melanotrichus*) *mimus*: Carvalho, Arq. Mus. Nac., 47: 116.
Distribution: Ariz.

Melanotrichus mistus (Knight), 1925
1925 *Orthotylus mistus* Knight, Can. Ent., 57: 91. [Alta].
1927 *Melanotrichus mistus*: Knight, Can. Ent., 59: 142.
1958 *Orthotylus* (*Melanotrichus*) *mistus*: Carvalho, Arq. Mus. Nac., 47: 116.
Distribution: Alta., B.C., Id., Wash.

Melanotrichus ovatus (Van Duzee), 1916
1916 *Orthotylus ovatus* Van Duzee, Proc. Cal. Acad. Sci., (4)6: 105. [Cal.].
1917 *Orthotylus* (*Orthotylus*) *ovatus*: Van Duzee, Univ. Cal. Publ. Ent., 2: 393.
1958 *Orthotylus* (*Melanotrichus*) *ovatus*: Carvalho, Arq. Mus. Nac., 47: 117.
Distribution: Cal.

Melanotrichus pallens Knight, 1968
1968 *Melanotrichus pallens* Knight, Brig. Young Univ. Sci. Bull., 9(3): 125. [Nev.].
Distribution: Nev.

Melanotrichus senectus (Van Duzee), 1916
1916 *Orthotylus senectus* Van Duzee, Proc. Cal. Acad. Sci., (4)6: 102. [Col.].
1917 *Orthotylus* (*Orthotylus*) *senectus*: Van Duzee, Univ. Cal. Publ. Ent., 2: 393.
1927 *Melanotrichus senectus*: Knight, Can. Ent., 59: 142.
1958 *Orthotylus* (*Melanotrichus*) *senectus*: Carvalho, Arq. Mus. Nac., 47: 117.
Distribution: Col.

Melanotrichus stanleyaea (Knight), 1968
1968 *Dichaetocoris stanleyaea* Knight, Brig. Young Univ. Sci. Bull., 9(3): 115. [Nev.].
1985 *Melanotrichus stanleyaea*: Polhemus, Pan-Pac. Ent., 61: 149.
Distribution: Nev.

Melanotrichus stitti Knight, 1968
1968 *Melanotrichus stitti* Knight, Brig. Young Univ. Sci. Bull., 9(3): 128. [Ariz.].
Distribution: Ariz.

Melanotrichus symphoricarpi (Knight), 1968
1968 *Dichaetocoris symphoricarpi* Knight, Brig. Young Univ. Sci. Bull., 9(3): 114. [Nev.].
1985 *Melanotrichus symphoricarpi*: Polhemus, Pan-Pac. Ent., 61: 149.
Distribution: Nev.

Melanotrichus uniformis Knight, 1968
1968 *Melanotrichus uniformis* Knight, Brig. Young Univ. Sci. Bull., 9(3): 128. [Ut.].
Distribution: S.D., Ut.

Melanotrichus utahensis Knight, 1968
1968 *Melanotrichus utahensis* Knight, Brig. Young Univ. Sci. Bull., 9(3): 127. [Ut.].
Distribution: Ut.

Melanotrichus virescens (Douglas and Scott), 1865
1865 *Litosoma virescens* Douglas and Scott, Brit. Hem., p. 339. [England].
1960 *Melanotrichus virescens*: Scudder, Proc. Ent. Soc. B.C., 57: 22.
1966 *Orthotylus virescens*: Waloff, J. Appl. Ecol., 3: 294.
Distribution: B.C., Cal. (Europe).

Note: Waloff (1966, J. Appl. Ecol., 3: 292-311) summarized the Pacific Northwest distribution and relative abundance of this introduced species that feeds on broom, *Cytisus scoparius* (L.) Wimm.

Melanotrichus viridulus (Knight), 1928
1928 *Labopidea viridula* Knight, Can. Ent., 60: 234. [Ut.].
1979 *Melanotrichus viridulus*: Kelton, Can. Ent., 111: 756.
Distribution: Ut.

Melanotrichus wallisi Kelton, 1980
1980 *Melanotrichus wallisi* Kelton, Can. Ent., 112: 337. [Sask.].
Distribution: Sask.

Melanotrichus wileyae Knight, 1927
1927 *Melanotrichus wileyae* Knight, Can. Ent., 59: 146. [Ut.].
Distribution: Ut.

Genus *Noctuocoris* Knight, 1923

1923 *Noctuocoris* Knight, Conn. Geol. Nat. Hist. Surv. Bull., 34: 523. Type-species: *Orthotylus fumidus* Van Duzee, 1916. Original designation.

Note: Schwartz and Stonedahl (1986, Pan-Pac. Ent., 62: 237-247) revised this genus and gave a key to species.

Noctuocoris conspurcatus Schwartz and Stonedahl, 1986
1986 *Noctuocoris conspurcatus* Schwartz and Stonedahl, Pan-Pac. Ent., 62: 241. [Ariz.].
Distribution: Ariz., Col.

Noctuocoris fumidus (Van Duzee), 1916
1916 *Orthotylus fumidus* Van Duzee, Proc. Cal. Acad. Sci., (4)6: 127. [Col.].
1917 *Orthotylus* (*Diommatus*) *fumidus*: Van Duzee, Univ. Cal. Publ. Ent., 2: 395.
1922 *Orthotylus fumatus* [*sic*]: Hussey, Occas. Pap. Mus. Zool., Univ. Mich., 118: 33.
1923 *Noctuocoris fumidus*: Knight, Conn. Geol. Nat. Hist. Surv. Bull., 34: 523.
Distribution: B.C., Col., Man., Mass., Mich., Minn., N.D., N.Y., Ok., Ont., Pa., Que., S.D., Ut.

Genus *Orthotylus* Fieber, 1858

1858 *Orthotylus* Fieber, Wien. Ent. Monat., 2: 315. Type-species: *Cimex nassatus* Fabricius, 1787. Designated by Kirkaldy, 1906, Trans. Am. Ent. Soc., 32: 127.

1858 *Tichorhinus* Fieber, Wien. Ent. Monat., 2: 314. Preoccupied. Type-species: *Cimex nassatus* Fabricius, 1787. Designated by Kirkaldy, 1906, Trans. Am. Ent. Soc., 32: 127. Synonymized by Reuter, 1883, Acta Soc. Sci. Fenn., 13(3): 342.

1887 *Diommatus* Uhler, Ent. Am., 3: 32. Type-species: *Diommatus congrex* Uhler, 1887, a junior synonym of *Lygus dorsalis* Provancher, 1872. Monotypic. Synonymized by Reuter, 1909, Acta Soc. Sci. Fenn., 36(2): 67.

Subgenus *Neomecomma* Southwood, 1953

1953 *Neomecomma* Southwood, Trans. R. Ent. Soc. London, 104: 443. Type-species: *Capsus bilineatus* Fallén, 1807. Original designation.

Orthotylus candidatus Van Duzee, 1916
1916 *Orthotylus candidatus* Van Duzee, Proc. Cal. Acad. Sci., (4)6: 124. [N.H.].
1917 *Orthotylus* (*Diommatus*) *candidatus*: Van Duzee, Univ. Cal. Publ. Ent., 2: 395.
1953 *Orthotylus* (*Neomecomma*) *candidatus*: Southwood, Trans. R. Ent. Soc. London, 104: 444.
Distribution: Alta., Col., Minn., N.H., N.Y., Que., Ut.

Subgenus *Orthotylus* Fieber, 1858

1858 *Orthotylus* Fieber, Wien. Ent. Monat., 2: 315. Type-species: *Cimex nassatus* Fabricius, 1787. Designated by Kirkaldy, 1906, Trans. Am. Ent. Soc., 32: 127.

Orthotylus affinis Van Duzee, 1916
1916 *Orthotylus affinis* Van Duzee, Proc. Cal. Acad. Sci., (4)6: 114. [Cal.].
1917 *Orthotylus* (*Diommatus*) *affinis*: Van Duzee, Univ. Cal. Publ. Ent., 2: 395.
1958 *Orthotylus* (*Orthotylus*) *affinis*: Carvalho, Arq. Mus. Nac., 47: 96.
Distribution: Cal.

Orthotylus alni Knight, 1923
1923 *Orthotylus alni* Knight, Conn. Geol. Nat. Hist. Surv. Bull., 34: 521. [N.Y.].
1958 *Orthotylus* (*Orthotylus*) *alni*: Carvalho, Arq. Mus. Nac., 47: 96.
Distribution: Alta., Man., Me., Minn., N.Y., Ont., Que.

Orthotylus angulatus (Uhler), 1895
1895 *Diommatus angulatus* Uhler, Bull. Col. Agr. Exp. Stn., 31: 44. [Col.].
1916 *Orthotylus angulatus*: Van Duzee, Proc. Cal. Acad. Sci., (4)6: 115.
1916 *Orthotylus* (*Diommatus*) *angulatus*: Van Duzee, Check List Hem., p. 45.
1958 *Orthotylus* (*Orthotylus*) *angulatus*: Carvalho, Arq. Mus. Nac., 47: 97.
Distribution: Alta., Col., Ut.

Orthotylus basicornis Knight, 1923
1923 *Orthotylus basicornis* Knight, Conn. Geol. Nat. Hist. Surv. Bull., 34: 515. [N.Y.].
1958 *Orthotylus* (*Orthotylus*) *basicornis*: Carvalho, Arq. Mus. Nac., 47: 97.
Distribution: Ia., Ill., Ind., Mich., Minn., Mo., N.Y., Que., S.D., Sask.

Orthotylus brunneus Van Duzee, 1916
1916 *Orthotylus angulatus brunneus* Van Duzee, Proc. Cal. Acad. Sci., (4)6: 116. [Cal.].
1917 *Orthotylus angulatus* var. *brunneus*: Van Duzee, Univ. Cal. Publ. Ent., 2: 395.
1958 *Orthotylus* (*Orthotylus*) *brunneus*: Carvalho, Arq. Mus. Nac., 47: 97.
Distribution: Cal.

Orthotylus celtidis Henry, 1979, New Subgeneric Combination
1979 *Orthotylus celtidis* Henry, Melsheimer Ent. Ser., 27: 2. [Tex.].
Distribution: Miss., Tex.
Note: Henry (1979, above) neglected to place this species in a subgenus.

Orthotylus contrastus Van Duzee, 1925
1925 *Orthotylus contrastus* Van Duzee, Proc. Cal. Acad. Sci., (4)14: 400. [Ut.].
1958 *Orthotylus* (*Orthotylus*) *contrastus*: Carvalho, Arq. Mus. Nac., 47: 99.
Distribution: Cal., Ut.

Orthotylus cruciatus Van Duzee, 1916
1916 *Orthotylus cruciatus* Van Duzee, Proc. Cal. Acad. Sci., (4)6: 119. [Me].
1917 *Orthotylus* (*Diommatus*) *cruciatus*: Van Duzee, Univ. Cal. Publ. Ent., 2: 395.
1958 *Orthotylus* (*Orthotylus*) *cruciatus*: Carvalho, Arq. Mus. Nac., 47: 99.

Distribution: Conn., Mass., Me., N.S., N.Y., Que., Vt.

Orthotylus cuneatus Van Duzee, 1916

1916 *Orthotylus cuneatus* Van Duzee, Proc. Cal. Acad. Sci., (4)6: 117. [Cal.].
1917 *Orthotylus* (*Diommatus*) *cuneatus*: Van Duzee, Univ. Cal. Publ. Ent., 2: 395.
1958 *Orthotylus* (*Orthotylus*) *cuneatus*: Carvalho, Arq. Mus. Nac., 47: 99.
Distribution: Cal.

Orthotylus dodgei Van Duzee, 1921

1921 *Orthotylus dodgei* Van Duzee, Proc. Cal. Acad. Sci., (4)11: 129. [Cal.].
1958 *Orthotylus* (*Orthotylus*) *dodgei*: Carvalho, Arq. Mus. Nac., 47: 99.
Distribution: Cal.

Orthotylus dorsalis (Provancher), 1872

1872 *Lygus dorsalis* Provancher, Nat. Can., 4: 104. [Quebec (?); no locality given]. Lectotype designated by Kelton, 1968, Nat. Can., 95: 1076.
1887 *Orthotylus dorsalis*: Provancher, Pet. Faune Ent. Can., 3: 138.
1887 *Diommatus congrex* Uhler, Ent. Am., 3: 33. ["Canada," Ill., Mass., Me., N.Y.]. Synonymized by Van Duzee, 1912, Can. Ent., 44: 322.
1916 *Orthotylus* (*Diommatus*) *dorsalis*: Van Duzee, Check List Hem., p. 45.
1953 *Orthotylus* (*Orthotylus*) *dorsalis*: Southwood, Trans. R. Ent. Soc. London, 104: 444.
Distribution: Alta., B.C., Col., Conn., Ia., Ill., Ks., Man., Mass., Me., Mich., Minn., Mo., N.J., N.S., N.Y., Oh., Ont., Que., Sask., Ut., Wis.

Orthotylus formosus Van Duzee, 1916

1916 *Orthotylus formosus* Van Duzee, Proc. Cal. Acad. Sci., (4)6: 108. [Cal.].
1917 *Orthotylus* (*Diommatus*) *formosus*: Van Duzee, Univ. Cal. Publ. Ent., 2: 394.
1958 *Orthotylus* (*Orthotylus*) *formosus*: Carvalho, Arq. Mus. Nac., 47: 101.
Distribution: B.C., Cal.

Orthotylus fraternus Van Duzee, 1916

1914 *Orthotylus inconspicuus*: Van Duzee, Trans. San Diego Soc. Nat. Hist., 2: 27.
1916 *Orthotylus fraternus* Van Duzee, Proc. Cal. Acad. Sci., (4)6: 99. [Cal.].
1917 *Orthotylus* (*Orthotylus*) *fraternus*: Van Duzee, Univ. Cal. Publ. Ent., 2: 393.
Distribution: Cal.

Orthotylus fuscicornis Knight, 1927

1927 *Orthotylus fuscicornis* Knight, Can. Ent., 59: 177. [Col.].
1958 *Orthotylus* (*Orthotylus*) *fuscicornis*: Carvalho, Arq. Mus. Nac., 47: 101.
Distribution: Alta., Col., Man., Sask., Ut.

Orthotylus hamatus Van Duzee, 1918

1918 *Orthotylus hamatus* Van Duzee, Proc. Cal. Acad. Sci., (4)8: 298. [Cal.].
1958 *Orthotylus* (*Orthotylus*) *hamatus*: Carvalho, Arq. Mus. Nac., 47: 101.
Distribution: Cal.

Orthotylus juglandis Henry, 1979, New Subgeneric Combination

1979 *Orthotylus juglandis* Henry, Melsheimer Ent. Ser., 27: 2. [Tex.].
Distribution: Miss., Tex.
Note: Henry (1979, above) neglected to give this species subgeneric placement.

Orthotylus katmai (Knight), 1921

1921 *Tichorhinus katmai* Knight, Oh. J. Sci., 21: 108. [Alk.]
1924 *Orthotylus katmai*: Downes, Proc. Ent. Soc. B.C., 21: 30.
1958 *Orthotylus* (*Orthotylus*) *katmai*: Carvalho, Arq. Mus. Nac., 47: 102.

Distribution: Alk., Alta., B.C., Man.

Orthotylus knighti Van Duzee, 1916
1916 *Orthotylus knighti* Van Duzee, Proc. Cal. Acad. Sci., (4)6: 121. [N.Y.].
1917 *Orthotylus (Diommatus) knighti*: Van Duzee, Univ. Cal. Publ. Ent., 2: 395.
1958 *Orthotylus (Orthotylus) knighti*: Carvalho, Arq. Mus. Nac., 47: 102.
Distribution: Ind., N.Y., Oh.

Orthotylus languidus Van Duzee, 1916
1916 *Orthotylus languidus* Van Duzee, Proc. Cal. Acad. Sci., (4)6: 107. [Cal.].
1917 *Orthotylus (Diommatus) languidus*: Van Duzee, Univ. Cal. Publ. Ent., 2: 395.
1958 *Orthotylus (Orthotylus) languidus*: Carvalho, Arq. Mus. Nac., 47: 102.
Distribution: Cal.

Orthotylus lateralis Van Duzee, 1916
1916 *Orthotylus lateralis* Van Duzee, Proc. Cal. Acad. Sci., (4)6: 120. [Ks.].
1917 *Orthotylus (Diommatus) lateralis*: Van Duzee, Univ. Cal. Publ. Ent., 2: 395.
1958 *Orthotylus (Orthotylus) lateralis*: Carvalho, Arq. Mus. Nac., 47: 102.
Distribution: Col., Ia., Ill., Ks., Minn., Mo., Oh., Ok.

Orthotylus marginatus (Uhler), 1895
1895 *Cyrtorrhinus* [*sic*] *marginatus* Uhler, Bull. Col. Agr. Exp. Stn., 31: 43. [Col.].
1916 *Orthotylus (Diommatus) marginatus*: Van Duzee, Check List Hem., p. 45.
1958 *Orthotylus (Orthotylus) marginatus*: Carvalho, Arq. Mus. Nac., 47: 103.
Distribution: Col., Me., N.Y., Ont., Que.

Orthotylus minuendus Knight, 1925
1925 *Orthotylus minuendus* Knight, Can. Ent., 57: 90. [Ariz.].
1958 *Orthotylus (Orthotylus) minuendus*: Carvalho, Arq. Mus. Nac., 47: 104.
Distribution: Ariz.

Orthotylus modestus Van Duzee, 1916
1916 *Orthotylus modestus* Van Duzee, Proc. Cal. Acad. Sci., (4)6: 109. [N.Y.].
1917 *Orthotylus (Diommatus) modestus*: Van Duzee, Univ. Cal. Publ. Ent., 2: 394.
1953 *Orthotylus (Orthotylus) modestus*: Southwood, Trans. R. Ent. Soc. London, 104: 444.
Distribution: Conn., D.C., Ga., Ia., Ill., Ind., Mich., Minn., Miss., Mo., N.C., N.J., N.Y., Oh., Ont., Pa., W.Va.

Orthotylus modestus immaculatus Knight, 1923
1923 *Orthotylus modestus* var. *immaculatus* Knight, Conn. Geol. Nat. Hist. Surv. Bull., 34: 520. [N.Y.].
1958 *Orthotylus (Orthotylus) modestus* var. *immaculatus*: Carvalho, Arq. Mus. Nac., 47: 104.
Distribution: Ga., N.Y., Ont.

Orthotylus modestus modestus Van Duzee, 1916
1916 *Orthotylus modestus* Van Duzee, Proc. Cal. Acad. Sci., (4)6: 109. [N.Y.].
1923 *Orthotylus modestus*: Knight, Conn. Geol. Nat. Hist. Surv. Bull., 34: 519.
Distribution: Same as for species.

Orthotylus molliculus Van Duzee, 1916
1916 *Orthotylus molliculus* Van Duzee, Proc. Cal. Acad. Sci., (4)6: 113. [Cal.].
1917 *Orthotylus (Diommatus) molliculus*: Van Duzee, Univ. Cal. Publ. Ent., 2: 395.
1958 *Orthotylus (Orthotylus) molliculus*: Carvalho, Arq. Mus. Nac., 47: 104.
Distribution: Cal.

Orthotylus nassatus (Fabricius), 1787
1787 *Cimex nassatus* Fabricius, Mant. Ins., 2: 304. [Europe].
1977 *Orthotylus nassatus*: Henry, U.S. Dept. Agr. Coop. Plant Pest Rept., 2: 605.
Distribution: N.S., P.Ed., Pa. (Europe).
Note: Kelton (1982, Can. Ent., 114: 283) first reported this species from Canada and gave host plants.

Orthotylus necopinus Van Duzee, 1916
1916 *Orthotylus necopinus* Van Duzee, Proc. Cal. Acad. Sci., (4)6: 125. [N.H.].
1917 *Orthotylus (Diommatus) necopinus*: Van Duzee, Univ. Cal. Publ. Ent., 2: 395.
1958 *Orthotylus (Orthotylus) necopinus*: Carvalho, Arq. Mus. Nac., 47: 105.
Distribution: Ill., Minn., N.H., N.Y., Ont., W.Va.

Orthotylus neglectus Knight, 1923
1923 *Orthotylus neglectus* Knight, Conn. Geol. Nat. Hist. Surv. Bull., 34: 515. [N.Y.].
1958 *Orthotylus (Orthotylus) neglectus*: Carvalho, Arq. Mus. Nac., 47: 105.
Distribution: Conn., Ill., Ind., Man., N.S., N.Y., Ont., Que., Sask.

Orthotylus notabilis Knight, 1927
1927 *Orthotylus notabilis* Knight, Can. Ent., 59: 176. [S.D.].
1958 *Orthotylus (Orthotylus) notabilis*: Carvalho, Arq. Mus. Nac., 47: 105.
Distribution: Alta., Ia., Ks., Man., Minn., S.D., Sask.

Orthotylus nyctalis Knight, 1927
1927 *Orthotylus nyctalis* Knight, Can. Ent., 59: 181. [Minn.].
1958 *Orthotylus (Orthotylus) nyctalis*: Carvalho, Arq. Mus. Nac., 47: 105.
Distribution: Ia., Ill., Minn., N.Y., Wis.

Orthotylus ornatus Van Duzee, 1916 [Fig. 67]
1916 *Orthotylus ornatus* Van Duzee, Proc. Cal. Acad. Sci., (4)6: 122. [N.Y.].
1917 *Orthotylus (Diommatus) ornatus*: Van Duzee, Univ. Cal. Publ. Ent., 2: 496.
1958 *Orthotylus (Orthotylus) ornatus*: Carvalho, Arq. Mus. Nac., 47: 106.
Distribution: Alta., Col., Conn., Del., Ga., Ia., Ill., Ind., La., Minn., Miss., Mo., N.Y., Oh., Ont., S.D., Tex., W.Va.

Orthotylus pacificus Van Duzee, 1919
1919 *Orthotylus pacificus* Van Duzee, *In* Parshley, Occas. Pap., Mus. Zool., Univ. Mich., 71: 33. [B.C.].
1958 *Orthotylus (Orthotylus) pacificus*: Carvalho, Arq. Mus. Nac., 47: 106.
Distribution: Alta., B.C.

Orthotylus pennsylvanicus Henry, 1979, New Subgeneric Combination
1979 *Orthotylus pennsylvanicus* Henry, Melsheimer Ent. Ser., 27: 4. [Pa.].
Distribution: Pa., W.Va.
Note: Henry (1979, above) neglected to give this species subgeneric placement.

Orthotylus plucheae Van Duzee, 1925
1925 *Orthotylus plucheae* Van Duzee, Proc. Cal. Acad. Sci., (4)14: 397. [Cal.].
1958 *Orthotylus (Orthotylus) plucheae*: Carvalho, Arq. Mus. Nac., 47: 107.
Distribution: Cal.

Orthotylus pullatus Van Duzee, 1916
1916 *Orthotylus pullatus* Van Duzee, Proc. Cal. Acad. Sci., (4)6: 118. [Cal.].
1917 *Orthotylus (Diommatus) pullatus*: Van Duzee, Univ. Cal. Publ. Ent., 2: 395.
1958 *Orthotylus (Orthotylus) pullatus*: Carvalho, Arq. Mus. Nac., 47: 108.
Distribution: Cal.

Orthotylus ramus Knight, 1927

1927 *Orthotylus ramus* Knight, Can. Ent., 59: 178. [Oh.].

1958 *Orthotylus* (*Orthotylus*) *ramus*: Carvalho, Arq. Mus. Nac., 47: 108.

Distribution: Fla., Ga., Ia., Ill., Mich., Miss., Mo., N.C., N.Y., Oh., Tex., W.Va. (Mexico).

Orthotylus robiniae Johnston, 1935

1935 *Orthotylus robiniae* Johnston, Bull. Brook. Ent. Soc., 30: 15. [Miss.].

1949 *Orthotylus robinae* [*sic*]: Froeschner, Am. Midl. Nat., 42: 170.

1958 *Orthotylus* (*Orthotylus*) *robiniae*: Carvalho, Arq. Mus. Nac., 47: 108.

Distribution: Ill., Miss., Mo.

Orthotylus rossi Knight, 1941

1941 *Orthotylus rossi* Knight, Ill. Nat. Hist. Surv. Bull., 22: 102. [Ill.].

1958 *Orthotylus* (*Orthotylus*) *rossi*: Carvalho, Arq. Mus. Nac., 47: 108.

Distribution: Ia., Ill.

Orthotylus serus Van Duzee, 1921

1921 *Orthotylus serus* Van Duzee, Proc. Cal. Acad. Sci., (4)11: 131. [N.Y.].

1958 *Orthotylus* (*Orthotylus*) *serus*: Carvalho, Arq. Mus. Nac., 47: 109.

Distribution: Ia., Ill., N.Y., W.Va.

Orthotylus spinosus Knight, 1925

1925 *Orthotylus spinosus* Knight, Bull. Brook. Ent. Soc., 20: 43. [Ariz.].

1958 *Orthotylus* (*Orthotylus*) *spinosus*: Carvalho, Arq. Mus. Nac., 47: 109.

Distribution: Ariz.

Orthotylus submarginatus (Say), 1832

1832 *Capsus submarginatus* Say, Descrip. Het. Hem. N. Am., p. 23. [U.S.].

1916 *Orthotylus submarginatus*: Van Duzee, Proc. Cal. Acad. Sci., (4)6: 123.

1917 *Orthotylus* (*Diommatus*) *submarginatus*: Van Duzee, Univ. Cal. Publ. Ent., 2: 395.

1958 *Orthotylus* (*Orthotylus*) *submarginatus*: Carvalho, Arq. Mus. Nac., 47: 109.

Distribution: Conn., Ill., Ind., Md., Mo., N.Y., Oh., Pa., R.I., Va., W.Va.

Orthotylus taxodii Knight, 1941, New Subgeneric Combination

1941 *Orthotylus taxodii* Knight, Ill. Nat. Hist. Surv. Bull., 22: 101. [Ill.].

Distribution: Ill., Miss.

Note: Knight (1941, above) did not give this species subgeneric placement.

Orthotylus uniformis Van Duzee, 1916

1916 *Orthotylus uniformis* Van Duzee, Proc. Cal. Acad. Sci., (4)6: 100. [Cal.].

1917 *Orthotylus* (*Orthotylus*) *uniformis*: Van Duzee, Univ. Cal. Publ. Ent., 2: 393.

Distribution: Ariz., Cal., Mass.(?), N.Y.(?), Que.(?)

Note: Eastern records for this species undoubtedly are in error and should be dropped from future lists.

Orthotylus ute Knight, 1927

1927 *Orthotylus ute* Knight, Can. Ent., 59: 179. [Col.].

1958 *Orthotylus* (*Orthotylus*) *ute*: Carvalho, Arq. Mus. Nac., 47: 110.

Distribution: Alta., Col., Id., Mont., Wyo.

Orthotylus vanduzeei Carvalho, 1955

1925 *Orthotylus cupressi* Van Duzee, Proc. Cal. Acad. Sci., (4)14: 399. [Cal.]. Preoccupied.

1955 *Orthotylus vanduzeei* Carvalho, Rev. Chil. Ent., 4: 225. New name for *Orthotylus cupressi* Van Duzee.

1958 *Orthotylus* (*Orthotylus*) *vanduzeei*: Carvalho, Arq. Mus. Nac., 47: 110.

Distribution: Cal.

Orthotylus viridinervis (Kirschbaum), 1856

1856 *Capsus viridinervis* Kirschbaum, Jahrb. Ver. Naturk. Nassau, 10: 238. [Europe].

1979 *Orthotylus viridinervis*: Henry and Wheeler, Proc. Ent. Soc. Wash., 81: 260.

Distribution: N.S., Ont. (Europe, northern Africa).

Note: Kelton (1982, Can. Ent., 114: 284-285) recorded Canadian hosts and probable prey of this predaceous species.

Orthotylus viridis Van Duzee, 1916

1916 *Orthotylus viridis* Van Duzee, Proc. Cal. Acad. Sci., (4)6: 103. [N.Y.].

1917 *Orthotylus* (*Orthotylus*) *viridis* Van Duzee, Univ. Cal. Publ. Ent., 2: 393.

Distribution: Alta., Conn., D.C., Ga., Ia., Ill., Md., Mich., Miss., Mo., N.C., N.Y., Oh., Pa., Que., S.D., Tenn, Va., W.Va.

Genus *Paraproba* Distant 1884

1884 *Paraproba* Distant, Biol. Centr.-Am., Rhyn., 1: 270. Type-species: *Paraproba fasciata* Distant, 1884. Designated by Kirkaldy, 1906, Trans. Am. Ent. Soc., 32: 138.

Note: Maldonado (1982, J. Agr. Univ. Puerto Rico, 66: 282-285) gave a key to all of the species of this genus, except *P. capitata*.

Paraproba capitata (Van Duzee), 1912

1912 *Diaphnidia capitata* Van Duzee, Bull. Buff. Soc. Nat. Sci., 10: 490. [N.Y.].

1950 *Diaphnida* [*sic*] *capitata*: Moore, Nat. Can., 77: 246.

1961 *Diaphnocoris capitata*: Kelton, Can. Ent., 93: 566.

1965 *Paraproba capitata*: Kelton, Can. Ent., 97: 1028.

Distribution: D.C., Ia., Ill., Ind., Mass., Me., Minn., Miss., N.B., N.S., N.Y., Oh., Ont., Pa., Que., W.Va.

Paraproba cincta Van Duzee, 1917

1917 *Paraproba cincta* Van Duzee, Proc. Cal. Acad. Sci., (4)7: 273. [Cal.].

Distribution: Cal.

Paraproba hamata (Van Duzee), 1912

1912 *Diaphnidia hamata* Van Duzee, Bull. Buff. Soc. Nat. Sci., 10: 489. [Cal.].

1916 *Paraproba hamata*: Van Duzee, Univ. Cal. Publ. Ent., 1: 241.

Distribution: Cal.

Paraproba nigrinervis Van Duzee, 1917

1917 *Paraproba nigrinervis* Van Duzee, Proc. Cal. Acad. Sci., (4)7: 274. [Cal.]

Distribution: B.C., Cal.

Paraproba pendula Van Duzee, 1914

1914 *Paraproba pendula* Van Duzee, Trans. San Diego Soc. Nat. Hist., 2: 25. [Cal.].

Distribution: Cal.

Genus *Parthenicus* Reuter, 1876

1876 *Parthenicus* Reuter, Öfv. K. Svens. Vet.-Akad. Förh., 32(9): 84. Type-species: *Parthenicus psalliodes* Reuter, 1876. Monotypic.

Note: Knight (1968, Brig. Young Univ. Sci. Bull., 9(3): 129-131, 142-144, 152) provided a key for many of the western U.S. species of *Parthenicus*; Henry (1982, Fla. Ent., 65: 355-356) keyed the eastern U.S. species.

Parthenicus accumulus Knight, 1968
1968 *Parthenicus accumulus* Knight, Brig. Young Univ. Sci. Bull., 9(3): 113. [Nev.].
Distribution: Nev.

Parthenicus albellus Knight, 1925
1925 *Parthenicus albellus* Knight, Oh. J. Sci., 25: 120. [Ariz.].
Distribution: Ariz.

Parthenicus aridus Knight, 1918
1918 *Parthenicus aridus* Knight, Bull. Brook. Ent. Soc., 13: 113. [N.M.].
Distribution: N.M.

Parthenicus atriplicis Knight, 1968
1968 *Parthenicus atriplicis* Knight, Brig. Young Univ. Sci. Bull., 9(3): 135. [Nev.].
Distribution: Nev.

Parthenicus aureosquamis Knight, 1925
1925 *Parthenicus aureosquamis* Knight, Oh. J. Sci., 25: 127. [Ariz.].
Distribution: Ariz., N.M., Tex.

Parthenicus baccharidis Knight, 1925
1925 *Parthenicus baccharidis* Knight, Oh. J. Sci., 25: 125. [Ariz.].
Distribution: Ariz.

Parthenicus basicornis Knight, 1968
1968 *Parthenicus basicornis* Knight, Brig. Young Univ. Sci. Bull., 9(3): 144. [Ut.]
Distribution: Ut.

Parthenicus becki Knight, 1968
1968 *Parthenicus becki* Knight, Brig. Young Univ. Sci. Bull., 9(3): 149. [Nev.].
Distribution: Nev.

*Parthenicus bouteloua*e Knight, 1968
1968 *Parthenicus boutelouae* Knight, Brig. Young Univ. Sci. Bull., 9(3): 154. [Ariz.].
Distribution: Ariz.

Parthenicus brevicornis Knight, 1968
1968 *Parthenicus brevicornis* Knight, Brig. Young Univ. Sci. Bull., 9(3): 141. [Nev.].
Distribution: Nev.

Parthenicus brindleyi Knight, 1968
1968 *Parthenicus brindleyi* Knight, Brig. Young Univ. Sci. Bull., 9(3): 141. [Id.].
Distribution: Id.

Parthenicus brooksi Kelton, 1980
1980 *Parthenicus brooksi* Kelton, Can. Ent., 112: 341. [Sask.].
Distribution: Sask.

Parthenicus brunneus Van Duzee, 1925
1925 *Parthenicus brunneus* Van Duzee, Proc. Cal. Acad. Sci., (4)14: 400. [Cal.].
Distribution: Cal.

Parthenicus candidus Van Duzee, 1918
1918 *Parthenicus candidus* Van Duzee, Proc. Cal. Acad. Sci., (4)8: 300. [Cal.].
Distribution: Cal. (Mexico).

Parthenicus cercocarpi Knight, 1968
1968 *Parthenicus cercocarpi* Knight, Brig. Young Univ. Sci. Bull., 9(3): 148. [Ariz.].
Distribution: Ariz.

Parthenicus condensus Knight, 1968
1968 *Parthenicus condensus* Knight, Brig. Young Univ. Sci. Bull., 9(3): 133. [Nev.].
Distribution: Nev.

Parthenicus consperus Knight, 1968
1968 *Parthenicus conspersus* Knight, Brig. Young Univ. Sci. Bull., 9(3): 139. [Ut.].
Distribution: Ut.

Parthenicus covilleae Van Duzee, 1918
1918 *Parthenicus covilleae* Van Duzee, Proc. Cal. Acad. Sci., (4)8: 300. [Cal.].
Distribution: Ariz., Cal., Nev., Tex. (Mexico).
Note: *Parthenicus percroceus* Van Duzee (1923, Proc. Cal. Acad. Sci., (4)12: 156), described from Mexico, was synonymized with *P. covilleae* by Knight (1968, Brig Young Univ. Sci. Bull., 9(3): 153).

Parthenicus cowaniae Knight, 1968
1968 *Parthenicus cowaniae* Knight, Brig. Young Univ. Sci. Bull., 9(3): 148. [Ariz.].
Distribution: Ariz.

Parthenicus cuneotinctus Knight, 1925
1925 *Parthenicus cuneotinctus* Knight, Oh. J. Sci., 25: 129. [Ariz.].
Distribution: Ariz., Nev., Tex., Ut. (Mexico).

Parthenicus davisi Knight, 1968
1968 *Parthenicus davisi* Knight, Brig. Young Univ. Sci. Bull., 9(3): 132. [Ut.].
Distribution: Ut.

Parthenicus deleticus Knight, 1968
1968 *Parthenicus deleticus* Knight, Brig. Young Univ. Sci. Bull., 9(3): 137. [Ariz.].
Distribution: Ariz.

Parthenicus desertus Knight, 1968
1968 *Parthenicus desertus* Knight, Brig. Young Univ. Sci. Bull., 9(3): 134. [Nev.].
Distribution: Nev.

Parthenicus discalis Van Duzee, 1925
1925 *Parthenicus discalis* Van Duzee, Proc. Cal. Acad. Sci., (4)14: 403. [Ariz.].
Distribution: Ariz., Cal.

Parthenicus femoratus (Van Duzee), 1916
1916 *Argyrocoris femoratus* Van Duzee, Univ. Cal. Publ. Ent., 1: 225. [Cal.].
1917 *Parthenicus femoratus*: Van Duzee, Univ. Cal. Publ. Ent., 2: 400.
Distribution: Cal.

Parthenicus furcatus Knight, 1968
1968 *Parthenicus furcatus* Knight, Brig. Young Univ. Sci. Bull., 9(3): 133. [Nev.].
Distribution: Nev.

Parthenicus fuscipilus Knight, 1968
1968 *Parthenicus fuscipilus* Knight, Brig. Young Univ. Sci. Bull., 9(3): 137. [Ariz.].
Distribution: Ariz.

Parthenicus fuscosus Knight, 1968
1968 *Parthenicus fuscosus* Knight, Brig. Young Univ. Sci. Bull., 9(3): 145. [Ariz.].
Distribution: Ariz.

Parthenicus giffardi Van Duzee, 1917
1917 *Parthenicus giffardi* Van Duzee, Proc. Cal. Acad. Sci., (4)7: 277. [Cal.].
Distribution: Cal.

Parthenicus grex Van Duzee, 1925
1925 *Parthenicus grex* Van Duzee, Proc. Cal. Acad. Sci., (4)14: 403. [Cal.].
Distribution: Cal.

Parthenicus incurvus Knight, 1968
1968 *Parthenicus incurvus* Knight, Brig. Young Univ. Sci. Bull., 9(3): 150. [Nev.].
Distribution: Nev.

Parthenicus irroratus Knight, 1925
1925 *Parthenicus irroratus* Knight, Oh. J. Sci., 25: 124. [Ariz.].
Distribution: Ariz.

Parthenicus juniperi (Heidemann), 1892
1892 *Psallus juniperi* Heidemann, Proc. Ent. Soc. Wash., 2: 225. [D.C., W.Va.]. Lectotype from D.C. designated by Wheeler and Henry, 1975, Trans. Am. Ent. Soc., 101: 361.
1905 *Psallus juniperi* Heidemann, J. N.Y. Ent. Soc., 13: 49 (as new). [D.C., Fla., Md., N.Y., Va.]. Preoccupied.
1914 *Apocremnus* (*Psallus*) *juniperi*: Barber, Bull. Am. Mus. Nat. Hist., 33: 500.
1923 *Parthenicus juniperi*: Knight, Conn. Geol. Nat. Hist. Surv. Bull., 34: 499.
1938 *Parthenicus junipera* [*sic*]: Brimley, Ins. N.C., p. 78.
Distribution: D.C., Fla., Ga., Ill., Ind., Mass., Md., Miss., Mo., N.C., N.Y., Neb., Oh., Ont., Va., W.Va., Wis.
Note: Wheeler and Henry (1975, above) clarified that Heidemann (1892, above) had validated the name *Psallus juniperi* before his (1905, above) formal description. Wheeler and Henry (1977, Trans. Am. Ent. Soc., 103: 637-639) studied biology and hosts and illustrated the fifth instar.

Parthenicus knighti Henry, 1982
1982 *Parthenicus knighti* Henry, Fla. Ent., 65: 359. [Fla.].
Distribution: Fla.

Parthenicus merinoi Knight, 1968
1968 *Parthenicus merinoi* Knight, Brig. Young Univ. Sci. Bull., 9(3): 135. [Nev.].
Distribution: Nev.

Parthenicus micans Knight, 1925
1925 *Parthenicus micans* Knight, Oh. J. Sci., 25: 123. [Ariz.].
Distribution: Ariz., Tex.

Parthenicus miniopunctatus Knight, 1968
1968 *Parthenicus miniopunctatus* Knight, Brig. Young Univ. Sci. Bull., 9(3): 139. [Nev.].
Distribution: Nev.

Parthenicus muchmorei Knight, 1968
1968 *Parthenicus muchmorei* Knight, Brig. Young Univ. Sci. Bull., 9(3): 144. [Cal.].
Distribution: Cal.

Parthenicus multipunctatus Knight, 1968
1968 *Parthenicus multipunctatus* Knight, Brig. Young Univ. Sci. Bull., 9(3): 135. [Ut.].
Distribution: Ut.

Parthenicus mundus Van Duzee, 1923
1923 *Parthenicus mundus* Van Duzee, Proc. Cal. Acad. Sci., (4)12: 155. [Cal.].
Distribution: Cal.

Parthenicus nevadensis Knight, 1968
1968 *Parthenicus nevadensis* Knight, Brig. Young Univ. Sci. Bull., 9(3): 146. [Nev.].

Distribution: Nev.

Parthenicus nicholellus Knight, 1968

1968 *Parthenicus nicholellus* Knight, Brig. Young Univ. Sci. Bull., 9(3): 145. [Ariz.].
Distribution: Ariz.

Parthenicus nicholi Knight, 1925

1925 *Parthenicus nicholi* Knight, Oh. J. Sci., 25: 119. [Ariz.].
Distribution: Ariz.

Parthenicus nigripunctus Knight, 1968

1968 *Parthenicus nigripunctus* Knight, Brig. Young Univ. Sci. Bull., 9(3): 132. [Nev.].
Distribution: Nev.

Parthenicus obsoletus Knight, 1968

1968 *Parthenicus obsoletus* Knight, Brig. Young Univ. Sci. Bull., 9(3): 147. [Ariz.].
Distribution: Ariz.

Parthenicus oreades Knight, 1925

1925 *Parthenicus oreades* Knight, Oh. J. Sci., 25: 122. [Ariz.].
Distribution: Ariz., Col.

Parthenicus pallidicollis Van Duzee, 1925

1925 *Parthenicus pallidicollis* Van Duzee, Proc. Cal. Acad. Sci., (4)14: 402. [Cal.].
Distribution: Cal.

Parthenicus pallipes Knight, 1968

1968 *Parthenicus pallipes* Knight, Brig. Young Univ. Sci. Bull., 9(3): 154. [Ariz.].
Distribution: Ariz.

Parthenicus peregrinus (Van Duzee), 1918

1918 *Atomoscelis peregrinus* Van Duzee, Proc. Cal. Acad. Sci., (4)8: 303. [Cal.].
1958 *Parthenicus peregrinus*: Carvalho, Arq. Mus. Nac., 47: 123.
1968 *Dichaetocoris peregrinus*: Knight, Brig. Univ. Young Sci. Bull., 9(3): 154.
Distribution: Cal., Nev.
Note: Polhemus (1985, Pan-Pac. Ent., 61: 149) discussed the generic placement of this species, concluding that it was best to keep *P. peregrinus* in the genus *Parthenicus*.

Parthenicus picicollis Van Duzee, 1916

1916 *Parthenicus picicollis* Van Duzee, Univ. Cal. Publ. Ent., 1: 226. [Cal.].
Distribution: Ariz., Cal. (Mexico).

Parthenicus pictus Knight, 1925

1925 *Parthenicus pictus* Knight, Oh. J. Sci., 25: 121. [Ariz.].
Distribution: Ariz., Nev.

Parthenicus pilipes Knight, 1968

1968 *Parthenicus pilipes* Knight, Brig. Young Univ. Sci. Bull., 9(3): 137. [Nev.].
Distribution: Nev.

Parthenicus pinicola Knight, 1968

1968 *Parthenicus pinicola* Knight, Brig. Young Univ. Sci. Bull., 9(3): 140. [Col.].
Distribution: Ariz., Col., Nev.

Parthenicus psalliodes Reuter, 1876

1876 *Parthenicus psalliodes* Reuter, Öfv. K. Svens. Vet.-Akad. Förh., 32(9): 85. [Tex.].
Distribution: Ala., Cal.(?), Fla., Tex. (Mexico).
Note: Henry (1982, Fla. Ent., 65: 361) questioned the Cal. record for this species.

Parthenicus ribesi Knight, 1968
1968 *Parthenicus ribesi* Knight, Brig. Young Univ. Sci. Bull., 9(3): 149. [Col.].
Distribution: Col.

Parthenicus ruber Van Duzee, 1917
1917 *Parthenicus ruber* Van Duzee, Proc. Cal. Acad. Sci., (4)7: 276. [Cal.].
Distribution: Cal. (Mexico).

Parthenicus rubrinervis Knight, 1925
1925 *Parthenicus rubrinervis* Knight, Oh. J. Sci., 25: 124. [Ariz.].
Distribution: Ariz.

Parthenicus rubromaculosus Knight, 1925
1925 *Parthenicus rubromaculosus* Knight, Oh. J. Sci., 25: 126. [Ariz.].
Distribution: Ariz.

Parthenicus rubropunctipes Knight, 1968
1968 *Parthenicus rubropunctipes* Knight, Brig. Young Univ. Sci. Bull., 9(3): 147. [Ariz.].
Distribution: Ariz.

Parthenicus rubrosignatus Knight, 1968
1968 *Parthenicus rubrosignatus* Knight, Brig. Young Univ. Sci. Bull., 9(3): 140. [Nev.].
Distribution: Nev.

Parthenicus rufiguttatus Knight, 1968
1968 *Parthenicus rufiguttatus* Knight, Brig. Young Univ. Sci. Bull., 9(3): 150. [Ariz.].
Distribution: Ariz.

Parthenicus rufivenosus Knight, 1925
1925 *Parthenicus rufivenosus* Knight, Oh. J. Sci., 25: 128. [Ariz.].
Distribution: Ariz.

Parthenicus rufus Henry, 1982
1982 *Parthenicus rufus* Henry, Fla. Ent., 65: 361. [Fla.].
Distribution: Fla., S.C.

Parthenicus rufusculus Knight, 1925
1925 *Parthenicus rufusculus* Knight, Oh. J. Sci., 25: 126. [Ariz.].
Distribution: Ariz., Nev.

Parthenicus sabulosus Van Duzee, 1925
1925 *Parthenicus sabulosus* Van Duzee, Proc. Cal. Acad. Sci., (4)14: 401. [Ut.].
Distribution: Ariz., Nev., Ut.

Parthenicus selectus Knight, 1925
1925 *Parthenicus selectus* Knight, Oh. J. Sci., 25: 121. [Ariz.].
Distribution: Ariz.

Parthenicus soror (Van Duzee) 1917
1917 *Psallus soror* Van Duzee, Proc. Cal. Acad. Sci., (4)7: 280. [Cal.].
1918 *Parthenicus soror*: Van Duzee, Proc. Cal. Acad. Sci., (4)8: 302.
Distribution: Cal., Ut.

Parthenicus taxodii Knight, 1941
1941 *Parthenicus taxodii* Knight, Ill. Nat. Hist. Surv. Bull, 22: 76. [Ill.].
Distribution: Ill., Miss., Mo.

Parthenicus tenuis Knight, 1968
1968 *Parthenicus tenuis* Knight, Brig. Young Univ. Sci. Bull., 9(3): 139. [Ut.].
Distribution: Nev., Ut.

Parthenicus trispinosus Knight, 1968

1968 *Parthenicus trispinosus* Knight, Brig. Young Univ. Sci. Bull., 9(3): 134. [Nev.].

Distribution: Nev.

Parthenicus utahensis Knight, 1968

1968 *Parthenicus utahensis* Knight, Brig. Young Univ. Sci. Bull., 9(3): 146. [Ut.].

Distribution: Nev., Ut.

Parthenicus vaccini (Van Duzee), 1915

1915 *Psallus vaccini* Van Duzee, Pomona J. Ent. Zool., 7: 117. [Mass.].

1916 *Parthenicus vaccini*: Van Duzee, Check List Hem., p. 45.

Distribution: Fla., Mass, Md., N.J., N.Y.

Note: Henry (1978, Proc. Ent. Soc. Wash., 80: 546) gave the host of this species and indicated that the specific name *vaccini* is a misnomer.

Parthenicus weemsi Henry, 1982

1982 *Parthenicus weemsi* Henry, Fla. Ent., 65: 364. [Fla.].

Distribution: Fla.

Genus *Phoradendrepulus* Polhemus and Polhemus, 1985

1985 *Phoradendrepulus* Polhemus and Polhemus, Pan-Pac. Ent., 61: 26. Type-species: *Phoradendrepulus myrmecomorphus* Polhemus and Polhemus, 1985. Original designation.

Phoradendrepulus myrmecomorphus Polhemus and Polhemus, 1985

1985 *Phoradendrepulus myrmecomorphus* Polhemus and Polhemus, Pan-Pac. Ent., 61: 26. [Ariz.].

Distribution: Ariz.

Genus *Pseudoloxops* Kirkaldy, 1905

1858 *Loxops* Fieber, Wien. Ent. Monat., 2: 314. Preoccupied. Type-species: *Capsus coccineus* Meyer-Dür, 1843. Monotypic.

1905 *Pseudoloxops* Kirkaldy, Wien. Ent. Zeit., 24: 268. New name for *Loxops* Fieber, 1858.

Pseudoloxops coccineus (Meyer-Dür), 1843

1843 *Capsus coccineus* Meyer-Dür, Verz. Schw. Rhyn., p. 75. [Europe].

1983 *Pseudoloxops coccineus*: Kelton, Can. Ent., 115: 107.

Distribution: Ont. (Europe, northern Africa).

Genus *Pseudopsallus* Van Duzee, 1916

1916 *Pseudopsallus* Van Duzee, Univ. Cal. Publ. Ent., 1: 224. Type-species: *Macrotylus angularis* Uhler, 1894. Original designation.

1930 *Bifidungulus* Knight, Bull. Brook. Ent. Soc., 25: 1. Type-species: *Bifidungulus viridicans* Knight, 1930. Original designation. Synonymized by Stonedahl and Schwartz, 1986, Am. Mus. Novit., 2842: 7.

1968 *Hesperocapsus* Knight, Brig. Young Univ. Sci. Bull., 9(3): 103. Type-species: *Pseudopsallus artemisicola* Knight, 1930. Original designation. Synonymized by Stonedahl and Schwartz, 1986, Am. Mus. Novit., 2842: 7.

Note: Stonedahl and Schwartz (1986, Am. Mus. Novit., 2842: 1-58) revised the genus and provided a key to species.

Pseudopsallus abroniae Knight, 1930

1930 *Pseudopsallus abroniae* Knight, Bull. Brook. Ent. Soc., 25: 5. [Col.].
1968 *Hesperocapsus abroniae*: Knight, Brig. Young Univ. Sci. Bull., 9(3): 107.
Distribution: Ariz., Col., N.M., Tex.

Pseudopsallus angularis (Uhler), 1894

1894 *Macrotylus angularis* Uhler, Proc. Cal. Acad. Sci., (2)4: 272. [Mexico].
1916 *Pseudopsallus angularis*: Van Duzee, Univ. Cal. Publ. Ent., 1: 224.
Distribution: Ariz., Cal., Nev. (Mexico).

Pseudopsallus anograe Knight, 1930

1930 *Pseudopsallus anograe* Knight, Bull. Brook. Ent. Soc., 25: 4. [Col.].
1968 *Hesperocapsus anograe*: Knight, Brig. Young Univ. Sci. Bull., 9(3): 107.
Distribution: Alta., Ariz., Cal., Col., Man., N.M., Ut.

Pseudopsallus artemisicola Knight, 1930

1930 *Pseudopsallus artemisicola* Knight, Bull. Brook. Ent. Soc., 25: 2. [Col.].
1968 *Hesperocapsus artemisicola*: Knight, Brig. Young Univ. Sci. Bull., 9(3): 103.
Distribution: Col., Ks., N.M., Neb., Wyo.

Pseudopsallus atriseta (Van Duzee), 1916

1914 *Oncotylus repertus*: Van Duzee, Trans. S. Diego Soc. Nat. Hist., 2: 29.
1916 *Labopidea atriseta* Van Duzee, Univ. Cal. Publ. Ent., 1: 221. [Cal.].
1930 *Pseudopsallus tanneri* Knight, Bull. Brook. Ent. Soc., 25: 5. [Col.]. Synonymized by Stonedahl and Schwartz, 1986, Am. Mus. Novit., 2842: 15.
1968 *Hesperocapsus tanneri*: Knight, Brig. Young Univ. Sci. Bull., 9(3): 103.
1969 *Hesperocapsus nigricornis* Knight, Ia. St. J. Sci., 44: 88. [Ut.]. Synonymized by Stonedahl and Schwartz, 1986, Am. Mus. Novit., 2842: 15.
1979 *Hesperocapsus atriseta*: Kelton, Can. Ent., 111: 756.
1986 *Pseudopsallus atriseta*: Stonedahl and Schwartz, Am. Mus. Novit., 2842: 15.
Distribution: Cal., Col., Mont., Nev., Ut. (Mexico).

Pseudopsallus demensus (Van Duzee), 1925

1925 *Orthotylus demensus* Van Duzee, Proc. Cal. Acad. Sci., (4)14: 398. [Ariz.].
1930 *Pseudopsallus demensus*: Knight, Bull. Brook. Ent. Soc., 25: 8.
1930 *Pseudopsallus nicholi* Knight, Bull. Brook. Ent. Soc., 25: 7. [Ariz.]. Synonymized by Stonedahl and Schwartz, 1986, Am. Mus. Novit., 2842: 17.
1968 *Hesperocapsus demensus*: Knight, Brig. Young Univ. Sci. Bull., 9(3): 107.
1968 *Hesperocapsus gaurae* Knight, Brig. Young Univ. Sci. Bull., 9(3): 107. [S.D.]. Synonymized by Stonedahl and Schwartz, 1986, Am. Mus. Novit., 2842: 17.
Distribution: Alta., Ariz., Col., Man., S.D., Sask., Tex., Wyo. (Mexico).

Pseudopsallus enceliae Stonedahl and Schwartz, 1986

1986 *Pseudopsallus enceliae* Stonedahl and Schwartz, Am. Mus. Novit., 2842: 18. [Cal.].
Distribution: Cal.

Pseudopsallus hixsoni (Knight), 1969

1930 *Bifidungulus viridicans*: Knight, Bull. Brook. Ent. Soc., 25: 1 (in part).
1969 *Hesperocapsus hixsoni* Knight, Ia. St. J. Sci., 44: 88. [Ok.].
1986 *Pseudopsallus hixsoni*: Stonedahl and Schwartz, Am. Mus. Novit., 2842: 19.
Distribution: Ks., N.M., Ok., Tex.

Pseudopsallus lajuntae Stonedahl and Schwartz, 1986

1986 *Pseudopsallus lajuntae* Stonedahl and Schwartz, Am. Mus. Novit., 2842: 19. [Col.].
Distribution: Ariz., Col.

Pseudopsallus lattini Stonedahl and Schwartz, 1986
1986 *Pseudopsallus lattini* Stonedahl and Schwartz, Am. Mus. Novit., 2842: 20. [Ore.].
Distribution: Ore.

Pseudopsallus major (Knight), 1969
1969 *Hesperocapsus major* Knight, Ia. St. J. Sci., 44: 87. [Ut.].
1986 *Pseudopsallus major*: Stonedahl and Schwartz, Am. Mus. Novit., 2842: 21.
Distribution: Cal., Col., Nev., Ut.

Pseudopsallus mojaviensis Stonedahl and Schwartz, 1986
1986 *Pseudopsallus mojaviensis* Stonedahl and Schwartz, Am. Mus. Novit., 2842: 22. [Cal.].
Distribution: Cal., Ut.

Pseudopsallus occidentalis Stonedahl and Schwartz, 1986
1986 *Pseudopsallus occidentalis* Stonedahl and Schwartz, Am. Mus. Novit., 2842: 23. [Cal.].
Distribution: Cal., Ore.

Pseudopsallus plagiatus (Knight), 1968
1968 *Hesperocapsus plagiatus* Knight, Brig. Young Univ. Sci. Bull., 9(3): 103. [Nev.].
1986 *Pseudopsallus plagiatus*: Stonedahl and Schwartz, Am. Mus. Novit., 2842: 24.
Distribution: Nev.

Pseudopsallus presidio Stonedahl and Schwartz, 1986
1986 *Pseudopsallus presidio* Stonedahl and Schwartz, Am. Mus. Novit., 2842: 25. [Tex.].
Distribution: Tex.

Pseudopsallus puberus (Uhler), 1894
1894 *Oncotylus puberus* Uhler, Proc. Cal. Acad. Sci., (2)4: 270. [Mexico].
1930 *Bifidungulus puberus*: Knight, Bull. Brook. Ent. Soc., 25: 2.
1973 *Bifidungulus puber*: Steyskal, Stud. Ent., 16: 207.
1986 *Pseudopsallus puberus*: Stonedahl and Schwartz, Am. Mus. Novit., 2842: 26.
Distribution: Ariz., Cal., Col., Nev., Ut. (Mexico).

Pseudopsallus repertus (Uhler), 1895
1895 *Oncotylus repertus* Uhler, Bull. Col. Agr. Exp. Stn., 31: 49. [Col.]. Synonymized and lectotype designated by Knight, 1968, Brig. Young Univ. Sci. Bull., 9(3): 105; resurrected by Stonedahl and Schwartz, 1986, Am. Mus. Novit., 2842: 27..
1930 *Pseudopsallus davisi* Knight, Bull. Brook. Ent. Soc., 25: 6. [Ut.]. Synonymized by Stonedahl and Schwartz, 1986, Am. Mus. Novit., 2842: 27.
1968 *Hesperocapsus davisi*: Knight, Brig. Young Univ. Sci. Bull., 9(3): 103.
1968 *Hesperocapsus utahensis* Knight, Brig. Young Univ. Sci. Bull., 9(3): 105. [Ut.]. Synonymized by Stonedahl and Schwartz, 1986, Am. Mus. Novit., 2842: 27.
1986 *Pseudopsallus repertus*: Stonedahl and Schwartz, Am. Mus. Novit., 2842: 27.
Distribution: Cal., Col., Id., Mont., Nev., Ore., Ut., Wyo.

Pseudopsallus sericatus (Uhler), 1895
1895 *Oncotylus sericatus* Uhler, Bull. Col. Agr. Exp. Stn., 31: 49. [Col.]. Lectotype designated by Knight, 1968, Brig. Young Univ. Sci. Bull., 9(3): 105.; lectotype reselected by Stonedahl and Schwartz, 1986, Am. Mus. Novit., 2842: 30.
1909 *Orthotylus sericatus*: Reuter, Acta Soc. Sci. Fenn., 36(2): 69.
1916 *Labopidea sericata*: Van Duzee, Check List Hem., p. 220.
1930 *Pseudopsallus sericatus*: Knight, Bull. Brook. Ent. Soc., 25: 3.
1958 *Pseudopsallus sericata*: Carvalho, Arq. Mus. Nac., 47: 129.

1968 *Hesperocapsus sericatus*: Knight, Brig. Young Univ. Sci. Bull., 9(3): 105.
Distribution: Alta., Ariz., Col., Mont., Nev., S.D., Sask.

Pseudopsallus stitti (Knight), 1968
1930 *Pseudopsallus abroniae*: Knight, Bull. Brook. Ent. Soc., 25: 5 (in part).
1968 *Hesperocapsus stitti* Knight, Brig. Young Univ. Sci. Bull., 9(3): 106. [Ariz.].
1969 *Hesperocapsus tinctus* Knight, Ia. St. J. Sci., 44: 89. [Ariz.]. Synonymized by Stonedahl and Schwartz, 1986, Am. Mus. Novit., 2842: 30.
1986 *Pseudopsallus stitti*: Stonedahl and Schwartz, Am. Mus. Novit., 2842: 30.
Distribution: Ariz., Cal., N.M., Nev.

Pseudopsallus viridicans (Knight), 1930
1930 *Bifidungulus viridicans* Knight, Bull. Brook. Ent. Soc., 25: 1. [Col.].
1986 *Pseudopsallus viridicans*: Stonedahl and Schwartz, Am. Mus. Novit., 2842: 31.
Distribution: Ariz., Col., Ks., Tex. (Mexico).

Genus *Pseudoxenetus* Reuter, 1909

1909 *Pseudoxenetus* Reuter, Acta Soc. Sci. Fenn., 36(2): 66. Type-species: *Xenetus regalis* Uhler, 1890. Designated by Van Duzee, 1916, Check List Hem., p. 44.

Pseudoxenetus regalis (Uhler), 1890 [Fig. 68]
1890 *Xenetus regalis* Uhler, Trans. Md. Acad. Sci., 1: 80. [Fla., Ill., Md., Pa., Tex.]. Lectotype designated by Henry, 1985, J. N.Y. Ent. Soc., 93: 1128.
1890 *Xenetus scutellatus* Uhler, Trans. Md. Acad. Sci., 1: 81. [Ont., Ill., and all states east]. Synonymized and lectotype designated by Henry, 1985, J. N.Y. Ent. Soc., 93: 1128.
1891 *Stenidea regalis*: Townsend, Proc. Ent. Soc. Wash., 2: 54.
1891 *Stenidea scutellata*: Townsend, Proc. Ent. Soc. Wash., 2: 54.
1909 *Pseudoxenetus scutellatus*: Reuter, Acta Soc. Sci. Fenn., 36(2): 67.
1909 *Pseudoxenetus regalis*: Reuter, Acta Soc. Sci. Fenn., 36(2): 67.
Distribution: Conn., Fla., Ga., Ill., Ind., Mass., Md., Mich., Minn., Miss., Mo., N.C., N.H., N.J., N.Y., Oh., Ok., Ont., Pa., Tex., Va., W.Va., Wis.

Genus *Renodaeus* Distant, 1893

1893 *Renodaeus* Distant, Biol. Centr.-Am., Rhyn., 1: 461. Type-species: *Renodaeus ficarius* Distant, 1893. Monotypic.

Note: Schuh (1974, Ent. Am., 47: 285) transferred this genus from Pilophorini to Orthotylini. Knight's (1926, Bull. Brook. Ent. Soc., 21: 57) tribe Renodaeini was placed in synonymy under Pilophorini by Carvalho (1958, Arq. Mus. Nac., 47: 135).

Renodaeus texanus Knight, 1926
1926 *Renodaeus texanus* Knight, Bull. Brook. Ent. Soc., 21: 56. [Tex.].
Distribution: Tex.

Genus *Reuteria* Puton, 1875

1875 *Reuteria* Puton, Pet. Nouv. Ent., 1: 519. Type-species: *Reuteria marqueti* Puton, 1875. Monotypic.

Note: Henry (1976, Ent. News, 87: 61-74) reviewed the genus and provided a key to species.

Reuteria bifurcata Knight, 1939

1904 *Malacocoris irroratus*: Wirtner, An. Carnegie Mus., 3: 201.
1939 *Reuteria bifurcata* Knight, Ia. St. Coll. J. Sci., 13: 130. [N.Y.].
Distribution: Ill., Md., Mo., N.C., N.Y., Ok., Pa.

Reuteria dobsoni Henry, 1976

1976 *Reuteria dobsoni* Henry, Ent. News, 87: 64. [Pa.].
Distribution: Pa.

Reuteria fuscicornis Knight, 1939

1939 *Reuteria fuscicornis* Knight, Ia. St. Coll. J. Sci., 13: 129. [N.Y.].
Distribution: D.C., Ia., Ill., Mass., Md., Minn., N.J., N.Y., Ont., Pa.

Reuteria irrorata (Say), 1832

1832 *Capsus irrorata* Say, Descrip. Het. Hem. N. Am., p. 25. [Ind.]. Neotype designated by Henry, 1976, Ent. News, 87: 70.
1878 *Malacocoris irroratus*: Uhler, Bull. U.S. Geol. Geogr. Surv. Terr., 4: 507.
1905 *Malorocoris* [*sic*] *irroratus*: Crevecoeur, Trans. Ks. Acad. Sci., 19: 233.
1908 *Reuteria irrorata*: Horvath, An. Hist.-Nat. Mus. Nat. Hung., 6: 10.
Distribution: D.C., Ia., Ill., Ind., Ks., Md., Mich., Minn., Mo., N.D., N.J., N.Y., Oh., Ont., Pa., Va., W.Va., Wis.
Note: The European records of *irrorata* pertain to *marqueti* Puton; the Mexican record given by Carvalho (1958, Arq. Mus. Nac., 47: 130) represents a misidentification; North Carolina was recorded in error by Henry (1976, Ent. News, 87: 66); other N.C. records for this species need verification.

Reuteria platani Knight, 1941

1941 *Reuteria platani* Knight, Ill. Nat. Hist. Surv. Bull., 22: 95. [Ill.].
Distribution: D.C., Ill., Pa.

Reuteria pollicaris Knight, 1939

1939 *Reuteria pollicaris* Knight, Ia. St. Coll. J. Sci., 13: 131. [Miss.].
Distribution: Miss.

Reuteria querci Knight, 1939

1939 *Reuteria querci* Knight, Ia. St. Coll. J. Sci., 13: 131. [Minn.].
Distribution: D.C., Ga., Ia., Ill., Man., Md., Minn., Mo., N.C., N.Y., Va.

Reuteria wheeleri Henry, 1976

1976 *Reuteria wheeleri* Henry, Ent. News, 87: 73. [N.C.].
Distribution: Ga., N.C.

Genus *Saileria* Hsiao, 1945

1945 *Saileria* Hsiao, Proc. Ent. Soc. Wash., 47: 27. Type-species: *Hyalochloria bella* Van Duzee, 1916. Original designation.
Note: Henry (1976, Ent. News, 87: 31) provided a key to species.

Saileria bella (Van Duzee), 1916

1916 *Hyalochloria bella* Van Duzee, Univ. Cal. Publ. Ent., 1: 217. [Cal.].
1945 *Saileria bella*: Hsiao, Proc. Ent. Soc. Wash., 47: 27.
Distribution: Cal.

Saileria compsus (Reuter), 1907

1907 *Orthotylus compsus* Reuter, Öfv. F. Vet.-Soc. Förh., 49(5): 14. [Jamaica].
1927 *Orthotylus compsus*: Knight, Can. Ent., 59: 181.

1958 *Orthotylus (Orthotylus) compsus*: Carvalho, Arq. Mus. Nac., 47: 98.
1976 *Saileria compsus*: Carvalho, Rev. Brasil. Biol., 36: 57.
Distribution: Tex. (Jamaica).
Note: Henry (1985, J. N.Y. Ent. Soc., 93: 1131) clarified the relationship of *S. compsus* to other species in the genus *Saileria*.

Saileria irrorata Henry, 1976
1976 *Saileria irrorata* Henry, Ent. News, 87: 29. [Ind.].
Distribution: Fla., Ind., Miss., Tex., W.Va.
Note: Henry (1980, Fla. Ent., 63: 490-493) figured male genitalia and gave new distribution records for this poorly known species.

Genus *Sericophanes* Reuter, 1876

1876 *Sericophanes* Reuter, Öfv. K. Svens. Vet-Akad. Förh., 32(9): 79. Type-species: *Sericophanes ocellatus* Reuter, 1876. Monotypic.

Sericophanes albomaculatus Knight, 1930
1930 *Sericophanes albomaculatus* Knight, Ent. News, 41: 320. [Tex.].
Distribution: Tex.

Sericophanes floridanus Knight, 1927
1927 *Sericophanes floridanus* Knight, Ent. News, 38: 305. [Fla.].
Distribution: Fla.

Sericophanes fuscicornis Knight, 1968
1968 *Sericophanes fuscicornis* Knight, Brig. Young Univ. Sci. Bull., 9(3): 160. [Ariz.].
Distribution: Ariz.

Sericophanes heidemanni Poppius, 1914
1898 *Sericophanes ocellatus*: Osborn, Proc. Ia. Acad. Sci., 5: 238.
1914 *Sericophanes heidemanni* Poppius, An. Soc. Ent. Belg., 58: 260. [N.Y.].
1917 *Sericophanes noctuans* Knight, Ent. News, 28: 4. [N.Y.]. Synonymized by Knight, 1923, Conn. Geol. Nat. Hist. Surv. Bull., 34: 545.
1979 *Serricophanes* [*sic*] *heidemanni*: Henry and Smith, J. Ga. Ent. Soc., 14: 215.
Distribution: Alta., Conn., Del., Fla., Ga., Ia., Ill., Ind., Mass., Md., Mich., Minn., Mont., N.C., N.Y., Oh., Ont., Pa., Que., S.D., Ut., W.Va., Wash., Wyo.

Sericophanes nevadensis Knight, 1968
1968 *Sericophanes nevadensis* Knight, Brig. Young Univ. Sci. Bull., 9(3): 160. [Nev.].
Distribution: Nev.

Sericophanes ocellatus Reuter, 1876
1876 *Sericophanes ocellatus* Reuter, Öfv. K. Svens. Vet.-Akad. Förh., 32(9): 79. [Tex.].
Distribution: Ia., Ks., Mass.(?), N.Y., Tex.

Sericophanes rubripes Knight, 1968
1968 *Sericophanes rubripes* Knight, Brig. Young Univ. Sci. Bull., 9(3): 161. [Cal.].
Distribution: Cal.

Sericophanes triangularis Knight, 1918
1918 *Sericophanes triangularis* Knight, Bull. Brook. Ent. Soc., 13: 81. [N.M.].
Distribution: Ariz., Col., N.D., N.M., Ok., S.D., Tex., Ut.

Sericophanes tumidifrons Knight, 1968
1968 *Sericophanes tumidifrons* Knight, Brig. Young Univ. Sci. Bull., 9(3): 160. [Cal.].
Distribution: Cal.

Genus *Scalponotatus* Kelton, 1969

1969 *Scalponotatus* Kelton, Can. Ent., 101: 15. Type-species: *Scalponotatus maturus* Kelton, 1969. Original designation.

Note: Kelton (1969, above) keyed the species and provided characters for separating this genus from *Slaterocoris* Wagner.

Scalponotatus albibasis (Knight), 1938

1938 *Strongylocoris albibasis* Knight, Iowa St. Coll. J. Sci., 13: 5. [Ariz.].
1969 *Scalponotatus albibasis*: Kelton, Can. Ent., 101: 16.
Distribution: Ariz.

Scalponotatus maturus Kelton, 1969

1969 *Scalponotatus maturus* Kelton, Can. Ent., 101: 17. [Cal.].
Distribution: Ariz., Cal.

Genus *Slaterocoris* Wagner, 1956

1956 *Slaterocoris* Wagner, Proc. Ent. Soc. Wash., 58: 280. Type-species: *Capsus stygicus* Say, 1832. Original designation.

Note: Although these shiny, black mirids superficially resemble some halticines and have been placed in that tribe, genitalia and other structures dictate that *Slaterocoris* and *Scalponotatus* belong in the Orthotylini, as commented on by Wagner (1956, Proc. Ent. Soc. Wash., 58: 280). Knight (1970, Ia. St. J. Sci., 45: 233-237) provided a key to species, updating Kelton's (1968, Can. Ent., 100: 1136-1137) key.

Slaterocoris alpinus Kelton, 1968

1968 *Slaterocoris alpinus* Kelton, Can. Ent., 100: 1128. [Col.].
Distribution: Col.

Slaterocoris ambrosiae (Knight), 1938

1938 *Strongylocoris ambrosiae* Knight, Ia. St. Coll. J. Sci., 13: 5. [S.D.].
1956 *Slaterocoris ambrosiae*: Wagner, Proc. Ent. Soc. Wash., 58: 280.
Distribution: Ariz., Ia., Ks., Mo., N.M., S.D., Tex. (Mexico).

Slaterocoris apache Kelton, 1968

1968 *Slaterocoris apache* Kelton, Can. Ent., 100: 1126. [Ariz.].
Distribution: Ariz., Col., Ut.

Slaterocoris arizonensis Knight, 1970

1970 *Slaterocoris arizonensis* Knight, Ia. St. J. Sci., 45: 248. [Ariz.].
Distribution: Ariz., N.M.

Slaterocoris atratus (Uhler), 1894

1894 *Stiphrosoma atrata* Uhler, Proc. Cal. Acad. Sci., (2)4: 268. [Cal.].
1956 *Slaterocoris atratus*: Wagner, Proc. Ent. Soc. Wash., 58: 280.
Distribution: Cal. (Mexico).
Note: The eastern U.S. record for this species by Smith (1910, Ins. N.J., p. 62) is in error according to Knight, 1970, Ia. St. J. Sci., 45: 246).

Slaterocoris atritibialis (Knight), 1938

1938 *Strongylocoris atritibialis* Knight, Ia. St. Coll. J. Sci., 13: 2. [N.Y.].
1953 *Strongylocoris atribialis* [*sic*]: Strickland, Can. Ent., 85: 200.
1956 *Slaterocoris atritibialis*: Wagner, Proc. Ent. Soc. Wash., 58: 280.
Distribution: Ala., Alta., B.C., Col., Ga., Ia., Ill., Ky., Man., Mass., Minn., Miss., N.C., N.D., N.J., N.Y., Oh., Ont., Pa., Que., S.D., Sask., Tenn., W.Va., Wash., Wyo.

Slaterocoris basicornis Knight, 1970
1970 *Slaterocoris basicornis* Knight, Ia. St. J. Sci., 45: 254. [Tex.].
Distribution: Tex.

Slaterocoris bifidus Knight, 1970
1970 *Slaterocoris bifidus* Knight, Ia. St. J. Sci., 45: 252. [Ut.].
Distribution: Ut.

Slaterocoris breviatus (Knight), 1938
1938 *Strongylocoris breviatus* Knight, Ia. St. Coll. J. Sci., 13: 1. [D.C.].
1946 *Stongylocoris* [*sic*] *breviatus*: Procter, Biol. Surv. Mt. Desert Reg., Ins. Fauna, p. 79.
1956 *Slaterocoris breviatus*: Wagner, Proc. Ent. Soc. Wash., 58: 280.
Distribution: Alta., B.C., D.C., Ill., Man., Mass., Md., Me., Minn., Miss., Mo., Mont., N.B., N.D., N.H., N.S., N.Y., Ont., Pa., Que., Sask., W.Va., Wyo.

Slaterocoris burkei Knight, 1970
1970 *Slaterocoris burkei* Knight, Ia. St. J. Sci., 45: 255. [Col.].
Distribution: Col.

Slaterocoris croceipes (Uhler), 1893
1893 *Stiphrosoma croceipes* Uhler, Proc. Ent. Soc. Wash., 2: 373. [Cal.].
1907 *Strongylocoris* (*Stiphrosoma*) *croceipes*: Tucker, Ks. Univ. Sci. Bull., 4: 60.
1914 *Strongylocoris croceipes*: Van Duzee, Trans. San Diego Soc. Nat. Hist., 2: 28.
1968 *Slaterocoris croceipes*: Knight, Brig. Young Univ. Sci. Bull., 9(3): 90.
Distribution: Ariz., Cal., Col., Ks., Nev., Ore., Ut.
Note: The N.Y. record listed by Carvalho (1958, Arq. Mus. Nac., 47: 341) probably represents a misidentification.

Slaterocoris custeri Knight, 1970
1970 *Slaterocoris custeri* Knight, Ia. St. J. Sci., 45: 254. [S.D.].
Distribution: S.D., Wyo.

Slaterocoris dakotae Knight, 1970
1970 *Slaterocoris dakotae* Knight, Ia. St. J. Sci., 45: 243. [S.D.].
Distribution: S.D.

Slaterocoris digitatus Knight, 1970
1970 *Slaterocoris digitatus* Knight, Ia. St. J. Sci., 45: 242. [Tex.].
Distribution: Tex.

Slaterocoris flavipes Kelton, 1968
1968 *Slaterocoris flavipes* Kelton, Can. Ent., 100: 1126. [Cal.].
Distribution: Cal.

Slaterocoris fuscicornis Knight, 1970
1970 *Slaterocoris fuscicornis* Knight, Ia. St. J. Sci., 45: 246. [Mo.].
Distribution: Mo.

Slaterocoris fuscomarginalis Knight, 1970
1970 *Slaterocoris fuscomarginalis* Knight, Ia. St. J. Sci., 45: 256. [Ut.].
Distribution: Id., Ut.

Slaterocoris getzendaneri Knight, 1970
1970 *Slaterocoris getzendaneri* Knight, Ia. St. J. Sci., 45: 251. [Wash.].
Distribution: Wash.

Slaterocoris hirtus (Knight), 1938
1938 *Strongylocoris hirtus* Knight, Ia. St. Coll. J. Sci., 13: 4. [Ia.].
1956 *Slaterocoris hirtus*: Wagner, Proc. Ent. Soc. Wash., 58: 280.
Distribution: Ia., Ill., Ks., Mo.

Slaterocoris knowltoni Knight, 1970
1970 *Slaterocoris knowltoni* Knight, Ia. St. J. Sci., 45: 240. [Ut.].
Distribution: Ut.

Slaterocoris longipennis Knight, 1968
1968 *Slaterocoris longipennis* Knight, Brig. Young Univ. Sci. Bull., 9(3): 90. [Nev.].
Distribution: Cal., Nev., Ut., Wyo.

Slaterocoris minimus Knight, 1970
1970 *Slaterocoris minimus* Knight, Ia. St. J. Sci., 45: 258. [Wis.].
Distribution: Wis.

Slaterocoris mohri (Knight), 1941
1941 *Strongylocoris mohri* Knight, Ill. Nat. Hist. Surv. Bull., 22: 78. [Ill.].
1968 *Slaterocoris mohri*: Kelton, Can. Ent., 100: 1130.
Distribution: Ill., Mo.

Slaterocoris nevadensis Knight, 1970
1970 *Slaterocoris nevadensis* Knight, Ia. St. J. Sci., 45: 239. [Nev.].
Distribution: Nev.

Slaterocoris nicholi Knight, 1970
1970 *Slaterocoris nicholi* Knight, Ia. St. J. Sci., 45: 243. [Ariz.].
Distribution: Ariz.

Slaterocoris ovatus Knight, 1970
1970 *Slaterocoris ovatus* Knight, Ia. St. J. Sci., 45: 247. [Cal.].
Distribution: Cal., Tex.

Slaterocoris pallidicornis (Knight), 1938
1938 *Strongylocoris pallidicornis* Knight, Ia. St. Coll. J. Sci., 13: 4. [S.D.].
1968 *Slaterocoris pallidicornis*: Kelton, Can. Ent., 100: 1123.
Distribution: Alta., Col., Man., Minn., N.D., N.S., S.D., Sask., Wyo.

Slaterocoris pallipes (Knight), 1926
1878 *Stiphrosoma stygica*: Uhler, Rept. U.S. Geol. Geogr. Surv. Terr., 4: 507 (in part).
1926 *Strongylocoris pallipes* Knight, Can. Ent., 58: 254. [Va.].
1956 *Slaterocoris pallipes*: Wagner, Proc. Ent. Soc. Wash., 58: 280.
Distribution: Md., N.C., N.J., N.Y., Va.
Note: Wheeler (1981, Proc. Ent. Soc. Wash., 83: 520-523) studied distribution, host, and seasonal history.

Slaterocoris pilosus Kelton, 1968
1968 *Slaterocoris pilosus* Kelton, Can. Ent., 100: 1127. [B.C.].
Distribution: B.C., Cal., Id., Wash.

Slaterocoris rarus Knight, 1970
1970 *Slaterocoris rarus* Knight, Ia. St. J. Sci., 45: 257. [N.Y.].
Distribution: N.Y.

Slaterocoris robustus (Uhler), 1895
1895 *Stiphrosoma robusta* Uhler, Bull. Col. Agr. Exp. Stn., 31: 45. [Col.].
1907 *Strongylocoris* (*Stiphrosoma*) *robustus*: Tucker, Ks. Univ. Sci. Bull., 4: 60.

1916 *Strongylocoris robustus*: Van Duzee, Check List Hem., p. 43.
1925 *Strongylocoris uniformis* Van Duzee, Proc. Cal. Acad. Sci., (4)14: 396. [Ut.]. Synonymized by Kelton, 1968, Can. Ent., 100: 1132.
1968 *Slaterocoris robustus*: Knight, Brig. Young Univ. Sci. Bull., 9(3): 90.
Distribution: B.C., Cal., Col., Id., Ks., Mont., N.M., Nev., Tex., Ut., Wyo.

Slaterocoris rubrofemoratus Knight, 1968
1968 *Slaterocoris rubrofemoratus* Knight, Brig. Young Univ. Sci. Bull., 9(3): 90. [Nev.].
Distribution: B.C., Cal., Col., Id., Mont., N.M., Nev., Ore., Tex., Ut., Wyo.

Slaterocoris schaffneri Knight, 1970
1970 *Slaterocoris schaffneri* Knight, Ia. St. J. Sci., 45: 240. [Tex.].
Distribution: N.D., Neb., S.D., Tex.

Slaterocoris sculleni Knight, 1970
1970 *Slaterocoris sculleni* Knight, Ia. St. J. Sci., 45: 257. [Ore.].
Distribution: Ore.

Slaterocoris severini Knight, 1970
1970 *Slaterocoris severini* Knight, Ia. St. J. Sci., 45: 250. [S.D.].
Distribution: S.D.

Slaterocoris sheridani Knight, 1968
1968 *Slaterocoris sheridani* Knight, Brig. Young Univ. Sci. Bull., 9(3): 92. [Wyo.].
Distribution: Ariz., Col., N.M., Wyo.

Slaterocoris solidaginis Kelton, 1968
1968 *Slaterocoris solidaginis* Kelton, Can. Ent., 100: 1131. [Cal.].
Distribution: Cal., Nev.

Slaterocoris sparsus Kelton, 1968
1968 *Slaterocoris sparsus* Kelton, Can. Ent., 100: 1125. [Cal.].
Distribution: Cal.

Slaterocoris stygicus (Say), 1832
1832 *Capsus stygicus* Say, Descript. Het. Hem. N. Am., p. 24. [Ind.].
1877 *Stiphrosoma stygicus*: Uhler, Wheeler's Rept. Chief Eng., p. 1328.
1889 *Stiphrosoma stygica*: Van Duzee, Can. Ent., 21: 4.
1890 *Strongylocoris stygicus*: Atkinson, J. Asiatic Soc. Bengal., 58(2): 120.
1900 *Styphrosoma* [*sic*] *stygica*: Osborn, Oh. Nat., 1: 12.
1907 *Strongylocoris* (*Stiphrosoma*) *stygicus*: Tucker, Ks. Univ. Sci. Bull., 4: 60.
1914 *Strongylocoris stygica*: Van Duzee, Trans. San Diego Soc. Nat. Hist., 2: 28.
1946 *Stongylocoris* [*sic*] *stygicus*: Procter, Biol. Surv. Mt. Desert Reg., Ins. Fauna, p. 79.
1956 *Slaterocoris stygicus*: Wagner, Proc. Ent. Soc. Wash., 58: 280.
Distribution: Alta., Ariz., B.C., Col., Conn., Ga., Ia., Id., Ill., Ind., Ks., Man., Mass., Md., Me., Mich., Miss., Mo., Mont., N.B., N.C., N.D., N.J., N.M., N.S., N.Y., Oh., Ont., Pa., Que., S.C., S.D., Sask., Tex., Ut., W.Va., Wyo.
Note: Leonard (1919, Can. Ent., 51: 178-180) described and illustrated the immature stages.

Slaterocoris texanus Knight, 1970
1970 *Slaterocoris texanus* Knight, Ia. St. J. Sci., 45: 252. [Tex.].
Distribution: Tex.

Slaterocoris tibialis Knight, 1970
1970 *Slaterocoris tibialis* Knight, Ia. St. J. Sci., 45: 257. [Ore.].

Distribution: Ore.

Slaterocoris utahensis Knight, 1968
1968 *Slaterocoris utahensis* Knight, Brig. Young Univ. Sci. Bull., 9(3): 92. [Ut.].
Distribution: Col., Id., Ut.

Slaterocoris woodgatei Knight, 1970
1970 *Slaterocoris woodgatei* Knight, Ia. St. J. Sci., 45: 254. [N.M.].
Distribution: N.M.

Genus *Squamocoris* Knight, 1968

1968 *Squamocoris* Knight, Brig. Young Univ. Sci. Bull., 9(3): 108. Type-species, *Squamocoris utahensis* Knight, 1968. Original designation.

Note: Stonedahl and Schuh (1986, Am. Mus. Novit., 2852: 1-26) revised this genus, gave a cladistic analysis, and provided a key to species.

Squamocoris arizonae Knight, 1968
1968 *Squamocoris arizonae* Knight, Brig. Young Univ. Sci. Bull., 9(3): 109. [Ariz.].
Distribution: Ariz., Cal.

Squamocoris fumosus Stonedahl and Schuh, 1986
1986 *Squamocoris fumosus* Stonedahl and Schuh, Am. Mus. Novit., 2852: 7. [Cal.].
Distribution: Cal.

Squamocoris latisquamus Stonedahl and Schuh, 1986
1986 *Squamocoris latisquamus* Stonedahl and Schuh, Am. Mus. Novit., 2852: 8. [Ariz.].
Distribution: Ariz., Nev.

Squamocoris pallidinervus Stonedahl and Schuh, 1986
1986 *Squamocoris pallidinervus* Stonedahl and Schuh, Am. Mus. Novit., 2852: 10. [Cal.].
Distribution: Cal., Nev., Ut.

Squamocoris purshiae Stonedahl and Schuh, 1986
1986 *Squamocoris purshiae* Stonedahl and Schuh, Am. Mus. Novit., 2852: 11. [Cal.].
Distribution: Cal.

Squamocoris utahensis Knight, 1968
1968 *Squamocoris utahensis* Knight, Brig. Young Univ. Sci. Bull., 9(3): 108. [Ut.].
Distribution: Cal., Id., Nev., Ut.

Genus *Texocoris* Schaffner, 1974

1974 *Texocoris* Schaffner, J. Ks. Ent. Soc., 47: 281. Type-species: *Texocoris secludis* Schaffner, 1974, a junior synonym of *Parthenicus nigrellus* Knight, 1939. Original designation.

Texocoris nigrellus (Knight), 1939
1939 *Parthenicus nigrellus* Knight, Bull. Brook. Ent. Soc., 34: 23. [Ia.].
1974 *Texocoris secludis* Schaffner, J. Ks. Ent. Soc., 47: 283. [Tex.]. Synonymized by Henry, 1982, Proc. Ent. Soc. Wash., 84: 340.
1982 *Texocoris nigrellus*: Henry, Proc. Ent. Soc. Wash., 84: 340.
Distribution: Ga., Ia., Ill., Mo., Tex., Wis.

Subfamily Phylinae Douglas and Scott, 1865

Note: Although not specifically treating the North American fauna, Schuh's Orthotylinae and Phylinae of South Africa (1974, Ent. Am., 47: 1-332) and Phylinae of the Indo-Pacific Region (1984, Bull. Am. Mus. Nat. Hist., 177: 1-476) are important works to consult on the classification of this subfamily.

Tribe Hallodapini Van Duzee, 1916

Genus *Coquillettia* Uhler, 1890

1890 *Coquillettia* Uhler, Trans. Md. Acad. Sci., 1: 78. Type-species: *Coquillettia insignis* Uhler, 1890. Monotypic.

Coquillettia ajo Knight, 1968
1968 *Coquillettia ajo* Knight, Brig. Young Univ. Sci. Bull., 9(3): 63. [Ariz.].
Distribution: Ariz.

Coquillettia albella Knight, 1968
1968 *Coquillettia albella* Knight, Brig. Young Univ. Sci. Bull., 9(3): 62. [Nev.].
Distribution: Nev.

Coquillettia albertae Kelton, 1980
1980 *Coquillettia albertae* Kelton, Can. Ent., 112: 285. [Alta.].
Distribution: Alta.

Coquillettia albiclava Knight, 1925
1925 *Coquillettia albiclava* Knight, Bull. Brook. Ent. Soc., 20: 36. [Ariz.].
Distribution: Ariz.

Coquillettia amoena (Uhler), 1877
1877 *Orectoderus amoenus* Uhler, Wheeler's Rept. Chief Eng., p. 1328. [Ill., Tex.].
1916 *Coquillettia amoena*: Van Duzee, Check List Hem., p. 42.
Distribution: Col., Fla., Ia., Ill., Ks., N.C., N.M., Tex., Wis.

Coquillettia aquila Bliven, 1962
1962 *Coquillettia aquila* Bliven, Occ. Ent., 1(6): 56. [Ariz.].
Distribution: Ariz.

Coquillettia attica Bliven, 1962
1962 *Coquillettia attica* Bliven, Occ. Ent., 1(6): 53. [Cal.].
Distribution: Cal.

Coquillettia balli Knight, 1918
1918 *Coquillettia balli* Knight, Bull. Brook. Ent. Soc., 13: 112. [Col.].
Distribution: Col.

Coquillettia foxi Van Duzee, 1921
1921 *Coquillettia foxi* Van Duzee, Proc. Cal. Acad. Sci., (4)11: 124. [Cal.].
Distribution: Cal.

Coquillettia granulata Knight, 1930
1930 *Coquillettia granulata* Knight, Ent. News, 41: 319. [Ut.].
Distribution: Ut.

Coquillettia insignis Uhler, 1890
1890 *Coquillettia insignis* Uhler, Trans. Md. Acad. Sci., 1: 79. [Western U.S.].

1953 *Coquilettia* [*sic*] *insignis*: Strickland, Can. Ent., 85: 200.
Distribution: Alta., Ariz., B.C., Cal., Col., "Dak.," Id., Ks., Mont., Sask., Tex., Ut., Wyo.

Coquillettia jessiana Knight, 1927
1927 *Coquillettia jessiana* Knight, Ent. News, 38: 303. [Col.].
Distribution: Ariz., Col.

Coquillettia luteiclava Knight, 1968
1968 *Coquillettia luteiclava* Knight, Brig. Young Univ. Sci. Bull., 9(3): 63. [Nev.].
Distribution: Nev.

Coquillettia mimetica Osborn, 1898
1898 *Coquillettia mimetica* Osborn, Proc. Ia. Acad. Sci., 5: 236. [Ia.].
Distribution: Ariz., Col., Fla., Ia., N.C., Tex., Wis.

Coquillettia mimetica floridana Knight, 1927
1927 *Coquillettia mimetica* var. *floridana* Knight, Ent. News, 38: 304. [Fla.].
Distribution: Fla.

Coquillettia mimetica laticeps Knight, 1927
1927 *Coquillettia mimetica* var. *laticeps* Knight, Ent. News, 38: 305. [Col.].
Distribution: Col.

Coquillettia mimetica mimetica Osborn, 1898
1898 *Coquillettia mimetica* Osborn, Proc. Ia. Acad. Sci., 5: 236. [Ia.].
Distribution: Same as for species.

Coquillettia nicholi Knight, 1925
1925 *Coquillettia nicholi* Knight, Bull. Brook. Ent. Soc., 20: 35. [Wyo.].
Distribution: Wyo.

Coquillettia nigrithorax Knight, 1930
1930 *Coquillettia nigrithorax* Knight, Ent. News, 41: 319. [Ariz.].
Distribution: Ariz.

Coquillettia numata Bliven, 1962
1962 *Coquillettia numata* Bliven, Occ. Ent., 1(6): 52. [Cal.].
Distribution: Cal.

Coquillettia saxetana Bliven, 1962
1962 *Coquillettia saxetana* Bliven, Occ. Ent., 1(6): 54. [Ariz.].
Distribution: Ariz.

Coquillettia soligena Bliven, 1962
1962 *Coquillettia soligena* Bliven, Occ. Ent., 1(6): 57. [Ariz.].
Distribution: Ariz.

Coquillettia terrosa Bliven, 1962
1962 *Coquillettia terrosa* Bliven, Occ. Ent., 1(6): 55. [Ariz.].
Distribution: Ariz.

Coquillettia uhleri Van Duzee, 1921
1921 *Coquillettia uhleri* Van Duzee, Proc. Cal. Acad. Sci., (4)11: 123. [Cal.].
Distribution: Cal.

Coquillettia virescens Knight, 1968
1968 *Coquillettia virescens* Knight, Brig. Young Univ. Sci. Bull., 9(3): 63. [Nev.].
Distribution: Nev.

Genus *Cyrtopeltocoris* Reuter, 1876

1876 *Cyrtopeltocoris* Reuter, Öfv. K. Svens. Vet.-Akad. Förh., 32(9): 81. Type-species: *Cyrtopeltocoris albofasciatus* Reuter, 1876. Monotypic.

Note: Schuh (1974, Ent. Am., 47: 301) discussed the tribal placement of this genus.

Cyrtopeltocoris ajo Knight, 1968
1968 *Cyrtopeltocoris ajo* Knight, Brig. Young Univ. Sci. Bull., 9(3): 163. [Ariz.].
Distribution: Ariz.

Cyrtopeltocoris albofasciatus Reuter, 1876
1876 *Cyrtopeltocoris albofasciatus* Reuter, Öfv. K. Vet.-Akad. Förh., 32(9): 81. [Tex.].
1906 *Cyrtopeltocoris albo-fasciatus* [*sic*]: Barber, Mus. Brook. Inst. Arts Sci., Sci. Bull., 1: 281.
1918 *Sericophanes transversus* Knight, Bull. Brook. Ent. Soc., 13: 82. [Tex.]. Synonymized by Knight, 1927, Can. Ent., 59: 41.
Distribution: Ariz., Cal., Col., N.M., Tex.

Cyrtopeltocoris arizonae Knight, 1968
1968 *Cyrtopeltocoris arizonae* Knight, Brig. Young Univ. Sci. Bull., 9(3): 163. [Ariz.].
Distribution: Ariz.

Cyrtopeltocoris balli Knight, 1968
1968 *Cyrtopeltocoris balli* Knight, Brig. Young Univ. Sci. Bull., 9(3): 164. [Col.].
Distribution: Col.

Cyrtopeltocoris barberi Knight, 1968
1968 *Cyrtopeltocoris barberi* Knight, Brig. Young Univ. Sci. Bull., 9(3): 164. [Ariz.].
Distribution: Ariz.

Cyrtopeltocoris conicatus Knight, 1968
1968 *Cyrtopeltocoris conicatus* Knight, Brig. Young Univ. Sci. Bull., 9(3): 163. [Ariz.].
Distribution: Ariz.

Cyrtopeltocoris gracilentis Knight, 1930
1927 *Cyrtopeltocoris albo-fasciatus* [*sic*]: Knight, Can. Ent., 59: 41.
1930 *Cyrtopeltocoris gracilentis* Knight, Ent. News, 41: 321. [Ala.].
Distribution: Ala., Tenn.

Cyrtopeltocoris huachucae Knight, 1968
1968 *Cyrtopeltocoris huachucae* Knight, Brig. Young Univ. Sci. Bull., 9(3): 162. [Ariz.].
Distribution: Ariz.

Cyrtopeltocoris illini Knight, 1941 [Fig. 70]
1941 *Cyrtopeltocoris illini* Knight, Ill. Nat. Hist. Surv. Bull., 22: 117. [Ill.].
Distribution: Ill., Mo.

Cyrtopeltocoris oklahomae Knight, 1968
1968 *Cyrtopeltocoris oklahomae* Knight, Brig. Young Univ. Sci. Bull., 9(3): 162. [Ok.].
Distribution: Ok.

Genus *Orectoderus* Uhler, 1876

1876 *Orectoderus* Uhler, Bull. U.S. Geol. Geogr. Surv. Terr., 1: 319. Type-species: *Orectoderus obliquus* Uhler, 1876. Monotypic.

Note: Knight (1968, Ia. St. J. Sci., 42: 311-312) provided a key to species.

Orectoderus arcuatus Knight, 1927
1927 *Orectoderus arcuatus* Knight, Ent. News, 38: 302. [Wash.].
Distribution: Wash.

Orectoderus bakeri Knight, 1968
1927 *Orectoderus arcuatus*: Knight, Ent. News, 38: 303.
1968 *Orectoderus bakeri* Knight, Ia. St. J. Sci., 42: 314. [Col.].
Distribution: Col.

Orectoderus cockerelli Knight, 1968
1968 *Orectoderus cockerelli* Knight, Ia. St. J. Sci., 42: 317. [Col.].
Distribution: Col.

Orectoderus longicollis Uhler, 1895
1895 *Orectoderus longicollis* Uhler, Bull. Col. Agr. Exp. Stn., 31: 47. [Col.].
Distribution: Col.

Orectoderus montanus Knight, 1968
1968 *Orectoderus montanus* Knight, Ia. St. J. Sci., 42: 315. [Wyo.].
Distribution: Alta., Id., Wyo.

Orectoderus obliquus Uhler, 1876
1876 *Orectoderus obliquus* Uhler, Bull. U.S. Geol. Geogr. Surv. Terr., 1: 320. [Col., Conn., Ill., Ks., Mass., Pa., "Lower Canada," "Washington Territory"].
Distribution: Alta., B.C., Col., Conn., Ill., Ks., Man., Mass., Me., Mich., Mont., N.B., N.D., N.J., N.M., N.Y., Ont., Ore., Pa., S.D., Sask., Ut., Wash., Wis.

Orectoderus obliquus obliquus Uhler, 1876
1876 *Orectoderus obliquus* Uhler, Bull. U.S. Geol. Geogr. Surv. Terr., 1: 320. [Col., Conn., Ill., Ks., Mass., Pa., "Lower Canada," Washington Territory"].
1923 *Orectoderus obliquus*: Knight, Conn. Geol. Nat. Hist. Surv. Bull., 34: 475.
Distribution: Same as for species.

Orectoderus obliquus ferrugineus Knight, 1923
1923 *Orectoderus obliquus* var. *ferrugineus* Knight, Conn. Geol. Nat. Hist. Surv. Bull., 34: 475. [N.Y.].
Distribution: N.Y.

Orectoderus ruckesi Knight, 1968
1968 *Orectoderus ruckesi* Knight, Ia. St. J. Sci., 42: 318. [Wyo.].
Distribution: Wyo.

Orectoderus salicis Knight, 1968
1968 *Orectoderus salicis* Knight, Ia. St. J. Sci., 42: 316. [Col.].
Distribution: Col.

Orectoderus schuhi Knight, 1964
1964 *Orectoderus schuhi* Knight, Ia. St. J. Sci., 39: 149. [Ore.].
Distribution: Nev., Ore.

Orectoderus utahensis Knight, 1968
1968 *Orectoderus utahensis* Knight, Ia. St. J. Sci., 42: 315. [Ut.].
Distribution: Col., Ut.

Genus *Teleorhinus* Uhler, 1890

1890 *Teleorhinus* Uhler, Trans. Md. Acad. Sci., 1: 74. Type-species: *Teleorhinus cyaneus* Uhler, 1890. Monotypic.

Teleorhinus brindleyi Knight, 1968
1968 *Teleorhinus brindleyi* Knight, Brig. Young Univ. Sci. Bull., 9(3): 65. [Id.].
Distribution: Id., Wyo.

Teleorhinus cyaneus Uhler, 1890
1890 *Teleorhinus cyaneus* Uhler, Trans. Md. Acad. Sci., 1: 75. [Cal.].
Distribution: Cal.

Teleorhinus floridanus Blatchley, 1926
1926 *Teleorhinus floridanus* Blatchley, Het. E. N. Am., p. 915 (as Knight ms. name). [Fla.].
1926 *Teleorhinus floridanus* Knight, Can. Ent., 58: 254. [Fla.]. Preoccupied. Synonymized by Knight, 1927, Bull. Brook. Ent. Soc., 22: 104.
Distribution: Fla.

Teleorhinus nigricornis Knight, 1968
1968 *Teleorhinus nigricornis* Knight, Brig. Young Univ. Sci. Bull., 9(3): 66. [Wash.].
Distribution: Cal., Wash.

Teleorhinus oregoni Knight, 1968
1968 *Teleorhinus oregoni* Knight, Brig. Young Univ. Sci. Bull., 9(3): 66. [Ore.].
Distribution: Ore.

Teleorhinus tephrosicola Knight, 1923 [Fig. 92]
1923 *Teleorhinus tephrosicola* Knight, Conn. Geol. Nat. Hist. Surv. Bull., 34: 476. [N.Y.].
1923 *Teleorhinus davisi*: Knight, Conn. Geol. Nat. Hist. Surv. Bull., 34:427 (error for *T. tephrosicola*).
Distribution: Mo., N.J., N.Y., Pa.

Teleorhinus utahensis Knight, 1968
1968 *Teleorhinus utahensis* Knight, Brig. Young Univ. Sci. Bull., 9(3): 65. [Ut.].
Distribution: N.M., Ut.

Tribe *Leucophoropterini* Schuh, 1974

Genus *Tytthus* Fieber, 1864

1864 *Tytthus* Fieber, Wien. Ent. Monat., 8: 82. Type-species: *Capsus geminus* Flor, 1860. Preoccupied(see below). Designated by Kirkaldy, 1906, Trans. Am. Ent. Soc., 32: 128.
Note: Carvalho and Southwood (1955, Bol. Mus. Goeldi, 11: 1-72) revised this genus and provided a key to species.

Tytthus alboornatus (Knight), 1931
1931 *Cyrtorhinus alboornatus* Knight, Bull. Brook. Ent. Soc., 26: 172. [Fla.].
1955 *Tytthus alboornatus*: Carvalho and Southwood, Bol. Mus. Goeldi, 11: 27.
Distribution: Fla., N.Y.

Tytthus balli (Knight), 1931
1931 *Cyrtorhinus balli* Knight, Bull. Brook. Ent. Soc., 26: 171. [Fla.].
1955 *Tytthus balli*: Carvalho and Southwood, Bol. Mus. Goeldi, 11: 30.
Distribution: Fla., Tex.

Tytthus pubescens (Knight), 1931 Revised Status
1860 *Capsus geminus* Flor, Rhyn. Livl., 1: 464. [Europe]. Preoccupied by *Capsus geminus* Say, 1832.

1931 *Cyrtorhinus pubescens* Knight, Bull. Brook. Ent. Soc., 26: 172. [Col.].
1955 *Tytthus geminus*: Carvalho and Southwood, Bol. Mus. Goeldi, 11: 28.
Distribution: Alk., Alta., B.C., Col., Sask., Yuk. (Europe).

Tytthus insperatus (Knight), 1925
1925 *Cyrtorhinus insperatus* Knight, Bull. Brook. Ent. Soc., 20: 42. [Ariz.].
1955 *Tytthus insperatus*: Carvalho and Southwood, Bol. Mus. Goeldi, 11: 31.
Distribution: Ariz., Cal.

Tytthus montanus Carvalho and Southwood, 1955
1955 *Tytthus montanus* Carvalho and Southwood, Bol. Mus. Goeldi, 11: 32. [Mont.].
Distribution: Mont.

Tytthus mundulus (Breddin), 1896
1896 *Periscopus mundulus* Breddin, Deut. Ent. Zeit., 1: 106. [Java].
1984 *Tytthus mundulus*: Nguyen et al., Fla. Dept. Agr. Cons. Serv., Ent. Circ. 265, p. 1.
Distribution: Fla.(?) (Australia, Fiji, Java, Philippine Islands; introduced into Hawaii).
Note: *T. mundulus* has been released in Fla. to help control the sugarcane delphacid, *Perkinsiella saccharicida* Kirkaldy (Nguyen et al., 1984, above), but its establishment is not yet confirmed.

Tytthus parviceps (Reuter), 1890
1890 *Cyrtorhinus parviceps* Reuter, Rev. d'Ent., 9: 258. [Egypt].
1893 *Cylloceps pellicia* Uhler, Proc. Zool. Soc. London, p. 712. [St. Vincent Island]. Synonymized by China, 1924, An. Mag. Nat. Hist., (9)14: 444.
1914 *Cylloceps pellicia*: Barber, Bull. Am. Mus. Nat. Hist., 33: 500.
1917 *Cyrtorhinus pygmaeus*: Van Duzee, Univ. Cal. Publ. Ent., 2: 824.
1939 *Cyrtorhinus pellicia*: Ingram et al., Int. Soc. Sugar Cane Tech., 6: 95.
1955 *Tytthus parviceps*: Carvalho and Southwood, Bol. Mus. Goeldi, 11: 21.
Distribution: Fla. (Africa, Europe, Venezuela, West Indies).

Tytthus pygmaeus (Zetterstedt), 1840
1840 *Capsus pygmaeus* Zetterstedt, Ins. Lapp. Descript., p. 279. [Europe].
1980 *Tytthus pygmaeus*: Kelton, Can. Ent., 112: 289.
Distribution: Alta., B.C., Nfld., Ont., Sask., Wyo., Yuk. (Europe, USSR).

Tytthus vagus (Knight), 1923
1923 *Cyrtorhinus caricis* var. *vagus* Knight, Conn. Geol. Nat. Hist. Surv. Bull., 34: 511. [N.Y.].
1926 *Cyrtorhinus caricis vagus*: Blatchley, Het. E. N. Am., p. 853.
1927 *Cyrtorhinus vagus*: Knight, Can. Ent., 59: 40.
1955 *Tytthus vagus*: Carvalho and Southwood, Bol. Mus. Geoldi, 11: 24.
Distribution: La., Md., N.C., N.J., N.Y., Va.

Tribe Phylini Douglas and Scott, 1865

Genus *Amblytylus* Fieber, 1858

1858 *Amblytylus* Fieber, Wien. Ent. Monat., 2: 325. Type-species: *Miris albidus* Hahn, 1830. Designated by Kirkaldy, 1906, Trans. Am. Ent. Soc., 32: 125.

Amblytylus nasutus (Kirschbaum), 1856
1856 *Lopus nasutus* Kirschbaum, Rhyn. Wiesb., 10: 281. [Germany].
1928 *Amblytylus vanduzeei* Blatchley, J. N.Y. Ent. Soc., 36: 15. [Ind.]. Synonymized

by Knight, 1930, Ent. News, 41: 256.
1930 *Amblytylus nasutus*: Knight, Ent. News, 41: 256
Distribution: Ind., Ky., Mass., Me., Mich., N.Y., Pa., W.Va., Wis. (Europe).
Note: Jewett and Townsend (1947, Ky. Agr. Exp. Stn. Bull., 508: 1-16) studied *A. nasutus* as a pest of Kentucky blue grass, *Poa pratensis* L.

Genus *Asciodema* Reuter, 1878

1878 *Asciodema* Reuter, Acta. Soc. Sci. Fenn., 13(1): 33. Type-species: *Tinocephalus obsoletus* Fieber, 1864. Designated by Kirkaldy, 1906, Trans. Am. Ent. Soc., 32: 122.

Asciodema obsoletum (Fieber), 1864
1864 *Tinocephalus obsoletus* Fieber, Wien. Ent. Monat., 8: 226. [Europe].
1966 *Asciodema obsoletum*: Waloff, J. Appl. Ecol., 3: 299.
Distribution: B.C. (northern Africa, Europe).

Genus *Atomoscelis* Reuter, 1875

1875 *Atomoscelis* Reuter, Bih. K. Vet.-Akad. Handl., 3(1): 57. Type-species: *Agalliastes onustus* Fieber, 1861. Designated by Kirkaldy, 1906, Trans. Am. Ent. Soc., 32: 123.

Atomoscelis modestus (Van Duzee), 1914
1914 *Tuponia modesta* Van Duzee, Trans. San Diego Soc. Nat. Hist., 2: 30. [Cal.].
1917 *Atomoscelis modestus*: Van Duzee, Univ. Cal. Publ. Ent., 2: 414.
Distribution: Alta., Ariz., Cal., Col., Man., N.M., S.D., Sask., Ut., Wash.

Genus *Atractotomus* Fieber, 1858

1858 *Atractotomus* Fieber, Wien. Ent. Monat., 2: 317. Type-species: *Capsus magnicornis* Fallén, 1807. Designated by Kirkaldy 1906, Trans. Am. Ent. Soc., 32: 124.
1872 *Dacota* Uhler, Prelim. Rept. U.S. Geol. Surv. Mont., p. 413. Type-species: *Dacota hesperia* Uhler, 1872. Monotypic. Synonymized by Knight, 1931, Bull. Brook. Ent. Soc., 26: 36.
Note: Froeschner (1963, Bull. Brook. Ent. Soc., 53: 1-5) reviewed this genus and provided a key to species.

Atractotomus acaciae Knight, 1925
1925 *Atractotomus acaciae* Knight, Bull. Brook. Ent. Soc., 20: 34. [Ariz.].
Distribution: Ariz.

Atractotomus albidicoxis Reuter, 1909
1909 *Atractotomus albidicoxis* Reuter, Acta Soc. Sci. Fenn., 36(2): 79. [Ariz.].
Distribution: Ariz.

Atractotomus balli Knight, 1931
1931 *Atractotomus balli* Knight, Bull. Brook. Ent. Soc., 26: 38. [Ariz.].
Distribution: Ariz.

Atractotomus cercocarpi Knight, 1931
1931 *Atractotomus cercocarpi* Knight, Bull. Brook. Ent. Soc., 26: 37. [Col.].
Distribution: Col., N.M.
Note: Froeschner's (1963, Bull. Brook. Ent. Soc., 53: 4) Ariz. record was a *lapsus* for Col.

Atractotomus crataegi Knight, 1931
1931 *Atractotomus crataegi* Knight, Bull. Brook. Ent. Soc., 26: 37. [Ia.].
Distribution: Ia.

Atractotomus hesperius (Uhler), 1872
1872 *Dacota hesperia* Uhler, Prelim. Rept. U.S. Geol. Surv. Mont., p. 413. ["Dakota"].
1931 *Atractotomus hesperius*: Knight, Bull. Brook. Ent. Soc., 26: 36.
Distribution: Alta., Ariz., Cal., Col., "Dakota," Mont., Sask., Wyo.

Atractotomus magnicornis (Fallén), 1807
1807 *Capsus magnicornis* Fallén, Monogr. Cimic. Suec., p. 99. [Sweden].
1923 *Atractotomus magnicornis*: Knight, Conn. Geol. Nat. Hist. Surv. Bull., 34: 461.
Distribution: Conn., N.Y., Pa., W.Va. (Europe).

Atractotomus magnicornis buenoi Knight, 1923
1923 *Atractotomus magnicornis* var. *buenoi* Knight, Conn. Geol. Nat. Hist. Surv. Bull., 34: 461. [N.Y.].
Distribution: N.Y.

Atractotomus magnicornis magnicornis (Fallén), 1807
1807 *Capsus magnicornis* Fallén, Monogr. Cimic. Suec., p. 99. [Sweden].
1923 *Atractotomus magnicornis*: Knight, Conn. Geol. Nat. Hist. Surv. Bull., 34: 461.
Distribution: Same as for species.

Atractotomus mali (Meyer-Dür), 1843
1843 *Capsus mali* Meyer-Dür, Verz. Schweiz. Rhyn., p. 63. [Switzerland].
1924 *Atractotomus mali*: Knight, Bull. Brook. Ent. Soc., 19: 65.
1949 *Criocoris saliens*: Lord, Can. Ent., 81: 217.
Distribution: Conn., N.B., N.S., P.Ed. (Europe).
Note: Sanford (1964, J. Econ. Ent., 57: 921-925) studied the life history of this species in Nova Scotia.

Atractotomus nicholi Knight, 1968
1968 *Atractotomus nicholi* Knight, Brig. Young Univ. Sci. Bull., 9(3): 58. [Ariz.].
Distribution: Ariz.

Atractotomus purshiae Froeschner, 1963
1963 *Atractotomus purshiae* Froeschner, Bull. Brook. Ent. Soc., 58: 1. [Id.].
Distribution: Id., Nev.

Atractotomus reuteri Knight, 1931
1909 *Atractotomus hesperius* Reuter, Acta Soc. Sci. Fenn., 36(2): 78. [Cal.]. Preoccupied.
1931 *Atractotomus reuteri* Knight, Bull. Brook. Ent. Soc., 26: 37. New name for *Atractotomus hesperius* Reuter.
Distribution: Ariz., Cal., N.M.

Genus *Beamerella* Knight, 1959

1959 *Beamerella* Knight, Ia. St. Coll. J. Sci., 33: 423. Type-species: *Beamerella personatus* Knight, 1959. Original designation.
1965 *Larinocerus* Froeschner, Ent. News, 76: 86. Type-species: *Larinocerus balius* Froeschner, 1965. Original designation. Synonymized by Henry and Schuh, 1979, Am. Mus. Novit., 2689: 2.
Note: Henry and Schuh (1979, above) provided a key to species and discussed the relationship of this genus to *Hambletoniola* Carvalho.

Beamerella balius (Froeschner), 1965
1965 *Larinocerus balius* Froeschner, Ent. News, 76: 88. [Cal.].
1979 *Beamerella balius*: Henry and Schuh, Am. Mus. Novit., 2689: 3.
Distribution: Cal., Nev.

Beamerella personatus Knight, 1959
1959 *Beamerella personatus* Knight, Ia. St. Coll. J. Sci., 33: 423. [Tex.].
Distribution: Tex.

Genus *Beckocoris* Knight, 1968

1968 *Beckocoris* Knight, Brig. Young Univ. Sci. Bull., 9(3): 35. Type-species: *Beckocoris laticephalus* Knight, 1968. Original designation.

Beckocoris laticephalus Knight, 1968
1968 *Beckocoris laticephalus* Knight, Brig. Young Univ. Sci. Bull., 9(3): 36. [Nev.].
Distribution: Nev.

Genus *Brachyceratocoris* Knight, 1968

1968 *Brachyceratocoris* Knight, Brig. Young Univ. Sci. Bull., 9(3): 61. Type-species: *Brachyceratocoris nevadensis* Knight, 1968. Original designation.

Brachyceratocoris nevadensis Knight, 1968
1968 *Brachyceratocoris nevadensis* Knight, Brig. Young Univ. Sci. Bull., 9(3): 61. [Nev.].
Distribution: Nev.

Genus *Campylomma* Reuter, 1878

1878 *Campylomma* Reuter, Acta Soc. Sci. Fenn., 13(1): 52. Type-species: *Campylomma nigronasuta* Reuter, 1878. Designated by Distant, 1904, Fauna Brit. India, Rhyn., 2: 483.

Campylomma verbasci (Meyer-Dür), 1843
1843 *Capsus verbasci* Meyer-Dür, Verz. Schw. Rhyn., p. 70. [Europe].
1886 *Agalliastes verbasci*: Uhler, Check-list Hem. Het., p. 21.
1890 *Campylomma verbasci*: Atkinson, J. Asiatic Soc. Bengal, 58(2): 177.
1933 *Campyloma* [*sic*] *verebasci*: Tate, J. Econ. Ent., 26: 1173.
1950 *Compylomma* [*sic*] *verbasci*: Moore, Nat. Can., 77: 249.
1982 *Campyloma* [*sic*] *verbacis* [*sic*]: Bouchard et al., An. Soc. Ent. Que., 27: 84.
Distribution: B.C., Col., Conn., Ia., Id., Ill., Ind., Mass., Me., Minn., N.B., N.S., N.Y., Ont., Ore., Pa., Que., Ut., W.Va., Wash., Wis. (Europe).
Note: Leonard (1915, J. N.Y. Ent. Soc., 23: 195-196) described the immature stages. McMullen and Jong (1970, Can. Ent., 102: 1390-1394) studied the biology in B.C.

Genus *Chaetophylidea* Knight, 1968

1968 *Chaetophylidea* Knight, Brig. Young Univ. Sci. Bull., 9(3): 33. Type-species: *Plagiognathus moerens* Reuter, 1909. Original designation.

Chaetophylidea moerens (Reuter), 1909
1909 *Plagiognathus moerens* Reuter, Acta Soc. Sci. Fenn., 36(2): 80. [Cal., Wyo.].
1955 *Hoplomachus moerens*: Carvalho, Rev. Chil. Ent., 4: 227.

1968 *Chaetophylidea moerens*: Knight, Brig. Young Univ. Sci. Bull., 9(3): 33.
Distribution: Alta., B.C., Cal., Col., Mich., Wash., Wyo.

Genus *Chlamydatus* Curtis, 1833

1833 *Chlamydatus* Curtis, Ent. Month. Mag., 1: 197. Type-species: *Chlamydatus marginatus* Curtis, 1833, a junior synonym of *Lygaeus saltitans* Fallén, 1807. Monotypic.

1858 *Agalliastes* Fieber, Wien. Ent. Monat., 2: 321. Type-species: *Lygaeus saltitans* Fallén, 1807. Designated by Kirkaldy, 1906, Trans. Am. Ent. Soc., 32: 123. Synonymy by virtue of common type-species.

Note: Kelton (1965, Can. Ent., 97: 1132-1144) reviewed the genus and provided a key to the North American species.

Chlamydatus arcuatus Knight, 1964
1964 *Chlamydatus arcuatus* Knight, Ia. St. J. Sci., 39: 139. [Col.].
Distribution: Ariz., Col., Id.

Chlamydatus artemisiae Kelton, 1965
1965 *Chlamydatus artemisiae* Kelton, Can. Ent., 97: 1138. [Alta.].
Distribution: Alta.

Chlamydatus associatus (Uhler), 1872
1872 *Agalliastes associatus* Uhler, Prelim. Rept. U.S. Geol. Surv. Mont., p. 419. [Ut.].
1890 *Chlamydatus associatus*: Atkinson, J. Asiatic Soc. Bengal, 58(2): 174.
1931 *Clamydatus* [*sic*] *associatus*: Knowlton, Ent. News, 42: 68.
Distribution: Alta., B.C., Cal., Col., Conn., Ia., Id., Ill., Ind., Ks., Man., Mass., Mich., Miss., Mo., Mont., N.C., N.H., N.J., N.M., N.Y., Oh., Ont., Ore., Pa., Que., S.D., Sask., Tex., Ut., W.Va., Wash., Wis., Wyo. (Mexico).

Chlamydatus auratus Kelton, 1965
1887 *Agalliastes pulicarius*: Van Duzee, Can. Ent., 19: 72.
1923 *Chlamydatus pulicarius*: Knight, Conn. Geol. Nat. Hist. Surv. Bull., 34: 430.
1964 *Chlamydatus arcuatus*: Knight, Ia. St. coll. J. Sci., 33: 139 (in part).
1965 *Chlamydatus auratus* Kelton, Can. Ent., 97: 1133. [Ont.].
Distribution: Alta., B.C., Conn., Man., Mich., Minn., Ont., Que., Sask., Wis.

Chlamydatus becki Knight, 1968
1968 *Chlamydatus becki* Knight, Brig. Young Univ. Sci. Bull., 9(3): 28. [Ut.].
Distribution: Nev., Ut.

Chlamydatus brevicornis Knight, 1964
1964 *Chlamydatus brevicornis* Knight, Ia. St. J. Sci., 39: 141. [Wyo].
Distribution: B.C., Col., Id., Ut., Wyo.

Chlamydatus fulvipes Knight, 1964
1964 *Chlamydatus fulvipes* Knight, Ia. St. J. Sci., 39: 139. [Minn.].
Distribution: Minn., N.Y.

Chlamydatus knighti Kelton, 1965
1965 *Chlamydatus knighti* Kelton, Can. Ent., 97: 1140. [Wash.].
Distribution: Ut., Wash.

Chlamydatus manzanitae Knight, 1964
1964 *Chlamydatus manzanitae* Knight, Ia. St. J. Sci., 39: 140. [Cal.].
Distribution: Cal.

Chlamydatus monilipes Van Duzee, 1921
1921 *Chlamydatus monilipes* Van Duzee, Proc. Cal. Acad. Sci., (4)11: 132. [Cal.].
Distribution: Cal., Nev.

Chlamydatus montanus Knight, 1964
1964 *Chlamydatus montanus* Knight, Ia. St. J. Sci., 39: 140. [Mont.].
Distribution: Alta., Mont., Sask.

Chlamydatus obliquus (Uhler), 1893
1893 *Agalliastes obliquus* Uhler, Proc. Ent. Soc. Wash., 2: 378. [Ut.].
1909 *Chlamydatus obliquus*: Reuter, Acta Soc. Sci. Fenn., 36(2): 83.
Distribution: Alta., B.C., Cal., Id., Mont., Ut., Wash.

Chlamydatus opacus (Zetterstedt), 1840
1840 *Capsus opacus* Zetterstedt, Ins. Lapp. Descr., p. 279. [Lapland].
1965 *Chlamydatus opacus*: Kelton, Can. Ent., 97: 1136.
Distribution: Man., N.T., Que. (Europe).

Chlamydatus pallidicornis Knight, 1964
1964 *Chlamydatus pallidicornis* Knight, Ia. St. J. Sci., 39: 138. [Wyo.].
Distribution: Alta., B.C., Col., Id., Wash., Wyo.

Chlamydatus pulicarius (Fallén), 1807
1807 *Lygaeus pulicarius* Fallén, Monogr. Cimic. Suec., p. 95. [Sweden].
1925 *Chlamydatus pulicarius*: Knight, Can. Ent., 57: 181.
Distribution: Alk., Alta., B.C. (Europe, USSR).
Note: There are a number of records of *C. pulicarius* in Carvalho's Catalog (1958, Arq. Mus. Nac., 45: 34), including Knight's 1925 report (above), but Kelton (1965, Can. Ent., 97: 1132) doubted that they apply to this species. Kelton (1980, Agr. Can. Publ., 1703: 302) further suggested that most previous records in North America probably refer to *C. pullus* (Reuter).

Chlamydatus pullus (Reuter), 1871
1871 *Agalliastes pullus* Reuter, Not. Förh. Sällsk. Fauna Flora Fenn., 11: 324. [Europe].
1925 *Chlamydatus pullus*: Knight, Can. Ent., 57: 181.
1964 *Chlamydatus arcuatus*: Knight, Ia. St. Coll. J. Sci., 39: 139 (in part).
Distribution: Alk., Alta., B.C., Col., Man., N.M., N.T., Nfld., Ont., Que., Sask., Wyo., Yuk. (northern Africa, Asia, Europe, Greenland).

Chlamydatus ruficornis Knight, 1959
1959 *Chlamydatus ruficornis* Knight, Ia. St. Coll. J. Sci., 33: 424. [Ia.].
Distribution: Alta., Col., Ia.

Chlamydatus schuhi Knight, 1964
1964 *Chlamydatus schuhi* Knight, Ia. St. J. Sci., 39: 141. [Ore.].
Distribution: B.C., Ore.

Chlamydatus suavis (Reuter), 1876
1876 *Agalliastes suavis* Reuter, Öfv. K. Svens. Vet.-Akad. Förh., 32(9): 92. [Tex.].
1890 *Chlamydatus suavis*: Atkinson, J. Asiatic Soc. Bengal, 58(2): 174.
1976 *Chlamydatus sauvis* [*sic*]: Goeden and Ricker, Environ. Ent., 5: 47.
1979 *Chlamydatus suevis* [*sic*]: Henry and Smith, J. Ga. Ent. Soc., 14: 213.
Distribution: Ala., Ariz., Cal., Conn., D.C., Fla., Ga., Ia., Ill., Ind., Ks., La., Mo., N.C., N.J., N.Y., Ont., Pa., Tex.

Chlamydatus uniformis (Uhler), 1893
1893 *Agalliastes uniformis* Uhler, Proc. Ent. Soc. Wash., 2: 379. [Ut.].

1917 *Chlamydatus uniformis*: Van Duzee, Univ. Cal. Publ. Ent., 2: 417.
Distribution: Cal., Ut.

Chlamydatus wilkinsoni (Douglas and Scott), 1866
1866 *Agalliastes wilkinsoni* Douglas and Scott, Ent. Month. Mag., 2: 273. [Europe].
1932 *Chlamydatus wilkinsoni*: Walley, Can. Ent., 64: 152.
Distribution: Alk., N.T., Que., Yuk. (Europe).

Genus *Compsidolon* Reuter, 1900

1900 *Compsidolon* Reuter, Öfv. F. Vet.-Soc. Förh., 42: 147. Type-species: *Compsidolon elegantulum* Reuter, 1900. Monotypic.

Subgenus *Coniortodes* Wagner, 1952

1952 *Psallus* (*Coniortodes*) Wagner, Tierw. Deut., 41: 172. Type-species: *Capsus salicellus* Herrich-Schaeffer, 1841 [author as Meyer-Dür, 1843, in error]. Original designation.

Compsidolon salicellum (Herrich-Schaeffer), 1841
1841 *Capsus salicellus* Herrich-Schaeffer, Wanz. Ins., 6: 47. [Europe].
1961 *Psallus* sp.: MacPhee and Sanford, Can. Ent., 93: 672.
1966 *Coniortodes salicellus*: Sanford and Herbert, Can. Ent., 98: 997.
1982 *Psallus salicellus*: Kelton, Can. Ent., 114: 172.
1984 *Compsidolon* (*Coniortodes*) *salicellum*: Lattin and Messing, J. N.Y. Ent. Soc., 92: 179.
Distribution: B.C., N.S., Ore., P.Ed., Wash. (Europe, Western Russia).
Note: Lattin and Messing (1984, above) reviewed the distribution, hosts, feeding habits, and literature of this introduced species. Messing and AliNiazee (1985, J. Ent. Soc. B.C., 82: 15) gave notes on the feeding habits and incorrectly claimed the first U.S. record.

Genus *Conostethus* Fieber, 1858

1858 *Conostethus* Fieber, Wien. Ent. Monat., 2: 318. Type-species: *Capsus roseus* Fallén, 1829. Monotypic.

Conostethus americanus Knight, 1939
1939 *Conostethus americanus* Knight, Ent. News, 50: 132. [Col.].
Distribution: Alta., Col., Mont., S.D., Sask.

Genus *Criocoris* Fieber, 1858

1858 *Criocoris* Fieber, Wien. Ent. Monat., 2: 319. Type-species: *Phytocoris crassicornis* Hahn, 1834 [as *Capsus crassicornis* Hahn]. Monotypic.
1876 *Strongylotes* Reuter, Öfv. K. Svens. Vet.-Akad. Förh., 32(9): 88. Preoccupied. Type-species: *Strongylotes saliens* Reuter, 1876. Monotypic.
1903 *Laodamia* Kirkaldy, Wien. Ent. Zeit., 22: 13. New name for *Strongylotes* Reuter. Synonymized by Reuter, 1910, Acta Soc. Sci. Fenn., 37(3): 167.
1910 *Laodamina* Banks, Cat. Nearc. Hem.-Het., p. 29. Unnecessary new name for *Laodamia* Kirkaldy.

Criocoris saliens (Reuter), 1876
1876 *Strongylotes saliens* Reuter, Öfv. K. Svens. Vet.-Akad. Förh., 32(9): 88. [Tex.].
1910 *Strongylotus* [*sic*] *saliens*: Smith, Cat. Ins. N.J., p. 160.
1910 *Laodamina saliens*: Banks, Cat. Nearc. Hem.-Het., p. 29.
1912 *Criocoris canadensis* Van Duzee, Bull. Buff. Soc. Nat. Sci., 10: 511. [Ont., Que.]. Synonymized by Blatchley, 1926, Het. E. N. Am., p. 961.
1914 *Criocoris saliens*: Van Duzee, Trans. San Diego Soc. Nat. Hist., 2: 29.
1939 *Atractotomus flavotarsus* Johnston, Bull. Brook. Ent. Soc., 34: 129. [Tex.]. Synonymized by Henry, 1985, J. N.Y. Ent. Soc., 93: 1121.
1984 *Crioceris* [*sic*] *saliens*: Larochelle, Fabreries, Suppl. 3, p. 318.
Distribution: Alta., Cal., Ga., Ia., Id., Ill., Ind., Ks., Man., Mass., Md., Minn., Mo., N.J., N.S., N.Y., Ont., Pa., Que., Sask., Tex., Va., W.Va., Wash., Wis.

Genus *Europiella* Reuter, 1909

1909 *Europiella* Reuter, Acta Soc. Sci. Fenn., 36(2): 83. Type-species: *Agalliastes stigmosus* Uhler, 1893. Original designation.
Note: Knight (1968, Brig. Young. Univ. Sci. Bull., 9(3): 37-39) provided a key to species of *Europiella*.

Europiella albata Knight, 1969
1969 *Europiella albata* Knight, Ia. St. J. Sci., 44: 85. [Ok.].
Distribution: Ok.

Europiella albipubescens Knight, 1968
1968 *Europiella albipubescens* Knight, Brig. Young Univ. Sci. Bull., 9(3): 46. [Nev.].
Distribution: Nev.

Europiella angulata (Uhler), 1895
1895 *Maurodactylus angulatus* Uhler, Bull. Col. Agr. Exp. Stn., 31: 53. [Col.].
1968 *Europiella angulata*: Knight, Brig. Young Univ. Sci. Bull., 9(3): 41.
Distribution: Col., Ut.

Europiella arizonae Knight, 1968
1968 *Europiella arizonae* Knight, Brig. Young Univ. Sci. Bull., 9(3): 45. [Ariz.].
Distribution: Ariz.

Europiella bakeri (Bergroth), 1898
1895 *Agalliastes signatus* Uhler, Bull. Col. Agr. Exp. Stn., 31: 55. [Col.]. Preoccupied.
1898 *Chlamydatus bakeri* Bergroth, Wien. Ent. Zeit., 17: 35. New name for *Agalliastes signatus* Uhler.
1909 *Chlamydatus uhlerianus* Kirkaldy, Can. Ent., 41: 390. Unnecessary new name for *Agalliastes signatus* Uhler.
1941 *Psallus bakeri*: Knight, Ill. Nat. Hist. Surv. Bull., 22: 43.
1955 *Europiella bakeri*: Carvalho, Rev. Chil. Ent., 4: 227.
Distribution: Alta., B.C., Col., Ill., Minn.

Europiella balli Knight, 1968
1968 *Europiella balli* Knight, Brig. Young Univ. Sci. Bull., 9(3): 44. [Ariz.].
Distribution: Ariz.

Europiella basicornis Knight, 1970
1970 *Europiella basicornis* Knight, Great Basin Nat., 30: 230. [Ut.].
Distribution: Ut. (Mexico).

Europiella brevicornis Knight, 1968
1968 *Europiella brevicornis* Knight, Brig. Young Univ. Sci. Bull., 9(3): 45. [Ariz.].
Distribution: Ariz.

Europiella concinna Reuter, 1909
1909 *Europiella concinna* Reuter, Acta Soc. Sci. Fenn., 36(2): 84. [Cal.].
Distribution: Ariz., Cal.

Europiella decolor (Uhler), 1893
1893 *Agalliastes decolor* Uhler, Proc. Ent. Soc. Wash., 2: 380. [Cal., Ut.].
1909 *Plagiognathus decolor*: Reuter, Acta Soc. Sci. Fenn., 36(2): 81.
1916 *Europiella decolor*: Van Duzee, Check List Hem., p. 47.
Distribution: Cal., Col., Id., Ks., Nev., Ut.

Europiella flavicornis Knight, 1969
1969 *Europiella flavicornis* Knight, Ia. St. J. Sci., 44: 82. [Col.].
Distribution: Cal., Col., Ks., Nev., Ut. (Mexico).

Europiella fuscicornis Knight, 1969
1969 *Europiella fusicornis* Knight, Ia. St. J. Sci., 44: 82. [Col.].
Distribution: Col.

Europiella grayiae Knight, 1968
1968 *Europiella grayiae* Knight, Brig. Young Univ. Sci. Bull., 9(3): 41. [Nev.].
Distribution: Nev.

Europiella humeralis (Van Duzee), 1923
1923 *Sthenarus humeralis* Van Duzee, Proc. Cal. Acad. Sci., (4)12: 162. [Mexico].
1968 *Europiella humeralis*: Knight, Brig. Young Univ. Sci. Bull., 9(3): 41.
Distribution: Ariz. (Mexico).

Europiella knowltoni Knight, 1970
1970 *Europiella knowltoni* Knight, Great Basin Nat., 30: 228. [Id.].
Distribution: Id.

Europiella lycii Knight, 1968
1968 *Europiella lycii* Knight, Brig. Young Univ. Sci. Bull., 9(3): 40. [Nev.].
Distribution: Alta., Nev., Sask.

Europiella montanae Knight, 1968
1968 *Europiella montanae* Knight, Brig. Young Univ. Sci. Bull., 9(3): 45. [Mont.].
Distribution: Mont., Sask.

Europiella monticola Knight, 1970
1970 *Europiella monticola* Knight, Great Basin Nat., 30: 230. [Col.].
Distribution: Col.

Europiella multipunctipes Knight, 1970
1970 *Europiella multipunctipes* Knight, Great Basin Nat., 30: 229. [Nev.].
Distribution: Nev.

Europiella nicholi Knight, 1968
1968 *Europiella nicholi* Knight, Brig. Young Univ. Sci. Bull., 9(3): 42. [Ariz.].
Distribution: Ariz.

Europiella nigricornis Knight, 1968
1968 *Europiella nigricornis* Knight, Brig. Young Univ. Sci. Bull., 9(3): 40. [Nev.].
Distribution: Nev.

Europiella nigrofemoratus Knight, 1968
1968 *Europiella nigrofemoratus* Knight, Brig. Young Univ. Sci. Bull., 9(3): 39. [Nev.].
Distribution: Nev.

Europiella pallida Knight, 1969
1969 *Europiella pallida* Knight, Ia. St. J. Sci., 44: 83. [Ut.].
Distribution: Ut.

Europiella pilosula (Uhler), 1893
1893 *Atomoscelis pilosulus* Uhler, Proc. Ent. Soc. Wash., 2: 377. [Ut.].
1917 *Psallus pilosulus*: Van Duzee, Univ. Cal. Publ. Ent., 2: 407.
1968 *Europiella pilosula*: Knight, Brig. Young Univ. Sci. Bull., 9(3): 44.
Distribution: Ut., Wyo.

Europiella punctipes Knight, 1968
1968 *Europiella punctipes* Knight, Brig. Young Univ. Sci. Bull., 9(3): 47. [Nev.].
Distribution: Nev.

Europiella rubricornis Knight, 1968
1968 *Europiella rubricornis* Knight, Brig. Young Univ. Sci. Bull., 9(3): 39. [Ut.].
Distribution: Ore., Ut.

Europiella rufiventris Knight, 1968
1968 *Europiella rufiventris* Knight, Brig. Young Univ. Sci. Bull., 9(3): 42. [Ariz.].
Distribution: Ariz.

Europiella sarcobati Knight, 1969
1969 *Europiella sarcobati* Knight, Ia. St. J. Sci., 44: 83. [Wash.].
Distribution: Ore., Wash.

Europiella signicornis Knight, 1969
1969 *Europiella signicornis* Knight, Ia. St. J. Sci., 44: 84. [Ariz.].
Distribution: Ariz., Nev.

Europiella similis Knight, 1969
1968 *Europiella stigmosa*: Knight, Brig. Young Univ. Sci. Bull., 9(3): 43.
1969 *Europiella similis* Knight, Ia. St. J. Sci., 44: 81. [Nev.].
Distribution: Ariz., Nev.

Europiella sparsa Van Duzee, 1918
1918 *Europiella sparsa* Van Duzee, Proc. Cal. Acad. Sci., (4)8: 305. [Cal.].
Distribution: Ariz., Cal., Col., Nev., Ut.

Europiella stigmosa (Uhler), 1893
1893 *Agalliastes stigmosus* Uhler, Proc. Ent. Soc. Wash., 2: 379. [Ut.].
1909 *Europiella stigmosa*: Reuter, Acta Soc. Sci. Fenn., 36(2): 84.
1909 *Europiella umbrina* Reuter, Acta Soc. Sci. Fenn., 36(2): 85. [Cal.]. Synonymized by Knight, 1968, Brig. Young Univ. Sci. Bull., 9(3): 43.
Distribution: Cal., Col., Nev., Ut.

Europiella stitti Knight, 1968
1968 *Europiella stitti* Knight, Brig. Young Univ. Sci. Bull., 9(3): 46. [Ariz.].
1969 *Europiella franseriae* Knight, Ia. St. J. Sci., 44: 85. [Ariz.]. Synonymized by Henry, 1985, J. N.Y. Ent. Soc., 93: 1124.
Distribution: Ariz.

Europiella unipuncta Knight, 1968
1968 *Europiella unipuncta* Knight, Brig. Young Univ. Sci. Bull., 9(3): 44. [Nev.].

Distribution: Nev., Ut.

Europiella viridiventris Knight, 1968
1968 *Europiella viridiventris* Knight, Brig. Young Univ. Sci. Bull., 9(3): 42. [Ariz.].
Distribution: Ariz.

Europiella yampae Knight, 1968
1968 *Europiella yampae* Knight, Brig. Young Univ. Sci. Bull., 9(3): 43. [Col.].
Distribution: Alta., Col., Man., Sask.

Genus *Hambletoniola* Carvalho, 1954

1954 *Hambletoniola* Carvalho, Ent. News, 15: 123. Type-species: *Hambletoniola antennata* Carvalho, 1954. Original designation.
Note: Henry and Schuh (1979, Am. Mus. Novit., 2689: 2) compared this genus to the closely related genus *Beamerella* Knight and provided a key to species.

Hambletoniola antennata Carvalho, 1954
1954 *Hambletoniola antennata* Carvalho, Ent. News, 15: 126. [Mexico].
1979 *Hambletoniola antennata*: Henry and Schuh, Am. Mus. Novit., 2689: 1.
Distribution: Tex. (Mexico).

Genus *Hoplomachidea* Reuter, 1909

1909 *Hoplomachidea* Reuter, Acta Soc. Sci. Fenn., 36(2): 75. Type-species: *Hoplomachus consors* Uhler, 1893. Monotypic.

Hoplomachidea consors (Uhler), 1893
1893 *Hoplomachus consors* Uhler, N. Am. Fauna, 7: 264. [Cal.].
1909 *Hoplomachidea consors*: Reuter, Acta Soc. Sci. Fenn., 36(2): 76.
Distribution: Cal., Nev.

Genus *Hoplomachus* Fieber, 1858

1858 *Hoplomachus* Fieber, Wien. Ent. Monat., 2: 324. Type-species: *Lygaeus thunbergi* Fallén, 1807. Designated by Kirkaldy, 1906, Trans. Am. Ent. Soc., 32: 125.

Hoplomachus affiguratus (Uhler), 1895
1895 *Macrotylus affiguratus* Uhler, Bull. Col. Agr. Exp. Stn., 31: 50. [Col.].
1912 *Plagiognathus affiguratus*: Reuter, Öfv. F. Vet.-Soc. Förh., 54A(7): 61.
1955 *Hoplomachus affiguratus*: Carvalho, Rev. Chil. Ent., 4: 226.
Distribution: Alta., Cal., Col., Id., Wash., Wyo.

Genus *Icodema* Reuter, 1875

1875 *Icodema* Reuter, Rev. Crit. Caps., 1: 97. Type-species: *Plagiognathus infuscatus* Fieber, 1861. Monotypic.

Icodema nigrolineatum (Knight), 1923
1904 *Macrotylus vestitus*: Wirtner, An. Carnegie Mus., 3: 201.
1923 *Plagiognathus nigrolineatus* Knight, Conn. Geol. Nat. Hist. Surv. Bull., 34: 443. [Minn.].
1955 *Icodema nigrolineata*: Carvalho, Rev. Chil. Ent., 4: 226.
1973 *Icodema nigrolineatum*: Steyskal, Stud. Ent., 16: 206.

Distribution: Conn., Fla., Ill., Man., Minn., Miss., Mo., Oh., Pa., Tex.
Note: Wheeler and Henry (1977, Great Lakes Ent., 10: 150-151) clarified the Wirtner (1904, above) record.

Genus *Keltonia* Knight, 1966

1966 *Keltonia* Knight, Can. Ent., 98: 590. Type-species: *Keltonia rubrofemorata* Knight, 1966. Original designation.
Note: Kelton (1966, Can. Ent., 98: 668-670) reviewed the genus and provided a key to species.

Keltonia balli (Knight), 1926
1926 *Psallus balli* Knight, Can. Ent., 58: 253. [Fla.].
1966 *Keltonia balli*: Knight, Can. Ent., 98: 591.
Distribution: Fla., Miss.

Keltonia clinopodii Kelton, 1966
1966 *Keltonia clinopodii* Kelton, Can. Ent., 98: 668. [Fla.].
Distribution: Fla.

Keltonia fuscipunctata Knight, 1966
1966 *Keltonia fuscipunctata* Knight, Can. Ent., 98: 591. [Fla.].
Distribution: Fla.

Keltonia rubrofemorata Knight, 1966
1966 *Keltonia rubrofemorata* Knight, Can. Ent., 98: 590. [Fla.].
Distribution: Fla.

Keltonia sulphurea (Reuter), 1907
1907 *Psallus sulphureus* Reuter, Öfv. F. Vet.-Soc. Förh., 49(5): 23. [Jamaica].
1911 *Pseudatomoscelus tuckeri* Poppius, An. Soc. Ent. Belg., 55: 86. [Tex.]. Synonymized by Carvalho, 1958, Arq. Mus. Nac., 45: 138.
1914 *Apocremnus* (*Psallus*) *sulphureus*: Barber, Bull. Am. Mus. Nat. Hist., 33: 500.
1916 *Psallus tuckeri*: Van Duzee, Check List Hem., p. 46.
1923 *Reuteroscopus sulphureus*: Knight, Conn. Geol. Nat. Hist. Surv. Bull., 34: 462.
1928 *Psallus conspurcatus* Blatchley, J. N.Y. Ent. Soc., 36: 16. [Fla.]. Synonymized by Knight, 1966, Can. Ent., 98: 591.
1938 *Reuteroscopus sulphurea*: Brimley, Ins. N.C., p. 81.
1966 *Keltonia sulphurea*: Knight, Can. Ent., 98: 591.
1979 *Keltonia sulphureus*: Henry and Smith, J. Ga. Ent. Soc., 14: 213.
Distribution: Ala., Ariz., Ark., Conn., Del., Fla., Ga., Ill., Ks., Mass., Miss., Mo., Oh., Tex., W.Va. (Mexico, West Indies).

Genus *Lepidopsallus* Knight, 1923

1923 *Lepidopsallus* Knight, Conn. Geol. Nat. Hist. Surv. Bull., 34: 470. Type-species: *Sthenarus rubidus* Uhler, 1895. Original designation.

Lepidopsallus arizonae Knight, 1968
1968 *Lepidopsallus arizonae* Knight, Brig. Young Univ. Sci. Bull., 9(3): 52. [Ariz.].
Distribution: Ariz.

Lepidopsallus australis Blatchley, 1926
1926 *Lepidopsallus australis* Blatchley, Het. E. N. Am., p. 953. [Fla.].

Distribution: Fla.

Lepidopsallus californicus Knight, 1968
1968 *Lepidopsallus californicus* Knight, Brig. Young Univ. Sci. Bull., 9(3): 52. [Cal.].
Distribution: Cal.

Lepidopsallus claricornis Knight, 1923
1923 *Lepidopsallus claricornis* Knight, Conn. Geol. Nat. Hist. Surv. Bull., 34: 471. [N.J.].
Distribution: N.J., W.Va.

Lepidopsallus hesperus Knight, 1968
1968 *Lepidopsallus hesperus* Knight, Brig. Young Univ. Sci. Bull., 9(3): 53. [Id.].
Distribution: Cal., Id., Wyo.

Lepidopsallus longirostris Knight, 1968
1968 *Lepidopsallus longirostris* Knight, Brig. Young Univ. Sci. Bull., 9(3): 54. [Col.].
Distribution: Alta., Ariz., Col., Sask., Wyo.

Lepidopsallus miniatus Knight, 1925
1925 *Lepidopsallus miniatus* Knight, Bull. Brook. Ent. Soc., 20: 226. [Fla.].
Distribution: Fla., Ill., Miss., Mo., S.C., Tex., W.Va.

Lepidopsallus minusculus Knight, 1923
1923 *Lepidopsallus minusculus* Knight, Conn. Geol. Nat. Hist. Surv. Bull., 34: 472. [N.Y.].
1982 *Lepidopsalus* [*sic*] *minisculus* [*sic*]: Bouchard et al., An. Soc. Ent. Que., 27: 84.
1982 *Lepidopsallus minisculus* [*sic*]: Kelton, Agr. Can. Monogr., 24: 136.
Distribution: N.Y., Ont., Que.

Lepidopsallus monticola Knight, 1968
1968 *Lepidopsallus monticola* Knight, Brig. Young Univ. Sci. Bull., 9(3): 54. [Col.].
Distribution: Ariz., Col.

Lepidopsallus nicholi Knight, 1968
1968 *Lepidopsallus nicholi* Knight, Brig. Young Univ. Sci. Bull., 9(3): 52. [Ariz.].
Distribution: Ariz.

Lepidopsallus nyssae Johnston, 1930
1930 *Lepidopsallus nyssae* Johnston, Bull. Brook. Ent. Soc., 25: 299. [Tex.].
Distribution: Ill., Tex.

Lepidopsallus olseni Knight, 1923
1923 *Lepidopsallus olseni* Knight, Conn. Geol. Nat. Hist. Surv. Bull., 34: 473. [Mass.].
Distribution: Mass.

Lepidopsallus ovatus Knight, 1925
1925 *Lepidopsallus ovatus* Knight, Bull. Brook. Ent. Soc., 20: 227. [Ariz.].
Distribution: Ariz., Cal.

Lepidopsallus pini Knight, 1968
1968 *Lepidopsallus pini* Knight, Brig. Young Univ. Sci. Bull., 9(3): 53. [Col.].
Distribution: Ariz., Col., Nev.

Lepidopsallus rostratus Knight, 1923
1923 *Lepidopsallus rostratus* Knight, Conn. Geol. Nat. Hist. Surv. Bull., 34: 470. [Minn.].
Distribution: Alta., Ia., Man., Minn., N.Y., Pa., Que., Sask., W.Va., Wis.

Lepidopsallus rubidus (Uhler), 1895
- 1889 *Psallus variabilis*: Van Duzee, Can. Ent., 21: 4.
- 1895 *Sthenarus rubidus* Uhler, Bull. Col. Agr. Exp. Stn., 31: 41. [Col., Fla., Ill., N.Y., Tex., Cuba, San Domingo, Mexico].
- 1909 *Europiella rubida*: Reuter, Acta Soc. Sci. Fenn., 36(2): 85.
- 1914 *Europiella rubidus*: Barber, Bull. Am. Mus. Nat. Hist., 33: 500.
- 1916 *Apocremnus variabilis*: Van Duzee, Check List Hem., p. 47.
- 1923 *Lepidopsallus rubidus*: Knight, Conn. Geol. Nat. Hist. Surv. Bull., 34: 471.
- 1923 *Lepidopsallus rubidus* var. *atricolor* Knight, Conn. Geol. Nat. Hist. Surv. Bull., 34: 472. [N.Y.]. Synonymized by Larochelle, 1984, Fabreries, Suppl. 3, p. 318.

Distribution: Alta., Cal., Col., Conn., Fla., Id., Ill., Ind., Ks., Man., Minn., N.Y., Oh., Ont.,Pa., Que., Sask., Tex., Ut., Vt., W.Va., Wash. (West Indies?, Mexico?).

Note: Records of this willow [*Salix* spp.]-feeding species from Mexico and the West Indies need verification.

Lepidopsallus tuthilli Knight, 1968
- 1968 *Lepidopsallus tuthilli* Knight, Brig. Young Univ. Sci. Bull., 9(3): 52. [Col.].

Distribution: Col.

Genus *Lopus* Hahn, 1831

1831 *Lopus* Hahn, Wanz. Ins., 1: 10. Type-species: *Lopus chrysanthemi* Hahn, 1831, a junior synonym of *Capsus decolor* Fallén, 1807. Designated by Kirkaldy, 1906, Can. Ent., 38: 376.

1879 *Onychumenus* Reuter, Acta Soc. Sci. Fenn., 13(2): 286. Type-species: *Capsus decolor* Fallén, 1807. Monotypic. Synonymy by virtue of common type-species..

Lopus decolor (Fallén), 1807
- 1807 *Capsus decolor* Fallén, Monogr. Cimic. Suec., p. 102. [Europe].
- 1872 *Lygus unicolor* Provancher, Nat. Can., 4: 105 (in part). [Que?]. Synonymized and lectotype designated by Kelton, 1968, Nat. Can., 95: 1074.
- 1887 *Oncotylus decolor*: Provancher, Pet. Faune Ent. Can., 3: 148.
- 1890 *Onychumenus decolor*: Atkinson, J. Asiatic Soc. Bengal, 58(2): 146.
- 1908 *Lopus decolor*: Horvath, An. Hist.-Nat. Mus. Nat. Hung., 6: 5.
- 1985 *Lupus* [*sic*] *decolor*: Messing and AliNiazee, J. Ent. Soc. B.C., 82: 15.

Distribution: B.C., Cal., Col., Conn., D.C., Mass., Md., Me., N.J., N.Y., Ont., Que., Va., W.Va. (northern Africa, Europe).

Genus *Macrotylus* Fieber, 1858

1858 *Macrotylus* Fieber, Wien. Ent. Monat., 2: 325. Type-species: *Macrotylus luniger* Fieber, 1895, a junior synonym of *Cimex quadrilineatus* Schrank, 1785. Monotypic.

Macrotylus amoenus Reuter, 1909 [Fig. 89]
- 1909 *Macrotylus amoenus* Reuter, Acta Soc. Sci. Fenn., 36(2): 75. [Conn.].

Distribution: Ark., Conn., Del., Fla., Mass., Me., Mo., N.C., N.Y., Pa.

Macrotylus dorsalis Van Duzee, 1916
- 1916 *Macrotylus dorsalis* Van Duzee, Pomona J. Ent. Zool., 8: 7. [Cal.].

Distribution: Cal.

Macrotylus essigi Van Duzee, 1916
1916 *Macrotylus essigi* Van Duzee, Pomona J. Ent. Zool., 8: 8. [Cal.].
Distribution: Cal.

Macrotylus geminus Knight, 1925
1925 *Macrotylus geminus* Knight, Bull. Brook. Ent. Soc., 20: 35. [Ariz.].
Distribution: Ariz.

Macrotylus infuscatus Van Duzee, 1916
1916 *Macrotylus infuscatus* Van Duzee, Pomona J. Ent. Zool., 8: 10. [Cal.].
Distribution: Cal., Id., Nev.

Macrotylus intermedius Van Duzee, 1917
1917 *Macrotylus intermedius* Van Duzee, Proc. Cal. Acad. Sci., (4)7: 278. [Cal.].
Distribution: Cal.

Macrotylus lineolatus Uhler, 1894
1894 *Macrotylus lineolatus* Uhler, Proc. Cal. Acad. Sci., (2)4: 270. [Cal., Mexico].
Distribution: Cal. (Mexico).

Macrotylus multipunctatus Van Duzee, 1916
1916 *Macrotylus multipunctatus* Van Duzee, Pomona J. Ent. Zool., 8: 7. [Cal.].
Distribution: B.C., Cal., Ore.

Macrotylus polymonii Knight, 1932
1932 *Macrotylus polymonii* Knight, Pan-Pac. Ent., 8: 79. [Wash.].
Distribution: Wash.

Macrotylus regalis Uhler, 1890
1890 *Macrotylus regalis* Uhler, Trans. Md. Acad. Sci., 1: 86. [Cal.].
Distribution: Cal.

Macrotylus salviae Knight, 1968
1968 *Macrotylus salviae* Knight, Brig. Young Univ. Sci. Bull., 9(3): 62. [Nev.].
Distribution: Nev.

Macrotylus sexguttatus (Provancher), 1887
1887 *Amblytylus sexguttatus* Provancher, Pet. Faune Ent. Can., 3: 150. [Que.]. Lectotype designated by Kelton, 1968, Nat. Can., 95: 1075.
1912 *Macrotylus sexguttatus*: Van Duzee, Ottawa Nat., 26: 68.
Distribution: Conn., Mass., Me., Mich., Minn., N.Y., Ont., Pa., Que., Wis.

Macrotylus tristis Uhler, 1890
1890 *Macrotylus tristis* Uhler, Trans. Md. Acad. Sci., 1: 87. [Cal.].
Distribution: Cal., Col.

Macrotylus vanduzeei Knight, 1932
1932 *Macrotylus vanduzeei* Knight, Pan-Pac. Ent., 8: 80. [Cal.].
Distribution: Cal.

Genus *Maurodactylus* Reuter, 1878

1878 *Maurodactylus* Reuter, Acta Soc. Sci. Fenn., 13(1): 27. Type-species: *Maurodactylus nigricornis* Reuter, 1878, a junior synonym of *Phytocoris albidus* Kolenati, 1845. Designated by Kirkaldy, 1906, Trans. Am. Ent. Soc., 32: 122.

Maurodactylus consors Uhler, 1895

1895 *Maurodactylus consors* Uhler, Col. Agr. Exp. Stn. Bull., 31: 53. [Col.].

Distribution: Col.

Maurodactylus semiustus Van Duzee, 1914

1914 *Maurodactylus semiustus* Van Duzee, Trans. San Diego Soc. Nat. Hist., 2: 31. [Cal.].

Distribution: Cal.

Note: Slater and Baranowski (1978, How to Know True Bugs, p. 160) suggested that *M. consors* and *M. semiustus* may belong in the genus *Lepidopsallus*; however, our examination of syntypes (USNM) suggests that these genera are not synonymous.

Genus *Megalocoleus* Reuter, 1890

1858 *Macrocoleus* Fieber, Wien. Ent. Monat., 2: 325. Preoccupied. Type-species: *Capsus exsanguis Herrich-Schaeffer, 1835*. Designated by Kirkaldy, 1906, Trans. Am. Ent. Soc., 32: 125.

1890 *Megalocoleus* Reuter, Rev. d'Ent., 9: 254. New name for *Macrocoleus* Fieber. Type-species: *Capsus exsanguis* Herrich-Schaeffer, 1835. Designated by Kirkaldy, 1906, Trans. Am. Ent. Soc., 32: 125.

Megalocoleus molliculus (Fallén), 1807

1807 *Lygaeus molliculus* Fallén, Monogr. Cimic. Suec., p. 77. [Europe].

1922 *Megalocoleus molliculus*: Knight, Can. Ent., 53: 283.

Distribution: Conn., Mass., N.J., Ont., Pa., W.Va., Wis. (Europe).

Note: Henry and Wheeler (1979, Proc. Ent. Soc. Wash., 81: 258-259) reviewed distribution and hosts and illustrated the adult and fifth instar.

Genus *Megalopsallus* Knight, 1927

1927 *Megalopsallus* Knight, An. Ent. Soc. Am., 20: 224. Type-species: *Megalopsallus atriplicis* Knight, 1927. Original designation.

Megalopsallus adustus Knight, 1927

1927 *Megalopsallus adustus* Knight, An. Ent. Soc. Am., 20: 227. [Tex.].

Distribution: Tex.

Megalopsallus atriplicis Knight, 1927

1927 *Megalopsallus atriplicis* Knight, An. Ent. Soc. Am., 20: 224. [Tex.].

Distribution: S.D., Tex.

Megalopsallus brittoni Knight, 1927

1927 *Megalopsallus brittoni* Knight, An. Ent. Soc. Am., 20: 227. [Conn.].

Distribution: Conn.

Megalopsallus diversipes Knight, 1927

1927 *Megalopsallus latifrons diversipes* Knight, An. Ent. Soc. Am., 20: 226. [Miss.].

1958 *Megalopsallus diversipes*: Carvalho, Arq. Mus. Nac., 45: 72.

Distribution: Miss.

Megalopsallus femoralis Kelton, 1980

1980 *Megalopsallus femoralis* Kelton, Can. Ent., 112: 285. [Sask.].

Distribution: Alta., Sask.

Megalopsallus latifrons Knight, 1927
1927 *Megalopsallus latifrons* Knight, An. Ent. Soc. Am., 20: 226. [Col.].
Distribution: Col., Tex., Ut.
Note: Wray's (1967, Ins. N.C., Suppl. 3, p. 27) record of this species from N.C. probably is based on a misidentification.

Megalopsallus marmoratus Knight, 1968
1968 *Megalopsallus marmoratus* Knight, Brig. Young Univ. Sci. Bull., 9(3): 27. [Ariz.].
Distribution: Ariz.

Megalopsallus nuperus (Van Duzee), 1923
1923 *Oncotylus nuperus* Van Duzee, Proc. Cal. Acad. Sci., (4)12: 157. [Cal.].
1968 *Megalopsallus nuperus*: Knight, Brig. Young Univ. Sci. Bull., 9(3): 26.
Distribution: Cal.

Megalopsallus rubropictipes Knight, 1927
1927 *Megalopsallus rubropictipes* Knight, An. Ent. Soc. Am., 20: 225. [Col.].
Distribution: Alta., Col., Nev., Sask., Ut.

Genus *Merinocapsus* Knight, 1968

1968 *Merinocapsus* Knight, Brig. Young Univ. Sci. Bull., 9(3): 34. Type-species: *Merinocapsus ephedrae* Knight, 1968. Original designation.
1968 *Ankylotylus* Knight, Brig. Young Univ. Sci. Bull., 9(3): 55. Type-species: *Ankylotylus pallipes* Knight, 1968. Original designation. Synonymized by Schuh, 1986, J. N.Y. Ent. Soc., 94: 217.

Merinocapsus ephedrae Knight, 1968
1968 *Merinocapsus ephedrae* Knight, Brig. Young Univ. Sci. Bull., 9(3): 34. [Nev.].
Distribution: Cal., Nev., Ut. (Mexico).

Merinocapsus froeschneri Schuh, 1986
1986 *Merinocapsus froeschneri* Schuh, J. N.Y. Ent. Soc., 94: 220. [Cal.].
Distribution: Cal., Nev., Ut.

Merinocapsus pallipes (Knight), 1968
1968 *Ankylotylus pallipes* Knight, Brig. Young Univ. Sci. Bull., 9(3): 56. [Nev.].
1986 *Merinocapsus pallipes*: Schuh, J. N.Y. Ent. Soc., 94: 224.
Distribution: Ariz., Nev., Ut.

Genus *Microphylellus* Reuter, 1909

1909 *Microphylellus* Reuter, Acta Soc. Sci. Fenn., 36(2): 76. Type-species: *Microphylellus modestus* Reuter, 1909. Designated by Reuter, 1912, Öfv. F. Vet.-Soc. Förh., 54A(7): 62.

1916 *Leptotylus* Van Duzee, Univ. Cal. Publ. Ent., 1: 215. Type-species: *Microphylellus longirostris* Knight, 1923. Designated and synonymized by Slater and Knight, 1954, Pan-Pac. Ent., 30: 143.

Microphylellus adustus Knight, 1929
1929 *Microphylellus adustus* Knight, Ent. News, 40: 40. [Col.].
Distribution: B.C., Col., Wash.

Microphylellus adustus adustus Knight, 1929
1929 *Microphylellus adustus*: Knight, Ent. News, 40: 40. [Col.].
Distribution: Col.

Microphylellus adustus binotatus Knight, 1929
1929 *Microphylellus adustus* var. *binotatus* Knight, Ent. News, 40: 40. [Wash.].
Distribution: B.C., Wash.

Microphylellus alpinus Van Duzee, 1916
1916 *Microphylellus alpinus* Van Duzee, Univ. Cal. Publ. Ent., 1: 242. [Cal.].
Distribution: Cal.

Microphylellus bicinctus (Van Duzee), 1914
1914 *Chlamydatus bicinctus* Van Duzee, Trans. San Diego Soc. Nat. Hist., 2: 30. [Cal.].
1916 *Microphylellus bicinctus*: Van Duzee, Univ. Cal. Publ. Ent., 1: 243.
Distribution: Cal.

Microphylellus brevicornis Knight, 1929
1929 *Microphylellus brevicornis* Knight, Ent. News, 40: 41. [Ariz.].
Distribution: Ariz.

Microphylellus elongatus Knight, 1923
1923 *Microphylellus elongatus* Knight, Conn. Geol. Nat. Hist. Surv. Bull., 34: 458. [N.Y.].
Distribution: Ill., Minn., N.Y., Que.

Microphylellus falcatus (Van Duzee), 1917
1917 *Reuteroscopus falcatus* Van Duzee, Proc. Cal. Acad. Sci., (4)7: 278. [Cal.].
1958 *Microphylellus falcatus*: Carvalho, Arq. Mus. Nac., 45: 73.
Distribution: Cal.

Microphylellus flavicollis Knight, 1929
1929 *Microphylellus flavicollis* Knight, Ent. News, 40: 43. [Wash.].
Distribution: Cal., Ore., Wash.

Microphylellus flavipes (Provancher), 1872
1872 *Capsus flavipes* Provancher, Nat. Can., 4: 104. [Que.]. Lectotype designated by Kelton, 1968, Nat. Can., 95: 1075.
1886 *Deraeocoris flavipes*: Uhler, Check-list Hem. Het., p. 19.
1917 *Capsus ater*: Van Duzee, Univ. Cal. Publ. Ent., 2: 338 (in part).
1923 *Microphylellus nigricornis* Knight, Conn. Geol. Nat. Hist. Surv. Bull., 34: 455. [N.Y.]. Synonymized by Kelton, 1968, Nat. Can., 95: 1075.
1968 *Microphylellus flavipes*: Kelton, Nat. Can., 95: 1075.
Distribution: Minn., N.Y., Ont., Que., Wis.
Note: Kelton (1968, above) clarified the confusion involving this species and noted that Van Duzee (1917, above) incorrectly considered *M. flavipes* a synonym of *Capsus ater*.

Microphylellus fuscicornis Knight, 1923
1923 *Microphylellus maculipennis* var. *fuscicornis* Knight, Conn. Geol. Nat. Hist. Surv. Bull., 34: 457. [Me.].
1980 *Microphylellus fuscicornis*: Kelton, Agr. Can. Publ. 1703: 311.
Distribution: Man., Me., Sask.

Microphylellus longirostris Knight, 1923

1923 *Microphylellus longirostris* Knight, Conn. Geol. Nat. Hist. Surv. Bull., 34: 458. [N.Y.].

Distribution: Conn., Ia., Ill., Man., Mass., Minn., N.Y., Oh., Que., Tex., Wis.

Microphylellus maculipennis Knight, 1923

1923 *Microphylellus maculipennis* Knight, Conn. Geol. Nat. Hist. Surv. Bull., 34: 456. [Minn.].

Distribution: Ill., Man., Me., Minn., Mo., N.Y., Tex., Wis.

Microphylellus mineus Knight, 1929

1929 *Microphylellus mineus* Knight, Ent. News, 40: 44. [Fla.].

Distribution: Fla.

Microphylellus minor Knight, 1929

1929 *Microphylellus minor* Knight, Ent. News, 40: 42. [Cal.].

Distribution: Cal.

Microphylellus minuendus Knight, 1927

1927 *Microphylellus minuendus* Knight, Proc. Biol. Soc. Wash., 40: 10. [Md.].

Distribution: D.C., Md.

Microphylellus modestus Reuter, 1912

1892 *Phylus modestus* Heidemann, Proc. Ent. Soc. Wash., 2: 226 (Uhler, ms. name). *Nomen nudum.*

1912 *Microphylellus modestus* Reuter, Öfv. F. Vet.-Soc. Förh., 54A(7): 62. [Pa.].

1938 *Microphylellus modesta*: Brimley, Ins. N.C., p. 81.

Distribution: Conn., Del., Ind., Man., Mass., Mo., N.C., N.J., N.Y., Oh., Ont., Pa., Que., Sask., Tex., Vt., W.Va., Wis.

Microphylellus nicholi Knight, 1929

1929 *Microphylellus nicholi* Knight, Ent. News, 40: 42. [Ariz.].

Distribution: Ariz.

Note: Carvalho (1958, Arq. Mus. Nac., 45: 74) incorrectly cited Knight (1941, Ill. Nat. Hist. Surv. Bull., 22: 21, 24, 41, 42) and the localities Minn. to Tex. and eastern states for this species. This distribution should be referred to *M. modestus.*

Microphylellus symphoricarpi Knight, 1968

1968 *Microphylellus symphoricarpi* Knight, Brig. Young Univ. Sci. Bull., 9(3): 30. [Nev.].

Distribution: Nev.

Microphylellus tsugae Knight, 1923

1923 *Microphylellus tsugae* Knight, Conn. Geol. Nat. Hist. Surv. Bull., 34: 456. [N.Y.].

Distribution: N.Y., Oh., Que., W.Va.

Microphylellus tumidifrons Knight, 1923

1923 *Microphylellus tumidifrons* Knight, Conn. Geol. Nat. Hist. Surv. Bull., 34: 455. [N.S.].

Distribution: Man., N.S., N.Y., Pa., Sask., W.Va.

Genus *Microphylidea* Knight, 1968

1968 *Microphylidea* Knight, Brig. Young Univ. Sci. Bull., 9(3): 29. Type-species: *Microphylidea prosopidis* Knight, 1968. Original designation.

Microphylidea pallens Knight, 1968
1968 *Microphylidea pallens* Knight, Brig. Young Univ. Sci. Bull., 9(3): 29. [Nev.].
Distribution: Nev.

Microphylidea prosopidis Knight, 1968
1968 *Microphylidea prosopidis* Knight, Brig. Young Univ. Sci. Bull., 9(3): 29. [Nev.].
Distribution: Ariz., Nev., Ut.

Genus *Mineocapsus* Knight, 1972

1972 *Mineocapsus* Knight, Ia. St. J. Sci., 46: 425. Type-species: *Mineocapsus minimus* Knight, 1972. Original designation.

Mineocapsus minimus Knight, 1972
1972 *Mineocapsus minimus* Knight, Ia. St. J. Sci., 46: 425. [Ut.].
Distribution: Ut.

Genus *Monosynamma* Scott, 1864

1864 *Monosynamma* Scott, Ent. An., p. 160, fig. 5. Type-species: *Microsynamma scotti* Fieber, 1864, a junior synonym of *Phytocoris bohemani* Fallén, 1829. Monotypic.
1864 *Microsynamma* Fieber, Wien. Ent. Monat., 8: 74. Type-species: *Microsynamma scotti* Fieber, 1864. Monotypic. Synonymized by Reuter, 1910, Acta Soc. Sci. Fenn., 37(3): 167.

Monosynamma bohemani (Fallén), 1829
1829 *Phytocoris bohemani* Fallén, Hem. Suec., p. 106. [Europe].
1908 *Neocoris bohemani*: Horvath, An. Hist.-Nat. Mus. Nat Hung., 6: 5.
1917 *Microsynamma bohemani*: Knight, Can. Ent., 49: 248.
1922 *Microsynamma bohemanni* [*sic*]: Hussey, Occas. Pap. Mus. Zool., Univ. Mich., 118: 33.
1958 *Monosynamma bohemani*: Carvalho, Arq. Mus. Nac., 45: 75.
1980 *Monosynamma bohemanni* [*sic*]: Kelton, Agr. Can. Publ., 1703: 315.
Distribution: Alta., B.C., Col., Conn., Ia., Ill., Ind., Man., Mich., N.J., N.Y., Nfld., Oh., Ore., Sask., Vt., Wash. (Europe, USSR).
Note: Knight (1917, above) gave the first definite North American record for this species; Horvath (1908, above) listed it as occurring in North America without citing an exact locality.

Genus *Myochroocoris* Reuter, 1909

1909 *Myochroocoris* Reuter, Acta Soc. Sci. Fenn., 36(2): 76. Type-species: *Myochroocoris griseolus* Reuter, 1909. Monotypic.

Myochroocoris griseolus Reuter, 1909
1909 *Myochroocoris griseolus* Reuter, Acta Soc. Sci. Fenn., 36(2): 77. [Tex.].
Distribution: Tex.

Genus *Nevadocoris* Knight, 1968

1968 *Nevadocoris* Knight, Brig. Young Univ. Sci. Bull., 9(3): 59. Type-species: *Nevadocoris becki* Knight, 1968. Original designation.

Nevadocoris becki Knight, 1968
1968 *Nevadocoris becki* Knight, Brig. Young Univ. Sci. Bull., 9(3): 59. [Nev.].
Distribution: Nev.

Nevadocoris bullatus Knight, 1968
1968 *Nevadocoris bullatus* Knight, Brig. Young Univ. Sci. Bull., 9(3): 60. [Nev.].
Distribution: Nev.

Nevadocoris pallidus Knight, 1968
1968 *Nevadocoris pallidus* Knight, Brig. Young Univ. Sci. Bull., 9(3): 60. [Nev.].
Distribution: Nev.

Genus *Nicholia* Knight, 1929

1929 *Nicholia* Knight, Can. Ent., 61: 215. Type-species: *Nicholia eriogoni* Knight, 1929. Original designation.

Nicholia eriogoni Knight, 1929
1929 *Nicholia eriogoni* Knight, Can. Ent., 61: 216. [Ariz.].
Distribution: Ariz.

Genus *Oncotylus* Fieber, 1858

1858 *Oncotylus* Fieber, Wien. Ent. Monat., 2: 318. Type-species: *Oncotylus tanaceti* Fieber, 1858, a junior homonym and synonym of *Capsus tanaceti* Herrich-Schaeffer, 1835 [not *Lygaeus tanaceti* Fallén, 1807, a synonym of *Oncotylus punctipes* Reuter, 1873]. Designated by Int. Comm. Zool. Nomen., 1950, Bull. Zool. Nomen., 4: 474.

Oncotylus guttulatus Uhler, 1894
1894 *Oncotylus guttulatus* Uhler, Proc. Cal. Acad. Sci., (2)4: 269. [Mexico].
1895 *Oncotylus guttulatus*: Uhler, Bull. Col. Agr. Exp. Stn., 31: 48.
Distribution: Ariz., Cal., Col., Nev., Tex. (Mexico).

Genus *Parapsallus* Wagner, 1952

1952 *Psallus* (*Parapsallus*) Wagner, Tierw. Deut., 4: 187. Type-species: *Capsus vitellinus* Scholtz, 1847. Original designation.

1964 *Parapsallus*: Kerzhner, Miridae Europ. USSR, 1: 760.

Parapsallus vitellinus (Scholtz), 1847
1847 *Capsus vitellinus* Scholtz, Arb. Ver. Schles. Kult., p. 130. [Europe].
1973 *Plagiognathus vitellinus*: Henry and Wheeler, Proc. Ent. Soc. Wash., 75: 480.
Distribution: N.Y., Ont., Pa., W.Va. (Europe).
Note: Henry and Wheeler (1973, above) gave an overview of the literature pertaining to this species, provided information on biology and hosts, and described the fifth-instar nymph.

Genus *Phoenicocoris* Reuter, 1875

1875 *Plagiognathus* (*Phoenicocoris*) Reuter, Rev. Crit. Caps., 1: 99. Type-species: *Capsus modestus* Meyer-Dür, 1843. Monotypic.

1962 *Phoenicocoris*: Kerzhner, Ent. Obozr., 41: 381.

Phoenicocoris dissimilis (Reuter), 1878

1878 *Sthenarus dissimilis* Reuter, Acta Soc. Sci. Fenn., 13(1): 174. [France].

1974 *Sthenarus dissimilis*: Henry and Wheeler, Proc. Ent. Soc. Wash., 76: 217.

1975 *Phoenicocoris dissimilis*: Wagner, Ent. Abh., 40: 104.

Distribution: N.Y., Pa. (Europe).

Note: Henry and Wheeler (1974, above) summarized the literature, gave hosts, and illustrated the adult and fifth-instar nymph.

Phoenicocoris strobicola (Knight), 1923

1923 *Psallus strobicola* Knight, Conn. Geol. Nat. Hist. Surv. Bull., 34: 467. [N.Y.].

1962 *Phoenicocoris strobicola*: Kerzhner, Ent. Obozr., 41: 381.

Distribution: Ill., Minn., N.Y., Oh., Ok., Pa., Que., W.Va.

Genus *Phyllopidea* Knight, 1919

1919 *Phyllopidea* Knight, Bull. Brook. Ent. Soc., 14: 127. Type-species: *Bolteria picta* Uhler, 1893. Original designation.

Note: Knight (1968, Brig. Young Univ. Sci. Bull., 9(3): 31-32) provided a key to the species of this genus.

Phyllopidea hirta (Van Duzee), 1916

1916 *Bolteria picta* var. *hirta* Van Duzee, Univ. Cal. Publ. Ent., 1: 244. [Cal.].

1919 *Phyllopidea hirta*: Knight, Bull. Brook. Ent. Soc., 14: 127.

Distribution: B.C., Cal., Nev.

Phyllopidea montana Knight, 1968

1968 *Phyllopidea montana* Knight, Brig. Young Univ. Sci. Bull., 9(3): 33. [Col.].

Distribution: Alta., Col., Id., Wyo.

Phyllopidea picta (Uhler), 1893

1893 *Bolteria picta* Uhler, Proc. Ent. Soc. Wash., 2: 373. [Ut.]. Lectotype designated by Knight, 1919, Bull. Brook. Ent. Soc., 14: 128.

1909 *Hyoidea picta*: Reuter, Acta Soc. Sci. Fenn., 36(2): 72.

1919 *Phyllopidea picta*: Knight, Bull. Brook. Ent. Soc., 14: 127.

Distribution: Cal., Col., Id., Nev., Ut.

Phyllopidea utahensis Knight, 1968

1968 *Phyllopidea utahensis* Knight, Brig. Young Univ. Sci. Bull., 9(3): 32. [Ut.].

Distribution: Ut.

Genus *Phylus* Hahn, 1831

1831 *Phylus* Hahn, Wanz. Ins., 1: 26. Type-species: *Phylus pallipes* Hahn, 1831, a junior synonym of *Cimex coryli* Linnaeus, 1758. Monotypic.

Phylus coryli (Linnaeus), 1758

1758 *Cimex coryli* Linnaeus, Syst. Nat., p. 451. [Europe].

1982 *Phylus coryli*: Kelton, Can. Ent., 114: 1127.

Distribution: B.C. (Europe).

Genus *Phymatopsallus* Knight, 1964

1964 *Phymatopsallus* Knight, Ia. St. J. Sci., 39: 127. Type-species: *Phymatopsallus strombocarpae* Knight, 1964. Original designation.

Note: Knight (1964, above) provided a key to the species of *Phymatopsallus*.

Phymatopsallus acaciae Knight, 1964
1964 *Phymatopsallus acaciae* Knight, Ia. St. J. Sci., 39: 131. [N.M.].
Distribution: Ariz., N.M.

Phymatopsallus chiricahuae Knight, 1964
1964 *Phymatopsallus chiricahuae* Knight, Ia. St. J. Sci., 39: 132. [Ariz.].
Distribution: Ariz.

Phymatopsallus croceguttatus Knight, 1964
1964 *Phymatopsallus croceguttatus* Knight, Ia. St. J. Sci., 39: 137. [Ariz.].
Distribution: Ariz., Cal.

Phymatopsallus croceus (Van Duzee), 1918
1918 *Psallus croceus* Van Duzee, Proc. Cal. Acad. Sci., (4)8: 302. [Cal.].
1964 *Phymatopsallus croceus*: Knight, Ia. St. J. Sci., 39: 137.
Distribution: Cal.

Phymatopsallus cuneopunctatus Knight, 1964
1964 *Phymatopsallus cuneopunctatus* Knight, Ia. St. J. Sci., 39: 134. [Ariz.].
Distribution: Ariz.

Phymatopsallus fulvipunctatus Knight, 1964
1964 *Phymatopsallus fulvipunctatus* Knight, Ia. St. J. Sci., 39: 137. [Ariz.].
Distribution: Ariz.

Phymatopsallus fuscipunctatus Knight, 1964
1964 *Phymatopsallus fuscipunctatus* Knight, Ia. St. J. Sci., 39: 139. [Ut.].
Distribution: Id., Ut.

Phymatopsallus huachucae Knight, 1964
1964 *Phymatopsallus huachucae* Knight, Ia. St. J. Sci., 39: 135. [Ariz.].
Distribution: Ariz.

Phymatopsallus longirostris Knight, 1964
1964 *Phymatopsallus longirostris* Knight, Ia. St. J. Sci., 39: 136. [Ariz.].
Distribution: Ariz.

Phymatopsallus nicholi Knight, 1964
1964 *Phymatopsallus nicholi* Knight, Ia. St. J. Sci., 39: 130. [Ariz.].
Distribution: Ariz.

Phymatopsallus pantherinus (Van Duzee), 1917
1917 *Psallus pantherinus* Van Duzee, Proc. Cal. Acad. Sci., (4)7: 279. [Cal.].
1964 *Phymatopsallus pantherinus*: Knight, Ia. St. J. Sci., 39: 138
Distribution: Cal., Col.

Phymatopsallus patagoniae Knight, 1964
1964 *Phymatopsallus patagoniae* Knight, Ia. St. J. Sci., 39: 132. [Ariz.].
Distribution: Ariz.

Phymatopsallus prosopidis Knight, 1968
1968 *Phymatopsallus prosopidis* Knight, Brig. Young Univ. Sci. Bull., 9(3): 49. [Nev.].
Distribution: Ariz., Nev.

Phymatopsallus ribesi Knight, 1968
1968 *Phymatopsallus ribesi* Knight, Brig. Young Univ. Sci. Bull., 9(3): 50. [Nev.].
Distribution: Nev.

Phymatopsallus rinconae Knight, 1964
1964 *Phymatopsallus rinconae* Knight, Ia. St. J. Sci., 39: 132. [Ariz.].
Distribution: Ariz.

Phymatopsallus rubropunctatus Knight, 1964
1964 *Phymatopsallus rubropunctatus* Knight, Ia. St. J. Sci., 39: 136. [Ariz.].
Distribution: Ariz.

Phymatopsallus strombocarpae Knight, 1964
1964 *Phymatopsallus strombocarpae* Knight, Ia. St. J. Sci., 39: 130. [Tex.].
Distribution: Ariz., Tex.

Phymatopsallus texanus Knight, 1964
1964 *Phymatopsallus texanus* Knight, Ia. St. J. Sci., 39: 134. [Tex.].
Distribution: Tex.

Phymatopsallus tuberculatus (Van Duzee), 1923
1923 *Psallus tuberculatus* Van Duzee, Proc. Cal. Acad. Sci., (4)12: 161. [Mexico].
1964 *Phymatopsallus tuberculatus*: Knight, Ia. St. J. Sci., 39: 133.
Distribution: Ariz. (Mexico).

Phymatopsallus viridescens Knight, 1964
1964 *Phymatopsallus viridescens* Knight, Ia. St. J. Sci., 39: 133. [Cal.].
Distribution: Cal.

Genus *Plagiognathus* Fieber, 1858

1858 *Plagiognathus* Fieber, Wien. Ent. Monat., 2: 320. Type-species: *Lygaeus arbustorum* Fabricius, 1794. Designated by Kirkaldy, 1906, Trans. Am. Ent. Soc., 32: 124.

1911 *Gerhardiella* Poppius, An. Soc. Ent. Belg., 55: 84. Type-species: *Gerhardiella rubida* Poppius, 1911. Monotypic. Synonymized by Carvalho, 1952, An. Acad. Brasil. Ci., 24: 65.

Plagiognathus albatus (Van Duzee), 1915
1915 *Psallus albatus* Van Duzee, Pomona J. Ent. Zool., 7: 116. [N.Y.].
1917 *Plagiognathus albatus*: Van Duzee, Univ. Cal. Publ. Ent., 2: 410.
1926 *Plagiognathus inopinus* Knight, Ent. News, 37: 11. [Pa.]. Synonymized by Henry, 1982, Proc. Ent. Soc. Wash., 84: 338.
Distribution: Cal., Conn., D.C., Fla., Ga., Ia., Ill., Ind., Mass., Mich., Minn., Mo., N.C., N.Y., Oh., Pa., Que., W.Va.
Note: Wheeler (1980, An. Ent. Soc. Am., 73: 354-356) presented information on life history and injury to ornamental *Platanus* species and described the fifth-instar nymph.

Plagiognathus albatus albatus (Van Duzee), 1915
1915 *Psallus albatus*: Van Duzee, Pomona J. Ent. Zool., 7: 116. [N.Y.].
1923 *Plagiognathus albatus*: Knight, Conn. Geol. Nat. Hist Surv. Bull., 34: 445.

Distribution: Same as for species.

Plagiognathus albatus vittiscutis Knight, 1923
1923 *Plagiognathus albatus* var. *vittiscutis* Knight, Conn. Geol. Nat. Hist. Surv. Bull., 34: 445. [N.Y.].
Distribution: Conn., La., Miss., N.Y., Pa., Que.

Plagiognathus albellus Knight, 1953
1953 *Plagiognathus albellus* Knight, Ia. St. Coll. J. Sci., 27: 509. [Mo.].
Distribution: Mo.

Plagiognathus albifacies Knight, 1927
1927 *Plagiognathus albifacies* Knight, Proc. Biol. Soc. Wash., 40: 11. [Ill.].
Distribution: D.C., Ga., Ill., Ind., Md., Minn.

Plagiognathus albonotatus Knight, 1923
1923 *Plagiognathus albonotatus* Knight, Conn. Geol. Nat. Hist. Surv. Bull., 34: 437. [N.Y.].
1923 *Plagiognathus albonotatus* var. *compar* Knight, Conn. Geol. Nat. Hist. Surv. Bull., 34: 438. [N.Y.]. Synonymized by Larochelle, 1984, Fabreries, Suppl. 3, p. 330.
Distribution: Alta., Col., Ill., Ind., Man., Me., Minn., N.D., N.Y., Oh., R.I., Sask.

Plagiognathus alboradialis Knight, 1923
1923 *Plagiognathus alboradialis* Knight, Conn. Geol. Nat. Hist. Surv. Bull., 34: 439. [N.Y.].
Distribution: Alta., B.C., Conn., Ind., Man., Me., N.B., N.H., N.J., N.Y., Nfld., Ont., Que., Sask., Vt.

Plagiognathus annulatus Uhler, 1895
1895 *Plagiognathus annulatus* Uhler, Bull. Col. Agr. Exp. Stn., 31: 51. [Col.].
Distribution: Alta., Col., Conn., Ind., Ks., Mass., Me., Mo., Mont., N.H., N.Y., Ont., Pa., Que., Sask., W.Va.

Plagiognathus annulatus annulatus Uhler, 1895
1895 *Plagiognathus annulatus* Uhler, Bull. Col. Agr. Exp. Stn., 31: 51. [Col.].
1923 *Plagiognathus annulatus*: Knight, Conn. Geol. Nat. Hist. Surv. Bull., 34: 443.
Distribution: Same as for species.

Plagiognathus annulatus nigrofemoratus Knight, 1923
1923 *Plagiognathus annulatus* var. *nigrofemoratus* Knight, Conn. Geol. Nat. Hist. Surv. Bull., 34: 443. [N.Y.].
Distribution: Me., N.Y.

Plagiognathus apiatus (Uhler), 1895
1895 *Agalliastes apiatus* Uhler, Bull. Col. Agr. Exp. Stn., 31: 53. [Col.].
1916 *Chlamydatus apiatus*: Van Duzee, Check List Hem., p. 47.
1965 *Plagiognathus apiatus*: Kelton, Can. Ent., 97: 1142.
Distribution: Col., Ks.

Plagiognathus arbustorum (Fabricius), 1794
1794 *Lygaeus arbustorum* Fabricius, Ent. Syst., 4: 175. [Europe].
1966 *Plagiognathus arbustorum*: Waloff, J. Appl. Ecol., 3: 300.
Distribution: B.C. (Europe, Siberia, China).
Note: Kelton (1982, Can. Ent., 114: 1128) summarized the N. Am. literature on this species, gave new records, and questioned Carvalho's (1958, Arq. Mus. Nac.,

45: 96) record from N.S.

Plagiognathus atricornis Knight, 1926
1926 *Plagiognathus atricornis* Knight, Ent. News, 37: 9. [Pa.].
Distribution: Ill., Mo., Pa., W.Va.

Plagiognathus blatchleyi Reuter, 1912
1912 *Plagiognathus blatchleyi* Reuter, Öfv. F. Vet.-Soc. Förh., 54A(7): 61. [N.Y.].
1923 *Plagiognathus blatchleyi* var. *nubilus* Knight, Conn. Geol. Nat. Hist. Surv. Bull., 34: 444. [N.Y.]. Synonymized by Larochelle, 1984, Fabreries, Suppl. 3, p. 327.
1926 *Plagiognathus nubilus*: Blatchley, Het. E. N. Am., p. 937.
Distribution: D.C., Ill., Ind., Man., Mass., Md., Me., Mo., N.J., N.Y., Oh., Pa., Va., W.Va.
Note: Blatchley (1926, above) argued that *P. nubilus* deserved specific rank, but Knight (1941, Ill. Nat. Hist. Surv. Bull., 22: 36) maintained its subspecific status. Larochelle (1984, above) considered it only a color form of *P. blatchleyi*, as we follow above.

Plagiognathus brevirostris Knight, 1923
1923 *Plagiognathus brevirostris* Knight, Conn. Geol. Nat. Hist. Surv. Bull., 34: 441. [N.Y.].
Distribution: Conn., Ill., Me., Mich., Minn., N.H., N.Y., Nfld., Que., W.Va.

Plagiognathus carinatus Knight, 1926
1926 *Plagiognathus carinatus* Knight, Ent. News, 37: 10. [Va.].
Distribution: Md., Va.

Plagiognathus carneolus Knight, 1927
1927 *Plagiognathus carneolus* Knight, Proc. Biol. Soc. Wash., 40: 10. [Md.].
Distribution: D.C., Md., Pa., Va., W.Va., Wis.

Plagiognathus caryae Knight, 1923
1923 *Plagiognathus caryae* Knight, Conn. Geol. Nat. Hist. Surv. Bull., 34: 448. [N.Y.].
Distribution: Miss., N.Y., Pa., Tex.

Plagiognathus chrysanthemi (Wolff), 1804
1804 *Miris chrysanthemi* Wolff, Icon. Cimic., 4: 157. [Europe].
1887 *Oncotylus punctipes*: Provancher, Pet. Faune Ent. Can., 3: 149.
1922 *Plagiognathus chrysanthemi*: Knight, Can. Ent., 53: 281.
1950 *Oncotylys* [*sic*] *punctipes*: Moore, Nat. Can., 77: 249.
1968. *Plagionathus* [*sic*] *chrysanthemi*: Kurczewski, J. Ks. Ent. Soc., 41: 198.
Distribution: B.C., Cal., Conn., Man., Mass., Me., N.B., N.H., N.S., N.Y., Nfld., Ont., P.Ed., Pa., Que., Vt., Wis. (Europe).
Note: Guppy (1963, An. Ent. Soc. Am., 56: 804-809) reported on the life history and damage to birdsfoot trefoil, *Lotus corniculatus* L., in Ontario.

Plagiognathus confusus Reuter, 1909
1909 *Plagiognathus confusus* Reuter, Acta Soc. Sci. Fenn., 36(2): 80. [Nev.].
Distribution: Cal., Nev.

Plagiognathus cornicola Knight, 1923
1923 *Plagiognathus cornicola* Knight, Conn. Geol. Nat. Hist. Surv. Bull., 34: 450. [N.Y.].
1979 *Plagiognathus conicola* [*sic*]: Carlson, Cat. Hymenop., 1: 707.
Distribution: Ill., Ind., Mass., N.Y., Va., W.Va., Wis.

Plagiognathus crocinus Knight, 1927
1927 *Plagiognathus crocinus* Knight, Proc. Biol. Soc. Wash., 40: 12. [Va.].
Distribution: Va.

Plagiognathus cuneatus Knight, 1923
1923 *Plagiognathus annulatus* var. *cuneatus* Knight, Conn. Geol. Nat. Hist. Surv. Bull., 34: 442. [N.Y.].
1941 *Plagiognathus cuneatus*: Knight, Ill. Nat. Hist. Surv. Bull., 22: 34.
Distribution: Ga., Ill., Mo., N.H., N.Y., Ont., Tex., Vt., W.Va., Wis.

Plagiognathus davisi Knight, 1923
1923 *Plagiognathus davisi* Knight, Conn. Geol. Nat. Hist. Surv. Bull., 34: 452. [N.Y.].
Distribution: Alta., Ia., Man., N.Y., Sask.

Plagiognathus delicatus (Uhler), 1887
1887 *Psallus delicatus* Uhler, Ent. Am., 3: 34. [Ga.].
1916 *Gerhardiella delicata*: Van Duzee, Univ. Cal. Publ. Ent., 1: 243.
1923 *Plagiognathus delicatus*: Knight, Conn. Geol. Nat. Hist. Surv. Bull., 34: 448.
Distribution: Cal.(?), Ga., Ia., Ill., Ind., Ks., Miss., Mo., N.J., N.Y., Oh., Pa., Va., W.Va., Wis. (Mexico).
Note: Van Duzee's (1916, above) California record of this species on *Quercus* undoubtedly is based on a misidentification. Wheeler and Henry (1976, An. Ent. Soc. Am., 69: 1095-1104) described the egg and fifth-instar nymph and discussed the seasonal history on honeylocust, *Gleditsia triacanthos* L.

Plagiognathus dispar Knight, 1923
1923 *Plagiognathus punctatipes* var. *dispar* Knight, Conn. Geol. Nat. Hist. Surv. Bull., 34: 451. [N.Y.].
1926 *Plagiognathus dispar*: Knight, Ent. News, 37: 11..
1929 *Plagiognathus crataegi* Knight, Ent. News, 40: 264. [Ia.]. Synonymized by Froeschner, 1949, Am. Midl. Nat., 42: 160.
1941 *Plagiognathus dispar* var. *crataegi*: Knight, Ill. Nat. Hist. Surv. Bull., 22: 39.
Distribution: Conn., Ia., Ill., Man., Md., Mich., Mo., N.Y., Pa., W.Va.

Plagiognathus diversus Van Duzee, 1917
1917 *Plagiognathus diversus* Van Duzee, Proc. Cal. Acad. Sci., (4)7: 283. [Cal.].
Distribution: Cal.

Plagiognathus diversus diversus Van Duzee, 1917
1917 *Plagiognathus diversus* Van Duzee, Proc. Cal. Acad. Sci., (4)7: 283. [Cal.].
Distribution: Cal.

Plagiognathus diversus cruralis Van Duzee, 1917
1917 *Plagiognathus diversus* var. *cruralis* Van Duzee, Proc. Cal. Acad. Sci., (4)7: 283. [Cal.].
Distribution: Cal.

Plagiognathus diversus pluto Van Duzee, 1917
1917 *Plagiognathus diversus* var. *pluto* Van Duzee, Proc. Cal. Acad. Sci., (4)7: 284. [Cal.].
Distribution: Cal.

Plagiognathus flavescens Knight, 1925
1925 *Plagiognathus flavescens* Knight, Bull. Brook. Ent. Soc., 20: 33. [Ariz.].
Distribution: Ariz., Col.

Plagiognathus flavicornis Knight, 1923
1923 *Plagiognathus flavicornis* Knight, Conn. Geol. Nat. Hist. Surv. Bull., 34: 436. [N.Y.].
Distribution: Ill., Mass., Minn., N.Y., Ont., W.Va.

Plagiognathus flavidus Knight, 1929
1929 *Plagiognathus shepherdiae* var. *flavidus* Knight, Ent. News, 40: 71. [S.D.].
1980 *Plagiognathus flavidus*: Kelton, Agr. Can. Publ., 1703: 318.
Distribution: Alta., Man., Que., S.D., Sask.

Plagiognathus flavoscutellatus Knight, 1923
1923 *Plagiognathus flavoscutellatus* Knight, Conn. Geol. Nat. Hist. Surv. Bull., 34: 440. [N.Y.].
Distribution: Ia., Ind., Mich., Minn., Mo., N.S., N.Y., Nfld., Oh., Pa., Que., Vt., Wis.

Plagiognathus flavus Knight, 1964
1964 *Plagiognathus flavus* Knight, Ia. St. J. Sci., 39: 146. [Ariz.].
Distribution: Ariz.

Plagiognathus fulvaceus Knight, 1964
1964 *Plagiognathus fulvaceus* Knight, Ia. St. J. Sci., 39: 144. [Col.].
Distribution: Col., Ut.

Plagiognathus fulvidus Knight, 1923
1923 *Plagiognathus fulvidus* Knight, Conn. Geol. Nat. Hist. Surv. Bull., 34: 447. [Conn.].
Distribution: Conn., Md., N.C., N.J., Oh., W.Va.

Plagiognathus fumidus (Uhler), 1875
1875 *Agalliastes fumidus* Uhler, Bull. Col. Agr. Exp. Stn., 31: 54. [Col.].
1917 *Chlamydatus fumidus*: Van Duzee, Univ. Cal. Publ. Ent., 2: 417.
1965 *Plagiognathus fumidus*: Kelton, Can. Ent., 97: 1142.
Distribution: Col.

Plagiognathus fusciflavus Knight, 1929
1929 *Plagiognathus fusciflavus* Knight, Ent. News, 40: 267. [Ore.].
Distribution: Ore.

Plagiognathus fuscipes Knight, 1929
1929 *Plagiognathus fuscipes* Knight, Ent. News, 40: 268. [Col.].
Distribution: Alta., Col.

Plagiognathus fuscosus (Provancher), 1872
1872 *Lygus fuscosus* Provancher, Nat. Can., 4: 105. [Que.]. Lectotype designated by Kelton, 1968, Nat. Can., 95: 1074.
1887 *Plagiognathus fuscosus*: Provancher, Pet. Faune Ent. Can., 3: 153.
1923 *Plagiognathus politus* var. *pallidicornis* Knight, Conn. Geol. Nat. Hist. Surv. Bull., 34: 435. [N.Y.]. Synonymized by Kelton, 1968, Nat. Can., 95: 1074.
1941 *Plagiognathus pallidicornis*: Knight, Ill. Nat. Hist. Surv. Bull., 22: 30.
Distribution: Alta., Conn., Ill., Man., Mass., Me., Minn., N.C., N.H., N.Y., Ont., Que., Sask., W.Va.

Plagiognathus fuscotibialis Knight, 1964
1964 *Plagiognathus fuscotibialis* Knight, Ia. St. J. Sci., 39: 143. [Wyo.].
Distribution: Wyo.

Plagiognathus geminatus Knight, 1929
1929 *Plagiognathus geminatus* Knight, Ent. News, 40: 265. [Tex.].
Distribution: Tex.

Plagiognathus geranii Knight, 1964
1964 *Plagiognathus geranii* Knight, Ia. St. J. Sci., 39: 142. [Col.].
Distribution: Col., Tex.

Plagiognathus gleditsiae Knight, 1929
1929 *Plagiognathus gleditsiae* Knight, Ent. News, 40: 265. [Tex.].
Distribution: Ill., Pa., Tex.
Note: Wheeler and Henry (1976, An. Ent. Soc. Am., 69: 1095-1104) described the egg and fifth-instar nymph and discussed the seasonal history on honeylocust, *Gleditsia triacanthos* L.

Plagiognathus grandis Reuter, 1876
1876 *Plagiognathus grandis* Reuter, Öfv. K. Svens. Vet.-Akad. Förh., 32(9): 91. [Tex.].
Distribution: Tex.

Plagiognathus guttatipes (Uhler), 1895
1895 *Lygus guttatipes* Uhler, Bull. Col. Agr. Exp. Stn., 31: 35. [Col.].
1917 *Plagiognathus guttatipes*: Knight, Cornell Univ. Agr. Exp. Stn. Bull., 391: 639.
Distribution: Alta., Col., Man., N.D., S.D., Sask., Wyo.

Plagiognathus guttulosus (Reuter), 1876
1876 *Psallus guttulosus* Reuter, Öfv. K. Svens. Vet.-Akad. Förh., 32(9): 89. [Tex.].
1914 *Apocremnus* (*Psallus*) *guttulosus*: Barber, Bull. Am. Mus. Nat. Hist., 33: 500.
1941 *Plagiognathus guttulosus*: Knight, Ill. Nat. Hist. Surv. Bull., 22: 40.
Distribution: Col., Fla., Ga., Ill., Miss., Mo., Tex. (Mexico).

Plagiognathus ilicis Knight, 1925
1925 *Plagiognathus ilicis* Knight, Ent. News, 36: 305. [N.Y.].
1971 *Plagiognathus illicis* [*sic*]: Khalaf, Fla. Ent., 54: 340.
Distribution: Miss., N.Y., Wis.

Plagiognathus intrusus Knight, 1926
1926 *Plagiognathus intrusus* Knight, Ent. News, 37: 12. [N.Y.].
Distribution: N.Y.

Plagiognathus laricicola Knight, 1923
1923 *Plagiognathus laricicola* Knight, Conn. Geol. Nat. Hist. Surv. Bull., 34: 452. [N.Y.].
Distribution: Alta., Conn., Ill., Man., Me., Minn., N.Y., Sask.

Plagiognathus lineatus Van Duzee, 1917
1917 *Plagiognathus lineatus* Van Duzee, Proc. Cal. Acad. Sci., (4)7: 282. [Cal.].
Distribution: Cal., Ore., Wash.

Plagiognathus longipennis (Uhler), 1895
1895 *Oncotylus longipennis* Uhler, Bull. Col. Agr. Exp. Stn., 31: 48. [Col.].
1958 *Plagiognathus longipennis*: Carvalho, Arq. Mus. Nac., 45: 103.
Distribution: Col., N.M.

Plagiognathus luteus Knight, 1929
1929 *Plagiognathus luteus* Knight, Ent. News, 40: 72. [Ariz.].
Distribution: Ariz.

Plagiognathus medicagus Arrand, 1958
1958 *Plagiognathus medicagus* Arrand, Can. Ent., 90: 498. [Sask.].
Distribution: Alta., B.C., Man., Mont., Sask.
Note: Arrand and McMahon (1974, Can. Ent., 106: 433-435) described the immature stages.

Plagiognathus mundus Van Duzee, 1917
1917 *Plagiognathus mundus* Van Duzee, Proc. Cal. Acad. Sci., (4)7: 281. [Cal.].
Distribution: Cal., Ore., Wash.

Plagiognathus negundinis Knight, 1929
1929 *Plagiognathus negundinis* Knight, Ent. News, 40: 263. [Ia.].
Distribution: Ia., Ill., Minn., Mo., Sask., Wis.

Plagiognathus negundinis fulvotinctus Knight, 1929
1929 *Plagiognathus negundinis* var. *fulvotinctus* Knight, Ent. News, 40: 264. [Minn.].
Distribution: Ia., Minn.

Plagiognathus negundinis negundinis Knight, 1929
1929 *Plagiognathus negundinis* Knight, Ent. News, 40: 263. [Ia.].
Distribution: Same as for species.

Plagiognathus nicholi Knight, 1964
1964 *Plagiognathus nicholi* Knight, Ia. St. J. Sci., 39: 147. [Minn.].
Distribution: Col., Minn., S.D.

Plagiognathus nigritibialis Knight, 1964
1964 *Plagiognathus nigritibialis* Knight, Ia. St. J. Sci., 39: 148. [Miss.].
Distribution: Miss.

Plagiognathus nigritus Knight, 1923
1923 *Plagiognathus nigritus* Knight, Conn. Geol. Nat. Hist. Surv. Bull., 34: 441. [Conn.].
Distribution: Alta., Col., Conn., Ind., Man., Mass., N.C., Oh., Sask.

Plagiognathus nigronitens Knight, 1923
1923 *Plagiognathus nigronitens* Knight, Conn. Geol. Nat. Hist. Surv. Bull., 34: 435. [N.Y.].
Distribution: Alta., Col., Ill., Man., Mass., Mich., Minn., Miss., N.J., N.Y., Oh., Ont., S.D., Sask., Wis.

Plagiognathus obscurus Uhler, 1872
1872 *Plagiognathus obscurus* Uhler, Prelim. Rept. U.S. Geol. Surv. Mont., p. 418. [Col., Ill., Mass., Md., Me., Mich., N.J., N.Y., Pa.].
1872 *Lygus brunneus* Provancher, Nat. Can., 4: 104. [Ont.]. Synonymized and lectotype designated by Kelton, 1968, Nat. Can. 95: 1073.
1887 *Pamerocoris brunneus*: Provancher, Pet. Faune Ent. Can., 3: 127.
1887 *Psallus delicatus*: Provancher, Pet. Faune Ent. Can., 3: 152.
1895 *Plagiognathus fraternus* Uhler, Bull. Col. Agr. Exp. Stn., 31: 51. [Col.]. Synonymized [reduced to subspecies] by Knight, 1941, Ill. Nat. Hist. Surv. Bull., 22: 33; subspecies synonymized by Larochelle, 1984, Fabreries, Suppl. 3, p. 322.
1895 *Fulvius brunneus*: Reuter, Ent. Tidskr., 16: 140.
1941 *Plagiognathus obscurus* var. *fraternus*: Knight, Ill. Nat. Hist. Surv. Bull., 22: 33.
Distribution: Alk., Alta., B.C., Cal., Col., Conn., Ga., Ill., Ks., Man., Mass., Md., Me., Mich., Minn., Mo., Mont., N.B., N.C., N.H., N.J., N.M., N.S., N.Y., Oh., Ont., Pa., Que., Sask., Tex., W.Va., Wis. (Chile).

Note: Beirne (1972, Mem. Can. Ent. Soc., 85: 28) commented that Canadian records of this species might be confused with those of *P. medicagus* Arrand.

Plagiognathus obscurus albocuneatus Knight, 1923

1923 *Plagiognathus obscurus* var. *albocuneatus* Knight, Conn. Geol. Nat. Hist. Surv. Bull., 34: 438. [N.Y.].

Distribution: Conn., Mass., N.H., N.Y., Vt.

Plagiognathus obscurus obscurus Uhler, 1872

1872 *Plagiognathus obscurus* Uhler, Prelim. Rept. U.S. Geol. Surv. Mont., p. 418. [Col., Ill., Mass., Md., Me., Mich., N.J., N.Y., Pa.].

Distribution: Same as for species.

Plagiognathus paddocki Knight, 1964

1964 *Plagiognathus paddocki* Knight, Ia. St. J. Sci., 39: 146. [Cal.].

Distribution: Cal.

Plagiognathus phorodendronae Knight, 1929

1929 *Plagiognathus phorodendronae* Knight, Ent. News, 40: 73. [Ariz.].

Distribution: Ariz.

Plagiognathus politus Uhler, 1895

1895 *Plagiognathus politus* Uhler, Bull. Col. Agr. Exp. Stn., 31: 52. [Col.].

1916 *Apocremnus politus*: Van Duzee, Univ. Cal. Publ. Ent., 1: 243.

1941 *Plagiognathus obscurus*: Knight, Ill. Nat. Hist. Surv. Bull., 22: 32, fig. 88.

1968 *Plagionathus* [*sic*] *politus*: Kurczewski, J. Ks. Ent. Soc., 41: 198.

Distribution: Cal., Col., Conn., Del., Fla., Ill., Ind., Ks., Man., Mass., Me., Mich., Mo., N.C., N.J., N.Y., Oh., Ont, Pa., Que., S.D., Ut., W.Va.

Note: Leonard (1915, J. N.Y. Ent. Soc., 23: 193-195) described the immature stages.

Plagiognathus politus flaveolus Knight, 1923

1923 *Plagiognathus politus flaveolus* Knight, Conn. Geol. Nat. Hist. Surv. Bull., 34: 434. [N.Y.].

Distribution: Conn., Ill., Mich., Mo., N.Y., Oh., Ont., Pa., Wis.

Plagiognathus politus politus Uhler, 1895

1895 *Plagiognathus politus* Uhler, Bull. Col. Agr. Exp. Stn. Ent., 31: 52. [Col.].

Distribution: Same as for species.

Plagiognathus punctatipes Knight, 1923

1923 *Plagiognathus punctatipes* Knight, Conn. Geol. Nat. Hist. Surv. Bull., 34: 450. [N.Y.].

Distribution: Conn., Ill., Mich., Mo., N.Y., Oh., Ont., Pa., W.Va.

Note: Wheeler and Hoebeke (1985, Proc. Ent. Soc. Wash., 87: 361) summarized the seasonal history and habits on ninebark, *Physocarpus opulifolius* (L.) Maxim.

Plagiognathus reinhardi Johnston, 1935

1935 *Plagiognathus reinhardi* Johnston, Bull. Brook. Ent. Soc., 30: 16. [Tex.].

Distribution: Tex.

Plagiognathus repetitus Knight, 1923

1923 *Plagiognathus repetitus* Knight, Conn. Geol. Nat. Hist. Surv. Bull., 34: 453. [N.Y.].

1967 *Plagiognathus repticus* [*sic*]: Ives, Can. Ent., 99: 607.

Distribution: Mass., Mich., N.J., N.S., N.Y., Que.

Plagiognathus repletus Knight, 1923
1923 *Plagiognathus repletus* Knight, Conn. Geol. Nat. Hist. Surv. Bull., 34: 449. [N.Y.].
Distribution: Conn., Ia., Ill., Mass., Mo., N.Y., Oh., Que., Va.

Plagiognathus repletus apicatus Knight, 1923
1923 *Plagiognathus repletus* var. *apicatus* Knight, Conn. Geol. Nat. Hist. Surv. Bull., 34: 449. [N.Y.].
Distribution: Conn., N.Y.

Plagiognathus repletus repletus Knight, 1923 [Fig. 88]
1923 *Plagiognathus repletus* Knight, Conn. Geol. Nat. Hist. Surv. Bull., 34: 449. [N.Y.].
Distribution: Same as for species.

Plagiognathus ribesi Kelton, 1982
1982 *Plagiognathus ribesi* Kelton, Can. Ent., 114: 169. [B.C.].
Distribution: B.C., Col.

Plagiognathus rolfsi Knight, 1964
1964 *Plagiognathus rolfsi* Knight, Ia. St. J. Sci., 39: 145. [Wash.].
Distribution: Wash.

Plagiognathus rosicola Knight, 1923
1923 *Plagiognathus rosicola* Knight, Conn. Geol. Nat. Hist. Surv. Bull., 34: 446. [Md.].
Distribution: Conn., Ill., Ks., Md., Mo., Que., W.Va.

Plagiognathus rubidus (Poppius), 1911
1911 *Gerhardiella rubida* Poppius, An. Soc. Ent. Belg., 55: 85. [Col.].
1912 *Poeciloscytus rosaceus* Van Duzee, Bull. Buff. Soc. Nat. Sci., 10: 488. [Col.]. Synonymized by Van Duzee, 1916, Check List Hem., p. 46 (as *Gerhardiella roseus; lapsus*).
1952 *Plagiognathus rubidus*: Carvalho, An. Acad. Brasil. Ci., 24: 67.
Distribution: Col.

Plagiognathus salicicola Knight, 1929
1929 *Plagiognathus salicicola* Knight, Ent. News, 40: 69. [Oh.].
Distribution: Ia., Ill., Ind., Mich., Minn., Oh., Que.

Plagiognathus salicicola depallens Knight, 1929
1929 *Plagiognathus salicicola* var. *depallens* Knight, Ent. News, 40: 70. [Minn.].
Distribution: Ia., Ind., Minn.

Plagiognathus salicicola salicicola Knight, 1929
1929 *Plagiognathus salicicola* Knight, Ent. News, 40: 69. [Oh.].
Distribution: Same as for species.

Plagiognathus salviae Knight, 1968
1968 *Plagiognathus salviae* Knight, Brig. Young Univ. Sci. Bull., 9(3): 30. [Nev.].
Distribution: Nev.

Plagiognathus shepherdiae Knight, 1929
1929 *Plagiognathus shepherdiae* Knight, Ent. News, 40: 70. [Col.].
Distribution: Col., S.D.

Plagiognathus shepherdiae shepherdiae Knight, 1929
1929 *Plagiognathus shepherdiae* Knight, Ent. News, 40: 70. [Col.].
Distribution: Same as for species.

Plagiognathus shepherdiae similatus Henry and Wheeler, New Name
1929 *Plagiognathus shepherdiae* var. *similis* Knight, Ent. News, 40: 71. [S.D.]. Preoccupied by *Plagiognathus similis* Knight, 1923.
Distribution: S.D.

Plagiognathus shoshonea Knight, 1964
1964 *Plagiognathus shoshonea* Knight, Ia. St. J. Sci., 39: 142. [Wyo.].
Distribution: Alta., B.C., Id., Mont., Ore., Wyo.

Plagiognathus similis Knight, 1923
1923 *Plagiognathus albatus* var. *similis* Knight, Conn. Geol. Nat. Hist. Surv. Bull., 34: 445. [Mich.].
1926 *Plagiognathus similis*: Blatchley, Het. E. N. Am., p. 944.
Distribution: Conn., Ill., Ind., Md., Mich., Mo., N.C., W.Va.

Plagiognathus similis furvus Knight, 1927
1927 *Plagiognathus similis* var. *furvus* Knight, Proc. Biol. Soc. Wash., 40: 12. [Md.].
Distribution: Md., N.C.

Plagiognathus similis similis Knight, 1923
1923 *Plagiognathus similis* Knight, Conn. Geol. Nat. Hist. Surv. Bull., 34: 445. [Mich.].
Distribution: Same as for species.

Plagiognathus stitti Knight, 1964
1964 *Plagiognathus stitti* Knight, Ia. St. J. Sci., 39: 145. [Ariz.].
Distribution: Ariz.

Plagiognathus subovatus Knight, 1929
1929 *Plagiognathus subovatus* Knight, Ent. News, 40: 266. [Minn.].
Distribution: Minn.

Plagiognathus suffuscipennis Knight, 1923
1923 *Plagiognathus suffuscipennis* Knight, Conn. Geol. Nat. Hist. Surv. Bull., 34: 454. [N.Y.].
Distribution: Alta., B.C., Ill., Man., Me., Minn., N.Y., Pa., Que., Sask., W.Va.

Plagiognathus syrticolae Knight, 1941
1941 *Plagiognathus syrticolae* Knight, Ill. Nat. Hist. Surv. Bull., 22: 31. [Ill.].
Distribution: Ill.

Plagiognathus tenellus Knight, 1929
1929 *Plagiognathus tenellus* Knight, Ent. News, 40: 73. [Ariz.].
Distribution: Ariz.

Plagiognathus tinctus Knight, 1923
1923 *Plagiognathus albonotatus* var. *tinctus* Knight, Conn. Geol. Nat. Hist. Surv. Bull., 34: 437. [Pa.].
1929 *Plagiognathus tinctus*: Knight, Ent. News, 40: 70.
Distribution: Ia., Ill., Ind., Md., Minn., Mo., N.C., N.Y., Oh., Pa., W.Va.

Plagiognathus tinctus debilis Blatchley, 1926
1926 *Plagiognathus debilis* Blatchley, Het. E. N. Am., p. 941. [Ind.].
1929 *Plagiognathus tinctus* var. *debilis*: Knight, Ent. News, 40: 70.
Distribution: Ind.

Plagiognathus tinctus tinctus Knight, 1929
1929 *Plagiognathus albonotatus* var. *tinctus* Knight, Conn. Geol. Nat. Hist. Surv. Bull.,

34: 437. [Pa.].
Distribution: Same as for species.

Plagiognathus urticae Knight, 1964
1964 *Plagiognathus urticae* Knight, Ia. St. J. Sci., 39: 148. [Cal.].
Distribution: Cal.

Plagiognathus verticalis (Uhler), 1894
1894 *Macrotylus verticalis* Uhler, Proc. Cal. Acad. Sci., (2)4: 272. [Cal.].
1914 *Plagiognathus verticalis*: Van Duzee, Trans. San Diego Soc. Nat. Hist., 2: 29.
1916 *Pseudopsallus verticalis*: Van Duzee, Univ. Cal. Publ. Ent., 1: 225
Distribution: Cal., Col., N.M. (Mexico).

Genus *Plesiodema* Reuter, 1875

1875 *Plesiodema* Reuter, Rev. Crit. Caps., 1: 97, 155. Type-species: *Phytocoris pinetellus* Zetterstedt, 1828. Monotypic.

Plesiodema sericeum (Heidemann), 1892
1892 *Psallus sericeus* Heidemann, Proc. Ent. Soc. Wash., 2: 226. [D.C.].
1926 *Plagiognathus tiliae* Knight, Can. Ent., 58: 252. [Minn.]. Synonymized by Knight, 1941, Ill. Nat. Hist. Surv. Bull., 22: 34.
1941 *Plagiognathus sericeus*: Knight, Ill. Nat. Hist. Surv. Bull., 22: 34.
1955 *Plesiodema sericeus*: Carvalho, Rev. Chil. Ent., 4: 226.
1973 *Plesiodema sericeum*: Steyskal, Stud. Ent., 16: 205.
Distribution: D.C., Ia., Ill., Mich., Minn., Mo., N.C., Pa., W.Va.

Genus *Pronotocrepis* Knight, 1929

1929 *Pronotocrepis* Knight, Can. Ent., 61: 217. Type-species: *Pronotocrepis clavicornis* Knight, 1929. Original designation.

Pronotocrepis clavicornis Knight, 1929
1929 *Pronotocrepis clavicornis* Knight, Can. Ent., 61: 217. [Col.].
Distribution: B.C., Col.

Pronotocrepis ribesi Knight, 1969
1969 *Pronotocrepis ribesi* Knight, Ia. St. J. Sci., 44: 79. [Wash.].
Distribution: Wash.

Pronotocrepis ruber Knight, 1969
1969 *Pronotocrepis ruber* Knight, Ia. St. J. Sci., 44: 80. [Ore.].
Distribution: Ore.

Genus *Psallus* Fieber, 1858

1858 *Psallus* Fieber, Wien. Ent. Monat., 2: 321. Type-species: *Cimex roseus* Fabricius, 1776. Designated by Distant, 1904, Fauna Brit. India, Rhyn., 2: 482.
1858 *Apocremnus* Fieber, Wien. Ent. Monat., 2: 320. Type-species: *Lygaeus ambiguus* Fallén, 1807. Designated by Kirkaldy, 1906, Trans. Am. Ent. Soc., 32: 124. Synonymized by Reuter, 1878, Acta Soc. Sci. Fenn., 13(1): 101.
1916 *Oligotylus* Van Duzee, Univ. Cal. Publ. Ent., 1: 216. Type-species: *Psallus brevitylus*

Slater and Knight, 1954. Designated and synonymized by Slater and Knight, 1954, Pan-Pac. Ent., 30: 143.

Psallus albipennis (Fallén), 1829
1829 *Phytocoris albipennis* Fallén, Hem. Suec., p. 107. [Europe].
1923 *Psallus waldeni* Knight, Conn. Geol. Nat. Hist. Surv. Bull., 34: 468. [Conn.]. Synonymized by Wheeler and Hoebeke, 1982, Proc. Ent. Soc. Wash., 84: 698.
1950 *Plagiognathus albipennis*: Moore, Nat. Can., 77: 248.
1982 *Psallus albipennis*: Wheeler and Hoebeke, Proc. Ent. Soc. Wash., 84: 696.
Distribution: Conn., Minn., N.Y., Que. (Europe).

Psallus alnicenatus Knight, 1923
1923 *Psallus alnicenatus* Knight, Conn. Geol. Nat. Hist. Surv. Bull., 34: 463. [N.Y.].
Distribution: Mass., Mich., Minn., N.S., N.Y.

Psallus amorphae Knight, 1930
1930 *Psallus amorphae* Knight, Can. Ent., 42: 125. [Minn.].
Distribution: Ia., Ill., Minn., Mo.

Psallus ancorifer (Fieber), 1858
1858 *Apocremnus ancorifer* Fieber, Wien. Ent. Monat., 2: 336. [Europe].
1908 *Psallus ancorifer*: Horvath, An. Hist.-Nat. Mus. Nat. Hung., 6: 5.
1916 *Apocremnus anchorifer* [*sic*]: Van Duzee, Univ. Cal. Publ. Ent., 1: 243.
Distribution: Cal., Conn., D.C., N.J., N.Y., Ore., Pa. (Asia Minor, Europe, northern Africa, USSR).

Psallus artemisicola Knight, 1964
1964 *Psallus artemisicola* Knight, Ia. St. J. Sci., 39: 149. [Minn.].
Distribution: Alta., Man., Minn., Mont., N.D., S.D., Sask., Wyo.

Psallus asperus Van Duzee, 1923
1923 *Psallus asperus* Van Duzee, Proc. Cal. Acad. Sci., (4) 12: 159. [Mexico; also from Cal.].
Distribution: Cal. (Mexico).

Psallus astericola Knight, 1930
1930 *Psallus astericola* Knight, Can. Ent., 62: 125. [Ia.].
Distribution: Ia.

Psallus atriplicis Knight, 1968
1968 *Psallus atriplicis* Knight, Brig. Young Univ. Sci. Bull., 9(3): 48. [Nev.].
Distribution: Nev.

Psallus atritibialis Knight, 1930
1930 *Psallus atritibialis* Knight, Can. Ent., 62: 129. [Ariz.].
Distribution: Ariz.

Psallus betuleti (Fallén), 1829
1829 *Phytocoris betuleti* Fallén, Hem. Suec., p. 97. [Europe].
1979 *Psallus betuleti*: Henry and Wheeler, Proc. Ent. Soc. Wash., 81: 261.
Distribution: Alk., Pa. (Palearctic).

Psallus biguttulatus Uhler, 1894
1894 *Psallus biguttulatus* Uhler, Proc. Cal. Acad. Sci., (2)4: 275. [Mexico].
1923 *Oncotylus biguttulatus*: Van Duzee, Proc. Cal. Acad. Sci., (4)12: 157.
Distribution: Cal., N.M., Tex. (Mexico).

Note: This species was returned to *Psallus* by Knight (1927, Can. Ent., 59: 35).

Psallus breviceps Reuter, 1909

1909 *Psallus breviceps* Reuter, Acta Soc. Sci. Fenn., 36(2): 78. [Tex.].

1916 *Plagiognathus breviceps*: Van Duzee, Check List Hem., p. 47.

Distribution: Cal., Tex.

Note: This species was returned to *Psallus* by Carvalho (1958, Arq. Mus. Nac., 45: 118).

Psallus brevitylus Slater and Knight, 1954

1954 *Psallus brevitylus* Slater and Knight, Pan-Pac. Ent., 30: 144. [Cal.].

Distribution: Cal.

Psallus carneatus Knight, 1930

1930 *Psallus carneatus* Knight, Can. Ent., 62: 128. [Cal.].

Distribution: Cal.

Psallus cercocarpicola Knight, 1930

1930 *Psallus cercocarpicola* Knight, Can. Ent., 62: 127. [Col.].

Distribution: Col., N.M.

Psallus drakei Knight, 1923

1923 *Psallus drakei* Knight, Conn. Geol. Nat. Hist. Surv. Bull., 34: 464. [N.Y.].

Distribution: Alta., Col., Man., N.Y., Sask.

Psallus falleni Reuter, 1883

1883 *Psallus falleni* Reuter, Acta Soc. Sci. Fenn., 13(3): 462. [Europe].

1922 *Psallus alnicola*: Knight, Can. Ent., 53: 283.

1983 *Psallus falleni*: Kelton, Can. Ent., 115: 325.

Distribution: Alta., B.C., Col., Id., Mich., Minn., N.H., N.Y., Ore., P.Ed., Que., W.Va., Wash. (Europe, USSR).

Note: Kelton (1983, Can. Ent., 115: 325) clarified that records of *Psallus alnicola* in N.Am. should be referred to *P. falleni*. *Psallus alnicola* Douglas and Scott, 1865, is a junior synonym of *Psallus scholtzi* Fieber, 1861 (Kerzhner, 1964, Miridae Europ. USSR, 1: 993).

Psallus flavellus Stichel, 1933

1933 *Psallus lepidus* var. *flavella* Stichel, Illus. Best. Deut. Wanz., 9: 268. [Europe].

1983 *Psallus flavellus*: Kelton, Can. Ent., 115: 327.

Distribution: N.S. (Europe).

Psallus flaviclavus Knight, 1930

1930 *Psallus flaviclavus* Knight, Can. Ent., 42: 130. [Col.].

Distribution: Col.

Psallus flora Van Duzee, 1923

1923 *Psallus flora* Van Duzee, Proc. Cal. Acad. Sci., (4)12: 158. [Mexico].

1961 *Psallus flora*: Bibby, J. Econ. Ent., 54: 329.

Distribution: Ariz. (Mexico).

Psallus fuscatus Knight, 1923

1923 *Psallus parshleyi* var. *fuscatus* Knight, Conn. Geol. Nat. Hist. Surv. Bull., 34: 466. [Minn.].

1941 *Psallus fuscatus*: Knight, Ill. Nat. Hist. Surv. Bull., 22: 44.

Distribution: Ill., Minn.

Psallus fuscopunctatus Knight, 1930
1930 *Psallus fuscopunctatus* Knight, Can. Ent., 62: 126. [Col.].
Distribution: Col.

Psallus lepidus Fieber, 1858
1858 *Psallus lepidus* Fieber, Wien. Ent. Monat., 2: 337. [Europe].
1983 *Psallus lepidus*: Kelton, Can. Ent., 115: 327.
Distribution: N.S., P.Ed. (Europe, northern Africa).

Psallus maculosus Knight, 1925
1925 *Psallus maculosus* Knight, Can. Ent., 57: 89. [Ariz.].
Distribution: Ariz.

Psallus merinoi Knight, 1968
1968 *Psallus merinoi* Knight, Brig. Young Univ. Sci. Bull., 9(3): 47. [Nev.].
Distribution: Nev.

Psallus morrisoni Knight, 1923
1923 *Psallus morrisoni* Knight, Conn. Geol. Nat. Hist. Surv. Bull., 34: 464. [Mass.].
Distribution: Mass., Me., Minn., N.Y.

Psallus nigerrimus (Van Duzee), 1916
1916 *Apocremnus nigerrimus* Van Duzee, Univ. Cal. Publ. Ent., 1: 243. [Cal.].
1958 *Psallus nigerrimus*: Carvalho, Arq. Mus. Nac., 45: 125.
Distribution: Cal.

Psallus nigrovirgatus Knight, 1930
1930 *Psallus nigrovirgatus* Knight, Can. Ent., 62: 130. [Col.].
Distribution: Col., N.M.

Psallus parshleyi Knight, 1923
1923 *Psallus parshleyi* Knight, Conn. Geol. Nat. Hist. Surv. Bull., 34: 465. [Mass.].
Distribution: Alta., B.C., Ill., Man., Mass., Minn., N.Y., Sask.

Psallus physocarpi Henry, 1981
1981 *Psallus physocarpi* Henry, Proc. Ent. Soc. Wash., 83: 399. [Pa.].
Distribution: N.Y., Pa.
Note: Wheeler and Hoebeke (1985, Proc. Ent. Soc. Wash., 87: 361) summarized the seasonal history and habits on ninebark, *Physocarpus opulifolius* (L.) Maxim.

Psallus piceicola Knight, 1923
1923 *Psallus piceicola* Knight, Conn. Geol. Nat. Hist. Surv. Bull., 34: 469. [N.Y.].
Distribution: Alta., Ill., Man., Minn., N.Y., Pa., Que., Sask.

Psallus pictipes (Van Duzee), 1918
1918 *Plagiognathus pictipes* Van Duzee, Proc. Cal. Acad. Sci., (4)8: 305. [Cal.].
1923 *Psallus pictipes*: Van Duzee, Proc. Cal. Acad. Sci., (4)12: 161.
1925 *Psallus suaedae* Knight, Bull. Brook. Ent. Soc., 20: 34. [Ariz.]. Synonymized by Carvalho, 1958, Arq. Mus. Nac., 45: 127.
Distribution: Ariz., Cal., Col., N.M., Tex. (Mexico).

Psallus purshiae Knight, 1968
1968 *Psallus purshiae* Knight, Brig. Young Univ. Sci. Bull., 9(3): 48. [Nev.].
Distribution: Nev.

Psallus roseus (Fabricius), 1777
1777 *Cimex roseus* Fabricius, Gen. Ins., p. 300. [Europe].

1983 *Psallus roseus*: Kelton, Can. Ent., 115: 325.
Distribution: B.C. (Europe, Siberia).

Psallus rubrofemoratus Knight, 1930
1930 *Psallus rubrofemoratus* Knight, Can. Ent., 62: 129. [Ariz.].
Distribution: Ariz.

Psallus vaccinicola Knight, 1930
1930 *Psallus vaccinicola* Knight, Can. Ent., 62: 128. [Ariz.].
Distribution: Ariz., Cal.

Psallus variabilis (Fallén), 1807
1807 *Lygaeus variabilis* Fallén, Monogr. Cimic. Suec., p. 88. [Europe].
1980 *Psallus variabilis*: Hoebeke, Coop. Plant Pest Rept., 5: 628.
Distribution: Conn., N.Y., Ont. (Europe, USSR).
Note: Knight (1927, Bull. Brook. Ent. Soc., 22: 104) discredited U.S. records for this species, stating that they referred to *Lepidopsallus rubidus atricolor* Knight. Wheeler and Hoebeke (1982, Proc. Ent. Soc. Wash., 84: 691) confirmed the establishment of *P. variabilis* in the U.S. and gave North American host plants.

Genus *Pseudatomoscelis* Poppius, 1911

1911 *Pseudatomoscelis* Poppius, An. Soc. Ent. Belg., 55: 85. Type-species: *Atomoscelis seriatus* Reuter, 1876. Original designation.

Pseudatomoscelis seriatus (Reuter), 1876
1876 *Atomoscelis seriatus* Reuter, Öfv. K. Svens. Vet.-Akad. Förh., 32(9): 91. [Tex.].
1898 *Psallus delicatus*: Howard, U.S. Dept. Agr., Div. Ent. Bull., new ser., 18: 101.
1911 *Atomoscelis serieatus* [*sic*]: Tucker, Can. Ent., 43: 30.
1911 *Pseudatomoscelis seriatus*: Poppius, An. Soc. Ent. Belg., 55: 86.
1915 *Atomoscelis seratus* [*sic*]: La Follette, J. Ent. Zool., 7: 128.
1916 *Psallus seriatus*: Van Duzee, Check List Hem., p. 46.
1979 *Pseudoatomoscelis* [*sic*] *seriatus*: Henry and Smith, J. Ga. Ent. Soc., 14: 213.
1980 *Gerhardiella delicatus*: Harrison et al., Ins. Involv. Transm. Bact. Path., p. 228.
Distribution: Ala., Ariz., Ark., Cal., Col., D.C., Fla., Ga., Ill., Ks., La., Md., Minn., Miss., Mo., N.J., N.M., Neb., Ok., S.C., Sask., Tex., Ut. (Mexico).
Note: Sterling and Dean (1977, Tex. Agr. Exp. Stn., MP-1342: 1-28) compiled a bibliography of the literature treating the cotton fleahopper, *P. seriatus*.

Genus *Ranzovius* Distant, 1893

1893 *Ranzovius* Distant, Biol. Centr.-Am., Rhyn., 1: 423. Type-species: *Ranzovius crinitus* Distant, 1893. Monotypic.

1905 *Nyctella* Reuter, Öfv. F. Vet.-Soc. Förh., 47(19): 35. Type-species: *Nyctella moerens* Reuter, 1905. Monotypic. Synonymized by Carvalho, 1952, An. Acad. Brasil. Ci., 24: 66.

Note: Henry (1984, Proc. Ent. Soc. Wash., 86: 53-67) revised this genus and provided a key to species.

Ranzovius agelenopsis Henry, 1984
1984 *Ranzovius agelenopsis* Henry, Proc. Ent. Soc. Wash., 86: 58. [Tenn.].
Distribution: Tenn.

Ranzovius californicus (Van Duzee), 1917

1917 *Excentricus californicus* Van Duzee, Proc. Cal. Acad. Sci., (4)7: 284. [Cal.]. Synonymized by Carvalho, 1955, Rev. Chil. Ent., 4: 224; resurrected by Henry, 1984, Proc. Ent. Soc. Wash., 86:60.

1955 *Ranzovius moerens*: Carvalho, Rev. Chil. Ent., 4: 224 (in part).

1984 *Ranzovius californicus*: Henry, Proc. Ent. Soc. Wash., 86: 60.

Distribution: Cal.

Note: Henry (1984, Proc. Ent. Soc. Wash., 86:60) clarified the status of this species. Davis and Russell (1969, Psyche, 76: 262-269) studied the habits [as *moerens* Reuter].

Ranzovius clavicornis (Knight), 1927

1927 *Psallus clavicornis* Knight, Proc. Biol. Soc. Wash., 40: 13. [Md.].

1926 *Excentricus mexicanus*: Blatchley, Het. E. N. Am., p. 962 (in part).

1984 *Ranzovius* (=*Exocentricus* [*sic*]) *clavicornis*: Mead, Fla. Dept. Agr. Cons. Serv. Tech. Rept., 23: 2.

1984 *Ranzovius contubernalis* Henry, Proc. Ent. Soc. Wash., 86: 61. [D.C.]. Synonymized by Henry, 1985, J. N.Y. Ent. Soc., 93: 1130.

1985 *Ranzovius clavicornis*: Henry, J. N.Y. Ent. Soc., 93: 1130.

Distribution: Ariz., Conn., D.C., Fla., La., Md., N.C., Tex., Va.

Note: Wheeler and McCaffrey (1984, Proc. Ent. Soc. Wash., 86: 68-81) studied the habits and seasonal history of *clavicornis* [as *contubernalis* Henry]. Mead (1984, above) incorrectly equated the genus *Excentricus* Reuter with *Ranzovius* Distant.

Genus *Reuteroscopus* Kirkaldy, 1905

1876 *Episcopus* Reuter, Öfv. K. Svens. Vet.-Akad. Förh., 32(9): 90. Preoccupied. Type-species: *Episcopus ornatus* Reuter, 1876. Monotypic.

1905 *Reuteroscopus* Kirkaldy, Wien. Ent. Zeit., 24: 268. New name for *Episcopus* Reuter.

1906 *Aristoreuteria* Kirkaldy, Trans. Am. Ent. Soc., 32: 125. Unnecessary new name for *Episcopus* Reuter. Synonymized by Kirkaldy, 1906, Can. Ent., 38: 373.

Note: Knight (1965, Ia. St. J. Sci., 40: 116-120) provided a key to species.

Reuteroscopus abroniae Knight, 1965

1965 *Reuteroscopus abroniae* Knight, Ia. St. J. Sci., 40: 105. [Col.].

Distribution: Col., Neb., Tex.

Reuteroscopus basicornis Knight, 1965

1965 *Reuteroscopus basicornis* Knight, Ia. St. J. Sci., 40: 104. [Ariz.].

Distribution: Ariz.

Reuteroscopus carolinae Knight, 1965

1965 *Reuteroscopus carolinae* Knight, Ia. St. J. Sci., 40: 106. [N.C].

Distribution: N.C.

Reuteroscopus dreisbachi Knight, 1965

1965 *Reuteroscopus dreisbachi* Knight, Ia. St. J. Sci., 40: 107. [Holotype from Mexico; also from Ariz.].

Distribution: Ariz. (Mexico).

Reuteroscopus femoralis Kelton, 1964
1964 *Reuteroscopus femoralis* Kelton, Can. Ent., 96: 1424. [Holotype from Mexico; also from Ariz.].
Distribution: Ariz., Tex. (Mexico).

Reuteroscopus froeschneri Knight, 1953
1953 *Reuteroscopus froeschneri* Knight, Ia. St. Coll. J. Sci., 27: 509. [Mo.].
Distribution: Mo.

Reuteroscopus hamatus Kelton, 1964
1964 *Reuteroscopus hamatus* Kelton, Can. Ent., 96: 1426. [Fla.].
Distribution: Fla. (Central America, West Indies).

Reuteroscopus longirostris Knight, 1925
1925 *Reuteroscopus longirostris* Knight, Can. Ent., 57: 90. [Ariz.].
Distribution: Ariz., Col.

Reuteroscopus nicholi (Knight), 1930
1930 *Psallus nicholi* Knight, Can. Ent., 62: 127. [Ariz.].
1964 *Reuteroscopus nicholi*: Kelton, Can. Ent., 96: 1432.
Distribution: Ariz. (Mexico).

Reuteroscopus ornatus (Reuter), 1876 [Fig. 90]
1876 *Episcopus ornatus* Reuter, Öfv. K. Svens. Vet.-Akad. Förh., 32(9): 90. [Tex.].
1905 *Reuteroscopus ornatus*: Reuter, Festschr. Palmén, 1: 46.
1905 *Reuteroscopus uvidus*: Van Duzee, Bull. Buff. Soc. Nat. Hist., 9: 182.
Distribution: Ariz., Ark., Cal., Col., Conn., Del., Fla., Ga., Ia., Ind., Ks., Ky., La., Mass., Mich., Miss., Mo., N.C., N.J., N.Y., Oh., Pa., Ont., S.C., Tenn., Tex., W.Va. (Mexico, Central America, West Indies).
Note: *Lygus uvidus* Distant, 1893, described from Mexico and Panama, was synonymized with *Reuteroscopus ornatus* by Reuter (1905, Festschr. Palmén, 1: 46), but later resurrected by Kelton (1964, Can. Ent., 96: 1432). Because *R. uvidus* is known only from Tex. southward, Van Duzee's (1905, above) record is listed under *R. ornatus*.

Reuteroscopus pallidiclavus Knight, 1965
1965 *Reuteroscopus pallidiclavus* Knight, Ia. St. J. Sci., 40: 104. [Ariz.].
Distribution: Ariz.

Reuteroscopus santaritae Knight, 1965
1965 *Reuteroscopus santaritae* Knight, Ia. St. J. Sci., 40: 106. [Ariz.].
Distribution: Ariz.

Reuteroscopus schaffneri Knight, 1965
1965 *Reuteroscopus schaffneri* Knight, Ia. St. J. Sci., 40: 108. [Tex.].
Distribution: Tex.

Reuteroscopus tinctipennis (Knight), 1925
1925 *Psallus tinctipennis* Knight, Can. Ent., 57: 89. [Ariz.].
1965 *Reuteroscopus tinctipennis*: Knight, Ia. St. J. Sci., 40: 103.
Distribution: Ariz., Tex.

Reuteroscopus uvidus (Distant), 1893
1893 *Lygus uvidus* Distant, Biol. Centr.-Am., Rhyn., 1: 433. [Mexico].
1964 *Reuteroscopus uvidus*: Kelton, Can. Ent., 96: 1431.

Distribution: Tex. (Mexico, Central America).

Genus *Rhinacloa* Reuter, 1876

1876 *Rhinacloa* Reuter, Öfv. K. Svens. Vet.-Akad. Förh., 32(9): 88. Type-species: *Rhinacloa forticornis* Reuter, 1876. Monotypic.

Note: Schuh and Schwartz (1985, Bull. Am. Mus. Nat. Hist., 179: 379-470) revised the genus and gave a key to species.

Rhinacloa basalis (Reuter), 1907

1907 *Sthenarus basalis* Reuter, Öfv. F. Vet.-Soc. Förh., 49(5): 26. [Jamaica].

1926 *Lepidopsallus pusillus* Knight, Bull. Brook. Ent. Soc., 20: 227. [Fla.]. Synonymized by Schuh and Schwartz, 1985, Bull. Am. Mus. Nat. Hist., 179: 395.

1955 *Rhinacloa pusilla*: Carvalho, Rev. Chil. Ent., 4: 226.

1985 *Rhinacloa basalis*: Schuh and Schwartz, Bull. Am. Mus. Nat. Hist., 179: 399.

Distribution: Fla. (Mexico to South America, West Indies).

Rhinacloa callicrates Herring, 1971

1971 *Rhinacloa callicrates* Herring, Proc. Ent. Soc. Wash., 73: 449. [Ariz.]

Distribution: Ariz., Cal. (Mexico).

Rhinacloa clavicornis (Reuter), 1905

1905 *Sthenarus clavicornis* Reuter, Öfv. F. Vet.-Soc. Förh., 47(19): 38. [Venezuela].

1926 *Rhinacloa subpallicornis* Knight, Bull. Brook. Ent. Soc., 20: 225. [Fla.]. Synonymized by Schuh and Schwartz, 1985, Bull. Am. Mus. Nat. Hist., 179: 399.

1964 *Rhinacloa subparallicornis* [*sic*]: Frost, Fla. Ent., 47: 136.

1985 *Rhinacloa clavicornis*: Schuh and Schwartz, Bull. Am. Mus. Nat. Hist., 179: 399.

Distribution: Fla. (South America, West Indies).

Rhinacloa forticornis Reuter, 1876

1876 *Rhinacloa forticornis* Reuter, Öfv. K. Svens. Vet.-Akad. Förh., 32(9): 88. [Tex.].

1887 *Rhinocloa* [*sic*] *forticomis* [*sic*]: Ashmead, Ent. Am., 3: 155.

1910 *Rhinocloa* [*sic*] *forticornis*: Smith, Cat. Ins. N.J., p. 132.

1915 *Rhinacola* [*sic*] *forticornis*: LaFollette, J. Ent. Zool., 7: 128.

1976 *Rhinacloa forticornus* [*sic*]: Goeden and Ricker, Environ. Ent. 5: 1171.

Distribution: Ariz., Cal., Col., Fla.(?), Ill.(?), Miss.(?), Mo., N.M., Nev., Tex., Ut. (Hawaii, Mexico to South America, West Indies).

Note: Records of *R. forticornis* from northeastern U.S. refer to *Spanagonicus albofasciatus* (Reuter). Schuh and Schwartz (1985, Bull. Am. Mus. Nat. Hist., 179: 418) considered the distribution of this species as western U.S. and Neotropical. Eastern U.S. records from Fla., Ill., and Miss. need verification.

Rhinacloa manleyi Schuh and Schwartz, 1985

1985 *Rhinacloa manleyi* Schuh and Schwartz, Bull. Am. Mus. Nat. Hist., 179: 421. [Holotype from Mexico; also from Tex.].

Distribution: Tex. (Mexico, Guatemala, Brazil).

Rhinacloa pallidipes Maldonado, 1969

1969 *Rhinacloa pallidipes* Maldonado, Univ. Puerto Rico, Agr. Exp. Stn. Tech. Pap., 45: 83. [Puerto Rico].

1984 *Rhinacloa pallidipes*: Henry, Proc. Ent. Soc. Wash., 86: 519.

Distribution: Fla. (Mexico to South America, West Indies).

Note: Schuh and Schwartz (1985, Bull. Am. Mus. Nat. Hist., 179: 407-410) discussed synonymy and expanded the known distribution of this species.

Rhinacloa rubroornata Schuh and Schwartz, 1985

1985 *Rhinacloa rubroornata* Schuh and Schwartz, Bull. Am. Mus. Nat. Hist., 179: 428. [Fla.].

Distribution: Fla.

Genus *Rhinocapsus* Uhler, 1890

1890 *Rhinocapsus* Uhler, Trans. Md. Acad. Sci., 1: 81. Type-species: *Rhinocapsus vanduzeei* Uhler, 1890. Monotypic.

Rhinocapsus rubricans (Provancher), 1887

1887 *Plagiognathus rubricans* Provancher, Pet. Faune Ent. Can., 3: 154. [Que.]. Lectotype designated by Kelton, 1968, Nat. Can., 95: 1075.

1912 *Rhinocapsus rubricans*: Van Duzee, Can. Ent., 44: 323.

1916 *Microphylellus rubricans*: Van Duzee, Check List Hem., p. 46.

1923 *Rhinocapsus miniatus* Knight, Conn. Geol. Nat. Hist. Surv. Bull., 34: 460. [N.J.]. Synonymized by Henry, 1982, Proc. Ent. Soc. Wash., 84: 338.

Distribution: Fla., Man., Mass., Me., N.C., N.J., Pa., Que., Sask.

Rhinocapsus vanduzeei Uhler, 1890

1890 *Rhinocapsus vanduzeei* Uhler, Trans. Md. Acad. Sci., 1: 82. [N.Y.].

1907 *Rhinocapsus vanduzei* [*sic*]: Moore, Can. Ent., 39: 163.

1923 *Rhinocapsus vanduzeii* [*sic*]: Knight, Conn. Geol. Nat. Hist. Surv. Bull., 34: 459.

Distribution: Conn., Fla., Ga., Man., Mass., Md., Me., Mich., Mo., N.C., N.Y., Ont., Pa., Que., Va., W.Va.

Note: Wheeler and Herring (1979, Quart. Bull. Am. Rhodo. Soc., 33: 12-14) discussed the habits of this species on cultivated azaleas, *Rhododendron* spp.

Genus *Semium* Reuter, 1876

1876 *Semium* Reuter, Öfv. K. Svens. Vet.-Akad. Förh., 32(9): 80. Type-species: *Semium hirtum* Reuter, 1876. Monotypic.

Note: Kelton (1959, Can. Ent., 91: 242) transferred this genus from Orthotylinae to Phylinae. Kelton (1973, Can. Ent., 105: 1584) provided a key to species.

Semium hirtum Reuter, 1876 [Fig. 64]

1876 *Semium hirtum* Reuter, Öfv. K. Svens. Vet.-Akad. Förh., 32(9): 80. [Tex.].

1885 *Eccritatarsus* [*sic*] *elegans* Popenoe, Trans. Ks. Acad. Sci., 9: 63. *Nomen nudum*.

1886 *Eccritotarsus elegans* Uhler, Check-list Hem. Het., p. 9. *Nomen nudum*.

1887 *Eccritotarsus elegans* Uhler, Ent. Am., 3: 149. [Cal., Ill., Ks., Tex.]. Synonymized by Reuter, 1909, Acta Soc. Sci. Fenn., 36(2): 73.

1971 *Seminum* [*sic*] *hirtum*: Khalaf, Fla. Ent., 54: 340.

Distribution: D.C., Fla., Ia., Ill., Ind., Ks., Miss., Mo., N.C., N.J., N.Y., Oh., Ont., Pa., S.D., Tex., W.Va. (Mexico).

Note: The Cal. record for this species undoubtedly should be applied to *S. subglabrum* Knight.

Semium subglabrum Knight, 1927
1927 *Semium subglaber* Knight, Bull. Brook. Ent. Soc., 22: 26. [Ariz.].
1973 *Semium subglabrum*: Steyskal, Stud. Ent., 16: 207.
Distribution: Ariz., Cal., Col., Nev.

Genus *Spanagonicus* Berg, 1883

1883 *Spanagonicus* Berg, An. Soc. Cient. Arg., 16: 78. Type-species: *Spanagonicus provincialis* Berg, 1883. Monotypic.

1907 *Leucopoecila* Reuter, Öfv. F. Vet.-Soc. Förh., 49(5): 24. Type-species: *Leucopoecila albofasciata* Reuter, 1907. Monotypic. Synonymized by Carvalho, 1952, An. Acad. Brasil. Ci., 24: 66.

Spanagonicus albofasciatus (Reuter), 1907
1904 *Rhinacloa forticornis*: Wirtner, An. Carn. Mus., 3: 202.
1907 *Leucopoecila albofasciata* Reuter, Öfv. F. Vet.-Soc. Förh., 49(5) 26. [Jamaica].
1914 *Leucopoecila ablofasciata* [*sic*]: Barber, Bull. Am. Mus. Nat. Hist., 33: 500.
1958 *Spanagonicus albofasciatus*: Carvalho, Arq. Mus. Nac., 45: 142.
1960 *Spanigonicus* [*sic*] *albifasciatus* [*sic*]: Knowlton, Proc. Ut. Acad. Sci., 37: 54.
1961 *Spanogonicus* [*sic*] *albofasciatus*: Bibby, J. Econ. Ent., 54: 329.
1976 *Spagnogicus* [*sic*] *albofasciatus*: Dietz et al., N.C. Agr. Res. Serv. Tech. Bull, 238: 151.
Distribution: Ala., Ariz., Ark., Cal., Fla., Ga., Ill., Miss., Mo., N.C., N.M., N.Y., Nev., Oh., Pa., S.C., Tenn., Tex., W.Va. (Hawaii, Mexico, West Indies).
Note: Butler (1965, J. Ks. Ent. Soc., 38: 70-75) and Butler and Stoner (1965, J. Econ. Ent., 58: 664-665) studied the biology. Neal et al. (1972, Fla. Ent., 55: 247-250) reported on predatory habits.

Genus *Sthenarus* Fieber, 1858

1858 *Sthenarus* Fieber, Wien. Ent. Monat., 2: 321. Type-species: *Capsus rotermundi* Scholtz, 1846. Designated by Kirkaldy 1906, Trans. Am. Ent. Soc., 32: 123.

Sthenarus cuneotinctus (Van Duzee), 1915
1915 *Psallus cuneotinctus* Van Duzee, Pomona J. Ent. Zool., 7: 118. [Cal.].
1916 *Sthenarus cuneotinctus*: Van Duzee, Check List Hem., p. 46.
Distribution: Cal.

Sthenarus mcateei Knight, 1927
1927 *Sthenarus mcateei* Knight, Proc. Biol. Soc. Wash., 40: 9. [Md.].
Distribution: Md., Miss.

Sthenarus pubescens (Reuter), 1876
1876 *Phoenicocoris pubescens* Reuter, Öfv. K. Svens. Vet.-Akad. Förh., 32(9): 90. [Tex.].
1890 *Sthenarus pubescens*: Atkinson, J. Asiatic Soc. Bengal, 58(2): 177.
Distribution: Tex.

Sthenarus rotermundi (Scholtz), 1847
1847 *Capsus rotermundi* Scholtz, Arb. Schles. Ges., p. 130. [Europe].
1979 *Sthenarus rotermundi*: Henry and Wheeler, Proc. Ent. Soc. Wash., 81: 263.
Distribution: Ont., Pa. (Europe, North Africa, USSR).

Note: Henry and Wheeler (1979, above) summarized the literature, gave hosts, and illustrated the adult and fifth-instar nymph.

Sthenarus viticola Johnston, 1935

1935 *Sthenarus viticola* Johnston, Bull. Brook. Ent. Soc., 30: 16. [Miss.].

Distribution: Miss.

Genus *Strophopoda* Van Duzee, 1916

1916 *Strophopoda* Van Duzee, Univ. Cal. Publ. Ent., 1: 216. Type-species: *Strophopoda aprica* Van Duzee, 1921. Designated by Van Duzee, 1921, Proc. Cal. Acad. Sci., (4)11: 132. Synonymized under *Chlamydatus* Curtis, by Carvalho, 1955, Rev. Chil. Ent., 4: 226; resurrected by Kelton, 1965, Can. Ent., 97: 1142.

Strophopoda aprica Van Duzee, 1921

1921 *Strophopoda aprica* Van Duzee, Proc. Cal. Acad. Sci., (4)11: 132. [Cal.].

1955 *Chlamydatus aprica*: Carvalho, Rev. Chil. Ent., 4: 226.

1973 *Chlamydatus apricus*: Steyskal, Stud. Ent., 16: 206.

Distribution: Cal.

Genus *Tannerocoris* Knight, 1970

1970 *Tannerocoris* Knight, Great Basin Nat., 30: 227. Type-species: *Tannerocoris sarcobati* Knight, 1970. Original designation.

Tannerocoris sarcobati Knight 1970

1970 *Tannerocoris sarcobati* Knight, Great Basin Nat., 30: 227. [Wash.].

Distribution: Col., Id., S.D., Ut., Wash.

Genus *Tuponia* Reuter, 1875

1875 *Megalodactylus* (*Tuponia*) Reuter, Rev. Crit. Caps., 1: 98. Type-species: *Megalodactylus lethierryi* Reuter, 1875. Designated by Kirkaldy, 1906, Trans. Am. Ent. Soc., 32: 122.

1878 *Tuponia*: Reuter, Acta Soc. Sci. Fenn., 13(1): 16.

Tuponia dubiosa Van Duzee, 1918

1918 *Tuponia dubiosa* Van Duzee, Proc. Cal. Acad. Sci., (4)8: 304. [Cal.].

Distribution: Cal.

Tuponia lucida Van Duzee, 1918

1918 *Tuponia lucida* Van Duzee, Proc. Cal. Acad. Sci., (4)8: 303. [Cal.].

Distribution: Cal.

Tuponia subnitida Uhler, 1895

1895 *Tuponia subnitida* Uhler, Bull. Col. Agr. Exp. Stn., 31: 45. [Col.].

Distribution: Col.

Tribe Pilophorini Douglas and Scott, 1876

Note: Schuh (1974, Ent. Am., 47: 311-313) transferred this tribe from Orthotylinae to Phylinae.

Genus *Alepidia* Reuter, 1909

1909 *Alepidia* Reuter, Acta Soc. Sci. Fenn., 36(2): 75. Type-species: *Pilophorus gracilis* Uhler, 1895. Monotypic.

Alepidia bellula Hussey, 1954
1954 *Alepidia bellula* Hussey, Proc. Ent. Soc. Wash., 56: 200. [Mich.].
Distribution: Conn., Mich.

Alepidia gracilis (Uhler), 1895
1895 *Pilophorus gracilis* Uhler, Bull. Col. Agr. Exp. Stn., 31: 42. [Col.].
1909 *Alepidia gracilis*: Reuter, Acta Soc. Sci. Fenn., 36(2): 75.
Distribution: Ala., Col., D.C., Fla., Ga., Ia., Ill., Ind., Mass., Md., Miss., N.J., N.Y., Va., W.Va.

Alepidia gracilis gracilis (Uhler), 1895
1895 *Pilophorus gracilis* Uhler, Bull. Col. Agr. Exp. Stn., 31: 42. [Col.].
1926 *Alepidia gracilis*: Knight, Bull. Brook. Ent. Soc., 21: 26.
Distribution: Same as for species.

Alepidia gracilis squamosa Knight, 1926
1926 *Alepidia gracilis* var. *squamosa* Knight, Bull. Brook. Ent. Soc., 21: 26. [Ala.].
Distribution: Ala., Fla., Ind., Md., Mich., N.Y., Va.

Genus *Alepidiella* Poppius, 1914

1914 *Alepidiella* Poppius, An. Soc. Ent. Belg., 58: 251. Type-species: *Alepidiella heidemanni* Poppius, 1914. Monotypic.

Alepidiella heidemanni Poppius, 1914
1914 *Alepidiella heidemanni* Poppius An. Soc. Ent. Belg., 58: 253. [D.C.].
Distribution: D.C., Ga., Md., Ok.

Genus *Paramixia* Reuter, 1900

1900 *Paramixia* Reuter, Öfv. F. Vet.-Soc. Förh., 42: 264. Type-species: *Paramixia suturalis* Reuter, 1900. Monotypic.

Note: Schuh (1974, Ent. Am., 47: 210) synonymized the genus *Orthotylellus* Knight, 1935, and transferred *Paramixia* to the tribe Pilophorini.

Paramixa polita (Uhler), 1894
1894 *Psallus politus* Uhler, Proc. Zool. Soc. London, p. 195. [Grenada]. Lectotype designated by Henry, 1985, J. N.Y. Ent. Soc., 93: 1128.
1907 *Sthenarus plebejus* Reuter, Öfv. F. Vet.-Soc. Förh., 49(5): 26. [Jamaica]. Synonymized by Henry, 1985, J. N.Y. Ent. Soc., 93: 1127.
1926 *Sthenarus plebejus*: Blatchley, Het. E. N. Am., p. 922.
1982 *Paramixia carmelitana*: Henry and Wheeler, Fla. Ent., 65: 236 (in part).
1985 *Paraxmixia polita*: Henry, J. N.Y. Ent. Soc., 93: 1127.
Distribution: Fla. (Cuba, Grenada, Jamaica, Puerto Rico).

Genus *Pilophorus* Hahn, 1826

1826 *Pilophorus* Hahn, Icon. Cimic., 1: 23. Type-species: *Cimex clavatus* Linnaeus, 1767. Designated by Int. Comm. Zool. Nomen., 1950, Bull. Zool. Nomen., 4: 474.

Note: Knight (1973, Ia. St. J. Res., 48: 129-133) provided a key to the U.S. species of *Pilophorus*.

Pilophorus americanus Poppius, 1914

1914 *Pilophorus crassipes* Poppius, An. Soc. Ent. Belg., 58: 242. [Col.]. Preoccupied.

1914 *Pilophorus americanus* Poppius, An. Soc. Ent. Belg., 58: 243. [Ariz.].

Distribution: Ariz., Cal., Col., Ut.

Note: This species previously was placed as a junior synonym of *P. crassipes* Poppius by Knight (1968, Brig. Young Univ. Sci. Bull., 9(3): 167), until Wheeler and Henry (1975, Trans. Am. Ent. Soc., 101: 358) showed that *P. crassipes* was preoccupied by *P. crassipes* Heidemann. *P. americanus* is the next available name. See note under *P. crassipes* Heidemann.

Pilophorus amoenus Uhler, 1887

1887 *Pilophorus amoenus* Uhler, Ent. Am., 3: 30. [Md.].

1887 *Pilophorus confusus*: Van Duzee, Can. Ent., 19: 72.

1889 *Pilophorus amoemus* [*sic*]: Van Duzee, Can. Ent., 21: 4.

1950 *Philophorus* [*sic*] *amoenus*: Wray, Ins. N.C., Suppl., p. 12.

Distribution: Cal.(?), Col., D.C., Ga., Ill., Ind., Man., Mass., Md., Me., N.C., N.H., N.J., N.M.(?), N.Y., Oh., Ont., Pa., Que., Va., W.Va.

Note: Records of this species from Cal. and N.M. undoubtedly are based on misidentifications and should be dropped from future lists.

Pilophorus australis Knight, 1926

1926 *Pilophorus australis* Knight, Bull. Brook. Ent. Soc., 21: 21. [La.].

Distribution: La., Pa.

Pilophorus balli Knight, 1968

1968 *Pilophorus balli* Knight, Brig. Young Univ. Sci. Bull., 9(3): 176. [Col.].

Distribution: Col.

Pilophorus banksianae Knight, 1973

1973 *Pilophorus banksianae* Knight, Ia. St. J. Res., 48: 134. [Minn.].

Distribution: Ia., Minn.

Pilophorus barberi Knight, 1968

1968 *Pilophorus barberi* Knight, Brig. Young Univ. Sci. Bull., 9(3): 171. [Ariz.].

Distribution: Ariz.

Pilophorus brunneus Poppius, 1914

1886 *Pilophorus confusus*: Uhler, Check-list Hem. Het., p. 20.

1910 *Pilophorus schwarzi*: Smith, Cat. Ins. N.J., p. 162.

1914 *Pilophorus brunneus* Poppius, An. Soc. Ent. Belg., 58: 244. [D.C., Md.].

1929 *Pilophorous* [*sic*] *brunneus*: Knight and McAtee, Proc. U.S. Nat. Mus., 75: 26.

Distribution: D.C., Ia., Ill., Ind., Md., Minn., Mo., N.J., N.Y., Oh., Ont., Pa.

Pilophorus buenoi Poppius, 1914

1914 *Pilophorus buenoi* Poppius, An. Soc. Ent. Belg., 58: 243. [N.C.].

Distribution: N.C.

Pilophorus chiricahuae Knight, 1968

1968 *Pilophorus chiricahuae* Knight, Brig. Young Univ. Sci. Bull., 9(3): 172. [Ariz.].

Distribution: Ariz.

Pilophorus clavatus (Linnaeus), 1767

1767 *Cimex clavatus* Linnaeus, Syst. Nat., ed. 12, p. 729. [Europe].

1857 *Capsus clavatus*: Fitch, Trans. N.Y. Agr. Soc., 17: 742.

1869 *Capsis* [*sic*] *clavatus*: Rathvon, Hist. Lancaster Co., Pa., p. 549.

1887 *Pilophorus bifasciatus*: Provancher, Pet. Faune Ent. Can., 3: 131.

1888 *Pilophorus clavatus*: Reuter, Acta Soc. Sci. Fenn. 15(1): 290.

Distribution: Alta., B.C., Cal., Col., Conn., Ia., Ill., Man., Mass., Mich. Minn., N.D., N.S., N.Y., Oh., Ont., Que., Sask. (Europe).

Pilophorus clavicornis Poppius, 1914

1914 *Pilophorus clavicornis* Poppius, An. Soc. Ent. Belg., 58: 248. [Ariz.].

Distribution: Ariz.

Pilophorus confusus Kirschbaum, 1856

1856 *Pilophorus confusus* Kirschbaum, Jahrb. Ver. Naturk. Nassau, 10: 232. [Europe].

1982 *Pilophorus confusus*: Kelton, Can. Ent., 114: 283.

Distribution: N.S. (Europe, Siberia).

Note: Records of *P. confusus* in N. Am. (e.g., Uhler, 1886, Check-list Hem. Het., p. 20) prior to Kelton (1982, above) are based on misidentifications.

Pilophorus crassipes Heidemann, 1892

1892 *Pilophorus crassipes* Heidemann, Proc. Ent. Soc. Wash., 2: 225. [D.C.].

1918 *Pilophorus crassipes* Van Duzee, Proc. Cal. Acad. Sci., (4)8: 293 (as new species). [D.C.]. Preoccupied.

1923 *Pilophorus vanduzeei* Knight, Conn. Geol. Nat. Hist. Surv. Bull., 34: 540. [N.Y.]. Synonymized by Wheeler and Henry 1975, Trans. Am. Ent. Soc., 101: 358.

Distribution: Alta., D.C., Ill., Ind., La., Mass., Md., Minn., N.J., N.Y., Oh., Pa., W.Va.

Note: Wheeler and Henry (1975, Trans. Am. Ent. Soc., 101: 358) showed that Heidemann (1892, above) validated the name *P. crassipes*, which later was described by Knight as *P. vanduzeei*. Western records for the preoccupied *P. crassipes* Poppius should be referred to *P. americanus* Poppius [which see].

Pilophorus depictus Knight, 1923

1923 *Pilophorus depictus* Knight, Conn. Geol. Nat. Hist. Surv. Bull., 34: 539. [D.C.].

Distribution: D.C., Md.

Pilophorus desertinus Knight, 1973

1973 *Pilophorus desertinus* Knight, Ia. St. J. Res., 48: 136. [Nev.].

Distribution: Nev.

Pilophorus diffusus Knight, 1968

1968 *Pilophorus diffusus* Knight, Brig. Young Univ. Sci. Bull., 9(3): 168. [Col.].

Distribution: Col., Wyo.

Pilophorus discretus Van Duzee, 1918

1918 *Pilophorus discretus* Van Duzee, Proc. Cal. Acad. Sci., (4)8: 290. [Cal.].

Distribution: Cal.

Pilophorus dislocatus Knight, 1968

1968 *Pilophorus dislocatus* Knight, Brig. Young Univ. Sci. Bull., 9(3): 171. [Col.].

Distribution: Ariz., Col.

Pilophorus exiguus Poppius, 1914
1914 *Pilophorus exiguus* Poppius, An. Soc. Ent. Belg., 58: 246. [Ariz.].
Distribution: Ariz.

Pilophorus floridanus Knight, 1973
1973 *Pilophorus floridanus* Knight, Ia. St. J. Res., 48: 140. [Fla.].
Distribution: Fla.

Pilophorus furvus Knight, 1923
1923 *Pilophorus furvus* Knight, Conn. Geol. Nat. Hist. Surv. Bull., 34: 539. [N.J.].
Distribution: Man.(?), N.J.
Note: The Manitoba record given by Bradley and Hinks (1968, Can. Ent., 100: 43) needs verification.

Pilophorus fuscipennis Knight, 1926
1926 *Pilophorus fuscipennis* Knight, Bull. Brook. Ent. Soc., 21: 23. [Col.].
Distribution: Ariz., Col.

Pilophorus geminus Knight, 1926
1926 *Pilophorus geminus* Knight, Bull. Brook. Ent. Soc., 21: 22. [Minn.].
Distribution: Man., Minn., Wis.

Pilophorus heidemanni Poppius, 1914
1914 *Pilophorus heidemanni* Poppius, An. Soc. Ent. Belg., 58: 240. [Va., W.Va.].
Distribution: Fla., N.C., Va., W.Va.

Pilophorus hesperus Knight, 1968
1968 *Pilophorus hesperus* Knight, Brig. Young Univ. Sci. Bull., 9(3): 169. [Col.].
Distribution: Ariz., Col., Ut., Wyo.

Pilophorus hirtus Knight, 1973
1973 *Pilophorus hirtus* Knight, Ia. St. J. Res., 48: 140. [Id.].
Distribution: Id.

Pilophorus jezzardi Knight, 1968
1968 *Pilophorus jezzardi* Knight, Brig. Young Univ. Sci. Bull., 9(3): 170. [Col.].
Distribution: Col.

Pilophorus juniperi Knight, 1923
1923 *Pilophorus juniperi* Knight, Conn. Geol. Nat. Hist. Surv. Bull., 34: 543. [N.Y.].
Distribution: D.C., Ia., Ill., Mass., Md., Minn., N.C., N.J., N.Y., Pa., S.D., W.Va., Wis.
Note: Wheeler and Henry (1977, Trans. Am. Ent. Soc., 103: 644-648) described the fifth-instar nymph and gave information on seasonal history and habits on *Juniperus* spp.

Pilophorus laetus Heidemann, 1892
1892 *Pilophorus laetus* Heidemann, Proc. Ent. Soc. Wash., 2: 225. [D.C.].
1918 *Pilophorus laetus* Van Duzee, Proc. Cal. Acad. Sci., (4)8: 294. [D.C.]. Preoccupied.
Distribution: Ala., D.C., Del., Ga., Man., Mass., Md., Miss., N.J., N.Y., Tenn., Va., W.Va.
Note: Wheeler and Henry (1975, Trans. Am. Ent. Soc., 101: 359-360) discussed Heidemann's (1892, above) validation of this species.

Pilophorus longisetosus Knight, 1968
1968 *Pilophorus longisetosus* Knight, Brig. Young Univ. Sci. Bull., 9(3): 174. [Col.].
Distribution: Col.

Pilophorus merinoi Knight, 1968

1968 *Pilophorus merinoi* Knight, Brig. Young Univ. Sci. Bull., 9(3): 175. [Nev.].
Distribution: Nev.

Pilophorus microsetosus Knight, 1968

1968 *Pilophorus microsetosus* Knight, Brig. Young Univ. Sci. Bull., 9(3): 169. [Nev.].
Distribution: Ariz., Col., Nev., Wyo.

Pilophorus minutus Knight, 1973

1973 *Pilophorus minutus* Knight, Ia. St. J. Res., 48: 139. [Tex.].
Distribution: Tex.

Pilophorus nasicus Knight, 1926

1926 *Pilophorus nasicus* Knight, Bull. Brook. Ent. Soc., 21: 18. [Fla.].
Distribution: Ala., Fla., Ga., Miss., N.C.

Pilophorus nevadensis Knight, 1968

1968 *Pilophorus nevadensis* Knight, Brig. Young Univ. Sci. Bull., 9(3): 172. [Nev.].
Distribution: Nev.

Pilophorus nicholi Knight, 1973

1973 *Pilophorus nicholi* Knight, Ia. St. J. Res., 48: 139. [Ariz.].
Distribution: Ariz.

Pilophorus opacus Knight, 1926

1926 *Pilophorus opacus* Knight, Bull. Brook. Ent. Soc., 21: 24. [Col.].
Distribution: Col.

Pilophorus perplexus Douglas and Scott, 1875

1875 *Pilophorus perplexus* Douglas and Scott, Ent. Month. Mag., 12: 101. [Europe].
1918 *Pilophorus walshii*: Fulton, An. Ent. Soc. Am., 11: 93.
1923 *Pilophorus perplexus*: Knight, Conn. Geol. Nat. Hist. Surv. Bull., 34: 544.
1964 *Philophorus* [*sic*] *perplexus*: Sanford, Can. Ent., 96: 1189.
Distribution: Conn., N.S., N.Y., Oh., Ont., Que. (Europe, North Africa).
Note: Fulton (1918, above) discussed this species as a predator of aphids on apple, *Malus sylvestris* Mill.

Pilophorus piceicola Knight, 1926

1926 *Pilophorus piceicola* Knight, Bull. Brook. Ent. Soc., 21: 19. [N.Y.].
Distribution: Ia., Man., N.Y., W.Va.

Pilophorus pinicola Knight, 1973

1973 *Pilophorus pinicola* Knight, Ia. St. J. Res., 48: 138. [Minn.].
Distribution: Alta., Man., Minn., Sask.

Pilophorus salicis Knight, 1968

1968 *Pilophorus salicis* Knight, Brig. Young Univ. Sci. Bull., 9(3): 173. [Col.].
Distribution: Alta., Col., Man., Nev., Sask.

Pilophorus schwarzi Reuter, 1909

1909 *Pilophorus schwarzi* Reuter, Acta. Soc. Sci. Fenn., 36(2): 74. [Cal.].
Distribution: Cal., Col., Ore.

Pilophorus setiger Knight, 1941

1941 *Pilophorus setiger* Knight, Ill. Nat. Hist. Surv. Bull., 22: 124. [Minn.].
Distribution: Ill., Man., Minn., S.D.

Pilophorus strobicola Knight, 1926
1922 *Pilophorus amoenus*: Drake, N.Y. St. Coll. For. Tech. Publ. 16, 22: 78.
1923 *Pilophorus crassipes*: Knight, Conn. Geol. Nat. Hist. Surv. Bull., 34: 542.
1926 *Pilophorus strobicola* Knight, Bull. Brook. Ent. Soc., 21: 19. [N.Y.].
Distribution: Ga., Ia., Ill., Ind., Minn., N.C., N.H., N.Y., Oh., Pa., W.Va.

Pilophorus tanneri Knight, 1968
1968 *Pilophorus tanneri* Knight, Brig. Young Univ. Sci. Bull., 9(3): 173. [Ut.].
Distribution: Ut.

Pilophorus taxodii Knight, 1941
1941 *Pilophorus taxodii* Knight, Ill. Nat. Hist. Surv. Bull., 22: 121. [Ill.].
Distribution: Ill., Miss.

Pilophorus tibialis Van Duzee, 1918
1918 *Pilophorus tibialis* Van Duzee, Proc. Cal. Acad. Sci., (4)8: 292. [Cal.].
Distribution: B.C., Cal., Col., Ore.

Pilophorus tomentosus Van Duzee, 1918
1918 *Pilophorus tomentosus* Van Duzee, Proc. Cal. Acad. Sci., (4)8: 291. [Cal.].
Distribution: Cal.

Pilophorus uhleri Knight, 1923
1923 *Pilophorus uhleri* Knight, Conn. Geol. Nat. Hist. Surv. Bull., 34: 541. [N.Y.].
Distribution: Alta., B.C., Ia., Ill., Man., Minn., N.J., N.Y., Ont., Sask., W.Va.

Pilophorus utahensis Knight, 1968
1968 *Pilophorus utahensis* Knight, Brig. Young Univ. Sci. Bull., 9(3): 175. [Ut.].
Distribution: Col., Ut.

Pilophorus vicarius Poppius, 1914
1914 *Pilophorus vicarius* Poppius, An. Soc. Ent. Belg., 58: 245. [Ariz.].
Distribution: Ariz., Col.

Pilophorus walshii Uhler, 1887
1887 *Pilophorus walshii* Uhler, Ent. Am., 3: 30. [Ill.].
1904 *Philophorus* [*sic*] *walshii*: Wirtner, An. Carnegie Mus., 3: 199.
1917 *Pilophorus walshi* [*sic*]: Van Duzee, Univ. Cal. Publ. Ent., 2: 380.
1928 *Philophorus* [*sic*] *walshi* [*sic*]: Watson, Oh. Biol. Surv. Bull., 16: 38.
Distribution: Col., D.C., Ga., Ia., Ill., Ind., Md., Mo., N.C., N.J., N.Y., Neb., Oh., Pa., Va., W.Va., Wis.
Note: Wheeler and Henry (1976, An. Ent. Soc. Am., 69: 1095-1104) described the egg and fifth-instar nymph and gave notes on seasonal history on honeylocust, *Gleditsia triacanthos L.*

Family Nabidae Costa, 1853

The Damsel Bugs

By Thomas J. Henry and John D. Lattin

The family Nabidae is a small group of usually dull yellowish-brown bugs. Most species measure 10 millimeters or less, and are generally characterized by a 4- or 5-segmented antenna, 2-lobed pronotum, and a 4-segmented rostrum arising from the apex of a slender, somewhat elongate, head. Adults may be fully winged or brachypterous with only short pads representing wings. In some species, like the shiny black *Nabicula subcoleoptrata* Kirby, winged forms are rare. Members of the genus *Carthasis* Champion are small, slender, and delicate, with the basal third of the hemelytra constricted, giving them a superficial resemblance to some Reduviidae.

All known nabids are predatory, feeding on co-existing insects and other arthropods on trees, shrubs, and weeds, but apparently some will also feed on plant material (Stoner, 1972, Environ. Ent., 1: 557-558). A few species live on the ground, at the bases of grasses, or under loose objects. Members of the highly specialized genus *Arachnocoris* Scott live in spider webs. Although some Nabidae resemble reduviids, they lack raptorial or grasping forelegs, and do not actively grasp their prey (Arnold, 1971, Can. J. Zool., 49: 131-132). A number of common species are important predators in agroecosystems (e.g., Richman et al., 1980, Environ. Ent., 9: 315-316; Burges, 1982, Can. Ent., 114: 763-764; Sloderbeck and Yeargan, 1983, Environ. Ent., 12: 161-165).

Much of the early literature on the Nabidae is found in scattered papers by Champion, Reuter, and Stål, which is nicely summarized in the Van Duzee Catalog (1917, Univ. Cal. Publ. Ent., 2: 274-283). Harris' monograph of the North American Nabidae (1928, Ent. Am., 9: 1-98) is the only comprehensive treatment of our fauna. Although considerably outdated, the descriptions and keys make it the most important reference to consult when identifying the nearctic species. Remane (1964, Zool. Beitr.,

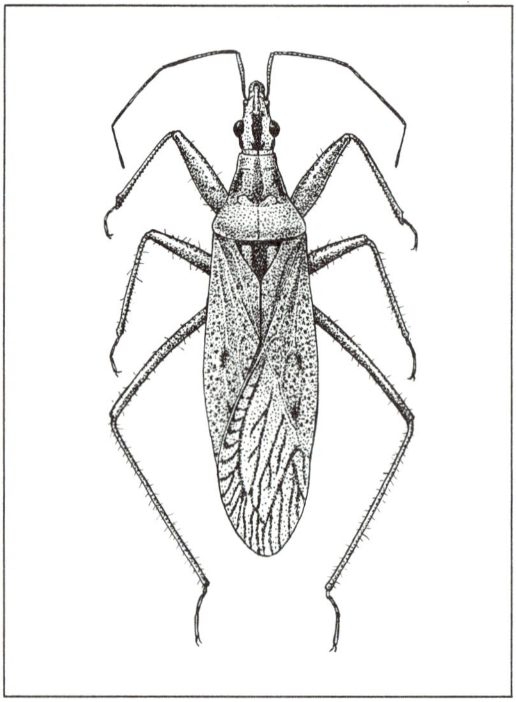

Fig. 94 *Nabis americoferus* [p. 517] (After Froeschner, 1944).

10: 253-314) reviewed the genus *Nabis* (*sensu lato*) and provided useful figures of male and female genitalia. Kerzhner's (1968, Ent. Obozr., 47: 848-863) review of the palearctic genera and (1981, Fauna U.S.S.R., Fam. Nabidae, 13: 1-324) monograph of the Nabidae of the USSR contain numerous changes in the generic and subgeneric classification of the family. Kerzhner's scheme of higher classification is followed in this catalog.

Several general works are useful in recognizing some of the genera and species of economic importance. Slater and Baranowski (1978, How To Know True Bugs, pp. 134-138) provide habitus illustrations and keys to the more common North American nabids. Hormchan et al. (1976, Tech. Bull. Miss. Agr. For. Exp. Stn., 76: 1-4) give a synoptic key to the species of agricultural importance, and Benedict and Cothran (1975, Pan-Pac. Ent., 51: 170-171) outline characters to help separate the often confused *Nabis alternatus* Parshley and *N. americoferus* Carayon. Elvin and Sloderbeck (1984, Fla. Ent., 67: 269-273) provided a key to four common species found in the southeastern United States.

Acknowledgment.--We are grateful to Dr. I. M. Kerzhner (Zoological Institute, Leningrad, U.S.S.R.) for his careful review and many useful comments on this manuscript.

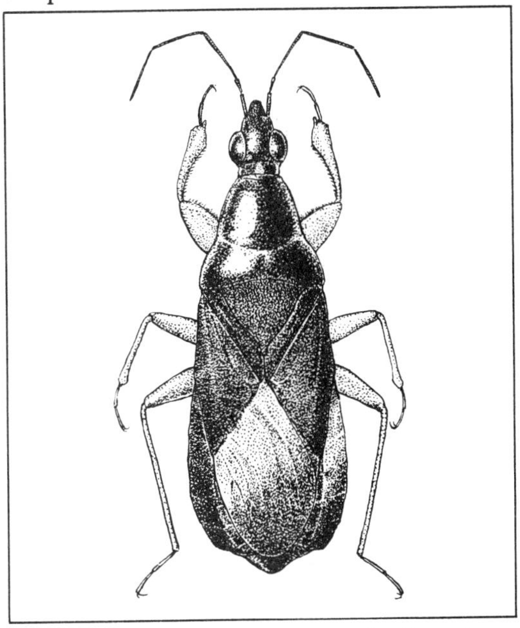

Fig. 95 *Pagasa fusca fusca* [p. 519] (After Froeschner, 1944).

Subfamily Nabinae Costa, 1853

Tribe Carthasini Blatchley, 1926

Genus *Carthasis* Champion, 1900

1900 *Carthasis* Champion, Biol. Centr.-Am., Rhyn., 2: 305. Type-species: *Carthasis rufonotatus* Champion, 1900. Monotypic.

1901 *Orthometrops* Uhler, Proc. Ent. Soc. Wash., 4: 508. Type-species: *Orthometrops decorata* Uhler, 1901. Monotypic. Synonymized by Banks, 1910, Cat. Nearc. Hem.-Het., p. 21.

Note: Blatchley's (1926, Het. E. N. Am., p. 538) placement of this genus in the family Reduviidae [as the subfamily Carthasinae] was refuted by Harris (1928, Bull. Brook. Ent. Soc., 23: 143-144 and 1928, Ent. Am., 9: 75) who returned it to Nabidae. This latter association has been used by all subsequent workers.

Carthasis decoratus (Uhler), 1901

1901 *Orthometrops decorata* Uhler, Proc. Ent. Soc. Wash., 4: 509. [Md., N.J., Pa.].

1908 *Carthasis contrarius* Reuter, Mém. Soc. Ent. Belg., 15: 97. [Md.]. Synonymized by Banks, 1910, Cat. Nearc. Hem.-Het., p. 21.

1910 *Carthasis decorata*: Banks, Cat. Nearc. Hem.-Het., p. 21.
1916 *Carthasis decoratus*: Van Duzee, Check List Hem., p. 33.
1916 *Carthasis rufo-notatus* [*sic*]: Barber, J. N.Y. Ent. Soc., 24: 308.
Distribution: D.C., Fla., La., Md., Miss., N.C., N.J., N.Y., Pa. ["It [*C. decoratus*] is now known to range throughout the eastern states." (Harris, 1928, Ent. Am., 9: 81)].
Note: Barber (1916, above) considered both *C. decoratus* and *C. contrarius* synonyms of *C. rufonotatus* Champion, but no subsequent worker has followed this opinion, including Harris in his 1928 monograph of the Nabidae.

Tribe Nabini Costa, 1853

Genus *Anaptus* Kerzhner, 1968

1968 *Anaptus* Kerzhner, Ent. Obozr., 47: 851. Type-species: *Nabis major* Costa, 1842. Original designation.

Anaptus major (Costa), 1842
1842 *Nabis major* Costa, Eserc. Accad. Aspir. Nat., Napoli, 2: 137. [Europe].
1932 *Nabis (Nabis) major*: Barber, Proc. Ent. Soc. Wash., 34: 66.
1961 *Stalia major*: Scudder, Proc. Ent. Soc. B.C., 58: 28.
1968 *Anaptus major*: Kerzhner, Ent. Obozr., 47: 851.
Distribution: B.C., Cal., N.Y., Ore., Pa., Wash. (N. Africa, Sw. Europe).
Note: Lattin (1966, Proc. Ent. Soc. Wash., 68: 314-318) summarized the North American distribution, illustrated the adult and genitalia, and suggested that this species was introduced into North America in ship ballasts. Wheeler (1976, Proc. Ent. Soc. Wash., 78: 382) summarized the N. Am. distribution.

Genus *Hoplistoscelis* Reuter, 1890

1890 *Nabis (Hoplistoscelis)* Reuter, Rev. d'Ent., 9: 295. Type-species: *Nabis sericans* Reuter, 1872. Designated by Van Duzee, 1916, Check List Hem., p. 32.
1963 *Hoplistoscelis*: Kerzhner, Acta Ent. Mus. Nat. Pragae, 35: 6.
Notes: Although Kirkaldy (1901, Wien. Ent. Zeit., 20: 223) previously designated as the type-species of this genus *N. nigriventris* Stål, which at the time was considered the senior synonym of *N. sericans* Reuter, he did not list *N. sericans* as such. Therefore, because *N. nigriventris* was not one of the originally included species of *Hoplistoscelis*, his designation is unavailable. Van Duzee's (1916, above) designation is considered valid.

Kerzhner (1963, above) raised *Hoplistoscelis* to generic status but did not list species; therefore, some combinations for species in this genus appear in print here for the first time.

Hoplistoscelis dentipes (Harris), 1928, New Combination
1872 *Nabis crassipes* Reuter, Öfv. K. Svens. Vet.-Akad. Förh., 29(6): 83. [Mexico]. Preoccupied.
1876 *Coriscus crassipes*: Uhler, Bull. U.S. Geol. Geogr. Surv. Terr., 1: 325.
1916 *Nabis (Hoplistoscelis) crassipes*: Van Duzee, Check List Hem., p. 32.
1928 *Nabis (Hoplistoscelis) dentipes* Harris, Ent. Am., 9: 43. New name for *Nabis crassipes* Reuter.
Distribution: Cal., Tex. (Mexico).
Note: Harris (1928, above, p. 44) noted that the Ga. record for this species should be referred to *Hoplistoscelis sericans* (Reuter).

Hoplistoscelis heidemanni (Reuter), 1908, New Combination
1908 *Reduviolus* (*Hoplistocelis* [*sic*]) *heidemanni* Reuter, Mém. Soc. Ent. Belg., 15: 100. [Cal.].
1916 *Nabis* (*Hoplistoscelis*) *heidemanni*: Van Duzee, Check List Hem., p. 32.
1944 *Nabis heidemanni*: Harris and Shull, Ia. St. Coll. J. Sci., 18: 206.
Distribution: Cal., Id.

Hoplistoscelis hubbelli (Hussey), 1953, New Combination
1953 *Nabis* (*Hoplistoscelis*) *hubbelli* Hussey, Occas. Pap. Mus. Zool., Univ. Mich., 550: 5. [Tenn.].
Distribution: Tenn.

Hoplistoscelis nigriventris (Stål), 1862, New Combination
1862 *Nabis nigriventris* Stål, Stett. Ent. Zeit., 23: 458. [Mexico].
1876 *Coriscus nigriventris*: Uhler, Bull. U.S. Geol. Geogr. Surv. Terr., 1: 326.
1890 *Nabis* (*Hoplistoscelis*) *nigriventris*: Reuter, Rev. d'Ent., 9: 296.
1899 *Nabis nigriventris*: Champion, Biol. Centr.-Am., Rhyn., 2: 303 (in part).
1908 *Reduviolus* (*Hoplistoscelis*) *nigriventris*: Reuter, Mém. Soc. Ent. Belg., 15: 99.
Distribution: Ariz., Col., N.M., Tex. (Mexico).

Hoplistoscelis sericans (Reuter), 1872
1872 *Nabis sericans* Reuter, Öfv. K. Svens. Vet.-Akad. Förh., 29(6): 83. [Tex.]. Synonymized with *H. nigriventris* (Stål) by Champion, 1899, Biol. Centr.-Am., Rhyn., 2: 303; resurrected by Kerzhner, 1986, J. N.Y. Ent. Soc., 94: 191.
1873 *Coriscus sericans*: Stål, K. Svens. Vet.-Akad. Handl., 11(2): 112.
1899 *Nabis nigriventris*: Champion, Biol. Centr.-Am., Rhyn., 2: 303 (in part).
1926 *Nabis crassipes*: Blatchley, Het. E. N. Am., p. 594 (in part).
1928 *Nabis* (*Hoplistoscelis*) *deceptivus* Harris, Ent. Am., 9: 34. [Tex.]. Synonymized by Kerzhner, 1986, J. N.Y. Ent. Soc., 94: 191.
1964 *Nabis deceptivus*: Dumas et al., J. Ks. Ent. Soc., 37: 198.
1976 *Hoplistocelis deceptivus*: Hormchan et al., Tech. Bull., Miss. Agr. For. Exp. Stn., 76: 2.
Distribution: Ala., Ark., Fla., Ga., Ill., Md., Miss., Mo., Tenn., Tex.
Note: Elvin and Sloderbeck (1984, Fla. Ent., 67: 270) illustrated nymphal instars 1-4 [as *H. deceptivus*].

Hoplistoscelis sordidus (Reuter), 1872
1872 *Nabis sordidus* Reuter, Öfv. K. Svens. Vet.-Akad. Förh., 29(6): 85. [Mexico].
1872 *Nabis pallescens* Reuter, Öfv. K. Svens. Vet.-Akad. Förh., 29(6): 85. [N.J., Pa., Wis.]. Synonymized by Champion, 1899, Biol. Centr.-Am., Rhyn., 2: 303.
1873 *Coriscus pallescens*: Stål, K. Svens. Vet.-Akad. Handl., 11(2): 112.
1890 *Nabis* (*Hoplistoscelis*) *pallescens*: Reuter, Rev. d'Ent., 9: 298.
1899 *Nabis* (*Hoplistoscelis*) *sordidus*: Champion, Biol. Centr.-Am., Rhyn., 2: 301.
1908 *Reduviolus* (*Hoplistoscelis*) *sordidus*: Reuter, Mém. Soc. Ent. Belg., 15: 100.
1916 *Reduviolus sordidus*: Weiss, Ent. News, 27: 10.
1974 *Hoplistocelis* [*sic*] *sordidus*: Slater, Mem. Conn. Ent. Soc., p. 157.
1976 *Hoplistoscelis sordidus*: Hormchan et al., Tech. Bull., Miss. Agr. For. Exp. Stn., 76: 2.
Distribution: D.C., Conn., Fla., Ia., Ill., Ind., La., Mass., Md., Me., Minn., Miss., Mo., N.J., N.Y., Neb., Oh., Pa., S.D., Tex., Wis. (Mexico to Panama, West Indies).

Genus *Lasiomerus* Reuter, 1890

1890 *Nabis* (*Lasiomerus*) Reuter, Rev. d'Ent., 9: 305. Type-species: *Nabis annulatus* Reuter, 1872. Original designation.

1963 *Lasiomerus*: Kerzhner, Acta Ent. Mus. Nat. Pragae, 35: 6.

Note: Kerzhner (1963, above, p. 6) raised *Lasiomerus* to generic status but did not list species; therefore, some combinations for species in this genus appear in print here for the first time.

Lasiomerus annulatus (Reuter), 1872

1872 *Nabis annulatus* Reuter, Öfv. K. Svens. Vet.-Akad. Förh., 29(6): 86. [Ill.].

1873 *Coriscus annulatus*: Stål, K. Svens. Vet.-Akad. Handl., 11(2): 112.

1890 *Nabis* (*Lasiomerus*) *annulatus*: Reuter, Rev. d'Ent., 9: 305.

1908 *Reduviolus* (*Lasiomerus*) *annulatus*: Reuter, Mém. Soc. Ent. Belg., 15: 103.

1978 *Lasiomerus annulatus*: Slater and Baranowski, How To Know True Bugs, p. 137.

Distribution: Conn., Ia., Ill., Ind., Mass., Minn., Miss., Mo., N.C., N.J., N.Y., Oh., Ont., Pa., Que., S.D.

Lasiomerus constrictus (Champion), 1899, New Combination

1899 *Nabis* (*Hoplistoscelis*) *constrictus* Champion, Biol. Centr.-Am., Rhyn., 2: 303. [Guatemala, Mexico, Panama].

1916 *Nabis constrictus*: Barber, J. N.Y. Ent. Soc., 24: 308.

1928 *Nabis* (*Lasiomerus*) *constrictus*: Harris, Ent. Am., 9: 51.

Distribution: D.C., Fla., Md., Va. (Mexico to Panama).

Lasiomerus spinicrus (Reuter), 1890, New Combination

1890 *Nabis* (*Acanthonabis*) *spinicrus* Reuter, Rev. d'Ent., 9: 305. [Brazil].

1916 *Nabis* (*Lasiomerus*) *spinicrus*: Van Duzee, Check List Hem., p. 32.

1926 *Nabis spinicrus*: Blatchley, Het. E. N. Am., p. 596.

Distribution: Fla. (Mexico to Brazil).

Note: Dr. I. M. Kerzhner (Zool. Inst., Leningrad, U.S.S.R) has informed us that the West Indian records for this species should be applied to *L. signatus* (Uhler), a species not known to occur in the U.S., and that specimens from Fla. identified as *L. spinicrus* may represent a new species.

Genus *Metatropiphorus* Reuter, 1872

1872 *Metatropiphorus* Reuter, Öfv. K. Svens. Vet.-Akad. Förh., 29(6): 93. Type-species: *Metatropiphorus belfragii* Reuter, 1872. Monotypic.

Metatropiphorus belfragii Reuter, 1872

1872 *Metatropiphorus belfragii* Reuter, Öfv. K. Svens. Vet.-Akad. Förh., 29(6): 94. [Tex.].

1876 *Metatropiphorus belfragei* [*sic*]: Uhler, Bull. U.S. Geol. Geogr. Surv. Terr., 1: 59.

Distribution: Conn., D.C., Fla., Ia., Ill., Mass., Md., Miss., Mo., N.C., N.J., N.Y., Tex., Va. (West Indies).

Genus *Nabicula* Kirby, 1837

1837 *Nabicula* Kirby, Fauna Bor.-Am., 4: 281. Type-species: *Nabicula subcoleoptrata* Kirby, 1837. Monotypic.

Note: This genus has been treated by most authors as a subgenus of *Nabis*; Kerzhner (1963, Acta Ent. Mus. Nat. Pragae, 35: 6) raised *Nabicula* to generic status and most recently recognized the following three subgenera.

Subgenus *Dolichonabis* Reuter, 1908

1908 *Reduviolus* (*Dolichonabis*) Reuter, Mém. Soc. Ent. Belg., 15: 104. Type-species: *Nabis limbatus* Dahlbom, 1851. Designated by Oshanin, 1912, Kat. Paläark. Hem., p. 55.

Note: Kerzhner (1963, Acta Ent. Mus. Nat. Pragae, 35: 6 and 1968, Ent. Obozr., 47: 852) treated *Dolichonabis* as a distinct genus, but later (Kerzhner, 1981, Fauna U.S.S.R., Fam. Nabidae, 13: 215) treated it as a subgenus.

Nabicula americolimbata (Carayon), 1961

1908 *Reduviolus* (*Dolichonabis*) *limbatus*: Reuter, Mém. Soc. Ent. Belg., 15: 107 (in part).
1908 *Reduviolus vicarius*: Van Duzee, Can. Ent., 40: 111.
1916 *Nabis* (*Dolichonabis*) *limbatus*: Van Duzee, Check List Hem., p. 33.
1917 *Nabis* (*Dolichonabis*) *limbatus*: Van Duzee, Univ. Cal. Publ. Ent., 2: 279 (in part).
1926 *Nabis limbatus*: Blatchley, Het. E. N. Am., p. 598.
1961 *Dolichonabis americolimbatus* Carayon, Bull. Mus. Nat. d'Hist. Nat., 33: 193. [N.Y.].
1981 *Nabicula* (*Dolichonabis*) *americolimbata*: Kerzhner, Fauna U.S.S.R., Fam. Nabidae, 13: 223.

Distribution: Alta., Me., Minn., N.Y., Nfld., Ont., Que. (Mongolia, USSR).

Note: This species was misidentified as the European *N. limbatus* (Dahlbom) by early workers, but Carayon (1961, above) showed that northern and eastern records of that species actually should be referred to a new species. Records of *N. limbatus* from Col. are referred to *N. nigrovittatus nearctica* Kerzhner.

Nabicula nigrovittata (Sahlberg), 1878

1878 *Nabis nigro-vittatus* [*sic*] Sahlberg, K. Svens Vet.-Akad. Handl., 16(4): 36. [Siberia].
1981 *Nabicula* (*Dolichonabis*) *nigrovittata*: Kerzhner, Fauna U.S.S.R., Fam. Nabidae, 13: 229.

Distribution: Alk., Col., Id., Man., N.Y., Ont., Que. (China, Mongolia, USSR).

Note: Kerzhner (1981, above) described three new subspecies for *N. nigrovittata*; only *N. nigrovittata nearctica* occurs in N. Am.

Nabicula nigrovittata nearctica Kerzhner, 1981

1908 *Reduviolus* (*Dolichonabis*) *limbatus*: Reuter, Mém. Soc. Ent. Belg., 15: 107 (in part).
1917 *Nabis* (*Dolichonabis*) *limbatus*: Van Duzee, Univ. Cal. Publ. Ent., 2: 279 (in part).
1928 *Nabis* (*Dolichonabis*) *nigro-vittatus* [*sic*]: Harris, Ent. Am., 9: 54 (in part).
1944 *Nabis nigrovittatus*: Harris and Shull, Ia. St. Coll. J. Sci., 18: 206.
1981 *Nabicula* (*Dolichonabis*) *nigrovittata nearctica* Kerzhner, Fauna U.S.S.R., Fam. Nabidae, 13: 234. [Alk.].

Distribution: Alk., Col., Id., Man., N.Y., Ont., Que.

Note: Early records of *N. nigrovittata* from N. Am. now are referred to this subspecies; records of *N. limbatus* from Col. also belong here.

Subgenus *Limnonabis* Kerzhner, 1968

1968 *Limnonabis* Kerzhner, Ent. Obozr., 47: 851. Type-species: *Nabis lineatus* Dahlbom, 1851. Original designation.

1981 *Nabicula* (*Limnonabis*): Kerzhner, Fauna U.S.S.R., Fam. Nabidae, 13: 204.

Nabicula propinqua (Reuter), 1872

1872 *Nabis propinquus* Reuter, Öfv. K. Svens. Vet.-Akad. Förh., 29(6): 87. [Wis.].

1872 *Nabis vicarius* Reuter, Öfv. K. Svens. Vet.-Akad. Förh., 29(6): 87. [Ill.]. Synonymized by Reuter, 1880, Öfv. F. Vet.-Soc. Förh., 22: p. 27.
1873 *Coriscus propinquus*: Stål, K. Svens. Vet.-Akad. Handl., 11(2): 113.
1873 *Coriscus vicarius*: Stål, K. Svens. Vet.-Akad. Handl., 11(2): 113.
1907 *Nabis elongatus* Hart, Bull. Ill. St. Lab. Nat. Hist., 7: 262. [Ill.]. Preoccupied. Synonymized by Van Duzee, 1916, Check List Hem., p. 33.
1908 *Reduviolus (Dolichonabis) propinquus*: Reuter, Mém. Soc. Ent. Belg., 15: 105.
1908 *Reduviolus propinquus*: Van Duzee, Can. Ent., 40: 111.
1916 *Nabis (Dolichonabis) propinquus*: Van Duzee, Check List Hem., p. 33.
1926 *Nabis propinquus* form *vicarius*: Blatchley, Het. E. N. Am., p. 598.
1966 *Dolichonabis propinquus*: Lattin, Proc. Ent. Soc. Wash., 68: 316.
1968 *Limnonabis propinquus*: Kerzhner, Ent. Obozr., 47: 851.
1981 *Nabicula (Limnonabis) propinqua*: Kerzhner, Fauna U.S.S.R., Fam. Nabidae, 13: 205.

Distribution: Alta., Ia., Ill., Mass., Md., Me., Mich., N.J., N.Y., Oh., Ont., Ore., Que., S.D., Wis.

Subgenus *Nabicula* Kirby, 1837

1837 *Nabicula* Kirby, Fauna Bor.-Am., 4: 281. Type-species: *Nabicula subcoleoptrata* Kirby, 1837. Monotypic.
1981 *Nabicula (Nabicula)*: Kerzhner, Fauna U.S.S.R., Fam. Nabidae, 13: 235.

Nabicula flavomarginata (Scholtz), 1847

1847 *Nabis flavomarginatus* Scholtz, Arb. Schles. Ges. Vat. Kultur., p. 114. [Europe].
1875 *Coriscus flavomarginatus* var. *sibiricus* Reuter, Pet. Nouv. Ent., 1: 545. [Siberia]. Synonymized by Kerzhner, 1981, Fauna U.S.S.R., Fam. Nabidae, 13: 236.
1895 *Coriscus flavomarginatus*: Heidemann, Proc. Ent. Soc. Wash., 3: 292.
1900 *Nabis flavo-marginatus* [*sic*]: Heidemann, Proc. Wash. Acad. Sci., 2: 506.
1900 *Nabis flavo-marginatus* [*sic*] var. *sibericus*: Heidemann, Proc. Wash. Acad. Sci., 2: 506.
1908 *Reduviolus (Reduviolus) flavomarginatus*: Reuter, Mém. Soc. Ent. Belg., 15: 111.
1916 *Nabis (Nabis) flavomarginatus*: Van Duzee, Check List Hem., p. 33.
1916 *Nabis (Nabis) flavomarginatus* var. *sibericus*: Van Duzee, Check List Hem., p. 33.
1921 *Nabis flavomarginatus*: Hickman, Bull. Brook. Ent. Soc., 16: 59.
1926 *Nabis flavomarginatus*: Blatchley, Het. E. N. Am., p. 600 (in part).
1968 *Nabicula flavomarginata*: Kerzhner, Ent. Obozr., 47: 852.
1981 *Nabicula (Nabicula) flavomarginata*: Kerzhner, Fauna U.S.S.R., Fam. Nabidae, 13: 236.

Distribution: Alk., Alta., N.B., N.S., Nfld., Que., (Asia, Europe, Greenland).

Nabicula subcoleoptrata Kirby, 1837

1837 *Nabicula subcoleoptrata* Kirby, Fauna Bor.-Am., 4: 282. ["New York to Cumberland House"].
1869 *Nabis canadensis* Provancher, Nat. Can., 1: 211. [No locality given; Quebec?]. Synonymized by Uhler, 1876, Bull. U.S. Geol. Geogr. Surv. Terr., 1: 325. Lectotype designated by Kelton, 1968, Nat. Can., 95: 1070.
1870 *Nabis marginatus* Riley, Second An. Rept. Nox. Ins. Mo., p. 32. *Nomen nudum*.
1872 *Nabis subcoleoptratus*: Reuter, Öfv. K. Svens. Vet.-Akad. Förh., 29(6): 81.
1873 *Coriscus subcoleoptratus*: Stål, K. Svens. Vet.-Akad. Handl., 11(2): 112.
1873 *Nabis xanthopus* Walker, Cat. Hem. Brit. Mus., 7: 143. *Nomen nudum*.

1887 *Coriscus propinquus*: Provancher, Pet. Faune Ent. Can., 3: 175.
1901 *Reduviolus subcoleoptratus*: Kirkaldy, Wien. Ent. Zeit., 20: 222.
1908 *Reduviolus (Reduviolus) subcoleoptratus*: Reuter, Mém. Soc. Ent. Belg., 15: 98.
1916 *Nabis (Nabicula) subcoleoptratus*: Van Duzee, Check List Hem., p. 32.
1981 *Nabicula (Nabicula) subcoleoptrata*: Kerzhner, Fauna U.S.S.R., Fam. Nabidae, 13: 235.
Distribution: Alta., B.C., Col., Conn., Ia., Id., Ill., Ind., Ks., Me., Mich., Minn., N.D., N.H., N.J., N.S., N.Y., Neb., Oh., Ont., Pa., Que., S.D., Tex., Wis.
Note: Harris (1928, Ent. Am., 9: 39) expressed doubt about this species occurring as far south as Tex.

Nabicula vanduzeei (Kirkaldy), 1901
1901 *Reduviolus vanduzeei* Kirkaldy, Wien. Ent. Zeit., 20: 223. [Col.].
1908 *Reduviolus (Reduviolus) flavomarginatus* var. *vanduzeei*: Reuter, Mém. Soc. Ent. Belg., 15: 111.
1916 *Nabis (Nabis) flavomarginatus* var. *vanduzeei*: Van Duzee, Check List Hem., p. 33.
1926 *Nabis vanduzeei*: Harris, Ent. News, 37: 287.
1926 *Nabis flavomarginatus*: Blatchley, Het. E. N. Am., p. 600 (in part).
1928 *Nabis (Nabis) vanduzeei*: Harris, Ent. Am., 9: 58.
1968 *Nabicula vanduzeei*: Kerzhner, Ent. Obozr., 47: 852.
1981 *Nabicula (Nabicula) vanduzeei*: Kerzhner, Fauna U.S.S.R., Fam. Nabidae, 13: 235.
Distribution: Alta., B.C., Col., Id., Mont., Ore., S.D., Wash., Wyo.

Genus *Nabis* Latreille, 1802

1802 *Nabis* Latreille, Hist. Nat., 3: 248. Type-species: *Cimex ferus* Linnaeus, 1758. Designated by Westwood, 1840, Intro. Modern Classif. Ins., Synopsis, p. 120; accepted by Int. Comm. Zool. Nomen., 1928, Opinion 104.
Note: The name *Coriscus* Schrank, 1801, used by several early authors has been applied to genera in the families Alydidae and Nabidae. China (1943, Gen. Names Brit. Hem.-Het., pp. 231-133) discussed the problems involved, including the identification of the type-species. The Int. Comm. Zool. Nomen (1950, Bull. Zool. Nomen., 4: 465-467) now has placed *Coriscus* Schrank on the *Official Index of Rejected and Invalid Names in Zoology.*

Subgenus *Nabis* Latreille, 1802

1802 *Nabis* Latreille, Hist. Nat., 3: 248. Type-species: *Cimex ferus* Linnaeus, 1758. Designated by Westwood, 1840, Intro. Modern Classif. Ins., Synopsis, p. 120.

Nabis edax Blatchley, 1929, New Subgeneric Combination
1929 *Nabis edax* Blatchley, Ent. News, 40: 75. [Cal.].
Distribution: Cal.
Note: Dr. I. M. Kerzhner (Zool. Inst., Leningrad, U.S.S.R.) has studied Blatchley's male holotype and informs us that this species belongs in the subgenus *Nabis*.

Nabis lovetti Harris, 1925
1925 *Nabis lovetti* Harris, Ent. News, 36: 205. [Cal.].
1928 *Nabis (Nabis) lovetti*: Harris, Ent. Am., 9: 59.
Distribution: B.C., Cal., Ore., Ut.(?), Wash.
Note: The record from Ut. needs verification.

Nabis roseipennis Reuter, 1872

1872 *Nabis roseipennis* Reuter, Öfv. K. Svens. Vet.-Akad. Förh., 29(6): 89. [Wis.].

1872 *Nabis punctipes* Reuter, Öfv. K. Svens. Vet.-Akad. Förh., 29(6): 89. [N.J., Wis.]. Synonymized by Reuter, 1880, Öfv. F. Vet.-Soc. Förh., 22:: 27.

1873 *Coriscus roseipennis*: Stål, K. Svens. Vet.-Akad. Handl., 11(2): 113.

1889 *Coriscus inscriptus*: Van Duzee, Can. Ent., 21: 5.

1890 *Nabis (Nabis) roseipennis*: Reuter, Rev. d'Ent., 9: 308.

1908 *Reduviolus roseipennis*: Reuter, Mém. Soc. Ent. Belg., 15: 118.

1908 *Nabis rudis* Reuter, Mém. Soc. Ent. Belg., 15: 118 (Uhler ms. name). *Nomen nudum.*

Distribution: Ala., Alta., B.C., Col., Conn., D.C., Del., Fla., Ia., Id., Ill., Ind., Ks., Mass., Me., Mich., Minn., Miss., Mo., N.C., N.J., N.S., N.Y., Neb., Oh., Ont., Pa., Que., S.D., Tenn., Va., Vt., W.Va., Wis.

Note: Mundinger (1923, Tech. Publ. No. 16 N.Y., St. Coll. For., 22: 149-160) studied the biology and described and illustrated the life stages. Elvin and Sloderbeck (1984, Fla. Ent., 67: 271) also illustrated the five nymphal instars. The combination *Nabis rudis* of Reuter (1908, above) was simply the reporting of a Uhler manuscript name.

Nabis rufusculus Reuter, 1872

1872 *Nabis rufusculus* Reuter, Öfv. K. Svens. Vet.-Akad. Förh., 29(6): 92. [Ill., N.Y.].

1873 *Coriscus rufusculus*: Stål, K. Svens. Vet.-Akad. Handl., 11(2): 113.

1878 *Coriscus assimilis* Uhler, Proc. Boston Soc. Nat. Hist., 19: 422. ["Canada," Md., Me.]. Synonymized by Reuter, 1908, Mém. Soc. Ent. Belg., 15: 119.

1901 *Reduviolus chewkeanus* Kirkaldy, Wien. Ent. Zeit., 20: 224. [N.C.]. Synonymized by Van Duzee, 1916, Check List Hem., p. 33. [Van Duzee (1917, Univ. Cal. Publ. Ent., 2: 283) noted that the spelling of this name should have been *cherokeanus*].

1908 *Reduviolus rufusculus*: Reuter, Mém. Soc. Ent. Belg., 15: 119.

1916 *Nabis (Nabis) rufusculus*: Van Duzee, Check List Hem., p. 33.

1928 *Nabis (Nabis) kalmii*: Harris, Ent. Am., 9: 64.

Distribution: Ala., Alta., B.C., Col., Conn., D.C., Ia., Id., Ill., Ind., Ks., Mass., Md., Me., Mich., Minn., Mo., N.C., N.D., N.Y., Oh., Ont., Ore., Pa., Que., Va., W.Va., Wash., Wis.

Note: Mundinger (1923, Tech. Publ. 16 N.Y. St. Coll. For., 22: 149, 160-164) studied its biology and described and illustrated the life stages. Kerzhner (1981, Fauna U.S.S.R., Fam. Nabidae, 13: 297) indicated that the *N. kalmii* record by Harris (1928, above) should be referred to *N. rufusculus.*

Subgenus *Reduviolus* Kirby, 1837

1837 *Reduviolus* Kirby, Fauna Bor.-Am., 4: 279. Type-species: *Reduviolus inscriptus* Kirby, 1837. Monotypic.

1981 *Nabis (Reduviolus)*: Kerzhner, Fauna U.S.S.R., Fam. Nabidae, 13: 297.

Nabis alternatus Parshley, 1922

1922 *Nabis alternatus* Parshley, S.D. St. Coll., Tech. Bull., 2: 12. [S.D.].

1928 *Nabis (Nabis) alternatus*: Harris, Ent. Am., 9: 66.

1935 *Nabis alternata* [*sic*]: Knowlton, Ent. News, 46: 110.

1968 *Reduviolus alternatus*: Kerzhner, Ent. Obozr., 47: 852.

1981 *Nabis (Reduviolus) alternatus*: Kerzhner, Fauna U.S.S.R., Fam. Nabidae, 13: 297.

Distribution: Alta., Ariz., Ark., B.C., Cal., Col., Fla., Ga., Ia., Id., Ind., Ks., Miss., Mo.,

Mont., N.D., N.M., N.Y., Neb., Nev., Ore., S.D., Tex., Ut., Wash., Wyo. (Mexico).

Nabis alternatus alternatus Parshley, 1922

1922 *Nabis alternatus* Parshley, S.D. St. Coll., Tech. Bull., 2: 12.

1928 *Nabis (Nabis) alternatus alternatus*: Harris, Ent. Am., 9: 67.

Distribution: Same as for species.

Nabis alternatus uniformis Harris, 1928

1928 *Nabis (Nabis) alternatus* var. *uniformis* Harris, Ent. Am., 9: 67. [Cal.].

1944 *Nabis alternatus uniformis*: Froeschner, Am. Midl. Nat., 31: 678.

Distribution: B.C., Cal., Col., Mo., N.M., Ore., S.D., Tex.

Nabis americanus Remane, 1964

1964 *Nabis americanus* Remane, Zool. Beitr., 10: 260. [Ala.].

1968 *Reduviolus americanus*: Kerzhner, Ent. Obozr., 47: 852.

1981 *Nabis (Reduviolus) americanus*: Kerzhner, Fauna U.S.S.R., Fam. Nabidae, 13: 297.

Distribution: Ala.

Nabis americoferus Carayon, 1961 [Fig. 94]

1926 *Nabis kalmii*: Blatchley, Het. E. N. Am., p. 604 (in part).

1928 *Nabis ferus*: Harris, Ent. Am., 9: 68.

1961 *Nabis americoferus* Carayon, Bull. Mus. Nat. d'Hist. Natur., 33: 190. [Mass.].

1968 *Reduviolus americoferus*: Kerzhner, Ent. Obozr., 47: 852.

1981 *Nabis (Reduviolus) americoferus*: Kerzhner, Fauna U.S.S.R., Fam. Nabidae, 13: 297.

Distribution: Alta., Ariz., B.C., Cal., Col., Conn., Ga., Ia., Id., Ill., Ind., Ks., Mass., Md., Me., Mich., Minn., Miss., Mo., Mont., N.D., N.C., N.J., N.M., N.Y., Neb., Oh., Ont., Ore., Pa., Que., S.D., Tenn., Tex., Ut., Wash., Wis. (Mexico).

Note: Carayon (1961, above) showed that all earlier nearctic records for the palearctic *N. ferus* (Linnaeus), in the combinations *Coriscus ferus*, *Nabis ferus*, and *Reduviolus ferus*, actually refer to his new species named above. Elvin and Sloderbeck (1984, Fla. Ent., 67: 272) illustrated the five nymphal instars.

Nabis inscriptus (Kirby), 1837

1837 *Reduviolus inscriptus* Kirby, Fauna Bor.-Am., 4: 280. ["Lat. 65 "; Canada].

1981 *Nabis (Reduviolus) inscriptus*: Kerzhner, Het. Fam. Nabidae, 13: 297.

Distribution: Alk., Alta., B.C., Col., Id. (Europe, USSR).

Note: Other distributions records reported for this species need verification.

Nabis kalmii Reuter, 1872

1872 *Nabis kalmii* Reuter, Öfv. K. Svens. Vet.-Akad. Förh., 29(6): 91. [Wis.].

1873 *Coriscus kalmii*: Stål, K. Svens. Vet.-Akad. Handl., 11(2): 113.

1901 *Reduviolus kalmii*: Kirkaldy, Wien. Ent. Zeit., 20: 225.

1916 *Nabis (Nabis) kalmii*: Van Duzee, Check List Hem., p. 33.

1928 *Nabis (Nabis) ferus* var. *pallidipennis* Harris, Ent. Am., 9: 69. [Ia.]. Synonymized by Kerzhner, 1981, Fauna U.S.S.R., Fam. Nabidae, 13: 297.

1964 *Nabis pallidipennis*: Remane, Zool. Beitr., 10: 275.

1968 *Reduviolus pallidipennis*: Kerzhner, Ent. Obozr., 47: 852.

1981 *Nabis (Reduviolus) kalmii*: Kerzhner, Fauna U.S.S.R., Fam. Nabidae, 13: 297.

Distribution: Ia., Minn., N.S., N.Y., Nfld., Oh., Pa., S.D., Wis.

Note: Many reported distribution records of this species are based on misidentifications. The above distribution follows Reuter (1872, above), Harris (1928, above, for *pallidipennis*), Remane (1964, above, for *pallidipennis*), and Kerzhner (1981, above).

Subgenus *Tropiconabis* Kerzhner, 1968

1968 *Tropiconabis* Kerzhner, Ent. Obozr., 47: 859. Type-species *Nabis capsiformis* Germar, 1838. Original designation.

1969 *Nabis (Tropiconabis)*: Benedek, Fauna Hung., 17: 17.

Nabis capsiformis Germar, 1838

1838 *Nabis capsiformis* Germar, Silber. Rev. Ent., 5: 132. [South Africa].
1872 *Nabis capsiformis*: Reuter, Öfv. K. Svens. Vet.-Akad. Förh., 29(6): 88.
1873 *Coriscus capsiformis*: Stål, K. Svens. Vet.-Akad. Handl., 11(2): 113.
1901 *Reduviolus capsiformis*: Kirkaldy, Wien. Ent. Zeit., 20: 223.
1912 *Nabis signatus*: Torre-Bueno, Ent. News, 23: 121.
1916 *Nabis (Nabis) capsiformis*: Van Duzee, Check List Hem., p. 33.
1968 *Tropiconabis capsiformis*: Kerzhner, Ent. Obozr., 47: 859.
1981 *Nabis (Tropiconabis) capsiformis*: Kerzhner, Fauna U.S.S.R., Fam. Nabidae, 13: 294.

Distribution: Ala., Ark., Fla., Ga., Miss., Mo., N.C., S.C., Tenn., Tex., Ut. (Africa, Asia, Europe, Mexico to South America, West Indies; "cosmopolitan throughout tropical and subtropical areas of the world").

Notes:Kerzhner (1981, Bull. Zool. Nomen., 38: 205-207) petitioned the Int. Comm. Zool. Nomen. to suppress the name *Nabis angustus* Spinola, 1837, which has priority over *N. capsiformis* Germar, 1838; the Commission (1985, Bull. Zool. Nomen., 42: 137, Opinion 1302) ruled to place *N. angustus* on the *Official Index of Rejected and Invalid Specific Names in Zoology*.

Elvin and Sloderbeck (1984, Fla. Ent., 67: 271) illustrated the five nymphal instars.

Torre-Bueno's (1912, Ent. News, 23: 121) Tex. record for *N. signatus* Uhler [previously synonymized with *L. spinicrus* Reuter by Champion (1899, Biol. Centr.-Am., Rhyn., 2: 304), but resurrected to species status by Kerzhner (1986, J. N.Y. Ent. Soc., 94: 191)] was considered a misidentification of *Nabis capsiformis* Germar by Harris (1928, above, p. 49).

Subfamily Prostemmatinae Reuter, 1890

Tribe Prostemmatini Reuter, 1890

Genus *Alloeorhynchus* Fieber, 1860

1860 *Alloeorhynchus* Fieber, Europ. Hem., 1: 43. Type-species: *Pirates flavipes* Fieber, 1836. First included species by Fieber, 1861, Europ. Hem., 2: 159.

Alloeorhynchus nigrolobus Barber, 1922

1922 *Alloeorhynchus nigrolobus* Barber, Proc. Ent. Soc. Wash., 24: 103. [Tex.].
1928 *Alloeorrhynchus* [*sic*] *nigrolobus*: Harris, Ent. Am., 9: 19.
Distribution: Ariz., N.M., Tex.

Genus *Pagasa* Stål, 1862

1862 *Pagasa* Stål, K. Svens. Vet.-Akad. Handl., 3(6): 60. Type-species: *Prostemma pallidiceps* Stål, 1860. Monotypic.

Subgenus *Lampropagasa* Reuter, 1909

1909 *Pagasa* (*Lampropagasa*) Reuter, *In* Reuter and Poppius, Acta Soc. Sci. Fenn., 37(2): 30. Type-species: *Prostemma fuscum* Stein, 1857. Original designation.

Pagasa fasciventris Harris, 1940

1940 *Pagasa* (*Lampropagasa*) *fasciventris* Harris, Ent. News, 51: 35. [Va.].

1953 *Pagasa fasciventris*: Hussey, Occas. Pap. Mus. Zool., Univ. Mich., 550: 3.

Distribution: Ks., Neb., Va.

Note: Harris (1942, Ent. News, 53: 36) described the male. Hussey (1953, above) suggested that *P. fasciventris* may belong to the subgenus *Parapagasa*.

Pagasa fusca (Stein), 1857

1857 *Prostemma fuscum* Stein, Berl. Ent. Zeit., 1: 90. [Pa.].

1873 *Pagasa nitida* Stål, K. Svens. Vet.-Akad. Handl., 11(2): 108. [Wis.]. Synonymized by Reuter, 1890, Rev. d'Ent., 9: 291.

1878 *Metastemma fusca*: Uhler, Proc. Bost. Soc. Nat. Sci., 19: 423.

1899 *Pagasa fusca*: Champion, Biol. Centr.-Am., Rhyn., 2: 299.

1909 *Pagasa* (*Lampropagasa*) *fusca*: Reuter and Poppius, Acta Soc. Sci. Fenn., 37(2): 31.

Distribution: Alta., Ariz., B.C., Cal., Col., Conn., D.C., La., Id., Ill., Ind., Ks., Man., Mass., Me., Minn., Miss., Mo., N.H., N.J., N.S., N.Y., Neb., Nfld., Ont., Pa., S.D., Tex., Ut., Wis. (Mexico to South America).

Pagasa fusca fusca (Stein), 1957 [Fig. 95]

1857 *Prostemma fuscum* Stein, Berl. Ent. Zeit., 1: 90. [Pa.].

1899 *Pagasa fusca*: Champion, Biol. Centr.-Am., Rhyn., 2: 299.

Distribution: Same as for species.

Pagasa fusca nigripes Harris, 1928

1928 *Pagasa fusca nigripes* Harris, Ent. Am., 9: 26. [Col.].

Distribution: Col.

Note: Also listed in key as *Pagasa fusca* var. *nigripes* (Harris, 1928, above, p. 21).

Subgenus *Pagasa* Stål, 1862

1862 *Pagasa* Stål, K. Svens. Vet.-Akad. Handl., 3(6): 60. Type-species: *Prostemma pallidiceps* Stål, 1860. Monotypic.

1909 *Pagasa* (*Pagasa*): Reuter and Poppius, Acta Soc. Sci. Fenn., 37(2): 26.

Pagasa pallipes Stål, 1873

1873 *Pagasa pallipes* Stål, K. Svens. Vet.-Akad. Handl., 11(2): 108. [Tex.].

1909 *Pagasa* (*Pagasa*) *pallipes*: Reuter and Poppius, Acta Soc. Sci. Fenn., 37(2): 29.

Distribution: Fla., Ks., Mich.(?), Tex., Ut. (Cuba, Panama).

Note: The Michigan record for this species reported by Hussey (1921, Psyche, 28: 10) was questioned by Harris (1928, Ent. Am., 9: 24).

Subgenus *Parapagasa* Hussey, 1953

1953 *Pagasa* (*Parapagasa*) Hussey, Occas. Pap. Mus. Zool., Univ. Mich., 550: 3. Type-species: *Pagasa insperata* Hussey, 1953. Original designation.

Pagasa insperata Hussey, 1953

1953 *Pagasa* (*Parapagasa*) *insperata* Hussey, Occas. Pap. Mus. Zool., Univ. Mich., 550: 3. [Mich.].

Distribution: Mich.

Tribe Phorticini Kerzhner, 1971

Genus *Phorticus* Stål, 1860

1860 *Phorticus* Stål, K. Svens. Vet.-Akad. Handl., 2(7): 69. Type-species: *Phorticus viduus* Stål, 1860. Designated by Distant, 1904, Fauna Brit. India, rhynch., 2(2): 395.

Phorticus collaris Stål, 1873

1873 *Phorticus collaris* Stål, K. Svens. Vet.-Akad. Handl, 11(2): 109. [Tex.].

Distribution: Tex. (Mexico).

Family Naucoridae Leach, 1815

The Creeping Water Bugs

By Dan A. Polhemus and John T. Polhemus

Creeping water bugs are distinguished by their ovate, strongly dorso-ventrally flattened bodies and highly modified raptorial forelegs. Nymphs and adults spend their lives in subsurface aquatic habitats, occurring in both lotic and lentic environments where they crawl amid the substrate or cling to aquatic plants. Eggs are glued to submerged pebbles, vegetation, and other objects. Adults and nymphs are predaceous on various aquatic insect larvae and mollusks, and can inflict a painful bite if mishandled.

The Naucoridae are a tropical group with a primarily southwestern distribution in North America. Only two species of *Pelocoris* Stål occur east of the Mississippi River, and the genera *Cryphocricos* Signoret and *Limnocoris* Stål are each represented by a single species in Texas. The genus *Ambrysus* Stål is well diversified in the western United States, and includes several Mexican species that range into southern Arizona and several others that have speciated in isolated desert hot springs

Major works on this family in North America are those of Usinger (1941, An. Ent. Soc. Am., 34: 5-16; 1946, Univ. Ks. Sci. Bull., 31: 185-210; and 1947, An. Ent. Soc. Am., 40: 329-343), La Rivers (1948, An. Ent. Soc. Am., 61: 371-376; 1951, Univ. Cal. Publ. Ent., 8: 277-338; 1971, Biol. Soc. Nev. Mem., 2: 1-120; 1974, Biol. Soc. Nev. Occas. Pap., 38: 1-17; and 1976, Biol. Soc. Nev. Occas. Pap., 41: 1-18), and J. Polhemus (1979, Bull. Cal. Ins. Surv., 21: 131-138; 1984, Aquatic Semiaquatic Hem., pp. 231-260); these provide keys to the species of our common genera, and the latter author gave an extensive overview of biology and ecology. In addition, several excellent studies exist dealing with morphology (Parsons, 1966, Trans. Royal Ent. Soc. London, 118: 119-151 and 1970, Zeit. Morph. Okol. Tiere, 66: 242-293; and Parsons and Hewson, 1975, Psyche, 81: 510-527) and ecology (Stout, 1978, Brenesia, 14-15: 1-

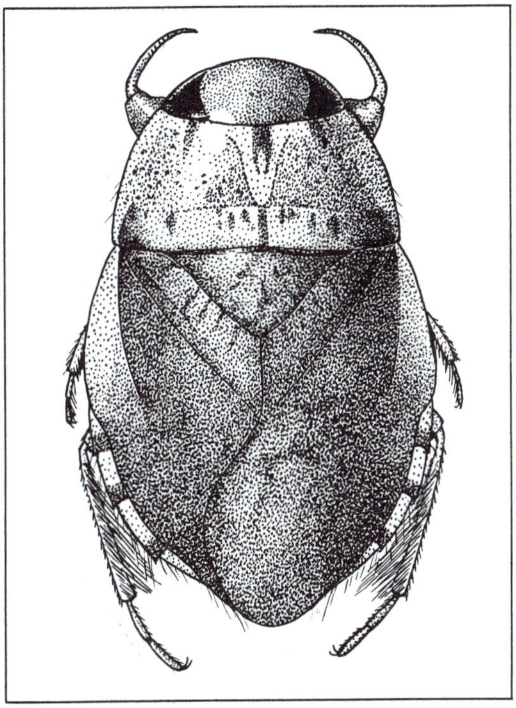

Fig. 96 *Pelocoris carolinensis* [p. 526] (After Froeschner, 1962).

11; 1981, Ecology, 62: 1170-1178).

Persistent records exist among the earlier literature regarding several Mexican *Ambrysus* species which do not occur in the United States. *Ambrysus signoreti* Stål and *A. guttatipennis* Stål have evidently been confused with a wide variety of other species occurring in our area. Records for *A. signoreti* probably refer to *A. mormon* Montandon or *A. occidentalis* La Rivers, and those for *A. guttatipennis* to *A. occidentalis* La Rivers or *A. arizonus* La Rivers. In most cases it is nearly impossible to determine which species were actually involved in the misidentifications short of first hand examination of the specimens in question, although in certain instances geographical considerations will indicate a likely choice.

Another species that has been erroneously reported from the United States is *Pelocoris poeyi* (Guérin-Méneville), which occurs in the Greater Antilles. Records of this species from our area represent misidentifications of *Pelocoris carolinensis* Torre-Bueno or *P. femoratus* (Palisot). Kirkaldy and Torre-Bueno (1909, Proc. Ent. Soc. Wash., 10: 187) inexplicably listed this species as *Ambrysus poeyi* (Amyot and Serville) (with North American records), and treated *Ambrysus signoreti* as a synonym of it; the generic combination, synonymy, and those records are incorrect.

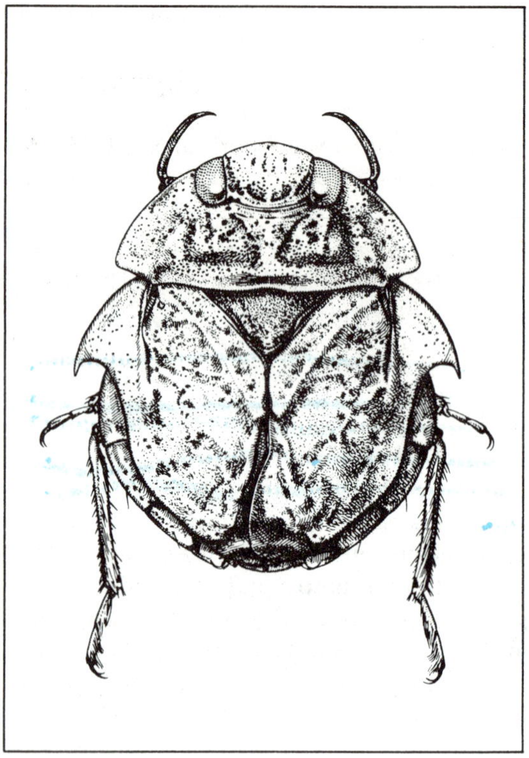

Fig. 97 *Usingerina moapensis* [p. 526] (After Usinger, 1956).

Subfamily Ambrysinae Usinger, 1941

Genus *Ambrysus* Stål, 1861

1862 *Ambrysus* Stål, Stett. Ent. Zeit., 23: 459. Type-species: *Ambrysus signoreti* Stål, 1862. Designated by Kirkaldy, 1906, Trans. Am. Ent. Soc., 32: 151.

Note: La Rivers (1951, Univ. Cal. Publ. Ent., 8: 291-292) provided a key to U.S. species, except for the subsequently described *A. amargosus* La Rivers and *A. thermarum* La Rivers.

Subgenus *Acyttarus* La Rivers, 1965

1965 *Acyttarus* La Rivers, Biol. Soc. Nev. Occas. Pap., 4: 4. Type-species: *Ambrysus funebris* La Rivers, 1949. Monotypic.

Ambrysus funebris La Rivers, 1949

1949 *Ambrysus funebris* La Rivers, Bull. So. Cal. Acad. Sci., 47: 103. [Cal.].

1965 *Ambrysus* (*Acyttarus*) *funebris*: La Rivers, Biol. Soc. Nev. Occas. Pap., 4: 4.
Distribution: Cal.
Note: Known only from isolated thermal spring outflows in the Furnace and Cow Creek drainages of Death Valley. La Rivers (1949, above; 1951, Univ. Cal. Publ. Ent., 8: 277-338) provided notes on ecology and biology.

Subgenus *Ambrysus* Stål, 1862

1862 *Ambrysus* Stål, Stett. Ent. Zeit. 23: 459. Type-species *Ambrysus signoreti* Stål, 1862. Designated by Kirkaldy, 1906, Trans. Am. Ent. Soc., 32: 151.
1965 *Ambrysus* (*Ambrysus*): La Rivers, Biol. Soc. Nev. Occas. Pap., 4: 4.

Ambrysus amargosus La Rivers, 1953
1953 *Ambrysus amargosus* La Rivers, Wasmann J. Biol., 11: 85. [Nev.].
Distribution: Nev.
Note: Known only from the thermal Point of Rocks Springs in Ash Meadows, Nev. La Rivers (1953, above) provided data on biology.

Ambrysus arizonus La Rivers, 1951
1951 *Ambrysus arizonus* La Rivers, Univ. Cal. Publ. Ent., 8: 320. [Ariz.].
Distribution: Ariz., N.M.

Ambrysus buenoi Usinger, 1946
1946 *Ambrysus buenoi* Usinger, Univ. Ks. Sci. Bull., 31: 199. [Tex.].
Distribution: Tex. (Mexico).
Note: Known in the U.S. only from the Big Bend Region of west Texas. Polhemus (1973, Great Basin Nat., 33: 113-119) provided notes on habits.

Ambrysus californicus Montandon, 1897
1897 *Ambrysus californicus* Montandon, Verh. Zool.-Bot. Ges. Wien, 47: 18. [Cal.].
Distribution: Cal.

Ambrysus californicus bohartorum Usinger, 1946
1946 *Ambrysus bohartorum* Usinger, Univ. Ks. Sci. Bull., 31: 195. [Cal.].
1951 *Ambrysus californicus bohartorum*: La Rivers, Univ. Cal. Publ. Ent., 8: 309.
Distribution: Cal.

Ambrysus californicus californicus Montandon, 1897
1897 *Ambrysus californicus* Montandon, Verh. Zool.-Bot. Ges. Wien, 47: 18. [Cal.].
1951 *Ambrysus californicus californicus*: La Rivers, Univ. Cal. Publ. Ent., 8: 306.
Distribution: Cal.

Ambrysus lunatus Usinger, 1946
1946 *Ambrysus lunatus* Usinger, Univ. Ks. Sci. Bull., 31: 203. [Tex.].
Distribution: N.M., Tex. (Mexico).

Ambrysus melanopterus Stål, 1862
1862 *Ambrysus melanopterus* Stål, Stett, Ent. Zeit., 23: 459. [Mexico].
1876 *Ambrysus melanopterus*: Uhler, Bull. U.S. Geol. Geogr. Surv. Terr, p. 337.
Distribution: Ariz. (Mexico, Guatemala).
Note: Known in the U.S. only from Pima Co. and Santa Cruz Co., Ariz., along the Mexican border. La Rivers (1951, Univ. Cal. Publ. Ent., 8: 295-297) was unable to confirm the early literature records for Tex.; these probably refer to *A. buenoi* Usinger or *A. puncticollis* Stål.

Ambrysus mormon Montandon, 1909
1909 *Ambrysus mormon* Montandon, Bull. Soc. Sci. Buc.-Roum., 18: 48. [Ut.].
Distribution: Ariz., Cal., Col., Id., N.M., Neb., Nev., Ore., S.D., Ut. (Mexico).
Note: Usinger (1946, Univ. Ks. Sci. Bull., 31: 185-210) provided notes on life history and illustrations of the nymphal instars.

Ambrysus mormon heidemanni Montandon, 1910
1910 *Ambrysus heidemanni* Montandon, Bull. Soc. Sci. Buc.-Roum., 18: 188. [Wyo.].
1951 *Ambrysus mormon heidemanni*: La Rivers, Univ. Cal., Publ. Ent., 8: 316.
Distribution: Wyo.
Note: Known only from the thermally heated waters of the Firehole River System in Yellowstone National Park.

Ambrysus mormon minor La Rivers, 1963
1963 *Ambrysus mormon minor* La Rivers, Biol. Soc. Nev. Occas. Pap., 1: 6. [Id.].
Distribution: Id.

Ambrysus mormon mormon Montandon, 1909
1909 *Ambrysus mormon* Montandon, Bull. Soc. Sci. Buc.-Roum., 18: 48. [Ut.].
1951 *Ambrysus mormon mormon*: La Rivers, Univ. Cal. Publ. Ent., 8: 313.
Distribution: Same as for species.

Ambrysus occidentalis La Rivers, 1951
1951 *Ambrysus occidentalis* La Rivers, Univ. Cal. Publ. Ent., 8: 322. [Ariz.].
Distribution: Ariz., Cal. (Mexico).
Note: Constantz (1974, Am. Midl. Nat., 92: 234-239) provided notes on mating behavior.

Ambrysus pudicus Stål, 1862
1862 *Ambrysus pudicus* Stål, Stett. Ent. Zeit., 23: 460. [Mexico].
Distribution: Tex. (Mexico, Guatemala).
Note: The nominate subspecies is not known from the U.S.

Ambrysus pudicus barberi Usinger, 1946
1946 *Ambrysus barberi* Usinger, Univ. Ks. Sci. Bull., 31: 189. [Mexico].
1953 *Ambrysus pudicus barberi*: La Rivers, Univ. Ks. Sci. Bull., 35: 1298.
Distribution: Tex. (Mexico).
Note: Known in the U.S. only from the vicinity of McAllen, Texas.

Ambrysus pulchellus Montandon, 1897
1897 *Ambrysus pulchellus* Montandon, Verh. Zool.-Bot. Ges. Wien, 47: 16. [Guatemala].
Distribution: Ariz., Tex. (Mexico, Guatemala).

Ambrysus pulchellus pallidulus Montandon, 1910
1910 *Ambrysus pulchellus* var. *pallidulus* Montandon, Bull. Soc. Sci. Buc.-Roum. Bull., 18: 189. [Tex.].
1946 *Ambrysus pallidulus*: Usinger, Univ. Ks. Sci. Bull., 31: 187. Synonymized by La Rivers, 1951, Univ. Cal. Publ. Ent., 8: 306.
1971 *Ambrysus pulchellus pallidulus*: La Rivers, Biol. Soc. Nev. Mem., 2: 68. Subspecies status reinstated.
Distribution: Tex.

Ambrysus pulchellus pulchellus Montandon, 1897
1897 *Ambrysus pulchellus* Montandon, Verh. Zool.-Bot. Ges. Wien, 47: 16. [Guatemala].
1971 *Ambrysus pulchellus pulchellus*: La Rivers, Biol. Soc. Nev. Mem., 2: 68.
Distribution: Same as for species.

Ambrysus puncticollis Stål, 1876
1876 *Ambrysus puncticollis* Stål, K. Svens. Vet.-Akad. Handl., 14(4): 143. [Tex.].
Distribution: Ariz., Cal., Tex. (Mexico).

Ambrysus thermarum La Rivers, 1953
1953 *Ambrysus thermarum* La Rivers, Proc. U.S. Nat. Mus., 103: 1. [N.M.].
Distribution: Ariz., Col., N.M.
Note: La Rivers (1953, above, p. 4) provided a modification of his (1951, Univ. Cal. Publ. Ent., 8: 291-292) key to accomodate *A. thermarum*. Polhemus (1973, Great Basin Nat., 33: 113-119) provided notes on habitat.

Ambrysus woodburyi Usinger, 1946
1946 *Ambrysus woodburyi* Usinger, Univ. Ks. Sci. Bull. 31: 194. [Ut.].
Distribution: Ariz., Nev., Tex., Ut.
Note: Records for this species from N.M. [Jemez Mts.] refer to the subsequently described *A. thermarum* La Rivers.

Subgenus *Syncollis* La Rivers, 1965

1965 *Syncollis* La Rivers, Biol. Soc. Nev. Occas. Pap., 4: 5. Type-species: *Ambrysus circumcinctus* Montandon, 1910. Original designation.

Ambrysus circumcinctus Montandon, 1910
1910 *Ambrysus circumcinctus* Montandon, Bull. Soc. Sci. Buc.-Roum., 19: 442. [Tex.].
Distribution: Ariz., N.M., Tex. (Mexico).

Ambrysus circumcinctus circumcinctus Montandon, 1910
1910 *Ambrysus circumcinctus* Montandon, Bull. Soc. Sci. Buc.-Roum., 19: 442. [Tex.].
1953 *Ambrysus circumcinctus circumcinctus*: La Rivers, Univ. Ks. Sci. Bull., 35: 1292.
Distribution: Same as for species.
Note: Three other subspecies occur from Mexico to Costa Rica (La Rivers, 1971, Biol. Soc. Nev. Mem., 2: 65).

Subfamily Cryphocricinae Montandon, 1897

Genus *Cryphocricos* Signoret, 1850

1850 *Cryphocricos* Signoret, Rev. Mag. Zool., 2(2): 290. Type-species: *Cryphocricos barozzi* Signoret, 1850. Monotypic.

Note: Usinger (1947, An. Ent. Soc. Am., 40: 329-343) reviewed nomenclatural problems in this genus and provided a key to species. The spellings *Cryphocricus* (Mayr, 1868, Reise Osterr. Freg. Nov. Jahr., p. 182) and *Cryptocricus* (Stål, 1876, K. Svens. Vet.-Akad. Handl., 14(4): 143) are incorrect emendations of the original name.

Cryphocricos hungerfordi Usinger, 1947
1947 *Cryphocricos hungerfordi* Usinger, An. Ent. Soc. Am., 40: 337. [Mexico].
Distribution: Tex. (Mexico).
Note: Known in the U.S. only from the Nueces, Frio, and Pecos Rivers in the vicinity of the Balcones Fault Zone.

Subfamily Limnocorinae Stål, 1876

Genus *Limnocoris* Stål, 1860

1860 *Limnocoris* Stål, K. Svens. Vet.-Akad. Handl., 2(7): 83. Type-species: *Limnocoris insignis* Stål, 1860. Monotypic.

Limnocoris lutzi La Rivers, 1957

1957 *Limnocoris lutzi* La Rivers, Pan-Pac. Ent., 33: 71. [Tex.].

Distribution: Tex. (Mexico).

Note: Known in the U.S. only from the Guadalupe and Pecos Rivers.

Genus *Usingerina* La Rivers, 1950

1950 *Usingerina* La Rivers, An. Ent. Soc. Am., 43: 368. Type-species: *Usingerina moapensis* La Rivers, 1950. Monotypic.

Usingerina moapensis La Rivers, 1950 [Fig. 97]

1950 *Usingerina moapensis* La Rivers, An. Ent. Soc. Am., 43: 368. [Nev.].

Distribution: Nev.

Note: Known only from the thermal Moapa Warm Springs. La Rivers (1950, above) provided data on biology and ecology.

Subfamily Naucorinae Stål, 1876

Genus *Pelocoris* Stål, 1876

1876 *Pelocoris* Stål, K. Svens. Vet.-Acad. Handl., 14(4): 144. Type-species: *Naucoris femorata* Palisot, 1820. Designated by Kirkaldy, 1906, Trans. Am. Ent. Soc., 32: 150.

Note: La Rivers (1948, An. Ent. Soc. Am., 61: 374) provided a key to U.S. species.

Pelocoris biimpressus Montandon, 1898

1898 *Pelocoris femorata* var. *biimpressus* Montandon, Bull. Soc. Sci. Buc.-Roum., 7: 285. [Guatemala, Mexico, and Uruguay].

1916 *Pelocoris biimpressus*: Van Duzee, Check List Hem., p. 52.

Distribution: Tex. (Mexico to Uruguay)

Note: This combination was based on a manuscript name by Stål, but was given species status by Van Duzee (1916, above and 1917, Univ. Cal. Publ. Ent., 2: 457) and La Rivers (1971, Biol. Soc. Nev. Mem., 2: 70).

Pelocoris carolinensis Torre-Bueno, 1907 [Fig. 96]

1907 *Pelocoris carolinensis* Torre-Bueno, Can. Ent., 39: 227. [N.C.].

Distribution: Fla., Miss., N.C., S.C.

Note: Hungerford's (1927, Bull. Brook. Ent. Soc., 22: 77-83) life history study of *P. carolinensis* has been referred to *P. femoratus* (Palisot).

Pelocoris femoratus (Palisot), 1820

1820 *Naucoris femorata* Palisot, Ins. Afr. Am., 14: 237. [U.S.].

1876 *Pelocoris femoratus*: Stål, K. Svens. Vet.-Acad. Handl., 14(4): 144.

1876 *Naucoris poeyi*: Uhler, Bull. U.S. Geol. Geogr. Surv. Terr., 1: 71.
1927 *Pelocoris carolinensis*: Hungerford, Bull. Brook. Ent. Soc., 22: 77.
Distribution: "Carolina," Conn., D.C., Fla., Ill., Ind., Ky., Ks., La., Mass., Mich., Minn., Miss., N.J., N.Y., Oh., Ont., Pa., R.I., S.D., Tenn., Wis., "Canada," (Mexico to Uruguay).
Note: Torre-Bueno (1903, J. N.Y. Ent. Soc., 11: 166-173) and Hungerford (1927, Bull. Brook. Ent. Soc., 22: 77-83 [as *P. carolinensis*]) provided life history information.

Pelocoris femoratus balius La Rivers, 1970
1970 *Pelocoris femoratus balius* La Rivers, Biol. Soc. Nev. Occas. Pap., 26: 1. [Fla.].
Distribution: Fla.

Pelocoris femoratus femoratus (Palisot), 1820
1820 *Naucoris femorata* Palisot, Ins. Afr. Am., 14: 237. [U.S.].
1970 *Pelocoris femoratus femoratus*: La Rivers, Biol. Soc. Nev. Occas. Pap., 26: 1.
Distribution: Same as for species.

Pelocoris shoshone La Rivers, 1948
1948 *Pelocoris shoshone* La Rivers, An. Ent. Soc. Am., 61: 371. [Nev.].
Distribution: Ariz., Cal., Nev. (Mexico).
Note: La Rivers (1948, above) discussed relations to other species of *Pelocoris* and gave notes on biology.

Pelocoris shoshone amargosus La Rivers, 1956
1956 *Pelocoris shoshone amargosus* La Rivers, Wasmann J. Biol., 14: 155. [Cal.].
Distribution: Ariz., Cal., Nev. (Mexico).

Pelocoris shoshone shoshone La Rivers, 1948
1948 *Pelocoris shoshone* La Rivers, An. Ent. Soc. Am., 61: 371. [Nev.].
1956 *Pelocoris shoshone shoshone*: La Rivers, Wasmann J. Biol., 14: 155.
Distribution: Nev.

Family Nepidae Latreille, 1802

The Water Scorpions

By Dan A. Polhemus

Water scorpions derive their common name from their elongate form, raptorial forelegs, and long posterior respiratory siphon. Species occur in both lotic and lentic habitats, preferring areas of still water where their brown, leaf- or sticklike bodies and legs render them extremely cryptic as they hang below the surface among tangles of submerged twigs, weeds, and debris that form their favored habitat.

The eggs of water scorpions are deposited either in floating vegetation (*Ranatra* Fabricius) or in mud near shore (*Nepa* Linnaeus, *Curicta* Stål) and bear respiratory horns that protrude into the open air; these egg structures were studied in detail by Hinton (1961, J. Ins. Phys., 7: 224-257; 1962, Proc. Roy. Ent. Soc. Lond., 37: 65-68). Total developmental time averages about two months. Nymphs and adults prey on mosquito larvae, tadpoles, and other aquatic organisms. Torre-Bueno (1903, J. N.Y. Ent. Soc., 11: 167) noted that these insects are sluggish and very susceptible to water mite attack. Adults are known to fly, and will feign death if captured and handled (Larsen, 1949, Lund. Univ. Ars., 45: 1-82).

Three of the twelve recognized genera, *Ranatra*, *Nepa*, and *Curicta*, occur in North America. *Ranatra* occurs worldwide, with a transcontinental range in the United States; *Nepa* is confined to the Northern Hemisphere with but a single species in the eastern U.S.; and *Curicta* is basically a neotropical genus that ranges through Central America and Mexico into the southern and southwestern states where two species occur. The major recent works on this family are one for North America by Hungerford (1922, Ks. Univ. Sci. Bull., 14: 425-469), which provides a key to our species of *Ranatra*, and a revision of the group

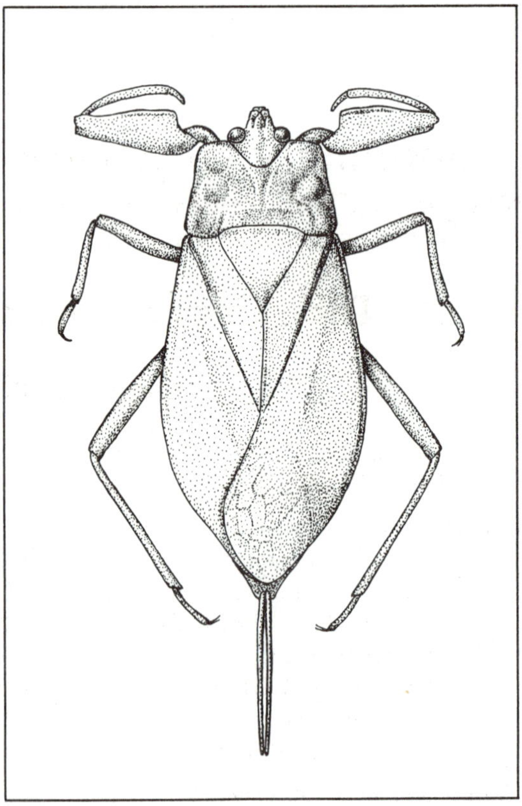

Fig. 98 *Nepa apiculata* [p. 530] (After Froeschner, 1962).

for the Americas by De Carlo, who treated each subfamily as a family: Nepidae (1951, Rev. Inst. Nac. Inv. Ci. Nac., Ci. Zool., 1: 385-421) and Ranatridae (1964, Rev. Mus. Arg. Ci. Nac., "Bernardino Rivadavia," Ent., 1: 133-215, pls. 1-6).

Van Duzee's (1917, Univ. Cal. Publ. Ent., 2: 463) listing of *Ranatra annulipes* Stål from Texas is incorrect since this species is known only from Brazil, Mesoamerica, and the West Indies. The manuscript name *R. grisea* Torre-Bueno cited in the same publication was used previously by Smith (1910, Cat. Ins. N.J., edit. 3, p. 158). Polhemus (1976, Pan-Pac. Ent., 53: 206) explained that the "Key West, Florida" specimen used by Kuitert as the holotype of his new species *Ranatra spatulata* (1949, J. Ks. Ent. Soc., 22: 32) was obviously a mislabeled African specimen and deleted that species name from our list.

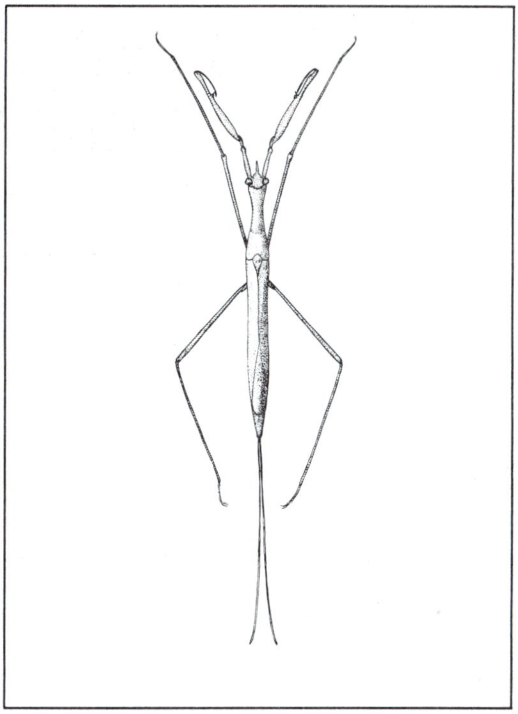

Fig. 99 *Ranatra fusca* [p. 531] (After Froeschner, 1962).

Subfamily Nepinae Douglas and Scott, 1865

Note: De Carlo (1951, Rev. Inst. Nac. Inv. Ci. Nac., Ci. Zool., 1: 385-421) revised this subfamily as the family Nepidae.

Tribe Curictini Menke and Stange, 1964

Genus *Curicta* Stål, 1861

1861 *Curicta* Stål, Öfv. K. Svens. Vet.-Akad. Förh., 18: 202. Type-species: *Curicta scorpio* Stål, 1861. Monotypic.

Curicta howardi Montandon, 1910

1910 *Curicta howardi* Montandon, Bull. Soc. Sci. Buc.-Roum., 18: 181. [Tex.].

1922 *Curicta drakei* Hungerford, Ks. Univ. Sci. Bull., 14: 432. [Tex.]. Synonymized by Kuitert, 1949, J. Ks. Ent. Soc., 22: 68.

Distribution: La., Tex. (Mexico).

Note: Wiley (1924, Ent. News, 35: 324-331) provided notes on life history and illustrations of the nymphal and adult stages, treating the species under the name *C. drakei* Hungerford.

Curicta pronotata Kuitert, 1949

1922 *Curicta howardii* [*sic*]: Hungerford, Ks. Univ. Sci. Bull., 14: 430.

1949 *Curicta pronotata* Kuitert, J. Ks. Ent. Soc., 22: 66. [Ariz.].

Distribution: Ariz. (Mexico).

Note: Known in the U.S. only from the Huachuca Mountains of southern Arizona.

Tribe Nepini Douglas and Scott, 1865

Genus *Nepa* Linnaeus, 1758

1758 *Nepa* Linnaeus, Syst. Nat., 10th ed., p. 440. Type-species: *Nepa cinerea* Linnaeus, 1758. Designated by Latreille, 1810, Consid. Gén., pp. 261, 434.

Nepa apiculata Uhler, 1862 [Fig. 98]

1862 *Nepa apiculata* Uhler, Rep. Ins. Inj. Veg., 3rd ed., p. 12. [E. N. Am.].

1888 *Nepa cinerea*: Ferrari, An. Nat. Hofmus., 3: 189 (in part).

Distribution: Conn., D.C., Ga., Ia., Ill., Ind., Ks., Man., Mass., Md., Mich., Minn., Mo., N.H., N.J., N.Y., Oh., Ont., Pa., Que., R.I., Wis. (Mexico).

Note: The original descrption of this species consists of figure 1 on plate I and a bracketed "[Nepa apiculata]" on p. 12, which in C. L. Flint's "Editor's Preface" is stated to be an addition to former editions, and is credited to Uhler. Uhler (1878, Proc. Bost. Soc. Nat. Hist., 19: 440) wrote that this species name was from a Thomas Say label on a specimen in the Harris collection and then (1884, Riverside Nat. Hist., 2: 253) provided a more formal description. Severin and Severin (1911, J. N.Y. Ent. Soc., 19: 99-108) provided notes on ecology.

Subfamily Ranatrinae Douglas and Scott, 1865

Note: De Carlo (1964, Rev. Mus. Arg. Ci. Nac., "Bernardino Rivadavia," Ent., 1: 133-215, pls. 1-6) revised this subfamily as the family Ranatridae.

Genus *Ranatra* Fabricius, 1790

1790 *Ranatra* Fabricius, Skriv. Nat., 1: 227. Type-species: *Nepa linearis* Linnaeus, 1758. Designated by Latreille, 1810, Consid. Gén., pp. 261, 434.

Note: Hungerford (1922, Ks. Univ. Sci. Bull., 14: 425-469) and De Carlo (1964, above under subfamily) gave keys to the genus. Neiswander (1925, Trans. Am. Ent. Soc., 51: 311-320) studied the morphology of certain species.

Ranatra australis Hungerford, 1922

1922 *Ranatra australis* Hungerford, Ks. Univ. Sci. Bull., 14: 449. [Tex.].

Distribution: Ala., Fla., Ga., Ks., La., Miss., N.C., Ok., S.C., Tex., Va.

Ranatra brevicollis Montandon, 1910

1910 *Ranatra brevicollis* Montandon, Bull. Soc. Sci. Buc.-Roum., 18: 184. [Cal.].

Distribution: Cal. (Mexico).

Note: Menke (1979, Bull. Cal. Ins. Surv., 21: 73) provided notes on range and habitat. A record for this species in B.C. is in error and refers to Baja California, Mexico.

Ranatra buenoi Hungerford, 1922

1905 *Ranatra fusca*: Torre-Bueno, Can. Ent., 37: 188.

1922 *Ranatra buenoi* Hungerford, Ks. Univ. Sci. Bull., 14: 442. [Tex.].

Distribution: Fla., Ga., La., Miss., Mo., N.C., S.C., Tex., Va.

Ranatra drakei Hungerford, 1922
1922 *Ranatra drakei* Hungerford, Ks. Univ. Sci. Bull., 14: 451. [Fla.].
Distribution: Fla., Ga., Miss., N.C.

Ranatra fusca Palisot, 1820 [Fig. 99]
1820 *Ranatra fusca* Palisot, Ins. Afr. Am., 14: 235. [U.S.].
1872 *Nepa 4-dentata* [*sic*]: Uhler, Prelim. Rept. U.S. Geol. Surv. Wyo., p. 471.
1876 *Ranatra quadridentata*: Uhler, Bull. U.S. Geol. Geogr. Surv. Terr., 1: 338 (in part).
1910 *Ranatra americana* Montandon, Bull. Soc. Sci. Buc.-Roum., 19: 65. [Fla., Mass., N.Y., Pa.]. Synonymized by Hungerford, 1922, Ks. Univ. Sci. Bull., 14: 436.
Distribution: Ark., B.C., Cal., Col., Conn., Fla., Ia., Ind., Ks., Me., Man., Mass., Md., Mich., Minn., N.B., N.C., N.D., N.J., N.S., Oh., Ont., Que., Ore., Tex., Wis.
Note: As pointed out by Torre-Bueno (1922, Bull. Brook. Ent. Soc., 17: 121), the combination *Ranatra quadrinotata* listed in synonymy under *R. americana* by Van Duzee (1917, Univ. Cal. Publ. Ent., 2: 463) was a *lapsus calami* for *Ranatra quadridentata;* all references under *R. quadrinotata* should be referred to *R. fusca*. Torre-Bueno (1906, Can. Ent., 38: 242-252) and Radinovsky (1964, Nat. Hist., 63: 16-25) provided information on ecology and life history.

Ranatra fusca edentula (Montandon), 1910
1910 *Ranatra americana edentula* Montandon, Bull. Soc. Sci. Buc.-Roum., 19: 66. [Pa., Tex.].
1922 *Ranatra fusca edentula*: Hungerford, Ks. Univ. Sci. Bull., 14: 440.
Distribution: La., Miss., Pa., Tex.

Ranatra fusca fusca Palisot, 1820
1820 *Ranatra fusca* Palisot, Ins. Afr. Am., 14: 235. [U.S.].
Distribution: Same as for species.

Ranatra kirkaldyi Torre-Bueno, 1905
1905 *Ranatra kirkaldyi* Torre-Bueno, Can. Ent., 37: 187. [Ill., N.Y.].
1910 *Ranatra fusca*: Montandon, Bull. Soc. Sci. Buc.-Roum., 19: 63.
Distribution: Fla., Ga., Ks., Ill., Ind., Mass., Mich., Minn., Miss., N.J., N.Y., N.C., Neb., Oh., Ont., S.C., S.D., Va., Wis.

Ranatra kirkaldyi hoffmanni Hungerford, 1922
1922 *Ranatra kirkaldyi hoffmanni* Hungerford, Ks. Univ. Sci. Bull., 14: 442. [Minn.].
Distribution: Minn.

Ranatra kirkaldyi kirkaldyi Torre-Bueno, 1905
1905 *Ranatra kirkaldyi* Torre-Bueno, Can. Ent., 37: 187. [Ill., N.Y.].
Distribution: Same as for species.

Ranatra montezuma Polhemus, 1976
1976 *Ranatra montezuma* Polhemus, Pan-Pac. Ent., 52: 204. [Ariz.].
Distribution: Ariz.
Note: Known only from a single insular population at Montezuma Well, a limestone sink in Arizona. Blinn et al. (1982, J. Ks. Ent. Soc., 55: 481-484) provided observations on ecology.

Ranatra nigra Herrich-Schaeffer, 1849
1849 *Ranatra nigra* Herrich-Schaeffer, Wanz. Ins., 9: 32. [N. Am.].
1876 *Ranatra fusca*: Uhler, Bull. U.S. Geol. Geogr. Surv. Terr., 1: 338.
1910 *Ranatra protensa* Montandon, Bull. Soc. Sci. Buc.-Roum., 18: 185. [N.Y.]. Synonymized by Hungerford, 1922, Ks. Univ. Sci. Bull., 14: 436.

Distribution: Ala., Ark., Fla., Ga., Ind., Ks., La., Md., Mass., Me., Minn., Miss., N.C., Oh., Ont., Pa., S.C., Tenn., Tex., Va., W.Va.

Ranatra quadridentata Stål, 1862

1862 *Ranatra quadridentata* Stål, Öfv. K. Svens. Vet.-Akad. Förh., 18: 204. [Mexico]. Lectotype designated by Menke, 1964, Bull. Brook. Ent. Soc., 58: 112.

1906 *Ranatra quadridentata*: Snow, Trans. Ks. Acad. Sci., 20: 180.

1917 *Ranatra americana*: Van Duzee, Univ. Cal. Publ Ent., 2: 463.

Distribution: Ariz., Cal., Tex. (Mexico).

Note: Uhler (1876, Bull. U.S. Geol. Geogr. Surv. Terr., 1: 338) and Van Duzee (1917, above [as *americana*]) listed *quadridentata* from many states and, while most of these records are based on misidentifications of other species, those from Arizona probably are correct but this has not been verified. The Snow (1906, above) record is probably the first valid U.S. record of this species; Hungerford (1922, Ks. Univ. Sci. Bull., 14: 452) verified the specimens (in the Univ. Ks. Ins. Colln.) as *quadridentata*; Menke (1979, Bull. Cal. Ins. Surv., 21: 75) questioned the Arizona and California records.

Ranatra texana Hungerford, 1930

1930 *Ranatra texana* Hungerford, Can. Ent., 62: 217. [Tex.].

Distribution: Tex. (Mexico).

Family Notonectidae Latreille, 1802

The Backswimmers

By John T. Polhemus and Dan A. Polhemus

Backswimmers are slender, elongate aquatic bugs, flattened ventrally and strongly convex dorsally, with large eyes and long oarlike hind legs modified for rowing. As their common name implies, they swim upside down. Notonectidae prefer relatively still water habitats, such as lakes, ponds, and pooled sections of streams, where they prey on various aquatic arthropods and small vertebrates. Swimming behavior varies among our three genera: *Notonecta* Linnaeus and *Buenoa* Kirkaldy species frequent quiet waters, the former hanging just beneath the surface film and the latter hovering in midwater zones, while *Martarega* White occur in areas of gentle current, where they school and hold position by rowing. Several species of a given genus may occur together, avoiding competition through habitat partitioning (Streams and Newfield, 1972, Univ. Conn. Occas. Pap., Biol. Sci. Ser., 2: 139-157). Eggs are laid in aquatic plants or on submerged debris, with development

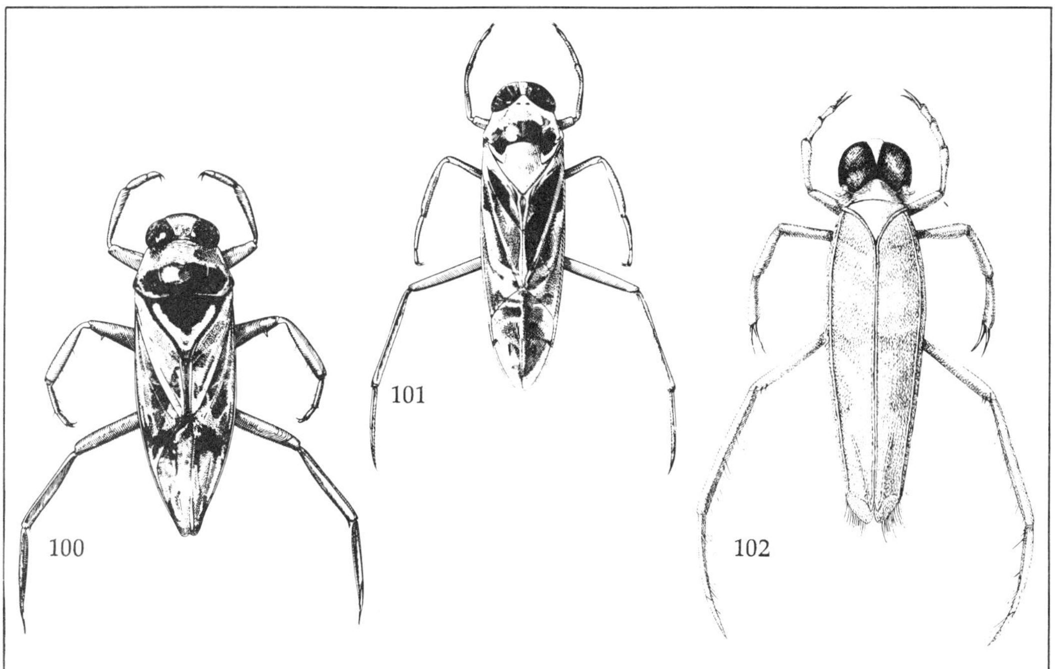

Figs. 100-102: 100, *Notonecta unifasciata* [p. 540] (After Usinger, 1956); 101, *Buenoa scimitra* [p. 535] (After Usinger, 1956); 102, *Martarega mexicana* [p.536] (After Truxal, 1979).

time being highly variable depending on the species and geographic locality but typically averaging 50 to 70 days (Rice, 1954, Am. Midl. Nat., 51: 105-132). In *Buenoa*, adults stridulate during courtship, the songs being species specific and apparently serving as premating reproductive isolating mechanisms (Hungerford, 1924, An. Ent. Soc. Am., 17: 223-226; Wilcox, 1975, Int. J. Ins. Morph. Embryo., 4: 169-182). Individuals of *Buenoa* contain haemoglobin in their tissues. This substance functions as a mechanism which allows them to maintain neutral buoyancy and remain stationary at any given depth (Miller, 1964, Proc. Roy. Ent. Soc. London, (A)39: 166-175; 1966, J. Exp. Biol., 44: 529-543). Backswimmers fly readily and often disperse into swimming pools where they can become an annoyance by inflicting a sharp, painful bite.

Notonectids are a cosmopolitan group found virtually throughout North America. The major works for the group in our region are generic revisions by Hungerford and Truxal listed under each of the three genera. A key to nymphs of West Coast species was provided by Voight and Garcia (1976, Pan-Pac. Ent., 52: 172-176). Several morphological studies provided by Parsons (1970, Zeit. Morph. Okol. Tiere, 66: 242-298; 1971, J. Morph., 133: 125-138; 1972, J. Morph., 138: 141-167) deal primarily with thoracic structures. An excellent review of biology was presented by Sanderson (1982, Aquatic Semiaquatic Het., pp. 642-648).

Subfamily Anisopinae Hutchinson, 1929

Genus *Buenoa* Kirkaldy, 1904

1904 *Buenoa* Kirkaldy, Wien. Ent. Zeit., 23: 120. Type-species: *Buenoa antigone* Kirkaldy, 1904. Original designation.

Note: Truxal (1953, Univ. Ks. Sci. Bull., 35: 1351-1523) revised the genus and provided a key to species.

Buenoa arida Truxal, 1953

1953 *Buenoa arida* Truxal, Univ. Ks. Sci. Bull., 35: 1435. [Ariz.].

Distribution: Ariz.

Buenoa arizonis Bare, 1928

1928 *Buenoa arizonis* Bare, Univ. Ks. Sci. Bull., 18: 342, pl. 54. [Ariz.].

Distribution: Ariz. (Mexico)

Note: Bare's (1928, above) figures of male genitalia [accompanied by a note that "a full description of this will follow"] served to adequately validate the name *B. arizonis*. His (1931, Pan-Pac. Ent., 7: 115) written description gives additional details of the species.

Buenoa artafrons Truxal, 1953

1953 *Buenoa artafrons* Truxal, Univ. Ks. Sci. Bull., 35: 1444. [Fla.].

Distribution: Fla., Ga., Miss.

Buenoa confusa Truxal, 1953

1905 *Anisops elegans*: Torre-Bueno, J. N.Y. Ent. Soc., 13: 46.

1908 *Buenoa elegans*: Torre-Bueno, J. N.Y. Ent. Soc., 16: 238.

1926 *Bueno* [*sic*] *elegans*: Blatchley, Het. E. N. Am., p. 1059.

1953 *Buenoa confusa* Truxal, Univ. Ks. Sci. Bull., 35: 1453. [Ks.].

Distribution: Ala., Alta., B.C., Conn., D.C., Fla., Ga., Ind., Ks., La., Man., Me., Mich., Minn., Miss., N.J., N.S., N.Y., Ont., P.Ed., Que. S.D., Sask., Tex., Va. (Mexico, West Indies).

Note: The type of *Anisops elegans* Fieber, 1851, was labelled "Brasil Coll. Germ." which led subsequent authors to identify American specimens under this name, e.g. Uhler (1894, above). All New World records of *B. elegans* (Fieber) should be transferred to *B. confusa.*

Buenoa hungerfordi Truxal, 1953

1953 *Buenoa hungerfordi* Truxal, Univ. Ks. Sci. Bull., 35: 1483. [Holotype from Mexico; also paratypes from Ariz.].

Distribution: Ariz. (Mexico).

Buenoa limnocastoris Hungerford, 1923

1923 *Buenoa limnocastoris* Hungerford, Ent. News, 34: 150. [Minn.].

Distribution: Fla., Ga., Me., Mich., Minn., N.J., N.S., Ont., P.Ed., Que., Va.

Note: Harris' (1943, J. Ks. Ent. Soc., 16: 153) S.D. record appears to be too distant from the continuous range of this species; it may belong to *B. confusa, B. macrotibialis,* or *B. margaritacea.*

Buenoa macrotibialis Hungerford, 1924

1924 *Buenoa macrotibialis* Hungerford, An. Ent. Soc. Am., 17: 225. [Mich.].

Distribution: B.C., Man., Mich., Minn., N.S., Ont., Que., S.D., Sask.

Buenoa margaritacea Torre-Bueno, 1908

1884 *Anisops platycnemis* Uhler, Stand. Nat. Hist., 2: 253. ["Maine to Cuba;...as far west as Illinois, and thence to Texas...to Tamaulipas, Mexico."]. Preoccupied. Neotype from Md. designated by Truxal, 1953, Univ. Ks. Sci. Bull., 35: 1400.

1908 *Buenoa margaritacea* Torre-Bueno, J. N.Y. Ent. Soc., 16: 238. New name for *Anisops platycnemis* Uhler.

Distribution: Ariz., Ark., Cal., Col., Fla., Ga., Ill., Ind., Ks., La., Man., Mass., Md., Mich., Minn., Miss., Mo., N.J., N.M., N.Y., Ok., Ont., Pa., Que., S.C., S.D., Tenn., Tex., Va., Vt. (Mexico).

Note: Hungerford (1917, Ent. News, 28: 174-183) provided notes on biology.

Buenoa marki Reichart, 1971

1971 *Buenoa marki* Reichart, Fla. Ent., 54: 311. [Fla.].

Distribution: Fla.

Buenoa omani Truxal, 1953

1953 *Buenoa omani* Truxal, Univ. Ks. Sci. Bull., 35: 1426. [Cal.].

Distribution: Cal. (Mexico).

Buenoa platycnemis (Fieber), 1851

1851 *Anisops platycnemis* Fieber, Abh. K. Boh. Ges. Wiss., (5)7: 485. [Puerto Rico].

1904 *Buenoa platycnemis*: Kirkaldy, Wien. Ent. Zeit., 23: 134.

Distribution: Fla., Tex. (Mexico to South America, West Indies).

Note: Truxal (1953, Univ. Ks. Sci. Bull., 35: 1423-1424) was able to verify the occurrence of this species in Fla. and Tex. and concluded that many other U.S. records are erroneous.

Buenoa scimitra Bare, 1925 [Fig. 101]

1925 *Buenoa scimitra* Bare, Ent. News, 36: 226. [Ks.].

Distribution: Ala., Ariz., Ark., Cal., Fla., Ga., Ks., La., Miss., N.C., N.M., Ok., S.C., Tenn., Tex., Ut., Va. (Mexico, West Indies).

Buenoa speciosa Truxal, 1953
1953 *Buenoa speciosa* Truxal, Univ. Ks. Sci. Bull., 35: 1437. [Holotype from Mexico; also from Tex.].
Distribution: Tex. (Mexico).

Buenoa uhleri Truxal, 1953
1953 *Buenoa uhleri* Truxal, Univ. Ks. Sci. Bull., 35: 1409. [Holotype from Mexico; also from U.S.].
Distribution: Cal., Tex. (Mexico).
Note: Truxal (1979, Bull. Cal. Ins. Surv., 21: 147) expressed belief that the Cal. record needs verification.

Subfamily Notonectinae Latreille, 1802

Tribe Nychini Hungerford, 1933

Genus *Martarega* White, 1879

1879 *Martarega* White, Trans. Ent. Soc. London, 4: 271. Type-species: *Martarega membranacea* White, 1879. Monotypic.
Note: Truxal (1949, J. Ks. Ent. Soc., 22: 1-24) provided a key to species.

Martarega mexicana Truxal, 1949 [Fig. 102]
1949 *Martarega mexicana* Truxal, J. Ks. Ent. Soc., 22: 11. [Mexico].
1966 *Martarega mexicana*: Polhemus, Proc. Ent. Soc. Wash., 68: 57.
Distribution: Ariz. (Mexico, Central America).
Note: Known in the U.S. only from the Verde River drainage in Arizona.

Tribe Notonectini Latreille, 1802

Genus *Notonecta* Linnaeus, 1758

1758 *Notonecta* Linnaeus, Syst. Nat., 10th ed., 1: 439. Type-species: *Notonecta glauca* Linneaus, 1758. Designated by Latreille, 1810, Cons. Gén., p. 434.
Notes:Hungerford (1934, Bull. Univ. Ks. Sci. Bull., 34: 5-195) revised the genus and provided a key to North American species. [The dates for all references to Hungerford's revision of the *Notonecta* of the World have been changed from 1933 as printed on the publication to 1934. This follows Zoological Record, 71: 71, which received it in 1934, and Hungerford (1950, J. Ks.Ent. Soc., 23: 93) who also used the 1934 date of publication for it.--R. C. Froeschner.]

Notonecta ochrothoe Kirkaldy, 1897, and *Notonecta impressa* Fieber, 1851, have been listed as species occurring in the United States. However, these cannot be confirmed. The records for *N. ochrothoe* almost certainly refer instead to *N. melaena* Kirkaldy, and it is quite possible that the two are synonyms. The type of *N. impressa* is a female from Mexico and, while Hungerford (1934 Univ. Ks. Sci. Bull., 21: 5-195) noted slight differences between this species and females of *N. montezuma* Kirkaldy, he also noted that they may be synonyms. The reference to "*Notonecta maculata* Olv." by Rathvon (1869, Hist. Lancaster Co., Pa., p. 550) could refer to any one of several species, except *N. maculata* Fabricius, which does not occur in the Americas.

Subgenus *Bichromonecta* Hungerford, 1934

1934 *Notonecta* (*Bichromonecta*) Hungerford, J. Ks. Ent. Sci., 7: 98. Type-species: *Notonecta shooterii* [*sic*] Uhler, 1894. Original designation.

Notonecta repanda Hungerford, 1934

1928 *Notonecta shooteri*: Hungerford, Can. Ent., 60: 76.
1934 *Notonecta repanda* Hungerford, Univ. Ks. Sci. Bull, 21: 100. [Ariz.].
Distribution: Ariz. (Mexico).

Notonecta shooteri Uhler, 1894

1894 *Notonecta shooterii* [*sic*] Uhler, Proc. Cal. Acad. Sci., (2) 4: 292. [Cal.].
1904 *Notonecta shooteri*: Torre-Bueno, J. N.Y. Ent. Soc., 12: 62.
1934 *Notonecta* (*Bichromonecta*) *shooteri*: Hungerford, Univ. Ks. Sci. Bull., 21: 24.
Distribution: Ariz., Cal., Ore. (Mexico).

Subgenus *Erythronecta* Hungerford, 1934

1934 *Notonecta* (*Erythronecta*) Hungerford, J. Ks. Ent. Soc., 7: 98. Type-species: *Notonecta mexicana* Amyot and Serville, 1843. Original designation.

Note: Hungerford (1934, Univ. Ks. Sci. Bull., 21: 24) erroneously gave *Notonecta lobata* Hungerford as the type-species of this subgenus.

Notonecta hoffmani Hungerford, 1925

1894 *Notonecta mexicana*: Uhler, Proc. Cal. Acad. Sci., (2)4: 292.
1925 *Notonecta hoffmani* Hungerford, Can. Ent., 57: 241. [Cal.].
1934 *Notonecta hoffmanni* [*sic*]: Hungerford, Univ. Ks. Sci. Bull., 21: 68.
1979 *Notonecta* (*Erythronecta*) *hoffmani*: Truxal, Bull. Cal. Ins. Surv., 21: 141-142.
Distribution: Ariz., Cal., N.M. (Mexico).

Note: Hungerford (1934, Univ. Ks. Sci. Bull., 21: 68) tabulated the *Notonecta mexicana* records that belong in the synonymy of *Notonecta hoffmani.* Because W. E. Hoffmann spelled his name with one and two n's [different papers in the same year] at the time of Hungerford's (1925, above) dedication, it is necessary to retain the original spelling of this species as *hoffmani*, as noted also by Truxal (1979, above, 21: 142).

Notonecta lobata Hungerford, 1925

1884 *Notonecta mexicana*: Uhler, Stand. Nat. Hist., 2: 252.
1925 *Notonecta lobata* Hungerford, Can. Ent., 57: 239. [Ariz.].
1934 *Notonecta* (*Erythronecta*) *lobata*: Hungerford, Univ. Ks. Sci. Bull., 32: 34.
Distribution: Ariz., N.M., Tex. (Mexico).

Note: Hungerford (1934, above, 21: 70) tabulated the *Notonecta mexicana* records that belong in the synonymy of *Notonecta lobata*.

Subgenus *Notonecta* Linnaeus, 1758

1758 *Notonecta* Linneaus, Syst. Nat., 10th ed., 1: 439. Type-species: *Notonecta glauca* Linnaeus, 1758. Designated by Lamarck, 1801, Sys. Nat. Anim., p. 296.

1929 *Notonecta* (*Notonecta*): Hutchinson, An. S. Afr. Mus., 25: 363.

Notonecta borealis Hussey, 1919

1897 *Notonecta insulata*: Kirkaldy, Trans. Ent. Soc. London, p. 405.
1904 *Notonecta lutea*: Torre-Bueno, Ent. News, 15: 220.
1919 *Notonecta borealis* Hussey, Occas. Pap. Mus. Zool., Univ. Mich., 75: 15. [Mich.].
1923 *Notonecta borealis* Torre-Bueno and Hussey, Bull. Brook. Ent. Soc., 18: 104. [B.C.]. Preoccupied.
Distribution: Alta., B.C., Man., Mich., Minn., N.T., Nfld., Ont., Que., Sask.
Note: Hussey (1919, above) validated this name when he included it in a key to species; he also foretold the early appearance of a description by Torre-Bueno and himself. The latter appeared in 1923 (above) and noted Hussey's earlier use as "without description"--but it had been described in the key. Hungerford (1934, Univ. Ks. Sci. Bull., 21: 96) tabulated the *Notonecta lutea* records that belong in the synonymy of *Notonecta borealis.*

Notonecta irrorata Uhler, 1879
1879 *Notonecta irrorata* Uhler, Proc. Boston Soc. Nat. Hist., 19: 443. [Mass.].
Distribution: Ala., Ariz., Ark., Conn., D.C., Fla., Ga., Ks., Ky., Ill., Ind., La., Man., Mass., Md., Me., Mich., Minn., Miss., N.B., N.C., N.J., N.S., N.Y., Oh., Ont., P.Ed., Pa., Que., R.I., Sask., Tenn., Va.

Subgenus *Paranecta* Hutchinson, 1929

1929 *Notonecta* (*Paranecta*) Hutchinson, An. S. Afr. Mus., 25: 363. Type-species: *Notonecta lactitans* Kirkaldy, 1897. Original designation.

Notonecta indica Linnaeus, 1771
1771 *Notonecta indica* Linnaeus, Mant. Plant., 2: 534. [West Indies].
1775 *Notonecta americana* Fabricius, Syst. Ent., p. 690. ["Americae aquis."]. Synonymized by Kirkaldy, 1904, Wien. Ent. Zeit., 23: 132.
1884 *Notonecta undulata*: Uhler, Stand. Nat. Hist., 2: 252 (in part).
1897 *Notonecta undulata* var. *charon*: Kirkaldy, Trans. Ent. Soc. London, p. 411 (in part). ["Most of U.S., Cuba, Mexico, Colombia, Peru"].
1905 *Notonecta howardii* Torre-Bueno, J. N.Y. Ent. Soc., 13: 151. [Ariz.]. Synonymized by Hungerford, 1934, Univ. Ks. Sci. Bull., 21: 114.
1979 *Notonecta* (*Paranecta*) *indica*: Truxal, Bull. Cal. Ins. Surv., 21: 141.
Distribution: Ariz., Ark., Cal., Fla., Ga., Ks., Ky., La., Md., Miss., N.C., N.M., S.C., Tenn., Tex. (Mexico to South America, West Indies).
Note: Hungerford (1934, above, 21: 113-114) tabulated the records of *Notonecta indica* as misidentified under other names.

Notonecta insulata Kirby, 1837
1837 *Notonecta insulata* Kirby, Fauna Bor.-Am., 4: 285. [Boreal America].
1851 *Notonecta rugosa* Fieber, Abh. K. Bohm. Ges. Wiss., (5)7: 476. [North America]. Synonymized by Uhler, 1878, Proc. Bost. Soc. Nat. Hist., 19: 442.
1851 *Notonecta rugosa* var. *bicolor* Fieber, Abh. K. Bohm. Ges. Wiss., (5)7: 476. Synonymized by Kirkaldy, 1904, Wien. Ent. Zeit, 23: 94.
1851 *Notonecta rugosa* var. *plagiata* Fieber, Abh. K. Bohm. Ges. Wiss., 7: 476. [Md., N.Y.]. Synonymized by Uhler, 1878, Proc. Bost. Soc. Nat. Hist., 19: 442.
1851 *Notonecta rugosa* var. *cordigera* Fieber, Abh. K. Bohm. Ges. Wiss., (5)7: 476. [Md., Pa.]. Synonymized by Uhler, 1878, Proc. Bost. Soc. Nat. Hist., 19: 442.
1897 *Notonecta impressa*: Kirkaldy, Trans. Ent. Soc. London, 1897: 403.
1897 *Notonecta insulata* var. *geala* Kirkaldy, Trans. Ent. Soc. London, p. 404. ["America"]. Synonymized by Hungerford, 1934, Univ. Ks. Sci. Bull., 21: 85.

1897 *Notonecta insulata* var. *odara* Kirkaldy, Trans. Ent. Soc. London, p. 404. ["America"]. Synonymized by Hungerford, 1934, Univ. Ks. Sci. Bull., 21: 85.

Distribution: Conn., Man., Mass., Md., Me., Mich., Minn., N.B., N.J., N.S., N.Y., Nfld., Oh., Ont., P.Ed., Que.

Note: Hungerford (1934, above, 21: 86) pointed out that the previous records for more western localities actually belong to *Notonecta kirbyi* Hungerford.

Notonecta kirbyi Hungerford, 1925

1875 *Notonecta insulata*: Uhler, Rept. U.S. Geol. Geogr. Surv. Terr., 5: 841.

1876 *Notonecta insulata*: Uhler, Bull. U.S. Geol. Geogr. Surv. Terr., 1(5): 339.

1897 *Notonecta insulata* var. *impressa*: Kirkaldy, Trans. Ent. Soc. London, p. 404.

1925 *Notonecta kirbyi* Hungerford, Can. Ent., 57: 241. [Ut.].

1979 *Notonecta* (*Paranecta*) *kirbyi*: Truxal, Bull. Cal. Ins. Surv., 21: 141.

Distribution: Alta., Ariz., B.C., Cal., Col., Id., Man., Mont., N.M., N.T., Neb., Nev., Ore., S.D., Sask., Tex., Ut., Wash., Wyo., Yuk.

Note: Hungerford (1934, above, 21: 88-89) tabulated the *Notonecta insulata* records that belong in the synonymy of *Notonecta kirbyi*.

Notonecta lunata Hungerford, 1926

1851 *Notonecta variabilis* var. *maculata* Fieber, Abh. K. Bohm. Ges. Wiss., (5)7: 477. [Md.]. Preoccupied.

1878 *Notonecta undulata*: Uhler, Proc. Bost. Soc. Nat. Hist., 19: 441 (in part).

1897 *Notonecta variabilis*: Kirkaldy, Trans. Ent. Soc. London, p. 44 (in part).

1900 *Notonecta americana*: Smith, Ins. N.J., p. 144.

1926 *Notonecta lunata* Hungerford, Psyche, 33: 12. New name for *Notonecta maculata* Fieber.

Distribution: Conn., Md., Me., Mich., N.B., N.S., N.Y., Ont., P.Ed., Que.

Note: Hungerford (1934, Univ. Ks. Sci. Bull., 21: 107) tabulated the *Notonecta lunata* records misidentified under other names.

Notonecta montezuma Kirkaldy, 1897

1897 *Notonecta montezuma* Kirkaldy, Trans. Ent. Soc. London, p. 402. [Mexico].

1928 *Notonecta montezuma*: Hungerford, An. Ent. Soc. Am., 21: 142.

Distribution: Tex. (Mexico).

Note: The earlier California records for this species were based on a misidentified specimen, according to Hungerford (1934, Univ. Ks. Sci. Bull., 21: 82).

Notonecta petrunkevitchi Hutchinson, 1945

1934 *Notonecta raleighi*: Hungerford, Univ. Ks. Sci. Bull., 21: 106 (in part).

1945 *Notonecta petrunkevitchi* Hutchinson, Trans. Conn. Acad. Arts Sci., 36: 599. [Conn.].

Distribution: Conn., Mass., Me., N.C., N.J.

Note: Hungerford (1950, J. Ks. Ent. Soc., 23: 93) provided notes on distribution.

Notonecta raleighi Torre-Bueno, 1907

1907 *Notonecta raleighi* Torre-Bueno, Can. Ent., 39: 225. [N.C.].

Distribution: D.C., Ill., Md., Miss., N.C., N.J.

Notonecta spinosa Hungerford, 1930

1930 *Notonecta spinosa* Hungerford, Can. Ent., 62: 217. [Ut.].

1979 *Notonecta* (*Paranecta*) *spinosa*: Truxal, Bull. Cal. Ins. Surv., 21: 141-142.

Distribution: B.C., Mont., Nev., Ore., Ut., Wyo.

Notonecta uhleri Kirkaldy, 1897

1897 *Notonecta uhleri* Kirkaldy, An. Mag. Nat. Hist., (6)20: 58. [Fla., Mass.].

Distribution: D.C., Fla., Ga., Ill., La., Mass., Md., Miss., Mo., N.C., N.J., N.Y., Pa., S.C., Va.

Notonecta undulata Say, 1832

1832 *Notonecta undulata* Say, Descrip. Het. Hem. N. Am., p. 39. [Ind., Mo.].

1851 *Notonecta punctata* Fieber, Abh. K. Bohm. Ges. Wiss., (5)7: 476. [Md.]. Synonymized by Kirkaldy, 1897, Trans. Ent. Soc. Lond., p. 410.

1897 *Notonecta undulata* var. *charon* Kirkaldy, Trans. Ent. Soc. London, p. 411 (in part). ["Most of U.S.A.; Cuba; Mexico; Colombia; Peru"]. Synonymized by Torre-Bueno, 1905, J. N.Y. Ent. Soc., 13: 154.

1979 *Notonecta* (*Paranecta*) *undulata*: Truxal, Bull. Cal. Ins. Surv., 21: 141.

Distribution: Alta., Ariz., Ark., B.C., Cal., Conn., D.C., Fla., Ia., Id., Ill., Ind., Ks., Ky., La., Man., Mass., Md., Me., Mich., Minn., Mo., Mont., N.B., N.C., N.J., N.M., N.S., N.T., N.Y., Nev., Nfld., Oh., Ok., Ont., Ore., P.Ed., Pa., Que., R.I., S.D., Sask., Tenn., Tex., Ut., Va., Wash. (Mexico).

Notonecta unifasciata Guérin-Méneville, 1857 [Fig. 100]

1857 *Notonecta unifasciata* Guérin-Méneville, Le Mon. Univ., 330: 1298. [Mexico].

1897 *Notonecta americana*: Kirkaldy, Trans. Ent. Soc. London, p. 409 (in part).

1900 *Notonecta undulata*: Champion, Biol. Centr.-Am., Rhyn., 2: 370 (in part).

1904 *Notonecta indica*: Kirkaldy, Wien. Ent. Zeit., 23: 95.

1979 *Notonecta* (*Paranecta*) *unifasciata*: Truxal, Bull. Cal., Ins. Surv., 21: 141-142.

Distribution: Ariz., B.C., Cal., N.M., Nev., Ore., Tex., Ut. (Mexico, South America).

Note: Hungerford (1934, Univ. Ks. Sci. Bull., 21: 109) tabulated the records of *Notonecta unifasciata* as misidentified under other names.

Notonecta unifasciata andersoni Hungerford, 1934

1934 *Notonecta unifasciata andersoni* Hungerford, Univ. Ks. Sci. Bull., 21: 110. [B.C.].

Distribution: Ariz., B.C., Nev., Ore., Ut.

Notonecta unifasciata cochisiana Hungerford, 1934

1934 *Notonecta unifasciata cochisiana* Hungerford, Univ. Ks. Sci. Bull., 21: 110. [Ariz.].

Distribution: Ariz.

Notonecta unifasciata unifasciata Guérin-Méneville, 1857

1857 *Notonecta unifasciata* Guérin-Méneville, Le Mon. Univ., 330: 1298. [Mexico].

1934 *Notonecta unifasciata unifasciata*: Hungerford, Univ. Ks. Sci. Bull., 21: 109.

Distribution: Cal., N.M. (Mexico).

Family Ochteridae Kirkaldy, 1906

The Velvety Shore Bugs

By Dan A. Polhemus and John T. Polhemus

Velvety shore bugs may be recognized by their compact, ovate form, exposed four-segmented antennae, and slender, nonraptorial legs with a 2:2:3 tarsal formula. Their common name is based on the softly textured appearance of the dorsum, which is generally dark with scattered light spots, frosted bluish areas, and set with tiny gold setae. Only a single genus, *Ochterus* Latreille, occurs in North America.

Ochterids are found in damp littoral habitats adjoining lakes and streams, where they prey on fly larvae, Collembola, and aphids. Eggs are laid singly on sand grains or other small objects, and the nymphs cover themselves with sand grains as an apparent means of concealment (Bobb, 1951, Bull. Brook. Ent. Soc., 46: 92). *Ochterus marginatus* (Latreille) completes its development within a month in Formosa (Takahashi, 1923, Bull. Brook. Ent. Soc., 18: 67), but North American species occurring in areas with cold winters diapause as fourth instars and consequently require nearly a year to reach maturity.

The Ochteridae are a small family, with only forty described species worldwide, of which six occur in North America. Our species are primarily southern and eastern in distribution, with only two species present in the West, both of them Mexican taxa ranging into the Colorado River drainage.

The major work on this family for the Western Hemisphere is that of Schell (1943, J. Ks. Ent. Soc., 16: 29), which contains a key including all but one of our North American species. Drake (1952, Fla. Ent., 35: 72) provided a checklist of New World species, and Menke (1979, Bull. Cal. Ins. Surv., 21: 124) gave a general review of the family. The name "Pelogoniidae" has been used by certain authors (e.g., China and Miller, 1955, Ann. Mag. Nat. Hist., 12: 257) as being based on the oldest family group name, but China and Miller (1959, Bull. Brit. Mus. (Nat. Hist.), Ent., 8: 16) used Ochteridae with the

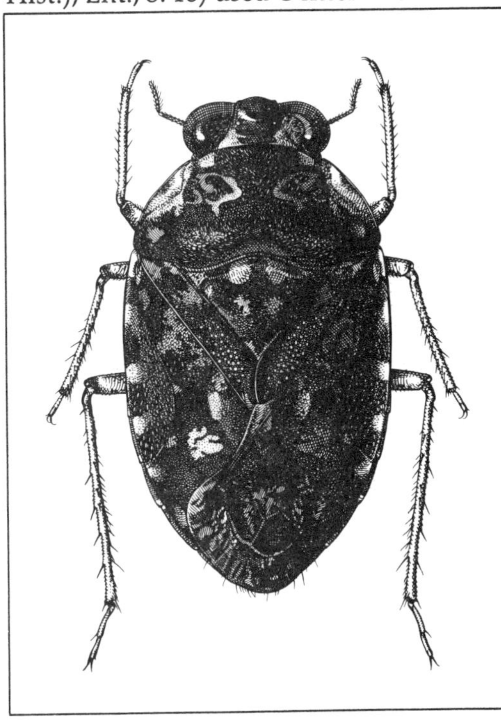

Fig. 103 *Ochterus barberi* [p. 542] (After Usinger, 1956).

explanation of 51 years of usage even though "Pelogoniidae" is the older name. Ochteridae has gained general acceptance among contemporary workers and is conserved under Article 40(b) of the 1985 Code.

Genus *Ochterus* Latreille, 1802

1807 *Ochterus* Latreille, Gen. Crus. Ins., 3: 142. Type-species: *Acanthia marginata* Latreille, 1804. Monotypic.

1809 *Pelogonus* Latreille, Gen. Crus. Ins., 4: 384. Unnecessary new name for *Ochterus* Latreille, 1807.

Note: Schell (1943, J. Ks. Ent. Soc., 16: 32) provided a key to species including all North American taxa, with the exception of the subsequently described *O. rotundus* Polhemus and Polhemus. The spellings *Pelegonus* (Laporte, 1832, Mag. Zool., 1, suppl., pp. 6, 13), *Ochtherus* (Agassiz, 1846, Nomen. Zool., p. 254, and Bergroth, 1890, Bull. Soc. Ent. Fr., p. lxvi), and *Octhera* (Bergroth, 1890, Bull. Soc. Ent. Fr., p. cxix) are all unnecessary emendations of the original generic names above, but these variant spellings have not appeared in the N. Am. literature.

Ochterus americanus (Uhler), 1876

1876 *Pelogonus americanus* Uhler, Bull. U.S. Geol. Geogr. Surv. Terr., 1: 335. [Ill., Mass., Pa., Tex., Cuba].

1906 *Pelogonus americatus* [*sic*]: Snow, Trans. Ks. Acad. Sci., 20: 180.

1907 *Ochterus americanus*: Torre-Bueno and Brimley, Ent. News, 18: 433.

1913 *Ochterus americanum* [*sic*]: Barber, Can. Ent., 45: 215.

Distribution: Ariz.(?), Ill., Ks., La., Mass., Md., Miss., Mo., N.C., N.J., N.Y., Neb., Pa., Tex., Va.

Note: The records for Florida (Drake and Chapman, 1958, An. Ent. Soc. Am., 51: 483-484) and South Carolina (Roback, 1958, Trans. Am. Ent. Soc., 84: 9) must be confirmed. The Florida specimens we have seen determined by Drake as *americanus* are actually *banksi*; the former has a northerly distribution, and the latter a more southerly one.

Ochterus banksi Barber, 1913

1913 *Ochterus banksi* Barber, Can. Ent., 45: 214. [Va.].

Distribution: Fla., Ind., Mass., Miss., N.J., N.Y., S.C., Tex., Va.

Note: Bobb (1951, Bull. Brook. Ent. Soc., 46: 92-100) provided data on biology and figures of the nymphal instars.

Ochterus barberi Schell, 1943 [Fig. 103]

1943 *Ochterus barberi* Schell, J. Ks. Ent. Soc., 16: 41. [Ariz.].

Distribution: Ariz., Cal., N.M. (Mexico).

Note: Known in the U.S. only from the Colorado River basin.

Ochterus flaviclavus Barber, 1913

1913 *Ochterus flaviclavus* Barber, Can. Ent., 45: 215. [Fla.].

Distribution: Fla., Ks., La., Tex.

Note: This species is more widely distributed than reported in the literature, and the La. and Tex. records are new.

Ochterus perbosci (Guérin-Méneville), 1843
1843 *Pelogonus perbosci* Guérin-Méneville, Rev. Zool., p. 113. [Mexico].
1909 *Ochterus perbosci*: Kirkaldy and Torre-Bueno, Proc. Ent. Soc. Wash., 10: 179.
Distribution: Ariz., Tex. (Mexico, West Indies, South America).

Ochterus rotundus Polhemus and Polhemus, 1976
1943 *Ochterus viridifrons*: Schell, J. Ks. Ent. Soc., 16: 37 (in part).
1976 *Ochterus rotundus* Polhemus and Polhemus, Great Basin Nat., 36: 223. [Mexico].
Distribution: Ariz. (Mexico).
Note: Known in the U.S. only from the Colorado River and its tributaries in the Grand Canyon.

Family Pentatomidae Leach, 1815

The Stink Bugs

By Richard C. Froeschner

The Pentatomidae comprise one of the largest families of Heteroptera. Its members usually exhibit one of two feeding habits, although these habits are not rigidly adhered to, even within a species. In general, members of the subfamily Asopinae are essentially predaceous, but the first-stage nymphs feed on plant juices; members of the other subfamilies are mostly phytophagous with a few species capable of incorporating insect haemolymph in their total diet. Very few, if any, Asopinae are host specific. The claim that *Perillus bioculatus* (Fabricius) confines its feeding to the Colorado potato beetle, *Leptinotarsa decimlineata* (Say), apparently results from the greater number of observations having been made under economic conditions; in contrast, for instance, Strickland (1953, Can. Ent., 85: 196) reports it as feeding "rather freely on larvae of *Lina scripta* [now *Chrysomela scripta* Fabricius] in addition to those of *Leptinotarsa decimlineata*"; and I have seen nymphs, as well as adults, feeding on various lepidopterous larvae. A good example of host specificity in the North American Pentatominae is to be found in the preference of the harlequin bug, *Murgantia histrionica* (Hahn), for cabbage and several other Cruciferae. A generalized outline of the biology of North American Pentatomidae includes one or two generations per year, depending upon the species involved. Under field conditions, the winter is passed in the adult stage, adults hibernating in the protection of crevices of plants and grass clumps or under bark, rocks, leaves and other objects, especially those on the ground. In spring, after leaving hibernation and successfully mating, the females begin laying eggs. The elongate or barrel-shaped eggs are usually laid in clusters of a few to a several dozen and individually glued to a host plant so that the opercular end is free. The eggs generally hatch within one week and the newly emerged first instar nymphs are often found clinging to the empty egg shells where they remain, apparently without feeding on the host, for about a week until their first molt. After this they move from the area of oviposition in search of food. The five nymphal instars generally span about a month to a month-and-a-half before becoming adults. Adults may be found during every month of the year, thus overlaping completely the nymphal period.

More detailed information on the bionomics and eggs of North American Pentatomidae can be found in two papers by Esselbaugh (1946, An. Ent. Soc. Am., 39: 667-691; 1949, Ent. Am., 28: 1-73) and a more recent one by McPherson (1982, Pent. Ne. N. Am., pp. 36-111).

The economic importance of these insects varies greatly from species to species and sometimes within a species according to as yet undetermined causes. Obviously the economic role of the predaceous forms must be judged by the insects attacked. In-

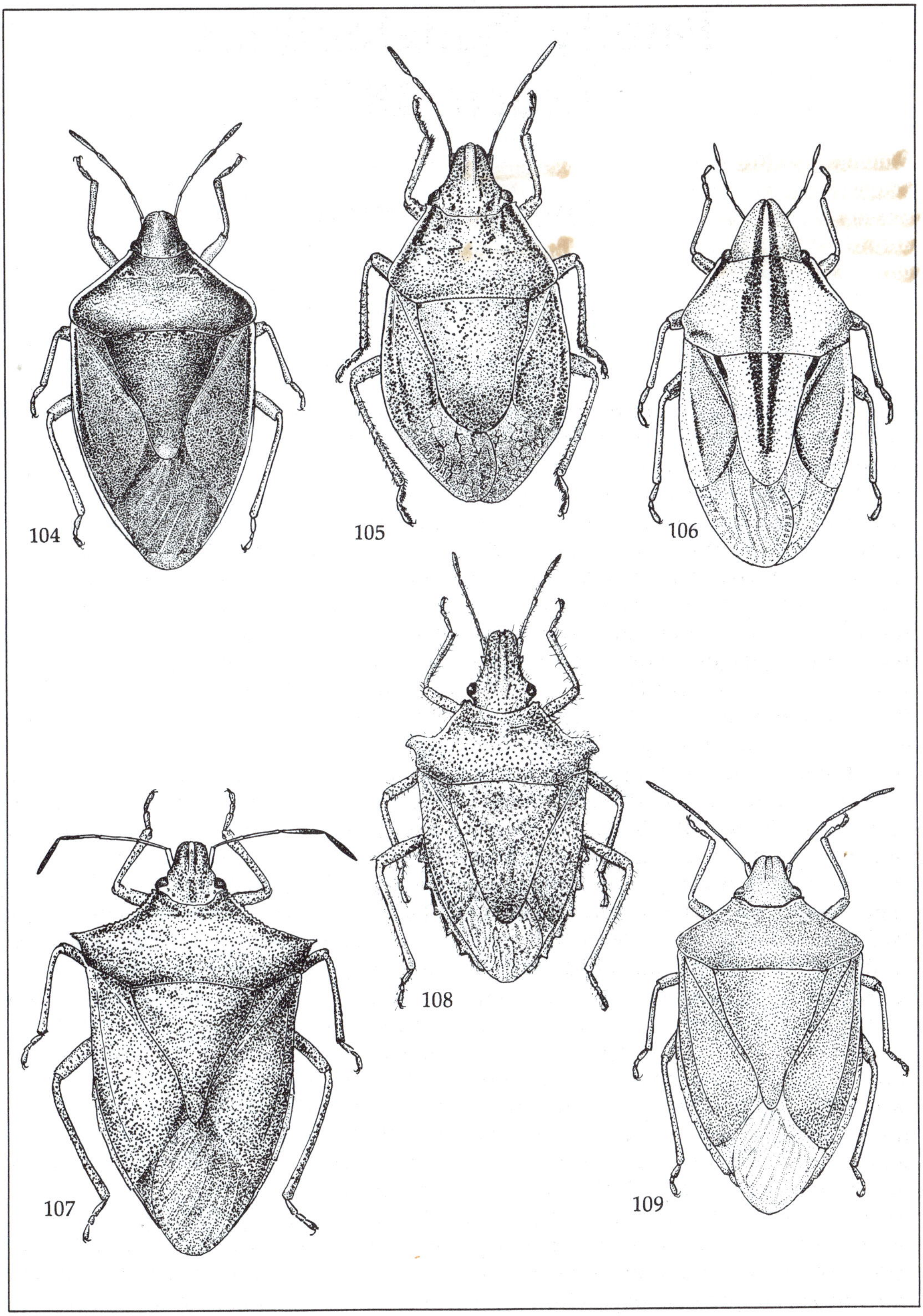

Figs. 104-109: 104, *Holcostethus limbolarius* [p. 581]; 105, *Coenus delius* [p. 573]; 106, *Aelia americana* [p. 565]; 107, *Euschistus variolarius* [p. 580]; 108, *Prionosoma podopioides* [p. 590]; 109, *Thyanta custator* [p.593] (After Froeschner, 1941).

dividuals feeding upon pest insects are considered beneficial, but the same or other individuals attacking insects desired by man--honey bees or parasites of objectionable insects--are considered harmful. Among the North American plant-feeding Pentatomidae, the most notorious and consistent pest species are the harlequin bug and the southern green stink bug, *Nezara viridula* (Linnaeus). Other species are weed-feeders and are seldom noticed by economic entomologists; yet numbers of individuals of these latter species may, for some inexplainable reason, move onto cultivated crops and cause serious damage for one season and then in the next and for many ensuing years confine their attention to their noneconomic plant hosts.

Stink bugs that annoy householders by appearing suddenly indoors are actually harmless, accidental intruders. During the fall months or after sunning themselves on warm winter or spring days, these bugs will enter crevices in which to pass the cold nights; if these crevices are under doors or around windows, the bugs may crawl into the house and wander around futilely trying to find their way out. Indoors they may bumble about flying into objects or resting on them, but they do not aggressively attack persons, house plants, buildings, or furnishings.

In North America the family Pentatomidae is probably the most collected and best known of the heteropteran families. Therefore, in the absence of recent reports other than those based on old literature records, the following two species should not be carried in our faunal list.

Dolycoris baccarum (Linnaeus). Oshanin's (1906, Verz. Pal. Hem., 1: 118) report of North American occurrence of the palearctic species appeared acceptable to Wu (1935, Cat. Ins. Sinensium, 2: 312) and China (1938, Zool. Ser., Field Mus. Nat. Hist., 20: 429) but not so to any American author.

Fabricius' (1794, Ent. Syst., 4: 134) original locality "America" for his pentatomid species *Lygaeus serratus*, now known under the combination *Mustha serrata*, appears erroneous as the species is in reality an Old World form.

The only publication attempting to provide keys for all species in our area was by Torre-Bueno (1939, Ent. Am., 19: 196-258). Numerous changes have been made in subsequent literature. The present catalog follows the suprageneric classification of the Western Hemisphere Pentatomidae as presented by Rolston and McDonald (1979, J. N.Y. Ent. Soc., 87: 189-207). Later articles by Rolston, Thomas, and McDonald (1980, J. N.Y. Ent. Soc., 88: 120-132), Rolston and McDonald (1981, J. N.Y. Ent. Soc., 88: 257-272; 1984, J. N.Y. Ent. Soc., 92: 69-86) presented keys to all North American and most Western Hemisphere genera of the tribe Pentatomini. McPherson (1982, Pent. Ne. N.Am., pp. 36-111) provided keys, an extensive literature review, and tabulations for the genera and species occurring in northeastern U.S.

Subfamily Asopinae Spinola, 1850

Genus *Alcaeorrhynchus* Bergroth, 1891

1862 *Mutyca* Stål, K. Svens. Vet.-Akad. Handl., 3(6): 58. Preoccupied. Type-species: *Canthecona grandis* Dallas, 1851. Monotypic.

1891 *Alcaeorrhynchus* Bergroth, Rev. d'Ent., 10: 235. New name for *Mutyca* Stål.

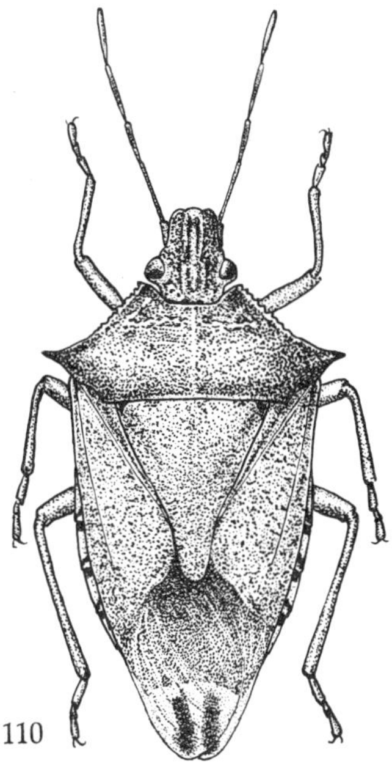

110

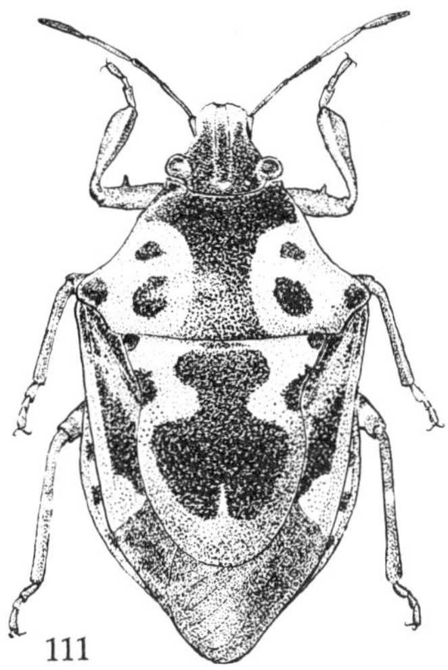

111

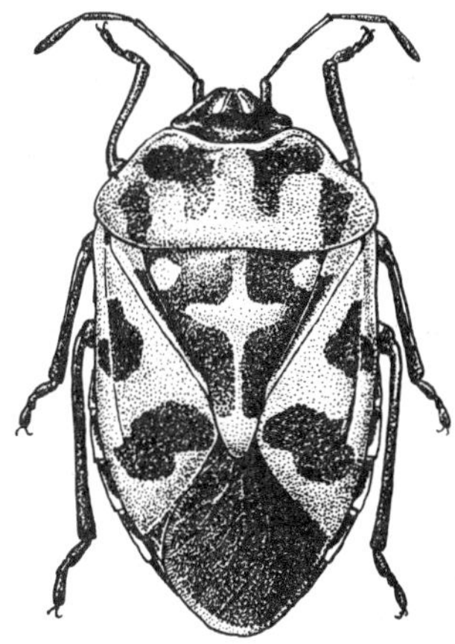

112

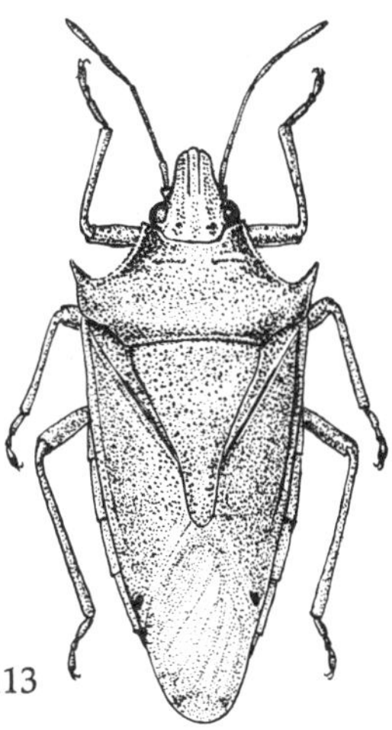

113

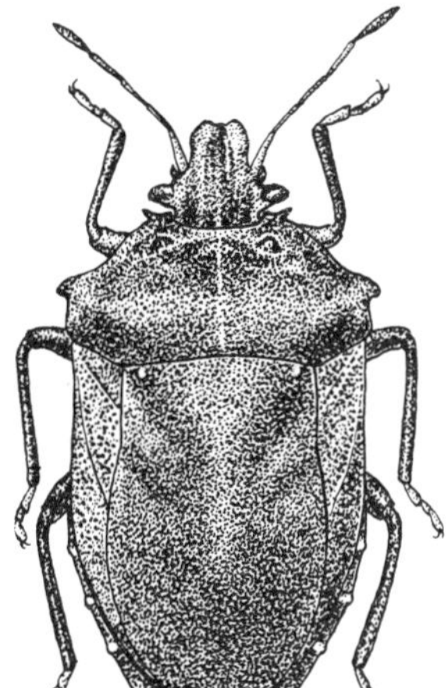

114

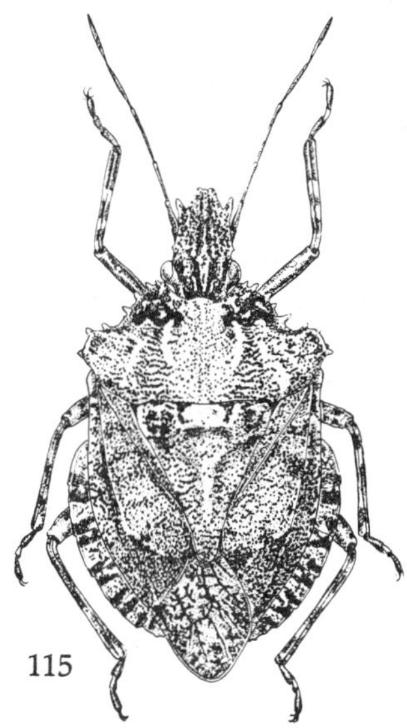

115

Figs. 110-115: 110, *Podisus maculiventris* [p. 555]; 111, *Stiretrus anchorago* "color var. *fimbriatus*" [p. 558]; 112, *Murgantia histrionica* [p. 585]; 113, *Oebalus pugnax* [p. 589]; 114, *Amaurochrous cinctipes* [p. 596]; 115, *Brochymena arborea* [p. 560] (After Froeschner, 1941).

Alcaeorrhynchus grandis (Dallas), 1851 [Fig. 116]

1851 *Canthecona grandis* Dallas, List Hem. Brit. Mus., 1: 91. [Colombia, Mexico].
1904 *Mutycha* [*sic*] *grandis*: Van Duzee, Trans. Am. Ent. Soc., 30: 72.
1907 *Alcaeorhynchus* [*sic*] *grandis*: Schouteden, Gen. Ins., 52: 32.
1909 *Alcaeorrhynchus grandis*: Kirkaldy, Cat. Hem., 1: 9.
1910 *Mutyca grandis*: Banks, Cat. Nearc. Hem.-Het., p. 94.
Distribution: Ark.(?), Fla., Tex. (Greater Antilles, Mexico south to Argentina).
Note: Barton and Lee (1981, Ark. Acad. Sci. Proc., 35: 20) reported two specimens from Ark. but suggested that they might be accidental introductions.

Alcaeorrhynchus phymatophorus (Palisot), 1811

1811 *Pentatoma phymatophora* Palisot, Ins. Rec. Afr. Am., p. 112, pl. 8, fig. 2. [Dominican Republic].
1904 *Mutycha* [*sic*] *phymatophora*: Van Duzee, Trans. Am. Ent. Soc., 30: 72.
1907 *Alcaeorhynchus* [*sic*] *phymatophora* [*sic*]: Schouteden, Gen. Ins., 52: 32.
1909 *Alcaeorrhynchus phymatophorus*: Kirkaldy, Cat. Hem., 1: 9.
1910 *Mutyca phymatophora*: Banks, Cat. Nearc. Hem.-Het., 1: 94.
1916 *Alcaeorrhynchus phymatophora* [*sic*]: Van Duzee, Check List Hem., p. 9.
Distribution: Fla. (Greater Antilles).

Genus Andrallus Bergroth, 1905

1862 *Audinetia* Ellenrieder, Nat. Tidschr. Ned. Indie, 24: 136. Preoccupied. Type-species: *Audinetia aculeata* Ellenrieder, 1862, a junior synonym of *Cimex spinidens* Fabricius, 1787. Monotypic.
1905 *Andrallus* Bergroth, An. Soc. Ent. Belg., 49: 370. New name for *Audinetia* Ellenrieder.

Andrallus spinidens (Fabricius), 1787

1787 *Cimex spinidens* Fabricius, Mant. Ins., 2: 285. [India].
1917 *Apateticus ludovicianus* Stoner, Ent. News, 28: 462. [La.]. Synonymized by Torre-Bueno, 1939, Bull. Brook. Ent. Soc., 34: 118.
1919 *Andrallus spinidens*: Hart, Ill. Nat. Hist. Surv. Bull., 13: 198.
Distribution: La., Tex. (Circumtropical).
Note: Manley (1982, Ent. News, 93: 19-24) studied the biology and life history and concluded that this species is a "Non-specific predator on Lepidoptera larvae."

Genus *Apateticus* Dallas, 1851

1851 *Apateticus* Dallas, List Hem. Brit. Mus., 1: 105. Type-species: *Halys lineolatus* Herrich-Schaeffer, 1838. Designated by Distant, 1902, Fauna Brit. India, Rhyn., 1: 253.

Subgenus *Apateticus* Dallas, 1851

1851 *Apateticus*: Dallas, List Hem. Brit. Mus., 1: 105. Type-species: *Halys lineolatus* Herrich-Schaeffer, 1838. Designated by Distant, 1902, Fauna Brit. India, Rhyn., 1: 253.
1907 *Apateticus* (*Apateticus*): Schouteden, Gen. Ins., 52: 70.

Apateticus lineolatus (Herrich-Schaeffer), 1840

1840 *Halys lineolatus* Herrich-Schaeffer, Wanz. Ins., 5: 69, fig. 514. [Mexico].
1906 *Podisus lineolatus*: Barber, Brook. Inst. Arts Sci., Sci. Bull., 1: 263.
1909 *Apateticus* (*Apateticus*) *lineolatus*: Van Duzee, Can. Ent., 41: 371.

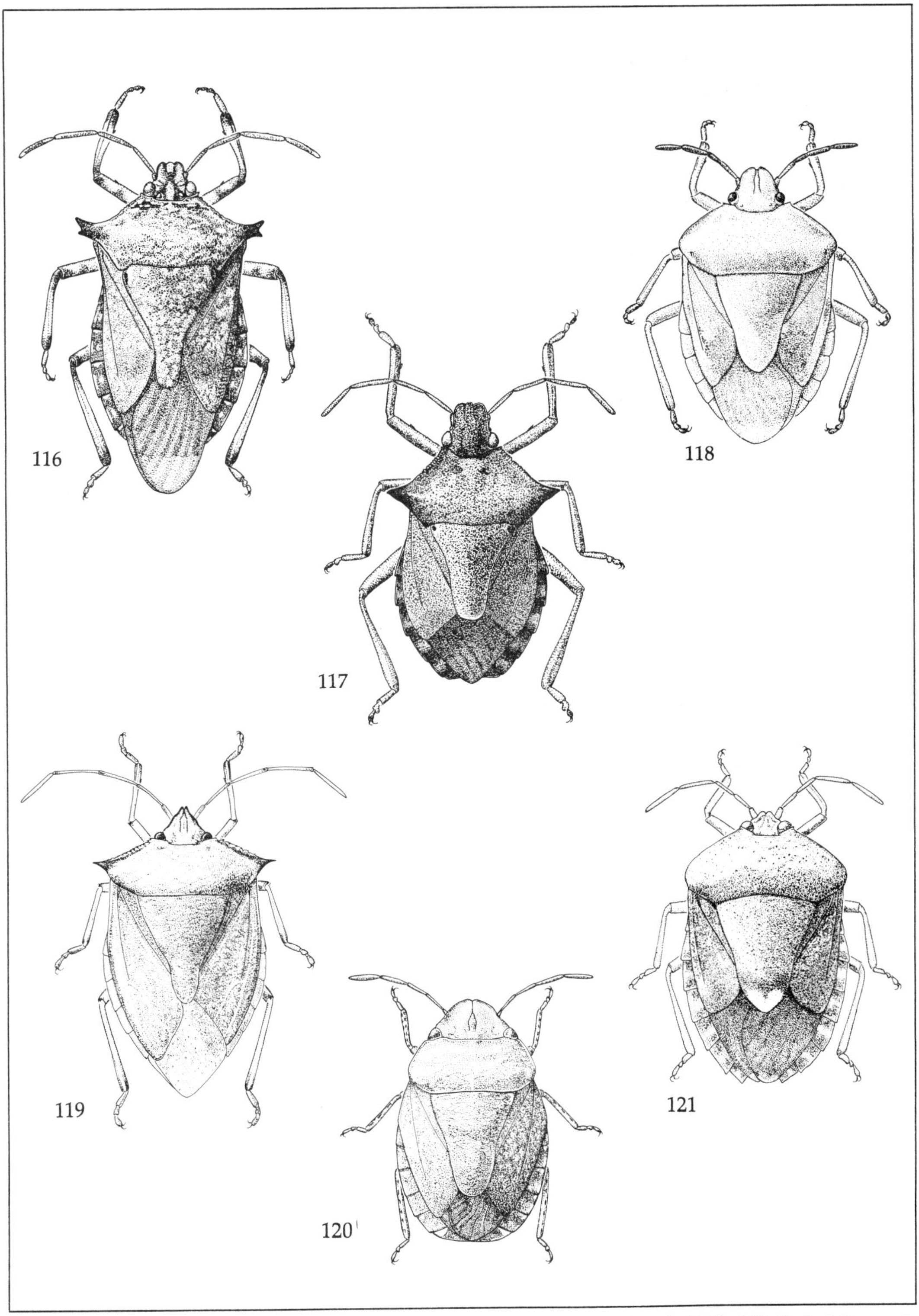

Figs. 116-121: 116, *Alcaeorrhynchus grandis* [p. 548]; 117, *Picromerus bidens* [p.554]; 118, *Codophila remota* [p.573]; 119, *Loxa flavicollis* [p. 583]; 120, *Lineostethus tenebricornis* [p. 559]; 121, *Edessa florida* [p.560] (Originals).

1910 *Apeteticus* [*sic*] *lineolatus*: Banks, Cat. Nearc. Hem.-Het., p. 93.
Distribution: Ariz., Fla., Tex. (Mexico to Venezuela).

Apateticus marginiventris (Stål), 1870
1870 *Podisus marginiventris* Stål, K. Svens. Vet.-Akad. Handl., 9(1): 49. [Mexico].
1895 *Podisus gillettei* Uhler, Bull. Col. Agr. Exp. Stn., 31: 12. [Col.]. Synonymized by Kirkaldy, 1909, Cat. Hem., 1: 22.
1904 *Podisus (Apateticus) gilletii* [*sic*]: Van Duzee, Trans. Am. Ent. Soc., 30: 69.
1907 *Apateticus (Apateticus) marginiventris*: Schouteden, Gen. Ins., Fasc. 52: 70.
1910 *Apeteticus* [*sic*] *gillettei*: Banks, Cat. Nearc. Hem.-Het., p. 93.
1926 *Apateticus marginiventris*: Barber, Ent. News, 37: 44.
Distribution: Ariz., Col., N.M., Neb. (Mexico).

Subgenus *Apoecilus* Stål, 1870

1870 *Podisus (Apoecilus)* Stål, K. Svens. Vet.-Akad. Handl., 9(1): 49. Type-species: *Pentatoma cynica* Say, 1832. Monotypic.
1907 *Apateticus (Apoecilus)*: Schouteden, Gen. Ins., 52: 70.
Note: Doubt about the authenticity of some earlier determinations in the subgenus casts question on those locality records--some of which, in the absence of refutation, are cited here.

Apateticus anatarius Van Duzee, 1935
1935 *Apateticus (Apoecilus) anatarius* Van Duzee, Pan-Pac. Ent., 11: 27. [Ariz.].
Distribution: Ariz.

Apateticus bracteatus (Fitch), 1856
1856 *Arma bracteata* Fitch, Trans. N.Y. St. Agr. Soc., 16: 336. [N.Y.].
1870 *Podisus bracteatus*: Stål, K. Svens. Vet.-Akad. Handl., 9(1): 54.
1904 *Podisus (Apoecilus) bracteatus*: Van Duzee, Trans. Am. Ent. Soc., 30: 70.
1907 *Apateticus (Apoecilus) bracteatus*: Schouteden, Gen. Ins., 52: 70.
1908 *Apateticus (Apoeicilus* [*sic*]) *bracteatus*: Torre-Bueno, J. N.Y. Ent. Soc., 16: 225.
1910 *Apeteticus* [*sic*] *bracteatus*: Banks, Cat. Nearc. Hem.-Het., p. 93 (in part).
1914 *Apateticus bracteatus*: Parshley, Psyche, 21: 149.
Distribution: Alta., B.C., Cal., Col., Conn., Id., Ill., Man., Mass., Mich., Neb., N.M.(?), N.S., N.Y., Oh., Ont., Pa., Que., S.D., Ut.
Note: The above listing for New Mexico, based on Ruckes (1937, Bull. Brook. Ent. Soc., 32: 35), needs verification.

Apateticus crocatus (Uhler), 1897
1879 *Podisus cynicus* var. *obscuripes* Riley, Rept. U.S. Dept. Agr. for 1878, p. 245 (Uhler ms. name). *Nomen nudum*.
1884 *Podiscus* [*sic*] *crocatus* Hagen, Can. Ent., 16: 40 (Uhler ms. name). *Nomen nudum*.
1886 *Podisus crocatus* Uhler, Check-list Hem. Het., p. 4. *Nomen nudum*.
1897 *Podisus crocatus* Uhler, Trans. Md. Acad. Sci., 1: 384. [B.C., Cal., Ore., "Washington Territories"].
1904 *Podisus (Apoecilus) crocatus*: Van Duzee, Trans. Am. Ent. Soc., 30: 68.
1907 *Apateticus (Apoecilus) crocatus*: Schouteden, Gen. Ins., 52: 70.
1909 *Apateticus (Apoecilus) bracteatus* var. *crocata* [*sic*]: Kirkaldy, Cat. Hem., 1: 22.
1910 *Apeteticus* [*sic*] *bracteatus*: Banks, Cat. Nearc. Hem.-Het., p. 93 (in part).
1921 *Apateticus crocatus*: Downes, Proc. Ent. Soc. B.C., 16: 21.
Distribution: Ariz., B.C., Cal., Col., Man., Mass., Mich., Minn., N.M., N.Y., Ore., Ut., Wash.

Apateticus cynicus (Say), 1832

1832 *Pentatoma cynica* Say, Descrip. Het. Hem. N. Am., p. 3. ["Missouri"].

1851 *Arma grandis* Dallas, List Hem. Brit. Mus., 1: 96. [N.Y.]. Synonymized by Uhler, 1878, Proc. Bost. Soc. Nat. Hist., 19: 370.

1870 *Podisus cynicus*: Stål, K. Svens. Vet.-Akad. Handl., 9(1): 54.

1870 *Podisus (Apoecilus) grandis*: Stål, K. Svens. Vet.-Akad. Handl., 9(1): 49.

1885 *Podisus cynica* [*sic*]: Lintner, 15th An. Rept. Ent. Soc. Ontario, p. 13.

1904 *Podisus (Apoecilus) cynicus*: Van Duzee, Trans. Am. Ent. Soc., 30:68.

1907 *Apateticus (Apoecilus) cynicus*: Schouteden, Gen. Ins., 52: 71.

1908 *Apateticus (Apoeicilus* [*sic*]) *cynicus*: Torre-Bueno, J. N.Y. Ent. Soc., 16: 225.

1910 *Apeteticus* [*sic*] *cynicus*: Banks, Cat. Nearc. Hem.-Het., p. 93.

1914 *Apateticus cynicus*: Parshley, Psyche, 21: 149.

Distribution: Ariz., Ark., Col., "Dakota," Fla., Ia., Ill., Ks., Man., Mass., Md., Mich., Mo., Neb., N.J., N.Y., Oh., Ont., Pa., Que., Tex., Wis.

Genus *Eocanthecona* Bergroth, 1915

1915 *Eocanthecona* Bergroth, An. Mag. Nat. Hist., ser. 8, 15: 484. Type-species: *Cimex furcellatus* Wolff, 1811. Original designation.

Eocanthecona furcellata (Wolff), 1811

1811 *Cimex furcellatus* Wolff, Icon. Cimic., 5: 182. [India].

Distribution: Fla.(?) (Asiatic countries and Pacific Islands).

Note: The late R. I. Sailer (1981, pers. comm.) supplied information that several thousands of these insects were released in Florida during the early 1980's as a potential biological control agent for combating the Colorado potato beetle, *Leptinotarsa declimlineata* (Say), and the eastern tent caterpillar, *Malacosoma americanum* (Fabricius). Recovery has not yet been reported.

Genus *Euthyrhynchus* Dallas, 1851

1851 *Euthyrhynchus* Dallas, List Hem. Brit. Mus., 1: 77, 104. Type-species: *Cimex floridanus* Linnaeus, 1767. Monotypic.

Euthyrhynchus floridanus (Linnaeus), 1767 [Frontispiece]

1767 *Cimex floridanus* Linnaeus, Syst. Nat., Ed. 12, 1, pt. 2: 719. ["Carolina"].

1831 *Pentatoma emarginata* Say, Descrip. Het. Hem. N. Am., p. 4. [Ga.]. Synonymized by Stål, 1870, K. Svens. Vet.-Akad. Handl., 9(1): 55.

1843 *Asopus floridanus*: Amyot and Serville, Hist. Nat. Ins. Hem., p. 84.

1870 *Euthyrhynchus floridanus*: Stål, K. Svens. Vet.-Akad. Handl., 9(1): 54.

Distribution: Ark., Fla., Ga., La., Miss., Mo., N.C., Pa., S.C., Tenn., Va. (Mexico to Brazil).

Note: Oetting and Yonke (1975, An. Ent. Soc. Am., 68: 659-662) and Ables (1975, J. Ga. Ent. Soc., 10: 353-356) provided notes on biology and immature stages.

Genus *Heteroscelis* Latreille, 1829

1829 *Heteroscelis* Latreille, Reg. Anim., 5: 194. Type-species: *Heteroscelis servillii* Laporte, 1833. First included species.

1907 *Heterosceloides* Schouteden, Gen. Ins., 52: 19. Type-species: *Phyllochirus lepida* Stål, 1862. Original designation. Synonymized by Piran, 1961, Rev. Inv. Agr., 15: 84.

Heteroscelis lepida (Stål), 1862
1862 *Phyllochirus lepida* Stål, Stett. Ent. Zeit., 23(1): 93. [Mexico].
1914 *Heteroscelis lepida*: Barber, J. N.Y. Ent. Soc., 22: 171.
1916 *Heterosceloides lepida*: Van Duzee, Check List Hem., p. 9.
Distribution: Tex. (Mexico to French Guiana).

Genus *Mineus* Stål, 1867

1867 *Mineus* Stål, Öfv. K. Vet.-Akad. Förh., 24(7): 498. Type-species: *Podisus strigipes* Herrich-Schaeffer, 1853. Monotypic.

Mineus strigipes (Herrich-Schaeffer), 1853
1853 *Podisus strigipes* Herrich-Schaeffer, Wanz. Ins., 9: 338. [No locality given].
1870 *Mineus strigipes*: Stål, K. Svens. Vet.-Akad. Handl., 9(1): 32.
Distribution: Ariz., Col., Conn., D.C., Fla., Ga., Ill., Ind., Ky., Mass. Md., Mich., Mo., N.C., N.J., N.M., N.Y., Oh., Ok., S.C., Tex., Va.

Genus *Oplomus* Spinola, 1840

1840 *Oplomus* Spinola, Essai Hem., p. 355. Type-species: *Cimex tripustulatus* Fabricius, 1803. Preoccupied. *Asopus salamandra Burmeister*, 1835, is next available name. Designated by Schouteden, 1907, Gen. Ins., 52: 34.

Subgenus *Oplomus* Spinola, 1840

1840 *Oplomus* Spinola, Essai Hem., p. 355. Type-species: *Cimex tripustulatus* Fabricius, 1803.
1870 *Oplomus* (*Oplomus*): Stål, K. Svens. Vet.-Akad. Handl., 9(1): 27.

Oplomus mundus Stål, 1862
1862 *Oplomus mundus* Stål, Stett. Ent. Zeit., 23(1): 87. [Mexico].
1906 *Oplomus mundus*: Barber, Brook. Inst. Arts Sci., Sci. Bull., 1: 263.
1909 *Oplomus* (*Oplomus*) *mundus*: Kirkaldy, Cat. Hem., 1: 8.
Distribution: Tex. (Mexico).

Oplomus salamandra (Burmeister), 1835
1803 *Cimex tripustulatus* Fabricius, Syst. Ent., p. 172. [South America]. Preoccupied.
1835 *Asopus salamandra* Burmeister, Handb. Ent., 2: 381. [Peru].
1916 *Oplomus* (*Oplomus*) *tripustulatus*: Van Duzee, Check List Hem., p. 9.
1926 *Oplomus tripustulatus*: Blatchley, Het. E. N. Am., p. 185.
Distribution: Fla. (Mexico to Paraguay).

Oplomus salamandra salamandra (Burmeister), 1835
1835 *Asopus salamandra* Burmeister, Handb. Ent., 2: 381. [Peru].
1916 *Oplomus* (*Oplomus*) *tripustulatus* var. *tripustulatus*: Van Duzee, Check List Hem., p. 9.
1926 *Oplomus tripustulatus*: Blatchley, Het. E. N. Am., p. 185.
1939 *Oplomus* (*Oplomus*) *tripustulatus*: Torre-Bueno, Ent. Am., 19: 249.
Distribution: Fla. (Mexico to Paraguay).
Note: Only nominate subspecies occurs in our area.

Subgenus *Polypoecilus* Stål, 1870

1870 *Oplomus* (*Polypoecilus*) Stål, K. Svens. Vet.-Akad. Handl., 9(1): 26. Type-species: *Asopus dichrous* Herrich-Schaeffer, 1839. Designated by Schouteden, 1907, Gen. Ins., 52: 33.

Oplomus dichrous (Herrich-Schaeffer), 1839

1839 *Asopus dichrous* Herrich-Schaeffer, Wanz. Ins., 4: 89, fig. 426. [Mexico].

1904 *Oplomus dichrous*: Van Duzee, Trans. Am. Ent. Soc., 30: 64.

1907 *Oplomus (Polypoecilus) dichroa* [*sic*]: Schouteden, Gen. Ins., 52.: 34.

1909 *Oplomus (Polypoecilus) dichrous*: Kirkaldy, Cat. Hem., 1: 9.

Distribution: Ariz. (Mexico).

Genus *Perillus* Stål, 1862

1862 *Oplomus (Perillus)* Stål, Stett. Ent. Zeit., 23(1): 88. Type-species: *Asopus confluens* Herrich-Schaeffer, 1839. Designated by Schouteden, 1907, Gen. Ins., 52: 11, 36.

1867 *Perillus*: Stål, Öfv. K. Svens. Vet.-Akad. Förh., 24(7): 496.

1907 *Perilloides* Schouteden, Gen. Ins., 52: 11, 37. Type-species: *Cimex bioculatus* Fabricius, 1775. Original designation. Synonymized by Van Duzee, 1916, Check List Hem., p. 9.

Note: Knight (1952, An. Ent. Soc. Am., 45: 229-232) reviewed *Perillus* and presented a key to species.

Perillus bioculatus (Fabricius), 1775

1775 *Cimex bioculatus* Fabricius, Syst. Ent., p. 715. ["America"].

1825 *Pentatoma clanda* Say, J. Acad. Nat. Sci. Phila., 4: 312. ["Missouri"]. Synonymized by Van Duzee, 1904, Trans. Am. Ent. Soc., 30: 66.

1869 *Zirona* [*sic*] *clauda* [*sic*]: Rathvon, Hist. Lancaster Co., Pa., p. 549.

1872 *Perillus bioculatus*: Stål, K. Svens. Vet.-Akad. Handl., 10(4): 129.

1872 *Perillus claudus* [*sic*]: Uhler, Prelim. Rept. Geol. Surv. Mont., p. 395.

1886 *Mineus bioculatus*: Uhler, Check-list Hem. Het., p. 4.

1906 *Perillus bioculatus claudus* [*sic*]: Snow, Trans. Ks. Acad. Sci., 20: 177.

1907 *Perilloides bioculatus*: Schouteden, Gen. Ins., 52: 37.

1917 *Perillus bioculatus* var. *bioculatus*: Van Duzee, Univ. Cal. Publ. Ent., 2: 74.

1917 *Perillus bioculatus* var. *clanda*: Van Duzee, Univ. Cal. Publ. Ent., 2: 74.

1926 *Perillus bioculatus* var. *clauda* [*sic*]: Stoner, Can. Ent., 58: 30.

Distribution: Alta., Ariz., Ark., Cal., Col., Fla., Ia., Id., Ill., Ks., Mich., Mo., Mont., Neb., Nev., N.J., N.M., N.Y., Oh., Ont., Ore., Pa., Que., Tex., Ut., Wyo. (Mexico).

Note: The "twospotted stink bug" is economically important as a predator of harmful insects, especially the Colorado potato beetle, and as such has been introduced in Europe. Experimental studies by Knight (1924, An. Ent. Am., 17: 258-272) showed temperature to be a controlling factor for the nonblack color of this insect. European entomologists and physiologists provided an extensive literature on development and physiology.

Perillus circumcinctus Stål, 1862

1862 *Perillus circumcinctus* Stål, Stett. Ent. Zeit., 23(1): 89. [Mexico].

1872 *Perillus marginatus* Provancher, Nat. Can., 4: 74. [No locality given]. Synonymized by Provancher, 1886, Pet. Faune Ent. Can., p. 32. Lectotype designated by Kelton, 1968, Nat. Can., 95: 1066.

1907 *Perilloides circumcinctus*: Schouteden, Gen. Ins., 52: 37.

Distribution: Ill., Man., Mass., Mich., Minn., Mo., Neb., N.H., N.J., N.Y., Oh., Ont., Sask., S.D.

Perillus confluens (Herrich-Schaeffer), 1840
1840 *Asopus confluens* Herrich-Schaeffer, Wanz. Ins., 5: 77, fig. 522. [Mexico].
1870 *Perillus confluens*: Stål, K. Svens. Vet.-Akad. Handl., 9(1): 32.
Distribution: Ariz., Col., N.M., Tex. (Mexico south to Costa Rica).

Perillus exaptus (Say), 1825
1825 *Pentatoma exapta* Say, J. Acad. Nat. Sci. Phila., 4: 313. ["Missouri"].
1837 *Pentatoma variegata* Kirby, Fauna Bor. Am., p. 267. ["New York to Cumberland-house"; "Boreali-Americana"]. Preoccupied. Synonymized by Stål, 1870, K. Svens. Vet.-Akad. Handl., 9(1): 32. See note below.
1851 *Zicrona marginella* Dallas, List Hem. Brit. Mus., 1: 109. [Hudson's Bay]. Synonymized by Stål, 1870, K. Svens. Vet.-Akad. Handl., 9(1): 32.
1861 *Zicrona exapta*: Uhler, Proc. Ent. Soc. Phila., 1: 23.
1870 *Perillus exaptus*: Stål, K. Svens. Vet.-Akad. Handl., 9(1): 32.
1907 *Perilloides exaptus*: Schouteden, Gen. Ins., 52: 38.
1909 *Perilloides exaptus* var. *variegatella* Kirkaldy, Cat. Hem., 1: 7. New name and status for *Pentatoma variegata* Kirby, 1837.
1921 *Perillus exaphus* [*sic*]: Barber, Can. Ent., 53: 146.
Distribution: Alta., B.C., Cal., Col., "Dakota," Ill., Ind., Mass., Mich., Mo., Mont., N.J., N.M., N.Y., N.S., Nfld., Oh., Ont., Que., Ut., Wash., Wyo.
Note: Kirkaldy (1909, above) raised Kirby's (1837, above) preoccupied "*Pentatoma variegata*" and gave it the new species name "*variegatella*." Subsequent authors have followed Stål's synonymizing and, apparently, overlooked Kirkaldy's replacement name.

Perillus lunatus Knight, 1952
1952 *Perillus lunatus* Knight, An. Ent. Soc. Am., 45: 230. [Col.].
Distribution: Col., Mont.

Perillus splendidus (Uhler), 1861
1861 *Zicrona splendida* Uhler, Proc. Ent. Soc. Phila., 1: 22. [Cal.]
1870 *Perillus splendidus*: Stål, K. Svens. Vet.-Akad. Handl., 9(1): 32.
1907 *Perilloides splendidus*: Schouteden, Gen. Ins., 52: 38.
1926 *Perillus splendens* [*sic*]: Barber, J. N.Y. Ent. Soc., 34: 219.
Distribution: Cal., Col., Tex. (Mexico).

Genus *Picromerus* Amyot and Serville, 1843

1843 *Picromerus* Amyot and Serville, Hist. Nat. Ins., Hem., p. 84. Type-species: *Cimex bidens* Linnaeus, 1758. Monotypic.

Picromerus bidens (Linnaeus), 1758 [Fig. 117]
1758 *Cimex bidens* Linnaeus, Syst. Nat., Edit. 10, p. 443. [Europe].
1967 *Picromerus bidens*: Cooper, Ent. News, 78: 36.
Distribution: Mass., Me., N.H., N.Y., Que., Vt. (Europe).
Note: The earliest North American specimen-record for this introduced European species is 1932. Kelton (1972, Can. Ent., 104: 1743-1744) presented a summary of this species in North America, comments on its habits, and a color photograph of the adult. Javahery (1986, Ent. News, 97: 87-98) studied the biology, ecology, and reproductive potential of North American populations.

Genus *Podisus* Herrich-Schaeffer, 1851

1851 *Podisus* Herrich-Schaeffer, Wanz. Ins., 9: 296, 337. Type-species: *Podisus vittipennis* Herrich-Schaeffer, 1853, a junior synonym of *Arma nigrispina* Dallas, 1851. Designated by Kirkaldy, 1908, Ent., 41: 124.

Note: Kirkland's (1898, 45th An. Rept. Sect. Mus. St. Bd. Agr. for 1897: 112-138) contains many notes on the habits and biology of species in this genus.

Subgenus *Podisus* Herrich-Schaeffer, 1851

1851 *Podisus*: Herrich-Schaeffer, Wanz. Ins., 9: 296, 337. Type-species: *Podisus vittipennis* Herrich-Schaeffer, 1853, a junior synonym of *Arma nigrispina* Dallas, 1851. Designated by Kirkaldy, 1908, Ent., 41: 124.

1870 *Podisus* (*Podisus*): Stål, K. Svens. Vet.-Akad. Handl., 9(1): 48.

Note: Evans (1985, Proc. Ent. Soc. Wash., 87: 94-97) presented a key to the nymphs of four species.

Podisus fretus Olsen, 1916

1906 *Podisus fuscescens*: Barber, Mus. Brook. Inst. Arts Sci., Sci. Bull. 1: 263.
1916 *Podisus fretus* Olsen, Bull. Brook. Ent. Soc., 11: 82. [N.J.].
1917 *Podisus* (*Podisus*) *fretus*: Van Duzee, Univ. Cal. Publ. Ent., 2: 79.
1939 *Podisus* (*Eupodisus*) *fretus*: Torre-Bueno, Ent. Am., 19: 256.
Distribution: Fla., Ind., Mass., Me., Mich., N.C., N.J., N.Y., Va.

Podisus fuscescens (Dallas), 1851

1851 *Arma fuscescens* Dallas, List Hem. Brit. Mus., 1: 102. [Mexico].
1909 *Apateticus* (*Podisus*) *fuscescens*: Kirkaldy, Cat. Hem., 1: 19.
1910 *Apeteticus* [*sic*] *fuscescens*: Banks, Cat. Nearc. Hem.-Het., p. 93.
1916 *Podisus* (*Podisus*) *fuscescens*: Van Duzee, Check List Hem., p. 10.
1939 *Podisus* (*Eupodisus*) *fuscescens*: Torre-Bueno, Ent. Am., 19: 256.
Distribution: Tex. (Mexico, Grenada).

Podisus maculiventris (Say), 1832 [Fig. 110]

1832 *Pentatoma maculiventris* Say, Ins. La., p. 11. [La.].
1851 *Arma spinosa* Dallas, List Hem. Brit. Mus., 1: 98. ["Trenton Falls"]. Synonymized by Van Duzee, 1904, Trans. Am. Ent. Soc., 30: 71.
1859 *Arma pallens* Stål, K. Svens. Freg. Eug. Resa Jorden, 3: 222. [Cal.]. Synonymized by Van Duzee, 1916, Check List Hem., p. 10.
1869 *Arma macula* [*sic*]: Rathvon, Hist. Lancaster Co., Pa., p. 549.
1872 *Podisus spinosus*: Uhler, Prelim. Rept. U.S. Geol. Surv. Mont., 5: 394.
1875 *Podisus pallens*: Uhler, Bull. U.S. Geol. Geogr. Surv. Terr., 1: 282.
1875 *Podisus* (*Arma*) *spinosus*: Glover, Rept. Dept. Agr. U.S., p. 118.
1879 *Arma* (*Podisus*) *spinosa*: Comstock, Rept. Cotton Ins., p. 166.
1903 *Podisus maculiventris*: Torre-Bueno, J. N.Y. Ent. Soc., 11: 128.
1904 *Podisus* (*Podisus*) *maculiventris*: Van Duzee, Trans. Am. Ent. Soc., 30: 71.
1907 *Apateticus* (*Eupodisus*) *maculiventris*: Schouteden, Gen. Ins., 52: 72.
1907 *Apateticus* (*Eupodisus*) *pallens*: Schouteden, Gen. Ins., 52: 72.
1909 *Apateticus* (*Podisus*) *maculiventris*: Kirkaldy, Cat. Hem., 1: 19.
1909 *Apateticus* (*Podisus*) *pallens*: Kirkaldy, Cat. Hem., 1: 20.
1910 *Apateticus maculiventris*: Kirkaldy, Proc. Haw. Ent. Soc., 2: 124.
1910 *Apeteticus* [*sic*] *maculiventris*: Banks, Cat. Nearc. Hem.-Het., p. 93.

Distribution: Alta., Ariz., Ark., B.C., Cal., Col., D.C., Fla., Ga., Ia., Ill., Ind., Ks., La., Man., Mass., Md., Mich., Miss., Mo., Mont., N.B., N.C., N.J., N.S., N.Y., Neb., Oh., Ok., Ont., Pa., Que., S.C., Tex., Ut., Va., Vt., Wis. (Introduced into Europe and Korea).

Note: Mukerji and LeRoux (1969, Can. Ent., 101: 314-327, 101: 387-403, 101: 449-460) discussed the predatory nature of this species.

Podisus modestus (Dallas), 1851

1851 *Arma modesta* Dallas, List Hem. Brit. Mus., 1: 101. ["N. America"; "Cincinnati"; "Trenton Falls"].

1870 *Podisus* (*Podisus*) *modestus*: Stål, K. Svens. Vet.-Akad. Handl., 9(1): 51.

1876 *Podisus modestus*: Uhler, Bull. U.S. Geol. Geogr. Surv. Terr, 1: 283.

1907 *Apateticus* (*Eupodisus*) *modestus*: Schouteden, Gen. Ins., 52: 72.

1909 *Apateticus* (*Podisus*) *modestus*: Kirkaldy, Cat. Hem., 1: 19.

1910 *Apeteticus* [*sic*] *modestus*: Banks, Cat. Nearc. Hem.-Het., p. 93.

1914 *Apateticus modestus*: Parshley, Psyche, 21: 149.

Distribution: Alta., B.C., Col., "Dakota," Ga., Ia., Id., Ill., Ind., Man., Mass, Me., Mich., Mont., N.C., N.H., N.M., N.S., N.Y., Neb., Oh., Ont., Que. (Greater Antilles, Mexico).

Podisus mucronatus Uhler, 1897

1897 *Podisus mucronatus* Uhler, Trans. Md. Acad. Sci., 1: 386. [Fla., Cuba].

1904 *Podisus* (*Podisus*) *mucronatus*: Van Duzee, Trans. Am. Ent. Soc., 30: 69, 72.

1907 *Apateticus* (*Eupodisus*) *mucronatus*: Schouteden, Gen. Ins., 52: 72.

1909 *Apateticus* (*Podisus*) *mucronatus*: Kirkaldy, Cat. Hem., 1: 20.

1910 *Apeteticus* [*sic*] *mucronatus*: Banks, Cat. Nearc. Hem.-Het., p. 93.

Distribution: Fla. (Cuba).

Podisus placidus Uhler, 1870

1870 *Podisus* (*Podisus*) *placidus* Uhler, Am. Ent., 2: 203. ["Canada"; "Washington Territory"; Mass.].

1869 *Steretrus* [*sic*] *fimbriatus*: Saunders, Can. Ent., 2: 15.

1870 *Arma placidum*: Saunders, Can. Ent., 2: 93.

1897 *Podisus placidus*: Kirkland, Can. Ent., 29: 116.

1904 *Podisus* (*Podisus*) *placidus*: Van Duzee, Trans. Am. Ent. Soc., 30: 71.

1907 *Apateticus* (*Eupodisus*) *placidus*: Schouteden, Gen. Ins., 52: 72.

1909 *Apateticus* (*Podisus*) *placidus*: Kirkaldy, Cat. Hem., 1: 20.

1910 *Apateticus placidus*: Smith, Rept. N.J. St. Mus., p. 138.

1910 *Apeteticus* [*sic*] *placidus*: Banks, Cat. Nearc. Hem.-Het., p. 94.

Distribution: Alta., Ark., B.C., Col., Ia., Id., Ill., Mass., Me., Mich., Mo., N.J., N.Y., Neb., Oh., Ont., Que., Ut.

Note: Biology and descriptions of immature stages provided by Oetting and Yonke (1971, Can. Ent., 103: 1506-1510).

Podisus sagitta (Fabricius), 1794

1794 *Cimex sagitta* Fabricius, Ent. Syst., 4: 99. ["Insular America meridionalis"].

1904 *Podisus* (*Podisus*) *sagitta*: Van Duzee, Trans. Am. Ent. Soc., 30: 69, 72.

1907 *Apateticus* (*Eupodisus*) *sagitta*: Schouteden, Gen. Ins., 52: 72.

1909 *Apateticus* (*Podisus*) *sagitta*: Kirkaldy, Cat. Hem., 1: 20.

1910 *Apeteticus* [*sic*] *sagitta*: Banks, Cat. Nearc. Hem.-Het., p. 94.

1926 *Podisus sagitta*: Blatchley, Het. E. N. Am., p. 196.

1939 *Podisus* (*Eupodisus*) *sagitta*: Torre-Bueno, Ent. Am., 19: 256.

Distribution: Fla., Tex. (Mexico to Brazil, West Indies).

Podisus serieventris Uhler, 1871

1871 *Podisus serieventris* Uhler, Proc. Boston Soc. Nat. Hist., 14: 94. [Mass., Me., Minn.].

1904 *Podisus (Podisus) serieventris*: Van Duzee, Trans. Am. Ent. Soc., 30: 69, 71.

1907 *Apateticus (Eupodisus) serieventris*: Schouteden, Gen. Ins., 52: 72.

1909 *Apateticus (Podisus) serieventris*: Kirkaldy, Cat. Hem., 1: 20.

1910 *Apeteticus* [*sic*] *serieventris*: Banks, Cat. Nearc. Hem.-Het., p. 94.

1914 *Apateticus serieventris*: Parshley, Psyche, 21: 149..

1940 *Podisus serviventris* [*sic*]: Clausen, Ent. Ins., p. 586.

Distribution: B.C., Col., Fla., Id., Ill., Ind., Mass., Me., Minn., Mont., N.C., N.H., N.M., N.S., N.Y., Nfld., Oh., Ont., Pa., Que., Ut.

Subgenus *Tylospilus* Stål, 1870

1870 *Podisus (Tylospilus)* Stål, K. Svens. Vet.-Akad. Handl., 9(1): 52. Type-species: *Arma chilensis* Spinola, 1852. Designated by Schouteden, 1907, Gen. Ins., 52: 70, 73.

Podisus acutissimus Stål, 1870

1870 *Podisus (Tylospilus) acutissimus* Stål, K. Svens. Vet.-Akad. Handl, 9(1): 53. [Mexico, Tex.].

1876 *Tylospilus acutissimus*: Uhler, Bull. U.S. Geol. Geogr. Surv. Terr., 1: 283.

1886 *Podisus acutissimus*: Uhler, Check-list Hem. Het., p. 4.

1907 *Apateticus (Tylospilus) acutissimus*: Schouteden, Gen. Ins., 52: 73.

1910 *Apeteticus* [*sic*] *acutissimus*: Banks, Cat. Nearc. Hem.-Het., p. 93.

Distribution: Ariz., Col., N.M., Tex. (Mexico, Guatemala).

Genus *Rhacognathus* Fieber, 1861

1861 *Rhacognathus* Fieber, Europ. Hem., pp. 81, 347. Type-species: *Cimex punctatus* Linnaeus, 1758. Monotypic.

Rhacognathus americanus Stål, 1870

1870 *Rhacognathus americanus* Stål, K. Svens. Vet.-Akad. Handl., 9(1): 33. [Ill.].

Distribution: Alta., Ill., Ind., Man., Mass., Mich., Minn., Neb., Oh.

Genus *Stiretrus* Laporte, 1833

1833 *Stiretrus* Laporte, Essai Class. Syst. Hem., p. 75. Type-species: *Scutellera decemguttata* Lepeletier and Serville, 1828. Designated by Schouteden, 1907, Gen. Ins., 52: 6.

Note: Rathvon (1869, Hist. Lancaster Co., Pa., p. 549) listed *Stiretus* [*sic*] *dentatus* and *Stiretus* [*sic*] *bidentatus* without authors but each followed by a question mark. These can be regarded only as manuscript names.

Subgenus *Oncogaster* Stål, 1870

1870 *Stiretrus (Oncogaster)* Stål, K. Svens. Vet.-Akad. Handl., 9(1): 24. Type-species: *Stiretrus caeruleus* Dallas, 1851. Designated by Schouteden, 1907, Gen. Ins., 52: 7, 8.

Stiretrus anchorago (Fabricius), 1781

1775 *Cimex anchorago* Fabricius, Syst. Ent., p. 699. ["America"].

1803 *Tetyra anchorago*: Fabricius, Syst. Rhyn., p. 137.

1837 *Pentatoma pulchella* Westwood, Hope Cat., 1: 42. [Ga.]. Synonymized by Stål, 1870, K. Svens. Vet.-Akad. Handl., 9(1): 24.

1837 *Pentatoma anchorago* Westwood, Hope Cat., p. 42 (*lapsus*, as new species). [Ga.].

1851 *Stiretrus anchorago*: Dallas, List Hem. Brit. Mus., 1: 80.

1870 *Stiretrus* (*Oncogaster*) *anchorago*: Stål, K. Svens. Vet.-Akad. Handl., 9(1): 24.

1912 *Stiretrus anchorago* subspecies *inflata* Schumacher, Sitz. Gesell. Natfors. Freude, p. 96. [U.S.]. Synonymized by Van Duzee, 1916, Check List Hem., p. 9.

Distribution: Ala., Ark., Conn., Fla., Ga., Ia., Ill., Ind., Ks., Ky., Mass., Md., Mich., Mo., N.C., N.J., N.M., N.Y., Ok., Pa., S.C., Tex., Va. (Mexico to Panama).

Notes:Records for Ont. and Cal. were based on misidentified nymphs--see Saunders (1870, Can. Ent., 2: 93) and Van Duzee (1917, Univ. Cal. Publ. Ent., 2: 203), respectively, for corrections to the present species.

The above scientific name here encompasses a range of color variation tending toward certain patterns which early entomologists named as distinct species. Taxonomic elucidation of these patterns needs the results of breeding experiments and an intensive taxonomic revision. The most frequently reported and geographically widespread pattern is that known under the name "*fimbriatus*," which form has been treated as a separate species or as a subspecies or "variety" of *S. anchorago*. In the following list the generally recognized "varieties" (not to be confused with subspecies here) are grouped for the convenience of workers who may wish to attempt solution of the puzzle. The total distribution is given under the species heading above, but none is given for the "varieties," the total distribution happens to be as extensive as the report for "*fimbriatus*," all the other "varieties" appear to be restricted to the southern part of the total range.

Stiretrus anchorago var. *anchorago* (Fabricius), 1775

1775 *Cimex anchorago* Fabricius, Syst. Ent., p. 699 ["America"].

1876 *Stiretrus anchorago* var. *anchorago*: Uhler, Bull. U.S. Geol. Geogr. Surv. Terr., 1: 281.

Stiretrus anchorago var. *diana* (Fabricius), 1803

1803 *Tetyra Diana* [*sic*] Fabricius, Syst. Rhyn., p. 137. ["Carolina"].

1835 *Asopus diana*: Burmeister, Handb. Ent., 2: 381.

1839 *Stiretrus Dianae* [*sic*]: Germar, Zeit. Ent., 1: 19.

1859 *Stiretrus Diana* [*sic*]: Dohrn, Cat. Hem., p. 7.

1876 *Stiretrus anchorago* var. *diana*: Uhler, Bull. U.S. Geol. Geogr. Surv. Terr., 1: 281.

Stiretrus anchorago var. *fimbriatus* (Say), 1828 [Fig. 111]

1828 *Tetyra fimbriata* Say, Am. Ent., 3: 43. [Pa.].

1839 *Stiretrus fimbriatus*: Germar, Zeit. Ent., 1: 16.

1838 *Asopus variegatus* Herrich-Schaeffer, Wanz. Ins., 4: 90, fig. 427. [America].

1869 *Strietrus* [*sic*] *fimbriatus*: Saunders, Can. Ent., 2: 15.

1869 *Steretrus* [*sic*] *fimbriatus*: Walsh, Can. Ent., 2: 33.

1876 *Stiretrus anchorago* var. *fimbriatus*: Uhler, Bull. U.S. Geol. Geogr. Surv. Terr., 1: 281.

Note: Biology and immature stages studied by Oetting and Yonke (1971, Can. Ent., 103: 1510-1515).

Stiretrus anchorago var. *personatus* Germar, 1839

1839 *Stiretrus personatus* Germar, Zeit. Ent., 1: 16. [Mexico].

1878 *Stiretrus anchorago* var. *personatus*: Uhler, Proc. Boston Soc. Nat. Hist., 19: 369.

Stiretrus anchorago var. *violaceus* (Say), 1828
1828 *Tetyra violaceus* Say, Am. Ent., 3: 43. [Fla.].
1839 *Stiretrus violaceus*: Germar, Zeit. Ent., 1: 12.
1876 *Stiretrus anchorago* var. *violaceus*: Uhler, Bull. U.S. Geol. Geogr. Surv. Terr., 1: 281
1907 *Stiretrus* (*Oncogaster*) *anchorago* var. *violaceus*: Schouteden, Gen. Ins., 52: 9.

Genus *Zicrona* Amyot and Serville, 1843

1843 *Zicrona* Amyot and Serville, Hist. Nat. Ins. Hem., p. 86. Type-species: *Cimex caeruleus* Linnaeus, 1758. Monotypic.

Zicrona caerulea (Linnaeus), 1758
1758 *Cimex caeruleus* Linnaeus, Syst. Nat., Ed. 10, p. 445. [Europe].
1851 *Zicrona cuprea* Dallas, Check List Brit. Hem., 1: 108. ["Hudson's Bay"]. Synonymized by Uhler, 1872, Prelim. Rept. U.S. Geol. Surv., 5: 395.
1904 *Zicrona caerulea*: Van Duzee, Trans. Am. Ent. Soc., 30: 68.
Distribution: Ariz., B.C., Cal., Col., Id., Me., Mich., N.H., N.M., Ont., Que., Tex., Ut.

Subfamily Discocephalinae Stål, 1867

Tribe Discocephalini Stål, 1867

Genus *Lineostethus* Ruckes, 1966

1966 *Lineostethus* Ruckes, Am. Mus. Novit., 2255: 10, 22. Type-species: *Discocephala clypeata* Stål, 1862. Original designation.

Note: Revision and key to species given by Ruckes (1966, above) and Hildebrand and Beck (1982, Rev. Brasil Biol., 42: 773-784).

Lineostethus marginellus (Stål), 1872
1872 *Discocephala* (*Platycarenus*) *marginella* Stål, K. Svens. Vet.-Akad. Handl., 10(4): 6. [Mexico].
1909 *Platycarenus* (*Discocephalessa*) *marginellus*: Kirkaldy, Cat. Hem., 1: 216.
1916 *Platycarenus marginellus*: Van Duzee, Check List Hem., p. 4.
1966 *Lineostethus clypeatus marginellus*: Ruckes, Am. Mus. Novit., 2255: 26.
1982 *Lineostethus marginellus*: Hildebrand and Becker, Rev. Brasil. Biol., 42: 780.
Distribution: Tex. (Mexico).

Lineostethus tenebricornis (Ruckes), 1957 [Fig. 120]
1957 *Platycarenus* (*Discocephalessa*) *tenebricornis* Ruckes, Bull. Brook. Ent. Soc., 52: 16. [Ariz.].
1966 *Lineostethus tenebricornis*: Ruckes, Am. Mus. Novit., 2255: 24.
Distribution: Ariz. (Mexico).

Subfamily Edessinae Amyot and Serville, 1843

Note: This group was elevated from tribal to subfamily status by Rolston and McDonald (1979, J. N.Y. Ent. Soc., 87: 197-200).

Genus *Edessa* Fabricius, 1803

1803 *Edessa* Fabricius, Syst. Rhyn., p. 145. Type-species: *Edessa cervus* Fabricius, 1787. Designated by Chenu, 1859, Encycl. Hist. Nat., 8: 217.

Edessa bifida (Say), 1832

1832 *Pentatoma* (*Ascra*) *bifida* Say, Ins. La., p. 7. [La.].

1844 *Edessa albirenis* Herrich-Schaeffer, Wanz. Ins., 7: 127. ["Carolina"]. Synonymized by Stål, 1872, K. Svens. Vet.-Akad. Handl., 10(4): 58.

1868 *Edessa obtusa*: Walker, Cat. Hem. Brit. Mus., 3: 420.

1872 *Edessa bifida*: Stål, K. Svens. Vet.-Akad. Handl., 10(4): 58.

Distribution: Ark., Fla., La., Md., Tex. (Mexico to Brazil, West Indies).

Edessa florida Barber, 1935 [Fig. 121]

1935 *Edessa florida* Barber, Proc. Ent. Soc. Wash., 37: 48. [Fla.].

Distribution: Fla., La., Md., N.C., S.C., Va.

Subfamily Pentatominae Leach, 1815

Tribe Halyini Amyot and Serville, 1843

Genus *Brochymena* Amyot and Serville, 1843

1843 *Brochymena* Amyot and Serville, Hist. Nat. Hem., p. 106. Type-species: *Halys serrata*: Amyot and Serville, 1843, a misidentification of *Cimex quadripustulata* Fabricius, 1775. Monotypic.

Note: Ruckes (1947, Ent. Am., 26: 143-238) revised the genus and gave keys to species.

Brochymena affinis Van Duzee, 1904

1904 *Brochymena affinis* Van Duzee, Trans. Am. Ent. Soc., 30: 29. [Cal., Id., Wash.].

Distribution: B.C., Cal., Col., Id., Nev., Ore., Ut., Wash.

Note: "Iowa" recorded by Torre-Bueno (1939, Ent. Am., 19: 206) apparently is a *lapsus* for Idaho, which he did not list even though it was one of three originally reported states.

Brochymena apiculata Van Duzee, 1923

1923 *Brochymena apiculata* Van Duzee, Proc. Cal. Acad. Sci., (4)12: 126. [Mexico].

Distribution: "Southwestern States" (Ruckes, 1947, Ent. Am., 26: 160) (Mexico).

Brochymena arborea (Say), 1825 [Fig. 115]

1825 *Pentatoma arborea* Say, J. Nat. Sci. Phila., 4: 311. ["Missouri"].

1840 *Halys erosa* Herrich-Schaeffer, Wanz. Ins., 5: 70. ["Nordamerika"]. Synonymized by Dallas, 1851, List Hem. Brit. Mus., 1: 188.

1851 *Brochymena arborea*: Dallas, List Hem. Brit. Mus., 1: 188.

1867 *Brochymena annulata*: Walker, Cat. Het.-Hem. Brit. Mus., 1: 230.
1869 *Brachymena* [*sic*] *arborea*: Rathvon, Hist. Lancaster Co., Pa., p. 549.
Distribution: Ark., Conn., D.C., Fla., Ga., Ill., Ind., Ks., La., Mass., Md., Me., Mich., Mo., N.C., N.H., N.J., N.M., N.Y., Oh., Ont., Pa., Que., S.D., Tex., Va. (Mexico).

Brochymena barberi Ruckes, 1939
1939 *Brochymena barberi* Ruckes, Bull. Brook. Ent. Soc., 34: 111. [Ariz.].
Distribution: Ariz., Tex.

Brochymena barberi barberi Ruckes, 1939
1939 *Brochymena barberi* Ruckes, Bull. Brook. Ent. Soc., 34: 111. [Ariz.].
Distribution: Ariz.

Brochymena barberi diluta Ruckes, 1939
1939 *Brochymena barberi* variety *diluta* Ruckes, Bull. Brook. Ent. Soc., 34: 113. [Tex.].
Distribution: Tex.

Brochymena cariosa Stål, 1872
1872 *Brochymena cariosa* Stål, K. Svens. Vet.-Akad. Handl., 10(4): 17. [Tex.].
Distribution: Ala., Ark., Fla., Ill., Ind., Ks., La., Miss., Mo., N.C., Neb., Oh., Tenn., Tex.
Note: Eikenbary and Raney (1968, J. Econ. Ent., 61: 1336) commented on predatory habits.

Brochymena carolinensis (Westwood), 1837
1775 *Cimex annulatus* Fabricius, Syst. Ent., p. 704. [Va.]. Preoccupied.
1803 *Halys annulata*: Fabricius, Syst. Rhyn., p. 182.
1811 *Halys serrata*: Wolff, Icones Cimic., 5: 179.
1837 *Halys carolinensis* Westwood, Hope Cat., 1: 22. [Carolina"].
1851 *Brochymena carolinensis*: Dallas, List Hem. Brit. Mus., 1: 189.
1859 *Brochymena annulata*: Dohrn, Cat. Hem. p. 13.
1869 *Brachymena* [*sic*] *carolinensis*: Rathvon, Hist. Lancaster Co., Pa., p. 549.
1871 *Brochymena Harrisii* [*sic*] Uhler, Proc. Boston Soc. Nat. Hist., 14: 95. [Pa., S.C.]. Synonymized by Van Duzee, 1916, Check List Hem., p. 4.
Distribution: Ala., Ark., Conn., D.C., Fla., Ga., Ind., Mass., Md., Mo., N.C., N.J., N.Y., Oh., Pa., Tex.

Brochymena chelonoides Ruckes, 1957
1957 *Brochymena chelonoides* Ruckes, Bull. Brook. Ent. Soc., 52: 22. [Tex.].
Distribution: Tex. (Mexico).

Brochymena dilata Ruckes, 1939
1939 *Brochymena dilata* Ruckes, Bull. Brook. Ent. Soc., 33: 239. [Ariz.].
Distribution: Ariz., N.M., Tex., Ut.

Brochymena exardentia Ruckes, 1961
1961 *Brochymena exardentia* Ruckes, J. N.Y. Ent. Soc., 68: 227. [Tex.].
Distribution: Tex.

Brochymena florida Ruckes, 1939
1939 *Brochymena florida* Ruckes, Bull. Brook. Ent. Soc., 34: 236. [Fla.].
Distribution: Fla.
Note: Ruckes (1939, above) suggested that the continental records of *Brochymena poeyi* "may" belong here.

Brochymena haedula Stål, 1862
1862 *Brochymena haedula* Stål, Stett. Ent. Zeit., 23: 99. [Mexico].
Distribution: Ariz., Tex. (Mexico).

Note: Ruckes (1947, Ent. Am., 26: 163) wrote "I have not seen an authentic specimen from the United States." This statement probably should delete the species from lists for the area north of the Rio Grande River.

Brochymena hoppingi Van Duzee, 1921
1921 *Brochymena hoppingi* Van Duzee, Proc. Cal. Acad. Sci., 11: 111. [Col.].
1937 *Brochymena myops*: Ruckes, Bull. Brook. Ent. Soc., 32: 33.
Distribution: Ariz., Cal., Col., N.M., Ut.

Brochymena laevigata Ruckes, 1957
1957 *Brochymena laevigata* Ruckes, Bull. Brook. Ent. Soc., 52: 19. [Tenn.].
Distribution: Tenn.

Brochymena lineata Ruckes, 1939
1939 *Brochymena lineata* Ruckes, Bull. Brook. Ent. Soc., 33: 236. [Ariz.].
Distribution: Ariz., N.M.

Brochymena marginella Stål, 1872
1872 *Brochymena annulata* variety *marginella* Stål, K. Svens. Vet.-Akad. Handl., 10(4): 16. [Tex.].
1893 *Brochymena marginella*: Lethierry and Severin, Gen. Cat. Hem., 1: 97.
Distribution: Fla., Tex.
Note: Van Duzee (1904, Trans. Am. Ent. Soc., 30: 31) pointed out that the Lethierry and Severin (1893, above) "Carolina" locality is in error.

Brochymena myops Stål, 1872
1844 *Halys quadripustulata*: Herrich-Schaeffer, Wanz. Ins., 7: 57.
1872 *Brochymena myops* Stål, K. Svens. Vet.-Akad. Handl., 10(4): 16. [La., Mexico].
1886 *Brochymena quadrinotata*: Uhler, Check-list Hem. Het., p. 5.
Distribution: Ala., La., Miss., N.C., N.M., Tex. (Mexico).
Note: The N.M. record was "accepted with reservations" by Ruckes (1947, Ent. Am., 26: 209).

Brochymena parva Ruckes, 1946
1840 *Halys obscura* Herrich-Schaeffer, Wanz. Ins., 5: 68. [Mexico]. Preoccupied. See note below about lectotype from Ariz.
1876 *Brochymena obscura*: Uhler, Bull. U.S. Geol. Geogr. Surv. Terr., 1: 283.
1909 *Brochymena tenebrosa*: Kirkaldy, Cat. Hem., 1: 192.
1946 *Brochymena parva* Ruckes, Bull. Brook. Ent. Soc., 41: 41. [Ariz.]. New name for *Halys obscura* Herrich-Schaeffer, 1840.
Distribution: Ariz., Cal., Col., Nev., N.M., Tex. (Mexico).
Note: The combination *Halys obscura*, used by Herrich-Schaeffer (1840, above) for a Mexican species, is preoccupied. All U.S. records for *B. tenebrosa* Walker apply to *B. parva*. Ruckes (1946, above) proposed the new name *Brochymena parva* and published data from two 1937 specimens from Arizona and labeled them as "lectotypes" for the species. By definition (Code, 72(a)) a lectotype must be part of the original series of specimens; Ruckes' specimens obviously could not have been part of that series; here the male of that series is considered the neotype of *B. parva* (specimen deposited in the American Museum of Natural History).

Brochymena pilatei Van Duzee, 1934
1934 *Brochymena pilatei* Van Duzee, Pan-Pac. Ent., 10: 22. [Cal.].
Distribution: Ariz., Cal., Ut. (Mexico).

Brochymena poeyi (Guérin-Méneville), 1857

1857 *Pentatoma poeyi* Guérin-Ménevile, Hist. Cuba Ins., 7: 365. [Cuba].

1904 *Brochymena poeyi*: Van Duzee, Trans. Am. Ent. Soc., 30: 28.

Distribution: Fla. (Greater Antilles).

Note: Ruckes (1947, Ent. Am., 26: 171) expressed doubts about the validity of Van Duzee's (1904, above) "Fla." record and wrote he "suspected" the specimens might be "a form of *B. arborea*" but stopped short of placing them there.

Brochymena punctata Van Duzee, 1909

1909 *Brochymena punctata* Van Duzee, Can. Ent., 41: 369. [Ga.].

Distribution: Ark., Fla., Ga., Ind., Va.

Note: Blatchley's (1926, Het. E. N. Am., p. 103) "Indiana" record was questioned by Ruckes (1947, Ent. Am., 26: 202).

Brochymena punctata punctata Van Duzee, 1909

1909 *Brochymena punctata* Van Duzee, Can. Ent., 41: 369. [Ga., Va.].

1947 *Brochymena punctata*: Ruckes, Ent. Am., 26: 182, 201.

Distribution: Fla., Ga., Ind., Va.

Brochymena punctata pallida Blatchley, 1926

1926 *Brochymena pallida* Blatchley, Het. E. N. Am., p. 101. [Fla.].

1947 *Brochymena punctata* variety *pallida*: Ruckes, Ent. Am., 26: 182, 202.

Distribution: Fla.

Brochymena quadripustulata (Fabricius), 1775

1775 *Cimex 4.pustulatus* [*sic*] Fabricius, Syst. Ent., p. 704. ["America"].

1803 *Halys 4pustulata* [*sic*]: Fabricius, Syst. Rhyng., p. 182.

1818 *Halys serrata*: Palisot, Ins. Rec. Afr. Am., p. 187.

1839 *Halys pupillata* Herrich-Schaeffer, Wanz. Ins., 4: 104. [Ga.]. Synonymized by Stål, 1872, K. Svens. Vet.-Akad. Handl., 10(4): 16.

1843 *Brochymena serrata*: Amyot and Serville, Hist. Nat. Ins., Hem., 107.

1872 *Brochymena quadripustulata*: Stål, K. Svens. Vet.-Akad. Handl., 10(4): 16.

1872 *Brochymena 4-notata* [*sic*] Provancher, Nat. Can., 4: 74. [Ga.]. Synonymized by Van Duzee, 1904, Trans. Am. Ent. Soc., 30: 29. Lectotype designated by Kelton, 1968, Nat. Can., 95: 1067.

1875 *Brochymena annulata*: Uhler, Rept. U.S. Geol. Geogr. Surv. Terr., 5: 283.

1930 *Brachymena* [*sic*] *quadripustulata*: Shaw, J. N.Y. Ent. Soc., 38: 463.

1957 *Brochymoena* [*sic*] *quadripustulata*: Rings, J. Econ. Ent., 50: 600.

Distribution: Alta., Ariz., Ark., B.C., Cal., Col., Conn., D.C., Fla., Ia., Ill., Ind., Ks., La., Mass., Md., Mich., Miss., Mo., N.C., N.H., N.J., N.M., N.Y., Neb., Oh., Ok., Ont., Ore., Pa., Que., S.C., Ut., Va. (Mexico; introduced into Hawaii).

Brochymena sulcata Van Duzee, 1918

1918 *Brochymena sulcata* Van Duzee, Proc. Cal. Acad. Sci., 8: 276. [Ariz., Cal.].

Distribution: Ariz., Cal., Col., Nev., N.M., Tex., Ut.

Tribe Mediceini Distant, 1902

Genus *Mecidea* Dallas, 1851

1851 *Mecidea* Dallas, List Hem. Brit. Mus., 1: 131, 139. Type-species: *Mecidea indica* Dallas, 1851. Designated by Distant, 1902, Fauna Brit. India, Rhyn., 1: 140.

Note: Sailer (1952, Proc. U.S. Nat. Mus., 102: 471-505) gave a generic review with key to species. N. Am. records for *M. longula* Stål appear to be misidentifications of the following two species.

Mecidea major Sailer, 1952

1876 *Mecidea longula*: Uhler, Bull. U.S. Geol. Geogr. Surv. Terr., 1: 283 (in part).

1952 *Mecida major* Sailer, Proc. U. S. Nat. Mus., 102: 478, 486. [Tex.].

Distribution: Ariz., Ark., Ill., Ks., Mo., Ok., Tex.

Mecidea minor Ruckes, 1946

1872 *Mecidea longula*: Stål, K. Svens. Vet.-Akad. Handl., 10(4): 17 (in part).

1946 *Mecidea minor* Ruckes, Bull. Brook. Ent. Soc., 41: 87. [N.M.].

Distribution: Ariz., Ark., Cal., Ia., Mo., N.M., Ok., S.D., Tex. (Mexico).

Tribe Pentatomini Leach, 1815

Genus *Acrosternum* Fieber, 1860

1860 *Acrosternum* Fieber, Europ. Hem., p. 79. Type-species: *Acrosterum heegeri* Fieber, 1861. First included species.

Note: Rolston (1983, J. N.Y. Ent. Soc., 91: 97-176) provided a revision and key for the 52 New World species of *Acrosternum*, all of which belong to the subgenus *Chinavia*.

Subgenus *Chinavia* Orian, 1965

1965 *Chinavia* Orian, Proc. R. Ent. Soc. London (B), 34: 25. Type-species: *Rhaphigaster pallidoconspersa* Stål, 1858. Original designation.

1983 *Acrosternum* (*Chinavia*): Rolston, J. N.Y. Ent. Soc., 91: 100.

Acrosternum hilare (Say), 1832

1832 *Pentatoma hilaris* Say, Ins. La., p. 9. [Ga., La., Mo., Pa.].

1851 *Rhaphigaster Sarpinus* [*sic*] Dallas, List Hem. Brit. Mus., 1: 276. [Oh.]. Synonymized by Uhler, 1878, Proc. Bost. Soc. Nat. Hist., 19: 380.

1856 *Rhaphigaster pennsylvanicus*: Fitch, Trans. N.Y. St. Agr. Soc., 16: 389.

1869 *Nizara* [*sic*] *hilaris*: Rathvon, Hist. Lancaster Co., Pa., p. 549.

1878 *Nezara hilaris*: Uhler, Proc. Boston Soc. Nat. Hist., 19: 380.

1879 *Raphigaster* [*sic*] (*Nezara*) *hilaris*: Comstock, Rept. Cotton Ins., p. 167.

1909 *Nezara* (*Acrosternum*) *hilaris*: Kirkaldy, Cat., Hem., 1: 119.

1915 *Acrosternum hilare*: Parshley, Psyche, 22: 175.

1917 *Acrosternum hilaris*: Van Duzee, Univ. Cal. Publ. Ent., 2: 60.

1983 *Acrosternum* (*Chinavia*) *hilare*: Rolston, J. N.Y. Ent. Soc., 91: 155.

Distribution: Ariz., Ark., Cal., Col., Conn., D.C., Fla., Ga., Ia., Ill., Ind., Ks., Ky., La., Mass., Md., Mich., Miss., Mo., Mont., N.C., N.Y., Neb., Oh., Ok., Ont., Pa., Que., R.I., S.C., S.D., Tex., Ut., Va., Vt.

Acrosterum marginatum (Palisot), 1811

1811 *Pentatoma marginatum* Palisot, Ins. Rec. Afr. Am., p. 147, pl. Hem. 10, fig. 1. [Dominican Republic].

1904 *Nezara marginata*: Van Duzee, Trans. Am. Ent. Soc., 30: 58.

1909 *Nezara* (*Acrosternum*) *marginata*: Kirkaldy, Cat. Hem., 1: 119.

1914 *Acrosternum marginatum*: Bergroth, An. Soc. Ent. Belg., 58: 25

1916 *Acrosternum marginata* [*sic*]: Van Duzee, Check List Hem., p. 8.

1983 *Acrosternum* (*Chinavia*) *marginatum*: Rolston, J. N.Y. Ent. Soc., 91: 152.

Distribution: Ariz., Cal., Fla., Tex. (Mexico to Ecuador, West Indies).

Acrosterum pennsylvanicum (Gmelin), 1790

1773 *Cimex viridis pensylvanicus* [*sic*] De Geer, Mem. Ins., 3: 330, pl. 34, fig. 5. [Pa.]. Unavailable trinomen.

1790 *Cimex pensylvanicus* [*sic*] Gmelin, Syst. Nat., 13: 2148. [Pa.].

1818 *Pentatoma pensylvanica* [*sic*]: Palisot, Ins. Rec. Afr. Am., p. 186.

1832 *Pentatoma abupta* Say, Descrip. Het. Hem. N. Am., p. 6. [Ga.]. Synonymized by Uhler, 1871, Proc. Bost. Soc. Nat. Hist., 14: 98.

1851 *Rhaphigaster parnisus* Dallas, List Hem. Brit. Mus., 1: 279. [No locality given]. Synonymized by Uhler, 1886, Check-list Hem., p. 8.

1869 *Raphigaster* [*sic*] *pensylvanica* [*sic*]: Rathvon, Hist. Lancaster Co., Pa., p. 548.

1871 *Rhaphigaster pennsylvanicum*: Uhler, Proc. Boston Soc. Nat. Hist., 4: 98.

1872 *Acrosternum pennsylvanicum*: Stål, K. Svens. Vet.-Akad. Handl., 10(4): 42.

1886 *Nezara pennsylvanica*: Uhler, Check-list Hem. Het., p. 8.

1909 *Nezara* (*Acrosterum*) *pensylvanica* [*sic*]: Kirkaldy, Cat. Hem., 1: 120.

1983 *Acrosternum* (*Chinavia*) *pennsylvanicum*: Rolston, J. N.Y. Ent. Soc., 91: 130.

Distribution: Conn., Fla., Ga., Ia., Ill., La., Mass., Mich., Mo., N.J., N.Y., Oh., Pa., Que.

Note: Gmelin (1790, above), who followed the Linnaean binomial system, should be given authorship of this species, not Palisot (1818, above) who has been given credit by previous authors.

Genus *Aelia* Fabricius, 1803

1803 *Aelia* Fabricius, Syst. Rhyn., p. 188. Type-species: *Cimex acuminata* Linnaeus, 1758. Designated by Curtis, 1838, Brit. Ent., pl. 704.

Aelia americana Dallas, 1851 [Fig. 106]

1851 *Aelia americana* Dallas, List Hem. Brit. Mus., 1: 223. ["N. America"].

Distribution: Ala., Alta., Ariz., Ark., B.C., Col., Ill., Ks., Man., Mich., Mont., N.M., Neb., Ok., S.D., Sask.

Note: Van Duzee (1912, Can. Ent., 44: 318) wrote that the earlier recording for Quebec was based on a misidentified specimen of *Neottiglossa undata* (Say).

Genus *Arvelius* Spinola, 1840

1840 *Arvelius* Spinola, Essai Hem., p. 344. Type-species: *Cimex gladiator* Fabricius, 1775. Designated by Kirdaldy, 1909, Cat. Hem., 1, 150, a junior synonym of *Cimex albopunctatus* De Geer, 1773.

Arvelius albopunctatus (De Geer), 1773

1773 *Cimex albopunctatus* De Geer, Mem. Ins., 3, 331, pl. 34, fig. 6. [Surinam].

1876 *Arvelius albopunctatus*: Uhler, Bull. U.S. Geol. Geogr. Surv. Terr., 1: 290.

Distribution: Ariz., Col., Fla., Tex. (Mexico to Argentina, West Indies).

Genus *Banasa* Stål, 1860

1860 *Banasa* Stål, K. Svens. Vet.-Akad. Handl., 2(7): 24. Type-species: *Banasa induta* Stål, 1860. Designated by Kirkaldy, 1909, Cat. Hem., 1: 115.

1871 *Atomosira* Uhler, Proc. Boston Soc. Nat. Hist., 14: 97. Type-species: *Atomosira sordida* Uhler, 1871. Monotypic. Synonymized by Stål, 1872, K. Svens. Vet.-Akad. Handl., 10(4): 43.

Note: Thomas and Yonke (1981, J. Ks. Ent. Soc., 54: 233-248) presented a review and key to the nearctic species of *Banasa* and gave (1985, An. Ent. Soc. Am., 78: 855-862) an analysis of certain evolutionary aspects of the species.

Banasa calva (Say), 1832

1832 *Pentatoma calva* Say, Descrip. Het. Hem. N. Am., p. 7. [Va.].

1851 *Rhaphigaster catinus* Dallas, List Hem. Brit. Mus., 1: 282. ["Canada"]. Synonymized by Uhler, 1878, Proc. Bost. Soc. Nat. Hist., 19: 379.

1876 *Banasa calva*: Uhler, Bull. U.S. Geol. Geogr. Surv. Terr., 1: 291.

1878 *Atomosira calva*: Uhler, Proc. Boston Soc. Nat. Hist., 19: 379.

1885 *Banasa dimidiata*: Provancher, Pet. Faune Ent. Can., 3: 46.

1909 *Nezara (Atomosira) calva*: Kirkaldy, Cat. Hem., 1: 122.

Distribution: B.C., Col., Ga., Ill., Me., Mich., Mont., N.C., N.J., N.Y., Ont., Ore., Pa., Que., Va. (Guatemala, Mexico).

Note: Life history studied by De Coursey (1963, An. Ent. Soc. Am., 56: 690-692).

Banasa dimiata (Say), 1832

1832 *Pentatoma dimiata* Say, Descrip. Het. Hem. N. Am., p. 7. [Fla., Ga.]. See note below.

1859 *Pentatoma dimidiata* [*sic*]: LeConte, Compl. Writ. T. Say, 1: 318. Unjustified emendation.

1872 *Banasa dimidiata* [*sic*]: Stål, K. Svens. Vet.-Akad. Handl., 10(4): 43.

1885 *Banasa calva*: Provancher, Pet. Faune Ent. Can., 3: 46.

1885 *Banasa euchlora*: Provancher, Pet. Faune Ent. Can., 3: 46.

1909 *Nezara (Banasa) dimiata*: Kirkaldy, Cat. Hem., 1: 122.

1958 *Banasa zenia* Bliven, Occ. Ent., 1: 8. [Cal.]. Synonymized by Thomas and Yonke, 1981, J. Ks. Ent. Soc., 54: 241.

1958 *Banasa samarana* Bliven, Occ. Ent., 1: 9. [Cal.]. Synonymized by Thomas and Yonke, 1981, J. Ks. Ent. Soc., 54: 241.

1958 *Banasa tempestiva* Bliven, Occ. Ent., 1:10. [Cal.]. Synonymized by Thomas and Yonke, 1981, J. Ks. Ent. Soc., 54: 241.

1958 *Banasa casterlini* Bliven, Occ. Ent., 1: 11. [Cal.]. Synonymized by Thomas and Yonke, 1981, J. Ks. Ent. Soc., 54: 241.

1958 *Banasa semigravis* Bliven, Occ. Ent., 1: 11. [Cal.]. Synonymized by Thomas and Yonke, 1981, J. Ks. Ent. Soc., 54: 241.

Distribution: Alta., Ark., B.C., Cal., Col., D.C., Fla., Ga., Ia., Ill., Ind., La., Man., Mass., Md., Mich., Mo., N.C., N.D., N.J., N.M., N.S., N.Y., Neb., Oh., Ok., Ont., Ore., Pa., Que., Tex., Ut., Va., W.Va., Wash., Wis.

Note: Say's (1832, above) original spelling of the species name was "*dimiata*." Apparently all subsequent authors, except Kirkaldy (1909, above), have followed the emended "*dimidiata*" given in LeConte's (1859, above) reprinting of Say's work, which is an unjustified emendation.

Note: Life history studied by De Coursey (1963, An. Ent. Soc. Am., 56: 687-689).

Banasa euchlora Stål, 1872

1872 *Banasa euchlora* Stål, K. Svens. Vet.-Akad. Handl., 10(4): 44. [S.C., Tex.].

1909 *Nezara (Atomosira) euchlora*: Kirkaldy, Cat. Hem., 1: 122.

Distribution: Ala., Ariz., Ark., Col., Fla., Ga., Ia., Id., Ill., Ind., Md., N.C., N.J., N.M., N.Y., Nev., Ok., S.C., Tex., Ut., Va.

Banasa grisea Ruckes, 1957

1957 *Banasa grisea* Ruckes, Bull. Brook. Ent. Soc., 52: 46. [Ariz.].

Distribution: Ariz.

Banasa herbacea (Stål), 1872
1872 *Piezodorus herbaceus* Stål, K. Svens. Vet.-Akad. Handl., 10(4): 44. [St. Thomas, Lesser Antilles].
1981 *Banasa herbacea*: Thomas and Yonke, J. Ks. Ent. Soc., 54: 246.
Distribution: Fla. (West Indies).

Banasa lenticularis Uhler, 1894
1894 *Banasa lenticularis* Uhler, Proc. Zool. Soc. London 1894, p. 174. [Grenada].
1909 *Banasa lenticularis*: Van Duzee, Bull. Buff. Soc. Nat. Sci., 9: 157.
Distribution: Fla., La., Tex. (Mexico to Panama, West Indies).

Banasa packardii Stål, 1872
1872 *Banasa Packardii* [*sic*] Stål, K. Svens. Vet.-Akad. Handl., 10(4): 43. [N.C.].
1886 *Banasa packardii*: Uhler, Check-list Hem. Het., p. 8.
1893 *Banasa packardi* [*sic*]: Lethierry and Severin, Cat. Hem., 1: 168.
1909 *Nezara* (*Atomosira*) *packardii*: Kirkaldy, Cat. Hem., 1: 123.
Distribution: Ariz., Fla., Ga., La., N.C., Va.

Banasa rolstoni Thomas and Yonke, 1981
1981 *Banasa rolstoni* Thomas and Yonke, J. Ks. Ent. Soc., 54: 237. [Nev.].
Distribution: Ariz., Cal., Nev., Ore., Tex., Ut.

Banasa sordida (Uhler), 1871
1871 *Atomosira sordida* Uhler, Proc. Bost. Soc. Nat. Hist., 14: 98. [No locality given].
1872 *Banasa sordida*: Stål, K. Svens. Vet.-Akad. Handl., 10(4): 44.
1909 *Nezara* (*Atomosira*) *sordida*: Kirkaldy, Cat. Hem., 1: 123.
Distribution: Ariz., B.C., Cal., Col., D.C., Ill., Mass. Md., N.J., N.M., Ont., Ut., Va., Wash.

Banasa subcarnea Van Duzee, 1935
1904 *Banasa varians*: Uhler, Proc. U.S. Nat. Mus., 27(1360): 351.
1909 *Banasa subrufescens*: Kirkaldy, Cat. Hem., 1: 122 (in part).
1935 *Banasa subcarnea* Van Duzee, Pan-Pac. Ent., 11: 26. [Ariz.].
Distribution: Ariz., Cal., Col., N.M., Ut.
Note: Thomas and Yonke (1981, J. Ks. Ent. Soc., 54: 234) noted that *B. subrufescens* (Walker) is a Brazilian species, not occurring in our area, and that most North American records for it belong to *B. subcarnea*.

Banasa tumidifrons Thomas and Yonke, 1981
1981 *Banasa tumidifrons* Thomas and Yonke, J. Ks. Ent. Soc., 54: 239. [Ore.].
Distribution: Cal., Id., Ore.

Genus *Brepholoxa* Van Duzee, 1904

1904 *Brepholoxa* Van Duzee, Trans. Am. Ent. Soc., 30: 78. Type-species: *Brepholoxa heidemanni* Van Duzee, 1904. Monotypic.

Brepholoxa heidemanni Van Duzee, 1904
1904 *Brepholoxa heidemanni* Van Duzee, Trans. Am. Ent. Soc., 30: 78. [Fla.].
Distribution: Fla.

Genus *Chlorochroa* Stål, 1872

1872 *Lioderma* (*Chlorochroa*) Stål, K. Svens. Vet.-Akad. Handl., 10(2): 33. Type-species: *Pentatoma ligata* Say, 1832. Designated by Kirkaldy, 1909, Cat. Hem., 1: 53.

1880 *Chlorochroa*: Distant, Biol. Centr.-Am., Rhyn., 1: 63.

1888 *Pitedia* Reuter, Acta Soc. Sci. Fenn., 15: 494. Type-species: *Cimex juniperinus* Linnaeus, 1758. Monotypic. Synonymized by Thomas, 1983, An. Ent. Soc. Am., 76: 217.

Notes:Thomas (1983, An. Ent. Soc. Am., 76: 215-224) treated *Chlorochroa* with two subgenera: *Rhytidolomia* Stål (as *Rhytidilomia* [*sic*]) and nominate *Chlorochroa*. He gave keys to the species of the subgenus *Rhytidolomia* and to the *opuntiae*-group of the subgenus *Chlorochroa*. Treatment and key to the species of the other species-group in subgenus *Chlorochroa*, the *sayi*-group, were given by Buxton et al. (1983, Cal. Dept. Agri. Food, Occas. Pap. Ent., 29: 1-14). The "Distribution" data given here is restricted to the records in those two papers. China (1943, Gen. Names Brit. Ins, 8: 225), following Reuter (1888, Acta Soc. Sci. Fenn., 15: 494), stated that *Chlorochroa* Sclater (1862, Athenaeum, 1834: 811) used in birds preoccupied Stål's usage. But examination of Sclater's paper showed that his generic name was without reference to an available species description or illustration and, hence, was a *nomen nudum* and still available to subsequent authors. Richmond (1917, Proc. U.S. Nat. Mus., 53(2221): 583) pointed out that the included species name was a *nomen nudum* and that the bird was described later (Sclater, 1863, Proc. Zool. Soc. London, 1862: 369) as *Vireo hypochryseus*. Reuter adopted *Pitedia* Amyot (1846, An. Soc. Ent. France, ser. 2, 3: 445) as the next available name, but China pointed out that it was not available there [having been described in a monomial system proposed as a system to replace Linnaeus' binomial system] and, hence, must take "Reuter, 1888" as its author and date of validity.

Buxton et al. (1983, Cal. Dept. Agr. Food, Occas. Pap. Ent., 29: 3) wrote of the species *Cimex albosparsus* Kushakevich (1867, Horae Ent. Soc. Ross., 4: 99), described from Cal. and generally assigned to this genus: "We must consider this form a *nomen dubium* until some of the types are found."

Subgenus *Chlorochroa* Stål, 1872

1872 *Lioderma* (*Chlorochroa*) Stål, K. Svens. Vet.-Akad. Handl., 10(4): 33. Type-species: *Pentatoma ligata* Say. Designated by Kirkaldy, 1909, Cat. Hem., 1: 53.

1983 *Chlorochroa* (*Chlorochroa*): Thomas, An. Ent. Soc. Am., 76: 218.

Chlorochroa congrua Uhler, 1876

1876 *Chlorochroa congrua* Uhler, Bull. U.S. Geol. Geogr. Surv. Terr., 1: 288. [Col.].

1886 *Lioderma congrua*: Uhler Check-list Hem. Het., p. 6 (in part).

1893 *Pentatoma congrua*: Lethierry and Severin, Cat. Hem., 1: 119.

1904 *Pentatoma* (*Chlorochroa*) *congrua*: Van Duzee, Trans. Am. Ent. Soc., 30: 36.

1909 *Rhytidolomia* (*Chlorochroa*) *congrua*: Kirkaldy, Gen. Cat. Hem., 1: 53.

Distribution: Col., Mont., Wyo.

Chlorochroa granulosa (Uhler), 1872

1872 *Pentatoma granulosa* Uhler, Prelim. Rept. Geol. Surv. Mont., p. 398. [Cal., Mont., Ut., "Western Territories"].

1886 *Lioderma congrua*: Uhler, Check-list Hem. Het., p. 6 (in part).

1904 *Pentatoma* (*Chlorochroa*) *sayi*: Van Duzee, Trans. Am. Ent. Soc., 30: 41 (in part).

1909 *Rhytidolomia* (*Chlorochroa*) *sayi*: Kirkaldy, Cat. Hem., 1: 54 (in part).

1916 *Chlorochroa sayi*: Van Duzee, Check List Hem., p. 5 (in part).

1983 *Chlorochroa granulosa*: Buxton et al., Cal. Dept. Agri. and Food, Occas. Papers Ent., 29: 12.

Distribution: Alta., Col., Id., Mont., Nev., Ut., Wash., Wyo.

Chlorochroa kanei Buxton and Thomas, 1983
1983 *Chlorochroa kanei* Buxton and Thomas, *In* Buxton et al., Cal. Dept. Agri. Food, Occas. Pap. Ent., 29: 15. [Cal.].
Distribution: Cal., Nev.

Chlorochroa ligata (Say), 1832
1832 *Pentatoma ligata* Say, Descrip. Het. Hem. N. Am., p. 5. [Mo.].
1867 *Cimex rubromarginatus* Kushakevich, Horae Soc. Ent. Ross., 4: 99. [Cal.]. Synonymized by Van Duzee, 1904, Trans. Am. Ent. Soc., 30: 41.
1867 *Pentatoma marginalis* Walker, Cat. Hem. Brit. Mus., 2: 288. ["North America," Mexico]. Preoccupied.
1872 *Lioderma (Chlorochroa) ligata*: Stål, K. Svens. Vet.-Akad. Handl., 10(4): 33.
1880 *Chlorochroa ligata*: Distant, Biol. Centr.-Am., Rhyn., 1: 64.
1886 *Lioderma ligata*: Uhler, Check-list Hem. Het., p. 6.
1904 *Pentatoma (Chlochroa) ligata*: Van Duzee, Trans. Am. Ent. Soc., 30: 41.
1909 *Rhytidolomia (Chlorochroa) ligata*: Kirkaldy, Cat. Hem., 1: 53.
1910 *Rhytidolomia ligata*: Kirkaldy, Proc. Haw. Ent. Soc., 2: 125.
1978 *Pitedia ligata*: Slater and Baranowski, How To Know True Bugs, p. 51.
Distribution: Alta., Ark., Ariz., B.C., Cal., Col., Id., Ks., Mo., Mont., N.M., Nev., Ore., S.D., Tex., Ut., Wash. (Mexico).

Chlorochroa lineata Thomas, 1983
1983 *Chlorochroa (Chlorochroa) lineata* Thomas, An. Ent. Soc. Am., 76: 218. [Ut.].
Distribution: Cal., Nev., Ut.

Chlorochroa norlandi Buxton and Thomas, 1983
1983 *Chlorochroa norlandi* Buxton and Thomas, *In* Buxton et al., Cal. Dept. Agri. Food, Occas. Pap. Ent., 29: 8. [Cal.].
Distribution: Cal.

Chlorochroa opuntiae Esselbaugh, 1948
1948 *Chlorochroa opuntiae* Esselbaugh, Bull. Brook. Ent. Soc., 42: 166. [Wash.].
1983 *Chlorochroa (Chlorochroa) opuntiae*: Thomas, An. Ent. Soc. Am., 76: 219.
Distribution: Alta., Ariz., Id., Mont., Ore., Tex., Wash., Wyo.

Chlorochroa persimilis Horvath, 1908
1885 *Pentatoma juniperina*: Provancher, Pet. Faune Ent. Can., 3: 36.
1904 *Pentatoma (Chlorochroa) juniperina*: Van Duzee, Trans. Am. Ent. Soc., 30: 39.
1908 *Chlorochroa persimilis* Horvath, An. Mus. Nat. Hung., 6: 555. [Canada, N.Y.].
1909 *Rhytidolomia (Chlorochroa) juniperina*: Kirkaldy, Cat. Hem., 1: 53 (in part).
1916 *Chlorochroa uhleri*: Van Duzee, Check List Hem., p. 5 (in part).
Distribution: Alta., Ark., Fla., Ill., Ind., Ks., Me., Mich., Miss., Mo., N.C., N.D., N.H., N.Y., Oh., Ont., Que., Va.

Chlorochroa rossiana Buxton and Thomas, 1983
1867 *Cimex flavomarginatus* Kushakevich, Horae Ent. Soc. Ross., 4: 99. [Cal.]. Preoccupied.
1893 *Chlorochroa rossiana* Buxton and Thomas, *In* Buxton et al., Cal. Dept. Agri. Food, Occas. Pap. Ent., 29: 10. New Name for *Cimex flavomarginatus* Kushakevich.
Distribution: B.C., Cal., Ore., Wash.

Chlorochroa sayi (Stål), 1872
1872 *Lioderma (Chlorochroa) sayi* Stål, K. Svens. Vet.-Akad. Handl., 10(4): 33. [Cal.].
1886 *Lioderma sayi*: Uhler, Check-list Hem. Het., p. 6.

1902 *Pentatoma (Lioderma) sayi*: Heidemann, Proc. Ent. Soc. Wash., 5: 80.
1904 *Pentatoma (Chlorochroa) sayi*: Van Duzee, Trans. Am. Ent. Soc., 30: 41 (in part).
1904 *Pentatoma sayi*: Snow, Ks. Univ. Sci. Bull., 2: 347.
1909 *Rhytidolomia (Chlorochroa) sayi*: Kirkaldy, Cat. Hem., 1: 54 (in part).
1910 *Rhytidolomia sayi*: Kirkaldy, Proc. Haw. Ent. Soc., 2: 125.
1916 *Chlorochroa sayi*: Van Duzee, Check List Hem., p. 5.
1935 *Chlorochroa sayii* [*sic*]: Patton and Mail, J. Econ. Ent., 28: 906.
1978 *Pitedia sayi*: Slater and Baranowski, How To Know True Bugs, p. 51.
Distribution: Ariz., Ark., Cal., Col., Id., Mont,. N.M., Nev., Tex., Ut., Wash., Wyo.

Chlorochroa uhleri (Stål), 1872
1872 *Lioderma (Chlorochroa) uhleri* Stål, K. Svens. Vet.-Akad. Handl., 10(4): 33. [Mexico].
1876 *Chlorochroa uhleri*: Uhler, Bull. U.S. Geol. Geogr. Surv. Terr., 1: 289.
1902 *Pentatoma (Lioderma) uhleri*: Heidemann, Proc. Ent. Soc. Wash., 5: 80.
1904 *Pentatoma (Chlorochroa) Uhleri* [*sic*]: Van Duzee, Trans. Am. Ent. Soc., 30: 39.
1909 *Rhytidolomia (Chlorochroa) uhleri*: Kirkaldy, Cat. Hem., 1: 54.
1912 *Liodermion (Chlorochroa) uhleri*: Zimmer, Univ. Neb., Contr. Dept. Ent., 4: 7.
Distribution: Alta., Ariz., B.C., Cal., Col., Id., Ks., Mont., N.D., N.M., Neb., Nev., Ore., S.D., Ut., Wash. (Mexico).

Subgenus *Rhytidolomia* Stål, 1872

1871 *Lioderma* Uhler, Proc. Bost. Soc. Nat. Hist., 14: 97. Preoccupied. Type-species: *Pentatoma saucia* Say, 1832. Designated by Kirkaldy, 1909, Cat. Hem., 1: 53.
1872 *Lioderma (Rhytidolomia)* Stål, K. Sven. Vet.-Akad. Handl., 10(4): 33. Type-species: *Pentatoma senilis* Say, 1832. Designated by Kirkaldy, 1900, Ent., 33: 240.
1904 *Liodermion* Kirkaldy, Ent., 37: 280. Unnecessary new name for *Lioderma* Uhler. Synonymized by Van Duzee, 1916, Check List Hem., p. 4.
1983 *Chlorochroa (Rhytidilomia* [*sic*]): Thomas, An. Ent. Soc. Am., 76: 219.

Chlorochroa belfragii (Stål), 1872
1872 *Lioderma (Rhytidolomia) Belfragii* [*sic*] Stål, K. Svens. Vet.-Akad. Handl., 10(4): 33. [Ill.].
1876 *Rhytidolomia Belfragii* [*sic*]: Uhler, Bull. U.S. Geol. Geogr. Surv. Terr., 1: 287.
1886 *Lioderma belfragii*: Uhler, Check-list Hem. Het., p. 6.
1904 *Pentatoma (Rhytidolomia) Belfragei* [*sic*]: Van Duzee, Trans. Am. Ent. Soc., 30: 37.
1909 *Rhytidolomia (Rhytidolomia) belfragii*: Kirkaldy, Cat. Hem., 1: 53.
1912 *Liodermion (Rhytidolomia) belfragii*: Zimmer, Univ. Neb., Contr. Dept. Ent., 4: 7.
1916 *Rhytidolomia belfragei* [*sic*]: Van Duzee, Check List Hem., p. 5.
1983 *Chlorochroa (Rhytidilomia* [*sic*]) *belfragei*: Thomas, An. Ent. Soc. Am., 76: 221.
Distribution: "Canada," Ia., Ill., Neb.

Chlorochroa dismalia Thomas, 1983
1983 *Chlorochroa (Rhytidilomia* [*sic*]) *dismalia* Thomas, An. Ent. Soc. Am., 76: 220. [Va.].
Distribution: Va.

Chlorochroa faceta (Say), 1825
1825 *Pentatoma faceta* Say, J. Acad. Sci. Phila., 4: 315. ["Missouri"].
1904 *Pentatoma (Rhytidolomia) faceta*: Van Duzee, Trans. Am. Ent. Soc., 30: 38.
1909 *Rhytidolomia (Rhytidolomia) faceta*: Kirkaldy, Cat. Hem., 1: 53.
1910 *Rhytidolomia faceta*: Kirkaldy, Proc. Haw. Ent. Soc., 2: 125.

1912 *Liodermion (Rhytidolomia) faceta* Zimmer, Univ. Neb., Contr. Dept. Ent., 4: 7.
1983 *Chlorochroa (Rhytidilomia [sic]) faceta*: Thomas, An. Ent. Soc. Am., 76: 221.
Distribution: Ariz., Cal., Col., Id., Ks., "Mo.," N.D., Neb., Sask.

Chlorochroa osborni (Van Duzee), 1904
1904 *Pentatoma (Rhytidolomia) Osborni* [*sic*] Van Duzee, Trans. Am. Ent. Soc., 30: 37. [Col., Tex.].
1906 *Pentatoma osborni*: Snow, Trans. Ks. Acad. Sci., 20 (pt. 1): 177.
1909 *Rhytidolomia (Rhytiodolomia) osborni*: Kirkaldy, Cat. Hem., 1: 53.
1916 *Rhytidolomia osborni*: Van Duzee, Check List Hem., p. 5.
1983 *Chlorochroa (Rhytidilomia [sic]) osborni*: Thomas, An. Ent. Soc. Am., 76: 220.
Distribution: Ariz., Col., Tex.

Chlorochroa rita (Van Duzee), 1934
1934 *Rhytidolomia rita* Van Duzee, Pan-Pac. Ent., 10: 96. [Ariz.].
1983 *Chlorochroa (Rhytidilomia [sic]) rita*: Thomas, An. Ent. Soc. Am., 76: 222.
Distribution: Ariz.

Chlorochroa saucia (Say), 1832
1832 *Pentatoma saucia* Say, Descrip. Het. Hem. N. Am., p. 6. [Fla., Va., "a third either in Pennsylvania or Indiana"]:
1871 *Lioderma saucia*: Uhler, Proc. Boston Soc. Nat. Hist., 14: 97.
1903 *Lioderma (Pentatoma) saucia*: Banks, J. N.Y. Ent. Soc., 11: 227.
1904 *Pentatoma (Lioderma) saucia*: Van Duzee, Trans. Am. Ent. Soc., 30: 36.
1904 *Liodermion saucia*: Kirkaldy, Ent., 37: 280.
1909 *Rhytidolomia (Liodermion) saucia*: Kirkaldy, Cat. Hem., 1: 53.
1915 *Rhytidolomia saucia*: Parshley: Psyche, 22: 174.
1927 *Rhytidolomia schotti* Barber, Bull. Brook. Ent. Soc., 22: 243. [Ala.]. Synonymized by Thomas, 1983, An. Ent. Soc. Am., 76: 222.
1983 *Chlorochroa (Rhytidilomia [sic]) saucia*: Thomas, An. Ent. Soc. Am., 76: 222.
Distribution: Ala., Conn., Fla., Mass., Md., N.C., N.J., N.Y., Pa., Tex., Va.

Chlorochroa senilis (Say), 1832
1832 *Pentatoma senilis* Say, Descrip. Het. Hem. N.Am., p. 5. ["U.S."].
1851 *Pentatoma grisea* Dallas, List Hem. Brit. Mus., 1: 246. ["N. America"]. Synonymized by Uhler, 1871, Proc. Bost. Soc. Nat. Hist., 14: 97.
1871 *Lioderma senilis*: Uhler, Proc. Boston Nat. Hist., 14: 97.
1903 *Pentatoma senilis*: Torre-Bueno, J. N.Y. Ent. Soc., 11: 129.
1904 *Pentatoma (Rhytidolomia) senilis*: Van Duzee, Trans. Am. Ent. Soc., 30: 37.
1909 *Rhytidolomia (Rhytidolomia) senilis*: Kirkaldy, Cat. Hem., 1: 53.
1910 *Chlorochroa senilis*: Smith, Rept. N.J. St. Mus., p. 134.
1983 *Chlorochroa (Rhytidilomia [sic]) senilis*: Thomas, An. Ent. Soc. Am., 76: 222.
Distribution: N.J., N.Y., Pa.

Chlorochroa viridicata (Walker), 1867
1867 *Hymenarcys viridicata* Walker, Cat. Hem. Brit. Mus., 2: 283 (Uhler ms. name). ["North America"].
1875 *Lioderma viridicata* Uhler, Rept. U.S. Geol. Geogr. Expl. Surv. 100th Mer., 5: 830. [Col.]. Preoccupied.
1893 *Pentatoma viridicata*: Lethierry and Severin, Gen. Cat. Hem., 1: 267.
1904 *Pentatoma (Lioderma) viridicata*: Van Duzee, Trans. Am. Ent. Soc., 30: 36.
1909 *Rhytidolomia (Liodermion) viridicata*: Kirkaldy, Cat. Hem., 1: 53.
1912 *Liodermion (Liodermion) viridicata*: Zimmer, Univ. Neb. Contr. Dept. Ent., 4: 7.

1916 *Rhytidolomia viridicata*: Van Duzee, Check List Hem., p. 5.
1983 *Chlorochroa* (*Rhytidilomia* [*sic*]) *viridicata*: Thomas, An. Ent. Soc. Am., 76: 221.
Distribution: Alta., Col., Mont., N.M., Neb. (Mexico).

Genus *Chlorocoris* Spinola, 1837

1837 *Chlorocoris* Spinola, Essai Hem., p. 228. Type-species: *Chlorocoris tau* Spinola, 1837. Monotypic.
Note: Thomas (1985, An. Ent. Soc. Am., 78: 674-690) revised this genus and provided a key to species.

Subgenus *Chlorocoris* Spinola, 1837

1837 *Chlorocoris* Spinola, Essai Hem., p. 228. Type-species: *Chlorocoris tau* Spinola, 1837. Monotypic.
1872 *Chlorocoris* (*Chlorocoris*): Stål, K. Svens. Vet.-Akad. Handl., 10(4): 35.

Chlorocoris distinctus Signoret, 1851
1851 *Chlorocoris distinctus* Signoret, An. Soc. Ent. France, ser. 2, 9: 330. [Mexico].
1862 *Chlorocoris atrispinus* Stål, Stett. Ent. Zeit., 23: 99. [Mexico]. Synonymized by Thomas, 1985, An. Ent. Soc. Am., 78: 680.
1910 *Chlorocoris atrispinus*: Barber, J. New York Ent. Soc., 18: 35.
1985 *Chlorocoris* (*Chlorocoris*) *distinctus*: Thomas, An. Ent. Soc. Am., 78: 680.
Distribution: Ariz., N.M. (Mexico to Ecuador).

Subgenus *Monochrocerus* Stål, 1872

1872 *Chlorocoris* (*Monochrocerus*) Stål, K. Svens. Vet.-Akad., Handl., 10(4): 35. Type-species: *Chlorocoris rufispinus* Dallas, 1851. Designated by Kirkaldy, 1909, Cat. Hem., 1: 95.

Chlorocoris flaviviridis Barber, 1914
1910 *Chlorocoris rufopictus*: Barber, J. N.Y. Ent. Soc., 18: 35.
1914 *Chlorocoris flaviviridis* Barber, J. N.Y. Ent. Soc., 22: 164. [Ariz.].
1916 *Chlorocoris* (*Monachrocerus* [*sic*]) *flaviviridis*: Van Duzee, Check List Hem., p. 7.
1917 *Chlorocoris* (*Monochrocerus*) *flaviviridis*: Van Duzee, Univ. Cal. Publ. Ent., 2: 54.
Distribution: Ariz.

Chlorocoris hebetatus Distant, 1890
1890 *Chlorocoris hebetatus* Distant, Biol. Centr.-Am., Rhyn., 1: 335. [Mexico].
1910 *Chlorocoris hebetatus*: Barber, J. N.Y. Ent. Soc., 18: 35.
1916 *Chlorocoris* (*Monachrocerus* [*sic*]) *hebetatus*: Van Duzee, Check List Hem., p. 7.
1917 *Chlorocoris* (*Monochrocerus*) *hebetatus*: Van Duzee, Univ. Cal. Publ. Ent., 2: 54.
Distribution: Ariz., Tex. (Mexico).

Chlorocoris subrugosus Stål, 1872
1872 *Chlorocoris subrugosus* Stål, K. Svens. Vet.-Akad. Handl., 10(4): 36. [Mexico].
1910 *Chlorocoris subrugosus*: Barber, J. N.Y. Ent. Soc., 18: 35.
1916 *Chlorocoris* (*Monachrocerus* [*sic*]) *subrugosus*: Van Duzee, Check List Hem., p. 7.
1917 *Chlorocoris* (*Monochrocerus*) *subrugosus*: Van Duzee, Univ. Cal. Publ. Ent., 2: 54.
Distribution: Ariz. (Mexico).

Genus *Codophila* Mulsant and Rey, 1866

1866 *Carpocoris* (*Codophila*) Mulsant and Rey, An. Soc. Linn. Lyon, ser. 2, 14: 160. Type-species: *Cimex varius* Fabricius, 1787. Designated by Distant, 1902, Fauna Brit. Ind., Rhyn., 1: 158.

1872 *Codophila*: Stål, Öfv. K. Vet.-Akad. Förh., 29(3): 38.

Note: The presence of *Codophila* instead of *Carpocoris* Kolenati in North America is explained by Thomas (1974, Pan-Pac. Ent., 50: 441-442). The "California" record for the Siberian form *Carpocoris lynx* var. *longiceps* Reuter, which Kirkaldy (1909, Cat. Hem., 1: 57) rejected as "in error," has not been located for the present catalog.

Subgenus *Antheminia* Mulsant and Rey, 1866

1866 *Carpocoris* (*Antheminia*) Mulsant and Rey, An. Soc. Linn. Lyon, Ser. 2, 14: 161. Type-species: *Cimex lynx* Fabricius, 1794, a junior synonym of *Cimex lunulatus* Goeze, 1778. Designated by Kirkaldy, 1909, Cat. Hem., 1: 56.

1958 *Codophila* (*Antheminia*): Tamanini, Mem. Mus. Civ. Stor. Nat. Hist., 6: 340.

Codophila remota (Horvath), 1907 [Fig. 118]

1872 *Carpocoris* (*Antheminia*) *lynx*: Stål, K. Svens. Vet.-Akad. Handl., 10(4): 33.

1872 *Carpocoris lynx*: Uhler, Prelim. Rept. U.S. Geol. Surv. Mont., 5: 398.

1907 *Carpocoris remotus* Horvath, An. Mus. Nat. Hung., 5: 296. [Col.].

1909 *Carpocoris* (*Antheminia*) *remotus*: Kirkaldy, Cat. Hem., 1: 57.

1974 *Codophila* (*Antheminia*) *remota*: Thomas, Pan-Pac. Ent., 50: 441.

Distribution: Alta., Ariz., B.C., Cal., Col., Id., Mont., N.D., N.M., N.T., Ut., Wash., Wyo.

Codophila sulcata (Van Duzee), 1918

1918 *Carpocoris sulcatus* Van Duzee, Proc. Cal. Acad. Sci., ser. 4, 8: 275. [Cal.].

1974 *Codophila* (*Antheminia*) *sulcata*: Thomas, Pan-Pac. Ent., 50: 442.

Distribution: Cal. (Mexico).

Genus *Coenus* Dallas, 1851

1851 *Coenus* Dallas, List Hem. Brit. Mus., 1: 194, 230. Type-species: *Coenus tarsalis* Dallas, 1851, a junior synonym of *Pentatoma delia* Say, 1832. Monotypic.

Coenus delius (Say), 1832 [Fig. 105]

1832 *Pentatoma delia* Say, Descrip. Het. Hem. N. Am., p. 8. ["Missouri," Mass.].

1843 *Hymenarcys aeruginosa* Amyot and Serville, Hist. Nat. Ins., Hem., p. 125. ["Amerique septentrionale"]. Synonymized by Stål, 1872, K. Svens. Vet.-Akad. Handl., 10(4): 30.

1851 *Coenus tarsalis* Dallas, List Hem. Brit. Mus., 1: 230. [No locality given; now considered to have been from N.Am.]. Synonymized by Stål, 1867, Öfv. K. Svens. Vet.-Akad. Förh., 24(7): 526.

1867 *Coenus delius*: Stål, Öfv. K. Svens. Vet.-Akad. Förh., 24(7): 526.

1868 *Coenus punctatissimus* Vollenhoven, Versl. Med. Kon. Akad. Weten. Amsterdam, ser. 2, 2: 183. [Wis.]. Synonymized by Stål, 1872, K. Svens. Vet.-Akad. Handl., 10(4): 30.

1869 *Coenus delia* [*sic*]: Rathvon, Hist. Lancaster Co., Pa., p. 548.

Distribution: Alta., Ark., B.C., Col., Conn., Fla., Ia., Id., Ill., Ind., Ks., Mass., Me.,

Mich., Mo., Mont., N.C., N.H., N.J., N.Y., Neb., Ont., Oh., Ok., Ont., Pa., Que., R.I., Tex., Ut., Wis.

Note: Biological information given by Oetting and Yonke (1971, J. Ks. Ent. Soc., 44: 447-449).

Coenus inermis Harris and Johnston, 1936

1936 *Coenus inermis* Harris and Johnston, Ia. St. Coll. J. Sci., 10: 378. [Ark.].

Distribution: Ark., Mo., Ok.

Genus *Cosmopepla* Stål, 1867

1867 *Cosmopepla* Stål, Öfv. K. Svens. Vet.-Akad. Förh., 24(7): 525. Type-species: *Cimex carnifex* Fabricius, 1798. Preoccupied. *Pentatoma bimaculata* Thomas, 1865, is next available name. Designated by Kirkaldy, 1909, Cat. Hem., 1: 80.

Note: McDonald (1986, J. N.Y. Ent. Soc., 94: 1-15) revised the genus and presented a key to species.

Cosmopepla bimaculata (Thomas), 1865

1798 *Cimex carnifex* Fabricius, Ent. Syst., Suppl., p. 535. ["America"]. Preoccupied.

1834 *Eysarcoris carnifex*: Hahn, Wanz. Ins., 2: 117.

1837 *Pentatoma carnifex*: Kirby, Fauna Bor.-Am., 4: 275.

1865 *Pentatoma bimaculata* Thomas, Trans. Ill. St. Agr. Soc., 5: 455. [Ill.].

1869 *Eusacoris* [*sic*] *carnifex*: Rathvon, Hist. Lancaster Co., Pa., p. 548.

1867 *Cosmopepla carnifex*: Popenoe, Trans. Ks. Acad. Sci., 9:62.

1909 *Cosmopepla lintneriana* Kirkaldy, Cat. Hem., 1:80. Unnecessary new name for *Cimex carnifex* Fabricius.

1912 *Cormopepla* [*sic*] *carnifex*: Olsen, J. N.Y. Ent. Soc., 19: 267.

Distribution: Alta., Ark., B.C., Col., Conn., D.C., Ga., Ill., Ks., Mass., Me., Mich., Miss., Mo., Mont., N.C., N.H., N.J., N.M., N.S., N.Y., Neb., Oh., Ok., Ont., Pa., Que., Tex., Ut., Va., Vt., Wash. (Mexico).

Note: Kirkaldy (1909, Cat. Hem., 1: 80) and McDonald (1986, J. N.Y. Ent. Soc., 94: 4) considered Thomas' combination *Pentatoma bimaculata* preoccupied by Westwood's (1837, Cat. Hem. Hope, pp. 8 and 35) use of that combination. Westwood's use is a *nomen nudum* because it has no descriptive comments, only in a list on p. 8, and the note on p. 35 that it should be deleted as a variation of the preceding species, *Pentatoma obscura*. Thus, Thomas' name is available for the North American species. McDonald (1968, Quaest. Ent., 4: 35-38) described the life history.

Cosmopepla binotata Distant, 1889

1889 *Cosmopepla binotata* Distant, Biol. Centr.-Am., Rhyn., 1: 327. [Mexico].

1910 *Cosmopepla binotata*: Barber, J. N.Y. Ent. Soc., 18: 35.

Distribution: Ariz., Tex. (Mexico).

Note: Montandon's (1893, Proc. U.S. Nat. Mus., 16: 49) "Wisconsin" record is here dropped as based on an adventive or mislabeled specimen.

Cosmopepla conspicillaris (Dallas), 1851 .

1851 *Eysarcoris conspicillaris* Dallas, List Hem. Brit. Mus., 1: 225. [Cal.].

1853 *Pentatoma conspicillaris*: Herrich-Schaeffer, Alpha-Syn. Verz. Wanz. Ins., p. 153.

1872 *Cosmopepla conspicillaris*: Stål, K. Svens. Vet.-Akad. Handl., 10(4): 19.

Distribution: B.C., Cal., Col., Mont., Ore., Ut., Wash. (Mexico).

Cosmopepla decorata (Hahn), 1834

1834 *Eysarcoris decoratus* Hahn, Wanz. Ins., 2: 117. [Mexico].

1844 *Pentatoma decorata*: Herrich-Schaeffer, Wanz. Ins., 7: 96.
1872 *Cosmopepla decorata*: Stål, K. Svens. Vet.-Akad. Handl., 10(4): 19.
Distribution: Ariz., Tex. (Mexico).

Cosmopepla intergressus (Uhler), 1893
1876 *Eysarcoris melanocephalus*: Uhler, Bull. U.S. Geol. Geogr. Surv. Terr., 1, pl. 9, fig. 7.
1893 *Eysarcoris intergressus* Uhler, Proc. Ent. Soc. Wash., 2: 368. [Ut].
1895 *Neottiglossa melanocephala*: Gillette and Baker, Bull. Col. Agr. Exp. Stn., 31: 14.
1955 *Cosmopepla humboldtensis* Bliven, Stud. Ins. Redwood Emp., 1: 8. [Cal.]. Synonymized by McDonald, 1986, J. N.Y. Ent. Soc., 94: 11.
1986 *Cosmopepla intergressa*: McDonald, J. N.Y. Ent. Soc., 94: 11.
Distribution: B.C., Cal., Col., Id., Ks., Mont., Ore., Ut., Wash.

Cosmopepla uhleri Montandon, 1893
1893 *Cosmopepla uhleri* Montandon, Proc. U.S. Nat. Mus., 16: 48. [Cal., Nev.].
Distribution: Ariz., Cal., Nev.
Note: Literature record for "Neb." was a *lapsus* for "Nev."

Genus *Cyptocephala* Berg, 1883

1883 *Cyptocephala* Berg, An. Soc. Cient. Arg., 32: 209. Type-species: *Cyptocephala cogitabunda* Berg, 1883. Monotypic.

Cyptocephala antiguensis (Westwood), 1837
1837 *Pentatoma antiguensis* Westwood, Hope Cat., 1: 36. [Antigua].
1872 *Thyanta taeniola*: Stål, K. Svens. Vet.-Akad. Handl., 10(4): 35.
1904 *Thyanta antiguensis*: Van Duzee, Trans. Am. Ent. Soc. 30: 54.
1984 *Cyptocephala antiguensis*: Rolston and McDonald, J. N.Y. Ent. Soc., 92: 77.
Distribution: Ariz., Cal., Fla., Tex. (Mexico to Venezuela, West Indies).

Cyptocephala bimini (Ruckes), 1952
1952 *Thyanta bimini* Ruckes, Bull. Brook. Ent. Soc., 47: 65. [Bahamas].
1952 *Thyanta bimini*: Ruckes, Am. Mus. Novit., 1591: 6.
1984 *Cyptocephala bimini*: Rolston and McDonald, J. N.Y. Ent. Soc., 92: 77.
Distribution: Fla. (Bahamas, Greater Antilles).

Cyptocepahala elegans (Malloch), 1919
1919 *Thyanta elegans* Malloch, Ill. Nat. Hist. Surv. Bull., 13: 217, 218. [Tex.].
1928 *Thyanta (Parathyanta) elegans*: Jensen-Haarup, Ent. Medd., 16: 186, 188.
1984 *Cyptocephala elegans*: Rolston and McDonald, J. N.Y. Ent. Soc., 92: 77.
Distribution: Fla., Tex.

Genus *Dendrocoris* Bergroth, 1891

1877 *Liotropis* Uhler, Bull. U.S. Geol. Geogr. Surv. Terr., 3: 399. Preoccupied. Type-species: *Liotropis humeralis* Uhler, 1877. Monotypic.
1891 *Dendrocoris* Bergroth, Rev. d'Ent., 10: 228. New Name for *Liotropis* Uhler.
Note: Revision with key to species was given by Nelson (1955, Proc. Ent. Soc. Wash., 57: 49-67).

Dendrocoris arizonensis Barber, 1911
1911 *Dendrocoris arizonensis* Barber, Ent. News, 22: 270. [Ariz.].
Distribution: Ariz., Cal., Col., Tex.

Dendrocoris contaminatus Uhler, 1897
1897 *Dendrocoris contaminatus* Uhler, Trans. Md. Acad. Sci., 1: 390. [Ariz.].
1904 *Liotropis contaminatus*: Van Duzee, Trans. Am. Ent. Soc., 30: 62.
1906 *Liopus* [*sic*] *contaminatus*: Snow, Trans. Ks. Acad. Sci., 20: 177.
Distribution: Ariz., Cal., N.M., Tex., Ut.

Dendrocoris fruticicola Bergroth, 1891
1891 *Dendrocoris fruticicola* Bergroth, Rev. d'Ent., 10: 228. [Fla.].
1904 *Liotropis fruticicola*: Van Duzee, Trans. Am. Ent. Soc., 30: 62.
Distribution: Ala., Fla., Ga., N.C.
Note: Kirkaldy (1910, Proc. Haw. Ent. Soc., 2: 126) included this species in his California list--certainly in error.

Dendrocoris humeralis (Uhler), 1877
1877 *Liotropis humeralis* Uhler, Bull. U.S. Geol. Geogr. Surv. Terr., 3: 400. [Col., Ga., Mass., Md., N.Y.].
1891 *Dendrocoris humeralis*: Bergroth, Rev. d'Ent., 10: 228, 229.
Distribution: Ark., Cal., Col., Conn., D.C., Ga., Id., Ill., Ind., Ks., Mass., Md., Mich., Me., N.C., N.H., N.J., N.M., N.Y., Oh., Pa., Vt., W.Va.

Dendrocoris parapini Nelson, 1957
1957 *Dendrocoris parapini* Nelson, Proc. Ent. Soc. Wash., 59: 198. [N.M.].
Distribution: N.M., Tex.

Dendrocoris pini Montandon, 1893
1893 *Dendrocoris pini* Montandon, Proc. U.S. Nat. Mus., 16: 51. [Cal.].
1904 *Liotropis pini*: Van Duzee, Trans. Am. Ent. Soc., 30: 63.
Distribution: Ariz., B.C., Cal., Col., N.M., Ore., Tex., Ut.

Dendrocoris reticulatus Barber, 1911
1904 *Dendrocoris fruticicola*: Uhler, Proc. U.S. Nat. Mus., 27: 351.
1911 *Dendrocoris reticulatus* Barber, Ent. News, 22: 270. [Ariz.].
Distribution: Ariz., Cal., N.M.

Genus *Euschistus* Dallas, 1851

1851 *Euschistus* Dallas, List. Hem. Brit. Mus., 1: 201. Type-species: *Pentatoma tristigma* Say, 1832. Designated by Stål, 1872, K. Svens. Vet.-Akad. Handl., 10(4): 26.
1909 *Euschistus* (*Paraschistus*) Kirkaldy, Gen. Cat., 1: 63. Type-species: *Euschistus integer* Stål, 1872. Monotypic. Synonymized by Rolston, 1974, Ent. Am., 48: 6.
Notes:A series of continuing experiments on laboratory breeding and interbreeding of populations of *Euschistus* published by Sailer (1952, U.S Dept. Agr., Bur. Ent. and Pl. Quarant., ET-303: 1-5; 1954, Year Book Am. Philosoph. Soc., 146-149; 1954, J. Econ. Ent., 47: 377-383; 1959, Proc. Ent. Soc. Wash., 61: 140-142; 1961, Germ Plasm Resources, pp. 295-303) indicated various abilities to interbreed and to backcross. While the results quite likely will importantly affect the taxonomic treatment of the species of *Euschistus*, too quick application of laboratory results may mask the existence of naturally occurring, visibly identifiable populations (admittedly intergrading in some areas) and hinder other experimental approaches and alternate explanations.

Rolston's (1974, Ent. Am., 48: 1-102) revision and keys to Middle American species of *Euschistus* is helpful in the study of our North American species.

Culliney (1985, Can. Ent., 117: 461-462) summarized the published reports of insect predation by members of the genus.

Subgenus *Euschistus* Dallas, 1851

1851 *Euschistus*: Dallas, List Hem. Brit. Mus., 1: 201. Type-species: *Pentatoma tristigma* Say, 1832. Designated by Stål, 1872, K. Svens. Vet.-Akad. Handl, 10(4): 26.

1872 *Euschistus* (*Euschistus*): Stål, K. Svens. Vet.-Akad. Handl., 10(4): 26.

Euschistus acuminatus Walker, 1867

1867 *Euschistus acuminatus* Walker, Cat. Hem. Brit. Mus., 2: 246. [Hispaniola].

1983 *Euschistus acuminatus*: Baranowski et al., Fla. Ent., 66: 287.

Distribution: Fla. (Greater Antilles).

Euschistus biformis Stål, 1862

1862 *Euschistus biformis* Stål, Stett. Ent. Zeit., 23: 100. [Mexico].

1872 *Euschistus* (*Euschistus*) *biformis*: Stål, K. Svens. Vet.-Akad. Handl., 10(4): 28.

Distribution: Ariz. (Mexico to Panama).

Euschistus comptus Walker, 1868

1868 *Euschistus comptus* Walker, Cat. Het. Brit. Mus., 3: 550. [Mexico].

1904 *Euschistus comptus*: Van Duzee, Trans. Am. Ent. Soc., 30: 48.

1916 *Euschistus* (*Euschistus*) *comptus*: Van Duzee, Check List Hem., p. 6.

Distribution: Tex. (Mexico to Guatemala).

Note: Rolston (1974, Ent. Am., 48: 33) omitted the Tex. record above without comment.

Euschistus conspersus Uhler, 1897

1897 *Euschistus conspersus* Uhler, Trans. Md. Acad. Sci., 1: 388. [Cal., Wash.].

1916 *Euschistus* (*Euschistus*) *conspersus*: Van Duzee, Check List Hem., p. 5.

Distribution: B.C., Cal., Wash.

Euschistus crassus Dallas, 1851

1851 *Euschistus crassus* Dallas, List Hem. Brit. Mus., 1: 205. [Fla.].

1916 *Euschistus* (*Euschistus*) *crassus*: Van Duzee, Check List Hem., p. 6.

Distribution: Fla., Ga., N.C. (Greater Antilles).

Note: Rolston (1974, Ent. Am., 48: 35) gave distribution as "Coastal plains of southern United States."

Euschistus crenator (Fabricius), 1794

1794 *Cimex crenator* Fabricius, Ent. Syst., 4: 101. ["Americae insulis"].

Distribution: Ariz., Cal., Fla., Tex. (Mexico to Brazil and Ecuador; West Indies).

Note: The nominate subspecies is not known from our area.

Euschistus crenator orbiculator Rolston, 1974

1974 *Euschistus crenator orbiculator* Rolston, Ent. Am., 48: 35. [Nicaragua].

1859 *Pentatoma crenata*: Dohrn, Cat. Hem., p. 16.

1872 *Euschistus* (*Euschistus*) *crenator*: Stål, K. Svens. Vet.-Akad. Handl., 10(4): 27.

Distribution: Ariz., Cal., Tex. (Mexico to South America).

Note: Although Rolston (1974, above) gave no United States records, he confirmed, in correspondence, that "*crenator*" records from our area belong to this subspecies.

Euschistus eggelstoni Rolston, 1974

1974 *Euschistus* (*Euschistus*) *eggelstoni* Rolston, Ent. Am., 48: 13, 76. [Holotype from Mexico; also paratypes from Ariz.].

Distribution: Ariz. (Mexico).

Euschistus ictericus (Linnaeus), 1763
1763 *Cimex ictericus* Linnaeus, Cent. Ins., p. 16. ["Carolina"].
1851 *Euschistus ictericus*: Dallas, List Hem. Brit. Mus., 1: 206.
1872 *Euschistus (Euschistus) ictericus*: Stål, K. Svens. Vet.-Akad. Handl., 10(4): 26.
Distribution: Ark., Col., Conn., Fla., Ia., Ill., Ind., La., Mass., Mich., N.C., Neb., N.J., N.Y., Oh., Ok., Ont., R.I., Tex., Ut., Vt., Wis.
Note: McPherson and Paskewitz (1984, J. N.Y. Ent. Soc. 92: 53-60) discussed the life cycle and described the immature stages of this species.

Euschistus inflatus Van Duzee, 1903
1903 *Euschistus inflatus* Van Duzee, Trans. Am. Ent. Soc., 29: 107. [Col., N.M.].
1916 *Euschistus (Euschistus) inflatus*: Van Duzee, Check List Hem., p. 5.
Distribution: Cal., Col., N.M., Ut.

Euschistus integer Stål, 1872
1872 *Euschistus integer* Stål, K. Svens. Vet.-Akad. Handl., 10(4): 28. [Mexico].
1909 *Euschistus (Paraschistus) integer*: Kirkaldy, Gen. Cat., p. 63.
Distribution: Ariz. (Mexico).
Note: Rolston (1974, Ent. Am., 48: 49) omitted the Ariz. record above without comment.

Euschistus latimarginatus Zimmer, 1910
1910 *Euschistus latimarginatus* Zimmer, Can. Ent., 42: 167. [Neb.].
1916 *Euschistus (Euschistus) latimarginatus*: Van Duzee, Check List Hem., p. 5.
Distribution: Col., Neb.

Euschistus obscurus (Palisot), 1817
1817 *Pentatoma obscura* Palisot, Ins. Rec. Afr. Am., p. 149. [Dominican Republic].
1904 *Euschistus bifibulus*: Van Duzee, Trans. Am. Ent. Soc., 30: 48.
1916 *Euschistus (Euschistus) bifibulus*: Van Duzee, Check List Hem., p. 5.
1927 *Euschistus atromaculosus* Barber, Bull. Brook. Ent. Soc., 22: 241. [Fla.]. Synonymized by Rolston, 1974, Ent. Am., 48: 55.
Distribution: Fla., Ga., Miss., Tex. (Mexico, Greater Antilles)
Note: Barber (1927, Bull. Brook. Ent. Soc., 22: 243) transferred earlier United States records of *Euschistus bifibulus* (Palisot) to this species.

Euschistus politus Uhler, 1897
1886 *Podisus politus* Uhler, Check-list Hem. Het., p. 4. *Nomen nudum.*
1897 *Euschistus politus* Uhler, Can. Ent., 29: 117. [D.C., Mass., Md., Pa.]
Distribution: Ark., Conn., D.C., Ill., Md., Mass., Md., Mich., Mo., N.H., N.J., N.Y., Oh., Pa., R.I., Tenn.
Note: McPherson (1974, An. Ent. Soc. Am., 67: 940-941) reported observations on the biology of this species.

Euschistus quadrator Rolston, 1974
1974 *Euschistus (Euschistus) quadrator* Rolston, Ent. Am., 48: 41. [Holotype from Mexico; also paratypes from La. and Tex.].
Distribution: La., Tex. (Mexico).

Euschistus servus (Say), 1832
1832 *Pentatoma serva* Say, Descrip. Het. Hem. N. Am, p. 4. [Fla., Pa.].
1837 *Pentatoma harrisii* Westwood, Hope Cat., 1: 41. [Ga.]. Synonymized by Van Duzee, 1916, Check List Hem., p. 5.
1872 *Euschistus (Euschistus) servus*: Stål, K. Svens. Vet.-Akad. Handl., 10(4): 26.
1872 *Euschistus (Euschistus) impictiventris* Stål, K. Svens. Vet.-Akad. Handl., 10(4): 26. [Tex.]. Synonymized by Rolston, 1974, Ent. Am., 48: 66.

1886 *Euschistus impictiventris*: Uhler, Check-list Hem. Het., p. 6.
1886 *Euschistus servus*: Uhler, Check-list Hem. Het., p. 6.
1888 *Euschistus jugalis* Provancher, Pet. Faune Ent. Can., 3: 204. [B.C.]. Synonymized under *Euschistus impictiventris* by Van Duzee, 1917, Univ. Cal. Publ. Ent., 2: 40; Rolston (1974, above) placed that name in synonymy under *Euschistus servus*. Lectotype designated by Kelton, 1968, Nat. Can., 95: 1067.
1919 *Euschistus subimpunctatus* McAtee, Ill. Nat. Hist. Surv. Bull., 13: 191. [Ill.]. Synonymized by Sailer, 1961, Germ Plasma Resources, p. 302.
1939 *Euschistus* (*Euschistus*) *subimpunctatus*: Torre-Bueno, Ent. Am., 19: 220.
Distribution: Ariz., Ark., Cal., Col., "Dakota," D.C., Fla., Ga., Ia., Ill., Ind., Ks., Ky., La., Mass., Md., Mich., Miss., Mo., N.J., N.C., N.M., N.Y., Oh., Pa., S.C., Tex., Ut., Va. (Mexico).
Note: Life history notes provided by Woodside (1946, J. Econ. Ent., 39: 161-163).

Euschistus servus euschistoides (Vollenhoven), 1868
1868 *Diceraeus euschistoides* Vollenhoven, Vers. Med. Kon. Weten. Amsterdam, 2: 180. [Wis.].
1869 *Euschitus* [*sic*] *pustulatus*: Rathvon, Hist. Lancaster Co., Pa., p. 549.
1871 *Euschistus fissilis* Uhler, Proc. Boston Soc. Nat. Hist., 14: 96. [No locality given]. Synonymized by Van Duzee, 1916, Check List Hem., p. 5.
1904 *Euschistus flissilis* [*sic*]: Van Duzee, Trans. Am. Ent. Soc., 30: 44.
1939 *Euschistus* (*Euschistus*) *euschistoides*: Torre-Bueno, Ent. Am., 19: 221.
1939 *Euschistus euschistoidea* [*sic*]: Beaulne, Can. Ent., 71: 120.
1946 *Euschistus servus euschistoides*: Sailer, J. Econ. Ent., 39: 158.
Distribution: Alta., B.C., Col., Fla., Ill., Ind., Ky., Mass., Mich., Mont., N.C., N.H., N.J., N.S., N.Y., Neb., Oh., Ont., Pa., Que.

Euschistus servus servus (Say), 1832
1832 *Pentatoma serva* Say, Descript. Het. Hem. N.Am., p. 4.
1872 *Euschistus* (*Euschistus*) *impictiventris* Stål, K. Svens. Vet.-Akad. Handl., 10(4): 26. [Tex.]. Synonymized by Rolston, 1974, Ent. Am., 48: 66.
1946 *Euschistus servus servus*: Sailer, J. Econ. Ent., 39: 158.
Distribution: Same as for species above.

Euschistus spurculus Stål, 1862
1862 *Euschistus spurculus* Stål, Stett. Ent. Zeit., 23: 100. [Mexico].
1906 *Euschistus spurculus*: Snow, Trans. Ks. Acad. Sci., 20: 177.
1916 *Euschistus* (*Euschistus*) *spurculus*: Van Duzee, Check List Hem., p. 6.
Distribution: Ariz. (Mexico).

Euschistus strenuus Stål, 1862
1862 *Euschistus strenuus* Stål, Stett. Ent. Zeit., 23: 100. [Mexico].
1893 *Euschistus zopilotensis* Distant, Biol. Centr.-Am., Rhyn., 1: 330. [Mexico]. Synonymized by Rolston, 1974, Ent. Am., 48: 79.
1904 *Euschistus zopilotensis*: Van Duzee, Trans. Am. Ent. Soc., 30: 48.
1910 *Euschistus zophilotensis* [*sic*]: Banks, Cat. Nearc. Hem.-Het., p. 86.
1916 *Euschistus* (*Euschistus*) *zopilotensis*: Van Duzee, Check List Hem., p. 6.
Distribution: Tex. (Mexico).

Euschistus tristigmus (Say), 1832
1832 *Pentatoma tristigma* Say, Descrip. Het. Hem. N.Am., p. 4. ["United States"].
1837 *Pentatoma inconspecta* Westwood, Hope Cat., 1: 42. [America meridionali]. Synonymized by Distant, 1903, Proc. Zool. Soc. Lond., 54: 812.

1844 *Pentatoma tristigma* Herrich-Schaeffer, Wanz. Ins., 7: 101. ["Nordamerika"]. Preoccupied.
1851 *Euschistus tristigma* [*sic*]: Dallas, List Hem. Brit. Mus., 1: 207.
1859 *Euschistus inconspectus*: Dohrn, Cat. Hem., p. 13.
1859 *Mormidea pyrrhocerra*: Dohrn, Cat. Hem., p. 14.
1861 *Euschistus tristigmus*: Uhler, Proc. Ent. Soc. Phila., 1: 23.
1869 *Euschitus* [*sic*] *inconspictus* [*sic*]: Rathvon, Hist. Lancaster Co., Pa., p. 549.
1925 *Euschistus custigmus* [*sic*]: Frost, J. N.Y. Ent. Soc., 32: 180.

Distribution: Col., D.C., Del., Fla., Ga., Ia., Ill., Ind., Ks., Ky., La., "Maritime Provinces" (Canada), Mass., Me., Mich., Minn., Mo., N.C., N.H., N.J., N.S., N.Y., Neb., Oh., Ok., Ont., Pa., Que., S.C., Va. (Mexico).

Note: The present catalog follows McPherson's (1982, Pent. Ne. N.Am., p. 63) summary of experimental and collections evidences that at approximately 41 degrees North Latitude, this species breaks into a northern subspecies *luridus* and the nominate subspecies ranging southward. The state records remotely north or south of that line of latitude can be placed easily by subspecies, but the records from close to either side need examination of specimens to establish their identities--therefore, no breakdown of distribution is given below with these two subspecies.

Euschistus tristigmus luridus Dallas, 1851

1851 *Euschistus luridus* Dallas, List Hem. Brit. Mus., 1: 207. [N.S., N.Y.].
1904 *Euschistus tristigmus tristigmus*: Van Duzee, Trans. Am. Ent. Soc., 30: 47.
1949 *Euschistus tristigmus luridus*: Sailer, Proc. Ent. Soc. Wash., 51: 163.

Euschistus tristigmus tristigmus (Say), 1832

1832 *Pentatoma tristigma* Say, Descrip. Het. Hem. N. Am., p. 4. ["United States"].
1841 *Cimex pyrrhocerus* Herrich-Schaeffer, Wanz. Ins., 6: 71. ["Nordamerika"]. Synonymized by Stål, 1872, K. Svens. Vet.-Akad. Handl., 10(4): 26.
1844 *Pentatoma tristigma* Herrich-Schaeffer, Wanz. Ins., 7: 101 (as new) ["Nordamerika"]. Preoccupied.
1851 *Euschistus tristigma* [*sic*]: Dallas, List Hem. Brit. Mus., 1: 107.
1859 *Mormidea pyrrhocera*: Dohrn, Cat. Hem., p. 14.
1872 *Euschistus* (*Euschistus*) *tristigmus*: Stål, K. Svens. Vet.-Akad. Handl., 10(4): 26.
1872 *Euschistus pyrrhocerus*: Uhler, Prelim. Rept. U.S. Geol. Surv. Mont., p. 369.
1904 *Euschistus tristigmus* var. *pyrrhocerus*: Van Duzee, Trans. Am. Ent. Soc., 30: 47.
1909 *Euschistus tricinctus*: Crevecoeur, Trans. Ks. Acad. Sci., 19: 232.
1925 *Euschistus custigmus* [*sic*]: Frost, J. N.Y. Ent. Soc., 32: 180.

Euschistus variolarius (Palisot), 1817 [Fig. 107]

1817 *Pentatoma variolaria* Palisot, Ins. Rec. Afr. Am., p. 149. [Dominican Republic].
1825 *Pentatoma punctipes* Say, J. Nat. Sci. Phila., 4: 314. [Mo., Pa.]. Preoccupied. Synonymized by Stål, 1872, K. Svens. Vet.-Akad. Handl, 10(4): 26.
1841 *Cimex sordidus* Herrich-Schaeffer, Wanz. Ins., 6:70. ["Nordamerika"]. Synonymized by Stål, 1872, K. Svens. Vet.-Akad. Handl., 10(4): 26.
1859 *Dendrocoris punctipes*: Dohrn, Cat. Hem., p. 14.
1869 *Euschitus* [*sic*] *punctipes*: Rathvon, Hist. Lancaster Co., Pa., p. 549.
1872 *Euschistus variolarius*: Stål, K. Svens. Vet.-Akad. Handl., 10(4): 26.
1912 *Euschistus varrolarius* [*sic*]: Olsen, J. N.Y. Ent. Soc., 20: 53.
1952 *Euschistis* [*sic*] *varialarius* [*sic*]: Ruckes, J. N.Y. Ent. Soc., 59: 249.

Distribution: Ark., B.C., Col., Conn., D.C., Fla., Ia., Id., Ill., Ind., Ks., Mass., Mich., Mo., N.C., N.H., N.J., N.Y., Neb., Oh., Ont., Ore., Pa., Que., Ut., Va. (West Indies).

Note: Parish (1934, An. Ent. Soc. Am., 27: 50-54) presented biological considerations of this species.

Genus *Holcostethus* Fieber, 1860

1806 *Holcostethus* Fieber, Europ. Hem., p. 79. Type-species: *Cimex sphacelatus* Fabricius, 1794. Designated by Kirkaldy, 1909, Cat. Hem., 1: 47.

1866 *Peribalus* Mulsant and Rey, An. Soc. Linn. Lyon, ser. 2, 14: 185. Type-species: *Cimex vernalis* Wolff, 1791. Designated by Kirkaldy, 1909, Cat. Hem., 1: xxiv. Synonymized by Kirkaldy, 1909, Cat. Hem., 1: 47.

1866 *Dryocoris* Mulsant and Rey, An. Soc. Linn. Lyon, ser. 2, 14: 190. Unnecessary new name for *Holcostethus*.

Note: McDonald revised the genus (1975, J. N.Y. Ent. Soc., 82: 245-258) and gave a key to the North American species (1982, J. N.Y. Ent. Soc., 90: 5-7).

Holcostethus abbreviatus Uhler, 1872

1872 *Holcostethus abbreviatus* Uhler, Prelim. Rept. U. S. Geol. Surv. Mont., 5: 397. [Cal.]. Lectotype from Cal. designated by McDonald, 1975, J. N.Y. Ent. Soc., 82: 250.

1904 *Peribalus abbreviatus*: Van Duzee, Trans. Am. Ent. Soc., 30: 33.

1960 *Peribalus eatoni* Bliven, Occid. Ent., 1: 36. [Cal.]. Synonymized by McDonald, 1975, J. N.Y. Ent. Soc., 82: 250.

Distribution: Alta., Ariz., B.C., Cal., Col., Ia., Id., Ks., Mont., N.D., Neb., Nev., N.M., Ore., S.C., Ut., Wash. (Mexico).

Note: McPherson (1982, Pent. N.E. N. Am., p. 49) pointed out that Michigan records were based on misidentified material of *H. fulvipes*.

Holcostethus fulvipes (Ruckes), 1957

1957 *Peribalus fulvipes* Ruckes, Bull. Brook. Ent. Soc., 52: 39. [N.Y.].

1975 *Holcostethus fulvipes*: McDonald, J. N.Y. Ent. Soc., 82: 252.

Distribution: Mich., N.H., N.Y.

Holcostethus hirtus (Van Duzee), 1937

1937 *Peribalus hirtus* Van Duzee, Pan-Pac. Ent., 13: 25. [Cal.].

1975 *Holcostethus hirtus*: McDonald, J. N.Y. Ent. Soc., 82: 254.

Distribution: Cal.

Holcostethus limbolarius (Stål), 1872 [Fig. 104]

1872 *Peribalus limbolarius* Stål, K. Svens. Vet.-Akad. Handl., 10(4): 34. [Tex.]. Lectotype designated by McDonald, 1975, J. N.Y. Ent. Soc., 82: 251.

1872 *Peribalus modestus* Uhler, Prelim. Rept. U. S. Geol. Surv. Mont., 5: 396. [Ariz., Col., Ks., New England "and generally throughout the States east of the Mississippi"]. Synonymized by Uhler, 1877, Bull. U.S. Geol. Geogr. Surv. Terr., 3: 403.

1909 *Holcostethus limbolarius*: Kirkaldy, Cat. Hem., 1: 48.

Distribution: Alta., Ariz., Ark., B.C., Cal., Col., Conn., Ga., Ia., Ill., Ind., Ks., Man., Mass., Me., Mich., Miss., Mo., Mont., N.C., N.J., N.M., N.Y., Neb., Oh., Ont., Ore., Pa., Tex., Ut., Wis. (Mexico).

Note: Biological information provided by Oetting and Yonke (1971, J. Ks. Ent. Soc., 44: 449-451).

Holcostethus piceus (Dallas), 1851

1851 *Pentatoma picea* Dallas, List Hem. Brit. Mus., 1: 236. ["Hudson's Bay"].

1886 *Peribalus piceus*: Uhler, Check-list Hem. Het., p. 7.
1909 *Holcostethus piceus*: Kirkaldy, Cat. Hem., 1: 48.
Distribution: Alta., Col., Ia., Ill., Man., Mich., Mont., Ont., Que., S.D., Sask.

Holcostethus ruckesi McDonald, 1975
1975 *Holcostethus ruckesi* McDonald, J. N.Y. Ent. Soc., 82: 254. [Ariz.].
Distribution: Ariz.

Holcostethus tristis (Van Duzee), 1904
1904 *Peribalus tristis* Van Duzee, Trans. Am. Ent. Soc., 30: 33. [B.C.].
1909 *Holcostethus tristis*: Kirkaldy, Cat. Hem., 1: 48.
Distribution: Alta., B.C., Cal., Id., Mont., Ore., Ut., Wash.

Genus *Hymenarcys* Amyot and Serville, 1843

1843 *Hymenarcys* Amyot and Serville, Hist. Nat. Ins., Hem., p. 124. Type-species: *Hymenarcys perpunctata* Amyot and Serville, 1843, a junior synonym of *Pentatoma nervosa* Say, 1832. Designated by Kirkaldy 1909, Cat. Hem., 1: 72.
Note: Rolston (1973, J. N.Y. Ent. Soc., 81: 111, 113) provided a key to species.

Hymenarcys aequalis (Say), 1832
1832 *Pentatoma aequalis* Say, Descrip. Het. Hem. N. Am., p. 7. [Ind.].
1842 *Cimex dentatus* Herrich-Schaeffer, Wanz. Ins., 5: 64. ["Nordamerika"]. Preoccupied. Synonymized by Stål, 1872, K. Svens. Vet.-Akad. Handl., 10(4): 30.
1851 *Pentatoma boxura* Dallas, List Hem. Brit. Mus., 1: 244. ["N. America"]. Synonymized by Stål, 1872, K. Svens. Vet.-Akad. Handl., 10(4): 30.
1872 *Hymenarcys aequalis*: Stål, K. Svens. Vet.-Akad. Handl., 10(4): 30.
Distribution: Ark., Col., D.C., Fla., Ga., Ia., Ill., Ind., Ks., Ky., Mass., Md., Mich., Miss., Mo., Mont., N.C., N.J., N.Y., Neb., Oh., Ok., Que., Sask., Tex. (Mexico).
Note: Immature stages and biology studied by Oetting and Yonke (1971 An. Ent. Soc. Am., 64: 1293-1296).

Hymenarcys crassa Uhler, 1897
1897 *Hymenarcys crassa* Uhler, Trans. Md. Acad. Sci., 1: 387. [Mexico].
Distribution: Ariz., Tex. (Mexico).
Note: Immature stages and biological notes given by Oetting and Yonke (1972, An. Ent. Soc. Am., 65: 474-478).

Hymenarcys nervosa (Say), 1832
1832 *Pentatoma nervosa* Say, Descrip. Het. Hem. N. Am., p. 9. [Ind.].
1837 *Pentatoma pennsylvaniae* Westwood, Hope Cat., 1: 35. [Pa.]. Synonymized by Stål, 1872, K. Svens. Vet-Akad. Handl., 10(4): 31.
1843 *Hymenarcys perpunctata* Amyot and Serville, Hist. Nat. Ins. Hem. p. 124. ["Amerique septentrionale"]. Synonymized by Stål, 1872, K. Svens. Vet.-Akad. Handl., 10(4): 31.
1867 *Hymenarcys nervosa*: Walker, Cat. Het., 2: 283.
1869 *Hymenarchi* [*sic*] *neriosa* [*sic*]: Rathvon, Hist. Lancaster Co., Pa., p. 548.
Distribution: Ark., D.C., "Dakota," Fla., Ga., Ia., Ill., Ind., Ky., Mass., Md., Mich., Mo., N.C., N.J., N.Y., Oh., Ok., Pa., Que., R.I., S.C., Tex., Va.
Note: Immature stages and biology studied Oetting and Yonke (1971, An. Ent. Soc. Am., 64: 1290-1293).

Hymenarcys reticulata Stål, 1872
1872 *Hymenarcys reticulata* Stål, K. Svens. Vet.-Akad. Handl., 10(4): 30. [Mexico].
1917 *Hymenarcys reticulata*: Van Duzee, Univ. Cal. Publ. Ent., 2: 46.
Distribution: Ariz. (Mexico).

Genus *Kermana* Rolston and McDonald, 1981

1981 *Kermana* Rolston and McDonald, J. N.Y. Ent. Soc., 88: 265. Type-species: *Rhaphigaster imbutus* Walker, 1867. Original designation.

Kermana imbuta (Walker), 1867
1867 *Rhaphigaster imbutus* Walker, Cat. Het., 2: 358. [Mexico].
1906 *Banasa imbuta*: Barber, Brook. Inst. Arts Sci., Sci. Bull., 1: 261.
1909 *Nezara (Atomosira) imbuta*: Kirkaldy, Cat. Hem., 1: 122.
1981 *Kermana imbuta*: Rolston and McDonald, J. N.Y. Ent. Soc., 88: 266.
Distribution: Tex. (Mexico to Costa Rica).

Genus *Loxa* Amyot and Serville, 1843

1843 *Loxa* Amyot and Serville, Hist. Nat. Ins. Hem., p. 137. Type-species: *Cimex flavicollis* Drury, 1773. Designated by Kirkaldy, 1909, Cat. Hem., 1: 96.
Note: Eger's (1978, J. N.Y. Ent. Soc., 86: 224-251) revision and key to the species of *Loxa* made significant changes in species concepts; earlier records need verification; here only the distribution as given by him is reported.

Loxa flavicollis (Drury), 1773 [Fig. 119]
1773 *Cimex flavicollis* Drury, Illus. Nat. Hist., 2: 67, pl. 36, fig. 4. [Jamaica].
1876 *Loxa flavicollis*: Uhler, Bull. U.S. Geol. Geogr. Surv. Terr., 1: 290 (in part).
1909 *Loxa florida* Van Duzee, Bull. Buff. Soc. Nat. Sci., 9: 156. [Fla.]. Synonymized by Eger, 1978, J. N.Y. Ent. Soc., 86: 229.
Distribution: Ala., Fla., N.M., Tex. (Greater Antilles).

Loxa viridis (Palisot), 1811
1811 *Pentatoma viridis* Palisot, Ins. Rec. Afr. Am., p. 111, pl. 8, fig. 3. ["Oware"].
1978 *Loxa viridis*: Eger, J. N.Y. Ent. Soc., 86: 243.
Distribution: Fla., Tex. (Mexico to Argentina).

Genus *Menecles* Stål, 1867

1867 *Menecles* Stål, Öfv. K. Svens. Vet.-Akad. Förh., 24(7): 527. Type-species: *Pentatoma inserta* Say, 1832. Monotypic.
Note: Revision and key given by Rolston (1973, J. N.Y. Ent. Soc., 80: 234-237).

Menecles insertus (Say), 1832
1832 *Pentatoma inserta* Say, Descrip. Het. Hem. N. Am., p. 6. [Ark., Mo.].
1872 *Menecles insertus*: Stål, K. Svens. Vet.-Akad. Handl., 10(4): 31.
1904 *Menecles incertus* [*sic*]: Van Duzee, Trans. Am. Ent. Soc., 30: 52.
1910 *Meneclas* [*sic*] *insertus*: Kirkaldy, Proc. Haw. Ent. Soc., 3: 125.
Distribution: Ariz., Ark., Cal., Conn., Ill., Ind., Ks., Ky., Mass., Mich., Mo., N.J., N.Y., Neb., Oh., Ont., Que., Pa., R.I., Ut.

Menecles portacrus Rolston, *1973*
1973 *Menecles portacrus* Rolston, J. N.Y. Ent. Soc., 80: 235. [Tex.].
Distribution: Tex.

Genus *Mormidea* Amyot and Serville, 1843

1843 *Mormidea* Amyot and Serville, Hist. Nat. Ins. Hem., p. 134. Type-species: *Cimex ypsilon* Linnaeus, 1758. Designated by Kirkaldy, 1903, Ent., 36: 231.

Note: A revision with a key to subgenera and species was provided by Rolston (1978, J. N.Y. Ent. Soc., 86: 161-219).

Subgenus *Melanochila* Stål, 1872

1872 *Mormidea* (*Melanochila*) Stål, K. Svens. Vet.-Akad. Handl., 10(4): 19. Type-species: *Cimex lugens* Fabricius, 1775. Monotypic.

Mormidea lugens (Fabricius), 1775

1775 *Cimex lugens* Fabricius, Syst. Ent., p. 716. ["America"].

1798 *Cimex albipes* Fabricius, Ent. Syst., Suppl., p. 535. ["Carolina"]. Preoccupied. Synonymized by Stål, 1872, K. Svens. Vet-Akad. Handl., 10(4): 19.

1803 *Cydnus lugens*: Fabricius, Syst. Rhyn., p. 187.

1803 *Cimex gamma* Fabricius, Syst. Rhyn., Suppl., Emend. [terminal, no p. number]. New name for *Cimex albipes* Fabricius.

1825 *Pentatoma punctipes* Say, J. Acad. Nat. Sci. Phila., 4: 313. ["United States"]. Preoccupied. Synonymized by Stål, 1872, K. Svens. Vet.-Akad. Handl., 10(4): 19.

1832 *Pentatoma gamma*: Say, Descrip. Het. Hem. N. Am, p. 9.

1851 *Pentatoma lugens*: Dallas, List Hem. Brit. Mus., 1: 248.

1868 *Mormidea lugens*: Walker, Cat. Hem. Brit. Mus., 3: 551.

1872 *Mormidea* (*Melanochila*) *lugens*: Stål, K. Svens. Vet.-Akad. Handl., 10(4): 19.

Distribution: Ala., Ark., Col., Conn., D.C., Fla., Ga., Ia., Ill., Ind., Ks., La., Man., Mass., Md., Me., Mich., Miss., N.C., N.D., N.H., N.J., N.S., N.Y., Neb., Oh., Ok., Ont., Pa., Que., R.I., S.C., S.D., Tenn., Tex., Va., Vt., W.Va., Wis., Wyo. (Mexico, West Indies).

Note: Biological information provided by Oetting and Yonke (1971, J. Ks. Ent. Soc., 44: 456-457) and McPherson (1974, An. Ent. Soc. Am., 67: 940-941).

Subgenus *Mormidea* Amyot and Serville, 1843

1843 *Mormidea*: Amyot and Serville, Hist. Nat. Ins., Hem., p. 135. Type-species: *Cimex ypsilon* Linnaeus, 1758. Designated by Kirkaldy, 1903, Ent., 36: 231.

1872 *Mormidea* (*Mormidea*): Stål, K. Svens. Vet.-Akad. Handl., 10(4): 19.

Mormidea cubrosa Dallas, 1851

1851 *Mormidea cubrosa* Dallas, List Hem. Brit. Mus., 1: 247. [Jamaica].

1867 *Eysarcoris punctifer* Walker, Cat. Hem. Brit. Mus., 2: 274. [Cal.]. Synonymized by Rolston, 1976, J. N.Y. Ent. Soc., 84: 5.

1872 *Mormidea* (*Mormidea*) *sordidula* Stål, K. Svens. Vet.-Akad. Handl., 10(4): 21. [Tex.]. Synonymized by Barber and Bruner, 1932, J. Dept. Agri. Puerto Rico, 16: 250. Lectotype designated by Rolston, 1978, J. N.Y. Ent. Soc., 86: 203.

1886 *Neottiglossa punctifer*: Uhler, Check-list Hem. Het., p. 5.

1886 *Mormidea sordidula*: Uhler, Check-list Hem. Het., p. 6.

Distribution: Cal., N.M., Tex. (Mexico to Colombia).

Mormidea pama Rolston, 1978

1978 *Mormidea* (*Mormidea*) *pama* Rolston, J. N.Y. Ent. Soc., 86: 205. [Honduras].

Distribution: Fla. (Mexico to Colombia, West Indies).

Mormidea pictiventris Stål, 1862

1862 *Mormidea pictiventris* Stål, Stett. Ent. Zeit., 23: 103. [Mexico].

1904 *Mormidea pictiventris*: Van Duzee, Trans. Am. Ent. Soc., 30: 42.

1916 *Mormidea* (*Mormidea*) *pictiventris*: Van Duzee, Check List Hem., p. 5.

Distribution: Tex. (Mexico to Colombia).

Note: Rolston (1978, J. N.Y. Ent. Soc., 86: 204-207) refers Fla. records of this species to *Mormidea pama* Rolston.

Genus *Moromorpha* Rolston, 1978

1978 *Moromorpha* Rolston, J. N.Y. Ent. Soc., 86: 163. Type-species: *Mormidea tetra* Walker, 1868. Monotypic.

Moromorpha tetra (Walker), 1868

1868 *Mormidea tetra* Walker, Cat. Het. Brit. Mus., 3: 551. [Mexico].

1904 *Mormidea tetra*: Van Duzee, Trans. Am. Ent. Soc., 30: 43.

1916 *Mormidea* (*Mormidea*) *tetra*: Van Duzee, Check List Hem., p. 5.

Distribution: Ariz., N.M., Tex. (Mexico).

Note: Present combination made by Rolston (1978, J. N.Y. Ent. Soc., 86: 163)

Genus *Murgantia* Stål, 1862

1862 *Murgantia* Stål, Stett. Ent. Zeit., 23: 105. Type-species: *Pentatoma tessellata* Amyot and Serville, 1843. Designated by Kirkaldy, 1909, Cat. Hem., 1: 106.

Murgantia angularis (Walker), 1867

1867 *Strachia angularis* Walker, Cat. Hem. Brit. Mus., 2: 315. [N. Am.].

1916 *Murgantia angularis*: Van Duzee, Check List Hem., p. 7.

Distribution: "N. Am."

Note: Torre-Bueno (1939, Ent. Am., 19: 235) omitted this species from his key to Pentatomidae because the North American record was "questioned" by Van Duzee (1917, Univ. Cal. Publ. Ent., 2: 56).

Murgantia histrionica (Hahn), 1834 [Fig. 112]

1834 *Strachia histrionica* Hahn, Wanz. Ins., 2: 116. [Mexico].

1868 *Strachia histrionicha* [*sic*]: Glover, U.S. Dept. Agr. Rept., p. 71.

1872 *Murgantia histrionica*: Stål, K. Svens. Vet.-Akad. Handl., 10(4): 37.

1903 *Murgantia histrionica* form *nigricans* Cockerell, Bull. S. Cal. Acad. Sci., 2: 85. [Cal.]. Synonymized by Kirkaldy, 1909, Cat. Hem., 1: 106.

Distribution: Ariz., Ark., Cal., Col., Conn., D.C., Del., Fla., Ga., Ia., Ill., Ind., Ks., La., Md., Mich., Minn., Miss., Mo., N.C., N.H., N.J., N.M., N.Y., Neb., Nev., Oh., Pa., S.D., Tex., Va., W.Va. (Mexico; introduced into Hawaii).

Note: Apparently the northern limits of established range for this species fluctuate markedly in response to the mildness or severity of the winters; during spring and summer it migrates northward.

Murgantia varicolor (Westwood), 1837

1837 *Pentatoma varicolor* Westwood, Hope Cat., 1: 37. [Brazil].

1851 *Stachia munda* Dallas, List Hem. Ins. Brit. Mus., 1: 264. [Colombia]. Synonymized by Distant, 1900, Proc. Zool. Soc. London, p. 812.

1886 *Murgantia munda*: Uhler, Check-list Hem. Het., p. 7.

1904 *Murgantia varicolor*: Van Duzee, Trans. Am. Ent. Soc., 30: 57.

Distribution: Col. (Mexico to Brazil).

Murgantia violascens (Westwood), 1837
1837 *Pentatoma violascens* Westwood, Hope Cat., 1: 34. ["Brasil?"].
1904 *Murgantia violascens*: Van Duzee, Trans. Am. Ent. Soc., 30: 57.
Distribution: Fla. (Brazil(?), West Indies).

Genus *Neopharnus* Van Duzee, 1910

1910 *Neopharnus* Van Duzee, Trans. Am. Ent. Soc., 36: 73. Type-species: *Neopharnus fimbriatus* Van Duzee, 1910. Monotypic.

Neopharnus fimbriatus Van Duzee, 1910
1910 *Neopharnus fimbriatus* Van Duzee, Trans. Am. Ent. Soc., 36: 73. [Fla.].
Distribution: Fla.

Genus *Neottiglossa* Kirby, 1837

1837 *Pentatoma* (*Neottiglossa*) Kirby, Fauna Bor.-Am., 4: 276. Type-species: *Pentatoma trilineata* Kirby, 1837. Monotypic.
1871 *Neottiglossa*: Uhler, Proc. Boston Soc. Nat. Hist., 14: 96.

Subgenus *Neottiglossa* Kirby 1837

1837 *Pentatoma* (*Neottiglossa*) Kirby, Fauna Bor.-Am., 4: 276. Type-species: *Pentatoma trilineata* Kirby, 1837. Monotypic.
1872 *Neottiglossa* (*Neottiglossa*): Stål, K. Svens. Vet.-Akad. Handl., 10(4): 18.

Neottiglossa trilineata (Kirby), 1837
1837 *Pentatoma* (*Neottiglossa*) *trilineata* Kirby, Fauna Bor.-Am., 4: 276. ["New York to Cumberland-house," Sask.].
1851 *Aelia trilineata*: Dallas, List Hem. Brit. Mus., 1: 224.
1871 *Neottiglossa undata*: Uhler, Proc. Boston Soc. Nat. Hist., 14: 395 (in part).
1877 *Neottiglossa trilineata*: Uhler, Bull. U.S. Geol. Geogr. Surv. Terr., 3: 401.
Distribution: Alta., B.C., Cal., Col., "Dakota," Man., Mich., N.S., Neb., Que., Sask.

Neottiglossa undata (Say), 1832
1832 *Pentatoma undata* Say, Descrip. Het. Hem. N. Am., p. 8. ["N. W. Territory"].
1871 *Neottiglossa undata*: Uhler, Proc. Boston Soc. Nat. Hist., 14: 395 (in part).
1877 *Neottiglossa undata*: Uhler, Bull. U. S. Geol. Geogr. Surv. Terr., 3: 401.
1886 *Aelia americana*: Provancher, Pet. Faune Ent. Can., 3: 85.
1903 *Mormidea undata*: MacGillivray and Houghton, Ent. News, 14: 263.
1909 *Neottiglossa* (*Neottiglossa*) *undata*: Kirkaldy, Cat. Hem., 1: 80.
1918 *Neotiglossa* [*sic*] *undata*: Torre-Bueno, Can. Ent., 50: 25.
Distribution: Alta., B.C., Cal., Col., Conn., Ia., Ill., Mass., Me., Mich., Minn., N.C., N.H., N.J., N.S., N.Y., Neb., Nfld., Oh., Ont., Pa., Que., Ut., Wis.

Subgenus *Texas* Kirkaldy, 1904

1872 *Melanostoma* Stål, K. Svens. Vet.-Akad. Handl., 10(4): 18. Preoccupied. Type-species: *Neottiglossa sulcifrons* Stål, 1872. Designated by Kirkaldy, 1909, Cat. Hem., 1: 80.
1904 *Texas* Kirkaldy, Ent., 37: 280. New name for *Melanostoma* Stål.
1909 *Neottiglossa* (*Texas*): Kirkaldy, Cat. Hem., 1: 80.

Neottiglossa californica Bliven, 1958, New Subgeneric Combination.
1958 *Neottiglossa californica* Bliven, Occ. Ent., 1: 12. [Cal.].
Distribution: Cal.
Note: Although this species was not placed subgenerically in original description, that description did mention the elongate ostiolar canal which allows its assignment to the subgenus *Texas*.

Neottiglossa cavifrons Stål, 1872
1872 *Neottiglossa (Melanostoma) cavifrons* Stål, K. Svens. Vet.-Akad. Handl., 10(4): 18. [Tex.].
1886 *Neottiglossa cavifrons*: Uhler, Check-list Hem. Het., p. 5.
1904 *Neottiglossa (Texas) cavifrons*: Kirkaldy, Ent., 37: 280.
Distribution: Ariz., Ark., Cal., Ga., Ill., Ind., Ky., Mo., N.C., Ore., Tex., Ut., Va.
Note: Biological information provided by Oetting and Yonke (1971, J. Ks. Ent. Soc., 44: 457-458). Downes (1935, Proc. Ent. Soc. B.C., 31: 46) showed that earlier records of this species for British Columbia should be applied to *N. tumidifrons* Downes.

Neottiglossa coronaciliata Ruckes, 1957
1957 *Neottiglossa corona-ciliata* [*sic*] Ruckes, Bull. Brook. Ent. Soc., 52: 41. [Tex.].
Distribution: Ark., La., Tex.

Neottiglossa sulcifrons Stål, 1872
1872 *Neottiglossa (Melanostoma) sulcifrons* Stål, K. Svens. Vet.-Akad. Handl, 10(4): 18. [Tex.].
1876 *Melanostoma sulcifrons*: Uhler, Bull. U.S. Geol. Geogr. Surv. Terr., 1: 284.
1886 *Neottiglossa sulcifrons*: Uhler, Check-list Hem. Het., p. 5.
1904 *Neottiglossa (Texas) sulcifrons*: Kirkaldy, Ent., 37: 280.
1905 *Neotiglossa* [*sic*] *sulcifrons*: Crevecoeur, Trans. Ks. Acad. Sci., 19: 232.
Distribution: Ark., Ga., Ia., Ill., Ind., Ks., Md., N.C., N.J., N.M., Neb., Tex., Ut., Va.
Note: Records for British Columbia and Michigan have been refuted respectively by Downes (1935, Proc. Ent. Soc. B.C., 31: 46) and Hussey (1921, Psyche, 28: 8) as misidentifications.

Neottiglossa tumidifrons Downes, 1928, New Subgeneric Combination
1928 *Neottiglossa tumidifrons* Downes, Can. Ent., 90: 90. [B.C.].
Distribution: B.C., Cal., Mont., Ore., Wash.
Note: Apparently *N. tumidifrons* was never formally assigned to the subgenus *Texas*, but its original description stresses the thickened dorsal submargins of the head characteristic of that subgenus.

Genus *Nezara* Amyot and Serville, 1843

1843 *Nezara* Amyot and Serville, Hist. Nat. Ins., Hem., p. 143. Type-species: *Cimex smaragdulus* Fabricius, 1775, a junior synonym of *Cimex viridula* Linnaeus, 1758. Designated by Kirkaldy, 1909, Cat. Hem., 1: 115.

Note: Freeman (1940, Trans. Royal Ent. Soc. London, 90: 351-374), in his revision of *Nezara*, discussed the several color patterns that may repeat in each of several species of this Old World genus. In the single species introduced into our area the color may be wholly green (*smaragdula* Fabricius) or have certain yellow areas dorsally (*smaragdula* or *torquata* Fabricius). Early workers used these conspicuous color differences to describe new taxa, but now such color differences are accorded no taxonomic value.

Nezara viridula (Linnaeus), 1758

1758 *Cimex viridulus* Linnaeus, Syst. Nat., Ed. 10: 444. [India].

1775 *Cimex torquatus* Fabricius, Syst. Ent., p. 710. [India]. Synonymized by Freeman, 1940, Trans. Roy. Ent. Soc. London, 90: 360.

1775 *Cimex smaragdulus* Fabricius, Syst. Ent., p. 711. [Madeira Islands]. Synonymized by Freeman, 1940, Trans. Roy. Ent. Soc. London, 90: 358.

1859 *Rhaphigaster viridulus*: Dohrn, Cat. Hem., p. 17.

1868 *Nezara viridula*: Stål, K. Svens. Vet.-Akad. Handl., 7(11): 31.

1916 *Nezara viridula* var. *smaragdula*: Van Duzee, Check List Hem., p. 7.

1916 *Nezara viridula* var. *torquata*: Van Duzee, Check List Hem., p. 7.

Distribution: Ark., Fla., Ga., Ill., La., Miss., N.C., Oh., Ok., S.C., Tex., Va. ("Cosmopolitan"; recorded from Afrotropical, Australian, Neotropical, Nearctic, Oriental, and Palearctic Regions).

Note: Cosmopolitan records for more northern states, like Oh., N.Y., and Va., probably are adventitious occurrences. Freeman (1940, above), after a careful and detailed study of the genus *Nezara*, decided that at least some of the included species exhibited various color manifestations. Feeling compelled to conform to the then-existing rules of nomenclature, he gave a "variety" name to each of certain color combinations but with the clearly stated understanding that each "variety" was nothing more than a color digression from the norm.

Genus *Odmalea* Bergroth, 1915

1915 *Odmalea* Bergroth, An. Soc. Ent. France, 83: 436. Type-species: *Odmalea quadripunctula* Bergroth, 1915. Original designation.

Note: Genus revised by Rolston (1978, J. N.Y. Ent. Soc., 86: 20-36).

Odmalea schaefferi (Barber), 1906

1906 *Dendrocoris* (*Liotropis*) *schaefferi* Barber, Brook. Inst. Arts Sci., Sci. Bull., 1: 262. [Tex.]. Lectotype designated by Rolston, 1978, J. N.Y. Ent. Soc., 86: 34.

1909 *Dendrocoris schaefferi*: Kirkaldy, Cat. Hem., 1: 151.

1915 *Odmalea schaefferi*: Bergroth, An. Soc. Ent. France, 83: 438, 439.

Distribution: Tex.

Genus *Oebalus* Stål, 1862

1862 *Oebalus* Stål, Stett. Ent. Zeit., 23: 102. Type-species: *Cimex typhoeus* Fabricius, 1803, a junior synonym of *Cimex pugnax* Fabricius, 1775. Designated by Kirkaldy, 1909, Cat. Hem., 1: 61.

1891 *Solubea* Bergroth, Rev. d'Ent., 10: 235. Unecessary new name for *Oebalus* Stål.

Note: Sailer (1944, Proc. Ent. Soc. Wash., 46: 105-127) revised this genus under the name *Solubea*; then in 1957 (Proc. Ent. Soc., 59: 41-42) presented a justification for replacing *Solubea* with the name *Oebalus*. Sailer (1944, p. 123, *op. cit.*), explained that the original locality given for *Oebalus poecilus* (Dallas) was from a mislabeled specimen. North American records as *Mormidea poecila* by Dohrn (1859, Cat. Hem., p. 14) and Kirkaldy (1909, Cat. Hem., 1: 60) given with a query, are undoubtedly based on Dallas' listing.

Oebalus grisescens (Sailer), 1944

1944 *Solubea grisescens* Sailer, Proc. Ent. Soc. Wash., 46: 118. [Argentina].

1983 *Oebalus grisescens*: Mead, Tri-Ology, 22(11): 4.
Distribution: Fla. (South America).

Oebalus insularis Stål, 1872

1872 *Oebalus insularis* Stål, K. Svens. Vet.-Akad. Handl., 10(4): 22. [Cuba].
1893 *Mormidea querini* Lethierry and Severin, Cat. Hem., 1: 123. New name for preoccupied *Pentatoma geographica* Guérin-Méneville, 1857. Synonymized by Sailer, 1944, Proc. Ent. Soc. wash., 46: 119.
1914 *Mormidea querini*: Barber, Bull. Am. Mus. Nat. Hist., 33: 522.
1932 *Solubea insularis*: Barber and Bruner, J. Dept. Agr. Puerto Rico, 16: 252 (in part).
Distribution: Fla. (Mexico to Colombia, West Indies).

Oebalus mexicanus (Sailer), 1944

1944 *Solubea mexicana* Sailer, Proc. Ent. Soc. Wash., 46: 114. [Holotype from Mexico; also paratypes from Ariz.].
1957 *Oebalus mexicanus*: Sailer, Proc. Ent. Soc. Wash., 59: 41.
Distribution: Ariz. (Mexico).

Oebalus pugnax (Fabricius), 1775 [Fig. 113]

1775 *Cimex pugnax* Fabricius, Syst. Ent., p. 704. ["Carolina"].
1803 *Cimex typhoeus* Fabricius, Syst. Rhyn., p. 162. ["Carolina"]. Synonymized by Stål, 1872, K. Svens. Vet.-Akad. Handl., 10(4): 22.
1832 *Pentatoma augur* Say, Descrip. Het. Hem. N. Am., p. 3. [Ga.]. Synonymized with *typhoeus* Fabricius by Stål, 1868, K. Svens. Vet.-Akad. Handl., 7(11): 27.
1835 *Cimex vitripennis* Burmeister, Handb. Ent., 2: 367. [N. Am.]. Synonymized by Stål, 1872, K. Svens. Vet.-Akad. Handl., 10(4): 22.
1851 *Mormidea typhoeus*: Dallas, List Hem. Brit. Mus., 1: 216.
1868 *Mormidea (Oebalus) typhea*: Glover, U.S. Dept. Agr. An. Rept., p. 71.
1868 *Oebalus typhoeus*: Stål, K. Svens. Vet.-Akad. Handl., 7(11): 27.
1872 *Oebalus pugnax*: Stål, K. Svens. Vet.-Akad. Handl., 10(4): 22.
1891 *Solubea pugnax*: Bergroth, Rev. d'Ent., 10: 235.
Distribution: Ariz., Ark., Col., Conn., D.C., Fla., Ga., Ia., Ill., Ind., Ks., Ky., La., Md., Mich., Minn., Miss., Mo., N.C., N.J., N.M., N.Y., Neb., Oh., Ok., Pa., S.C., Tenn., Tex., Va. (Mexico to Colombia, West Indies).
Note: Sailer (1944, Proc. Ent. Soc. Wash., 46: 113) expressed belief that at least some Central American records for this species may belong to *O. mexicana* (Sailer). Only the nominate subspecies occurs in our area.

Oebalus pugnax pugnax (Fabricius), 1775

1775 *Cimex pugnax* Fabricius, Syst. Ent., p. 704.
1944 *Solubea pugnax pugnax*: Sailer, Proc. Ent. Soc. Wash., 46: 109.
Distribution: Same range as for species, except Colombia, which applies to the subspecies *O. pugnax torrida* (Sailer).

Genus *Padaeus* Stål, 1862

1862 *Padaeus* Stål, Stett. Ent. Zeit., 23, 101. Type-species: *Cimex irroratus* Herrich-Schaeffer, 1837 [p. 19, female; see *P. trivittatus* Stål]. Preoccupied. Next available name *Mormidea vidua* Vollenhoven, 1868. Designated by Kirkaldy, 1909, Cat. Hem., 1, 68.

Padaeus trivittatus Stål, 1872

1837 *Cimex irroratus*: Herrich-Schaeffer, Wanz. Ins., 4: 20 (in part, male).
1872 *Padaeus trivittatus* Stål, K. Svens. Vet.-Akad. Handl., 10(4): 29. New name for

male of *Cimex irroratus* Herrich-Schaeffer. [Mexico].
1904 *Padaeus irroratus*: Van Duzee, Trans. Am. Ent. Soc., 30: 78.
1978 *Padaeus trivittatus*: Froeschner, Proc. Ent. Soc. Wash., 80: 131.
Distribution: Ariz. (Mexico to Guatemala).
Note: Fla. records for this species are in error.

Genus *Pellaea* Stål, 1872

1872 *Nezara* (*Pellaea*) Stål, K. Svens. Vet.-Akad. Handl., 10(4): 40. Type-species: *Rhaphigaster sticticus* Dallas, 1851. Monotypic.
1914 *Pellaea*: Bergroth, An. Soc. Ent. Belg., 58: 25..

Pellaea stictica (Dallas), 1851
1851 *Rhaphigaster sticticus* Dallas, List Hem. Brit. Mus., 1: 281. [Mexico, Colombia, British Guiana].
1984 *Pellaea stictica*: Henry, Proc. Ent. Soc. Wash., 86: 519.
Distribution: Tex. (Mexico to Argentina).

Genus *Piezodorus* Fieber, 1860

1860 *Piezodorus* Fieber, Europ. Hem., p. 78. Type-species: *Piezodorus degeeri* Fieber, 1861, a junior synonym of *Cimex lituratus* Fabricius, 1794. Monotypic.

Piezodorus guildinii (Westwood), 1837
1837 *Rhaphigaster guildinii* Westwood, Hope Cat., p. 31. [St. Vincent Island].
1894 *Piezodorus guildingi* [*sic*]: Uhler, Proc. Zool. Soc. London, p. 175.
1904 *Piezodorus Guildingi* [*sic*]: Van Duzee, Trans. Am. Ent. Soc., 30, 61.
1916 *Piezodorus guildinii*: Van Duzee, Check List Hem., p. 8.
Distribution: Fla., Ga., N.M. (West Indies, South America).
Note: Panizzi and Slansky (1985, Fla. Ent., 68: 215-216) summarized the foodplants and economic importance of this species.

Piezodorus lituratus (Fabricius), 1794
1794 *Cimex lituratus* Fabricius, Ent. Syst., 4: 114. [Italy].
1821 *Pentatoma incarnata* Germar, Fauna Ins. Europ., 4: 23. [Europe]. Synonymized by Reuter, 1885, Berl. Ent. Zeit., 29: 42.
1837 *Rhaphigaster punctulatus* Westwood, Hope Cat. p. 31. [N. Am.]. Synonymized by Reuter, 1888, Acta Soc. Sci. Fenn., 15: 497.
1904 *Piezodorus incarnatus*: Van Duzee, Trans. Am. Ent. Soc., 30: 61.
1916 *Piezodorus lituratus*: Van Duzee, Check List Hem., p. 8.
Distribution: Fla. (Africa, Asia, Europe).
Note: Van Duzee (1904, above) expressed doubt that this species is a member of our fauna.

Genus *Prionosoma* Uhler, 1863

1863 *Prionosoma* Uhler, Proc. Ent. Soc. Phila., 2: 363. Type-species: *Prionosoma podopioides* Uhler, 1863. Monotypic.
1960 *Neurohalys* Bliven, Occ. Ent., 1: 34. Type-species: *Neurohalys bucculatus* Bliven, 1960. Monotypic. Synonymized by Rolston and McDonald, 1984, J. N.Y. Ent. Soc., 92: 83.

Prionosoma podopioides Uhler, 1863 [Fig. 108]

1863 *Prionosoma podopioides* Uhler, Proc. Ent. Soc. Phila., 2: 364. [Cal.].
1888 *Prionosoma villosum* Provancher, Pet. Faune Ent. Can., 3: 204. [B.C.]. Synonymized by Van Duzee, 1904, Trans. Am. Ent. Soc., 30: 52. Lectotype designated by Kelton, 1968, Nat. Can., 95: 1067.
1960 *Neurohalys bucculatus* Bliven, Occ. Ent., 1: 34. [Cal.]. Synonymized by Rolston and McDonald, 1984, J. N.Y. Ent. Soc., 92: 83.
Distribution: Ariz., Ark., B.C., Cal., Col., Ia., Id., Ill., Mich., N.M., Neb., Nev., Ut. (Mexico).

Genus *Proxys* Spinola, 1840

1840 *Proxys* Spinola, Essai Hem., p. 325. Type-species: *Cimex victor* Fabricius, 1775. Monotypic.

Proxys albopunctulatus (Palisot), 1811
1811 *Pentatoma albopunctulata* Palisot, Ins. Rec. Afr. Am., p. 130. [Saint-Dominique].
1886 *Proxys albo-punctatus* [*sic*]: Uhler, Check-list Hem. Het., p. 6.
1916 *Proxys albopunctulatus*: Van Duzee, Check List Hem., p. 6.
Distribution: "Southern States" (Uhler, 1886, above) (Brazil, Panama, West Indies).
Note: Van Duzee (1904, Trans. Am. Ent. Soc., 30: 49) questioned Uhler's (1886, above) record for the United States, but Kirkaldy (1909, Cat. Hem., 1: 71) accepted it without comment.

Proxys punctulatus (Palisot), 1818
1818 *Halys punctulatus* Palisot, Ins. Rec. Afr. Am., p. 188. [No locality given].
1832 *Pentatoma tenebrosa* Say, Ins. La., p. 10. [La.]. Synonymized by Stål, 1872, K. Svens. Vet.-Akad. Handl., 10(4): 29.
1872 *Proxys punctulatus*: Stål, K. Svens. Vet.-Akad. Handl., 10(4): 29.
Distribution: Ariz., Ark., Fla., Ga., Ill., Ind., La., Md., Mo., N.C., Ok., Pa., Tex., Va.
Note: Vangeison and McPherson (1975, An. Ent. Soc. Am., 68: 25-30) described the life history, immature stages, and laboratory rearing of this species.

Genus *Runibia* Stål, 1861

1861 *Runibia* Stål, Stett. Ent. Zeit., 22: 140. Type-species: *Cimex perspicua* Fabricius, 1798. Designated by Kirkaldy, 1909, Cat. Hem., 1: xxx, 110.

Runibia proxima (Dallas), 1851
1851 *Pentatoma proxima* Dallas, List Hem. Brit. Mus., 1: 255. [Jamaica].
1906 *Runibia proxima*: Snow, Trans. Ks. Acad. Sci., 20: 151.
Distribution: Tex. (West Indies).

Genus *Tepa* Rolston and McDonald, 1984

1984 *Tepa* Rolston and McDonald, J. N.Y. Ent. Soc., 92: 77. Type-species: *Pentatoma rugulosa* Say, 1832. Original designation.
Note: Rolston (1973, J. Ga. Ent. Soc., 7: 278-285) treated and keyed the North American species of *Tepa* under the designation of "small *Thyanta* species." Rider (1986, J. N.Y. Ent. Soc., 94: 552-557) reviewed this genus and presented a key to its 6 species.

Tepa brevis (Van Duzee), 1904
1904 *Thyanta brevis* Van Duzee, Trans. Am. Ent. Soc., 30: 56. [Col.].
1984 *Tepa brevis*: Rolston and McDonald, J. N.Y. Ent. Soc., 92: 77.

Distribution: Ariz., Cal., Col., Id., N.M., Nev., Ore., Tex., Ut. (Mexico).

Tepa jugosa (Van Duzee), 1923

1923 *Thyanta jugosa* Van Duzee, Proc. Cal. Acad. Sci., ser. 4, 12: 129. [Isla Raza, Gulf of Cal.].

1956 *Thyanta coloradensis* Bliven, New. Hem. W. St., p. 5. [Cal.]. Synonymized by Rider, 1986, J. N.Y. Ent. Soc., 94: 553.

1984 *Tepa jugosa*: Rolston and McDonald, J. N.Y. Ent. Soc., 92: 77.

Distribution: Ariz., Cal., Tex., (Mexico).

Tepa panda (Van Duzee), 1923

1923 *Thyanta panda* Van Duzee, Proc. Cal. Acad. Sci., ser. 4, 12: 128. [Mexico].

1972 *Thyanta panda*: Rolston, J. Ga. Ent. Soc., 7: 283.

1984 *Tepa panda*: Rolston and Mcdonald, J. N.Y. Ent. Soc., 92: 78.

Distribution: Fla. (Curaçāo, Mexico).

Tepa rugulosa (Say), 1832

1832 *Pentatoma rugulosa* Say, Descrip. Het. Hem. N. Am., p. 7. ["N.W. Territory"].

1872 *Thyanta rugulosa*: Stål, K. Svens. Vet.-Akad. Handl., 10(4): 35.

1904 *Thyanta punctiventris* Van Duzee, Trans. Am. Ent. Soc., 30: 55. [Cal., Col., N.D., Ut.]. Synonymized by Rider, 1986, J. N.Y. Ent. Soc., 94: 554.

1986 *Tepa rugulosa*: Rider, J. N.Y. Ent. Soc., 94: 554.

Distribution: Alta., Ariz., B.C., Cal., Col., Id., Mo., Mont., N.D., N.M., Nev., Ore., S.D., Tex., Ut., Wash. (Cuba, Mexico).

Tepa vanduzeei Rider, 1986

1919 *Thyanta punctiventris*: Hart, Ill. Nat. Hist. Surv. Bull., 13: 217.

1984 *Tepa punctiventris*: Rolston and McDonald, J. N.Y. Ent. Soc., 92: 78.

1986 *Tepa vanduzeei* Rider, J. N.Y. Ent. Soc., 94: 555. [Tex.].

Distribution: Ariz., Cal., Col., Ks., Mo., N.M., Neb., Ok., S.D., Tex. (Mexico).

Tepa yerma (Rolston), 1972

1972 *Thyanta yerma* Rolston, J. Ga. Ent. Soc., 7: 278. [Ut.].

1984 *Tepa yerma*: Rolston and McDonald, J. N.Y. Ent. Soc., 92: 78.

Distribution: Ariz., Cal., Col., Id., Nev., Ore., Ut., Wyo.

Genus *Thyanta* Stål, 1860

1860 *Thyanta* Stål, K. Svens. Vet.-Akad. Handl., 2(7): 58. Type-species: *Cimex perditor* Fabricius, 1794. Designated by Kirkaldy, 1909, Cat. Hem., 1: 94.

Notes:Rolston and McDonald (1984, J. NY. Ent. Soc., 92: 76-80) modified the concept of *Thyanta* by transferring some of its species to the genus *Cyptocephala* Berg and some to their new genus *Tepa;* these three genera are separated in their key. Ruckes' study (1957, Am. Mus. Novit., p. 1-22), as modified by genetic evidence presented by Ueshima (1963, Pan-Pac. Ent., 39: 149-154), can still be used to identify nearly all the species now remaining in *Thyanta.*

Thyanta graminicolor Uhler (1871, Rept. U.S. Geol. Survey Wyo., p. 471)) is a *nomen nudum* and is not treated further here.

Thyanta accerra McAtee, 1919

1919 *Thyanta custator* var. *accerra* McAtee, Bull. Brook. Ent. Soc., 14: 16. [Ala.].

1926 *Thyanta accerra*: Blatchley, Het. E. N. Am., p. 118.

1951 *Thyanta custator* "color phase" *accera* [*sic*]: Sailer, Proc. Ent. Soc. Wash., 53: 42.

1957 *Thyanta pallidovirens accerra*: Ruckes, Am. Mus. Novit., 1824: 15.

1971 *Thyanta custator*: Oetting and Yonke, J. Ks. Ent. Soc., 44: 455.

Distribution: Ala., Ark., Col., Fla., Ga., Id., Ill., Ind., Ks., Mich., Miss., Mo., Mont., N.C., N.D., N.M., N.Y., Que., S.D., Tex., Ut., Wyo., "South Central Canada" (introduced into Hawaii).

Note: Ueshima (1963, Pan-Pac. Ent., 39: 149-154) presented genetic evidence for specific status of this taxon. McPherson (1982, Pent. Ne. N.Am., p. 78) gave evidence that the biological notes given under the combination *Thyanta custator* by Oetting and Yonke (1971, above) actually belong to *Thyanta accerra.*

Thyanta calceata (Say), 1832

1832 *Pentatoma calceata* Say, Descrip. Het. Hem. N. Am., p. 8. [U.S.; N.J.].

1911 *Thyanta calceata*: Barber, J. N.Y. Ent. Soc., 19: 108.

Distribution: Ala., Ark., Fla., Ga., Ill., N.C., S.C., Tex., Va.

Note: Biological information was provided by Oetting and Yonke (1971, J. Ks. Ent. Soc., 44: 453-455). The "New England" occurrence given by Blatchley (1926, Het. E. N. Am., p. 117) was not confirmed by Ruckes (1957, Am. Mus. Nov., 1824: 21) studies.

Thyanta casta Stål, 1862

1862 *Thyanta casta* Stål, Stett. Ent. Zeit., 23: 104. [Mexico].

1894 *Thyanta casta*: Uhler, Proc. Cal. Acad. Sci., ser. 2, 4: 231.

Distribution: Ariz., Cal., Fla., N.M., Tex. (Mexico, West Indies).

Thyanta custator (Fabricius), 1803 [Fig. 109]

1803 *Cimex custator* Fabricius, Syst. Rhyn., p. 164. ["Carolina"].

1851 *Pentatoma custator*: Dallas, List Hem. Brit. Mus., 1: 251.

1860 *Thyanta custator*: Stål, K. Svens. Vet.-Akad. Handl., 2(7): 58.

Distribution: Ark., Conn., Del., Fla., La., Mass., Miss., N.C., N.H., N.J., N.Y., Oh., Pa., S.C., Tex., Va.

Thyanta pallidovirens (Stål), 1859

1859 *Pentatoma pallido-virens* [*sic*] Stål, K. Svens. Freg. Eug. Resa Jorden, p. 227. [Cal.].

1860 *Thyanta pallido-virens* [*sic*]: Stål, K. Svens. Vet.-Akad. Handl., 2(7): 58.

1978 *Thyanta pallidovirens*: Slater and Baranowski, How To Know True Bugs, p. 51.

Distribution: Ariz., B.C., Cal., Id., N.M., Nev., Ore., Tex., Ut., Wash. (Mexico).

Thyanta pallidovirens pallidovirens (Stål), 1859

1859 *Pentatoma pallido-virens* [*sic*] Stål, K. Svens. Freg. Eug. Resa Jorden, p. 227. [Cal.].

1957 *Thyanta pallido-virens* [*sic*] *pallido-virens* [*sic*]: Ruckes, Am. Mus. Novit., 1824: 7.

Distribution: Cal., Id., Nev., Ut., Wash.

Thyanta pallidovirens setosa Ruckes, 1957

1957 *Thyanta pallido-virens* [*sic*] *setosa* Ruckes, Am. Mus. Novit., 1824: 17. [Wash.].

Distribution: B.C., Cal., Id., Nev., Ore., Wash.

Thyanta pallidovirens spinosa Ruckes, 1957

1957 *Thyanta pallido-virens spinosa* Ruckes, Am. Mus. Novit., 1824: 18. [Ariz.].

Distribution: Ariz., N.M., Nev., Tex., Ut. (Mexico).

Thyanta perditor (Fabricius), 1794

1794 *Cimex perditor* Fabricius, Ent. Syst., 4: 102. [Insular America].

1872 *Thyanta perditor*: Uhler, Prelim. Rept. U.S. Geol. Surv. Mont., p. 399.

Distribution: Ariz., Fla. (Mexico to South America, West Indies).

Note: Panizzi and Herzog (1984, An. Ent. Soc. Am., 77: 646-650) studied the biology of this species.

Thyanta pseudocasta Blatchley, 1926
1926 *Thyanta pseudocasta* Blatchley, Het. E. N. Am., p. 120. [Fla.].
Distribution: Fla.

Genus *Trichopepla* Stål, 1867

1867 *Trichopepla* Stål, Öfv. K. Svens. Vet.-Akad. Förh., 24: 528. Type-species: *Peribalus pilipes* Dallas, 1851, a junior synonym of *Pentatoma semivittata* Say, 1832. Monotypic.
Note: Genus revised by McDonald (1976, J. N.Y. Ent. Soc., 84: 9-22).

Trichopepla atricornis Stål, 1872
1872 *Trichopepla atricornis* Stål, K. Svens. Vet-Akad. Handl., 10(4): 34. [Ill.]. Lectotype designated by McDonald, 1976, J. N.Y. Ent. Soc., 84: 13.
Distribution: Alk., Alta., B.C., Cal., Id., Ill., Mich., Mont., Oh., Ore., Ut., Wis.

Trichopepla aurora Van Duzee, 1918
1918 *Trichopepla aurora* Van Duzee, Proc. Cal. Acad. Sci., 8: 273. [Cal.].
Distribution: B.C., Cal., Col., Id., Mont., Ore., Wash.

Trichopepla dubia (Dallas), 1851
1851 *Pentatoma dubia* Dallas, List Hem. Brit. Mus., 1: 237. ["N. America"].
1904 *Peribalus dubius*: Van Duzee, Trans. Am. Ent. Soc., 30: 34.
1909 *Holocostethus dubius*: Kirkaldy, Cat. Hem., 1: 48.
1918 *Trichopepla californica* Van Duzee, Proc. Cal. Acad. Sci., 8: 272. [Cal.]. Synonymized by McDonald, 1976, J. N.Y. Ent. Soc., 84: 15.
1937 *Trichopepla klotsi* Ruckes, Am. Mus. Novit., 935: 2. [Wyo.]. Synonymized by McDonald, 1976, J. N.Y. Ent. Soc., 84: 15.
1976 *Trichopepla dubia*: McDonald, J. N.Y. Ent. Soc., 84: 15.
Distribution: Ariz., B.C., Cal., Col., Id., Ore., Ut., Wash., Wyo. (Mexico).

Trichopepla semivittata (Say), 1832
1832 *Pentatoma semivittata* Say, Descrip. Het. Hem. N. Am., p. 9. [Ind.].
1844 *Pentatoma semivittatum* Herrich-Schaeffer, Wanz. Ins., 7: 101 (as new species). [Pa.]. Preoccupied.
1851 *Pentatoma pilipes* Dallas, List Hem. Brit. Mus., 1: 247. [Fla.]. Synonymized by Uhler, 1871, Proc. Bost. Soc. Nat. Hist., 14: 96.
1871 *Trichopepla semivittata*: Uhler, Proc. Bost. Soc. Nat. Hist., 14: 96.
Distribution: Ark., Col., D.C., Del., Fla., Ill., Ind., Ks., La., Mich., Mo., N.C., N.J., N.Y., Neb., Oh., Ont., Pa., Que., Tex., Va. (Mexico).

Trichopepla vandykei Van Duzee, 1918
1918 *Trichopepla vandykei* Van Duzee, Proc. Cal. Acad. Sci., 8: 271. [Cal.].
Distribution: Cal.

Genus *Vulsirea* Spinola, 1840

1840 *Vulsirea* Spinola, Essai Hem., p. 350. Type-species: *Vulsirea nigrorubra* Spinola, 1837. Designated by Kirkaldy, 1909, Cat. Hem., 1: 112.

Vulsirea violacea (Fabricius), 1803
1803 *Cimex violaceus* Fabricius, Syst. Rhyn., p. 167. [South America].
1904 *Vulsirea violacea*: Van Duzee, Trans. Am. Ent. Soc., 30: 57.
Distribution: Fla. (Mexico to Brazil, West Indies).

Tribe Sciocorini Amyot and Serville, 1843

Genus *Sciocoris* Fallén, 1829

1829 *Sciocoris* Fallén, Hem. Suec. Cimic., p. 20. Type-species: *Cydnus umbrinus* Fallén, 1807. Preoccupied. Next available name is *Naucoris cursitans* Fabricius, 1794. Monotypic.

Sciocoris longifrons Barber, 1933

1933 *Sciocoris longifrons* Barber, Proc. Ent. Soc. Wash., 34: 149. [Tex.].
Distribution: Tex.

Sciocoris microphthalmus Flor, 1860.

1860 *Sciocoris microphthalmus* Flor, Rhyn. Livlands, 1: 114. ["Livlands"].
1904 *Sciocoris microphthalmus*: Van Duzee, Trans. Am. Ent. Soc., 30: 32.
Distribution: Alta., B.C., Conn., Ia., Mass., Me., Mich., Minn., N.D., N.H., N.Y., Ok., Ont., Que., Sask. (Europe to China).

Subfamily Podopinae Amyot and Serville, 1843

Note: The North American fauna of this subfamily was revised, with keys and numerous illustrations, by Barber and Sailer (1953, J. Wash. Acad. Sci., 43(5): 150-162) who treated it as the "Tribe Podopini"; subsequently, the genus *Neapodops* Slater and Baranowski was described.

Genus *Allopodops* Harris and Johnston, 1936

1936 *Allopodops* Harris and Johnston, Ia. St. Coll. J. Sci., 10: 377. Type-species: *Allopodops mississippiensis* Harris and Johnston, 1936. Monotypic.

Allopodops mississippiensis Harris and Johnston, 1936

1936 *Allopodops mississippiensis* Harris and Johnston, Ia. St. Coll. J. Sci., 10: 378. [Miss.].
Distribution: Miss., S.C., Va.

Genus *Amaurochrous* Stål, 1872

1872 *Podops* (*Amaurochrous*) Stål, K. Svens. Vet.-Akad. Handl., 10(4): 15. Type-species: *Scutellera dubia* Palisot, 1805. Designated by Schouteden, 1905, Gen. Ins., 30: 33.
1905 *Amaurochrous*: Schouteden, Gen. Ins., 30: 32.
Note: Kirkaldy (1909, Cat. Hem., 1: xxxiii) credits Schouteden with designating *dubia* as type-species, but on p. 237 gives "type *cinctipes*," the latter apparently a *lapsus*.

Amaurochrous brevitylus Barber and Sailer, 1953

1839 *Podops dubius*: Germar, Zeit. Ent., 1: 64 (in part).
1869 *Padops* [*sic*] *dubius*: Rathvon, Hist. Lancaster Co., Pa., p. 548.
1904 *Podops parvulus*: Van Duzee, Trans. Am. Ent. Soc., 30: 22 (in part).
1905 *Amaurochrous parvulus*: Schouteden, Gen. Ins., 30: 33.
1912 *Amaurochrous dubius*: Olsen, J. N.Y. Ent. Soc., 20: 50.
1953 *Amaurochrous brevitylus* Barber and Sailer, J. Wash. Acad. Sci., 43: 160. [Mass.].
Distribution: Ariz., Ia., Ill., Ks., Mass., Mich., Minn., N.J., N.Y., Neb., Pa., Que., Wis.

Amaurochrous cinctipes (Say), 1828 [Fig. 114]

1828 *Tetyra cinctipes* Say, Am. Ent., 3: 49. ["Middle States"].

1859 *Podops dubius*: Dohrn, Cat. Hem., p. 5 (in part).

1869 *Scutellera cinctipes*: Rathvon, Hist. Lancaster Co., Pa., p. 548.

1872 *Podops (Amaurochrous) cinctipes*: Stål, K. Svens. Vet.-Akad. Handl., 10(4): 15.

1886 *Podops cinctipes*: Uhler, Check-list Hem. Het., p. 5.

1905 *Amaurochrous cinctipes*: Schouteden, Gen. Ins., 30: 33.

1939 *Podops peninsularis*: Torre-Bueno, Bull. Brook. Ent. Soc., 34: 214.

Distribution: Ark., D.C., Del., Ia., Ill., Ind., Ks., La., Mass., Mich., Minn., Mo., N.C., N.J., N.Y., Neb., Ont., Pa., Que., S.C., Tex.

Note: McPherson and Paskewitz (1984, J. N.Y. Ent. Soc., 92: 61-68) reared this species in the laboratory and described the immature stages.

Amaurochrous dubius (Palisot), 1805

1805 *Scutellera dubia* Palisot, Ins. Rec. Afr. Am., p. 33. [San Domingo].

1859 *Podops dubius*: Dohrn, Cat. Hem., p. 5 (in part).

1886 *Podops dubius*: Uhler, Check-list Hem. Het., p. 5.

1905 *Amaurochrous dubius*: Schouteden, Gen. Ins., 30: 33.

1924 *Podops peninsularis* Blatchley, Ent. News, 35: 87. [Fla.]. Synonymized by Barber and Sailer, 1953, J. Wash. Acad. Sci., 43: 159.

Distribution: Fla., La. (West Indies, Central America).

Amaurochrous magnus Barber and Sailer, 1953

1953 *Amaurochrous magnus* Barber and Sailer, J. Wash. Acad. Sci., 43: 162. [Fla.].

Distribution: Fla., La.

Amaurochrous ovalis Barber and Sailer, 1953

1953 *Amaurochrous ovalis* Barber and Sailer, J. Wash. Acad. Sci., 43: 162. [S.C.].

Distribution: N.C., S.C.

Amaurochrous vanduzeei Barber and Sailer, 1953

1953 *Amaurochrous vanduzeei* Barber and Sailer, J. Wash. Acad. Sci., 43: 160. [Cal.].

Distribution: Cal.

Genus *Neapodops* Slater and Baranowski, 1970

1970 *Neapodops* Slater and Baranowski, Fla. Ent., 53: 139. Type-species: *Neapodops floridanus* Slater and Baranowski, 1970. Original designation.

Neapodops floridanus Slater and Baranowski, 1970

1970 *Neapodops floridanus* Slater and Baranowski, Fla. Ent., 53: 141. [Fla.].

Distribution: Fla.

Genus *Notopodops* Barber and Sailer, 1953

1953 *Notopodops* Barber and Sailer, J. Wash. Acad. Sci., 43: 152. Type-species: *Notopodops omani* Barber and Sailer, 1953. Original designation.

Notopodops omani Barber and Sailer, 1953.

1906 *Oncozygia clavicornis*: Barber, Brook. Inst. Arts Sci., Sci. Bull., 1: 256 (in part).

1953 *Notopodops omani* Barber and Sailer, J. Wash. Acad. Sci., 43: 154. [Tex.].

Distribution: Tex.

Genus *Oncozygia* Stål, 1872

1872 *Oncozygia* Stål, K. Svens. Vet.-Akad. Handl., 10(4): 15. Type-species: *Oncozygia clavicornis* Stål, 1872. Monotypic.

Oncozygia clavicornis Stål, 1872

1872 *Oncozygia clavicornis* Stål, K. Svens. Vet.-Akad. Handl., 10(4): 15. [Tex.].

Distribution: Fla., Miss., S.C., Tex., Va.

Note: The B.C. record by Hart (1919, Bull. Ill. Nat. Hist. Surv., 13: 171) "is probably based on a misidentification" (Barber and Sailer, 1953, J. Wash. Acad. Sci., 43: 152).

Genus *Weda* Schouteden, 1905

1905 *Weda* Schouteden, An. Soc. Ent. Belg., 49: 150. Type-species: *Weda horvathi* Schouteden, 1905, a junior synonym of *Podops parvulus* Van Duzee, 1904. Monotypic.

Weda parvula (Van Duzee) 1904

1904 *Podops parvulus* Van Duzee, Trans. Am. Ent. Soc., 30: 22 (in part). [Col.]. Lectotype designated by Barber and Sailer, 1953, J. Wash. Acad. Sci., 43: 155.

1905 *Weda horvathi* Schouteden, An. Soc. Ent. Belg., 49: 145. [Col.] Synonymized Barber and Sailer, 1953, J. Wash. Acad. Sci., 43: 155.

1953 *Weda parvula*: Barber and Sailer, J. Wash. Acad. Sci., 43: 155.

Distribution: Col., Ut.

Weda stylata Barber and Sailer, 1953

1953 *Weda stylata* Barber and Sailer, J. Wash. Acad. Sci., 43: 156. [Cal.].

Distribution: Cal.

Weda tumidifrons Barber and Sailer, 1953

1906 *Oncozygia clavicornis*: Barber, Brook. Inst. Arts Sci., Sci. Bull., 1: 256 (in part).

1953 *Weda tumidifrons* Barber and Sailer, J. Wash. Acad. Sci., 43: 156. [Tex.].

Distribution: Col., Tex.

Family Phymatidae Laporte, 1832

The Ambush Bugs

By Richard C. Froeschner

The common name for members of this family was earned by the insects' habit of lying concealed, generally in flowers, and waiting until some appropriate-sized insect comes within reach of the grasping front legs. In a flash the prey is captured, subdued by a lethal and perhaps digesting injection, and then leisurely drained of its body fluids. The prey, varying from tiny frail flies through bees up to medium-sized butterflies and moths, is selected as appropriate for each developmental stage of the ambush bugs. Eggs, laid during summer and fall in small, froth-covered batches, hatch the following spring. The new generation begins to reach adulthood by summer. For North American species extensive details of biology are known for only one species, *Phymata americana* Melin, which was studied and reported on by Balduf (1941, An. Ent. Soc. Am., 34: 204-214), published under the combination *Phymata pennsylvanica americana*). The following year Balduf (1942, J. Econ. Ent., 34: 445-448) evaluated the economic importance of the same species, basing his conclusions on observa-

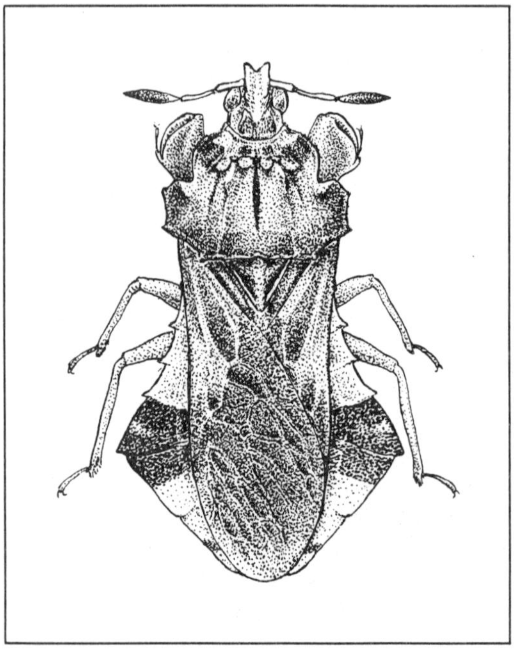

Fig. 122 *Phymata pennsylvanica* [p. 603] (After Froeschner, 1944).

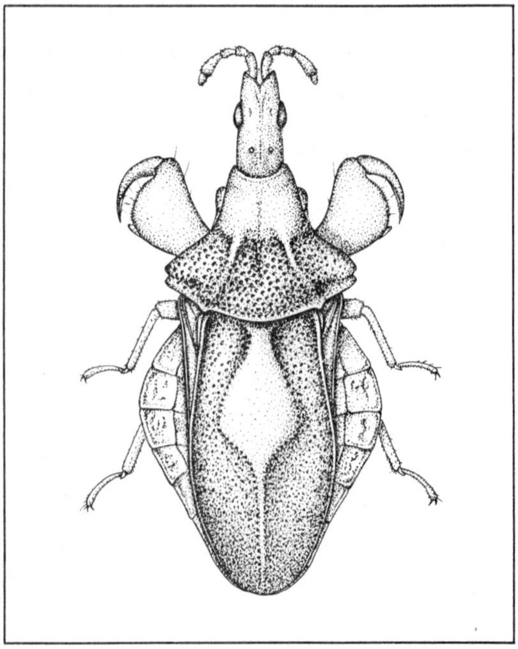

Fig. 123. *Macrocephalus notatus* [p. 600] (Original).

tions of over 800 feedings on seven orders of insects; he decided the ambush bug "should neither be destroyed nor promoted because its injuries appear to have been almost entirely cancelled by its good services."

The exact taxonomic level to be accorded this group is not yet settled, either in the long held position of a family, or as a subfamily of Reduviidae as proposed by Carayon et al. (1958, Rev. Zool. Bot. Afr., 57: 256-281). The matter rests on evaluation of the Old World genus *Themonocoris* Carayon et al., which some authors accept as a connecting link between the traditional Phymatidae and Reduviidae, while others consider it simply a primitive subfamily of Phymatidae. The conservative view of retaining the Phymatidae as a family is being followed here.

The basic classification for study of the 26 species (in three genera) occurring north of the Rio Grande follows Handlirsch (1897, An. Nat. Hofmus., 12: 127-230), Melin (1930, Ark. Zool., 22: 1-40), Evans (1931, An. Ent. Soc. Am., 24: 711-738), and Kormilev (1962, Philip. J. Sci., 89: 287-486). Only a few species have been described subsequent to those works.

The opinion of Curtis W. Sabrosky, Past President of the International Commission on Zoological Nomenclature, on certain nomenclatorial problems is gratefully acknowledged.

Subfamily Macrocephalinae Amyot and Serville, 1843

Note: Kormilev (1984, J. Nat. Hist., 18: 624) furnished keys to the genera of the world by region, including the three in the Americas.

Genus *Lophoscutus* Kormilev, 1951

1951 *Macrocephalus* (*Lophoscutus*) Kormilev, Rev. Inst. Nac. Invest. Ci. Nat., 2: 101. Type-species: *Macrocephalus affinis* Guérin-Méneville, 1838. Designated by Maa and Lin, 1956, Quart. J. Taiwan Mus., 9: 116.

1956 *Lophoscutus*: Maa and Lin, Quart. J. Taiwan Mus., 9: 116.

Notes: Kormilev (1981, Sociobiology, 6: 214) returned *Lophoscutus* to subgeneric status under *Macrocephalus* but in his later work (1984, J. Nat. Hist., 18: 623-637) keyed and treated it as a genus.

Dohrn's (1859, Cat. Hem., p. 41) "Am. bor." record for "*Macrocephalus macilentus*" Westwood [*now Lophoscutus macilentus* (Westwood)] resulted from a typographical error which placed "id." under the distribution record for the previous species. It is a South American species and not included in the present catalog.

Lophoscutus prehensilis (Fabricius), 1803, New Combination

1803 *Syrtis prehensilis* Fabricius, Syst. Rhyn., p. 123. ["Carolina"].

1841 *Macrocephalus pallidus* Westwood, Trans. Ent. Soc. London, 3: 27. [Ga.]. Synonymized by Stål, 1876, K. Svens. Vet.-Akad. Handl., 14(4): 135.

1843 *Macrocephalus prehensilis*: Amyot and Serville, Hist. Nat. Ins., Hem., p. 293.

1954 *Macrocephalus* (*Lophoscutus*) *prehensilis*: Kormilev, J. Ks. Ent. Soc., 27: 159.

Distribution: Ala., Ark., Fla., Ga., Ks., Ky., Miss., Mo., N.C., N.M., Ok., Tenn., Tex. (Mexico).

Note: Although this species was not included among those named when *Lophoscutus* was raised to generic rank, it was assigned to that taxon by Kormilev (1954, above).

Lophoscutus prehensilis minor (Kormilev), 1954
1954 *Macrocephalus prehensilis minor* Kormilev, J. Ks. Ent. Soc., 27: 159. [Fla.].
Distribution: Fla.

Lophoscutus prehensilis prehensilis (Fabricius), 1803
1803 *Syrtis prehensilis* Fabricius, Syst. Rhyn., p. 123. ["Carolina"].
1954 *Macrocephalus prehensilis prehensilis*: Kormilev, J. Ks. Ent. Soc., 27: 159.
Distribution: Ala., Ark., Ga., Ks., Ky., Miss., Mo., N.C., N.M., Ok., Tenn., Tex. (Mexico).

Genus *Macrocephalus* Swederus, 1787

1787 *Macrocephalus* Swederus, Vet. Akad. Handl., 8: 183. Type-species: *Macrocephalus cimicoides* Swederus, 1787. Monotypic.

Note: With the elevation of *Lophoscutus* to generic status, *Macrocephalus* no longer divides into subgenera.

Macrocephalus arizonicus Cockerell, 1900
1900 *Macrocephalus arizonicus* Cockerell, Ent., 33: 66. [Ariz.].
Distribution: Ariz.
Note: On the basis of Ashmead's comparison of the holotype of *arizonicus* to the description of *M. uhleri* Handlirsch, Cockerell (1909, Ent., 33: 201) synonymized his form with Handlirsch's. Authors followed this until Evans (1931, An. Ent. Soc. Am., 24: 733) resurrected Cockerell's species in a key and gave descriptive comments.

Macrocephalus barberi Evans, 1931
1931 *Macrocephalus barberi* Evans, An. Ent. Soc. Am., 24: 733. [Cal.].
Distribution: Cal.

Macrocephalus cimicoides Swederus, 1787
1787 *Macrocephalus cimicoides* Swederus, Vet. Akad. Handl., 8: 185. [Ga.].
1835 *Macrocephalus manicatus*: Burmeister, Handb. Ent., 2: 252.
Distribution: Cal., "Canada," "Carolina," Fla., Tex.

Macrocephalus dorannae Evans, 1931
1931 *Macrocephalus dorannae* Evans, An. Ent. Soc. Am., 24: 731. [Ariz.].
Distribution: Ariz.

Macrocephalus gracilis Handlirsch, 1897
1897 *Macrocephalus gracilis* Handlirsch, An. Nat. Hofmus., 12: 193. [N. Am.].
Distribution: N. Am.
Note: This species apparently has not been recognized since its original description.

Macrocephalus manicatus (Fabricius), 1803
1803 *Syrtis manicata* Fabricius, Syst. Rhyn., p. 123. ["Carolina"].
1847 *Macrocephalus manicatus*: Herrich-Schaffer, Wanz. Ins., 8: 107.
Distribution: "Carolina," Ga., Tex. (Brazil).

Macrocephalus notatus (Westwood), 1841 [Fig. 123]
1841 *Syrtis notata* Westwood, Trans. Ent. Soc. London, 3: 24. [Colombia].
1910 *Macrocephalus notatus*: Banks, Cat. Nearct. Hem.-Het., p. 51.
Distribution: Tex. (Mexico to Argentina).

Macrocephalus similis Kormilev, 1972
1972 *Macrocephalus similis* Kormilev, Bull. S. Cal. Acad. Sci., 71: 93. [Ariz.].
Distribution: Ariz.

Macrocephalus uhleri Handlirsch, 1898
1898 *Macrocephalus uhleri* Handlirsch, Verh. Zool.-Bot. Ges. Wien, 48: 383. [Ariz.].
Distribution: Ariz.

Subfamily Phymatinae Laporte, 1832

Genus *Phymata* Latreille, 1802

1802 *Phymata* Latreille, Hist. Nat. Crust. Ins., p. 247. Type-species: *Acanthia crassipes* Fabricius, 1775. Monotypic.
1803 *Syrtis* Fabricius, Syst. Rhyn., p. 121. Type-species: *Cimex erosus* Linnaeus, 1758. Designated by Kirkaldy, 1900, Ent., 33: 263. Synonymized by Amyot and Serville, 1843, Hist. Nat. Ins. Hem., p. 288.
Note: Many of the pre-1962 records for species of the genus *Phymata* were confused as to identification and nomenclature and still remain relatively unsorted. Therefore, many of them (particularly for "*erosa*" which belongs to a European species) are being omitted here; in essence the records given below begin with Evan's (1931, An. Ent. Soc. Am., 24: 712-730) study of that family in North America as redefined and extended in Kormilev's (1962, Philip. J. Sci., 89: 287-486) "Revision of the Phymatinae." Publications after the latter work are taken at face value.

Subgenus *Phymata* Latreille, 1802

1802 *Phymata* Latreille, Hist. Nat. Crust. Ins., p. 247. Type-species: *Acanthia crassipes* Fabricius, 1775. Monotypic.
1962 *Phymata* (*Phymata*): Kormilev, Philip. J. Sci., 89: 308.

Phymata albopicta Handlirsch, 1897
1897 *Phymata albopicta* Handlirsch, Ann. Nat. Hofmus., 12: 151. [Ga., Mexico].
Distribution: D.C., Ga., N.C., S.C., Tex., Va. (Mexico to Guatemala).

Phymata americana Melin, 1930
1930 *Phymata americana* Melin, Ark. Zool., 22: 6. [Col., N.Y., Ont., Tex., Wis., Mexico].
Distribution: Ark., Ariz., B.C., Col., D.C., Del., Ia., Ill., Ind., Ks., Mass., Md., Mich., Minn., Mo., Mont., N.C., N.D., N.J., N.M., N.Y., Nev., Oh., Ont., Pa., S.D., Tex., Va., Vt., Wash., Wis. (Mexico).

Phymata americana americana Melin, 1930
1930 *Phymata americana* Melin, Ark. Zool., 22: 6. [Col., N.Y., Ont., Tex., Wis., Mexico].
1930 *Phymata americana wisconsina* Melin, Ark. Zool., 22: 6. [Wis., Tex., Mexico]. Synonymized by Evans, 1931, An. Ent. Soc. Am., 24: 715.
1930 *Phymata americana ottawensis* Melin, Ark. Zool., 22: 7. [Ont.]. Synonymized by Evans, 1931, An. Ent. Soc. Am., 24: 714.
1931 *Phymata pennsylvanica americana*: Evans, An. Ent. Soc. Am., 24: 715.
1953 *Phymata americana americana*: Kormilev, Mis. Estud. Pat. Reg. Argentina, 24: 66.
Distribution: Ark., Ariz., Col., D.C., Del., Ia., Ill., Ind., Ks., Mass., Md., Mich., Minn., Mo., Mont., N.C., N.D., N.J., N.M., N.Y., Neb., Oh., Ont., Pa., S.D., Tex., Va., Vt., Wis. (Mexico).

Note: Melin (1930, above) proposed the new species *P. americana* comprised of four subspecies, but not including the nominate subspecies, *P. americana americana*, as is required by the International Rules of Zoological Nomenclature. Evans (1931, above) recognized this situation and selected the subspecies wisconsina to be the nominate *P. americana americana*, which he considered a subspecies of *P. pennsylvanica*. Kormilev (1953, above) reinstated *P. americana* to species status.

Phymata americana coloradensis Melin, 1930

1930 *Phymata americana coloradensis* Melin, Ark. Zool., 22: 7. [Col.].

1931 *Phymata pennsylvanica coloradensis*: Evans, An. Ent. Soc. Am., 24: 716.

Distribution: Ariz., Col., Ks., N.M., Neb., Ok., Ore., Tex., Ut., Wash. (Mexico).

Note: Kormilev (1953, Mis. Estud. Pat. Reg. Argentina, 24: 66) reestablished the combination *P. americana coloradensis*.

Phymata americana metcalfi Evans, 1931

1930 *Phymata metcalfi* Evans, An. Ent. Soc. Am., 24: 723. [B.C., Ore.].

1962 *Phymata americana metcalfi*: Kormilev, Philip. J. Sci., 89: 414.

Distribution: B.C., Nev., Ore., Ut.

Phymata americana obscura Kormilev, 1957

1957 *Phymata americana obscura* Kormilev, Rev. Brasil. Biol., 17: 136. [Id.].

Distribution: Id.

Phymata arctostaphylae Van Duzee, 1914

1914 *Phymata erosa arctostaphylae* Van Duzee, Trans. San Diego Soc. Nat. Hist., 2: 11. [Cal.].

1931 *Phymata arctostaphylae*: Evans, An. Ent. Soc. Am., 24: 719.

Distribution: Cal.

Phymata borica Evans, 1931

1931 *Phymata borica* Evans, An. Ent. Soc. Am., 24: 721. [Ut.].

Distribution: Ariz., Col., Id., Ut.

Phymata fasciata (Gray), 1832

1832 *Syrtis fasciatus* Gray, Animal Kingd., 15: 252. [N. Am.].

1876 *Phymata wolffii* Stål, K. Svens. Vet.-Akad. Handl., 14(4): 133. ["Carolina," N.Y., Tex., Wis., Mexico]. Synonymized by Bergroth, 1892, Rev. d'Ent., 11: 264.

1917 *Phymata erosa wolffii*: Van Duzee, Univ. Cal. Publ. Ent., 2: 228.

1930 *Phymata fasciata*: Melin, Arkiv Zool., 22: 9.

Distribution: Ariz., "Carolina," Fla., Ga., La., Miss., Tex. (Mexico to Panama).

Note: Melin (1930, Ark. Zool., 22: 9-12) treated *P. fasciata* Gray as being composed of two subspecies: *georgiensis* (from the United States) and *mexicana* (from Mexico). Evans (1931, An. Ent. Soc. Am., 24: 716) and Kormilev (1962, Philip. J. Sci., 89: 407) differed in their treatments of those two subspecies. Evans noted that neither subspecies repeated the species name--as required by the "Entomological Code" (and now by the International Code)--so he made the subspecies *mexicana* the nominate subspecies and reduced it as a junior synonym of *P. fasciata fasciata* Gray. Kormilev, in contrast and without comment on Evans' actions, designated one of Melin's cotypes of *P. fasciata georgiensis* as neotype of *Phymata fasciata* (Gray) and thereby sank *P. fasciata georgiensis* Melin as a junior synonym. The latter action, involving types, is binding and followed here.

Phymata fasciata fasciata (Gray), 1832
1832 *Syrtis fasciatus* Gray, Animal Kingd., 15: 242. [N. Am.].
1930 *Phymata fasciata georgiensis* Melin, Ark. Zool., 22: 9. ["Carolina," Ga., Tex.]. Synonymized by Kormilev, 1962, Philip. J. Sci., 89: 407.
1962 *Phymata fasciata fasciata*: Kormilev, Philip. J. Sci., 89: 398.
Distribution: Ariz., "Carolina," Ga., La., Miss., Tex. (Mexico).

Phymata fasciata mystica Evans, 1931
1931 *Phymata mystica* Evans, An. Ent. Soc. Am., 24: 717. [Fla.].
1962 *Phymata fasciata mystica*: Kormilev, Philip. J. Sci., 89: 409.
Distribution: Fla., Ga.

Phymata granulosa Handlirsch, 1897
1897 *Phymata erosa granulosa* Handlirsch, An. Nat. Hofmus., 12: 163. [Mexico]. Lectotype designated by Kormilev, 1962, Philip. J. Sci., 89: 398.
1930 *Phymata granulosa*: Melin, Ark. Zool., 22: 15.
Distribution: Tex. (Mexico to Cost Rica).

Phymata granulosa texasana Kormilev, 1957
1957 *Phymata granulosa texasana* Kormilev, Rev. Brasil. Biol., 17: 134. [Tex.].
Distribution: Tex.

Phymata luteomarginata Kormilev, 1957
1957 *Phymata luteomarginata* Kormilev, Rev. Brasil. Biol., 17: 130. [Nev.].
Distribution: Nev.

Phymata luxa Evans, 1931
1931 *Phymata luxa* Evans, An. Ent. Soc. Am., 24: 728. [Tex.].
Distribution: Col., Tex.

Phymata maculata Kormilev, 1957
1957 *Phymata maculata* Kormilev, Rev. Brasil. Biol., 17: 124. [Ut.].
Distribution: Ariz., Ut.

Phymata noualhieri Handlirsch, 1897
1897 *Phymata noualhieri* Handlirsch, An. Nat. Hofmus., 12: 153. [Ga., Pa., Mexico, Guatemala].
Distribution: Ariz., Fla., Ga., N.C., Pa. (Mexico to Guatemala).

Phymata pacifica Evans, 1931
1931 *Phymata pacifica* Evans, An. Ent. Soc. Am., 24: 725. [Cal.].
Distribution: Cal.

Phymata pacifica pacifica Evans, 1931
1931 *Phymata pacifica* Evans, An. Ent. Soc. Am., 24: 725. [Cal.].
Distribution: Cal.

Phymata pacifica stanfordi Evans, 1931
1931 *Phymata pacifica stanfordi* Evans, An. Ent. Soc. Am., 24: 726. [Cal.].
Distribution: Cal.

Phymata pallida Kormilev, 1957
1957 *Phymata pallida* Kormilev, Rev. Brasil. Biol., 17: 128. [Ariz.].
Distribution: Ariz.

Phymata pennsylvanica Handlirsch, 1897 [Fig. 122]
1897 *Phymata erosa pennsylvanica* Handlirsch, An. Nat. Hofmus., 12: 163. [Conn., Fla., Ill., Ind., Mo., N.Y., Pa.].

1930 *Phymata americana newyorkensis* Melin, Ark. Zool., 22: 7. [N.Y., Oh.]. Synonymized by Evans, 1931, An. Ent. Soc. Am., 24: 714.
1931 *Phymata pennsylvanica pennsylvanica*: Evans, An. Ent. Soc. Am., 24: 715.
1953 *Phymata pennsylvanica*: Kormilev, Mis. Estud. Pat. Reg. Argentina, 24: 63.
Distribution: Ark., Col., Conn., D.C., Del., Fla., Ga., Ia., Ill., Ind., Ks., Ky., Mass., Md., Mich., Minn., Mo., Mont., N.C., N.D., N.J., N.Y., Neb., Oh., Pa., S.D., Va., W.Va., Wis.

Phymata rossi Evans, 1931
1931 *Phymata rossi* Evans, An. Ent. Soc. Am., 24: 720. [Ariz.].
Distribution: Ariz., Ks., N.M., Ut.

Phymata saileri Kormilev, 1957
1957 *Phymata saileri* Kormilev, Rev. Brasil. Biol., 17: 133. [Ariz.].
Distribution: Ariz.

Phymata salicis Cockerell, 1900
1900 *Phymata erosa salicis* Cockerell, Ent., 33: 66. [Ariz., Cal.].
1931 *Phymata salicis*: Evans, An. Ent. Soc. Am., 24: 723.
Distribution: Ariz., Cal.

Phymata vicina Handlirsch, 1897
1897 *Phymata vicina* Handlirsch, An. Nat. Hofmus., 12: 150. [Fla.].
Distribution: Alta., Ariz., Col., Conn., D.C., Fla., Ill., Ind., Ks., Mass., N.J., N.Y., Neb., Pa., R.I., S.D., Tex., Ut., Va.

Phymata vicina parvula Kormilev, 1957
1957 *Phymata vicina parvula* Kormilev, Rev. Brasil. Biol., 17: 126. [Tex.].
Distribution: Tex.

Phymata vicina vicina Handlirsch, 1897
1897 *Phymata vicina* Handlirsch, An. Nat. Hofmus., 12: 150. [Fla.].
1957 *Phymata vicina vicina*: Kormilev, Rev. Brasil. Biol., 17: 126.
Distribution: Alta., Ariz., Col., Conn., D.C., Fla., Ill., Ind., Ks., Mass., N.J., N.Y., Neb., Pa., R.I., S.D., Tex., Ut., Va.

Family Piesmatidae Spinola Amyot and Serville, 1843

The Piesmatids

By Richard C. Froeschner

All members of this small family (a world total of 37 species in 3 genera) obtain their nourishment from plant fluids. When feeding they may cause the leaves to curl and whiten; and, as summarized by Drake and Davis (1958, An. Ent. Soc. Am., 51: 574-575), when adults of certain members of the nearly worldwide genus *Piesma* Lepeletier and Serville imbibe virus-contaminated sap of sugar beets they can, even after hibernating, serve as a vector of that virus when feeding on other sugar beet plants. In North America *Piesma cinereum* (Say) is a vector of the virus causing Sugar Beet Savoy, which stunts and distorts the plants and reduces sugar yield. In Europe another species of *Piesma* is a vector of another virus disease of sugar beets. Our common American species probably has two or more generations each year. Egg laying occurs throughout much of the spring and summer with nymphs of the last generation maturing into adults which hibernate after crawling into protective shelter. Most North American members of this family feed on plants of the families Amaranthaceae and Chenopodiaceae.

Earlier entomologists considered the reticulated upper surface and 2-segmented tarsi to ally the family Piesmatidae to the lace bug family Tingidae, but several technical characters, such as the presence of abdominal trichobothria, mark the family as remote from the Tingidae and much closer to the Lygaeidae.

The most significant treatments of the family in North America are McAtee's (1919, Bull. Brook. Ent. Soc., 14: 80-93) key and Drake and Davis' (1958, An. Ent. Soc. Am., 5: 567-581) monograph containing a discussion ofthe morphology and classification and keys to the three known world genera (only *Piesma* occurs in North America) and the seven American species.

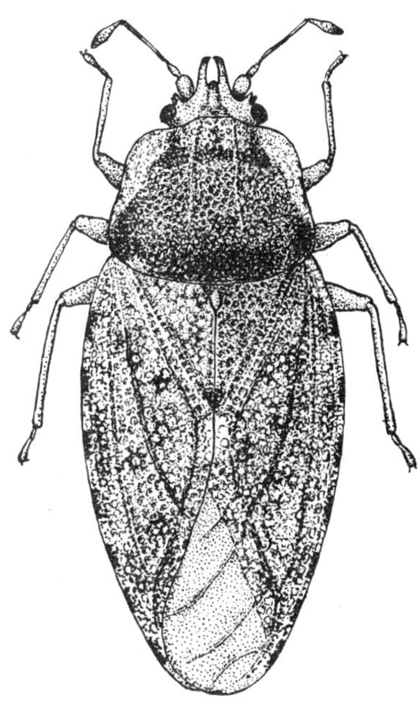

Fig. 124. *Piesma cinereum* [p. 606] (After Froeschner, 1944).

Subfamily Piesmatinae Amyot and Serville, 1843

Genus *Piesma* Lepeletier and Serville, 1825

1825 *Piesma* Lepeletier and Serville, Encycl. Meth., 10: 652. Type-species: *Tingis capitata* Wolff, 1804. Designated by Brullé, 1835, Hist. Nat. Ins., 9:342.

1833 *Zosmenus* Laporte, Mag. Zool., 2: 49. Type-species: *Zosmenus maculatus* Laporte, 1833. Monotypic. Synonymized by Westwood, 1840, Introd. Mod. Class. Ins., Synop., p. 120.

1895 *Agrammodes* Uhler, Bull. Col. Agr. Exp. Stn., 31: 56. Type-species: *Agrammodes costatus* Uhler, 1895. Monotypic. Synonymized by McAtee, 1919, Bull. Brook. Ent. Soc., 14: 82.

Note: As pointed out by Drake and Davis (1958, An. Ent. Soc. Am., 51: 567), the name *Piesma* is neuter in gender and all included species must end in a neuter suffix.

Piesma brachiale McAtee, 1919

1919 *Piesma brachialis* McAtee, Bull. Brook. Ent. Soc., 14: 88. [Ariz., Ut.].

1958 *Piesma brachiale*: Drake and Davis, An. Ent. Soc. Am., 51: 577.

Distribution: Ariz., Ut.

Note: Types collected on *Isocoma* (*Bigelovia*) *hartwegi*.

Piesma ceramicum McAtee, 1919

1919 *Piesma ceramica* McAtee, Bull. Brook. Ent. Soc., 14: 89. [Ut.].

1919 *Piesma rugulosa* McAtee, Bull. Brook. Ent. Soc., 14: 89. [Col.]. Synonymized by Drake and Davis, 1958, An. Ent. Soc. Am., 51: 578.

1958 *Piesma ceramicum*: Drake and Davis, An. Ent. Soc. Am., 51: 578.

Distribution: Col., Ut.

Note: Types taken on "greasewood."

Piesma cinereum (Say), 1832 [Fig. 124]

1832 *Tingis cinerea* Say, Descrip. Het. Hem. N. Am., p. 27. ["U.S.].

1869 *Zosmenus cinereus*: Rathvon, Hist. Lancaster Co., Pa., p. 549.

1873 *Piesma cinerea*: Stål, K. Svens. Vet.-Akad. Handl., 11(2): 116.

1885 *Piesma pusilla*: Popenoe, Trans. Ks. Acad. Sci., 9: 63.

1905 *Piesma cenerea* [*sic*]: Crevecoeur, Trans. K. Acad. Sci., 19: 234.

1919 *Piesma cinerea* var. *inornata* McAtee, Bull. Brook. Ent. Soc., 14: 88. [Cal.]. Synonymized by Drake and Davis, 1958, An. Ent. Soc. Am. 51: 576.

1926 *Piesma cinerea* var. *floridana* Blatchley, Het. E. N. Am., p. 484. [Fla.]. Synonymized by Drake and Davis, 1958, An. Ent. Soc. Am., 51: 576.

1958 *Piesma cinereum:* Drake and Davis, An. Ent. Soc. Am., 51: 576.

Distribution: Alta., Ariz., B.C., Cal., Col., Conn., D.C., Fla., Ia., Ill., Ind., Ks., Mass., Me., Mich., Mo., N.C., N.J., N.Y., Oh., Ont., Ore., Pa., R.I., Tex., Ut., Wash., Wis. (Mexico to Argentina, West Indies).

Note: Primary hosts belong to the plant genera *Amaranthus, Atriplex, Chenopodium,* and *Salsola*.

Piesma costatum (Uhler), 1895

1895 *Agrammodes costatus* Uhler, Bull. Col. Agr. Exp. Stn., 31: 56. [Ariz., Col.].

1919 *Piesma costata*: McAtee, Bull. Brook. Ent. Soc., 14: 90.

1919 *Piesma costata* var. *defecta* McAtee, Bull. Brook. Ent. Soc., 14: 91. [Cal.]. Synonymized by Drake and Davis, 1958, An. Ent. Soc. Am., 51: 577.

1958 *Piesma costatum*: Drake and Davis, An. Ent. Soc. Am., 51: 577.
Distribution: Ariz., Cal., Col.

Piesma explanatum McAtee, 1919
1919 *Piesma explanata* McAtee, Bull. Brook. Ent. Soc., 14: 91. [Ut.].
1958 *Piesma explanatum*: Drake and Davis, An. Ent. Soc. Am., 51: 578.
Distribution: B.C., Ut.

Piesma patruele McAtee, 1919
1919 *Piesma patruela* McAtee, Bull. Brook. Ent. Soc., 14: 86. [Ariz., Tex.].
1958 *Piesma patruele* : Drake and Davis, An. Ent. Soc. Am., 51: 577.
Distribution: Ariz., Tex.

Piesma proteum McAtee, 1919
1919 *Piesma protea* McAtee, Bull. Brook. Ent. Soc., 14: 89. [Ariz., Neb.].
1919 *Piesma depressa* McAtee, Bull. Brook. Ent. Soc., 14: 91. [Ariz.]. Synonymized by Drake and Davis, 1958, An. Ent. Soc. Am., 51: 578.
1919 *Piesma incisa* McAtee, Bull. Brook. Ent. Soc., 14: 91. [Ariz., Ut.]. Synonymized by Drake and Davis, 1958, An. Ent. Soc. Am., 51: 578.
1958 *Piesma proteum*: Drake and Davis, An. Ent. Soc. Am., 51: 578.
Distribution: Ariz., Col., Neb., Nev., Ut.
Note: Found on *Atriplex* sp. and breeding on *Suaeda torreyana*.

Family Pleidae Fieber, 1851

The Pygmy Backswimmers

By Dan A. Polhemus

The pygmy backswimmers are distinguished by their strongly convex form, small size, coleopteriform hemelytra, and hind legs set with long hairs for swimming. All species are truly aquatic, frequenting weedy ponds and other slow-water habitats where they prey on ostracods, *Daphnia*, and mosquito larvae. Pleids swim upside down, propelling themselves in a rowing fashion similar to that of Notonectidae, but unlike certain members of the latter group they are apparently unable to maintain the hydrostatic equilibrium necessary to hold station at a given depth, and instead frequently grasp onto or crawl amidst aquatic vegetation. Total developmental time from eggs, which are laid in the tissues of various aquatic weeds, to adults averages about seventy days (Rice, 1954, Am. Midl. Nat., 51; 105). All stages are predaceous, despite reports that some species are herbivorous (Torre-Bueno, 1905, J. N.Y. Ent. Soc., 8: 29). Populations reach their highest densities in small ponds and shallow portions of larger lakes, where they gain protection from predation by fish (Gittelman, 1974, J. Ks. Ent. Soc., 47: 491).

Pleids are a cosmopolitan group, with their greatest diversity in the tropics. Of the three described genera two, *Neoplea* Esaki and China and *Paraplea* Esaki and China, occur in North America. Our species are mostly southeastern in distribution, although *Neoplea striola* (Fieber) is transcontinental. With the exception of a single uncertain record from Utah, this family is unknown in the interior West or the Colorado and Rio Grande drainages.

Family status for Pleidae was first proposed by Fieber (1851, Abh. Bohem. Ges. Wiss. Prag., (5)7: 17) under the name Pleae, but for many years the group was regarded as a subfamily of Notonectidae and was treated as such in most of the early litera-

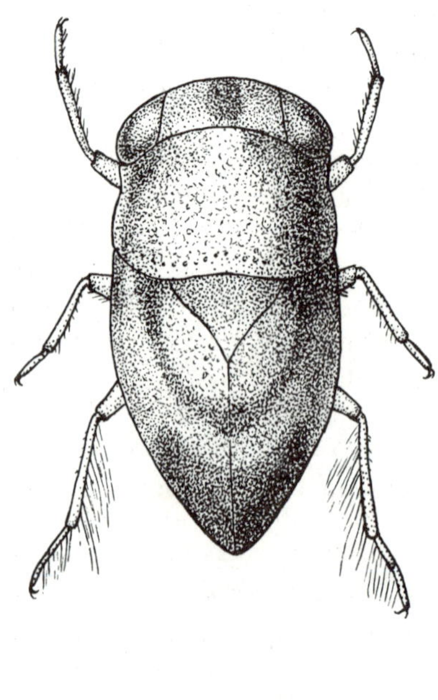

Fig. 125. *Neoplea striola* [p. 609] (After Froeschner, 1962).

ture. Esaki and China (1928, Rev. Esp. Ent., 4: 129) established the family status after detailed morphological studies, concluding that Pleidae were most closely allied to the pantropical Helotrephidae. In this same study Esaki and China recognized three subgenera, which were subsequently raised to generic status by Drake and Maldonado (1956, Bull. Brook. Ent. Soc., 51: 53). Fieber's spelling *Ploa* (1844, Abh. Bohem. Ges. Wiss. Prag., (5)3: 296) is an unjustified emendation of the name *Plea* Leach, 1817.

Pleidae are badly in need of revision in North America, and no key to our species exists. Species concepts are in certain cases confused due to differences in color pattern and hemelytral sculpturing correlated with varying degrees of wing development. The best recent American work is that of Drake and Chapman (1953, Proc. Biol. Soc. Wash., 66: 53-60), which provides morphological descriptions and distributional data.

Genus *Neoplea* Esaki and China, 1928

1928 *Plea* (*Neoplea*) Esaki and China, Rev. Esp. Ent., 4: 166. Type-species: *Plea striola* Fieber, 1844. Monotypic.

1956 *Neoplea*: Drake and Maldonado, Bull. Brook. Ent. Soc., 51: 53.

Neoplea apopkana (Drake and Chapman), 1953

1953 *Plea apopkana* Drake and Chapman, Proc. Biol. Soc. Wash., 66: 59. [Fla.].

1956 *Neoplea apopkana*: Drake and Maldonado, Bull. Brook. Ent. Soc., 51: 53.

Distribution: Fla., Miss.

Neoplea notana (Drake and Chapman), 1953

1953 *Plea notana* Drake and Chapman, Proc. Biol. Soc. Wash., 66: 59. [Fla.].

1956 *Neoplea notana*: Drake and Maldonado, Bull. Brook. Ent. Soc., 51: 53.

Distribution: Fla., Miss.

Neoplea striola (Fieber), 1844 [Fig. 125]

1844 *Ploa* [*sic*] *striola* Fieber, Abh. Bohem. Ges. Wiss. Prag., (5)3: 296. [North America].

1884 *Plea striola*: Uhler, Stand. Nat. Hist., 2: 253.

1922 *Plea harnedi* Drake, Oh. J. Sci., 22: 114. [Miss.]. Synonymized by Ellis, 1950, Proc. Ent. Soc. Wash., 52: 104; resurrected by Drake and Chapman, 1953, Proc. Biol. Soc. Wash., 66: 57; synonymy reinstated by Ellis, 1965, Fla. Ent., 48: 77..

1928 *Plea* (*Neoplea*) *striola*: Esaki and China, Rev. Esp. Ent., 4: 166.

1956 *Neoplea striola*: Drake and Maldonado, Bull. Brook. Ent. Soc., 51: 53.

Distribution: Ala., Ark., Cal., Conn., Fla., Ia., Ill., Ind., Ks., Man., Mass., Md., Mich., Minn., Miss, Mo., N.C., N.J., N.Y., Neb., Oh., Ont., Pa., Que., Tex., Ut.(?), Wash., W.Va., Va. (Mexico).

Note: Data on biology and ecology were provided by Bare (1926, An. Ent. Soc. Am., 19: 93), Rice (1954, Am. Midl. Nat., 51: 105), and Gittelman (1974, J. Ks. Ent. Soc., 47: 491; 1975, An. Ent. Soc. Am., 68: 1011). Ellis (1965, Fla. Ent., 48: 77) reported this species from swiftly flowing sections of the Tombigbee River in Mississippi. This species has been introduced into the San Joaquin Valley of California for mosquito control.

Genus *Paraplea* Esaki and China, 1928

1928 *Plea* (*Paraplea*) Esaki and China, Rev. Esp. Ent., 4: 166. Type-species: *Plea pallescens*

Distant, 1906. Original designation.
1956 *Paraplea*: Drake and Maldonado, Bull. Brook. Ent. Soc., 51: 53.

Paraplea nilionis (Drake and Chapman), 1953
1953 *Plea* (*Paraplea*) *nilionis* Drake and Chapman, Proc. Biol. Soc. Wash., 66: 55. [Fla.].
1956 *Paraplea nilionis*: Drake and Maldonado, Bull. Brook. Ent. Soc., 51: 53.
Distribution: Fla., Miss.

Paraplea puella (Barber), 1923
1923 *Plea* (*Paraplea*) *puella* Barber, Am. Mus. Novit., 75: 11. [Puerto Rico].
1953 *Plea* (*Paraplea*) *puella*: Drake and Chapman, Proc. Biol. Soc. Wash., 66: 54.
1956 *Paraplea puella*: Drake and Maldonado, Bull. Brook. Ent. Soc., 51: 53.
Distribution: Fla., La., Miss., Tex. (Mexico, Central America, West Indies).

Family Polyctenidae Westwood, 1874

The Bat Bugs

By Richard C. Froeschner

The 23 known species of this family are referred to as bat bugs, so-called because they live as ectoparasites among the hairs on bats. Adaptations to this life developed the following specializations: a flattened body; elongate legs; an anteriorly narrowed triangular head without compound eyes; the much shortened wings (not even hinged at the base) that expose the apical half or more of the upper side of the abdomen; and rows of stout setae, called "combs," on the head and certain other parts of the body. Two genera of free-tailed bats (family Molassidae) are utilized as hosts by North American Polyctenidae.

The reproductive process of the Polyctenidae, markedly modified from the general insectan type exhibited by most heteropterans, is referred to as "pseudoplacental viviparity." The method of insemination is not yet known, but Hagan (1931, J. Morph. Physiol., 51: 33-38) reports the presence of spermatozoa in the female's haemocoel. Whether they had been introduced into the reproductive tract whence they escaped into the haemocoel, as occurs in certain Nabidae, or had been introduced directly into the haemocoel by traumatic insemination through the body wall, as happens among the Cimicidae, remains to be discovered. From the haemocoel the sperm must enter the reproductive tract to fertilize the egg. The fertilized egg, unlike that of other heteropterans in having no yolk or "shell" (chorion), remains in the mother's ovariole where the outer wall of the developing embryo lies against the wall of the ovariole as a "pseudoplacenta" that absorbs nourishment from the female's tissues. The nymphs, issuing singly and at intervals during the life of the parent, leave the reproductive tract in an advanced stage and pass through only two postnatal molts before transforming into the adult stage. Hagan further noted embryoes in postnatal

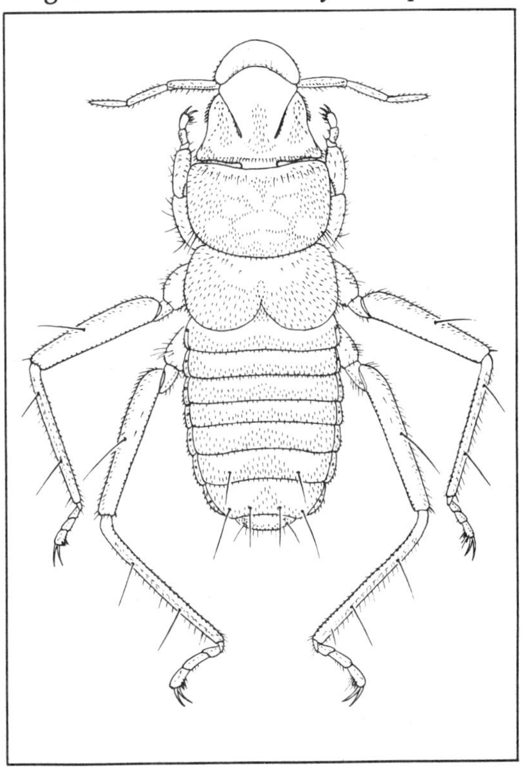

Fig. 126. *Hesperoctenes eumops* [p.612] (Original).

nymphs, which suggests insemination may take place before the female reaches adulthood.

The literature most useful for study of this group of insects includes work by Ferris and Usinger (1939, Microent., 4: 1-50), who review the morphology and classification of this group from a global viewpoint, a world catalog by Usinger (1946, Gen. Cat. Hem., Fasc. 5: 1-18), and a new checklist and bibliography by Ryckman and Casdin (1977, Cal. Vector Views, 24: 25-31). In North America the family Polyctenidae is represented by only two species, which are keyed in Ferris and Usinger (1939, *op. cit.*).

Genus *Hesperoctenes* Kirkaldy, 1906

1906 *Hesperoctenes* Kirkaldy, Can. Ent., 38: 375. Type-species: *Polyctenes fumarius* Westwood, 1874. Original designation.

Hesperoctenes eumops Ferris and Usinger, 1939 [Fig. 126]

1920 Hesperoctenes longiceps: Ferris, J. N.Y. Ent. Soc., 27: 261.
1939 Hesperoctenes eumops Ferris and Usinger, Microent., 4: 18, 19. [Cal.].
Distribution: Cal. (Brazil).
Note: Host is a free-tailed bat, *Eumops californicus*.

Hesperoctenes hermsi Ferris and Usinger, 1939

1939 Hesperoctenes hermsi Ferris and Usinger, Microent., 4: 17, 20. [Tex.].
Distribution: Tex.
Note: Host is a free-tailed bat, *Tadarida macrotis*.

Family Pyrrhocoridae Fieber, 1860

The Cotton Stainers

By Thomas J. Henry

The cotton stainers or red bugs are primarily a tropical group with only a few species ranging into temperate North America. In our region, they are medium sized and aposematically colored red, yellow, and white. The lateral margins of the pronotum are flattened and usually reflexed, the rostrum is relatively long, extending well onto the abdomen, and the body is glabrous. Only the genus *Dysdercus* Guérin-Méneville is represented in the Western Hemisphere.

Little is known about the biology of most species, but several are considered serious pests of cotton and other crops. Members of this family are commonly referred to as cotton stainers because of their habit of introducing a staining fungus into cotton bolls while feeding. Also, when abundant, the gregarious nymphs are sometimes ground up when cotton is mechanically picked, leaving a red stain from their brightly colored bodies in the cotton fiber (Henry, 1984, U.S. Dept. Agr., APHIS, PPQ Info. Circ., 39: 1-6).

Pyrrhocoris apterus Linnaeus, reported in the U.S. by Barber (1911, J. N.Y. Ent. Soc., 19: 111-112) from New Jersey based on two specimens, has not been collected in this country since 1896. It, therefore, is not considered established in North America and is excluded from this catalog.

The major works treating the Pyrrhocoridae on a world basis are Lethierry and Severin (1894, Cat. Gen. Hem., 2: 243-255, 261) and Hussey (1929, Gen. Cat. Hem., 3: 1-144). Doesburg (1968, Mus. Leiden Zool. Ver., 97: 1-215) revised the New World species of *Dysdercus* and provided a summary of papers treating biology and systematics. Freeman (1947, Trans. Roy. Ent. Soc. London, 98: 373-424) revised the Old World species of *Dysdercus*.

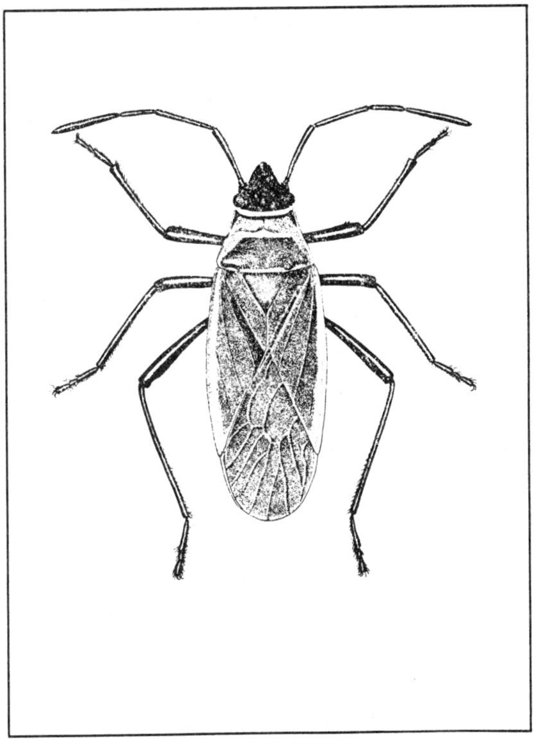

Fig. 127. *Dysdercus mimulus* [p. 614] (After T. C. Barber, 1925).

Subfamily Pyrrhocorinae Fieber, 1860

Genus *Dysdercus* Guérin-Méneville, 1831

1931 *Dysdercus* Guérin-Méneville, Voy. Coq., Atlas, pl. 12, fig. 16. Type-species: *Dysdercus peruvianus* Guérin-Méneville, 1831. Monotypic.

Notes:Doesburg (1968, Mus. Leiden Zool. Ver., 95: 1-215) revised the New World species and gave distributions based on specimens examined, but did not specifically treat all records mentioned in the U.S. literature. Thus, three species previously recorded from our region were excluded without comment, viz., *D. peruvianus* (Guérin-Méneville), *D. obliquus* (Herrich-Schaeffer), and *D. obscuratus* Distant. Herein, the *D. peruvianus* record, considering the verified distribution of this neotropical species, is omitted because its presence probably was based on a port interception or a misidentification, whereas *D. obliquus* and *D. obscuratus* are retained, provisionally, as part of the southwestern fauna.

A fourth species, *D. ocreatus* (Say), although described from Georgia, was considered by Doesburg (1968, above) a West Indian species. He concluded that Say's type locality was an error or the specimen on which he based his description was imported on cotton from Hispaniola. This species, therefore, is excluded from the catalog.

Subgenus *Dysdercus* Guérin-Méneville, 1831

1831 *Dysdercus* Guérin-Méneville, Voy. Coq., Atlas, p. 12, fig. 16. Type-species: *Dysdercus peruvianus* Guérin-Méneville, 1831. Monotypic.

Note: This is the only subgenus of *Dysdercus* represented in our fauna.

Dysdercus andreae (Linnaeus), 1758

1758 *Cimex andreae* Linnaeus, Syst. Nat., Ed. 10, 1: 448. [Jamaica].
1775 *Cimex suturalis* Fabricius, Syst. Ent., p. 721. ["America"]. Synonymized by Stål, 1868, K. Svens. Vet.-Akad. Handl., 7(11): 85.
1850 *Dysdercus suturalis*: Herrich-Schaeffer, Wanz. Ins., 9: 177.
1907 *Dysdercus andrae* [*sic*]: Van Duzee, Bull. Buff. Soc. Nat. Sci., 8: 18.
1914 *Dysdercus andreae*: Barber, Bull. Am. Mus. Nat. Hist., 33: 509.
Distribution: Fla. (West Indies).

Dysdercus bimaculatus (Stål), 1854

1854 *Dysderus* [*sic*] *bimaculatus* Stål, Öfv. K. Svens. Vet.-Akad. Förh., 11(8): 236. [Panama].
1968 *Dysdercus bimaculatus*: Doesburg, Mus. Leiden Zool. Ver., 97: 75.
Distribution: Ariz. (Mexico to Venezuela and Colombia).

Dysdercus concinnus Stål, 1861

1861 *Dysdercus concinnus* Stål, Öfv. K. Svens. Vet.-Akad. Förh., 18(4): 198. [Mexico].
1883 *Dysdercus splendidus* Distant, Biol. Centr.-Am., Rhyn., 1: 231. [Panama]. Synonymized by Doesburg, 1968, Mus. Leiden Zool. Ver., 97: 68.
1906 *Dysdercus concinnus*: Barber, Mus. Brook. Inst. Arts Sci., Sci. Bull., 1: 277.
1929 *Dysdercus mimus* var. *splendidus*: Hussey, Gen. Cat. Hem., 3: 96.
Distribution: Tex. (Mexico to South America, Galapagos Islands).

Dysdercus mimulus Hussey, 1929 [Fig. 127]

1832 *Capsus mimus*: Say, Descrip. Het. Hem. N. Am., p. 20 (in part; var. A & B).
1862 *Dysdercus mimus*: Stål, Stett. Ent. Zeit., 23: 316.
1925 *Dysdercus obscuratus*: Barber, J. Agr. Res., 31: 1137.

1929 *Dysdercus mimulus* Hussey, Gen. Cat. Hem., 3: 95. New name for *Dysdercus mimus* Say (var. A & B), 1832, and *D. mimus* Stål, 1862.
Distribution: Ariz., Cal., Fla., Tex. (Mexico, Central America, West Indies).
Note: In the early literature, this species frequently has been confused with *Dysdercus mimus* (Say) and other species. Doesburg (1968, Mus. Leiden Zool. Ver., 97: 104-105) provided a list of synonymy and sorted out many of the misidentifications. Early records of *mimus* should be carefully evaluated. The record of *D. obscuratus* by McAtee (1926, Ent. News, 37: 14) was based on specimens reported by T. C. Barber (1925, above).

Dysdercus mimulus luteus Doesburg, 1968
1968 *Dysdercus mimulus luteus* Doesburg, Mus. Leiden Zool. Ver., 97: 110. [Ariz.].
Distribution:Distribution: Ariz. (Mexico).

Dysdercus mimulus mimulus Hussey, 1929
1929 *Dysdercus mimulus* Hussey, Gen. Cat. Hem., 3: 5.
1968 *Dysdercus mimulus mimulus*: Doesburg, Mus. Leiden Zool. Ver., 97: 104.
Distribution: Same as for species.

Dysdercus mimus (Say), 1832
1832 *Capsus mimus* Say, Descrip. Het. Hem. N. Am., p. 20. [Mexico].
1854 *Dysderus* [*sic*] *albidiventris* Stål, Öfv. K. Svens. Vet.-Akad. Förh., 11(8); 236. [Mexico]. Synonymized by Hussey, 1929, Gen. Cat. Hem., 3: 96.
1876 *Dysdercus albidiventris*: Uhler, Bull. U.S. Geol. Geogr. Surv. Terr., 1: 314.
1876 *Dysdercus mimus*: Uhler, Bull. U.S. Geol. Geogr. Surv. Terr., 1: 314 (in part).
1927 *Dysdercus splendidus*: Hussey, Bull. Brook. Ent. Soc., 22: 235.
Distribution: Ariz., Cal., Tex. (Mexico to South America).

Dysdercus obliquus (Herrich-Schaeffer), 1843
1843 *Pyrrhocoris obliquus* Herrich-Schaeffer, Wanz. Ins., 7: 19. [Mexico].
1876 *Dysdercus obliquus*: Uhler, Bull. U.S. Geol. Geogr. Surv. Terr., 1: 314.
Distribution: Ariz., Cal., Tex. (Mexico to Colombia and Venezuela).

Dysdercus obscuratus Distant, 1883
1883 *Dysdercus obscuratus* Distant, Biol. Centr.-Am., Rhyn., 1: 230. [Costa Rica, Guatemala]. Lectotype from Costa Rica designated by Doesburg, 1968, Mus. Leiden Zool. Ver., 97: 98.
Distribution: Tex. (Mexico to Bolivia and Peru).
Note: The nominate subspecies has not been reported from the U.S.

Dysdercus obscuratus incertus Distant, 1883
1883 *Dysdercus incertus* Distant, Biol. Centr.-Am., Rhyn., 1: 230. [Costa Rica].
1912 *Dysdercus obscuratus*: Torre-Bueno, Ent. News, 23: 121.
1940 *Dysdercus incertus*: Torre-Bueno, Bull. Brook. Ent. Soc., 35: 12.
1968 *Dysdercus obscuratus incertus*: Doesburg, Mus. Leiden Zool. Ver., 97: 103.
Distribution: Tex. (Mexico to Bolivia and Peru).
Note: Doesburg (1968, above), in omitting this subspecies from his U.S. list, questioned, but did not transfer, Torre-Bueno's (1940, above and 1941, Ent. Am., 21: 116)) records from Tex.

Dysdercus suturellus (Herrich-Schaeffer), 1842
1842 *Pyrrhocoris suturellus* Herrich-Schaeffer, Wanz. Ins., 6: 76. [N. Am.].
1870 *Dysdercus suturellus*: Stål, K. Svens. Vet.-Akad. Handl., 9(1): 123.
1918 *Dysdercus saturellus* [*sic*]: Watson, Fla. Buggist, 2: 88.
Distribution: Ala., Fla., Ga., S.C. (West Indies).

Family Reduviidae Latreille, 1807

The Assassin Bugs

By Richard C. Froeschner

The Reduviidae comprise one of the larger families of Heteroptera but are proportionally less represented north of the Rio Grande River than are some of the other heteropteran families. The following list enumerates 159 species in 46 genera.

All forms, except for a few reports of early instars inserting their beaks in plant tissue, are predaceous. Most feed on insects and other terrestrial arthropods but members of the subfamily Triatominae have specialized in feeding regularly on the blood of birds, mammals--including man--, or lizards. The prey-seeking habit often involves stealth and chance encounter, but some forms seek out and frequent living spaces of their prey, whether it be the burrows of rodents, the buildings of man, flowers visited by flies, or the webs of spiders on which they travel without being ensnared. This latter habit was independently developed in several subfamilies. Schumacher (1917, Zeit. Wiss. Ins.-Biol., 13: 217-218) reported one form inside the webs of a psocid (order Psocoptera) on which it feeds. Attack on the prey is accompanied by an injection of a stupefying fluid, which apparently also facilitates sucking of its body juices.

The seasonal history of North American Reduviidae varies considerably. According to species, hibernation may be in the egg, immature, or adult stage. Most species have a single generation per year and none are known to require more than one year for a complete life cycle. Two life cycles per year have been described for a few species--but perhaps in the southern states this phenomenon might be more common than has been reported. Habits of nymphs and adults appear quite similar except that some nymphs are unlike their adults in bearing on their body surface sticky hairs and the small flecks of material from their environment which adhere to these and conceal the body outline. Adults and nymphs are capable of making a rasping sound--stridulation--by moving the tip of their beak back and forth in the cross-striated groove on their prosternum.

The economic value of the assassin bugs may or may not be easy to assess depending on the species involved and the interest of the observer. Those forms that feed on terrestrial arthropods are rather indiscriminate feeders and capture whatever is available whether it be a harmful caterpillar, the valuable honey bee, or insects of no economic concern to man. While these assassin bugs do exert a certain population pressure on these other terrestrial arthropods, they cannot be credited with bringing about appreciable population reductions of particular insects that attack man's crops.

Reduviidae bite humans for one of two reasons. Some, especially those of the subfamily Peiratinae, are quick to bite in defense against deliberate or accidental attack. Others, particularly of the subfamily Triatominae, bite to obtain blood. The bite of the latter, when well adapted to deriving a blood meal, causes no noticeable sensation in the victim and the bug can eat unmolested. In tropical America the Triatominae are vectors of American trypanosomiasis or Chagas' disease. When the causative organism, *Trypanosoma cruzi* Chagas, completes its development in the bug's digestive tract, it passes therefrom with the excrement; when dropped on the skin of a host it can enter the host's body through breaks in the skin or through the mucous membranes. There is as yet no cure for the disease in humans, and its prevention is complicated by the fact that the triatomines colonize buildings, hiding in crevices, etc., and bring with them the causative *Trypanosoma* from a reservoir in the blood of wild and domestic animals. While the causative agent *Trypanosoma cruzi* is known to infest a number of wild hosts in the United States, the absence of house-infesting triatomines--although some of the sylvatic (or naturally occurring) species do come to artificial light--probably accounts for the absence of human cases of Chagas' disease in this country. The wild hosts most commonly infested in the United States are several species of wood rats (*Neotoma* species) and the opossum (*Didelphis virginiana* Kerr).

Literature of greatest use for the study of this family in the area north of Mexico includes the following: A checklist of the species of the Western Hemisphere by Wygodzinsky (1949, Univ. Nac. Tucumán, Monogr., 1: 1-102); two efforts to key all the species of the area, one by Fracker (1912, Proc. Ia. Acad. Sci., 19: 217-252) and one by Readio (1927, Univ. Ks. Sci. Bull., 17: 5-291); a key for the generic identification of nearctic reduviid nymphs by Fracker and Usinger (1949, An. Ent. Soc. Am., 42: 273-278); Readio's (1927, above) assemblage of published natural history information offered with his original observations; and Usinger's (1941, Bull. Brook. Ent. Soc., 36: 206-208) summary of Reduviidae dwelling in spider webs. References for works treating infra-familial taxa will be presented at appropriate places in the text.

Under the family Reduviidae, Rathvon (1869, Hist. Lancaster Co., Pa., p. 550) includes the combination "*Leptopus* Latr. *alternatus* Har." This is apparently a Harris manuscript name for which there is no other published usage. Uhler (1871 and 1878) in his two papers on the Harris collection did not report such a label on any of the specimens. In modern usage *Leptopus* belongs to the family Leptopodidae, which is represented in North America by the single species, *Patapius spinosus* (Rossi), recently introduced in California.

Walker's (1873, Cat. Hem. Brit. Mus., 7: 97) record of the South American *Phorastes femoratus* (DeGeer) [using the combination *Pirates femoralus* [*sic*]] for North America must have been in error. No other author listed it for our region and no specimens are available.

Subfamily Apiomerinae Amyot and Serville, 1843

Note: Davis (1969, An. Ent. Soc. Am., 62: 84) treated this group as a tribe, Apiomerini, under the subfamily Harpactorinae.

Genus *Apiomerus* Hahn, 1831

1831 *Apiomerus* Hahn, Wanz. Ins., 1: 29. Type-species: *Reduvius hirtipes* Fabricius, 1787. Monotypic.

Note: Costa Lima et al. (1951, Mem. Inst. Oswaldo Cruz, 49: 273-442) revised this genus

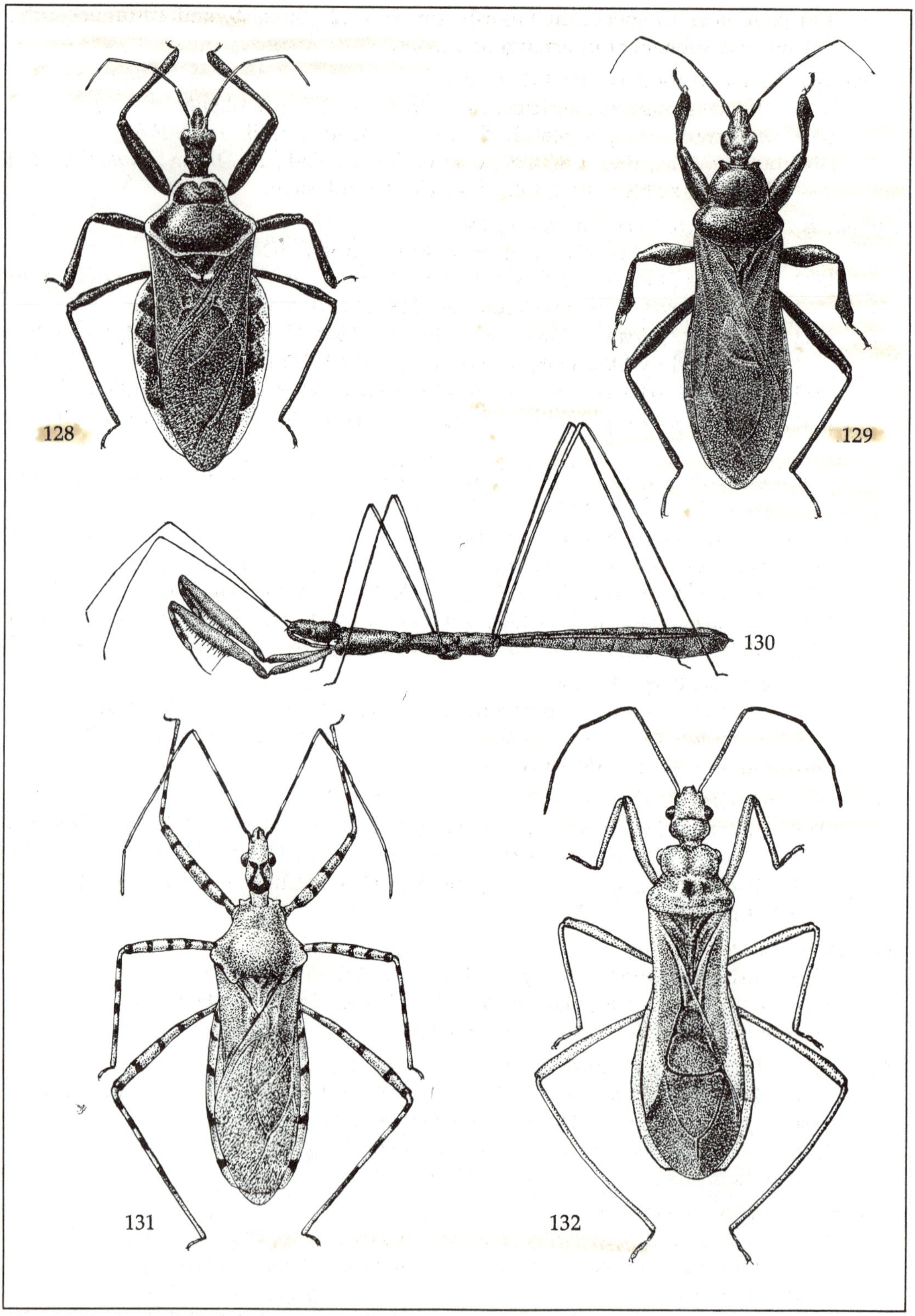

128 129 130 131 132

Figs. 128-132: 128, *Apiomerus crassipes* [p.619]; 129, *Melanolestes picipes* [p. 641]; 130, *Barce uhleri* [p.626]; 131, *Pselliopus barberi* [p. 633]; 132, *Oncerotrachelus acuminatus* [p. 644] (After Froeschner, 1944).

and gave keys to species in English (pp. 299-312), Spanish, and Portuguese; they did not use subgenera in arranging species.

Apiomerus crassipes (Fabricius), 1803 [Fig. 128]
1803 *Reduvius crassipes* Fabricius, Syst. Rhyn., p. 273. [Carolina].
1868 *Apiomerus crassipes*: Stål, K. Svens. Vet.-Akad. Handl., 7(11): 117.
Distribution: Ariz., B.C., Ca., Col., Conn., Fla., Ill., Ind., Ks., Minn., Mo., N.C., N.J., N.M., N.Y., Neb., Oh., Ok., Tex., Ut., Va. (Mexico).

Apiomerus crasspipes crassipes (Fabricius), 1803
1803 *Reduvius crassipes* Fabricius, Syst. Rhyn., p. 273. [Carolina].
1832 *Reduvius linitaris* Say, Descrip. Het. Hem. N. Am., p. 31. [Ind.]. Synonymized by Stål, 1868, K. Svens. Vet.-Akad. Handl., 7(11): 117.
1843 *Herega rubrolimbata* Amyot and Serville, Hist. Nat. Hem., p. 354. [U.S.]. Synonymized by Stål, 1868, K. Svens. Vet.-Akad. Handl., 7(11): 117.
1872 *Apiomerus* (*Herega*) *crassipes*: Stål, K. Svens. Vet.-Akad. Handl., 10(4): 98.
1951 *Apiomerus crassipes crassipes* Costa Lima et al., Mem. Inst. Oswaldo Cruz, 49: 350.
Distribution: Ariz., B.C., Cal., Col., Conn., Fla., Ill., Ind., Ks., Minn., Mo., N.C., N.J., N.M., N.Y., Neb., Oh., Ok., Tex., Ut., Va.

Apiomerus flaviventris Herrich-Schaeffer, 1846
1846 *Apiomerus flaviventris* Herrich-Schaeffer, Wanz. Ins., 8: 77. [Mexico].
1872 *Apiomerus* (*Herega*) *flaviventris*: Stål, K. Svens. Vet.-Akad. Hand., 10(4): 98.
1876 *Apiomerus flaviventris*: Uhler, Bull. U.S. Geol. Geogr. Surv. Terr., ser. 2, 5: 328.
Distribution: Ariz., Cal., Col., Fla., N.M., Tex. (Mexico).

Apiomerus immundus Bergroth, 1898
1898 *Apiomerus immundus* Bergroth, Bull. Soc. Ent. France, p. 307. [Mexico].
1916 *Apiomerus* (*Apiomerus*) *immundus*: Van Duzee, Check List Hem., p. 30.
Distribution: Cal., Tex. (Mexico).

Apiomerus longispinis Champion, 1899
1899 *Apiomerus longispinis* Champion, Biol. Centr.-Am., Rhyn., 2: 233, 239. [Mexico].
1907 *Apiomerus longispunus* [*sic*]: Snow, Trans. Ks. Acad. Sci., 20: 160.
1916 *Apiomerus* (*Apiomerus*) *longispinis*: Van Duzee, Check List Hem., p. 30.
Distribution: Ariz., Tex. (Mexico).

Apiomerus moestus Stål, 1862
1862 *Apiomerus moestus* Stål, Stett. Ent. Zeit., 23: 455. [Mexico].
1916 *Apiomerus* (*Apiomerus*) *moestus*: Van Duzee, Check List Hem., p. 30.
Distribution: Ariz. (Mexico to Guatemala).

Apiomerus pictipes Herrich-Schaeffer, 1846
1846 *Apiomerus pictipes* Herrich-Schaeffer, Wanz. Ins., 8: 75. [Central America].
1886 *Apiomerus pictipes*: Uhler, Check-list Hem. Het., p. 24.
1916 *Apiomerus* (*Herega*) *pictipes*: Van Duzee, Check List Hem., p. 30.
Distribution: Col., N.M. (Mexico to Brazil).

Apiomerus repletus Uhler, 1876
1876 *Apiomerus repletus* Uhler, U.S. Geol. Geogr. Surv. Terr., ser. 2, 5: 329. [Cal.].
1916 *Apiomerus* (*Apiomerus*) *repletus*: Van Duzee, Check List Hem., p. 30.
Distribution: Cal.
Note: Champion, (1899, Biol. Centr.-Am., Rhyn., 2: 236) refers to a species "*occidentalis* Glover" from California in synonymy under this species. The name "*oc-*

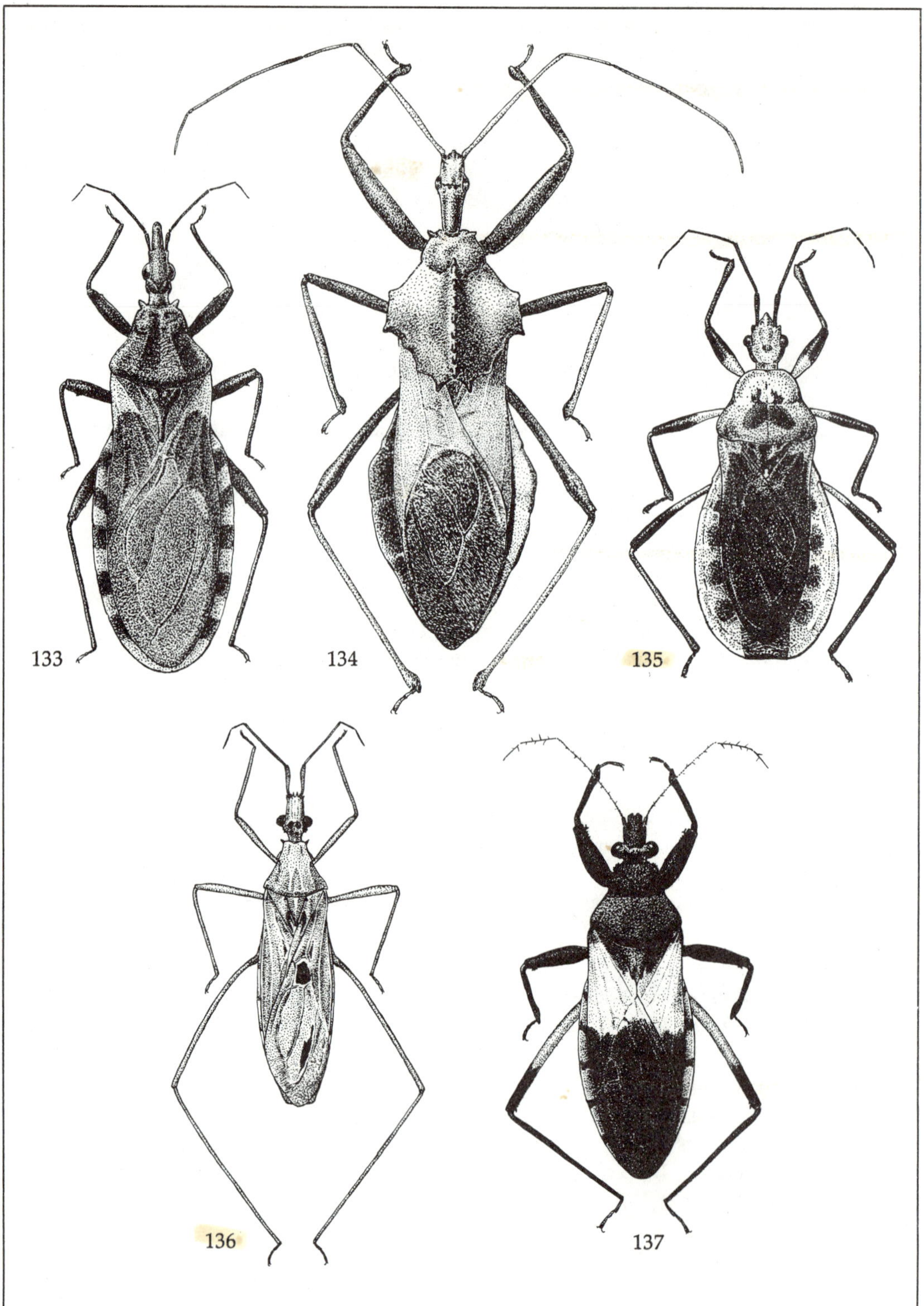

Figs. 133-137: 133, *Triatoma sanguisuga* [p. 651]; 134, *Arilus cristatus* [p. 631]; 135, *Rhiginia cruciata* [p. 621]; 136, *Narvesus carolinensis* [p. 646]; 137, *Microtomus purcis* [p. 640] (After Froeschner, 1944).

cidentalis" appears to be a manuscript name as no previously published source for it was found. Wygodzinsky (1949, Inst. Med. Reg., Tucumán, Monogr., 1: 17) treated *A. repletus* as a junior synonym of *A. elatus* Stål, but Costa Lima et al. (1951, Mem. Inst. Oswaldo Cruz, 49: 395) treated it as a distinct species.

Apiomerus spissipes (Say), 1825

1825 *Reduvius spissipes* Say, J. Acad. Nat. Sci. Phila., 4: 328. [Ark.].
1872 *Herega spissipes*: Uhler, Prelim. Rept. U.S. Geol. Surv. Mont., p. 420.
1872 *Apiomerus* (*Herega*) *spissipes*: Stål, K. Svens. Vet.-Akad. Handl., 10(4): 98.
1876 *Apiomerus spissipes*: Uhler, U.S. Geol. Geogr. Surv. Terr., ser. 2, 5: 328.
1922 *Apimerous* [*sic*] *spissipes*: Parshley, S.D. St. Coll., Tech. Bull., 2: 12.
Distribution: Ala., Ariz., Ark., Col., Fla., Ks., Mo., N.C., Ok., S.D., Tex., Ut. (Mexico to Costa Rica).

Apiomerus subpiceus Stål, 1862

1862 *Apiomerus subpiceus* Stål, Stett. Ent. Zeit., 23: 454. [Mexico].
1906 *Apiomerus subpiceus*: Snow, Trans. Ks. Acad. Sci., 20: 180.
1916 *Apiomerus* (*Apiomerus*) *subpiceus*: Van Duzee, Check List Hem., p. 30.
Distribution: Ariz. (Mexico to Costa Rica).

Subfamily Ectrichodiinae Amyot and Serville, 1843

Note: Stål (1864, An. Soc. Ent. France, 4: 59) described *Pothea "aeneo-nitens"* from "America borealis." All subsequent records appear to be based on the original description, and several authors point out the locality is probably erroneous. In light of this, the species is here considered not a member of this fauna.

Genus *Rhiginia* Stål, 1859

1859 *Rhiginia* Stål, Öfv. K Svens. Vet.-Akad. Förh., 16: 176, 181. Type-species: *Reduvius lateralis* Lepeletier and Serville, 1872. Monotypic.

Rhiginia cinctiventris (Stål), 1872

1872 *Ectrichodia cinctiventris* Stål, K. Svens. Vet.-Akad. Handl., 10(4): 103. [Tex.].
1916 *Rhiginia cinctiventris*: Van Duzee, Check List Hem., p. 29.
Distribution: N.M., Tex. (Mexico).

Rhiginia cruciata (Say), 1832 [Fig. 135]

1832 *Petalocheirus cruciatus* Say, Descrip. Het. Hem. N. Am., p. 33. [Ga., Ind., Mo.].
1846 *Ectrychotes bicolor* Herrich-Schaeffer, Wanz. Ins., 8: 53. [Md.]. Synonymized by Stål, 1872, K. Svens. Vet.-Akad. Handl., 10(4): 103.
1859 *Physorhynchus bicolor*: Dohrn, Cat. Hem., p. 49.
1872 *Ectrichodia cruciata*: Stål, K. Svens. Vet.-Akad. Handl., 10(4): 103.
1873 *Ectrichodia media* Walker, Cat. Hem. Brit. Mus., 8: 62. [La.]. Synonymized by Van Duzee, 1916, Check List Hem., p. 29.
1916 *Rhiginia cruciata*: Van Duzee, Check List Hem., p. 29.
Distribution: Fla., Ga., Ill., Ind., La., Md., Mo., N.C., N.J., Ok., Pa., Tex., Va. (Greater Antilles, Mexico).

Subfamily Emesinae Amyot and Serville, 1843

Note: This subfamily was monographed in great detail by Wygodzinsky (1966, Bull. Am. Mus. Nat. Hist., 133: 1-614, pls. 1-4). His paper is essential for any studies in the group.

Tribe Emesini Amyot and Serville, 1843

Genus *Emesa* Fabricius, 1803

1803 *Emesa* Fabricius, Syst. Rhyn., p. 263. Type-species: *Gerris mantis* Fabricius, 1794. Designated by Laporte, 1833, Essai Hem., p. 84.

Note: McAtee and Malloch (1925, Proc. U.S. Nat. Mus., 67(1): 40) explained why the name must be treated as masculine.

Emesa annulatus (Dohrn), 1860

1860 *Westermannia annulata* Dohrn, Linnaea Ent., 14: 251. [Mexico].
1925 *Emesa (Emesa) annulatus*: McAtee and Malloch, Proc. U.S. Nat. Mus., 67(1): 40.
1966 *Emesa annulata*: [*sic*] Wygodzinsky, Bull. Am. Mus. Nat. Hist., 133: 239.
Distribution: Ariz. [*sic*] (Mexico to Panama).

Genus *Gardena* Dohrn 1859

1859 *Gardena* Dohrn, Cat. Hem., p. 252. *Nomen nudum*.

1860 *Gardena* Dohrn, Linnaea Ent., 14: 214. Type-species: *Gardena melinarthrum* Dohrn, 1860. Monotypic.

Gardena elkinsi Wygodzinsky, 1966

1925 *Lutevopsis longimanus*: McAtee and Malloch, Proc. U.S. Nat. Mus., 67(1): 37.
1927 *Luteviopsis* [*sic*] *longimanus*: Readio, Univ. Ks. Sci. Bull., 17: 49.
1951 *Lutevopsis* species: Elkins, Tex. J. Sci., 3: 408. [Tex.].
1966 *Gardena elkinsi* Wygodzinsky, Bull. Am. Mus. Nat. Hist., 133: 257. [Mexico].
Distribution: Ariz., Cal., Fla., Tex. (Mexico).

Gardena poppaea McAtee and Malloch, 1925

1925 *Gardena poppaea* McAtee and Malloch, Proc. U.S. Nat. Mus., 67(1): 74. [Tex.].
1925 *Gardena messalina* McAtee and Malloch, Proc. U.S. Nat. Mus., 67(1): 72. [Tex.]. Synonymized by Elkins, 1953, J. Ks. Ent. Soc., 26: 139.
Distribution: Fla., Tex. (Mexico).

Genus *Stenolemoides* McAtee and Malloch, 1925

1925 *Stenolemus (Stenolemoides)* McAtee and Malloch, Proc. U.S. Nat. Mus., 67(1): 28. Type-species: *Luteva arizonensis* Banks, 1909. Monotypic.

1947 *Stenolemoides*: Wygodzinsky, Rev. Brasil. Biol., 6: 514.

Stenolemoides arizonensis (Banks), 1909

1909 *Luteva arizonensis* Banks, Psyche, 16: 45. [Ariz.].
1916 *Ploiaria arizonensis*: Van Duzee, Check List Hem., p. 28.
1922 *Stenolemus arizonensis*: McAtee and Malloch, Proc. Biol. Soc. Wash., 35: 95.

1922 *Westermannia arizonensis*: Barber, J. N.Y. Ent. Soc., 30: 130.
1925 *Stenolemus* (*Stenolemoides*) *arizonensis*: McAtee and Malloch, Proc. U.S. Nat. Mus., 67(1): 28.
1947 *Stenolemoides arizonensis*: Wygodzinsky, Rev. Brasil. Biol., 6: 515.
Distribution: Ariz., Cal., N.M., Nev., Tex., Ut. (Mexico).

Genus *Stenolemus* Signoret, 1858

1858 *Stenolemus* Signoret, An. Soc. Ent. France, ser. 3, 6: 251. Type-species: *Stenolemus spiniventris* Signoret, 1858. Monotypic.

Stenolemus lanipes Wygodzinsky, 1949
1916 *Stenolemus spiniventris*: Van Duzee, Check List Hem., p. 27.
1925 *Stenolemus* (*Stenolemus*) *hirtipes* McAtee and Malloch, Proc. U.S. Nat. Mus., 67(1): 32. [Miss.]. Preoccupied.
1949 *Stenolemus lanipes* Wygodzinsky, Inst. Med. Reg. Tucumán, Monogr., 1: 33. New name for *Stenolemus hirtipes* McAtee and Malloch.
Distribution: Fla., Miss., N.C., S.C.

Stenolemus longicornis (Blatchley), 1925
1925 *Malacopus longicornis* Blatchley, Ent. News, 36: 46. [Fla.].
1925 *Stenolemus* (*Stenolemus*) *pristinus* McAtee and Malloch, Proc. U.S. Nat. Mus., 67(1): 29. [Fla.]. Synonymized by Blatchley, 1926, Het. E. N. Am., p. 514.
1926 *Stenolemus longicornis*: Blatchley, Het. E. N. Am., p. 514.
Distribution: Fla.

Stenolemus pallidipennis McAtee and Malloch, 1925
1925 *Stenolemus* (*Stenolemus*) *pallidipennis* McAtee and Malloch, Proc. U.S. Nat. Mus., 67(1): 30. [Ariz.].
1927 *Stenolemus pallidipennis*: Readio, Univ. Ks. Sci. Bull., 17: 47.
Distribution: Ariz.

Stenolemus spiniventris Signoret, 1858
1858 *Stenolemus spiniventris* Signoret, An. Soc. Ent. France, ser. 3, 6: 253. [Mexico].
1916 *Stenolemus spinifrons*: Van Duzee, Check List Hem., p. 27.
1925 *Stenolemus* (*Stenolemus*) *spiniger* McAtee and Malloch, Proc. U.S. Nat. Mus., 67(1): 33. [Tex.]. Synonymized by Wygodzinsky, 1948, Rev. Brasil. Biol., 8: 220.
Distribution: Ariz., Fla., Tex. (Mexico to Guatemala).

Tribe Leistarchini Stål, 1862

Genus *Ploiaria* Scopoli, 1786

1786 *Ploiaria* Scopoli, Delic. Florae Faunae Insub., 1: 60. Type-species: *Ploiaria domestica* Scopoli, 1786. Monotypic.

Ploiaria aptera McAtee and Malloch, 1925
1925 *Ploiaria* (*Ploiaria*) *aptera* McAtee and Malloch, Proc. U.S. Nat. Mus., 67(1): 66. [Ariz.].
1966 *Ploiaria aptera*: Wygodzinsky, Bull. Am. Mus. Nat. Hist., 133: 169.
Distribution: Ariz.

Ploiaria californiensis Baker, 1910

1910 *Ploiaria californiensis* Baker, Pomona Coll. J. Ent., 2: 226. [Cal.]

Distribution: Cal.

Note: McAtee and Malloch (1925, Proc. U.S. Nat. Mus., 67(1): 52) and Wygodzinsky (1966, Bull. Am. Mus. Nat. Hist., 133: 173) both expressed doubts about the identity of this species, but for different reasons.

Ploiaria carolina (Herrich-Schaeffer), 1850

1850 *Emesodema carolina* Herrich-Schaeffer, Wanz. Ins., 9: 115. ["Carolina"].

1886 *Luteva carolina*: Uhler, Check-list Hem. Het., p. 26.

1873 *Cerascopus carolinus*: Walker, Cat. Hem. Brit. Mus., 8: 148.

1909 *Ploiaria carolina*: Banks, Psyche, 16: 44 (in part).

1925 *Ploiaria (ploiaria) carolina*: McAtee and Malloch, Proc. U.S. Nat Mus., 67(1): 58.

Distribution: Fla., Ga., N.C., S.C.

Ploiaria chilensis (Philippi), 1862

1862 *Stenolemus chilensis* Philippi, An. Univ. Chile, 21: 387. [Chile].

1951 *Ploiaria chilensis*: Kuschel, *In* Wygodzinsky, Rev. Chil. Ent., 1: 113.

Distribution: Cal. (Europe, Colombia to Argentina and Chile).

Ploiaria denticauda McAtee and Malloch, 1925

1925 *Ploiaria (Ploiaria) denticauda* McAtee and Malloch, Proc. U.S. Nat. Mus., 67(1): 63. [Ariz.].

Distribution: Ariz., Cal., Tex.

Ploiaria floridana (Bergroth), 1922

1922 *Luteva floridana* Bergroth, Konowia, 1: 218. [Fla.].

1925 *Ploiaria (Ploiria) floridana*: McAtee and Malloch, Proc. U.S. Nat. Mus., 67(1): 59.

1926 *Ploiaria floridana*: Blatchley, Het. E. N. Am., p. 526.

Distribution: Fla.

Ploiaria hirticornis (Banks), 1909

1909 *Ploiariopsis hirticornis* Banks, Psyche, 16: 44. [N.C.].

1909 *Ploiaria carolina*: Banks, Psyche, 16: 44 (in part).

1925 *Ploiaria (Ploiaria) hirticornis*: McAtee and Malloch, Proc. U.S. Nat. Mus., 67(1): 64.

Distribution: D.C., Fla., La., N.C.

Ploiaria pilicornis McAtee and Malloch, 1925

1925 *Ploiaria (Ploiaria) pilicornis* McAtee and Malloch, Proc. U.S. Nat. Mus., 67(1): 61. [Ariz.].

1966 *Ploiaria pilicornis*: Wygodzinsky, Bull. Am. Mus. Nat. Hist., 133: 196.

Distribution: Ariz.

Ploiaria reticulata (Baker), 1910

1910 *Ploiariopsis reticulata* Baker, Pomona Coll. J. Ent., 2: 225. [Cal.].

1925 *Ploiaria (Ploiaria) reticulata*: McAtee and Malloch, Proc. U.S. Nat. Mus., 67(1): 63.

1966 *Ploiaria reticulata*: Wygodzinsky, Bull. Am. Mus. Nat. Hist., 133: 199.

Distribution: Cal., Tex. (Mexico).

Ploiaria setulifera McAtee and Malloch, 1925

1925 *Ploiaria (Lutvea) setulifera* McAtee and Malloch, Proc. U.S. Nat. Mus., 67(1): 55. [Fla.].

1966 *Ploiaria setulifera*: Wygodzinsky, Bull. Am. Mus. nat. Hist., 133: 200.

Distribution: Fla. (Cuba).

Ploiaria similis McAtee and Malloch, 1925

1925 *Ploiaria* (*Ploiaria*) *similis* McAtee and Malloch, Proc. U.S. Nat. Mus., 67(1): 62. [Tex.].

1953 *Ploiaria similis*: Elkins, J. Ks. Ent. Soc. 26: 139.

Distribution: Tex. (Mexico).

Ploiaria sonoraensis (Van Duzee), 1923

1923 *Ploiariopsis sonoraensis* Van Duzee, Proc. Cal. Acad. Sci., ser. 4, 12: 144. [Cal.].

1925 *Ploiaria sonoraensis*: McAtee and Malloch, Proc. U.S. Nat. Mus., 67(1): 52.

Distribution: Cal.

Ploiaria texana Banks, 1909

1909 *Ploiaria texana* Banks, Psyche, 16: 44. [Tex.].

Distribution: Tex.

Ploiaria uniseriata McAtee and Malloch, 1925

1925 *Ploiaria* (*Ploiaria*) *uniseriata* McAtee and Malloch, Proc. U.S. Nat. Mus., 67(1): 61. [Tex.].

1966 *Ploiaria uniseriata*: Wygodzinsky, Bull. Am. Mus. Nat. Hist., 133: 201.

Distribution: Tex. (Mexico, Guatemala).

Tribe Metapterini Stål, 1859

Genus *Barce* Stål, 1866

1866 *Barce* Stål, Hem. Africana, 3: 163. Type-species: *Barce annulipes* Stål, 1867, a junior synonym of *Ploiaria fraterna* Say, 1832. Monotypic.

Barce aberrans (McAtee and Malloch), 1925

1925 *Metapterus aberrans* McAtee and Malloch, Proc. U.S. Nat. Mus., 67(1): 86. [Tex.].

1966 *Barce aberrans*: Wygodzinsky, Bull. Am. Mus. Nat. Hist., 133: 441.

Distribution: Tex.

Barce fraterna (Say), 1832

1832 *Ploiaria fraterna* Say, Descrip. Het. Hem. N. Am., p. 33. [La.].

1859 *Emesa fraterna*: Dohrn, Cat. Hem., p. 52.

1909 *Barce fraterna*: Banks, Psyche, 16: 47.

1916 *Ploiaria simplicipes*: Van Duzee, Check List Hem., p. 28.

1925 *Metapterus fraternus*: McAtee and Malloch, Proc. U.S. Nat. Mus., 67(1): 89.

Distribution: Cal., Fla., Ill., Ind., Ks., La., Man., Mass., Md., Me., Mo., N.C., N.J., N.Y., Oh., Ont., Pa., Que., Tex., Va., Wis. (Greater Antilles, Mexico to Ecuador).

Barce fraterna annulipes Stål, 1867

1867 *Barce annulipes* Stål, Berl. Ent. Zeit., 10: 168. [Wis.].

1869 *Zelus simplicipes* Rathvon, Hist. Lancaster Co., Pa., p. 550. *Nomen nudum*.

1873 *Emesa annulipes*: Walker, Cat. Hem. Brit. Mus., 8: 147.

1878 "*G.*" *simplicipes* Uhler, Proc. Bost. Soc. Nat. Hist., 19: 430. [Mass.]. Synonymized by Weiss, 1916, Ent. News, 127: 10.

1916 *Ploiaria simplicipes*: Van Duzee, Check List Hem., p. 28.

1925 *Metapterus annulipes*: McAtee and Malloch, Proc. U.S. Nat. Mus., 67(1): 88.

1966 *Barce fraterna annulipes*: Wygodzinsky, Bull. Am. Mus. Nt. Hist., 133: 441.

Distribution: Conn., Fla., Ill., Ind., La., Man., Mass., Md., Me., Mo., N.C., N.H., N.J.,

N.Y., Ok., Pa., Que., Tex.

Note: The letter "G." used in Uhler (1878, above) appears to be a typographical error because the species clearly is treated within the genus *Emesodema*.

Barce fraterna banksii Baker, 1910

1910 *Barce banksii* Baker, Pomona Coll. J. Ent., 2: 227. [Cal.].

1925 *Metapterus banksii*: McAtee and Malloch, Proc. U.S. Nat. Mus., 67(1): 87.

1927 *Metapterus banksi* [*sic*]: Readio, Univ. Ks. Sci. Bull., 17: 69.

1966 *Barce fraterna banksii*: Wygodzinsky, Bull. Am. Mus. Nat. Hist., 133: 44.

1951 *Metapterus normae* Elkins, Field and Lab., 19: 90. [Tex.]. Synonymized by Wygodzinsky, 1966, Bull. Am. Mus. Nat. Hist., 133: 44.

Distribution: Cal., Tex. (Greater Antilles, Mexico to Ecuador).

Barce fraterna fraterna (Say), 1832

1832 *Ploiaria fraterna* Say, Descrip. Hem. Het. N. Am., p. 33.

1966 *Barce fraterna fraterna*: Wygodzinsky, Bull. Am. Mus. Nat. Hist., 133: 441.

Distribution: Fla., Ill., Ks., La., Mass., Md., Mo., N.C., N.Y., Ok., Que., Tex., Va. (Greater Antilles).

Barce husseyi Wygodzinsky, 1966

1925 *Ischnonyctes* species: McAtee and Malloch, Proc. U.S. Nat. Mus., 67(1): 11.

1966 *Barce husseyi* Wygodzinsky, Bull. Am. Mus. Nat. Hist., 133: 443. [La.].

Distribution: La.

Barce neglecta (McAtee and Malloch), 1925

1925 *Metapterus neglectus* McAtee and Malloch, Proc. U.S. Nat. Mus., 67(1): 87. [N.J.].

1966 *Barce neglecta*: Wygodzinsky, Bull. Am. Mus. Nat. Hist., 133: 445.

Distribution: Mass., N.J., N.Y.

Barce uhleri Banks, 1909 [Fig. 130]

1909 *Barce uhleri* Banks, Psyche, 16: 47. [N.C.].

1909 *Barce uhleri* var. *brunnea* Banks, Psyche, 16: 47. [N.Y., Va.]. Synonymized by McAtee and Malloch, 1925, Proc. U.S. Nat. Mus., 67(1): 87.

1925 *Metapterus uhleri*: McAtee and Malloch, Proc. U.S. Nat. Mus., 67(1): 86.

Distribution: Alta., Ia., Ind., Ks., Mass., Mo., N.C., N.J., N.Y., Ok., S.D., Sask., Va.

Barce werneri Wygodzinsky, 1966

1966 *Barce werneri* Wygodzinsky, Bull. Am. Nat. Hist., 133: 445. [La.].

Distribution: La.

Genus *Emesaya* McAtee and Malloch, 1925

1925 *Emesaya* McAtee and Malloch, Proc. U.S. Nat. Mus., 67(1): 74. Type-species: *Ploiaria brevipennis* Say, 1832. Original designation.

Emesaya banksi McAtee and Malloch, 1925

1925 *Emesaya banksi* McAtee and Malloch, Proc. U.S. Nat. Mus., 67(1): 77 [Tex.].

Distribution: Cal., Tex. (Mexico).

Emesaya brevicoxa (Banks), 1909

1909 *Emesa brevicoxa* Banks, Psyche, 16: 48. [Cal.].

1925 *Emesaya brevicoxa*: McAtee and Malloch, Proc. U.S. Nat. Mus., 67(1): 77.

Distribution: Cal.

Emesaya brevipennis (Say), 1832
1773 *Cimex longipes* De Geer, Mem. Hist. Ins., 3: 352. [Pa.]. Preoccupied.
1828 *Ploiaria brevipennis* Say, Am. Ent., p. 105. [Pa.].
1859 *Emesa brevipennis*: Dohrn, Cat. Hem., p. 52.
1925 *Emesaya brevipennis*: McAtee and Malloch, Proc. U.S. Nat. Mus., 67(1): 78.
Distribution: Cal., Conn., Fla., Ga., Ia., Ill., Ind., Mass., Md., Mo., N.C., N.J., N.Y., Ok., Pa., R.I., Tex. (Mexico to Brazil).

Emesaya brevipennis australis McAtee and Malloch, 1925
1925 *Emesaya brevipennis australis* McAtee and Malloch, Proc. U.S. Nat. Mus., 67(1): 79. [Panama; also U.S.].
Distribution: Fla., Ga., Tex. (Mexico to Brazil).

Emesaya brevipennis brevipennis (Say), 1828
1773 *Cimex longipes* De Geer, Mem. Hist. Ins., 3: 352. [Pa.]. Preoccupied.
1828 *Ploiaria brevipennis* Say, Am. Ent., 3, pl. 47. [Pa.].
1832 *Emesa filum*: Gray, Anim. Kingdom, 15: 789.
1843 *Emesa pia* Amyot and Serville, Hist. Nat. Ins. Hem., p. 394. [Pa.]. Synonymized by Uhler, 1871, Proc. Bost. Soc. Nat. Hist., 14: 107.
1859 *Emesa longipes*: Dohrn, Cat. Hem., p. 32.
1869 *Ploiarius* [*sic*] *brevipennis*: Rathvon, Hist. Lancaster Co., Pa., p. 550.
1925 *Emesaya brevipennis brevipennis*: McAtee and Malloch, Proc. U.S. Nat. Mus., 67(1): 79.
Distribution: Cal., Fla., Ia., Ill., Ind., Ks., Mass., Md., Mo., N.C., N.J., N.Y., Pa., R.I., Tex. (Mexico).

Emesaya brevipennis occidentalis McAtee and Malloch, 1925
1925 *Emesaya brevipennis occidentalis* McAtee and Malloch, Proc. U.S. Nat. Mus., 67(1): 80. [Holotype from Mexico; also from U.S.].
Distribution: Cal., Fla. (Mexico).

Emesaya incisa McAtee and Malloch, 1925
1925 *Emesaya incisa* McAtee and Malloch, Proc. U.S. Nat. Mus., 67(1): 78. [Cal.].
Distribution: Ariz., Cal. (Mexico).

Emesaya lineata McAtee and Malloch, 1925
1925 *Emesaya lineata* McAtee and Malloch, Proc. U.S. Nat. Mus., 67(1): 81. [Fla.].
Distribution: Fla.

Genus *Ghinallelia* Wygodzinsky, 1966

1966 *Ghinallelia* Wygodzinsky, Bull. Am. Mus. Nat. Hist., 133: 485. Type-species: *Ghilianella globifera* Bergroth, 1906. Original designation.

Ghinallelia productilis (Barber), 1914
1914 *Ghilianella productilis* Barber, Bull. Am. Mus. Nat. Hist., 33: 502. [Fla.].
1966 *Ghinallelia productilis*: Wygodzinsky, Bull. Am. Mus. Nat. Hist., 133: 492.
Distribution: Ala., Fla. (West Indies).

Genus *Pseudometapterus* Wygodzinsky, 1966

1966 *Pseudometapterus* Wygodzinsky, Bull. Am. Mus. Nat. Hist., 133: 547. Type-species: *Ghilianella argentina* Berg, 1900. Original designation.

Pseudometapterus butleri Wygodzinsky, 1966
1966 *Pseudometapterus butleri* Wygodzinsky, Bull. Am. Mus. Nat. Hist., 133: 551. [Ariz.].
Distribution: Ariz.

Pseudometapterus umbrosus (Blatchley), 1926
1926 *Metapterus umbrosus* Blatchley, Het. E. N. Am., p. 553. [Fla.].
1966 *Pseudometapterus umbrosus*: Wygodzinsky, Bull. Am. Mus. Nat. Hist., 133: 557.
Distribution: Fla.

Pseudometapterus wygodzinskyi (Elkins), 1953
1953 *Metapterus wygodzinskyi* Elkins, J. Ks. Ent. Soc., 26: 137. [Tex.].
1966 *Pseudometapterus wygodzinskyi*: Wygodzinsky, Bull. Am. Mus. Nat. Hist., 133: 558.
Distribution: Tex. (Mexico).

Tribe Ploiariolini Van Duzee, 1916

Genus *Emesopsis* Uhler, 1893

1893 *Emesopsis* Uhler, Proc. Zool. Soc. London, p. 718. Type-species: *Emesopsis nubilus* Uhler, 1893. Monotypic.

Emesopsis nubilus Uhler, 1893
1893 *Emesopsis nubilus* Uhler, Proc. Zool. Soc. London, p. 718. [St. Vincent, West Indies; Cuba].
1966 *Emesopsis nubilus*: Wygodzinsky, Bull. Am. Mus. Nat. Hist., 133: 365.
Distribution: Fla. (Africa, Asia, Hawaii, Mexico, South America, West Indies).

Genus *Empicoris* Wolff, 1811

1811 *Empicoris* Wolff, Icon. Cimic., 5: 197. Type-species: *Cimex vagabundus* Linnaeus, 1758. Monotypic.

Empicoris armatus (Champion), 1898
1898 *Ploiariodes armata* Champion, Biol. Centr.-Am., Rhyn., 2: 165. [Guatemala, Panama].
1922 *Ploeariola mansueta* Bergroth, Not. Ent., 2: 80. [Fla., Jamaica]. Synonymized by McAtee and Malloch, 1925, Proc. U.S. Nat. Mus., 67(1): 20.
1925 *Empicoris armatus*: McAtee and Malloch, Proc. U.S. Nat. Mus., 67(1): 20.
Distribution: Fla. (Greater Antilles, Guatemala, Panama).

Empicoris barberi (McAtee and Malloch), 1923
1923 *Ploiariodes barberi* McAtee and Malloch, Am. Mus. Novit., 75: 7. [Puerto Rico].
1925 *Empicoris barberi*: McAtee and Malloch, Proc. U.S. Nat. Mus., 67(1): 19.
Distribution: Fla. (Puerto Rico).

Empicoris culiciformis (De Geer), 1773
1773 *Cimex culiciformis* De Geer, Mem. Hist. Ins., 3: 223. [France].
1847 *Ploiaria maculata* Haldeman, Proc. Acad. Nat. Sci. Phila., 3: 151. [Pa.]. Synonymized by Uhler, 1871, Proc. Bost. Soc. Nat. Hist., 14: 107.
1859 *Emesa longipes*: Dohrn, Cat. Hem., p. 52.
1886 *Cerascopus errabundus*: Uhler, Check-list Hem. Het., p. 26.
1909 *Ploiaria errabunda*: Banks, Psyche, 16: 46.
Distribution: Conn., Md., Ore., Va. (Africa, Europe, South America).

Empicoris errabundus (Say), 1832

1832 *Ploiaria errabunda* Say, Descrip. Het. Hem. N. Am., p. 34. [N. Am.].
1871 *Plocaria* [*sic*] *errabunda*: Uhler, Proc. Bost. Soc. Nat. Hist., 14: 107.
1891 *Cerascopus errabundus*: Townsend, Proc. Ent. Soc. Wash., 2: 55.
1896 *Ploiariola errabunda*: Lethierry and Severin, Cat. Gen. Am., 3: 69.
1909 *Ploiariodes tuberculata* Banks, Psyche, 16: 46. [N.Y., Va.]. Synonymized by McAtee and Malloch, 1923, Proc. Biol. Soc. Wash., 36: 161.
1922 *Ploeariola* [*sic*] *tuberculata*: Bergroth, Not. Ent., 2: 79.
1925 *Empicoris errabundus*: McAtee and Malloch, Proc. U.S. Nat. Mus., 67(1): 24.
1929 *Ploiariodes errabundus*: Torre-Bueno, Bull. Brook. Ent. Soc., 24: 34.
Distribution: Fla., Ill., Ks., Mass., Md., Me., Mich., Mo., N.H., N.Y., Ont., Ore., Pa., Que., Tex., Va., W.Va., Wash. (Jamaica, Mexico).

Empicoris incredibilis Wygodzinsky, 1966

1966 *Empicoris incredibilis* Wygodzinsky, Bull. Am. Mus. Nat. Hist., 133: 375. [Ariz.].
Distribution: Ariz.

Empicoris nudus McAtee and Malloch, 1925

1925 *Empicoris nudus* McAtee and Malloch, Proc. U.S. Nat. Mus., 67(1): 22. [Fla.].
Distribution: Fla.

Empicoris orthoneuron McAtee and Malloch, 1925

1925 *Empicoris orthoneuron* McAtee and Malloch, Proc. U.S. Nat. Mus., 67(1): 18. [Cal.].
1925 *Empicoris reticulatus* McAtee and Malloch, Proc. U.S. Nat. Mus., 67(1): 20. [Holotype from Mexico; also from U.S.]. Synonymized by Wygodzinsky, 1966, Bull. Am. Mus. Nat. Hist., 133: 381.
Distribution: Ariz., B.C., Cal., Mass., Md., Miss., Nev., Ore., Tex., Va. (Mexico to Argentina).

Empicoris palmensis Blatchley, 1926

1926 *Empicoris palmensis* Blatchley, Het. E. N. Am., p. 522. [Fla.].
Distribution: Fla.

Empicoris parshleyi (Bergroth), 1922

1922 *Ploeariola parshleyi* Bergroth, Not. Ent., 2: 79. [Va.].
1925 *Empicoris parshleyi*: McAtee and Malloch, Proc. U.S. Nat. Mus., 67(1): 22.
Distribution: Mass., Md., Mich., N.H., Va.

Empicoris pilosus (Fieber), 1861

1861 *Ploearia* [*sic*] *pilosus* Fieber, Europ. Hem., p. 149. [Germany].
1912 *Ploiariodes hirtipes* Banks, Psyche, 19: 97. [Vt.]. Synonymized by McAtee and Malloch, 1922, Proc. Biol. Soc. Wash., 35: 95.
1916 *Ploiariola hirtipes*: Van Duzee, Check List Hem., p. 27.
1925 *Empicoris vagabundus* var. *pilosus*: McAtee and Malloch, Proc. U.S. Nat. Mus., 67(1): 18.
1926 *Empicoris pilosus*: Blatchley, Het. E. N. Am., p. 523.
Distribution: B.C., Mass., Mich., Pa., Que., Vt., Wis. (Europe).

Empicoris rubromaculatus (Blackburn), 1889

1889 *Ploiariodes rubromaculatus* Blackburn, Proc. Linn. Soc. New South Wales, ser. 2, 3: 349. [Hawaii].
1909 *Ploiariodes californica* Banks, Psyche, 16: 46. [Cal.]. Synonymized by McAtee and Malloch, 1922, Proc. Biol. Soc. Wash., 35: 95.
1925 *Empicoris rubromaculatus*: McAtee and Malloch, Proc. U.S. Nat. Mus. 67(1): 16.
Distribution: B.C., Cal., Fla., Ind., Miss., N.C., Tex., Va. (Cosmopolitan).

Empicoris subparallelus McAtee and Malloch, 1925

1925 *Empicoris subparallelus* McAtee and Malloch, Proc. U.S. Nat. Mus., 67(1): 21. [Holotype from Cuba; also from Tex.].

Distribution: Tex. (Cuba).

Empicoris vagabundus (Linnaeus), 1758

1758 *Cimex vagabundus* Linnaeus, Syst. Nat., 10: 450. [Europe].

1919 *Ploiariola canadensis* Parshley, Occas. Pap. Mus. Zool., Univ. Mich., 71: 25. [B.C.]. Synonymized by McAtee and Malloch, 1922, Proc. Biol. Soc. Wash., 35: 95.

1925 *Empicoris vagabundus vagabundus*: McAtee and Malloch, Proc. U.S. Nat. Mus., 67(1): 18.

1929 *Ploearia* [*sic*] *canadensis*: China and Myers, An. Mag. Nat. Hist., ser. 10, 3: 123.

1966 *Empicoris vagabundus*: Wygodzinsky, Bull. Am. Mus. Nat. Hist., 133: 385.

Distribution: B.C., D.C., Wash. (Europe).

Empicoris winnemana McAtee and Malloch, 1925

1925 *Empicoris winnemana* McAtee and Malloch, Proc. U.S. Nat. Mus., 67(1): 19. [Md.].

Distribution: Conn., Md., Va.

Subfamily Harpactorinae Amyot and Serville, 1843

Genus *Acholla* Stål, 1862

1862 *Acholla* Stål, Stett. Ent. Zeit., 23: 445. Type-species: *Reduvius sexspinosus* Wolff, 1802, a junior synonym of *Cimex multispinosus* De Geer, 1773. Monotypic.

Acholla ampliata Stål, 1872

1872 *Acholla ampliata* Stål, K. Svens. Vet.-Akad. Handl., 10(4): 72. [Mexico].

1904 *Acholla ampliata*: Uhler, Proc. U.S. Nat. Mus., 27: 364.

Distribution: Ariz., N.M., Ut. (Mexico).

Acholla multispinosa (De Geer), 1773

1773 *Cimex multispinosus* De Geer, Memoirs, 3: 348. [Pa.].

1802 *Reduvius sexspinosus* Wolff, Icon. Cimic., 3: 124. [N. Am.]. Synonymized by Stål, 1872, K. Svens. Vet.-Akad. Handl., 10(4): 72.

1846 *Harpactor subarmatus* Herrich-Schaeffer, Wanz. Ins., 8: 83. [N. Am.]. Synonymized by Stål, 1872, K. Svens, Vet.-Akad. Handl., 10(4): 72.

1854 *Reduvius* species Emmons, Agr. Nat. Hist. N.Y., 5: 168. [N.Y.].

1854 *Sinea stimulatrix* Emmons, Agr. Nat. Hist. N.Y., pl. 29, fig. 8. [N.Y.]. Synonymized by Van Duzee, 1916, Check List. Hem., p. 31.

1872 *Acholla multispinosa*: Stål, K. Svens. Vet.-Akad. Handl., 10(4): 72.

1873 *Sinea sexspinosa*: Walker, Cat. Hem. Brit. Mus., 8: 140.

Distribution: Ariz., Col., Ia., Ill., Ind., Ks., Mass., Me., Mich., Mo., N.C., N.J., N.Y., Neb., Oh., Ont., S.D., Pa., Tex., Va., W.Va., Wis.

Acholla tabida (Stål), 1862

1862 *Ascra tabida* Stål, Stett. Ent. Zeit., 23: 446. [Mexico].

1876 *Acholla tabida*: Uhler, Bull. U.S. Geol. Geogr. Surv. Terr., 1: 326.
Distribution: Cal. (Mexico).

Genus *Arilus* Hahn, 1831

1831 *Arilus* Hahn, Wanz. Ins., 1: 33. Type-species: *Cimex serratus* Fabricius, 1775, a junior synonym of *Cimex carinatus* Forster, 1771. Monotypic.

Arilus cristatus (Linnaeus), 1763 [Fig. 134]

1763 *Cimex cristatus* Linnaeus, Cent. Ins. Rar., p. 16. ["Carolina"].
1825 *Reduvius novenarius* Say, Am. Ent., 2: 31. ["Pennsylvania to the southern boundary" of the U. S. A.]. Synonymized by Stål, 1872, K. Svens. Vet.-Akad. Handl., 10(4): 72.
1832 *Nabis novenarius*: Say, Descrip. Hem. Het. N. Am., p. 33.
1837 *Arilus denticulatus* Westwood, *In* Drury, Illus. Exot. Ins., edition 2, p. 73. [Pa.]. Synonymized by Uhler, 1878, Proc. Bost. Soc. Nat. Hist., 19: 426.
1859 *Prionotus novenarius*: Stål, Öfv. K. Svens. Vet.-Akad. Förh., 16: 196.
1872 *Prionotus cristatus*: Stål, K. Svens. Vet.-Akad. Handl., 10(4): 72.
1899 *Arilus cristatus*: Champion, Biol. Centr.-Am., Rhyn., 2: 288.
Distribution: D.C., Del., Fla., Ia., Ill., Ind., Ks., Md., Mo., N.C., N.J., N.M., N.Y., Ok., Ont., Pa., S.C., Tex. (Guatemala, Mexico).

Genus *Atrachelus* Amyot and Serville, 1843

1843 *Atrachelus* Amyot and Serville, Hist. Nat. Hem., p. 374. Type-species: *Atrachelus heterogeneus* Amyot and Serville, 1843, a junior synonym of *Reduvius cinereus* Fabricius, 1796. Monotypic.

Note: Elkins (1954, Proc. Ent. Soc. Wash., 56: 97-120) presented a "Synopsis of *Atrachelus*" with a key to species.

Subgenus *Atrachelus* Amyot and Serville, 1843

1843 *Atrachelus* Amyot and Serville, Hist. Nat. Hem., p. 374. Type-species: *Atrachelus heterogeneus* Amyot and Serville, 1843, a junior synonym of *Reduvious cinereus* Fabricius, 1796. Monotypic.

Atrachelus cinereus (Fabricius), 1798

1798 *Reduvius cinereus* Fabricius, Ent. Syst., Suppl., p. 545. ["Carolina"].
1872 *Atrachelus cinereus*: Stål, K. Svens. Vet.-Akad. Handl., 10(4): 78.
Distribution: Ala., Ariz., Cal., Fla., Ga., La., Mich., Miss., N.C., N.M., Pa., S.C., Tex. (Cuba, Mexico to Argentina).

Atrachelus cinereus cinereus (Fabricius), 1798

1798 *Reduvius cinereus* Fabricius, Ent. Syst., Suppl., p. 545. ["Carolina"].
1803 *Zelus cinereus*: Fabricius, Syst. Rhyn., p. 287.
1843 *Atrachelus heterogeneus* Amyot and Serville, Hist. Nat. Hem., p. 374. [Pa.]. Synonymized by Stål, 1872, K. Svens. Vet.-Akad. Handl., 10(4): 78.
1859 *Sinea cinerea*: Dohrn, Cat. Hem., p. 46.
1872 *Atrachelus cinereus*: Stål, K. Svens. Vet.-Akad. Handl., 10(4): 78.
1895 *Heza clavata*: Ashmead, Ins. Life, 7: 321.
1954 *Atrachelus* (*Atrachelus*) *cinereus cinereus*: Elkins, Proc. Ent. Soc. Wash., 56: 101.

Distribution: Ala., Fla., Ga., La., Mich., Miss., Mo., N.C., Pa., S.C., Tex. (Cuba, Guatemala, Mexico).

Note: Van Duzee (1917, Univ. Cal. Publ. Ent., 2: 267) commented that Ashmead's (1895) record of *H. clavata* probably was a "wrong determination of *Atrachelus cinereus* or an allied form."

Atrachelus cinereus wygodzinskyi Elkins, 1954

1954 *Atrachelus cinereus wygodzinskyi* Elkins, Proc. Ent. Soc. Wash., 56: 104. [Ariz.].

Distribution: Ariz., Cal., N.M., Tex. (Mexico).

Subgenus *Phorobura* Stål, 1859

1859 *Phorobura* Stål, Öfv. K. Svens. Vet.-Akad. Förh., 16: 368. Type-species: *Phorobura ignobilis* Stål, 1859. Designated by Wygodzinsky, 1949, Univ. Nac. Tucuman, Monogr., 16: 6.

Atrachelus mucosa (Champion), 1899

1899 *Repipta mucosa* Champion, Biol. Centr.-Am., Rhyn., 2: 271. [Panama].

1916 *Repipta mucosa*: Van Duzee, Check List Hem., p. 31.

1956 *Atrachelus* (*Phorobura*) *mucosa*: Elkins, Fla. Ent., 39: 44.

Distribution: Tex. (Mexico to Panama).

Genus *Castolus* Stål, 1858

1858 *Castolus* Stål, Öfv. K. Svens. Vet.-Akad. Förh., 15: 447. Type-species: *Castolus plagiaticollis* Stål, 1858. Monotypic.

Note: Maldonado (1976, Proc. Ent. Soc. Wash., 78: 436-437) provided a key to the species of *Castolus*.

Castolus ferox (Banks), 1910

1910 *Zelus ferox* Banks, Ent. News, 21: 325. [Ariz.].

1913 *Castolus ferox*: Bergroth, Ent. News, 24: 265.

Distribution: Ariz., Tex.

Castolus subinermis (Stål), 1862

1862 *Repipta subinermis* Stål, Stett. Ent. Zeit., 23: 447. [Mexico].

1906 *Castolus subinermis*: Snow, Trans. Ks. Acad. Sci., 20: 180.

Distribution: Ariz. (Mexico).

Genus *Doldina* Stål, 1859

1859 *Doldina* Stål, Öfv. K. Svens. Vet.-Akad. Förh., 16: 366, 368. Type-species: *Doldina carinulata* Stål, 1859. Monotypic.

Note: Hussey and Elkins (1956, Quart. J. Fla. Acad. Sci., 18: 261-278) reviewed this genus and gave a key to the species.

Doldina interjungens Bergroth, 1913

1910 *Hygromystes* species: Torre-Bueno and Engelhardt, Can. Ent., 42: 150.

1913 *Doldina interjungens* Bergroth, Ent. News, 24: 263. [N.C.].

1913 *Doldina praetermissa* Bergroth, Ent. News, 24: 264. [Fla., British Honduras]. Synonymized by Hussey and Elkins, 1956, Quart. J. Fla. Acad. Sci., 18: 273.

Distribution: Fla., Md., N.C., Tex. (Middle America).

Genus *Fitchia* Stål, 1859

1859 *Fitchia* Stål, Öfv. K. Svens. Vet.-Akad. Förh., 16: 367, 370. Type-species: *Fitchia aptera* Stål, 1859. Monotypic.

Fitchia aptera Stål, 1859

1859 *Fitchia aptera* Stål, Öfv. K. Svens. Vet.-Akad. Förh., 16: 371. [N. Am.].

1866 *Fitchia nigrovittata* Stål, Öfv. K. Svens. Vet.-Akad. Förh., 23: 296. Unnecessary new name for *Fitchia aptera* Stål.

Distribution: Alta., Col., D.C., Ill., Ind., Ks., Mass., Me., Mo., N.C., N.J., N.Y., Ok., Pa., Que., S.C., S.D., Tex., Ut.

Fitchia spinosula Stål, 1872

1872 *Fitchia spinosula* Stål, K. Svens. Vet.-Akad. Handl. 10(4): 79. [Tex.].

Distribution: Col., Fla., Ga., N.C., Tex.

Genus *Heza* Amyot and Serville, 1843

1843 *Heza* Amyot and Serville, Hist. Nat. Hem., p. 374. Type-species: *Reduvius binotatus* Lepeletier and Serville, 1825. Monotypic.

Note: Maldonado (1976, J. Agr. Univ. Puerto Rico, 60: 403-433) revised the genus.

Heza similis Stål, 1859

1859 *Heza similis* Stål, Öfv. K. Svens. Vet.-Akad. Förh., 16: 199. [Colombia].

1914 *Heza similis*: Van Duzee, Trans. San Diego Soc. Nat. Hist., 2: 13.

Distribution: Cal., Fla., Tex. (Greater Antilles, Mexico to Paraguay).

Genus *Pselliopus* Bergroth, 1905

1862 *Milyas* Stål, K. Svens. Vet.-Akad., Handl., 3(6): 61. Preoccupied. Type-species: *Hiranetis ornaticeps* Stål, 1862. Monotypic.

1905 *Pselliopus* Bergroth, Rev. d'Ent., 24: 112. New name for *Milyas* Stål.

Note: A Key to the U.S. species was given by Barber (1924, Proc Ent. Soc. Wash., 26: 212-213).

Pselliopus barberi Davis, 1912 [Fig. 131]

1912 *Pselliopus barberi* Davis, Psyche, 19: 21. [Md., Mo., Va., Tex.].

Distribution: Ill., Ks., La., Md., Mo., N.C., Oh., Ok., S.C., Tex., Va.

Pselliopus cinctus (Fabricius), 1776

1776 *Reduvius cinctus* Fabricius, Gen. Ins., p. 302. [N. Am.].

1790 *Cimex praecinctus* Gmelin, Syst. Nat., edition 13, 1(4): 2198. [N. Am.]. Synonymized by Stål, 1872, K. Svens. Vet.-Akad. Handl., 10(4): 87.

1846 *Harpactor cincta*: Herrich-Schaeffer, Wanz. Ins., 8: 83.

1859 *Sinea cincta*: Dohrn, Cat. Hem., p. 46.

1868 *Milyas cinctus*: Stål, K. Svens. Vet.-Akad. Handl., 7(11): 106.

1873 *Harpactor cinctus*: Walker, Cat. Hem. Brit. Mus., 8: 97.

1891 *Milyas zebra*: Townsend, Proc. Ent. Soc. Wash., 2: 55.

1909 *Pselliopus cinctus*: Van Duzee, Bull. Buff. Soc. Nat. Hist., 9: 177.

Distribution: Col., Conn., Fla., Ga., Ill., Ind., Ks., Mass., Md., Mich., Mo., N.C., N.J., N.Y., Oh., Ok., Pa., R.I., Tex., Va., Wis., Wyo.

Note: Yonke and Medler (1970, J. Ks. Ent. Soc., 43: 441) reported the tachinid fly *Xanthomelanodes arcuata* (Say) hatching from a fifth-instar nymph of *P. cinctus*.

Pselliopus inermis (Champion), 1899

1899 *Milyas inermis* Champion, Biol. Centr.-Amer., Rhyn., 2: 246. [Mexico].
1910 *Milyas inermis*: Barber, J. N.Y. Ent. Soc., 18: 39.
1916 *Pselliopus inermis*: Van Duzee, Check List Hem., p. 30.
Distribution: Ariz. (Mexico).

Pselliopus latifasciatus Barber, 1924

1924 *Pselliopus latifasciatus* Barber, Proc. Ent. Soc. Wash., 26: 211. [La.].
Distribution: Col., Fla., Md., La., Mich., Mo., Ok., Tex., Va.

Pselliopus spinicollis (Champion), 1899

1899 *Milyas spinicollis* Champion, Biol. Centr.-Am., Rhyn., 2: 245. [Mexico].
1908 *Milyas spinicollis*: Barber, J. N.Y. Ent. Soc., 16: 193.
1913 *Pselliopus spinicollis*: Fracker, Proc. Ia. Acad. Sci., 19: 238.
Distribution: Ariz., B.C., Cal., Col., Id. (Mexico).

Pselliopus zebra (Stål), 1862

1862 *Milyas zebra* Stål, Stett. Ent. Zeit., 23: 448. [Mexico].
1906 *Milyas zebra*: Snow, Trans. Ks. Acad. Sci., 20: 180
1913 *Pselliopus zebra*: Fracker, Proc. Ia. Acad. Sci., 19: 237.
Distribution: Ariz., Cal. (Mexico).

Genus *Repipta* Stål, 1859

1859 *Repipta* Stål, Öfv. K. Svens. Vet.-Akad. Förh., 16: 369. Type-species: *Zelus taurus* Fabricius, 1803. Designated by Van Duzee, 1916, Check List Hem., p. 31.

Repipta flavicans (Amyot and Serville), 1843

1843 *Zelus flavicans* Amyot and Serville, Hist. Nat. Hem., p. 374. [French Guiana].
1916 *Repipta flavicans*: Van Duzee, Check List Hem., p. 31.
Distribution: Tex. (Mexico to Argentina).
Note: Fracker's (1913, Proc. Ia. Acad. Sci., 19: 242) Illinois record of this species appears to be a typographical error and the sentence "Mr. Hart has found this species in Illinois" should undoubtedly be shifted to the paragraph dealing with *R. taurus*.

Repipta taurus (Fabricius), 1803

1803 *Zelus taurus* Fabricius, Syst. Rhyn., p. 291. ["Carolina"].
1843 *Zelus lineatus* Amyot and Serville, Hist. Nat. Hem., p. 373. ["Pa."]. Synonymized by Stål, 1872, K. Svens. Vet.-Akad. Handl., 10(4): 80.
1872 *Repipta taurus*: Stål, K. Svens. Vet.-Akad. Handl., 10(4): 80.
1913 *Repipta flavicans*: Fracker, Proc. Ia. Acad. Sci., 19: 242.
Distribution: "Carolina," Col., Fla., Ill., La., Pa., Tex. (Mexico to Panama).

Genus *Rhynocoris* Hahn, 1834

1834 *Rhynocoris* Hahn, Wanz. Ins., 2: 20. Type-species: *Reduvius cruentus* Fabricius, 1787, a junior synonym of *Cimex iracundus* Poda, 1761. Designated by Kirkaldy, 1900, Ent., 33: 242.

Rhynocoris leucospilus (Stål), 1859

1859 *Reduvius leucospilus* Stål, Öfv. K. Svens. Vet.-Akad. Förh., 16: 203. [Siberia].
1896 *Harpactor leucospilus*: Lethierry and Severin, Cat. Gen. Hem., 3: 160.
1916 *Rhynocoris leucospilus*: Van Duzee, Check List Hem., p. 30.

Distribution: Alk. (Palearctic Asia, Transcaucasia).

Rhynocoris ventralis (Say), 1832
1832 *Reduvius ventralis* Say, Descrip. Het. Hem. N. Am., p. 31. [Mo.].
1914 *Rhynocoris ventralis*: Van Duzee, Trans. San Diego Soc. Nat. Hist., 2: 13.
Distribution: Alta., B.C., Cal., Col., Ill., Ind., Mass., Me., Mo., N.D., Neb., Ok., Sask., Ut.

Rhynocoris ventralis americanus (Bergroth), 1897
1897 *Harpactor americanus* Bergroth, Ent. News, 8: 96. [Cal.].
1914 *Rhynocoris ventralis americanus*: Van Duzee, Trans. San Diego Soc. Nat. Hist., 2: 13.
Distribution: Cal.

Rhynocoris ventralis annulipes Van Duzee, 1914
1914 *Rhynocoris ventralis annulipes* Van Duzee, Trans. San Diego Soc. Nat. Hist., 2: 13. [Cal.].
Distribution: Cal.

Rhynocoris ventralis femoralis Van Duzee, 1914
1914 *Rhynocoris ventralis femoralis* Van Duzee, Trans. San Diego Soc. Nat. Hist., 2: 13. [Cal.].
Distribution: Cal.

Rhynocoris ventralis ventralis (Say), 1832
1832 *Reduvius ventralis* Say, Descrip. Het. Hem. N. Am., p. 31. [Mo.].
1876 *Apiomerus ventralis*: Uhler, Bull. U.S. Geol. Geogr. Surv. Terr., 5: 328.
1914 *Rhynocoris ventralis ventralis*: Van Duzee, Trans San Diego Soc. Nat. Hist., 2: 13.
Distribution: Alta., B.C., Cal., Col., Ill., Ind., Mass., Me., Mo., N.D., Neb., Sask., Ut.

Genus *Rocconota* Stål, 1859

1859 *Rocconota* Stål, Öfv. K. Svens. Vet.-Akad. Förh., 16: 366, 370. Type-species: *Rocconota sextuberculata* Stål, 1859. Designated by Van Duzee, 1916, Check List Hem., p. 31.

Rocconota annulicornis (Stål), 1872
1872 *Heza annulicornis* Stål, K. Svens. Vet.-Akad. Handl., 10(4): 77. [Mexico, Tex.].
1899 *Rocconota annulicornis*: Champion, Biol. Centr.-Am., Rhyn., 2: 273.
1906 *Rocconata* [*sic*] *annulicornis*: Snow, Trans. Ks. Acad. Sci., 20: 153.
Distribution: Ala., Fla., Ky., Ind., N.C., N.J., N.Y., Tex. (Mexico).

Genus *Sinea* Amyot and Serville, 1843

1843 *Sinea* Amyot and Serville, Hist. Nat. Hem., p. 375. Type-species: *Sinea multispinosa* Amyot and Serville, 1843, a junior synonym of *Reduvius diadema* Fabricius, 1776. Designated by Caudell, 1901, J. N.Y. Ent. Soc. 9: 2.

Note: Caudell (1901, J. N.Y. Ent. Soc., 9: 1-11) revised and keyed the species of *Sinea*.

Sinea anacantha Hussey, 1953
1953 *Sinea anacantha* Hussey, J. Ks. Ent. Soc., 26: 63. [Ut.].
Distribution: Ariz., Ut.

Sinea complexa Caudell, 1900
1900 *Sinea complexa* Caudell, Can. Ent., 32: 67. [Cal.].

Distribution: Ariz., Cal., Col., Mo., Tex., Ut.
Note: Caudell (1901, J. N.Y. Ent. Soc., 9: 7) rejects accuracy of an "Alabama" label.

Sinea confusa Caudell, 1901
1886 *Sinea integra*: Uhler, Check-list Hem. Het., p. 23.
1901 *Sinea confusa* Caudell, J. N.Y. Ent. Soc., 9: 6. [Ariz.].
Distribution: Ariz., Cal., Tex., Ut. (Mexico and Central America).

Sinea coronata Stål, 1862
1862 *Sinea coronata* Stål, Stett. Ent. Zeit., 23: 444. [Mexico].
1876 *Sinea coronata*: Uhler, Bull. U.S. Geol. Geogr. Surv. Terr., 5: 326.
Distribution: Cal., Tex. (Mexico, Guatemala).

Sinea defecta Stål, 1862
1862 *Sinea defecta* Stål, Stett. Ent. Zeit., 23: 445. [Mexico].
1906 *Sinea defecta*: Snow, Trans. Ks. Acad. Sci., 20: 180.
Distribution: Ariz., Tex. (Mexico to Panama).

Sinea diadema (Fabricius), 1776
1773 *Cimex multispinosus*: De Geer, Mem. Ins., 3: 348 (in part).
1776 *Reduvius diadema* Fabricius, Gen. Ins., p. 302. [N. Am.].
1778 *Cimex diadema*: Goeze, Ent. Beytr., 2: 252.
1783 *Cimex hispidus* Thunberg, Nova Ins. Spec., 2: 33. [India-in error!]. Synonymized by Stål, 1872, K. Svens. Vet.-Akad. Handl., 10(4): 71.
1790 *Cimex setosus*: Gmelin, *In* Linnaeus, Syst. Nat., Edit. 13, 1(4): 2144 (in part).
1803 *Zelus diadema*: Fabricius, Syst. Rhyn., p. 286.
1825 *Reduvius raptatorius* Say, J. Acad. Nat. Sci. Phila., 4: 327. [Mo., Pa.]. Synonymized by Stål, 1872, K. Svens. Vet.-Akad. Handl., 10(4): 71.
1843 *Sinea multispinosa*: Amyot and Serville, Hist. Nat. Hem., p. 375.
1869 *Reduvius multispinosus*: Rathvon, Hist. Lancaster Co., Pa., p. 549.
1872 *Sinea diadema*: Stål, K. Svens. Vet.-Akad. Handl., 10(4): 70.
Distribution: Alta., Ark., B.C., Cal., Col., Conn., D.C., Fla., Ga., Id., Ill., Ind., Ks., Mass., Md., Me., Mich., Mo., N.C., N.D., N.H., N.J., N.S., N.Y., Neb., Nev., Ok., Ont., Pa., Que., R.I., S.C., S.D., Ut., Va., Vt., Wash., Wis.

Sinea raptoria Stål, 1862
1862 *Sinea raptoria* Stål, Stett. Ent. Zeit., 23: 444. [Mexico].
1876 *Sinea raptoria*: Uhler, Bull. U.S. Geol. Geogr. Surv. Terr., 5: 61.
Distribution: Ariz., Cal., Tex. (Mexico to Columbia).

Sinea rileyi Montandon, 1893
1893 *Sinea rileyi* Montandon, Proc. U.S. Nat. Mus., 16: 51. [Cal.].
1893 *Ginea* [*sic*] *rileyi*: Riley, N. Am. Fauna, 7: 250.
Distribution: Ariz., Cal., Fla., N.C., Tex., Ut. (Introduced into England).

Sinea sanguisuga Stål, 1862
1862 *Sinea sanguisuga* Stål, Stett. Ent. Zeit., 23: 444. [Mexico].
1901 *Sinea sanguisuga*: Caudell, J. N.Y. Ent. Soc., 9: 9.
Distribution: Fla., Tex. (Greater Antilles, Mexico, Guatemala).

Sinea spinipes (Herrich-Schaeffer), 1846
1846 *Harpactor spinipes* Herrich-Schaeffer, Wanz. Ins., 8: 82. [South America--probably in error].
1886 *Sinea spinipes*: Uhler, Check-list Hem. Het., p. 23.
Distribution: Ark., Col., Fla., Ga., Ill., Ind., Ks., Ky., La., Mo., N.C., N.Y., Ok., S.D., Tex., Va., Wis. (Mexico).

Sinea undulata Uhler, 1894

1894 *Sinea undulata* Uhler, Proc. Cal. Acad. Sci., series 2, 4: 282. [Cal.].

Distribution: Cal. (Mexico).

Genus *Zelus* Fabricius, 1803

1803 *Zelus* Fabricius, Syst. Rhyn., p. 281. Type-species: *Cimex longipes* Linnaeus, 1758. Designated by Latreille, 1810, Consid. Gén., p. 443.

Note: Hart (1986, An. Ent. Soc. Am., 79: 535-548) presented a revision of the species from northern Mexico northward with a key to the 9 species; he did not use subgeneric categories but his synonymizing of names did not cross the subgenera used earlier.

Subgenus *Diplacodus* Kirkaldy, 1900

1843 *Diplodus* Amyot and Serville, Hist. Nat. Hem., p. 370. Preoccupied. Type-species: *Reduvius armillatus* Lepeletier and Serville, 1825. Designated by Van Duzee, 1916, Check List Hem., p. 90.

1862 *Zelus* (*Diplodus*): Stål, Stett. Ent. Zeit., 23: 450.

1900 *Zelus* (*Diplacodus*): Kirkaldy, Ent., 33: 242.

Zelus exsanguis Stål, 1862

1862 *Zelus* (*Diplodus*) *exsanguis* Stål, Stett. Ent. Zeit., 23: 452. [Mexico].

1986 *Zelus exsanguis*: Hart, An. Ent. Soc. Am., 79: 539.

Distribution: Tex. (Mexico, Central America).

Note: Hart (1986, above) wrote "Almost without exception, past references to any *Z. exsanguis* from the United States are actually to *Z. luridus*." In the present catalog such combinations are placed under *Z. luridus* Stål. West and Delong (1955, Proc. Ent. Soc. Ont., 86: 97-101) gave notes on biology of *Z. exsanguis*. Yonke and Medler (1970, J. Ks. Ent. Soc., 43: 441-443) reported the hatching of a parasitic scelionid wasp *Telenomus* sp. from egg masses of this species.

Zelus nugax Stål, 1862

1862 *Zelus* (*Diplodus*) *nugax* Stål, Stett. Ent. Zeit., 23: 450. [Mexico].

1986 *Zelus nugax*: Hart, An. Ent. Soc. Am., 79: 540.

Distribution: Tex. (Mexico to Paraguay).

Zelus renardii Kolenati, 1856

1856 *Zelus renardii* Kolenati, Bull. Soc. Imp. Nat. Moscou, 29: 460. [Cal.].

1872 *Zelus* (*Diplodus*) *renardii*: Stål, K. Svens. Vet.-Akad. Handl., 10(4): 91.

1876 *Diplodus renardii*: Uhler, Bull. Geol. Geogr. Surv. Terr., 5: 328.

1899 *Zelus* (*Zelus*) *laevicollis* Champion, Biol. Centr.-Am., Rhyn., 2: 260. [Mexico]. Synonymized by Hart, 1986, An. Ent. Soc. Am., 79: 540.

1910 *Zelus laevicollis*: Banks, Cat. Nearc. Hem.-Het., p. 16.

1916 *Zelus* (*Diplocodus* [*sic*]) *laevicollis*: Van Duzee, Check List Hem., p. 30.

1916 *Zelus* (*Diplocodus* [*sic*]) *renardii*: Van Duzee, Check List Hem., p. 30.

Distribution: Cal., Ks., La., Tex. (Mexico to Guatemala, Jamaica; introduced into Midway Atoll, Hawaiian and Philippine Islands, and Samoa).

Subgenus *Pindus* Stål, 1862

1862 *Zelus* (*Pindus*) Stål, Stett. Ent. Zeit., 23: 454. Type-species: *Zelus tetracanthus* Stål, 1862. Monotypic.

Zelus tetracanthus Stål, 1862

1862 *Zelus* (*Pindus*) *tetracanthus* Stål, Stett. Ent. Zeit., 23: 454. [Mexico].
1872 *Pindus socius* Uhler, Prelim. Rept. U.S. Geol. Surv. Mont., p. 420. [Ariz., "Dakota," Ks.]. Synonymized by Hart, 1986, An. Ent. Soc. Am., 79: 536.
1886 *Diplodus socius*: Uhler, Check-list Hem. Het., p. 24.
1910 *Zelus audax* Banks, Ent. News, 21: 325. [N.Y., Va.]. Synonymized by Hart, 1986, An. Ent. Soc. Am., 79: 536.
1913 *Zelus* (*Pindus*) *socius*: Fracker, Proc. Ia. Acad. Sci., 19: 240.
1913 *Zelus* (*Pindus*) *occiduus* Torre-Bueno, Ent. News, 24: 22. [Cal.]. Synonymized by Hart, 1986, An. Ent. Soc. Am., 79: 536.
1915 *Zelus socius*: Torre-Bueno, Ent. News, 26: 277.
1916 *Zelus* (*Pindus*) *audax*: Van Duzee, Check List Hem., p. 30.
1925 *Zelus* (*Pindus*) *angustatus* Hussey, J. N.Y. Ent. Soc., 33: 66. [Fla.]. Synonymized by Hart, 1986, An. Ent. Soc. Am., 79: 536.
1928 *Zelus angustatus*: Blatchley, J. N.Y. Ent. Soc., 36: 6.
1951 *Zelus occiduus*: Elkins, Tex. J. Sci., 3: 410.
1986 *Zelus tetracanthus*: Hart, An. Ent. Soc. Am., 79: 536.
Distribution: Alta., Ariz., B.C., Cal., Col., Conn., Fla., Id., Ill., Ind., Ks., Mass., Me., Mich., Mo., N.C., N.J., N.M., N.Y., Ok., Ont., Que., S.D., Tex., Ut., Va. (Mexico to Paraguay).

Subgenus *Zelus* Fabricius, 1803

1803 *Zelus* Fabricius, Syst. Ent., p. 281. Type-species: *Cimex longipes* Linnaeus, 1758. Designated by Latreille, 1810, Consid. Gén., p. 443.
1862 *Zelus* (*Zelus*): Stål, Stett. Ent. Zeit., 23: 449.

Zelus cervicalis Stål, 1872

1872 *Zelus* (*Zelus*) *cervicalis* Stål, K. Svens. Vet.-Akad. Handl., 10(4): 90. ["Carolina," Tex., Mexico].
1876 *Zelus cervicalis*: Uhler, Bull. U.S. Geol. Geogr. Surv. Terr., 5: 327.
1887 *Evagoras marginata* Provancher, Pet. Faune Ent. Can., 3: 182. [Ont.]. Synonymized by Van Duzee, 1912, Can. Ent., 44: 324.
1899 *Zelus* (*Zelus*) *pictipes* Champion, Biol. Centr.-Am., Rhyn., 2: 255. [Mexico to Colombia]. Synonymized by Hart, 1986, An. Ent. Soc. Am., 79: 542.
1906 *Zelus pictipes*: Snow, Trans. Ks. Acad. Sci., 20: 180.
Distribution: Ariz., Cal., Fla., La., Mo., N.C., Ok., S.C., Tex., Va.
Note: The Ontario specimen on which Provancher (1887, above) based his *Evagoras marginata* record was probably an accidental introduction.

Zelus longipes (Linnaeus), 1767

1767 *Cimex longipes* Linnaeus, Syst. Nat., Edit. duodecima, 1(2): 724. [St. Thomas, West Indies].
1832 *Zelus bilobus* Say, Descrip. Het. Hem. N. Am., p. 2. [Ga., La.]. Synonymized by Hart, 1986, An. Ent. Soc. Am., 79: 543.
1872 *Zelus* (*Zelus*) *longipes*: Stål, K. Svens. Vet.-Akad. Handl., 10(4): 88.
1872 *Zelus* (*Zelus*) *bilobus*: Stål, K. Svens. Vet.-Akad. Handl., 10(4) 88.
1873 *Euagoras longipes*: Walker, Cat. Hem. Brit. Mus., 8: 117.
Distribution: Alta., B.C., Conn., Fla., Ga., La., Mass., Me., Miss., N.C., N.Y., Ont., S.C., Tex., Va. (Mexico to Argentina, West Indies).
Note: Stål (1862, Stett. Ent. Zeit., 23: 449), Walker (1873, above) and Uhler (1878, Proc.

Boston Soc. Nat. Hist., 19: 427) all treated Linnaeus' species as a senior synonym of Say's. Subsequent authors, including Stål (1872, above), generally treated Say's species as distinct from Linnaeus' until Hart (1986, An. Ent. Soc. Am., 79: 543) again treated Say's species as a junior synonym with the explanation "new synonymy."

Zelus luridus Stål, 1862

1854 *Reduvius* species: Emmons, Agr. Nat. Hist. N.Y., 5: 168, pl. 7, fig. 3 [a nymph].

1862 *Zelus luridus* Stål, Stett. Ent. Zeit., 23: 454. ["Carolina"].

1869 *Zelus acanthogonius* Rathvon, Hist. Lancaster Co., Pa., p. 550. *Nomen nudum.* See note below.

1872 *Zelus (Diplodus) luridus*: Stål, K. Svens. Vet.-Akad. Handl., 19(4): 91.

1872 *Darbanus georgiae* Provancher, Nat. Can., 4: 106. [Fla.]. Synonymized by Hart, 1986, An. Ent. Soc. Am., 79: 537.

1873 *Diplodus luridus*: Walker, Cat. Hem. Brit. Mus., 8: 124.

1887 *Darbanus palliatus* Provancher, Pet. Faune Ent. Can., 3: 182. [Ontario]. Synonymized by Hart, 1986, An. Ent. Soc. Am., 79: 537.

1907 *Diplodus exsanguis*: Snow, Trans. Ks. Acad. Sci., 20: 159.

1914 *Diplocodus* [*sic*] *exsanguis*: Van Duzee, Trans. San Diego Soc. Nat. Hist., 2: 13.

1916 *Zelus* (*Diplocodus* [*sic*]) *exsanguis*: Van Duzee, Check List Hem., p. 30.

Distribution: Ala., Ariz., Ark., Cal., Col., Conn., Del., Fla., Ga., Ia., Ill., Ind., Ks., Ky., La., Man., Md., Me., Mich., Minn., Miss., Mo., N.C., N.H., N.J., N.M., N.Y., Neb., Ok., Ont., Pa., Que., R.I., S.C., S.D., Tenn., Tex., Va., Vt., W.Va., Wis., Wyo. (Mexico).

Note: See note under *Zelus exsanguis*. Uhler (1878, Proc. Boston Soc. Nat. Hist., 19: 427) cited the combination "*Zelus acanthogonius*" as being a manuscript name used by Thomas Say in identifying specimens of this species in the Harris Collection. Both of Provancher's (above) species, *Darbanus georgiae* and *D. palliatus*, were first assigned to synonymy under "*Diplocodus luridus*" by Van Duzee (1912, Can. Ent., 44: 324), but were transferred to synonymy under *Zelus exsanguis* when Parshley (1914, Psyche, 21: 144) made that name the senior synonym of *Z. luridus*. Hart (1986, An. Ent. Soc. Am., 79: 537) again raised *Z. luridus* to specific status and noted transfer of those synonyms by the designation "new synonymy."

Subfamily Microtominae Schumacher, 1924

Genus *Homalocoris* Perty, 1833

1833 *Platycoris* Perty, Delec. Anim. Articul., p. 174. Preoccupied. Type-species: *Platycoris varius* Perty, 1833. Monotypic.

1833 *Homalocoris* Perty, Delec. Anim. Articul., p. 216. New name for *Platycoris* Perty.

Homalocoris guttatus (Walker), 1873

1873 *Reduvius guttatus* Walker, Cat. Hem. Brit. Mus., 7: 181. [Mexico].

1910 *Homalocoris guttatus*: Barber, J. N.Y. Ent. Soc., 18: 38.

Distribution: Ariz. (Mexico).

Homalocoris maculicollis Stål, 1872

1872 *Homalocoris maculicollis* Stål, K. Svens. Vet.-Akad. Handl., 10(4): 101. [Mexico].
1886 *Homalocoris maculicollis*: Uhler, Check-list Hem. Het., p. 25.
Distribution: Ariz.(?) (Mexico, Central America).

Homalocoris minutus (Mayr), 1865

1865 *Hammatocerus minutus* Mayr, Ver. Zool.-Bot. Gesell. Wien, 15: 439. [Locality unknown].
1907 *Homalocoris minutus*: Snow, Trans. Ks. Acad. Sci., 20: 160.
Distribution: Ariz.

Genus *Microtomus* Illiger, 1807

1807 *Microtomus* Illiger, Fauna Etrusca, 2: 240. Type-species: *Cimex purcis* Drury, 1782. Monotypic.

Note: Revisions of *Microtomus* provided by Stichel (1926, Deut. Ent. Zeit., pp. 179-190, 1 pl.) and Costa Lima (1935, An. Acad. Brasil. Sci., 7: 315-322, 1 pl.).

Microtomus luctuosus (Stål), 1854

1854 *Hammatocerus luctuosus* Stål, Öfv. K. Svens. Vet.-Akad. Förh., 11(8): 237. [Mexico].
1916 *Hammacerus luctuosus*: Van Duzee, Check List Hem., p. 29.
1906 *Hammatocerus luctuosus*: Barber, Brook. Inst. Arts Sci., Sci. Bull., 1: 285.
1926 *Microtomus luctuosus*: Stichel, Deut. Ent. Zeit., p.: 182.
Distribution: Tex. (Mexico to Panama).

Microtomus purcis (Drury), 1782 [Fig. 137]

1782 *Cimex purcis* Drury, Illust. Exot. Ent., 3: 63. [Va.].
1800 *Reduvius nychthemerus* Illiger, Arch. Zool. Zoot, 1: 147. [Ga.]. Synonymized by Amyot and Serville, 1843, Hist. Nat. Hem., p. 346.
1807 *Microtomus purcis*: Illiger, Fauna Etrusca, 2: 240.
1832 *Nabis purcis*: Say, Descrip. Het. Hem. N. Am., p. 33.
1833 *Hammacerus furcis* [*sic*]: Laporte, Essai Classif. Syst. Hem., p. 79.
1835 *Hammatocerus nychthemerus*: Burmeister, Handb. Ent., 2: 236.
1854 *Hammacerus purcis*: Emmons, Agr. Nat. Hist. N.Y. Rept., 5: 168.
1872 *Hammatocerus purcis*: Stål, K. Svens. Vet.-Akad. Handl., 10(4): 100.
Distribution: Ala., Col., Fla., Ga., Ind., La., N.C., Ok., S.C., Tex., Va.

Subfamily Peiratinae Amyot and Serville, 1843

Notes:The long-used spelling "Piratinae" must give way to the above spelling because, as pointed out by Kerzhner (1974, Ent. Rev., 35: 92), the International Rules of Nomenclature make it necessary to follow the original spelling of the type-genus, *Peirates* Serville.

Accidental transference of certain type in reassembling the galley for Coscaron's article on Peiratinae (1983, Rev. Soc. Ent. Arg., 42: 369-382) resulted in erroneous distribution data. The heading *Melanolestes abdominalis* (Herrich-Schaeffer), Fig. 2 and the first three lines under it should be moved from page 370 to page 372 and inserted ahead of the indented line reading "214; Barber, 1906, p. 284; Banks, 1910, p. Fracker, 1912, p. 233; Wygodzins-." This will correctly eliminate the North American distributional data from under *Tydides rufus* (Serville). Then from page 380 transfer the heading "*Thymbreus crocinopterus* Stål, Fig. 1" into the space on page 370 from which *Melanolestes abdominalis* was removed; and further from page 380 the two lines (lines 10 and 11) of distribution beginning with "Concepcion: Horqueta..." must be transferred to page 372 and added to the end of the distribution data for *Tydides rufus*--after the word "PARAGUAY."

Genus *Melanolestes* Stål, 1866

1866 *Melanolestes* Stål, Öfv. K. Svens. Vet.-Akad. Förh., 23: 251, 259. Type-species: *Pirates* [*sic*] *picipes* Herrich-Schaeffer, 1848. Designated by Van Duzee, 1916, Check List Hem., p. 29.

Melanolestes abdominalis (Herrich-Schaeffer), 1846

1846 *Pirates abdominalis* Herrich-Schaeffer, Wanz. Ins., 8: 63. [N. Am.].

1872 *Melanolestes picipes* variety *abdominalis*: Stål, K. Svens. Vet.-Akad. Handl., 10(4): 107.

1872 *Melanolestes abdominalis*: Uhler, Prelim. Rept. U.S. Geol. Surv. Mont., p.: 421.

1983 *Tydides rufus*: Coscaron, Rev. Soc. Ent. Arg., 42: 372 (in part).

Distribution: Cal., Conn., D.C., Ill., Ind., Mass., Md., Miss., Mo., N.C., N.J., N.M., N.Y., Ok., Pa., R.I., S.D., Tex. (Mexico, Hawaii-introduced).

Note: The above record (Coscaron, 1983) for *Tydides rufus* (Serville) for North America resulted from erroneous placement of a species-heading during conversion from galley proof to page format; *T. rufus* is a species of the American tropics.

Melanolestes morio (Erichson), 1848

1848 *Pirates morio* Erichson, Reisen British-Guiana 1840-1844, p. 613. [Guyana].

1873 *Pirates picipes*: Walker, Cat. Hem. Brit. Mus., 7: 97 (in part).

1983 *Melanolestes morio*: Coscaron, Rev. Soc. Ent. Arg., 42: 374.

Distribution: Fla., Ill. (Greater Antilles, Mexico to Brazil).

Melanolestes picipes (Herrich-Schaeffer), 1846 [Fig. 129]

1846 *Pirates picipes* Herrich-Schaeffer, Wanz. Ins., 8: 62. [N. Am.].

1856 *Reduvius pungens* Leconte, Proc. Acad. Nat. Sci. Phila., 7: 404. [Ga.]. Synonymized by Uhler, 1878, Proc. Bost. Soc. Nat. Hist., 19: 424.

1866 *Melanolestes picipes*: Stål, Öfv. K. Svens. Vet.-Akad. Förh., 23: 259.

Distribution: Ala., Cal., Conn., Del., Fla., Ill., Ind., Ks., La., Mass., Me., Minn., Mo., N.C., N.H., N.J., N.M., N.Y., Ok., Pa., Que., R.I., S.D., Tex., Ut. (Mexico to Brazil).

Genus *Rasahus* Amyot and Serville, 1843

1843 *Rasahus* Amyot and Serville, Hist. Nat. Hem., p. 325. Type-species: *Peirates sulcicollis* Serville, 1831. Designated by Van Duzee, 1916, Check List Hem., p. 29.

Note: Coscaron (1983, Rev. Mus. La Plata, new ser., 13: 75-138, key to 25 species, pp. 79-82) presented a revision of *Rasahus*.

Rasahus biguttatus (Say), 1832

1832 *Petalocheirus biguttatus* Say, Descrip. Het. Hem. N. Am., p. 13. [La.].

1859 *Reduvius mutillarius*: Uhler, *In* Leconte, Say's Compl. Writings, 1: 307.
1866 *Callisphodrus biguttatus*: Stål, Öfv. K. Svens. Vet.-Akad. Förh., 23: 258.
1872 *Rasahus biguttatus*: Stål, K. Svens. Vet.-Akad. Handl., 10(4): 106.
1877 *Rasahus liguttatus* [*sic*]: Uhler, An. Rept. Geogr. Surv. 100th Merid., p. 1330.
Distribution: Ariz., Fla., La., N.C., Tex. (Greater Antilles, Mexico to Brazil).

Rasahus hamatus (Fabricius), 1781
1781 *Reduvius hamatus* Fabricius, Spec. Ins., 2: 381. [Central America].
1869 *Reduvius humeralis*: Rathvon, Hist. Lancaster Co., Pa., p. 550.
1906 *Rasahus hamatus*: Barber, Mus. Brook. Inst. Arts Sci., Sci. Bull. 1: 285.
Distribution: Fla., Mo., Ok., Tex. (Mexico to Argentina, West Indies).

Rasahus scutellaris (Fabricius), 1787
1787 *Reduvius scutellaris* Fabricius, Mant. Ins., 2: 313. [French Guiana].
1983 *Rasahus scutellaris*: Coscaron, Rev. Mus. La Plata, new ser., 13: 111.
Distribution: Tex. (Costa Rica to Brazil).

Rasahus thoracicus Stål, 1872
1872 *Rasahus thoracicus* Stål, K. Svens. Vet.-Akad. Handl., 10(4): 106. [Mexico].
1906 *Rasahus thoracicus*: Snow, Trans. Ks. Acad. Sci., 20: 180.
Distribution: Ariz., Cal., Nev., Tex. (Greater Antilles, Mexico).

Genus *Sirthenea* Spinola, 1837

1837 *Sirthenea* Spinola, Essai Hem., p. 100. Type-species: *Reduvius carinatus* Fabricius, 1798. Monotypic.

Sirthenea carinata (Fabricius), 1798
1798 *Reduvius carinatus* Fabricius, Ent. Sys. Suppl. p. 545. ["Carolina"].
1831 *Peirates carinatus*: Serville, An. Sci. Nat., 23: 221.
1837 *Sirthenea carinata*: Spinola, Essai Hem., p. 100.
1843 *Rasahus carinatus*: Amyot and Serville, Hist. Nat., Hem., p. 326.
1873 *Pirates* [*sic*] *carinatus*: Walker, Cat. Hem. Brit. Mus., 7: 97.
1899 *Sirthenea stria*: Champion, Biol. Centr.-Am., Rhyn., 2: 220 (in part).
Distribution: Fla., Ga., Ill., Ind., La., Mich., Mo., N.C., N.J., Oh., S.C., Tex., Va. (Mexico to Brazil).
Note: All North American records of *S. stria* (Fabricius) can be traced to those made when *S. carinata* was considered a synonym of it.

Subfamily Reduviinae Latreille, 1807

Note: A key to the American genera of Reduviinae was given by Lent and Wygodzinsky (1947, Rev. Brasil. Biol., 7: 342-343).

Genus *Pseudozelurus* Lent and Wygodzinsky, 1947

1947 *Pseudozelurus* Lent and Wygodzinsky, Rev. Brasil. Biol., 7: 344. Type-species: *Spiniger arizonicus* Banks, 1910. Original designation.

Pseudozelurus arizonicus (Banks), 1910
1910 *Spiniger arizonicus* Banks, Ent. News, 21: 324. [Ariz.].

1947 *Pseudozelurus arizonicus*: Lent and Wygodzinsky, Rev. Brasil. Biol., 7: 347.
Distribution: Ariz.

Genus *Reduvius* Fabricius, 1775

1775 *Reduvius* Fabricius, Syst. Ent., p. 729. Type-species: *Cimex personatus* Linnaeus, 1758. Designated by Laporte, 1832, Mag. Zool., p. 7.

Note: Wygodzinsky and Usinger (1964, Am. Mus. Novit., 2175: 4-5) presented a key to the species of *Reduvius* in North America.

Reduvius personatus (Linnaeus), 1758

1758 *Cimex personatus* Linnaeus, Syst. Nat., ed. 10, 1: 446. [Europe].

1855 *Reduvius pungens* Leconte, Proc. Acad. Sci. Phila., 7: 404. [Ga.]. Synonymized by Uhler, 1878, Proc. Bost. Soc. Nat. Hist., 19: 422.

1869 *Reduvius personatus*: Rathvon, Hist. Lancaster Co., Pa., p. 549.

1872 *Reduvius albosignatus* Provancher, Nat. Can., 4: 105. [Ga.]. Synonymized by Provancher, 1887, Pet. Faune Ent. Can., 3: 184.

1872 *Opsicoetus pungens*: Stål, K. Svens. Vet.-Akad. Handl., 10(4): 119.

1887 *Opsicoetus personatus*: Provancher, Nat. Can., 3: 184.

Distribution: Ala., B.C., Conn., Fla., Ill., Ind., Ks., Mass., Md., Me., Mich., Mo., N.C., N.H., N.J., N.M., N.Y., Ok., Ont., Pa., Que., S.D., Tex., Ut., Vt. (Cosmopolitan).

Reduvius senilis Van Duzee, 1906

1906 *Reduvius* (*Opsicoetus*) *senilis* Van Duzee, Ent. News, 17: 390. [Ariz.].

1916 *Reduvius senilis*: Van Duzee, Check List Hem., p. 29.

1927 *Reduvius senilus* [*sic*]: Readio, Univ. Ks. Sci. Bull., 17: 112.

Distribution: Ariz., Tex., Ut. (Mexico).

Note: There appear to be no valid Cal. records subsequent to Wygodzinsky and Usinger's (1964, see below) transfer of Usinger's (1933, see below) record of *R. senilis* to *R. vanduzeei*.

Reduvius sonoraensis Usinger, 1942

1942 *Reduvius sonoraensis* Usinger, Ent. News, 53: 198. [Mexico].

1964 *Reduvius sonoraensis*: Wygodzinsky and Usinger, Am. Mus. Novit., 2175: 5.

Distribution: Ariz., Cal., Tex. (Mexico).

Reduvius vanduzeei Wygodzinsky and Usinger, 1964

1933 *Reduvius senilis*: Usinger, Pan-Pac. Ent., 9: 171 (in part).

1954 *Reduvius senilus* [*sic*]: Ryckman, Bull. So. Cal. Acad. Sci., 53: 88.

1964 *Reduvius vanduzeei* Wygodzinsky and Usinger, Am. Mus. Novit., 2175: 11. [Ut.].

Distribution: Cal., Tex., Ut.

Genus *Zeluroides* Lent and Wygodzinsky, 1948

1948 *Zeluroides* Lent and Wygodzinsky, Rev. Brasil. Biol., 8: 49. Type-species: *Zeluroides mexicanus* Lent and Wygodzinsky, 1948. Original designation.

Note: Lent and Wygodzinsky (1959, Rev. Brasil. Biol., 19: 360) offered a key (in Spanish) to the species and subspecies of *Zeluroides*.

Zeluroides americanus Lent and Wygodzinsky, 1948

1948 *Zeluroides americanus* Lent and Wygodzinsky, Rev. Brasil. Biol., 8: 53. [Ariz.].

Distribution: Ariz. (Mexico).

Zeluroides americanus americanus Lent and Wygodzinsky, 1948

1948 *Zeluroides americanus* Lent and Wygodzinsky, Rev. Brasil. Biol., 8: 53. [Ariz.].
1959 *Zeluroides americanus americanus* Lent and Wygodzinsky, Rev. Brasil. Biol., 8: 53.
Distribution: Ariz.

Genus *Zelurus* Hahn, 1826

1826 *Zelurus* Hahn, Icones Monogr. Cimic., pl. 6. Type-species: *Zelurus ocellatus* Hahn, 1826, a junior synonym of *Reduvius eburneus* Lepeletier and Serville, 1825. Monotypic.

Zelurus bicolor (Stål), 1859

1859 *Spiniger bicolor* Stål, Stett. Ent. Zeit., 20: 396. [Brazil].
1910 *Spiniger bicolor*: Banks, Cat. Nearc. Hem.-Het., p. 18.
1951 *Zelurus bicolor*: Elkins, Tex. J. Sci., 3: 411.
Distribution: Ariz., Tex. (Brazil).

Subfamily Saicinae

Note: The generic and specific names "*Saicodes annulatus* Uhl." reported from the western states by Uhler (1886, Check-list Hem. Het., p. 26) are *nomina nuda*. McAtee and Malloch (1923, An. Ent. Soc. Am., 16: 253) reported Uhler's only specimen associated with this name was actually from Cuba. Maldonado (1981, J. Agr. Univ. Puerto Rico, 65: 406-407) provided a key to the six American genera of Saicinae.

Genus *Oncerotrachelus* Stål, 1868

1868 *Oncerotrachelus* Stål, K. Svens. Vet.-Akad. Handl., 7(11): 130. Type-species *Reduvius acuminatus* Say, 1832. Monotypic.

Oncerotrachelus acuminatus (Say), 1832 [Fig. 132]

1832 *Reduvius acuminatus* Say, Descrip. Het. Hem. N. Am., p. 32. [Ind.].
1872 *Oncerotrachelus acuminatus*: Stål, K. Svens. Vet.-Akad. Handl., 10(4): 124.
Distribution: Ala., D.C., Del., Fla., Ill., Ind., Ks., Mass., Md., Minn., Mo., N.C., N.J., N.Y., Ok., Pa., S.C., Tex. (Greater Antilles).

Oncerotrachelus pallidus Barber, 1922

1922 *Oncerotrachelus pallidus* Barber, Proc. Ent. Soc. Wash., 24: 104. [Tex.].
Distribution: Ok., Tex.

Genus *Saica* Amyot and Serville, 1843

1843 *Saica* Amyot and Serville, Hist. Nat., Hem., p. 371. Type-species: *Saica rubella* Amyot and Serville, 1843, a junior synonym of *Zelus recurvatus* Fabricius, 1803. Monotypic.

Saica apicalis Osborn and Drake, 1915

1915 *Saica apicalis* Osborn and Drake, Oh. Nat., 15: 530. [Guatemala].
1951 *Saica apicalis*: Elkins, Tex. J. Sci., 3: 411.
Distribution: Tex. (Guatemala, Panama).

Saica florida Barber, 1914

1914 *Saica fusco-vittata* [*sic*] Barber, Bull. Am. Mus. Nat. Hist., 33: 504. [Fla.]. Preoccupied.

1953 *Saica florida* Barber, Proc. Ent. Soc. Wash., 55: 142. New name for *Saica fuscovittata* Barber.

Distribution: Fla., Va.

Subfamily Stenopodainae Amyot and Serville, 1843

Note: The New World species of this subfamily were revised by Barber (1930, Ent. Am., 10: 149-238). The above spelling of the subfamily name follows Opinion 868 (1969) of the International Commission on Zoological Nomenclature which proposed it to solve a conflict of the hemipteran Stenopodinae with the crustacean family name Stenopodidae.

Genus *Ctenotrachelus* Stål, 1868

1868 *Ctenotrachelus* Stål, K. Svens. Vet.-Akad. Handl., 7(11): 127. Type-species: *Ctenotrachelus macilentus* Stål, 1872. Monotypic.

Ctenotrachelus mexicanus (Champion), 1898

1898 *Schumannia mexicana* Champion, Biol. Centr.-Am., Rhyn. 2: 185. [Mexico].

1951 *Ctenotrachelus mexicanus*: Elkins, Tex. J. Sci., 3: 411 (in part).

Distribution: Tex. (Mexico to Bolivia).

Ctenotrachelus shermani Barber, 1930

1904 *Schumannia mexicana*: Heidemann, Proc. Ent. Soc. Wash., 6: 12.

1930 *Ctenotrachelus shermani* Barber, Ent. Am., 10: 185. [N.C.].

1951 *Ctenotrachelus mexicanus*: Elkins, Texas J. Sci., 3: 411 (in part).

Distribution: N.C.

Genus *Diaditus* Stål, 1859

1859 *Diaditus* Stål, Öfv. K. Svens. Vet.-Akad. Förh., 16: 383. Type-species: *Diaditus semicolon* Stål, 1859. Monotypic.

Note: Giacchi (1982, Physis, Secc. C., 41(100): 9-27) revised this genus and provided a key to species.

Diaditus pictipes Champion, 1898

1898 *Diaditus pictipes* Champion, Biol. Centr.-Am., Rhyn., 2: 189. [Guatemala, Mexico].

1903 *Diaditus pictipes*: Heidemann, Proc. Ent. Soc. Wash., 5: 102.

Distribution: Tex. (Mexico, Guatemala).

Diaditus tejanus Giacchi, 1980

1980 *Diaditus tejanus* Giacchi, Rev. Soc. Ent. Arg., 39: 1. [Tex.].

Distribution: Tex. (Mexico).

Genus *Gnathobleda* Stål, 1859

1859 *Gnathobleda* Stål, Öfv. K. Svens. Vet.-Akad. Förh., 16: 380. Type-species: *Gnathobleda fraudulenta* Stål, 1859. Monotypic.
Note: Giacchi (1977, Physis, 37 Secc. C., (93): 261-267) revised this genus and provided a key to species.

Gnathobleda litigiosa Stål, 1862
1862 *Gnathobleda litigiosa* Stål, Stett. Ent. Zeit., 23: 442. [Mexico].
1872 *Gnathobleda tumidula* Stål, K. Svens. Vet.-Akad. Handl., 10(4): 121. [Tex., Cuba]. Synonymized by Barber, 1930, Ent. Am., 10: 179.
Distribution: Fla., Ga., Tex. (Greater Antilles, Mexico).

Genus *Narvesus* Stål, 1859

1859 *Narvesus* Stål, Öfv. K. Svens. Vet.-Akad. Förh., 16: 383, 384. Type-species: *Narvesus carolinensis* Stål, 1859. Monotypic.

Narvesus carolinensis Stål, 1859 [Fig. 136]
1859 *Narvesus carolinensis* Stål, Öfv. K. Svens. Vet.-Akad. Förh., 16: 385. ["Carolina"].
1873 *Stenopoda carolinensis*: Walker, Cat. Hem. Brit. Mus., 8: 31.
Distribution: Ariz., Fla., Ill., Mo., N.C., N.J., Ok., S.C., Tex., Va. (Antilles, Mexico, Nicaragua).

Genus *Oncocephalus* Klug, 1830

1830 *Reduvius* (*Oncocephalus*) Klug, Symb. Physicae, Ins., 2: pl. 19, figs. 1, 2. Type-species: *Reduvius notatus* Klug, 1830. Designated by Distant, 1903, Fauna Brit. India, 2: 227.
1860 *Oncocephalus*: Fieber, Europ. Hem., p. 42.
Note: Giacchi (1984, Physis, Secc. C., 42(103): 39-62) revised this genus and provided a key to species.

Oncocephalus apiculatus Reuter, 1882
1882 *Oncocephalus apiculatus* Reuter, Acta Soc. Sci. Fenn., 12: 728. [N. Am.].
Distribution: Mo., Tex.

Oncocephalus geniculatus (Stål), 1872
1872 *Spilalonius geniculatus* Stål, K. Svens. Vet.-Akad. Handl., 10(4): 123. [Tex.].
1882 *Oncocephalus geniculatus*: Reuter, Acta Soc. Sci. Fenn., 12: 727.
Distribution: Col., Fla., Ks., Mo., N.C., Ok., Tex.

Oncocephalus nubilus Van Duzee, 1914
1914 *Oncocephalus nubilus* Van Duzee, Trans. San Diego Soc. Nat. Hist., 2: 12. [Ariz.].
Distribution: Ariz., Cal., Tex.

Genus *Pnirontis* Stål, 1859

1859 *Pnirontis* Stål, Öfv. K. Svens. Vet.-Akad. Förh., 16: 183. Type-species: *Pnirontis scutellaris* Stål, 1859. Designated by Van Duzee, 1916, Check List Hem., p. 28.
Note: In his revision of the Stenopodainae, Barber (1930, Ent. Am., 10: 153-175) did not use subgenera under *Pnirontis*.

Pnirontis brimleyi Blatchley, 1926
1926 *Pnirontis brimleyi* Blatchley, Het. E. N. Am., p. 545. [N.C.].
Distribution: N.C., Tex.

Pnirontis granulosa Barber, 1930
1930 *Pnirontis granulosa* Barber, Ent. Am., 10: 163. [Fla.].
Distribution: Fla.

Pnirontis infirma Stål, 1859
1859 *Pnirontis infirma* Stål, Öfv. K. Svens. Vet.-Akad. Förh., 16: 382. ["Carolina"].
1872 *Pnirontis (Centromelus) infirma*: Stål, K. Svens. Vet.-Akad. Handl., 10(4): 120.
1873 *Pygolampis infirma*: Walker, Cat. Hem. Brit. Mus., 8: 35.
Distribution: "Carolina," Fla., Ga., Ill., N.J., Tex. (Greater Antilles, Honduras to Brazil).
Note: Barber (1930, Ent. Am., 10: 171) noted Mayr's opinion that Kolenati's (1856, Bull. Soc. Imp. Nat. Moscou, 29: 469) *Pygolampis spinosissima* was erroneously described from Caucasus and is probably synonymous with *Pnirontis infirma*; however, he took no further action. Examination of the type-specimen is needed to solve this matter.

Pnirontis languida Stål, 1859
1859 *Pnirontis languida* Stål, Öfv. K. Svens. Vet.-Akad. Förh., 16: 381, 382. [Brazil, Carolina].
1873 *Pygolampis languida*: Walker, Cat. Hem. Brit. Mus., 8: 35.
1876 *Centromelus languidus*: Uhler, Bull. U. S. Geol. Geogr. Surv. Terr., 1: 331.
1868 *Pnirontis (Centromelus) languida*: Stål, K. Svens. Vet.-Akad. Handl., 7(11): 128.
Distribution: Fla., Ga., N.C., Ok., S.C., Tex. (Greater Antilles, Mexico to Brazil).

Pnirontis modesta Banks, 1910
1910 *Pnirontis modesta* Banks, Ent. News, 21: 324. [Va.].
1916 *Pnirontis (Centromelus) modesta*: Van Duzee, Check List Hem., p. 28.
Distribution: D.C., Fla., Ind., Mo., N.C., Tex., Va. (Greater Antilles).

Genus *Pygolampis* Germar, 1825

1825 *Pygolampis* Germar, Fauna Europ. Hem., 8: 22. Type-species: *Acanthia denticulata* Rossi, 1790, a junior synonym of *Cimex bidentatus* Goeze, 1778. Monotypic.
Note: Van Duzee (1916, Check List Hem., p. 28) reported the neotropical *Pygolampis spurca* Stål for the "S. St." Because I cannot locate the source of Van Duzee's quoted record and because the species has not been reported for North America on subsequent finds, I am here following Barber (1930, Ent. Am., 10: 177) who ignored the northern record when he listed *P. spurca* from Panama south to Brazil and Bolivia.

Pygolampis pectoralis (Say), 1832
1832 *Reduvius pectoralis* Say, Descrip. Het. Hem. N. Am., p. 33. [Fla., Ind., La.].
1859 *Pnirontis fuscipennis* Stål, Öfv. K. Svens. Vet.-Akad. Förh., 16: 379. ["Carolina"]. Synonymized by Stål, 1872, K. Svens. Vet.-Akad. Handl. 10(4): 121.
1869 *Nabis pectoralis*: Rathvon, Hist. Lancaster Co., Pa., p. 550.
1871 *Pygolampis pectoralis*: Uhler, Proc. Bost. Soc. Nat. Hist., 14: 106.
1873 *Pygolampis fuscipennis*: Walker, Cat. Hem. Brit. Mus., 8: 35.
Distribution: Ariz., Cal., Col., Del., Fla., Ill., Ind., La., Mass., Md., Me., Mo., N.C., N.J., N.Y., Ok., S.C., Tex. (Greater Antilles).

Pygolampis sericea Stål, 1859
1859 *Pygolampis sericea* Stål, Öfv. K. Svens. Vet.-Akad. Förh., 16: 378, 380. [Pa.].
Distribution: B.C., Ill., Ind., Mass., Md., N.C., N.Y., Pa., S.C., Tex., Va.

Genus *Stenopoda* Laporte, 1832

1833 *Stenopoda* Laporte, Mag. Zool., p. 26. Type-species: *Stenopoda cinerea* Laporte, 1873. Monotypic.

Note: Giacchi (1969, Physis Secc. C., 29: 1-26) presented a revision of *Stenopoda*.

Stenopoda cinerea Laporte, 1833

1775 *Cimex culiciformis* Fabricius, Syst. Ent., p. 728. ["America"]. Preoccupied.
1803 *Gerris culiciformis*: Fabricius, Ent. Syst., 4: 189
1833 *Stenopoda cinerea* Laporte, Mag. Zool., 2: 26. [Cuba].
1872 *Stenopoda culiciformis*: Stål, K. Svens. Vet.-Akad. Handl., 10(4): 122.
1908 *Stenopoda culicis* [*sic*]: Torre-Bueno, J. N.Y. Ent. Soc., 16: 235.

Distribution: Ala., Ark., Fla., Ill., Ks., La., Miss., Mo., N.B., N.C., N.J., N.Y., Ok., Tenn., Tex. (Antilles, Mexico to Argentina).

Subfamily Triatominae Jeannel, 1919

Note: Lent and Wygodzinsky (1979, Bull. Am. Mus. Nat. Hist., 163: 123-520) provided a comprehensive, fully illustrated revision of the subfamily Triatominae with descriptions, keys (in English, Spanish, and Portuguese), distribution, and various other information. Only one of the six known tribes occurs in our area.

Tribe Triatomini Jeannel, 1919

Notes:The species *Dipetalogaster maximus* (Uhler) was originally described from "Lower California" in Mexico and subsequently listed from the United States in combination with the generic names *Conorhinus* (Snow, 1907, Trans. Ks. Acad. Sci., 20: 160), *Triatoma* (Van Duzee, 1916, Check List Hem., p. 29, and 1917, Univ. Cal. Publ. Ent., 2: 248), and *Eutriatoma* (Neiva and Lent, 1936, Rev. Brasil. Biol., 6: 172, 183). More recent studies by Usinger (1944, U.S. Publ. Health Bull., 288: 41) and Lent and Wygodzinsky (1979, Bull. Am. Mus. Nat. Hist., 163: 360) reveal that the species is restricted to the southern part of Baja California, Mexico--hence the species is here dropped from our list.

Population analyses of the intensely studied blood-sucking Triatomini have revealed localized patterns of deviation; these are often given subspecific names. The present work follows Lent and Wygodzinsky (1979, above) and cites those names in the synonymy under their appropriate species; in other words, the species are here considered to be "variable."

Genus *Paratriatoma* Barber, 1938

1938 *Paratriatoma* Barber, Proc. Ent. Soc. Wash., 40: 104. Type-species: *Paratriatoma hirsuta* Barber, 1938. Original Designation.

Paratriatoma hirsuta Barber, 1938

1938 *Paratriatoma hirsuta* Barber, Proc. Ent. Soc. Wash., 40: 104. [Cal.].
1967 *Paratriatoma hirsuta hirsuta*: Ryckman, Bull. Pan-Am. Res. Inst., 1: 3.
1967 *Paratriatoma hirsuta kamiensis* Ryckman, Bull. Pan-Am. Res. Inst., 1: 2. [Cal.]. Synonymized by Lent and Wygodzinsky, 1979, Bull. Am. Mus. Nat. Hist., 163: 401.

1967 *Paratriatoma hirsuta pimae* Ryckman, Bull. Pan-Am. Res. Inst., 1: 2. [Ariz.]. Synonymized by Lent and Wygodzinsky, 1979, Bull. Am. Mus. Nat. Hist., 163: 401.
Distribution: Ariz., Cal., Nev. (Mexico).

Genus *Triatoma* Laporte, 1832

1832 *Triatoma* Laporte, Mag. Zool., 2: 6, 11. Type-species: *Nabis gigas* Fabricius, 1775, a junior synonym of *Cimex rubrofasciatus* De Geer, 1773. Monotypic.

Note: Uhler (1876, Bull. Geol. Geogr. Surv. Terr., 5: 64-65) reported the conspicuously large "*Meccus phyllosoma*" from "near San Diego" [California]. That species, which currently goes under the combination *Triatoma phyllosoma* (Burmeister), has been reported by several subsequent authors to be known only from the state of Oaxaca, Mexico. In the absence of the Uhler specimen and because of the subsequent literature, this species is here excluded from our list.

Triatoma gerstaeckeri (Stål), 1859
1859 *Conorhinus gerstaeckeri* Stål, Berl. Ent. Zeit., 3: 111. [Tex.].
1914 *Triatoma gerstaekeri* [*sic*]: Neiva, Rev. *Triatoma*, p. 40.
1916 *Triatoma gerstaeckeri*: Van Duzee, Check List Hem., p. 29.
Distribution: Tex. (Mexico).

Triatoma incrassata Usinger, 1939
1939 *Triatoma incrassata* Usinger, Univ. Cal. Publ. Ent., 7: 45. [Mexico].
1962 *Triatoma incrassata apachensis* Ryckman, Univ. Cal. Publ. Ent., 27: 109, 119. [Ariz.]. Synonymized by Lent and Wygodzinsky, 1979, Bull. Am. Mus. Nat. Hist., 163: 243.
Distribution: Ariz. (Mexico).

Triatoma indictiva Neiva, 1912
1912 *Triatoma indictiva* Neiva, Brasil-Medico, 26: 22. [Tex.].
1944 *Triatoma sanguisuga indictiva*: Usinger, U.S. Public Health Bull., 288: 68.
Distribution: Ariz., Tex. (Mexico).

Triatoma lecticularia (Stål), 1859
1859 *Conorhinus lecticularius* Stål, Berl. Ent. Zeit., 3: 107. ["India orientalis," apparently in error!]. Lectotype designated by Lent and Wygodzinsky, 1979, Bull. Am. Mus. Nat. Hist., 163: 253.
1872 *Conorhinus variegatus*: Stål, K. Svens. Vet.-Akad. Handl., 10(4): 111 (in part).
1911 *Triatoma heidemanni* Neiva, Brasil-Medico, 25: 441. [Tex.]. Synonymized by Lent and Wygodzinsky, 1979, Bull. Am. Mus. Nat. Hist., 163: 250.
1911 *Triatoma occulta* Neiva, Brasil-Medico, 25: 441. [Tex.]. Synonymized by Lent and Wygodzinsky, 1979, Bull. Am. Mus. Nat. Hist., 163: 250.
1914 *Triatoma sanguisuga*: Neiva, Rev. *Triatoma*, p. 63 (in part).
1944 *Triatoma lecticularia occulta*: Usinger, U.S. Public Health Bull., 288: 63.
1944 *Triatoma lecticularia floridana* Usinger, U.S. Public Health Bull., 288: 63. [Fla.]. Synonymized by Lent and Wygodzinsky, 1979, Bull. Am. Mus. Nat. Hist., 163: 250.
Distribution: Ariz., Cal., Fla., Ga., Ill., Ks., La., Md., Mo., N.C., N.M., Ok., Pa., S.C., Tenn., Tex. (Mexico).

Triatoma neotomae Neiva, 1911
1911 *Triatoma neotomae* Neiva, Brasil-Medico, 25: 422. [Ariz., Tex.].
Distribution: Ariz., Cal., N.M., Tex.

Triatoma peninsularis Usinger, 1940

1940 *Triatoma peninsularis* Usinger, Pan-Pac. Ent., 16: 73. [Mexico].

1955 *Triatoma protracta peninsularis*: Ryckman et al., J. Econ. Ent., 48: 330.

Distribution: "Southwestern United States" (Mexico).

Note: *T. peninsularis* is included in our list on the basis of Ryckman et al. (1955, above) who wrote it occurred in the "southwestern United States."

Triatoma protracta (Uhler), 1894

1894 *Conorhinus protractus* Uhler, Proc. Cal. Acad. Sci., ser. 2, 4: 284. [Cal., Mexico].

1939 *Triatoma protracta woodi* Usinger, Univ. Cal. Publ. Ent., 7: 42. [N.M.]. Synonymized by Lent and Wygodzinsky, 1979, Bull. Am. Mus. Nat. Hist., 163: 300.

1962 *Triatoma protracta protracta*: Ryckman, Univ. Cal. Publ. Ent., 27: 110.

1962 *Triatoma protracta navajoensis* Ryckman, Univ. Cal. Publ. Ent., 27: 114. [N.M.]. Synonymized by Lent and Wygodzinsky, 1979, Bull. Am. Mus. Nat. Hist., 163: 300.

Distribution: Ariz., Cal., Col., N.M., Nev., Tex., Ut. (Mexico).

Triatoma recurva (Stål), 1868

1868 *Conorhinus recurvus* Stål, K. Svens. Vet.-Akad. Handl., 7(11): 124. [Brazil].

1937 *Triatoma longipes* Barber, Proc. Ent. Soc. Wash., 39: 86. [Ariz.]. Synonymized by Lent, 1951, Ciencia, 11: 156.

1938 *Eutriatoma longipes*: Pinto, Zoo-Parasit. Med. Vet., p. 94.

1951 *Triatoma recurva*: Lent, Ciencias, 11: 156.

Distribution: Ariz. (Mexico to Brazil).

Triatoma rubida (Uhler), 1894

1894 *Conorhinus rubidus* Uhler, Proc. Cal. Acad. Sci., ser. 2, 4: 285. [Mexico].

1914 *Triatoma rubida*: Neiva, Rev. Triatoma, p. 59.

Distribution: Ariz., Cal., Col., N.M., Nev., Tex. (Mexico).

Note: Lent and Wygodzinsky (1979, Bull. Am. Mus. Nat. Hist., 163: 309) listed five infraspecific names, including the nominate one, in synonymy under *T. rubida* but at no place did they call them identical. After discussing the size, color patterns, and distribution of these several infraspecific taxa (which Ryckman, 1976, Bull. Pan-Am. Res. Inst., 1: 1-3 differentiated in a tabular key to "subspecies") they concluded much more study was needed to understand the biosystematics of this species. The only form reported north of the U.S.-Mexican border is *T. uhleri* Neiva.

Triatoma rubida uhleri Neiva, 1911

1899 *Conorhinus rubrofasciatus*: Champion, Biol. Centr.-Am., Rhyn., 2: 208 (in part).

1907 *Conorhinus maximus*: Snow, Trans. Ks. Acad. Sci., 20: 160.

1911 *Triatoma uhleri* Neiva, Brasil-Medico, 25: 421. [Ariz., Cal., N.M., Tex.].

1914 *Triatoma ocellata* Neiva, Rev. *Triatoma*, p. 55. [Ariz.]. Synonymized by Usinger, 1944, Public Health Bull., 288: 66.

1916 *Triatoma maxima*: Van Duzee, Check List Hem., p. 29.

1931 *Eutriatoma uhleri*: Pinto, Bol. Biol., 19: 102.

1944 *Triatoma rubida uhleri*: Usinger, U.S. Public Health Bull., 288: 66.

Distribution: Same as for species.

Tiatoma rubrofasciata (De Geer), 1773

1773 *Cimex rubro-fasciatus* [*sic*] De Geer, Mem. Hist. Ins., 3: 349. [India].

1859 *Conorhinus rubro-fasciatus* [*sic*]: Dohrn, Cat. Hem., p. 103

1873 *Conorhinus rubrofasciatus*: Walker, Cat. Hem. Brit. Mus., 8: 15.

1916 *Triatoma rubrofasciata*: Van Duzee, Check List Hem., p. 29.
Distribution: Fla. (Almost cosmopolitan).

Triatoma sanguisuga (Leconte), 1856 [Fig. 133]
1855 *Conorhinus sanguisuga* Leconte, Proc. Acad. Nat. Sci. Phila., 7: 404. [Ga.].
1859 *Conorhinus lateralis* Stål, Berl. Ent. Zeit., 3: 107. [Ga.]. Synonymized by Stål, 1872, K. Svens. Vet.-Akad. Handl., 10(4): 111.
1872 *Conorhinus variegatus*: Stål, K. Svens. Vet.-Akad. Handl., 10(4): 111 (in part).
1885 *Conorhinus sanguisuga*: Popenoe, Trans. Ks. Acad. Sci., 9: 63.
1911 *Triatoma sanguisuga* var. *ambigua* Neiva, Brasil-Medico, 25: 422. [Fla.]. Synonymized by Hussey, 1922, Psyche, 29: 120.
1914 *Triatoma sanguisuga*: Neiva, Rev. *Triatoma*, p. 31.
1926 *Triatoma pintoi* Larrousse, An. Parasit. Hum. Comp., 4: 138. [Fla.]. Synonymized by Neiva and Lent, 1941, Rev. Ent., 12: 87.
1938 *Triatoma sanguesuga* [*sic*]: Pinto, Zoo-Parasit. Med. Vet., p. 95.
1943 *Triatoma ambigua*: Davis et al., U.S. Public Health Serv. Rept., 58: 353.
1944 *Triatoma sanguisuga texana* Usinger, U.S. Public Health Bull., 288: 69. [Tex.]. Synonymized by Lent and Wygodzinsky, 1979, Bull. Am. Mus. Nat. Hist., 163: 321.
Distribution: Ala., Ariz., Ark., Fla., Ga., Ill., Ind., Ks., Ky., La., Md., Miss., Mo., N.C., Oh., Ok., Pa., S.C., Tenn., Tex., Va.

Family Rhopalidae Amyot and Serville, 1843

(= Corizidae Douglas and Scott, 1865)

The Scentless Plant Bugs

By Thomas J. Henry

The Rhopalidae or scentless plant bugs are a small, difficult-to-identify group of coreoid bugs. They are small to medium sized, usually heavily punctate and pubescent and, often, possess tubercles and small spines on the head, pronotum, and legs. They are best recognized by the numerous veins on the wing membrane, closed cells on the corium, and the greatly reduced ostiole or scent-gland opening. From the latter character comes the common name scentless plant bugs.

Most members of the family feed on fruits and seeds of herbaceous plants growing in old fields, along roadsides, and other disturbed or harsh habitats. A few species are arboreal. The common boxelder bug, *Boisea trivittata* (Say) [formerly known in the genus *Leptocoris* Hahn], feeds on the seeds of boxelder, *Acer negundo* L., and other trees of the maple family. It is this bug that gives the rhopalids notoriety. In the fall or spring when boxelder bugs seek or leave overwintering shelters, they often congregate in large masses in sunny locations, frequently by the thousands. Many manage to enter buildings through small openings found in and around foundations, siding,

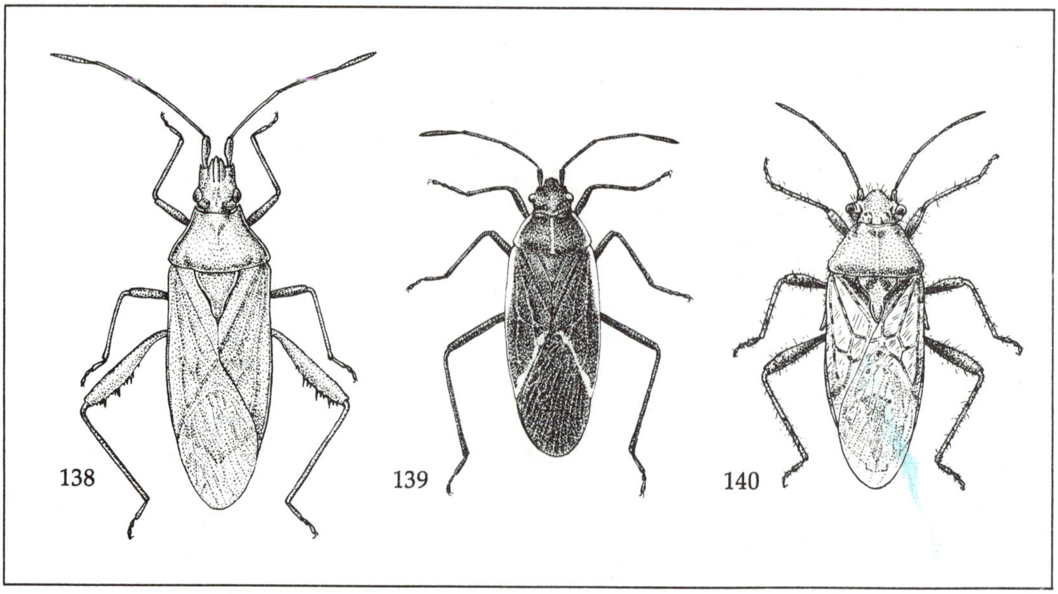

Figs. 138-140: 138, *Harmostes reflexulus* [p.655]; 139, *Boisea trivittata* [p. 662]; 140, *Arhyssus lateralis* [p. 656] (After Froeschner, 1942).

and window sills, causing homeowners much anxiety. Although harmless to humans, the bugs sometimes defecate on curtains, furniture and other objects, leaving unsightly spots and, when crushed, giving off an unpleasant odor. Wheeler (1982, Bed Bugs, pp. 319-351) gave a thorough review of the problems associated with this nuisance pest.

The taxonomy of the family, until recently, has been relatively unstable. Most of the early work and keys to species were based on size and color, making many of the older records in the literature unreliable. Recent revisions of the genera *Arhyssus* Stål, *Harmostes* Burmeister, *Leptocoris* Hahn, *Liorhyssus* Stål, *Niesthrea* Spinola, *Stictopleurus* Stål, and others should be consulted for correct species relationships [see respective genera for references].

The major literature for this family was compiled by Göllner-Scheiding (1983, Mitt. Zool. Mus. Berlin, 59: 37-189) in her world catalog of the Rhopalidae. Chopra's "Higher Classification of the family Rhopalidae" (1967, Trans. Royal Ent. Soc. London, 119: 363-399), containing keys to subfamilies, tribes, and genera, is followed in this catalog. Slater and Baranowski (1978, How To Know True Bugs, pp. 66-70) provided a key to the North American genera [excepting the genera *Rhopalus* Schilling and *Xenogenus* Berg, each represented by single species]. Schaefer and Chopra (1982, An. Ent. Soc. Am., 75: 224-233) presented a cladistic analysis and list of food plants for the family. Hoebeke and Wheeler (1982, Proc. Ent. Soc. Wash., 84: 213-218) gave a good key to the species found in eastern North America.

Subfamily Rhopalinae Amyot and Serville, 1843

Tribe Chorosomini Douglas and Scott, 1865

Genus *Xenogenus* Berg, 1883

1883 *Xenogenus* Berg, An. Soc. Cient. Arg., 15: 252. Type-species: *Xenogenus picturatum* Berg, 1883. Monotypic.

Xenogenus picturatum Berg, 1883

1883 *Xenogenus picturatum* Berg, An. Soc. Cient. Arg., 15: 253. [Uruguay]. Lectotype designated by Göllner-Scheiding, 1980, Mitt. Zool. Mus. Berlin, 56: 120.

1893 *Xenogenus extensum* Distant, Biol. Centr.-Am., Rhyn., 1: 461. [Mexico]. Synonymized by Göllner-Scheiding, 1980, Mitt. Zool. Mus. Berlin, 56: 120.

1910 *Xenogenus extensum*: Barber, J. N.Y. Ent. Soc., 18: 37.

1980 *Xenogenus picturatum*: Göllner-Scheiding, Mitt. Zool. Mus. Berlin, 56: 120.

Distribution: Ariz. (Mexico to Argentina).

Note: Göllner-Scheiding (1980, above) considered *X. extensum* a synonym of *X. picturatum*, but Brailovsky and Soria (1981, An. Inst. Biol. Univ. Nal. Aut. Méx., 51: 153) treated it as a separate species. Because Göllner-Scheiding studied type material and Brailovsky and Soria apparently did not, *X. extensum* is here considered a junior synonym of *X. picturatum*.

Tribe Harmostini Stål, 1873

Genus *Aufeius* Stål, 1870

1870 *Aufeius* Stål, K. Svens. Vet.-Akad. Handl., 9(1): 221. Type-species: *Aufeius impressicollis* Stål, 1870. Monotypic.

Aufeius impressicollis Stål, 1870.

1870 *Aufeius impressicollis* Stål, K. Svens. Vet.-Akad. Handl., 9(1): 222. [Tex.].

1881 *Harmostes propinquus* Distant, Biol. Centr.-Am., Rhyn., 1: 168. [Guatemala]. Synonymized by Van Duzee, 1916, Check List Hem., p. 14.

1906 *Harmostes propinquus*: Barber, Mus. Brook. Inst. Arts Sci., Sci. Bull., 1: 271.

Distribution: Ariz., Ark., Cal., Col., Ia., Id., Ind., Ks., Md., Mo., N.M., Neb., Oh., S.D., Tex., Ut., Wash. (Mexico to Guatemala).

Note: Wheeler (1984, Proc. Ent. Soc. Wash., 92: 174-178) reviewed distribution and gave biological notes and host information.

Genus *Harmostes* Burmeister, 1835

1835 *Harmostes* Burmeister, Handb. Ent., 2: 307. Type-species: *Harmostes dorsalis* Burmeister, 1835. Monotypic.

Note: Göllner-Scheiding (1978, Mitt. Zool. Mus. Berlin, 54: 257-311) revised this genus and provided a key to subgenera and species. Malloch (1921, Bull. Brook. Ent. Soc., 16: 56) reported *H. prolixus* Stål new to the U.S. from Brownsville, Tex., but Göllner-Scheiding (1978, above) showed that this is a South American species, and Brailovsky and Soria (1981, An. Inst. Biol. Univ. Aut. Méx., 51: 152) did not list it from Mexico, indicating that this record undoubtedly is based on a misidentification.

Subgenus *Harmostes* Burmeister, 1835

1835 *Harmostes* Burmeister, Handb. Ent., 2: 307. Type-species: *Harmostes dorsalis* Burmeister, 1835. Monotypic.

Note: This is the only subgenus recorded from the U.S. or Canada; *Neoharmostes* Göllner-Scheiding, 1978, the only other subgenus, occurs in the Neotropics.

Harmostes angustatus Van Duzee, 1918

1918 *Harmostes angustatus* Van Duzee, Proc. Cal. Acad. Sci., (4)8: 277. [Cal.].

Distribution: Ariz., Cal., N.M., Tex., Ut. (Mexico).

Harmostes dorsalis Burmeister, 1835

1835 *Harmostes dorsalis* Burmeister, Handb. Ent., 2: 307. [Mexico]. Lectotype designated by Göllner-Scheiding, 1978, Mitt. Zool. Mitt. Berlin, 54: 274.

1881 *Harmostes subrufus* Distant, Biol. Centr.-Am., Rhyn., 1: 167. [Guatemala]. Synonymized and lectotype designated by Göllner-Scheiding, 1978, Mitt. Zool. Mus. Berlin, 54: 273-274.

1910 *Harmostes subrufus*: Barber, J. N.Y. Ent. Soc., 28: 37.

1917 *Harmostes croceus* Gibson, Ent. News, 28: 445. [Cal.]. Synonymized by Göllner-Scheiding, 1978, Mitt. Zool. Mus. Berlin, 54: 273.

1956 *Harmostes chemebuevi* Bliven, N. Hem. W. St., p. 9. [Cal.]. Synonymized by Göllner-Scheiding, 1978, Mitt. Zool. Mus. Berlin, 54: 273.

Distribution: Ariz., B.C., Cal., Ore., Tex. (Mexico to Argentina and Peru).

Note: Brailovsky and Soria (1981, Ann. Inst. Biol. Univ. Nal. Aut. Méx., 51: 149) recognized *H. subrufus* as a valid species, but because Göllner-Scheiding (1978, above) studied type material, and Brailovsky and Soria apparently did not, *H. subrufus* is here considered a junior synonym of *H. dorsalis*.

Harmostes formosus Distant, 1881

1881 *Harmostes formosus* Distant, Biol. Centr.-Am., Rhyn., 1: 167. [Mexico].

1941 *Harmostes formosus*: Torre-Bueno, Bull. Brook. Ent. Soc., 36: 84.

Distribution: Tex. (Mexico).

Harmostes fraterculus (Say), 1832

1832 *Syromates* [*sic*] *fraterculus* Say, Descrip. Het. Hem. N. Am., p. 10. [Ind., Ga.].

1859 *Syromastes fraterculus*: LeConte, Compl. Writ. Say, p. 324.

1870 *Harmostes fraterculus*: Stål, K. Svens. Vet.-Akad. Handl., 9(1): 221.

Distribution: Ariz., Cal., D.C., Fla., Ga., Ill., Ind., Md., Minn., Mo., N.C., N.J., Ok., Pa., Tex., Va. (Central America, Mexico, West Indies).

Note: Wheeler and Miller (1983, Proc. Ent. Soc. Wash., 85: 426-434) studied biology and hosts and described and illustrated the immature stages.

Harmostes reflexulus (Say), 1832 [Fig. 138]

1832 *Syromates* [*sic*] *reflexulus* Say, Descrip. Het. Hem. N. Am., p. 10. [Pa.].

1859 *Syromastes reflexulus*: LeConte, Compl. Writ. Say, p. 323.

1851 *Harmostes costalis* Herrich-Schaeffer, Wanz. Ins., p. 9. [Pa.]. Synonymized by Stål, 1870, K. Svens. Vet.-Akad. Handl., 9(1): 220.

1852 *Harmostes virescens* Dallas, List Hem. Brit. Mus., 2: 520. [Ga.]. Synonymized by Stål, 1870, K. Svens. Vet.-Akad. Handl., 9(1): 220.

1869 *Hamostes* [*sic*] *reflexus* [*sic*]: Rathvon, Hist. Lancaster Co., Pa., p. 549.

1870 *Harmostes reflexulus*: Stål, K. Svens. Vet.-Akad. Handl., 9(1): 220.

1881 *Harmostes bicolor* Distant, Biol. Centr.-Am., Rhyn., 1: 167. [Mexico]. Synonymized by Göllner-Scheiding, 1978, Mitt. Zool. Mus. Berlin, 54: 285.

1913 *Harmostes bruesi* Bergroth, Ent. News, 24: 266. [Tex.]. Synonymized by Van Duzee, 1916, Check List Hem., p. 14.

1917 *Harmostes bicolor*: Gibson, Ent. News, 28: 447.

1956 *Harmostes mirificus* Bliven, N. Hem. W. St., p. 8. [Cal.]. Synonymized by Göllner-Scheiding, 1978, Mitt. Zool. Mus. Berlin, 54: 285.

1956 *Harmostes caurus* Bliven, N. Hem. W. St., p. 9. [Cal.]. Synonymized by Göllner-Scheiding, 1978, Mitt. Zool. Mus. Berlin, 54: 285.

Distribution: Recorded from all U.S. states; B.C., Que. (Mexico, Cuba).

Note: Yonke and Walker (1970, J. Ks. Ent. Soc., 43: 444-450; 1970, An. Ent. Soc. Am., 63: 1749-1754) studied biology and described and illustrated the immature stages.

Harmostes obliquus (Say), 1832

1832 *Syromates* [*sic*] *obliquus* Say, Descrip. Het. Hem. N. Am., p. 11. [U.S.].

1859 *Syromastes obliquus*: LeConte, Compl. Writ. Say, p. 324.

1886 *Harmostes obliquus*: Uhler, Check-list Hem. Het., p. 13.

Distribution: "U.S."

Note: This name has not been associated with a species in the U.S. fauna since the original description. Blatchley (1926, Het. E. N. Am., p. 274) commented, based on Say's original description, "it must also have been a common species of wide range." Uhler (1886, above) first placed this species in *Harmostes* with a question. Göllner-Scheiding (1983, Mitt. Zool. Mus. Berlin, 59: 124) listed it as a "*Species dubia.*"

Harmostes serratus (Fabricius), 1775
1775 *Acanthia serrata* Fabricius, Syst. Ent., p. 695. [No locality given].
1852 *Harmostes affinis* Dallas, List Hem. Ins. Brit. Mus., 2: 522. [No locality given]. Synonymized by Göllner-Scheiding, 1978, Mitt. Zool. Mus. Berlin, 54: 287.
1868 *Harmostes serratus*: Stål, K. Svens. Vet.-Akad. Handl., 7(11): 67.
1907 *Harmostes affinis*: Van Duzee, Bull. Buff. Soc. Nat. Sci., 8: 13.
Distribution: Fla., Tex. (Mexico to South America).
Note: Early records, such as Gibson's (1917, Ent. News, 28: 444) for "... practically the entire United States," are in error.

Tribe Niesthreini Chopra, 1967

Genus *Arhyssus* Stål, 1870

1870 *Corizus* (*Arhyssus*) Stål, K. Svens. Vet.-Akad, Handl., 9(1): 223. Type-species: *Corizus bohemani* Signoret, 1859, a junior synonym of *Corizus nigristernum* Signoret, 1859. Designated by Van Duzee, 1916, Check List Hem., p. 15.
1908 *Arhyssus*: Baker, Can. Ent., 40: 244.
Note: Chopra (1968, An. Ent. Soc. Am., 61: 629-655) revised this genus and provided a key to species.

Arhyssus barberi Harris, 1942
1908 *Corizus scutatus*: Hambleton, An. Ent. Soc. Am., 1: 139.
1919 *Corizus scutatus*: Gibson, Can. Ent., 51: 91.
1941 *Corizus scutatus*: Torre-Bueno, Ent. Am., 21: 97.
1942 *Arhyssus barberi* Harris, J. Ks. Ent. Soc., 15: 101. [Id.].
Distribution: B.C., Cal., Id., Ore., Ut., Wash.
Note: Chopra (1968, An. Ent. Soc. Am., 61: 633) clarified the misidentifications listed above.

Arhyssus confusus Chopra, 1968
1968 *Arhyssus confusus* Chopra, An. Ent. Soc. Am., 61: 644. [Ariz.].
Distribution: Ariz., Cal., Col., N.M., Nev., Tex. (Mexico).

Arhyssus crassus Harris, 1942
1942 *Arhyssus crassus* Harris, J. Ks. Ent. Soc., 15: 103. [Id.].
Distribution: Ariz., Cal., Id., Ore., Ut., Wash., Wyo.

Arhyssus distinctus Chopra, 1968
1968 *Arhyssus distinctus* Chopra, An. Ent. Soc. Am., 61: 639. [Ore.].
Distribution: B.C., Cal., Ore.

Arhyssus hirtus (Torre-Bueno), 1912
1912 *Corizus hirtus* Torre-Bueno, Ent. News, 23: 217. [N.Y.]. Lectotype designated by Chopra, 1968, An. Ent. Soc. Am., 61: 650.
1913 *Rhopalus hirtus*: Bergroth, Mém. Soc. Ent. Belg., 22: 163.
1916 *Corizus* (*Arhyssus*) *hirtus*: Van Duzee, Check List Hem., p. 15.
1968 *Arhyssus hirtus*: Chopra, An. Ent. Soc. Am., 61: 649.
Distribution: Mass., Md., N.J., N.Y.
Note: Wheeler and Henry (1984, Fla. Ent., 67: 525-527) studied hosts and distribution and illustrated the fifth instar.

Arhyssus lateralis (Say), 1825 [Fig. 140]
1825 *Coreus lateralis* Say, J. Acad. Nat. Sci. Phila., 4: 320. [Neb., Pa.]. Neotype from Neb. designated by Chopra, 1968, An. Ent. Soc. Am., 61: 645.

1852 *Rhopalus punctipennis* Dallas, List Hem. Brit. Mus., 2: 526. [N. Am.]. Synonymized by Hambleton, 1908, An. Ent. Soc. Am., 1: 140.
1870 *Corizus punctipennis*: Stål, K. Svens. Vet.-Akad. Handl., 9(1): 224.
1885 *Corisus* [*sic*] *lateralis*: Provancher, Pet. Faune Ent. Can., 3: 60.
1885 *Corizus lateralis*: Popenoe, Trans. Ks. Acad. Sci., 9: 63.
1908 *Niesthrea lateralis*: Baker, Can. Ent., 40: 244.
1908 *Niesthrea lateralis* form *roseus* Baker, Can. Ent., 40: 244. [Cal.]. Synonymized by Chopra, 1968, An. Ent. Soc. Am., 61: 645.
1913 *Rhopalus lateralis*: Bergroth, Mém. Soc. Ent. Belg., 22: 163 (in part.).
1916 *Corizus* (*Arhyssus*) *lateralis*: Van Duzee, Check List Hem., p. 15.
1916 *Corizus* (*Arhyssus*) *lateralis* var. *roseus*: Van Duzee, Check List Hem., p. 15.
1968 *Arhyssus lateralis*: Chopra, An. Ent. Soc. Am., 61: 645.
Distribution: Recorded from every U.S. state (except Id., Me., Mont.); Ont., Que., Sask. (Mexico to Guatemala).
Note: Even though Göllner-Scheiding (1983, Mitt. Zool. Mus. Berlin, 59: 49) listed "*C. lateralis* form *roseus* Baker" as a variety in her Catalog, Chopra's (1968, above) revisionary treatment of it as a synonym of *A. lateralis* is followed here. Hambleton (1909, An. Ent. Soc. Am., 2: 272-276) studied the life history.

Arhyssus longirostratus Chopra, 1968
1968 *Arhyssus longirostratus* Chopra, An. Ent. Soc. Am., 61: 648. [Wash.].
Distribution: Cal., Nev., Wash.

Arhyssus nigristernum (Signoret), 1859
1859 *Corizus nigristernum* Signoret, An. Soc. Ent. France, 7: 100. [N.Y.].
1859 *Corizus bohemani* Signoret, An. Soc. Ent. France, 7: 100. [Carolina]. Synonymized by Hambleton, 1908, An. Ent. Soc. Am., 1: 143.
1870 *Corizus* (*Arhyssus*) *bohemani*: Stål, K. Svens. Vet.-Akad. Handl., 9(1): 223.
1908 *Niesthrea lateralis* form *nigristernum*: Baker, Can. Ent., 40: 243.
1913 *Rhopalus lateralis*: Bergroth, Mém. Soc. Ent. Belg., 22: 163 (in part).
1916 *Corizus* (*Arhyssus*) *lateralis*: Van Duzee, Check List Hem., p. 15 (in part).
1918 *Corizus bohemanni* [*sic*]: Torre-Bueno, Can. Ent., 50: 25.
1919 *Corizus bohemanii* [*sic*]: Gibson, Can. Ent., 51: 90.
1919 *Corizus bohoemanie* [*sic*]: Torre-Bueno, Bull. Brook. Ent. Soc., 14: 125.
1934 *Corizus bohoemannii* [*sic*]: Brown, Can. Ent., 66: 223.
1941 *Corizus* (*Arhyssus*) *bohemanii* [*sic*]: Torre-Bueno, Ent. Am., 21: 96.
1968 *Arhyssus nigristernum*: Chopra, An. Ent. Soc. Am., 61: 647.
Distribution: Ark., Conn., D.C., Fla., Ind., Ks., Ky., La., Mass., Md., Me., Mich., Miss., Mo., N.C., N.J., N.Y., Oh., Ont., Pa., Que., R.I., Tenn., Tex., Va., W.Va.
Note: Hambleton (1908, above) included Ariz. and Cal. in the distribution of this species, but Chopra (1968, An. Ent. Soc. Am., 61: 648) excluded these far western records.

Arhyssus parvicornis (Signoret), 1859
1859 *Corizus parvicornis* Signoret, An. Soc. Ent. France, 7: 101. [Mexico].
1908 *Corizus parvicornis*: Hambleton, An. Ent. Soc. Am., 1: 143.
1916 *Corizus* (*Arhyssus*) *parvicornis*: Van Duzee, Check List Hem., p. 15.
1968 *Arhyssus parvicornis*: Chopra, An. Ent. Soc. Am., 61: 650.
Distribution: Ariz., Cal., Col., Ks., N.M., Nev., Ok., Tex., Ut., Wash.(?) (Mexico, Costa Rica).
Note: The Wash. record given by Hambleton (1908, above) may have been based on a misidentification.

Arhyssus punctatus (Signoret), 1859
- 1859 *Corizus punctatus* Signoret, An. Soc. Ent. France, 7: 81. [Mexico].
- 1908 *Niesthrea punctatus*: Baker, Can. Ent., 40: 244.
- 1916 *Corizus* (*Arhyssus*) *punctatus*: Van Duzee, Check List Hem., p. 15.
- 1968 *Arhyssus punctatus*: Chopra, An. Ent. Soc. Am., 61: 646.

Distribution: Ariz., Cal., Col., Fla.(?), Ok., Ore.(?), N.M., Tex. (Mexico, Guatemala, Honduras, Haiti?).

Note: Records in the literature from Fla. and Ore. may have been based on misidentifications.

Arhyssus schaeferi Chopra, 1968
- 1968 *Arhyssus schaeferi* Chopra, An. Ent. Soc. Am., 61: 637. [Cal.].

Distribution: Cal.

Arhyssus scutatus (Stål), 1859
- 1859 *Rhopalus scutatus* Stål, K. Svens. Freg. Eug. Resa. Jorden, 3: 239. [Cal.]. Chopra, 1968, An. Ent. Soc. Am., 61: 637. Holotype recognized in error by Chopra, 1968, An. Ent. Soc. Am., 61:1637.
- 1859 *Corizus jactatus* Signoret, An. Soc. Ent. France, 7: 81. [Cal.]. Synonymized by Hambleton, 1908, An. Ent. Soc. Am., 1: 139.
- 1870 *Corizus* (*Arhyssus*) *scutatus*: Stål, K. Svens. Vet.-Akad. Handl., 9(1): 224.
- 1908 *Niesthrea scutatus*: Baker, Can. Ent., 40: 243.
- 1908 *Corizus indentatus* Hambleton, An. Ent. Soc. Am., 1: 139. [B.C., Cal., Col., Ore., Wash., Wyo.]. Synonymized and lectotype from Wyo. designated by Chopra, 1968, An. Ent. Soc. Am., 61: 637.
- 1908 *Corizus tuberculatus* Hambleton, An. Ent. Soc. Am., 1: 140. [Wash.]. Synonymized and lectotype designated by Chopra, 1968, An. Ent. Soc. Am., 61: 635.
- 1913 *Rhopalus indentatus*: Bergroth, Mém. Soc. Ent. Belg., 22: 163.
- 1913 *Rhopalus tuberculatus*: Bergroth, Mém. Soc. Ent. Belg., 22: 163.
- 1916 *Corizus* (*Arhyssus*) *indentatus*: Van Duzee, Check List Hem., p. 15.
- 1942 *Arhyssus scutatus*: Harris, J. Ks. Ent. Soc., 15: 101.
- 1942 *Arhyssus brevipilis* Harris, J. Ks. Ent. Soc., 15: 104. [Id.]. Synonymized by Chopra, 1968, An. Ent. Soc. Am., 61: 635.
- 1942 *Arhyssus indentatus*: Harris, J. Ks. Ent. Soc., 15: 104.

Distribution: Alta., B.C., Cal., Id., Nev., Ore., Ut., Wash., Wyo.

Note: Gibson's (1919, Can. Ent., 51: 91) record for Mexico and Torre-Bueno's (1941, Ent. Am., 21: 97) for Col. and Ks. were probably based on misidentifications.

Arhyssus usingeri Harris, 1942
- 1942 *Arhyssus usingeri* Harris, J. Ks. Ent. Soc., 15: 102. [Cal.].

Distribution: Cal.

Arhyssus validus (Uhler), 1893
- 1893 *Corizus validus* Uhler, Proc. Ent. Soc. Wash., 2: 370. [Cal., Ore., Ut.]. Lectotype from Ut. designated by Chopra, 1968, An. Ent. Soc. Am., 61: 643.
- 1908 *Niesthrea validus*: Baker, Can. Ent., 40: 243.
- 1916 *Corizus* (*Arhyssus*) *validus*: Van Duzee, Check List Hem., p. 15
- 1919 *Corizus lateralis* var. *validus*: Gibson, Can. Ent., 51: 91.
- 1968 *Arhyssus validus*: Chopra, An. Ent. Soc. Am., 61: 643.

Distribution: Ariz., B.C., Cal., Col., Id., Ore., Ut., Wash. (Mexico).

Genus *Niesthrea* Spinola, 1837

1837 *Corizus* (*Niesthrea*) Spinola, Essai Hém., p. 245. Type-species: *Lygaeus sidae* Fabricius, 1794. Monotypic.
1908 *Niesthrea*: Baker, Can. Ent., 40: 243.
Note: Chopra (1973, J. Nat. Hist., 7: 441-459) revised this genus and gave figures of male genitalia to help separate species.

Niesthrea louisianica Sailer, 1961
1876 *Corizus sidae*: Uhler, Bull. U.S. Geol. Geogr. Surv. Terr., 1: 301 (in part).
1893 *Corizus pictipes*: Uhler, Proc. Ent. Soc. Wash., 2: 369.
1916 *Corizus* (*Niesthrea*) *sidae* var. *pictipes*: Van Duzee, Check List Hem., p. 15 (in part).
1961 *Niesthrea louisianica* Sailer, Proc. Ent. Soc. Wash., 63: 297. [La.].
Distribution: Ala., Ariz., Ark., Fla., Ga., Ia., Ks., La., Md., Miss., Mo., N.C., N.J., N.M., N.Y., Ok., S.C., Tenn., Tex., Ut., Va.
Note: Early U.S. records of *N. sidae* outside Fla., Ga., and Tex. refer to this species (see Sailer, 1961, Proc. Ent. Soc. Wash., 63: 293-299 and Wheeler, 1977, An. Ent. Soc. Am., 70: 631). Readio (1928, An. Ent. Soc. Am., 21: 189-201) made observations on the life cycle [as *sidae*] and illustrated the adult and immatures. Wheeler (1977, An. Ent. Soc. Am., 70: 631-634) studied the life history of this species on rose of Sharon, *Hibiscus syriacus* L., in N.C. Jones et al. (1985, An. Ent. Soc. Am., 78: 326-330) studied biology on *Abutilon theophrasti* Medik. in Miss.

Niesthrea sidae (Fabricius), 1794
1794 *Lygaeus sidae* Fabricius, Ent. Syst., 4: 169. ["Amer. Insul."]. Lectotype designated by Sailer, 1961, Proc. Ent. Soc. Wash., 63: 296.
1876 *Corizus sidae*: Uhler, Bull. U.S. Geol. Geogr. Surv. Terr., 1: 301 (in part).
1913 *Rhopalus sidae*: Bergroth, Mém. Soc. Ent. Belg., 22: 163 (in part).
1908 *Niesthrea sidae*: Baker, Can. Ent., 40: 243 (in part).
1916 *Corizus* (*Niesthrea*) *sidae*: Van Duzee, Check List Hem., p. 15 (in part).
Distribution: Fla., Ga., Tex. (Mexico to Colombia and Venezuela, West Indies).
Note: Sailer (1961, above) showed that the U.S. records for *N. sidae,* other than most from Fla., Ga., and Tex., should apply to the new species he described as *N. louisianica.*

Niesthrea ventralis (Signoret), 1859
1859 *Corizus ventralis* Signoret, An. Soc. Ent. France, 7: 92. [Mexico].
1908 *Niesthrea sidae*: Baker, Can. Ent., 40: 243 (in part).
1913 *Rhopalus ventralis*: Bergroth, Mém. Soc. Ent. Belg., 22: 163.
Distribution: Tex. (Mexico to Guatemala).
Note: This species was long considered a synonym of *N. sidae* (e.g. Baker, 1908, above; Hambleton, 1908, An. Ent. Soc. Am., 1: 142) but Chopra (1973, J. Nat. Hist., 7: 452) gave it species status, after studying Signoret's cotypes.

Tribe Rhopalini Amyot and Serville, 1843

Genus *Liorhyssus* Stål, 1870

1870 *Corizus* (*Liorhyssus*) Stål, K. Svens. Vet.-Akad. Handl., 9(1): 222. Type-species: *Lygaeus hyalinus* Fabricius, 1794. Designated by Reuter, 1888, Acta Soc. Sci. Fenn., 15: 763.

1872 *Liorhyssus*: Stål, Öfv. K. Svens. Vet.-Akad. Förh., 29(6): 55.

Note: Göllner-Scheiding (1976, Deut. Ent. Zeit., 23: 181-206) revised the genus and provided a key to species.

Liorhyssus hyalinus (Fabricius), 1794

1794 *Lygaeus hyalinus* Fabricius, Ent. Syst., 4: 168. ["Am. Insul."].

1872 *Corizus viridicatus*: Uhler, Prelim. Rept. Geol. Surv. Mont., p. 404 (in part). See note below.

1876 *Corizus hyalinus*: Uhler, Bull. U.S. Geol. Geogr. Surv. Terr., 1: 300 (in part).

1877 *Corizus hyalinus* var. *viridicatus*: Uhler, Bull. U.S. Geol. Geogr. Surv. Terr., 3: 407 (in part).

1907 *Rhopalus hyalinus*: Kirkaldy, Proc. Haw. Ent. Soc., 1: 146.

1908 *Liorhyssus hyalinus*: Baker, Can. Ent., 40: 243.

1913 *Rhopalus viridicatus*: Bergroth, Mém. Soc. Ent. Belg., 22: 163 (in part).

1916 *Corizus (Liorhyssus) hyalinus*: Van Duzee, Check List Hem., p. 14.

1967 *Liorhyssus hyalinatus* [*sic*]: Kurczewski, J. Ks. Ent. Soc., 40: 207.

Distribution: Ark., Ariz., B.C., Cal., Col., Conn., Fla., Ia., Ind., Ks., La., Man., Mass., Md., Miss., Mo., N.C., N.M., Neb., Nev., Oh., Ont., S.C., S.D., Tex., Ut., Va., Wyo. (Africa, Asia, Australia, Hawaii, Midway Atoll, Europe, Mexico to South America, West Indies).

Note: Harris (1944, Ia. St. J. Sci., 19: 103) showed that Uhler's description of *Corizus viridicatus* was in part for the species *L. hyalinus* and in part for *Stictopleurus viridicatus* (which see). Readio (1928, An. Ent. Soc. Am., 21: 189-198) provided notes on biology and illustrated the adult and immatures. Carlson (1959, J. Econ. Ent., 52: 242-244) studied the effects of this "hyaline grass bug" on lettuce seed production.

Genus *Stictopleurus* Stål, 1872

1872 *Stictopleurus* Stål, Öfv. K. Svens. Vet.-Akad. Förh., 29(6): 55. Type-species: *Cimex crassicornis* Linnaeus, 1758. Designated by Oshanin, 1912, Kat. Paläark. Hem., p. 25.

Note: Göllner-Scheiding (1975, Deut. Ent. Zeit., 22: 1-60) revised this genus and provided a key to species.

Stictopleurus intermedius (Baker), 1908

1908 *Corizus novaeboracensis* form *intermedia* Baker, Can. Ent., 40: 243. [Nev.].

1916 *Corizus (Stictopleurus) crassicornis* var. *intermedius*: Van Duzee, Check List Hem., p. 15.

1944 *Stictopleurus intermedia*: [*sic*] Harris, Ia. St. Coll. J. Sci., 19: 105.

1975 *Stictopleurus intermedius*: Göllner-Scheiding, Deut. Ent. Zeit., 22: 22.

Distribution: Col., Mont., Nev., Ore., Ut., Wash.

Stictopleurus knighti Harris, 1942

1942 *Stictopleurus knighti* Harris, J. Ks. Ent. Soc., 15: 100. [Minn.].

Distribution: Mich., Minn., Wyo. (Mexico?).

Note: Mexico was recorded with a question by Göllner-Scheiding (1983, Mitt. Zool. Mus. Berlin, 59: 145).

Stictopleurus plutonius (Baker), 1908

1908 *Corizus novaeboracensis* form *plutonius* Baker, Can. Ent., 40: 243. [No locality given].

1916 *Corizus (Stictopleurus) crassicornis* var. *plutonius*: Van Duzee, Check List Hem., p. 15.

1944 *Stictopleurus plutonius*: Harris and Shull, Ia. St. Coll. J. Sci., 18: 202.
Distribution: Ariz., B.C., Cal., Col., Id., N.M., Nev., Ore., Ut., Wash., Wyo.

Stictopleurus punctiventris (Dallas), 1852
1852 *Rhopalus punctiventris* Dallas, List Hem. Brit. Mus., 2: 526. [N. Am.].
1859 *Corizus novaeboracensis* Signoret, An. Soc. Ent. France, 7: 97. [N.Y.]. Synonymized by Hambleton, 1908, An. Ent. Soc. Am., 1: 137.
1861 *Corizus borealis* Uhler, Proc. Acad. Nat. Sci., Phila., p. 284. [Col.]. Synonymized by Uhler, 1876, Bull. U.S. Geol. Geogr. Surv. Terr., 1: 301.
1870 *Corizus* (*Arhyssus*) *punctiventris*: Stål, K. Svens. Vet.-Akad. Handl., 9(1): 223.
1872 *Corizus punctiventris*: Uhler, Prelim. Rept. U.S. Geol. Surv. Mont., p. 403.
1908 *Corizus crassicornis*: Van Duzee, Can. Ent., 40: 110.
1908 *Corizus novaeboracensis* form *novaeboracensis*: Baker, Can. Ent., 40: 243.
1908 *Corizus novaeboracensis* form *occidentalis* Baker, Can. Ent., 40: 243. [Col. to Nev.]. Synonymized by Harris, 1944, Ia. St. Coll. J. Sci., 19: 102.
1916 *Corizus (Stictopleurus) crassicornis*: Van Duzee, Check List Hem., p. 15.
1944 *Stictopleurus punctiventris*: Harris, Ia. St. J. Sci., 19: 101.
Distribution: Alta., Ariz., B.C., Cal., Col., Id., Ind., Mass., Me., Mich., Minn., Mont., N.H., N.J., N.M., N.S., N.Y., Neb., Oh., Ore., Pa., Que., S.D., Sask., Tex., Ut., Wash., Yuk. (Mexico).
Note: For many years this species went under the name *S. crassicornis*, a European species. Göllner-Scheiding (1975, Deut. Ent. Zeit., 22: 37-39) showed that all North American records apply to *S. punctiventris*. Yonke and Medler (1967, Proc. No. Centr. Branch, Ent. Soc. Am., 22: 74-75) presented life history notes [as *crassicornis*].

Stictopleurus viridicatus (Uhler), 1872
1872 *Corizus viridicatus* Uhler, Prelim. Rept. U.S. Geol. Surv. Mont., p. 404. [Col., "Dak.," Neb.]. See note below.
1876 *Corizus hyalinus*: Uhler, Bull. U.S. Geol. Geogr. Surv. Terr., 1: 300 (in part).
1903 *Corizus hyalinus* var. *viridicatus*: Van Duzee, Trans. Am. Ent. Soc., 29: 109.
1908 *Liorhyssus hyalinus* form *viridicatus*: Baker, Can. Ent., 40: 243.
1908 *Corizus novaeboracensis* form *pallidus* Baker, Can. Ent., 40: 243. [Col. to Nev.]. Preoccupied. Synonymized by Harris, 1944, Ia. St. Coll. J. Sci., 19: 103.
1913 *Rhopalus viridicatus*: Bergroth, Mem. Soc. Ent. Belg., 22: 163 (in part).
1916 *Corizus* (*Liorhyssus*) *viridicatus*: Van Duzee, Check List Hem., p. 15.
1941 *Corizus* (*Stictopleurus*) *viridicatus*: Torre-Bueno, An. Ent. Soc. Am., 34: 285.
1944 *Stictopleurus viridicatus*: Harris, Ia. St. Coll. J. Sci., 19: 103.
Distribution: Alta., Ariz., B.C., Cal., Col., D.C., Ia., Id., Ks., Minn., Mont., N.D., N.M., Neb., Nev., S.D., Sask., Ut., Wash., Wyo.
Note: Harris (1944, Ia. St. Coll. J. Sci., 19: 103) showed that part of Uhler's (1872, above) syntype series of *S. viridicata* contained specimens of *Liorhyssus hyalinus* (which see).

Genus Rhopalus Schilling, 1827

1827 *Rhopalus* Schilling, Uebers. Arb. Schles. Ges. Kult., p. 22. Type-species: *Lygaeus capitatus* Fabricius, 1794, a junior synonym of *Cimex subrufus* Gmelin, 1790. Designated by Westwood, 1840, Intr. Class. Ins., 2(B): 123.
Note: The name "*Rhopalus maculigerus* Uhler" listed from Illinois by Walker (1872, Cat. Hem. Brit. Mus., 5: 21) is a *nomen nudum*. Göllner-Scheiding (1978, Mitt. Zool. Mus. Berlin, 54: 313-331) revised this genus and provided keys to subgenera and species.

Subgenus *Brachycarenus* Fieber, 1861

1861 *Brachycarenus* Fieber, Europ. Hem., p. 236. Type-species: *Rhopalus tigrinus* Schilling, 1829. Monotypic.
1978 *Rhopalus* (*Brachycarenus*): Göllner-Scheiding, Mitt. Zool. Mus. Berlin, 54: 327.

Rhopalus tigrinus Schilling, 1829

1829 *Rhopalus tigrinus* Schilling, Beit. Ent. Schles. Fauna, 1: 53. [Europe].
1977 *Brachycarenus tigrinus*: Hoebeke, Coop. Plant Pest Rept., 2: 802.
1982 *Rhopalus* (*Brachycarenus*) *tigrinus*: Hoebeke and Wheeler, Proc. Ent. Soc. Wash., 84: 213.
Distribution: Md., N.J., N.Y., Pa. (Asia Minor, Europe).
Note: Hoebeke and Wheeler (1982, above) redescribed the adult, reviewed host plants and literature, and gave the North American distribution of this recently discovered immigrant.

Subfamily Serinethinae Stål, 1873

Genus *Boisea* Kirkaldy, 1910

1910 *Leptocoris* (*Boisea*) Kirkaldy, Proc. Haw. Ent. Soc., 3: 123. Type-species: *Lygaeus trivittatus* Say, 1825. Original designation.
1980 *Boisea*: Göllner-Scheiding, Deut. Ent. Zeit., 27: 144.
Note: Schaefer (1975, An. Ent. Soc. Am., 68: 537-541) studied species relationships and distribution of *B. rubrolineata* and *B. trivittata*. Schaefer and Chopra (1982, An. Ent. Soc. Am., 75: 231) questioned Göllner-Scheiding's (1980, above) decision to give *Boisea* generic status.

Boisea rubrolineata (Barber), 1956

1956 *Leptocoris rubrolineatus* Barber, Pan-Pac. Ent., 32: 9. [Cal.].
1980 *Boisea rubrolineata*: Göllner-Scheiding, Deut. Ent. Zeit., 27: 147.
1986 *Boisea rubrolineatus* [*sic*]: Schowalter, Environ. Ent., 15: 1055.
Distribution: Ariz., B.C., Cal., Id., Nev., Ore., Tex., Wash.
Note: Early western records of *L. trivittatus* refer to *B. rubrolineata*, as noted by Barber (1956, above). Schowalter (1986, above) presented information on overwintering habits.

Boisea trivittata (Say), 1825 [Fig. 139]

1825 *Lygaeus trivittatus* Say, J. Acad. Sci. Phila., 4: 322. [Mo.].
1870 *Leptocoris trivittatus*: Stål, K. Svens. Vet.-Akad. Handl., 9(1): 226.
1872 *Lygaeus californicus* Walker, Cat. Het. Brit. Mus., 5: 42. [Cal. in error]. Synonymized by Barber, 1956, Pan-Pac. Ent., 32: 9.
1910 *Leptocoris* (*Boisea*) *vittatus* [*sic*]: Kirkaldy, Proc. Haw. Ent. Soc., 3: 123.
1980 *Boisea trivittata*: Göllner-Scheiding, Deut. Ent. Zeit., 27: 148.
Distribution: Alta., Ariz., B.C.(?), Col., Conn., D.C., Fla., Ia., Ill., Ind., Ks., Mass., Md., Minn., Miss., Mo., Mont., N.D., N.H., N.J., N.M., N.Y., Neb., Oh., Ok., Ont., Pa., R.I., S.D., Sask., Tenn., Tex., Ut., Va., W.Va., Wis.
Note: Smith and Shepherd (1937, Trans. Ks. Acad. Sci., 40: 143-159) and Tinker (1952, Ecology, 33: 407-414) studied the life history. Slater and Schaefer (1963, Bull.

Brook. Ent. Soc., 58: 114-117) gave an interesting discussion on distribution and the eastern spread in New England. Wheeler (1982, Bed Bugs, pp. 337-341) provided an extensive review of literature on economic importance and life history.

Genus *Jadera* Stål, 1862

1862 *Jadera* Stål, K. Svens. Vet.-Akad. Handl., 3(6): 59. Type-species: *Leptocoris coturnix* Burmeister, 1835. Designated by Van Duzee, 1916, Check List Hem., p. 15.

Note: Göllner-Scheiding (1979, Deut. Ent. Zeit., 26: 47-75) revised this genus and provided a key to species. Mead (1985, Fla. Dept. Agr. Cons. Serv., Ent. Circ. No. 277, 2 pp.) reviewed the species occurring in Florida.

Jadera antica (Walker), 1872

1872 *Lygaeus anticus* Walker, Cat. Het. Brit. Mus., 5: 46. [West Indies]. Lectotype designated by Göllner-Scheiding, 1979, Deut. Ent. Zeit., 26: 52.

1979 *Jadera antica*: Göllner-Scheiding, Deut. Ent. Zeit., 26: 52.

Distribution: Fla. (West Indies).

Note: Göllner-Scheiding (1979, above) has indicated that records of *J. sanguinolenta* (Fabricius) from Fla. are misidentifications of *J. antica*. *J. sanguinolenta* is restricted to the West Indies.

Jadera coturnix (Burmeister), 1835, Revised Status

1835 *Leptocoris coturnix* Burmeister, Handb. Ent., 2: 305. [Brazil]. Lectotype designated by Göllner-Scheiding, 1979, Deut. Ent. Zeit., 26: 49.

1852 *Serinetha aeola* Dallas, List Hem. Brit. Mus., 2: 463. [Mexico]. Synonymized by Göllner-Scheiding, 1979, Deut. Ent. Zeit., 26: 49.

1870 *Jadera aeola*: Stål, K. Svens. Vet.-Akad. Handl., 9(1): 227.

Distribution: Tex. (Mexico to South America).

Note: Stål (1868, K. Svens. Vet.-Akad. Handl., 7(11): 67-68) considered *J. coturnix* (Burmeister), 1835, a junior synonym of *J. sanquinolenta* (Fabricius), 1775, but Göllner-Scheiding (1979, above, 26: 48-49) re-evaluated the species and placed *J. coturnix* in synonymy with *J. aeola*, not *J. sanquinolenta*. However, she incorrectly argued to maintain the name *J. aeola*, based on its long usage, rather than *J. coturnix*, which has priority. [It should be noted also that *Pyrrhotes cinerea* Westwood, 1842 and *Therapha cinerea* Amyot and Serville, 1843, were considered synonymous with *J. aeola* by Göllner-Scheiding (1979, above), and, likewise, have priority in order after *J. coturnix*.]. Tex. records of *J. sanguinolenta* (Fabricius) refer to this species.

Jadera haematoloma (Herrich-Schaeffer), 1847

1847 *Leptocoris haematoloma* Herrich-Schaeffer, Wanz. Ins., 8: 103. [Mexico].

1870 *Jadera haematoloma*: Stål, K. Svens. Vet.-Akad. Handl., 9(1): 226.

1918 *Pyrrhotes haematoloma*: Malloch, Ent. News, 29: 284.

Distribution: Ala., Ariz., Cal., Col., Fla., Ill., Ks., Mo., N.C., Tex. (Mexico to Colombia and Venezuela, West Indies).

Jadera hinnulea Göllner-Scheiding, 1979

1979 *Jadera hinnulea* Göllner-Scheiding, Deut. Ent. Zeit., 26: 60. [Mexico; also paratypes from Tex.].

1982 *Jadera bayardae* Göllner-Scheiding, Deut. Ent. Zeit., 29: 459. Unnecessary new name for *J. hinnulea* Göllner-Scheiding, 1979.

Distribution: Tex. (Mexico to Panama).

Note: Göllner-Scheiding (1979, above) published the description of this species using a manuscript name given by E. B. Bayard (1943, M.S. Thesis, Ia. St. Coll.) and, later (Göllner-Scheiding, 1982, above), proposed the new name *J. bayardae*. Her earlier validation of the name *hinnulea* has priority (Göllner-Scheiding, 1984, Deut. Ent. Zeit., new ser., 31: 337).

Family Saldidae Amyot and Serville, 1843

The Shore Bugs

By John T. Polhemus

The saldids are a cosmopolitan group ranging over all of North America as far north as the shores of the Arctic Ocean. They are oval or elongate-oval bugs with moderately long legs fitted for running or jumping. The long antennae are four segmented. The hemelytra are divided into a clavus, corium, and membrane, the latter having veins which delimit 4 or 5 cells. There are 23 described genera worldwide of which 12 occur in North America.

As their common name implies, most saldids are littoral, inhabiting lake shores, beaches and stream banks. However, the variety of habitats utilized by the Saldidae in North America is much wider, from intertidal salt marshes to almost dry land far from water. The more secretive species living in dense grasses, on large stones, or on vertical rock faces blend in with the substrate and may hide in holes or crevices. Most Saldidae are quick to jump or fly when disturbed and are quite difficult to capture. These bugs are not often taken by the general collector and are poorly represented in most collections.

The eggs are generally laid at the base of grass blades or inserted into them, with a developmental time to imago of about a month. Saldids are predators and scavengers, feeding on organisms found on the surface or in the damp surface layers of the substrate.

Some species overwinter as adults and others in the egg stage. Biological studies of several species occurring in North America have been published by Wiley (1922, Univ. Ks. Sci. Bull., 14: 301-311) and Stock and Lattin (1976, J. Ks. Ent. Soc., 49: 313-326). A review of previous works and much original data on biology, ecology, morphology, and phylogeny was given by Polhemus (1985, Shore Bugs, 252 pp.).

Taxonomically the group is very difficult; the literature is voluminous and replete with misidentifications. The phylogeny of the Saldidae and closely related families (the Leptopodomorpha) has been reviewed by Schuh and Polhemus (1980, Syst. Zool., 29: 1-26) and Polhemus (1985, Shore Bugs, 252 pp.). The latter contains a generic review, one new North American genus, and retains but revises the genera established by Reuter (1912, Öfv. F. Vet.-Soc. Förh, 54 (A, 12): 1-24). The higher order classification was revised by Cobben (1959, Zool. Meded. Leid., 36: 303-316). The last comprehensive revision of North American Saldidae was by Uhler (1877, Bull. U.S. Geol. Geogr. Surv. Terr., 3: 428-475), but regional treatments containing keys may be found in Chapman (1962, Pan-Pac. Ent., 38: 147-159; Nevada), Schuh (1967, Contr. Am. Ent. Inst., 2: 1- 35; Great Lakes Region), Brooks and Kelton (1967, Mem. Ent. Soc. Can., 51: 1-92; Canada, Prairie Provinces), Polhemus and Chapman (1979, Bull. Cal. Ins. Surv., 21: 16-33; California), and Polhemus (1985, Shore Bugs; Great Basin southward).

Because of the many misidentifications in the literature, much of the distributional data contained in this section is derived

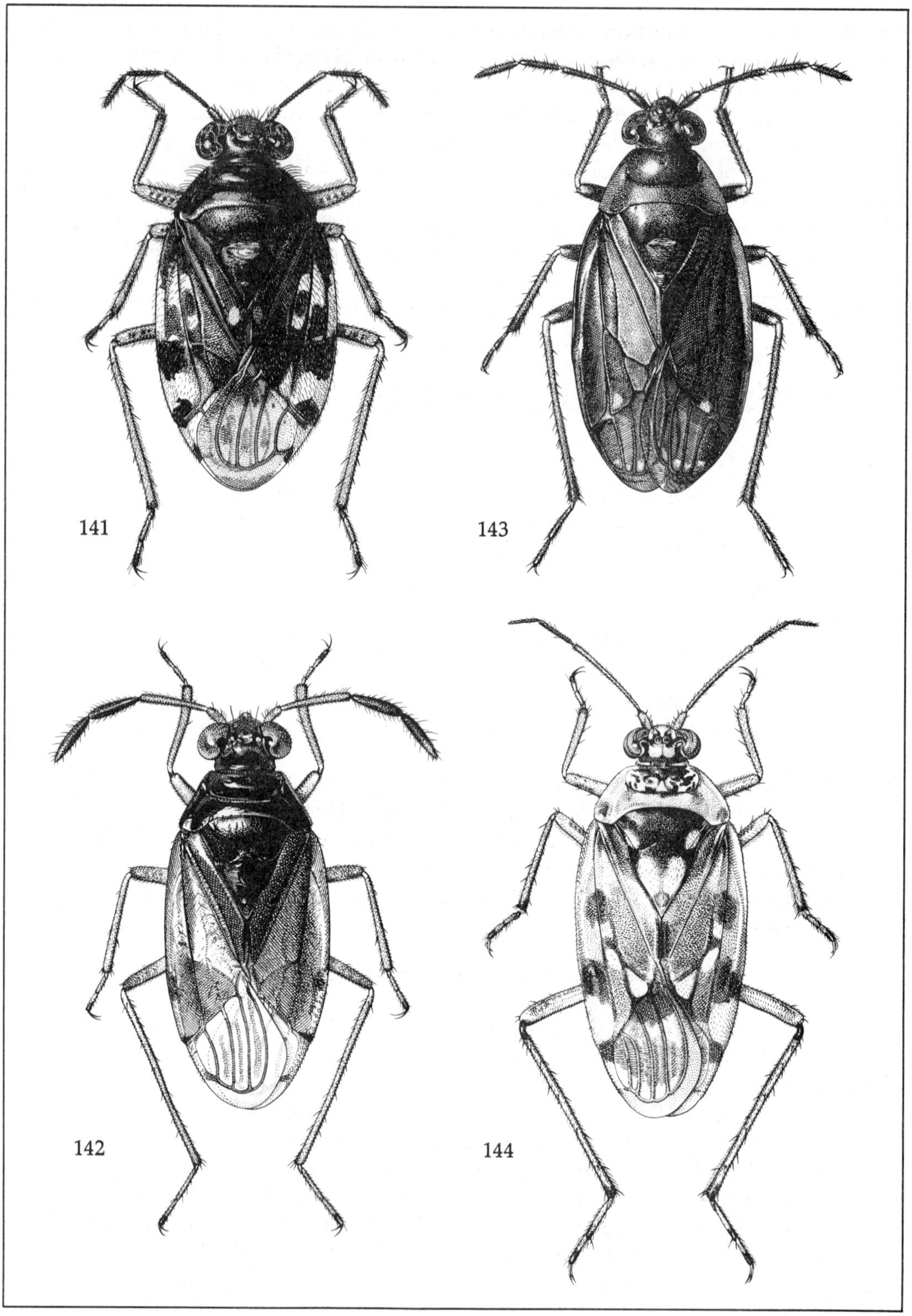

Figs. 141-144: 141, *Saldula comatula* [p.677]; 142, *Ioscytus politus* [p.673]; 143, *Salda buenoi* [p.669]; 144, *Pentacora signoreti* [p.668] (After Usinger, 1956).

from the Polhemus Collection. Literature sources were used if they were considered to be reliable. Lethierry and Severin (1896, Cat. Gen. Hem., 3: 223) considered Uhler's (1877, Bull. U.S. Geol. Geogr. Surv. Terr., 3: 450) U.S. records of "*Salda littoralis*" to be misidentifications of "*Salda scotica*" (Curtis) and thus could list the latter species for "America Bor." In this they were followed by Banks (1910, Cat. Nearc. Hem.-Het., p. 12); Van Duzee (1916, Check List Hem., p. 50 and 1917, Univ. Cal. Publ. Ent., 2: 445). Blatchley (1926, Het. E. N. Am., p. 1008) questioned the occurrence of *S. scotica* in North America, and Drake (1952, Proc. Ent. Soc. Wash., 54: 147) pointed out the American record was in error. Polhemus (1985, Proc. Ent. Soc. Wash., 87: 893) removed *Saldula c-album* (Fieber) from the list of American species, stating that the records pertain to *Saldula saltatoria* (Linnaeus) or an undescribed species.

Subfamily Chiloxanthinae Cobben, 1959

Genus *Chiloxanthus* Reuter, 1891

1891 *Chiloxanthus* Reuter, Medd. Soc. Fauna Flora Fenn., 17: 145. Type-species: *Salda pilosa* Fallén, 1807. Designated by Reuter, 1912, Öfv. F. Vet.-Soc. Förh., 54(A)12: 12.

Chiloxanthus arcticus (Sahlberg), 1878

1878 *Salda arctica* Sahlberg, K. Svens. Vet.-Akad. Handl., 16(4): 33. [N.W. Siberia].
1960 *Chiloxanthus arcticus*: Usinger, Pan-Pac. Ent., 36: 190.
Distribution: Alk. (Scandinavia, Siberia).
Note: Usinger (1960, above) gave notes on the biology of this species in Alk.

Chiloxanthus stellatus (Curtis), 1835

1835 *Acanthia stellata* Curtis, Descrip. Ins. App. Narr. Second Voyage N.W. Passage, p. 75. [Alk.].
1877 *Salda stellata*: Uhler, Bull. U.S. Geol. Geogr. Surv. Terr., 3: 442.
1916 *Chiloxanthus stellata*: Van Duzee, Check List Hem., p. 50.
1919 *Chiloxanthus stellatus*: Van Duzee, Rept. Can. Arctic Exp., 3(F): 47.
Distribution: Alk., N.T. (N. Europe, Siberia).
Note: Usinger (1960, Pan-Pac. Ent., 36: 189-190) gave notes on the biology of this species in Alk.

Genus *Pentacora* Reuter, 1912

1912 *Pentacora* Reuter, Öfv. F. Vet.-Soc. Förh., 54(A)12: 10. Type-species: *Salda signoreti* Guérin-Méneville 1857. Original designation.

Pentacora hirta (Say), 1832

1832 *Acanthia hirta* Say, Descrip. Het. Hem. N. Am., p. 34. [Ind.]. Neotype from Fla. designated by Polhemus, 1985, Shore Bugs, p. 115.
1877 *Salda pellita* Uhler, Bull. U.S. Geol. Geogr. Surv. Terr., 3: 433. [Mass.]. Synonymized by Torre-Bueno, 1923, Conn. Geol. Nat. Hist. Surv. Bull. 34, Add. et Corrig., p. 3. Lectotype designated by Polhemus, 1985, Shore Bugs, p. 115.
1896 *Salda hirta*: Lethierry and Severin, 1896, Cat. Gen. Hem., 3: 218.
1906 *Salda vagator* Snow, Trans. Ks. Acad. Sci., 20: 153 (Uhler ms. name). *Nomen nudum*.
1912 *Pentacora hirta*: Reuter, Öfv. F. Vet.-Soc. Förh., 54(A)12: 11.
1916 *Saldula pellita*: Van Duzee, Check List Hem., p. 50.

1923 *Pentacora pellita*: Torre-Bueno, Conn. Geol. Nat. Hist. Surv. Bull., 34: 411.
Distribution: Ala., Conn., Fla., Ia., Ind., Mass., Md., Me., Miss., N.C., N.Y., Pa., Que., Tex. (Mexico, West Indies).

Pentacora ligata (Say), 1832
1832 *Acanthia ligata* Say, Descrip. Het. Hem. N. Am., p. 34. [Ind.].
1872 *Salda variegata* Provancher, Nat. Can., 4: 107. [Que.]. Synonymized by Provancher, 1888, Pet. Faune Ent. Can., 3: 189. Lectotype designated by Kelton and Lattin, 1968, Nat. Can., 95: 663.
1876 *Salda ligata*: Uhler, Bull. U.S. Geol. Geogr. Surv. Terr., 1: 333.
1912 *Pentacora ligata*: Reuter, Öfv. F. Vet.-Soc. Förh., 54(A)12: 11.
Distribution: Col., Conn., Ga., Ia., Ill., Ind., Ks., La., Man., Mass., Md., Me., Mich., Minn., Miss., Mo., N.H., N.J., N.C., N.Y., Nebr., Nfld., Oh., Ok., Ont., Pa., Que., S.C., Tenn., Tex., Va., Wis. (Mexico).
Note: Uhler (1878, Proc. Bost. Soc. Nat. Hist., 19: 431) reported a Harris manuscript name "*Acanthia maritima*" for this species.

Pentacora saratogae Cobben, 1965
1965 *Pentacora saratogae* Cobben, Pan-Pac. Ent., 41: 180. [Cal.].
Distribution: Cal.
Note: Described from Death Valley and known only from inland salty biotopes.

Pentacora signoreti (Guérin-Méneville), 1857 [Fig. 144]
1857 *Salda signoretii* [*sic*] Guérin-Méneville, Hist. Phys. Polit. Nat. Cuba, p. 401. [Cuba].
1873 *Acanthia signoretii* [*sic*]: Stål, K. Svens. Vet.-Akad. Handl., 11(2): 148.
1912 *Pentacora signoreti*: Reuter, Öfv. F. Vet.-Soc. Förh., 54(A)12: 11.
1949 *Pentacora signoretii* [*sic*]: Hodgden, J. Ks. Ent. Soc., 22: 150.
Distribution: Ala., Cal., Col., Conn., Fla., Ga., Ia., Ks., Man., Mass., Md., Mo., N.C., N.H., N.J., N.M., N.J., Neb., S.C., S.D., Sask., Tex., Ut. (Mexico, West Indies).
Note: The original -ii ending of this species has been emended by several authors to -i. This change is accepted here in order to be consistent with the world Leptopodomorpha catalog (R. T. Schuh, B. Galil, and J. T. Polhemus, 1987, Bull. Am. Mus. Nat. Hist., 185: 272) and usage elsewhere in this catalog (e.g. *Ambrysus signoreti* Stål). This species occurs along seacoasts and inland on salty biotopes.

Pentacora signoreti signoreti (Guérin-Méneville), 1857
1857 *Salda signoretii* [*sic*] Guérin-Méneville, Hist. Phys. Polit. Nat. Cuba, p. 401. [Cuba].
1949 *Pentacora signoretii* [*sic*] *signoretii* [*sic*]: Hodgden, J. Ks. Ent. Soc., 22: 150.
1985 *Pentacora signoreti signoreti*: Polhemus, Shore Bugs, p. 119.
Distribution: Same as for species.

Pentacora sphacelata (Uhler), 1877
1877 *Salda sphacelata* Uhler, Bull. U.S. Geol. Geogr. Surv. Terr., 3: 434. [Mass., Md.]. Lectotype from Md. designated by Polhemus, 1969, An. Ent. Soc. Am., 62: 1208.
1909 *Acanthia sphacelata*: Kirkaldy and Torre-Bueno, Proc. Ent. Soc. Wash., 10: 178.
1916 *Saldula sphacelata*: Van Duzee, Check List Hem., p. 50.
1923 *Pentacora sphacelata*: Torre-Bueno, Conn. Geol. Nat. Hist. Surv. Bull., 34: 412.
Distribution: Cal., Conn., Del., Fla., Mass., Md., Me., Miss., Nfld., N.J., N.Y., R.I., Tex., Ut.(?) (Europe, Galapagos, Mexico to South America, West Indies).

Note: The Missouri record for this species has been shown by Froeschner (1962, Am. Midl. Nat., 67: 229) to pertain instead to *P. ligata* (Say). The Utah record, not confirmed in modern times despite extensive collecting, is suspect.

Subfamily Saldinae Amyot and Serville, 1843

Tribe Saldini Amyot and Serville, 1843

Genus *Lampracanthia* Reuter, 1912

1912 *Lampracanthia* Reuter, Öfv. F. Vet.-Soc. Förh., 54(A)12: 21. Type-species: *Salda crassicornis* Uhler, 1877. Monotypic

1959 *Salda (Lampracanthia)*: Cobben, Zool. Meded. Leid., 36: 307.

Note: Polhemus and Chapman, (1979, Bull. Cal. Ins. Surv., 21: 19) restored *Lampracanthia* to generic status.

Lampracanthia crassicornis (Uhler), 1877

1877 *Salda crassicornis* Uhler, Bull. U.S. Geol. Geogr. Surv. Terr., 3: 438. [Sask.].

1909 *Acanthia crassicornis*: Kirkaldy and Torre-Bueno, Proc. Ent. Soc. Wash., 10: 176.

1912 *Lampracanthia crassicornis*: Reuter, Öfv. F. Vet.-Soc. Förh., 54(A)12: 21.

1918 *Lampracanthia crassicornia* [*sic*]: Hungerford, J. N.Y. Ent. Soc., 26: 15 (in part).

1967 *Salda (Lampracanthia) crassicornis*: Schuh, Contr. Am. Ent. Inst., 2: 26.

Distribution: Alk., B.C., Col., Ind., N.H., N.Y., Neb., Nfld., Que., Sask.

Genus *Salda* Fabricius, 1803

1803 *Salda* Fabricius, Syst. Rhyn., p. 113. Type-species: *Cimex littoralis* Linnaeus, 1758. Designated by Int. Comm. Zool. Nomen., 1954, Opinion 245.

1843 *Sciodopterus* Amyot and Serville, Hist. Nat. Ins. Hem., p. 404. Type-species: *Salda flavipes* Fabricius, 1803. Monotypic. Synonymized by Fieber, 1859, Wien. Ent. Monat., 3: 231.

Note: The history of type fixing for the genus *Salda* Fabricius is described by China (1947, Bull. Zool. Nomen., 1: 276).

Salda alta Polhemus, 1967

1967 *Salda alta* Polhemus, Proc. Ent. Soc. Wash., 69: 29. [Col.].

Distribution: Col.

Salda anthracina Uhler, 1877

1877 *Salda anthracina* Uhler, Bull. U.S. Geol. Geogr. Surv. Terr., 3: 438. [U.S.].

1909 *Acanthia anthracina*: Kirkaldy and Torre-Bueno, Proc. Ent. Soc. Wash., 10: 175.

1923 *Lampracanthia anthracina*: Torre-Bueno, Conn. Geol. Nat. Hist. Surv. Bull., 34: 416.

1967 *Salda (Salda) anthracina*: Schuh, Contr. Am. Ent. Inst., 2: 22.

Distribution: Ala., Alk., B.C., Ks., Mich., N.H., N.J., N.Y., Neb., Oh., Pa., Tenn.

Salda buenoi (McDunnough), 1925 [Fig. 143]

1925 *Saldula buenoi* McDunnough, Can. Ent., 57: 259. [Alta.].

1943 *Salda buenoi*: Harris, J. Ks. Ent. Soc., 16: 152.

1953 *Salda lugubris*: Strickland, Can. Ent., 85: 203.

1967 *Salda (Salda) buenoi*: Schuh, Contr. Am. Ent. Inst., 2: 24.
Distribution: Alta., Ariz., B.C., Cal., Col., Ia., Id., Ill., Man., Mass., Mich., Mont., N.D., N.H., N.M., Neb., Nev., Nfld., Ont., Ore., S.D., Sask., Ut., Wash., Wis., Wyo.

Salda coloradensis Polhemus, 1967
1967 *Salda coloradensis* Polhemus, Proc. Ent. Soc. Wash., 69: 27. [Col.].
Distribution: Col.

Salda littoralis (Linnaeus), 1758
1758 *Cimex littoralis* Linnaeus, Syst. Nat., Ed. 10, p. 442. [Europe].
1893 *Salda littoralis*: Uhler, Proc. Ent. Soc. Wash., 2: 382.
1909 *Acanthia littoralis*: Kirkaldy and Torre-Bueno, Proc. Ent. Soc. Wash., 10: 177.
Distribution: Alk., Cal., Col., Ia.(?), Ill.(?), Ind.(?), Man., Me., Mont., N.D., N.T., N.Y., Nfld., Que., Ut.(?) (Europe, Iceland, Siberia).
Note: Records for this species from the 48 contiguous United States may well refer to *Salda buenoi* (McDunnough), *Salda coloradensis* Polhemus, or *Salda provancheri* Kelton and Lattin; the specimens reported from Alaska by Drake (1952, Proc. Ent. Soc. Wash., 54: 145) were reexamined and are of *S. littoralis*.

Salda lugubris (Say), 1832
1832 *Acanthia lugubris* Say, Descrip. Het. Hem. N. Am., p. 34. [Mo.]. Neotype from Ia. designated by Polhemus, 1985, Shore Bugs, p. 131.
1872 *Salda major* Provancher, Nat. Can., 4: 107. [Que.]. Synonymized by Drake and Hottes, 1950, Great Basin Nat., 10: 55. Lectotype designated by Kelton and Lattin, 1968, Nat. Can., 95: 662.
1876 *Salda lugubris*: Uhler, Bull. U.S. Geol. Geogr. Surv. Terr., 1: 333.
1877 *Salda deplanata* Uhler, Bull. U.S. Geol. Geogr. Surv. Terr., 3: 442. [Ill., Mass., Md., Me., Mich., Minn., Mo., N.M., N.T., N.Y., Ont., Tex.]. Synonymized with *S. major* by Van Duzee, 1912, Can. Ent., 44: 324. Lectotype from Mass. designated by Polhemus, 1985, Shore Bugs, p. 131.
1909 *Acanthia deplanata*: Kirkaldy and Torre-Bueno, Proc. Ent. Soc. Wash., 10: 176.
1909 *Acanthia major*: Kirkaldy and Torre-Bueno, Proc. Ent. Soc. Wash., 10: 177.
1916 *Saldula lugubris*: Van Duzee, Check List Hem., p. 50.
1916 *Saldula major*: Van Duzee, Check List Hem., p. 50.
Distribution: Alta., Ariz., B.C., Col., Fla., Ga., Ia., Ill., Ind., Ks., Man., Mass., Md., Me., Mich., Minn., Mo., Mont., N.C., N.D., N.H., N.J., N.M., N.T., N.Y., Nfld., Oh., Ont., Pa., Que., R.I., S.C., S.D., Sask., Tex., Ut., Va., Wis., Wyo., Yuk. (Mexico).
Note: Some old records listed as *Salda coriacea* Uhler may pertain to this species.

Salda obscura Provancher, 1872
1872 *Salda obscura* Provancher, Nat. Can., 4: 107. [Canada].
1909 *Acanthia obscura*: Kirkaldy and Torre-Bueno, Proc. Ent. Soc. Wash., 10: 177.
1916 *Salda littoralis*: Van Duzee, Check List Hem., p. 50 (in part).
1925 *Saldula (Acanthia) obscura*: McDunnough, Can. Ent., 57: 257.
1935 *Lampracanthia obscura*: Downes, Proc. Ent. Soc. B.C., 31: 47.
1967 *Salda (Salda) obscura*: Schuh, Contr. Am. Ent. Inst., 2: 25.
Distribution: Alk., Alta., Col., Man., Mich., Nev., Nfld., Ore., Que., Sask., Wash.

Salda provancheri Kelton and Lattin, 1968
1869 *Acanthia coriacea* Rathvon, Hist. Lancaster Co., Pa., p. 550. *Nomen nudum*.
1872 *Salda coriacea* Uhler, Prelim. Rept. Geol. Surv. Mont., 4: 421. ["British America," Ill., "New England," Ut.]. Preoccupied. "Type" from Ut. designated by Uhler, 1877, Bull. U.S. Geol. Geogr. Surv. Terr., 3: 438. Same specimen designated as

lectotype by Kelton and Lattin, 1968, Nat. Can., 95: 664.
1888 *Sciodopterus bouchervillei*: Provancher, Pet. Faune Ent. Can., 3: 192. See note below.
1909 *Acanthia bouchervillei*: Kirkaldy and Torre-Bueno, Proc. Ent. Soc. Wash., 10: 175.
1909 *Acanthia coriacea*: Kirkaldy and Torre-Bueno, Proc. Ent. Soc. Wash., 10: 176 (in part).
1923 *Lampracanthia coriacea*: Torre-Bueno, Bull. Brook. Ent. Soc., 18: 150.
1930 *Salda ceriacea* [*sic*]: Johnson, Publ. Nantucket Maria Mitchell Assoc., 3: 31.
1948 *Salda bouchervillei*: Hussey, Bull. Brook. Ent. Soc., 43: 153.
1967 *Salda* (*Salda*) *bouchervillei*: Schuh, Contr. Am. Ent. Inst., 2: 23.
1968 *Salda provancheri* Kelton and Lattin, Nat. Can., 95: 664. New name for *Salda coriacea* Uhler, 1872.

Distribution: Alk., Alta., Ariz., B.C., Cal., Col., Ga., Ill., Ind., Ks., Man., Mass., Me., Mich., Minn., Mo., N.D., N.J., N.S., N.T., N.Y., Neb., Nfld., Ont., Ore., Pa., Que., Sask., Tenn., Ut., Wis.

Note: *Salda provancheri* Kelton and Lattin was known for many years under the name *Salda bouchervillei* Provancher until Kelton and Lattin (1968, above) found the Provancher type (No. 73) to be a *Saldula* species. Provancher labeled another specimen (No. 125), a *Salda* species, as *bouchervillei* and this specimen was correctly synonymized with *S. coriacea* Uhler by Van Duzee (1912, above). [Note by R. C. Froeschner] Froeschner (1986, Proc. Ent. Soc. Wash., 88: 394) recently proposed to resurrect the preoccupied name *Salda coriacea* Uhler, arguing that it was a secondary homonym not noted until after 1960, at which time it became permissible to restore such names when they are no longer in conflict as homonyms. However, Froeschner overlooked Hussey's (1948, Bull. Brook. Ent. Soc., 43: 153) rejection that permanently invalidated *S. coriacea*. The next available name for this species is *S. provancheri*, proposed by Kelton and Lattin (1968, above).

Genus *Teloleuca* Reuter, 1912

1912 *Teloleuca* Reuter, Öfv. F. Vet.-Soc. Förh., 54(A)12: 17. Type-species: *Acanthia pellucens* Fabricius, 1779. Original designation.
1923 *Chartolampra* Torre-Bueno, Bull. Brook. Ent. Soc., 18: 154. Type-species: *Chartoscirta cursitans* Torre-Bueno, 1912, a junior synonym of *Acanthia pellucens* Fabricius, 1779. Monotypic. Synonymy by virture of common type-species.
1959 *Salda* (*Teloleuca*): Cobben, Zool. Meded. Leid., 36: 307.

Note: Polhemus and Chapman (1979, Bull. Cal. Ins. Surv., 21: 33) restored *Teloleuca* to generic status. Polhemus (1985, Shore Bugs, p. 130) further supported the generic validity of *Teloleuca* using cladistic analysis.

Teloleuca bifasciata (Thomson), 1871

1871 *Salda bifasciata* Thomson, Opusc. Ent., 20: 404. [Lapland, N. Europe].
1924 *Acanthia bellatrix* Torre-Bueno, Can. Ent., 56: 298. [Alta.]. Synonymized by Drake and Hottes, 1950, Great Basin Nat., 10: 52.
1950 *Teloleuca bifasciata*: Drake and Hottes, Great Basin Nat., 10: 52.
1967 *Salda* (*Teloleuca*) *bifasciata*: Schuh, Contr. Am. Ent. Soc., 2: 26.

Distribution: Alk., Alta., B.C., Cal.(?), Col., Mont., Que., Wash., Wyo., Yuk. (Europe, Siberia).

Teloleuca pellucens (Fabricius), 1779

1779 *Acanthia pellucens* Fabricius, Reise Norwegen Nat. Oekon., p. 234. [Norway].

1877 *Salda elongata* Uhler, Bull. U.S. Geol. Geogr. Surv. Terr., 3: 448. [B.C.]. Synonymized by Drake, 1952, Proc. Ent. Soc. Wash., 54: 147.
1909 *Acanthia elongata*: Kirkaldy and Torre-Bueno, Proc. Ent. Soc. Wash., 10: 176.
1923 *Chartoscirta (Chartolampra) cursitans* Torre-Bueno, Bull. Brook. Ent. Soc., 18: 151. [N.Y.]. Synonymized by Drake and Hottes, 1950, Great Basin Nat., 10: 52.
1924 *Acanthia celeripedis* Torre-Bueno, Can. Ent., 56: 296. [Alta.]. Synonymized by Drake and Hottes, 1950, Great Basin Nat., 10: 52.
1950 *Teloleuca pellucens*: Drake and Hottes, Great Basin Nat., 10: 52.
1953 *Salda celeripedis*: Strickland, Can. Ent., 86: 202.
1967 *Salda (Teloleuca) pellucens*: Schuh, Contr. Am. Ent. Inst., 2: 26.
Distribution: Alk., Alta., B.C., Col., Man., Mont., N.T., N.Y., Nfld., Ont., Ore., Sask., Wash., Wis. (N. Europe, Siberia).

Tribe Saldoidini Reuter, 1912

Genus *Calacanthia* Reuter, 1891

1891 *Acanthia (Calacanthia)* Reuter, Medd. Soc. Fauna Flora Fenn., 17: 145. Type-species: *Salda trybomi* Sahlberg, 1878. Designated by Reuter, 1895, Acta Soc. Sci. Fenn., 21(2): 8.
1912 *Calacanthia*: Reuter, Öfv. F. Vet.-Soc. Förh., 54(A)12: 19.

Calacanthia trybomi (Sahlberg), 1878
1878 *Salda trybomi* Sahlberg, K. Svens. Vet.-Akad. Handl., 16(4): 35. [Siberia].
1919 *Calacanthia trybomi*: Van Duzee, Rept. Can. Arctic Exp., 3(F): 4.
Distribution: Alk., Man., N.T. (Siberia).

Genus *Ioscytus* Reuter, 1912

1912 *Ioscytus* Reuter, Öfv. F. Vet.-Soc. Förh., 54(A)12: 19. Type-species: *Salda polita* Uhler, 1877. Original designation.
Note: This genus was revised by McKinnon and Polhemus (1986, J. N.Y. Ent. Soc., 94: 434-441) who furnished ecological notes and a key to species.

Ioscytus beameri (Hodgden), 1949
1949 *Salda beameri* Hodgden, J. Ks. Ent. Soc., 22: 150. [N.M.].
1951 *Ioscytus beameri*: Drake and Hoberlandt, Acta Ent. Mus. Nat. Prague, 26(376): 4.
Distribution: N.M.
Note: This species is known only from the type series taken at Las Cruces, New Mexico.

Ioscytus chapmani McKinnon and Polhemus, 1986
1986 *Ioscytus chapmani* McKinnon and Polhemus, J. N.Y. Ent. Soc., 94: 434. [Oh.].
Distribution: Ky., Oh., Va.

Ioscytus cobbeni Polhemus, 1964
1964 *Ioscytus cobbeni* Polhemus, Proc. Ent. Soc. Wash., 66: 253. [Col.].
Distribution: Ariz., Col., N.M., Tex. (Mexico).

Ioscytus franciscanus (Drake), 1949
1949 *Saldula franciscana* Drake, Psyche, 56: 192. [Cal.].
1951 *Ioscytus franciscana* [*sic*]: Drake and Hoberlandt, Acta Ent. Mus. Nat. Pragae,

26(376): 4.
1979 *Ioscytus franciscanus*: Polhemus and Chapman, Bull. Cal. Ins. Surv., 21: 22.
Distribution: Cal., Ore.

Ioscytus nasti Drake and Hottes, 1955
1951 *Ioscytus nasti* Drake and Hoberlandt, Acta Ent. Mus. Nat. Prague, 26(376): 9. *Nomen nudum.*
1955 *Ioscytus nasti* Drake and Hottes, Bol. Ent. Venezolana, 11: 3. [Cal.].
Distribution: Cal., Nev.

Ioscytus politus (Uhler), 1877 [Fig. 142]
1877 *Salda polita* Uhler, Bull. U.S. Geol. Geogr. Surv. Terr., 3: 442. [Cal.]. Lectotype designated by Polhemus, 1969, An. Ent. Soc. Am., 62: 1208.
1909 *Acanthia polita*: Kirkaldy and Torre-Bueno, Proc. Ent. Soc. Wash., 10: 177.
1912 *Ioscytus polita*: Reuter, Öfv. F. Vet.-Soc. Förh., 51(A)12: 21.
1914 *Ioscytus politus*: Van Duzee, Trans. San Diego Soc. Nat. Hist., 2: 32.
1968 *Saldula politus*: Cobben., Evol. Trends Het., 1: 8.
Distribution: Ariz., B.C., Cal., Col., N.M., Nev., Ore., Ut.

Ioscytus politus flavicosta Reuter, 1912
1912 *Ioscytus politus* var. *flavicosta* Reuter, Öfv. F. Vet.-Soc. Förh., 54(A)12: 21. [Ut.].
Distribution: Col., Nev., Ut.

Ioscytus politus politus (Uhler), 1877
1877 *Salda polita* Uhler, Bull. U.S. Geol. Geogr. Surv. Terr., 3: 442.
1979 *Ioscytus politus politus*: Polhemus and Chapman, Bull. Cal. Ins. Surv., 16: 22.
Distribution: Same as for species.

Ioscytus tepidarius (Hodgden), 1949
1949 *Salda tepidaria* Hodgden, J. Ks. Ent. Soc., 22: 161. [Mexico].
1985 *Ioscytus tepidarius*: Polhemus, Shore Bugs, p. 139.
Distribution: Ariz. (Mexico).

Genus *Macrosaldula* Leston and Southwood, 1964

1964 *Saldula* (*Macrosaldula*) Leston and Southwood, Ent. Month. Mag., 100: 80. Type-species: *Acanthia scotica* Curtis, 1833. Original designation.
1968 *Macrosaldula*: Wroblewski, Polsk. Tow. Ent., 58: 13.

Macrosaldula monae (Drake), 1952
1952 *Saldula monae* Drake, Proc. Ent. Soc. Wash., 54: 146. [Alk.].
1985 *Macrosaldula monae*: Cobben, Tijd. Ent., 128: 254.
Distribution: Alk.
Note: Cobben (1985, above) provisionally placed *monae*, which is known only from the Aleutian Islands, in the genus *Macrosaldula* which is otherwise Palearctic. He noted that this and two other species from the far eastern Palearctic deviate from typical *Macrosaldula* species.

Genus *Micracanthia* Reuter, 1912

1912 *Micracanthia* Reuter, Öfv. F. Vet.-Soc. Förh., 54(A)12: 16. Type-species: *Salda marginalis* Fallén, 1807. Original designation.
Note: Slater and Baranowski (1978, How To Know True Bugs, p. 224) commented on *Micracanthia* and wrote, "apparently in the future this will be reduced to subgeneric status" and so used it in listing species. To date, subsequent authors have kept

Micracanthia as a genus.

Micracanthia bergrothi Jakovlev, 1893

1893 *Salda bergrothi* Jakovlev, Horae Soc. Ent. Rossicae, 27: 304. [Siberia].

1957 *Micracanthia ripula* Drake, Bull. So. Cal. Acad. Sci., 56: 142. [Man.]. Synonymized by Polhemus, 1985, Proc. Ent. Soc. Wash., 87: 893.

Distribution: Alta., B.C., Man., Mich., Mont., N.Y.(?), Nfld., Sask., Wash., Wyo., Yuk. (Siberia).

Note: Records of *Micracanthia marginalis* (Fallén) from the New World (e.g. Drake and Chapman, 1953, Bull. Brook. Ent. Soc., 48: 64) probably pertain to *M. bergrothi* Jakovlev, which it resembles.

Micracanthia fennica (Reuter), 1884

1884 *Salda fennica* Reuter, Ent. Tidskr., 5: 171. [Finland].

1951 *Micracanthia fennica*: Drake and Hoberlandt, Acta Ent. Mus. Nat. Pragae, 26(376): 5.

Distribution: Alk., "Canada," Col., Ks., Mass., N.Y., Neb., Ore. (Europe, Siberia).

Micracanthia floridana Drake and Chapman, 1953

1953 *Micracanthia floridana* Drake and Chapman, Bull. Brook. Ent. Soc., 48: 64. [Fla.].

Distribution: Col., Fla., Ill., Ks., Mich., Miss., Mo., N.J., Pa., Tex.

Micracanthia humilis (Say), 1832

1832 *Acanthia humilis* Say, Descrip. Het. Hem. N. Am., p. 35. [Fla.]. Neotype designated by Polhemus, 1985, Shore Bugs, p. 143.

1877 *Salda humilis*: Uhler, Bull. U.S. Geol. Geogr. Surv. Terr., 3: 451.

1912 *Micracanthia humilis*: Reuter, Öfv. F. Vet.-Soc. Förh., 52(A) 16.

1978 *Saldula* (*Micracanthia*) *humilis*: Slater and Baranowski, How To Know True Bugs, p. 224.

Distribution: All 48 contiguous United States; Alta., B.C., Man., N.T., Que. (Mexico to Brazil, West Indies).

Micracanthia hungerfordi (Hodgden), 1949

1949 *Salda hungerfordi* Hodgden, J. Ks. Ent. Soc., 22: 156. [N.Y.].

1951 *Micracanthia hungerfordi*: Drake and Hoberlandt, Acta Ent. Mus. Nat. Pragae, 26(376): 5.

Distribution: Conn., Fla., Mass., Md., N.J., N.Y.

Micracanthia husseyi Drake and Chapman, 1952

1952 *Micracanthia husseyi* Drake and Chapman, Fla. Ent., 35: 148. [Fla.].

Distribution: Fla., La., Miss., N.J., Tex. (Belize, Mexico, West Indies).

Micracanthia pumpila Blatchley, 1928

1928 *Micranthia* [*sic*] *pumpila* Blatchley, J. N.Y. Ent. Soc., 36: 22. [Fla.].

1930 *Micranthia* [*sic*] *pumila* [*sic*]: Blatchley, Blatchleyana, p. 66.

1951 *Micracanthia pumpila*: Drake and Hoberlandt, Acta Ent. Mus. Nat. Pragae, 26(376): 5.

Distribution: Ala., Fla., Md., Tex., Va. (West Indies).

Note: According to the rules of the Int. Code of Zool. Nomen. (1985), Blatchley (1930, above) improperly emended the misspelling "*pumpila*" in the original description, which resulted from a printer's error.

Micracanthia quadrimaculata (Champion), 1900

1900 *Salda quadrimaculata* Champion, Biol. Centr.-Am., Rhyn., 2: 342. [Panama]. Lectotype designated by Polhemus, 1985, Shore Bugs, p. 149.

1914 *Micranthia* [*sic*] *pusilla* Van Duzee, Trans. San Diego Soc. Nat. Hist., 2: 32. [Cal.].

Synonymized by Drake and Hottes, 1950, Great Basin Nat., 10: 52.
1917 *Micracanthia pusilla*: Van Duzee, Univ. Cal. Publ. Ent., 2: 447.
1950 *Micracanthia quadrimaculata*: Drake and Hottes, Great Basin Nat., 10: 51.
1978 *Saldula (Micracanthia) quadrimaculata*: Slater and Baranowski, How To Know True Bugs, p. 224.
Distribution: B.C., Cal., Col., Fla.(?), Id., N.M., Nev., Ore., S.D., Wash., Wyo. (Central America, Mexico).

Micracanthia schuhi Lattin, 1968
1968 *Micracanthia schuhi* Lattin, Proc. Ent. Soc. Wash., 70: 165. [Ore.].
Distribution: Ore.

Micracanthia utahensis Drake and Hottes, 1955
1951 *Micracanthia utahensis* Drake and Hoberlandt, Acta Ent. Mus. Nat. Prague, 26(376): 5. *Nomen nudum*.
1955 *Micracanthia utahensis* Drake and Hottes, Bol. Ent. Venezolana, 11: 1. [Ut.].
Distribution: Col., N.M., Nev., Ut. (Mexico).

Genus *Rupisalda* Polhemus, 1985

1985 *Rupisalda* Polhemus, Shore Bugs, p. 151. Type-species: *Rupisalda petricola* Polhemus, 1985. Original designation.

Rupisalda dewsi (Hodgden), 1949
1949 *Salda dewsi* Hodgden, J. Ks. Ent. Soc., 22: 153. [Costa Rica].
1966 *Saldula dewsi*: Polhemus, Proc. Ent. Soc. Wash., 68: 57.
1985 *Rupisalda dewsi*: Polhemus, Shore Bugs, p. 156.
Distribution: Ariz. (Costa Rica, Mexico, Venezuela).

Rupisalda saxicola (Polhemus), 1972
1972 *Saldula saxicola* Polhemus, Great Basin Nat., 32: 143. [Ariz.].
1985 *Rupisalda saxicola*: Polhemus, Shore Bugs, p. 162.
Distribution: Ariz. (Central America, Mexico).

Rupisalda teretis (Drake) 1950
1950 *Saldula teretis* Drake, Bull. Brook. Ent. Soc., 45: 1. [Id.].
1985 *Rupisalda teretis*: Polhemus, Shore Bugs, p. 152.
Distribution: Id., Nev.

Genus *Saldoida* Osborn, 1901

1901 *Saldoida* Osborn, Can. Ent., 33: 181. Type-species: *Saldoida slossonae* Osborn, 1901. Designated by Kirkaldy, 1906, Trans. Am. Ent. Soc., 32: 148.
Note: A review of the genus was given by Drake and Chapman (1958, An. Ent. Soc. Am., 51: 480-485).

Saldoida cornuta Osborn, 1901
1901 *Saldoida cornuta* Osborn, Can. Ent., 33: 182. [Fla.].
Distribution: Fla., Miss.

Saldoida slossonae Osborn, 1901
1901 *Saldoida slossoni* [*sic*] Osborn, Can. Ent., 33: 181. [Fla.].
1958 *Saldoida slossonae*: Drake and Chapman, An. Ent. Soc. Am., 51: 483.
Distribution: Fla., Ga., Ks., La., Miss., N.J., Tex., Va.

Saldoida slossonae slossonae Osborn, 1901

1901 *Saldoida slossoni* [*sic*] Osborn, Can. Ent., 33: 181. [Fla.].

1958 *Saldoida slossonae*: Drake and Chapman, An. Ent. Soc. Am., 51: 483.

Distribution: Fla., Ga., La., Miss., N.J., Va.

Saldoida slossonae wileyae Hungerford, 1922

1922 *Saldoida slossoni* [*sic*] *wileyi* [*sic*] Hungerford, Bull. Brook. Ent. Soc., 17: 64. [Tex.]. Synonymized by Usinger, 1945, Bull. Brook. Ent. Soc., 40: 116.

1982 *Saldoida slossonae wileyae*: A. Slater, 1982, Tech. Publ. St. Biol. Surv. Ks., 12: 41.

Distribution: Ks., Tex.

Note: Usinger (1945, above) synonymized the subspecies *S. slossonae wileyae*, stating that it was within the limits of variability of *S. slossonae*; A. Slater (1982, Tech. Publ. St. Biol. Surv. Ks., 12: 41) listed it as a subspecies without comment.

Saldoida turbaria Schuh, 1967

1967 *Saldoida turbaria* Schuh, Contr. Am. Ent. Inst., 2: 12. [Mich.].

Distribution: Mich.

Genus *Saldula* Van Duzee, 1914

1912 *Acanthia*: Reuter, Öfv. F. Vet.-Soc. Förh., 54(A)12: 14.

1914 *Saldula* Van Duzee, Trans. San Diego Soc. Nat. Hist., 2: 32. New name for *Acanthia* of Reuter, 1912. Type-species: *Cimex saltatoria* Linnaeus, 1758. Original designation.

Note: Reuter (1912, above) redefined and restricted the genus *Acanthia* Fabricius, 1775, and accepted *Acanthia saltatoria* (Linnaeus) as type-species. Van Duzee (1914, above) rejected Reuter's usage and proposed the name *Saldula* to replace it; later (1914, Can. Ent., 46: 386-387) he repeated his proposal along with his explanation of the history of *Acanthia*. In 1924 the International Commission on Zoological Nomenclature (Opinion 81, Smithson. Misc. Coll., 73(2): 19) named *Cimex lectularius* Linnaeus the type-species of both *Cimex* Linnaeus and *Acanthia* Fabricius and thus made the latter a junior objective synonym. In that Commission's "Official Index of Rejected and Invalid Generic Names in Zoology" the name *Acanthia* Fabiricius, 1775, is listed as name number 904, based on Directive 63.

Saldula ablusa Drake and Hottes, 1954

1951 *Saldula ablusa* Drake and Hoberlandt, Acta Ent. Mus. Nat. Pragae, 26(376): 6. [E. U.S.]. *Nomen nudum.*

1954 *Saldula ablusa* Drake and Hottes, Occas. Pap. Mus. Zool., Univ. Mich., 553: 2. [Mich.].

Distribution: Ill., Ind., Mich., N.J., N.Y., Oh., Ont., S.D.(?), Wis.

Note: References to *Saldula xanthochila* (Fieber) (1859, Wien. Ent. Monat., 3: 234) and its subspecies *limbosa* (Horvath) (1891, Rev. d'Ent., 10: 80) from the northeastern United States almost certainly refer to this species or *S. pallipes* (which see).

Saldula andrei Drake, 1949

1949 *Saldula andrei* Drake, Arkiv Zool., 42B: 3 (July). [N.M.].

1949 *Salda laviniae* Hodgden, J. Ks. Ent. Soc., 22: 158 (Oct.). [Tex.]. Synonymized by Drake and Hottes, 1950, Great Basin Nat., 10: 60.

Distribution: Ariz., B.C., Cal., Col., Id., Mont., N.M., Nev., Tex., Ut., Wash., Wyo. (Mexico).

Saldula andrei andrei Drake, 1949

1949 *Saldula andrei* Drake, Arkiv Zool., 42B: 3.

1985 *Saldula andrei andrei*: Polhemus, Shore Bugs, p. 167.
Distribution: Same as for species.

Saldula andrei azteca Drake and Hottes, 1949
1949 *Saldula azteca* Drake and Hottes, Proc. Biol. Soc. Wash., 62: 177. [N.M.].
1985 *Saldula andrei azteca*: Polhemus, Shore Bugs, p. 168.
Distribution: Alta., B.C., Col., Id., N.M., Ut., Wash., Wyo. (Mexico).

Saldula balli Drake, 1950
1950 *Saldula balli* Drake, Bull. Brook. Ent. Soc., 45: 6. [Ariz.].
1950 *Saldula varionis* Drake and Hottes, Great Basin Nat., 10: 57. [Col.]. Synonymized by Polhemus, 1985, Shore Bugs, p. 171.
Distribution: Ariz., Cal., Col., Ks., N.M., Neb., Nev., Ut.

Saldula basingeri Drake, 1949
1949 *Saldula bassingeri* [*sic*] Drake, Psyche, 56: 190. [Cal.].
1979 *Saldula basingeri*: Polhemus and Chapman, Bull. Cal. Ins. Surv., 21: 26.
Distribution: Cal., Col.
Note: The records from Canada and Minnesota given by Brooks and Kelton (1967, Mem. Ent. Soc. Can., 51: 68) refer to a form of the *Saldula pallipes* (Fabricius) complex, but probably not *S. basingeri* Drake.

Saldula bouchervillei (Provancher), 1872
1872 *Sciodopterus bouchervillei* Provancher, Nat. Can., 4: 106. [Que.]. Lectotype designated by Kelton and Lattin, 1968, Nat. Can., 95: 664.
1878 *Salda separata* Uhler, Proc. Boston Soc. Nat. Hist., 19: 432. [U.S.]. Synonymized with *Saldula pallipes* (Fabricius), 1794, by Drake and Hoberlandt, 1951, Acta Ent. Mus. Nat. Pragae, 26(376): 9; restored to specific status by Polhemus, 1967, Proc. Ent. Soc. Wash., 69: 24; synonymized with *bouchervillei* by Kelton and Lattin, 1968, Nat. Can., 95: 664.
1909 *Acanthia bouchervillei*: Kirkaldy and Torre-Bueno, Proc. Ent. Soc. Wash., 10: 175.
1909 *Acanthia separata*: Kirkaldy and Torre-Bueno, Proc. Ent. Soc. Wash., 10: 178 (in part).
1916 *Saldula separata*: Van Duzee, Check List Hem., p. 50.
1949 *Saldula illinoiensis* Drake, Arkiv Zool., 42B: 2. [Ill.]. Synonymized with *S. separata* by Polhemus, 1967, Proc. Ent. Soc. Wash., 69: 24; synonymized with *S. bouchervillei* by Kelton and Lattin, 1968, Nat. Can., 95: 664.
1952 *Salda bouchervillei*: Drake, 1952, Proc. Ent. Soc. Wash., 54: 145.
Distribution: Alk., Alta., Col., Ia., Ill., Ind., Man., Mass., Mich., Minn., N.D., N.H., N.J., N.M.(?), N.Y., Neb., Nfld., Ont., Pa., Que., Sask., Tenn., Vt., Wis.

Saldula comatula Parshley, 1923 [Fig. 141]
1921 *Saldula comata* Parshley, Proc. Ent. Soc. B.C., 18: 21. [B.C.]. Preoccupied.
1922 *Saldula comatula* Parshley, Ent. News, 33: 71. New name for *Saldula comata* Parshley.
Distribution: Alta., Ariz., B.C., Cal., Col., Id., Ill., Ks., Mont., N.M., Neb., Nev., S.D., Sask., Ut., Wash., Wyo. (Mexico).

Saldula confluenta (Say), 1832
1832 *Acanthia confluenta* Say, Descrip. Het. Hem. N. Am., p. 35. [U.S.].
1859 *Acanthia confluens*: Leconte, Compl. Writings of Thomas Say, 1: 361.
1877 *Salda confluens*: Uhler, Bull. U.S. Geol. Geogr. Surv. Terr., 3: 433.
1916 *Saldula confluenta*: Van Duzee, Check List Hem., p. 50.
1917 *Saldula confluens*: Parshley, Occas. Pap. Boston Soc. Nat. Hist., 7: 110.

Distribution: Col., Conn., Ia., Ill., Ind., Ks., Man., Mich., Minn., Mont., N.C., N.J., N.Y., Oh., Ont., Que., S.C., S.D., Tenn., Tex., Va., Wis.

Saldula coxalis (Stål), 1873

1873 *Acanthia coxalis* Stål, K. Svens. Vet.-Akad. Handl., 11(2): 149. [Cuba].

1879 *Salda argentina* Berg, An. Soc. Cient. Arg., 9: 293. [Argentina]. Synonymized by Drake and Hottes, 1950, Great Basin Nat., 10: 57. Lectotype designated by Drake and Carvalho, 1948, Rev. Ent., 19: 474.

1914 *Acanthia xanthochila* var. *limbosa*: Barber, Bull. Am. Mus. Nat. Hist., 33: 499.

1950 *Saldula argentina*: Drake, Bull. Brook. Ent. Soc., 45: 3.

1950 *Saldula coxalis*: Drake and Hottes, Great Basin Nat., 10: 57.

Distribution: Fla., Tex. (Mexico, South America, West Indies).

Saldula dispersa (Uhler), 1893

1893 *Salda dispersa* Uhler, Proc. Ent. Soc. Wash., 2: 383. [Ut.]. Lectotype from Ut. designated by Polhemus, 1985, Proc. Ent. Soc. Wash., 87: 893.

1909 *Acanthia dispersa*: Kirkaldy and Torre-Bueno, Proc. Ent. Soc. Wash., 10: 176.

1916 *Saldula dispersa*: Van Duzee, Check List Hem., p. 50.

1949 *Saldula ourayi* Drake and Hottes, Proc. Biol. Soc. Wash., 62: 179. [Wash.]. Synonymized by Polhemus, 1985, Proc. Ent. Soc. Wash., 87. 893.

1955 *Saldula arenicola*: Drake and Hottes, Bol. Ent. Venezolana, 11: 12 (in part).

Distribution: B.C., Cal., Col.(?), Id., Nev., Ore., Ut., Wash., Wyo.(?).

Note: Polhemus (1985, above) reestablished the validity of this species and considered U.S. records for *Saldula arenicola* (Scholz) to be based on misidentifications and refer mainly to this species.

Saldula explanata (Uhler), 1893

1893 *Salda explanata* Uhler, Proc. Ent. Soc. Wash., 2: 383. [Ut.]. Lectotype designated by Polhemus, 1985, Shore Bugs, p. 181.

1909 *Acanthia explanata*: Kirkaldy and Torre-Bueno, Proc. Ent. Soc. Wash., 10: 176.

1916 *Saldula explanata*: Van Duzee, Check List Hem., p. 50.

Distribution: Alta., Ariz., B.C., Cal., Col., Id., Man., Mont., Nev., Ore., Sask., Ut., Wash., Wyo. (Mexico).

Saldula laticollis (Reuter), 1875

1875 *Acanthia (Salda) laticollis* Reuter, Pet. Nouv. Ent., 1: 544. [Alk., Siberia]. Synonymized with *Saldula pallipes* (Fabricius), 1794, by Drake and Hoberlandt, 1951, Acta Ent. Mus. Nat. Pragae, 26(376): 8; resurrected by Lindskog, 1981, Pan-Pac. Ent., 57: 322. Lectotype from Alk. designated by Lindskog, 1982, Pan-Pac. Ent., 57: 322.

1890 *Acanthia pallipes* var. *laticollis*: Reuter, Rev. Ent., 9: 251.

1896 *Salda laticollis*: Lethierry and Severin, Cat. Gen. Hem., 3: 219.

1909 *Acanthia laticollis*: Kirkaldy and Torre-Bueno, Proc. Ent. Soc. Wash., 10: 176.

1916 *Saldula laticollis*: Van Duzee, Check List Hem., p. 50.

1949 *Saldula fernaldi* Drake, Psyche, 56: 191. [Nfld.]. Synonymized with *Saldula palustris* (Douglas), 1874, by Drake, 1962, Proc. Biol. Soc. Wash., 75: 116; synonymized with *S. laticollis* by Lindskog, 1982, Pan-Pac. Ent., 57: 322.

1950 *Saldula notalis* [*sic*] Drake, Bull. Brook. Ent. Soc. 45: 4. [Cal.]. Synonymized by Lindskog, 1982, Pan-Pac. Ent., 57: 322.

1950 *Saldula notabilis* Drake, Bull. Brook. Ent. Soc., 45: 5. [Drake and Hoberlandt (1951, Acta Ent. Mus. Nat. Pragae, 26(376): 8) chose *notabilis* instead of *notalis* as correct name.].

1951 *Saldula pallipes*: Drake and Hoberlandt, Acta Ent. Mus. Nat. Pragae, 26(376): 8 (in part).
Distribution: Alk., Alta., B.C., Cal., Man., Nfld., Ore., Sask., Wash. (Mexico).
Note: Lindskog (1982, Pan-Pac. Ent., 57: 321-326) clarified the identity and many occurrences of this species and is being followed here, even to accepting his synonymy of *Saldula fernaldi* rather than Polhemus' (1985, Shore Bugs, p. 183--in press in 1981; see foreward of same for explanation) proposal to restore it as a separate species. Stock and Lattin (1976, J. Ks. Ent. Soc., 49: 313-316) reported on the biology [as *S. palustris*].

Saldula lattini Chapman and Polhemus, 1965
1965 *Saldula lattini* Chapman and Polhemus, J. Ks. Ent. Soc., 38: 359. [Ore.].
Distribution: Ore.

Saldula lomata Polhemus, 1985
1960 *Saldula palustris*: Cobben, Stud. Fauna Curação Caribb. Isl., 11: 50.
1985 *Saldula lomata* Polhemus, Shore Bugs, p. 186. [Tex.].
Distribution: Fla., Miss., N.J., S.C., Tex. (Central America, Mexico, West Indies).

Saldula luctuosa (Stål), 1859
1859 *Salda luctuosa* Stål, K. Svens. Freg. Eug. Resa, p. 263. [Cal.].
1873 *Acanthia luctuosa*: Stål, K. Svens. Vet.-Akad. Handl, 11(2): 149.
1916 *Saldula luctuosa*: Van Duzee, Check List Hem., p. 51.
Distribution: Cal.
Note: This species is known only from coastal salt marshes from La Jolla to Tomales Bay in California.

Saldula nigrita Parshley, 1921
1921 *Saldula nigrita* Parshley, Proc. Ent. Soc. B.C., 18: 23. [B.C.].
Distribution: Alta., B.C., Cal., Col., Man., Mich., Minn., Mont., N.H., N.M., N.S., Nev., Nfld., Ore., Que., Sask., Wash., Yuk. (Mexico).

Saldula opacula (Zetterstedt), 1840
1840 *Salda opacula* Zetterstedt, Ins. Lapp., p. 268. [Lapland].
1895 *Acanthia opacula*: Reuter, Acta Soc. Sci. Fenn., 21(2): 46.
1916 *Saldula opacula*: Van Duzee, Check List Hem., p. 50.
Distribution: Alk., Alta., B.C., Cal., Col., Fla., Ia., Ind., Man., Mass., Md., Me., Mich., Minn., Mont., N.J., N.Y., Neb., Nev., Nfld., Ore., Pa., Que., Sask., Ut., Wash., Wis. (N. Europe, Siberia).

Saldula opiparia Drake and Hottes, 1955
1916 *Saldula xanthochila*: Van Duzee, Check List Hem., p. 50 (in part, western records).
1950 *Saldula xanthochila limbosa*: Drake, Bull. Brook. Ent. Soc., 45: 3.
1951 *Saldula opiparia* Drake and Hoberlandt, Acta Ent. Mus. Nat. Prague, 26(376): 9. *Nomen nudum*.
1955 *Saldula opiparia* Drake and Hottes, Bol. Ent. Venezolana, 11: 9. [Ut.].
1962 *Saldula coxalis*: Chapman, Pan-Pac. Ent., 38: 154.
Distribution: Alta., Ariz., B.C., Cal., Col., Id., Man., Mont., N.T., Nev., Ont.(?), Ore., Sask., Ut., Wash., Wyo. (Mexico).
Note: Polhemus (1985, Shore Bugs, p. 194) redefined *Saldula opiparia* Drake and Hottes, 1955, and showed that western records attributed to other species with a light stripe on the lateral pronotal margin refer instead to this species.

Saldula orbiculata (Uhler), 1877

1877 *Salda orbiculata* Uhler, Bull. U.S. Geol. Geogr. Surv. Terr., 3: 450. [Cal., Ill., Mass., Pa., N.Y., Tex.]. Lectotype from Mass. designated by Polhemus, 1969, An. Ent. Soc. Am., 62: 1208.

1909 *Acanthia orbiculata*: Kirkaldy and Torre-Bueno, Proc. Ent. Soc. Wash., 10: 177.

1916 *Saldula orbiculata*: Van Duzee, Check List Hem., p. 50.

Distribution: Cal., Col.(?), Ia., Ill., Ind., Ks., Man., Mass., Mich., Minn., N.J., N.S., N.Y., Neb., Nev., Nfld., Oh., Pa., Que., Tex. (Guatemala, Mexico).

Saldula pallipes (Fabricius), 1794

1794 *Acanthia pallipes* Fabricius, Ent. Syst. Emend., 4: 71. [Denmark].

1873 *Acanthia pallipes*: Stål, K. Svens. Vet.-Akad. Handl., 11(2): 149.

1825 *Acanthia interstitialis* Say, J. Acad. Nat. Sci. Phila., 4: 324. [Mo.]. Synonymized by Drake, 1950, Bull. Brook. Ent. Soc., 45: 3. Neotype designated by Polhemus, 1985, Shore Bugs, p. 196.

1871 *Salda interstitialis*: Uhler, Am. J. Sci. Arts, 1: 105.

1877 *Salda pallipes*: Uhler, Bull. U.S. Geol. Geogr. Surv. Terr., 3: 446.

1877 *Salda reperta* Uhler, Bull. U.S. Geol. Geogr. Surv. Terr., 3: 447. [Mass.]. Synonymized by Drake and Hoberlandt, 1951, Acta Ent. Mus. Nat. Pragae, 26(376): 9. Lectotype designated by Polhemus, 1985, Shore Bugs, p. 196.

1914 *Saldula interstitialis*: Van Duzee, Trans. San Diego Soc. Nat. Hist., 2: 32.

1916 *Saldula pallipes*: Van Duzee, Check List Hem., p. 50.

1916 *Saldula reperta*: Van Duzee, Check List Hem., p. 50.

1955 *Saldula chipetae* Drake and Hottes, Bol. Ent. Venezolana, 11: 6. [Col.]. Synonymized by Polhemus, 1985, Shore Bugs, p. 197.

Distribution: All 48 contiguous United States; Alk., Alta., B.C., Man., Nfld., Ont., Que. (Africa, Central America, Europe, Mexico, South America(?), West Indies).

Note: This species and the subspecies *dimidiata* (Curtis) are quite variable and their limits are not yet well understood for North America. Reference to *Saldula xanthochila* (Fieber) and its subspecies *limbosa* (Horvath) from northeastern U.S. almost certainly refers to this species or *S. ablusa* Drake and Hottes (which see).

Saldula pallipes dimidiata (Curtis), 1835

1835 *Acanthia dimidiata* Curtis, Brit. Ent., Vol. 7, pl. 548. [Britain].

1951 *Saldula pallipes* var. *dimidiata*: Drake and Hoberlandt, Acta Ent. Mus. Nat. Pragae, 26(376): 9.

Distribution: Cal., Col., Id., Nev., S.D., Ut., Wash. (Asia, Europe).

Saldula pallipes pallipes (Fabricius), 1794

1794 *Acanthia pallipes* Fabricius, Ent. Syst. Emend., 4: 71. [Denmark].

Distribution: Same as for species.

Saldula palustris (Douglas), 1874

1874 *Salda palustris* Douglas, Ent. Month. Mag., (1)11: 10. [England].

1962 *Saldula palustris*: Chapman, Pan-Pac. Ent., 38: 157.

Distribution: Alta., B.C., Cal., Col., Mich., N.M., N.Y., Nev., Ore., Sask., Ut. (Europe).

Note: The American populations from inland localities referred to above may not be conspecific with the Old World species. Coastal populations from western North America and Newfoundland previously identified as *palustris* have been shown to be *laticollis* Reuter (see Lindskog, 1982, Pan-Pac. Ent. 57: 322). Populations from the West Indies, Mexico and the Eastern and Southeastern United

States, compared with *palustris* by Cobben (1960, Stud. Fauna Curaçâo Caribb. Isl., 11: 50), have been described as the new species *Saldula lomata* by Polhemus, 1985. Thus only inland records are presently placed under *palustris*.

Saldula pexa Drake, 1950

1950 *Saldula hirsuta*: Drake, Bull. Brook. Ent. Soc., 45: 5.

1950 *Saldula hirsuta pexa* Drake, Bull. Brook. Ent. Soc., 45: 5. [Cal.].

1962 *Saldula pexa*: Drake, Proc. Biol. Soc. Wash., 75: 116.

Distribution: Ariz., Cal., Col., Ks., N.M., Nev., Tex., Ut. (Mexico).

Saldula saltatoria (Linnaeus), 1758

1758 *Cimex saltatoria* Linnaeus, Syst. Nat., Ed. 10, 1: 448. [Europe].

1873 *Acanthia saltatoria*: Stål, K. Svens. Vet.-Akad. Handl., 11(2): 149.

1876 *Salda saltatoria*: Uhler, Bull. U.S. Geol. Geogr. Surv. Terr., 1: 334.

1916 *Saldula saltatoria*: Van Duzee, Check List Hem., p. 51.

1950 *Saldula c-album*: Drake and Hottes, Great Basin Nat., 10: 59 (in part).

Distribution: Alk., Alta, B.C., Cal., Col., Conn., D.C., Ia., Ill., Ind., Man., Mass., Me., Mich., Minn., N.J., N.T., N.Y., Neb., Nev., Ont., Ore., Que., R.I., Sask., Tenn., Va., Wash., Wis. (Asia, Colombia(?), Europe).

Note: American records for *Saldula c-album* (Fieber) refer mainly to *Saldula saltatoria* (Linnaeus). A few may pertain to undescribed species.

Saldula severini Harris, 1943

1943 *Saldula severini* Harris, J. Ks. Ent. Soc., 16: 152. [S.D.]. Synonymized with *orbiculata* by Drake and Hottes, 1954, Occas. Pap. Mus. Zool. Univ. Mich., 553: 1; restored to specific status by Polhemus, 1967, Proc. Ent. Soc. Wash., 69: 24.

Distribution: Cal., Col., Ia., Ks., N.M., Nev., S.D., Tenn., Wyo.

Saldula sulcicollis (Champion), 1900

1900 *Salda sulcicollis* Champion, Biol. Centr.-Amer., Rhyn., 2: 340. [Guatemala, Mexico, Panama]. Lectotype from Guatemala designated by Polhemus, 1985, Shore Bugs, p. 201.

1966 *Saldula sulcicollis*: Polhemus, Proc. Ent. Soc. Wash., 68: 57.

Distribution: Ariz. (Central America, Mexico).

Note: In the United States known only from Aravaipa Canyon, Ariz.

Saldula usingeri Polhemus, 1967

1967 *Saldula usingeri* Polhemus, Proc. Ent. Soc. Wash., 69: 346. [Cal.].

Distribution: Cal.

Note: This species is known only from mineral springs in California. Resh and Sorg (1983, Environ. Ent., 12: 1628-1634) have shown that the occurrence of *S. usingeri* can be predicted using water chemistry parameters.

Saldula villosa (Hodgden), 1949

1949 *Salda villosa* Hodgden, J. Ks. Ent. Soc., 22: 161. [Cal.].

1951 *Saldula villosa*: Drake and Hoberlandt, Acta Ent. Mus. Nat. Pragae, 26(376): 11.

Distribution: Cal., Ore.

Family Schizopteridae Reuter, 1891

The Schizopterids

By Thomas J. Henry

Schizopteridae are tiny heteropterans ranging in length from 0.8 to 2.0 millimeters. Small size and cryptic habits account for their being overlooked by the general collector. Although their habits are poorly known, most are thought to be predatory. They can be collected by sifting forest litter or in pitfall traps, and occasionally are taken at lights. Only four species are known from the United States. Emsley's (1962, Animal Kingdom, 45: 114-116) popular title "5000 bugs--a year's work would fit in a teaspoon" playfully illustrates the trials and tribulations of schizopterid work.

Members of this family are convex with sclerotized, beetlelike forewings. The length of the 1st and 2nd antennal segment is subequal. The ventrally enlarged propleura surrounds the procoxae and undersurface of the head, and the hindlegs, in conjunction with the metasternum, are modified for jumping. Males have three tarsal segments on each leg, and females have only two on the pro- and mesotarsi and three on the metatarsi. Nymphs have one pair of dorsal abdominal scent glands.

The major references for the family include Reuter's (1891) "Monographia Ceratocomidarum"; McAtee and Malloch's (1925) "Revision of the bugs in the family Cryptostematidae...."; and Emsley's (1969) monograph of "The Schizopteridae...." Slater and Baranowski (1978) provided a good synopsis of the North American species, furnishing keys to the known genera and species.

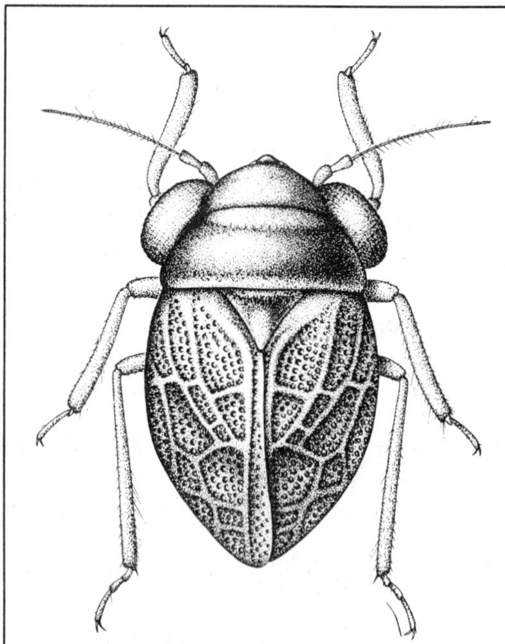

Fig. 145 *Glyptocombus saltator* [p. 683] (Orignal).

Subfamily Hypselosomatinae Esaki and Miyamoto, 1959

Genus *Glyptocombus* Heidemann, 1906

1906 *Glyptocombus* Heidemann, Proc. Ent. Soc. Wash., 7: 192. Type-species: *Glyptocombus saltator* Heidemann, 1906. Monotypic.

Glyptocombus saltator Heidemann, 1906 [Fig. 145]
1906 *Glyptocombus saltator* Heidemann, Proc. Ent. Soc. Wash., 7: 194. [Md.].
Distribution: D.C., Md., Mich., Tenn., Wash.

Subfamily Schizopterinae Reuter, 1891

Genus *Corixidea* Reuter, 1891

1891 *Schizoptera* (*Corixidea*) Reuter, Acta Soc. Sci. Fenn., 19(6): 18. Type-species: *Schizoptera lunigera* Reuter, 1891. Monotypic.
1912 *Corixidea*: Reuter, Öfv. F. Vet.-Soc. Förh., 54(7): 66.

Corixidea major McAtee and Malloch, 1925
1925 *Corixidea major* McAtee and Malloch, Proc. U.S. Nat. Mus., 67: 26. [Tenn.]
Distribution: Tenn.

Genus *Nannocoris* Reuter, 1891

1891 *Schizoptera* (*Nannocoris*) Reuter, Acta Soc. Sci. Fenn., 19(6): 18. Type-species: *Schizoptera nebulifera* Reuter, 1891. Designated by Kirkaldy, 1906, Trans. Am. Ent. Soc., 32: 148.
1925 *Nannocoris*: McAtee and Malloch, Proc. U.S. Nat. Mus., 67: 28.

Nannocoris arenarius Blatchley, 1926
1926 *Nannocoris arenaria* Blatchley, Het. E. N. Am., p. 651. [Fla.].
Distribution: Fla.

Genus *Schizoptera* Fieber, 1860

1860 *Schizoptera* Fieber, Wien. Ent. Monat., 4: 269. Type-species: *Schizoptera cicadina* Fieber, 1860. Monotypic.
Note: No representatives of the nominate subgenus *Schizoptera* occur in our region.

Subgenus *Lophopleurum* McAtee and Malloch, 1925

1925 *Schizoptera* (*Lophopleurum*) McAtee and Malloch, Proc. U.S. Nat. Mus., 67: 14. Type-species: *Schizoptera sulcata* McAtee and Malloch, 1925. Original designation.

Schizoptera bispina McAtee and Malloch, 1925
1925 *Schizoptera* (*Lophopleurum*) *bispina* McAtee and Malloch, Proc. U.S. Nat. Mus., 67: 24. [Guatemala, Mexico].
1926 *Schizoptera bispina* Blatchley, Het. E. N. Am., p. 649.
Distribution: Fla. (Guatemala, Mexico).

Family Scutelleridae Leach, 1815

The Shield Bugs

By Richard C. Froeschner

The shield bugs, which have been reported from every zoogeographic region of the world, are phytophagous, a life-requirement that makes them potential enemies of plants. In fact, on lands in the Old World at the eastern end of the Mediterranean Sea members of the genus *Eurygaster* Laporte are significant pests of cultivated small grains. Fortunately the North American members of the family, even those of the same genus *Eurygaster*, have not transferred to introduced cultivated crops in economic numbers; however, where native host plants have become important sources of products, the associated natural fauna dependent upon that plant for substance has been termed enemies of man's interests. For example, the native pine-frequenting shield bug, *Tetyra bipunctata* (Herrich-Schaeffer), is now so judged because it feeds on the developing seeds of pines--a habit it probably had long before Europeans migrated to the Americas.

Present knowledge of our species suggests they generally pass through one or two generations a year and hibernate in the adult stage. Eggs are glued upright to the surface of the host plants and are generally arranged in parallel rows in the same fashion used by members of the related Pentatomidae.

The classificatory level assigned to this group has varied between full family, as assigned here, and subfamily under the Pentatomidae. Taxonomic treatments for the family in North America were given by Van Duzee (1904, Trans. Am. Ent. Soc., 30: 11-21) and by Torre-Bueno (1939, Ent. Am., 19: 164-175). From a world aspect, Schouteden (1904, Gen. Ins., 24: 1-98) presented keys to tribes and genera and included a list of species; and Leston (1952, Public. Cult. Comp. Diam. Angola, 16: 11-26, and 1953, Ent. Gaz. 4: 18) revised the suprageneric classification of the group. McPherson (1982, Pent. Ne. N. Am., p. 9-17) presented keys and summaries of biologies for the shield bugs of northeastern North America.

Subfamily Eurygastrinae Amyot and Serville, 1843

Genus *Euptychodera* Bergroth 1908

1906 *Ptychodera* Reuter, Acta Soc. Sci. Fenn., 33(8): 44. Preoccupied. Type-species: *Phimodera corrugata* Van Duzee, 1904. Monotypic.

1908 *Euptychodera* Bergroth, Mém. Soc. Ent. Belg., 15: 143. New name for *Ptychodera* Reuter.

Euptychodera corrugata (Van Duzee), 1904

1904 *Phimodera corrugata* Van Duzee, Trans. Am. Ent. Soc., 30: 16. [Col.].

1906 *Ptychodera corrugata*: Reuter, Acta Soc. Sci. Fenn., 33(8): 45.

1908 *Euptychodera corrugata*: Bergroth, Mém. Soc. Ent. Belg., 15: 143.

Distribution: Ariz., Cal., Col., S.D., Tex., Ut.

Genus *Eurygaster* Laporte, 1832

1832 *Eurygaster* Laporte, Mag. Zool., 1: 68. Type-species: *Cimex hottentotta* Fabricius, 1775. Designated by Reuter, 1888, Acta Soc. Sci. Fenn., 15(2): 761.

Note: Vojadni (1961, Pan-Pac. Ent., 37: 97-107) revised the nearctic species of *Eurygaster*.

Eurygaster alternata (Say), 1828

1828 *Tetyra alternata* Say, Am. Ent., 3: 94. ["Middle States"].

1851 *Eurygaster alternata*: Dallas, List Hem. Brit. Mus., 1: 47.

1869 *Scutellera alternata*: Rathvon, Hist. Lancaster Co., Pa., p. 548.

1870 *Eurygaster alternatus* Stål, K. Svens. Vet.-Akad. Handl., 9(1): 18.

1872 *Eurygaster nicoletanensis* Provancher, Nat. Can., 4: 73. [Que.]. Synonymized by Provancher, 1885, Pet. Faune Can., 3: 23.

Distribution: Alta., B.C., Cal., Col., Conn., Ia., Ill., Ind., Man., Mass., Md., Me., Mich., Mont., N.C., N.D., N.H., N.J., N.M., N.S., N.Y., N.T., Neb., Oh., Ore., Que., Sask., S.D., Ut., Va., Wis.

Eurygaster amerinda Bliven, 1956

1956 *Eurygaster amerinda* Bliven, New Hem. W. St., p. 4. [Cal.].

1956 *Eurygaster greggii* Bliven, New Hem. W. St., p. 2. [Cal.]. Synonymized by Vojdani, 1961, Pan-Pac. Ent., 37: 103.

1956 *Eurygaster macclellani* Bliven, New Hem. W. St., p. 2. [Cal.]. Synonymized by Vojdani, 1961, Pan-Pac. Ent., 37: 103.

1956 *Eurygaster ukiah* Bliven, New Hem. W. St., p. 3., [Cal.]. Synonymized by Vojdani, 1961, Pan-Pac Ent., 37: 103.

Distribution: Cal., Ill.

Eurygaster minidoka Bliven, 1956

1956 *Eurygaster minidoka* Bliven, New Hem. W. St., p. 4. [Id.].

Distribution: Cal.(?), Col., Id., Nev., Ore., Wash., Wyo.

Note: Bliven (1962, Occ. Ent., 1: 66) contended that the California record for this species is in error.

Eurygaster paderewskii Bliven, 1962

1961 *Eurygaster minidoka*: Vojdani, An. Ent. Soc. Am., 54: 567.

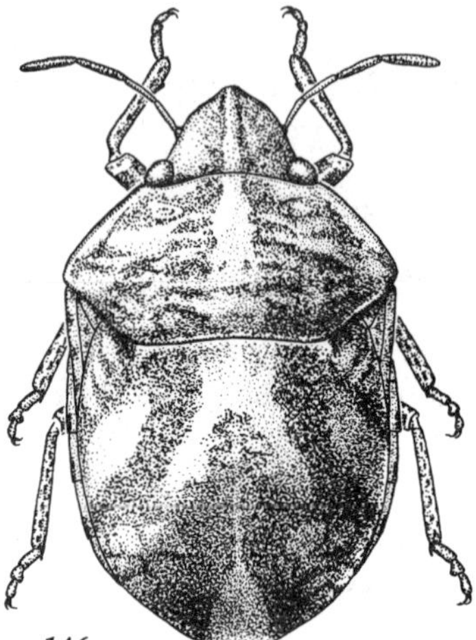

146

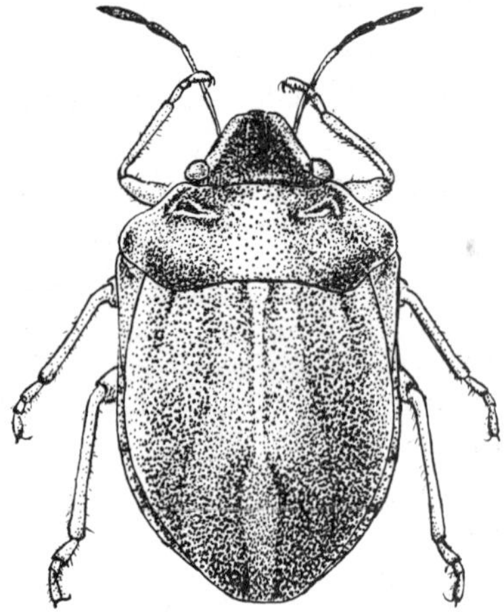

147

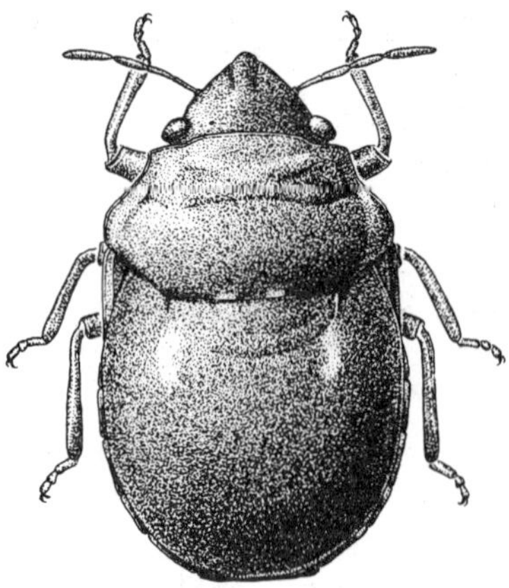

148

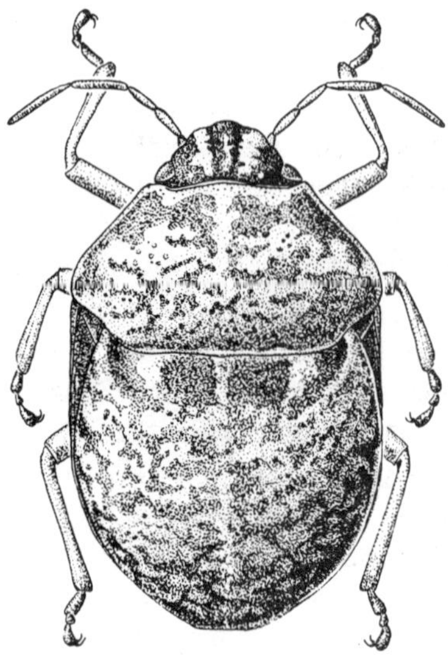

149

Figs. 146-149: 146, *Sphyrocoris obliquus* [p. 691]; 147, *Homaemus parvulus* [p. 690]; 148, *Camiris porosus* [p.689]; 149, *Chelysomidea guttata* [p.689] (All originals, except fig. 147, after Froeschner, 1941).

1962 *Eurygaster paderewskii* Bliven, Occ. Ent., 1: 66. [Cal.].
Distribution: Cal.

Eurygaster shoshone Kirkaldy, 1909
1904 *Eurygaster carinatus* Van Duzee, Trans. Am. Ent. Soc., 30: 18. [Id., Nev., Ut.]. Preoccupied.
1909 *Eurygaster shoshone* Kirkaldy, Cat. Hem., 1: 273. New name for *Eurygaster carinatus* Van Duzee.
Distribution: B.C., Cal., Id., Nev., Ore., Ut., Wash.

Genus *Fokkeria* Schouteden, 1904

1904 *Fokkeria* Schouteden, An. Soc. Ent. Belg., 48: 301. Type-species: *Fokkeria crassa* Schouteden, 1904, a junior synonym of *Odontoscellis producta* Van Duzee, 1904. Monotypic.

Fokkeria producta (Van Duzee), 1904
1904 *Odontoscelis producta* Van Duzee, Trans. Ent. Soc. Am., 30: 20. [Col.].
1904 *Fokkeria crassa* Schouteden, An. Soc. Ent. Belg., 48: 301. [Col.]. Synonymized by Schouteden, 1904, Gen. Ins., 24: 84.
1904 *Fokkeria producta*: Schouteden, Gen. Ins., 24: 84.
Distribution: Alta., Col.

Genus *Phimodera* Germar, 1839

1839 *Phimodera* Germar, Zeit. Ent., 1: 60. Type-species: *Podops galgulina* Herrich-Schaeffer, 1837. Designated by Kirkaldy, 1903, Ent., 36: 233.
Note: Reuter (1906, Acta Soc. Sci. Fenn., 33(8): 1-51) monographed *Phimodera*.

Phimodera binotata (Say), 1824
1824 *Scutellera binotata* Say, Long's Second Exped., 2: 298. ["North-West Territory"].
1886 *Phimodera binotata*: Uhler, Check-list Hem. Het., p. 2.
1893 *Tetyra binotata*: Lethierry and Severin, Gen. Cat. Hem., 1: 31.
Distribution: Alta., Ariz., B.C., Cal., Col., Ill., Man., Mich., N.M., N.T., Neb., Ore., S.D., Wash.

Phimodera torpida Walker, 1867
1867 *Phimodera torpida* Walker, Cat. Hem. Brit. Mus., 1: 75. [Sask.].
Distribution: Alta., B.C., Col., Mich., Sask., Ut.

Phimodera torrida Reuter, 1906
1906 *Phimodera torrida* Reuter, Acta Soc. Sci. Fenn., 33(8): 16. [Nev.].
Distribution: Nev.

Genus *Vanduzeeina* Schouteden, 1904

1904 *Vanduzeeina* Schouteden, Gen. Ins., 24: 85. Type-species: *Odontoscelis balli* Van Duzee, 1904. Original designation.

Vanduzeeina balli (Van Duzee), 1904
1904 *Odontoscelis balli* Van Duzee, Trans. Am. Ent. Soc., 30: 19. [Cal., Col., N.M., Wyo.].
1904 *Vanduzeeina balli*: Schouteden, Gen. Ins., 24: 86.
Distribution: B.C., Cal., Col., N.M., Nev., Wyo.

Vanduzeeina borealis Van Duzee, 1925
1925 *Vanduzeeina borealis* Van Duzee, Proc. Cal. Acad. Sci., ser. 4, 14: 391. [Cal.].
Distribution: Alta., B.C., Cal., Ill., Man., S.D., Yuk.

Vanduzeeina californica Van Duzee, 1925
1925 *Vanduzeeina californica* Van Duzee, Proc. Cal. Acad. Sci., ser. 4, 14: 392. [Cal.].
Distribution: Cal.

Vanduzeeina senescens Usinger, 1930
1930 *Vanduzeeina senescens* Usinger, Pan-Pac. Ent., 6: 132. [Cal.].
Distribution: Cal.

Vanduzeeina slevini Usinger, 1930
1930 *Vanduzeeina slevini* Usinger, Pan-Pac. Ent., 6: 132. [Cal.].
Distribution: Cal.

Subfamily Pachycorinae Amyot and Serville, 1843

Genus *Acantholomidea* Sailer, 1945

1867 *Acantholoma* Stål, Öfv. K. Svens. Vet.-Akad. Förh., 24(7): 491. Preoccupied. Type-species: *Acantholoma denticulata*, Stål, 1870. First included species by Stål, 1870, K. Svens. Vet.-Akad. Handl., 9(1): 17.

1945 *Acantholomidea* Sailer, Proc. Ent. Soc. Wash., 47: 135. New name for *Acantholoma* Stål, 1867.

Acantholomidea denticulata (Stål), 1870
1870 *Acantholoma denticulata* Stål, K. Svens. Vet.-Akad. Handl., 9(1): 17. [Ill.].
1926 *Acantholoma denticulata* var. *kansiana* Blatchley, Het. E. N. Am., p. 50. [Ks.]. Synonymized by Harris and Andre, 1934, An. Ent. Soc. Am., 27: 5.
1970 *Acantholomidea denticulata*: McPherson, Mich. Ent., 3: 36.
Distribution: Ia., Ill., Ind., Ks., Ky., Mich., Minn., N.J., N.Y., Ont., Va., Wis.
Note: Biological data given by Harris and Andre (1934, An. Ent. Soc. Am., 27: 5-12, pls. 1-2).

Genus *Camirus* Stål, 1862

1851 *Zophoessa* Dallas, List Hem. Brit. Mus., 1: 43. Preoccupied. Type-species: *Pachycoris porosus* Germar, 1839. Monotypic.

1862 *Camirus* Stål, K. Svens. Vet.-Akad. Handl., 3(6): 57. Type-species: *Pachycoris conicus* Germar, 1839. Designated by Kirkaldy, 1909, Cat. Hem., 1:xxiv.

Camirus consocius (Uhler), 1876
1876 *Zophoessa consocius* Uhler, Bull. U.S. Geol. Geogr. Surv. Terr., 1: 274. [Ariz.].
1886 *Camirus consocius*: Uhler, Check-list Hem. Het., p. 2.
Distribution: Ariz., Cal., Tex.

Camirus moestus (Stål), 1862
1862 *Zophoessa moesta* Stål, Stett. Ent. Zeit., 23: 83. [Mexico].

1961 *Camirus moestus*: Bibby, J. Econ. Ent., 54: 329.
Distribution: Ariz. (Mexico).

Camirus porosus (Germar), 1839 [Fig. 148]
1839 *Pachycoris porosus* Germar, Zeit. Ent., 1: 108. [Cal.].
1851 *Zophoessa porosa*: Dallas, List Hem. Brit. Mus., 1: 43.
1886 *Camirus porosus*: Uhler, Check-list Hem. Het., p. 2.
1982 *Acantholomidea porosa*: McPherson, Pent. Ne. N. Am., p. 12.
Distribution: B.C., Cal., Col., Fla., N.C., Tex., Va. (Mexico to Colombia).
Note: McPherson's (1982, above) placed this species in *Acantholomidea* without explanation.

Genus *Chelysomidea* Lattin, New Name

1867 *Orsilochus* Stål, Öfv. K. Svens. Vet.-Akad. Förh., 24: 493. Preoccupied. Type-species: *Pachycoris variabilis* Herrich-Schaeffer, 1837. Designated by Schouteden, 1904, Gen. Ins., 24: 52.
1891 *Chelysoma* Bergroth, Rev. d'Ent., 10: 235. New name for *Orsilochus* Stål. Preoccupied.
Note: Herein, J.D. Lattin (Oregon State University, Corvallis, Ore.) proposes the replacement name *Chelysomidea* for *Chelysoma* Bergroth, which is preoccupied by *Chelysoma* Gravenhorst, 1843 (tunicates).

Chelysomidea guttata (Herrich-Schaeffer), 1839, New Combination [Fig. 149]
1839 *Pachycoris guttatus* Herrich-Schaeffer, Wanz. Ins., 4: 4. [Ga.].
1870 *Orsilochus guttatus*: Stål, K. Svens. Vet.-Akad. Handl., 9(1): 9.
1904 *Chelysoma guttatum*: Schouteden, Gen. Ins., 24: 52.
Distribution: Fla., Ga., N.C., S.C.

Genus *Diolcus* Mayr, 1864

1864 *Diolcus* Mayr, Verh. Zool.-Bot. Gesell. Wien, 14: 904. Type-species: *Cimex irroratus* Fabricius, 1775. Designated by Schouteden, 1904, Gen. Ins., 24: 56.

Diolcus chrysorrhoeus (Fabricius), 1803
1803 *Tetyra chrysorrhoeus* Fabricius, Syst. Rhyn., p. 138. ["Carolina"].
1832 *Scutellera viridipunctata* Say, Descrip. Het. Hem. N. Am., p. 2. [Fla.]. Synonymized by Uhler, 1859, *In* Le Conte, Complete Writings Thomas Say, 1: 310.
1839 *Pachycoris chrysorrhoeus*: Germar, Zeit. Ent., 1: 95.
1870 *Diolcus chrysorrhoeus*: Stål, K. Svens. Vet.-Akad. Handl., 9(1): 11.
1911 *Diolcus chryssorrheus* [*sic*]: Rosenfeld, J. Econ. Ent., 4: 408.
Distribution: Ala., Ark., Fla., Ga., Ill.(?), La., Miss., N.C., S.C., Tex.
Note: McPherson (1982, Pent. Ne. N. Am., p. 12) discussed and questioned the Ill. record.

Diolcus irroratus (Fabricius), 1775
1775 *Cimex irroratus* Fabricius, Syst. Ent., p. 699. ["America bor."].
1904 *Diolcus irroratus*: Van Duzee, Trans. Am. Ent. Soc., 30: 12.
Distribution: Fla. (West Indies).

Genus *Homaemus* Dallas, 1851

1851 *Homaemus* Dallas, List Hem. Brit. Mus., 1: 36. Type-species: *Pachycoris exilis* Herrich-Schaeffer, 1839, a junior synonym of *Scutellera aeneifrons* Say, 1824. Designated by

Schouteden, 1904, Gen. Ins., 24: 59.

Homaemus aeneifrons (Say), 1824

1824 *Scutellera aeneifrons* Say, Long's Second Exped., 2: 299. ["North-West Territory"].

1839 *Pachycoris exilis* Herrich-Schaeffer, Wanz. Ins., 4: 5. [N. Am.]. Synonymized by Stål, 1870, K. Svens. Vet.-Akad. Handl, 9(1): 15.

1859 *Homaemus exilis*: Dohrn, Cat. Hem., p. 3.

1870 *Homaemus aeneifrons*: Stål, K. Svens. Vet.-Akad. Handl., 9(1): 15.

1904 *Homaemus aenifrons* [*sic*]: Schouteden, Gen. Ins., 24: 59.

1905 *Homoemus* [*sic*] *aenifrons* [*sic*]: Crevecoeur, Trans. Ks. Acad. Sci., 19: 232.

1970 *Homaemus aeneifrors* [*sic*]: McPherson, Mich. Ent., 3: 36.

Distribution: Alk., Alta., B.C., Cal., Col., Man., Mass., Md., Me., Mich., Minn., Mont., N.B., N.C., N.H., N.J., N.M., N.S., N.T., N.Y., Neb., Nfld., Ont., Pa., Que., Sask., S.D., Ut., Va., Wis. (Mexico).

Homaemus aeneifrons aeneifrons (Say), 1824

1824 *Scutellera aeneifrons* Say, Long's Second Exped., 2: 299. [North-west Territory].

1929 *Homaemus aeneifrons aeneifrons*: Walley, Can. Ent., 61: 254.

Distribution: Same as for species.

Homaemus aeneifrons extensus Walley, 1929

1929 *Homaemus aeneifrons extensus* Walley, Can. Ent., 61: 254. [B.C.].

Distribution: Alk.(?), Alta., B.C., Col., Mont., N.T.

Note: Froeschner and Halpin (1981, Proc. Biol. Soc. Wash., 94: 425) reported a female and one nymph from Alk., but in the absence of males could only surmize they represented this western subspecies.

Homaemus bijugis Uhler, 1872

1872 *Homaemus bijugis* Uhler, Prelim. Rept. U.S. Geol. Surv. Mont., 5: 393. [Col., Neb.].

1876 *Homaemus consors* Uhler, Bull. U.S. Geol. Geogr. Surv. Terr., 1: 272. ["west of the Mississippi River"]. Synonymized by Van Duzee, 1916, Check List Hem., p. 1.

Distribution: Alta., Ariz., Cal., Col., Fla., Ia., Id., Ill., Ks., Minn., Mont., N.M., Neb., Nev., Sask., S.D., Ut., Wyo.

Homaemus parvulus (Germar), 1839 [Fig. 147]

1839 *Pachycoris parvulus* Germar, Zeit. Ent., 1: 107. ["Carolina"].

1851 *Homaemus parvulus*: Dallas, List Hem. Brit. Mus., 1: 36.

1886 *Homaemus grammicus*: Uhler, Check-list Hem. Het., p. 2.

1905 *Homoemus* [*sic*] *grammicus*: Crevecoeur, Trans. Ks. Acad. Sci., 19: 232.

Distribution: Ark., Cal., Col., Fla., Ia., Ill., Ind., Ks., La., Md., Mo., N.C., N.M., Ore., Tex. (Mexico).

Homaemus proteus Stål, 1862

1862 *Homaemus proteus* Stål, Stett. Ent. Zeit., 23: 82. [Mexico].

1904 *Homaemus proteus*: Uhler, Proc. U.S. Nat. Mus., 27: 349.

Distribution: Cal., Fla., N.M., Tex. (Mexico to Colombia).

Note: Van Duzee (1904, Trans. Am. Ent. Soc., 30: 15) pointed out that Osborn's (1892, Proc. Ia. Acad. Sci., 1: 122) Ia. record for this species was in error.

Homaemus variegatus Van Duzee, 1914

1914 *Homaemus variegatus* Van Duzee, Trans. San Diego Soc. Nat. Hist., 2: 3. [Cal.].

Distribution: Cal., Ut.

Genus *Pachycoris* Burmeister, 1835

1835 *Pachycoris* Burmeister, Handb. Ent., 2: 391. Type-species: *Tetrya fabricii* Burmeister, 1835, a misidentification of *Cimex torridus* Scopoli, 1772. Designated by Kirkaldy, 1903, Ent., 36: 213.

Pachycoris torridus (Scopoli), 1772

1772 *Cimex torridus* Scopoli, Annus Hist. Nat., p. 110. ["America torrida"].
1876 *Pentatoma fabricii*: Uhler, Bull. U. S. Geol. Geogr. Surv. Terr., 1: 273.
1880 *Pachycoris torridus*: Distant, Biol. Centr.-Am., Rhyn., 1: 16.
Distribution: Cal. (Mexico to Argentina).
Note: Van Duzee (1904, Trans. Am. Ent. Soc., 30: 12) deliberately excluded this species from the fauna north of Mexico, but in his 1916 Check List Hem. (p. 1) and 1917 Catalog (Univ. Publ. Cal. Ent., 2: 5) listed it for California without expressed reservation. All subsequent reports of natural occurrence in North America appear to be based on this record.

Genus *Sphyrocoris* Mayr, 1864

1864 *Sphyrocoris* Mayr, Verh. Zool.-Bot. Ges. Wien, 14: 904. Type-species: *Pachycoris obliquus* Germar, 1839. Monotypic.

Sphyrocoris obliquus (Germar), 1839 [Fig. 146]

1839 *Pachycoris obliquus* Germar, Zeit. Ent., 1: 94. [Martinique].
1876 *Sphyrocoris obliquus*: Uhler, Bull. U.S. Geol. Geogr. Surv. Terr., 1: 273.
1910 *Sphyrocoris obilquus* [*sic*]: Kirkaldy, Proc. Haw. Ent. Soc., 2: 126.
Distribution: Ariz., Cal., Fla., N.M., Tex. (West Indies, Mexico to Colombia).

Sphyrocoris punctellus (Stål), 1862

1862 *Homaemus punctellus* Stål, Stett. Ent. Zeit., 23: 81. [Mexico].
1907 *Sphyrocoris punctellus*: Snow, Trans. Ks. Acad. Sci., 20: 158.
Distribution: Ariz., Cal. (Mexico).

Genus *Stethaulax* Bergroth, 1891

1871 *Aulacostethus* Uhler, Proc. Boston Soc. Nat. Hist., 14: 93. Preoccupied. Type-species: *Tetyra marmorata* Say, 1832. Monotypic.
1891 *Stethaulax* Bergroth, Rev. d'Ent., 10: 235. New name for *Aulacostethus* Uhler.
Note: Barber and Bruner (1932, J. Dept. Agr. Puerto Rico, 16: 243) concluded *Stethaulax* Bergroth was a junior synonym of *Symphylus* Dallas; however, subsequent authors have continued, without comment, to use *Stethaulax* and *Symphylus* as distinct genera, without refutation of the Barber and Bruner study.

Stethaulax marmoratus (Say), 1832

1832 *Tetyra marmorata* Say, Descrip. Het. Hem. N. Am., p. 2. [N.J.].
1871 *Aulacostethus marmoratus*: Uhler, Proc. Boston Soc. Nat. Hist., 14: 94.
1904 *Stethaulax marmoratus*: Van Duzee, Trans. Am. Ent. Soc., 30: 12, 76.
1926 *Symphylus deplanatus*: Blatchley, Het. E. N. Am., p. 44 (in part).
Distribution: Cal., Fla., Ga., Ill., Ky., Md., Miss., Mo., N.C., N.J., N.Y., Ore., Tex.
Note: Biological notes were given by McPherson and Walt (1971, Trans. Ill. St. Acad. Sci., 64: 198-200). Walt and McPherson (1973, An. Ent. Soc. Am., 66: 1103-1107) described and illustrated the immature stages.

Stethaulax simulans (Uhler), 1876

1876 *Aulacostethus simulans* Uhler, Bull. U.S. Geol. Geogr. Surv. Terr., 1: 272. [Cal.].

1904 *Stethaulax marmoratus*: Van Duzee, Trans. Am. Ent. Soc., 30: 76.

1910 *Stethaulax simulans*: Banks, Cat. Nearc. Hem.-Het., p. 98.

Distribution: Ariz., Cal.

Note: Blatchley (1926, Het. E. N. Am., p. 45) resurrected this species and defined it, but Torre-Bueno (1939, Ent. Am., 19: 169) treated it without comment as a junior synonym of *S. marmoratus*, possibly following Van Duzee's Catalogue (1917, Univ. Cal. Publ. Ent., 2: 8).

Genus *Symphylus* Dallas 1851

1851 *Symphylus* Dallas, List Hem. Brit. Mus., 1: 5, 37. Type-species: *Symphylus obtusus* Dallas, 1851. Designated by Schouteden, 1904, Gen. Ins., 24: 60.

Symphylus caribbeanus Kirkaldy 1909

1857 *Scutellera* (*Pachycoris*) *obliqua* Guérin-Méneville, Hist. l'Ile Cuba, p. 362. [Cuba]. Preoccupied.

1909 *Symphylus caribbeanus* Kirkaldy, Cat. Hem., 1: 280. New name for *Scutellera obliqua* Guérin-Méneville, 1857.

1926 *Symphylus deplanatus*: Blatchley, Het. E. N. Am., p. 43 (in part).

Distribution: Fla., Ga., Tex. (Cuba).

Genus *Tetyra* Fabricius, 1803

1803 *Tetyra* Fabricius, Syst. Rhyn., p. 128. Type-species: *Tetyra antillarum* Kirkaldy, 1909. Designated by Int. Comm. Zool. Nomen., 1954, Opinion 255, Bull. Zool. Nomen., 4: 471.

Tetyra antillarum Kirkaldy, 1909

1842 *Tetyra arcuata* Schiödte, Nat. Tidskr., 4: 292. ["Americae meridionalis insulis"]. Preoccupied.

1904 *Tetyra arcuata*: Van Duzee, Trans. Am. Ent. Soc., 30: 11.

1909 *Tetyra antillarum* Kirkaldy, Cat. Hem., 1: 284. New name for *Tetyra arcuata* Schiödte, 1842.

Distribution: Fla. (Mexico, West Indies).

Tetyra bipunctata (Herrich-Schaeffer), 1839

1839 *Pachycoris bipunctatus* Herrich-Schaeffer, Wanz. Ins., 4: 87. [No locality given].

1861 *Macraulax tristis* Uhler, Proc. Ent. Soc. Phila., 1: 21. [Md.]. Synonymized by Stål, 1870, K. Svens. Vet.-Akad. Handl., 9(1): 4.

1868 *Macraulax tristris* [*sic*]: Walker, Cat. Hem. Brit. Mus., 3: 515 [spelling of species name corrected to *tristis* on p. 578].

1870 *Tetyra bipunctata*: Stål, K. Svens. Vet.-Akad. Handl., 9(1): 4.

Distribution: Ala., Conn., D.C., Fla., Ga., Ind., La., Md., Mich., Minn., Miss., N.C., N.J., N.Y., Ont., Tenn., Tex., Va., Wis. (Mexico).

Note: Gilbert et al. (1967, An. Ent. Soc. Am., 60: 698-701) presented bionomics of the species when associated with pine.

Tetyra robusta Uhler, 1897

1897 *Tetyra robusta* Uhler, Trans. Md. Acad. Sci., 1: 383. [Ariz., Mexico].

Distribution: Ariz., Nev., Ut. (Mexico).

Subfamily Scutellerinae Leach, 1815

Genus *Augocoris* Burmeister, 1835

1835 *Augocoris* Burmeister, Handb. Ent., 2: 396. Type-species: *Augocoris gomesii* Burmeister, 1835. Designated by Kirkaldy, 1903, Ent., 36: 213.

Augocoris illustris (Fabricius), 1781

1781 *Cimex illustris* Fabricius, Species Ins., 2: 339. ["Cap. bon. Sp."]
1956 *Augocoris illustris*: Hussey, Fla. Ent., 39: 88.
Distribution: Fla. (Mexico to South America, West Indies).

Family Tessaratomidae Stål, 1864

The Tessaratomids

By Richard C. Froeschner

The family Tessaratomidae is essentially an Old World group, only two species in one genus being known from the New World where they occur principally in the Neotropics. Of these two species, one has been tentatively reported from North America north of Mexico. Also included in this family is the infamous Australian "bronze orange bug," *Musgraveia sulciventris* (Stål).

The biology of members of the Tessaratomidae closely resembles that of the common Pentatomidae and Scutelleridae, the eggs here also being glued upright in clusters on the host plant.

As is true of several of the higher level taxonomic groups within the superfamily Pentatomoidea, this group has been assigned various hierarchical positions from a tribal level up to a family. For North American studies, the above-mentioned lone species is keyed in Torre-Bueno (1939, Ent. Am., 19: 242). Several authors have reviewed and modified the classification of this family during the last few decades; interested persons may consult their works: Leston (1954, Publ. Cult. Comp. Diamantes Angola, 24: 11-22; 1955, Proc. Roy. Ent. Soc. London, B, 24: 62-68); Kumar (1969, An. Ent. Soc. Am., 62: 681-695; 1969, Austr. J. Zool., 17:553-606); and Kumar and Ghauri, 1970, Deut. Ent. Zeit., 17: 1-32).

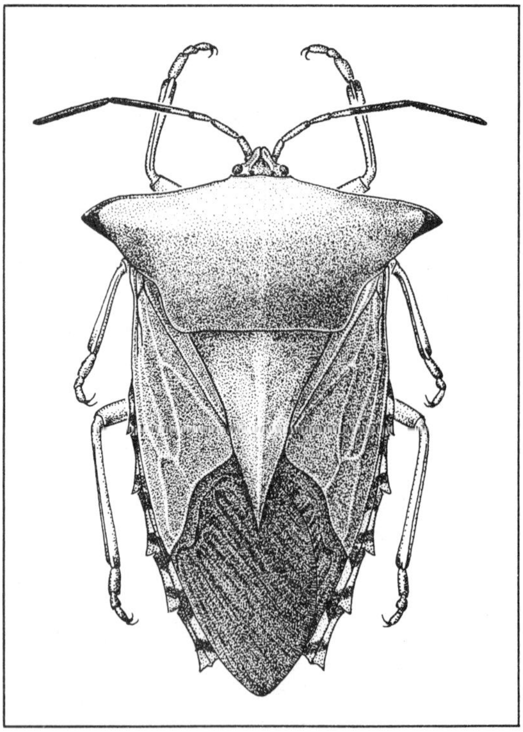

Fig. 150 *Piezosternum subulatum* [p. 695] (Original).

Subfamily Oncomerinae Stål, 1870

Tribe Piezosternini Leston, 1955

Genus *Piezosternum* Amyot and Serville, 1843

1843 *Piezosternum* Amyot and Serville, Hist. Nat. Ins.. Hem., p. 161. Type-species: *Pentatoma mucronata* Palisot, 1805, a junior synonym of *Cimex subulatus* Thunberg, 1783. Monotypic.

Piezosternum subulatum (Thunberg), 1783 [Fig. 150]

1783 *Cimex subulatus* Thunberg, Diss. Ent. Nov. Ins. Sp., 2: 41. [No locality given].

1939 *Piezosternum subulatum*: Torre-Bueno, Ent. Am., 19: 242.

Distribution: Tex.(?) (Mexico to Brazil; West Indies).

Note: Torre-Bueno (1939, above) claimed an earlier but unlocated record for Tex.

Family Thaumastocoridae Kirkaldy, 1907

The Thaumastocorid Bugs or Royal Palm Bugs

By Richard C. Froeschner

Only fragmentary information is available on the members of this family of fewer than two dozen species. All were described from widely scattered places south of the Tropic of Cancer; only one has been introduced and established north of that line: A Cuban species, *Xylastodoris luteolus* Barber, now known from Florida where it occurs on another Cuban introduction, the Royal Palm, *Roystonea regia* Humboldt, Bonplan, and Kunth. According to biological observations by Baranowski (1958, An. Ent. Soc. Am., 51: 547-551), adults and nymphs feed in the newly opening fronds on which their feeding causes yellow spots, sometimes to the extent that "severe damage results." Apparently their populations build slowly because only the older palm trees have high numbers of individuals, a fact that can be attributed, at least in part, to females producing only one or two eggs per day. This association with palms seems to be a feature of the subfamily Xylastodoridinae; members of the subfamily Thaumastocorinae "appear to have a more diverse range of host plants," according to the observations of Slater (1973, J. Austr. Ent. Soc., 12: 155-156).

The fundamental taxonomic papers for study in this group are those by Drake and Slater (1957, An. Ent. Soc. Am., 50: 353-370) and Slater and Drake (1958, Proc. 10th Int. Congr. Ent., 1: 321-323).. The most complete coverage of the biology of our only species is Baranowski's (1958, above) report. Viana and Carpintero's (1983, Comun. Mus. Argent. Cienc. Nat., Ent., 1: 63-74) treatment of both subfamilies as families was refuted by Slater and Brailovsky (1983, Proc. Ent. Soc. Wash., 85: 560-561). The latter authors also explained why the subfamily name should be spelled "Xylastodoridinae," rather than "Xylastodorinae."

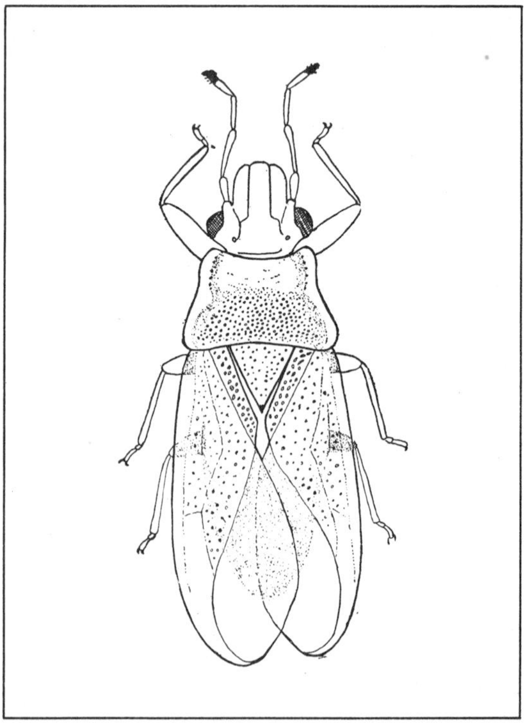

Fig. 151. *Xylastodoris luteolus* [p. 697] (After Barber, 1920).

Subfamily Xylastodoridinae Barber, 1920

Genus *Xylastodoris* Barber, 1920

1920 *Xylastodoris* Barber, 1920, Bull. Brook. Ent. Soc., 15: 100. Type-species: *Xylastodoris luteolus* Barber, 1920. Monotypic.

Xylastodoris luteolus Barber, 1920 [Fig. 151]

1920 *Xylastodoris luteolus* Barber, Bull. Brook. Ent. Soc., 15: 101. [Cuba].
1921 *Xylastodoris luteolus*: Moznette, Quart. Bull. St. Plant Board Fla., 6: 10.
Distribution: Fla. (Cuba).

Family Thyreocoridae Amyot and Serville, 1843

The Negro Bugs

By Richard C. Froeschner

These insects occur abundantly on grasses, weeds, and shrubs from which they obtain their sustenance, apparently concentrating their feeding on the flowers and developing seeds. Sometimes, especially during the winter months, the adults will conceal themselves among plant debris or under objects on the ground. During spring and early summer, eggs are glued to the plants on which the immature individuals feed in company with the adults. Each year a single generation is completed. Besides scattered reports of perceptible feeding damage to crops and ornamental plants, frequent reference is made to a disagreeable taste imparted to cultivated berries on which these insects occur.

The taxonomic study essential to work on this family was presented by McAtee and Malloch (1933, An. Carnegie Mus., 21: 191-411, i-v, pls. 4-17); Torre-Bueno (1939, Ent. Am., 19: 184-196) abstracted therefrom a key to the North American forms. Noteworthy subsequent contributions to the taxonomy of this family in North America were made by Sailer (1940, J. Ks. Ent. Soc.,

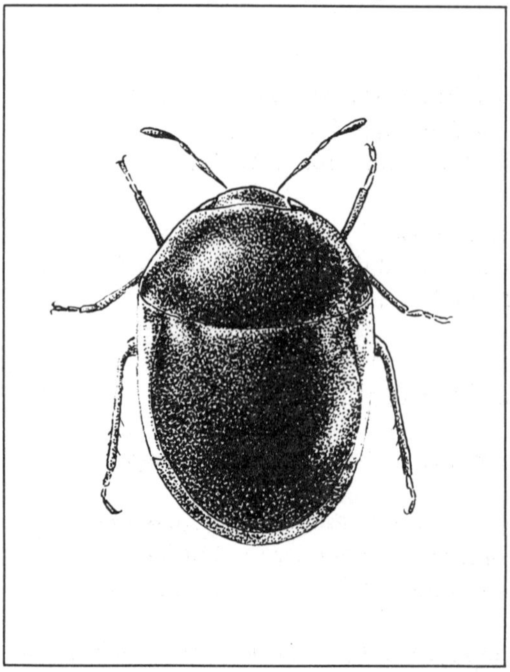

Fig. 152 *Corimelaena pulicaria* [p. 702] (After Froeschner, 1941).

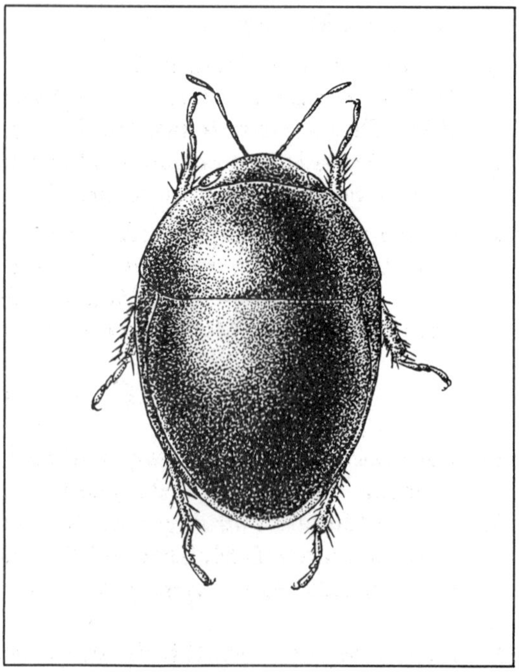

Fig. 153: *Galgupha carinata* [p. 705] (After Froeschner, 1941).

13: 62-63 and 1945, Proc. Ent. Soc. Wash., 47: 129-135); and McPherson and Sailer (1978, J. Ks. Ent. Soc., 51: 516-520).

Many American authors accepted Uhler's (1871, Prelim. Rept. U.S. Geol. Surv. Wyo., p. 471) proposal of the name Corimelaenidae for this family. The *International Code of Zoological Nomenclature* (1985, Art. 23 and 36), however, requires acceptance of the oldest group-name (regardless of original ending) based on the name of an included genus, the group-name taking the authorship and date of its first use. Thus Amyot and Serville's (1843, Hist. Nat. Ins., Hem., p. 60) name "Thyreocorides," almost thirty years older than Uhler's "Corimelaenidae," takes priority and must be accepted as the proper name for this family group.

Two North American species described under the names *Scutellera unicolor* Palisot (1805, Ins. Afr. Am., p. 32, pl. 5, fig. 5) and *Tetyra helopiodes* Wolff (1811, Icon. Cimic., p. 174, pl. 27, fig. 168) were considered unidentifiable by McAtee and Malloch (1933, above) because of the generalized nature of the original descriptions and unavailability of type specimens. Although both were described as black and hence probably belong to in the genus *Galgupha* Amyot and Serville, they are not placed in the list below.

Rathvon (1869, Hist. Lancaster Co., Pa., p. 548), in his Pennsylvania list, recorded a species under the double misspelling "*Odontocellis scaraboides*," with both parts of the combination queried. He quite probably intended the combination to be *Odontoscelis scarabaeoides* as used by Burmeister (1835, Handb. Ent., 2: 385). Because this species, now known as *Thyreocoris scarabaeoides* (Linnaeus), has otherwise been reported only from Europe and Asia, no further listing of it as a member of our fauna is warranted.

Genus *Amyssonotum* Horvath, 1919

1919 *Amyssonotum* Horvath, An. Mus. Nat. Hung., 17: 212. Type-species: *Corimelaena rastrata* Stål, 1862. Monotypic.

Amyssonotum rastratum (Stål), 1862

1862 *Corimelaena rastrata* Stål, K. Svens. Vet.-Akad. Handl., 2(7): 8. [Brazil].
1917 *Thyreocoris rastratus*: Van Duzee, Univ. Cal. Publ. Ent., 2: 15.
1919 *Amyssonotum rastratum*: Horvath, An. Mus. Nat. Hung., 17: 213.
Distribution: Tex. (Mexico to Brazil).

Amyssonotum rastratum flexum McAtee and Malloch, 1933

1933 *Amyssonotum rastratum flexum* McAtee and Malloch, An. Carnegie Mus., 21: 351. [Holotype from Panama; also from Tex. and other countries].
Distribution: Tex. (Mexico to Colombia).

Genus *Corimelaena* White, 1839

1839 *Corimelaena* White, An. Mag. Nat. Hist., new ser., 3: 539. Type-species: *Tetyra lateralis* Fabricius, 1803. Original designation.

1865 *Eucoria* Mulsant and Rey, Hist. Nat. Punaises, Scutellerides, p. 12. Type-species: *Eucoria marginipennis* Mulsant and Rey, 1865, a junior synonym of *Odontoscelis pulicaria* Germar, 1839. Monotypic. Synonymized by Sailer, 1945, Proc. Ent. Soc. Wash., 47: 132.

1933 *Allocoris* McAtee and Malloch, An. Carnegie Mus., 21: 203, 358. Type-species: *Corimelaena gillettii* Van Duzee, 1904, a junior synonym of *Tetyra lateralis* Fabricius, 1803. Original designation.

Subgenus *Corimelaena* White, 1839

1839 *Corimelaena* White, An. Mag. Nat. Hist., new ser., 3: 539. Type-species: *Tetyra lateralis* Fabricius, 1803. Original designation.

1933 *Allocoris* (*Allocoris*): McAtee and Malloch, An. Carnegie Mus., 21: 203, 358.

1939 *Corimelaena* (*Corimelaena*): Torre-Bueno, Ent. Am., 19: 191.

Corimelaena agrella McAtee, 1919

1919 *Corimelaena agrella* McAtee, Bull. Ill. Nat. Hist. Surv., 13: 216. [Md.].
1933 *Allocoris* (*Allocoris*) *agrella*: McAtee and Malloch, An. Carnegie Mus., 21: 362.
1939 *Corimelaena* (*Corimelaena*) *agrella*: Torre-Bueno, Ent. Am., 19: 194.
Distribution: D.C., Ia., Ill., Ky., Md., Tex.

Corimelaena alpina (McAtee and Malloch), 1933

1933 *Allocoris* (*Allocoris*) *alpinus* McAtee and Malloch, An. Carnegie Mus., 21: 363. [N.Y.].
1939 *Corimelaena* (*Corimelaena*) *alpina*: Torre-Bueno, Ent. Am., 19: 193.
Distribution: Ill., Minn., N.Y.

Corimelaena barberi (McAtee and Malloch), 1933

1933 *Allocoris* (*Allocoris*) *barberi* McAtee and Malloch, An. Carnegie Mus., 21: 365. [Tex.].
1939 *Corimelaena* (*Corimelaena*) *barberi*: Torre-Bueno, Ent. Am., 19: 194.
Distribution: Tex. (Mexico to Costa Rica).

Corimelaena contrasta (McAtee and Malloch), 1933

1933 *Allocoris* (*Allocoris*) *contrasta* McAtee and Malloch, An. Carnegie Mus., 21: 366. [Ariz.].
1939 *Corimelaena* (*Corimelaena*) *contrasta*: Torre-Bueno, Ent. Am., 19: 193.
Distribution: Ariz. (Mexico).

Corimelaena feminea (McAtee and Malloch), 1933

1933 *Allocoris* (*Allocoris*) *feminea* McAtee and Malloch, An. Carnegie Mus., 21: 368. [Tex.].
1939 *Corimelaena* (*Corimelaena*) *feminea*: Torre-Bueno, Ent. Am., 19: 192.
Distribution: Tex.

Corimelaena harti Malloch, 1919

1919 *Corimelaena harti* Malloch, Bull. Ill. Nat. Hist. Surv., 13: 215. [Ill.].
1933 *Allocoris* (*Allocoris*) *harti*: McAtee and Malloch, An. Carnegie Mus., 21: 371.
1939 *Corimelaena* (*Corimelaena*) *harti*: Torre-Bueno, Ent. Am., 19: 194.
Distribution: D.C., Ga., Ill., Md., Miss., Mo., N.C., N.Y., Va.

Corimelaena interrupta Malloch, 1919

1919 *Corimelaena interrupta* Malloch, Bull. Ill. Nat. Hist. Surv., 13: 214. [Tex.].
1933 *Allocoris* (*Allocoris*) *interrupta*: McAtee and Malloch, An. Carnegie Mus., 21: 372.
1939 *Corimelaena* (*Corimelaena*) *interrupta*: Torre-Bueno, Ent. Am., 19: 193.
Distribution: Tex. (Mexico to Costa Rica).

Corimelaena lateralis (Fabricius), 1803

1803 *Tetyra lateralis* Fabricius, Syst. Rhyn., p. 142. ["Carolina"].
1933 *Allocoris* (*Allocoris*) *gillettii*: McAtee and Malloch, Ann. Carnegie Mus., 21: 369.
Distribution: B.C., Conn., D.C., Fla., Ia., Ill., Ks., Ky., Mass., Md., Mich., Mo., N.C., N.D., N.J., Neb., Oh., Ok., Que., R.I., Tex., Ut., Va., Wash. (Mexico, Guatemala).

Corimelaena lateralis lateralis (Fabricius), 1803

1803 *Tetyra lateralis* Fabricius, Syst. Rhyn., p. 142. ["Carolina"].

1839 *Odontoscelis lateralis*: Germar, Zeit. Ent., 1: 39.

1904 *Corimelaena Gillettii* [*sic*] Van Duzee, Trans. Am. Ent. Soc., 30: 8. [D.C., Md., N.J., Tex., "Canada," "Indian Territory"]. Synonymized by Sailer, 1945, Proc. Ent. Soc. Wash., 47: 131.

1908 *Thyreocoris Gillettei* [*sic*]: Bergroth, Mém. Soc. Ent. Belg., 15: 139.

1910 *Thyreocoris lateralis*: Banks, Cat. Nearc. Hem.-Het., p. 102.

1919 *Corimelaena lateralis*: Malloch, Bull. Ill. Nat. Hist. Surv., 13: 212, 213.

1933 *Allocoris (Allocoris) gillettii*: McAtee and Malloch, An. Carnegie Mus., 21: 360, 369.

1939 *Corimelaena* (*Corimelaena*) *gillettii*: Torre-Bueno, Ent. Am., 19: 192.

1972 *Corimelaena lateralis lateralis*: McPherson, An. Ent. Soc. Am., 63: 906.

1982 *Corimelaena* (*Corimelaena*) *lateralis lateralis*: McPherson, Pent. Ne. N. Am., p. 24

Distribution: B.C., Conn., D.C., Fla., Ia., Ill., Ky., Mass., Mich., Mo., N.C., N.D., N.J., Neb., Oh., Ok., Que., R.I., Tex., Ut., Va., Wash.

Note: Life history and description of immature stages were provided by McPherson (1972, above, pp. 906-911).

Corimelaena marginella Dallas, 1851

1851 *Corimelaena marginella* Dallas, List Hem. Brit. Mus., 1: 59. [N.Y.].

1904 *Corimelaena marginalis* [*sic*]: Van Duzee, Trans. Am. Ent. Soc., 30: 3.

1919 *Corimelaena nanella* McAtee, Bull. Ill. Nat. Hist. Surv., 13: 212, 215. [Md.]. Synonymized by Van Duzee, 1923, Ent. News, 34: 303.

1933 *Allocoris* (*Allocoris*) *marginella*: McAtee and Malloch, An. Carnegie Mus., 21: 373.

1939 *Corimelaena* (*Corimelaena*) *marginella*: Torre-Bueno, Ent. Am., 19: 194.

Distribution: Fla., Ind., Ks., Md., N.J., N.Y., R.I., Tex., Va.

Corimelaena minuta Uhler, 1863

1863 *Corimelaena minuta* Uhler, Proc. Ent. Soc. Phila., 2: 155. [Cuba].

1904 *Corimelaena minuta*: Van Duzee, Trans. Am. Ent. Soc., 30: 76.

1910 *Thyreocoris minuta*: Banks, Cat. Nearc. Hem.-Het., p. 102.

1916 *Thyreocoris minutus*: Van Duzee, Check List Hem., p. 2.

1939 *Corimelaena* (*Corimelaena*) *minuta*: Torre-Bueno, Ent. Am., 19: 194.

Distribution: Fla., Tex. (West Indies).

Note: The Fla. and Tex. records above were rejected by Blatchley (1926, Het. E. N. Am., p. 68) and McAtee (1927, Bull. Brook. Ent. Soc., 22: 274) but accepted by Torre-Bueno (1939, Ent. Am., 19: 194).

Corimelaena nigra Dallas, 1851

1851 *Corimelaena nigra* Dallas, List Hem. Brit. Mus., 1: 57. ["Hudson's Bay"].

1876 *Corimelaena anthracina* Uhler, Bull. U.S. Geol. Geogr. Surv. Terr., 5: 270. [Cal.]. Synonymized by Torre-Bueno, 1939, Ent. Am., 19: 192.

1910 *Thyreocoris anthracina*: Banks, Cat. Nearc. Hem.-Het., p. 101.

1910 *Thyreocoris nigra*: Banks, Cat. Nearc. Hem.-Het., p. 102.

1916 *Thyreocoris niger* [*sic*]: Van Duzee, Check List Hem., p. 2.

1919 *Galgupha nigra*: Hart, Bull. Ill. Nat. Hist. Surv., 13: 210.

1919 *Corimelaena anthracina*: Malloch, Bull. Ill. Nat. Hist. Surv., 13: 213.

1933 *Allocoris* (*Allocoris*) *nigra*: McAtee and Malloch, An. Carnegie Mus., 21: 376.

1939 *Corimelaena* (*Corimelaena*) *nigra*: Torre-Bueno, Ent. Am., 19: 192.

Distribution: Alta., Ariz., B.C., Cal., Col., "Hudson's Bay," Mich., N.M., N.Y., Nev.,

Ont., Ore., Que., S.D., Sask., Ut., Wash., Yuk. (Mexico).
Note: McPherson (1982, Pent. Ne. N. Am., p. 25) found that the Illinois specimens reported for this species by Hart (1919, above) were misidentified.

Corimelaena obscura McPherson and Sailer, 1978
1978 *Corimelaena* (*Corimelaena*) *obscura* McPherson and Sailer, J. Ks. Ent. Soc., 51: 516. [Ia.].
Distribution: Ala., D.C., Ga., Ia., Ill., Ks., Ky., La., Md., Mich., Mo., Pa., Tenn., Va.

Corimelaena polita Malloch, 1919
1919 *Corimelaena polita* Malloch, Bull. Ill. Nat. Hist. Surv., 13: 211, 213. [Tex.].
1933 *Allocoris* (*Allocoris*) *polita* [*sic*]: McAtee and Malloch, An. Carnegie Mus., 21: 378.
1939 *Corimelaena* (*Corimelaena*) *polita*: Torre-Bueno, Ent. Am., 19: 193.
Distribution: Tex.

Corimelaena pulicaria (Germar), 1839 [Fig. 152]
1839 *Odontoscelis pulicarius* Germar, Zeit. Ent., 1: 39. [Pa.].
1851 *Corimelaena pulicaria*: Dallas, List. Hem. Brit. Mus., 1: 59.
1865 *Galgupha flavo-marginata* [*sic*] Thomas, Trans. Ill. St. Agr. Soc., 5: 455. [Ill.]. Synonymized by Van Duzee, 1904, Trans. Am. Ent. Soc., 30: 9.
1886 *Corimelaena tibialis*: Uhler, Check-list Hem. Het., p. 2.
1910 *Thyreocoris pulicaria*: Banks, Cat. Nearc. Hem.-Het., p. 102.
1916 *Thyreocoris pulicarius*: Van Duzee, Check List Hem., p. 2.
1933 *Allocoris* (*Allocoris*) *pulicarius*: McAtee and Malloch, An. Carnegie Mus., 21: 378.
1939 *Corimelaena* (*Corimelaena*) *pulicaria*: Torre-Bueno, Ent. Am., 19: 193.
Distribution: Alta., Ark., B.C., Cal., "Carolina," Col., Conn., "Dakota," D.C., Fla., Ga., Ia., Ill., Ks., La., Mass., Md., Me., Mich., Minn., Miss., Mo., N.J., N.Y., Neb., Ont., Ore., Pa., Que., R.I., S.D., Tex., Va., Vt. (Guatemala, Mexico).

Subgenus *Parapora* McAtee and Malloch, 1933

1933 *Allocoris* (*Parapora*) McAtee and Malloch, An. Carnegie Mus., 21: 382. Type-species: *Corimelaena extensa* Uhler, 1863. Original designation.
1945 *Corimelaena* (*Parapora*): Sailer, Proc. Ent. Soc. Wash., 47: 131.

Corimelaena californica Van Duzee, 1929
1929 *Corimelaena californica* Van Duzee, Pan-Pac. Ent., 6: 10. [Cal.].
1933 *Allocoris* (*Parapora*) *californica*: McAtee and Malloch, An. Carnegie Mus., 21: 383.
1939 *Corimelaena* (*Parapora*) *californica*: Torre-Bueno, Ent. Am., 19: 195.
Distribution: Cal.

Corimelaena cognata (Van Duzee), 1907.
1907 *Thyreocoris cognatus* Van Duzee, Bull. Buff. Soc. Nat. Sci., 8: 6. [Jamaica].
1933 *Allocoris* (*Parapora*) *cognata*: McAtee and Malloch, An. Carnegie Mus., 21: 384.
1939 *Corimelaena* (*Parapora*) *cognata*: Torre-Bueno, Ent. Am., 19: 195.
Distribution: Ariz. (Mexico to Colombia).

Corimelaena extensa Uhler, 1863
1863 *Corimelaena extensa* Uhler, Proc. Ent. Soc. Phila., 2: 155. ["near Fort Benton"].
1909 *Thyreocoris montanus* Van Duzee, Ent. News, 20: 231. [Ut.]. Synonymized by McAtee and Malloch, 1933, An. Carnegie Mus., 21: 385.
1910 *Thyreocoris extensa*: Banks, Cat. Nearc. Hem.-Het., p. 101.
1916 *Thyreocoris extensus*: Van Duzee, Check List Hem., p. 2.

1933 *Allocoris* (*Parapora*) *extensus*: McAtee and Malloch, An. Carnegie Mus., 21: 385.
1939 *Corimelaena* (*Parapora*) *extensa*: Torre-Bueno, Ent. Am., 19: 195.
Distribution: Ariz., B.C., Cal., Ia., Mont., Nev., Ore., S.D., Ut., Wash. (Mexico).

Corimelaena incognita (McAtee and Malloch), 1933
1933 *Allocoris* (*Parapora*) *incognita* McAtee and Malloch, An. Carnegie Mus., 21: 386. [Ariz.].
1939 *Corimelaena* (*Parapora*) *incognita*: Torre-Bueno, Ent. Am., 19: 196.
Distribution: Ariz., B.C., Cal., Col., Tex. (Mexico).

Corimelaena virilis (McAtee and Malloch), 1933
1933 *Allocoris* (*Parapora*) *virilis* McAtee and Malloch, An. Carnegie Mus., 21: 387. [Ariz.].
1939 *Corimelaena* (*Parapora*) *virilis*: Torre-Bueno, Ent. Am., 19: 196.
Distribution: Ariz., Cal., Id., Ut., Wash.

Subgenus *Termapora* McAtee and Malloch, 1933

1933 *Allocoris* (*Termapora*) McAtee and Malloch, An. Carnegie Mus., 21: 381. Type-species: *Corimelaena minutissima* Malloch, 1919. Monotypic.
1939 *Corimelaena* (*Termapora*): Torre-Bueno, Ent. Am., 19: 195.

Corimelaena minutissima Malloch, 1919
1919 *Corimelaena minutissima* Malloch, Bull. Ill. Nat. Hist. Surv., 13: 214. [Tex.].
1933 *Allocoris* (*Termapora*) *minutissima*: McAtee and Malloch, An. Carnegie Mus., 21: 381.
1939 *Corimelaena* (*Termapora*) *minutissima*: Torre-Bueno, Ent. Am., 19: 195.
Distribution: Miss., Tex. (Mexico).

Genus *Cydnoides* Malloch, 1919

1919 *Cydnoides* Malloch, Bull. Ill. Nat. Hist. Surv., 13: 208. Type-species: *Corimelaena ciliata* Uhler, 1863. Original designation.

Subgenus *Cydnoides* Malloch, 1919

1919 *Cydnoides* Malloch, Bull. Ill. Nat. Hist. Surv., 13: 208. Type-species: *Corimelaena ciliata* Uhler, 1863. Original designation.
1933 *Cydnoides* (*Cydnoides*): McAtee and Malloch, An. Carnegie Mus., 21: 338.

Cydnoides ciliatus (Uhler), 1863
1863 *Corimelina ciliata* Uhler, Proc. Ent. Soc. Phila., 2: 156. [Cal.].
1910 *Thyreocoris ciliata*: Banks, Cat. Nearc. Hem.-Het., p. 101.
1916 *Thyreocoris ciliatus*: Van Duzee, Check List Hem., p. 2.
1919 *Cydnoides ciliatus*: Malloch, Bull. Ill. Nat. Hist. Surv., 13: 208.
1933 *Cydnoides* (*Cydnoides*) *ciliatus*: McAtee and Malloch, An. Carnegie Mus., 21: 339.
Distribution: Ariz., Cal., Col., Fla., Ill., Ks., Mich., Minn., Mo., N.C., N.M., Neb., Nev., Ore., Tex., Ut. (Mexico).

Cydnoides ciliatus ciliatus (Uhler), 1863
1863 *Corimelaena ciliata* Uhler, Proc. Ent. Soc. Phila., 2: 156. [Cal.].
1933 *Cydnoides* (*Cydnoides*) *ciliatus ciliatus*: McAtee and Malloch, An. Carnegie Mus., 21: 339.
Distribution: Ariz., Cal., N.M., Nev., Ore., Tex., Ut. (Mexico).

Cydnoides ciliatus orientis McAtee and Malloch, 1933

1933 *Cydnoides* (*Cydnoides*) *ciliatus orientis* McAtee and Malloch, An. Carnegie Mus., 21: 340. [Fla.].

Distribution: Col., Fla., Ill., Ks., Mich., Minn., Mo., N.C., Neb., Tex.

Cydnoides confusus McAtee and Malloch, 1933

1933 *Cydnoides* (*Cydnoides*) *confusus* McAtee and Malloch, An. Carnegie Mus., 21: 341. [N.M.].

Distribution: Ariz., N.M., Tex. (Mexico).

Cydnoides renormatus (Uhler), 1895

1895 *Corimelaena renormata* Uhler, Bull. Col. Agr. Exp. Stn., 31: 11. [Col.].

1908 *Thyreocoris renormatus*: Bergroth, Mém. Soc. Ent. Belg., 15: 139.

1910 *Thyreocoris renormata*: Banks, Cat. Nearc. Hem.-Het., p. 102.

1919 *Cydnoides renormatus*: Malloch, Bull. Ill. Nat. Hist. Surv., 13: 208.

1933 *Cydnoides* (*Cydnoides*) *renormatus*: McAtee and Malloch, An. Carnegie Mus., 21: 339, 341.

Distribution: Ariz., Col., Id., Ill., Minn., N.M., Tex.

Subgenus *Sayocoris* McAtee and Malloch, 1933

1933 *Cydnoides* (*Sayocoris*) McAtee and Malloch, An. Carnegie Mus., 21: 342. Type-species: *Thyreocoris albipennis* Say, 1832. Original designation.

Cydnoides albipennis (Say), 1859

1832 *Thyreocoris albipennis* Say, Descrip. Het. Hem. N. Am., p. 2. ["Missouri River"].

1886 *Corimelaena albipennis*: Uhler, Check-list Hem. Het., p. 2.

1904 *Corimelaena sayi* Van Duzee, Trans. Am. Ent. Soc., 30: 10. Unnecessary new name for *Thyreocoris albipennis* Say.

1908 *Thyreocoris Sayi* [*sic*]: Bergroth, Mém, Soc. Ent. Belg., 15: 139.

1919 *Cydnoides sayi*: Malloch, Bull. Ill. Nat. Hist. Surv., 13: 209.

1927 *Cydnoides albipennis*: Knight, Ent. News, 38: 40.

1933 *Cydnoides* (*Sayocoris*) *albipennis*: McAtee and Malloch, An. Carnegie Mus., 13: 343.

1943 *Cydnoides albidipennis* [*sic*]: Harris, J. Ks. Ent. Soc., 16: 150.

Distribution: Col., Ks., Ia., Neb., S.D.

Cydnoides obtusus (Uhler), 1894

1894 *Corimelaena obtusus* Uhler, Proc. Cal. Acad. Sci., ser. 2, 4: 225. [Mexico].

1916 *Thyreocoris obtusus*: Van Duzee, Check List Hem., p.2

1919 *Cydnoides obtusa*: Malloch, Bull. Ill. Nat. Hist. Surv., 13: 208.

1933 *Cydnoides* (*Sayocoris*) *obtusus*: McAtee and Malloch, An. Carnegie Mus., 21: 344.

Distribution: Cal., Tex. (Mexico).

Genus *Galgupha* Amyot and Serville, 1843

1843 *Galgupha* Amyot and Serville, Hist. Nat. Ins., Hem., p. 68. Type-species: *Galgupha atra* Amyot and Serville, 1843. Designated by Horvath, 1919, An. Mus. Nat. Hung., 17: 216.

Note: Horvath's type designation actually read "*Scutellera unicolor* P.B. (*Galgupha atra* A.S.)," indicating that he was naming one of the originally included species, *G. atra*, as type-species but considered it a junior synonym of *Scutellera unicolor*. McAtee and Malloch (1933, An. Carnegie Mus., 21: 391) considered *Scutellera unicolor* as an unidentifiable species and (on their page 203) credited Horvath with designation of *G. atra* as the type-species.

Subgenus *Galgupha* Amyot and Serville, 1843

1843 *Galgupha* Amyot and Serville, Hist. Nat. Ins., Hem., p. 68. Type-species: *Galgupha atra* Amyot and Serville, 1843. Designated by Horvath, 1919, An. Mus. Nat. Hung., 17: 216.
1933 *Galgupha* (*Galgupha*): McAtee and Malloch, An. Carnegie Mus., 21: 279.

Galgupha aterrima Malloch, 1919
1919 *Galgupha aterrima* Malloch, Bull. Ill. Nat. Hist. Surv., 13: 211. [Ill.].
1933 *Galgupha* (*Galgupha*) *aterrima*: McAtee and Malloch, An. Carnegie Mus., 21: 280.
Distribution: Ala., Conn., Ill., Ind., La., Mass., Md., Me., Mich., Miss., Mo., N.H., N.J., N.Y., Ok., Pa., Que., R.I., S.C., Tex., Va., Wis.

Galgupha atra Amyot and Serville, 1843
1843 *Galgupha atra* Amyot and Serville, Hist. Nat. Ins., Hem., p. 68. ["Mexique; Amérique septentrionale"].
1878 *Corimelaena ater* [*sic*]: Uhler, Proc. Boston Soc. Nat. Hist., 19: 366.
1916 *Thyreocoris ater* [*sic*]: Van Duzee, Check List Hem., p. 2.
1919 *Galgupha atra*: Malloch, Bull. Ill. Nat. Hist. Surv., 13: 210.
1933 *Galgupha* (*Galgupha*) *atra*: McAtee and Malloch, An. Carnegie Mus., 21: 281.
Distribution: Ala., Alta., Ariz., B.C., Col., Conn., D.C., Fla., Ga., Ia., Ill., Ind., Ks., Man., Mass., Md., Me., Mich., Mo., Mont., N.C., N.D., N.H., N.J., N.S., N.Y., Neb., Nfld., Oh., Ont., Pa., Que., R.I., S.D., Tenn., Tex., Va., Vt., Wash., Wis. (Mexico).

Galgupha carinata McAtee and Malloch, 1933 [Fig. 153]
1933 *Galgupha* (*Galgupha*) *carinata* McAtee and Malloch, An. Carnegie Mus., 21: 282. [La.].
Distribution: Ala., Ga., Ill., La., Md., Mich., Miss., Mo., Ok., Tenn., Tex., Va.

Galgupha denudata (Uhler), 1863
1863 *Corimelaena denudata* Uhler, Proc. Ent. Soc. Phila., 2: 157. [La.].
1916 *Thyreocoris denudatus*: Van Duzee, Check List Hem., p. 2.
1919 *Galgupha denudata*: Malloch, Bull. Ill. Nat. Hist. Surv., 13: 210.
1933 *Galgupha* (*Galgupha*) *denudata*: McAtee and Malloch, An. Carnegie Mus., 21: 284.
Distribution: Ala., D.C., Fla., Ga., Ill., La., Miss., Tex., Va.

Galgupha hesperia McAtee and Malloch, 1933
1933 *Galgupha* (*Galgupha*) *hesperia* McAtee and Malloch, An. Carnegie Mus., 21: 284. [Cal.].
Distribution: Cal.

Galgupha loboprostethia Sailer, 1940
1919 *Galgupha aterrima*: Malloch, Bull. Ill. Nat. Hist. Surv., 13: 211 (in part).
1940 *Galgupha loboprostethia* Sailer, J. Ks. Ent. Soc., 13: 62. [Ks.].
1943 *Galgupha lobostethia* [*sic*]: Harris, J. Ks. Ent. Soc., 16: 150.
1970 *Galgupha laboprostethia* [*sic*]: McPherson, Mich. Ent., 3: 38.
Distribution: Ark., Ia., Ill., Ks., Mich., Mo., Ok., S.D., Tenn., Va.

Galgupha ovalis Hussey, 1925
1925 *Galgupha ovalis* Hussey, J. N.Y. Ent. Soc., 33: 62. [Ga.].

1933 *Galgupha* (*Galgupha*) *ovalis*: McAtee and Malloch, An. Carnegie Mus., 21: 285.
Distribution: Ala., Ariz., Ark., Col., Conn., D.C., Del., Fla., Ga., Ia., Ill., Ind., Ks., Ky., La., Mass., Md., Mich., Mo., Mont., N.C., N.J., N.M., N.Y., Neb., Ok., Pa., S.C., S.D., Tex., Va., Wyo. (Guatemala, Mexico).

Subgenus *Gyrocnemis* McAtee and Malloch, 1928

1928 *Galgupha* (*Gyrocnemis*): McAtee and Malloch, An. Mus. Zool. Polin., 7: 33. Type-species: *Odontoscelis maculipennis* Germar, 1839. Original designation.

Galgupha diminuta (Van Duzee), 1923
1923 *Euryscytus diminutus* Van Duzee, Ent. News, 34: 305. [Cal.].
1933 *Galgupha* (*Gyrocnemis*) *diminuta*: McAtee and Malloch, An. Carnegie Mus., 21: 252.
Distribution: Ariz., Cal.

Galgupha guttiger (Stål), 1862
1862 *Thyreocoris guttiger* Stål, Stett. Ent. Zeit., 23: 94. [Mexico].
1933 *Galgupha* (*Gyrocnemis*) *guttiger*: McAtee and Malloch, An. Carnegie Mus., 21: 256.
Distribution: Tex. (Greater Antilles, Mexico to Bolivia).

Galgupha punctifer McAtee and Malloch, 1933
1933 *Galgupha* (*Gyrocnemis*) *punctifer* McAtee and Malloch, An. Carnegie Mus., 21: 264. [Holotype from Guatemala; also from Tex.].
Distribution: Tex. (Mexico to Panama).

Galgupha texana McAtee and Malloch, 1933
1933 *Galgupha* (*Gyrocnemis*) *texana* McAtee and Malloch, An. Carnegie Mus., 21: 269. [Tex.].
Distribution: Tex.

Subgenus *Nothocoris* McAtee and Malloch, 1928

1928 *Galgupha* (*Nothocoris*) McAtee and Malloch, An. Mus. Zool. Polin., 7: 40. Type-species: *Odontoscelis brunnipennis* Germar, 1839. Original designation.

Galgupha bakeri McAtee and Malloch, 1933
1933 *Galgupha* (*Nothocoris*) *anomala* McAtee and Malloch, An. Carnegie Mus., 21: 296 [not McAtee and Malloch, 1933, above, p. 272, f. "Errata" on p. vii]. [Col.]. Preoccupied.
1933 *Galgupha* (*Nothocoris*) *bakeri* McAtee and Malloch, An. Carnegie Mus., 21: vii. New name for *Galgupha anomala* McAtee and Malloch, 1933, p. 296 [not p. 272].
Distribution: Col.

Galgupha eas McAtee and Malloch, 1933
1933 *Galgupha* (*Nothocoris*) *eas* McAtee and Malloch, An. Carnegie Mus., 21: 301. [Tex.].
Distribution: Tex.

Galgupha nitiduloides (Wolff), 1802
1802 *Cimex nitiduloides* Wolff, Icon. Cimic., 3: 98. [Pa.].
1859 *Corimelaena nitiduloides*: Dohrn, Cat. Hem., p. 5.
1933 *Galgupha* (*Nothocoris*) *nitiduloides*: McAtee and Malloch, An. Carnegie Mus., 21: 305.

Distribution: Alta., Ariz., Cal., Col., Conn., D.C., Ia., Id., Ill., Ind., Ks., Mass., Mich., Minn., Miss., Mo., Mont., N.C., N.D., N.J., N.M., N.S., N.Y., Neb., Oh., Ont., Ore., Pa., Que., Tex., Wash. (Mexico to Costa Rica).

Note: A Brazilian record for this species was questioned by McAtee and Malloch (1933, above, p. 308).

Galgupha nitiduloides coerulescens (Stål), 1862

1862 *Thyreocoris coerulescens* Stål, Stett. Ent. Zeit., 23: 94. [Mexico].

1863 *Corimelaena cyanea* Uhler, Proc. Ent. Soc. Phila., 2: 157. [Cal.]. Synonymized by Distant, 1880, Biol. Centr.-Am., Rhyn., 1: 10.

1886 *Corimelaena caerulescens* [*sic*]: Uhler, Check-list Hem. Het., p. 2.

1893 *Corimelaena coerulescens*: Lethierry and Severin, Gen. Cat. Hem., p. 11.

1910 *Thyreocoris caerulescens* [*sic*]: Kirkaldy, Proc. Haw. Ent. Soc., 2: 126.

1923 *Galgupha coerulescens*: Van Duzee, Ent. News, 34: 305.

1933 *Galgupha* (*Northocoris*) *nitiduloides caerulescens* [*sic*]: McAtee and McAtee, An. Carnegie Mus., 21: 307.

Distribution: Ariz., Cal., Tex. (Mexico to Costa Rica).

Note: A Brazilian record for this species was questioned by McAtee and Malloch (1933, above).

Galgupha nitiduloides nitiduloides (Wolff), 1802

1802 *Cimex nitiduloides* Wolff, Icon. Cimic., 3: 98. [Pa.].

1832 *Thyreocoris histeroides* Say, Descrip. Het. Hem. N. Am., p. 2. [Ark.]. Synonymized by Uhler, 1878, Proc. Bost. Soc. Nat. Hist., 19: 366.

1839 *Odontoscelis nitiduloides*: Herrich-Schaeffer, Wanz. Ins., 5: 12.

1851 *Corimelaena nitiduloides*: Dallas, List Hem. Brit. Mus., 1: 56.

1885 *Corimelaena caerulescens* [*sic*]: Popenoe, Trans. Ks. Acad. Sci., 9: 62.

1910 *Thyreocoris nitiduloides*: Banks, Cat. Nearc. Hem.-Het., p. 102.

1919 *Galgupha nitiduloides*: Malloch, Bull. Ill. Nat. Hist. Surv., 13: 210.

1933 *Galgupha* (*Nothocoris*) *nitiduloides nitiduloides*: McAtee and Malloch, An. Carnegie Mus., 21: 309.

Distribution: Alta., Ariz., Ark., Col., Conn., D.C., Ia., Id., Ill., Ind., Ks., Mass., Mich., Minn., Miss., Mo., Mont., N.C., N.D., N.J., N.M., N.S., N.Y., Neb., Oh., Ont., Ore., Pa., Que., Tex., Wash.

Galgupha nitiduloides texensis McAtee and Malloch, 1933

1933 *Galgupha* (*Nothocoris*) *nitiduloides texensis* McAtee and Malloch, An. Carnegie Mus., 21: 308. [Tex.].

Distribution: Miss., Tex.

Subgenus *Orocoris* McAtee and Malloch, 1933

1933 *Galgupha* (*Orocoris*) McAtee and Malloch, An. Carnegie Mus., 21: 272. Type-species: *Cydnoides arizonensis* Van Duzee, 1923. Original designation.

Galgupha arizonensis (Van Duzee), 1923

1923 *Cydnoides arizonensis* Van Duzee, Ent. News, 34: 304. [Ariz.].

1933 *Galgupha* (*Orocoris*) *arizonensis*: McAtee and Malloch, An. Carnegie Mus., 21: 273.

Distribution: Ariz.

Family Tingidae Laporte, 1807

(= Tingididae; Tingitidae)

The Lace Bugs

By Richard C. Froeschner

The beautiful, graceful, and delicate network of veins that led to the common name of lace bugs for these small insects (2-5 mm) can be properly appreciated only under magnification. All included species are phytophagous. Their normal habitat is on the surface of plants from mosses to enormous trees; some appear restricted to one species of host plant, whereas others will thrive on a variety of plants in several families. When separated from their plant host, lace bugs will probe damp surfaces and, if they sample perspiring human skin, the resulting sharp sensation leads some victims to the erroneous conclusion that the lace bugs are seeking blood.

Lace bugs complete one or more life cycles during each growing season. Adults thrust their eggs into plant parts. The five immature stages following the egg, as well as the resulting adults, are quite sedentary and locally may accumulate large numbers in a limited space. Such overcrowding populations may cause considerable damage or death to the plant or plant part on which they are feeding. Most native species ordinarily are of little economic importance except under unusual circumstances. But several may attack and damage cultivated plants; among these are *Gargaphia solani* Heidemann on egg plant and *Corythucha cydoniae* (Fitch) on rosaceous shrubs and trees. Severe and consistent damage is caused to azaleas, rhododendrons, and Japanese andromeda by the introduced species *Stephanitis pyrioides* (Scott), *S. rhododendri* Horvath, and *S. takeyai* Drake and Maa. Sometimes the lace bugs' ability to destroy plants is turned to man's advantage when it can be used to help control certain weeds. Populations of the lantana lace bug, *Teleonemia scrupulosa* Stål, which freely attacks lantana plants "at home" in the southwestern

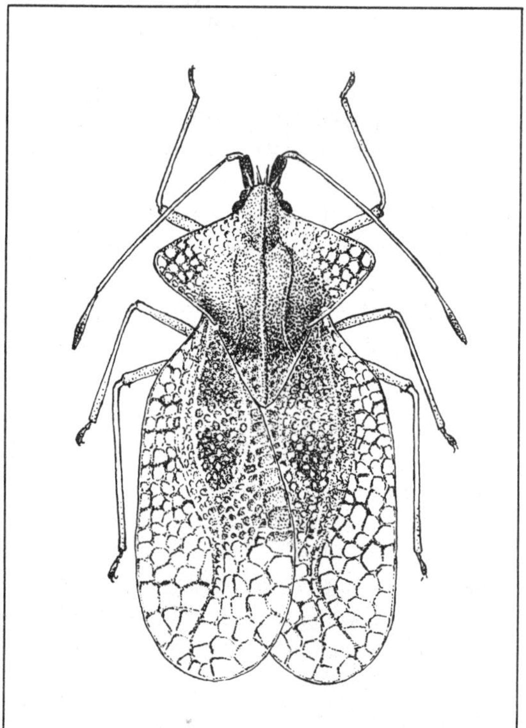

Fig. 154 *Gargaphia solani* [p. 723] (After Froeschner, 1944).

United States and tropical America, have been transferred to the Old World to combat that plant where it appears as an introduced weed.

Landmarks in the literature of this group for North America were by Stål (1873, K. Svens. Vet.-Akad. Handl., 11(2): 115-134); Van Duzee (1917, Univ. Cal. Publ. Ent., 2: 209-224); Hurd (1946, Ia. St. Coll. J. Sci., 20: 429-492); and Drake and Ruhoff (1965, U.S. Nat. Mus. Bull., 243: 17-634), a world catalog including host plants, references to biologies, etc. Several publications dealing with state or regional faunas make significant contributions, often with useful keys to species: Provancher (1886, Pet. Faune Ent. Can., 3: 156-161); Osborn and Drake (1916, Oh. Biol. Surv. Bull., 8: 217-251); Barber and Weiss, (1922, N.J. Dept. Agr., 54: 3-15); Drake, (1922, N.Y. St. Coll. Forest. Tech. Publ., 16: 64-66); McAtee (1923, Proc. Ent. Soc. Wash., 25: 143-151); Parshley (1923, Conn. Geol. Nat. Hist. Surv. Bull., 34: 695-707); Blatchley (1926, Het. E. N. Am., pp. 448-501); Downes (1927, Proc. Ent. Soc. B.C., 23: 1-22); Froeschner (1944, Am. Midl. Nat., 31: 646-649, 667-670); and Bailey (1951, Ent. Am., 31: 1-140).

Subfamily Tinginae Laporte, 1807

Tribe Tingini Laporte, 1807

Genus *Acalypta* Westwood, 1840

1840 *Acalypta* Westwood, Synop. Gen. Brit. Ins., p. 121. Type-species: *Tingis carinata* Panzer, 1806. Monotypic.

1916 *Fenestrella* Osborn and Drake, Oh. Biol. Surv. Bull., 8: 222. Preoccupied. Type-species: *Fenestrella ovata* Osborn and Drake, 1916; preoccupied in *Acalypta*; next available name *Acalypta duryi* Drake, 1930. Monotypic.

1922 *Drakella* Bergroth, An. Soc. Ent. Belg., 62: 152. New name for *Fenestrella* Osborn and Drake. Synonymized by Drake, 1928, Bull. Brook. Ent. Soc., 23: 2.

Note: Drake and Lattin (1963, Proc. U.S. Nat. Mus., 115: 331-345) reviewed the American species of this genus. Their inclusion of the Japanese species *A. sauteri* Drake among the "American species" was based on a specimen intercepted at a port-of-entry at New York City, not on an established population. Froeschner (1976, Am. Midl. Nat., 96: 257-269) discussed the zoogeography of the American species.

Acalypta cooleyi Drake, 1917

1917 *Acalypta cooleyi* Drake, Oh. J. Sci., 17: 213. [Mont.]. Distribution: Ariz., Cal., Mont., Ore. (Asia).

Acalypta duryi Drake, 1930

1916 *Fenestrella ovata* Osborn and Drake, Oh. Biol. Surv. Bull., 8: 223. [Oh.]. Preoccupied.

1922 *Drakella ovata*: Bergroth, An. Soc. Ent. Belg., 62: 152.

1928 *Acalypta ovata*: Drake, Bull. Brook. Ent. Soc., 23: 3.

1930 *Acalypta duryi* Drake, Bull. Brook. Ent. Soc., 25: 268. [Oh.].

Distribution: Ga., N.C., Oh., Tenn.

Acalypta elegans Horvath, 1906

1906 *Acalypta elegans* Horvath, An. Mus. Nat. Hung., 4: 32. [Siberia].

1928 *Acalypta nyctalis* Drake, Bull. Brook. Ent. Soc., 23: 5. [N.H.]. Synonymized by

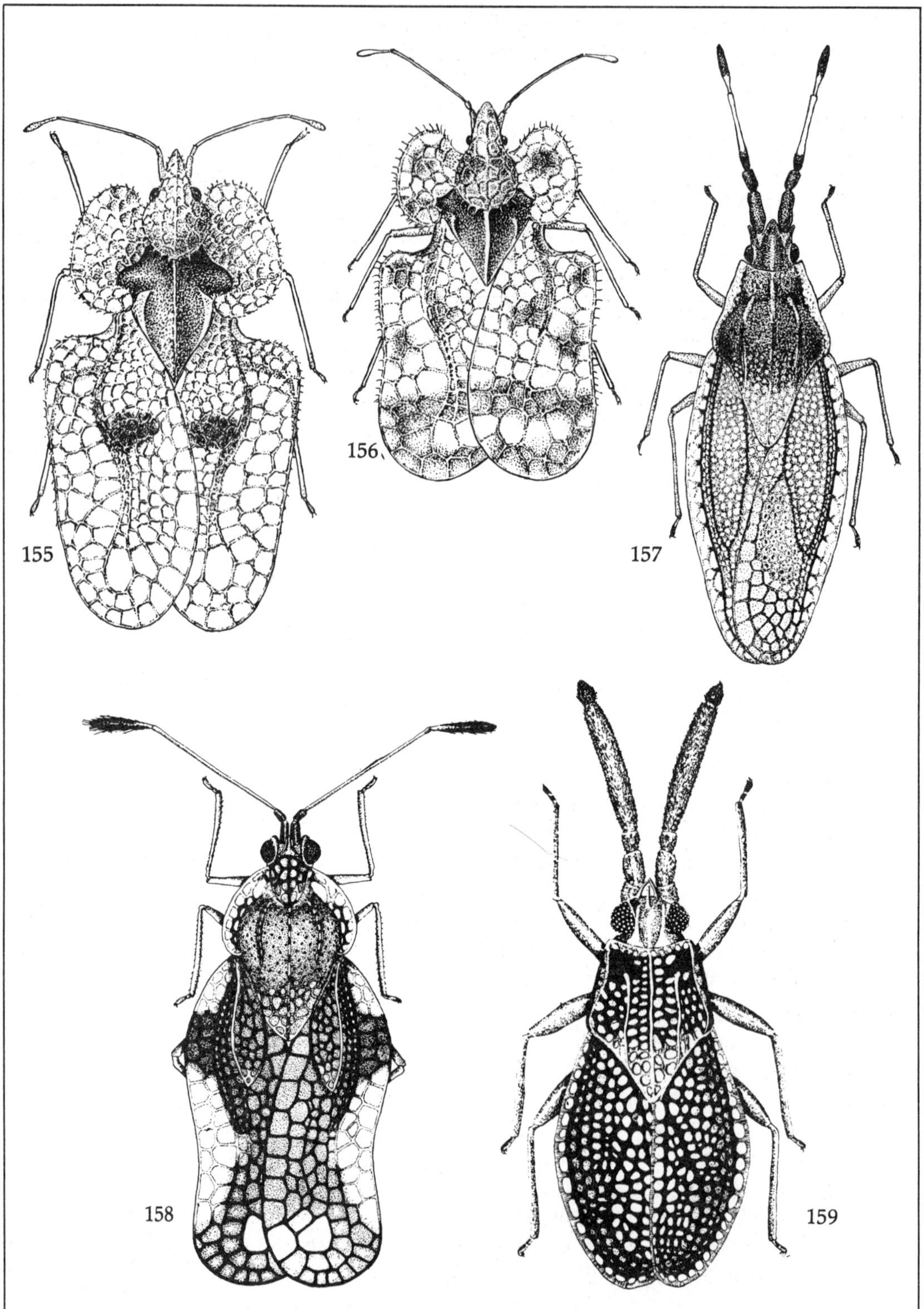

Figs.155-159: 155, *Corythucha ciliata* [p. 715]; 156, *Corythucha marmorata* [p. 718]; 157, *Atheas mimeticus* [p.712]; 158, *Leptopharsa clitoriae* [p.725]; 159, *Alveotingis grossocerata* [p. 712] (All after Froeschner, 1944, except 158-159, after Osborn and Drake, 1916).

Péricart, 1978, An. Ent. Soc. France, n. ser., 14: 697.
1935 *Acalypta nyctilis* [*sic*]: Downes, Proc. Ent. Soc. B.C., 31: 46.
Distribution: Alk., Alta., B.C., N.H., Nfld. (Russia).

Acalypta lillianis Torre-Bueno, 1916
1916 *Acalypta lillianis* Torre-Bueno, Bull. Brook. Ent. Soc., 11: 39. [N.Y.]. Lectotype designated by Drake and Lattin, 1963, Proc. U.S. Nat. Mus., 115(3486): 342.
1916 *Acalypta ovata* Osborn and Drake, Oh. J. Sci., 17: 9. [N.H.]. Synonymized by Parshley, 1917, Psyche, 24: 14.
1917 *Acalypta grisea* Heidemann, Proc. Ent. Soc. Wash., 18: 218. [Mass.]. Synonymized by Drake, 1928, Bull. Brook. Ent. Soc., 23: 6.
1921 *Acalypta modesta* Parshley, Proc. Ent. Soc. B.C., 18: 16. [B.C.]. Synonymized by Hurd, 1946, Ia. St. J. Sci., 20: 463.
1921 *Acalypta lilianis* [*sic*]: Parshley, Proc. Ent. Soc. B.C., 18: 16.
1960 *Acalypta lilliana* [*sic*]: Drake and Ruhoff, Bull. Brook. Ent. Soc., 54: 138.
Distribution: Alk., B.C., Conn., D.C., Ga., Ia., Id., Ill., Ind., Mass., Md., Me., Mich., Minn., N.C., N.D., N.H., N.J., N.Y., Neb., Nfld., Oh., Ont., Pa., Que., R.I., Tenn., Va., Vt., Wis.

Acalypta parvula (Fallén), 1807
1807 *Tingis parvula* Fallén, Monogr. Cimic. Svec., p. 53. [Austria, Czechoslovakia].
1934 *Acalypta barberi* Drake, Bull. Brook. Ent. Soc., 29: 196. [N.Y.]. Synonymized by Golub, 1973, Rev. d'Ent., URSS, 52: 631.
1973 *Acalypta parvula*: Golub, Rev. d'Ent., URSS, 52: 631.
Distribution: B.C., N.B., N.Y., Ont., Ore., Pa., Wash. (Europe).
Note: Golub (1973, above) suggested that this species was introduced into N. Am.

Acalypta saundersi (Downes), 1927
1927 *Drakella saundersi* Downes, Can. Ent., 59: 60. [B.C.].
1928 *Acalypta saundersi*: Drake, Bull. Brook. Ent. Soc., 23: 4.
Distribution: B.C., Cal., Ore., Wash.

Acalypta thomsonii Stål, 1873
1873 *Acalypta thomsonii* Stål, K. Svens. Vet.-Akad. Handl., ll(2): 122. ["Carolina"].
1926 *Acalypta thomsoni* [*sic*]: Drake, An. Carnegie Mus., 16: 377.
1926 *Acalypta madelainae* Torre-Bueno, Bull. Brook. Ent. Soc., 21: 117. [Mass.]. Synonymized by Drake, 1928, Bull. Soc., 23: 4.
Distribution: D.C., Fla., Ga., Mass., Md., N.C., N.J., R.I., S.C., Va.

Acalypta vanduzeei Drake, 1928
1928 *Acalypta vanduzeei* Drake, Bull. Brook. Ent. Soc., 23: 8. [Cal.].
Distribution: Cal.

Acalypta vandykei Drake, 1928
1928 *Acalypta vandykei* Drake, Bull. Brook. Ent. Soc., 23: 8. [Cal.].
Distribution: Cal.
Note: Drake and Ruhoff's (1965, U.S. Nat. Mus. Bull., 243: 256) record for Oregon is dropped here because the paper prepared by Drake and Lattin (1963, Proc. U.S. Nat. Mus., 115: 331-345), while the 1965 paper was in press, omitted it; no Oregon specimen of *vandykei* is at hand.

Genus *Acanthocheila* Stål, 1858

1858 *Monanthia* (*Acanthocheila*) Stål, K. Svens. Vet.-Akad. Handl., 2(7): 61. Type-species:

Monanthia armigera Stål, 1858. Designated by Van Duzee, 1916, Check List Hem., p. 26.

1916 *Acanthocheila*: Van Duzee, Check List Hem. p. 26.

Acanthocheila armigera (Stål), 1858

1858 *Monanthia (Acanthocheila) armigera* Stål, K. Svens. Vet.-Akad. Handl., 2(7): 61. [Brazil].

1937 *Acanthocheila armigera*: Drake and Poor, Mem. Carnegie Mus., 11: 306.

Distribution: Tex. (Central and South America, Greater Antilles).

Acanthocheila exquisita Uhler, 1889

1889 *Acanthocheila exquisita* Uhler, Proc. Ent. Soc. Wash., l: 143. [Fla.].

Distribution: Fla. (Bahama Islands).

Genus *Alveotingis* Osborn and Drake, 1916

1916 *Alveotingis* Osborn and Drake, Oh. Biol. Surv. Bull., 8: 245. Type-species: *Alveotingis grossocerata* Osborn and Drake, 1916. Monotypic.

Alveotingis brevicornis Osborn and Drake, 1917

1917 *Alveotingis brevicornis* Osborn and Drake, Oh. J. Sci., 17: 306. [Ia.].

Distribution: Ia., Minn., Mo.

Alveotingis grossocerata Osborn and Drake, 1916 [Fig. 159]

1916 *Alveotingis grossocerata* Osborn and Drake, Oh. Biol. Surv. Bull., 8: 245. [Me.].

Distribution: Conn., Ks., Mass., Me., N.H., N.Y., Pa.

Note: Parshley (1920, Ent. News, 31: 274) mentioned a Uhler manuscript name, "*Rhombodea areolata*" on a specimen in the Slosson collection.

Alveotingis minor Osborn and Drake, 1917

1917 *Alveotingis minor* Osborn and Drake, Oh. J. Sci., 17: 305. [Ia.].

Distribution: Ia.

Genus *Atheas* Champion, 1898

1898 *Atheas* Champion, Biol. Centr.-Am., Rhyn., 2: 44. Type-species: *Atheas nigricornis* Champion, 1898. Designated by Van Duzee, 1916, Check List Hem., p. 26.

Atheas austroriparius Heidemann, 1909

1909 *Atheas austroriparius* Heidemann, Bull. Buff. Soc. Nat. Sci., 9: 235. [Fla.].

1925 *Athaes* [*sic*] *angustroriparius* [*sic*]: Drake, Fla. Ent., 9: 39.

Distribution: Fla., Ga., Miss., Mo., S.C., Tex. (Mexico).

Atheas exiguus Heidemann, 1909

1909 *Atheas exiguus* Heidemann, Bull. Buff. Soc. Nat. Sci., 9: 233. [Fla.].

Distribution: Fla., Miss., Tex.

Atheas insignis Heidemann, 1909

1909 *Atheas insignis* Heidemann, Bull. Buff. Soc. Nat. Sci., 9: 232. [Md.].

Distribution: D.C., Md., Miss., Va.

Atheas mimeticus Heidemann, 1909 [Fig. 157]

1909 *Atheas mimeticus* Heidemann, Bull. Buff. Soc. Nat. Sci., 9: 234. [Ks.].

1917 *Atheas annulatus* Osborn and Drake, Oh. J. Sci., 17: 295. [Ark.]. Synonymized by Hurd, 1946, Ia. St. Coll. J. Sci., 20: 460.

1917 *Atheas sordidus* Osborn and Drake, Oh. J. Sci., 17: 296. [Ia.]. Synonymized by Hurd, 1946, Ia. St. Coll. J. Sci., 20: 460.

Distribution: Ark., Col., Fla., Ia., Ks., La., Minn., Miss., Mo., N.M., Neb., Wis., Wyo.

Atheas nigricornis Champion, 1898
1898 *Atheas nigricornis* Champion, Biol. Centr.-Am., Rhyn., 2: 45. [Guatemala].
1917 *Atheas nigricornis*: Osborn and Drake, Oh. J. Sci., 17: 296.
Distribution: Ariz., Tex. (Mexico to Ecuador).

Genus *Calotingis* Drake, 1918

1918 *Calotingis* Drake, Bull. Brook. Ent. Soc., 13: 86. Type-species: *Calotingis knighti* Drake, 1918. Monotypic.

Calotingis knighti Drake, 1918
1918 *Calotingis knighti* Drake, Bull. Brook. Ent. Soc., 13: 87. [Tex.].
Distribution: Tex. (Mexico).

Genus *Corythaica* Stål, 1873

1873 *Corythaica* Stål, K. Svens. Vet.-Akad. Handl., 11(2): 119, 122. Type-species: *Tingis monacha* Stål, 1858. Monotypic.
1898 *Dolichocysta* Champion, Trans. Ent. Soc. London, l: 56. Type-species:*Dolichocysta venusta* Champion, 1898. Monotypic. Synonymized by Hurd, 1945, Ia. St. Coll. J. Sci., 20: 80.

Corythaica acuta (Drake), 1917
1917 *Dolichocysta acuta* Drake, Oh. J. Sci., 17: 214. [Col.].
1919 *Corythaica acuta*: Gibson, Proc. Biol. Soc. Wash., 32: 100.
Distribution: Col., Mont., Nev.

Corythaica bellula Torre-Bueno, 1917
1917 *Corythaica bellula* Torre-Bueno, Bull. Brook. Ent. Soc., 12: 19. [N.Y.].
1926 *Corythaica floridana* Blatchley, Het. E. N. Am., p. 472. [Fla.]. Synonymized by Drake, 1930, Bull. Brook. Ent. Soc., 25: 268.
Distribution: Conn., Fla., Ga., Mass., Me., N.H., N.Y., Neb., R.I.

Corythaica carinata Uhler, 1894
1894 *Corythaica carinata* Uhler, Proc. Zool. Soc. London, Ser. 2, 4: 203. [Grenada].
1917 *Corythaica constricta* Osborn and Drake, Oh. J. Sci., 17: 304. [Col.]. Synonymized by Drake and Bruner, 1924, Mem. Soc. Cubana Hist. Nat., 6: 151.
1919 *Dolichocysta constricta*: Gibson, Proc. Biol. Soc. Wash., 32: 102.
Distribution: Col., Fla., Tex. (Mexico to Honduras, West Indies).

Corythaica venusta (Champion), 1898
1898 *Dolichocysta venusta* Champion, Trans. Ent. Soc. London, 1: 57. [Mexico].
1899 *Dolichocysta venusta*: Heidemann, Proc. Ent. Soc. Wash., 4: 339.
1919 *Dolichocysta magna* Gibson, Proc. Biol. Soc. Wash., 32: 102. [Col.] Synonymized by Hurd, 1945, Ia. St. Coll. J. Sci., 20: 90.
1919 *Dolichocysta densata* Gibson, Proc. Biol. Soc. Wash., 32: 102. [Tex]. Synonymized by Hurd, 1945, Ia. St. Coll. J. Sci., 20: 90.
1923 *Dolichocysta obscura* Van Duzee, Proc. Cal. Acad. Sci., ser. 4, 12: 140. [Ariz.]. Synonymized by Hurd, 1945, Ia. St. Coll. J. Sci., 20: 90.
1945 *Corythaica venusta*: Hurd, Ia. St. Coll. J. Sci., 20: 90.
Distribution: Ariz., B.C., Cal., Col., Ks., Mont., N.M., Neb., S.D., Tex. (Mexico).

Genus *Corythucha* Stål, 1873

1873 *Corythucha* Stål, K. Svens. Vet.-Akad. Handl., 11(2): 119, 122. Type-species: *Tingis fuscigera* Stål, 1862. Designated by Van Duzee, 1916, Check List Hem., p. 25.

Corythucha aesculi Osborn and Drake, 1916.

1916 *Corythucha aesculi* Osborn and Drake, Oh. Biol. Surv. Bull., 8: 232. [Oh.].
1918 *Corythucha fuscigera*: Gibson, Trans. Am. Ent. Soc., 44: 78 (in part).
1926 *Corythuca* [*sic*] *aesculi*: Blatchley, Het. E. N. Am., p. 465.
Distribution: Ark., Conn., D.C., Fla., Ia., Ill., Ind., La., Md., Mo., N.C., N.Y., Oh., Va., W.Va.

Corythucha arcuata (Say), 1832

1832 *Tingis arcuata* Say, Descrip. Het. Hem. N. Am., p. 27. ["United States"].
1873 *Corythucha arcuata*: Stål, K. Svens. Vet.-Akad. Handl., ll(2): 123.
1884 *Corythuca* [*sic*] *arcuata*: Uhler, Stand. Nat. Hist., 2: 284.
1886 *Corythuca* [*sic*] *ciliata*: Provancher, Pet. Faune Ent. Can., 3: 158.
1886 *Corythuca* [*sic*] *juglandis*: Provancher, Pet. Faune Ent. Can., 3: 158.
1888 *Corythucha polygrapha* Lintner, Rept. St. Ent. N.Y. 1887, 41: 109. [Md., N.Y.]. Synonymized by Drake and Ruhoff, 1965, U.S. Nat. Mus. Bull., 243: 142.
1889 *Corythuca* [*sic*] *arquata* [*sic*]: Van Duzee, Can. Ent., 21: 5.
1903 *Corythucha arcuata arcuata*: Morrill, Psyche, 10: 133.
1910 *Corythuca* [*sic*] *polygraphia* [*sic*]: Banks, Cat. Nearc. Hem.-Het., p. 56.
1916 *Corythucha juglandis*: Van Duzee, Check List Hem., p. 25 (in part).
1918 *Corythucha mali* Gibson, Trans. Am. Ent. Soc., 44: 98. [N.J.]. Synonymized by Froeschner, 1944, Am. Midl. Nat., 31: 668.
1918 *Corythucha piercei* Gibson, Trans. Am. Ent. Soc., 44: 85. [Ariz.]. Synonymized by Drake and Ruhoff, 1965, U.S. Nat. Mus. Bull., 243: 142.
1926 *Corythuca* [*sic*] *arcuata* var. *mali*: Blatchley, Het. E. N. Am., p. 465.
Distribution: Ala., Ariz., Col., Conn., D.C., Ga., Ia., Ill., Ind., Ks., Mass., Md., Me., Mich., Minn., Miss., Mo., N.C., N.D., N.H., N.J., N.M., N.Y., Neb., Oh., Ont., Pa., Que., R.I., S.C., S.D., Tex., Ut., Va., Vt., Wis.

Corythucha associata Osborn and Drake, 1916

1916 *Corythucha associata* Osborn and Drake, Oh. J. Sci., 17: 14. [Tenn.].
1918 *Corythucha spinulosa* Gibson, Trans. Am. Ent. Soc., 44: 79. [N.J.]. Synonymized by Drake, 1921, Fla. Ent., 4: 54.
1918 *Corythucha fuscigera*: Gibson, Trans. Am. Ent. Soc., 44: 78 (in part).
1926 *Corythuca* [*sic*] *associata*: Blatchley, Het. E. N. Am., p. 456.
Distribution: Ala., Conn., D.C., Ga., Ia., Ill., Ind., Md., Miss., N.J., N.Y., Oh., Pa., R.I., S.C., Tenn.

Corythucha baccharidis Drake, 1922

1922 *Corythucha baccharidis* Drake, Fla. Ent., 5: 37. [Fla.].
1926 *Corythuca* [*sic*] *baccharidis*: Blatchley, Het. E. N. Am., p. 463.
Distribution: Fla., Miss., Tex.

Corythucha bellula Gibson, 1918

1918 *Corythucha bellula* Gibson, Trans. Am. Ent. Soc., 44: 93. [Oh.].
1926 *Corythuca* [*sic*] *bellula*: Blatchley, Het. E. N. Am., p. 458.
Distribution: Me., N.H., N.Y., Oh., Vt.

Corythucha brunnea Gibson, 1918

1918 *Corythucha brunnea* Gibson, Trans. Am. Ent. Soc., 44: 93. [Oh.].
Distribution: Ill., La., Miss., Tex.

Corythucha bulbosa Osborn and Drake, 1916
1916 *Corythucha bulbosa* Osborn and Drake, Oh. Biol. Surv. Bull., 8: 232. [Oh.].
1919 *Corythuca* [*sic*] *bulbosa*: Weiss, Oh. J. Sci., 20: 17.
Distribution: D.C., Ill., Ind., Md., Mo., N.J., Oh., Pa., Va.

Corythucha caelata Uhler, 1884
1884 *Corythuca* [*sic*] *caelata* Uhler, Stand. Nat. Hist., 2: 279. [Mexico].
1910 *Corythuca* [*sic*] *caelata*: Banks, Cat. Nearc. Hem.-Het., p. 55.
1918 *Corythucha caelata*: Gibson, Trans. Am. Ent. Soc., 44: 101 (in part).
1944 *Corythucha compta* Drake and Hambleton, Proc. Ent. Soc. Wash., 46: 96. [Cal.]. Synonymized by Drake and Ruhoff, 1965, Bull. U.S. Nat. Mus., 243: 144.
Distribution: Cal. (Mexico).

Corythucha caryae Bailey, 1951
1951 *Corythucha caryae* Bailey, Ent. Am., 31: 68. [Mass.].
Distribution: Mass., N.H., R.I.

Corythucha celtidis Osborn and Drake, 1916
1916 *Corythucha celtidis* Osborn and Drake, Oh. Biol. Surv. Bull., 8: 227. [Oh.].
1926 *Corythuca* [*sic*] *celtidis*: Blatchley, Het. E. N. Am., p. 468.
Distribution: Ia., Ill., Ind., Ga., Ky., Md., N.C., N.J., Oh., S.C., Tenn.

Corythucha celtidis celtidis Osborn and Drake, 1916
1916 *Corythucha celtidis* Osborn and Drake, Oh. Biol. Surv. Bull., 8: 227.
1925 *Corythucha celtidis* var. *celtidis*: Drake, Fla. Ent., 9: 36.
Distribution: Ia., Ill., Ind., Ga., Ky., Md., N.C., N.J., Oh., S.C., Tenn.

Corythucha celtidis mississippiensis Drake, 1925
1925 *Corythucha celtidis* var. *mississippiensis* Drake, Fla. Ent., 9: 36. [Miss.].
1926 *Corythuca* [*sic*] *celtidis* var. *mississippiensis*: Blatchley, Het. E. N. Am., p. 468.
Distribution: Ga., Miss., S.C., Tenn.

Corythucha cerasi Drake, 1948
1948 *Corythucha cerasi* Drake, Rev. Ent., 19: 435. [Wash.].
Distribution: Ind., Wash.

Corythucha championi Drake and Cobben, 1960
1906 *Corythucha decens*: Snow, Trans. Ks. Acad. Sci., 20: 153.
1910 *Corythuca* [*sic*] *decnes* [*sic*]: Banks, Cat. Nearc. Hem.-Het., p. 56.
1929 *Corythucha decens*: Lutz, Ent. News, 40: 233.
1960 *Corythucha championi* Drake and Cobben, Stud. Fauna Curaçāo Caribbean Is., 10(54): 93. [Curaçao].
1965 *Corythucha championi*: Drake and Ruhoff, U.S. Nat. Mus. Bull., 243: 145.
Distribution: Pa.(?), Tex. (Mexico to Colombia, Netherlands Antilles).
Note: The geographic remoteness of the Pa. record suggests that it needs verification. The label on the documentary specimen in the J.C. Lutz Collection (now in the USNM) shows it was collected by Lutz himself.

Corythucha ciliata (Say), 1832 [Fig. 155]
1832 *Tingis ciliata* Say, Descrip. Het. N. Am., p. 27. ["United States"].
1840 *Tingis hyalina* Herrich-Schaeffer, Wanz. Ins., 5: 84. [Carolina]. Synonymized by Uhler, 1878, Proc. Boston Soc. Nat. Hist., 19: 414.
1873 *Corythucha ciliata*: Stål, K. Svens. Vet.-Akad. Handl., ll(2): 123.
1873 *Corythucha hyalina*: Stål, K. Svens. Vet.-Akad. Handl., ll(2): 123.
1886 *Corythuca* [*sic*] *ciliata*: Uhler, Check-list Hem. Het., p. 22.

1891 *Corythuca* [*sic*] *ciliati* [*sic*]: Summers, Bull. Agr. Exp. Stn., Univ. Tenn., 4: 90.
1910 *Corythuca* [*sic*] *hyalina*: Banks, Cat. Nearc. Hem.-Het., p. 56.
Distribution: Ala., Col., Conn., D.C., Fla., Ga., Ia., Ill., Ind., Ks., Mass., Md., Me., Mich., Mo., N.C., N.J., N.Y., Oh., Ok., Ont., Pa., Que., R.I., Tenn., Tex., Va., Vt.

Corythucha confraterna Gibson, 1918
1918 *Corythucha confraterna* Gibson, Trans. Am. Ent. Soc., 44: 102. [Cal.].
Distribution: Ariz., Cal. (Mexico).

Corythucha coryli Osborn and Drake, 1917
1917 *Corythucha coryli* Osborn and Drake, Oh. J. Sci., 17: 299. [Md.].
1926 *Corythuca* [*sic*] *coryli*: Blatchley, Het. E. N. Am., p. 455.
Distribution: Conn., D.C., Ia., Ill., Ind., Ks., Mass., Md., N.H., N.J., Neb., R.I., Va.

Corythucha cydoniae (Fitch), 1861
1861 *Tingis cydoniae* Fitch, Country Gentleman, 17: 114. [Mass.].
1880 *Corythuca* [*sic*] *arcuata*: Comstock, U.S. Dept. Agr., Rept. Ent., p. 221.
1903 *Corythuca* [*sic*] *arcuata crataegi* Morril, Psyche, 10: 132. [Mass.]. Synonymized by Gibson, 1918, Trans. Am. Ent. Soc., 44: 88.
1910 *Corythuca* [*sic*] *crataegi*: Banks, Cat. Nearc. Hem.-Het., p. 56.
1910 *Corythuca* [*sic*] *cydoniae*: Banks, Cat. Nearc. Hem.-Het., p. 56.
1916 *Corythucha crataegi*: Osborn and Drake, Oh. Biol. Surv. Bull., 8: 229.
1918 *Corythucha cydoniae*: Gibson, Trans. Am. Ent. Soc., 44: 87.
1918 *Corythucha occidentalis* Drake, Trans. Am. Ent. Soc., 44: 91. [Cal.]. Synonymized by Drake and Ruhoff, 1965, Bull. U.S. Nat. Mus., 243: 147.
Distribution: Ala., Ark., Cal., Col., Conn., D.C., Del., Fla., Ga., Ia., Id., Ind., Man., Mass., Md., Me., Mich., Miss., Mo., N.C., N.D., N.J., N.Y., Neb., Oh., Ok., Ont., Ore., Pa., R.I., S.C., S.D., Tex., Ut., Va., Vt., W.Va., Wash., Wis. (Mexico).
Note: Wheeler (1981, Great Lakes Ent., 14: 37-43) reviewed hosts of the hawthorn lace bug.

Corythucha distincta Osborn and Drake, 1916
1893 *Corythucha* species: Cockerell, Trans. Am. Ent. Soc., 20: 364.
1895 *Corythuca* [*sic*] *fuscigera*: Gillette and Baker, Col. Agr. Exp. Stn., Bull., 31: 364.
1916 *Corythucha distincta* Osborn and Drake, Oh. J. Sci., 17: 13. [Col.].
Distribution: B.C., Cal., Col., Mont., S.D., Ut., Wash., Wyo.

Corythucha distincta distincta Osborn and Drake, 1916
1916 *Corythucha distincta* Osborn and Drake, Oh. J. Sci., 17: 13.
1917 *Corythucha distincta distincta*: Osborn and Drake, Oh. J. Sci., 17: 301. [Mont.].
Distribution: Same as for species.

Corythucha distincta spinata Osborn and Drake, 1917
1917 *Corythucha distincta* var. *spinata* Osborn and Drake, Oh. J. Sci., 17: 301. [Mont.].
Distribution: Mont.

Corythucha elegans Drake, 1918
1918 *Corythucha elegans* Drake, Trans. Am. Ent. Soc., 44: 89. [Col.].
1926 *Corythuca* [*sic*] *elegans*: Blatchley, Het. E. N. Am., p. 456.
Distribution: B.C., Col., Man., Mass., Mich., N.Y., Ont., Sask., Wis.

Corythucha eriodictyonae Osborn and Drake, 1917
1917 *Corythucha eriodictyonae* Osborn and Drake, Oh. J. Sci., 17: 302. [Cal.].
Distribution: Cal., Nev.

Corythucha floridana Heidemann, 1909
1909 *Corythuca* [*sic*] *floridana* Heidemann, Bull. Buff. Soc. Nat. Sci., 9: 236. [Fla.].
1917 *Corythucha floridana*: Osborn and Drake, Oh. J. Sci., 17: 300.
Distribution: Fla.

Corythucha fuscigera (Stål), 1862
1862 *Tingis fuscigera* Stål, Stett. Ent. Zeit., 23: 323. [Mexico].
1873 *Monathia lucida* Walker, Cat. Hem. Het. Brit. Mus., 6: 191. [Mexico]. Synonymized by Champion, Biol. Centr.-Am., Rhyn., 2: 7.
1897 *Corythucha fuscigera*: Champion, Biol. Centr.-Am., Rhyn., 2: 7.
1886 *Leptostyla lucida*: Uhler, Check-list Hem. Het., p. 22.
1886 *Corythuca* [*sic*] *fuscigera*: Uhler, Check-list Hem. Het., p. 22.
Distribution: Ariz., Cal. (Mexico to Guatemala).

Corythucha gossypii (Fabricius), 1794
1794 *Acanthia gossypii* Fabricius, Ent. Syst., 4: 78. ["Americae meridionalis Insulis"].
1862 *Tingis decens* Stål, Stett. Ent. Zeit., 23: 324. [Mexico]. Synonymized by Drake and Cobben, 1960, Stud. Fauna Curação, 11: 91.
1886 *Corythuca* [*sic*] *gossypii*: Uhler, Check-list Hem. Het., p. 22.
1914 *Corythucha gossypii*: Barber, Bull. Am. Mus. Nat. Hist., 33: 507.
1929 *Corythucha decens*: Lutz, Ent. News, 40: 233.
Distribution: Fla., N.M., Pa., Tex. (Mexico to Ecuador, West Indies).

Corythucha heidemanni Drake, 1918
1918 *Corythucha heidemanni* Drake, Trans. Am. Ent. Soc., 44: 87. [N.Y.].
1918 *Corythucha borealis* Parshley, Trans. Am. Ent. Soc., 44: 92. [Me.]. Synonymized by Parshley, 1923, Conn. Geol. Nat. Hist. Surv. Bull., 34: 701.
1926 *Corythuca* [*sic*] *heidemanni*: Blatchley, Het. E. N. Am., p. 457.
Distribution: Conn., Mass., Me., N.B., N.H., N.S., N.Y., Nfld., Ont., Vt.

Corythucha hewitti Drake, 1919
1919 *Corythucha hewitti* Drake, Can. Ent., 51: 159. [B.C.].
1919 *Corythucha hesperia* Parshley, Occas. Pap. Mus. Zool., Univ. Mich., 71: 23. [B.C.]. Synonymized by Parshley, 1921, Proc. Ent. Soc. B.C., 18: 14.
Distribution: B.C., Col., Conn., Ia., Man., Minn., Ont., Pa., Wis.

Corythucha hispida Uhler, 1894
1894 *Corythuca* [*sic*] *hispida* Uhler, Proc. Cal. Acad. Sci., ser. 2, 4: 279. [Mexico].
1910 *Corythuca* [*sic*] *hispida*: Banks, Cat. Nearc. Hem.-Het., p. 56.
1918 *Corythucha hispida*: Gibson, Trans. Am., Ent. Soc., 44: 104 (in part).
Distribution: Cal. (Mexico).

Corythucha immaculata Osborn and Drake, 1916
1916 *Corythucha immaculata* Osborn and Drake, Oh. J. Sci., 17: 11. [Cal.].
1917 *Corythuca* [*sic*] *pura* Gibson, Ent. News, 28: 258. [Wash.]. Synonymized by Drake, 1918, Bull. Brook. Ent. Soc., 13: 86.
1918 *Corythucha pura*: Gibson, Trans. Am. Ent. Soc., 44: 103.
Distribution: B.C., Cal., Col., Id., Mont., Ore., Ut., Wash.

Corythucha incurvata Uhler, 1894
1894 *Corythuca* [*sic*] *incurvata* Uhler, Proc. Cal. Acad. Sci., ser. 2, 4: 280. [Cal.]
1911 *Corythuca* [*sic*] *arcuata*: Pemberton, J. Econ. Ent., 4: 339.
1917 *Corythucha bullata* Van Duzee, Proc. Cal. Acad. Sci., ser. 4, 7: 258. [Cal.]. Synonymized by Drake and Ruhoff, 1962, Bull. So. Cal. Acad. Sci., 60: 156.
1917 *Corythucha incurvata*: Van Duzee, Univ. Cal. Publ. Ent., 2: 213.

1920 *Corythucha heteromelecola* Drake, Oh. J. Sci., 20: 50. [Cal.]. Synonymized by Drake and Ruhoff, 1962, Bull. So. Cal. Acad. Sci., 60: 156.
Distribution: Ariz., Cal., Ore. (Mexico).

Corythucha juglandis (Fitch), 1856
1856 *Tingis juglandis* Fitch, Trans. N.Y. St. Agr. Soc., 16: 446. [N.Y.].
1873 *Corythucha juglandis*: Stål, K. Svens. Vet.-Akad. Handl., 11(2): 123.
1878 *Tingis arcuata*: Uhler, Proc. Boston Soc. Nat. Hist., 19: 415 (in part).
1884 *Corythuca* [*sic*] *juglandis*: Uhler, Stand. Nat. Hist., 2: 285.
1910 *Corythuca* [*sic*] *arcuata*: Smith, An. Rept. N.J. St. Mus., p. 148.
1916 *Corythucha contracta* Osborn and Drake, Oh. Biol. Surv. Bull., 8: 230. [Oh.]. Synonymized by Bailey, 1951, Ent. Am., 31: 84.
1918 *Corythucha parshleyi* Gibson, Trans. Am. Ent. Soc., 44: 82. [N.J.]. Synonymized by Drake, 1921, Fla. Ent., 4: 51.
1926 *Corythuca* [*sic*] *contracta*: Blatchley, Het. E. N. Am., p. 467.
Distribution: B.C., Conn., Ga., Ia., Ill., Ind., Ks., Mass., Md., Me., Miss., N.C., N.H., N.J., N.Y., Oh., R.I., S.C., Tenn., Tex., Va., Vt., Wis.
Note: Two specimens were taken from the stomach of a frog (Frost, 1925, J. N.Y. Ent. Soc., 32: 180).

Corythucha lowryi Drake, 1948
1948 *Corythucha lowryi* Drake, Rev. Ent., 19: 434. [Wis.].
Distribution: Wis.

Corythucha marmorata (Uhler), 1878 [Fig. 156]
1878 *Tingis marmorata* Uhler, Proc. Boston Soc. Nat. Hist., 19: 415. [N.C.].
1889 *Corythuca* [*sic*] *marmorata*: Van Duzee, Can. Ent., 21: 5.
1898 *Corythuca* [*sic*] *irrorata* Howard, U.S. Dept. Agr., Div. Ent. Bull. 10, n. ser., p. 99 (Riley ms. name). [Ala.]. *Nomen nudum*. Listed as synonym by Van Duzee, Univ. Cal. Publ. Ent., 2: 214.
1908 *Corythuca* [*sic*] *gossypii*: Torre-Bueno, J. N.Y. Ent. Soc., 16: 232.
1914 *Corythucha marmorata*: Van Duzee, Trans. San Diego Soc. Nat. Hist., 2: 11.
1917 *Corythucha decens*: Van Duzee, Univ. Cal. Publ. Ent., 2: 214 (in part).
1918 *Corythucha lactea* Drake, Trans. Am. Ent. Soc., 44: 94. [Ut.]. Synonymized by Drake and Ruhoff, 1965, U.S. Nat. Mus. Bull., 243: 154.
1919 *Corythucha marmorata* var. *informis* Parshley, Occas. Pap. Mus. Zool., Univ. Mich., 71: 20. [B.C.]. Synonymized by Drake and Ruhoff, 1965, U.S. Nat. Mus. Bull., 243: 154.
1920 *Corythucha marmorata* var. *minutissima* Drake, Oh. J. Sci., 20: 50. [Cal.]. Synonymized by Drake and Ruhoff, 1965, U.S. Nat. Mus. Bull., 243: 154.
Distribution: Ala., Ariz., B.C., Cal., Conn., Col., D.C., Fla., Ga., Ia., Id., Ill., Ind., Ks., La., Mass., Md., Me., Mich., Miss., Mo., Oh., Ore., N.C., N.D., N.H., N.J., N.M., N.Y., Neb., Pa., S.D., Tenn., Tex., Ut., Va., Vt., Wis., W.Va.

Corythucha mollicula Osborn and Drake, 1916
1916 *Corythucha mollicula* Osborn and Drake, Oh. J. Sci., 17: 12. [Mich.].
1917 *Corythucha salicis* Osborn and Drake, Oh. J. Sci., 17: 298. [Mass., Mont., Wis.]. Synonymized by Drake, 1921, Fla. Ent., 4: 53.
1918 *Corythucha molliculata* [*sic*]: Gibson, Trans. Am. Ent. Soc., 44: 90.
1919 *Corythucha canadensis* Parshley, Occas. Pap. Mus. Zool., Univ. Mich., 71: 18. [B.C.]. Synonymized by Parshley, 1921, Proc. Ent. Soc. B.C., 18: 14.
1926 *Corythuca* [*sic*] *mollicula*: Blatchley, Het. E. N. Am., p. 461.

Distribution: B.C., Conn., Fla., Ga., Id., Ill., Ind., Ks., Man., Mass., Md., Me., Mich., Mo., Mont., Ont., Ore., N.D., N.H., N.J., N.Y., Neb., Pa., R.I., S.C., S.D., Va., Wash., Wis.

Corythucha montivaga Drake, 1919

1919 *Corythucha montivaga* Drake, Oh. J. Sci., 19: 417. [Mont.].

Distribution: Mont., Ut., Wyo.

Corythucha morrilli Osborn and Drake, 1917

1917 *Corythucha morrilli* Osborn and Drake, Oh. J. Sci., 17: 298. [Col.].

1917 *Corythucha decens*: Van Duzee, Univ. Cal. Publ. Ent., 2: 214 (in part).

1933 *Corythucha morelli* [*sic*]: Knowlton, Ent. News, 44: 263.

Distribution: Ariz., B.C., Cal., Col., Fla., N.M., Tex., Ut. (Hawaii; Greater, Lesser, and Netherlands Antilles; Mexico to Guatemala).

Corythucha nicholi Drake, 1928

1928 *Corythucha nicholi* Drake, Fla. Ent., 12: 3. [Ariz.].

Distribution: Ariz.

Corythucha obliqua Osborn and Drake, 1916

1916 *Corythucha obliqua* Osborn and Drake, Oh. J. Sci., 17: 11. [Cal.].

1917 *Corythucha maculata* Van Duzee, Proc. Cal. Acad. Sci., ser. 4, 7: 257. [Cal.]. Synonymized by Drake, 1919, Oh. J. Sci., 19: 418.

1917 *Corythucha coelata* [*sic*]: Van Duzee, Proc. Cal. Acad. Sci., ser. 4, 7: 259.

1917 *Corythucha fuscigera*: Van Duzee, Univ. Cal. Publ. Ent., 2: 213 (in part).

1918 *Corythucha contaminata* Gibson, Trans. Am. Ent. Soc., 44: 82. [Id.]. Synonymized by Drake and Ruhoff, 1965, U.S. Nat. Mus. Bull., 243: 156.

Distribution: Cal., Col., Id., Ore., Ut., Wash.

Corythucha omani Drake, 1941

1941 *Corythucha omani* Drake, J. Wash. Acad. Sci., 31: 144. [Ariz.].

Distribution: Ariz.

Corythucha padi Drake, 1917

1917 *Corythucha padi* Drake, Oh. J. Sci., 17: 215. [Mont.].

1918 *Corythucha coloradensis* Gibson, Trans. Am. Ent. Soc., 44: 89. [Col.]. Synonymized by Drake and Ruhoff, 1965, U.S. Nat. Mus. Bull., 243: 156.

1918 *Corythucha fuscigera*: Gibson, Trans. Am. Ent. Soc., 44: 78 (in part).

Distribution: B.C., Col., Id., Mont., Neb., Ore., Ut., Wash.

Corythucha pallida Osborn and Drake, 1916

1916 *Corythucha pallida* Osborn and Drake, Oh. Biol. Surv. Bull., 8: 230. [Oh.].

1927 *Corythuca* [*sic*] *pallida*: Blatchley, Het. E. N. Am., p. 459.

Distribution: Ala., Ariz., D.C., Ga., Ill., Ind., Md., Me., Miss., Mo., N.C., Oh., S.C., Tenn., Va., W.Va.

Corythucha pallipes Parshley, 1918

1917 *Corythucha juglandis*: Van Duzee, Univ. Cal. Publ. Ent., 2: 215 (in part).

1918 *Corythucha pallipes* Parshley, Trans. Am. Ent. Soc., 44: 82. [Conn.].

1918 *Corythucha betulae* Drake, Trans. Am. Ent. Soc., 44: 86. [N.Y.]. Synonymized by Parshley, 1920, Psyche, 31: 273.

1918 *Corythucha cyrta* Parshley, Trans. Am. Ent. Soc., 44: 86. [Me.]. Synonymized by Drake, 1922, N.Y. St. Coll. Forest., Syracuse Univ. Tech. Publ., 16: 111.

1922 *Corythucha paleipes* [*sic*]: Barber, St. N.J. Dept. Agr., Bur. Stat. Inspect. Circ., 54: 16.

1926 *Corythuca* [*sic*] *pallipes*: Blatchley, Het. E. N. Am., p. 457.
Distribution: Conn., Man., Mass., Me., Mich., N.B., N.H., N.J., N.Y., Nfld., Ont., Ore., Que., Vt., Wash., Wis.

Corythucha pergandei Heidemann, 1906
1906 *Corythuca* [*sic*] *pergandei* Heidemann, Proc. Ent. Soc. Wash., 8: 10. [D.C.].
1916 *Corythucha pergandei*: Osborn and Drake, Oh. Biol. Surv. Bull., 8: 228.
Distribution: B.C., Conn., D.C., Ga., Ia., Ks., Mass., Me., Minn., Mo., N.H., N.J., N.S., N.Y., Neb., Ok., Ont., Tex., Vt., Wis.

Corythucha pruni Osborn and Drake, 1916
1916 *Corythucha pruni* Osborn and Drake, Oh. Biol. Surv. Bull., 8: 231. [D.C.].
1917 *Corythucha hoodiana* Osborn and Drake, Oh. J. Sci., 17: 302. [Ore.]. Synonymized by Drake and Ruhoff, 1965, U.S. Nat. Mus. Bull., 243: 159.
1918 *Corythucha exigua* Drake, Trans. Am. Ent. Soc., 44: 83. [N.C.]. Synonymized by Drake and Ruhoff, 1965, U.S. Nat. Mus. Bull., 243: 159.
1920 *Corythucha pyriformis* Parshley, Can. Ent., 52: 81. [Me.]. Synonymized by Parshley, 1923, Conn. Geol. Nat. Hist. Surv. Bull., 34: 701.
1926 *Corythuca* [*sic*] *pruni*: Blatchley, Het. E. N. Am., p. 465.
1926 *Corythuca* [*sic*] *exigua*: Blatchley, Het. E. N. Am. p. 467.
Distribution: Conn., D.C., Ga., Ia., Ill., Ind., Mass., Md., Me., Mich., Mo., N.C., N.H., N.Y., Neb., Oh., Ore., R.I., Ut., Va., Vt.

Corythucha sagillata Drake, 1932
1932 *Corythucha sagillata* Drake, Psyche, 39: 101. [Ariz.].
Distribution: Ariz.

Corythucha salicata Gibson, 1918
1918 *Corythucha salicata* Gibson, Trans. Am. Ent. Soc., 44: 90. [Ore.].
1918 *Corythucha drakei* Gibson, Trans. Am. Ent. Soc., 44: 98. [Cal.]. Synonymized by Drake, 1921, Fla. Ent., 4: 53.
1918 *Corythucha essigi* Drake, J. Econ. Ent., 11: 385. [Cal.]. Synonymized by Drake and Ruhoff, 1965, U.S. Nat. Mus. Bull., 243: 159.
1920 *Corythucha platani* Drake, Oh. J. Sci., 20: 49. [Cal.]. Synonymized by Drake and Ruhoff, 1965, U.S. Nat. Mus. Bull., 243: 159.
Distribution: B.C., Cal., Id., Man., Ore., Ut., Wash.

Corythucha scitula Drake, 1948
1948 *Corythucha scitula* Drake, Bol. Ent. Venez., 7: 24. [Ore.].
Distribution: Ore.

Corythucha setosa Champion, 1897
1897 *Corythucha setosa* Champion, Biol. Centr.-Am., Rhyn., 2: 8. [Guatemala].
1917 *Corythucha setosa*: Van Duzee, Univ. Cal. Publ. Ent., 2: 215.
Distribution: Cal. (Guatemala, Mexico).

Corythucha sphaeralceae Drake, 1920
1917 *Corythucha hispida*: Van Duzee, Univ. Cal. Publ. Ent., 2: 214 (in part).
1920 *Corythucha sphaeralceae* Drake, Oh. J. Sci., 20: 51. [Cal.].
1920 *Corythucha pacifica* Drake, Oh. J. Sci., 20: 52. [Wash.]. Synonymized by Drake and Ruhoff, 1965, U.S. Nat. Mus. Bull., 243: 160.
Distribution: Ariz., Cal., Ut., Wash.

Corythucha spinosa (Duges), 1889
1889 *Tingis spinosa* Duges, La Naturaleza, Soc. Mex. Hist. Nat., ser. 2, 1: 207. [Mexico].

1917 *Corythucha spinosa*: Van Duzee, Univ. Cal. Publ. Ent., 2: 214.
1918 *Corythucha unifasciata*: Gibson, Trans. Am. Ent. Soc., 44: 97.
Distribution: Cal. (Costa Rica, Cuba, Mexico, Trinidad).

Corythucha tuthilli Drake, 1940
1940 *Corythucha tuthilli* Drake, Ent. News., 51: 172. [Col.].
Distribution: Col.

Corythucha ulmi Osborn and Drake, 1916
1916 *Corythucha pallida* var. *ulmi* Osborn and Drake, Oh. Biol. Surv. Bull., 8: 231. [Oh.].
1918 *Corythucha ulmi*: Gibson, Trans. Am. Ent. Soc., 44: 79.
1926 *Corythuca* [*sic*] *ulmi*: Blatchley, Het. E. N. Am., p. 464.
Distribution: Conn., Ga., Ill., Ind., Mass., Md., Mich., Minn., Miss., Mo., N.H., N.Y., Neb., Oh., S.C., S.D., Va., Vt., Wis.

Genus *Dichocysta* Champion, 1898

1898 *Dichocysta* Champion, Biol. Centr.-Am., Rhyn., 2: 33. Type-species: *Dichocysta pictipes* Champion, 1898. Monotypic.

Dichocysta pictipes Champion, 1898
1898 *Dichocysta pictipes* Champion, Biol. Centr.-Am., Rhyn., 2: 34. [Panama].
1899 *Dichocysta pictipes*: Heidemann, Proc. Ent. Soc. Wash., 4: 339.
1910 *Dichroysta* [*sic*] *pictipes*: Banks, Cat. Nearc. Hem.-Het., p. 56.
Distribution: Ariz., Fla., Tex. (Mexico to Panama).

Genus *Dictyla* Stål, 1874

1874 *Dictyla* Stål, Öfv. K. Svens. Vet.-Akad. Förh., 31(3): 57. Type-species: *Monanthia platyoma* Fieber, 1861. Monotypic.

Dictyla coloradensis (Drake), 1917
1917 *Monanthia coloradensis* Drake, Bull. Brook. Ent. Soc., 12: 51. [Col.].
1960 *Dictyla coloradensis:* Drake and Ruhoff, Proc. U.S. Nat. Mus., 112: 51.
Distribution: Col., Ore., Tex.

Dictyla echii (Schrank), 1782
1782 *Cimex echii* Schrank, Fuess. Neues Mag. Liebh. Ent., 1: 276. [Austria].
1962 *Dictyla echii*: Udine, Coop. Econ. Ins. Rept., 12: 778.
Distribution: Md., N.Y., Oh., Ont., Pa., Va., W.Va. (throughout Europe and around Mediterranean Sea to Morocco).
Note: Introduced European species; see review by Hambleton (1968, Coop. Econ. Ins. Rept., 18: 658) who reports its U.S. host as the naturalized plant *Echium vulgare* L., also from Europe. Wheeler and Hoebeke (1985, J. N.Y. Ent. Soc., 93: 1057-1063) studied seasonal history and the North American distribution; they explained that the Fla. record for this species was based on a port interception.

Dictyla ehrethiae (Gibson), 1917
1917 *Monanthia ehrethiae* Gibson, Bull. Brook. Ent. Soc., 12: 50. [Tex.].
1960 *Dictyla ehrethiae*: Drake and Ruhoff, Proc. U.S. Nat. Mus., 112: 52.
Distribution: Tex. (Mexico).

Dictyla labeculata (Uhler) 1893

1893 *Monanthia labeculata* Uhler, N. Am. Fauna, 7: 264. [Cal.].

1916 *Monanthia labecula* [*sic*]: Van Duzee, Check List Hem., p. 26.

1943 *Monanthia lobeculata* [*sic*]: Drake, Proc. Ent. Soc. Wash., 45: 142.

1960 *Dictyla labeculata*: Drake and Ruhoff, Proc. U.S. Nat. Mus., 112: 51.

Distribution: Ariz., B.C., Cal., Col., N.M.

Genus *Galeatus* Curtis, 1833

1833 *Galeatus* Curtis, Ent. Mag., 1: 196. Type-species: *Tingis spinifrons* Fallén, 1807. Monotypic.

Galeatus affinis (Herrich-Schaeffer), 1835

1835 *Tingis affinis* Herrich-Schaeffer, Nomen. Ent., 1: 58. [Germany].

1923 *Galeatus uhleri* Horvath, An. Carnegie Mus., 15: 108. [N.M.]. Synonymized by Péricart, 1982, An. Soc. Ent. France, n. ser., 18: 355.

1962 *Galeatus angusticollis uhleri*: Drake and Ruhoff, Bull. So. Cal. Acad. Sci., 60: 163.

1982 *Galeatus affinis*: Péricart, An. Soc. Ent. France, n. ser., 18: 355.

Distribution: N.M. (Finland and Italy to China and Japan).

Galeatus spinifrons (Fallén), 1807

1807 *Tingis spinifrons* Fallén, Monogr. Cimic. Svec., p. 38. [Sweden].

1887 *Sphaerista peckhami* Ashmead, Ent. Am., 3: 156. [Wis.]. Synonymized under *G. angusticollis* Reuter by Drake and Ruhoff, 1962, Bull. So. Cal. Acad. Sci., 60: 163, a name cataloged by Drake and Ruhoff, 1965, U.S. Nat. Mus. Bull., 243: 220, as a junior synonym of *G. spinifrons*.

1889 *Galeatus peckhami*: Van Duzee, Can. Ent., 21: 5.

1962 *Galeatus spinifrons*: Drake and Ruhoff, Stud. Ent., 5: 491.

Distribution: Alta., Col., Man., Mass., Me., Mich., Minn., N.H., N.Y., Ont., Ut., Wis.

Note: The N.M. record was referred to *G. affinis* when Péricart (1982, An. Soc. Ent. France, n. ser., 18: 355) transferred *G. uhleri* to synonymy under that species.

Genus *Gargaphia* Stål, 1862

1862 Monanthia (*Gargaphia*) Stål, Stett. Ent. Zeit., 23: 324. Type-species: *Monanthia patricia* Stål, 1862. Designated by Van Duzee, 1916, Check List Hem., p. 25.

Gargaphia albescens Drake, 1917

1917 *Gargaphia albescens* Drake, Ent. News, 28: 228. [Cal.].

Distribution: Cal.

Gargaphia amorphae (Walsh), 1864

1864 *Tingis amorphae* Walsh, Proc. Ent. Soc. Phila., 3: 409. [Ill. assumed].

1886 *Gargaphia amorphae*: Uhler, Check-list Hem. Het., p. 22.

1917 *Gargaphia tiliae* var. *amorphae*: Van Duzee, Univ. Cal. Publ. Ent., 2: 217.

Distribution: Ia., Ill., Ind., Ks., Miss., Mo., N.C., N.J., N.Y., Oh., Pa., Va., W.Va.

Gargaphia angulata Heidemann, 1899

1899 *Gargaphia angulata* Heidemann, Can. Ent., 31: 301. [Ala.].

1901 *Gargaphia undulata* Heidemann, Proc. Ent. Soc. Wash., 4: 493. [D.C., Md.]. *Nomen nudum.*

1916 *Garzaphia* [*sic*] *angulata*: Torre-Bueno, Ent. News, 27: 10.

Distribution: Ala., Ariz., Col., Conn., D.C., Ga., Ks., Mass., Md., Mich., Minn., Miss.,

Mo., N.C., N.J., N.Y., Oh., Pa., S.C., Tenn., Tex., Va.

Gargaphia arizonica Drake and Carvalho, 1944
1944 *Gargaphia arizonica* Drake and Carvalho, Bull. Brook. Ent. Soc., 39: 43. [Ariz.].
Distribution: Ariz., N.M., Tex. (Mexico).

Gargaphia balli Drake and Carvalho, 1944
1944 *Gargaphia balli* Drake and Carvalho, Bull. Brook. Ent. Soc., 39: 42. [Ariz.].
Distribution: Ariz., N.M.

Gargaphia bimaculata Parshley, 1920
1920 *Gargaphia bimaculata* Parshley, Ent. News, 31: 271. [Fla.].
1925 *Gargaphia binotata* [*sic*]: Drake, Fla. Ent., 9: 37.
Distribution: Fla.

Gargaphia condensa Gibson, 1919
1919 *Gargaphia condensa* Gibson, Trans. Am. Ent. Soc., 45: 197. [Ariz.].
1919 *Gargaphia carinata* Gibson, Trans. Am. Ent. Soc., 45: 199. [Ariz.]. Synonymized by Drake, 1922, Fla. Ent., 5: 41.
Distribution: Ariz.

Gargaphia iridescens Champion, 1897
1897 *Gargaphia iridescens* Champion, Biol. Centr.-Am., Rhyn., 2: 10. [Mexico].
1910 *Gargaphia iridescens*: Banks, Cat. Nearc. Hem.-Het., p. 56.
Distribution: Ariz., Cal., Col., N.M., Tex. (Mexico).

Gargaphia mexicana Drake, 1922
1922 *Gargaphia mexicana* Drake, Fla. Ent., 5: 40. [Mexico].
1946 *Gargaphia mexicana*: Hurd, Ia. St. Coll. J. Sci., 20: 480.
Distribution: Tex. (Mexico).

Gargaphia opacula Uhler, 1893
1893 *Gargaphia opacula* Uhler, N. Am. Fauna, 7: 263. [Cal.].
Distribution: B.C., Cal., Ks., N.M., Ut., Wash. (Mexico).

Gargaphia oregona Drake and Hurd, 1945
1945 *Gargaphia oregona* Drake and Hurd, Bol. Ent. Venez., 4: 131. [Ore.].
Distribution: Ore.

Gargaphia solani Heidemann, 1914 [Fig. 154]
1914 *Gargaphia solani* Heidemann, Proc. Ent. Soc. Wash., 16: 136. [Mo.].
Distribution: Ala., Ariz., Ark., B.C., Conn., D.C., Ga., Ia., Ill., Ind., Ks., Md., Ms., Miss., Mo., N.C., N.J., Oh., Ok., Pa., S.C., Tenn., Tex., Va. (Mexico).

Gargaphia sororia Hussey, 1957
1957 *Gargaphia sororia* Hussey, Proc. Ent. Soc. Wash., 59: 175. [Fla.].
Distribution: Fla.

Gargaphia tiliae (Walsh), 1864
1864 *Tingis tiliae* Walsh, Proc. Ent. Soc. Phila., 3: 408. [Ill. assumed].
1873 *Gargaphia fasciata* Stål, K. Svens. Vet.-Adad. Handl., ll(2): 125. [Ill.]. Synonymized by Drake, 1922, Fla. Ent., 5: 41.
1886 *Gargaphia tiliae*: Uhler, Check-list Hem. Het., p. 22.
Distribution: Ala., Ariz., Col., Conn., Fla., Ga., Ia., Ill., Ind., Ks., Ky., Mass., Md., Me., Mich., Minn., Miss., Mo., N.B., N.C., N.H., N.J., N.Y., Neb., Oh., Ont., Pa., Que., S.C., S.D., Tenn., Va., Vt., W.Va., Wis. (Mexico).

Gargaphia tuthilli Drake and Carvalho, 1944
1944 *Gargaphia tuthilli* Drake and Carvalho, Bull. Brook. Ent. Soc., 39: 42. [Col.].
Distribution: Col.

Genus *Hesperotingis* Parshley, 1917

1917 *Hesperotingis* Parshley, Psyche, 24: 21. Type-species: *Hesperotingis antennata* Parshley, 1917. Original designation.

Hesperotingis antennata Parshley, 1917
1917 *Hesperotingis antennata* Parshley, Psyche, 24: 21. [N.J.].
Distribution: Conn., D.C., Del., Fla., Mass., Mo., N.H., N.J., N.Y., Pa.

Hesperotingis antennata antennata Parshley, 1917
1917 *Hesperotingis antennata* var. *antennata* Parshley, Psyche, 24: 21. [N.J.].
Distribution: Conn., D.C., Del., Fla., Mass., Mo., N.H., N.J., N.Y., Pa.

Hesperotingis antennata borealis Parshley, 1917
1917 *Hesperotingis antennata* var. *borealis* Parshley, Psyche, 24: 24. [N.H.].
Distribution: D.C., Mo., N.H.

Hesperotingis duryi (Osborn and Drake), 1916
1916 *Melanorhopala duryi* Osborn and Drake, Oh. J. Sci., 17: 15. [Tex.].
1946 *Hesperotingis duryi*: Hurd, Ia. St. Coll. J. Sci., 20: 447.
Distribution: Fla., Tex.

Hesperotingis duryi confusa Drake, 1922
1922 *Hesperotingis duryi* var. *confusa* Drake, Fla. Ent., 5: 48. [Tex.].
Distribution: Tex.

Hesperotingis duryi duryi (Osborn and Drake), 1916
1916 *Melanorhopala duryi* Osborn and Drake, Oh. J. Sci., 17: 15. [Tex.].
1922 *Hesperotingis duryi* var. *duryi*: Drake, Fla. Ent., 5: 48.
Distribution: Fla., Tex.

Hesperotingis floridana Drake, 1928
1928 *Hesperotingis floridana* Drake, Fla. Ent., 12: 4. [Fla.].
Distribution: Fla.

Hesperotingis fuscata Parshley, 1917
1917 *Hesperotingis fuscata* Parshley, Psyche, 24: 24. [Col.].
Distribution: Col., Ks.

Hesperotingis illinoiensis Drake, 1918
1918 *Hesperotingis illinoiensis* Drake, Bull. Brook. Ent. Soc., 13: 88. [Ill.].
Distribution: Conn., Ill., Ind.

Hesperotingis mississippiensis Drake, 1928
1928 *Hesperotingis mississippiensis* Drake, Fla. Ent., 12: 4. [Miss.].
Distribution: Fla., Ga., Miss., S.C.

Hesperotingis occidentalis Drake, 1922
1922 *Hesperotingis occidentalis* Drake, Fla. Ent., 5: 49. [Col.]
Distribution: Cal., Col., Id.

Genus *Leptodictya* Stål, 1873

1873 *Leptodictya* Stål, K. Svens. Vet.-Akad. Handl., ll(2): 121, 127. Type-species: *Monanthia ochropa* Stål, 1858. Designated by Oshanin, 1912, Kat. Paläark. Hem. p. 45.

Subgenus *Hanuala* Kirkaldy, 1905

1905 *Hanuala* Kirkaldy, Bull. Soc. Ent. France., 15: 217. Type-species: *Hanuala leinahoni* Kirkaldy, 1905. Monotypic.
1931 *Leptodictya* (*Hanuala*): Drake, Bull. Mus. Nac., 7: 119.

Leptodictya bambusae Drake, 1918
1918 *Leptodictya bambusae* Drake, Oh. J. Sci., 18: 175. [Puerto Rico].
1945 *Leptodictya* (*Hanuala*) *bambusae*: Drake and Hambleton, J. Wash. Acad. Sci., 35: 362.
Distribution: Tex. (Mexico to Peru, Cuba, Haiti, Puerto Rico, Greater Antilles).

Leptodictya nicholi Drake, 1926
1926 *Leptodictya nicholi* Drake, Bull. Brook. Ent. Soc., 21: 126. [Ariz.].
1931 *Leptodictya* (*Hanuala*) *nicholi*: Drake, Bull. Mus. Nac., 7: 121.
Distribution: Ariz.

Leptodictya plana Heidemann, 1913
1913 *Leptodictya plana* Heidemann, Proc. Ent. Soc. Wash., 15: 1. [Ok.].
1931 *Leptodictya* (*Hanuala*) *plana*: Drake, Bol. Mus. Nac., 7: 121.
Distribution: Ala., Ariz., Fla., Ks., Miss., Ok., Tex. (Mexico).

Leptodictyla simulans Heidemann, 1913
1913 *Leptodictya simulans* Heidemann, Proc. Ent. Soc. Wash., 15: 3. [Va.].
1931 *Leptodictya* (*Hanuala*) *simulans*: Drake, Bol. Mus. Nac., 7: 120.
Distribution: Ala., Fla., Ga., Miss., N.C., S.C., Tex., Va.

Leptodictya tabida (Herrich-Schaeffer), 1840
1840 *Monanthia tabida* Herrich-Schaeffer, Wanz. Ins., 5: 86. [Mexico].
1886 *Leptostyla tabida*: Uhler Check-list Hem. Het., p. 22.
1925 *Leptodictya tabida*: Drake, Fla. Ent., 9: 38.
1965 *Leptodictya* (*Hanuala*) *tabida*: Drake and Ruhoff, U.S. Nat. Mus. Bull., 243: 268.
Distribution: Tex. (Mexico to Venezuela).

Genus *Leptopharsa* Stål, 1873

1873 *Leptopharsa* Stål, K. Svens. Vet.-Akad. Handl., ll(2): 122, 126. Type-species: *Leptopharsa elegantula* Stål, 1873. Designated by Drake, 1922, Mem. Carnegie Mus., 9: 370.
1873 *Leptostyla* Stål, K. Svens. Vet.-Akad. Handl., 11(2): 120, 125. Preoccupied. Type-species: *Tingis oblonga* Say, 1825. Designated by Drake, 1922, Mem. Carnegie Mus., 9: 372. Synonymized by Drake, 1928, Proc. Biol. Soc. Wash., 41: 21.
1904 *Gelchossa* Kirkaldy, Ent., 37: 280. New name for *Leptostyla* Stål. Synonymized by Drake, 1928, Proc. Biol. Soc. Wash., 41: 21.

Leptopharsa clitoriae (Heidemann), 1911 [Fig. 158]
1911 *Leptostyla clitoriae* Heidemann, Proc. Ent. Soc. Wash., 13: 137, 180. [D.C.].
1911 *Leptostyla clitoriae* Heidemann, Proc. Ent. Soc. Wash., 13: 180. [D.C.]. Preoccupied.
1916 *Leptostyla costofasciata* Drake, Oh. J. Sci., 16: 326. [Tenn.]. Synonymized by Osborn and Drake, 1916, Oh. Biol. Surv. Bull., 8: 239.
1923 *Gelchossa clitoriae*: McAtee, Proc. Ent. Soc. Wash., 25: 146.
1928 *Leptopharsa clitoriae*: Drake, Proc. Biol. Soc. Wash., 41: 21.
Distribution: Ark., D.C., Ga., Ia., Ill., Ind., Ks., Mass., Md., Me., Mich., Miss., Mo., N.C., Neb., N.Y., Oh., Pa., S.C., Tenn., Tex., Va., Wis., W.Va.

Note: Heidemann proposed "*Leptostyla clitoriae* new species" twice in two separate articles in the same issue (number 3) of the 13th volume of Proc. Ent. Soc. Wash.: First on p. 137 where he described the egg in an article entitled "Some Remarks on the Eggs of North American Species of Hemiptera-Heteroptera" (pp. 128-140); then on p. 180, he described the adult in an article entitled "A New Species of North American Tingitidae." Neither article referred to the other.

Leptopharsa heidemanni (Osborn and Drake), 1916

1916 *Leptostyla heidemanni* Osborn and Drake, Oh. Biol. Surv. Bull., 8: 238. [Ark.].

1923 *Gelchossa heidemanni*: McAtee, Proc. Ent. Soc. Wash., 25: 146.

1944 *Leptopharsa heidemanni*: Froeschner, Am. Midl. Nat., 31: 669.

Distribution: Ark., Conn., D.C., Ga., Ia., Ill., Ind., La., Mass., Md., Miss., Mo., N.J., N.Y., Oh., Pa., R.I., Va.

Leptopharsa hintoni Drake, 1938

1938 *Leptopharsa hintoni* Drake, Pan-Pac. Ent., 14: 71. [Mexico].

1946 *Leptopharsa hintoni*: Hurd, Ia. St. Coll. J. Sci., 20: 466.

Distribution: Ariz., Tex. (Mexico).

Leptopharsa machalana Drake and Hambleton, 1946

1946 *Leptopharsa machalana* Drake and Hambleton, Proc. Biol. Soc. Wash., 59: 13. [Ecuador].

Distribution: Fla. (Mexico to Ecuador).

Note: The nominate subspecies does not occur in our area.

Leptopharsa machalana vinnula Drake and Hambleton, 1946

1946 *Leptopharsa machalana* var. *vinnula* Drake and Hambleton, Proc. Biol. Soc. Wash., 59: 13. [Fla.].

Distribution: Fla. (Guatemala, Mexico).

Note: Stahler (1946, J. Econ. Ent., 39: 545-546) reported on the life history, habits, ecology, and laboratory experiments with this subspecies.

Leptopharsa oblonga (Say), 1825

1825 *Tingis oblonga* Say, J. Acad. Nat. Sci. Phila., 4: 325. [Mo.].

1873 *Leptostyla oblonga*: Stål, K. Svens. Vet.-Akad. Handl., ll(2): 126.

1923 *Gelchossa oblonga*: McAtee, Proc. Ent. Soc. Wash., 25: 147.

1944 *Leptopharsa oblonga*: Froeschner, Am. Midl. Nat., 31: 669.

Distribution: Ark., "Canada," D.C., Ga., Ia., Ill., Ind., Ks., Md., Mo., N.J., Neb., S.D., Va., Wis.

Leptopharsa papella Drake, 1941

1941 *Leptopharsa papella* Drake, J. Wash. Acad. Sci., 31: 143. [Md.].

Distribution: Ind., Md.

Leptopharsa velifer (McAtee), 1917

1917 *Leptostyla velifer* McAtee, Bull. Brook. Ent. Soc., 12: 60. [Ariz.].

1946 *Leptopharsa velifer*: Hurd, Ia. St. Coll. J. Sci., 20: 466.

Distribution: Ariz.

Genus *Leptoypha* Stål, 1873

1873 *Leptoypha* Stål, K. Svens. Vet.-Akad. Handl., 11(2): 129. Type-species: *Tingis mutica* Say, 1832. Monotypic.

Leptoypha barberi Drake and Ruhoff, 1960
1910 *Leptoypha brevicornis*: Barber, J. N.Y. Ent. Soc., 18: 38.
1919 *Leptoypha minor*: McAtee, Bull. Brook. Ent. Soc., 14: 142 (in part).
1960 *Leptoypha barberi* Drake and Ruhoff, J. Ks. Ent. Soc., 33: 152. [Ariz.].
Distribution: Ariz.

Leptoypha costata Parshley, 1917
1917 *Leptoypha costata* Parshley, Psyche, 24: 16. [Md.].
1917 *Leptoypha distinguenda* Heidemann, Proc. Ent. Soc. Wash., 18: 218. [D.C.]. Synonymized by Van Duzee, 1917, Univ. Cal. Publ. Ent., 2: 817.
Distribution: Ark., Col., Conn., D.C., Ill., La., Md., Miss., Va.

Leptoypha drakei McAtee, 1919
1893 *Leptoypha mutica*: Uhler, N. Am. Fauna, 7: 264.
1917 *Leptoypha brevicornis*: McAtee, Bull. Brook. Ent. Soc., 12: 59.
1919 *Leptoypha drakei* McAtee, Bull. Brook. Ent. Soc., 14: 143. [Cal.].
Distribution: Ariz., Cal., Tex.

Leptoypha elliptica McAtee, 1917
1917 *Leptoypha elliptica* McAtee, Bull. Brook. Ent. Soc., 12: 57. [Tex.].
Distribution: Fla., Ga., Ind., Mo., Tenn., Tex.

Leptoypha ilicis Drake, 1919
1919 *Leptoypha ilicis* Drake, Oh. J. Sci., 19: 420. [Ga.].
Distribution: Ga., Fla., Ok., Tex.

Leptoypha mcateei Drake, 1921
1921 *Leptoypha mcateei* Drake, Fla. Ent., 4: 49. [Fla.].
1925 *Leptoypha mcatella* [*sic*]: Drake, Fla. Ent., 9: 38.
Distribution: Fla.

Leptoypha minor McAtee, 1917
1917 *Leptoypha minor* McAtee, Bull. Brook. Ent. Soc., 12: 56. [Cal.].
1941 *Leptoypha nubilis* Drake, Pan-Pac. Ent., 17: 141. [Cal.]. Synonymized by Drake and Ruhoff, 1962, Bull. So. Cal. Acad. Sci., 60: 156.
Distribution: Cal.

Leptoypha morrisoni Drake, 1922
1922 *Leptoypha morrisoni* Drake, Fla. Ent., 5: 43. [Dominican Republic].
1946 *Leptoypha morrisoni*: Hurd, Ia. St. Coll. J. Sci., 20: 457.
Distribution: Fla. (Panama, Greater Antilles).

Leptoypha mutica (Say), 1832
1832 *Tingis mutica* Say, Descrip. Het. Hem. N. Am., p. 27. [Ind.].
1873 *Monanthia mutica*: Walker, Cat. Hem. Brit. Mus., 6: 190.
1873 *Leptoypha mutica*: Stål, K. Svens. Vet.-Akad. Handl., ll(2): 129.
1916 *Leptophya* [*sic*] *mutica*: Osborn and Drake, Oh. Biol. Surv. Bull., 8: 241.
Distribution: Conn., D.C., Fla., Ga., Ia., Ill., Ind., Ks., Mass., Md., Me., Mich., Minn., Mo., N.C., N.D., N.H., N.J., N.Y., Neb., Oh., Ont., Pa., Que., S.C., S.D., Tenn., Tex., Va., Wis. (Mexico).

Genus *Melanorhopala* Stål, 1873

1873 *Tingis* (*Melanorhopala*) Stål, K. Svens. Vet.-Akad. Handl., 11(2): 130. Type-species: *Tingis clavata* Stål, 1873. Designated by Van Duzee, 1916, Check List Hem., p. 26.

1908 *Melanorhopala*: Horvath, An. Mus. Nat. Hung., 6: 564.

Note: Henry and Wheeler (1986, J. N.Y. Ent. Soc., 94: 236) provided a key to the species of *Melanorhopala*.

Melanorhopala balli Drake, 1928

1928 *Melanorhopala balli* Drake, Fla. Ent., 12: 3. [Col.].

Distribution: Col.

Melanorhopala clavata (Stål), 1873

1873 *Tingis* (*Melanorhopala*) *clavata* Stål, K. Svens. Vet.-Akad. Handl., 11(2): 131. [Ill.].

1873 *Tingis* (*Melanorhopala*) *lurida* Stål, K. Svens. Vet.-Akad. Handl., 11(2): 131. [Ill.]. Synonymized by Parshley, 1919, Bull. Brook. Ent. Soc., 14: 102.

1873 *Tingis* (*Melanorhopala*) *uniformis* Stål, K. Svens. Vet.-Akad. Handl., 11(2): 131. [Ill.]. Synonymized by Drake, 1926, An. Carnegie Mus., 16: 376.

1886 *Cantacader henshawi* Ashmead, Can. Ent., 18: 20. [Mass.]. Synonymized by Horvath, 1908, An. Mus. Nat. Hung., 6: 564.

1896 *Lasiacantha clavata*: Lethierry and Severin, Gen. Cat. Hem., 3: 19.

1896 *Lasiacantha lurida*: Lethierry and Severin, Gen. Cat. Hem., 3: 19.

1896 *Lasiacantha uniformis*: Lethierry and Severin, Gen. Cat. Hem. 3: 19.

1905 *Tingis clavata*: Crevecoeur, Trans. Ks. Acad. Sci., 19: 233.

1908 *Melanorhopala clavata*: Horvath, An. Mus. Nat. Hung., 6: 564.

1915 *Tingis clavata*: Osborn and Drake, Oh. Nat., 15: 506.

1916 *Melanorhopala lurida*: Osborn and Drake, Oh. Biol. Surv. Bull., 8: 244.

1916 *Melanorhopala uniformis*: Osborn and Drake, Oh. Biol. Surv. Bull., 8: 245.

1916 *Melanorhopala obscura* Parshley, Psyche, 23: 167. [Mass.]. Synonymized by Parshley, 1919, Bull. Brook. Ent. Soc., 14: 102.

1926 *Melanorhopala reflexa* Blatchley, Het. E. N. Am., p. 492. [Ind.]. Synonymized by Drake, 1930, Bull. Brook. Ent. Soc., 25: 269.

Distribution: Col., Conn., Fla., Ia., Ill., Ind., Ks., Man., Mass., Me., Mo., N.D., N.H., N.J., N.Y., Neb., Oh., R.I., S.D., Wis., Wyo.

Melanorhopala froeschneri Henry and Wheeler, 1986

1986 *Melanorhopala froeschneri* Henry and Wheeler, J. N.Y. Ent. Soc., 94: 236. [Tenn.].

Distribution: Del., Ky., Md., N.C., Tenn.

Melanorhopala infuscata Parshley, 1917

1917 *Melanorhopala infuscata* Parshley, Psyche, 24: 19. [Va.].

Distribution: D.C., Md., N.C., Va.

Genus *Phymacysta* Monte, 1942

1942 *Phymacysta* Monte, Papeis Avulsos Dept. Zool., 2: 106. Type-species: *Leptostyla tumida* Champion, 1897. Monotypic.

Phymacysta tumida (Champion), 1897

1897 *Leptostyla tumida* Champion, Biol. Centr.-Am., Rhyn., 2: 14. [Guatemala].

1960 *Phymacysta tumida*: Drake and Cobben, Stud. Fauna Curação Caribbean Is., 11(54): 85.

Distribution: Tex. (Mexico to Brazil and Peru; Greater, Lesser, and Netherlands Antilles).

Genus *Physatocheila* Fieber, 1844

1844 *Monanthia* (*Physatocheila*) Fieber, Ent. Monogr., p. 80. Type-species: *Acanthia quadri-*

maculata Wolff, 1804, a junior synonym of *Acanthia costata* Fabricius, 1794. Designated by Oshanin, 1912, Kat. Paläark. Hem., 1: 45.

Physatocheila brevirostris Osborn and Drake, 1916
1916 *Physatochila* [*sic*] *brevirostris* Osborn and Drake, Oh. Biol. Surv. Bull., 8: 243. [Oh.].
1917 *Physatocheila brevirostris*: Osborn and Drake, Psyche, 24: 155.
Distribution: Conn., Ia., Ill., Ind., Mass., Md., N.J., N.Y., Oh., Pa., Que., Va.

Physatocheila major Osborn and Drake, 1917
1917 *Physatocheila major* Osborn and Drake, Psyche, 24: 158. [Ill.].
Distribution: D.C.. Ill., Ind., Md., Va.

Physatocheila plexa (Say), 1832
1832 *Tingis plexus* Say, Descrip. Het. Hem. N. Am., p. 27. ["U.S."].
1873 *Physatochila* [*sic*] *plexa*: Stål, K. Svens. Vet.-Akad. Handl., 11(2): 129.
1873 *Monanthia plexa*: Walker, Cat. Hem. Brit. Mus., 6: 190.
1910 *Physatocheila plexa*: Smith, An. Rept. N.J. St. Mus., p. 149.
1917 *Physatocheila parshleyi* Osborn and Drake, Psyche, 24: 155, 156. [Ia., Id., Ill., Ks., Mass., Mich., N.Y., Neb., Ore., Va., Wis.]. Synonymized by Barber, 1922, N.J. Dept. Agr., Bur. Stat. Insp. Circ., 54: 17.
1924 *Physatocheila ornata*: Downes, Proc. Ent. Soc. B.C., 21: 28.
Distribution: B.C., Conn., D.C., Ia., Ill., Ind., Mass., Md., Me., Mich., N.D., N.H., N.J., N.S., N.Y., Neb., Ont., Pa., R.I., Va., W.Va., Wis.

Physatocheila variegata Parshley, 1916
1916 *Physatocheila variegata* Parshley, Psyche, 23: 166. [N.Y.].
1916 *Physatocheila plexa*: Osborn and Drake, Oh. Biol. Surv. Bull., 8: 242.
1928 *Physatocheila plexa* var. *variegata*: Drake, Cornell Univ. Agr. Exp. Sta. Mem., 191: 102.
Distribution: Cal., Conn., Ia., Id., Ill., Ind., Mass., Me., Mich., Mo., N.H., N.J., N.Y., Oh., Ont., Ore., Va., W.Va.
Note: Froeschner (1944, Am. Midl. Nat., 31: 669) and subsequent authors treat this form as at the species level.

Physatocheila variegata ornata Van Duzee, 1917
1914 *Physatochila* [*sic*] *plexa*: Van Duzee, Trans. San Diego Soc. Nat. Hist., 2: 11.
1917 *Physatocheila ornata* Van Duzee, Proc. Cal. Acad. Sci., ser. 4, 7: 259.
1917 *Physatocheila parshleyi*: Osborn and Drake, Psyche, 24: 156 (in part).
1917 *Physatocheila plexa*: Osborn and Drake, Pysche, 24: 156 (in part).
1946 *Physatocheila variegata* var. *ornata*: Hurd, Ia. St. Coll. J. Sci., 20: 452.
Distribution: Cal., Id., Ore.

Physatocheila variegata variegata Parshley, 1916
1916 *Physatocheila variegata* Parshley, Psyche, 23: 166. [N.Y.].
1946 *Physatocheila variegata* var. *variegata*: Hurd, Ia. St. Coll. J. Sci., 20: 452.
Distribution: Conn., Ia., Ill., Ind., Mass., Me., Mich., Mo., N.H., N.J., N.Y., Ont., Va., W.Va.

Genus *Pseudacysta* Blatchley, 1926

1926 *Pseudacysta* Blatchley, Het. E. N. Am., p. 497. Type-species: *Acysta perseae* Heidemann, 1908. Monotypic.

Pseudacysta perseae (Heidemann), 1908
1908 *Acysta perseae* Heidemann, Proc. Ent. Soc. Wash., 10: 103. [Fla.].
1926 *Pseudacysta persea* [*sic*]: Blatchley, Het. E. N. Am., p. 497.
1946 *Pseudacysta perseae*: Hurd, Ia. St. Coll. J. Sci., 20: 459.
Distribution: Fla., Ga., La., Tex. (Mexico).

Genus *Stephanitis* Stål, 1873

1873 *Stephanitis* Stål, K. Svens. Vet.-Akad. Handl., ll(2): 123. Type-species: *Acanthia pyri* Fabricius, 1775. Designated by Oshanin, 1912, Kat. Paläark. Hem., 1: 130.

Subgenus *Stephanitis* Stål, 1873

1873 *Stephanitis* Stål, K. Svens. Vet.-Akad. Handl., 11(2): 123. Type-species: *Acanthia pyri* Fabricius, 1775. Designated by Oshanin, 1912, Kat. Paläark. Hem., 1: 130.
1912 *Stephanitis* (*Stephanitis*): Horvath, An. Mus. Nat. Hung., 10: 319.

Stephanitis blatchleyi Drake, 1925
1925 *Stephanitis blatchleyi* Drake, Fla. Ent., 9: 37. [Fla.].
1926 *Leptobyrsa blatchleyi*: Blatchley, Het. E. N. Am., p. 470.
1965 *Stephanitis* (*Stephanitis*) *blatchleyi*: Drake and Ruhoff, U.S. Nat. Mus. Bull., 243: 354.
Distribution: Fla.

Stephanitis pyrioides (Scott), 1874
1874 *Tingis pyrioides* Scott, An. Mag. Nat. Hist., ser. 4, 14: 291, 440. [Japan].
1905 *Stephanitis azaleae* Horvath, An. Mus. Hung., 3: 565. Unnecessary new name for *Tingis pyrioides* Scott, 1874.
1912 *Stephanitis* (*Stephanitis*) *azaleae*: Horvath, An. Mus. Nat. Hung., 10: 333.
1917 *Stephanitis pyrioides*: Dickerson and Weiss, Ent. News, 28: 101.
1922 *Stephanitis pyriodes* [*sic*]: Barber, St. N.J. Dept. Agr., Bur. Stat. Insp. Circ., 54: 16.
1965 *Stephanitis* (*Stephanitis*) *pyrioides*: Drake and Ruhoff, U.S. Nat. Mus. Bull., 243: 361.
Distribution: Conn., D.C., Fla., Ga., Mass., Mo., N.J., N.Y., Pa., R.I. (Argentina, Australia, China, England, Germany, Korea, Japan, Morocco, Netherlands, Taiwan).

Stephanitis rhododendri Horvath, 1905
1904 *Leptobyrsa* "Species nova?": Wirtner, An. Carnegie Mus., 3: 202.
1905 *Stephanitis rhododendri* Horvath, An. Mus. Nat. Hung., 3: 567. [Netherlands].
1908 *Leptobyrsa explanata* Heidemann, Proc. Ent. Soc. Wash., 10: 105. [D.C.]. Synonymized by Champion, 1916, Ent. Month. Mag., 52: 507.
1912 *Stephanitis* (*Stephanitis*) *Rhododendri* [*sic*]: Horvath: An. Mus. Nat. Hung., 10: 329.
1917 *Leptobyrsa rhododendri*: Parshley, Psyche, 24: 15.
Distribution: B.C., Conn., D.C., Fla., Ga., Mass., Md., Me., N.C., N.H., N.J., N.Y., Nfld., Oh., Ore., Pa., R.I., S.C., Va., Vt., W.Va., Wash. (Europe, New Zealand, South Africa).
Note: The North American distribution consists of two widely separated ranges: One coastwise on the Atlantic Coast from Nfld. to Fla., plus W.Va.; the other on the Pacific Coast from B.C to Ore. Crosby and Hadley's (1915, J. Econ. Ent., 8: 409-414) discussion of this species and illustrations of all stages--only four

nymphal instars are present-- was expanded by Dickerson (1917, J. N.Y. Ent. Soc., 25: 105-112.).

Stephanitis takeyai Drake and Maa, 1955

1905 *Tingis globulifera* Matsumura, Thous. Ins. Japan, 2: 98. [Japan]. Preoccupied.

1950 *Stephanitis globulifera*: Bailey, Psyche, 57: 143.

1955 *Stephanitis takeyai* Drake and Maa, Quart. J. Taiwan Mus., 8: 10. New name for *Tingis globulifera* Matsumura, 1905.

1964 *Stephanitis* (*Stephanitis*) *takeyai*: Drake and Ruhoff, U.S. Nat. Mus. Bull., 243: 364.

Distribution: Conn., Del., Mass., Md., N.J., N.Y., Oh., Pa., R.I., Va.

Note: Wheeler (1977, Proc. Ent. Soc. Wash., 79: 168-171) summarized information on this species as it occurs in North America.

Genus *Teleonemia* Costa, 1864

1864 *Teleonemia* Costa, An. Mus. Zool. R. Univ. Napoli, 2: 144. Type-species: *Teleonemia funerea* Costa, 1864. Monotypic.

Teleonemia barberi Drake, 1918

1918 *Teleonemia barberi* Drake, Oh. J. Sci., 18: 325, 328. [Ariz.].

Distribution: Ariz., Tex.

Teleonemia belfragii Stål, 1873

1873 *Teleonemia belfragii* Stål, K. Svens. Vet.-Akad. Handl., 11(2): 132. [Tex.].

1896 *Teleonemia Belfragei* [*sic*]: Lethierry and Severin, Cat. Gen. Hem., 3: 22.

1914 *Telconemia* [*sic*] *belfragei* [*sic*]: Barber, Bull. Am. Mus. Nat. Hist., 33: 507.

Distribution: Ala., Fla., Miss., Tex.

Teleonemia consors Drake, 1918

1918 *Teleonemia consors* Drake, Oh. J. Sci., 18: 324, 327. [Ariz.].

Distribution: Ariz.

Teleonemia cylindricornis Champion, 1898

1898 *Teleonemia cylindricornis* Champion, Biol. Centr.-Am., Rhyn., 2: 41. [Guatemala].

1925 *Teleonemia cylindricornis*: Drake, Fla. Ent., 9: 38.

Distribution: Ill., Miss. (Mexico to British Honduras, Jamaica).

Teleonemia huachucae Drake, 1941

1941 *Teleonemia huachucae* Drake, Pan-Pac. Ent., 17: 140. [Ariz.].

Distribution: Ariz.

Teleonemia monile Van Duzee, 1918

1918 *Telenemia monile* Van Duzee, Proc. Cal. Acad. Sci., ser. 4, 8: 279. [Cal.].

Distribution: Cal.

Teleonemia montivaga Drake, 1920

1920 *Teleonemia montivaga* Drake, Oh. J. Sci., 20: 52. [Cal.].

Distribution: Cal.

Teleonemia nigrina Champion, 1898

1886 *Teleonemia elongata* Uhler, Check-list Hem. Het., p. 22. ["U.S."]. *Nomen nudum*.

1898 *Teleonemia nigrina* Champion, Biol. Centr.-Am., Rhyn., 2: 41. [Guatemala].

1905 *Taleonemia* [*sic*] *elongata*: Crevecoeur, Trans. Ks. Acad. Sci., 19: 233.

Distribution: Ariz., Ark., Cal, Ga., Ks., Mo., N.C., N.J., N.M., Ok., S.C., Tex., Ut., Va. (Mexico to Guatemala).

Teleonemia novicia Drake, 1920
1920 *Teleonemia novicia* Drake, Oh. J. Sci., 20: 53. [Cal.].
Distribution: Ariz., Cal.

Teleonemia sacchari (Fabricius), 1794
1794 *Acanthia sacchari* Fabricius, Ent. Syst., 4: 77. ["Americae meridionalis Insulis"].
1910 *Teleonemia sacchari*: Banks, Cat. Nearc. Hem.-Het., p. 57.
Distribution: Cal., Fla. (Greater, Lesser, and Netherlands Antilles, Mexico).

Teleonemia schwarzi Drake, 1918
1918 *Teleonemia schwarzi* Drake, Oh. J. Sci., 18: 326. [Cal.].
Distribution: Ariz., Cal. (Mexico).

Teleonemia scrupulosa Stål, 1873
1873 *Teleonemia scrupulosa* Stål, K. Svens. Vet.-Akad. Handl., 11(2): 132. [Brazil, Colombia].
1906 *Teleonemia scrupulosa*: Barber, Mus. Brook. Inst. Arts Sci., Sci. Bull., 1: 281.
1919 *Teleonemia vanduzeei* Drake, Fla. Buggist, 3: 24. [Fla., Jamaica]. Synonymized by Drake and Frick, 1939, Proc. Haw. Ent. Soc., 10: 201.
Distribution: Fla., Ga., Tex. (Mexico to Brazil, Colombia, and French Guiana; Greater, Lesser, and the Netherlands Antilles; introduced into many other parts of the world as a possible biological control agent for *Lantana* weeds.

Teleonemia variegata Champion, 1898
1898 *Teleonemia variegata* Champion, Biol. Centr.-Am., Rhyn., 2: 42. [Guatemala].
1910 *Teleonemia variegata*: Barber, J. N.Y. Ent. Soc., 18: 38.
Distribution: Ariz. (Mexico to Honduras).

Teleonemia vidua Van Duzee, 1918
1918 *Teleonemia vidua* Van Duzee, Proc. Cal. Acad. Sci., ser. 4, 8: 278. [Cal.].
Distribution: Cal.

Genus *Tingis* Fabricius, 1803

1803 *Tingis* Fabricius, Syst. Rhyn., 124. Type-species: *Cimex cardui* Linnaeus, 1758. Designated by Latreille, 1810, Consid. Gén. Crust. Arach. Ins. p. 433; confirmed by Int. Comm. Zool. Nomen., Direction 4, 1954.

Subgenus *Tingis* Fabricius, 1803

1803 *Tingis* Fabricius, Syst. Rhyn., 124. Type-species: *Cimex cardui* Linneaus, 1758. Designated by Latreille, 1810, Consid. Gén. Crust. Arach. Ins., p. 433; confirmed by Int. Comm. Zool. Nomen., Direction 4, 1954.
1873 *Tingis* (*Tingis*): Stål, K. Svens. Vet.-Akad. Handl., 11(2): 130.

Tingis auriculata (Costa), 1847
1847 *Catoplatus auriculatus* Costa, Cimic. Reg. Neapol., p. 20. [Italy].
1919 *Monanthia necopina* Drake, Oh. J. Sci., 19: 420. [Md.]. Synonymized by Drake and Ruhoff, 1962, Bull. So. Cal. Acad. Sci., 60: 156.
1923 *Tingis necopina*: McAtee, Proc. Ent. Soc. Wash., 25: 145.
1965 *Tingis auriculata*: Drake and Ruhoff, U.S. Nat. Mus. Bull., 243: 393.
Distribution: Md. (Europe, North Africa).
Note: N. Am. listing based on lone specimen collected in 1890; no subsequent collections.

Tribe Ypsotingini Drake and Ruhoff, 1965

Genus *Dictyonota* Curtis, 1827

1827 *Dictyonota* Curtis, Brit. Ent., pl. 154. Type-species: *Dictyonota strichnocera* Fieber, 1844. Designated by Int. Comm. Zool. Nomen., Opinion 251, 1954.

Subgenus *Dictyonota* Curtis, 1827

1827 *Dictyonota* Curtis, Brit. Ent., pl. 154. Type-species: *Dictyonota strichnocera* Fieber, 1844. Designated by Int. Comm. Zool. Nomen., Opinion 251, 1954.

1906 *Dictyonota* (*Dictyonota*): Horvath, Ann. Mus. Nat. Hung., 4: 39.

Dictyonota fuliginosa Costa, 1855

1855 *Dyctionota* [*sic*] *fuliginosa* Costa, Cimic. Reg. Neapol., p. 10. [Italy].

1960 *Dictyonota fuliginosa*: Scudder, Proc. Ent. Soc. B.C., 57: 22.

Distribution: B.C. (Europe).

Note: Probably introduced into Canada along with its host Scotch broom, *Cytisus scoparius* (Linnaeus). See Scudder (1960, above).

Subgenus *Alcletha* Kirkaldy, 1900

1900 *Alcletha* Kirkaldy, Ent., 33: 241. Type-species: *Acanthia tricornis* Schrank, 1801. Original designation.

1906 *Dictyonota* (*Alcletha*): Horvath, An. Mus. Nat. Hung., 4: 39.

Dictyonota tricornis (Schrank), 1801

1801 *Acanthia tricornis* Schrank, Fauna Boica, p. 67. [Europe].

1916 *Dictyonota tricornis* var. *americana* Parshley, Psyche, 23: 164. [Me.]. Synonymized by Drake and Ruhoff, 1962, Bull. So. Cal. Acad. Sci., 61: 142.

1965 *Dictyonota tricornis*: Drake and Ruhoff, U.S. Nat. Mus. Bull., 243: 437.

Distribution: Me., N.B., N.S. (Europe, North Africa).

Family Veliidae Amyot and Serville, 1843

The Small Water Striders

By Cecil L. Smith

The veliids, also commonly known as riffle bugs, small water striders, or water crickets, are primarily inhabitants of water surfaces but some crawl over adjacent mudflats and wet rocks. While most of our forms occur on fresh waters, members of a few genera (for instance *Trochopus* Carpenter and *Husseyella* Herring) are found on salt or brackish waters along coasts.

Locomotion over the water's surface varies in different components of the family. The majority simply walk or run over quiet waters, especially those among emergent vegetation. Some frequent open or running water (subfamily Rhagoveliinae) where their movement over the surface is facilitated by an expanded tuft of hairs in a deep cleft in the last segment of the middle tarsus; this tuft can be fanned out to gain greater push against the water and thus gain speed and strength to overcome the effects of moving water. A third type of locomotion results when a salivary secretion emitted through the beak causes a reduction in the water's surface tension, which literally pulls the insect onto this area of lowered surface tension.

In much of our area adults hibernate and begin reproduction shortly after resuming activities in spring. Eggs are laid in batches and glued to objects close to the

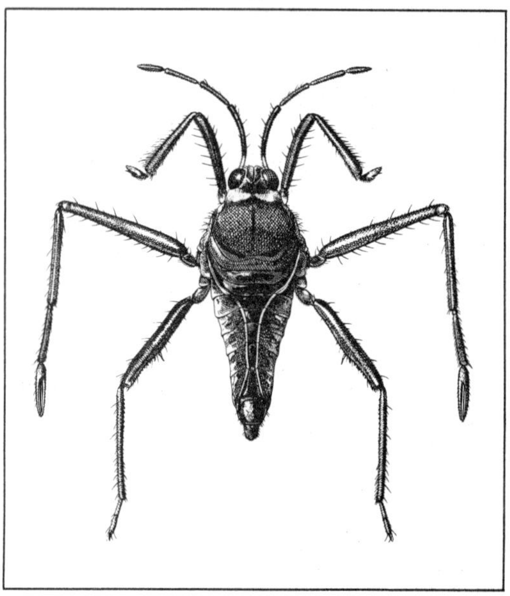

Fig. 160 *Rhagovelia distincta* [p.740] (After Usinger, 1956).

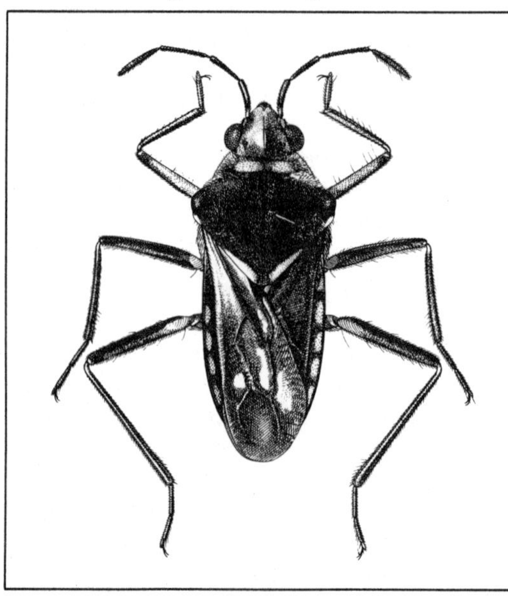

Fig.161 *Microvelia beameri* [p. 736] (After Usinger, 1956).

water. The number of instars in the nymphal stage varies from 4-5. In some species the reduced number appears to be a reaction to conditions; in others it appears to be constant and normal. Flight capability may vary among individuals of a species. While some species are always macropterous, some are dimorphic and may have either macropterous and apterous or macropterous and micropterous individuals.

The higher classification of the family was treated by China and Usinger (1949, An. Mag. Nat. Hist., 2: 343-354) and later revised by Štys (1976, Acta Ent. Boh., 73: 399-403). Smith and Polhemus (1978, Proc. Ent. Soc. Wash., 80: 56-68) published keys and checklists for the five North American genera and 35 species. An overview of the family, including notes on biology, ecology, taxonomy, morphology, as well as a key to the Californian species, was provided by Polhemus and Chapman (1979, Bull. Cal. Ins. Surv., 21: 49-57).

A similar treatment can be found in Bennett and Cook's (1981, Univ. Minn. Agr. Exp. Stn. Tech. Bull., 332: 24-28) review of the Minnesota species.

Subfamily Microveliinae China and Usinger, 1949

Tribe Microveliini China and Usinger, 1949

Genus *Husseyella* Herring, 1955

1955 *Husseyella* Herring, Fla. Ent., 38: 21. Type-species: *Microvelia turmalis* Drake and Harris, 1933. Original designation.

Husseyella turmalis (Drake and Harris), 1933

1933 *Microvelia turmalis* Drake and Harris, Proc. Biol. Soc. Wash., 46: 53. [British Honduras].

1955 *Husseyella turmalis*: Herring, Fla. Ent., 38: 25.

Distribution: Fla. (British Honduras, West Indies).

Genus *Microvelia* Westwood, 1834

1834 *Velia* (*Microvelia*) Westwood, An. Soc. Ent. France, 3: 647. Type-species: *Microvelia pulchella* Westwood, 1834. Monotypic.

1834 *Microvelia* Westwood, An. Soc. Ent. France, 3: pl. 6, fig. 5.

1835 *Hydroessa* Burmeister, Handb. Ent., 2: 213. Type-species: *Hydroessa reticulata* Burmeister, 1835, a junior synonym of *Velia pygmaea* Dufour, 1833. Monotypic. Synonymized by Brullé, 1835, Hist. Nat. Ins., 9: 295.

Note: Torre-Bueno (1924, Bull. Brook. Ent. Soc., 19: 186-194) reviewed species of the Western Hemisphere; Drake and Hussey (1955, Fla. Ent., 38: 95-115) later presented a synoptic list of the New World species. Smith and Polhemus (1978, Proc. Ent. Soc. Wash., 80: 58-62) provided a key to the apterous species of America north of Mexico.

Subgenus *Kirkaldya* Torre-Bueno, 1910

1910 *Microvelia* (*Kirkaldya*) Torre-Bueno, Can. Ent., 42: 186. Type-species: *Hebrus americanus* Uhler, 1884. Monotypic.

Note: *Kirkaldya* was restored from synonymy to subgeneric status by Polhemus (1970, Proc. Ent. Soc. Wash., 72: 443).

Microvelia americana (Uhler), 1884

1884 *Hebrus americanus* Uhler, Stand. Nat. Hist., 2: 274. ["Middle States," Me.]. Lectotype from N.H. designated by Polhemus, 1977, Proc. Ent. Soc. Wash., 79: 638.

1910 *Microvelia (Kirkaldya) americana*: Torre-Bueno, Can. Ent., 42: 186.

Distribution: Ark., Conn., D.C., Fla., Ga., Ia., Ill., Ind., Ks., Ky., Mass., Md., Me., Mich., Minn., Miss., Mo., N.C., N.H., N.J., N.S., N.Y., Neb., Oh., Ok., Ont., Pa., Que., R.I., S.C., S.D., Tenn., Tex., Va., Vt., W.Va., Wis.

Note: Torre-Bueno (1910, Can. Ent., 42: 176-186) studied the life history.

Microvelia beameri McKinstry, 1937 [Fig. 161]

1937 *Microvelia beameri* McKinstry, J. Ks. Ent. Soc., 10: 30. [Ariz.].

1978 *Microvelia (Kirkaldya) beameri*: Smith and Polhemus, Proc. Ent. Soc. Wash., 80: 61.

Distribution: Ariz., Cal., Col., N.M., Nev., Tex., Ut. (Jamaica, Mexico).

Microvelia californiensis McKinstry, 1937

1937 *Microvelia californiensis* McKinstry, J. Ks. Ent. Soc., 10: 39. [Cal.].

1978 *Microvelia (Kirkaldya) californiensis*: Smith and Polhemus, Proc. Ent. Soc. Wash., 80: 61.

Distribution: Cal., Ore. (Mexico).

Microvelia fasciculifera McKinstry, 1937

1937 *Microvelia fasciculifera* McKinstry, J. Ks. Ent. Soc., 10: 31. [Ariz.].

1978 *Microvelia (Kirkaldya) fasciculifera*: Smith and Polhemus, Proc. Ent. Soc. Wash., 80: 61.

Distribution: Ariz., N.M., Tex. (Mexico).

Microvelia gerhardi Hussey, 1924

1895 *Microvelia americana* Uhler, Bull. Col. Agr. Exp. Stn., 31: 61. [Col.]. Preoccupied. Lectotype designated by Polhemus, 1977, Proc. Ent. Soc. Wash., 79: 639.

1924 *Microvelia gerhardi* Hussey, Bull. Brook. Ent. Soc., 19: 164. [Neb.].

1978 *Microvelia (Kirkaldya) gerhardi*: Smith and Polhemus, Proc. Ent. Soc. Wash., 80: 61.

Distribution: Ariz., Cal., Col., N.M., Neb., Nev., S.D., Tex., Ut., Wyo. (Mexico).

Microvelia paludicola Champion, 1898

1898 *Microvelia paludicola* Champion, Biol. Centr.-Am., Rhyn., 2: 127. [Guatemala]. Lectotype designated by Polhemus, 1977, Proc. Ent. Soc. Wash., 79: 640.

1924 *Microvelia paludicola*: Torre-Bueno, Bull. Brook. Ent. Soc., 19: 194.

1950 *Microvelia alachuana* Hussey and Herring, Fla. Ent., 33: 117. [Fla.]. Synonymized by Drake and Hussey, 1954, Fla. Ent., 37: 137.

1978 *Microvelia (Kirkaldya) paludicola*: Smith and Polhemus, Proc. Ent. Soc. Wash., 80: 60.

Distribution: Ala., Ariz., Cal., Col., Fla., Ga., Ks., Ky., La., Miss., Mo., N.M., Ok., Tenn., Tex. (Greater Antilles, Mexico to Central America).

Microvelia torquata Champion, 1898

1898 *Microvelia torquata* Champion, Biol. Centr.-Am., Rhyn., 2: 128. [Guatemala]. Lectotype designated by Polhemus, 1977, Proc. Ent. Soc. Wash., 79: 643.

1955 *Microvelia paludicola*: Drake and Hussey, Fla. Ent., 38: 114 (in part).

1977 *Microvelia torquata*: Polhemus, Proc. Ent. Soc. Wash., 79: 643.

1978 *Microvelia (Kirkaldya) torquata*: Smith and Polhemus, Proc. Ent. Soc. Wash., 80: 59.

Distribution: Ariz., Cal., Col., N.M., Nev., Tex., Ut. (Mexico to Central America).

Note: Polhemus and Chapman (1979, Bull. Cal. Ins. Surv., 21: 55) pointed out that most southwestern U.S. records [except Tex.] of *M. paludicola* belong to this species.

Subgenus *Microvelia* Westwood, 1834

1834 *Velia (Microvelia)* Westwood, An. Soc. Ent. France, 3: 647. Type-species: *Microvelia pulchella* Westwood, 1834. Monotypic.

1910 *Microvelia (Microvelia)*: Torre-Bueno, Can. Ent., 42: 186.

Microvelia albonotata Champion, 1898

1898 *Microvelia albonotata* Champion, Biol. Centr.-Am., Rhyn., 2: 129. [Guatemala].

1907 *Microvelia albonotata*: Van Duzee, Bull. Buff. Soc. Nat. Sci., 8: 24.

1908 *Rhagovelia capitata*: Torre-Bueno, J. N.Y. Ent. Soc., 16: 233.

1916 *Microvelia (Microvelia) albonotata*: Van Duzee, Check List Hem., p. 49.

Distribution: Conn., Fla., Ga., Ia., Ill., Ind., Mass., Md., Mich., Minn., Miss., N.C., N.J., N.Y., Pa., R.I., S.C., Tex. (Mexico to South America, West Indies).

Microvelia atrata Torre-Bueno, 1916

1916 *Microvelia atrata* Torre-Bueno, Bull. Brook. Ent. Soc., 11: 60. [Ga.].

1917 *Microvelia (Microvelia) atrata*: Van Duzee, Univ. Cal. Publ. Ent., 2: 434.

Distribution: Ala., Fla., Ga., La., Miss.

Microvelia austrina Torre-Bueno, 1924

1924 *Microvelia austrina* Torre-Bueno, Bull. Brook. Ent. Soc., 19: 191. [N.C.].

1925 *Microvelia parallela* Blatchley, Ent. News, 36: 48. [N.C.]. Synonymized by Blatchley, 1926, Het. E. N. Am., p. 992.

1978 *Microvelia (Microvelia) austrina* Smith and Polhemus, Proc. Ent. Soc. Wash., 80: 59.

Distribution: Ala., D.C., Ga., Ind., Md., Miss., N.C., S.C., Tenn., Va. (Mexico).

Microvelia buenoi Drake, 1920

1920 *Microvelia buenoi* Drake, Bull. Brook. Ent. Soc., 15: 20. [N.Y.].

1978 *Microvelia (Microvelia) buenoi*: Smith and Polhemus, Proc. Ent. Soc. Wash., 80: 62.

Distribution: Alk., Alta., B.C., Cal., Fla., Ia., Ill., Ind., Man., Mass., Me., Mich., Minn., Miss., Mont., N.J., N.T., Nev., Ont., Ore., Que., S.C., Sask., Ut., Wash., Wis.

Note: Hoffmann (1925, Bull. Brook. Ent. Soc., 20: 93-94) gave life history notes.

Microvelia cerifera McKinstry, 1937

1937 *Microvelia cerifera* McKinstry, J. Ks. Ent. Soc., 10: 37. [Ks.].

1978 *Microvelia (Microvelia) cerifera*: Smith and Polhemus, Proc. Ent. Soc. Wash., 80: 59.

Distribution: Ariz., Cal., Col., Ia., Ks., N.M., Neb., Nev., Ut., Wyo.

Microvelia cubana Drake, 1951

1951 *Microvelia cubana* Drake, Great Basin Nat., 11: 41. [Cuba].

1955 *Microvelia portoricensis*: Hussey, Quart. J. Fla. Acad. Sci., 18: 120.

1978 *Microvelia (Microvelia) cubana*: Smith and Polhemus, Proc. Ent. Soc. Wash., 80: 62.

Distribution: Fla. (Cuba, Dominican Republic).

Note: Herring (1958, Pan-Pac. Ent., 34: 174-176) discussed the dispersal of this species [as *M. portoricensis*] by hurricanes.

Microvelia fontinalis Torre-Bueno, 1916

1916 *Microvelia fontinalis* Torre-Bueno, Bull. Brook. Ent. Soc., 11: 58. [N.Y.].

1917 *Microvelia (Microvelia) fontinalis*: Van Duzee, Univ. Cal. Publ. Ent., 2: 434.

Distribution: Conn., D.C., Ga., Ia., Ill., Ind., Mass., Md., Mich., Minn., Miss., N.J., N.Y., Oh., Pa., R.I., Tenn., Va., W.Va., Wis.

Microvelia glabrosulcata Polhemus, 1974

1974 *Microvelia glabrosulcata* Polhemus, Great Basin Nat., 34: 212. [Mexico].

1974 *Microvelia glabrosulcuta* [*sic*]: Polhemus, Great Basin Nat., 34: 209.

1978 *Microvelia (Microvelia) glabrosulcata*: Smith and Polhemus, Proc. Ent. Soc. Wash., 80: 59.

Distribution: Ariz. (Mexico).

Microvelia hinei Drake, 1920

1920 *Microvelia hinei* Drake, Oh. J. Sci., 20: 207. [Oh.].

1920 *Microvelia borealis*: Hungerford, Univ. Ks. Sci. Bull., 21: 137.

1978 *Microvelia (Microvelia) hinei*: Smith and Polhemus, Proc. Ent. Soc. Wash., 80: 62.

Distribution: Ala., Ariz., Ark., Cal., Col., Conn., D.C., Fla., Ga., Ia., Ill., Ks., Ky., La., Mass., Md., Mich., Minn., Miss., Mo., N.J., N.M., N.Y., Nev., Oh., Ore., Pa., Tenn., Tex., Va., Wyo. (Mexico to Argentina, West Indies).

Microvelia marginata Uhler, 1893

1893 *Microvelia marginata* Uhler, Proc. Zool. Soc. Lond., p. 719. [St. Vincent, West Indies].

1955 *Microvelia marginata*: Drake and Hussey, Fla. Ent., 38: 106.

Distribution: Fla. (West Indies).

Note: Several early literature reports of *marginata* from across the U.S. are based on misidentifications. The only confirmed record is a Fla. specimen listed by Drake and Hussey (1955, above), which may represent a hurricane introduction and not an established population.

Microvelia pulchella Westwood, 1834

1834 *Microvelia pulchella* Westwood, An. Soc. Ent. France, 3: Pl. 6, fig. 5.

1834 *Velia (Microvelia) pulchella* Westwood, An. Soc. Ent. France, 3: 647. [West Indies].

1857 *Microvelia capitata* Guérin-Méneville, Sagra's Hist. Cuba, 7: 174. [Cuba]. Tentatively synonymized [as *species inquirenda*] by Drake and Hussey, 1955, Fla. Ent., 38: 107; followed by Polhemus and Chapman, 1979, Bull. Cal. Ins. Surv., 21: 53.

1890 *Ragovelia incerta* Kirby, J. Linn. Soc. London, 20: 548. [Brazil]. Synonymized by Polhemus and Chapman, 1979, Bull. Cal. Ins. Surv., 21: 53.

1894 *Microvelia robusta* Uhler, Proc. Zool. Soc. London, p. 219. [Grenada]. Synonymized by Drake and Maldonado, 1954, Proc. Biol. Soc. Wash., 67: 220 [as synonym of *M. incerta* Kirby].

1899 *Microvelia incerta*: Kirkaldy, Rev. d'Ent., 18: 95.

1907 *Microvelia pulchella*: Van Duzee, Bull. Buff. Soc. Nat. Sci., 8: 23.

1908 *Rhagovelia capitata*: Torre-Bueno, J. N.Y. Ent. Soc., 16: 233.

1916 *Microvelia robusta*: Van Duzee, Check List Hem., p. 49.

1916 *Microvelia borealis* Torre-Bueno, Bull. Brook. Ent. Soc., 11: 59. [N.J.]. Synonymized by Drake and Hussey, 1955, Fla. Ent., 38: 104 [as synonym of *Microvelia pulchella incerta* (Kirby)].

1916 *Microvelia (Microvelia) pulchella*: Van Duzee, Check List Hem., p. 49.

1916 *Microvelia (Microvelia) capitata*: Van Duzee, Check List Hem., p. 49.

Distribution: Ala., Alk., Ariz., Ark., B.C., Cal., Conn., D.C., Fla., Ga., Ia., Ill., Ind., Ks., La., Man., Mass., Md., Mich., Minn., Miss., Mo., N.C., N.J., N.Y., Oh., Ont., Ore., Pa., Que., R.I., S.C., Tex., W.Va., Wis., Wyo., Va. (Mexico to South America, West Indies).

Note: Blatchley (1926, Het. E. N. Am., p. 988) dismissed Van Duzee's (1917, Univ. Cal. Publ. Ent., 2: 219) Fla. record for *M. robusta* [now a junior synonym of *M. pulchella*] by stating "Since he cites only the original description, where it is mentioned only from Grenada, it also is not further considered." Torre-Bueno (1917, Ent. News, 28: 354-359) presented the life history [as *M. borealis*].

Microvelia signata Uhler, 1894

1894 *Microvelia signata* Uhler, Proc. Cal. Acad. Sci., 4: 288. [Mexico].

1928 *Microvelia oreades* Drake and Harris, Oh. J. Sci., 28: 274. [Col.]. Synonymized by Drake, 1951, Great Basin Nat., 11: 37.

1916 *Microvelia (Microvelia) signata*: Van Duzee, Check List Hem., p. 49.

Distribution: Ariz., Cal., Col., N.M., Tex., Ut. (Mexico to Central America).

Subfamily Rhagoveliinae China and Usinger, 1949

Genus *Rhagovelia* Mayr, 1865

1865 *Rhagovelia* Mayr, Verh. Zool.- Bot. Ges. Wien, 15: 445. Type-species: *Velia nigricans* Burmeister, 1835. Designated by Kirkaldy, 1901, Ent., 34: 286.

Note: Bacon (1956, Univ. Ks. Sci. Bull., 38: 695-913) revised the New World species, and Matsuda (1956, Univ. Ks. Sci. Bull., 38: 915-1018) analysed the subgeneric arrangement. Only the nominate subgenus occurs in the United States and Canada. Smith and Polhemus (1978, Proc. Ent. Soc. Wash., 80: 62-62) provided a key to the species of America north of Mexico.

Subgenus *Rhagovelia* Mayr, 1865

1865 *Rhagovelia* Mayr, Verh. Zool.- Bot. Ges. Wien, 15: 445. Type-species: *Velia nigricans* Burmeister, 1835. Designated by Kirkaldy, 1901, Ent., 34: 286.

1956 *Rhagovelia (Rhagovelia)*: Matsuda, Univ. Ks. Sci. Bull., 38: 989.

Rhagovelia becki Drake and Harris, 1936

1936 *Rhagovelia becki* Drake and Harris, Proc. Biol. Soc. Wash., 49: 106. [Mexico].

1956 *Rhagovelia (Rhagovelia) becki*: Matsuda, Univ. Ks. Sci. Bull., 38: 989.

Distribution: Nev., Tex. (Mexico).

Rhagovelia choreutes Hussey, 1925

1925 *Rhagovelia choreutes* Hussey, J. N.Y. Ent. Soc., 33: 67. [Fla.].

1956 *Rhagovelia (Rhagovelia) choreutes*: Matsuda, Univ. Ks. Sci. Bull., 38: 989.

Distribution: Cal., Fla., Miss., N.M., Ok., Tex. (Mexico to Central America).

Rhagovelia distincta Champion, 1898 [Fig. 160]

1877 *Rhagovelia mexicana* Signoret, Bull. Soc. Ent. France, 2: 53. [Mexico]. *Nomen nudum.*

1898 *Rhagovelia distincta* Champion, Biol. Centr.-Am., Rhyn., 2: 135. [Mexico]. Lectotype designated by Polhemus, 1977, Proc. Ent. Soc. Wash., 79: 644.

1914 *Rhagovelia obesa*: Van Duzee, Trans. San Diego Soc. Nat. Hist., 2: 32.

1927 *Rhagovelia excellentis* Drake and Harris, Proc. Biol. Soc. Wash., 40: 134. [Col.]. Synonymized by Gould, 1931, Univ. Ks. Sci. Bull., 20: 26.

1931 *Rhagovelia distincta distincta*: Gould, Univ. Ks. Sci. Bull., 20: 25.

1931 *Rhagovelia distincta* var. *arizonensis* Gould, Univ. Ks. Sci. Bull., 20: 26. [Ariz.]. Synonymized by Bacon, 1956, Univ. Ks. Sci. Bull., 38: 841.

1931 *Rhagovelia distincta* var. *cadyi* Gould, Univ. Ks. Sci. Bull., 20: 27. [Wyo.]. Synonymized by Bacon, 1956, Univ. Ks. Sci. Bull., 38: 841.

1931 *Rhagovelia distincta* var. *harmonia* Gould, Univ. Ks. Sci. Bull., 20: 28. [Ariz.]. Synonymized by Bacon, 1956, Univ. Ks. Sci. Bull., 38: 841.

1931 *Rhagovelia distincta* var. *proxima* Gould, Univ. Ks. Sci. Bull., 20: 29. [Col.]. Synonymized by Bacon, 1956, Univ. Ks. Sci. Bull., 38: 841.

1931 *Rhagovelia distincta* var. *valentina* Gould, Univ. Ks. Sci. Bull., 20: 29. [Tex.]. Synonymized by Bacon, 1956, Univ. Ks. Sci. Bull., 38: 841.

1956 *Rhagovelia* (*Rhagovelia*) *distincta*: Matsuda, Univ. Ks. Sci. Bull., 38: 989.

Distribution: Ariz., Cal., Col., Id., N.M., Tex., Ut., Wyo. (Mexico to Central America).

Note: Blatchley (1926, Het. E. N. Am., p. 996) rejected the "Ind." record as probably an example of mislabelling.

Rhagovelia knighti Drake and Harris, 1927

1927 *Rhagovelia knighti* Drake and Harris, Proc. Biol. Soc. Wash., 40: 133. [Mo.].

1956 *Rhagovelia* (*Rhagovelia*) *knighti*: Matsuda, Univ. Ks. Sci. Bull., 38: 989.

Distribution: Ark., Mo., Ok.

Rhagovelia obesa Uhler, 1871

1871 *Rhagovelia obesa* Uhler, Proc. Boston Soc. Nat. Hist., 14: 107. [Mass., Md.].

1924 *Rhagovelia arctoa* Torre-Bueno, Trans. Am. Ent. Soc., 50: 250. [Minn.]. Synonymized by Bacon, 1956, Univ. Ks. Sci. Bull., 38: 849.

1924 *Rhagovelia flavicincta* Torre-Bueno, Trans. Am. Ent. Soc., 50: 249. [N.C.]. Synonymized by Bacon, 1956, Univ. Ks. Sci. Bull., 38: 849.

1956 *Rhagovelia* (*Rhagovelia*) *obesa*: Matsuda, Univ. Ks. Sci. Bull., 38: 989.

Distribution: Ala., D.C., Ga., Ill., Ind., Me., Man., Md., Mass., Mich., Minn., Miss., N.C., N.H., N.J., N.Y., Oh., Ont., Pa., S.C., Tenn., Va., Vt.

Note: Signoret (1877, Bull. Soc. Ent. France, 2: 65) listed his manuscript combination *Rhagovelia aeneipes* from "Niagara." It is generally associated with *obesa*, the only species of *Rhagovelia* occurring in that area. The early records of this eastern species for Cal. and Ut. undoubtedly were based on misdetermined specimens of *R. distincta*. Cheng and Fernando (1971, Can. J. Zool., 49: 435-442) studied the life history and biology.

Rhagovelia oriander Parshley, 1922

1922 *Rhagovelia oriander* Parshley, S.D. St. Coll. Tech. Bull., 2: 19. [S.D.].

1956 *Rhagovelia* (*Rhagovelia*) *oriander*: Matsuda, Univ. Ks. Sci. Bull., 38: 989.

Distribution: Ia., Ill., Ind., Ks., Mich., Minn., Mo., Oh., S.D.

Rhagovelia rivale Torre-Bueno, 1924

1924 *Rhagovelia rivale* Torre-Bueno, Trans. Am. Ent. Soc., 50: 247. [Ks.].

1956 *Rhagovelia* (*Rhagovelia*) *rivale*: Matsuda, Univ. Ks. Sci. Bull., 38: 989.
Distribution: Col., Ia., Ks., Mo., Neb., Ok., S.D., Tex.

Rhagovelia torreyana Drake and Hussey, 1957
1957 *Rhagovelia torreyana* Drake and Hussey, Occas. Pap. Mus. Zool., Univ. Mich., 580: 2. [Fla.].
Distribution: Fla.
Note: The description of this new species was in press when the subgenera of *Rhagovelia* were proposed; no subsequent assignment to subgenus has been made. Here it is arbitrarily assigned to the only subgenus occurring in our area.

Rhagovelia varipes Champion, 1898
1898 *Rhagovelia varipes* Champion, Biol. Centr.-Am., Rhyn. 2: 133. [Mexico].
1931 *Rhagovelia beameri* Gould, Univ. Ks. Sci. Bull., 20: 18. [Ariz.]. Synonymized by Drake and Harris, 1936, Proc. Biol. Soc. Wash., 49: 106.
1956 *Rhagovelia* (*Rhagovelia*) *varipes*: Matsuda, Univ. Ks. Sci. Bull., 38: 989.
Distribution: Ariz., N.M. (Mexico).

Genus *Trochopus* Carpenter, 1898

1898 *Trochopus* Carpenter, Ent. Month. Mag., 34: 78. Type-species: *Trochopus marinus* Carpenter, 1898. Monotypic.
Note: Drake and Doesburg (1966, Stud. Fauna Suriname, 8: 65-76) revised the three species of this genus and presented a table for their separation.

Trochopus plumbeus (Uhler), 1894
1894 *Rhagovelia plumbea* Uhler, Proc. Zool. Soc. Lond., p. 217. [Fla., Grenada, St. Vincent].
1898 *Trochopus marinus* Carpenter, Ent. Month. Mag., 34: 79. [Jamaica]. Synonymized by Carpenter, 1898, Ent. Month. Mag., 34: 109.
1916 *Rhagovelia* (*Trochopus*) *plumbea*: Torre-Bueno, Bull. Brook. Ent. Soc., 11: 56.
1966 *Trochopus plumbeus*: Drake and Doesburg, Stud. Fauna Suriname, 8: 68.
Distribution: Fla. (West Indies, Mexico to Venezuela).

Subfamily Veliinae Amyot and Serville, 1843

Genus *Paravelia* Breddin, 1898

1898 *Paravelia* Breddin, Jahr. Natur. Ver. Magdeburg, p. 159. Type-species: *Velia basalis* Spinola, 1837. Original designation.
Note: Our species of this genus were considered members of the genus *Velia* Latreille, until Polhemus (1976, J. Ks. Ent. Soc., 49: 509-513) resurrected *Paravelia*, reviewed the genus, and provided a checklist of the species. Smith and Polhemus (1978, Proc. Ent. Soc. Wash., 80: 62) provided a key to the species of America north of Mexico.

Paravelia beameri (Hungerford), 1929
1929 *Velia beameri* Hungerford, An. Ent. Soc. Am., 22: 759. [Ariz.].
1976 *Paravelia beameri*: Polhemus, J. Ks. Ent. Soc., 49: 512.
Distribution: Ariz.

Paravelia brachialis (Stål), 1860

1860 *Velia brachialis* Stål, K. Svens. Vet.-Akad. Handl., 2(7): 82. [Brazil].

1916 *Velia australis* Torre-Bueno, Bull. Brook. Ent. Soc., 11: 54. [Fla.]. Synonymized by Blatchley, 1926, Het. E. N. Am., p. 1002.

1976 *Paravelia brachialis*: Polhemus, J. Ks. Ent. Soc., 49: 512.

Distribution: Ariz., D.C., Fla., Ga., La., Miss., N.C., Oh., Pa., Tex. (Mexico to Brazil, West Indies)

Paravelia stagnalis (Burmeister), 1835

1835 *Velia stagnalis* Burmeister, Handb. Ent., 2: 212. [Pa.].

1919 *Velia watsoni* Drake, Fla. Buggist, 3: 123. [Fla.]. Synonymized by Polhemus, 1976, J. Ks. Ent. Soc., 49: 511.

1953 *Velia paulineae* Wilson, Fla. Ent., 36: 27. [Miss.]. Synonymized by Polhemus, 1976, J. Ks. Ent. Soc., 49: 511.

1976 *Paravelia stagnalis*: Polhemus, J. Ks. Ent., Soc., 49: 513.

Distribution: Ala., D.C., Fla., Ga., Ks., La., Md., Miss., Mo., N.C., N.J., Oh., Pa., Tex., Va. (Cuba).

Paravelia summersi (Drake), 1951

1951 *Velia summersi* Drake, Rev. Ent., 22: 371. [Ariz.].

1976 *Paravelia summersi*: Polhemus, J. Ks. Ent. Soc., 99: 513.

Distribution: Ariz., N.M. (Mexico).

Literature Cited

Abbott, J. F.
1912. A new type of Corixidae (*Ramphocorixa balanodis*, n. gen., et sp.) with an account of its life history. Canadian Entomologist, 44: 113-120, plate IV.
1912. A new genus of Corixidae (Hemip.). Entomological News, 23: 337-339, plate 18.
1913. A new species of Corixidae (Heteroptera). Canadian Entomologist, 45: 113-115.
1913. Corixidae of Georgia. Bulletin of the Brooklyn Entomological Society, 8: 81-91.
1916. New species of Corixidae. Entomological News, 27: 340-342.
1923. Family Corixidae. Pages 386-390. *In* Britton, W. E. (editor). The Hemiptera or sucking insects of Connecticut. Connecticut State Geological and Natural History Survey Bulletin, 34: 1-807.

Ables, J. R.
1975. Notes on the biology of the predaceous pentatomid *Euthyrhynchus floridanus* (L.). Journal of the Georgia Entomological Society, 10: 353-356.

Ables, J. R., S. L. Jones, and D. W. McCommas, Jr.
1978. Response of selected predator species to different densities of *Aphis gossypii* and *Heliothis virescens* eggs. Environmental Entomology, 7: 402-404.

Adams, C. C.
1909. An ecological survey of Isle Royale, Lake Superior. Michigan Biological Survey, *In* Report to the Board of the Geological Survey. 468 pages.

Adkins, W. S.
1917. Some Ohio Heteroptera records. Ohio Journal of Science, 18: 58-61.

Adler, P. H. and A. G. Wheeler, Jr.
1984. Extra-phytophagous food sources of Hemiptera-Heteroptera: Bird droppings, dung, and carrion. Journal of the Kansas Entomological Society, 57: 21-27.

Agassiz, J. L. R.
1846. Nomenclatoris Zoologici. 2. Index Universalis. Jent et Grassmann, Soloduri. 393 pages.

Agnew, C. W., W. L. Sterling, and D. A. Dean
1982. Influence of cotton nectar on red imported fire ants and other predators. Environmental Entomology, 11: 629-634.

Ahmad, I.
1965. The Leptocorisinae (Heteroptera: Alydidae) of the World. Bulletin of the British Museum (Natural History), Entomolgy, Supplement 5: 1-156.

Ahmad, T. R., S. D. Kindler, and K. P. Pruess
1984. Recovery of two sorghum varieties from sublethal infestations of chinch bug *Blissus leucopterus leucopterus* (Say) (Hemiptera: Lygaeidae). Journal of Economic En-tomology, 77: 151-152.

Ahmad, T. R., K. P. Pruess, and S. D. Kindler
1984. Non-crop grasses as hosts for the chinch bug, *Blissus leucopterus leucopterus* (Say) (Hemiptera: Lygaeidae). Journal of the Kansas Entomological Society, 57: 17-20.

Akingbohungbe, A. E.
1972. Two previously undescribed species of Miridae in North America (Hemiptera: Heteroptera). Annals of the Entomological Society of America, 65: 840-842.

Akingbohungbe, A. E. and T. J. Henry
1984. A review of the taxonomic characters and the higher classification of the Isometopinae (Hemiptera: Miridae). XVII International Congress of Entomology, Hamburg, Germany, Abstract Volume, Page 13.

Akingbohungbe, A. E., J. L. Libby, and R. D. Shenefelt
1972. Miridae of Wisconsin (Hemiptera: Heteroptera). University of Wisconsin Research Bulletin R 2396, 24 pages.
1973. Nymphs of Wisconsin Miridae (Hemiptera: Heteroptera). University of Wisonsin Research Bulletin R 2561, 25 pages.

Alayo, P. D.
1973. Los Hemipteros de Cuba. Parte XI. Familia Lygaeidae. Torreia, (New Series) 25: 1-79.

Alden, B., H. Dingle, and B. Possidente
1983. Diel organization of behavior in milkweed bugs, *Oncopeltus* spp. Physiological Entomology, 8: 223-230.

Aldrich, J. M.

1915. Results of twenty-five years' collecting in the Tachinidae, with notes on some common species. Annals of the Entomological Society of America, 8: 79-84.

Ali A.-S. A. and T. F. Watson

1982. Efficacy of dipel and *Geocoris punctipes* (Hemiptera: Lygaeidae) against the tobacco budworm (Lepidoptera: Noctuidae) on cotton. Journal of Economic Entomology, 75: 1002-1004.

Allen, R. C.

1969. A revision of the genus *Leptoglossus* Guerin (Hemiptera: Coreidae). Entomologica Americana, 45: 35-140.

Aller, T. and R. L. Caldwell

1979. An investigation of the possible presence of an aggregation pheromone in the milkweed bugs, *Oncopeltus fasciatus* and *Lygaeus kalmii*. Physiological Entomology, 4: 287-290.

Aller, T., D. Hirai, and R. L. Caldwell

1979. A comparison of the correcting behavior exhibited by two milkweed bugs, *Oncopeltus fasciatus* and *Lygaeus kalmii*. Physiological Entomology, 4: 99-102.

Altieri, M. A., J. W. Todd, E. W. Hauser, M. Patterson, G. A. Buchanan, and R. H. Walker

1981. Some effects of weed management and row spacing on insect abundance in soybean fields. Protective Ecology, 3: 339-344.

Altieri, M. A. and W. H. Whitcomb

1980. Predaceous and herbivorous arthropods associated with camphorweed (*Heterotheca subaxillaris* Lamb.) in north Florida. Journal of the Georgia Entomological Society, 15: 290-299.

Amyot, C. J. B. and J. G. A. Serville

1843. Histoire naturelle des insectes Hémiptères. *In* "Suites à Buffon." Fain et Thunot, Paris. LXXVI + 675 + 8 pages [Atlas], 12 plates.

1845-1846. Methode mononymique. Annales de la Société Entomologique de France, Series 2, 1845, 3: 369-492, plates 8-9; 1846, 4: 73-193, 359-452, plate 10.

Andersen, N. M.

1975. The *Limnogonus* and *Neogerris* of the Old World with character analysis and a reclassification of the Gerrinae (Hemiptera: Gerridae). Entomologica Scandinavica, (Supplementum) 7: 1-96.

1976. A comparative study of locomotion on the water surface in semiaquatic bugs (Insecta, Hemiptera, Gerromorpha). Videnskabelige Meddelelser fra Dansk Naturhistorisk Forening, 139: 337-396.

1977. A new and primitive genus and species of Hydrometridae (Hemiptera, Gerromorpha) with a cladistic analysis of relationships within the family. Entomo-logica Scandinavica, 8: 301-316.

1978. A new family of semiaquatic bugs for *Paraphrynovelia* Poisson with a cladistic analysis of relationships (Insecta, Hemiptera, Gerromorpha). Steenstrupia [Zoological Museum, University of Copenhagen], 4: 211-225.

1979. Phylogenetic inference as applied to the study of evolutionary diversification of semiaquatic bugs (Hemiptera: Gerromorpha). Systematic Zoology, 28: 554-578.

1981. Semiaquatic bugs: phylogeny and classification of the Hebridae (Heteroptera: Gerromorpha) with revisions of *Timasius*, *Neotimasius* and *Hyrcanus*. Systematic Entomology, 6: 377-412.

1982. The Semiaquatic Bugs (Hemiptera, Gerromorpha). Phylogeny, Adaptations, Biogeography and Classification. Entomonograph Volume 3. Scandinavian Science Press, Klampenborg, Denmark. 455 pages.

Andersen, N. M. and J. T. Polhemus

1980. Four new genera of Mesoveliidae (Hemiptera, Gerromorpha) and the phylogeny and classification of the family. Entomologica Scandinavica, 11: 369-392.

Anderson, L. D.

1932. Three new species of *Metrobates* (= *Trepobatopsis* Champion). Journal of the Kansas Entomological Society, 5: 56-60.

1932. A monograph of the genus *Metrobates* (Hemiptera, Gerridae). University of Kansas Science Bulletin, 20: 297-311.

Anderson, N. H.

1962. Anthocoridae of the Pacific Northwest with notes on distributions, life histories, and habits (Heteroptera). Canadian Entomologist, 94: 1325-1334.

Anderson, N. H. and L. A. Kelton

1963. A new species of *Anthocoris* from Canada, with distribution records for three other species (Heteroptera: Anhocoridae). Canadian Entomologist, 95: 439-442.

Andre, F.

1934. Notes on the biology of *Oncopeltus fasciatus* (Dallas). Iowa State College Journal of Science, 9: 73-87.

1937. An undescribed chinch bug from Iowa. Iowa State College Journal of Science, 11: 165-167.

Anonymous

1969. Chinch bugs. Illinois Cooperative Extension Service Fact Sheet NHE 35. 3 pages.

Apstein, C.

1916. Erscheinungsjahre von Gmelin (Linné), Systema Naturae, ed. 13. Zoologischer Anzeiger, 47: 32.

Arbogast, R. T.

1975. Population growth of *Xylocoris flavipes*: Influence of temperature and humidity. Environmental Entomology, 4: 825-831.

1976. Suppression of *Oryzaephilus surninamensis* (L.) (Coleoptera, Curcujidae) on shelled corn by the predator *Xylocoris flavipes* (Reuter) (Hemiptera, Anthocoridae). Journal of the Georgia Entomological Society, 11: 67-71.

1978. The biology and impact of the predatory bug *Xylocoris flavipes* (Reuter). Proceedings of the Second International Working Conference on Stored-Product Entomology, Nigeria, pages 91-105.

1979. Cannibalism in *Xylocoris flavipes* (Hemiptera: Anthocoridae), a predator of stored-product insects. Entomologica Experimentalis et Applicata, Netherlands, 25: 128-135.

Arbogast, R. T., M. Carthon, and J. R. Roberts, Jr.

1971. Developmental stages of *Xylocoris flavipes* (Hemiptera: Anthocoridae), a predator of stored-product insects. Annals of the Entomological Society of America, 64: 1131-1134.

Arbogast, R. T., B. R. Flaherty, R. V. Byrd, and J. W. Press

1985. Developmental stages of *Xylocoris sordidus* (Hemiptera: Anthocoridae). Entomological News, 96: 53-58.

Arbogast, R. T., B. R. Flaherty, and J. W. Press

1983. Demography of the predaceous bug *Xylocoris sordidus* (Reuter). American Midland Naturalist, 109: 398-405.

Arbogast, R. T., G. L. LeCato, and M. Carthon

1977. Longevity of fed and starved *Xylocoris flavipes* (Reuter) (Hemiptera, Anthocoridae) under laboratory conditions. Journal of the Georgia Entomological Society, 12: 58-64.

Arnaud, P. H., Jr.

1978. A host-parasite catalog of North American Tachinidae (Diptera). United States Department of Agriculture Miscellaneous Publication, No. 1319. 860 pages.

Arnold, J. W.

1971. Feeding behavior of a predaceous bug (Hemiptera: Nabidae). Canadian Journal of Zoology, 49: 131-132.

Arrand, J. C.

1958. A new species of *Plagiognathus* from alfalfa in western Canada (Hemiptera: Miridae). Canadian Entomologist, 90: 497-500.

1960. *Liocoris* spp. collected on alfalfa in central and northern British Columbia. Proceedings of the Entomological Society of British Columbia, 57: 60.

Arrand, J. C. and H. McMahon

1974. *Plagiognathus medicagus* (Hemiptera: Miridae): Descriptions of egg and five nymphal instars. Canadian Entomologist, 106: 433-435.

Ashlock, P. D.

1957. An investigation of the taxonomic value of the phallus in the Lygaeidae (Hemiptera-Heteroptera). Annals of the Entomological Society of America, 50: 407-426.

1958. A revision of the genus *Malezonotus* (Hemiptera-Heteroptera: Lygaeidae). Annals of the Entomological Society of America, 51: 199-208.

1960a. H. G. Barber: Bibliography and list of names proposed. Proceedings of the Entomological Society of Washington, 62: 129-138.

1960b. New synonymies and generic changes in the Lygaeidae (Hemiptera-Heteroptera). Proceedings of the Biological Society of Washington, 73: 235-238.

1961. A review of the genus *Arphnus* Stål with a new species from Mexico (Hemiptera: Lygaeidae). Pan-Pacific Entomologist, 37: 17-22.

1963. A new species of the genus *Malezonotus* from California (Hemiptera-Heteroptera: Lygaeidae). Pan-Pacific Entomologist, 39: 264-266.

Ashlock, P. D. (continued)
1964. Two new tribes of Rhyparochrominae: A re-evaluation of the Lethaeini (Hemiptera-Heteroptera: Lygaeidae). Annals of the Entomological Society of America, 57: 414-422.
1967. A generic classification of the Orsillinae of the world (Hemiptera-Heteroptera: Lygaeidae). University of California Publications in Entomology, 48: vi + 82 pages.
1969. Robert L. Usinger bibliography and list of names proposed. Pan-Pacific Entomologist, 45: 185-203.
1975. Toward a classification of North American Lygaeinae (Hemiptera-Heteroptera: Lygaeidae). Journal of the Kansas Entomological Society, 48: 27-32.
1977. New records and name changes of North American Lygaeidae (Hemiptera: Heteroptera: Lygaeidae). Proceedings of the Entomological Society of Washington, 79: 575-582.
1979. A new *Eremocoris* from California with a key to North American genera of Drymini (Hemiptera-Heteroptera: Lygaeidae). Pan-Pacific Entomologist, 55: 149-154.

Ashlock, P. D. and J. D. Lattin
1963. Stridulatory mechanisms in the Lygaeidae, with a new American genus of Orsillinae (Hemiptera: Heteroptera). Annals of the Entomological Society of America, 56: 693-703.

Ashlock, P. D. and C. W. O'Brien
1964. *Catherosia lustrans*, a tachinid parasite of some drymine Lygaeidae (Diptera and Hemiptera-Heteroptera). Pan-Pacific Entomologist, 40: 98-100.

Ashlock, P. D. and J. A. Slater
1982. A review of the genera of Western Hemisphere Ozophorini with two new genera from Central America (Hemiptera-Heteroptera: Lygaeidae). Journal of the Kansas Entomological Society, 55: 737-750.

Ashmead, W. H.
1886. On two new Hemiptera-Heteroptera. Canadian Entomologist, 18: 18-20.
1887. Hemipterological contributions. (No. I.). Entomologica Americana, 3: 155-156.
1892. Notes on the Genus *Enicocephalus* Westwood, and description of a new species from Utah. Proceedings of the Entomological Society of Washington, 2: 328-330.
1895. Notes on cotton insects found in Mississippi. Insect Life, 7: 320-326.
1897. A new water-bug from Canada. Canadian Entomologist, 29: 56.

Atim, A. B. and H. M. Graham
1983. Parasites of *Geocoris* spp. near Tucson Arizona. Southwestern Entomologist, 8: 210-215.
1984. Predation of *Geocoris punctipes* and *Nabis alternatus*. Southwestern Entomologist, 9: 227-231.

Atkins, I. M., M. E. McDaniel, and J. H. Gardenhire
1969. Growing oats in Texas. Texas Agricultural Experiment Station Bulletin, 1091: 3-28.

Atkinson, E. T.
1890. Catalogue of the Insecta. No. 2. Order Rhynchota. Suborder Hemiptera-Heteroptera. Family Capsidae. Journal of the Asiatic Society of Bengal, 58(2): 25-200.

Ayala, S. C., O. B. Quintero, and P. Barretto
1975. Trypanosomatidos de plantas laticiferas y sus insectos transmisores en Colombia y Costa Rica. Revista de Biologia Tropical, 23: 5-15.

Ayers, G. S., T. M. Brown, and R. E. Monroe
1974. Studies on D-alanine in the Heteroptera. Journal of Entomology (A), 49: 1-5.

Bacheler, J. S. and R. M. Baranowski
1975. *Paratriphelips laeviusculus*, a phytophagous anthocorid new to the United States (Hemiptera: Anthocoridae). Florida Entomologist, 58: 157-163.

Bacon, J. A.
1956. A taxonomic study of the genus *Rhagovelia* (Hemiptera, Veliidae) of the Western Hemisphere. University of Kansas Science Bulletin, 38: 695-913.

Baerensprung, F. von
1857. *Myrmedobia* und *Lichenobia*, zwei neue einheimische rhynchoten-Gattungen. Berliner Entomologishe Zeitschrift, 1: 161-168.
1860. Catalogus Hemipterorum Europae. Hemiptera Heteroptera Europaea systematice disposita. Berliner Entomologische Zeitschrift, 4 (appendix): 1-25.

Bailey, N.S.
1950. An Asiatic tingid new to North America (Heteroptera). Psyche, 57: 143-145.

1951. The Tingoidea of New England and their biology. Entomologica Americana, New Series, 31: 1-140.

Baker, C. F.

1906. Notes on the *Nysius* and *Ortholomus* of America. Invertebrata Pacifica, 1: 133-140.

1908. Preliminary remarks on American Corizini (Hemiptera). Canadian Entomologist, 40: 241-244.

1910. Californian Emesidae (Hemiptera). Pomona Journal of Entomology, 2: 225-227.

Baker, P. B. and R. H. Ratcliffe

1977. Evaluation of blue grasses for tolerance to *Blissus leucopterus hirtus* (Hemiptera: Lygaeidae). Journal of the New York Entomological Society, 85: 165-166.

Baker, P. B., R. H. Ratcliffe, and A. L. Steinhauer

1981a. Tolerance to hairy chinch bug feeding in Kentucky blue grass. Environmental Entomology, 10: 153-157.

1981b. Laboratory rearing of the hairy chinch bug. Environmental Entomology, 10: 226-229b.

Baker, R. S., M. L. Laster, and W. F. Kitten

1985. Effects of the herbicide monosodium methanearsonate on insect and spider populations in cotton fields. Journal of Economic Entomology, 78: 1481-1484.

Balduf, W. V.

1941. Life history of *Phymata pennsylvanica americana* Melin (Phymatidae, Hemiptera). Annals of the Entomological Society of America, 34: 204-214.

1942. Evaluating the economic status of *Phymata*. Journal of Economic Entomology, 35: 445-448.

1964. Numbers of ovarioles in the Heteroptera (Insecta). Proceedings of the Entomological Society of Washington, 66: 2-5.

Banks, N.

1903. Additions to the list of New York Pentatomidae. Journal of the New York Entomological Society, 11: 227.

1909. Notes on our species of Emesidae. Psyche, 16: 43-48.

1910. Catalogue of the Nearctic Hemiptera-Heteroptera. American Entomological Society, Philadelphia. viii + 103 pages.

1910. Four new Reduviidae. Entomological News, 21: 324-325.

1912. A new species of Emesidae from Vermont. Psyche, 19: 97.

1912. At the *Ceanothus* in Virginia. Entomological News, 23: 102-110.

Baranowski, R. M.

1958. Notes on the biology of the royal palm bug, *Xylastodoris luteolus* Barber (Hemiptera, Thaumastocoridae). Annals of the Entomological Society of America, 51: 547-551.

1979a. Notes on the biology of *Ischnodemus oblongus* and *Ischnodemus fulvipes* with descriptions of the immature stages (Hemiptera: Heteroptera: Lygaeidae). Annals of the Entomological Society of America, 72: 655-658.

Baranowski, R. M., R. I. Sailer, and F. W. Mead

1983. *Euschistus acuminatus*, a pentatomid new to the United States (Hemiptera: Pentatomidae). Florida Entomologist, 66: 287-291.

Baranowski, R. M. and J. A. Slater

1975. The life history of *Craspeduchus pulchellus*, a lygaeid new to the United States (Hemiptera: Lygaeidae). Florida Entomologist, 58: 297-302.

1983. The *Ozophora pallescens* complex in the West Indies with the description of four new species (Hemiptera: Lygaeidae). Florida Entomologist, 66: 440-463.

Barber, G. W.

1921. Collecting about Walden Pond. Canadian Entomologist. 53: 145-146.

1926. A short list of Scutelleroidea collected in New Mexico in 1916. Entomological News, 37: 43-44.

Barber, H. G.

1906. Hemiptera from southwestern Texas. Museum of the Brooklyn Institute of Arts and Sciences, Science Bulletin, 1: 255-289.

1908. [Notes in minutes of meeting]. Journal of the New York Entomological Society, 16: 193.

1908. [Note in minutes of meeting]. Journal of the New York Entomological Society, 16: 248.

1909. [Notes in minutes of meeting]. Journal of the New York Entomological Society, 17: 137-138.

1910. Some Mexican Hemiptera-Heteroptera new to the fauna of the United States. Journal of the New York Entomological Society, 18: 34-39.

1911. Descriptions of some new Hemiptera-

Barber, H. G. (continued)
Heteroptera. Journal of the New York Entomological Society, 19: 23-31.

1911. The resurrection of *Thyanta calceata* Say from synonymy. Journal of the New York Entomological Society, 19: 108-111.

1911. *Pyrrhocoris apterus* Linn. in the United States. Journal of the New York Entomological Society, 19: 111-112.

1911. Arrangement of the species of *Dendrocoris* Bergr., with the descriptions of two new species (Hemip.). Entomological News, 22: 268-271.

1912. A preliminary report on the Hemiptera-Heteroptera of Porto Rico collected by the American Museum of Natural History. American Museum Novitates, 75: 1-13.

1913. Description of two new species of *Ochterus* Latr. (Hemiptera) with an arrangement of the North American species. Canadian Entomologist, 45: 213-215.

1914. Insects of Florida. II. Hemiptera. Bulletin of the American Museum of Natural History, 33: 495-535.

1914. New Hemiptera-Heteroptera, with comments upon the distribution of certain known species. Journal of the New York Entomological Society, 22: 164-171.

1914. Hemiptera from Wilmington and Southport, North Carolina. Journal of the New York Entomological Society, 22: 269.

1916. [Notes on Nabidae]. Journal of the New York Entomological Society, 24: 308.

1917. Synoptic keys to the Lygaeidae (Hemiptera) of the United States. Part I. Psyche, 24: 128-135.

1918a. Corrections to "New York Scoloposthethi" (Family Lygaeidae: Heter.). Entomological News, 29: 51-52.

1918b. Concerning Lygaeidae.- No. I. Journal of the New York Entomological Society, 26: 44-46.

1918c. Concerning Lygaeidae.- No. 2. Journal of the New York Entomological Society, 26: 49-66.

1918d. Synoptic keys to the Lygaeidae (Hemiptera) of the United States. Part II. Rhyparochrominae. Psyche, 25: 71-88.

1918e. A new species of *Leptoglossus*: A new *Blissus* and varieties. Bulletin of the Brooklyn Entomological Society, 13: 35-39.

1918f. The genus *Plinthisus* Latr. [Lygaeidae-Hemiptera] in the United States. Proceedings of the Entomological Society of Washington, 20: 108-111.

1920. A new member of the family Thaumastocoridae. Bulletin of the Brooklyn Entomological Society, 15: 98-105.

1921a. Revision of the genus *Lygaeus* Fab. (Hemiptera-Heteroptera). Proceedings of the Entomological Society of Washington, 23: 63-68.

1921b. Revision of the genus *Ligyrocoris* Stål (Hemiptera, Lygaeidae). Journal of the New York Entomological Society, 29: 100-114.

1922. Collecting Hemiptera in the Adirondacks. Journal of the New York Entomological Society, 30: 110-111.

1922. List of the superfamily Tingidoideae of New Jersey with synonymy and food plants. State of New Jersey Department of Agriculture, Bureau of Statistics and Inspection Circular, 54: 16-17.

1922. Note on *Luteva carolina* H.S. (Hemiptera-Heteroptera. Fam. Reduviidae). Journal of the New York Entomological Society, 30: 130.

1922. Two new species of Reduviidae from the United States (Hem.). Proceedings of the Entomological Society of Washington, 24: 103-104.

1923. Family Lygaeidae. Pages 708-737. *In* W. E. Britton (editor). Guide to the insects of Connecticut. Part IV. The Hemiptera or sucking insects of Connecticut. Connecticut Geological Natural History Survey Bulletin, 34: 1-807, plates I-XX.

1923. A preliminary report of the Hemiptera-Heteroptera of Porto Rico collected by the American Museum of Natural History. American Museum Novitates, 75: 1-13.

1923. Report on certain families of Hemiptera-Heteroptera collected by the Barbados-Antigua Expedition from the University of Iowa in 1918. University of Iowa Studies in Natural History, 10: 17-29.

1924a. Two new species of *Cymus* (Hemiptera-Lygaeidae). Bulletin of the Brooklyn Entomological Society, 19: 87-90.

1924b. Corrections and comments Hemiptera-Heteroptera. Journal of the New York Entomological Society, 32: 133-137.

1924. A new species of *Pselliopus* (Hemiptera: Reduviidae). Proceedings of the Entomological Society of Washington, 26: 211-213.

1924. The genus *Arhaphe* in the United States (Hemiptera-Pyrrhocoridae). Canadian Entomologist, 56: 227-228.

1926. Notes on Coreidae in the collection of the U. S. National Museum with description of a new *Catorhintha* (Hemiptera-Heteroptera). Journal of the New York Entomological Society, 34: 209-216.

1926a. A new *Geocoris* from Illinois (Hemiptera, Lygaeidae). Bulletin of the Brooklyn Entomological Society, 21: 38-39.

1927. Two new species of Pentatomidae from the southern United States (Hemiptera-Heteroptera). Bulletin of the Brooklyn Entomological Society, 22: 241-244.

1928. A new genus and species of Coreidae from the Western States (Hemiptera-Heteroptera). Journal of the New York Entomological Society, 36: 25-28.

1928a The genus *Eremocoris* in the eastern United States, with description of a new species and a new variety (Hemiptera-Lygaeidae). Proceedings of the Entomological Society of Washington, 30: 59-60.

1928b. Revision of the genus *Ptochiomera* Say (Hemiptera, Lygaeidae). Journal of the New York Entomological Society, 36: 175-177.

1928c. Order Hemiptera. Pages 74-142. *In* Leonard, M. D.(editor). A list of the insects of New York with a list of the spiders and certain other allied groups. Cornell University Agricultural Experiment Station Memoir 101. 1121 pages.

1928d. Some quantitative results in collecting Hemiptera. Entomological News, 39: 193-194.

1928e. *Ptochiomera* Say or *Plochiomera* Stål? Bulletin of the Brooklyn Entomological Society, 23: 153.

1928f. Two new Lygaeidae from the western United States (Hemiptera-Lygaeidae). Bulletin of the Brooklyn Entomological Society, 23: 264-268.

1930. Essay on the subfamily Stenopodinae of the New World. Entomologica Americana, New Series, 10: 149-238, plate 11.

1932. Two palearctic Hemiptera in the nearctic fauna (Heteroptera- Pentatomidae: Nabidae). Proceedings of the Entomological Society of Washington, 34: 65-66.

1932. Three new species of *Exptochiomera* from the United States (Hemiptera-Lygaeidae). Journal of the New York Entomological Society, 40: 357-363.

1933. A new *Sciocoris* from Texas (Hemiptera-Heteroptera: Pentatomidae). Proceedings of the Entomological Society of Washington, 34: 149-150 (1932).

1935a. New *Geocoris* from the United States, with key to species (Lygaeidae: Geocorinae). Journal of the New York Entomological Society, 43: 131-137.

1935. A new *Edessa* from Florida (Hemiptera-Heteroptera: Pentatomidae). Proceedings of the Entomological Society of Washington, 37: 48-49.

1937. Descriptions of six new species of *Blissus* (Hemiptera-Heteroptera: Lygaeidae). Proceedings of the Entomological Society of Washington, 39: 81-86.

1937. A new species of *Triatoma* from Arizona (Hemiptera-Heteroptera: Reduviidae). Proceedings of the Entomological Society of Washington, 39: 86-87.

1938a. A new species of *Cistalia* (Hemiptera-Heteroptera: Lygaeidae). Proceedings of the Entomological Society of Washington, 40: 87-88.

1938b. A review of the genus *Crophius* Stål, with descriptions of three new species (Hemiptera-Heteroptera: Lygaeidae). Journal of the New York Entomological Society, 46: 313-319.

1938. A new genus and species of the subfamily Triatominae (Reduviidae: Hemiptera). Proceedings of the Entomological Society of Washington, 40: 104-105.

1939. A new species of *Heterogaster* from the southern part of the United States (Hemiptera: Lygaeidae). Proceedings of the Entomological Society of Washington, 41: 173-174.

1939. Insects of Porto Rico and the Virgin Islands-- Hemiptera-Heteroptera (excepting the Miridae and Corixidae). Scientific Survey of Porto Rico and the Virgin Islands. New York Academy of Sciences, 14: 263-441.

1939. A new bat bug from the eastern United States (Hemiptera-Heteroptera: Cimicidae). Proceedings of the Entomological Society of Washington, 41: 243-246.

1941. Description of a new bat bug (Hemiptera-Heteroptera: Cimicidae). Journal of the Washington Academy of Sciences, 31: 315-317.

1947a. Revision of the genus *Nysius* in the United States and Canada (Hemiptera Heteroptera: Lygaeidae). Journal of the

Barber, H. G. (continued)
Washington Academy of Sciences, 37: 354-366.
1947b. The family Lygaeidae (Hemiptera-Heteroptera) of the island of Cuba and the Isle of Pines. Part I. Memorias de la Sociedad Cubana de Historia Natural, 19: 55-75.
1947c. Records of the species of *Nysius* occurring in the Dominion of Canada (Hemiptera: Lygaeidae). Canadian Entomologist, 79: 194.
1948. A case of synonymy in the family Neididae (Hemiptera-Heteroptera). Bulletin of the Brooklyn Entomological Society, 43: 21.
1948. Lygaeidae collected in western Texas, with a new *Lygaeospilus* from California. Ohio Journal of Science, 48: 66-68.
1948. New records for *Stygnocoris rusticus* Fallén. Bulletin of the Brooklyn Entomological Society, 43: 31.
1948. Concerning *Esuris* Barber (not Stål) and *Neosuris* Barber, with a new subspecies from Idaho (Hemiptera-Heteroptera: Lygaeidae). Psyche, 55: 84-86.
1948. The genus *Cligenes* in the United States (Hemiptera, Lygaeidae). Proceedings of the Entomological Society of Washington, 50: 157-158.
1949. Some new Lygaeidae chiefly from the United States (Hemiptera: Heteroptera). Pan-Pacific Entomologist, 24: 201-206 (1948).
1949. A new genus in the subfamily Blissinae from Mexico and a new *Nysius* from the north west (Lygaeidae; Hemiptera-Heteroptera). Bulletin of the Brooklyn Entomological Society, 44: 141-144.
1949. Some synonyms in the family Lygaeidae (Hemiptera). Proceedings of the Entomological Society of Washington, 51: 273-276.
1952a. The genus *Antillocoris* Kirk. in the United States (Hemiptera-Lygaeidae). Bulletin of the Brooklyn Entomological Society, 47: 85-87.
1952b. The genus *Pachybrachius* in the United States and Canada with the description of two new species (Hemiptera: Lygaeidae). Journal of the New York Entomological Society, 60: 211-220.
1953a. A second revision of the genus *Ptochiomera* Say and its allies (Hemiptera, Lygaeidae). Proceedings of the Entomological Society of Washington, 55: 19-27.
1953. A change of name in the family Reduviidae. Proceedings of the Entomological Society of Washington, 55: 142.
1953b. A revision of the genus *Kleidocerys* Stephens in the United States (Hem-iptera, Lygaeidae). Proceedings of the Entomological Society of Washington, 55: 273-283.
1953. A new subfamily, genus, and species belonging to the family Enicocephalidae (Hemiptera, Heteroptera). American Museum Novitates, 1614: 1-4.
1954. The family Lygaeidae (Hemiptera-Heteroptera) of the island of Cuba and the Isle of Pines - Part II. Memorias de la Sociedad Cubana de Historia Natural, 22: 335-353.
1954. A report on the Hemiptera Heteroptera from the Bimini Islands, Bahamas, British West Indies. American Museum Nov-itates, 1682: 1-18.
1955. The genus *Cryphula* Stål, with the descriptions of two new species (Heteroptera: Lygaeidae). Journal of the New York Entomological Society, 63: 135-137.
1956. A new arrangement in the subfamily Cyminae (Hemiptera, Lygaeidae). Proceedings of the Entomological Society of Washington, 58: 282.
1956. Lectotype designated for *Kleidocerys franciscanus* (Stål) 1859 (Hemiptera: Lygaeidae). Entomological News, 67: 264-265.
1956. A new species of *Leptocoris* (Coreidae: Leptocorini). Pan-Pacific Entomologist, 32: 9-11.
1958. A new species of *Nysius* from Alaska and Alberta, Canada (Hemiptera, Lygaeidae). Proceedings of the Entomological Society of Washington, 60: 70.
1958b. Heteroptera: Lygaeidae. Pages 173-218. *In* Insects of Micronesia. Volume 7. Bernice P. Bishop Museum, Honolulu.

Barber, H. G. and P. D. Ashlock
1960. The Lygaeidae of the Van Voast-- American Museum of Natural History expedition to the Bahama Islands, 1953 (Hemiptera: Heteroptera). Proceedings of the Entomological Society of Washington, 62: 117-124.

Barber, H. G. and S. C. Bruner
1932. The Cydnidae and Pentatomidae of Cuba. Journal of the Department of Agri-

culture of Puerto Rico, 16: 231-284, plates 24-26.

Barber, H. G. and R. I. Sailer
1953. A revision of the turtle bugs of North America (Hemiptera: Pentatomidae). Journal of the Washington Academy of Sciences, 43: 150-162.

Barber, H. G. and Weiss, H. B.
1922. The lace bugs of New Jersey. State of New Jersey Department of Agriculture, Bureau of Statistics and Inspection Circular, 54: 3-15.

Barber, T. C.
1925. Preliminary observations on an insect of the cotton stainer group new to the United States. Journal of Agricultural Research, 31: 1137-1147.

Bare, C. O.
1925. A new species of *Buenoa* (Hemiptera, Notonectidae). Entomological News, 36: 225-228.
1926. Life histories of some Kansas "backswimmers." Annals of the Entomological Society of America, 19: 93-101.
1928. Haemoglobin cells and other studies of the genus *Buenoa* (Hemiptera, Notonectidae). University of Kansas Science Bulletin, 18: 265-349.
1931. A *Buenoa* of southwest United States and Mexico (Hemiptera). Pan-Pacific Ento-mologist, 7: 115-118.

Barnes, M. M.
1970. Genesis of a pest: *Nysius raphanus* and *Sisymbrium irio* in vineyards. Journal of Economic Entomology, 63: 1462-1463.

Barry, R. M.
1973. A note on species composition of predators in Missouri soybeans. Journal of the Georgia Entomological Society, 8: 284-286.

Barton, H. E. and L. A. Lee
1981. The Pentatomidae of Arkansas. Arkansas Academy of Science Proceedings, 35: 20-25.

Bay, E, C.
1974. Predator-prey relationships among aquatic insects. Annual Review of Entomology, 19: 441-453.

Beauline, J. I.
1939. Parasites and predators reared at Quebec. Canadian Entomologist, 71: 120.

Becker, M.
1967. Estudos sôbre a subfamília Scaptocorinae na Região Neotropical (Hemiptera: Cydnidae). Arquivos de Zoologia, São Paulo, 15: 291-325.

Béique, R. and A. Robert
1963. Les lygéides de la Province de Quebec (Hétéroptères). Annals of the Entomological Society of Quebec, 8: 70-96.
1964. Les lygéides de la Province de Quebec (Hétéroptères) (2e partie). Annals of the Entomological Society of Quebec, 9: 72-102.

Beirne, B. P.
1972. Pest insects of annual crop plants in Canada. Part IV, Hemiptera-Homoptera; part V, Orthoptera; part VI, other groups. Memoirs of the Entomological Society of Canada, 85: 1-73.

Bell, K. D., Jr. and W. H. Whitcomb
1962. Efficiency of egg predators of the bollworm. Arkansas Farm Research, 11: 9.
1964. Field studies on egg predators of the bollworm, *Heliothis zea* (Boddie). Florida Entomologist, 47: 171-180.

Benedek, P.
1969. Magyarors zág állat Heteroptera (Fauna Hungariae) (No. 94). 17(7): 1-86.

Benedict, J. H. and W. R. Cothran
1975a. A faunistic survey of the Hemiptera-Heteroptera found in northern California hay alfalfa. Annals of the Entomological Society of America, 68: 897-900.
1975. Identification of the damsel bugs, *Nabis alternatus* Parshley and *N. americoferus* Carayon (Heteroptera: Nabidae). Pan-Pacific Entomologist, 51: 170-171.

Bennett, D. V. and E. F. Cook
1981. The semiaquatic Hemiptera of Minnesota (Hemiptera: Heteroptera). University of Minnesota Agricultural Experiment Station Technical Bulletin, 332: 3-59.

Bequaert, J.
1950. A bibliographic note on Say's two tracts of March, 1831, and January, 1832. Psyche, 57: 146.

Berg, C.
1879. Hemiptera Argentina enumeravit speciesque novas. P. E. Coni, Bonairiae. 316 pages
1883-1884. Addenda et emendand ad Hemiptera Argentinae (2). Anales de la Sociedad

Berg, C. (continued)
Cientifica Argentina, 1883, 15: 193-217, 241-269; 1883, 16: 5-32, 73-87, 105-125, 180-191, 231-241, 285-294; 1884, 17: 97-118, 166-176 [Also separate, 1884, P.E. Coni, Bonariae, 213 pages].
1892. Nova Hemiptera faunarum Argentinae et Uruguayensis. Lygaeidae. Annales Sociedad Cientifica Argentina, 33: 151-165. [Reprint-1892, Pauli E. Coni et Filiorum, Bonariae, 112 pages].
1901. Substitution d'un nom generique d'Hemiptera. Communicaciones del Museo Nacional, Buenos Aires, 1: 281.

Bergroth, E.
1886. Ueber einige amerikanische Aradiden. Wiener Entomologische Zeitung, 5: 97-98.
1886. Zur Kenntniss der Aradiden. Verhandlungen der Kaiserlich-Koniglichen Zoologisch-Botanischen Gesellschaft in Wien, 36: 53-60, plate II.
1887. Synopsis of the genus *Neuroctenus* Fieb. Öfversigt af Finska Vetenskaps-Societatens Fördhandlingar, 29: 173-189.
1887. Sur Quelques Aradides Nouveaux ou Péu Connus. Revue d'Entomologie, 6: 244-247.
1889. Sur Quelques Aradides Appartenant au Musée Royal d'Histoire Naturelle de Bruxelles. Annales de la Société Entomologique de Belgique, 33: clxxx-clxxxii.
1890. [Notes hémiptérologiques addressées de Forssa (Finlande)]. Bulletin de la Société Entomologique de France, Series 6, 10: lxv-lxvi, cxix.
1891. Contributions a l'etude des Pentatomides. Revue d'Entomologie, 10: 200-235.
1892. Note on the water-bug, found by Rev. J. L. Zabriskie. Insect Life, 4: 321.
1892. Notes on the nearctic Aradidae. Proceedings of the Entomological Society of Washington, 2: 332-337.
1892. Notes synonymiques. Revue d'Entomologie, 11: 262-264.
1893. Notes Hémiptérologiques. Revue d'Entomologie, 12: 153-155.
1895. Le genre *Cligenes* Dist. (Lygaeidae). Revue d'Entomologie, 14: 142-145.
1897. On two remarkable Californian Hemiptera. Entomological News, 8: 95-96.
1898. Ueber einige amerikanische Capsiden. Wiener Entomologische Zeitung, 17: 33-35.
1898. Diagnoses of some new Aradidae. Entomologist's Monthly Magazine, 34: 100-101.
1898. Aradidae americanae novae vel minus cognitae. Wiener Entomologische Zeitung, 17: 26-27.
1898. Description de deaux Reduviides nouveaux [Hémipt.]. Bulletin de la Société Entomologique de France, pages 307-308.
1905. Rhynchota Aethiopica. IV. Fam. Pentatomidae. Annales de la Société Entomologique de Belgique, 49: 368-378.
1905. Rhynchota Neotropica. Revue d'Entomologie, 24: 104-112.
1906. Notes on American Hemiptera. I. Canadian Entomologist, 38: 198-202.
1907. Notes on American Hemiptera. III. The Acanthosomatinae of North America. Entomological News, 18: 48-50.
1908. Enumeratio Pentatomidarum post Catalogum bruxellensem descriptarum. Mémoires de la Société Entomologique Belgique, 15: 131-200.
1910. On some Miridae from French Guiana. Annales de la Société Entomologique de Belgique, 54: 60-68.
1913. On some North American Hemiptera. Entomological News, 24: 263-267.
1913. Supplementum catalogi Heteropterorum bruxellensis. II. Coreidae, Pyrrhocoridae, Colobathristidae, Neididae. Mémoires de la Société Entomologique de Belgique, 22: 125-183.
1913. Notes on American Hemiptera. II. Canadian Entomologist, 45: 1-8.
1914. Zwei neue paläarktische Hemipteren, nebst synonymischen Mitteilungen. Wiener Entomologische Zeitung, 33: 177-184.
1914. Notes on some genera of Heteroptera. Annales de la Société Entomologique de Belgique, 58: 23-28.
1915. Ein neuer Ameisengast aus Südafrika (Hem. Heteropt.). Wiener Entomologische Zeitung, 34: 291-292.
1915. A new species of *Rheumatobates* Bergr. (Hem., Gerridae). Bulletin of the Brooklyn Entomological Society, 10: 62-64.
1915. Pentatomides nouveaux de la Guyane Française. Annales de la Société Entomologique de France, 83: 423-441, plate 11 (1914).
1915. New Oriental Pentatomoidea. Annals and Magazine of Natural History, series 8, 15: 481-493.

1919. Die Erscheinungsdata zweier hemipterologischen Werke. Entomologische Mitteilungen, 8: 188-191.
1920. List of the Cylapinae (Hem., Miridae) with descriptions of new Philippine forms. Annales de la Société Entomologique de Belgique, 55: 67-83.
1922. On some neotropical Tingidae (Hem.). Annales Société Entomologique de Belgique, 62: 149-152.
1922. Two new American Ploeariinae (Hem., Reduviidae). Konowia, 1: 218-220.
1922. The American species of *Ploeariola* Reut. (Hem., Reduviidae). Notulae Entomologicae, 2: 49-51, 77-81.
1922. New neotropical Miridae (Hem.). Arkiv för Zoologie, 14: 1-14.
1924. On the Isometopidae (Hem. Het.) of North America. Notulae Entomologicae, 4: 3-9.
1925. On the "annectant bugs" of Messrs. McAtee and Malloch. Bulletin of the Brooklyn Entomological Society, 20: 159-164.

Beyer, A. H.
1921. Garden flea-hopper in alfalfa and its control. United States Department of Agriculture Bulletin, 964: 1-27.

Bibby, F. F.
1961. Notes on miscellaneous insects of Arizona. Journal of Economic Entomology, 54: 324-333.

Bird, R. D. and A. V. Mitchener
1953. Insects of the season 1952 in Manitoba. Canadian Insect Pest Review, 31: 40-52.

Bisabri-Ershadi, B. and L. E. Ehler
1981. Natural biological control of western yellow-striped armyworm in hay alfalfa in northern California. Hilgardia, 49: 1-23.

Blackburn, T.
1889. Notes on the Hemiptera of the Hawaiian Islands. Proceedings of the Linnaean Society of New South Wales, Series 2, 3: 343-354.

Blackman, M. W.
1918. On insect visitors to the blossoms of wild blackberry and wild *Spiraea* -- a study in seasonal distribution. New York State College of Forestry, Syracuse, Technical Publication 10, 28: 119-144.

Blakley, N. R.
1980. Divergence in seed resource use among neotropical milkweed bugs (*Oncopeltus*). Oikos, 35: 8-15.
1981. Life history significance of size-triggered metamorphosis in milkweed bugs (*Oncopeltus*). Ecology, 62: 57-64.

Blakley, N. R. and H. Dingle
1978. Competition, butterflies eliminate milkweed bugs from a Caribbean island. Oecologia. 37: 133-136.

Blakley, N. R. and S. R. Goodner
1978. Size dependent timing of metamorphosis in milkweed bugs (*Oncopeltus*) and its life history implications. Biological Bulletin, 155: 499-510.

Blanchard, E.
1840. Histoire naturelle des insectes. Orthoptères, Neuroptères, Hémiptères, Lép-idop-tères, et Diptères. 3 volumes. P. Dumeril, Paris. (Lygaeidae, volume 3, pages 85-218)
1840. Hemipteres. *In* Laporte, F. L. (editor). Histoire Naturelle des Animaux Articules, Annelides, Crustaces, Arachnides, Myriapodes, et Insectes. P. Dumeril, Paris. 3: 85-218, plates 1-15.
1844. *Cydnides* and *Cydnus*. *In* d'Orbigny, C. Dictionnaire Universel d'Histoire Naturelle, 4: 505.
1852. Orden VII: Hemipteros. *In* Gay, C. (editor). Historia fisica y politica de Chile. Zoologia, 7: 113-320.

Blanford, W. T.
1903. Sokotra [book review]. Nature, 69: 199-201.

Blatchley, W. S.
1924. Some apparently new Heteroptera from Florida. Entomological News, 35: 85-90.
1925. Some additional new species of Heteroptera from the southern United States, with characterization of a new genus. Entomological News, 36: 45-52.
1925. Two changes of names (Heteroptera, Coleoptera). Entomological News, 36: 245.
1926. Heteroptera or True Bugs of Eastern North America, with Especial Reference to the Faunas of Indiana and Florida. Nature Publishing Company, Indianapolis. 1116 pages.
1926. Some new Miridae from the eastern United States. Entomological News, 37: 163-169.
1928. Notes on the Heteroptera of eastern North America with descriptions of new species, I. Journal of the New York Entomological Society, 36: 1-23.

Blatchley, W. S. (continued)
1928. The Mexican chicken bug in Florida. Florida Entomologist, 12(3): 43-44.
1928. Two new Anthocoridae and a new microphysid from Florida (Heteroptera). Entomological News, 39: 85-88.
1929. Two new Heteroptera from southern California (Cydnidae, Nabidae). Entomological News, 40: 74-76.
1930. Blatchleyana. Nature Publishing Company, Indianapolis. 77 pages.
1934. Notes on a collection of Heteroptera taken in winter in the vicinity of Los Angeles, California. Transactions of the American Entomological Society, 60: 1-16.

Blinn, D. W., C. Pinney, and M. W. Sanderson
1982. Nocturnal planktonic behavior of *Ranatra montezuma* Polhemus (Nepidae: Hemiptera) in Montezuma Well, Arizona. Journal of the Kansas Entomological Society, 55: 481-484.

Bliven, B. P.
1954. New Hemiptera from redwood (Miridae, Cicadellidae). Bulletin of the Brooklyn Entomological Society, 49: 109-119.
1955. Studies on Insects of the Redwood Empire. I. New phytophagous Hemiptera from Coast Range Mountains (Pentatomidae, Miridae, Cicadellidae, Psyllidae). Published by author, Eureka, California. Pages 8-14.
1956. New Hemiptera from the western states with illustrations of previously described species and new synonymy in the Psyllidae. Privately published by author, Eureka, California. 27 pages.
1957. Some Californian mirids and leafhoppers, including two new genera and four new species. Occidental Entomologist, l(1): 1-7.
1958. Studies on insects of the Redwood Empire II: New Hemiptera and further notes on the *Colladonus* complex. Occidental Entomologist, l(2): 8-24.
1959. New Pyrrhocoridae and Miridae from the western United States (Hemiptera). Occidental Entomologist, l(3): 25-33.
1960. Studies on insects of the Redwood Empire III: New Hemiptera with notes on others. Occidental Entomologist, l(4): 34-42.
1961. New species of *Irbisia* from California. Occidental Entomologist, l(5): 45-51.
1962. New species of antlike Miridae with notes on others (Hemiptera). Occidental Entomologist, l(6): 52-61.
1962. Contributions to a knowledge of the Scutelleroidea I: On the identity of *Eurygaster minidoka*. Occidental Entomologist, 1(6): 66-67.
1963. New species of *Irbisia* from California II. Occidental Entomologist, l(7): 68-86.
1966. New Miridae from western North America. Occidental Entomologist, l(9): 115-122.
1973. A third paper of Hemiptera associated with Pyrrhocoridae. Occidental Entomologist, 1(10): 123-133.
1973. Studies on insects of the Redwood Empire IV. Miridae. Occidental Entomologist, l(10): 134-142.

Blum, M. S.
1978. Biochemical defenses of insects. Pages 465-513. *In* Rockstein, M. (editor). Biochemistry of Insects. Academic Press, New York. 649 pages.

Bobb, M. L.
1951. Life history of *Ochterus banksi* Barber (Hemiptera: Ochteridae). Bulletin of the Brooklyn Entomological Society, 46: 92-100.
1953. Observations on the life history of *Hesperocorixa interrupta* (Say) (Hemiptera: Corixidae). Virginia Journal of Science, 4: 111-115.
1974. The aquatic and semi-aquatic Hemiptera of Virginia. The insects of Virginia: No. 7. Research Division Bulletin, Virginia Polytechnic Institute and State University, Blacksburg, 87: 1-195.

Böcher, J.
1972. Feeding biology of *Nysius groelandicus* (Zett.) (Heteroptera: Lygaeidae) in Greenland. With a note on oviposition in relation to food-source and dispersal of the species. Meddelelser om Gronland, 191: 1-41, plates 1-5.
1975a. Notes on the reproductive biology and egg diapause in *Nysius groenlandicus* (Zett.) (Heteroptera: Lygaeidae). Videnskabelige Meddelelser fra Dansk Naturhistorisk Forening I Kobenhavn, 138: 21-38.
1975b. Spredningen af den grönlandske frötäge, *Nysius groenlandicus* (Zett.) (Heteroptera: Lygaeidae). (With a summary: Dispersal of *Nysius groenlandicus* (Zett.) in Greenland). Entomologiske Meddelelser, 43: 105-109.
1976. Population studies on *Nysius groelandicus*

(Zett.) (Heteroptera: Lygaeidae) in Greenland with particular references to climatic factors, especially the snow cover. Videnskabelige Meddelelser fra Dansk Naturhistorisk Forening I København, 139: 61-89.
1978. Biology and ecology of the arctic alpine bug *Nysius groenlandicus* (Heteroptera: Lygaeidae) in Greenland. Norwegian Journal of Entomology, 25: 72.

Boheman, C. H.
1852. Entomologiska Anteckningar under en resa i Södra Sverige 1852. Öfversigt af Kongliga Svenska Vetenskaps-Akademiens Förhandlingar, pages 55-210.

Boitard, M.
1827. Manuel d'Histoire Naturelle, Volume 1. Roret, Paris.

Boivin, G.
1983. Bibliography of the pear plant bug, *Lygocoris communis* (Knight) (Hemiptera: Miridae). Bibliographies of the Entomological Society of America, 2: 1-9.

Bolkan, H. A., J. M. Ogawa, R. E. Rice, R. M. Bostock, and J. C. Crane
1984. Leaffooted bug (Hemiptera: Coreidae) and epicarp lesion of pistachio fruits. Journal of Economic Entomology, 77: 1163-1165.

Bonhag, P. F. and J. R. Wick
1953. The functional anatomy of the male and female reproductive systems of the milkweed bug, *Oncopeltus fasciatus* (Dallas) (Heteroptera: Lygaeidae). Journal of Morphology, 3: 177-284.

Borror, D. J. and R. E. White
1970. A Field Guide to the Insects of America North of Mexico. Houghton Mifflin, Boston. 404 pages.

Bouchard, D., J.C. Tourneur, and R. O. Paradis
1982. Le complexe entomophage limitant les populations d'*Aphis pomi* de Geer (Homoptera: Aphididae) dans le sud-ouest du Québec. Données préléminaires. Annales de la Société Entomologique du Québec, 27:80-93.

Bowers, W. S., T. Ohta, J. S. Cleere, and P. A. Marsella
1976. Discovery of insect anti-juvenile hormones in plants. Science, 193: 542-547.

Bradley, G. A. and J. D. Hinks
1968. Ants, aphids, and jack pine in Manitoba. Canadian Entomologist, 100: 40-50.

Bradshaw, G. V. R. and A. Ross
1961. Ectoparasites of Arizona Bats. Journal of the Arizona Academy of Sciences, 1: 109-112.

Brailovsky, H.
1975a. Contribución al estudio de los Hemiptera-Heteroptera de México: VII. Distribución y diagnosis de las especies del género *Melanopleurus* Stål (Lygaeidae-Lygaeinae) y descripción de dos nuevas especies. Anales de Instituto de Biologia Universidad Nacional Autónoma México, 46 (Serie Zoologia 1): 53-62.
1975. Distribucion de las especies de *Narnia* Stal (Coreidae- Coreinae-Anisoscelini) y descripcion de una nueva especie. Revista de la Sociedad Mexicana de Historia Natural, 36: 169-176.
1976. Contribución al estudio de los Hemiptera-Heteroptera de México: X. Una nueva especie del género *Kleidocerys* Stephens (Lygaeidae-Ischnorhynchinae) y datos de la distribución geográfica de las especies mexicanas del género. Anales de Instituto de Biologia Universidad Nacional Autónoma México, 47 (Serie Zoologia 2): 43-48.
1977a. Una nueva especie del género *Neacoryphus* (Hemiptera-Heteroptera- Lygaeidae -Lygaeinae) de Costa Rica. Anales de Instituto de Biologia Universidad Nacional Autónoma México, 48 (Serie Zoologia 1): 93-96.
1977b. Contribución al estudio de los Hemiptera-Heteroptera de México. XII. El género *Neacoryphus* Scudder (Lygaeidae-Lygaeinae) y descripción de tres nuevas especies. Anales de Instituto de Biologia Universidad Nacional Autónoma México, 48 (Serie Zoologia 1): 97-122.
1977c. Una nueva especie del género *Melanopleurus* Stâl (Hemiptera-Heteroptera-Lygaeidae-Lygaeinae) de California, Estados Unidos de Norteamérica. Anales de Instituto de Biologia Universidad Nacional Autónoma México, 48 (Serie Zoologia 1): 129-132.
1978a. Un nuevo género y una nueva especie de Lygaeinae (Hemiptera-Heteroptera-Lygaeidae) del Peru. Anales de Instituto de Biologia Universidad Nacional Autónoma México, 49 (Serie Zoologia 1): 115-122.
1978b. Estudio del género *Lygaeus* Fabricius 1794, del nuevo mundo, con descripción

Brailovsky, H. (continued)
de cinco nuevas especies (Hemiptera-Heteroptera-Lygaeidae-Lygaeinae). Anales de Instituto de Biologia Universidad Nacional Autónoma México, 49 (Serie Zoologia 1): 123-166.

1978c. Contribución al estudio de los Hemiptera-Heteroptera de México: XIV. Una nueva especie de *Neosuris* Barber (Lygaeidae: Udeocorini). Anales de Instituto de Biologia Universidad Nacional Autonóma México, 49 (Serie Zoologia 1): 167-170.

1979a. Seis nuevas especies del género *Melanopleurus* Stål (Hemiptera-Heteroptera-Lygaeidae-Lygaeinae). Annales de Instituto de Biologia Universidad Nacional Autónoma México, 50 (Serie Zoologia 1): 193-204.

1979b. Revision del genero *Craspeduchus* Stål, con descripción de dos nuevas especies (Hemiptera-Heteroptera-Lygaeidae-Lygaeinae). Anales de Instituto de Biologia Universidad Nacional Autónoma México, 50 (Serie Zoologia 1): 205-226.

1981. Descripción de dos nuevas especies de la tribu Myodochini (Heteroptera-Rhyparochrominae) del continente américano. Anales de Instituto de Biologia Universidad Nacional Autónoma México, 51, (1980) Serie Zoologia (1): 217-226.

1982. Revision del complejo *Ochrimnus*, con descripcion de nuevos especies y nuevos generos (Hemiptera, Heteroptera, Lygaeidae, Lygaeinae). Folia Entomológica Méxicana, 51: 1-163.

1985. Revision del genero *Anasa* Amyot-Serville (Hemiptera-Heteroptera-Coreidae-Coreinae-Coreini). Monografias del Instituto de Biologia Nacional Autónoma México, 2: 1-266.

Brailovsky, H. and E. Barrera

1979. Contribución al estudio de los Hemiptera-Heteroptera de Mexico. XVI. La subfamilia Oxycareninae (Lygaeidae), con descripcion de una nueva especie. Folia Entomólogica Méxicana, 41: 81-93.

1981. Hemiptera-Heteroptera de México: XIX. Revisión de la tribu Drymini Stål (Lygaeidae-Rhyparochrominae) y descriptión de seis nuevas especies. Anales de Instituto de Biologia Universidad Nacional Autónoma México, 51, (1980) Serie Zoologia (1): 169-204.

1984. Una nueva especie del género *Ereminellus* Harrington y algunas observaciones acerca de Myodochini américanos (Hemiptera-Heteroptera-Lygaeidae-Rhyparochrominae). Anales de Instituto de Biologia Universidad Nacional Autónoma México , 54 (1983), Serie Zoologia (1): 53-62.

Brailovsky, H. and F. Soria

1981. Contribución al estudio de los Hemiptera-Heteroptera de México: XVIII. Revisión de la tribu Harmostini Stål (Rhopalidae) y descripción de una nueva especie. Anales de Instituto de Biologia Universidad Nacional Autónoma de México, 51(1980), Serie Zoologia (1): 123-168.

Braimah, S. A., L. A. Kelton, and R. K. Stewart

1982. The predaceous and phytophagous plant bugs (Heteroptera: Miridae) found on apple trees in Quebéc. Naturaliste Canadien, 109: 153-180.

Braman, S. K., K. E. Godfrey, and K. V. Yeargan

1985. Rates of development of a Kentucky population of *Geocoris uliginosus*. Journal of Agricultural Entomology, 2: 185-191.

Breddin, G.

1896. Javanische Zuckerrohrschädlinge aus der Familie der Rhynchoten. Deutsche Entomologische Zeitschrift, 1: 105-110.

1898. Studia hemipterologica. IV. Jahresbericht und Abhandlungen des Naturwissenschaftlichen Vereins in Magdeburg, pages 149-163.

1907. Berytiden und Myodochiden von Ceylon aus der Sammelausbeute von Dr. W. Horn (Rhynch. Het.). Deutsche Entomologische Zeitschrift, 1907: 34-47.

Brimley, C. S.

1938. The insects of North Carolina, being a list of the insects of North Carolina and their close relatives. North Carolina Department of Agriculture, Division of Entomology, Raleigh. 560 pages.

Brittain, W. H.

1919. Notes on *Lygus campestris* Linn. in Nova Scotia. Proceedings of the Nova Scotia Entomological Society for 1918, 4: 76-81, plate V.

Britton, W. E.

1920. Check-list of the insects of Connecticut. Connecticut Geological and Natural History Survey Bulletin, 31: 1-397.

Brodie, W. and J. E. White
1883. Checklist of insects of the Dominion of Canada. C. Blackett Robinson Company, Toronto. 67 pages.

Brooks, A. R.
1959. A new *Palmacorixa* from western Canada (Hemiptera, Corixidae). Proceedings of the Entomological Society of Washington, 61: 179-181.

Brooks, A. R. and L. A. Kelton
1967. Aquatic and semiaquatic Heteroptera of Alberta, Saskatchewan, and Manitoba (Hemiptera). Memoirs of the Entomological Society of Canada, 51: 1-92.

Brown, A. W. A.
1934. A contribution to the insect fauna of Timagami. Canadian Entomologist, 66: 220-231.

Browne, W. N.
1916. A comparative study of the chromosomes of six species of *Notonecta*. Journal of Morphology, 27: 119-162, plates 1-7.

Brullé, A.
1835. Histoire naturelle des insects (Coleopteres, Orthopteres et Hemipteres), traitant de leur organisation et de leurs moeurs en general part V. Audouin; et comprenant leur classification et la description des especes part Brulle. 9: 1-415, 56 plates.

Bruner, S. C.
1934. Notes on Cuban Dicyphinae (Hemiptera, Miridae). Memorias de la Sociedad Poey, Habana, 8: 35-49.

Bruton, B. D., J. A. Reinert, and R. W. Toler
1979. Effects of the southern chinch bug, *Blissus insularis*, and the St. Augustine decline strain of Panicum mosaic virus on seventeen accessions and two cultivars of St. Augustine grass. Phytopathology, 69: 525-526.

Bruton, B. D., R. W. Toler, and J. A. Reinert
1983. Combined resistance in St. Augustinegrass to the southern chinch bug and the St. Augustine decline strain of panicum mosaic virus. Plant Diseases, 67: 171-172.

Bueno, J. R. de la Torre [see Torre-Bueno, J. R. de la]

Bull, D. L.
1973. Effects of juvenile hormone analogues on certain species of insects associated with cotton. Folia Entomólogica Mexicana, 25/ 26: 95-96.

Bull, D. L., R. L. Ridgway, W. E. Buxkemper, M. Schwarz, T. P. McGovern, and R. Sarmiento
1973. Effects of synthetic juvenile hormone analogues on certain injurious and beneficial arthropods associated with cotton. Journal of Economic Entomology, 66: 623-626.

Burdick, D. J.
1968. Distributional information on *Hydrometra martini* Kirkaldy. Pan-Pacific Entomolog-ist, 44: 81.

Burgess, L.
1977. *Geocoris bullatus*, an occasional predator on flea beetles (Hemiptera: Lygaeidae). Canadian Entomologist, 109: 1519-1520.
1982. Predation on adults of the flea beetle *Phyllotreta cruciferae* by the western damsel bug, *Nabis alternatus* (Hemiptera: Nabidae). Canadian Entomologist, 114: 763-764.

Burgess, L., J. Dueck, and D. L. McKenzie
1983. Insect vectors of the yeast *Nematospora coryli* in mustard *Brassica juncea* crops in southern Saskatchewan. Canadian Entomologist, 115: 25-30.

Burgess, L. and H. H. Weeger
1986. A method for rearing *Nysius ericae* (Hemiptera: Lygaeidae), the false chinch bug. Canadian Entomologist, 118: 1059-1061.

Burmeister, H. C. C.
1835-1839. Handbuch der Entomologie. Tome 2. T. Enslin, Berlin. 1835, Abtheil 1: i-xii, 1-400; 1838, Abtheil 2 (Erste Halfte): 397-756 [page numbers overlap with Abtheil 1]; 1839, Abtheil 2 (Zweite Halfte): 757-1050.

Buschman, L. L., H. N. Pitre, and H. F. Hodges
1984. Soybean cultural practices: Effects on populations of geocorids, nabids, and other soybean arthropods. Environmental Entomology, 13: 305-317.

Buschman, L. L., W. H. Whitcomb, R. C. Hemenway, D. L. Mays, Nguyen Ru, N. C. Leppla, and B. J. Smittle
1977. Predators of velvetbean caterpillar eggs in Florida soybeans. Environmental Entomology, 6: 403-407.

Butler, E. A.
1923. A Biology of the British Hemiptera Heteroptera. H. F. & G. Witherby, London. viii + 682 pages, 5 plates.

Butler, G. D., Jr.
1965. *Spanogonicus albofasciatus* as an insect and mite predator (Hemiptera: Miridae). Journal of the Kansas Entomological Society, 38: 70-75.
1966a. Insect predators of bollworm eggs. Progressive Agriculture in Arizona, 18: 26-27.
1966b. Development of several predaceous Hemiptera in relation to temperature. Journal of Economic Entomology, 59: 1306-1307.
1967. Big-eyed bugs as predators of lygus bugs. Progressive Agriculture in Arizona, 19: 13.

Butler, G. D., Jr. and A. Stoner
1965. The biology of *Spanogonicus albofasciatus*. Journal of Economic Entomology, 58: 664-665.

Buxton, G. M., D. B. Thomas and R. C. Froeschner
1983. Revision of the species of the *sayi*-group of Chlorochroa Stal (Hemiptera: Pentatomidae). California Department of Food and Agriculture, Occasional Papers in Entomology, 29: 1-25.

Byerly, K. F., A. P. Gutierrez, R. E. Jones, and R. F. Luck
1978. A comparison of sampling methods for some arthropod populations in cotton. Hilgardia, 46: 257-282.

Byers, G. W.
1973. A mating aggregation of *Nysius raphanus* (Hemiptera: Lygaeidae). Journal of the Kansas Entomological Society, 46: 281-282.

Calabrese, D. M.
1974. Population and subspecific variation in *Gerris remigis* Say. Entomological News, 85: 27-28.
1974. Keys to the adults and nymphs of the species of *Gerris* Fabricius occurring in Connecticut. Memoirs of the Connecticut Entomological Society. Pages 227-266.
1980. Zoogeography and cladistic analysis of the Gerridae (Hemiptera: Heteroptera). Miscellaneous Publications of the Entomological Society of America, 11: 1-119.

Caldwell, R. L.
1968. The effects of reproduction on flight in a migrating and a non-migrating species of milkweed bug. American Zoologist, 8: 743.
1970. A comparison of the dispersal strategies of two milkweed bugs, *Oncopeltus fasciatus* and *Lygaeus kalmii*. Dissertation Abstracts International (B), 30: 3430.
1974. A comparison of the migratory strategies of two milkweed bugs, *Oncopeltus fasciatus* and *Lygaeus kalmii*. Pages 304-316. *In* Browne, L. B. (editor). Experimental Analysis of Insect Behavior. Springer-Verlag, New York.

Caldwell, R. L. and J. P. Hegmann
1969. Heritability of flight duration in the milkweed bug *Lygaeus kalmii*. Nature, 223 (5201): 91-92.

Callahan, P. S., A. N. Sparks, J. W. Snow, and W. W. Copeland
1972. Corn earworm moth: Vertical distribution in nocturnal flight. Environmental Entomology, 1: 497-503.

Capriles, J. Maldonado [see Maldonado Capriles, J.]

Carayon, J.
1958. études sur les Hémiptères Cimicoidea. l.--Position des genres *Bilia, Biliola, Bilianella* et *Wollastoniella* dans une tribu nouvelle (Oriini) des Anthocoridae; différences entre ces derniers et les Miridae Isometopinae (Heteroptera). Mémoires du Muséum National d'Histoire Naturelle, Paris., Serie A, Zoologie, 16: 141-172.
1961. Valeur systématique des voies ectodermiques de l'appareil génital femelle chez les Hémiptères Nabidae. Bulletin du Muséum National d'Histoire Naturelle, 33: 183-196.
1972. Caractères systématiques et classification des Anthocoridae [Hemipt.]. Annales de la Société Entomologique France (New Series), 8: 309-349.
1972. Le genre *Xylocoris*: Subdivision et espèces nouvelles [Hem. Anthocoridae]. Annales de la Société Entomologique France (New Series), 8: 579-606.

Carayon, J., R. L. Usinger, and P. Wygodzinsky
1958. Notes on the higher classification of the Reduviidae, with the description of a new tribe of the Phymatinae (Hemiptera-Heteroptera). Revue de Zoologie et de Botanique Africaines, 57: 256-281.

Carillo, J. L.
1967a. A mechanism for egg dispersal in *Nysius tenellus* Barber (Hemiptera: Lygaeidae). Pan-Pacific Entomologist, 43: 80.
1967b. Larval stages in *Solierella blaisdelli* (Bridwell) and *S. peckhami* (Ashmead). Pan-Pacific Entomologist, 43: 201-203.

Carillo, J. L. and L. E. Cattagirone
1970. Observations on the biology of *Solierella peckhami, S. blaisdelli* (Sphecidae), and two species of Chrysididae (Hymenoptera). Annals of the Entomological Society of America, 63: 672-681.

Carlson, E. C.
1959. The effect of lygus and hyaline grass bugs on lettuce seed production. Journal of Economic Entomology, 52: 242-244.

Carlson, R. W.
1979. Family Ichneumonidae. Pages 315-740. *In* Krombein, K. V., P. D. Hurd, Jr., D. R. Smith, and B. D. Burks (editors). Catalog of Hymenoptera in America North of Mexico. Volume 1. Smithsonian Institution Press, Washington, D.C.

Carpenter, G. H.
1898. A new marine hydrometrid. Entomologist's Monthly Magazine, 34: 78-81.
1898. *Trochopus* and *Rhagovelia*. Entomologist's Monthly Magazine, 34: 109-112.

Carroll, D. P. and S. C. Hoyt
1984. Natural enemies and their effects on apple aphid, *Aphis pomi* DeGeer (Homoptera: Aphididae), colonies on young apple trees in central Washington. Environmental Entomology, 13: 469-481.

Carvalho, J. C. M.
1945. Mirídeos neotropicais: XVI- Revisão do gênero *Garganus* Stäl (Hemiptera). Boletim do Museu Nacional (Nova Serie) (Zoologia), 45: 1-15.
1947. Mirídeos neotropicais, XXVII: Gêneros *Porpomiris* Berg, *Lampethusa* Distant, *Cyrtopeltis* Fieber e *Dicyphus* Fieber (Hemiptera). Boletim do Museu Nacional (Nova Serie)(Zoologia), 77: 1-23, 19 plates.
1951. Mirídeos neotropicais, XXXIII: Espécies da coleção do American Museum of Natural History, inclusive a descrição de uma espécie Neártica (Hemiptera). Arquivos do Museu Nacional, Rio de Janeiro, 42: 153-157.
1952. On the major classification of the Miridae (Hemiptera). (With keys to subfamilies and tribes and a catalogue of the world genera). Anais da Academia Brasileira de Ciências, 24: 31-110.
1952. Neotropical Miridae, L: On the present generic assignment of the species in the Biologia Centrali Americana (Hemiptera). Boletim do Museu Nacional, (Nova Serie)(Zoologia)118: 1-17.
1952. Neotropical Miridae. LI: On the present generic assignment of the species in "Bidrag till Rio de Janeiro-Traktens Hemipter-Fauna" (Hemiptera). Revista Brasileira de Biologia, 12: 215-217.
1952. Neotropical Miridae. XLVII.-- Notes on the Blanchard, Spinola and Signoret types in the Paris Museum. Revue Française d'Entomologique, 19: 181-188.
1953. A new species of *Bothynotus* from Florida (Hemiptera: Miridae). Florida Entomologist, 36: 161-163.
1954. Neotropical Miridae, LXXI: Genus *Cyrtocapsus* Reuter with descriptions of four new species (Hemiptera). Bulletin of the Brooklyn Entomological Society, 49: 12-17.
1954. Neotropical Miridae, LXVII: Genus *Ranzovius* Distant, predacious on eggs of *Theridion* (Araneida) in Trinidad (Hemiptera). Annals and Magazine of Natural History, (12)7: 92-96.
1954. Neotropical Miridae, LXIX: A remarkable new genus of Phylini (Hemiptera). Entomological News, 65: 123-126.
1954. Neotropical Miridae, LXXVII: Miscellaneous observations in some European museums (Hemiptera). Anais da Academia Brasileira de Ciências, 26: 423-427.
1955. Analecta miridologica: Miscellaneous observations in some American museums and bibliography (Hemiptera). Revista de Chilena Entomologia, 4: 221-227.
1955. Analecta miridologica: Einige nomenklat-orische Berichtigungen für die paläarktische Fauna (Hemiptera: Heteroptera). Beiträge zur Entomologie, 5: 333-336.
1957-1960. Catalogue of the Miridae of the world. Arquivos do Museu Nacional, Rio de Janeiro. Part I. Cylapinae, Deraeocorinae, Bryocorinae, 44(1): 1-158 (1957); Part II. Phylinae, 45(2): 1-216 (1958); Part III. Orthotylinae, 47(3): 1-161 (1958); Part IV. Mirinae, 48(4): 1-384 (1959); Part V. Bibliography & Index, 51(5): 1-194 (1960).
1975. Neotropical Miridae, CLXXXVIII: On the genera *Dolichomiris* Reuter, *Megaloceroea* Fieber, *Stenodema* Laporte, *Trigonotylisca* n. gen. and *Trigonotylus* Fieber (Hemiptera). Revista Brasileira de Biologia, 35: 121-140.

Carvalho, J. C. M. (continued)
1975. Mirídeos neotropicais, CXCII: Descrição de dois subgêneros e espécies novas do gênero *Notholopus* Bergroth (Hemiptera). Revista Brasileira de Biologia, 35: 369-378.
1976. Analecta miridologica: Concerning changes of taxonomic position of some genera and species (Hemiptera). Revista Brasileira de Biologia, 36: 49-59.

Carvalho, J. C. M. and C. J. Drake
1943. A new genus and two new species of neotropical Dicyphinae (Hemiptera, Miridae). Revista Brasileira de Biologia, 3: 87-89.

Carvalho, J. C. M. and A. V. Fontes
1973. Mirídeos neotropicais, CLI: Estudos sobre o gênero *Prepops* Reuter-VII (Hemiptera). Revista Brasileira de Biologia, 33: 539-546.
1981. Mirídeos neotropicais CCXXV: Revisão do gênero *Collaria* Provancher no continente Americano (Hemiptera). Experientiae, 27: 11-46.
1983. Mirídeos neotropicais, CCXXXIII: Gênero *Dagbertus* Distant - Descrições de espécies e revisão das que ocorrem na região (Hemiptera). Revista Brasileira de Biologia, 43: 157-176.

Carvalho, J. C. M., A. V. Fontes, and T. J. Henry
1983. Taxonomy of the South American species of *Ceratocapsus*, with descriptions of 45 new species (Hemiptera: Miridae). United States Department of Agriculture Technical Bulletin, 1676: 1-58.

Carvalho, J. C. M. and I. P. Gomes
1980. Mirídeos neotropicais, CCXVIII; Revisão do gênero «*Derophthalma*» Berg, 1883 - (Hemiptera). Experientiae, 26: 93-146.

Carvalho, J. C. M. and R. F. Hussey
1954. On a collection of Miridae (Hemiptera) from Paraguay, with descriptions of three new species. Occasional Papers of the Museum of Zoology, University of Michigan, 552: 1-1l.

Carvalho, J. C. M. and J. Jurberg
1974. Neotropical Miridae, CLXXX: On the *Horcias* complex (Hemiptera). Revista Brasileira de Biologia, 34: 49-65.

Carvalho, J. C. M., H. H. Knight, and R. L. Usinger
1961. *Lygus* Hahn, 1833 (Insecta, Hemiptera); proposed designation under the plenary powers of a type-species in harmony with accustomed usage. Z.N.(S.) 1062. Bulletin of Zoological Nomenclature, 18: 281-284.

Carvalho, J. C. M. and J. C. Schaffner
1974. Neotropical Miridae, CLIII: *Sixeonotopsis*, new genus and other new or little known Bryocorini (Hemiptera). Revista Brasileira de Biologia, 33 (suplementum): 11-16 (1973).
1974. Neotropical Miridae, CLXI: *Apachemiris[,] Jornandinus[,] and Oaxacaenus*, new genera of Orthotylini (Hemiptera). Revista Brasileira de Biologia, 33 (suplementum): 65-73 (1973).
1975. Neotropical Miridae, CXCI: Descriptions of two new genera and new species of Herdoniini with a note on synonymy (Hemiptera). Revista de Brasileira Biologia, 35: 349-358.
1976. Neotropical Miridae, CXCVIII: Review of the genera *Callichila* Reuter and *Platytylus* Fieber (Hemiptera). Revista Brasileira de Biologia, 35: 705-736 (1975).

Carvalho, J. C. M. and T. R. E. Southwood
1955. Revisão do complexo *Cyrtorhinus* Fieber-*Mecomma* Fieber (Hemiptera-Heteroptera, Miridae). Boletim do Museu Paraense Emilio Goeldi, 11: 7-72, 25 figures.

Carvalho, J. C. M. and R. L. Usinger
1957. A new genus and two new species of myrmecomorphic Miridae from North America (Hemiptera). Wasmann Journal of Biology, 15: 1-13.

Carvalho, J. C. M. and E. Wagner
1957. A world revision of the genus *Trigonotylus* Fieber (Hemiptera-Heteroptera, Miridae). Arquivos do Museu Nacional, Rio de Janeiro, 43: 121-155.

Caudell, A. N.
1900. A new species of *Sinea*. Canadian Entomologist, 32: 67-68.
1901. The genus *Sinea* of Amyot & Serville. Journal of the New York Entomological Society, 9: 1-11, plates 1-2.

Champion, G. C.
1897-1901. Insecta: Rhynchota (Hemiptera-Heteroptera). Volume II. *In* Goodwin and Salvin (editors). Biologia Centrali-Americana. London. xvi + 416 pages, 22 plates. [1897: 1-32; 1898: 33-192; 1899: 193-304; 1900: 305-344; 1901: i-xvi + 345-416].

1898. Notes on American and other Tingitidae, with descriptions of two new genera and four species. Transactions of the Entomological Society of London, 1: 55-64, 2 pls.
1916. [Taxonomic notes]. Entomologist's Monthly Magazine, 52: 207-208.

Champlain, R. A. and L. L. Sholdt
1966. Rearing *Geocoris punctipes*, a lygus bug predator, in the laboratory. Journal of Economic Entomology, 59: 1301.
1967a. Life history of *Geocoris punctipes* (Hemiptera: Lygaeidae) in the laboratory. Annals of the Entomological Society of America, 60: 881-883.
1967b. Temperature range for development of immature stages of *Geocoris punctipes* (Hemiptera: Lygaeidae). Annals of the Entomological Society of America, 60: 883-885.

Chaplin, S. B. and S. J. Chaplin
1981. Comparative growth energetics of a migratory and non-migratory insect: the milkweed bugs. Journal of Animal Ecology, 50: 407-420.

Chaplin, S. J.
1973. Reproductive isolation between two sympatric species of *Oncopeltus* (Hemiptera: Lygaeidae) in the tropics. Annals of the Entomological Society of America, 66: 997-1000.

Chapman, H. C.
1959. Distribution and ecological records for some aquatic and semi-aquatic Heteroptera of New Jersey. Bulletin of the Brooklyn Entomological Society, 54: 8-12.
1962. The Saldidae of Nevada (Hemiptera). Pan-Pacific Entomologist, 38: 147-159.

Chapman, H. C. and Polhemus, J. T.
1965. A new shore bug from Oregon (Hemiptera: Saldidae). Journal of the Kansas Entomological Society, 38: 359-361.

Cheng, L. and C. H. Fernando
1971. Life history and biology of the riffle bug *Rhagovelia obesa* Uhler in southern Ontario. Canadian Journal of Zoology, 49: 435-442.

Chenu, J. C.
1859. Hemiptera. *In* Encyclopedie d'Histoire Naturelle, 8: 182-219.

Childs, L.
1914. Insect notes. Monthly Bulletin of the California State Commission of Horticulture, 3: 220.

China, W. E.
1924. The Hemiptera-Heteroptera of Rodriquez, together with the description of a new species of *Cicada* from that island. Annals and Magazine of Natural History, Series 9, 14: 427-453.
1938. Notes on the nomenclature of British Corixidae. Entomologist's Monthly Magazine, 74: 34-39.
1941. Genotype fixations in Hemiptera Heteroptera. Proceedings of the Royal Entomological Society of London, Series B, 10: 130.
1941. A new subgeneric name for *Lygus* Reuter 1875 nec Hahn 1833 (Hemipt.-Heteropt.). Proceedings of the Royal Entomological Society of London, Series B, 10: 60.
1943. The Generic Names of British Insects. Pt. 8. The Generic Names of the British Hemiptera-Heteroptera, with a Check List of the British Species. Royal Entomological Society of London, pages 211-342.
1946. Opinion 705. *Blissus* Burmeister, 1835 (Insecta, Hemiptera): Added to the official list of generic names. Bulletin of Zoological Nomenclature, 21: 198-201.
1947. On the status of the name *Salda* Fabricius, 1803 (Class Insecta, Order Hemiptera). Bulletin of Zoological Nomenclature, 1: 276.
1954. Notes on the nomenclature of the Pyrrhocoridae (Hemiptera Heteroptera). Entomologist's Monthly Magazine, 90: 188-189.

China, W. E. and J. C. M. Carvalho
1952. The *"Cyrtopeltis-Engytatus"* complex (Hemiptera, Miridae, Dicyphini). Annals and Magazine of Natural History, Series 12, 5: 158-166.

China, W. E. and N. C. E. Miller
1955. Check-list of family and subfamily names in Hemiptera-Heteroptera. Annals and Magazine of Natural History, Series 12, 8: 257-267.
1959. Check-list and keys to the families and subfamilies of the Hemiptera Heteroptera. Bulletin of the British Museum (Natural History) Entomology, Series 8, 1: 1-45.

China, W. E. and J. G. Myers
1929. A reconsideration of the classification of the cimicoid families (Heteroptera),

China, W. E. and J. G. Myers (continued)
with the description of two new spider-web bugs. Annals and Magazine of Natural History, Series 10, 3: 97-125.

China, W. E. and R. L. Usinger
1949. A new genus of Hydrometridae from the Belgian Congo, with a new subfamily and a key to the genera. Revue de Zoologie et de Botanique Africans, 41: 314-319.
1949. Classification of the Veliidae (Hemiptera) with a new genus from South Africa. Annals and Magazine of Natural History, 2: 343-354.

Chiravathanapong, S. N. and H. N. Pitre
1980. Effects of *Heliothis virescens* larval size on predation by *Geocoris punctipes*. Florida Entomologist, 63: 146-151.

Chittenden, F. H.
1898. The European bat bug in America. United States Department of Agriculture, Division of Entomology, Technical Bulletin (new series), 18: 97.

Choban, R. G. and A. P. Gupta
1972. Meiosis and early embryology of *Blissus leucopterus hirtus* Montandon (Heteroptera: Lygaeidae). International Journal of Insect Morphology and Embryology, 1: 301-314.

Chopra, N. P.
1967. The higher classification of the family Rhopalidae (Hemiptera). Transactions of the Royal Entomological Society of London, 119: 363-399.
1968. A revision of the genus *Arhyssus* Stål. Annals of the Entomological Society of America, 61: 629-655.
1973. A revision of the genus *Niesthrea* Spinola (Rhopalidae: Hemiptera). Journal of Natural History, 7: 441-459.

Chow, T., G. E. Long, and G. Tamaki
1983. Effects of temperature and hunger on the functional response of *Geocoris bullatus* (Say) (Hemiptera: Lygaeidae) to *Lygus* spp. (Hemiptera: Miridae) density. Environmental Entomology, 12: 1332-1338.

Clancy, C. A. and H. D. Pierce.
1966. Natural enemies of some lygus bugs. Journal of Economic Entomology, 59: 853-858.

Clancy, D. W.
1946. Natural enemies of some Arizona cotton insects. Journal of Economic Entomology, 39: 326-327.

Clausen, C. P.
1940. Entomophagous Insects. New York & London. Pages i-ix, 1-688.

Cobben, R. H.
1959. Notes on the classification of Saldidae with the description of a new species from Spain. Zoologische Mededelingen uitgegeven door het Rijksmuseum van Natuurlijke Historie te Leiden, 36: 303-316.
1960. The Heteroptera of the Netherlands Antilles - III. Saldidae (Shore Bugs). *In* Studies on the fauna of Curaçāo and other Caribbean Islands, 11(52): 44-61.
1965. A new shore-bug from Death Valley, California (Heteroptera: Saldidae). Pan-Pacific Entomologist, 41: 180-185.
1968. Evolutionary trends in Heteroptera. Part I. Eggs, architecture of the shell, gross embryology and eclosion. Mededeling no. 151 of the Laboratory of Entomology of the Agricultural University, Wageningen, Nederland. Centre for Agricultural Publishing and Documentation, Wageningen. 475 pages.
1978. Evolutionary trends in Heteroptera. Part II. Mouthpart-structures and feeding strategies. Mededelingen Landbouwhogeschool Wageningen 78-5. J. Veenman & Zonen B. V., Wageningen, Nederland. 407 pages.
1985. Additions to the Eurasian saldid fauna, with a description of fourteen new species (Heteroptera, Saldidae). Tijdschrift voor Entomologie, 128: 215-270.

Cockerell, T. D. A.
1893. The entomology of the mid-alpine zone of Custer County, Colorado. Transactions of the American Entomological Society, 20: 305-370.
1900. New insects from Arizona, and a new bee from Mexico. Entomologist, 33: 61-66.
1900. *Macrocephalus arizonicus* = *uhleri*. Entomologist, 33: 201.
1903. New bees from southern California and other records. Bulletin of the Southern California Academy of Sciences, 2: 84-85.
1910. Some insects from Steamboat Springs, Colo.--II. Canadian Entomologist, 42: 366-370.
1915. Sunflower insects. Canadian Entomologist, 47: 280-282.

Cohen, A. C.
1981. An artificial diet for *Geocoris punctipes* (Say) (Hemiptera, Lygaeidae). Southwestern Entomologist, 6: 109-113.
1982. Water and temperature relations of two hemipteran members of a predator-prey complex. Environmental Entomology, 11: 715-719.
1983. Improved method of encapsulating artificial diet for rearing predators of harmful insects. Journal of Economic Entomology, 76: 957-959.
1984. Food consumption, food utilization, and metabolic rates of *Geocoris punctipes* [Het.: Lygaeidae] fed *Heliothis virescens* [Lep.: Noctuidae] eggs. Entomophaga, 29: 361-367.
1985a. Metabolic rates of two hemipteran members of a predator-prey complex. Comparative Biochemistry and Physiology A: Comparative Physiology, 81: 833-836.
1985b. Simple method for rearing the insect predator *Geocoris punctipes* (Heteroptera: Lygaeidae) on a meat diet. Journal of Economic Entomology, 78: 1173-1175.

Cohen, A. C. and J. W. Debolt
1983. Rearing *Geocoris punctipes* on insect eggs. Southwestern Entomologist, 8: 61-64.

Cohen, A. C. and N. M. Urias
1986. Meat-based artificial diets for *Geocoris punctipes*. Southwestern Entomologist, 11: 171-176.

Comstock, J. H.
1880. Report upon cotton insects. United States Department of Agriculture, Division of Entomology. 511 pages, 3 plates (1879). [1880 publication date given by Comstock, 1881, U.S. Dept. Agr. Ent. Rept. for 1880, p. 275].
1880. The hawthorn *Tingis* (*Corythucha arcuata*, Say, var.), order Hemiptera; family Tingidae. United States Department of Agriculture, Report of the Entomologist for 1879, pages 221-222.

Condit, B. P. and J. R. Cate
1982. Determination of host range in relation to systematics for *Peristenus stygicus* [Hym.: Braconidae], a parasitoid of Miridae. Entomophaga, 27: 203-210.

Constantz, G.
1974. The mating behavior of a creeping water bug, *Ambrysus occidentalis*. American Midland Naturalist, 92: 234-239.

Cook, A. J.
1891. Kerosene emulsion and its uses. Bulletin of the Michigan Agricultural College Experiment Station, 76: 1-16.

Cooper, J. C.
1870. The cliff swallows. Ornithology, 1: 104-106.

Cooper, K. W.
1967. *Picromerus bidens* (Linn.), a beneficial, predatory European bug discovered in Vermont (Heteroptera: Pentatomidae). Entomological News, 78: 36-40.

Cory, E. N. and P. A. McConnell
1927. The phlox plant bug. University of Maryland Agricultural Experiment Station Bulletin, 292: 15-22.

Coscaron, M. D. C.
1983. Revision del genero *Rasahus* (Insecta, Heteroptera, Reduviidae). Revista del Museo de La Plata, new series, 13: 75-138.
1983. Nuevas citas de distribucion geografica para la subfamilia Peiratinae (Insecta, Heteroptera, Reduviidae). Revista de la Sociedad Entomologica Argentina, 42: 369-382.

Cosper, R. D., M. J. Gaylor, and J. C. Williams
1983. Intraplant distribution of three insect predators on cotton, and seasonal effects of their distribution on vacuum sampler efficiency. Environmental Entomology, 12: 1568-1571.

Costa, A.
1842. Ragguaglio delle specie iu interessanti di Emitteri Eterotteri raccolti in Sicilia, e descrizione di alcuni nuove specie dei contorni di Palermo. Pages 129-147. *In* Esercitazioni accademiche degli aspiranti naturalisti diretti dal Dotter O.-G. Costa, t. 2, part 2. Napoli (1840).
1843. Saggio d'una monografia delle speciede l genere *Ophthalmicus* (Emittere Eterotteri) indigene al regno di Napoli conhote su talune altre di Europa. Annali dell Accademia degli Aspiranti Naturalisti Napoli., 1: 293-316.
1844-1862. Cimicum regni Neapolitani centuria. Atti del Reale Istituto d'Incorra giamento alle Scienze Naturali, parts 1-5. [1844 (separate), part 1, 7: 143-216, plate (Figs. 1-12) (Centuria prima, journal 1847); 1847 (separate), part 2-3, 7: 239-279, plates 1-2, 7: 365-405, plates 3-4 (Centuria secunda, journal 1847); 1853

Costa, A. (continued)
(separate), part 4, 8: 225-299 (Centuria tertia et quartoe, journal 1855); 1862 (separate) part 5, 10: 329-367 (journal 1863).] [Dates for parts follow Kerzhner, 1974.]
1864. Descrizione di taluni insetti stranieri all'Europa. Annuario del Museo Zoologico dell Regia Universita di Napoli, 2: 139-153, plates I-II.

Costa Lima, A. M. da
1935. Genero *Microtomus* Illiger, 1807 (Reduviidae: Microtominae). Annales da Academia Brasileira de Sciencias, 7: 315-322, 1 plate.
1936. Terceiro catalogo dos insectos que vivem nas plantas do Brasil. Ministerio da Agricultura Departamento Nacional da Producção Vegetal Escola Nacional de Agronomia, Directoria de Estatistica da Producção, Rio de Janeiro. 460 pages.

Costa Lima, A. M. da, C. A. Campos Seabra, and C. R. Hathaway
1951. Estudo dos Apiêmeros (Hemiptera: Reduviidae). Memórias do Instituto Oswaldo Cruz, 49: 273-442.

Coulianos, C. C. and O. Kugelberg
1973. A simple method for rearing terrestrial Heteroptera, with special reference to seed-bugs (Het., Lygaeidae). Entomologica Scandinavica, 4: 105-110.

Craig, C. H.
1963. The alfalfa plant bug, *Adelphocoris lineolatus* (Goeze) in northern Saskatchewan. Canadian Entomologist, 95: 6-13.

Crevecoeur, F. F.
1905. Additions to the list of the hemipterous fauna of Kansas. Transactions of the Kansas Academy of Science, 19: 232-237.

Crocker, R. L. and C. L. Simpson
1981. Pesticide screening test for the southern chinch bug. Journal of Ecomomic Entomology, 74: 730-731.

Crocker, R. L., R. W. Toler, and C. L. Simpson
1982. Bioassay of St. Augustinegrass lines for resistance to southern chinch bug (Hemiptera: Lygaeidae) and to St. Augustine decline virus. Journal of Economic Entomology, 75: 515-516.

Crocker, R. L. and W. H. Whitcomb
1980. Feeding niches of the big-eyed bugs *Geocoris bullatus*, *Geocoris punctipes*, and *Geocoris uliginosus* (Hemiptera: Heteroptera: Lygaeidae: Geocorinae). Environmental Entomology, 9: 508-513.

Crocker, R. L., W. H. Whitcomb, and R. M. May
1975. Effects of sex, developmental stage, and temperature on predation by *Geocoris punctipes*. Environmental Entomology, 4: 531-534.

Crosby, C. R. and C. H. Hadley, Jr.
1915. The rhododendron lace-bug, *Leptobyrsa explanata* Heidemann (Tingitidae, Hemiptera). Journal of Economic Entomology, 8: 409-414, plate 22.

Culliney, T. W.
1985. Predation on the imported cabbageworm *Pieris rapae* by the stink bug *Euschistus servus euschistoides* (Hemiptera: Pentatomidae). Canadian Entomologist, 117: 641-642.

Cummings, C.
1933. The giant water bugs (Belostomatidae-Hemiptera). University of Kansas Science Bulletin, 21: 197-219.

Curtis, J.
1824-1840. British entomology: being illustrations and descriptions of the genera of insects found in Great Britain and Ireland; containing coloured figures from nature of the most rare and beautiful species, and in many instances of the plants upon which they are found. London. 16 Volumes + 769 plate-sheets [Plate sheets issued and numbered without regard to orders or families; author's suggestions included assembling plates of Hemiptera as volume 7 when binding]. 1824, 1: 1-5; 1825, 2: 51-98; 1826, 3: 99-146; 1827, 4: 147-194; 1828, 5: 195-241; 1829, 6: 242-289; 1830, 7: 290-337; 1831, 8: 338-383; 1832, 9: 384-433; 1833, 10: 434-481; 1834, 11: 482-529; 1835, 12: 530-577; 1836, 13: 578-625; 1837, 14: 626-673; 1838, 15: 674-721; 1839, 16: 722-769; 1840, Index. [Dates follow Horn, W. and S. Schenkling, 1928, Index Litteratureae Entomologicae. Volume 1. Berlin-Dahlem.].
1831. A guide to an arrangement of British Insects printed on one side for labelling cabinets; being a catalog of all the named species hitherto discovered in Great Britain and Ireland. Westley, London. 6 pages, 248 columns.
1833. Characters of some undescribed genera and species, indicated in the "Guide to and arrangement of British Insects." Entomologist's Monthly Magazine, 1:

186-199.
1835. Insects. Descriptions of the insects brought home by Commander James Clark Ross, R.N., F.R.S., & C. *In* Ross, J. C. Appendix to the narrative of a second voyage in search of a north-west passage and of a residence in arctic regions during the years 1829, 1830, 1831, 1832, 1833. A.W. Webster, London. Pages lvii-lxxx.
1837. Guides to an Arrangement of British Insects; Being A Catalog of All the Named Species Hitherto Discovered in Great Britain and Ireland. 2nd ed. J. Pigot and Co, London. 13: 578-625.

Dahlbom, A. G.
1851. Anteckningar ofver Inseckter, som blifvit observerade pa Gottland och i en del af Calmare San, under Sommaren 1850. Öfversigt af Kongliga Svenska Vetenskaps-Akademiens Förhandlingar, 1850: 155-229.

Dailey, P. J.
1977. Insects frequenting the common milkweed *Asclepias syriaca*. Part 1. Coleoptera and Hemiptera. American Zoologist, 17: 924.

Dailey, P. J., R. C. Graves, and J. L. Herring
1978. Survey of Hemiptera collected on common milkweed, *Asclepias syriaca*, at one site in Ohio. Entomological News, 89: 157-162.

Dallas, W. S.
1851-1852. List of the Specimens of Hemipterous Insects in the Collection of the British Museum. Taylor & Francis Incorporated, London. 1851, 1: 1-368, plates 1-11; 1852, 2: 369-592, plates 12-15.

Daly, H. V., J. T. Doyen, and P. R. Ehrlich
1978. Introduction to insect biology and diversity. McGraw-Hill, New York. 564 pages.

Davis, D. J., T. McGregor, and T. De Shazo
1943. *Triatoma sanguisuga* (LeConte) and *Triatoma ambigua* Neiva as natural carriers of *Trypanosoma cruzi* in Texas. U.S. Public Health Service Report, 58: 353-354.

Davis, L. V. and I. E. Gray
1966. Zonal and seasonal distribution of insects in North Carolina salt marshes. Ecological Monographs, 36: 275-295.

Davis, N. T.
1961. Morphology and phylogeny of the Reduvioidea (Hemiptera: Heteroptera). Part II. Wing venation. Annals of the Entomological Society of America, 54: 340-354.
1969. Contribution to the morphology and phylogeny of the Reduvioidea. Part IV. The harpactoroid complex. Annals of the Entomological Society of America, 62: 74-94.

Davis, R. M. and M. P. Russell
1969. Commensalism between *Ranzovius moerens* (Reuter) (Hemiptera: Miridae) and *Hololena curta* (McCook) (Araneida: Agelenidae). Psyche, 76: 262-269.

Davis, W. T.
1912. A new species of *Pselliopus* (*Milyas*). Psyche, 19: 20-21.

Davis, W. T. and C. W. Leng
1912. Insects on a recently felled tree. Journal of the New York Entomological Society, 20: 119-121.

Deakle, J. P. and J. R. Bradley, Jr.
1982. Effects of early season applications of diflubenzuron and azinphosmethylon population levels of certain arthropods in cottonfields. Journal of the Georgia Entomological Society, 17: 200-204.

Deay, H. O. and G. E. Gould
1936. Hemiptera unrecorded from Indiana. Proceedings of the Indiana Academy of Science, 45: 305-309.

Decker, G. C. and F. Andre
1938. Biological notes on *Blissus iowensis* Andre (Hemiptera-Lygaeidae). Annals of the Entomological Society of America, 31: 457-466.

De Carlo, J. A.
1932. Nuevas especies de Belostomidos (Hemiptera). Revista de la Sociedad Entomologica Argentina, 5: 121-126, 1 plate.
1938. I) Dos nuevas especies del genero *Abedus* Stål. II) Nuevas consideraciones sobre *Belostoma costa-limai* De Carlo y *Lethocerus truncatus* Cummings (Hemiptera-Belostomatidae). Revista de la Sociedad Entomologica Argentina, 10: 41-45, 2 plates.
1938. Los Belostomidos Americanos (Hemiptera). Anales del Museu Argentino de Ciéncias Naturales "Bernadino Rivadavia," 39: 189-252.
1948. Revisión del género *Abedus* Stäl (Ins. Hemipt. Belostom.). Communicaciones del Museo Argentino de Ciencias Naturales, 5: 1-24, 3 plates.

De Carlo, J. A. (continued)

1951. I) Nueva agrupacion en subgeneros de las especies del genero *Abedus* Stål (Hemipt. Belostom.). Revista de la Socieded Entomologica Argentina, 15: 69-72.

1951. Népidos de América (Hemiptera-Nepidae). Revista del Instituto Nacional de Investigacion de las Ciencias Naturales, Ciencias Zoológicas, 1: 385-421.

1963. Especies del genero *Abedus* Stal consideradas erroneamente sinonimas de otras (Hemiptera-Belostomatidae). Anales Sociedad Cientifica Argentina, 175: 69-78.

1964. Los «Ranatridae» de America (Hemiptera). Revista del Museo Argentino de Ciencias Naturales "Bernardino Rivadavia," Entomologia, 1: 133-215, plates 1-6.

1964. Género *Lethocerus* Mayr (Hemiptera-Belostomatidae). Physis [Revista de la Asociacion Argentina de Ciencias Naturales Argentina], 24: 337-350.

DeCoursey, R. M.

1963. The life histories of *Banasa dimidiata* and *Banasa calva* (Hemiptera: Pentatomidae). Annals of the Entomological Society of America, 56: 687-693.

1971. Keys to the families and subfamilies of the nymphs of North America Hemiptera-Heteroptera. Proceedings of the Entomological Society of Washington, 73: 413-428.

DeGeer, C.

1773. Mémoires pour servir à l'Histoire des Insects. Volume III. P. Hosselberg, Stockholm. ii + 696 pages, 44 plates.

Delcourt, A.

1909. Recherches sur la variabilite du genre *Notonecta*. Bulletin Scientifique de la France et de la Belgique, 43: 373-461.

De Lima, J. O. G. and T. F. Leigh

1984. Effect of cotton genotypes on the western bigeyed bug (Heteroptera: Miridae [*sic*]). Journal of Economic Entomology, 77: 898-902.

DeLoach, C. J. and J. C. Peters

1972. Effect of strip-planting vs. solid-planting on predators of cotton insects in southeastern Missouri, 1969. Environmental Entomology, 1: 94-102.

Deonier, D. L., S. P. Kincaid, and J. F. Scheiring

1976. Substrate and moisture preferences in common toad bug, *Gelastocoris oculatus* (Hemiptera: Gelastocoridae). Entomological News, 87: 257-264.

Desjardins, J. F.

1837. Notice sur un insecte nouveau faisant partie de la faune de l'ile Maurice. Annals de la Société Entomologique de France, 6: 239-243.

Dickerson, E. L.

1917. Notes on *Leptobyrsa rhododendri* Horv. Journal of the New York Entomological Society, 25: 105-112, plate 8.

Dickerson, E. L. and H. B. Weiss

1916. The ash leaf bug, *Neoborus amoenus* Reut. (Hem.). Journal of the New York Entomological Society, 24: 302-306, plate 16.

1917. The azalea lace-bug, *Stephanitis pyrioides* Scott (Tingitidae, Hemiptera). Entomological News, 28: 101-105, plate 9.

Dietz, L. L., J. W. Van Duyn, J. R. Bradley, Jr., R. L. Rabb, W. M. Brooks, and R. E. Stinner

1976a. A guide to the identification and biology of soybean arthropods in North Carolina. North Carolina Agricultural Experiment Station Technical Bulletin, 238: 88-107.

1976. A guide to the identification and biology of soybean arthropods in North Carolina. North Carolina Agricultural Research Service Technical Bulletin, 238: 1-264.

Dimmock, G.

1886. Belostomatidae and some other fish-destroying bugs. Annual Report of the Fish and Game Commission of Massachusetts, 25: 67-74.

Dingle, H.

1972. Migration strategies of insects. Science, 175: 1327-1335.

1974. The experimental analysis of migration and life-history strategies in insects. Page 329-342. *In* Brown, L. B. (editor). Experimental Analysis of Insect Behavior. Springer-Verlag, New York.

1978. Migration and diapause in tropical, temperature, and island milkweed bugs. Page 254-276. *In* Dingle, H. (editor). Evolution of Insect Migration and Diapause. Springer-Verlag, New York.

Dingle, H., B. M. Alden, N. R. Blakley, D. Kopec, and E. R. Miller

1980. Variation in photoperiodic response within and among species of milkweed bugs (*Oncopeltus*). Evolution, 34: 356-370.

Dingle, H., N. R. Blakley, and E. R. Miller
1980. Variation in body size and flight performance in milkweed bugs (*Oncopeltus*). Evolution, 34: 371-385.

Dingle, H. and R. L. Caldwell
1971. Temperature and reproductive success in *Oncopeltus fasciatus*, *Oncopeltus unifasciatellus*, *Lygaeus kalmii* and *Lygaeus turcicus*. Annals of the Entomological Society of America, 64: 1171-1172.

Dinkins, R. L., J. R. Brazzel, and C. A. Wilson
1970a. Species and relative abundance of *Chrysopa*, *Geocoris*, and *Nabis* in Mississippi cotton fields. Journal of Economic Entomology, 63: 660-661.
1970b. Seasonal incidence of major predaceous arthropods in Mississippi cotton fields. Journal of Economic Entomology, 63: 814-817.

Distant, W. L.
1880-1893. Insecta. Rhynchota. Hemiptera-Heteroptera. Volume I. *In* Goodman and Salvin (editors). Biologia Centrali-Americana. London. x + 462 pages, 39 plates. [1880: 1-88; 1881: 89-168; 1882: 169-224; 1883: 225-264; 1884: 265-304; 1889: 305-328; 1893: i-xx + 329-462].
1900. Revision of the Rhynchota belonging to the family Pentatomidae in the Hope Collection of Oxford. Proceedings of the Zoological Society of London, 54: 807-824, plates 52-53.
1900. Rhynchotal notes.--IV. Heteroptera: Pentatomidae. Annals and Magazine of Natural History, Series 7, 5: 386-397.
1901. Rhynchotal notes.--X. Heteroptera: Fam. Lygaeidae. Annals and Magazine of Natural History, Series 7, 7: 531-541.
1901. Rhynchotal notes.--XI. Heteroptera: Fam. Lygaeidae. Annals and Magazine of Natural History, Series 7, 8: 464-486, 497-510.
1902. The Fauna of British India Including Burma. Rhynchota- Vol. I (Heteroptera). Taylor and Francis, London. xxxviii + 438 pages.
1903-1904. The Fauna of British India, including Ceylon and Burma. Rhynchota. Volume II. (Heteroptera). Taylor and Francis, London. xvii + 503 pages. [1903, xvii + 242 pages ; 1904, pages 243-503].
1904. Rhynchotal notes.--XX. Heteroptera. Fam. Capsidae. (Part I). Annals and Magazine of Natural History, Series 7, 13: 103-114.
1904. Rhynchotal notes.--XXI. Heteroptera. Fam. Capsidae. (Part II). Annals and Magazine of Natural History, Series 7, 13: 194-206.
1904. Rhynchotal notes.--XXII. Heteroptera from North Queensland. Annals and Magazine of Natural History, Series 7, 13: 263-276.
1904. Rhynchotal notes.--XXV. Heteroptera. Fam. Anthocoridae. Annals and Magazine Natural History, Series 7, 14: 219-222.
1909. Oriental Rhynchota Heteroptera. Annals and Magazine of Natural History, Series 8, 3: 491-507.
1910. The Fauna of British India, including Ceylon and Burma. Rhynchota. Vol. V. Heteroptera: Appendix. Taylor and Francis, London. xii + 362 pages.

Dodson, V. E.
1975. Life histories of three species of Corixidae (Hemiptera: Heteroptera) from western Colorado. American Midland Naturalist, 94: 257-266.

Doesburg, P. H. van, Jr.
1968. A revision of the New World species of *Dysdercus* Guérin Méneville (Heteroptera, Pyrrhocoridae). Zoologische Verhandelingen, 97: 1-215, 16 plates.

Dohrn, F. A.
1859. Catalogus Hemipterorum. Herrcke und Lebeling, Stettin. 112 pages.
1860. Beiträge zu einer monographischen Bearbeitung der Familie der Emesina. Linnaea Entomologica, 14: 206-255, 1 plate.

Dolling, W. R.
1972. A new species of *Dicyphus* Fieber (Hem.,M iridae) from southern England. Entomologist's Monthly Magazine, 107: 244-245 (1971).

Dolling, W. R. and T. R. Yonke
1976. The genus *Coriomeris* in North America. Annals of the Entomological Society of America, 69: 1147-1152.

Douglas, J. W.
1874. British Hemiptera--additional species. Entomologist's Monthly Magazine, lst series, 11: 9-12, 142-144.

Douglas, J. W. and J. Scott
1865. The British Hemiptera. Vol. I. Hemiptera-Heteroptera. R. Hardwicke, London. xii + 627 pages, 21 plates.

Douglas, J. W. and J. Scott (continued)

1866. Additions to the British fauna (Hemiptera). Entomologist's Monthly Magazine, 2: 217-220, 272-276; 3: 13-16.

1867. British Hemiptera: Additions and corrections. Entomologist's Monthly Magazine, 4: 1-6, 45-52, 93-100.

1869. British Hemiptera: Additions and corrections. Entomologist's Monthly Magazine, 5: 259-268.

1871. British Hemiptera. Additions and corrections. Entomologist's Monthly Magazine, 8: 23-29, 60-63.

1875. British Hemiptera.-- Additions and corrections- Section Capsina. Entomologist's Monthly Magazine, 12: 100-102.

Dowdy, W. W.

1947. An ecological study of the Arthropoda of an oak-hickory forest with reference to stratification. Ecology, 28: 418-439.

Downes, W.

1921. The Life History of *Apateticus crocatus* Uhl. Proceedings of the Entomological Society of British Columbia, 16: 21-27.

1924. New records of Hemiptera from British Columbia. Proceedings of the Entomological Society of British Columbia, 21: 27-33.

1927. A premliminary list of the Hemiptera and Homoptera of British Columbia. Proceedings of the Entomological Society of British Columbia, 23: 1-22.

1927. A new species of *Drakella* (Heteroptera-Tingitidae). Canadian Entomologist, 59: 60.

1935. Additions to the list of B.C. Hemiptera. Proceedings of the Entomological Society of British Columbia, 31: 46-48.

1957. Notes on some Hemiptera which have been introduced into British Columbia. Proceedings of the Entomological Society of British Columbia, 54: 11-13.

Drake, C. J.

1916. A new tingid from Tennessee. Ohio Journal of Science, 16: 326-328.

1917. The North American species of *Monanthia* (Tingidae). Bulletin of the Brooklyn Entomological Society, 12: 49-52.

1917. A survey of the North American species of *Merragata*. Ohio Journal of Sciences, 17: 101-105.

1917. New and noteworthy Tingidae from the United States. Ohio Journal of Science, 17: 213-216.

1917. Key to the nearctic species of *Gargaphia* with the description of a new species (Hem., Het.). Entomological News, 28: 227-228.

1918. [New species]. *In* Gibson, E. H. The genus *Corythucha* Stal (Tingidae: Heteroptera). Transactions of the American Entomological Society, 44: 69-104.

1918. Notes on North American Tingidae (Hem.-Het.). Bulletin of the Brooklyn Entomological Society, 13: 86-88.

1918. Two new tingids from the West Indies (Hem.-Heter.). Ohio Journal of Science, 18: 175-177.

1918. The North American species of *Teleonemia* occurring north of Mexico. Ohio Journal of Science, 18: 323-332.

1918. A new corn insect from California (Heteroptera). Journal of Economic Entomology, 11: 385.

1919. An undescribed *Teleonemia* from Florida and Jamaica (Hemip.). Florida Buggist, 3: 24.

1919. A new species of *Velia* from Florida (Hem. Het.). Florida Buggist, 3: 122-123.

1919. On some Tingidae new to the fauna of Canada (Hemip.). Canadian Entomologist, 51: 159-160.

1919. On some North American Tingidae (Hemip.). Ohio Journal of Science, 19: 417-421.

1920. An undescribed water-strider from the Adirondacks. Bulletin of the Brooklyn Entomological Society, 15: 19-21.

1920. Descriptions of new North American Tingidae. Ohio Journal of Science, 20: 49-54.

1920. Water striders new to the fauna of Ohio, including the description of a new species. Ohio Journal of Science, 20: 205-208.

1921. Notes on some American Tingidae, with descriptions of new species. Florida Entomologist, 4: 49-54.

1922. Heteroptera in the vicinity of Cranberry Lake. Technical Publication No. 16, New York State College of Forestry, Syracuse University, 22: 54-86.

1922. On some North and South American Tingidae (Hemip.). Florida Entomologist, 5: 37-43, 48-50.

1922. A new species of *Plea* (Hemiptera-Notonectidae). Ohio Journal of Science, 22: 114-116.

1922. Neotropical Tingitidae with descriptions of three new genera and thirty-two new

species and varieties (Hemiptera). Memoirs of the Carnegie Museum, 9: 351-378, 1 plate.
1925. Concerning some Tingitidae from the Gulf States (Heteroptera). Florida Entomologist, 9: 36-39.
1926. Notes on some Tingitidae from Cuba (Hemiptera). Psyche, 33: 86-88.
1926. An undescribed tingitid from Arizona (Hemiptera). Bulletin of the Brooklyn Entomological Society, 21: 126-127.
1926. The North American Tingitidae (Heteroptera) described by Stål. Annals of the Carnegie Museum, 16: 375-380, 1 plate.
1928. Four undescribed tingitids from United States. Florida Entomologist, 12: 3-5.
1928. A synopsis of the American species of *Acalypta* (Hemip.-Tingitidae). Bulletin of the Brooklyn Entomological Society, 23: 1-9.
1928. Synonymical notes on tingitid genera with the descriptions of two new species from Haiti (Hemip.). Proceedings of the Biological Society of Washington, 41: 21-24.
1928. Family Tingitidae Costa. Pages 100-103. *In* Leonard, M. D. A list of the insects of New York. Cornell University Agricultural Experiment Station, Memoir, 101: 1-1121.
1930. Notes on American Tingitidae (Hemiptera). Bulletin of the Brooklyn Entomological Society, 25: 268-272.
1931. Concerning the genus "*Leptodictya*" Stal (Hemiptera, Tingitidea). Boletim do Museu Nacional, Rio de Janeiro, 7: 119-122.
1932. Notes on some American Tingitidae (Hemiptera). Psyche, 39: 100-102.
1934. An Undescribed *Acalypta* from New York (Tingitidae: Hemiptera). Bulletin of the Brook. Entomological Society, 29: 196.
1938. Mexican Tingitidae (Hemiptera). Pan-Pacific Entomologist, 14: 70-72.
1940a. Dos nuevas especies del género *Blissus* Klug. de la Argentina. Notas del Museo de la Plata, 5 (Zoologia 41): 223-227.
1940. An undescribed *Corythucha* (Tingitidae-Hemip.) from Colorado. Entomological News, 51: 172.
1941. New American Tingitidae (Hemiptera). Journal of the Washington Academy Sciences, 31: 141-145.
1941. Three new American Tingitidae (Hemiptera). Pan-Pacific Entomologist, 17: 139-141.
1943. A list of species of *Monanthia* Lep. & Serv. of the Western Hemisphere, including description of a new species (Hemiptera: Tingitidae). Proceedings of the Entomological Society of Washington, 45: 141-142.
1948. New American Tingitidae (Hemiptera). Boletin de Entomologia Venezolana, 7: 20-25.
1948. Two new Mesoveliidae, with check list of American species (Hemiptera). Boletin de Entomologia Venezolana, 7: 145-147.
1948. American Tingidae (Hemiptera). Revista de Entomologia, Rio de Janeiro, 19: 429-436.
1949. Concerning North American Saldidae (Hemiptera). Arkiv för Zoologi Experimentale et Generale, 42B: 1-4.
1949. Some American Saldidae (Hemiptera). Psyche, 56: 187-193.
1950. Concerning North American Saldidae (Hemiptera). Bulletin of the Brooklyn Entomological Society, 45: 1-7.
1951. New American chinch bugs (Hemiptera: Lygaeidae). Journal of the Washington Academy of Sciences, 41: 319-323.
1951. New neotropcial water-striders (Hemiptera-Veliidae). Great Basin Naturalist, 11: 37-42.
1951. New water-striders from the Americas (Hemiptera: Veliidae). Revista de Entomologia, 22: 371-378.
1952. Concerning American Ochteridae (Hemiptera). Florida Entomologist, 35: 72-75.
1952. A new tropical hebrid (Hemiptera). Pan-Pacific Entomologist, 28: 194.
1952. Alaskan Saldidae (Hemiptera). Proceedings of Entomological Society of Washington, 54: 145-148.
1955. A new *Metrobates* from the Gulf States (Hemiptera: Gerridae). Journal of the Kansas Entomological Society, 28: 130-131.
1956. New neotropical Hydrometridae (Hemiptera). Proceedings of the Biological Society of Washington, 69: 153-156.
1957. New apterous Aradidae (Hemiptera). Proceedings of the Biological Society of Washington, 70: 35-42.
1957. An undescribed shore-bug from Manitoba (Hemiptera, Saldidae). Bulletin of the Southern California Academy of Sciences, 56: 142-143.

Drake, C. J. (continued)
1962. Synonymic data and two new genera of shore-bugs (Hemiptera: Saldidae). Proceedings of the Biological Society of Washington, 75: 115-124.

Drake, C. J. and S. C. Bruner
1924. Concerning some Tingitidae occurring in the West Indies (Hemip). Memorias de la Sociedad Cubana de Historia Natural "Felipe Poey," 6: 144-154.

Drake, C. J. and J. C. M. Carvalho
1944. Four new American Tingitidae (Hemiptera). Bulletin of the Brooklyn Entomological Society, 39: 41-44.
1948. Concerning South American Saldidae (Hemiptera). Revista do Entomologia Rio de Janeiro, 19: 473-479.

Drake, C. J. and H. C. Chapman
1952. A new species of *Micracanthia* from Florida (Hemiptera: Saldidae). Florida Entomologist, 35: 147-150.
1953. An undescribed saldid from the Gulf states (Hemiptera: Saldidae). Bulletin of the Brooklyn Entomological Society, 48: 64-66.
1953. A new species of *Trepobates* Uhler from Florida (Hemiptera: Gerridae). Florida Entomologist, 36: 109-112.
1953. Distributional data and description of a new hebrid (Hemiptera). Great Basin Naturalist, 13: 9-11.
1953. Preliminary report on the Pleidae (Hemiptera) of the Americas. Proceedings of the Biological Society of Washington, 66: 53-60.
1954. New American waterstriders (Hemiptera). Florida Entomologist, 37: 151-155.
1958. New neotropical Hebridae, including a catalogue of the American species (Hemiptera). Journal of the Washington Academy of Sciences, 48: 317-326.
1958. The subfamily Saldoidinae (Hemiptera: Saldidae). Annals of the Entomological Society of America, 51: 480-485.
1963. A new genus and species of water-strider from California (Hemiptera: Macroveliidae). Proceedings of the Biological Society of Washington, 76: 227-234.

Drake, C. J. and Cobben, R. H.
1960. The Heteroptera of the Netherlands Antilles - V, Tingidae (lace bugs). Studies on the Fauna of Curaçāo and other Caribbean Islands, 11(54): 67-97.

Drake, C. J. and N. T. Davis
1958. The morphology and systematics of the Piesmatidae (Hemiptera), with keys to world genera and American species. Annals of the Entomological Society of America, 51: 567-581.

Drake, C. J. and P. H. Doesburg, van
1966. Water-striders of the American genus *Trochopus* (Hemiptera: Veliidae). Studies of the Fauna of Suriname and other Guyanas, 8: 65-76.

Drake, C. J. and Frick, D. M.
1939. Synonymy and distribution of the lantana lace bug (Hemiptera: Tingitidae). Proceedings of the Hawaiian Entomological Society, 10: 199-202.

Drake, C. J. and Hambleton, E. J.
1944. Four new American Tingitidae (Hemiptera). Proceedings of the Entomological Society of Washington, 46: 94-96.
1945. Concerning neotropical Tingitidae (Hemiptera). Journal of the Washington Academy of Sciences, 35: 356-367.
1946. New species and new genera of American Tingitidae (Hemiptera). Proceedings of the Biological Society of Washington, 59: 9-16.

Drake, C. J. and H. M. Harris
1926. Notes on American Anthocoridae with descriptions of new forms. Proceedings of the Biological Society of Washington, 39: 33-46.
1927. Three new species of Enicocephalidae. Ohio Journal of Science, 27: 102-103.
1927. Notes on the genus *Rhagovelia*, with descriptions of six new species. Proceedings of the Biological Society of Washington, 40: 131-138.
1928. Three new gerrids from North America (Hemip.). Proceedings of the Biological Society of Washington, 41: 25-29.
1928. Two undescribed water-striders from Grenada (Hemiptera). Florida Entomologist, 12: 7-8.
1928. *Tetraphleps canadensis* Provancher, a true *Tetraphleps* (Hemip.). Canadian Entomo logist, 60: 50.
1928. Concerning some North American water-striders with descriptions of three new species. Ohio Journal of Science, 28: 269-276.
1930. A wrongly identified American water-strider. Bulletin of the Brooklyn Entomological Society, 25: 145-146.

1932. A survey of the species of *Trepobates* Uhler (Hemiptera, Gerridae). Bulletin of the Brooklyn Entomological Society, 27: 113-123.

1932. An undescribed water-strider from Honduras (Hemiptera, Gerridae). Pan-Pacific Entomologist, 8: 157-158.

1932. Some miscellaneous Gerridae in the collection of the Museum of Comparative Zoology (Hemiptera). Psyche, 39: 107-112.

1933. New American Veliidae (Hemiptera). Proceedings of the Biological Society of Washington, 46: 45-53.

1934. The Gerrinae of the Western Hemisphere (Hemiptera). Annals of the Carnegie Museum, 23: 179-240.

1936. Notes on American water-striders. Proceedings of the Biological Society of Washington, 49: 105-108.

1938. Concerning Mexican Gerridae (Hemiptera). Pan-Pacific Entomologist, 14: 73-75.

1942. Notas sôbre *Rheumatobates*, com descrição de uma nova espécie (Hemiptera, Gerridae). Revista Brasileira Biologia, 2: 399-402.

1943. Notas sobre Hebridae del hemisferio occidental (Hemiptera). Notas del Museo de La Plata, Zoologia, 8: 41-58.

1945. Concerning the genus *Metrobates* Uhler (Hemiptera, Gerridae). Revista Brasileira Biologia, 5: 179-180.

Drake, C. J. and J. L. Herring

1964. The genus *Nidicola* (Hemiptera: Anthocoridae). Proceedings of the Biological Society of Washington, 77: 53-64.

Drake, C. J. and L. Hoberlandt

1951. Check-list and distributional records of Leptopodidae (Hemiptera). Acta Entomologica Musei Nationalis Pragae, 26: 1-5 (1950).

1951. Catalogue of genera and species of Saldidae (Hemiptera). Acta Entomologica Musei Nationalis Pragae, 26(376): 1-12 (1950).

Drake, C. J. and F. C. Hottes

1925. Four undescribed species of water-striders (Hemip.-Gerridae). Ohio Journal of Science, 25: 46-50.

1925. Five new species and a new variety of water-striders from North America (Hemiptera-Gerridae). Proceedings of the Biological Society of Washington, 38: 69-74.

1949. Two new species of Saldidae (Hemiptera) from western United States. Proceedings of the Biological Society of Washington, 62: 177-184.

1950. Saldidae of the Americas (Hemiptera). Great Basin Naturalist, 10: 51-61.

1951. Notes on the genus *Rheumatobates* Bergroth (Hemiptera: Heteroptera). Proceedings of the Biological Society of Washington, 64: 147-158.

1952. Genus *Trepobates* Herrich-Schaeffer (Hemiptera; Gerridae). Great Basin Naturalist, 12: 35-38.

1952. Distributional and synonymical data and descriptions of two new *Hydrometra* (Hemiptera: Hydrometridae). Journal of the Kansas Entomological Society, 25: 106-110.

1954. Synonymic data and description of a new saldid (Hemiptera). Occasional Papers of the Museum of Zoology, University of Michigan, 553: 1-5.

1955. Concerning Saldidae (Hemiptera) of the Western Hemisphere. Boletín de Entomología Venezolana, 11: 1-12.

Drake, C. J. and M. P. Hurd

1945. New American Tingitidae (Hemiptera). Boletín de Entomologia Venezolana, 4: 127-133.

Drake, C. J. and R. F. Hussey

1951. Concerning some American *Microvelia* (Hemiptera: Veliidae). Florida Entomologist, 34: 137-145.

1954. Notes on some American Veliidae(Hemiptera) with the description of two new *Microvelia* from Jamaica. Florida Entomologist, 37: 133-138.

1955. Concerning the genus *Microvelia* Westwood, with descriptions of new species and a checklist of the American forms (Hemiptera: Veliidae). Florida Entomologist, 38: 95-115.

1957. Notes on some American *Rhagovelia*, with descriptions of two new species (Hemiptera: Veliidae). Occasional Papers of the Museum Zoology, University Michigan, 580: 1-6.

Drake, C. J. and Lattin, J. D.

1963. American species of the lacebug genus *Acalypta* (Hemiptera: Tingidae). Proceedings of the United States National Museum, 115: 331-345, 15 plates.

1959. Descriptions, synonymy, and check-list

Drake, C. J. and Lattin, J. D. (continued)
of American Hydrometridae (Hemiptera: Heteroptera). Great Basin Naturalist, 19: 43-52.

Drake, C. J. and T.-C. Maa
1955. Chinese and other Oriental Tingoidea (Hemiptera). III. Quarterly Journal of the Taiwan Museum, 8: 1-11.

Drake, C. J. and J. Maldonado Capriles
1954. Puerto Rican water-striders (Hemiptera). Proceedings of the Biological Society of Washington, 67: 219-222.
1956. Some pleids and water-striders from the Dominican Republic (Hemiptera). Bulletin of the Brooklyn Entomological Society, 51: 53-56.

Drake, C. J. and Poor, M. E.
1937. The South American Tingitidae (Hemiptera) described by Stål. Memoirs of the Carnegie Museum, 11: 301-314, 1 plate.

Drake, C. J. and Ruhoff, F. A.
1960. A new moss-feeding tingid from Mexico (Hemiptera). Bulletin of the Brooklyn Entomological Society, 54: 136-139 (1959).
1960. Lace-bug genera of the world (Hemiptera: Tingidae). Proceedings of the United States National Museum, 112: 1-105, 9 plates.
1960. An undescribed tingid from Arizona (Hemiptera). Journal of the Kansas Entomological Society, 33: 152-154.
1962. Synonymic charges and four new species of Tingidae (Hemiptera). Bulletin of the Southern California Academy of Sciences, 60: 156-164, 3 plates (1961).
1962. Synonymic notes and descriptions of new Tingidae (Hemiptera). Studia Entomologica, 5: 489-506.
1962. Taxonomic changes and descriptions of new Tingidae (Hemiptera). Bulletin of the Southern California Academy of Sciences, 61: 133-142.
1965. Lacebugs of the World: A Catalog (Hemiptera: Tingidae). United States National Museum Bulletin, 243: i-viii, 1-634, frontispiece, plates 1-56.

Drake, C. J. and J. A. Slater
1957. The phylogeny and systematics of the family Thaumastocoridae (Hemiptera: Heteroptera). Annals of the Entomological Society of America, 50: 353-370.

Drury, D.
1770-1782. Illustrations of Natural History Wherein Are Exhibited Upwards of Two Hundred and Forty Figures of Exotic Insects According to Their Different Genera. Volumes 1-3. B. White, London. [1770, 1: i-xxvii, 1-130, plates 1-50; 1773, Index [giving scientific names]; Volume 2, 1773; pages i-vii, 1-90, plates 1-50, Index [giving scientific names]; 1782, 3: i-xii, 1-76, plates 1-50, Index [giving scientific names]]. [The 1773 date for the "Index" giving scientific names in Volume I was established in Westwood's 1837 "New Edition" of Drury's above work].

Duarte, J. and D. M. Calabrese
1982. Is the binomen *Lygaeus kalmii* Stål (Hemiptera: Heteroptera: Lygaeidae) applied to sibling species? Proceedings of the Entomological Society of Washington, 84: 301-303.

Duffey, S. S.
1970. Cardiac glycosides and distastefulness: Some observations on the palatability spectrum of butterflies. Science, 169: 78-79.

Duffey, S. S. and G. G. E. Scudder
1972. Cardiac glycosides in North American Asclepidaceae, a basis for unpalatability in brightly coloured Hemiptera and Coleoptera. Journal of Insect Physiology, 18: 63-78.

Dufour, L.
1831. Description et figure du *Xylocoris rufipennis*, Hémiptère nouveau. Annales des Sciences Naturelles, Paris. 22: 423-426.
1833. Mémoire sur les genres *Xylocoris, Leptopus* et *Velia*. Annales de la Société Entomologique de France, 2: 104-118, 1 plate.
1863. Essai monographique sur les Bélostomides. Annales de la Société Entomologique de France, Series 4, 3: 373-400.

Duges, A.
1889. *Tingis spinosa*. La Naturaleza, Series 2, 1: 207-209, 1 plate.
1892. *Acanthia inodora*, A. Dug. (chinches de gallos). La Naturaleza, Series 2, 2: 169-170, plate 8.

Dumas, B. A., W. P. Boyer, and W. H. Whitcomb
1962. Effect of time of day on surveys of predaceous insects in field crops. Florida Entomologist, 45: 121-128.

Dunbar, D. M.
1971. Big-eyed bugs in Connecticut lawns. Connecticut Agricultural Experiment Station Circular 244. 6 pages.

1972. Notes on the mating behavior of *Geocoris punctipes* (Hemiptera: Lygaeidae). Annals of the Entomological Society of America, 65: 764-765.

Dunbar, D. M. and O. G. Bacon

1972a. Influence of temperature on development and reproduction of *Geocoris atricolor, G. pallens,* and *G. punctipes* (Heteroptera: Lygaeidae) from California. Environmental Entomology, 1: 596-599.

1972b. Feeding, development, and reproduction of *Geocoris punctipes* (Heteroptera: Lygaeidae) on eight diets. Annals of the Entomological Society of America, 65: 892-895.

Dunn, C. E.

1979. A revision and phylogenetic study of the genus *Hesperocorixa* Kirkaldy (Hemiptera: Corixidae). Proceedings of the Academy of Natural Sciences of Philadelphia, 131: 158-190.

Dupuis, C.

1953. Prioritéde quelques noms d'Hétéroptères de Guérin Méneville (1831). Bulletin de la Société Zoologique, 77: 447-454 (1952).

Egbert, A. M.

1946. A new corixid from Georgia (Hemiptera, Corixidae). Journal Kansas Entomological Society, 19: 133-135.

Eger, J. E. [also as Eger, J. E., II]

1978. Revision of the genus *Loxa* (Hemiptera: Pentatomidae). Journal of the New York Entomological Society, 86: 224-259.

Ehler, L. E.

1977. Natural enemies of cabbage looper on cotton in the San Joaquin Valley. Hilgardia, 45: 73-106.

Ehler, L. E., K. G. Eveleens and R. van den Bosch

1973. An evaluation of some natural enemies of cabbage looper on cotton in Califonia. Environmental Entomology, 2: 1009-1015.

Ehler, L. E. and J. C. Miller

1978. Biological control in temporary agroecosystems. Entomophaga, 23: 207-212.

Eikenbary, R. D. and H. G. Raney

1968. Population trends of insect predators of the elm leaf beetle. Journal of Economic Entomology, 61: 1336-1339

Eichler, W.

1942. *Oeciacus gerdheinrichi* nov. spec., eine Salanganenwanze (Heteroptera, Cimicidae) von *Collocalia spodiopygia sororum* Stresemann 1931 aus Celebes. Mittheilungen aus dem Zoologischen Museum in Berlin, 25: 292-299.

Elkins, J. C.

1951. The Reduviidae of Texas. Texas Journal of Science, 3: 407-412.

1951. A new species of *Metapterus* (Hemiptera, Reduviidae). Field and Laboratory, 19: 90-93.

1953. Notes on some Texas Reduviidae (Hemiptera, Reduviidae). Journal of the Kansas Entomological Society, 26: 137-140.

1954. A synopsis of *Atrachelus* (Hemiptera, Reduviidae). Proceedings of the Entomological Society of Washington, 56: 97-120.

1956. A synonymic note on *Atrachelus* A. and S. (Hemiptera: Reduviidae). Florida Entomologist, 39: 44.

Ellenrieder, C. A. M.

1862. Eerste Bijdrage Tot de Kennis der Hemipteren van den Indischen Archipel. Natuurkundige Tijdschrift voor Nederlandsche Indie, 24: 130-174, plates 1-6.

Ellington, J., M. Cardenas, K. Kiser, L. Guerra, V. Salguero, and G. Ferguson

1984. Approach to the evaluation of some factors affecting insect resistance in one 'Acala' and seven sister genotypes of Stoneville cotton in New Mexico. Journal of Economic Entomology, 77: 612-618.

Ellington, J., K. Kiser, G. Ferguson, and M. Cardenas

1984. A comparison of sweepnet, absolute, and insectavac sampling methods in cotton ecosystems. Journal of Economic Entomology, 77: 599-605.

Ellis, L. L.

1950. The status of *Plea striola* and *harnedi* (Hemiptera, Pleidae). Proceedings of the Entomological Society of Washington, 52: 104-105.

1952. The aquatic Hemiptera of southeastern Louisiana (exclusive of the Corixidae). American Midland Naturalist, 48: 302-329.

1965. An unusual habitat for *Plea striola* (Hemiptera: Pleidae). Florida Entomologist, 48: 77.

Elsey, K. D.
1972. Predation of eggs of *Heliothis* spp. on tobacco. Environmental Entomology, 1: 433-438.
1973. *Jalysus spinosus*: Effect of insecticide treatments on this predator of tobacco pests. Environmental Entomology, 2: 240-243.

Elvin, M. K. and P. E. Sloderbeck
1984. A key to nymphs of selected species of Nabidae (Hemiptera) in the south-eastern USA. Florida Entomologist, 67: 269-273.

Emmons, E.
1854. Natural History of New York Agriculture. Volume 5. Insects. Albany, New York. 272 pages, 3 + 47 plates.
1854. The more common and injurious species of insects. Agriculture Natural History of New York Report, 5: i-viii, 1-272, plates 1-47.

Emsley, M. G.
1962. 5000 bugs-- a year's work would fit in a teaspoon. Animal Kingdom, 45: 114-116.
1969. The Schizopteridae (Hemiptera: Heteroptera) with the description of new species from Trinidad. Memoirs of the American Entomological Society, 25: 1-154.

Enderlein, G.
1912. Die Insekten des Antarkto-Archiplata-Gebietes. Kongliga Svenska Vetenskaps-Akademiens Handlingar, 48(3): 111-118, 154-155. [Heteropteran parts].

Englehardt, G. P.
1918. [In minutes of meeting]. Bulletin of the Brooklyn Entomological Society, 13: 40.

Erichson, W. F.
1848. Insecten. 3: 553-617. *In* Reisen in British-Guiana in den Jahren 1840-1844. Im Auftrag Sr. Mäjestat des Königs von Preussen, Ausgeführt von Richard Schomburgk. J. J. Weber, Leipzig.

Esaki, T.
1924. On the genus *Halobates* from Janpanese and Formosan coasts (Hemiptera: Gerridae). Psyche, 31: 112-118, 1 plate.
1926. The water-striders of the subfamily Halobatinae in the Hungarian National Museum. Annales Historico-Naturales Musei Nationalis Hungarici, 23: 117-164.
1928. A monograph of the Helotrephidae, subfamily Helotrephinae (Hem. Heteroptera). EOS, Revista Española de Entomología. 4: 129-172.

Eschscholtz, J. F.
1822 Entomographien. Volume 1 (erste Lieferung). G. Reimer, Berlin. 128 pages.

Esselbaugh, C. O.
1946 A study of the eggs of the Pentatomidae (Hemipters). Annals of the Entomological Society of America, 39: 667-691.
1948. Some remarks on the genus *Chlorochroa* (Hemiptera, Pentatomidae) and a new species. Bulletin of the Brooklyn Entomological Society. 42: 164-169.
1949. Notes on the bionomics of some midwestern pentatomidae. Entomologica Americana, 28: 1-73 (1948).
1949. A bionomic note on the taxonomic status of the form *pyrrhocerus* of Euschistus tristigmus Say (Hemiptera, Pentatomidae). Proceedings of the Entomological Society of Washington, 51: 160-163.

Essig, E. O.
1915. Injurious and beneficial insects of California. 2nd edition, Monthly Bulle-tin of the California State Commission of Horticulture, Supplement 4, 541 pages.
1926. Insects of Western North America. Macmillan Co., New York.

Essig, E. O. and R. L. Usinger
1940. The life nd works of Edward Payson Van Duzee. Pan-Pacific Entomology, 16: 145-177.

Estrada, J.
1976. Los insectos plaga del algodonero del Valle de Juarez, D. B., Chih., y sus enimigos naturales. Folia Entomologica Mexicana, 36: 21.

Evans, E. W.
1983. The influence of neighboring hosts on colonization of prairie milkweeds by a seed-feeding bug. Ecology, 64: 648-653.
1985. A key to nymphs of four species of the genus *Podisus* (Hemiptera: Pentatomidae) of northeastern North America. Proceedings of the Entomological Society of Washington, 87: 94-97.

Evans, J. H.
1931. A preliminary revision of the ambush bugs of North America, (Hemiptera, Phymatidae). Annals of the Entomological Society of America, 24: 711-738.

Eveleens, K. G.
1974. De inductie van sekundaire plagen in de katoenteelt van California door chem-

ische insektenbestrijding. Entomologische Berichten, 34: 4.

Eveleens, K. G., R. van den Bosch, and L. E. Ehler
1973. Secondary outbreak induction of beet armyworm by experimental insecticide applications in cotton in California. Environmental Entomology, 2: 497-503.

Eyles, A. C.
1964. Feeding habits of some Rhyparochrominae (Heteroptera: Lygaeidae) with particular reference to the value of natural foods. Transactions of the Royal Entomological Society of London, 116: 89-114.
1969. The validity of *Dieuches occidentalis* from Arizona. New Zealand Journal of Science, 12: 728-731.
1971. List of Isometopidae (Heteroptera: Cimicoidea). New Zealand Journal of Science, 14: 940-944.
1972. Supplement to list of Isometopidae (Heteroptera: Cimicoidea). New Zealand Journal of Science, 15: 463-464.
1975. Further new genera and other new combinations for species previously assigned to *Megaloceroea* (Heteroptera: Miridae: Stenodemini). Journal of Natural History, 9: 153-167.

Eyles, A. C. and J. C. M. Carvalho
1975. Revision of the genus *Dolichomiris*, with a revised key to the genera of Stenodemini (Heteroptera: Miridae). Journal of Natural History, 9: 257-269.

Fabricius, J. C.
1775. Systema entomologiae, sistens insectorum classes, ordines, genera, species, adjectis synonymis, locis, descriptionibus, observationibus. Flensburgi et Lipsiae, Korte. xxvii + 832 pages.
1776. Genera insectorum eorumque characteres naturales secundum numerum, figuram, situm, et proportionem omnium partium oris adjecta mantissa specierum nuper detectarum. Friedr. Bartschii, Chilonii. xiv + 310 pages.
1779. Reise nach Norwegen mit Bemerkungen aus der Naturhistorie und Oekonomie. C. E. Sohn, Hamburg. 388 pages. + index.
1781. Species insectorum exhibentes eorum differentias specificas, synonyma auctorum, loca natali, metamorphosin adjectis observationibus, descriptionibus. Hamburgi et Kilonii, Bohn.1: 1-552; 2: 1-517.
1787. Mantissa insectorum sistens eorum species nuper dectas: Adjectis characteribus genericis, differentiis specificis, emendationibus, observationibus. Kiel. 1: 1-348; 2: 1-382. [Classis VII: Ryngota. 2: 266-320, 382].
1794. Entomologia systematica emendata et aucta, secundum classes, ordines, genera, species, adjectis synonymis, locis, observationibus. C. G. Proft, Hafniae. 4: i-vi, 1-472. [Classis X. Ryngota. 4: 1-229].
1798. Supplementum entomologiae systematicae. Proft et Storch, Hafniae. 572 pages. [Classis XII. Rhyngota. Pages 511-546].
1803. Systema Rhyngotorum secundum ordines, genera, species adjectis synonymis, locis, observationibus, descriptionibus. Carolum Reichard, Brunsvigae. x + 314 pages + 1 [emendanda].

Falcon, L. A., R. van den Bosch, C. A. Ferris, L. K. Stromberg, L. K. Etzel, R. E. Stinner, and T. F. Leigh
1968. A comparison of season-long cotton-pest-control programs in California during 1966. Journal of Economic Entomology, 61: 633-642.

Fallén, C. F.
1807. Monographia Cimicum Sueciae. C. G. Proft, Hafniae. 123 pages.
1814. Specimen novam Hemiptera disponendi methodum exhibens. Lundae, Berlinguianus. 26 pages.
1828-1829. Hemiptera Sueciae. Cimicides eorumque familiae affines. 1828, pages i-iv, 1-16; 1829, pages 17-186 + 4.

Farlow, R. A. and H. N. Pitre
1981. Effect of herbicides on a hemipteran predator. Journal of the Mississippi Academy of Sciences, 26: 5.
1983. Bioactivity of the postemergent herbicides acifluorfen and bentazon on *Geocoris punctipes* (Say) (Hemiptera: Lygaeidae). Journal of Economic Entomology, 76: 200-203.

Ferguson, H. J. and R. M. McPherson
1982. Effects of selected insecticides on soybean pest and predator species. virginia Journal of Science, 33: 72.

Ferguson, H. J., R. M. McPherson, and W. A. Allen
1984. Effect of four soybean cropping systems on the abundance of foliage-inhabiting

Ferguson, H. J., R. M. McPherson, and W. A. Allen (continued)
insect predators. Environmental Entomology, 13: 1105-1112.

Ferrari, E. von
1888. Die Hemipteren-Gattung *Nepa* Latr. (sens. natur.). Annalen des Kaiserlich-Koniglichen Naturhistorischen Hofmuseum, 3: 161-194, plates 8-9.

Ferris, G. F.
1920. Some records of Polyctenidae (Hemiptera). Journal of the New York Entomological Society, 27: 261-263, plate 24 (1919).

Ferris, G. F. and R. L. Usinger
1939. The family Polyctenidae (Hemiptera; Heteroptera). Microentomology, 4: 1-50.

Fieber, F. X.
1837. Beiträge zur Kenntniss der Schnabelkerfe, (Rhynchota). Beiträge zur gesammten Natur-und Heilwissenschaft (Von Weilenweber). Prague. 1: 97-111, 337-355.
1844. Entomologische Monographien. Abhandlung der Königlische Böhmischen Gesellschaft der Wissenschaften in Prag, Series 5, 3: 279-416. [separate: 1844, Barth, Leipzig. 138 pages].
1848. Synopsis aller bisher in Europa entdeckten Arten der Gattung *Corisa*. Bulletin de la Société Impériale des Naturalistes de Moscou, 21: 505-539, plate 10.
1851. Species generis *Corisa* monographice despositae. Abhandlung der Königlische Böhemischen Gesellschaft der Wissenschaften in Prag, Series 5, 7: 213-260, 2 plates. [Separate: Actis Regiae Bohemicae Societatis Scientiarum, Pragae, 1: 1-48, 2 plates].
1851. Genera Hydrocoridum secundum ordinem naturalem in Familias disposita. Abhandlung der Königlische Böhemischen gesellschaft der Wissenschaften in Prag, Series 5, 7: 181-212, 4 plates. [Separate: Actis Regiae Bohemicae Societatis Scientiarum, Pragae, 1: 1-30, 4 plates].
1851. Rhynchotographieen. Abhandlungen Kongliga Bohmischen Gesellschaft Wissenschaften in Prag, Series 5, 7: 425-486.
1858. Criterien zur generischen Theilung der Phytocoriden (Capsini aut.). Wiener Entomologische Monatschrift, 2: 289-327, 329-347.
1859. Die europäischen Arten der Gattung *Salda* Fab. Wiener Entomologische Monatschrift, 3: 230-241.
1860. Exegesen in Hemipteren. Wiener Entomologische Zeitung, 4: 257-272.
1860-1861. Die europäischen Hemiptera. Halbflügler (Rhynchota Heteroptera). Nach der analytischen Methode bearbeitet. Gerold, Wien. 1860, i-vi, 1-112; 1861, 113-444, 2 plates. [Above dating follows Horn and Schenkling, 1929, Index Litturaturae Entomologicae, 4: 1354, except for "i-vi," which they do not mention].
1861. Die Gattung *Ophthalmicus*, monographisch nach der analylischen methode bearbeitet. Wiener Entomologische Monatschrift, 5(9): 266-285.
1864. Erörterungen zur Nomenclatur der Rhynchoten (Hemiptera) Livland's. Weiner Entomologische Monatschrift, 7: 53-62 (1863).
1864. Neuere Entdeckungen in europäischen Hemipteren. Wiener Entomologische Monatschrift, 8(3): 65-86, 205-236, 321-336.
1870. Dodecas neuer Gattungen und neuer Arten europäischer Hemiptera. Verhandlungen der Kaiserlich-Koniglischen Zoologisch-Botanischen Gesellschaft in Wien, 20: 243-264, 2 plates.

Fitch, A.
1855a. The chinch bug. Cultivator, (3)3: 191.
1855b. Comments on the above [The chinch bug, by E. C. Smith] by Dr. A. Fitch. Cultivator, (3)3: 238-239.
1855c. Chinch bug. Pages 277-297. *In* First and second report on the noxious, beneficial and other insects of the state of New York. [Journal publication in Transactions of the New York State Agricultural Society, 15: 409-559 in 1856.]
1855. Second report on the noxious, beneficial and other insects, of the state of New York. Transactions of the New York State Agricultural Society, 15: 409-559.
1856. Third report on the noxious, beneficial and other insects, of the state of New York. Transactions of the New York State Agricultural Society, 16: 315-490.
1857. Fourth report on the noxious and other insects of the state of New York. Transactions of the New York Agricultural Society, 17: 687-814.
1861. The quince tingis. Country Gentleman, 17: 114.
1870. Thirteeth report on the noxious and other insects of the state of New York.

Transactions of the New York Agricultural Society, 29: 495-566.

Fleischer, S. J., M. J. Gaylor, and J. V. Edelson
1985. Estimating absolute density from relative sampling of *Lygus lineolaris* (Heteroptera: Miridae) and selected predators in early to mid-season cotton. Environmental Entomology, 14: 709-717.

Fletcher, J.
1885. [Note on *Podisus modestus*]. 15th Annual Report of the Entomological Society of Ontario for 1884, page 22.

Flor, G.
1860-1861. Die Rhynchoten Livlands in systematischer Folge bescreiben. Erster Theil: Rhynchota frontirostria Zett. (Hemiptera Heteroptera Aut.). Archiv für die Naturkunde Liv-, Ehst-, und Kurlands. II Serie, Biologische Naturkunde. Carl Schulz, Dorpat. Band I: 1-826 (1860); band II: 1-638 (1861).

Fontes, A. V.
1981. Estudos comparativos da genitália da fêmea no gênero *Notholopus* Bergroth, 1922 (Hemiptera: Miridae). Arquivos Museu Nacional, Rio de Janeiro, 56: 137-183.

Forbes, S. A.
1876. List of Illinois Crustacea, with descriptions of new species. Bulletin of the Illinois Museum of Natural History, 1: 4-5.
1885. Some insect enemies of the soft maple (*Acer dasycarpum*). Page 111. *In* Fourteenth Report of the State Entomologist of Illinois.
1900. Report of the state entomologist on the noxious and beneficial insects of the state of Illinois. Report of the Illinois Entomologist, 21: 1-184.
1900. The economic entomology of the sugar beet. Pages 49-186. *In* Twenty-First Report of the State Entomologist of Illinois.

Forster, J. R.
1771. Novae species insectorum; Centuria 1. London. viii + 100 pages.

Fourcroy, A. F.
1785. Entomologia parisiensis, sive catalogus insectorum quae in Agro parisiensi reperiuntur; secundum methodum Geoffroeanam in sectiones, genera et species distributus; cui addita sunt nomina trivialia et fere trecentae novae species. Paris. 2 volumes. Pages 1-231, 233-544.

Fracker, S. B.
1912. A systematic outline of the Reduviidae of North America. Proceedings of the Iowa Academy of Science, 19: 217-252. [Often cited as 1913 but the journal is dated by the printer as 1912, the Zoological Record (50(12): 43) lists it as 1912, and the Smithsonian Institution Library copy is dated as being received July 14, 1912].
1918. The Alydinae of the United States. Annals of the Entomological Society of America, 11: 255-280, plates 24-25.
1919. *Chariesterus* and its neotropical relatives (Coreidae Heteroptera). Annals of the Entomological Society of America, 12: 227-230.
1923. A review of the North American Coreini (Heteroptera). Annals of the Entomological Society of America, 16: 165-174.

Fracker, S. B. and R. L. Usinger
1949. The generic identification of nearctic reduviid nymphs (Hemiptera). Annals of the Entomological Society of America, 42: 273-278.

Franz, J. M.
1961. Biological control of pest insects in Europe. Annual Review of Entomology, 6: 183-200.

Freeman, P.
1947. A revision of the genus *Dysdercus* Boisduval (Hemiptera, Pyrrhocoridae), excluding the American species. Transacions of the Royal Entomological Society of London, 98: 373-424.

Froeschner, R. C.
1941. Contributions to a synopsis of the Hemiptera of Missouri, pt. I. Scutellaridae, Podopidae, Pentatomidae, Cydnidae, Thyreocoridae. American Midland Naturalist, 26: 122-146.
1942. Contributions to a synopsis of the Hemiptera of Missouri, pt. II. Coreidae, Aradidae, Neididae. The American Midland Naturalist, 27: 591-609.
1944. Contributions to a synopsis of the Hemiptera of Missouri, pt. III. Lygaeidae, Pyrrhocoridae, Piesmidae, Tingididae, Enicocephalidae, Phymatidae, Ploiariidae, Reduviidae, Nabidae. American Midland Naturalist, 31: 638-683.
1949. Contributions to a synopsis of the Hemiptera of Missouri, pt. IV. Hebridae, Mesoveliidae, Cimicidae, Anthocoridae, Cryptostemmatidae, Isometopidae,

Froeschner, R. C. (continued)

Miridae. American Midland Naturalist, 42: 123-188.

1960. Cydnidae of the Western Hemisphere. Proceedings of the United States National Museum, 111(3430): 337-680, plates 1-13.

1962. Contributions to a synopsis of the Hemiptera of Missouri, pt. V. Hydrometridae, Gerridae, Veliidae, Saldidae, Octeridae, Gelastocoridae, Naucoridae, Belostomatidae, Nepidae, Notonectidae, Pleidae, Corixidae. American Midland Naturalist, 67: 208-240.

1963. The genus *Ceraleptus* Costa in the Western Hemisphere (Hemiptera: Coreidae). Journal of the Kansas Entomological Society, 36: 109-113.

1963. Review of the genus *Atractotomus* Fieber in North America with notes, key, and description of one new species (Hemiptera: Miridae). Bulletin of the Brooklyn Entomological Society, 58: 1-5.

1965. *Larinocerus balius*, a new genus and new species of plant bug from the United States (Hemiptera: Miridae). Entomological News, 76: 85-89.

1967. Revision of the cactus plant bug genus *Hesperolabops* Kirkaldy (Hemiptera: Miridae). Proceedings of the United States National Museum, 123(3614): 1-11.

1976. Zoogeographic notes on the lace bug genus *Acalypta* Westwood in the Americas with description of a new species from Mexico (Hemiptera: Tingidae). American Midland Naturalist, 96: 257-269.

1978. The stink bug *Padaeus trivittatus* Stål and not *Padaeus viduus* (Vollenhoven) in the United States (Heteroptera: Pentatomidae). Proceedings of the Entomological Society of Washington, 80: 131-132.

1985. Synopsis of the Heteroptera or true bugs of the Galápagos Islands. Smithsonian Contributions to Zoology, 407: 1-84.

1986. Restoration of the species name *Salda coriacea* Uhler, 1872 (Hemiptera: Saldidae). Proceedings of the Entomological Society of Washington, 88: 394.

Froeschner, R. C. and R. M. Baranowski

1970. First United States records for a West Indian burrower bug, *Amnestus trimaculatus* (Hemiptera: Cydnidae). Florida Entomologist, 53: 15.

Froeschner, R. C. and Q. L. Chapman

1963. A South American cydnid, *Scaptocoris castaneus* Perty, established in the United States (Hemiptera: Cydnidae). Entomological News, 74: 95-98.

Froeschner, R. C., E. V. Coan, and R. E. Ryckman

1981. *Oeciacus vicarius* Horvath, 1912 (Insecta, Hemiptera, Cimicidae): Proposed conservation under the Plenary Powers. Z.N.(S.) 2358. Bulletin of Zoological Nomenclature, 40: 65-66.

Froeschner, R. C. and L. Halpin

1981. Heteroptera recently collected in the Ray Mountains in Alaska. Proceedings of the Biological Society of Washington, 94: 423-426.

Froeschner, R. C. and L. E. Peña

1985. First South American record for the Circum-Mediterranean *Patapius spinosus* (Rossi) (Heteroptera: Leptopodidae). Revista Chilena de Entomologia, 12: 223.

Froeschner, R. C. and W. E. Steiner, Jr.

1983. Second record of South American burrowing bug, *Scaptocoris castaneus* Perty (Hemiptera: Cydnidae), in the United States. Entomological News, 94: 176.

Frost, S. W.

1925. Frogs as insect collectors. Journal of the New York Entomological Society, 32: 174-185, plate 14 (1924).

1952. Miridae from light traps. Journal of the New York Entomological Society, 60: 237-240.

1964. Insects taken in light traps at the Archbold Biological Station, Highlands County, Florida. Florida Entomologist, 47: 129-161.

Fuessly, J. C.

1775. Verzeichniss der ihm bekannten Schweizerishen Insecten. Zurich und Winterthur, 4: 1-62.

Fulton, B. B.

1918. Observations on the life history and habits of *Pilophorus walshii* Uhler. Annals of the Entomological Society of America, ll: 93-96.

Furth, D., R. Albrecht, K. Muraszko, and G. E. Hutchinson

1978. Scanning electron microscope study of the palar pegs of three species of Corixidae (Hemiptera). Systematic Entomology, 3: 147-152.

Fye, R. E.
1971. Grain sorghum - a source of insect predators for insects on cotton. Progressive Agriculture in Arizona, 23: 12-13.
1974. Population defined and approaches to measuring population density, dispersal, and dispersion. Pp. 46-61. *In* F. G. Maxwell and F. A. Harris (eds.). Proceedings of the summer institute on biological control of plant insects and diseases. University Press of Mississippi, Jackson. 647 pp.

Fye, R. E. and R. L. Carranza
1972. Movement of insect predators from grain sorghum to cotton. Environmental Entomology, 1: 790-791.

Fyles, T. W.
1903. *Aratus* [*sic*] *luteolus*, n. sp. Canadian Entomologist, 35: 75-76.

Gagné, W. C. and F. G. Howarth
1975. The cavernicolous fauna of Hawaiian lava tubes, 6. Mesoveliidae or water treaders (Heteroptera). Pacific Insects, 16: 399-413.

Galbreath, J. E.
1973. Diapause in *Mesovelia mulsanti* (Hemiptera: Mesoveliidae). Journal of the Kansas Entomological Society, 46: 224-233.
1975. Thoracic polymorphism in *Mesovelia mulsanti* (Hemiptera: Mesoveliidae). University of Kansas Science Bulletin, 50: 459-482.
1976. The effect of the age of the female on diapause in *Mesovelia mulsanti* (Hemiptera: Mesoveliidae). Journal of the Kansas Entomological Society, 49: 27-31.
1977. Reproduction in *Mesovelia mulsanti*: (Hemiptera: Mesoveliidae). Transactions of the Illinois Academy of Science, 69: 91-99.

Germar, E. F.
1817-1847.Fauna Insectorum Europae. Kümmel, Halae. Fasicles 3-25; 1817 Fascicle 3; 1821 Fascicle 4; 1822 Fascicle 5-7; 1824 Fascicle 8-10; 1825 Fascicle 11-12; 1827 Fascicle 13; 1831 Fascicle 14-15; 1837 Fascicle 16-17; 1836 Fascicle 18-19; 1838 Fascicle 20; 1839 Fascicle 21; 1842 Fascicle 22; 1845 Fascicle 23; 1847 Fascicle 24. [from Horn & Schenkling, 1938, Index Liter. Ent., page 412].
1837a. Hemiptera Heteroptera promontorii bonae spei, nundum descripta, quae collegit C. F. Drége. Revue entomologique publiée par Silbermann, 5: 121-192.
1839. Beiträge zu einer Monographie der Schildwanzen. Zeitschrift für die Entomologie, 1: 1-146, plate 1.

Ghauri, M. S. K. and F. Y. K. Ghauri
1983. A new genus and new species of Isometopidae from North India, with a key to world genera (Heteroptera). Reichenbachia, 21: 19-25.

Giacchi, J. C.
1969. Revision del genero *Stenopoda* Laporte, 1833 (Hemiptera, Reduviidae, Stenopodainae). Physis, 29: 1-26.
1977. Revision de los Stenopodainos americanos. IV. El genero *Gnathobleda* Stal, 1859 (Hemiptera-Reduviidae). Physis, Secc. C., 37(93): 261-274.
1980. Una nueva especie para el genero *Diaditus* Stal, 1859 (Stenopodainae-Reduviidae). Revista de la Sociedad Entomologia Argentina, 39: 1-4.
1982. Revision de los Stenopodainos americanos. V. El género *Diaditus* Stal, 1859 (Heteroptera, Reduviidae). Physis, Secc. C., 41(100): 9-27.
1984. Revision de los Stenopodainos americanos. VI. Los especies americanos del genero *Oncocephalus* Klug, 1830 (Heteroptera-Reduviidae). Physis, Secc. C., 42(103): 39-62.

Giard, M. A.
1892. Sur un Hémiptère Hétéroptère (*Halticus* minutus Reuter) qui ravage les arachides en Cochinchine. Comptes Rendus des Séances de la SociétéBiologie Serie 9, 4: 79-82.

Gibson, A.
1911. The entomological records, 1910. 41st Report Entomological Society of Ontario, 1910: 101-120.
1913. The entomological record, 1912. Forty-Third Annual Report of the Entomological Society of Ontario, 1912: 112-140.

Gibson, E. H.
1917. [*Monanthia ehrethiae* Gibson, sp. nov.]. *In* Drake, C. J. The North American species of *Monanthia* (Tingidae). Bulletin of the Brooklyn Entomological Society, 12: 49-52.
1917. The family Isometopidae Fieb. as represented in North America (Heteroptera.). Bulletin of the Brooklyn Entomological Society, 12: 73-77.
1917. A new species of *Corythuca* from the

Gibson, E. H. (continued)
Northwest (Heterop., Tingitidae). Entomological News, 28: 258.
1917. The genus *Harmostes* Burm. (Coreidae, Heterop.). Entomological News, 28: 439-450.
1918. The genus *Hadronema* Uhl. (Miridae; Heteroptera). Canadian Entomologist, 50: 81-84.
1918. The genus *Corythucha* Stal (Tingidae; Heteroptera). Transactions of the American Entomological Society, 44: 69-104.
1919. Notes on the North American species of *Corizus* (Coreidae, Heteroptera). Canadian Entomologist, 51: 89-92.
1919. The genera *Corythaica* Stål and *Dolichocysta* Champion (Tingidae: Heteroptera). Proceedings of the Biological Society of Washington, 32: 97-104.
1919. The genus *Gargaphia* Stål (Tingidae; Heteroptera). Transactions of the American Entomological Society, 45: 187-201.

Gibson, E. H. and A. Holdridge
1918. Notes on the North and Central American species of *Acanthocephala* Lap. (Fam. Coreidae: Heteroptera). Canadian Entomologist, 50: 237-241.

Gilbert, B. L., S. J. Barras, and D. M. Norris
1967. Bionomics of *Tetyra bipunctata* (Hemiptera: Pentatomidae: Scutellerinae) as associated with *Pinus banksiana* in Wisconsin. Annals of the Entomological Society of America, 60: 698-701.

Gillette, C. P.
1890. *Abcanthia* [*sic*] *papistrilla* in nests of the barn swallow. Entomological News, 1: 26-27.

Gillette, C. P. and C. F. Baker
1895. A preliminary list of the Hemiptera of Colorado. Bulletin of the Colorado Agricultural Experiment Station, 31(Technical Series 1): 1-137.

Girard, G. L.
1937. Life history, habits and food of the sage grouse, *Centrocercus urphasianus* Bonaparte. University of Wyoming Publication, 3: 1-56.

Girault, A. A.
1905. The bedbug, *Clinocoris* (=*Cimex* = *Acanthia* = *Klinophilos*) *lectularia* Linnaeus. Part I. Life-history at Paris, Texas, with biological notes, and some considerations on the present state of our knowledge concerning it. Psyche, 12: 61-74.

Gistl, J. N. F. X.
1837. Systematische Übersicht der Wanzen und Cicaden der Umgegund von München. Faunus Germany, 1(2): 98-111.
1847. *In* Gistel, J. and J. & F. Bromme. Handbuch der Naturgeschichte aller drei Reiche, für Lehrer and Lernende, für Schule und Haus. Hoffmann, Stuttgart. 1037 pages, 48 plates (1850). [See Menke, A. S., 1976, Ent. News, 87: 167-170, for clarification of dates].
1848. Naturgeschichte des Thierreichs. Für höhere Schulen. Hoffman, Stuttgart. XVI + 216 + 4 pages, 32 plates.

Gittelman, S. H.
1974. The habitat preference and immature stages of *Neoplea striola* (Hemiptera: Pleidae). Journal of the Kansas Entomological Society, 47: 491-503.
1975. Physical gill efficiency and winter dormancy in the pygmy backswimmer, *Neoplea striola* (Hemiptera: Pleidae). Annals of the Entomological Society of America, 68: 1011-1017.

Glenn, P. A.
1923. The onion capsid, *Orthotylus translucens* Tucker. Journal of Economic Entomology, 16: 79-81, 1 plate.

Glick, P. A.
1939. The distribution of insects, spiders and mites in the air. United States Department of Agriculture Technical Bulletin, 673: 1-150.
1957. Collecting insects by airplane in southern Texas. United States Department of Agriculture Technical Bulletin, 1158: 1-28.

Glick, P. A. and W. B. Lattimore, Jr.
1954. The relation of insecticides to insect populations in cotton fields. Journal of Economic Entomology, 47: 681-684.

Glover, T.
1868. Report of the Entomologist. *In* United States Department of Agriculture, Report of the Comissioner of Agriculture for the year 1867, pages 58-76.
1875. Report of the Entomologist. Heteroptera or plant-bugs. Report of the Department of Agriculture of the United States, page 125.

1876. Manuscript notes from my journal, or illustrations of insects, native and foreign. Order Hemiptera, suborder Heteroptera, or plant bugs. Washington, D.C. II + 132 pages, 10 plates [+ 10 captions].

Gmelin, J. F.

1790. Hemiptera. *In* Caroli a Linné Systema Naturae. Edition 13 [editre by J. F. Gmelin]. 1 (4): 1520, 1523, 2041-2224. [Although title page is dated 1788 and Sherborn (1902, Index Animalium) dated the species in the above work as 1789, the present list follows Hopkinson (1907, Proc. Zool. Soc. Lond., pp. 1035-1036), Apstein (1916, Zool. Anz., 47: 32) and most subsequent authors who use the date 1790].

Goble, H. W.

1972. Insects of the season 1972 related to fruit, vegetables, field crops and ornamentals. Proceedings of the Entomological Society of Ontario, 103: 1-2.

Goeden, R. D. and D. W. Ricker

1976. The phytophagous insect fauna of the ragweed, *Ambrosia dumosa*, in southern California. Environmental Entomology, 5: 45-50.

1976. The phytophagous insect fauna of the ragweed, *Ambrosia psilostachya*, in southern California. Environmental Entomology, 5: 1169-1177.

Goel, S. C. and C. W. Schaefer

1970. The structure of the pulvillus and its taxonomic value in the land Heteroptera (Hemiptera). Annals of the Entomological Society of America, 63: 307-313.

Goeze, J. A. E.

1777-1783. Entomologische Beyträge zu des Ritter Linne zwolften Ausgabe des Natursystems. 3 volumes. Weidmann, Leipzig. [1777, 1: i-xvi + 1-736; 1778, 2: i-lxxii + 1-352; 1779, 3(1): i-xl + 1-390; 1780, 3(2): i-xxiv + 1-350;, 1781, 3(3): i-xlviii + 1-439; 1783, 3(4): i-xx + 1-178].

Göllner-Scheiding, U.

1975. Revision der Gattung *Stictopleurus* Stål, 1872 (Heteroptera, Rhopalidae). Deutsche Entomologische Zeitschrift, 22: 1-60.

1976. Revision der Gattung *Liorhyssus* Stål, 1870 (Heteroptera, Rhopalidae). Deutsche Entomologische Zeitschrift, 23: 181-206.

1978. Revision der Gattung *Harmostes* Burm., 1835 (Heteroptera, Rhopalidae) und einige Bemerkungen zu den Rhopalinae. Mitteilungen aus dem Zoologischen Museum in Berlin, 54: 257-311.

1978. Bemerkungen zu der Gattung *Rhopalus* Schilling einschliesslich *Brachycarenus* Fieber (Heteroptera, Rhopalidae). Mitteilungen aus dem Zoologischen Museum in Berlin, 54: 313-331.

1979. Die Gattung *Jadera* Stål, 1862 (Heteroptera, Rhopalidae). Deutsche Entomologische Zeitschrift, 26: 47-75.

1980. Revision der afrikanischen Arten sowie Bemerkungen zu weiteren Arten der Gattungen *Leptocoris* Hahn, 1833, und *Boisea* Kirkaldy, 1910. Deutsche Entomologische Zeitschrift, 27: 103-148.

1980. Einige Bemerkungen zu den Gattungen *Corizus* Fallén, 1814, und *Xenogenus* Berg, 1883 (Heteroptera, Rhopalidae). Mitteilungen Zoologischen aus dem Zoologischen Museum in Berlin, 56: 111-121.

1982. Ergänzungen zu Gattungen der Rhopalidae (Heteroptera). Deutsche Entomologische Zeitschrift, 29: 459-467.

1983. General-Katalog der Familie Rhopalidae (Heteroptera). Mitteilungen aus dem Zoologischen Museum in Berlin, 59: 37-189.

1984. Korrektur zu: Dt. Entom. Z. 26 (1979): 60 und 29 (1982): 459. Deutsche Entomologische Zeitschrift, 31: 337.

Golub, V. B.

1973. On the systematics the palearctic lace bugs of the genus *Acalypta* Westw. (Heteroptera, Tingidae). Revue d'Entomologie de l'URSS, 52: 628-632. [English translation in Entomological Review, 52: 417-419].

Gonsoulin, G. J.

1974. Seven families of aquatic and semiaquatic Hemiptera in Louisiana. Part IV. Family Gerridae. Transactions of the American Entomological Society, 100: 513-549.

Gonzalez, D., D. A. Ramsey, T. F. Leigh, B. S. Ekbom, and R. van den Bosch

1977. A comparison of vacuum and whole-plant methods for sampling predaceous arthropods on cotton. Environmental Entomology, 6: 750-760.

Gonzales, D. and L. T. Wilson

1982. A food-web approach to economic thresholds: A sequence of pests/predaceous arthropods on California cotton. Entomophaga, 27(Special Issue): 31-43.

Gordh, G. and R. A. Coker
1973. A new species of *Telenomus* parasitic on *Geocoris* (Hymenoptera: Proctotrupoidea: Hemiptera: Lygaeidae) in California. Canadian Entomologist, 105: 1407-1411.

Gould, G. E.
1931. The *Rhagovelia* of the Western Hemisphere, with notes on world distribution (Hemiptera, Veliidae). University of Kansas Science Bulletin, 20: 5-61.

Gracen, V. E., Jr.
1986. Host plant resistance for insect control in some important crop plants. Critical Reviews in Plant Science, 4: 277-291.

Graham, H. M., A. A. Negm, and L. R. Ertle
1984. Worldwide literature of the *Lygus* complex (Hemiptera: Miridae), 1900-1980. United States Department of Agriculture, Bibliography and Literature of Agriculture, No. 30, 205 pages.

Gravena, S. and W. L. Sterling
1983. Natural predation on the cotton leafworm (Lepidoptera: Noctuidae). Journal of Economic Entomology, 76: 779-784.

Gray, G. R.
1832. [New genera and species]. *In* Griffith, E. The Animal Kingdom arranged in accordance with its organization by the Baron Cuvier Member of the Institute of France. Volume 15, part 2.

Greene, G. L., W. H. Whitcomb, and R. Baker
1974. Minimum rates of insecticide on soybeans. *Geocoris* and *Nabis* populations following treatment. Florida Entomologist, 57: 114.

Griffin, F. J.
1937. A further note on "Palisot de Beauvois, Insectes Rec. Afr. Amér." 1805-1821. Journal of the Society for the Bibliography of Natural History, 1: 121-122.

Griffith, M. E.
1945. The environment, life history and structure of the water boatman, *Ramphocorixa acuminata* (Uhler) (Hemiptera, Corixidae). University of Kansas Science Bulletin, 30: 241-365.

Griffith, R.
1980. Transmission of microorganisms associated with cedros wilt disease of coconuts. Journal of the Agricultural Society of Trinidad and Tobago, 80: 303-310.

Gross, H. R., Jr., S. D. Pair, and R. D. Jackson
1985. Behavioral responses of primary entomophagous predators to larval homogenates of *Heliothis zea* and *Spodoptera frugiperda* (Lepidoptera: Noctuidae) in whorl-stage corn. Environmental Entomology, 14: 360-364.

Gruetzmacher, M. C. and J. C. Schaffner
1977. Modification of the concept of the genus *Hadronema* Uhler as necessitated by a new species from Texas (Heteroptera, Miridae). Southwestern Entomologist, 2: 53-56.

Guérin-Méneville, F. E.
1831. Hémiptères. *In* L. J. Duperrey (editor). Voyage autour du monde, Zoologies, 2: Atlas, plates 10-12. [Text published in 1838].
1831-1838.Septième ordre. Les Hémiptères, *In* Iconographie du régne animal de G. Cuvier, ou représentation d'après nature de l'une des espéces les plus remarquables, et souvent non encore figurées, de chaque genre d'animaux. Avec un texte descriptif mis au courant de la science. Insectes, pages 343-841, plates 55-59. [Atlas published in 1831, text in 1838. See Dupuis, 1953, Bulletin de la Société-Zoologique de France, 77: 447-454].
1843. Note sur la *Naucoris rugosa* de J. Desjardins, formant un nouveau genre d'-Hémiptères et description de plusieurs espèces des genres *Pelogonus* et *Mononyx*. Revue Zoologique Travaux Inedits, 1843: 112-114.
1857. Ordre des Hémiptères, Latr. Première section. Hétéroptères, Latr. *In* M. R. Sagra's Historie Physique, Politique et Naturelle de l'Ile de Cuba. Arthus Bertrand, Paris. 7: 359-424.
1857. Entomologie appliquée Hautle-Pain D'Insectes. Le Moniteur Universel, J. Officiel de l'Empire Français. 330: 1298.

Gulde, J.
1936. Die Wanzen Mitteleuropas. Hemiptera Heteroptera Mitteleuropas. 7. Familie: Lygaeidae. Frankfort. 5(1): 1-104.
1937. Die Wanzen Mitteleuropas. Hemiptera Heteroptera Mitteleuropas. 7. Familie: Lygaeidae. Frankfort. 5(2): 105-222.

Guppy, J. C.
1963. Observations on the biology of *Plagiognathus chrysanthemi* (Hemiptera: Miridae), a pest of birdsfoot trefoil in On-

tario. Annals of the Entomological Society of America, 56: 804-809.

Gutierrez, A. P., D. W. Demichele, Y. Wang, G. L. Curry, R. Skeith, and L. G. Brown
1980. The systems approach to research and decision making for cotton pest control. Pages 155-186. *In* Huffaker, C. B. (editor). Environmental science and technology: new technology of pest control. Wiley, New York. 500 pp.

Hagan, H. R.
1931. The embryogeny of the polyctenid, *Hesperoctenes fumarius* Westwood, with reference to viviparity in insects. Journal of Morphology and Physiology, 51: 3-117, 1 plate.

Hagen, H. A.
1884. Enemies of *Pieris menapia*. Canadian Entomologist, 16: 40.

Hagen, K. S., S. Bombosch, and J. A. McMurty
1976. Chapter 5. The biology and impact of predators. Pages 93-142. *In* C. B. Huffaker and P. S. Messenger (editors). Theory and practice of biological control. Academic Press, New York. 788 pages.

Hagen, K. S. and R. Hale
1974. Increasing natural enemies through use of supplementary feeding and non-target prey. Pages 170-181. *In* Maxwell, F. G. and F. A. Harris (editors). Proceedings of the summer institute on biological control of plant insects and diseases. University Press of Mississippi, Jackson. 647 pages.

Hagen, K. S., R. van den Bosch, and D. L. Dahlsten
1971. The importance of naturally-occurring biological control in the western United States. Pages 253-293. *In* Huffaker, C. B. (editor). Biological control. Plenum Press, New York and London.

Haglund, C. J. E.
1868. Hemiptera nova. Stettiner Entomologische Zeitung, 29: 150-163.

Hahn, C. W.
1826. Icones and monographiam Cimicum. Lechner, Nürnburg. 1 page, 24 plates.
1831-1836. Die Wanzenartigen Insecten. C. H. Zeh'schen Buchhandlung, Nürnburg. 1: 1-36 (1831); 1: 37-118 (1832); 1: 119-236 (1833); 2: 1-32 (1833); 2: 33-120 (1834); 2: 121-142 (1835); 3: 1-16 (1835); 3: 17-34 (1836). [for subsequent parts see Herrich-Schaeffer, G. A. W. 1836-1853].

Haldeman, S. S.
1847. Descriptions of several new species and one new genus of insects. Proceedings of the Academy of Natural Sciences of Philadelphia, 3: 149-151.
1852. Appendix C-Insects. Pages 366-378, 2 plates. *In* H. Stansbury, Exploration and Survey of the Valley of the Great Salt Lake of Utah including a reconnoissance of a new route through the Rocky Mountains - Zoology. U.S. Government Printing Office, Washington, D.C.
1853. Descriptions of some new species of insects, with observations on described species. Proceedings of the Academy of Natural Sciences of Philadelphia, 6: 361-365.

Halstead, T. F.
1970. A new species of the genus *Largus* Hahn with a key to the species of the genus in the southwestern United States (Hemiptera: Largidae). Pan-Pacific Entomologist, 46: 45-46.
1972. A new species of *Largus* (Hemiptera: Largidae). Canadian Entomologist, 104: 959.
1972. A review of the genus *Arhaphe* Herrich-Schäffer (Hemiptera: Largidae). Pan-Pacific Entomologist, 48: 1-7.
1972. Notes and synonymy in *Largus* Hahn with a key to United States species (Hemiptera: Largidae). Pan-Pacific Entomologist, 48: 246-248.

Hambleton, E. J.
1968. New state records for a lace bug (*Dictyla echii*). United States Department of Agriculture, Cooperative Economic Insect Report, 18: 658.

Hambleton, J. C.
1908. The genus *Corizus* with a review of the North and Middle American species. Annals of the Entomological Society of America, 1: 133-147, plates 8-11.
1909. Life history of *Corizus lateralis* Say. Annals of the Entomological Society of America, 2: 272-276, 1 plate.

Hamid, A.
1971. The life cycles of three species of *Cymus* (Hemiptera: Lygaeidae) in Connecticut. University of Connecticut Occasional Papers, Biological Science Series, 2: 21-28.

Hamid, A. (continued)

1975. A systematic revision of the Cyminae (Heteroptera: Lygaeidae) of the world with a discussion of the morphology, biology, phylogeny and zoogeography. Entomological Society of Nigeria Occasional Publication, 14: 1-179.

Hamilton, S. W.

1983. *Neortholomus*, a new genus of Orsillini (Hemiptera-Heteroptera: Lygaeidae: Orsillinae). University of Kansas Science Bulletin, 52: 197-234.

Hamlin, J. C.

1923. New cactus bugs of the genus *Chelinidea* (Hemiptera). Proceedings of the Royal Society of Queensland, 35: 43-45.

1924. A review of the genus *Chelinidea* (Hemiptera-Heteroptera) with biological data. Annals of the Entomological Society of America, 17: 193-208.

Handlirsch, A.

1897. Monographie der Phymatiden. Annalen des Kaiserlich-Koniglichen Naturhistorischen Hofsmuseums, 12: 127-230, plates IV-IX.

1898. Zwei neue Phymatiden. Verhandlungen der Kaiserlich-Koniglichen Zoologisch-Botanischen Gesellschaft in Wien, 48: 382-283.

1925. Handbuch der Entomologie. Band III. Geschichte, Literatur, Technik, Paleontologie, Phylogenie, Systematik. G. Fisher, Jena. Pages 276-277, 1071-1075.

Hantsbarger, W. M.

1957. *Nysius* of South Dakota (Lygaeidae-Hemiptera). Journal of the Kansas Entomological Society, 30: 156-159.

Hardee, D. D., H. Y. Forsythe, Jr., and G. G. Gyrisco

1963. A survey of the Hemiptera and Homoptera infesting grasses (Grammineae) in New York. Journal of Economic Entomology, 56: 555-559.

Harrington, B. J.

1980. A generic level revision and cladistic analysis of the Myodochini of the world (Hemiptera, Lygaeidae, Rhyparochrominae). Bulletin of the American Museum of Natural History, 167: 45-116.

Harrington, J. E.

1972. Notes on the biology of *Ischnodemus* species of America north of Mexico (Hemiptera: Lygaeidae: Blissinae). University of Connecticut Occasional Papers, Biological Science Series, 2: 47-56.

Harrington, W. H.

1892. Fauna Ottawaensis, Hemiptera. Ottawa Naturalist, 6: 25-32.

Harris, H. M.

1925. A new species of Nabidae (Costa) from the western United States (Hemiptera). Entomological News, 36: 205-206.

1926. Notes on some American Nabidae (Hemiptera). Entomological News, 37: 287.

1928. A monographic study of the hemipterous family Nabidae as it occurs in North America. Entomologica Americana, 9 (new series): 1-98.

1937. Contributions to the South Dakota list of Hemiptera. Iowa State College Journal of Science, 11: 169-176.

1940. A new *Pagasa* from the United States (Hemiptera, Nabidae). Entomological News, 51: 35-37.

1941. Concerning Neididae, with new species and new records for North America. Bulletin of the Brooklyn Entomological Society, 36: 105-109.

1942. Some new American Rhopalidae (Hemip-tera). Journal of the Kansas Entomological Society, 15: 100-105.

1942. The male of *Pagasa fasciventris* H. M. Harris (Hemiptera, Nabidae). Entomological News, 53: 36.

1942. *Hebrus* Curtis antedates *Naeogeus* LaPorte (Hemiptera, Hebridae). Pan-Pacific Entomologist, 18: 124.

1943a. Additions to the South Dakota list of Hemiptera. Journal of the Kansas Entomological Society, 16: 150-153.

1944. Concerning American Rhopalini (Hemiptera, Rhopalidae). Iowa State College Journal of Science, 19: 99-109.

Harris, H. M. and Andre, F.

1934. Notes on the biology of *Acantholoma denticulata* Stal (Hemiptera, Scutelleridae). Annals of the Entomological Society of America, 27: 5-15.

Harris, H. M. and C. J. Drake

1941. A new genus and species of Anthocoridae (Hemiptera). Iowa State College Journal of Science, 15: 343-344.

1944. New apterous Aradidae from the Western Hemisphere (Hemiptera). Proceedings of the Entomological Society of Washington, 46: 128-132.

Harris, H. M. and H. G. Johnston
1936. A new genus and species of Podopidae and a new *Coenus* (Hemiptera: Scutelleroideae). Iowa State College Journal of Science, 10: 377-380.

Harris, H. M. and W. E. Shull
1944. A preliminary list of Hemiptera of Idaho. Iowa State College Journal of Science, 18: 199-208.

Harris, T. W.
1833. Insects. Pages 566-595. *In* Hitchock, E. Report on the Geology, Minerology, Botany, and Zoology of Massachusetts. Amherst. xii + 700 pages. [1st edition 1833; 2nd edition 1835].
1841. A report on the insects of Massachusetts injurious to vegetation. Cambridge. 459 pages.
1862. A treatise on some of the insects injurious to vegetation. 3rd edition. William White, Boston. xi + 640 pages [2nd edition, 1852, viii + 513 pages].

Harris, V. E. and J. R. Phillips
1986. The effect of mowing spring weed hosts of *Heliothis* spp. on predatory arthropods. Journal of Agricultural Entomology, 3: 77-86.

Harrison, M. D., J. W. Brewer, and L. D. Merrill
1980. Insect involvement in the transmission of bacterial pathogens. Pages 201-292. *In* Harris, K. F. and K. Maramorosch (editors). Vectors of Plant Pathogens. Academic Press, New York. 467 pages.

Harrison, R. G.
1980. Dispersal polymorphism in insects. Annual Review of Ecology and Systematics, 11: 98-118.

Hart, C. A.
1907. On the biology of the sand areas of Illinois. Part III. Zoological studies in the Sand Regions of the Illinois and Mississippi River Valleys. Bulletin of the Illinois State Laboratory of Natural History, 7: 195-267.
1919. The Pentatomoidea of Illinois with keys to the nearctic genera. Bulletin of the Illinois Natural History Survey, 13: 155-223, plates 16-21.

Hawley, I. M.
1917. The hop redbug (*Paracalocoris hawleyi* Knight). Journal of Economic Entomology, 10: 545-552, plate 28.

Hayes, R. O., D. B. Francy, J. S. Lazuick, G. C. Smith, and E. P. J. Gibbs
1977. Role of the cliff swallow bug (*Oeciacus vicarius*) in the natural cycle of a western equine encephalitis-related alphavirus. Journal of Medical Entomology, 14: 257-262.

Hayward, C. L.
1948. Biotic communities of the Wasatch Chaparral, Utah. Ecological Monographs, 18: 473-506.

Heer, O.
1853. Die Insectenfauna der Tieriärgebilde von Oeningen und Radoboj in Croatien. Engelmann, Leipzig. Volume 3. 4 + 138 pages + 15 plates.

Heidemann, O.
1891. Note on the occurrence of a rare capsid, near Washington, D. C. Proceedings of the Entomological Society of Washington, 2: 68-69.
1892. Note on the food-plants of some Capsidae from the vicinity of Washington, D. C. Proceedings of the Entomological of Society Washington, 2: 224-226.
1895. [In notes of meeting]. Proceedings of the Entomological Society of Washington, 3: 292.
1895. [Records of Hemiptera from D.C. exhibited at monthly meeting]. Proceedings of the Entomological Society of Washington, 3: 143.
1899. A new species of Tingitidae. Canadian Entomologist, 31: 301-302.
1899. [Notes in minutes of meeting]. Proceedings of the Entomological Society of Washington, 4: 339.
1900. Papers from the Harriman Alaska Expedition. XIII. Entomological results (7): Heteroptera. Proceedings of the Washington Academy of Sciences, 2: 503-506.
1901. [*Gargaphia undulata*; in minutes of meeting]. Proceedings of the Entomological Society of Washington, 4: 493.
1902. Hemiptera. Pages 80-82. *In* Caudell, A. M. Some insects from the summit of Pike's Peak, found on snow. Proceedings of the Entomological Society of Washington, 5: 74-82.
1903. [Notes in minutes of meeting]. Proceedings of the Entomological Society of Washington, 5: 102.
1903. Remarks on *Ligyrocoris constrictus* Say and description of *Perigenes fallax*, a

Heidemann, O. (continued)
new species. Proceedings of the Entomological Society of Washington, 5: 155-157.
1903. [Notes in Minutes of Meeting]. Proceedings of the Entomological Society of Washington, 5: 309-310.
1904. [Notes in minutes of meeting]. Proceedings of the Entomological Society of Washington, 6: 11-12.
1904. Notes of North American Aradidae, with descriptions of two new species. Proceedings of the Entomological Society of Washington, 6: 161-165.
1904. Notes on a few Aradidae occurring north of the Mexican Boundary. Proceedings of the Entomological Society of Washington, 6: 229-233.
1905. A list of capsids from the state of New York, with the description of a new species. Journal of the New York Entomological Society, 13: 48-50.
1905. Description of a new *Anasa* from North America. Proceedings of the Entomological Society of Washington, 7: 11-12.
1906. Account of a new tingitid. Proceedings of the Entomological Society of Washington, 8: 10-13.
1906. A new genus and species of the hemipterous family Ceratocombidae from the United States. Proceedings of the Entomological Society of Washington, 7: 192-194.
1907. Three new species of North American Aradidae. Proceedings 'of the Entomological Society of Washington, 8: 68-71.
1908. Two new species of North American Tingitidae [Hemiptera-Heteroptera]. Proceedings of the Entomological Society of Washington, 10: 103-108, plate IV.
1908. Notes on *Heidemannia cixiiformis* Uhler and other species of Isometopinae. Proceedings of the Entomological Society of Washington, 9: 126-130.
1909. New species of Tingitidae and description of a new *Leptoglossus* (Hemiptera-Heteroptera). Bulletin of the Buffalo Society of Natural Sciences, 9: 231-238.
1909. Two new species of North American Aradidae. Proceedings of the Entomological Society of Washington, 11: 189-191.
1910. New species of *Leptoglossus* from North America. Proceedings of the Entomological Society of Washington 12: 191-197, plates VII and VIII [captions of figures omitted above, but given in volume 13: 83].
1910. [Notes in minutes of meeting]. Proceedings of the Entomological Society of Washington, 12: 45-47.
1910. Description of a new capsid. Proceedings of the Entomological Society of Washington, 12: 200-201.
1911. Some remarks on the eggs of North American species of Hemiptera-Heteroptera. Proceedings of the Entomological Society of Washington, 13: 128-140, plates 9-12.
1911. A new species of North American Tingitidae. Proceedings of the Entomological Society of Washington, 13: 180-181.
1913. Description of two new species of North American Tingitidae. Proceedings of the Entomological Society of Washington, 15: 1-4.
1914. A new species of North American Tingitidae. Proceedings of the Entomological Society of Washington, 16: 136-137.
1917. Two new species of lace-bugs (Heteroptera: Tingidae). Proceedings of the Entomological Society of Washington, 18: 217-220 (1916).

Heiss, E.
1980. Nomenklatorische Änderungen und Di fferenzierung von *Aradus crenatus* Say, 1831, und *Aradus cinnamomeus* Panzer, 1806, aus Europa und USA (Insecta: Heteroptera, Aradidae). Bericht des Naturwissenschaftlich-Medizinischen Vereins in Innsbruck, 67: 103-116.

Hemming, F. and D. Noakes
1958. Official list of books approved as available for Zoological Nomenclature, First Installment: Names 1-38. London, xi + 12 pages.
1958. Official index of rejected and invalid generic names in Zoology. International Trust for Zoological Nomenclature, London. Direction 63, pages 93-94.

Henneberry, T. J., L. A. Bariola and D. L. Kittock
1977. Nectariless cotton: Effect on cotton leafperforator and other cotton insects in Arizona. Journal of Economic Entomology, 70: 797-799.

Henneberry, T. J. and R. E. Clayton
1985. Consumption of pink bollworm (Lepidoptera: Gelechiidae) and tobacco budworm (Lepidoptera: Noctuidae) eggs by some predators commonly found in cotton fields. Environmental Entomology,

14: 416-419.

Henry, T. J.

1974. Two new pine-inhabiting *Phytocoris* from Pennsylvania (Hemiptera: Miridae). Entomological News, 85: 187-191.

1976. A new *Saileria* from eastern United States (Hemiptera: Miridae). Entomological News, 87: 29-31.

1976. Review of the genus *Reuteria* Puton 1875, with descriptions of two new species (Hemiptera: Miridae). Entomological News, 87: 61-74.

1977. *Orthotylus nassatus*, a European plant bug new to North America (Heteroptera: Miridae). United States Department of Agriculture Cooperative Plant Pest Report, 2: 605-608.

1977. *Teratodia* Bergroth, new synonym of *Diphleps* Bergroth with descriptions of two new species (Heteroptera: Miridae: Isometopinae). Florida Entomologist, 60: 201-210.

1978. Review of the neotropical genus *Hyalochloria*, with descriptions of ten new species (Hemiptera: Miridae). Transactions of the American Entomological Society, 104: 69-90.

1978. Two new *Ceratocapsus* Reuter 1876, from the eastern United States (Hemiptera: Miridae). Proceedings of the Entomological Society of Washington, 80: 383-387.

1978. Description of a new *Polymerus*, with notes on two other little known mirids from the New Jersey Pine-Barrens (Hemiptera: Miridae). Proceedings of the Entomological Society of Washington, 80: 543-547.

1979. Review of the New World species of *Bothynotus* Fieber (Hemiptera: Miridae). Florida Entomologist, 62: 232-244.

1979. Review of the *Ceratocapsus lutescens* group, with descriptions of seven new species from the eastern United States (Hemiptera: Miridae). Proceedings of the Entomological Society of Washington, 81: 401-423.

1979. Review of the New World species of *Myiomma* Puton with descriptions of eight new species (Hemiptera: Miridae: Isometopinae). Proceedings of the Entomological Society of Washington, 81: 552-569.

1979. Descriptions and notes on five new species of Miridae from North America (Hemiptera). Melsheimer Entomological Series, 27: 1-10.

1980. Review of *Lidopus* Gibson and *Wetmorea* McAtee and Malloch, descriptions of three new genera and two new species, and key to New World genera (Hemiptera: Miridae: Isometopidae). Proceedings of the Entomological Society of Washington, 82: 178-194.

1980. New records for *Saileria irrorata* and *Tropidosteptes adustus* (Hemiptera: Miridae). Florida Entomologist, 63: 490-493.

1981. A new eastern United States *Psallus* Fieber (Heteroptera: Miridae) from *Physocarpus* (Rosaceae). Proceedings of the Entomological Society of Washington, 83: 399-402.

1982. The onion plant bug genus *Labopidicola* (Hemiptera: Miridae): Economic implications, taxonomic review, and description of a new species. Proceedings of the Entomological Society of Washington, 84: 1-15.

1982. New synonymies and a new combination in the North American Miridae (Hemiptera). Proceedings of the Entomological Society of Washington, 84: 337-341.

1982. Genus *Parthenicus* in the eastern United States, with descriptions of new species (Hemiptera: Miridae). Florida Entomologist, 65: 354-366.

1983 The garden fleahopper genus *Halticus* (Hemiptera: Miridae): Resurrection of an old name and key to species of the Western Hemisphere. Proceedings of the Entmological Society of Washington, 85: 607-611.

1984. Revision of the spider-commensal plant bug genus *Ranzovius* Distant (Heteroptera: Miridae). Proceedings of the Entomological Society of Washington, 86: 53-67.

1984. Pests not known to occur in the United States or of limited distribution. No. 39: Peruvian cotton stainer. United States Department of Agriculture, APHIS, PPQ Information Circular. 6 pages.

1984. New species of Isometopinae (Hemiptera: Miridae) from Mexico, with new records for previously described North American species. Proceedings of the Entomological Society of Washington, 86: 337-345.

1984. New United States records for two Het-

Henry, T. J. (continued)
eroptera: *Pellaea stictica* (Pentatomidae) and *Rhinacloa pallidipes* (Miridae). Proceedings of the Entomological Society of Washington, 86: 519-520.
1985. Newly recognized synonyms, homonyms, and combinations in the North American Miridae (Heteroptera). Journal of the New York Entomological Society, 93: 1121-1136.
1985. Two new species of *Ceratocapsus* from North America (Heteroptera: Miridae). Proceedings of the Entomological Society of Washington, 87: 387-391.
1985. What is *Capsus frontifer* Walker, 1873 (Heteroptera: Miridae)? Proceedings of the Entomological Society of Washington, 87: 679.
1985. *Caulotops distanti* (Miridae: Heteroptera), a potential *Yucca* pest newly discovered in the United States. Florida Entomologist, 68: 320-323.

Henry, T. J. and J. L. Herring
1979. Review of the genus *Corticoris* with descriptions of two new species from Mexico (Hemiptera: Miridae: Isomeotopinae). Proceedings of the Entomological Society of Washington, 81: 82-96.

Henry, T. J. and L. A. Kelton
1986. *Orthocephalus saltator* Hahn (Heteroptera: Miridae): Corrections of misidentifications and the first authentic report for North America. Journal of the New York Entomological Society, 94: 51-55.

Henry, T. J. and K. C. Kim
1984. Genus *Neurocolpus* Reuter (Heteroptera: Miridae): Taxonomy, economic implications, hosts, and phylogenetic review. Transactions of the American Entomological Society, 110: 1-75.

Henry, T. J., J. W. Neal, Jr., and K. M. Gott
1986. *Stethoconus japonicus* (Heteroptera: Miridae): A predator of *Stephanitis* lace bugs newly discovered in the United States, promising in the biocontrol of azalea lace bug (Heteroptera: Tingidae). Proceedings of the Entomological Society of Washington, 88: 722-730.

Henry, T. J. and R. T. Schuh
1979. Redescription of *Beamerella* Knight and *Hambletoniola* Carvalho and included species (Hemiptera, Miridae), with review of their relationships. American Museum Novitates, No. 2689. 13 pages.

Henry, T. J. and C. L. Smith
1979. An annotated list of the Miridae of Georgia (Hemiptera-Heteroptera). Journal of the Georgia Entomological Society, 14: 212-220.

Henry, T. J. and G. M. Stonedahl
1984. Type designations and new synonymies for nearctic species of *Phytocoris* Fallen (Hemiptera: Miridae). Journal of the New York Entomological Society, 91: 442-465.

Henry, T. J. and A. G. Wheeler, Jr.
1973. *Plagiognathus vitellinus* (Scholtz), a conifer-feeding mirid new to North America (Hemiptera: Miridae). Proceedings of the Entomological Society of Washington, 75: 480-485.
1974. *Sthenarus dissimilis* and *Orthops rubricatus*: Conifer-feeding mirids new to North America (Hemiptera: Miridae). Proceedings of the Entomological Society of Washington, 76: 217-224.
1976. *Dicyphus rhododendri* Dolling, first records from North America (Hemiptera: Miridae). Proceedings of the Entomological Society of Washington, 78: 108-109.
1979. *Orthotylus translucens*: Taxonomic status and corrections of published misidentifications (Hemiptera: Miridae). Proceedings of the Entomological Society of Washington, 81: 60-63.
1979. Palearctic Miridae in North America: Records of newly discovered and little-known species (Hemiptera: Heteroptera). Proceedings of the Entomological Society of Washington, 81: 257-268.
1982. New United States records for six neotropical Miridae (Hemiptera) in southern Florida. Florida Entomologist, 65: 233-241.
1986. *Melanorhopala froeschneri* (Heteroptera: Tingidae): A new lace bug from eastern United States, with notes on host plant and habits, description of fifth instar, and key to species of the genus. Journal of the New York Entomological Society, 94: 235-244.

Herbert, D. A. and J. D. Harper
1986. Bioassays of a betaexotoxin of *Bacillus thuringiensis* against *Geocoris punctipes* (Hemiptera: Lygaeidae). Journal of Economic Entomology, 79: 592-595.

Herrich-Schaeffer, G. A. W.

1835. Nomenclator entomologicus. Verzeichniss der europäischen Insecten; zur Erleichterung des Tauschverkehrs mit Preisen versehen. I. Lepidoptera und Hemiptera. Friedrich Pustet, Regensburg. [Hemiptera, 1: 35-116].

1835. Hemiptera. Part 135 (II). *In* Panzer, G. W. F. Faunae Insectorum Germanicae initia oder Deutschlands Insecten. Regensburg.

1836-1853. Die Wanzenartigen Insecten. C. H. Zeh'schen Buchhandlung, Nürnburg. 3: 33-114 (1836); 4: 1-32 (1837); 4: 33-92 (1838); 4: 93-108 (1839); 5: 1-60 (1839); 5: 61-108 (1840); 6: 1-36 (1840); 6: 37-72 (1841); 6: 73-118 (1842); 7: 1-16 (1842); 7: 17-40 (1843); 7: 41-134 (1844); 8: 1-48 (1845); 8: 49-100 (1846); 8: 101-130 (1847); 9: 1-44 (1849); 9: 45-256 (1850); 9: 257-348 (1851); 9 ["historischer übersicht" and "index"]: 1-210 (1853). [Dates and pagination follow Bergroth, 1919, Ent. Mitt., 8: 188-189].

1900. Bibliographical and nomenclatorial notes on the Rhynchota. No. 1. Entomologist, 33: 238-243.

Herring, J. L.

1949. Taxonomic and distributional notes on the Hydrometridae of Florida (Hemiptera). Florida Entomologist, 31: 112-116 (1948).

1949. A new species of *Rheumatobates* from Florida (Hemiptera, Gerridae). Florida Entomologist, 32: 160-165.

1950. The aquatic and semiaquatic Hemiptera of northern Florida. Part 1: Gerridae. Florida Entomologist, 33: 23-32.

1950. The aquatic and semiaquatic Hemiptera of northern Florida. Part 2: Veliidae and Mesoveliidae. Florida Entomologist, 33: 145-150.

1951. The aquatic and semiaquatic Hemiptera of northern Florida. Part 3: Nepidae, Belostomatidae, Notonectidae, Pleidae and Corixidae. Florida Entomologist, 34: 17-29.

1955. A new American genus of Veliidae (Hem-iptera). Florida Entomologist, 38: 21-25.

1958. Evidence for hurricane transport and dispersal of aquatic Hemiptera. Pan-Pacific Entomologist, 34: 174-175.

1961. The genus *Halobates* (Hemiptera: Gerridae). Pacific Insects, 3: 223-305.

1965. The status of *Amphiareus* Distant, *Buchananiella* Reuter and *Poronotellus* Kirkaldy (Hemiptera: Anthocoridae). Proceedings of the Entomological Society Washington, 67: 202-203.

1966. The Anthocoridae of the Galápagos and Cocos Islands (Hemiptera). Proceedings of the Entomological Society Washington, 68: 127-130.

1966. The genus *Orius* of the Western Hemisphere (Hemiptera: Anthocoridae). Annals of the Entomological Society of America, 59: 1093-1109.

1971. A new species of *Rhinacloa* from palo verde and ocotillo in the western U. S. (Hemiptera: Miridae). Proceedings of the Entomological Society of Washington, 73: 449.

1972. A new species of *Neoborella* from dwarf mistletoe in Colorado (Hemiptera: Miridae). Proceedings of the Entomological Society of Washington, 74: 9-10.

1976. Keys to genera of Anthocoridae of America north of Mexico, with description of a new genus (Hemiptera: Heteroptera). Florida Entomologist, 59: 143-150.

1980. A review of the cactus bugs of the genus *Chelinidea* with the description of a new species (Hemiptera: Coreidae). Proceedings of the Entomological Society of Washington 82: 237-251.

Hesse, A. J.

1947. A remarkable new dimorphic isometopid and two other new species of Hemiptera predacious upon the red scale of citrus. Journal of the Entomological Society of South Africa, 10: 31-45.

Hickman, D. J.

1921. Illustrations of the male hooks in *Nabis* (Nabidae, Hemiptera). Bulletin of the New York Entomological Society, 16: 58-59.

Hidalgo, J., Jr.

1935. The genus *Abedus* Stal. (Hemiptera, Belostomatidae). University of Kansas Science Bulletin, 22: 493-519.

Hill, A. R.

1957. A key to the North American members of the genus *Anthocoris* Fallen (Hemiptera: Anthocoridae). Pan-Pacific Entomologist, 33: 171-174.

Hinton, H. E.

1961. The structure and function of the eggshell in the Nepidae (Hemiptera). Journal of Insect Physiology, 7: 224-257.

Hinton, H. E.(continued)

1962. A key to the eggs of the Nepidae (Hemiptera). Proceedings of the Royal Entomological Society of London, Series A, 37: 65-68.

Hiura, I.

1960. Contribution to the knowledge of Anthocoridae from Japan and its adjacent territories (Hemiptera-Heteroptera). Bulletin of the Osaka Museum of Natural History, 12: 43-55.

Hodgden, B. B.

1949. New Saldidae from Western Hemisphere (Hemiptera). Journal of the Kansas Entomological Society, 22: 149-165.

Hoebeke, E. R.

1977. A rhopalid bug (*Brachycarenus tigrinus* (Schilling). United States Department of Agriculture Cooperative Plant Pest Report, 2: 802.

1978. (Note on *Aethus nigritus* (Fabricius). United States Department of Agriculture Cooperative Plant Pest Report, 3: 376.

1980. A mirid bug (*Psallus variabilis* (Fallen)). United States Department of Agriculture Cooperative Plant Pest Report, 5: 628.

Hoebeke, E. R. and A. G. Wheeler, Jr.

1982. *Rhopalus* (*Brachycarenus*) *tigrinus*, recently established in North America, with a key to the genera and species of Rhopalidae in eastern North America (Hemiptera: Heteroptera). Proceedings of the Entomological Society of Washington, 84: 213-218.

1982. *Catorhintha mendica*, a Great Plains coreid now established on the Atlantic Coast (Hemiptera: Coreidae). Entomological News, 93: 29-31.

1984. *Aethus nigritus* (F.), a palearctic burrower bug established in eastern North America (Hemiptera - Heteroptera: Cydnidae). Proceedings of the Entomological Society of Washington, 86: 738-744.

Hoffmann, C. H.

1932. The biology of the three North American species of *Mesovelia* (Hemiptera-Mesoveliidae). Canadian Entomologist, 64: 88-94, 113-120, 126-133.

1949. Field studies on the effects of airplane applications of DDT on forest invertebrates. Ecological Monographs, 19: 1-46.

Hoffmann, W. E. [also spelled Hoffman]

1925. Some aquatic Hemiptera having only four nymphal stages. Bulletin of the Brooklyn Entomological Society, 20: 93-94.

1941. Catalogue of aquatic Hemiptera of China, Indo-China, Formosa, and Korea. Lingnan Science Journal, 20: 1-78E + 5 pages.

Hogue, C. L.

1974. The insects of the Los Angeles basin. Natural History Museum Los Angeles County Science, Series 27. 173 pages.

Holmquist, A. M.

1926. Studies in arthropod hibernation. I. Ecological survey of hibernating species from forest environments of the Chicago Region. Annals of the Entomological Society of America, 19: 395-428.

Hopkinson, F. L. S.

1907. Dates of publication of the separate parts of Gmelin's edition (13th) of the "Systema Naturae" of Linnaeus. Proceedings of the Zoological Society of London, pages 1035-1036.

Hormchan, P., L. W. Hepner, and M. F. Schuster

1976. Predacious damsel bugs: Identification and distribution of the subfamily Nabinae in Mississippi. Mississippi Agricultural & Forestry Experiment Station Technical Bulletin, 76: 1-4.

Horn, G. C.

1962. Chinch bugs and fertilizer, is there a relationship? Florida Turf-Grass Association Bulletin, 9: 3, 5.

Horn, W. and S. Schenkling

1928-1929. Index Litteraturae Entomologicae. Serie I: Die Welt-Literatur über die gesamte Entomologie bis inklusive 1863. 4 volumes. 1928, 1: 1-352, 2: 353-704, 3: 705-1056; 1929, 4: I-XXI, 1057-1426.

Horning, D. S., Jr. and W. F. Barr

1970. Insects of Craters of the Moon National Monument Idaho. University of Idaho College of Agriculture Miscellaneous Series, 8: 1-118.

Horvath, G.

1875. Monographia Lygaeidarum Hungariae. Budapest. 109 pages, 1 color plate.

1890. Synopsis des *Nysius* paléarctiques. Revue d'Entomologie, 9: 185-191.

1891. Hémiptères recueillis dans l'Arménie Russe avec la description d'espécies et

variétès nouvells. Revue d'Entomologie, 10: 68-81.
1893. Les *Scolopostethus* américains. Revue d'Entomologie, 12: 238-241.
1899. Remarques synonymiques sur les Hémiptères Paléarctiques. Revue d'Entomologie, 17: 275-281.
1905. Tingitidae novae vel minus cognitae e regione Palaearctica. Annales Historico-Naturales Musei Nationalis Hungarici, 3: 556-572.
1906. Synopsis Tingitidarum regionis palaearcticae. Annales Historico-Naturales Musei Nationalis Hungarici, 4: 1-118, 1 plate.
1907. Hemiptera nova vel minus cognita e regione palaearctica. Annales Historico-Naturales Musei Nationalis Hungarici, 5: 289-323.
1908. Les relations entre les faunes Hémiptèrologiques de l'Europe et de l'Amérique du Nord. Annales Historico-Naturales Musei Nationalis Hungarici, 6: 1-14.
1908. Remarques sur quelques Hémiptères de l'Amérique du Nord. Annales Historico-Naturales Musei Nationalis Hungarici, 6: 555-569.
1910. Description of a new bat-bug from British Columbia. Entomologist's Monthly Magazine, Series 2, 21: 12-13.
1911. Révision des Leptopodides. Annales Musei Nationalis Hungarici, 9: 358-370.
1912. Revision of the American Cimicidae. Annales Musei Nationalis Hungarici, 10: 257-262.
1912. Miscellanea Hemipterologica. Annales Musei Nationalis Hungarici, 10: 599-609.
1912. Species generis tingidarum *Stephanitis*. Annales Musei Nationalis Hungarici, 10: 319-339.
1915. Monographie des Mésovéliides. Annales Musei Nationalis Hungarici, 8: 535-556.
1919. Analecta ad cognitionem Cydnidarum. Annales Musei Nationalis Hungarici, 17: 205-273.
1923. A new species of *Galeatus* from New Mexico (Hemiptera - Tingitidae). Annales of the Carnegie Museum, 15: 108-109.
1929. General Catalog of the Hemiptera. Fascicle II. Mesoveliidae. Smith College, Northampton, Massachusetts. 15 pages.

Howard, L. O.
1898. The so-called "Cotton Flea." Page 101. Notes from correspondence. *In* United States Department of Agriculture, Division of Entomology Bulletin, 18: 99-101.
1898. Injury to chrysanthemums by *Corythuca irrorata*. United States Department of Agriculture, Division of Entomology Bulletin, New Series, 10: 99.
1907. The Insect Book. Doubleday, Page and Company, New York. 429 pages, 48 plates.

Howard, W. R.
1872. The radish bug - a new insect (*Nysius raphanus*, n. sp.). Canadian Entomologist, 4: 219-220. [Also: Phillips' Southern Farmer (Planter?), 1872; and Country Gentleman 37: 507.].

Hsiao, T-Y.
1942. A new mirid from Oregon (Hemiptera). Pan-Pacific Entomologist, 18: 160-161.
1945. A new plant bug from Peru, with note on a new genus from North America (Miridae: Hemiptera). Proceedings of the Entomological Society of Washington, 47: 24-27.

Huckaba, R. M., J. R. Bradley, and J. W. Van Duyn
1983. Effects of herbicidal applications of toxaphene on the soybean thrips, certain predators and corn earworm in soybeans. Journal of the Georgia Entomological Society, 18: 200-207.

Hungerford, H. B.
1917. Food habits of corixids. Journal of the New York Entomological Society, 25: 1-5.
1917. The life-history of *Mesovelia mulsanti* White. Psyche, 24: 73-84.
1917. Life history of a boatman. Journal of the New York Entomological Society, 25: 112-122, 1 plate.
1917. The egg laying habits of a back-swimmer (Hem.), *Buenoa margaritacea* Bueno, and other biological notes concerning it. Entomological News, 18: 174-183.
1917. The life history of the backswimmer, *Notonecta undulata* Say (Hem., Het.). Entomological News, 28: 267-278, 2 plates.
1918. Concerning the oviposition of Notonectae [*sic*] (Hem.). Entomological News, 29: 241-245, 2 plates.
1918. Notes on the oviposition of some semiaquatic Hemiptera (*Hebrus, Salda, Lampracanthia*). Journal of the New York Entomological Society, 26: 12-18, 1 plate.
1920. The biology and ecology of aquatic and

Hungerford, H. B. (continued)
semiaquatic Hemiptera. University of Kansas Science Bulletin, 11: 1-328 (1919).
1922. *Saldoida slossoni* Osb. var. *wileyi*, new var., taken in Texas. Bulletin of the Brooklyn Entomological Society, 17: 64.
1922. The life history of the toad bug. *Gelastocoris oculatus* Fabr. (Gelastocoridae). University of Kansas Science Bulletin, 24: 145-171.
1922. The Nepidae of North America (Further studies in aquatic Hemiptera). Kansas University Science Bulletin, 14: 425-469.
1922. Oxyhaemoglobin present in backswimmer *Buenoa margaritacea* Bueno (Hemiptera). Canadian Entomologist, 54: 262-263.
1923. A new species of the genus *Buenoa* (Hemiptera, Notonectidae). Entomological News, 34: 149-152.
1923. Some studies on the genus *Hydrometra* in America north of Mexico with a description of a new species (Hydrometridae, Hemip.). Canadian Entomologist, 55: 54-58.
1924. A new *Mesovelia* with some biological notes regarding it (Hemiptera - Mesoveliidae). Canadian Entomologist, 56: 142-144.
1924. A second new *Mesovelia* from the Douglas Lake, Michigan Region (Hemiptera-Mesoveliidae). Annals of the Entomological Society of America, 17: 453-456.
1924. Stridulation of *Buenoa limnocastoris* Hungerford and systematic notes on the *Buenoa* of the Douglas Lake Region of Michigan, with the description of a new form (Notonectidae-Hemiptera). Annals of the Entomological Society of America, 17: 223-226, 1 plate.
1925. Study of the *Notonecta mexicana* A. & S. series with descriptions of new species (Hemiptera-Notonectidae). Canadian Entomologist, 57: 238-241.
1925. Notes on some North American corixids from the southwest. Bulletin of the Brooklyn Entomological Society, 20: 17-25.
1925. A study of the *interrupta-harrisii* group of the genus *Arctocorixa* with descriptions of new species (Hemiptera-Corixidae). Bulletin of the Brooklyn Entomological Society, 20: 141-145.
1926. Some new Corixidae from the north. Canadian Entomologist, 58: 268-272.
1926. Some new records of aquatic Hemiptera from northern Michigan with the description of seven new Corixidae. Bulletin of the Brooklyn Entomological Society, 21: 194-200, 1 plate.
1926. Some undescribed Corixidae from Alaska. Annals of the Entomological Society of America, 19: 461-463.
1926. Some *Notonecta* from South America. Psyche, 33: 11-16.
1927. A new species of *Hydrometra* from North America. Annals of the Entomological Society of America, 20: 262.
1927. A new *Ramphocorixa* from Haiti (Hemiptera-Corixidae). American Museum Novitates, 278: 1 + 1 plate.
1927. *Arctocorixa atopodonta*, new name for *Arctocorixa dubia* Abbott. Bulletin of the Brooklyn Entomological Society, 22: 35.
1927. The life history of the creeping water bug, *Pelocoris carolinensis* Bueno (Naucoridae). Bulletin of the Brooklyn Entomological Society, 22: 77-83.
1928. Melanchroism in *Notonecta borealis* Bueno and Hussey. Canadian Entomologist, 60: 76.
1928. Aquatic Hemiptera from New Mexico and Georgia, including a new species of Corixidae. Entomological News, 39: 156-157.
1928. Some recent studies in aquatic Hemiptera (including a new subgenus and a new species). Annals of the Entomological Society of America, 21: 139-146.
1928. Some Corixidae of the northern states and Canada. Canadian Entomologist, 60: 226-230.
1929. A new *Velia* from Arizona with notes on other species (Hemiptera-Veliidae). Annals of the Entomological Society of America, 22: 759-761.
1930. Two new water bugs from the western U.S.A. (Nepidae and Notonectidae). Canadian Entomologist, 62: 216-218.
1930. New Corixidae from western North America (Hemiptera). Pan-Pacific Entomologist, 7: 22-26.
1934. The genus *Notonecta* of the world (Notonectidae-Hemiptera). University of Kansas Science Bulletin, 21: 5-195 (1933). [Because this work contains two species described in 1934--see *meinertzhageni* Poisson on p. 44 and *hintoni* Hungerford on p. 72--it could not have appeared in 1933].

1934. A new *Notonecta* from Mexico. Journal of the Kansas Entomological Society, 7: 97-98.
1936. Aquatic and semiaquatic Hemiptera collected in Yucatan and Campeche. Carnegie Institute of Washington Publication No. 457, pp. 145-150.
1938. Some new *Graptocorixa* from Mexico and other notes (Corixidae-Hemiptera). Journal of the Kansas Entomological Society, 11: 134-141.
1939. Two new Corixidae from Mexico. Journal of the Kansas Entomological Society, 12: 123-125.
1939. A corixid from deep water. Annals of the Entomological Society of America, 32: 585-586.
1939. Report on some water bugs from Costa Rica, C. A. Annals of the Entmological Society of America, 32: 587-588.
1940. New Corixidate [*sic*] from China, Manchuria and Formosa. Journal of the Kansas Entomological Society, 13: 8-14.
1942. Three new Corixidae from the southern states. Bulletin of the Brooklyn Entomological Society, 37: 127-131.
1948. The Corixidae of the Western Hemisphere (Hemiptera). University of Kansas Science Bulletin, 32: 1-827.
1950. On the distribution of *Notonecta petrunkevitchi* Hutchinson (Hemiptera: Notonectidae). Journal of the Kansas Entomological Society, 23: 93.
1953. Concerning *Mesovelia douglasensis* Hungerford. Journal of the Kansas Entomological Society, 26: 76-77.
1954. First Florida record for *Hydrometra consimilis* Barber. Journal of the Kansas Entomological Society, 27: 80.
1954. The genus *Rheumatobates* Bergroth (Hemiptera-Gerridae). University of Kansas Science Bulletin, 36 : 529-588.
1956. A new *Cenocorixa* from the northwestern United States (Hemiptera- Corixidae). Journal Kansas Entomological Society, 29: 39-41.

Hungerford, H. B. and N. E. Evans
1934. The Hydrometridae of the Hungarian National Museum and other studies in the family (Hemiptera). Annales Musei Nationalis Hungarici, 28: 31-112.

Hungerford, H. B. and R. F. Hussey
1957. A new corixid (Hemiptera) from Georgia. Quarterly Journal of the Florida Academy of Sciences, 20: 89-92.

Hungerford, H. B. and R. Matsuda
1959. Concerning the Genus *Limnogonus* and a new subgenus (Heteroptera: Gerridae). Journal of the Kansas Entomological Society, 32: 40-41.
1961. A review of the subgenus *Neogerris* Matsumura (Hemiptera: Gerridae). Insecta Matsumurana, 24: 112-114.

Hungerford, H. B. and R. I. Sailer
1943. A new corixid from Minnesota. Bulletin of the Brooklyn Entomological Society, 37: 179-180 (1942).

Hunt, L.-M.
1979. Observations of the habits of *Lygaeus kalmii angustomarginatus* (Hemiptera: Lygaeidae) in southern Michigan. Great Lakes Entomologist, 12: 31-33.
1979. Comparison of effects of juvenile hormone mimic and synthetic juvenile hormone I on the bug *Lygaeus kalmii*. Physiological Entomology, 4: 135-138.

Hunter, W. D.
1912. Some notes on insect abundance in Texas in 1911. Proceedings of the Entomological Society of Washington, 14: 62-66.

Hunter, W. D., F. C. Pratt, and J. D. Mitchell
1912. The principal cactus insects of the United States. United States Department of Agriculture, Bureau of Entomology Bulletin, 113: 1-71.

Hurd, M. P.
1945. A monograph of the genus *Corythaica* Stål (Hemiptera: Tingidae). Iowa State College Journal of Science, 20: 79-99
1946. Generic classification of North American Tingoidea (Hemiptera-Heteroptera). Iowa State College Journal of Science, 20: 429-493.

Hurd, P. D. and E. G. Linsley
1975. Some insects other than bees associated with *Larrea tridentata* in the southwestern United States. Proceedings of the Entomological Society of Washington, 77: 100-120.

Hussain, M.
1975. Predators of the alfalfa weevil, *Hypera postica*, in western Nevada - a greenhouse study (Coleoptera: Curculionidae). Journal of the New York Entomological Society, 83: 226-228.

Hussey, P. B.

1926. Studies on the pleuropodia of *Belostoma flumineum* Say and *Ranatra fusca* Palisot de Beauvois, with a discussion of these organs in other insects. Entomologica Americana, 7: 1-81.

Hussey, R. F.

1919. The waterbugs (Hemiptera) of the Douglas Lake Region, Michigan. Occasional Papers of the Museum of Zoology, University of Michigan, 75: 1-23.

1920. An American species of *Cymatia* (Corixidae, Hemiptera). Bulletin of the Brooklyn Entomological Society, 15: 80-83, 1 plate.

1921. Distributional notes on Hemiptera, with the description of a new *Gerris*. Psyche, 28: 8-15.

1922. Hemipterological notes. Psyche, 29: 229-233.

1922. On some Hemiptera from North Dakota. Occasional Papers of the Museum of Zoology, University of Michigan, 115: 1-23.

1922. Hemiptera from Berrien County, Michigan. Occasional Papers of the Museum of Zoology, University of Michigan, 118: 1-39.

1922. Ecological notes on *Cymatia americana* (Corixidae, Hemiptera). Bulletin of the Brooklyn Entomological Society, 16: 131-136 (1921).

1922. A bibliographical notice on the reduviid genus *Triatoma* (Hemip.). Psyche, 29: 109-123.

1924. A new North American species of *Microvelia* (Hem.). Bulletin of the Brooklyn Entomological Society, 19: 164-165.

1924. A change of name (Hemiptera, Miridae). Bulletin of the Brooklyn Entomological Society, 19: 165.

1925. Some new or little-known Hemiptera from Florida and Georgia. Journal of the New York Entomological Society, 33: 61-69.

1927. On some American Pyrrhocoridae [Hemiptera]. Bulletin of the Brooklyn Entomological Society, 22: 227-235.

1929. Pyrrhocoridae. Fascicle III. Pages 1-144. *In* Horvath, G. and H. M. Parshley (editors). General Catalogue of the Hemiptera. Smith College, Northampton, Massachusetts.

1948. A new *Metrobates* from Florida (Hemiptera, Gerridae). Florida Entomologist, 31: 123-124.

1948. A necessary change of name (Hemiptera: Saldidae). Bulletin of the Brooklyn Entomological Society, 43: 153.

1950. Bilateral abnormality of the antennae in *Ptochiomera nodosa* (Hemiptera, Lygaeidae). Bulletin of the Brooklyn Entomological Society, 45: 27.

1950. Two synonymic notes (Hemiptera: Coreidae, Corixidae). Entomological News, 61: 12-13.

1951. *Leptocorixa filiformis* in the United States (Hemiptera: Coreidae). Florida Entomologist, 33: 150-154 (1950).

1952. Food plants and new records for some Hemiptera in Florida. Florida Entomologist, 35: 117-118.

1953. Concerning some North American Coreidae (Hemiptera). Bulletin of the Brooklyn Entomological Society, 48: 29-34.

1953. Some new and little-known American Hemiptera. Occasional Papers of the Museum of Zoology, University of Michigan, 550: 1-12.

1953. Two new species of *Sinea* (Hemiptera, Reduviidae). Journal of the Kansas Entomological Society, 26: 61-64.

1954. Concerning the Floridian species of *Fulvius* (Hemiptera, Miridae). Florida Entomologist, 37: 19-22.

1954. Some new or little-known Miridae from the northeastern United States (Hemiptera). Proceedings of the Entomological Society of Washington, 56: 196-202.

1955. Some records of Hemiptera new to Florida. Quarterly Journal of the Florida Academy of Science, 18: 120-122.

1956. Additions to the United States list of Hemiptera. Florida Entomologist, 39: 88.

1957. Two changes of name in Hemiptera (Aneuridae and Miridae). Florida Entomologist, 40: 80.

1957. A new *Gargaphia* from Florida (Hemiptera: Tingidae). Proceedings of the Entomological Society of Washington, 59: 175-176.

1958. A new North American *Mozena* (Hemiptera: Coreidae). Florida Entomologist, 41: 142-143.

1960. A lygaeid new to the United States list (Hemiptera). Florida Entomologist, 43: 93.

Hussey, R. F. and J. C. Elkins

1956. Review of the genus *Doldina* Stål (Hem-

Hussey, R. F. and J. C. Elkins (continued)
iptera: Reduviidae). Quarterly Journal of the Florida Academy of Sciences, 18: 261-278 (1955).

Hussey, R. F. and J. L. Herring
1949. Notes on the variation of the *Metrobates* of Florida (Hemiptera, Gerridae). Florida Entomologist, 32: 166-170.
1950. A remarkable new belostomatid (Hemiptera) from Florida and Georgia. Florida Entomologist, 33: 84-89.
1950. A new *Microvelia* from Florida (Hemiptera, Veliidae). Florida Entomologist, 33: 117-120.
1950. Rediscovery of a belostomatid named by Thomas Say (Hemiptera). Florida Entomologist, 33: 154-256.

Hutchinson, G. E.
1929. A revision of the Notonectidae and Corixidae of South Africa. Annals of the South African Museum, 25: 359-474, 15 plates.
1934. Report on terrestrial families of Hemiptera-Heteroptera. Yale North India Expedition. Memoirs of the Connecticut Academy of Arts and Sciences, 10: 119-146, 3 plates.
1940. A revision of the Family Corixidae of India and adjacent regions. Transactions of the Connecticut Academy of Arts and Sciences, 33: 339-476, 36 plates.
1945. On the species of *Notonecta* (Hemiptera-Heteroptera) inhabiting New England. Transactions of the Connecticut Academy of Arts and Sciences, 36: 599-605.

Hutchison, W. D. and H. N. Pitre
1982. Diurnal variation in sweep net estimates of *Geocoris punctipes* (Say) (Hemiptera: Lygaeidae) density in cotton. Florida Entomologist, 65: 578-579.
1983. Predation of *Heliothis virescens* (Lepidoptera: Noctuidae) eggs by *Geocoris punctipes* (Hemiptera: Lygaeidae) adults on cotton. Environmental Entomology, 12: 1652-1656.

Ignatowicz, S. and J. Boczek
1980. Insect hormones and the possibilities of their use for control of plant pests. Kosmos, Series A. Biology, 29(2): 183-192.

Illiger, J. C. W.
1800. Vierzig neue Insekten aus der Lellwigischen Sammlung in Braunschweig. Arkiv für Zoologie und Zootomie (Wiedmann), 1(2): 103-150, 2 plates.
1807. [Notes]. *In* Rossi, P. Classis septima. Ryngota. Fauna Etrusca, sistens Insecta quae in Provinciis Florentina et Pisana Praesertim collegit. Iterum edita et annotatis perpetuis acta a D. Carlo Illiger. 2: 212-267.

Ingram, J. W., H. A. Jaynes, and R. N. Lobdell
1939. Sugarcane pests in Florida. International Society of Sugar Cane Technologists, Proceedings of the Sixth Congress, Baton Rouge, Louisiana, 6: 89-98.

International Commission on Zoological Nomenclature
1924. The genotype of *Cimex, Acanthia, Clinocoris,* and *Klinophilos*. Smithsonian Miscellaneous Collections, 73: 19-32.

Irwin, M. E., R. W. Gill and D. Gonzalez
1974. Field-cage studies of native egg predators of the pink bollworm in southern California cotton. Journal of Economic Entomology, 67: 193-196.

Irwin, M. E. and M. Shepard
1980. Sampling predaceous Hemiptera on soybeans. Pages 505-531. *In* Kogan, M. and D. C. Herzog (editors). Sampling Methods in Soybean Entomology. Springer-Verleg, New York. 587 pages.

Isenhour, D. J. and J. W. Todd
1984. Toxicity of the experimental insectcide SD-52618 to the predator *Geocoris punctipes* and selected lepidopterous species in soybean. Journal of Agricultural Entomology, 1: 376-379.

Isman, M. B.
1979. Cardenolide content of lygaeid bugs on *Asclepias curassavica* in Costa Rica. Biotropica, 11: 78-79.

Isman, M. B., S. S. Duffey, and G. G. E. Scudder
1977a. Variation in cardenolide content of the lygaeid bugs *Oncopeltus fasciatus* and *Lygaeus kalmii kalmii* and of their milkweed hosts (*Asclepias* spp.) in central California. Journal of Chemical Ecology, 3: 613-624.
1977b. Cardenolide content of some leaf- and stem-feeding insects on temperate North American milkweeds (*Asclepias* spp.). Canadian Journal of Zoology, 55: 1024-1028.

Ives, W. G. H.
1967. Relations between invertebrate predators and prey associated with larch sawfly eggs and larvae on tamarack.

Ives, W. G. H. (continued)
Canadian Entomologist, 99: 607-622.

Jaczewski, T.
1930. Notes on the American species of the genus *Mesovelia* Muls. (Heteroptera, Mesoveliidae). Annales Musei Zoologici Polonici, 9: 3-12, 3 plates.
1931. Die Corixiden (Corixidae, Heteroptera) des Zoologischen Staatsinstituts und Zoologischen Museums in Hamburg. II. Archiv für Hydrobiologie, 23: 507-519.
1931. Studies on Mexican Corixidae. Annales Musei Zoologici Polonici, 9: 187-230, 5 plates.
1932. Notes on Corixidae. VIII-XI. Annales Musei Zoologici Polonici, 9: 147-154, 1 plate.
1936. Notes on Corixidae (Hem.). Proceedings of the Royal Entomological Society of London, Series B, 5: 34-43.

Jakovlev, V. E.
1877. [New heteropterous insects (Hemipt. Het.) of the Astrakan fauna]. Second supplement (II). Bulletin de la Société-Impériale des Naturalistes de Moscou, 52: 269-300.
1883. Hemiptera-Heteroptera des Gouverments Irkutsk. Horae Societas Entomologica Rossicae, 17(3-4): 282-310. [Text in Russian; descriptions in German].

Jansson, A.
1972. Mechanisms of sound production and morphology of the stridulatory apparatus in the genus *Cenocorixa*. Annales Zoologici Fennici, 9: 120-129.
1972. Systematic notes and new synonymy in the genus *Cenocorixa* (Hemiptera: Corixidae). Canadian Entomologist, 104: 449-459.
1973. Stridulation and its significance in the genus *Cenocorixa*. Behavior, 46: 1-36.
1975. Comparison of *Sigara saileri* Wilson and *S. johnstoni* Hungerford. Journal of the Kansas Entomological Society, 48: 1-3.
1976. Audiospectographic analysis of stridulatory signals of some North American Corixidae (Hemiptera). Annales Zoologici Fennici, 13: 48-62.
1979. A new species of *Callicorixa* from northwestern North America. Pan-Pacific Entomologist, 54: 261-266 (1978).

Jansson, A. and G. G. E. Scudder
1972. Corixidae (Hemiptera) as predators: Rearing on frozen brine shrimp. Journal of the Entomological Society of British Columbia, 69: 44-45.

Javahery, M.
1986. Biology and ecology of *Picromerus bidens* (Hemiptera: Pentatomidae) in southeastern Canada. Entomological News, 97: 87-98.

Jay, E., R. Davis, and S. Brown
1968. Studies on the predacious habits of *Xylocoris flavipes* (Reuter) (Hemiptera: Anthocoridae). Journal of the Georgia Entomological Society, 3: 126-130.

Jeannel, R.
1942. Les Hénicocéphalides: Monographie d'un groupe d'Hémiptères Hématophages. Annales de la Société Entomologique de France, 110: 273-368. [dated 1941, but Jeannel, 1943, Bulletin de la Société Entomologique de France, 48: 125, footnote, says it was published 20 March 1942].
1943. Nouveaux Hénicocéphalides sudaméricains. Bulletin de la Société Entomologique de France, 48: 125-128.

Jensen-Haarup, A. C.
1928. Hemipterological notes and descriptions V. Entomologiske Meddelelser, 16: 185-202.

Jewett, H. H. and L. H. Townsend
1947. *Miris dolobratus* (Linn.) and *Amblytylus nasutus* (Kirschbaum). Two destructive insect pests of Kentucky bluegrass. Kentucky Agricultural Experiment Station Bulletin, 508: 1-16.

Jimenez, J. G.
1975. Evalucion de estimulantes alimenticios para incrementar la fauna benefica en algodonero. Folia Entomológica Méxicana, 33: 57.
1976. Cultivos intercalados algodon/maiz, algodon/alfalfa para proteger e incrementa la fauna benéfica como medida de control del complejo *Heliothis* en algodonero en la region de Ceballos, Dgo. Folia Entomológica Méxicana, 36: 29-30.

Johannsen, O. A.
1909. North American Enicocephalidae.-Plate I. Psyche, 16: 1-4, l plate.

Johansen, C. A.
1972. Spray additives for insecticidal selectivity to injurious vs. beneficial insects. Environmental Entomology, 1: 51-54.

Johansen, C. A. and J. D. Eves
1972. Acidified sprays, pollinator safety, and integrated pest control on alfalfa grown for seed. Journal of Economic Entomology, 65: 546-555.
1973. Development of a pest management program on alfalfa grown for seed. Environmental Entomology, 2: 515-517.

Johnson, C. and R. Ledig
1918. Tentative list of Hemiptera from the Clarement-Laguna Region. Journal of Entomology and Zoology, 10: 3-8.

Johnson, C. G.
1969. The migration and dispersal of insects by flight. Methuen, London.

Johnson, C. W.
1927. The insect fauna with reference to the flora and other biological features. Pages 9-247. *In* Proctor, W. Biological Survey of the Mount Desert Region. Part I. Wistar Institute of Anatomy and Biology, Philadelphia, Pennsylvania. 247 pages.
1930. A list of the insect fauna of Nantucket, Massachusetts. Publications of the Nantucket Maria Mitchell Association, 3(2): 1-174 + 1 (errata and addendum) + i-xviii (index).

Johnson, E. K., J. H. Young, D. R. Molnar, and R. D. Morrison
1976. Effects of three insect control schemes on populations of cotton insects and spiders, fruit damage, and yield of Westburn 70 cotton. Environmental Entomology, 5: 508-510.

Johnson-Cicalese, J. M., S. Ahmad, and C. R. Funk
1985. Endophyte-enhanced billbug resistance in perennial ryegrass. Bulletin of the New Jersey Academy of Science, 30: 54.

Johnston, H. G.
1928. A partial list of Miridae from Texas (Order Hemiptera). Bulletin of the Brooklyn Entomological Society, 24: 217-219.
1930. Four new species of Miridae from Texas (Hemiptera). Bulletin of the Brooklyn Entomological Society, 25: 295-300.
1930. *Dicyphus minimus* Uhler, a pest on tomatoes (Hemiptera: Miridae). Journal of Economic Entomology, 23: 642.
1935. Five new species of Miridae (Hemiptera). Bulletin of the Brooklyn Entomological Society, 30: 15-19.
1939. Five new species of Miridae from Texas (Hemiptera). Bulletin of the Brooklyn Entomological Society, 34: 129-133.

Jones, M. P.
1935. A peculiar insect situation along a seashore. Proceedings of the Entomological Society of Washington, 37: 150-151.

Jones, W. A., Jr. and J. E. McPherson
1980. The first report of the occurrence of acanthosomatids in South Carolina. Journal of the Georgia Entomological Society, 15: 286-289.

Jones, W. A., H. E. Walker, P. C. Quimby, and J. D. Outz
1985. Biology of *Niesthrea louisianica* (Hemiptera: Rhopalidae) on selected plants, and its potential for biocontrol of velvetleaf, *Abutilon theophrasti* (Malvaceae). Annals of the Entomological Society of America, 78: 326-330.

Jordon, K. H. C.
1933. Beiträge zur biologie Heimischer Wanzen (Heteropt.). Entomologische Zeitung (Herausgegeben von dem entomologischen Vereine zu Stettin), 94: 212-236.
1935. Beitrag zur lebensweise der Wanzen auf feuchten Boden (Heteropt.). Entomologische Zeitung (Herausgegeben von dem entomologischen Vereine zu Stettin), 96: 1-26.

Jotwani, M. G.
1981. Insect resistance in sorghum plants. Insect Science and Its Application, 2: 93-98.

Jubb, G. L., Jr., E. C. Masteller, and A. G. Wheeler, Jr.
1979. Survey of the arthropods in vineyards of Erie County, Pennsylvania: Hemiptera: Heteroptera. Environmental Entomology, 8: 982-986.

Judd, W. W.
1963. Studies of the Byron Bog in southwestern Ontario XVII. Seasonal distribution of Hemiptera (Corixidae, Notonectidae, Belostomatidae, Nepidae) in Redmond's Pond. Canadian Entomologist, 95: 1109-1111.
1970. Insects associated with flowering wild carrot, *Daucus carota* L., in southern Ontario. Proceedings of the Entomological Society of Ontario, 100: 176-181.

Keever, D. W., J. R. Bradley, Jr., and M. C. Ganyard
1977. Effects of diflubenzuron (Dimilin) on

Keever, D. W., J. R. Bradley, Jr., and M. C. Ganyard (continued)
selected beneficial arthropods in cotton fields. Environmental Entomology, 6: 732-736.

Kellen, W. R.
1953. A quantitative sampler for aquatic insects. Journal of Economic Entomology, 46: 913-914.

Kelso, L. H.
1937. See Girard, G. L. 1937.

Kelton, L. A.
1955. Genera and subgenera of the *Lygus* complex (Hemiptera: Miridae). Canadian Entomologist, 87: 277-301.
1955. New species of *Liocoris* from North America (Hemiptera: Miridae). Canadian Entomologist, 87: 484-490.
1955. Species of *Lygus, Liocoris,* and their allies in the Prairie Provinces of Canada (Hemiptera: Miridae). Canadian Entomologist, 87: 531-556.
1959. Synopsis of the genus *Semium,* and description of a new species from Mexico (Hemiptera: Miridae). Canadian Entomologist, 91: 242-246.
1959. Male genitalia as taxonomic characters in the Miridae (Hemiptera). Canadian Entomologist, 91(Supplement 11): 1-72.
1960. A new species of *Mecomma* Fieber from Canada, with reference to *M. mimetica* Carv. and South. and *M. dispar* (Boh.) in North America (Hemiptera: Miridae). Canadian Entomologist, 92: 570-573.
1961. A new species of *Gerris* F. from Yukon and Alaska (Hemiptera: Gerridae). Canadian Entomologist, 93: 663-665.
1961. Synopsis of the nearctic species of *Stenodema* Laporte, and description of a new species from western Canada (Hemiptera: Miridae). Canadian Entomologist, 93: 450-455.
1961. A new nearctic genus of Miridae, with notes on *Diaphnidia* Uhler 1895 and *Brachynotocoris* Reuter 1880 (Hemiptera). Canadian Entomologist, 93: 566-568.
1963. Synopsis of the genus *Orius* Wolff in America north of Mexico (Heteroptera: Anthocoridae). Canadian Entomologist, 95: 631-636.
1964. Revision of the genus *Reuteroscopus* Kirkaldy 1905 with descriptions of eleven new species (Hemiptera: Miridae). Canadian Entomologist, 96: 1421-1433.
1965. *Diaphnidia* Uhler and *Diaphnocoris* Kelton in North America (Hemiptera: Miridae). Canadian Entomologist, 97: 1025-1030.
1965. *Chlamydatus* Curtis in North America (Hemiptera: Miridae). Canadian Entomologist, 97: 1132-1144.
1966. Synopsis of the genus *Tetraphleps* Fieber in North America (Hemiptera: Anthocoridae). Canadian Entomologist, 98: 199-204.
1966. Two new species of *Keltonia* Knight, with a key to known species (Hemiptera: Miridae). Canadian Entomologist, 98: 668-670.
1966. Review of the species of *Teratocoris* Fieber, with description of a new species from the Nearctic Region (Hemiptera: Miridae). Canadian Entomologist, 98: 1265-1271.
1966. *Pithanus maerkeli* (Herrich-Schaeffer) and *Actitocoris signatus* Reuter in North America (Hemiptera: Miridae). Canadian Entomologist, 98: 1305-1307.
1967. Synopsis of the genus *Lyctocoris* in North America and description of a new species from Quebec (Heteroptera: Anthocoridae). Canadian Entomologist, 99: 807-814.
1968a. On the Heteroptera in the Provancher Collection (Hemiptera). Naturaliste Canadien, 95: 1065-1080.
1968. Revision of the North American species of *Slaterocoris* (Heteroptera: Miridae). Canadian Entomologist, 100: 1121-1137.
1969. *Scalponotatus,* a new genus near *Slaterocoris,* with descriptions of new species (Heteroptera: Miridae). Canadian Entomologist, 101: 15-23.
1970. Four new species of *Trigonotylus* from North America (Heteroptera: Miridae). Canadian Entomologist, 102: 334-338.
1971. Revision of the species of *Trigonotylus* in North America (Heteroptera: Miridae). Canadian Entomologist, 103: 685-705.
1971. Four new species of *Lygocoris* from Canada (Heteroptera: Miridae). Canadian Entomologist, 103: 1107-1110.
1971. Review of *Lygocoris* species found in Canada and Alaska (Heteroptera: Miridae). Memoirs of the Entomological Society of Canada, 83: 1-87.
1972. *Picromerus bidens* in Canada (Heteroptera: Pentatomidae). Canadian Entomologist, 104: 1743-1744, unnumbered color plate.

1972. Four new species of *Bolteria*, with a key to the North American species and a note on the species found in Canada (Heteroptera: Miridae). Canadian Entomologist, 104: 627-640.

1972. Species of *Dichrooscytus* found in Canada, with descriptions of four new species (Heteroptera: Miridae). Canadian Entomologist, 104: 1033-1049.

1972. A note on *Dichrooscytus elegans*, with descriptions of four new species from New Mexico and Texas (Heteroptera: Miridae). Canadian Entomologist, 104: 1439-1444.

1972. Descriptions of nine new species of *Dichrooscytus* from North America (Heteroptera: Miridae). Canadian Entomologist, 104: 1457-1464.

1973. *Knightomiris* new genus, for *Lygus distinctus* (Heteroptera: Miridae). Canadian Entomologist, 105: 1417-1420.

1973. Two species of *Lygus* from North America, and a note on the status of *Lygus abroniae* (Heteroptera: Miridae). Canadian Entomologist, 105: 1545-1548.

1973. A new species of *Semium* (Heteroptera: Miridae) from Mexico, with new records on distribution for two other species. Canadian Entomologist, 105: 1583-1584.

1974. On the status of seven nearctic species currently included in the genus *Lygus* Hahn (Heteroptera: Miridae). Canadian Entomologist, 106: 377-380.

1975. The lygus bugs (genus *Lygus* Hahn) of North America (Heteroptera: Miridae). Memoirs of the Entomological Society of Canada, 95: 1-101.

1976. Three new species of *Xylocoris* from North America, and a note on the status of species in the genus *Scoloposcelis* Fieber (Heteroptera: Anthocoridae). Canad-ian Entomologist, 108: 193-198.

1976. The genus *Elatophilus* Reuter in North America with descriptions of two new species (Heteroptera: Anthocoridae). Canadian Entomologist, 108: 631-634.

1977. Species of the genus *Pinalitus* Kelton found in North America (Heteroptera: Miridae). Canadian Entomologist, 109: 1549-1554.

1977. New species of *Cardiastethus* Fieber and *Melanocoris* Champion, and new records of European *Acompocoris* Reuter and *Temnostethus* Fieber in Canada (Heteroptera: Anthocoridae). Canadian Entomologist, 109: 243-248.

1977. A new species of *Elatophilus* Reuter from Ontario, and new synonymy for *Piezostethus flaccidus* Van Duzee (Heteroptera: Anthocoridae). Canadian Entomologist, 109: 1017-1018.

1978. The Anthocoridae of Canada and Alaska. Heteroptera Anthocoridae. The Insects and Arachnids of Canada. Part 4. Agriculture Canada Research Publication, No. 1639. 101 pages.

1978. *Xenoborus* Reuter (1908): A new synonym of *Tropidosteptes* Uhler (1878), with description of a new species (Heteroptera: Miridae). Canadian Entomologist, 110: 471-473.

1979. Two new species of *Phytocoris* from western Canada (Heteroptera: Miridae). Canadian Entomologist, 111: 689-692.

1979. *Labopidea* Uhler in North America, with descriptions of new species and a new genus (Heteroptera: Miridae). Canadian Entomologist, 111: 753-758.

1979. A new genus *Brooksella*, near *Ilnacora* Reuter, with new synonymy and new combinations for 15 species currently placed in *Melanotrichus* Reuter (Heteroptera: Miridae). Canadian Entomologist, 111: 949-954.

1979. Replacement name for *Brooksella* Kelton (Heteroptera: Miridae). Canadian Entomologist, 111: 1423.

1980. Descriptions of three new species of Miridae from the Prairie Provinces and a new record of European Phylini in the Nearctic Region (Heteroptera). Canadian Entomologist, 112: 285-292.

1980. Two new species of *Melanotrichus* Reuter from western Canada and a description of the male of *M. atriplicis* (Heteroptera: Miridae). Canadian Entomologist, 112: 337-339.

1980. Description of a new species of *Parthenicus* Reuter, new records of holarctic Orthotylini in Canada, and new synonymy for *Diaphnocoris pellucida* (Heteroptera: Miridae). Canadian Entomologist, 112: 341-344.

1980. Lectotype designation for *Idolocoris agilis*, and descriptions of three new species of *Dicyphus* Fieber from North America (Heteroptera: Miridae). Canadian Entomologist, 112: 387-392.

1980. The Plant Bugs of the Prairie Provinces of Canada. Heteroptera: Miridae. The

Kelton, L. A. (continued)
Insects and Arachnids of Canada. Part 8. Agriculture Canada Research Publication, No. 1703. Ottawa. 408 pages.
1980. First record of a European bug, *Loricula pselaphiformis*, in the Nearctic Region (Heteroptera: Microphysidae). Canadian Entomologist, 112: 1085-1087.
1981. First record of a European bug, *Myrmedobia exilis* (Heteroptera: Microphysidae), in the Nearctic Region. Canadian Entomologist, 113: 1125-1127.
1982. Description of a new species of *Plagiognathus* Fieber, and additional records of European *Psallus salicellus* in the Nearctic Region (Heteroptera: Miridae). Canadian Entomologist, 114: 169-172.
1982. New records of European *Pilophorus* and *Orthotylus* in Canada (Heteroptera: Miridae). Canadian Entomologist, 114: 283-287.
1982. New and additional records of palearctic *Phylus* Hahn and *Plagiognathus* Fieber in North America (Heteroptera: Miridae). Canadian Entomologist, 114: 1127-1128.
1982. Plant bugs on fruit crops in Canada. Heteroptera: Miridae. Agriculture Canada Monograph, 24: 1-201.
1983. European *Pseudoloxops coccineus* found in Canada, and additional records of *Camptozygum aequale* in the Nearctic Region (Heteroptera: Miridae). Canadian Entomologist, 115: 107-108.
1983. Four European species of *Psallus* Fieber found in Canada (Heteroptera: Miridae). Canadian Entomologist, 115: 325-328.
1985. Species of the genus *Fulvius* Stål found in Canada (Heteroptera: Miridae: Cylapinae). Canadian Entomologist, 117: 1071-1073.

Kelton, L. A. and N. H. Anderson
1962. New Anthocoridae from North America, with notes on the status of some genera and species (Heteroptera). Canad-ian Entomologist, 94: 1302-1309.

Kelton, L. A. and J. L. Herring
1978. Two new species of *Neoborella* Knight (Heteroptera: Miridae) found on dwarf mistletoe, *Arceuthobium* spp. Canadian Entomologist, 110: 779-780.

Kelton, L. A. and H. H. Knight
1959. A new species of *Paradacerla* from Mexico, and synopsis of the genus in North America (Hemiptera: Miridae). Canadian Entomologist, 91: 122-126.
1962. *Mecomma* Fieber in North America (Hemiptera: Miridae). Canadian Entomologist, 94: 1296-1302.
1970. Revision of the genus *Platylygus*, with descriptions of 26 new species (Hemiptera: Miridae). Canadian Entomologist, 102: 1429-1460.

Kelton, L. A. and J. D. Lattin
1968. On the Saldidae types in the Provancher Collection, and a new name for *Salda coriacea* Uhler (Heteroptera). Naturliste Canadien, 95: 661-666.

Kelton, L. A. and J. C. Schaffner
1972. A note on *Dichrooscytus elegans*, with descriptions of four new species from New Mexico and Texas (Heteroptera: Miridae). Canadian Entomologist, 104: 1439-1444.

Kerr, S. H.
1962. Lawn insect studies--1962. Proceedings University Florida Turf Management Conference, 10: 201-208.
1966. Biology of the lawn chinch bug, *Blissus insularis*. Florida Entomologist, 49: 9-18.

Kerzhner, I. M.
1962. Materials on the taxonomy of capsid bugs (Heteroptera, Miridae) in the fauna of the USSR. Entomologischeskoe Obozrenie, 41: 372-387 [English translation in Entomological Review, 41: 226-235].
1963. Beitrag zur Kenntnis der unterfamilie Nabinae (Heteroptera: Nabidae). Acta Entomologica Musei Nationalis Pragae, 35: 5-61.
1964. Family Miridae (Capsidae). Pages 913-1003. *In* Bei-Bienko, G. Ya. (editor). Keys to the insects of the European USSR. I [English translation, 1967, Israel Program for Scientific Translations, Jerusalem].
1968. New and little known palearctic bugs of the family Nabidae (Heteroptera). Entomologicheskoe Obozrenie, 47: 848-863. [English translation in Entomological Review, 47: 517-525].
1974. On the date of publication of A. Costa's "Cimicum Regni Neapolitani Centuria" and of the family-group names (Heteroptera) contained in it. Revue d'Entomologie de l'URRS, 63: 854-860. [English

translation in Entomological Review, 35: 90-93.]

1981. Heteroptera of the Family Nabidae. *In* The Fauna of the U.S.S.R. Academy of Sciences USSR, Zoological Institute, New Series, No. 124, 13(2): 1-324. [In Russian].

1981. *Nabis capsiformis* Germar, [1838] (Insecta, Heteroptera, Nabidae): Proposed conservation under the plenary powers. Z.N.(S.)2147. Bulletin of Zoological Nomenclature, 38: 205-207.

Kessel, R. G. and H. W. Beams

1963. Micropinocytosis and yolk formation in oocytes of the small milkweed bug. Experimental Cell Research, 30: 440-443.

Khalaf, K. T.

1971. Miridae from Louisiana and Mississippi (Hemiptera). Florida Entomologist, 54: 339-342.

Kiman, Z. B. and K. V. Yeargan

1985. Development and reproduction of the predator *Orius insidiosus* (Hemiptera: Anthocoridae) reared on diets of selected plant material and arthropod prey. Annals of the Entomological Society of America, 78: 464-467.

King, G. B.

1897. Some ants and myrmecophilous insects from Toronto. Canadian Entomologist, 29: 100-103.

Kinzelbach, R. K.

1970. *Loania canadensis* n. gen., n. sp. und die Untergliederung der Callipharixenidae (Insecta: Strepsiptera). Senckenburgiana Biologica, 51: 99-107.

Kinzer, R. E., C. B. Cowan, R. L. Ridgway, J. W. Davis, Jr., J. R. Coppedge, and S. L. Jones

1977. Populations of arthropod predators and *Heliothis* spp. after applications of aldicarb and monocrotophos to cotton. Environmental Entomology, 6: 13-16.

Kirby, W.

1837. Order Hemiptera Latr. Pages 275-285. *In* Richardson's Fauna Boreali-Americana or the Zoology of the Northern Parts of British America. J. Fletcher, Norwich. 4: i-xxxix + 1-325, plates 1-8.

Kiritshenko, A. N. and G. G. E. Scudder

1973. Some new genera and species of Rhyparochrominae (Hemiptera: Lygaeidae) from the Soviet Union, with a key to the genera of Gonianotini. Journal of Natural History, 7: 133-151.

Kirkaldy, G. W.

1897. Revision of the Notonectidae. Part I. Introduction, and systematic revision of the genus Notonecta. Transactions of the Entomological Society of London, 1897: 393-426.

1897. Synonymic notes on aquatic Rhynchota. Entomologist, 30: 258-260.

1897. Aquatic Rhynchota: Descriptions and notes. No. 1. Annals and Magazine of Natural History, Series 6, 20: 52-60.

1898. Notes on aquatic Rhynchota. No. 1. Entomologist, 31: 2-4.

1898. On the nomenclature of the European subgenera of *Corixa*, Geoffr. (Rhynchota). Entomologist, 32: 252-253.

1899. On some aquatic Rhynchota from Jamaica. Entomologist, 32: 28-30.

1899. On the nomenclature of the Rhynchota.- part 1. Entomologist, 32: 217-221.

1899. Sur quelques Hémiptères aquatiques nouveaux ou peu connus. Revue d'Entomologie, 18: 85-96, 7 figures.

1900. Bibliographical and nomenclatorial notes on the Rhynchota. No. 1. Entomologist, 33: 238-243.

1900. Recent notes on *Hydrometra martini*, Kirk. = *lineata*, Say. Entomologist, 33: 175-176.

1900-1906. On the nomenclature of the genera of the Rhynchota, Heteroptera and auchenorrhynchous Homoptera. Entomologist, 1900, 33: 25-28, 262-265; 1901, 34: 176-179, 218-219; 1903, 36: 213-216, 230-233; 1906, 39: 253-257.

1901. Anmerkungen über bemerkenswerte Nabinen (Rhynchota). Weiner Entomologische Zeitung, 20: 219-225.

1901. The stridulating organs of waterbugs, especially Corixidae. Journal of the Quekett Microscopical Club, Series 2, 8: 33-46, plates 3-4.

1901. Miscellanea Rhynchotalia. Entomologist, 34: 5-6.

1901. Notes on the division Veliiaria. Rhynchota (= Subfam. Velidae, Leth. & Sev.). Entomologist, 34: 285-286, 308-310.

1902. Hemiptera. Fauna Hawaiiensis, 3: 93-174, plates IV-V.

1902. Memoirs on Oriental Rhynchota. Journal of the Bombay Natural History Society, 14: 294-309.

1902. Miscellanea Rhynchotalia.--No. 3. Entomologist, 35: 136-138.

Kirkaldy, G. W. (continued)

1902. Memoir upon the rhynchotal family Capsidae Auctt. Transactions of the Entomological Society of London, pages 243-272, 2 plates.

1903. Einige neue und wenig bekannte Rhynchoten. Wiener Entomologische Zeitung, 22: 13-17.

1904. Über Notonectiden (Hemiptera). Wiener Entomologische Zeitung, 23: 93-135.

1904. Bibliographical and nomenclatoral notes on the Hemiptera.--No. 3. Entomologist, 37: 254-258, 279-283.

1905. Quelques Tingides nouveaux ou peu connus [Hém.]. Bulletin de la Société-Entomologique de France, 15: 216-217.

1905. Neue und wenig bekannte Hemiptera. Wiener Entomologische Zeitung, 24: 266-268.

1906. List of the genera of the pagiopodous Hemiptera-Heteroptera, with their type species, from 1758 to 1904 (and also of the aquatic and semi-aquatic Trochalopoda). Transactions of the American Entomological Society, 32: 117-156, 156a, 156b.

1906. Notes on the classification and nomenclature of the hemipterous superfamily Miroidea. Canadian Entomologist, 38: 369-376.

1908. Notes on Corixidae No. 1 [Hem.]. Canadian Entomologist, 40: 117-120.

1908. Bibliographical and nomenclatorial notes on the Hemiptera.- No. 8. The Entomologist, 41: 123-124.

1909. Hemiptera: New and old.-No. 1. Canadian Entomologist, 41: 30-32.

1909. Hemiptera, old and new, No. 2. Canadian Entomologist, 41: 388-392.

1909. A list of the Hemiptera (excluding Sternorrhyncha) of the Maorian Subregion, with notes on a few of the species. Transactions of the New Zealand Institute, 41: 22-29.

1909. Catalogue of the Hemiptera (Heteroptera) with Biological and Anatomical References, Lists of Foodplants and Parasites, etc. Felix L. Dames, Berlin. 1: I-XL, 1-392.

1910. Further notes on Hemiptera, chiefly Hawaiian. Proceedings of the Hawaiian Entomological Society, 2: 118-123.

1910. A preliminary list of the Hemiptera of California, pt. 1. Proceedings of the Hawaiian Entomological Society, 3: 123-126.

1911. A new species of *Gerris* (Hemip.). Entomological News, 22: 246.

Kirkaldy, G. W. and S. Edwards

1902. Anmerkungen über bemerkenswerte Pyrrhocorinen (Rynchota). Weiner Entomologische Zeitung, 21: 161-172.

Kirkaldy, G. W. and J. R. de la Torre-Bueno

1909. A catalogue of American aquatic and semiaquatic Hemiptera. Proceedings of the Entomological Society of Washington, 10: 173-221 (1908).

Kirkland, A. H.

1897. Notes on predaceous Heteroptera, with Prof. Uhler's description of two species. Canadian Entomologist, 29: 115-118.

1898. The species of *Podisus* occurring in the United States. Forty-fifth Annual Report of the Secretary of the Massachusetts State Board of Agriculture for 1897, pages 112-138.

Kirschbaum, C. L.

1856. Rhynchotographische Beiträge. I. Die Capsinen der Gegend von Wiesbaden. Jahrbücher des Vereins für Naturkunde im Herzogthum Nassau, Wiesbaden, 10: 161-348 (1855).

Kittle, P. D.

1977. A revision of the genus *Trepobates* Uhler (Hemiptera: Gerridae). Ph.D. Dissertation, University of Arkansas, xii + 255 pages.

1980. The water striders (Hemiptera: Gerridae) of Arkansas. Proceedings of the Arkansas Academy of Science, 34: 68-71.

1982. Two new species of water striders of the genus *Trepobates* Uhler (Hemiptera: Gerridae). Proceedings of the Entomological Society of Washington, 84: 157-164.

Klug, J. C. F.

1829-1845. Symbolae physicae, seu icones et descriptiones insectorum, quae ex itinere per Africam borealem et Asiam F. G. Hemprich et C. H. Ehrenberg studio novae aut illustratae redierunt. Parts 1-5. Mittler, Berlin. [1829, 1: a-i, colored plates 1-10; 1830, 2: a-f, colored plates 11-20; 1832, 3: a-1, colored plates 21-30; 1834, 4: a-1, colored plates 31-40; 1845, 5: a-1, 10 colored plates].

Knausenberger, J. G. and W. A. Allen

1977. Seasonal abundance of predators of *Heliothis zea* in soybeans in southeastern Virginia. Virginia Journal of Science, 28: 52.

Knight, H. H.

1916. Remarks on *Lygus invitus* Say, with descriptions of a new species and variety of *Lygus* (Hemiptera Miridae). Canadian Entomologist, 48: 345-349.

1917. New and noteworthy forms of North American Miridae (Hemip.). Entomological News, 28: 3-8.

1917. Notes of species of Miridae inhabiting ash trees (*Fraxinus*) with the description of a new species (Hemip.). Bulletin of the Brooklyn Entomological Society, 12: 80-82.

1917. Records of European Miridae occurring in North America (Hemiptera, Miridae). Canadian Entomologist, 49: 248-252.

1917. New species of *Lopidea* (Miridae, Hemip.). Entomological News, 28: 455-461.

1917. A revision of the genus *Lygus* as it occurs in America north of Mexico, with biological data on the species from New York. Cornell University Agricultural Experiment Station Bulletin, 391: 555-645.

1918. New species of *Platylygus* with a note on the male of *Largidea grossa* Van Duzee (Hemip. Miridae). Bulletin of the Brooklyn Entomological Society, 13: 16-18, 1 plate.

1918. Synoptic key to the subfamilies of Miridae (Hemiptera-Heteroptera). Journal of the New York Entomological Society, 26: 40-44.

1918. Additional data on the distribution and food plants of *Lygus* with descriptions of a new species and variety (Hemip. Miridae). Bulletin of the Brooklyn Entomological Society, 13: 42-45.

1918. The genus *Sericophanes* with descriptions of two new species (Miridae, Hemip.). Bulletin of the Brooklyn Entomological Society, 13: 80-83.

1918. New species of *Lopidea* from Arizona (Hemip. Miridae). Entomological News, 29: 172-176, 1 plate.

1918. Old and new species of *Lopidea* from the United States (Hemip., Miridae). Entomological News, 29: 210-216.

1918. Interesting new species of Miridae from the United States, with a note on *Orthocephalus mutabilis* (Fallen) (Hemip. Miridae). Bulletin of the Brooklyn Entomological Society, 13: 111-116.

1919. The male of *Lygus univittatus* with the description of a new *Lygus* (Hemip. Miridae). Bulletin of the Brooklyn Entomological Society, 14: 21-22.

1919. The genus *Bolteria* Uhler (Hemiptera-Miridae). Bulletin of the Brooklyn Entomological Society, 14: 126-128.

1920. New and little-known species of *Phytocoris* from the eastern United States (Heteroptera-Miridae). Bulletin of the Brooklyn Entomological Society, 15: 49-66.

1921. Monograph of the North American species of *Deraeocoris* (Heteroptera, Miridae). 18th Report of the State Entomologist of Minnesota. pages 75-210 (1920). [also 1921, University of Minnesota Agricultural Experiment Station Technical Bulletin, 1: 75-210].

1921. A new species of *Bolteria* (Heteroptera, Miridae). Bulletin of the Brooklyn Entomological Society, 16: 73-74.

1921. Scientific results of the Katmai Expeditions of the National Geographic Society. XIV. Hemiptera of the family Miridae. Ohio Journal of Science, 21: 107-112.

1922. Nearctic records for species of Miridae known heretofore only from the Palearctic Region (Heterop.). Canadian Entomologist, 53: 281-288 (1921).

1922. The North American species of *Labops* (Heteroptera-Miridae). Canadian Entomologist, 54: 258-261.

1923. A new species of *Labopidea* on garlic (Heteroptera-Miridae). Bulletin of the Brooklyn Entomological Society, 18: 31.

1923. A new *Peritropis* from the eastern United States (Heteroptera-Miridae). Entomological News, 34: 50-52.

1923. A fourth paper on the species of *Lopidea* (Heteroptera, Miridae). Entomological News, 34: 65-72, 1 plate.

1923. Family Miridae (Capsidae). Pages 422-658. *In* Britton, W. E. (editor). The Hemiptera or sucking insects of Connecticut. Connecticut Geological and Natural History Survey Bulletin, 34: 1-807.

1924. *Atractotomus mali* (Meyer) found in Nova Scotia (Heteroptera, Miridae). Bulletin of the Brooklyn Entomological Society, 19: 65.

1925. Descriptions of thirty new species and two new genera of North American Miridae (Hemiptera). Bulletin of the Brooklyn Entomological Society, 20: 33-58.

Knight, H. H. (continued)

1925. *Neocapsus cuneatus* Distant in Arizona and Texas, with a variety described from Mississippi and North Carolina (Heteroptera, Miridae). Entomological News, 36: 78-79.

1925. Descriptions of a new genus and eleven new species of North American Miridae (Hemiptera). Canadian Entomologist, 57: 89-97.

1925. Descriptions of fourteen undescribed species of *Parthenicus* (Hemiptera, Miridae). Ohio Journal of Science, 25: 119-129.

1925. A list of Miridae and Anthocoridae from Alberta, Canada (Hemiptera). Canadian Entomologist, 57: 181-182.

1925. Descriptions of twelve new species of *Polymerus* from North America (Hemiptera, Miridae). Canadian Entomologist, 57: 244-253.

1925. Description of a new species of *Plagiognathus* from the eastern United States (Hem., Miridae). Entomological News, 36: 305-306.

1925. A new *Rhinacloa* and three new species of *Lepidopsallus* (Hemiptera, Miridae). Bulletin of the Brooklyn Entomological Society, 20: 225-228.

1926. Descriptions of four new species of *Plagiognathus* from the eastern United States (Hem., Miridae). Entomological News, 37: 9-12.

1926. *Capsus simulans* (Stal) and *Labops burmeisteri* Stal recognized from the Nearctic Region (Hemiptera, Miridae). Canadian Entomologist, 58: 59-60.

1926. Descriptions of seven new species of *Pilophorus* (Hemiptera, Miridae). Bulletin of the Brooklyn Entomological Society, 21: 18-26.

1926. Description of a new *Renodaeus* from Texas (Hemiptera, Miridae). Bulletin of the Brooklyn Entomological Society, 21: 56-57.

1926. Descriptions of nine new species of Bryocorinae (Hemiptera, Miridae). Bulletin of the Brooklyn Entomological Society, 21: 101-108.

1926. Descriptions of eleven new species of *Phytocoris* from eastern North America (Hemiptera, Miridae). Bulletin of the Brooklyn Entomological Society, 21: 158-168.

1926. Descriptions of seven new *Paracalocoris* with keys to the nearctic species and varieties (Hemiptera, Miridae). Annals of the Entomological Society of America, 19: 367-377.

1926. Descriptions of four new species of *Eustictus* (Hemiptera, Miridae). Psyche, 33: 121-123.

1926. Notes on species of *Polymerus* with descriptions of four new species and two new varieties (Hemiptera, Miridae). Canadian Entomologist, 58: 164-168.

1926. Descriptions of six new species of Miridae from eastern North America (Hemiptera-Miridae). Canadian Entomologist, 58: 252-256.

1926. *Capsus externus* Herrich-Schaeffer is a *Paracalocoris* (Hemiptera, Miridae). Entomological News, 37: 258-262.

1926. A key to the North American species of *Macrolophus* with descriptions of two new species (Hem.: Miridae). Entomological News, 37: 313-316.

1927. Descriptions of twelve new species of Miridae from the District of Columbia and vicinity (Hemiptera). Proceedings of the Biological Society of Washington, 40: 9-18.

1927. Notes on the distribution and host plants of some North American Miridae (Hemiptera). Canadian Entomologist, 59: 34-44.

1927. A new *Semium* from Arizona and Colorado (Hemiptera, Miridae). Bulletin of the Brooklyn Entomological Society, 22: 26-27.

1927. Notes on the collecting of Say's mulatto-bug (*Cydnoides albipennis* Say) (Hemiptera, Cydnidae). Entomological News, 38: 40-42.

1927. On the Miridae in Blatchley's "Heteroptera of Eastern North America." Bulletin of the Brooklyn Entomological Society, 22: 98-105.

1927. Descriptions of fifteen new species of *Ceratocapsus* (Hemiptera, Miridae). Ohio Journal of Science, 27: 143-154.

1927. New species and a new genus of Deraeocorinae from North America (Hemiptera, Miridae). Bulletin of the Brooklyn Entomological Society, 22: 136-143.

1927. Descriptions of nine new species of *Melanotrichus* Reuter from North America (Hemiptera, Miridae). Canadian Entomologist, 59: 142-147.

1927. Descriptions of seven new species of the

genus *Orthotylus* Fieber (Hemiptera, Miridae). Canadian Entomologist, 59: 176-181.

1927. *Acetropis americana*, a new species of Miridae from Oregon (Hemiptera). Entomological News, 38: 206-207.

1927. *Megalopsallus*, a new genus of Miridae with five new species from North America (Hemiptera). Annals of the Entomological Society of America, 20: 224-228.

1927. New species of mimetic Miridae from North America (Hemiptera). Entomological News, 38: 302-307.

1927. *Dacerla downesi*, a new species of Miridae from Oregon (Hemiptera). Entomological News, 38: 314-315.

1928. Key to species of *Clivinema* with descriptions of seven new species (Hemiptera, Miridae). Proceedings of the Biological Society of Washington, 41: 31-36.

1928. *Hesperolabops periscopis*, a new periscopic bug from Salvador, with a note on the Texas member of this genus (Hemiptera, Miridae). Proceedings of the Entomological Society of Washington, 30: 67-68.

1928. New species of *Phytocoris* from North America (Hemiptera, Miridae). Bulletin of the Brooklyn Entomological Society, 23: 28-46.

1928. A new key for *Bolteria* with descriptions of two new species (Hemiptera, Miridae). Bulletin of the Brooklyn Entomological Society, 23: 129-132.

1928. Key to the species of *Hadronema* Uhler with descriptions of five new species (Hemiptera, Miridae). Canadian Entomologist, 60: 177-182.

1928. Key to the species of *Oncerometopus* with descriptions of five new species (Hemiptera, Miridae). Journal of the New York Entomological Society, 36: 189-194.

1928. New species of *Labopidea* and *Macrotyloides* (Hemiptera, Miridae). Canadian Entomologist, 60: 233-236.

1928. Descriptions of four new North American species of *Megaloceroea* (Hemip.: Miridae). Entomological News, 39: 247-251.

1928. Families Miridae and Isometopidae. Pages 110-135. *In* Leonard, M. D. (editor). A List of the Insects of New York. New York Agricultural Experiment Station (Cornell) Memoir, 101: 1-1121.

1929. New species of *Halticotoma* and *Sixeonotus* (Hemiptera, Miridae). Bulletin of the Brooklyn Entomological Society, 23: 241-249(1928).

1929. New species of *Neoborus* and *Xenoborus* (Hemiptera, Miridae). Bulletin of the Brooklyn Entomological Society, 24: 1-11.

1929. Descriptions of six new species of *Microphylellus* (Hemip., Miridae). Entomological News, 40: 40-43.

1929. Descriptions of five new species of *Plagiognathus* from North America (Hemip., Miridae). Entomological News, 40: 69-74.

1929. Rectifications for Blatchley's "Heteroptera" with the description of a new species (Hemiptera). Bulletin of the Brooklyn Entomological Society, 24: 143-154.

1929. *Labops verae*, new species with *Labopella*, *Nicholia*, and *Pronotocrepis*, new genera of North American Miridae (Hemiptera). Canadian Entomologist, 61: 214-218.

1929. New species and varieties of *Platytylellus* from North America (Hemiptera: Miridae). Entomological News, 40: 189-192.

1929. The fourth paper on new species of *Plagiognathus* (Hemiptera, Miridae). Entomological News, 40: 263-268.

1930. New species of *Pseudopsallus* Van D. with an allied new genus described (Hemiptera, Miridae). Bulletin of the Brooklyn Entomological Society, 25: 1-8.

1930. New species of *Ceratocapsus* (Hemiptera, Miridae). Bulletin of the Brooklyn Entomological Society, 25: 187-198.

1930. An European plant-bug (*Amblytylus nasutus* Kirschbaum) recognized from Massachusetts (Hemiptera, Miridae). Entomological News, 41: 256-258.

1930. Descriptions of four new species of mimetic Miridae (Hemiptera). Entomological News, 41: 319-321.

1930. New species of *Psallus* Fieb. (Hemiptera, Miridae). Canadian Entomologist, 62: 125-131.

1930. A new key to *Paracalocoris* with descriptions of eight new species (Hemiptera, Miridae). Annals of the Entomological Society of America, 23: 810-827.

1931. *Dacota hesperia* Uhler referred to *Atractotomus*, also descriptions of three new species (Hemiptera, Miridae). Bulletin of the Brooklyn Entomological Society, 26: 36-38.

1931. Three new species of *Cyrtorhinus* from North America (Hemiptera, Miridae).

Knight, H. H. (continued)
Bulletin of the Brooklyn Entomological Society, 26: 171-173.

1932. Two new species of *Macrotylus* Fieber from the western United States (Hemiptera, Miridae). Pan-Pacific Entomologist, 8: 79-80.

1933. *Bothynotus* Fieber: Descriptions of two new species from North America (Hemip., Miridae). Entomological News, 44: 132-135.

1933. *Calocorisca californica* n. sp., an additonal genus for the United States (Hemiptera, Miridae). Pan-Pacific Entomologist, 9: 69-70.

1933. *Lampethusa nicholi*, a new species from Arizona and Texas (Hemiptera, Miridae). Pan-Pacific Entomologist, 9: 71-72.

1934. *Phytocoris* Fallen: Twelve new species from the western United States (Hemiptera-Miridae). Bulletin of the Brooklyn Entomological Society, 29: 1-16.

1934. *Neurocolpus* Reuter: Key with five new species (Hemiptera, Miridae). Bulletin of the Brooklyn Entomological Society, 29: 162-167.

1938. *Stronglyocoris* Blanchard: Six new species from North America (Hemiptera, Miridae). Iowa State College Journal of Science, 13: 1-7.

1939. Three new species of Miridae from North America (Hemiptera). Bulletin of the Brooklyn Entomological Society, 34: 21-23.

1939. *Conostethus americanus* new species from Colorado, Montana and South Dakota (Hemiptera, Miridae). Entomological News, 50: 132-133.

1939. *Reuteria* Puton; four new species from the United States (Hemiptera, Miridae). Iowa State College Journal of Science, 13: 129-133.

1941. The plant bugs, or Miridae, of Illinois. Illinois Natural History Survey Bulletin, 22: 1-234.

1941. New species of *Irbisia* Reuter (Hemiptera, Miridae). Bulletin of the Brooklyn Entomological Society, 36: 75-79.

1941. New species of *Lygus* from the western United States (Hemiptera, Miridae). Iowa State College Journal of Science, 15: 269-273.

1942. *Stittocapsus* new genus and *Calocoris texanus* new speies from the United States (Hemiptera, Miridae). Entomological News, 53: 156-158.

1943. Five new species of *Dicyphus* from western North America and one new *Cyrtopeltis* (Hemiptera, Miridae). Pan-Pacific Entomologist, 19: 53-58.

1943. New species of *Polymerus* (Westwood) from the United States (Hemiptera, Miridae). Canadian Entomologist, 75: 179-182.

1944. *Lygus* Hahn; six new species from western North America (Hemiptera, Miridae). Iowa State College Journal of Science, 18: 471-477.

1952. Review of the genus *Perillus* with description of a new species (Hemiptera, Pentatomidae). Annals of the Entomological Society of America, 45: 229-232.

1953. New species of Miridae from Missouri (Hemiptera). Iowa State College Journal of Science, 27: 509-518.

1959. New genera and species of North American Miridae (Hemiptera). Iowa State College Journal of Science, 33: 421-426.

1961. Ten new species of *Phytocoris* from North America. Iowa State Journal of Science, 35: 473-484.

1962. Ten new and six old species of *Lopidea* from North America (Hemiptera, Miridae). Iowa State Journal of Science, 37: 29-41.

1963. Review of the genus *Ilnacora* Reuter with descriptions of ten new species (Hemiptera, Miridae). Iowa State Journal of Science, 38: 161-178.

1964. *Phymatopsallus* new genus, and new species of Phylinae from North America (Hemiptera, Miridae). Iowa State Journal of Science, 39: 127-152.

1965. Old and new species of *Lopidea* Uhler and *Lopidella* Knight (Hemiptera, Miridae). Iowa State Journal of Science, 40: 1-26.

1965. A new key to species of *Reuteroscopus* Kirk. with descriptions of new species (Hemiptera, Miridae). Iowa State Journal of Science, 40: 101-120.

1966. *Schaffneria*, a new genus of ground dwelling plant bugs (Hemiptera, Miridae). Iowa State Journal of Science, 41: 1-6.

1966. *Keltonia*, a new genus near *Reuteroscopus* Kirk., with descriptions of new species (Hemiptera: Miridae). Canadian Entomologist, 98: 590-591.

1968. Review of genus *Orectoderus* Uhler with

a key to the species (Hemiptera, Miridae). Iowa State Journal of Science, 42: 311-318.

1968. Taxonomic review: Miridae of the Nevada Test Site and the western United States. Brigham Young University Science Bulletin, 9(3): 1-282.

1969. New species of *Pronotocrepis* Kngt., *Europiella* Reut., and *Hesperocapsus* Kngt., from western United States (Miridae, Hemiptera). Iowa State Journal of Science, 44: 79-91.

1970. *Tannerocoris* new genus, and new species of Miridae (Hemiptera) from the western United States. Great Basin Naturalist, 30: 227-231.

1970. Review of the genus *Slaterocoris* Wagner, with a key and descriptions of new species (Hemiptera, Miridae). Iowa State Journal of Science, 45: 233-267.

1971. A key to the species of *Bolteria* Uhler with descriptions of six new species (Hemiptera, Miridae). Iowa State Journal of Science, 46: 87-94.

1972. *Mineocapsus minimus* new genus and new species; with a length of one millimeter this insect is a candidate for the smallest species in family Miridae (Hemiptera). Iowa State Journal of Science, 46: 425-427.

1973. A key to the North American species of *Pilophorus* Hahn with descriptions of new species (Hemiptera, Miridae). Iowa State Journal of Research, 48: 129-145.

1974. A key to species of *Phytocoris* Fallen belonging to the *Phytocoris junceus* Kngt. group of species (Hemiptera, Miridae). Iowa State Journal of Research, 49: 123-135.

Knight, H. H. and W. L. McAtee

1929. Bugs of the family Miridae of the District of Columbia and vicinity. Proceedings of the United States National Museum, 75: 1-27.

Knight, H. H. and J. C. Schaffner

1968. *Lopidea* Uhler: New species and records from Mexico and southwestern United States, with *Mayamiris*, related new genus from Mexico (Hemiptera, Miridae). Iowa State Journal of Science, 43: 71-81.

1972. New species of *Lopidea* Uhler from Mexico and the western United States (Hemiptera, Miridae). Iowa State Journal of Research, 47: 107-115.

1976. New and old species of the genus *Ilnacora* Reuter (Hemiptera, Miridae). Iowa State Journal of Research, 50: 399-407.

Knowlton, G. F.

1931. Notes on Utah Heteroptera and Homoptera. Entomological News, 42: 68-72.

1933. Notes on Utah Heteroptera. Entomological News, 44: 261-264.

1935a. Beet leafhopper insect predator studies. Proceedings of the Utah Academy of Sciences, Arts, and Letters, 12: 255-260.

1935. Further notes of Utah Heteroptera and Homoptera. Entomological News, 46: 108-112.

1936. The insect fauna of Utah (from the standpoint of an economic entomologist). Proceedings of the Utah Academy of Sciences, Arts, and Letters, 13: 249-262.

1937. Biological control of the beet leafhopper in Utah. Proceedings of the Utah Academy of Sciences, Arts, and Letters, 14: 113-139.

1945. *Labops* damage to range grasses. Journal of Economic Entomology, 38: 707-708.

1960. An unusual flight of Hemiptera in southern Utah. Proceedings of the Utah Academy of Sciences, Arts, and Letters, 37: 53-54.

Kolenati, F. A.

1845. Meletemata entomologica. Fascicle II. Hemiptera Caucasi Tessaratomidae monographice dispositae. Imperialis Academiae Scientiarum, Petropoli. 132 pages, 8 plates.

1846. Meletemata entomologica. Fascicle IV. Hemiptera Caucasi Pentatomidae monographice dispositae. Imperialis Academiae Scientiarum, Petropoli. 72 pages, 2 plates.

1856. Meletemata entomologica hemipterorum heteropterorum Caucasi. Harpagocorisae, monographice dispositae. Fascicle VI. Imperialis Academie Scientiarum, Petropoli. Pages 419-502, 1 plate. [Also separate in Bulletin de la Société, Impériale des Naturalistes de Moscou, 29: 419-502].

Komarek, E. V.

1971. Insect control - fire for habitat management. Proceedings Tall Timbers Conference on Ecological Animal Control by Habitat Management, 2: 157-171.

Kormilev, N. A.

1951. Phymatidae Argentinas (Hemiptera), con Observaciones sobre Phymatidae en General. Revista del Instituto Nacional de Investigacion de las Ciencias Naturales, Ciencias Zoologicas, 2: 45-110, plates 1-14.

1953. Notas sobre Phymatidae nearcticas I (Hemiptera). Misiín de Estudios de Patología Regional Argentina, 24: 63-67.

1954. Notes on nearctic Phymatidae II, (Hemiptera). A new subspecies of *Macrocephalus prehensilis* (F.) from Florida. Journal of the Kansas Entomological Society, 27: 158-159.

1957. Notes on American "Phymatidae" (Hemiptera). Revista Brasilieira de Biologia, 17: 123-138.

1959. Notas sobre Aradidae neotropicales, V, (Hemiptera). Sobre el género *Pictinus* Stal, 1873. Revista de la Sociedad Uruguaya de Entomologia, 3: 21-33.

1962. Revision of the Phymatinae (Hemiptera, Phymatidae). Philippine Journal of Science, 89(3-4): 287-486, plates 1-19 (1960).

1964. Notes on the Aradidae in the Naturhistoriska Riksmusem, Stockholm. Hemiptera-Heteroptera. Arkiv för Zoologi, Series 2, 16: 463-479.

1966. Notes on Aradidae in the U.S. National Museum, IV (Hemiptera-Heteroptera). Proceedings of the United States National Museum 119(3548): 1-25.

1966. Two new American Aradidae (Hemiptera-Heteroptera). Psyche, 73: 26-29.

1967. On some Aradidae from Brasil, Argentina and Laos (Hemiptera, Heteroptera). Opuscula Zoologica, 100: 1-10.

1968. North and Central American species of *Aneurus* (Hemiptera: Aradidae). Proceedings of the United States National Museum, 125(3657): 1-12.

1968. Notes on neotropical Aradidae XIX. (Hemiptera-Heteroptera). Revista de la Facultad de Agronomia, Universidad Central de Venezuela, 5: 43-56.

1971. Key to American species of the genus *Mezira* (Hemiptera: Aradidae). Proceedings of the Entomological Society of Washington, 73: 282-292.

1972. A new species of ambush bug from Arizona (Hemiptera: Phymatidae). Bulletin of the Southern California Academy of Sciences, 71: 93-95.

1978. Two new species of American Aradidae (Hemiptera). Proceedings of the Entomological Society of Washington, 80: 228-233.

1980. Notes on American Aradidae (Hemiptera: Aradidae). Proceedings of the Entomological Society of Washington, 82: 99-107.

1982. Records and descriptions of North American and Oriental Aradidae (Hemiptera). The Wasmann Journal of Biology, 40: 1-17.

1982. On some Aradidae in the collections of the University of Georgia, Athens, Georgia, U. S. A. (Hemiptera). Journal of the Georgia Entomological Society, 17: 333-337.

1982. On *Mezira granulata* (Say) group (Hemiptera: Aradidae). Journal of Natural History, 16: 775-779.

1984. Keys to the genera and descriptions of new taxa of macrocephaline ambush bugs (Heteroptera: Phymatidae). Journal of Natural History, 18: 623-637.

Kormilev, N. A. and E. Heiss

1979. A new species of the genus *Aradus* F., 1803, from Alaska (Insecta: Heteroptera). Bericht des Naturwissenschaftlich-Medizinischen Vereins in Innsbruck, 66: 47-52.

Kouchakevitch, A. A.

1867. Some New Species of Hemiptera. Horae Societatis Entomologicae Rossicae, 4(2): 97-101.

Kraus, B.

1985. Oviposition on the back of female giant water bugs, *Abedus indentatus*: The consequence of a shortage in male back space? (Hemiptera: Belostomatidae). Pan-Pacific Entomologist, 61: 54-57.

Kritsky, G. R.

1977. Observations on the morphology and behavior of the Enicocephalidae (Hemiptera). Entomological News, 88: 105-110.

1977. Two new genera of Enicocephalidae (Hemiptera). Entomological News, 88: 161-168.

1978. The North American and Caribbean species of *Systelloderes* (Hemiptera: Enicocephalidae). Entomological News, 89: 65-73.

1978. A new species of *Hymenocoris* from Mexico (Hemiptera: Enicocephalidae). Ento-

mological News, 89: 74-76.
1981. Two new species of *Alienates* (Hemiptera: Enicocephalidae). Entomological News, 92: 130-132.

Krombein, K. V.
1969. The generic placement of two Nearctic *Holopyga* with biological notes (Hymenoptera: Chrysididae). Proceedings of the Entomological Society of Washington, 71: 351-361.

Kuitert, L. C.
1942. Gerrinae in the University of Kansas collections. University of Kansas Science Bulletin, 28:113-143.
1949. Some new species of Nepidae (Hemiptera). Journal of the Kansas Entomological Society, 22: 60-68.

Kumar, R.
1969. Morphology and relationships of the Pentatomoidea (Heteroptera). III. Natalicolinae and some Tessaratomidae of uncertain position. Annals of the Entomological Society of America, 62: 681-695.
1969. Morphology and relationships of the Pentatomoidea (Heteroptera). IV. Oncomerinae (Tessaratomidae). Australian Journal of Zoology, 17: 553-606.
1974. A revision of world Acanthosomatidae (Heteroptera: Pentatomoidea): Keys to and descriptions of subfamilies, tribes and genera, with designation of types. Australian Journal of Zoology, Supplemental Series, 34: 1-60.

Kumar, R. and M. S. K. Ghauri
1970. Morphology and relationships of the Pentatomoidea (Heteroptera) 2--World Genera of Tessaratomini (Tessaratomidae). Deutsche Entomologische Zeitschrift, 17: 1-32.

Kumar, R., R. J. Lavigne, J. E. Lloyd, and R. E. Pfadt
1976. Insects of the Central Plains Experiment Range, Pawnee National Grassland. University of Wyoming Agricultural Experiment Station Science Monograph, 32: 1-74.

Kurczewski, F. E.
1967. A note on the nesting behavior of *Solierella inermis* (Hymenoptera: Sphecidae, Larrinae). Journal of the Kansas Entomological Society, 40: 203-208.
1972. Observations on the nesting behavior of *Diplopectron peglowi* Krombein (Hymenoptera: Sphecidae). Proceedings of the Entomological Society of Washington, 74: 385-396.

Kushakevich, A.
1867. Some new species of Hemiptera [In Russian]. Horae Societatis Entomologicae Rossicae, 4: 97-101, plate 2.

Labrador, S., J. Ramon, and E. Rubio Espina
1967. Preliminary studies on the biology and control of *Blissus leucopterus* pest of pangola pastures in Zulia State. Anon. Reunion Latinamericana de Fitotecnia Resumenes de Los Trabahos Cientificos VII. Pp. 70-71.

LaFollette, R. A.
1915. Preliminary list of common Heteroptera from the Claremont-Laguna Region. Journal of Entomology and Zoology, 7: 123-129.

Lamarck, J. B.
1801. Système des animaux sans vertèbres ou tableau général des classes, des ordres et des genres de ces animaux; présentant leurs charactères essentiels et leur distribution, d'après la consideration de leurs rapports naturels et de leur organization, et suivant l'arrangement éstabli dans les galeries du Muséum d'Histoire Naturelle parmi leurs dépouilles conservées, précédédu discours d'ouverture du cours de Zoologie donnéedans le Mus. d'hist. nat. Paris. VIII + 432 pages.

Lamp, W. O. and T. O. Holtzer
1980. Distribution of overwintering chinch bugs, *Blissus leucopterus leucopterus* (Hemiptera Heteroptera Lygaeidae). Journal of the Kansas Entomological Society, 53: 320-324.

Lampman, R. L. and N. J. Fashing
1978. The effect of temperature and feeding on longevity, fecundity, and intrinsic rate of increase of *Gryon parkeri*, an egg parasite of the large and small milkweed bugs. Virginia Journal of Science, 29: 63.

Lanciani, C. A.
1971. Host exploitation and synchronous development in a water mite parasite of the marsh treader *Hydrometra myrae* (Hemiptera: Hydrometridae). Annals of the Entomological Society America, 64: 1254-1259.

Lansbury, I.

1960. The Corixidae (Hemiptera-Heteroptera) of British Columbia. Proceedings of the Entomological Society of British Columbia, 57: 34-43.

1960. Note on Corixidae occurring at black light traps (Hemiptera: Heteroptera). Entomological News, 71: 244.

Laporte, F. L.

1832-1833. Essai d'une classification systématique de l'ordre des Hémiptères (Hémiptères Heteroptères, Latr.). Magazin de Zoologie (Guerin), 1832, 2: 1-16; 1833, 2: 17-75, Supplement 76-88, plates 51-55 [This dating follows the International Commission of Zoological Nomenclature's Direction 63 (1956)].

La Rivers, I.

1948. A new species of *Pelocoris* from Nevada, with notes on the genus in the United States (Hemiptera: Naucoridae). Annals of the Entomological Society of America, 41: 371-376.

1949. A new species of *Ambrysus* from Death Valley, with notes on the genus in the United States (Hemiptera: Naucoridae). Bulletin of the Southern California Academy of Sciences, 47: 103-110 (1948).

1950. A new naucorid genus and species from Nevada (Hemiptera). Annals of the Entomological Society of America, 43: 368-373.

1951. A revision of the genus *Ambrysus* in the United States (Hemiptera: Naucoridae). University of California Publications in Entomology, 8: 277-338.

1953. New gelastocorid and naucorid records and miscellaneous notes, with a description of the new species, *Ambrysus amargosus* (Hemiptera: Naucoridae). Wasmann Journal of Biology, 11: 83-96.

1953. Two new naucorid bugs of the genus *Ambrysus*. Proceedings of the United States National Museum, 103(3311): 1-7.

1953. The *Ambrysus* of Mexico (Hemiptera, Naucoridae). University of Kansas Science Bulletin, 35: 1279-1349.

1956. A new subspecies of *Pelocoris shoshone* from the Death Valley drainage (Naucoridae: Hemiptera). Wasmann Journal of Biology, 14: 155-158.

1957. A *Limnocoris* for the United States (Hemiptera: Naucoridae). Pan-Pacific Entomologist, 33: 71-75.

1963. Two new ambrysi (Hemiptera: Naucoridae). Biological Society of Nevada Occasional Papers, 1: 1-7.

1965. The subgenera of the genus *Ambrysus* (Hemiptera, Naucoridae). Biological Society of Nevada Occasional Papers, 4: 1-8.

1970. A new subspecies of *Pelocoris femoratus* (Palisot-Beauvois) from Florida (Hemiptera: Naucoridae). Biological Society of Nevada Occasional Papers, 26: 1-4.

1971. Studies of Naucoridae (Hemiptera). Biological Society of Nevada Memoire, 2: 1-120.

1974. Catalogue of taxa described in the family Naucoridae (Hemiptera) supplement no. 1: Corrections, emendations and additions, with descriptions of new species. Biological Society of Nevada Occasional Papers, 38: 1-17.

1976. Supplement no. 2 to the catalogue of taxa described in the family Naucoridae (Hemiptera), with descriptions of new species. Biological Society of Nevada Occasional Papers, 41: 1-18.

Larochelle, A.

1984. Les Punaises Terrestres (Heteropteres: Geocorises) du Quebec. Fabreries, Supplément 3. 513 pages.

Larrousse, F.

1926. Description de deux espèces nouvelles du genre *Triatoma*: *T. carrioni* n. sp., et *T. pintoi* n. sp. Annales de Parasitologie Humaine et Comparée, 4: 136-139.

Larsen, O.

1949. Die ortsbewegungen von *Ranatra linearis* L. ein beitrag zur vergleichende physiologie der lokomotionsorgane der insekten. Lund Universitatis Arsskrift N. F. Avd. 2, 45: 1-82.

Laster, M. L. and J. R. Brazzel

1968. A comparison of predator populations in cotton under different control programs in Mississippi. Journal of Economic Entomology, 61: 714-719.

Latreille, P. A.

1796. Précis des caractères génériques des Insectes, disposés dans un ordre Naturel. Bordeaux. xii + 201 pages.

1802. Histoire naturelle, générale et particuliere des Crustacés et des Insectes. Dufart, Paris. 3 : i-xii, 13-468; 4: 1-387.

1806-1809. Genera crustaciorum et insectorum secundum ordinem naturalem in familias disposita, inconibus exemplisque plurimis explicata. Amand Koenig, Paris. 4 volumes. 1806, 1: i-xviii, 1-302; 1807, 2: 1-280; 1807, 3: 1-258; 1809, 4: 1-399.

1810. Ordre III: Hémiptères. Hemiptera. Section Première. Hétéroptères. Heteroptera. Pages 250-251, 254-261, 421, 433-434. *In* Considérations générales sur l'ordre naturel des animaux composant les classes des crustacés, des arachnides, et des insectes, avec un tableau méthodique de leurs genres, disposés en familles. Schoell, Paris. 444 pages.

1817. Les crustacés, les arachnides et les insects, Tome 3. Les Hémiptères. (Rhygota Fab.). Pages 385-417. *In* Cuvier, G. L.. Le règne animal distribue d'après son organization, pour servir de base a l'histoire naturelle des animaux et d'introduction a l'anatomie comparée. Deterville, Paris. xxiv + 653 pages.

1817. Insectes de l'Amérique Equinoxiale, recueillis pendant le voyage de MM. de Humboldt et Bonpland. Secunde Partie. *In* Humboldt, F. and A. Bonpland. Voyage aux Regions Equinoxiales du Nouveau Continent, faiten 1799-1804, Part II, Zoology. 10: 97-144.

Lattin, J. D.

1966. *Stalia major* (Costa) in North America (Hemiptera: Nabidae). Proceedings of the Entomological Society of Washington, 68: 314-318.

1968. A new species of *Micracanthia* Reuter from Oregon (Heteroptera: Saldidae). Proceedings of the Entomological Society of Washington, 70: 165-169.

Lattin, J. D. and R. H. Messing

1984. *Compsidolon* (*Coniortodes*) *salicellum* (Hemiptera: Miridae): A predaceous plant bug, new to the United States, found on filbert. Journal of the New York Entomological Society, 92: 179-183.

Lattin, J. D. and G. M. Stonedahl

1984. *Campyloneura virgula*, a predacious Miridae not previously recorded from the United States (Hemiptera). Pan-Pacific Entomologist, 60: 4-7.

Lauck, D. R.

1959. Three new species of *Belostoma* from Mexico and Central America (Hemiptera: Belostomatidae), with a list of North American species. Bulletin of the Chicago Academy of Sciences, 11: 1-9.

1962. A monograph of the genus *Belostoma* (Hemiptera). Part I. Introduction and *B. dentatum* and *subspinosum* groups. Bulletin of the Chicago Academy of Sciences, 11: 34-81.

1963. A monograph of the genus *Belostoma* (Hemiptera). Part II. *B. aurivillianum, stollii, testaceopallidum, dilatatum,* and *discretum* groups. Bulletin of the Chicago Academy of Sciences. 11: 82-101.

1964. A monograph of the genus *Belostoma* (Hemiptera). Part III. *B. triangulum, bergi, minor, bifoveolatum* and *flumineum* groups. Bulletin of the Chicago Academy of Sciences, 11: 102-154.

1966. A new species of *Sigara* from California (Corixidae, Hemiptera). Pan-Pacific Entomologist, 42: 168-172.

1979. Family Corixidae/water boatmen. Pages 87-123. *In* Menke, A. S. (editor). Aquatic and Semiaquatic Hemiptera of California. Bulletin of the California Insect Survey, 21: 1-166.

Lauck, D. R. and A. S. Menke

1961. The higher classification of the Belostomatidae (Hemiptera). Annals of the Entomological Society of America, 54: 644-657.

Lauck, D. R. and W. G. Wheatcroft

1958. Notes on a new habitat for *Nerthra* (Gelastocoridae-Hemiptera). Entomological News, 69: 20.

Lavigne, R. J.

1976. Rangeland insect-plant associations on the Pawnee site. Annals of the Entomological Society of America, 69: 753-763.

Lawrence, R. K. and T. F. Watson

1979. Predator-prey relationships of *Geocoris punctipes* and *Heliothis virescens*. Environmental Entomology, 8: 245-248.

Lawson, F. A.

1959a. Identification of the nymphs of common families of Hemiptera. Journal of the Kansas Entomological Society, 32: 88-92.

Leach, W. E.

1815. Hemiptera. Page 123. *In* Brewster's Edinburgh Encyclopedia. Volume 9. Edinburgh, Scotland. Pages 57-192.

LeBaron, W.

1850. The chinch bug. Prairie Farmer, 10: 280-281.

LeCato, G. L. and R. Davies

1973. Preferences of the predator *Xylocoris flavipes* (Hemiptera: Anthocoridae) for species and instars of stored-product insects. Florida Entomologist, 56: 57-59.

LeConte, J. L., Jr.

1855. Remarks on two species of American cimex. Proceedings of the Academy of Natural Sciences of Philadelphia, 7: 404.

1859. The Complete Writings of Thomas Say on the Entomology of North America. Bailliere Brothers, New York. 2 volumes. [1: i-xxi, 1-412, plates 1-54; 2: 1-814].

Lee, R. D. and R. E. Ryckman

1955. First precise locality record of *Hesperocimex coloradensis* List from Mexico (Hemiptera, Cimicidae). Proceedings of the Entomological Society of Washington, 57: 164.

Leidy, J.

1847. History and anatomy of the hemipterous genus *Belostoma*. Journal of the Academy of Natural Sciences of Philadelphia, 2nd Series, 1: 57-67, 1 plate.

Leigh, T. F.

1961. Insecticide susceptibility of *Nysius raphanus*, a pest of cotton. Journal of Economic Entomology, 54: 120-122.

Leigh, T. F., J. H. Black, E. C. Jackson, and V. E. Burton

1966. Insecticides and beneficial insects in cotton fields. California Agriculture, 20: 4-6.

Leigh, T. F. and D. Gonzalez

1976. Field cage evaluation of predators for control of *Lygus hesperus* on cotton. Environmental Entomology, 5: 948-952.

Leigh, T. F., D. W. Grimes, W. L. Dickens, and C. E. Jackson

1974. Planting pattern, plant population, irrigation, and insect interactions in cotton. Environmental Entomology, 3: 492-496.

Leigh, T. F., D. W. Grimes, H. Yamada, J. R. Stockton, and D. Bassett.

1969. Arthropod abundance in cotton in relation to some cultural management variables. Proceedings of the Tall Timbers Conference on Ecological Animal Control by Habitat Management, 1: 71-83.

Lent, H.

1951. *Triatoma longipes* Barber, 1937, um sinonimo de *Triatoma recurva* (Stal, 1868) (Hemipt., Reduv.). Ciencia, 11: 156-158.

Lent, H. and P. Wygodzinsky

1947. Contribuição ao conhecimento dos "Reduviinae" Americanos (Reduviidae, Hemiptera). Revista Brasileira de Biologia, 7: 341-368.

1948. On two new genera of American "Reduviinae," with a key and notes on others (Reduviidae, Hemiptera). Revista Brasileira de Biologia, 8: 43-55.

1959. Sobre los géneros "*Pseudozelurus*" Lent & Wygod. y "*Zeluroides*" Lent & Wygod. (Reduviinae, Reduviidae, Hemiptera). Revista Brasileira de Biologia, 19: 351-365.

1979. Revision of the Triatominae (Hemiptera, Reduviidae), and their significance as vectors of Chagas' Disease. Bulletin of the American Museum of Natural History, 163: 123-520.

Lentz, G. L., A. Y. Chambers, and R. M. Hayes

1983. Effects of systemic insecticide-nematicides on midseason pest and predator populations in soybean. Journal of Economic Entomology, 76: 836-840.

Leonard, D. E.

1966. Biosystematics of the "*leucopterus* complex" of the genus *Blissus* (Heteropera: Lygaeidae). Connecticut Agricultural Experiment Station Bulletin, 677: 1-47.

1968a. A revision of the genus *Blissus* (Heteroptera: Lygaeidae) in eastern North America. Annals of the Entomological Society of America, 61: 239-250.

1968b. Three new species of *Blissus* from the Antilles (Heteroptera: Lygaeidae). Proceedings of the Entomological Society of Washington, 70: 150-153.

1970. A new North American species of *Blissus* (Heteroptera: Lygaeidae). Canadian Entomologist, 102: 1531-1533.

Leonard, M. D.

1915. The immature stages of *Plagiognathus politus* Uhler and *Campylomma verbasci* Herrich-Schaeffer (Capsidae, Hemiptera). Journal of the New York Entomological Society, 23: 193-197.

1916. The immature stages of *Tropidosteptes cardinalis* Uhler (Capsidae, Hemiptera). Psyche, 23: 1-3.

1916. The immature stages of two Hemiptera *Empoasca obtusa* Walsh (Typhlocybidae) and *Lopidea robiniae* Uhler (Capsidae). Entomological News, 27: 49-54, 2 plates.

1916. A tachinid parasite reared from an adult

capsid (Dip., Hom.). Entomological News, 27: 236.
1919. The immature stages of the goldenrod leaf-bug, *Strongylocoris stygica* Say (Miridae, Heterop.). Canadian Entomologist, 51: 178-180.

Lepeletier, A. L. M. and J. G. A. Serville.
1825. Hemiptera Heteroptera. *In* Olivier, G. A. (editor). Encyclopédie Méthodique. Agasse, Paris. 10: 1-833.

Le Quesne, W. J.
1956. An examination of the British species of *Drymus* Fieber (Hem., Lygaeidae) with a new subgenus and an addition and a generic reassignment in the British list. Entomologist's Monthly Magazine, 92: 337-341.

LeRoux, E. J.
1960. Effect of "modified" and "commercial" spray programs on the fauna of apple orchards in Quebec. Annals of the Entomological Society of Quebec, 6: 87-121.

Leslie, J. F. and H. Dingle
1983. Interspecific hybridization and genetic divergence in milkweed bugs (*Oncopeltus*: Hemiptera: Lygaeidae). Evolution, 37: 583-591.

Leston, D.
1952. Notes on the Ethiopian Pentatomoidea (Hemiptera): VIII, Scutellerinae Leach of Angola, with remarks on the male genitalia and classification of the subfamily. Publicações Culturais da Companhia de Diamantes de Angola, 16: 9-26.
1952. On certain subgenera of *Lygus* Hahn 1833 (Hem., Miridae), with a review of the British species. Entomologist's Gazette, 3: 213-230.
1953. The suprageneric nomenclature of the British Pentatomoidea (Hemiptera). Entomologist's Gazette, 4: 13-25.
1954. Notes on the Ethiopian Pentatomoidea (Hemiptera): XVII, Tessaratominae, Dinidorinae and Phyllocephalinae of Angola. Publicações Culturais da Companhia de Diamantes de Angola, 24: 11-22.
1955. A key to the genera of Oncomerini Stal (Heteroptera: Pentatomidae, Tessaratominae), with description of a new genus and species from Australia and new synonymy. Proceedings of the Royal Entomological Society of London, Series B Taxonomy, 24: 62-68.
1979. The species of *Dagbertus* (Hemiptera: Miridae) associated with avocado in Florida. Florida Entomologist, 62: 376-379.

Leston, D. and T. R. E. Southwood
1964. A new subgenus of *Saldula* Van Duzee 1914 (Hem., Saldidae). Entomologist's Monthly Magazine, 100: 80.

Lethierry, L. F.
1869. Catalogue des Hémiptères du départment du Nord. Mémoires de la Sociétédes Sciences d' Agriculture de des Arts à Lille., (3)6: 307-374. [Separate L. Danel, Lille. 1869. 70 pages.].
1874. Catalogue des Hémiptères du départment du Nord. 2nd edition. L. Danel, Lille. 108 pages. [Serial publication. 1876. Mémoires de la Sociétédes Sciences d'Agriculture et des Arts à Lille., (4)1: 205-312].
1888. Liste des Hémiptères recueillis à Sumatra et dans l'ile Nias par M.E. Modigliani. Annali del Museo Civico di Storia Naturale di Genova, (2)6: 460-470.

Lethierry, L. F. and G. Severin
1893-1896. Catalogue Général des Hémiptères. R. Friedlander and Fils, Bruxelles and Berlin. [Pentatomidae, 1893, 1: i-x, 1-286; Coreidae, Berytidae, Lygaeidae, Pyrrhocoridae, 1894, 2: 1-277, i-iii; Tingidae, Phymatidae, Aradidae, Hebridae, Hydrometridae, Henicocephalidae, Saldidae, Aepophilidae, Ceratocombidae, Cimicidae, Anthocoridae, 1896, 3: 1-275.].

Lindberg, H.
1935. Wissenschaftliche Ergebnisse der Niederlandischen Expeditionen in den Karakorus und die Angrenzenden Gebiete, 1922, 1925, & 1929-30. Zoologie; Hemiptera. Kommission bei F. A. Brockhaus, Leipzig, 1925. Pages 415-424.
1958. Hemiptera Heteroptera from Newfoundland collected by the Swedish-Finnish expedition of 1949 and 1951. Acta Zoologica Fennica, 96: 1-25.

Lindroth, C. H.
1931. Die Insektenfauna Islands & ihre Problems. Zoologiska Bidrag Fran Uppsala, 13: 105-599.

Lindroth, C. H. and G. E. Ball
1969. An annotated list of invertebrates of the Kodiak Island Refugium. Boreal Institute

Lindroth, C. H. and G. E. Ball (continued)
of the University of Alberta. Pages 122-155.

Lindskog, P.
1982. On the identity of *Saldula laticollis* (Reuter) (Heteroptera: Saldidae). Pan-Pacific Entomologist, 57: 321-326 (1981).

Lingren, P. D. and R. L. Ridgway
1967. Toxicity of five insecticides to several insect predators. Journal of Economic Entomology, 60: 1639-1641.

Lingren, P. D., R. L. Ridgway, C. B. Cowan, Jr., J. W. Davis, and W. C. Watkins
1968. Biological control of bollworm and tobacco budworm by arthropod predators affected by insecticides. Journal of Economic Entomology, 61: 1521-1525.

Lingren, P. D., R. L. Ridgway, and S. L. Jones
1968. Consumption by several arthropod predators of eggs and larvae of two *Heliothis* species that attack cotton. Annals of the Entomological Society of America, 61: 613-618.

Linnaeus, C.
1758. Systema naturae per regna tria naturae, secundum classes, ordines, genera, species, cum characteribus, differentiis, synonymis, locis. Editio decima, reformata. Laurentii Salvii, Holmiae. 1: 1-823, i-iii. [Hemiptera. Pages 434-457].
1761. Fauna Svecica sistens animalia Sveciae regni: Mammalia, Aves, Amphibia, Pices, Insecta, Vermes. Distributa per classes et ordines, genera et species, cum, differentiis specierum, synonymis auctorum, nominibus incolarum, locis natalium, descriptionibus insectorum. Laurentii Salvii, Stockholmiae. 578 pages.
1763. Centuria Insectorum Rariorum. Upsaliae. vi + 32 pages.
1767. Systemma Naturae per regna tria naturae secundum classes, ordines, genera, species cum characteribus, differentiis, synonymis, locis. Editio duodecima, reformata. Hemiptera, 1(2): 687-743.
1771. Mantissa Plantarum altera Generum. Edition VI, 7 Special edition II. Regni animalis appendix, 521-552.

Linnavuori, R.
1951. Studies on the family Cryptostemmatidae. Annales Entomologici Fennici, 17(3): 92-103.
1953. Hemipterological studies. 1. *Chiloxanthus arcticus* J. Sahlberg (Het.: Saldidae) a valid species. Suomen Hyonteistieteellinen Aikakauskirja [Annales Entomologici Fennici], 19: 107.

Linsley, E. G. and R. L. Usinger
1936. Insect collecting in California--II. Foothills regions. Pan-Pacific Entomologist, 12: 49-55.

Lintner, J. A.
1885. [Note on predatory Pentatomidae]. 15th Annual Report of the Entomological Society of Ontario for 1884: 13.
1888. Fourth report on the injurious and other insects of the State of New York. New York State Museum of Natural History, No. 41. Report of the State Entomologist of New York for 1887. 237 pages.

Lipsey, R. L.
1970. The hosts of *Neurocolpus nubilus* (Say), the clouded plant bug (Hemiptera, Miridae). Entomological News, 81: 213-219.
1970. The life history of *Neurocolpus nubilus* (Say), the clouded plant bug (Hemiptera, Miridae). Entomological News, 81: 257-262.

List, G. M.
1925. Three new genera and three new species of Cimicidae from North America. Proceedings of the Biological Society of Washington, 38: 103-110.

Little, V. A.
1972. General and Applied Entomology. 3rd edition. Harper and Row, New York. 527 pages.

Liu, H. J. and F. L. McEwen
1977. *Nosema blissi* sp. n. (Microsporidea: Nosematidae), a pathogen of the chinch bug, *Blissus leucopterus hirtus* (Hemiptera: Lygaeidae). Journal of Invertebrate Pathology, 29: 141-146.
1979. The use of temperature accumulations and sequential sampling in predicting damaging populations of *Blissus leucopterus hirtus*. Environmental Entomology, 8: 512-515.

Livingston, J. M., W. C. Yearian and S. Y. Young
1978. Effect of insecticides, fungicides, and insecticide-fungicide combinations on development of lepidopterous larval populations in Soybean. Environmental Entomology, 7: 823-828.

Loan, C. C.
1980. Plant bug hosts (Heteroptera: Miridae) of some euphorine parasites (Hymenoptera: Braconidae) near Belleville, Ontario, Canada. Naturaliste Canadien, 107: 87-93.

Lopez, E. G. and G. L Teetes
1976. Selected predators of aphids in grain sorghum and their relation to cotton. Journal of Economic Entomology, 69: 198-204.

Lopez, J. D., Jr., R. L. Ridgway, and R. E. Pinnell
1976. Comparative efficacy of four insect predators of the bollworm and tobacco budworm. Environmental Entomology, 5: 1160-1164.

Lord, F. T.
1949. The influence of spray programs on the fauna of apple orchards in Nova Scotia. III. Mites and their predators. Canadian Entomologist, 81: 217-230.

Lundblad, O.
1927. Studien über schwedische Corixiden. VI. Entomologisk Tidskrift, 48: 57-97, plates 1-7.
1928. Monographie der bis jetzt bekannten Arten der neotropischen Corixidengattung *Heterocorixa* B. White. Entomologisk Tidskrift, 49: 66-83.
1928. Drei neue Corixidengattungen. Zoologischen Anzeiger, 79: 148-163, 17 figures.
1928. Beitrag zur Kenntnis der Corixiden. Entomologisk Tidskrift, 49: 219-243, plates IV-VI.
1929. Über einige Corixiden des Berliner Zoologischen Museums. Archiv für Hydrobiologie, 20: 296-311, plates 10-12.
1929. Beitrag zur Kenntnis der Corixiden. II. Entomologisk Tidskrift, 50: 17-48, plates II-V.
1929. Zur Kenntnis dreier wenig bekanter amerikanischer Corixiden. Zoologischer Anzeiger, 80: 193-204.
1931. Über die Corixiden des zoologischen Museums in Halle, nebst einer Ubersicht der Gattung *Trichocorixa*. Zoologischer Anzeiger, 96: 85-95.

Lutz, J. C.
1929. *Corythucha decens* Stal in Pennsylvania (Heteropt.: Tingidae). Entomological News, 40: 233.

Lynch, R. E. and J. W. Garner
1980. Insects associated with sunflower, *Helianthus annuus*, in southern Georgia. Journal of the Georgia Entomological Society, 15: 182-189.

Maa, T.-C. and K.-S. Lin
1956. A synopsis of the Old World Phymatidae (Hem.). Quarterly Journal of the Taiwan Museum, 9: 109-154, plates I-IV.

MacCreary, D.
1965. Flight range observations on *Lygus lineolarius* and certain other Hemiptera. Journal of Economic Entomology, 58: 1004-1005.

MacGillivray, A. D. and C. O. Houghton
1903. A list of insects taken in the Adirondack Mountains, N.Y.--III. Entomological News, 14: 262-265.

Mackey, H. E., Jr.
1972. A life history survey of *Gelastocoris oculatus* in eastern Tennessee. Journal of the Tennessee Academy of Science, 47: 153-155.

MacPhee, A. W. and K. H. Sanford
1961. The influence of spray programs on the fauna of apple orchards in Nova Scotia. XII. Second supplement to VII. Effects on beneficial arthropods. Canadian Entomologist, 93: 671-673.

Mailloux, G. and H. T. Streu
1979. A sampling technique for estimating hairy chinch bug (*Blissus leucopterus hirtus* Montandon: Hemiptera: Lygaeidae)

Mailloux, G. and H. T. Streu (continued)
populations and other arthropods from turf grass. Annals of the Entomological Society of Quebec, 24: 139-143.
1981. Population biology of the hairy chinch bug (*Blissus leucopterus hirtus* Montandon: Hemiptera: Lygaeidae). Annals of the Entomological Society of Quebec, 26: 51-90.
1982. Spatial distribution pattern of hairy chinch bug (*Blissus leucopterus hirtus* Montandon: Hemiptera: Lygaeidae) populations in turfgrass. Annals of the Entomological Society of Quebec, 27: 111-131.

Maldonado-Capriles, J.
1969. The Miridae of Puerto Rico (Insecta, Hemiptera). University of Puerto Rico Agricultural Experiment Station Technical Paper, 45: 1-133.
1974. The neotropical genus *Melanopleurus* (Hemiptera: Lygaeidae). Proceedings of the Entomological Society of Washington, 76: 22-30.

Maldonado-Capriles, J. (continued)
1976. The genus *Heza* (Hemiptera: Reduviidae). The Journal of Agriculture of the University of Puerto Rico, 60: 403-433.
1976. Three new species of *Castolus* and a key to the species (Hemiptera: Reduviidae). Proceedings of the Entomological Society of Washington, 78: 435-446.
1980. The genus *Jobertus* Distant, 1884 (Hemiptera: Miridae: Orthotylinae). Journal of Agriculture of the University of Puerto Rico, 64: 304-309.
1981. A new *Ghilianella* and a new saicine genus *Buninotus* (Hemiptera: Reduviidae) from Panama. The Journal of Agriculture of the University of Puerto Rico, 65: 401-407.
1982. Designation of the senior types of *Paraproba pendula* Van Duzee, 1914, and a key to the North American species of *Paraproba* (Hemiptera: Miridae). Journal of Agriculture of the University of Puerto Rico, 66: 282-285.

Malipatil, M. B.
1978. Revision of the Myodochini (Hemiptera: Lygaeidae: Rhyparochrominae) of the Australian Region. Australian Journal of Zoology, Supplement Series, 56: 1-178.

Malloch, J. R.
1918. *Pyrrhotes haematoloma* H. S., and *Leptocoris trivittatus* Say in Illinois (Hemiptera, Coreidae). Entomological News, 29: 284.
1919. [Editorial additions]. *In* Hart, C. A. The Pentatomoidea of Illinois with keys to the nearctic genera. Bulletin of the Illinois Natural History Survey, 13: 155-223, plates 16-21.
1919. Thyreocorinae. Pages 206-216. *In* Hart, C. A. The Pentatomoidea of Illinois with Keys to the Nearctic Genera. Bulletin of the Illinois State Laboratory of Natural History Survey, 13: 157-223, plates XVI-XXI.
1921. Systematic notes on Hemiptera Heteroptera. Bulletin of the Brooklyn Entomological Society, 16: 54-56.

Manley, G. V.
1982. Biology and life history of the rice field predator *Andrallus spinidens* F. (Hemiptera: Pentatomidae). Entomological News, 93: 19-24.

Mann, J.
1969. Cactus-feeding insects and mites. Bulletin of the United States National Museum, 256: 1-158.

Marlatt, C. L.
1896. [Notes on insects in minutes of meeting]. Proceedings of the Entomological Society of Washington, 4: 44-46.

Marston, N. L. and G. D. Thomas
1979. Seasonal cycles of soybean arthropods in Missouri: Effect of pesticidal and cultural practices. Environmental Entomology, 8: 165-173.

Marti, O. G., Jr. and J. J. Hamm
1986. Effect of *Vairimorpha* sp. on survival of *Geocoris punctipes* in the laboratory. Journal of Entomological Science, 20: 354-358.

Martin, C. H.
1929. An exploratory survey of characters of specific value in the genus *Gelastocoris* Kirkaldy, and some new species. University of Kansas Science Bulletin, 18: 351-369 (1928).

Martin, J. O.
1900. A study of *Hydrometra lineata*. Canadian Entomologist, 32: 70-76.

Martin, P. B., P. D. Lingen, G. L. Greene, and A. H. Baumhover
1981. Seasonal occurrence of *Rachiplusia ou*, *Autographa biloba*, and associated entomophages in clover. Journal of the Georgia Entomological Society, 16: 288-295.

Martin, P. B., R. L. Ridgway, and C. E. Schuetze
1978. Physical and biological evaluations of an encapsulated diet for rearing *Chrysopa carnea*. Florida Entomologist, 61: 145-152.

Matsuda, R.
1956. A supplementary taxonomic study of the genus *Rhagovelia* (Hemiptera, Veliidae) of the Western Hemisphere. A deductive method. University of Kansas Science Bulletin, 38: 915-1017.
1960. Morphology, evolution and a classification of the Gerridae (Hemiptera-Heteroptera). University of Kansas Science Bulletin, 41: 25-632.
1971. A new species of *Aradus* (Aradidae: Hemiptera). Canadian Entomologist, 103: 1195-1196.
1977. The Aradidae of Canada. Hemiptera: Aradidae. The Insects and Arachnids of Canada. Part 3. Research Branch, Canada Department of Agriculture, Publication 1634. 116 pages.

1980. Description of a new species of *Aradus* (Hemiptera: Heteroptera) from Canada. Canadian Entomologist, 112: 855-856.

Matsumura, S.

1913. Thousand insects of Japan. Additamenta. I. 184 pages + plates XI-XV.

Mayr, G. L.

1853. Zwei neue Wanzen aus Kordofan. Verhandlungen des Kaiserlich Koniglichen Zoologisch-Botanischen Vereins in Wien, 2: 14-18.

1863. Hemipterologische studien. Die Belostomiden. Verhandlungen der Kaiserlich Koniglichen Zoologisch-Botanischen Gesellschaft in Wien, 13: 339-364.

1864. Diagnosen neuer Hemipteren. Verhandlungen der Kaiserlich-Konliglichen Zoologisches-Botanischen Gesellschaft in Wien, 14: 903-914.

1865. Diagnosen neuer Hemipteren. 2. Verhandlungen der Kaiserlich-Koniglichen Zoologisch-Botanischen Gesellschaft in Wien, 15: 429-446.

1866. Diagnosen neuer Hemipteren. Verhandlungen der Kaiserlich-Koniglichen Zoologish-Botanischen Gesellschaft in Wien, 15: 429-430.

1868a. Reise der Österreichischen Fregatte Novara um die Erde in den Jahren 1857, 1858, 1859 unter den refehlen des commodore B. Von Wüllerstorf-Urbair. Kaiserlich-Koniglichen Hof-und Staatsdruckerei, Wien, Zoologische Theil 2, Abtheilung 1, B. 2. Hemiptera. 204 pages, 5 tables. [B. Karl Gerold's Sohn, Wien.]

1871. Die Belostomiden. Monographisch Bearbeitet. Verhandlungen der Kaiserlich-Koniglichen Zoologisch-Botanischen Gesellschaft in Wien, 21: 399-440.

McAtee, W. L.

1914. Key to the nearctic genera and species of Geocorinae (Heteroptera; Lygaeidae). Proceedings of the Biological Society of Washington, 27: 125-136.

1917. Key to the nearctic species of *Leptoypha* and *Leptostyla* (Heteroptera Tingidae). Bulletin of the Brooklyn Entomological Society, 12: 55-64.

1917. Key to the nearctic species of *Paracalocoris* (Heteroptera; Miridae). Annals of the Entomological Society of America, 9: 366-390 (1916)

1919. Notes on nearctic Heteroptera. Coreidae. Bulletin of the Brooklyn Entomological Society, 14: 8-16.

1919. Corrections and additions to an article on *Leptoypha* and *Leptostyla* (Heteroptera; Tingidae). Bulletin of the Brooklyn Entomological Society, 14: 142-144.

1919. Notes on two Miridae, *Camptobrochis* and *Paracalocoris* (Heteroptera). Entomological News, 30: 246.

1919. Key to the nearctic genera and species of Berytidae (Heteroptera). Journal of the New York Entomological Society, 27: 79-92.

1919. Key to the nearctic species of Piesmidae (Heteroptera). Bulletin of the Brooklyn Entomological Society, 14: 80-93.

1919. [New species]. Pages 215-216. *In* Malloch, J. R., Thyreocorinae, *In* Hart, C. A. The Pentatomoidea of Illinois with Keys to the Nearctic Genera. Bulletin of the Illinois State Natural History Survey, 13: 157-223, plates XVI-XXI.

1923. Tingitoidea of the vicinity of Washington, D.C. (Heteroptera). Proceedings of the Entomological Society of Washington, 25: 143-151.

1926. Notes on nearctic Hemiptera. Entomological News, 37: 13-16.

1927. Notes on "Heteroptera or True Bugs of Eastern North America." Bulletin of the Brooklyn Entomological Society, 22: 267-281.

McAtee, W. L. and J. R. Malloch

1922. Changes in names of America Rhynchota chiefly Emesinae. Proceedings of the Biological Society of Washington, 35: 95-96.

1923. Notes on American Bactrodinae and Saicinae (Heteroptera: Reduviidae). Annals of the Entomological Society of America, 16: 247-255.

1923. Further notes on names of Emesinae and other Rhynchota. Proceedings of the Biological Society of Washington, 36: 161-164.

1923. *Ploiariodes barberi*, new species. Pages 7-8. *In* Barber, H. G. A preliminary report on the Hemiptera-Heteroptera of Porto Rico collected by the American Museum of Natural History. American Museum Novitates, 75: 1-13.

1924. Some annectant bugs of the superfamily Cimicoideae (Heteroptera). Bulletin of the Brooklyn Entomological Society, 19: 69-82, 1 plate.

1925. Another annectant genus (Hemiptera; Cimicoidea). Proceedings of the Biological

McAtee, W. L. and J. R. Malloch (continued)
Society of Washington, 38: 145-148.
1925. Revision of the American bugs of the reduviid subfamily Ploiariinae. Proceedings of the United States National Museum, 67(1): 1-135, 9 plates.
1925. Revison of bugs of the family Cryptostemmatidae in the collection of the United States National Museum. Proceedings of the United States National Museum, 67(13): 1-42, 4 plates.
1928. Thyreocorinae from the state of Paranà, Brazil (Hemiptera: Pentatomidae). Annales Musei Zoologici Polonici, 7: 32-44.
1932. Notes on genera of Isometopinae (Heteroptera). Stylops, l: 62-70.
1933. Revision of the subfamily Thyreocorinae of the family Pentatomidae (Hemiptera- Heteroptera). Annals of the Carnegie Museum, 21: 191-411, plates 4-17, i-vii.

McCarty, M. T., M. Shepard, and S. G. Turnipseed
1980. Identification of predaceous arthropods in soybeans by using autoradiography. Environmental Entomology, 9: 199-203.

McClure, H. E.
1943. Aspection in the biotic communities of the Churchill area. Ecological Monographs, 13: 1-35.
1943. Further notes on aero-plankton of Kentucky. Entomological News, 43: 5, 11, 37-45.

McDaniel, S. G. and W. L. Sterling
1979. Predator determination and efficiency on *Heliothis virescens* eggs in cotton using 32P2. Environmental Entomology, 8: 1083-1087.

McDaniel, S. G., W. L. Sterling, and D. A. Dean
1981. Predators of tobacco budworm in larvae in Texas cotton. Southwestern Entomologist, 6: 102-108.

McDonald, F. J. D.
1968. The life history of *Cosmopepla bimaculata* (Thomas)(Heteroptera: Pentatomidae) in Alberta. Quaestiones Entomologicae, 4: 35-38.
1975. Revision of the genus *Holcostethus* in North America (Hemiptera: Pentatomidae). Journal of the New York Entomological Society, 82: 245-258 (1974).
1976. Revision of the genus *Trichopepla* (Hemiptera: Pentatomidae) in N. America. Journal of the New York Entomological Society, 84: 9-22.
1982. Description of the male genitalia of *Holcostethus hirtus* (Van Duzee) with a revised key to North American species (Hemiptera: Pentatomidae). Journal of the New York Entomological Society, 90: 5-7.
1986. Revision of *Cosmopepla* Stål (Hemiptera: Pentatomidae). Journal of the New York Entomological Society, 94: 1-15.

McDunnough, J.
1925. Notes on *Saldula obscura* Prov. with description of a new species (Hemiptera). Canadian Entomologist, 57: 257-260.

McEwen, F. L.
1973. Insects of the season in Ontario 1973 related to fruit, vegetables, field crops and ornamentals. Proceedings of the Entomological Society of Ontario, 104: 102.

McGavin, G. C.
1982. A new genus of Miridae (Hem.: Heteroptera). Entomologist's Monthly Magazine, 118: 79-86.

McGhee, R. B.
1972. First record of *Oncopeltus sandarachatus* (Hemiptera: Heteroptera: Lygaeidae) in the eastern United States. Journal of the Georgia Entomological Society, 7: 36-37.

McGhee, R. B. and F. J. Postell
1982. Transmission of the trypanosomatid flagellate *Phytomonas davidi*, a symbiont of the Euphorbiaceae, by the hemipteran bug *Pachybrachius bilobata scutellatus*. Journal of Protozoology, 29: 445-448.

McGregor, E. A. and F. L. McDonough
1917. The red spider on cotton. United States Department of Agriculture Bulletin, 416: 1-72.

McKinstry, A. P.
1937. Some new species of *Microvelia* (Veliidae, Hemiptera). Journal of the Kansas Entomological Society, 10: 30-36, 37-41.

McLain, D. K.
1984a. Host plant density and territorial behavior of the seed bug, *Neacoryphus bicrucis* (Hemiptera: Lygaeidae). Behavioral Ecology and Sociobiology, 14: 181-188.
1984b. Coevolution: Müllerian mimicry between a plant bug (Miridae) and a seed bug (Lygaeidae) and the relationship between host plant choice and unpalatability. Oikos, 43: 143-148.

1986. Resource patchiness and variation in the intensity of sexual selection in a resource defending polygynous insect species. Oikos, 47: 19-25.

McLain, D. K. and D. J. Shure

1985. Host plant toxins and unpalatability of *Neacoryphus bicrucis* (Hemiptera: Lygaeidae). Ecological Entomology, 10: 291-298.

McMullen, R. D. and C. Jong

1967. New records and discussion of predators of the pear psylla, *Psylla pyricola* Forster, in British Columbia. Journal of the Entomological Society of British Columbia, 64: 35-40.

1970. The biology and influence of pesticides on *Campylomma verbasci* (Heteroptera: Miridae). Canadian Entomologist, 102: 1390-1394.

McPherson, J. E.

1965. Notes on the life history of *Notonecta hoffmanni* (Hemiptera: Notonectidae). Pan-Pacific Entomologist, 41: 86-89.

1970. A key and annotated list of the Scutelleroidea of Michigan (Hemiptera). Michigan Entomologist, 3: 34-63.

1972. Life history of *Corimelaena lateralis lateralis* (Hemiptera: Thyreocoridae) with descriptions of immature stages and list of other species of Scutelleroidea found with it on wild carrot. Annals of the Entomological Society of America, 65: 906-911.

1974. Notes on the biology of *Mormidea lugens* and *Euschistus politus* (Hemiptera: Pentatomidae) in southern Illinois. Annals of the Entomological Society of America, 67: 940-942.

1982. The Pentatomoidea (Hemiptera) of Northeastern North America with Emphasis on the Fauna of Illinois. Southern Illinois University Press, Carbondale. ix + 241 pages.

McPherson, J. E. and S. M. Paskewitz

1984. Life history and laboratory rearing of *Euschistus ictericus* (Hemiptera: Pentatomidae), with descriptions of immature stages. Journal of the New York Entomological Society, 92: 53-60.

1984. Laboratory rearing of *Amaurochrous cinctipes* (Hemiptera: Pentatomidae: Podopinae) with descriptions of immature stages. Journal of the New York Entomological Society, 92: 61-68.

McPherson, J. E. and R. I. Sailer.

1978. A new species of *Corimelaena* (Hemiptera: Thyreocoridae) from America north of Mexico. Journal of the Kansas Entomological Society, 51: 516-520.

McPherson, J. E. and J. F. Walt

1971. The first record in Illinois of a population of *Stethaulax marmoratus* (Say) (Hemiptera: Scutelleridae) with information on life history. Transactions of the Illinois State Academy of Science, 64: 198-200.

McPherson, R. M., J. C. Smith, and W. A. Allen

1982. Incidence of arthropod predators in different soybean cropping systems. Environmental Entomology, 11: 685-689.

Mead, F. W.

1972. Key to the species of big eyed bugs, *Geocoris* spp., in Florida. Hemiptera: Lygaeidae. Tri-Ology, Florida Department of Agriculture & Consumer Services, Division of Plant Industry, Entomology Circular No. 121, 2 pages.

1983. A stink bug, *Oebalus grisescens* (Sailer). Tri-Ology, Florida Department of Agriculture & Consumer Services, Division of Plant Industry, 22(11): 4.

1984. A plant bug, *Ranzovius* (= *Exocentricus*) clavicornis (Knight). Tri-Ology, Florida Department of Agriculture & Consumer Services, Division of Plant Industry, 23(4): 2.

1985. *Jadera* scentless plant bugs in Florida (Hemiptera: Rhopalidae). Florida Department of Agriculture & Consumer Services, Division of Plant Industry, Entomology Circular No. 277, 2 pages.

Meinert, F. V.

1895. *Rheumatobates Bergrothi* n. sp. Entomologische Meddelelser udgivne of Entomologisk Forening, 5: 1-9, 2 plates.

Melber, A., L. Hoelscher, and G. H. Schmidt

1980. Further studies on the social behavior and its ecological significance in *Elasmucha grisea* (Hem.-Het.: Acanthosomatidae). Zoologischer Anzeiger, 205: 27-38.

Melin, D.

1929. Hemiptera from South and Central America. I. (Revision of the genus *Gelastocoris* and the American species of *Mononyx*). Zoologiska Bidrag Fran Uppsala, 12: 151-198.

1930. Hemiptera from South and Central America. II. (Contributions to a revision

Melin, D. (continued)
of the genus *Phymata*). Arkiv för Zoologi, 22: 1-40, plates 1-7.

Menhinick, E. F.
1963. Insect species in the herb stratum of a *Sericea lespedeza* stand, AEC Savannah River Project, Aiken, South Carolina. United States Atomic Energy Commission, Division of Technical Information, TID-19136: 1-47.

Menke, A. S.
1958. A synopsis of the genus *Belostoma* Latreille, of America north of Mexico, with the description of a new species (Hemiptera; Belostomatidae). Bulletin, Southern California Academy of Sciences, 57: 154-174.
1960. Lectotype designation for *Lethocerus angustipes* (Mayr). (Hemiptera: Belostomatidae). Pan-Pacific Entomologist, 36: 104.
1960. A taxonomic study of the genus *Abedus* Stål (Hemiptera, Belostomatidae). University of California Publications in Entomology, 16: 393-440.
1962. Notes on species of *Lethocerus* Mayr and *Hydrocyrus* Spinola described by Guerin-Meneville, L. Dufour, A. L. Montandon, and G. A. W. Herrich-Schäffer (Belostomatidae; Hemiptera). Proceedings of the Biological Society of Washington, 75: 61-66.
1963. The dates of publication of Palisot de Beauvois' "Insectes Recueilles en Afrique et en Amérique." Annals and Magazine of Natural History, Series 13, 5: 701-702.
1963. A review of the genus *Lethocerus* in North and Central America, including the West Indies (Hemiptera: Belostomatidae). Annals of the Entomological Society of America, 56: 261-267.
1964. Lectotype designation for *Ranatra quadridentata* (Stål) (Hemiptera: Nepidae). Bulletin of the Brooklyn Entomological Society, 58: 112-113 (1963).
1966. A new toe biter from Mexico (Belostomatidae, Hemiptera). Contributions in Science, Los Angelos County Museum, 118: 1-6.
1977. Synonymical notes and new distribution records in *Abedus* (Hemiptera, Belostomatidae). Southwestern Naturalist, 22: 115-223.
1979. Family Nepidae/water scorpions. Pages 70-75. *In* Menke, A. S. (editor). The Semiaquatic and Aquatic Hemiptera of California (Heteroptera: Hemiptera). Bulletin of the California Insect Survey, 21: xi + 1-166.
1979. Family Belostomatidae/giant water bugs, electric light bugs, toe biters. Pages 76-86. *In* Menke, A. S. (editor). The Semiaquatic and Aquatic Hemiptera of California (Heteroptera: Hemiptera). Bulletin of the California Insect Survey, 21: xi + 1-166.
1979. Family Ochteridae/ velvety shore bugs. Pages 124-125. *In* Menke, A. S. (editor). The Semiaquatic and Aquatic Hemiptera of California (Heteroptera: Hemiptera). Bulletin of the California Insect Survey, 21: xi + 1-166.
1979. Family Gelastocoridae/toad bugs. Pages 126-130. *In* Menke, A. S. (editor). The Semiaquatic and Aquatic Hemiptera of California (Heteroptera: Hemiptera). Bulletin of the California Insect Survey, 21: xi + 1-166.
1979. *Lethocerus* Mayr, 1853 (Insecta, Hemiptera, Belostomatidae), proposed conservation in place of *Iliastus* Gistel [1847]. Z.N.(S.) 2161. Bulletin of Zoological Nomenclature, 35: 236-238.

Menke, A. S. and J. T. Polhemus
1973. Lectotype designation for *Gerris buenoi* Kirkaldy (Heteroptera: Gerridae). Pan-Pacific Entomologist, 49: 257.

Menke, W. W. and G. L. Greene
1976. Experimental validation of a pest management model. Florida Entomologist, 59: 135-142.

Merkle, O. G., K. J. Starks, and A. J. Casady
1983. Registration of pearl millet germplasm lines with chinch bug resistance. Crop Science, 23: 601.

Messenger, P. S., E. Biliotti, and R. van den Bosch
1976. Chapter 22. The importance of natural enemies in integrated control. Pages 543-563. *In* Huffaker, C. B. and P. S. Messenger (editors). Theory and Practice of Biological Control. Academic Press, New York. 788 pages.

Messina, F. J.
1978. Mirid fauna associated with old-field goldenrods (*Solidago*: Compositae) in Ithaca, NY. Journal of the New York Entomological Society, 86: 137-143.

Messing, R. H. and M. T. AliNiazee
1985. Natural enemies of *Myzocallis coryli* (Hom.: Aphididae) in Oregon hazelnut orchards. Journal of the Entomological Society of British Columbia, 82: 14-18.

Meyer-Dür, L. R.
1843. Verzeichniss der in der Schweiz einheimischen Rhynchoten (Hemiptera Linn.). Erstes Heft, die Familie der Capsini. Lent und Grassmann, Solothurn. i-x, 116 + 4 pages, 7 plates.

Miller, N. C. E.
1971. The Biology of the Heteroptera. Second (revised) Edition. E.W. Classey, Hampton. 206 pages.

Miller, P. L.
1964. The possible role of haemoglobin in *Anisops* and *Buenoa* (Hemiptera: Notonectidae). Proceedings of the Royal Entomological Society of London, 39: 166-175.
1966. The function of haemoglobin in relation to the maintenance of neutral buoyancy in *Anisops pellucens*. Journal of Experimental Biology, 44: 529-543.

Miller, W. O.
1971. Recent developments with Dursban insecticide for chinch bug control. Down to Earth, 27: 1-5.

Miller, W. O. and C. E. Miller
1979. Plant bug reduction through the use of Premerge 3 dinitri amine herbicides as a directed spray in cotton. Down to Earth, 35: 14-15.

Milliken, F. B.
1918. *Nysius ericae*, the false chinch bug. Journal of Agricultural Research, 13: 571-578, 2 plates.

Milliken, F. B. and F. M. Wadley
1913. *Phasia* (*Phorantha*) *occidentis* Walker, an internal parasite of the false chinch bug. Bulletin of the Brooklyn Entomological Society, 18: 28-31.

Millspaugh, D. D.
1939. Bionomics of the aquatic and semiaquatic Hemiptera of Dallas County, Texas. Field and Lab, 7: 67-87.

Milne, L. J. and M. J. Milne
1944. Selection of colored lights by night-flying insects. Part I. Analysis of the experiment. Entomologica Americana, 24: 21-57.
1980. The Audubon Society Field Guide to North American Insects and Spiders. Alfred A. Knopf, New York. 989 pages.

Mize, T. W. and G. Wilde
1986a. New resistant germplasm to the chinch bug (Heteroptera: Lygaeidae) in grain sorghum: contribution of tolerance and antixenosis as resistance mechanisms. Journal of Economic Entomology, 79: 42-45.
1986b. New grain sorghum sources of antibiosis to the chinch bug (Heteroptera: Lygaeidae). Journal of Economic Entomology, 79: 176-180.
1986c. Reproduction of the chinch bug (Heteroptera: Lygaeidae) on new resistance sources in grain sorghum. Journal of Economic Entomology, 79: 664-667.

Mize, T., G. Wilde, and M. T. Smith
1980. Chemical control of chinch bug and greenbug on seedling sorghum with seed, soil, and foliar treatments. Journal of Economic Entomology, 73: 544-547.

Moffett, J. O., L. S. Smith, C. C. Burkhardt, and C. W. Shipman
1976. Insect visitors to cotton flowers. Journal of the Arizona Academy of Science, 11: 47-48.

Montandon, A. L.
1893. Lygaeides exotiques. Notes et descriptions d'espèces nouvelles. Annales de la Société Entomologique de Belgique, 37: 399-406.
1893. Notes on American Hemiptera Heteroptera. Proceedings of the United States National Museum, 16(924): 45-52.
1896. Hemiptères-hétéroptères exotiques. Notes et descriptions. II. Fam. Belostomatidae. Annales de la Société Entomologique de Belgique, 40: 508-520.
1897. Hemiptera cryptocerata. Fam. Naucoridae, Sous-fam. Cryptocricinae. Verhandlungen der Kaiserlich-Koniglichen Zoologisch-Botanischen Gesellschaft in Wien, 47: 6-23.
1898. Hemiptera cryptocerata. Notes et descriptions d'espèces nouvelles. Bulletin de la Société des Sciences de Bucarest-Roumanie, 7: 282-290.
1899. Hemiptera Heteroptera. Fam. Coreidae. Notes et Descriptions de trois nouvelles espèces américaines. Bulletin de la Société des Sciences de Bucarest-Roumanie, 8: 190-195.
1900. Notes sur quelques genres de la Fam: Belostomatidae. Bulletin de la Société

Montandon, A. L. (continued)
des Sciences de Bucarest-Roumanie, 9: 264-273.
1903. Hemipteres aquatiques, notes synonymiques et geographiques, descriptions d'especes nouvelles. Bulletin de la Sociétédes Sciences de Bucarest-Roumanie, 12: 97-121.
1907. Nouveaux genres et espèces du groupe des Geocorinae. Annales Historico-Naturales Musei Nationalis Hungarici, 5: 89-97.
1908. Espèces nouvelles ou peu connues du genre *Geocoris* Fall. Bulletin de la Société Sciences de Bucarest-Roumanie, 16: 214-234.
1909. Naucoridae. Descriptions d'espèces nouvelles. Bulletin de la Société des Sciences de Bucarest-Roumanie, 18: 43-61.
1910. Trois espèces nouvelles de la famille Naucoridae. Bulletin de la Société des Sciences de Bucarest-Roumanie, 19: 438-444.
1910. Hydrocorises de l'Amérique du Nord. Notes et descriptions d'espèces nouvelles. Bulletin de la Société des Sciences de Bucarest-Roumanie, 18: 180-191 (1909).
1910. Notes sur quelques formes Nord-Américains du genre *Ranatra* (Hem.). Bulletin de la Société des Sciences de Bucarest-Roumanie, 19: 62-67.
1913. Nouvelles études sur les Geocorinae (Hemipt.). Note présentée dans la séances du 20 Mai 1913. Bulletin de la Section Scientifique de l'Academie Roumaine, 2: 48-60.
1913. Nepidae et Belostomatidae. Descriptions de deux especes nouvelles. Bulletin de la Société des Sciences de Bucarest-Roumanie, 22: 122-125.
1914. Formes peu connues et nouvelles varietes du genre *Geocoris*. Bulletin de la Société des Sciences Bucarest, 23: 234-243.

Monte, O.
1942. Crítica sobre alguns gêneros e espécies de Tingítideos. Papéis Avulsos do Departamento de Zoologia, Sao Paulo, 2: 103-115.

Montgomery, T. H., Jr.
1902. A list of the Hemiptera Heteroptera of the vicinity of Wood's Hole, Massachusetts. Entomological News, 13: 12-13.

Moore, G. A.
1907. List of Hemiptera taken at Como, Quebec. Canadian Entomologist, 39: 161-163, 189-191.
1950. Check-list of Hemiptera of the province of Quebec. Naturaliste Canadien, 77: 233-271. [also published as Contributions de l'Institut de Biologie de l'Universitéde Montréal, 26: 1-49.].
1950. Catalogue des Hémiptères de la Province de Québec. Naturaliste Canadien, 77: 233-271.

Moore, T. B., R. Stevens, and E. D. McArthur
1982. Preliminary study of some insects associated with rangeland shrubs with emphasis on *Kochia prostrata*. Journal of Range Management, 35: 128-130.

Moore, T. E.
1955. A new species of *Agnocoris* from Illinois, and a synopsis of the genus in North America. Proceedings of the Entomological Society of Washington, 57: 175-180.
1956. Two new species of *Lopidea* Uhler from Illinois (Hemiptera, Miridae). Entomological News, 67: 39-42.
1956. *Agnocoris rubicunda* in North America (Hemiptera, Miridae). Journal of the Kansas Entomological Society, 29: 37-39.

Morgan, D. L., G. W. Frankie and M. J. Gaylor
1978. Potential for developing insect-resistant plant materials for use in urban environments. Pages 267-294. *In* G. W. Frankie and C. S. Koehler (editors). Perspectives in Urban Entomology. Academic Press, New York. 417 pages.

Morrill, A. W.
1903. Notes on the immature stages of some tingitids of the genus *Corythuca*. Psyche, 10: 127-134, plate 3.
1910. Plant-bugs injurious to cotton bolls. United States Department of Agriculture, Bureau of Entomology Bulletin, 86: 1-110.

Morris, D. A. and J. C. Smith
1981. Effect of row spacing on insect populations, using twelve commercial varieties of soybean. Virginia Journal of Science, 32: 78.

Morrison, D. E., J. J. Bradley, Jr., and J. W. Van Duyn
1979. Populations of corn earworm and associated predators after applications of certain soil-applied pesticides to soy-

Morrison, D. E., J. J. Bradley, Jr., and J. W. Van Duyn (continued)
beans. Journal of Economic Entomology, 72: 97-100.

Moser, J. C., B. C. Thatcher, and L. S. Pickard
1971. Relative abundance of southern pine beetle associates in east Texas. Annals of the Entomological Society of America, 64: 72-77.

Moznette, G. F.
1921. Notes on the royal palm bug. Quarterly Bulletin of the State Plant Board of Florida, 6: 10-15.

Mukerji, M. K. and E. J. LeRoux
1969. The effect of predator age on the functional response of *Podisus maculiventris* to the prey size of *Galleria mellonella.* Canadian Entomologist, 101: 314-327.
1969. A quantitative study of food consumption and growth of *Podisus maculiventris* (Hemiptera: Pentatomidae). Canadian Entomologist, 101: 387-403.
1969. A study on energetics of *Podisus maculiventris* (Hemiptera: Pentatomidae). Canadian Entomologist, 101: 449-460.

Müller, O. F.
1766. Manipulus insectorum Taurinensium. Mélanges de Philosophie et de Mathématique Société Royale, Turin, 3(7): 185-198.

Mulsant, E. and C. Rey
1852. Description de quelques Hémiptères Hétéroptères nouveaux ou peu connus. Annales de la Société Linnéenne de Lyon. Pages 76-141.
1885-1871. Histoire naturelle des punaises de France. Scutellerides, 1865, 1: 1-112, 1 plate; Pentatomides, 1866, 2: 1-372, 2 plates; Coreides, Alydides, Berytides, Stenocephalides, 1871, 3: 1-434 + 3, 2 plates. F. Savy & Deyrolle, Paris. [also published in Annales de la Société Linnéene de Lyon].

Muminov, N. N.
1986. Middle-Asiatic species of the genus *Deraeocoris* Kbm. (Heteroptera, Miridae). Izvesliya Akademii Nauk Tadzhikskoi S.S.R. Otdelenie Biologicheskikh Nauk, 101: 39-41 (1985). (In Russian).

Mundinger, F. G.
1923. The life history of two species of Nabidae (Hemip. Heterop.). *Nabis roseipennis* Reut. and *Nabis rufusculus* Reut. Technical Publication No. 16 of New York State College of Forestry at Syracuse University, 22: 149-167 + 3 pages + plates (1922).

Mychajliw, S.
1961. Four new species of *Hydrometra* from the New World (Hemiptera: Hydrometridae). Journal of the Kansas Entomological Society, 34: 27-33.

Myers, J. G.
1927. Ethological observations on some Pyrrhocoridae of Cuba (Hemiptera-Heteroptera). Annals of the Entomological Society of America, 20: 279-300.

Nadgauda, D. and H. N. Pitre
1978. *Reduviolus roseipennis* feeding behavior: Acceptability of tobacco budworm larvae. Journal of the Georgia Entomological Society, 13: 304-308.

Naranjo, S. E. and J. L. Stimac
1985. Development, survival, and reproduction of *Geocoris punctipes* (Hemiptera: Lygaeidae): Effects of plant feeding of soybean and associated weeds. Environmental Entomology, 14: 523-530.

Neal, T. M., G. L. Greene, F. W. Mead, and W. H. Whitcomb
1972. *Spanogonicus albofasciatus* (Hemiptera: Miridae): A predator in Florida soybeans. Florida Entomologist, 55: 247-250.

Neering, T.
1954. Morphological variations in *Mesovelia mulsanti* (Hemiptera, Mesoveliidae). University of Kansas Science Bulletin, 36: 125-148.

Negron, J. F. and T. J. Riley
1985. Effect of chinch bug (Heteroptera: Lygaeidae) feeding in seedling field corn. Journal of Economic Entomology, 78: 1370-1372.

Neiswander, C. R.
1925. On the anatomy of the head and thorax in *Ranatra* (Heteroptera). Transactions of the American Entomological Society, 51: 311-320.
1931. The sources of American corn insects. Ohio Agricultural Experiment Station Bulletin, 473: 1-98.

Neiva, A.
1911. Notas de entomologia medica. Duas novas especies norte-americanas de

Neiva, A. (continued)

hemipteros hematophagos. Brasil-Medico, 25: 421-422.

1911. Notas de entomologia medica. Tres novas especies de reduvidas norte-americanos. Brasil-Medico, 25: 441.

1912. Notas de entomologia medica e descripção de duas novas especies de triatomas norte-americanos. Brasil-Medico, 25: 21-22.

1914. Revisão do genero *Triatoma* Lap. Rodrigues & C., Rio de Janeiro. ii + 80 pages.

Neiva, A. and H. Lent

1936. Notas e commentarios sobre triatomideos. Lista de especies e sua distribução geographica. Revista Brasileira de Biologia, 6: 153-190.

1941. Sinopse dos triatomideos. Revista de Entomologia, 12: 61-92.

Nelson, G. H.

1955. A revision of the genus *Dendrocoris* and its generic relationships (Hemiptera, Pentatomidae). Proceedings of the Entomological Society of Washington, 57: 49-67.

1957. A new species of *Dendrocoris* and a new combination of *Atizies* (Hemiptera, Pentatomidae). Proceedings of the Entomological Society of Washington, 59: 197-199.

Newsom, L. D., R. F. Smith, and W. H. Whitcomb

1976. Chapter 23. Selective pesticides and selective use of pesticides. Pages 565-591. *In* Huffaker, C. B. and P. S. Messenger (editors). Theory and Practice of Biological Control. Academic Press, New York. 788 pages.

Nguyen, R., O. Sosa, and F. W. Mead

1984. Sugarcane delphacid, *Perkinsiella saccharicida* Kirkaldy 1903 (Homoptera: Delphacidae). Florida Department of Agriculture & Consumer Services, Division of Plant Industry, Entomology Circular 265, 2 pages.

Nixon, P. L. and J. E. McPherson

1977. An annotated list of phytophagous insects collected on immature black walnut trees in southern Illinois. Great Lakes Entomologist, 10: 211-222.

Nokkala, S. and C. Nokkala

1985. Mitotic and meiotic behavior of axial core structure of holokinetic chromosomes. Hereditas, 103: 107-110.

Oatman, E. R., F. E. Gilstrap, R. L. Hale, and V. Voth

1974. Effect of phorate on the twospotted spider mite, associated predators, and aphids on strawverry in southern California. Environmental Entomology, 3: 642-644.

Oatman, E. R. and J. A. McMurty

1966. Biological control of the two-spotted spider mite on strawberry in southern California. Journal of Economic Entomology, 59: 433-429.

Oatman, E. R., J. A. McMurty, H. H. Shorey, and V. Voth

1967. Studies on integrating *Phytoseiulus persimilis* releases, chemical applications, cultural manipulations, and natural predation for control of the two-spotted spider mite on strawberry in southern California. Journal of Economic Entomology, 60: 1344-1351.

Oatman, E. R. and V. Voth

1972. An ecological study of the two-spotted spider mite on strawberry in southern California. Environmental Entomology, 1: 34-39.

Oatman, E. R., J. A. Wyman, H. W. Browning, and V. Voth

1981. Effects of releases and varying infestation levels of the two-spotted spider mite on strawberry yield in southern California. Journal of Economic Entomology, 74: 112-115.

Oetting, R. D. and T. R. Yonke

1971. Biology of some Missouri stink bugs. Journal of the Kansas Entomological Society, 44: 446-459.

1971. Immature stages and biology of *Hymenarcys nervosa* and *H. aequalis* (Hemiptera.: Pentatomidae). Annals of the Entomological Society of America, 64: 1289-1296.

1971. Immature stages and biology of *Podisus placidus* and *Stiretrus fimbriatus* (Hemiptera: Pentatomidae). Canadian Entomologist, 103: 1505-1516.

1972. Immature stages and notes on the biology of *Hymenarcys crassa* (Hemiptera: Pentatomidae). Annals of the Entomological Society of America, 65: 474-478.

1975. Immature stages and notes on the biology of *Euthyrhynchus floridanus* (L.) (Hemiptera: Pentatomidae). Annals of the Entomological Society of America,

68: 659-662.

Oldham, T. W.
1978. New records of aquatic and semi-aquatic Hemiptera in Kansas. Technical Publications of the State Biological Survey of Kansas, 6: 59-69.

Oliver, A. D. and K. M. Komblas
1968. Controlling chinch bugs in St. Augustine grass. Louisiana Agriculture, 11: 3-16.

Olivier, A. G.
1811. Encyclopedie methodique. Historie naturelle. Hemiptera. Volume 8. Agasse, Paris. 722 pages.

Olsen, C. E.
1912. [Notes in minutes of meeting]. Journal of the New York Entomological Society, 19: 267 (1911).
1912. Contribution to an annotated list of Long Island insects. Journal of the New York Entomological Society, 20: 48-58.
1915. A capsid new to our fauna. Bulletin of the Brooklyn Entomological Society, 10: 34-35.
1916. A new pentatomid. Bulletin of the Brooklyn Entomological Society, 11: 82-83.

Orian, A. J. E.
1965. A new genus of Pentatomidae from Africa, Madagascar and Mauritius (Hemiptera). Proceedings of the Royal Entomological Society of London, Series B, 34: 25-29.

O'Rourke, F. A.
1980. Hybridization in milkweed bugs of the genus *Oncopeltus* (Hemiptera: Lygaeidae). Evolution, 33: 1098-1113 (1979).

Orphanides, G. M., D. Gonzalez, and B. R. Bartlett
1971. Identification and evaluation of pink bollworm predators in southern California. Journal of Economic Entomology, 64: 421-424.

Osborn, H.
1886. [Notes on minutes of Brooklyn Entomological Society for January 6, 1886]. Entomologica Americana, 1: 220.
1892. Notes on the species of *Acanthia*. Canadian Entomologist, 24: 262-265.
1892. Catalogue of the Hemiptera of Iowa. Proceedings of the Iowa Academy of Science, 1: 120-131.
1893. Notes on the distribution of Hemiptera. Proceedings of the Iowa Academy of Sciences, 1: 120-123.
1893. Additions and corrections to catalogue of Hemiptera. Proceedings of the Iowa Academy of Science, 1: 103-104.
1894. Notes on the distribution of Hemiptera. Proceedings of the Iowa Academy of Science, 1: 120-123.
1898. Additions to the list of Hemiptera of Iowa, with descriptions of new species. Proceedings of the Iowa Academy of Science, 5: 232-247.
1900. A list of the Hemiptera collected in the vicinity of Bellaire, Ohio. Ohio Naturalist, 1: 11-12.
1900. Remarks on the hemipterous fauna of Ohio with a preliminary record of species. 8th Annual Report of the Ohio Academy of Sciences, Pp. 60-79.
1901. New genus including two new species of Saldidae. Canadian Entomologist, 33: 181-182.
1903. Aradidae of Ohio. Ohio Naturalist, 4: 36-42.
1918. A meadow plant bug, *Miris dolabratus.* Journal of Agricultural Research, 15: 175-200, 1 plate.
1919. The meadow plant bug. Maine Agricultural Experiment Station Bulletin, 276: 1-16, 1 plate.

Osborn, H. and C. J. Drake
1915. Additions and notes on the Hemiptera-Heteroptera of Ohio. Ohio Naturalist, 15: 501-508.
1915. Records of Guatemalan Hemiptera-Heteroptera with descriptions of new species. Ohio Naturalist, 15: 529-541.
1916. The Tingitoidea or "lace bugs" of Ohio. Ohio Biological Survey Bulletin, 8: 217-251.
1916. Some new species of nearctic Tingidae. Ohio Journal of Science, 17: 9-15.
1917. Notes on Tingidae. Psyche, 24: 155-161, plate X.
1917. Notes on American Tingidae with descriptions of new species. Ohio Journal of Science, 17: 295-307.

Oshanin, B. F.
1906-1909. Verzeichnis der Palaearktischen Hemipteren mit besonderer Berücksichtigung ihrer Verteilung im russischen Reiche. Beilage zum, Annuaire Musée Zoologique de l'Académie Impériale des Sciences. St. Petersburg. Volume I. Part I: 1-393 (1906); Part II: 395-586 (1908); Part III: 587-1087 (1909).

Oshanin, B. F. (continued)
1912. Katalog der paläarktischen Hemipteren (Heteroptera, Homoptera - Auchenorhyncha und Psylloideae). R. Friedlander und Sohn, Berlin. xvi + 187 pages.

O'Shea, R.
1979. Redescriptions of three neotropical coreid genera of uncertain tribal placement (Heteroptera). Entomological News, 90: 45-50.
1980. A generic revision of the Acanthocerini (Hemiptera: Coreidae: Coreinae). Studies on Neotropical Fauna and Environment, 15: 57-80.
1980. A generic revision of the Nematopodini (Heteroptera: Coreidae: Coreinae). Studies on Neotropical Fauna and Environment, 15: 197-227.

O'Shea, R. and C. W. Schaefer
1978. The Micitini are not monophyletic (Hemiptera: Coreidae: Coreinae). Annals of the Entomological Society of America, 71: 776-784.

Overgaard, N. A.
1968. Insects associated with the southern pine beetle in Texas, Louisiana, and Mississippi. Journal of Economic Entomology, 61: 1197-1201.

Palisot de Beauvois, A. M. F. J.
1805-1821. Insectes recueillis en Afrique et en Amérique, dans les royaumes d'Oware et de Benin, à Saint-Dominique et dans les Etats-Unis, pendant les années. 1786-1797, 15 parts. Paris. [Griffin (1973, Journal of the Society Bibliography of Natural History, 1: 121-122) and Menke (1963, Annals Magazine of Natural History, series 13, 5: 701-702] agree upon the dates of the parts containing Heteroptera as follows: 1805, (1-2):1-40; 1806, (3):41-56; 1807, (4-5): 57-72, 73-88; 1811, (7-8): 101-120, 121-136; 1817, (9-10): 137-156, 157-172; 1818, (11-12): 173-192, 193-208; 1820, (13-14): 209-224, 225-240; 1821, (15): 241-276].

Palmer, W. A.
1986. Host specificity of *Ochrimnus mimulus* (Stål) (Hemiptera: Lygaeidae) with notes on its phenology. Proceedings of the Entomological Society of Washington, 88: 451-454.

Panizzi, A. R. and D. C. Herzog
1984. Biology of *Thyanta perditor* (Hemiptera: Pentomidae). Annals of the Entomological Society of America, 77: 646-650.

Panizzi, A. R. and F. Slansky, Jr.
1985. New host plant records for the stink bug *Piezodorus guildinii* in Florida (Hemiptera: Pentatomidae). Florida Entomologist, 68: 215-216.

Panzer, G. W. F.
1793-1813. Faunae Insectorum Germanicae initia oder Deutschlands Insecten, Hemiptera. 109 volumes. Felssecker, Nürnburg. [cited in catalog--40: 20 (1797)].

Pape, D. J. and L. A. Crowder
1981. Toxicity of methyl parathion and toxaphene to several insect predators in central Arizona. Southwestern Entomologist, 6: 44-48.

Parker, F. W. and N. M. Randolph
1972. Mass rearing the chinch bug in the laboratory. Journal of Economic Entomology, 65: 894-895.

Parish, H. E.
1934. Biology of *Euschistus variolarius* P. de B. (family Pentatomidae; order Hemiptera). Annals of the Entomological Society of America, 27: 50-54.

Parshley, H. M.
1914. List of the Hemiptera-Heteroptera of Maine. Psyche, 21: 139-149.
1915. Hemiptera-Heteroptera of Maine. Corrections and additions. Psyche, 22: 22-23.
1915. Systematic papers on New England Hemiptera. II. Synopsis of the Pentatomidae. Psyche, 22: 170-177.
1916. A new list of North American Hemiptera. Psyche, 23: 128-129.
1916. New and noteworthy Hemiptera from New England. Entomological News, 27: 103-106.
1916. On some Tingidae from New England. Psyche, 23: 163-168.
1917. Notes on North American Tingidae (Hemiptera). Psyche, 24: 13-25.
1917. Fauna of New England. 14. List of the Hemiptera - Heteroptera. Occasional Papers of the Boston Society of Natural History, 7: 1-125.
1917. Insects in ocean drift. I. Hemiptera Heteroptera. Canadian Entomologist, 49: 45-48.
1917. A species of *Macrotracheliella* found in New England (Hemip., Anthocoridae). Entomological News, 28: 37-38.

1918. [New species]. *In* Gibson, E. H. The genus *Corythucha* (Tingidae: Heteroptera). Transactions of the American Entomological Society, 44: 69-104.
1918. Three species of *Anasa* injurious in the North (Hemiptera, Coreidae). Journal of Economic Entomology, 11: 471-472.
1919. On some Hemiptera from western Canada. Occasional Papers of the Museum of Zoology, University of Michigan, 71: 1-35.
1919. Note on the sexes of the tingid *Melanorhopala clavata* Stal (Hemiptera). Bulletin of the Brooklyn Entomological Society, 14: 102-103.
1920. Hemiptera from Peaks Island, Maine, collected by Mr. G. A. Moore. Canadian Entomologist, 52: 80-87.
1920. Hemiptera collected in western New England, chiefly from mountains. Psyche, 27: 139-143.
1920. Hemipterological notices I. (Tingidae). Entomological News, 31: 271-274.
1921. A report on some Hemiptera from British Columbia. Proceedings of the Entomological Society of British Columbia, Systematic Series, 18: 13-24.
1921. Distributional notes on Hemiptera, with the description of a new *Gerris*. Psyche, 28: 8-15.
1921. Essay on the American species of *Aradus* (Hemiptera). Transaction of the American Entomological Society, 47: 1-106, 7 plates.
1922. Report on a collection of Hemiptera-Heteroptera from South Dakota. South Dakota State College Technical Bulletin, 2: 1-22.
1922. A change of name in *Ischnodemus* (Hemiptera, Lygaeidae). Bulletin of the Brooklyn Entomological Society, 17: 123.
1922 Hemipterological notices.--II. Entomological News, 33: 41-43.
1922. A change of name in the Saldidae (Hemiptera). Entomological News, 33: 71.
1923. The distribution and forms of *Lygaeus kalmii* Stal, with remarks on insect zoogeography (Hemiptera, Lygaeidae). Canadian Entomologist, 55: 81-84.
1923. Hemipterological notices.--III. (Miridae, Lygaeidae). Entomological News, 34: 21-22.
1923. Records of Nova Scotian Hemiptera-Heteroptera. Proceedings of the Acadian Entomological Society for 1922, 8: 102-108.
1923. Family Anthocoridae. Pages 665-668. *In* Britton, W. E. (editor). Guide to the Insects of Connecticut. Part IV. The Hemiptera or sucking insects of Connecticut. Connecticut Geological and Natural History Survey Bulletin, 34: 1-807, 20 plates.
1923. Family Enicocephalidae. Pages 693-694. *In* Britton, W. E. (editor). Guide to the Insects of Connecticut. Part IV. The Hemiptera or Sucking Insects of Connecticut. Connecticut Geological and Natural History Survey Bulletin 34: 1-807, 20 plates.
1923. Family Tingidae. Pages 695-707. *In* Britton, W. E. (editor). Guide to the Insects of Connecticut. Part IV. The Hemiptera or Sucking Insects of Connecticut. Connecticut State Geological and Natural History Survey of Connecticut Bulletin, 34: 1-807, 20 plates.
1923. Family Neididae. Pages 737-738. *In* Britton, W. E. (editor). Guide to the Insects of Connecticut. Part IV. The Hemiptera or Sucking Insects of Connecticut. Connecticut Geological and Natural History Survey Bulletin, 34: 1-807, 20 plates.
1925. A Bibliography of the North American Hemiptera-Heteroptera. Smith College, Northampton, Massachusetts. 252 pages.
1929. New species and new records of *Aradus* (Aradidae, Hemiptera). Canadian Entomologist, 61: 243-246.

Parsons, M. C.
1959. Skeleton and musculature of the head of *Gelastocoris oculatus* (Fabricius) (Hemiptera-Heteroptera). Bulletin of the Museum of Comparative Zoology at Harvard College, 122: 1-53.
1960. Skeleton and musculature of the thorax of *Gelastocoris oculatus* (Fabricius) (Hemiptera-Heteroptera). Bulletin of the Museum of Comparative Zoology at Harvard College, 122: 299-357.
1966. Studies on the cephalic anatomy of Naucoridae. Transactions of the Royal Entomological Society of London, 118: 119-151.
1970. Respiratory significance of the thoracic and abdominal morphology of three aquatic bugs *Ambrysus, Notonecta,* and *Hesperocorixa* (Insecta, Heteroptera).

Parsons, M. C. (continued)
Zeitschrift für Morphologie und Okologie der Tiere, 66: 242-298.

1971. Respiratory significance of the external morphology of adults and fifth instar nymphs of *Notonecta undulata* Say (Aquatic Heteroptera; Notonectidae). Journal of Morphology, 133: 125-138.

1972. Fine structure of the triturating devices in the food pump of *Notonecta* (Heteroptera: Notonectidae). Journal of Morphology, 138: 141-168.

Parsons, M. C. and R. J. Hewson

1975. Plastral respiratory devices in adult *Cryphocricos* (Naucoridae: Heteroptera). Psyche, 81: 510-527 (1974).

Patton, R. L. and G. A. Mail

1935. The grain bug (*Chlorochroa sayii* Stal) in Montana with special reference to the effects of cold weather. Journal of Economic Entomology, 28: 906-913.

Payne, J. A., F. W. Mead, and E. W. King

1968. Hemiptera associated with pig carrion. Annals of the Entomological Society of America, 61: 565-567.

Peet, W. B., Jr.

1973. Biological studies on *Nidicola marginata* (Hemiptera: Anthocoridae). Annals of the Entomological Society of America, 66: 344-348.

1979. Description and biology of *Nidicola jaegeri*, n. sp., from Southern California (Hemiptera: Anthocoridae). Annals of the Entomological Society of America, 72: 430-437.

Pemberton, C.

1911. The California Christmas-berry tingis. Journal of Economic Entomology, 4: 339-346.

Pemberton, R. W. and E. M. Hoover

1980. Insects associated with wild plants in Europe and the Middle East. Biological control of weeds surveys. United States Department of Agriculture Miscellaneous Publication, 1382: i-iv, 1-33.

Péricart, J.

1971. Observations diverses et nouvelles synonymies concernant les Anthocoridae et Microphysidae Palearctiques (Heteroptera). Bulletin de la Société Linnéenne De Lyon, 40: 93-114.

1972. Hémiptères. Anthocoridae, Cimicidae et Microphysidae de L'Quest-Paléarctique. Faune de L'Europe et du Bassin Méditerranéen. Masson et Cie, Paris. 7: 1-402.

1978. Révision systématique des Tingidae ouest-paléarctiques. 5. Contribution a la connaissance du genre *Acalypta* Westwood (Hemiptera). Annales de la Société Entomologique de France, new series, 14: 683-701.

1982. Révision systématique des Tingidae ouest-paléarctiques (Hemiptera). 9. Compléments et corrections. Annales de la Société Entomologique de France, new series, 18: 349-372.

Perty, J. A. M.

1830-1834. Delectus animalium articulatorum, quae in itinere per Braziliam annis MDCCCXVII-MDCCCXX [1817-1820], jussu et auspiciis Maximiliani Josephi Bavariae regis augustissimi peracto, collegerunt Dr. J. B. de Spix et Dr. C. F. Ph. de Martius. [1830: i-iii, 1-44, 1-60, plates 1-12; 1832: 61-124, plates 13-24; 1833: 125-224, plates 25-40].

Peters, L. L.

1982. Susceptibility of chinch bugs to selected insecticides -- laboratory study (Hemiptera: Lygaeidae). Journal of the Kansas Entomological Society, 55: 317-322.

1983. Chinch bug (Heteroptera: Lygaeidae) control with insecticides on wheat, field corn, and grain sorghum, 1981. Journal of Economic Entomology, 76: 178-181.

Peters, W. and J. Spurgeon

1971. Biology of the water-boatman *Krizousacorixa femorata* (Heteroptera: Corixidae). American Midland Naturalist, 86: 197-207.

Pfaler, E. V.

1936. Lebenszyklen der Lygaeiden (Hem.). Notulae Entomologicae, 16: 65-85.

Philippi, R. A.

1862. Viaje a los baños i al nuevo volcan de Chillan por don Rodolpho Philippi.--Comunicacion del mismo a la Facultad de Ciencias Físicas en su sesion del presente mes. Anales de la Universidad de Chile. 21: 377-389.

Picchi, V. D.

1977. A systematic review of the genus *Aneurus* of North and Middle America and the West Indies (Hemiptera: Aradidae). Quaestiones Entomologicae,

13: 255-308.

Pielou, D. P.
1950. The effect of insecticide applications on the insect fauna and seed yield of alsike clover in southern Ontario. Canadian Entomologist. 82: 141-160.
1966. The fauna of *Polyporus betulinus* (Builliard) Fries (Basidiomycetes: Polyporaceae) in Gatineau Park, Quebec. Canadian Entomologist, 98: 1233-1237.

Pieters, E. P.
1978. Comparison of sample-unit sizes for D-vac sampling of cotton arthropods in Mississippi. Journal of Economic Entomology, 71: 107-108.

Pieters, E. P. and W. L. Sterling
1973. Inferences on the dispersion of cotton arthropods in Texas. Environmental Entomology, 2: 863-867.
1973. Comparison of sampling techniques for cotton arthropods in Texas. Texas Agricultural Experiment Station Miscellaneous Publication, 1120. 8 pages.
1974. Aggregation indices of cotton arthropods in Texas. Environmental Entomology, 3: 598-600.

Pimental, D. and A. G. Wheeler, Jr.
1973. Species and diversity of arthropods in the alfalfa community. Environmental Entomology, 2: 659-668.

Pinto, C.
1927. *Ornithocoris toledoi*, novo genero e nova especie de percevejo de ave (Hemiptera-Fam. Cimicidae). Revista de Biologia e Hygiene, 1: 17-22.
1931. Valor do rostro e antenas na caracterização dos generos de Triatomideos. Hemiptera. Reduvidioidea. Boletim Biologico, 19: 45-136, 1 fold-out table.
1938. Zoo-Parasitas de Interesse Medico e Veterinario. Pimenta de Mello & Cia, Rio de Janeiro. 376 pages.

Pinto, J. D. and S. I. Frommer
1980. A survey of the arthropods on jojoba (*Simmondsia chinensis*). Environmental Entomology, 9: 137-143.

Piran, A. A.
1961. Sinopsis del género *Heteroscelis* Latreille 1829, (Hem. Pentatomidae) con la descripción de cinco especies nuevas. Revista de Investigaciones Agricolas, Instituto Nacional de Tecnologia Agro-pecuaria, Buenos Aires, 15: 83-99.

Pitre, H., T. L. Hillhouse, M. C. Donahue and H. C. Kinard
1978. Beneficial arthropods on soybeans and cotton in different ecosystems in Mississippi. Mississippi Agricultural and Forestry Experiment Station Technical Bulletin, 90: 1-9.

Pitts, D. L. and E. P. Pieters
1982. Toxicity of chlordimeform and methomyl to predators of *Heliothus* spp. on cotton. Journal of Economic Entomology, 75: 353-355.

Polhemus, D. A.
1984. A new species of *Dichaetocoris* Knight from the western United States, with notes on other species (Hemiptera: Miridae). Pan-Pacific Entomologist, 60: 33-36.
1985. A review of *Dichaetocoris* Knight (Heteroptera: Miridae): New species, new combinations, and additional distribution records. Pan-Pacific Entomologist, 61: 146-151.

Polhemus, D. A. and J. T. Polhemus
1984. *Ephedrodoma*, a new genus of orthotyline Miridae (Hemiptera) from western United States. Proceedings of the Entomological Society of Washington, 86: 550-554.
1985. Myrmecomorphic Miridae (Hemiptera) on mistletoe: *Phoradendrepulus myrmecomorphus* n. gen., n. sp., and a redescription of *Pilophoropsis brachypterus* Poppius. Pan-Pacific Entomologist, 61: 26-31.

Polhemus, J. T.
1964. A new species of *Ioscytus* from the western United States (Hemiptera: Saldidae). Proceedings of the Entomological Society of Washington, 66: 253-255.
1966. Some Hemiptera new to the United States (Notonectidae, Saldidae). Proceedings of the Entomological Society of Washington, 68: 57.
1967. Notes on North American Saldidae (Hemiptera). Proceedings of the Entomological Society of Washington, 69: 24-30.
1967. A new saldid from California (Hemiptera: Saldidae). Proceedings of the Entomological Society of Washington, 69: 346-348.
1969. Lectotype designations for some Saldidae (Hemiptera) described by P. R. Uhler. Annals of the Entomological Society of America, 62: 1207-1208.

Polhemus, J. T. (continued)

1970. A new genus of Veliidae from Mexico (Hemiptera). Proceedings of the Entomological Society of Washington, 72: 443-448.

1972. Notes concerning Mexican Saldidae, including the description of two new species (Hemiptera). Great Basin Naturalist, 32: 138-153.

1972. Notes on the genus *Nerthra*, including the description of a new species (Hemiptera: Gelastocoridae). Proceedings of the Entomological Society of Washington, 74: 306-309.

1973. Notes on aquatic and semiaquatic Hemiptera from the southwestern United States (Insecta: Hemiptera). Great Basin Naturalist, 33: 113-119.

1974. The *austrina* group of the genus *Microvelia* (Hemiptera; Veliidae). Great Basin Naturalist, 34: 207-217.

1975. Lectotype designation for *Hebrus sobrinus* Uhler (Heteroptera: Hebridae). Proceedings of the Entomological Society of Washington, 77: 128.

1975. New estuarine and intertidal water striders from Mexico and Costa Rica (Hemiptera: Gerridae, Mesoveliidae). Pan-Pacific Entomologist, 51: 243-247.

1976. Notes on North American Nepidae (Hemiptera: Heteroptera). Pan-Pacific Entomologist, 52: 204-208.

1976. A reconsideration of the status of the genus *Paravelia* Breddin with other notes and a check list of species (Veliidae: Heteroptera). Journal of the Kansas Entomological Society, 49: 509-513.

1977. Type designations and other notes concerning Veliidae (Insecta: Hemiptera). Proceedings of the Entomological Society of Washington, 79: 637-648.

1977. Neotype designation for *Hebrus sobrinus* Uhler (Heteroptera: Hebridae). Proceedings of the Entomological Society of Washington, 79: 237.

1979. Family Naucoridae/creeping water bugs, saucer bugs. Pages 131-138. *In* Menke, A. S. (editor). The Semiaquatic and Aquatic Hemiptera of California (Heteroptera: Hemiptera). Bulletin of the California Insect Survey, 21: xi + 1-166.

1982. Hemiptera. Hydrometridae. Pages 313-314. *In*: Hurlbert, S. H. and A. Villalobos-Figueroa (editors). Aquatic Biota of Mexico, Central America and the West Indies. San Diego State University, San Diego, California.

1984. Aquatic and semiaquatic Hemiptera. Pages 231-260. *In* Merritt, R. W. and K. W. Cummins (editors). An Introduction to the Aquatic Insects of North America. 2nd edition. Kendall Hunt, Dubuque, Iowa.

1985. Shore Bugs (Heteroptera Hemiptera; Saldidae). A World Overview and Taxonomic Treatment of Middle American Forms. The Different Drummer, Englewood, Colorado. 252 pages.

1985. Nomenclatural changes for North American Saldidae. Proceedings of the Entomological Society of Washington, 87: 893.

Polhemus, J. T. and H. C. Chapman

1966. Notes on some Hebridae from the United States with the description of a new species (Hemiptera). Proceedings of the Entomological Society of Washington, 68: 209-211.

1970. Some notes concerning American Hebridae, with the description of a new species and subspecies (Hemiptera). Proceedings of the Entomological Society of Washington, 72: 51-54.

1979. Family Salididae/shore bugs. Pages 16-33. *In* Menke, A. S. (editor). The Semiaquatic and Aquatic Hemiptera of California (Heteroptera: Hemiptera). Bulletin of the California Insect Survey, 21: xi + 1-166.

1979. Family Hebridae/velvet water bugs. Pages 34-38. *In* Menke, A. S. (editor). The Semiaquatic and Aquatic Hemiptera of California (Heteroptera: Hemiptera). Bulletin of the California Insect Survey, 21: xi + 1-166.

1979. Family Mesoveliidae/water treaters. Pages 39-42. *In* Menke, A. S. (editor). The Semiaquatic and Aquatic Hemiptera of California (Heteroptera: Hemiptera). Bulletin of the California Insect Survey, 21: xi + 1-166.

1979. Family Hydrometridae/marsh treaders, water measurers. Pages 43-45. *In* Menke, A. S. (editor). The Semiaquatic and Aquatic Hemiptera of California (Heteroptera: Hemiptera). Bulletin of the California Insect Survey, 21: xi + 1-166.

1979. Family Macroveliidae. Pages 46-48. *In* Menke, A. S. (editor). The Semiaquatic and Aquatic Hemiptera of California

Polhemus, J. T. and H. C. Chapman (continued)
(Heteroptera: Hemiptera). Bulletin of the California Insect Survey, 21: xi + 1-166.
1979. Family Veliidae/small water striders, water crickets, riffle bugs. Pages 49-57. *In* Menke, A. S. (editor). The semiaquatic and aquatic Hemiptera of California (Heteroptera: Hemiptera). Bulletin of the California Insect Survey, 21: xi + 1-166.
1979. Family Gerridae/water striders, pond skaters, wherrymen. Pages 58-69. *In* Menke, A. S. (editor). The Semiaquatic and Aquatic Hemiptera of California (Heteroptera: Hemiptera). Bulletin of the California Insect Survey, 21: xi + 1-166.

Polhemus, J. T. and J. R. Hendrickson
1974. The occurrence of *Trichocorixa reticulata* in the Gulf of California. Pan-Pacific Entomologist, 50: 52.

Polhemus, J. T. and C. N. McKinnon
1983. Notes on the Hebridae of the Western Hemisphere with descriptions of two new species (Heteroptera: Hemiptera). Proceedings of the Washington Entomological Society, 85: 110-115.

Polhemus, J. T. and M. S. Polhemus
1976. Aquatic and semiaquatic Heteroptera of the Grand Canyon (Insecta: Hemiptera). Great Basin Naturalist, 36: 221-226.

Polivka, J. B.
1963. Control of hairy chinch bug, *Blissus leucopterus hirtus* Mont., in Ohio. Ohio Agricultural Experiment Station Research Circular, 122. 8 pages.

Polivka, J. B. and F. Irons
1966. Experimental control of chinch bugs on corn. Journal of Economic Entomology, 59: 759.

Popenoe, E. A.
1885. Contributions to a knowledge of the Hemiptera-fauna of Kansas. Transactions of the Kansas Academy of Science, 9: 62-64.
1890. Some insects injurious to the bean. Second Report of the Kansas Experiment Station for 1889. Pages 206-212.
1909. Contributions to a knowledge of the Hemiptera-fauna of Kansas. Transactions of the Kansas Academy of Science, 9: 62-64.

Poppius, B.
1909. Zur Kenntnis der Miriden-Unterfamilie Cylapina Reut. Acta Societatis Scientiarum Fennicae, 37(4): 1-45, 1 plate.
1909. Beiträge zur Kenntnis der Anthocoriden. Acta Societatis Scientiarum Fennicae, 38(9): 1-43.
1911. Zwei neue nearktische Miriden-Gattungen. Annales de la Société Entomologique de Belgique, 55: 84-87.
1912. Die Miriden der Äthiopischen Region. I. Mirina, Cylapina, Bryocorina. Acta Societatis Scientiarum Fennicae, 41(3): 1-203, 1 plate.
1913. Beiträge zur Anthocoriden-fauna von Central und Nord Amerika. Annales de la Société Entomologique de Belgique, 57: 11-15.
1914. Übersicht der *Pilophorus-* Arten nebst Beschreibung verwandter Gattungen (Hem. Het.). Annales de la Société Entomologique de Belgique, 58: 237-254.
1914. Einige neue Miriden-Gattungen und Arten aus Nord-Amerika und Cuba. Annales de la Société Entomologique de Belgique, 58: 255-261.
1914. Die Miriden der Äthiopischen Region. II. Macrolophinae, Heterotominae, Phylinae. Acta Societatis Scientiarum Fennicae, 44(3): 1-138.

Porter, T. W.
1952. Three new species of Hebridae (Hebridae) from the Western Hemisphere. Journal of the Kansas Entomological Society, 25: 9-12.
1952. A new species of Hebridae (Hemiptera) from the Southwest. Journal of the Kansas Entomological Society, 25: 147-149.

Prendergast, B.
1943. Observations on the sand dune chinch bug, *Blissus mixtus* Barber (Lygaeidae Hemiptera). Pan-Pacific Entomologist, 19: 59-60.

Press, J. W., B. R. Flaherty, and R. T. Arbogast
1975. Control of the red flour beetle, *Tribolium castaneum*, in a warehouse by a predaceous bug, *Xylocoris flavipes*. Journal of the Georgia Entomological Society, 10: 76-78.

Price, P. W.
1975. Insect Ecology. John Wiley and Sons, New York. 514 pages.

Price, P. W. and M. F. Willson
1979. Abundance of herbivores on six milkweed species in Illinois. American Midland Naturalist, 101: 76-86.

Procter, W.
1946. Biological Survey of the Mount Desert Region. Part VII. Being a Revision of Parts I and VI with the Addition of 1100 Species. The Insect Fauna with Reference to Methods of Capture, Food Plants, the Flora and Other Biological Features. Wistar Institute of Anatomy and Biology, Philadelphia. 566 pages.

Provancher, L.
1871. Liste des Hémiptères pris a Québec. Naturaliste Canadien, 3(4): 136-139.
1869. Description d'un nouvel Hémiptère. Naturaliste Canadien, 1: 211-212.
1872. Description de plusieurs Hémiptères nouveaux. Hétéroptères. Naturaliste Canadien, 4: 73-79, 103-108, 319-320, 350-352, 376-379.
1885-1890. Petite Faune Entomologique du Canada et Particulierement de la Province de Quebec. Volume 3. Cinquieme Ordre les Hemipteres. Naturaliste Canadien, 1885, 1-64; 1886, 65-112; 1887, 113-184; 1888, 185-204; 1889, 205-282; 1890, 283-354, plates 1-5.

Puchkov, V. G.
1956. Basic trophic groups of phytophagus hemipterous insects and changes in the character of their feeding during the process of development. Zoologicheskii Zhurnal, 35: 32-44. [In Russian].

Puton, A.
1872. *Myiomma fieberi* Put. Petites Nouvelles Entomologiques, 44: 177.
1875. Description de deux genres nouveaux d'Hémiptères de la famille des Capsides. Petites Nouvelles Entomologiques, 1: 519.
1878. Synopsis des Hémiptères-Hétéroptères de France de la famille des Lygaeides. Mémoires de la Sociétédes Sciences de l'Agriculture et des Arts a Lille, (4)6: 273-354. [Separate published 1878, periodical 879].
1886. Catalogue des Hémiptères (Hétéroptères, Cicadines et Psyllides) de la faune Paléarctique. 3rd edition. Caen. Pages 3-100.
1887. Hémiptères nouveaux ou peu connus de la Faune Paléarctique. Revue d'Entomologie, 6: 96-105.

Pyke, B., W. Sterling, and A. Hartstack
1980. Beat and shake bucket sampling of cotton terminals for cotton fleahoppers, other pests and predators. Environmental Entomology, 9: 572-576.

Quaintance, A. L.
1897. Some strawberry insects. Florida Agricultural Experiment Station Bulletin, 42: 551-600.
1898. A preliminary report upon the insect enemies of tobacco in Florida. Florida Agricultural Experiment Station Bulletin, 48: 150-188.

Radinovsky, S.
1964. Cannibal of the pond. Natural History, 73: 16-25.

Rakickas, R. J. and T. F. Watson
1974. Population trends of *Lygus* spp. and selected predators in strip cut alfalfa. Environmental Entomology, 3: 781-784.

Rambur, J. P.
1839. Faune entomologique d l'Andalousie. Tome II. Arthus Bertrand, Libraire, Paris. 336 pages. [Hemiptera, pages 177-212].

Ramoska, W. A.
1984. The influence of relative humidity on *Beauveria bassiana* infectivity and replication in the chinch bug, *Blissus leucopterus*. Journal of Invertebrate Pathology, 43: 389-394.

Ramoska, W. A. and T. Todd
1985. Variation in efficacy and viability of *Beauveria bassiana* in the chinch bug (Hemiptera: Lygaeidae) as a result of feeding activity on selected host plants. Environmental Entomology, 14: 146-148.

Randolph, N. M. and G. L. Teetes
1971. Control of the chinch bug on grain sorghum. Texas Agricultural Experiment Station Progress Report, 2863-2876: 32-34.

Raney, H. G. and K. V. Yeargan
1977. Seasonal abundance of common phytophagous and predaceous insects in Kentucky soybeans. Transactions of the Kentucky Academy of Science, 38: 83-87.

Rathvon, S. S.
1869. Entomology. Pages 521-572. *In* Mombert, J. I. An Authentic History of Lancaster County in the State of Pennsylvania. J. E. Barr and Co., Lancaster, Pennsylvania. [Heteroptera, pages 548-552].

Rau, P.
1943. The neon-sign dance of the water-boatman, *Tricocorixa verticalis* Fieb. (Hemiptera). Entomological News, 54: 258-259.

Rauf, A., D. M. Benjamin, and R. A. Cecich
1984. Bionomics of *Platylygus luridus* (Hemiptera: Miridae) in Wisconsin jack pine seed orchards. Canadian Entomologist, 116: 1219-1225.

Rauf, A., R. A. Cecich, and D. M. Benjamin
1984. Conelet abortion in jack pine caused by *Platylygus luridus* (Hemiptera: Miridae). Canadian Entomologist, 116: 1213-1218.

Razafimahatratra, V. and J. D. Lattin
1982. Five new species and new synonymies for the genus *Deraeocoris* (Heteroptera: Miridae) from western North America. Pan-Pacific Entomologist, 58: 352-364.

Readio, J. and M. H. Sweet
1982. A review of the Geocorinae of the United States east of the 100th meridian (Hemiptera: Lygaeidae). Miscellaneous Publications of the Entomological Society of America, 12: 1-91.

Readio, P. A.
1927. Studies on the biology of the Reduviidae of America north of Mexico. University of Kansas Science Bulletin, 17: 5-291.
1928. Studies on the biology of the genus *Corizus* (Coreidae, Hemiptera). Annals of the Entomological Society of America, 21: 189-201.

Reed, T., M. Shepard, and S. G. Turnipseed
1984. Assessment of the impact of arthropod predators on noctuid larvae in cages in soybean fields. Environmental Entomology, 13: 954-961.

Rees, A. and R. Offord
1969. Studies on the protease and other enzymes from the venom of *Lethocerus cordofanus*. Nature, 221: 675-677.

Reichart, C. V.
1966. A new species of *Zeridoneus* from Utah (Hemiptera: Lygaeidae). Ohio Journal of Science, 66: 347-348.
1971. A new *Buenoa* from Florida (Hemiptera: Notonectidae). Florida Entomologist, 54: 311-313.

Reid, D. G., C. C. Loan, and R. Harmsen
1976. The mirid (Hemiptera) fauna of *Solidago canadensis* (Asteracea) in south-eastern Ontario. Canadian Entomologist, 108: 561-567.

Reinert, J. A.
1972a. New distribution and host record for the parasitoid *Eumicrosoma benefica*. Florida Entomologist, 55: 143-144.
1972b. Control of the southern chinch bug, *Blissus insularis*, in south Florida. Florida Entomologist, 55: 231-235.
1972c. Turf-grass insect research. Proceedings of the Florida Turf-Grass Management Conference, 20: 79-84.
1974. Tropical sod webworm and southern chinch bug control in Florida. Florida Entomologist, 57: 275-279.
1978a. Antibiosis to the southern chinch bug by St. Augustinegrass accessions. Journal of Economic Entomology, 71: 21-24.
1978b. Natural enemy complex of the southern chinch bug in Florida. Annals of the Entomological Society of America, 71: 728-731.
1982. Carbamate and synthetic pyrethroid insecticides for control of organophosphate-resistant southern chinch bugs (Heteroptera: Lygaeidae). Journal of Economic Entomology, 75: 716-718.

Reinert, J. A., B. D. Bruton, and R. W. Toler
1980. Resistance of St. Augustinegrass to southern chinch bug and St. Augustine decline strain of *Panicum* mosaic virus. Journal of Economic Entomology, 73: 602-604.

Reinert, J. A., P. Busey, and F. G. Bilz
1986. Old World St. Augustine grasses resistant to the southern chinch bug(Heteroptera:Lygaeidae). Journal of Economic Entomology, 79: 1073-1075.

Reinert, J. A., and A. E. Dudeck
1974. Southern chinch bug resistance in St. Augustinegrass. Journal of Economic Entomology, 67: 275-277.

Reinert, J. A. and S. H. Kerr
1973. Bionomics and control of lawn chinch bugs. Bulletin of the Entomological Society of America, 19: 91-92.

Reinert, J. A. and K. M. Portier
1983. Distribution and characterization of oranophosphate-resistant southern chinch bugs (Heteroptera: Lygaeidae) in Florida. Journal of Economic Entomology, 76: 1187-1190.

Reinert, J. A., R. W. Toler, B. D. Bruton, and P. Busey
1981. Retention of resistance by mutants of floratam St. Augustinegrass to the southern chinch bug and St. Augustine decline. Crop Science, 21: 464-466.

Reis, P. R., A. Costa, Jr., and L. Lobato
1976. *Blissus leucopterus* (Say, 1832) (Hemiptera, Lygaeidae), nova praga de Gramineas, introduzida no estado de Minas Gerais. Anais de Sociedade Entomologica do Brasil, 5: 241-242.

Remane, R.
1964. Wietere Beiträge zur Kenntnis der Gattung *Nabis* Latr. (Hemiptera Heteroptera, Nabidae). Zoologische Beiträge, 10: 253-314.

Remold, H.
1962. Über die biologische Bedeutung der Duftdrüsen bei den Landwanzen (Geocorisae). Zeitschrift für Vergleichende Physiologie, 45: 636-694.

Resh, V. H. and K. L. Sorg
1983. Distribution of the Wilbur Springs shore bug (Hemiptera: Saldidae): Predicting occurrence using water chemistry parameters. Environmental Entomology, 12: 1628-1635.

Reuter, O. M.
1871. Skandinaviens och Finlands Acanthiider beskrifne. Öfversight af Kongliga Svenska Vetenskaps-Akademiens Förhandlinger, 28(3): 403-429.
1871. Acanthiidae americanae, descriptae. Öfversight af Kongliga Svenska Vetenskaps-Akademiens Förhandlinger, 28(5): 557-568, 1 plate.
1871. Pargas sockens Heteroptera, förtechnade. Notiser ur Förhandlingar Sallskapets pro Fauna et Flora Fennicae, 11: 309-326 (1870).
1872. Nabidae novae et minus cognitae. Bidrag till Nabidernas kännedom af. Öfversigt Kongliga Svenska Vetenskaps-Akademiens Förhandlingar, 29(6): 79-96, 1 plate.
1874. Remarques synonymiques sur quelques Hétéroptères. Annales de la Société Entomologique de France, Series 5, 4: 559-566.
1875. Genera Cimicadarum Europae. Bihang till Kongliga Svenska Vetenskaps-Akademiens Handlingar, 3(1): 1-66.
1875. Hemipteres nouveaux. Petites Nouvelles Entomologiques, l(136): 544-545.
1875. Remarques sur le catalogue des Hémiptères d'Europe et du bassin de la Méditerrranée par les Dr. A. Puton. Petites Nouvelles Entomologiques, l(137): 547-548.
1875. Revisio critica Capsinarum, praecipue Scandinaviae et Fenniae. Akademisk Afhandling, Helsingfors, 1(1): 1-101; 1(2): 1-190. [taxonomic descriptions in 2nd part].
1876. Diagnoses praecursioriae Hemipterorum - Heteropterorum. Petites Nouvelles Entomologiques, 2(147): 33-34.
1876. Capsinae ex America boreali in Museo Holmiensi asservatae, descriptae. Öfversigt af Kongliga Svenska Vetenskaps-Akademiens Förhandlingar, 32(9): 59-92 (1875).
1878. *Actinocoris*, novum Hemipterorum genus e Fennia australi. Meddelanden af Societas pro Fauna et Flora Fennicae, 2: 194-197.
1878-1896. Hemiptera Gymnocerata Europae. Hémiptères Gymnocèrates d'Europe, du bassin de la Méditerranée et l'Asie Russe. I. Acta Societatis Scientiarum Fennicae, 13(1): 1-188 (1878); 13(2): 193-312 (1879); 13(3): 313-496 and suppl. 13(3): 497-568 (1883); 23(4): 1-179 (1891); 33(5): 1-392 (1896).
1879. De Hemipteris e Sibiria orientali nonnullis adnotationes criticae. Öfversigt af Finska Vetenskaps-Societatens Förhandlingar, 21: 42-63.
1879. Remarks on some British Hemiptera-Heteroptera. Entomologist's Monthly Magazine, 15: 66-67.
1880. Diagnoses Hemipterorum novorum. II. Öfversigt af Finska Vetenskaps-Societatens Förhandlingar, 22: 9-24.
1880. Anteckningar om *Coriscus lineatus* Dahlb. Öfversigt. Finska af Vetenskaps-Societatens Förhandlingar, 22: 25-32.
1880. Nya bidrag till Åbo och Ålands skärgards Hemipter-fauna. Meddelanden af Societas pro Fauna et Flora Fennica, 5: 160-236.
1881. Ad cognitionem Reduviidarum mundi antiqui. Acta Societatis Scientarium Fennicae, 12: 269-339.
1882. Monographia generis *Oncocephalus* Klug proximeque affinium. Acta Societatis Scientiarum Fennicae, 12: 673-758, plates 1-3.
1882. Ad cognitionem Heteropterorum Africae occidentalis. Öfversigt af Finska Vetenskaps-Societatens Förhandlingar, 25: 1-43.
1884. Genera nova hemipterorum. IV. Terma-

tophylina, nova familia Anthocoridarum ex Aegypto. Wiener Entomologische Zeitung, 3: 218-219.

1884. Finlands och den Skandinaviska Halföns Hemiptera Heteroptera. Entomologische Tidskrift, 5: 173-184.

1884. Monographia Anthocoridarum orbis terrestris. Acta Societatis Scientiarum Fennicae, 14: 555-758.

1884. Entomologiska meddelanden från societas' pro fauna et flora Fennica sammantraden Åren 1882 och 1883. Entomologisk Tidskrift, 5: 163-171.

1885. Ad cognitionem Lygaeidarum palaearcticarum. Revue d'Entomologie, 4: 199-233.

1885. Synonymische Bemerkungen über Hemipteren. Berliner Entomologische Zeitschrift, 29: 39-47.

1886. Notes synonymiques. Revue d'Entomologie, 5: 120-122.

1888. Revisio synonymica Heteropterorum palaearcticorum quae descripserunt auctores vetustiores (Linné 1758-Latreille 1806). Acta Societatis Scientiarum Fennicae, 15(1): 241-315; 15(2): 443-812. [also published separately in 1888 with continuous pagination, 1-448].

1890. Ad Cognitionem Nabidarum. Revue d'Entomologie, 9: 289-309 (1889).

1890. Adnotationes Hempterologicae. Revue d'Entomologie, 9: 248-254. (Corrigenda, 10: 27, 1891).

1890. Capsidae novae ex Africa boreali. Revue d'Entomologie, 9: 255-260.

1891. Monographia Ceratocombidarum orbis terrestris. Acta Societatis Scientiarum Fennicae, 19(6): 1-27, 1 plate.

1891. De skandinaviskt-finska *Acanthia-*(*Salda-*) *arterna* af *saltatoria*-gruppen. Meddelanden af Societas pro Fauna et Flora Fennica, 17: 144-160.

1895. Ad cognitionem Capsidarum. III. Capsidae ex Africa boreali. Revue d'Entomologie, 14: 131-142.

1895. Zur Kenntnis der Capsiden-Gattung *Fulvius* Stål. Entomologisk Tidskrift, 16: 129-154.

1895. *Fulvius heidemanni,* eine Berichtigung. Entomologisk Tidskrift, 16: 254.

1895. Species palaearticae generis *Acanthia* Fabr., Latr. Acta Societatis Scientiarium Fennicae, 21(2): 1-58.

1896. Dispositio generum palearcticorum divisionis Capsaria familiae Capsidae. Öfversigt af Finska Vetenskaps-Societatens Förhandlingar, 38: 156-171.

1900. De Finska aterna af *Aradus lugubris*-gruppen. Meddelanden af Societas pro Fauna et Flora Fennica, 26: 131-139, 221-222.

1900. Capsidae novae mediterraneae, descriptae. I. Öfversigt af Finska Vetenskaps-Societatens Förhandlingar, 42: 131-160.

1900. Capsidae novae mediterraneae descripsit. II. Öfversigt af Finska Vetenskaps-Societatens Förhandlingar, 42: 259-267.

1901. Synonymiska notiser rorande nagra finska Hemiptera Heteroptera. Meddelanden af Societas pro Fauna et Flora Fennica, 27: 61-62.

1904. Uebersicht der paläarktischen *Stenodema*-Arten. Öfversigt af Finska Vetenkaps-Societatens Förhandlingar, 46(15): 1-21.

1905. Über die Verwendung des Gattungsnamen *Lopus* (Heteroptera, Capsidae). Wiener Entomologische Zeitung, 25: 216.

1905. Capsidae Stålianae secundum specimina typica redescriptae, I, II. Öfversigt af Finska Vetenskaps-Societatens Förhandlingar, 47(12): 1-20.

1905. Capsidae in Venezuela a D:o D:re Fr. Meinert collectae enumeratae novaeque species descriptae. Öfversigt af Finska Vetenskaps-Societatens Förhandlingar, 47(19): 1-39.

1905. Hemipterologische Spekulationen, I. Die Klassifikation der Capsiden. Festschrift für Palmén. I. Helsingfors. 58 pages.

1906. Monographia Generis Heteropterorum *Phimodera* Germ. Acta Societatis Scientarum Fennicae, 33(8): 1-51, 2 plates.

1907. Capsidae novae in insula Jamaica mense Aprilis 1906 a D. E. P. Van Duzee collectae. Öfversigt af Finska Vetenskaps-Societatens Förhandlingar, 49(5): 1-27.

1907. Capsidae in Brasilia collectae in Museo I. R. Vindobonensi asservatae. Annalen des Kaiserlich-Konglichen Naturhistorischen Hofmuseums Wien, 22: 33-80.

1908. Capsidae Mexicanae a Do. Bilimek collectae in museo i. r. Vindobonensi asservatae. Annalen des Kaiserlich-Konglichen Naturhistorischen Hofmuseums Wien, 22: 150-179 (1907).

1908. *Hemisphaerodella mirabilis* n. gen. et sp., eine merkwürdige Capsiden-Gattung

Reuter, O. M. (continued)
aus den Antillen. Wiener Entomologische Zeitung, 27: 297-298.
1908. Bermerkungen über Nabiden nebst Beschreibung neuer Arten. Mémoires de la Société Entomologique de Belgique, 15: 87-130.
1908. Enumeratio Pentatomidarum post Calalogum bruxellensem descriptarum. Mémoires Société Entomologique de Belgique, 15: 131-200.
1909. Capsidae Argentinae. Kritische und neue argentinische Capsiden. Öfversigt af Finska Vetenskaps-Societatens Förhandlingar, 51A(13): 1-20.
1909. Bemerkungen über nearktische Capsiden nebst Beschreibung neurer Arten. Acta Societatis Scientiarum Fennicae, 36(2): 1-86.
1910. Neue Beiträge zur Phylogenie und Systematik der Miriden nebst einleitenden Bemerkungen über die Phylogenie der Heteropteren-Familien. Acta Societatis Scientiarum Fennicae, 37(3): 1-167.
1912. Bemerkungen über mein neues Heteropterensystem. Öfversight af Finska Vetenskaps-Societatens Förhandlingar, 54A(6): 1-62.
1912. Hemipterologische Miscellen. Öfversigt af Finska Vetenskaps-Societatens Förhandlingar, 54A(7): 1-76.
1912. Zur generischen Teilung der paläarktischen und nearktischen Acanthiaden. Öfversigt af Finska Vetenskaps-Societatens Fördhandlangar, 52A(12): 1-24.
1913. Amerikanische Miriden. Öfversigt af Finska Vetenskaps-Societatens Förhandlingar, 55A(18): 1-64, 1 plate.
1913. Über *Sixeonotus luteiceps* Reut. und Beschreibung einer neuen Bryocorine (Hem. Het.). Annales de la Société Entomologique de Belgique, 57: 278-279.

Reuter, O. M. and B. R. Poppius
1912. Zur Kenntnis der Termatophyliden. Öfversigt af Finska Vetenskaps-Societatens Förhandlingar, 54A(1): 1-17.

Rice, L. A.
1954. Observations on the biology of ten notonectid species found in the Douglas Lake, Michigan region. American Midland Naturalist, 51: 105-132.

Richman, D. B., R. C. Hemenway, Jr., and W. H. Whitcomb
1980. Field cage evaluation of predators of the soybean looper, *Pseudoplusia includens* (Lepidoptera: Noctuidae). Environmental Entomology, 9: 315-317.

Richmond, C. W.
1917. Generic names applied to birds during the years 1906 to 1915, inclusive, with additions and corrections to Waterhouse's "Index Generum Avium." Proceedings of the United States National Museum, 53(2221): 565-636.

Rider, D. A.
1986. A new species and new synonymy in the genus *Tepa* Rolston and McDonald (Hemiptera: Pentatomidae). Journal of the New York Entomological Society, 94: 552-558.

Ridgway, R. L. and S. L. Jones
1968. Plant feeding by *Geocoris pallens* and *Nabis americoferus*. Annals of the Entomological Society of America, 61: 232-233.

Ridgway, R. L. and P. D. Lingren
1972. Predaceous and parasitic arthropods as regulators of *Heliothis* populations. United States Department of Agriculture Southern Cooperative Service Bulletin, 169. 92 pages.

Ridgway, R. L., P. D. Lingren, C. B. Cowan, Jr., and J. W. Davis
1967. Populations of arthropod predators and *Heliothis* spp. after applications of systematic insecticides to cotton. Journal of Economic Entomology, 60: 1012-1016.

Riley, C. V.
1870. The chinch bug -- *Micropus leucopterus*, Say. (Heteroptera, Lygaeidae). Pages 15-37. Second Annual Report of the Noxious, Beneficial and Other Insects of the State of Missouri.
1870. Beneficial Insects- The glassy winged soldier-bug --*Campyloneura vitripennis*, Say. Pages 137-139. Third Annual Report on the Noxious, Beneficial and other Insects of the State of Missouri.
1873. Fifth annual report on the noxious, beneficial, and other insects of the state of Missouri, made to the State Board of Agriculture, pursuant to an appropriation for this purpose from the Legislature of the state. 8th Annual Report of the State Board of Agriculture for 1872. 160 + 8 pages, 75 figures.
1875. The chinch bug - *Micropus leucopterus*,

Say. Seventh Report of Noxious and Beneficial Insects of Missouri. Pages 19-71.

1879. The thick-thighed walking stick. Pages 241-245, plate 3. Report of the United States Department of Agriculture for 1879.

1888. Some recent entomological matters of international concern: The *Icerya* or fluted scale. Insect Life, 1: 126-137.

1893. Report on a small collection of insects made during the Death Valley Expeditions. North American Fauna, 7: 235-238, 249-250, 260-265.

Rings, R. W.

1957. Types and seasonal incidence of stink bug injury to peaches. Journal of Economic Entomology, 50: 599-604.

1958. Types and seasonal incidence of plant bug injury to peaches. Journal of Economic Entomology, 51: 27-32.

Roach, S. H.

1980. Arthropod predators on cotton, corn, tobacco and soybeans in South Carolina. Journal of the Georgia Entomological Society, 15: 131-138.

Roach, S. H. and A. R. Hopkins

1981. Reduction in arthropod predator populations in cotton fields treated with insecticides for *Heliothis* spp. control. Journal of Economic Entomology, 74: 454-457.

Roback, S. S.

1958. New records of aquatic Heteroptera from the United States and Canada. Transactions of the American Entomological Society, 84: 1-11.

Rogers, D. J. and M. J. Sullivan.

1986. Nymphal performance of *Geocoris punctipes* (Hemiptera: Lygaeidae) on pest-resistant soybeans. Environmental Entomology, 15: 1032-1036.

Rogers, L. E. and R. J. Lavigne

1972. Asilidae of the Pawnee National Grasslands in northeastern Colorado. Wyoming Agricultural Experiment Station Monograph, 25: 1-35.

Rolston, L. H.

1972. The small *Thyanta* species of North America (Hemiptera: Pentatomidae). Journal of the Georgia Entomological Society, 7: 278-285.

1973. The genus *Menecles* Stål (Hemiptera; Pentatomidae). Journal of the New York Entomological Society, 80: 234-237 (1972).

1973. A review of *Hymenarcys* (Hemiptera: Pentatomidae). Journal of the New York Entomological Society, 81: 111-117.

1974. Revision of the genus *Euschistus* in Middle America (Hemiptera, Pentatomidae, Pentatomini). Entomologica Americana, 48: 1-102.

1976. A evaluation of the generic assignment of some American Pentatomini (Hemiptera: Pentatomidae). Journal of the New York Entomological Society, 84: 2-8.

1978. A revision of the genus *Odmalea* Bergroth (Hemiptera: Pentatomidae). Journal of the New York Entomological Society, 86: 20-36.

1978. A revision of the genus *Mormidea* (Hemiptera: Pentatomidae). Journal of the New York Entomological Society, 86: 161-219.

1983. A revision of the genus *Acrosternum* Fieber, subgenus *Chinavia* Orian, in the Western Hemisphere (Hemiptera: Pentatomidae). Journal of the New York Entomological Society, 91: 97-106

Rolston, L. H. and Kumar, R.

1975. Two new genera and two new species of Acanthosomatidae (Hemiptera) from South America, with a key to the genera of the Western Hemisphere. Journal of the New York Entomological Society, 82: 271-278 (1974).

Rolston, L. H. and F. J. D. McDonald

1979. Keys and diagnoses for the families of Western Hemisphere Pentatomoidea, subfamilies of Pentatomidae and tribes of Pentatominae (Hemiptera). Journal of the New York Entomological Society, 87: 189-207.

1981. Conspectus of Pentatomini genera of the Western Hemisphere--Part 2 (Hemiptera: Pentatomidae). Journal of the New York Entomological Society, 88: 257-272 (1980).

Rolston, L. H., F. J. D. McDonald and D. B. Thomas, Jr.

1980. A Conspectus of Pentatomini genera of the Western Hemisphere. Part I (Hemiptera: Pentatomidae). Journal of the New York Entomological Society, 88: 120-132.

Root, R. B.

1986. The life of a Californian population of the facultative milkweed bug, *Lygaeus*

Root, R. B. (continued)
kalmii (Heteroptera: Lygaeidae). Proceedings of the Entomological Society of Washington, 88: 201-214.

Root, R. B. and S. J. Chaplin
1976. The life-styles of tropical milkweed bugs, *Oncopeltus* (Hemiptera: Lygaeidae) utilizing the same hosts. Ecology, 57: 132-140.

Rosenfeld, A. H.
1911. Insects and spiders in Spanish moss. Journal of Economic Entomology, 4: 398-409.

Rossi, P.
1790. Fauna Etrusca, sistens Insecta, quae in provinciis Florentina et Pisana praesertim collegit. Liburni. 2: 1-272.

Ruckes, H.
1937. An annotated list of some pentatomids (Heteroptera) from New Mexico. Bulletin of the Brooklyn Entomological Society, 32: 32-36.
1937. *Trichopepla klotsi*, a new species of pentatomid from Wyoming (Heteroptera). American Museum Novitates, 935: 1-2.
1939. Two new species of *Brochymena* (Pentatomidae, Heteroptera) from Arizona. Bulletin of the Brooklyn Entomological Society, 33: 236-242 (1938).
1939. Three new species of *Brochymena* (Pentatomidae) from the United States and Mexico. Bulletin of the Brooklyn Entomological Society, 34: 111-119.
1939. *Brochymena florida*, a new species of pentatomid from Florida. Bulletin of the Brooklyn Entomological Society, 34: 236-239.
1946. *Brochymena obscura* (H.-S.), *Brochymena tenebrosa* Walker, and *Brochymena parva*, a new name. Bulletin of the Brooklyn Entomological Society, 41: 41-44.
1946. *Mecidea minor*, a new species of pentatomid from New Mexico. Bulletin of the Brooklyn Entomological Society, 41: 86-88.
1947. Notes and keys on the genus *Brochymena* (Pentatomidae, Heteroptera). Entomologica Americana, 26 (New Series): 143-238 (1946).
1952. [Notes in minutes of meeting]. Journal of the New York Entomological Society, 59: 249 (1951).
1955. The genus *Chariesterus* de Laporte (Heteroptera, Coreidae). American Museum Novitates, 1721: 1-16.
1955. Three new species and a subspecies of the genus *Mozena* (Heteroptera, Coreidae). American Museum Novitates, 1702: 1-8.
1957. The taxonomic status and distribution of *Thyanta custator* (Fabricius) and *Thyanta pallido-virens* (Stål) (Heteroptera, Pentatomidae). American Museum Novitates, 1824: 1-23.
1957. New species of Pentatomidae from North and South America (Heteroptera) I. Bulletin of the Brooklyn Entomological Society, 52: 16-24.
1957. New species of Pentatomidae from North and South America (Heteroptera) II. Bulletin of the Brooklyn Entomological Society, 52: 39-47.
1961. Three New World halyine pentatomids (Hemiptera; Pentatomidae). Journal of the New York Entomological Society, 68: 225-231 (1960).
1966. An analysis and a breakdown of the genus *Platycarenus* Fieber (Heteroptera, Pentatomidae, Discocephalinae). American Museum Novitates, 2255: 1-42.

Rudolph, R.
1971. The locomotion of *Hydrometra*. Forma et Functia, 4: 454-464.

Rummel, D. R. and R. E. Reeves
1971. Response of bollworm and predaceous arthropod populations to aldicarb treatments in cotton. Journal of Economic Entomology, 64: 907-911.

Rush, W. A., D. B. Francy, G. C. Smith, and C. B. Cropp
1980. Transmission of an arbovirus by a member of the family Cimicidae. Annals of the Entomological Society of America, 73: 315-322.

Ryckman, R. E.
1954. *Reduvius senilus* [*sic*] Van Duzee from the lodges of *Neotoma* in San Juan County, Utah (Hemiptera: Reduviidae). Bulletin of the Southern California Academy of Sciences, 53: 88.
1958. Description and biology of *Hesperocimex sonorensis*, new species, an ectoparasite of the purple martin (Hemiptera, Cimicidae). Annals of the Entomological Society of America, 51: 33-47.
1962. Biosystematics and hosts of the *Triatoma protracta* complex in North America (Hemiptera: Reduviidae) (Rodentia:

Cricetidae). University of California Publications in Entomology, 27: 93-239.
1967. Six new populations of Triatominae from western North America (Hemiptera: Reduviidae). Bulletin of the Pan-American Research Institute, 1: 1-3.

Ryckman, R. E., D. G. Bentley, and E. F. Archbold
1981. The Cimicidae of the Americas and oceanic islands, a checklist and bibliography. Bulletin of the Society of Vector Ecologists, 6: 93-142.

Ryckman, R. E. and M. A. Casdin
1977. The Polyctenidae of the world, a checklist with bibliography. California Vector Views, 24: 25-31.

Ryckman, C. V., C. P. Christianson, and D. Spencer
1955. *Triatoma recurva* collected from its natural host in Sonora, Mexico. Journal of Economic Entomology, 48: 330-332.

Ryerson, S. A. and J. D. Stone
1979. A selected bibliography of two species of *Orius*: the minute pirate bug, *Orius tristicolor, and Orius insidiosus* (Heteroptera: Anthocoridae). Bulletin of the Entomological Society of America , 25: 131-135.

Saad, A. A. B. and G. W. Bishop
1976. Attraction of insects to potato plants through use of artificial honeydews and aphid juice. Entomophaga, 21: 49-57.

Sahlberg, J. R.
1870. Hemiptera Heteroptera samlade under en resa i ryska Karalen sommaren 1869. Notiser ur Förhandlingar Sällskapets pro Fauna et Flora Fennica, 11: 277-307.
1878. Bidrag till Nordvestra Sibiriens insektfauna. Hemiptera Heteroptera insamlade under expeditionerna till obi och jenesej 1876 och 1877. Öfversigts af Kongliga Svenska Vetenskaps-Akademiens Handlingar, 16(4): 1-39.

Sahlberg, R. F.
1842. Nova species generis *Phytocoris* (Fallén) ex ordine Hemipterorum descripta. Acta Societatis Scientiarum Fennicae, l(2): 411-412.
1848. Monographia Geocorisarum Fenniae. Frenckelliana, Helsingforsiae. 154 pages.

Sailer, R. I.
1940. A new species of Thyreocorinae. Journal of the Kansas Entomological Society, 13: 62-63.
1944. The genus *Solubea* (Heteroptera: Pentatomidae). Proceedings of the Entomological Society of Washington, 46: 105-127.
1945. The status of *Corimelaena* White, 1839, *Eucoria* Mulsant and Rey, 1865, *Allocoris* McAtee and Malloch, 1933 (Heteroptera: Pentatomidae). Proceedings of the Washington Entomological Society, 47: 129-135.
1945. A new name for *Acantholoma* Stal (Hemiptera: Scutelleridae). Proceedings of the Entomological Society of Washington, 47: 135.
1946. [Taxonomic status of *Euschistus servus euschistoides* (Vollenhoven)] *In* Woodside, A. M. Cat-facing and Dimpling in Peaches. Journal of Economic Entomology, 39: 158-161.
1946. The synonymy and distribution of *Trichocorixa reticulata* (Guérin-Méneville). Hemiptera: Corixidae. Proceedings of the Hawaiian Entomological Society, 12: 617-620.
1948. [The genus *Trichocorixa* (Corixidae, Hemiptera)]. Pages 289-407. *In* Hungerford, H. B. The Corixidae of the Western Hemisphere (Hemiptera). University of Kansas Science Bulletin, 32: 1-827.
1949. [Taxonomic comment]. *In* Esselbaugh, C. O. A bionomic note on the taxonomic status of the form *pyrrhocerus* of *Euschistus tristigmus* Say (Hemiptera, Pentatomidae). Proceedings of the Entomological Society of Washington, 51: 160-163.
1951. The status of *Thyanta accera* McAtee (Hemiptera, Pentatomidae). Proceedings of the Entomological Society of Washington, 53: 42.
1952. A review of the stink bugs of the genus *Mecidea*. Proceedings of the United States National Museum, 102(3309): 471-505, plates 47-48.
1952. A technique for rearing certain Hemiptera. United States Department of Agriculture, Bureau of Entomology and Plant Quarantine, ET-303: 1-5.
1954. Significance of hybridization among stink bugs of the genus *Euschistus*. Yearbook of the American Philosophical Society, pages 146-149 (1953).
1954. Interspecific hybridization among insects with a report on crossbreeding experiments with stink bugs. Journal of Economic Entomology, 47: 377-383.
1957. *Solubea* Bergroth, 1891, a synonym of *Oebalus* Stål, 1862, and a note concerning

Sailer, R. I. (continued)
the distribution of *O. ornatus* (Sailer) (Hemiptera, Pentatomidae). Proceedings of the Entomological Society of Washington, 59: 41-42.
1959. Experimental systematics. Proceedings of the Entomological Society of Washington, 61: 140-142.
1961. Possibilities for genetic improvement of beneficial insects. Germ Plasm Resources, pages 295-303.
1961. The identity of *Lygaeus sidae* Fabricius, type species of the genus *Niesthrea* (Hemiptera: Coreidae). Proceedings of the Entomological Society of Washington, 63: 293-299.

Sailer, R. I. and S. E. Lienk
1954. Insect predators of mosquito larvae and pupae in Alaska. Mosquito News, 14: 14-16.

Sanderson, E. D.
1906. Report on miscellaneous cotton insects in Texas. United States Department of Agriculture, Bureau of Entomology Bulletin, 57: 1-63.

Sanderson, M. W.
1982. Aquatic and Semiaquatic Hemiptera. Pages 6.1-6.94. *In* Brigham, A., W. Brigham, and A. Gnilka (editors). Aquatic Insects and Oligochaetes of North and South Carolina. Midwest Aquatic Enterprises, Mahomet, Illinois. xi + 817 pages.

Sanford, K. H.
1964. Life history and control of *Atractotomus mali*, a new pest of apple in Nova Scotia (Miridae: Hemiptera). Journal of Economic Entomology, 57: 921-925.
1964. Eggs and oviposition sites of some predacious mirids on apple trees (Miridae: Hemiptera). Canadian Entomologist, 96: 1185-1189.

Sanford, K. H. and H. J. Herbert
1966. The influence of spray programs on the fauna of apple orchards in Nova Scotia. XV. Chemical controls for winter moth, *Operophtera brumata* (L.), and their effects on phytophagous mite and predator populations. Canadian Entomologist, 98: 991-999.
1967. The influence of spray programs on the fauna of apple orchards in Nova Scotia XVIII. Predator and prey populations in relation to miticides. Canadian Entomologist, 99: 689-696.

Sauer, D. and D. Feir
1972. Field observations of predation on the large milkweed bug, *Oncopeltus fasciatus*. Environmental Entomology, 1: 268.

Saunders, W.
1869. Notes and experiments on currant worms. Canadian Entomologist, 2: 13-17.
1870. An insect friend. *Arma placidum*, Ulke. Canadian Entomologist, 2: 93-94.
1883. Insects injurious to fruits. J. B. Lippincott & Co., Philadelphia. 436 pages.

Say, T. [see LeConte, 1859, for reprinted Say papers]
1824-1928. American Entomology. I. Philadelphia, Pennsylvania. 112 pages, plates 1-18. 3 volumes. 1824, 1: 8 + 112 pages, plates 1-18; 1825, 2: plates 19-36; 1828, 3: plates 37-54.
1824. [Descriptions of insects.]. *In* Keatings' Narrative of an expedition to the source of St. Peter's River, Lake Winnepeck, Lake of the Woods, under command of Major Long, 1823, 2: 268-378.
1825a. Descriptions of new hemipterous insects collected in the expedition to the Rocky Mountains, performed by order of Mr. Calhoun, Secretary of War, under command of Major Long. Journal of the Academy of Natural Sciences of Philadelphia, 4: 07-345.
1831. Descriptions of new species of North American insects found in Louisiana by Joseph Barabino. New Harmony, Indiana. 19 pages. [See Scudder, S. H., 1899, Psyche, 8: 306-307 and Bequaert, J., 1950, Psyche, 57: 146 for dating and clarification of this and Say's 1832 paper with a similar title].
1831-1832. Descriptions of new species of heteropterous Hemiptera of North America. New Harmony, Indiana. 39 pages.. [The T. W. Harris copy in the Houghton Library at Harvard, Concord, Massachusetts, has at the bottom of page 5 a footnote referring to "New Sp. N. Am. Ins. found by J. Barabino, 1832, p. 9" -- thus, no more than the first 4 pages could have appeared in 1831].
1832. New species of North American insects, found by Joseph Barabino, chiefly in Louisiana. New Harmony, Indiana. 16 pages.

Scarbrough, A. G. and B. E. Sraver
1979. Predatory behavior and prey of *Atomosia*

puella (Diptera: Asilidae). Proceedings of the Entomological Society of Washington, 81: 630-639.

Schaefer, C. W.

1963a. Remarks on Scudder's classification of the lygaeoid-coreoid complex of the Heteroptera. Canadian Journal of Zoology, 41: 1174-1175.

1964a. Book review: A Catalogue of the Lygaeidae of the World, by James Alexander Slater, 1964. Bulletin of the Entomological Society of America, 10: 234.

1964. The morphology and higher classification of the Coreoidea (Hemiptera-Heteroptera): Parts I and II. Annals of the Entomological Society of America., 57: 670-684.

1965. The morphology and higher classification of the Coreoidea (Hemiptera: Heteroptera). Part III. The families Rhopalidae, Alydidae, and Coreidae. Miscellaneous Publications, Entomological Society of America, 5: 1-76.

1966. Some notes on heteropteran trichobothria. Michigan Entomologist, 1: 85-90.

1966. The morphology and higher systematics of the Idiostolinae (Hemiptera: Lygaeidae). Annals of the Entomological Society of America, 59: 602-613.

1968. Book review: Evolutionary Trends in Heteroptera. Part I. Eggs, Architecture of the Shell, Gross Embryology, and Eclosion, by R. H. Cobben, 1968. Bulletin of the Entomological Society of America, 14: 293-294.

1968. The morphology and higher classification of the Coreoidea (Hemiptera-Heteroptera). Part IV. The *Acanthocephala*-group and the position of the *Stenoscelidea* Westwood (Coreidae). University of Connecticut Occasional Papers, Biological Science Series, 1: 153-199.

1974. Rise and fall of the apple redbugs. Memoirs of the Connecticut Entomological Society, pages 101-116.

1975. Heteropteran trichobothria (Hemiptera: Heteroptera). International Journal of Insect Morphology and Embryology, 4: 193-264.

1975. A re-assessment of North American *Leptocoris* (Hemiptera-Heteroptera: Rhopalidae). Annals of the Entomological Society of America, 68: 537-541.

1980. The host plants of the Alydinae, with notes on heterotypic feeding aggregations (Hemiptera: Coreoidea: Alydidae). Journal of the Kansas Entomological Society, 53: 115-122.

Schaefer, C. W. and N. P. Chopra

1982. Cladistic analysis of the Rhopalidae, with a list of food plants. Annals of the Entomological Society of America, 75: 224-233.

Schaefer, K. F. and W. A. Drew

1968. The aquatic and semiaquatic Hemiptera of Oklahoma. Part III. Gerridae and Veliidae. Proceedings of the Oklahoma Academy of Science, 47: 125-134.

1969a. The Lygaeidae (Hemiptera) of Oklahoma. Proceedings of the Oklahoma Academy of Science, 48: 83-104.

Schaefer, C. W. and A. Hamid

1971. An unreported abdominal structure in some lygaeids. Journal of the Kansas Entomological Society, 44: 301-304.

Schaefer, C. W. and P. L. Mitchell

1983. Food plants of the Coreoidea (Hemiptera: Heteroptera). Annals of the Entomological Society of America, 76: 591-615.

Schaefer, C. W. and R. J. Pupedis

1981. A stridulatory device in certain Alydinae (Hemiptera: Heteroptera: Alydidae). Journal of the Kansas Entomological Society, 54: 143-152.

Schaffner, J. C.

1974. *Texocoris secludis*, a new genus and species of Orthotylinae from Texas (Heteroptera: Miridae). Journal of the Kansas Entomological Society, 47: 281-284.

1977. *Acaciaocoris*, a new genus of Orthotylini occurring in Mexico and southwestern United States (Hemiptera: Miridae). Folia Entomológica Méxicana, 38: 5-12.

Schechter, R. B. and W. E. Brickley

1959. Insects associated with milkweed. Proceedings of the Entomological Society of Washington, 61: 248.

Schell, D. V.

1943. The Ochteridae (Hemiptera) of the Western Hemisphere. Journal of the Kansas Entomological Society, 16: 29-47.

Schellenberg, J. R.

1800. Das Geschlecht der Land- und Wasserwanzen, nach Familien geordnet mit Abbildunge. Orell, Zurich. 32 pages.

Schilling, P. S.

1827. g. Aus der Ordnung der Hemiptera. Uebersicht der Arbeiten und Veränderungen der schlesichen Gesellschaft für Vaterländische Kultur, Breslau (1826). Page 22.

1829. Hemiptera Heteroptera Silesiae systematice disposuit. Beiträge zur Entomologie, 1: 34-92.

1834. Über eine geffügelte Hauswanze, *Cimex domestica*. Isis (Oder Encyclopädische Zeitung), pages 738-739.

Schmidt, E.

1931. Zur Kenntnis der Familie Pyrrhocoridae Fieber (Hemiptera-Heteroptera). Stettiner Entomologische Zeitung (Herausgegeben vom Entomologischen Verein zu Stettin), 92: 1-51.

Schiödte, J. M. C. [also Schijdte, J. M. C.]

1842. Revisio critica specierum generis Tetyrae Fabricii, quarum in Museo Regio Hafniensi exempla typica. Naturhistorisk Tidskrift, 4: 279-312, 346-348.

Scholtz, H.

1847. Prodromus zu einer Rhynchoten-Fauna von Schlesien. Uebersicht der Arbeiten und Veränderungen de Schlesischen Gesellschaft für Vaterländische Kultur, Breslau (1846). Pages 104-164.

Schouteden, H.

1904. Descriptions de Scutellériens nouveaux ou peu connus (Pentatomides). Annales de la Société Entomologique de Belgique, 48: 296-303.

1904. Heteroptera. Fam. Pentatomidae. Subfam. Scutellerinae. Genera Insectorum, 24: 1-98, plates 1-5.

1905-1906. Heteroptera. Fam. Pentatomidae. Subfam. Graphosomatinae. Genera Insectorum. 1905, Fasc. 30: 1-46, plates 1-3; 1906, Addenda et Corrigenda, page 47.

1905. Description de Graphosomiens nouveaux. Annales de la Société Entomologique de Belgique, 49: 141-147.

1907. Heteroptera. Fam. Pentatomidae, Subfam. Asopinae (Amyoteinae). Genera Insec-t-orum, 52: 1-82, plates 1-5.

Schowalter, T. D.

1986. Overwintering aggregation of *Boisea rubrolineatus* (Heteroptera: Rhopalidae) in western Oregon. Environmental Entomology, 15: 1055-1056.

Schrank, F. P.

1796. Sammlung naturhistorischer und physikalischer Aufsatze. Raspe, Nürnburg. xvi + 485 pages, 7 plates.

1782. Kritische Revision des österreichischen Insectenverzeichnisses. Füessly Neues Magazin für die Liebhaber der Entomologie. 1: 135-168, 263-306.

1798-1804. Fauna Boica. Nürnburg. 1798, 1: 1-720; 1801, 2(1): 1-374; 2(2): 1-412; 1803, 3(1): 1-272; 1804, 3(2): 1-372.

Schread, J. C.

1963. The chinch bug and its control. Connecticut Agricultural Experiment Station Circular 223. 4 pages.

1970. Chinch bug control. Connecticut Agricultural Experiment Station Circular 233. 6 pages.

Schuh, R. T.

1967. The shore bugs (Hemiptera: Saldidae) of the Great Lakes Region. Contributions of the American Entomological Institute, 2: 1-35.

1974. The Orthotylinae and Phylinae (Hemiptera: Miridae) of South Africa with phylogenetic analysis of the ant-mimetic tribes of the two subfamilies for the world. Entomologica Americana, 47: 1-322.

1975. The structure, distribution, and taxonomic importance of trichobothria in the Miridae (Hemiptera). American Museum Novitates, 2585: 1-26.

1976. Pretarsal structure in the Miridae (Hemiptera) with a cladistic analysis of relationships within the family. American Museum Novitates, 2601: 1-39.

1984. Revision of the Phylinae (Hemiptera, Miridae) of the Indo-Pacific. Bulletin of the American Museum of Natural History, 177: 1-476.

1986. *Merinocapsus froeschneri*, a new species of phyline Miridae from western North America, with notes on the genus (Heteroptera). Journal of the New York Entomological Society, 94: 217-225.

1986. The influence of cladistics on heteropteran classification. Annual Review of Entomology, 31: 67-93.

Schuh, R. T., B. Galil, and J. T. Polhemus

1987. Catalog and bibliography of Leptopodomorpha (Heteroptera). Bulletin of the American Museum of Natural History, 185(3): 243-406.

Schuh, R. T. and J. D. Lattin

1980. *Myrmecophes oregonensis*, a new species

of Halticini (Hemiptera, Miridae) from the western United States. American Museum Novitates, 2697: 1-11.

Schuh, R. T. and J. T. Polhemus

1980. Analysis of taxonomic congruence among morphological, ecological, and biogeographic data sets for the Leptopodomorpha (Hemiptera). Systematic Zoology, 29: 1-26.

1980. *Saldolepta kistnerorum*, new genus and new species from Ecuador (Hemiptera, Leptopodomorpha), the sister group of *Leptosalda chiapensis*. American Museum Novitates, 2698: 1-5.

Schuh, R. T. and M. D. Schwartz

1985. Revision of the plant bug genus *Rhinacloa* Reuter with a phylogenetic analysis (Hemiptera: Miridae). Bulletin of the American Museum of Natural History, 179: 379-470.

Schultz, J. C., D. Otte, and F. Enders

1977. *Larrea* as a habitat for desert arthropods. Pages 176-208. *In* Mabry, T. J., J. H. Hunziker and D. R. Difeo, Jr. (editors). Creosote Bush: Biology and Chemistry of *Larrea* in New World Deserts. US'IBP Synthesis Series, volume 6. Dowden, Hutchinson and Ross, Inc., Stroudsburg. 284 pages.

Schumacher, F.

1912. Neue amerikanische Formen aus der Unterfamilie der Asopinen (Hem. Het. Pent.). Sitzungsberichte der Gesselschaft Naturforschender Freude zu Berlin, pages 91-98.

1917. Über die Gattung *Stethoconus* Flor. (Hem. Het. Caps.). Sitzungsbericht Gesellschaft Naturforscheuder Freunde zu Berlin, 9: 344-346 (1916).

1917. Ueber Psocidenfeinde aus der Ordnung der Hemipteren. Zeitschrift für wissenschaftliche Insektenbiologie, 13: 217-218.

Schummel, T. E.

1832. Versuch einer genauen Beschreibung der in Schlesien einheimischen Arten der Familie der Ruderwangen (Ploteres, Latr.). E. Pelz, Breslau. 56 pages.

Schuster, M. F., M. J. Lukefahr, and F. G. Maxwell

1976. Impact of nectariless cotton on plant bugs and natural enemies. Journal of Economic Entomology, 69: 400-402.

Schwartz, M. D.

1984. A revision of the black grass bug genus *Irbisia* Reuter (Heteroptera: Miridae). Journal of the New York Entomological Society, 92: 193-306.

Schwartz, M. D. and J. D. Lattin

1984. *Irbisia knighti*, a new mirine plant bug (Heteroptera: Miridae) from the Pacific Northwest. Journal of the New York Entomological Society, 91: 413-417. (1983)

Schwartz, M. D. and G. M. Stonedahl

1986. Revision of the plant bugs genus *Noctuocoris* Knight (Heteroptera: Miridae: Orthotylinae). Pan-Pacific Entomologist, 62: 237-242.

Schwarz, E. A.

1888. On the insects found on *Uniola paniculata* in southeastern Florida. Proceedings of the Entomological Society of Washington, 1: 104-107.

1899. List of insects hitherto known from the Pribilof Islands. Pages 547-554. *In* The Fur Seals and Fur-Seal Islands of the North Pacific Ocean. Part III. Washington, D.C.

1901. [Notes on minutes of meeting]. Proceedings of the Entomological Society of Washington, 4: 391.

Sclater, P. E.

1862. [Note in minutes of meeting]. Athenaeum, 1834: 811.

1863. On some birds to be added to the avifauna of Mexico. Proceedings of the Zoological Society of London, 1862: 367-369.

Scopoli, I. A.

1763. Entomologia Carniolica exhibens insecta Carnioliae indigena et distributa in ordines, genera, species, varietates. Methodo Linneana. Trattner, Vindobonae. A2-A5 + 19 [index] + 420 pages.

1772. Annus V. Historico-Naturalis. C. G. Hilscheri, Lipsiae. 128 pages.

1786-1788. Deliciae faunae et florae insubricae, seu novae aut minus cognitae species plantarum et animalium quas in Insubria austriaca tum spontaceas quum exoticas vidit descripsit et aeri incindi curavit. 3 volumes. St. Salvator, Ticini. [1786, 1: ix + 85, 25 plates; 2: ii + 115, 25 plates; 1788, 3: ii + 87, 25 plates].

Scott, J.

1864. Additions to the fauna of Great Britain, and descriptions of two new species. Entomologist's Annual for 1864, pages 154-162.

Scott, J. (continued)
1874. On a collection of Hemiptera Heteroptera from Japan. Descriptions of various new genera and species. Annals and Magazine of Natural History, Series 4, 14: 289-304, 360-365, 426-452.

Scott, W. P., J. W. Smith, and C. R. Parencia, Jr.
1983. Effect of boll weevil (Coleoptera: Curculionidae) diapause control insecticide treatments on predaceous arthropod populations in cotton fields. Journal of Economic Entomology, 76: 87-90.

Scudder, G. G. E.
1957a. The higher classification of the Rhyparochrominae (Hem., Lygaeidae). Entomologist's Monthly Magazine, 93: 152-156.
1957b. A revision of Ninini (Hemiptera-Heteroptera, Lygaeidae) including the description of a new species from Angola. Publicações Culturais da Companhia de Diamantes de Angola, 34: 91-108.
1957c. The systematic position of *Pachymerus distinguendus* Flor, with a revised key to the British species of *Peritrechus* (Hem., Lygaeidae). Entomologist's Monthly Magazine, 93: 244-245.
1958. Review of Stichel "Illustrierte Bestimmungstabellen des Wanzen Europa II." Entomologist's Monthly Magazine, 94: xix.
1960. *Dictyonota fuliginosa* Costa (Hemiptera: Tingidae) in the Nearctic. Proceedings of the Entomological Society of British Columbia, 57: 22.
1961. Some Heteroptera new to British Columbia. Proceedings of the Entomological Society of British Columbia, 58: 26-29.
1962a. The Ischnorhynchinae of the world (Hemiptera: Lygaeidae). Transactions of the Royal Entomological Society of London, 114: 163-194.
1962b. The world Rhyparochrominae (Hemiptera: Lygaeidae). I. New Synonymy and generic changes. Canadian Entomologist, 94: 764-773.
1962c. New Heterogastrinae (Hemiptera) with a key to the genera of the world. Opuscula Entomologica, 27: 117-127.
1962d. Results of the Royal Society expedition to southern Chile, 1958-59: Lygaeidae (Hemiptera), with the description of a new subfamily. Canadian Entomologist, 94: 1064-1075.
1963a. Adult abdominal characters in the lygaeoid-coreoid complex of the Heteroptera, and the classification of the group. Canadian Journal of Zoology, 41: 1-14.
1963b. Heteroptera stranded at high altitudes in the Pacific Northwest. Proceedings of the Entomological Society of British Columbia, 60: 41-44.
1965. *Neacoryphus* Scudder, a new genus of Lygaeinae (Hemiptera: Lygaeidae). Proceedings of the Entomological Society of British Columbia, 62: 34-37.
1966a. *Hyalomya aldrichii* Townsend (Dipt., Tachinidae), a parasite of *Emblethis vicarius* Horvath (Hem., Lygaeidae). Entomologist's Monthly Magazine, 101: 286.
1966. The immature stages of *Cenocorixa bifida* (Hung.) and *C. expleta* (Uhler) (Hemiptera: Corixidae). Journal of the Entomological Society of British Columbia, 63: 33-40.
1967. Rhyparochrominae types in the British Museum (Natural History) (Hemiptera: Lygaeidae). Bulletin of the British Museum (Natural History), Entomology, 20: 253-285.
1968. Air-borne Lygaeidae (Hemiptera) trapped over the Atlantic, Indian and Pacific Oceans, with the description of a new species of *Appolonius* Distant. Pacific Insects, 10: 155-160.
1970a. The world Rhyparochrominae (Hemiptera: Lygaeidae). X. Further systematic changes. Canadian Entomologist, 102: 98-104.
1970b. The world Rhyparochrominae (Hemiptera, Lygaeidae). XI. The Horváth types. Annales Historico-Naturales Musei Nationalis Hungarici, 62: 197-206.
1971. The Gerridae (Hemiptera) of British Columbia. Journal of the Entomological Society of British Columbia, 68: 3-10.
1976. Water-boatmen of saline waters (Hemiptera: Corixidae). Pages 263-289. *In* Cheng, L. (editor). Marine Insects. North Holland/ American Elsevier, Amsterdam, Oxford, New York. 581 pages.
1977. The world Rhyparochrominae types (Hemiptera: Lygaeidae) XIII. The Stal types. Entomologia Scandinavica, 8: 29-35.
1979a. Present patterns in the fauna and flora of Canada. Pages 87-179. *In* Danks, H. V.

(editor). Canada and its insect fauna. Memoirs of the Entomological Society of Canada, 108: 1-573.

1979b. Hemiptera. Pages 329-348. *In* Danks, H. V. (editor). Canada and its insect fauna. Memoirs of the Entomological Society of Canada, 108: 1-573.

1981. Two new species of Lygaeinae (Hemiptera: Lygaeidae) from Canada. Canadian Entomologist, 113: 747-753.

1984. Two new genera of Rhyparochrominae (Hemiptera: Lygaeidae) from North America. Canadian Entomologist, 116: 1293-1300.

Scudder, G. G. E. and S. S. Duffey

1972. Cardiac glycosides in the Lygaeinae (Hemiptera: Lygaeidae). Canadian Journal of Zoology, 50: 35-42.

Scudder, G. G. E. and G. S. Jamieson

1972. The immature stages of *Gerris* in British Columbia. Journal of the Entomological Society of British Columbia, 69: 72-79.

Scudder, S. H.

1890. The Tertiary Insects of North America. Report of the United States Geological Survey of the Territories, Government Printing Office, Washington, D.C. 734 pages.

1899. An unknown tract on American insects by Thomas Say. Psyche, 8: 306-307.

Sears, M. K., F. L. McEwen, G. Ritcey, and R. R. McGraw

1980. Evaluation of insecticides for the control of the hairy chinch bug (Hemiptera: Lygaeidae) in Ontario lawns. Proceedings of the Entomological Society of Ontario, 111: 13-20.

Seidenstücker, G.

1950. Über *Myrmedobia* Bärensprung (Heteropt. Microphysidae). Senckenbergiana Biologica, 31: 287-296.

Seiss, C. F.

1897. [Hemiptera taken on sunflower]. Entomological News, 8: 67.

Semtner, P. J.

1979. Insect predators and pests on tobacco following applications of systemic insecticides. Environmental Entomology, 8: 1095-1098.

Serville, J. G. A.

1831. Description du genre Peirate, de l'ordre des Hemiptères, famille des Géocorises, tribu des Nudicolles. Annales des Sciences Naturelles, 23: 213-221.

Severin, H. H. P. and H. C. Severin

1911. An experimental study on the death-feigning of *Belostoma* (= *Zaitha* auctt.) *flumineum* Say and *Nepa apiculata* Uhler. Behavior Monographs, 1: 1-44, 1 plate.

1911. Habits of *Belostoma* (= *Zaitha*) *flumineium* Say and *Nepa apiculata* Uhler, with observations on other closely related aquatic Hemiptera. Journal of the New York Entomological Society, 19: 99-108.

Shanks, C. H., Jr. and B. Finnigan

1972. Population dynamics of the strawberry aphid in southwestern Washington. Environmental Entomology, 1: 81-89.

Shaw, E. L.

1930. Insects from *Lactuca* stems. Journal of the New York Entomological Society, 38: 463-468.

Shelton, A. M., J. A. Wyman, and A. J. Mayor

1981. Effects of commonly used insecticides on the potato tuberworm and its associated parasites and predators in potatoes. Journal of the New York Entomological Society, 74: 303-308.

Shepard, M.

1980. Sequential sampling plans for soybean arthropods. Pages 79-93. *In* Kogan, M. and D. C. Herzog (editors). Sampling Methods in Soybean Entomology. Springer Verlag, New York. 587 pages.

Shepard, M, G. R. Carner, and S. G. Turnipseed

1974. A comparison of three sampling methods for arthropods in soybeans. Environmental Entomology, 3: 227-232.

1974. Seasonal abundance of predaceous arthropods in soybeans. Environmental Entomology, 3: 985-988.

1977. Colonization and resurgence of insect pests of soybean in response to insecticides and field isolation. Environmental Entomology, 6: 501-506.

Shepard, M., W. Sterling, and J. K. Walker, Jr.

1972. Abundance of beneficial arthropods on cotton genotypes. Environmental Entomology, 1: 117-121.

Shepard, M., V. Wadill, and S. G. Turnipseed

1974. Dispersal of *Geocoris* spp. in soybeans. Journal of the Georgia Entomological Society, 9: 120-126.

Sherman, F., Jr.

1905. The flat-bugs (Aradidae) of North Carolina. Entomological News, 16: 7-9.

Sherman, F., Jr. (continued)
1948. Coreidae of South Carolina in comparison with North Carolina (Hemiptera). Entomological News, 59: 15-17.

Short, D. E. and P. G. Koehler
1979. A sampling technique for mole crickets and other pests in turf grass and pasture. Florida Entomologist, 62: 282-283.

Shull, W. E.
1933. An investigation of the *Lygus* species which are pests of beans (Hemiptera, Miridae). Idaho Agricultural Experiment Station Research Bulletin, 11: 1-42.
1933. The identity of two *Lygus* pests (Hemiptera, Miridae). Journal of Economic Entomology, 26: 1076-1079.

Shull, W. E. and C. Wakeland
1931. Tarnished plant bug injury to beans. Journal of Economic Entomology, 24: 326-327.

Sibley, C. K. ["and others"]
1926. Insects and some other invertebrates. Pages 87-184. *In* A preliminary biological survey of The Lloyd-Cornell Reservation. Bulletin of the Lloyd Library, 27(Entomological Series S): 1-247.

Signoret, V. A.
1850. Description d'un genre nouveau de l'ordre des Hémiptères Hétéroptères, et de la section des Hydrocoryses. Revue et Magasin de Zoologie pure et appliquee (etc.) par Guérin-Méneville, Paris, Series 2, 2: 289-291.
1852. Notice sur quelques Hémiptères nouveaux ou peu connus. Annales de la Société Entomologique de France, Series 2, 10: 539-544, plate 16.
1857. Essai monographique du genre *Micropus* Spinola. Annales de la Société Entomologique de France, Series 3, 5: 23-32, 1 plate.
1858. Description d'un nouveau genre de la tribu des Longicoxes, Amyot and Serville, groupe des Emésides. Annales de la Société Entomologique de France, Series 3, 6: 251-253, plate 6.
1859. Monographie du genre *Corizus*. Annales de la Société Entomologique de France, Series 3, 7: 75-105.
1877. [Note on the Veliidae]. Bulletin de la Société Entomologique de France, Series 5, 2: LIV-LV (or 1877, 2: 64-65).
1879. [Notes and descriptions]. Bulletin de la Société Entomologique de France, Series 5, 9: clxii-clxiii.
1880. [Notes and descriptions]. Bulletin de la Société Entomologique de France, Series 5, 10: vii-viii, xliv.
1881. [*Germatus* [*sic*] *violaceans*]. Bulletin de la Société Entomologique de France, page 50.
1881. [Notes in meeting]. Bulletin des Séances de la Société Entomologique de France, Series 6, I: clvi-clviii [also reprinted on pages 218-220].
1881-1884. Révision du Groupe des Cydnides de la Famille des Pentatomoides. Annales de la Société Entomologique de France, Series 6, 1881, 1: 25-52, 193-218, 319-332, 423-436, plates 1, 2, 6-12; 1882, 2: 23-42, 145-168, 241-266, 465-484, plates 1, 2, 6-9, 13, 14; 1883, 3: 33-60, 207-220, 357-374, 517-534, plates 2-5, 9, 10, 15, 16; 1884, 4: 45-62, 117-128, plates 2-3.

Simanton, W. A. and F. Andre
1936. A biological study of *Lygaeus kalmii* Stål (Hemiptera-Lygaeidae). Bulletin of the Brooklyn Entomological Society, 31: 99-107.

Simmands, F. J., J. M. Franz, and R. I. Sailer
1976. Chapter 2. History of biological control. Pages 17-39. *In* Huffaker, C. B. and P. S. Messenger (editors). Theory and Practice of Biological Control. Academic Press, New York. 788 pages.

Slater, Alex
1981. Aquatic and semiaquatic Heteroptera in the collection of the State Biological Survey of Kansas. Technical Publication of the State Biological Survey of Kansas, 10: 71-88.
1981. Hemiptera (True Bugs). Chapter 15, pages 119-134. *In* Guide to the freshwater invertebrates of the midwest. Technical Publication of the State Biological Survey of Kansas, 11: v + 1-221.
1982. Aquatic and semiaquatic Heteroptera in the collection of the State Biological Survey of Kansas. Addendum. Technical Publications of the State Biological Survey of Kansas, 12: 39-42.
1983. On the biology and food plants of *Lygaeus turcicus* (Fabr.) (Hemiptera: Lygaeidae). Journal of the New York Entomological Society, 91: 48-56.

Slater, J. A.
1948. Notes on *Uhleriola floralis* (Uhl.) in Illinois (Heteroptera, Lygaeidae). Bulletin

of the Brooklyn Entomological Society, 43: 69-71.

1950. Notes and new records of Iowa Hemiptera (Heteroptera). Proceedings of the Iowa Academy of Science, 57: 519-521.

1950. An investigation of the female genitalia as taxonomic characters in the Miridae (Hemiptera). Iowa State College Journal of Science, 25: 1-81.

1951. A key to the nymphs of midwestern Lygaeidae (Hemiptera: Heteroptera). Bulletin of the Brooklyn Entomological Society, 46: 42-48.

1952a. A contribution to the biology of the subfamily Cyminae (Heteroptera: Lygaeidae). Annals of the Entomological Society of America, 45: 315-326.

1952b. An annotated list of the Lygaeidae of Iowa and Illinois (Hemiptera: Heteroptera). Proceedings of the Iowa Academy of Science, 59: 521-540.

1952c. The immature stages of American Pachygronthinae (Hemiptera: Lygaeidae). Proceedings of the Iowa Academy of Science, 58: 553-561.

1954. Notes on the genus *Labops*, Burmeister in North America, with the descriptions of three new species (Hemiptera: Miridae). Bulletin of the Brooklyn Entomological Society, 49: 57-65, 89-94.

1955. A revision of the subfamily Pachygronthinae of the world (Hemiptera: Lygaeidae). Philippine Journal of Science, 84: 1-160, plates 1-4.

1956a. Neotropical Pachygronthinae in the American Museum of Natural History (Hemiptera, Lygaeidae). American Museum Novitates, 1769: 1-5.

1956. *Megaloceraea recticornis* (Geoffr.), a mirid new to the eastern United States, with the description of a new genus of Stenodemini (Hemiptera, Miridae). Proceedings of the Entomological Soceity of Washington, 58: 116-120.

1957. Nomenclatorial considerations in the family Lygaeidae (Hemiptera: Heteroptera). Bulletin of the Brooklyn Entomological Society, 52: 35-38.

1960. The responsibilities of the insect taxonomist. Bulletin of the Entomological Society of America, 6: 17-19.

1963. Immature stages of the subfamilies Cyminae and Ischnorhynchinae (Hemiptera: Lygaeidae). Journal of the Kansas Entomological Society, 36: 84-93.

1964a. A Catalogue of the Lygaeidae of the World. 2 volumes. University of Connecticut, Storrs. 1688 pages.

1964b. Hemiptera (Heteroptera): Lygaeidae. South African Animal Life, 10: 15-228.

1966. A further contribution to our knowledge of the Pachygronthinae (Hemiptera: Lygaeidae). Journal of the Entomological Society of Queensland, 5: 51-65.

1971. The first neotropical records of the genus *Plinthisus* with the description of three new species. Journal of the Kansas Entomological Society, 44: 377-384.

1972a. Lygaeid bugs (Hemiptera: Lygaeidae) as seed predators of figs. Biotropica, 4: 145-151.

1972b. An analysis of the genus *Migdilybs* with descriptions of three new species and comments on the distribution of the Ozophorini (Hemiptera: Lygaeidae). Journal of the Entomological Society of South Africa, 35: 159-169.

1973. Book review. Robert Leslie Usinger, Autobiography of an Entomologist. Bulletin of the Entomological Society of America, 19: 60-61.

1973. A contribution to the biology and taxonomy of Australian Thaumastocoridae with the description of a new species (Hemiptera: Heteroptera). Journal of the Australian Entomological Society, 12: 151-156.

1974. A preliminary analysis of the derivation of the Heteroptera fauna of the northeastern United States with special reference to the fauna of Connecticut. Memoirs of the Connecticut Entomological Society, pages 145-213.

1975a. On the biology and zoogeography of Australian Lygaeidae (Hemiptera: Heteroptera) with special reference to the southwest fauna. Journal of the Australian Entomological Society, 14: 47-64.

1975b. The Pachygronthinae of the West Indies, with the description of a new species of *Pachygrontha* from Cuba (Hemiptera: Lygaeidae). Florida Entomologist, 58: 65-74.

1977. The incidence and evolutionary significance of wing polymorphism in lygaeid bugs with particular reference to those of South Africa. Biotropica, 9: 217-229.

1978. Monocots and chinch bugs: A study of host plant relationships in the lygaeid subfamily Blissinae (Hemiptera, Lygae-

Slater, J. A. (continued)
idae). Biotropica, 8: 143-165.
1979. The systematics, phylogeny, and zoogeography of the Blissinae of the world (Hemiptera, Lygaeidae). Bulletin of the American Museum of Natural History, 165(1): 1-180.
1980. Systematic relationships of the Antillocorini of the Western Hemisphere (Hemiptera: Lygaeidae). Systematic Entomology, 5: 199-226.
1982. Hemiptera. Pages 417-447. *In* Parker, S. P. (editor). Synopsis and Classification of Living Organisms. McGraw Hill Book Company, Inc., New York.
1983. On the biology and food plants of *Lygaeus turcicus* (Fabr.) (Hemiptera: Lygaeidae). Journal of the New York Entomological Society, 91: 48-56.
1986. A synopsis of the zoogeography of the Rhyparochrominae (Heteroptera: Lygaeidae). Journal of the New Entomological Society, 94: 262-280.

Slater, J. A. and P. D. Ashlock
1966. *Atrazonotus*, a new genus of Gonianotini from North America (Hemiptera: Lygaeidae). Proceedings of the Entomological Society of Washington, 68: 152-156.
1976. The phylogenetic position of *Praetorblissus* Slater with the description of two new species (Hemiptera: Lygaeidae). Journal of the Kansas Entomological Society, 49: 567-579.

Slater, J. A., P. D. Ashlock, and D. B. Wilcox
1969. The Blissinae of Thailand and Indochina (Hemiptera: Lygaeidae). Pacific Insects, 11: 671-733.

Slater, J. A. and R. M. Baranowski
1970. A new genus and species of turtle bug from southern Florida (Hemiptera: Pentatomidae). Florida Entomologist, 53: 139-142.
1973. A review of the genus *Cistalia* Stål (Hemiptera: Lygaeidae). Florida Entomologist, 56: 263-272.
1978. How to Know the True Bugs (Hemiptera-Heteroptera). Wm. C. Brown Company Publishers, Dubuque, Iowa. 256 pages.
1983. The genus *Ozophora* in Florida (Hemiptera: Lygaeidae). Florida Entomologist, 66: 416-440.

Slater, J. A., H. G. Barber, and R. I. Sailer
1959. Nomenclatoral considerations relative to the genus *Myodocha* Latreille, 1807 (Hemiptera). Entomological News, 70: 185-189.
1961. *Myodocha* Latreille, 1807 (Hemiptera); proposed designation of a type species under the plenary powers. Z.N.(S.) 1431. Bulletin of Zoological Nomenclature, 18: 287-288.

Slater, J. A. and H. Brailovsky
1983a. Review of the neotropical genus *Toonglasa* (Hemiptera: Lygaeidae). Annals of the Entomological Society of America, 76: 523-535.
1983. The systematic status of the family Thaumastocoridae with the description of a new species of *Discocoris* from Venezuela (Hemiptera: Heteroptera). Proceedings of the Entomological Society of Washington, 85: 560-563.

Slater, J. A. and J. Carayon
1963. Ethiopean Lygaeidae IV: A new predatory lygaeid with a discussion of its biology and morphology (Hemiptera: Heteroptera). Proceedings of the Royal Entomological Society of London, (A), 38: 1-11.

Slater, J. A. and W. E. China
1961. *Pamera* Say, 1831; proposed suppression under the plenary powers and addition of *Rhyparochromus* Hahn, 1826, and *Megalonotus* Fieber, [1860] to the official list (Class Insecta, Order Hemiptera). Bulletin of Zoological Nomenclature, 18: 342-345.
1961. *Blissus* Burmeister, 1835 (Insecta, Hemiptera): Proposed designation of a type-species under the plenary powers. Z.N.(S.) 1471. Bulletin of Zoological Nomenclature, 18: 346-348.
1961. Heterogastrinae Stål, 1872 (Insecta, Hemiptera): Proposed validation under the plenary powers. Z.N.(S.) 1474. Bulletin of Zoological Nomenclature, 18: 349-350.
1961. *Scolopostethus* Fieber, [1860] (Insecta, Hemiptera): Proposed validation under the plenary powers. Z.N.(S.) 1475. Bulletin of Zoological Nomenclature, 18: 351-352.

Slater, J. A. and C. J. Drake
1958. The systematic position of the family Thaumastocoridae (Hemiptera: Heteroptera). Proceedings of the Tenth International Congress of Entomology, 1: 321-323.

Slater, J. A. and H. W. Hurlbutt

1957. A comparative study of the metathoracic wing in the family Lygaeidae (Hemiptera: Heteroptera). Proceedings of the Entomological Society of Washington, 59: 67-79.

Slater, J. A. and H. H. Knight

1954. The taxonomic status of *Oligotylus* Van Duzee and *Leptotylus* Van Duzee, with the description of a new species of *Psallus*. Pan-Pacific Entomologist, 30: 143-145.

Slater, J. A. and N. F. Knop

1969. Geographic variation in the North American milkweed bugs of the *Lygaeus kalmii* complex. Annals of the Entomological Society of America, 62: 1221-1232.

Slater, J. A. and J. D. Lattin

1965. *Lachnestes singalensis* (Dohrn), a lygaeid new to the Western Hemisphere (Hemiptera). Pan-Pacific Entomologist, 41: 58-60.

Slater, J. A. and S. Miyamoto

1963. A revision of the sugar cane bugs of the genus *Cavelerius* (Lygaeidae: Blissinae). Mushi, 37: 139-154, plates 6-9.

Slater, J. A. and J. E. O'Donnell

1979. An analysis of the *Ozophora laticephala*-complex with the description of eight new species (Hemiptera: Lygaeidae). Journal of the Kansas Entomological Society, 52: 154-179.

Slater, J. A. and C. W. Schaefer

1963. *Leptocoris trivittatus* (Say) and *Coriomeris humilis* Uhl. in New England (Hemiptera: Coreidae). Bulletin of the Brooklyn Entomological Society, 58: 114-117.

Slater, J. A. and B. Sperry

1973. The biology and distribution of the South African Lygaeinae, with descriptions of new species (Hemiptera: Lygaeidae). Annals of the Transvaal Museum, 28: 117-201, 1 plate.

Slater, J. A. and M. H. Sweet

1958. The occurrence of *Megalonotus chiragra* (F.) in the eastern United States with notes on its biology and ecology (Hemiptera: Lygaeidae). Bulletin of the Brooklyn Entomological Society, 53: 102-107.

1961. A contribution to the higher classification of the Megalonotinae (Hemiptera: Lygaeidae). Annals of the Entomological Society of America, 54: 203-209.

1965. The systematic position of the Psamminae (Heteroptera: Lygaeidae). Proceedings of the Entomological Society of Washington, 67: 255-262.

1970. The systematics and ecology of new genera and species of primitive Stygnocorini from South Africa, Madagascar and Tasmania (Hemiptera: Lygaeidae). Annals of the Natal Museum, 20: 257-292.

Slater, J. A., M. H. Sweet, and R. M. Baranowski

1977. The systematics and biology of the genus *Bathydema* Uhler (Hemiptera: Lygaeidae). Annals of the Entomological Society of America, 70: 343-358.

Slater, J. A. and E. Wagner

1955. Neuer Beitrag zur Systematik der Gattung *Trigonotylus* Fieb. (Hem. Heterop. Miridae). Deutsche Entomologische Zeitschrift, 2: 101-105.

Slater, J. A. and D. B. Wilcox

1966. An analysis of three new genera of neotropical Blissinae (Hemiptera: Lygaeidae). Annals of the Entomological Society of America, 59: 61-76.

1969. A revision of the genus *Ischnodemus* in the Neotropical Region (Hemiptera: Lygaeidae; Blissinae). Miscellaneous Publications of the Entomological Society of America, 6: 197-238.

Slater, J. A. and T. E. Woodward

1982. Lilliputocorini, a new tribe with six new species of *Lilliputocoris*, and a cladistic analysis of the Rhyparochrominae (Hemiptera, Lygaeidae). American Museum Novitates, 2754: 1-23.

Slater, J. A., T. E. Woodward, and M. H. Sweet

1962. A contribution to the classification of the Lygaeidae, with the description of a new genus from New Zealand (Hemiptera: Heteroptera). Annals of the Entomological Society of America, 55: 597-605.

Slifer, E. H. and S. S. Sekhon

1963. Sense organs on the antennal flagellum of the small milkweed bug, *Lygaeus kalmii* Stal (Hemiptera: Lygaeidae). Journal of Morphology, 112: 165-193.

Slingerland, M. V.

1909. A red bug on apple. *In* Proceedings of the Fifty-fourth Annual Meeting of the Western New York Horticultural Society, pages 90-91.

Sloderbeck, P. E. and K. V. Yeargan

1983. Comparison of *Nabis americoferus* and *Nabis roseipennis* (Hemiptera: Nabidae) as predators of the green cloverworm (Lepidoptera: Noctuidae). Environmental Entomology, 12: 161-165.

Slosson, A. T.

1898. Additional list of insects taken in alpine region of Mt. Washington. Entomological News, 9: 251-253.

1900. Additional list of insects taken in alpine region of Mt. Washington. Entomological News, 11: 319-323.

1901. On a Florida beach. Entomological News, 12: 10-12.

1902. Additional list of insects taken in alpine region of Mt. Washington. Entomological News, 13: 4-8.

1906. Additional list of insects taken in alpine region of Mt. Washington. Entomological News, 17: 323-326.

Smith, C. L. and J. T. Polhemus

1978. The Veliidae (Heteroptera) of America north of Mexico - keys and checklist. Proceedings of the Entomological Society of Washington, 80: 56-68.

Smith, E. C.

1855. The chinch bug. Cultivator, (3)3: 237-238.

Smith, J. B.

1890. Catalogue of insects found in New Jersey. Pages 1-486. *In* Geological Survey of New Jersey. Final report of the State Geologist. Volume II. Part 2. Zoology. 486 pages. [Heteroptera, pages 416-436].

1900. Insects of New Jersey. A List of the Species Occurring in New Jersey, with Notes on Those of Economic Importance. Supplement to the Twenty-Seventh Annual Report of the State Board of Agriculture (1899). 755 pages. [Heteroptera, pages 115-145].

1910. A report on the insects of New Jersey. Report of the New Jersey State Museum, 1909. 888 pages. [Heteroptera, pages 131-170].

Smith, L. W., Jr. and W. R. Enns

1969. Insects associated with oxidation lagoons in Missouri. Journal of the Kansas Entomological Society, 42: 409-412.

Smith, M. T., G. Wilde, and T. Mize

1981. Chinch bug: Damage and effects of host plant and photoperiod. Environmental Entomology, 10: 122-124.

Smith, R. C. and B. L. Shepherd

1937. The life history and control of the boxelder bug in Kansas. Transactions of the Kansas Academy of Science, 40: 143-159.

Smith, R. L.

1974. Life history of *Abedus herberti* in central Arizona (Hemiptera: Belostomatidae). Psyche, 81: 272-283.

1975. Surface molting behavior and its possible respiratory significance for a giant water bug *Abedus herberti* Hidalgo (Hemiptera: Belostomatidae). Pan-Pacific Entomologist, 51: 259-267.

1976. Brooding behavior of a male water bug *Belostoma flumineum* (Hemiptera: Belostomatidae). Journal of the Kansas Entomological Society, 49: 333-343.

1976. Male brooding behavior of the water bug *Abedus herberti* (Hemiptera: Belostomatidae). Annals of the Entomological Society of America, 69: 740-747.

Snodgrass, G. L., T. J. Henry, and W. P. Scott

1984. An annotated list of the Miridae (Heteroptera) found in the Yazoo-Mississippi Delta and associated areas in Arkansas and Louisiana. Proceedings of the Entomological Society of Washington, 86: 845-860.

Snow, F. H.

1904. Lists of Coleoptera, Lepidoptera, Diptera and Hemiptera collected in Arizona bythe entomological expeditions of the University of Kansas in 1902 and 1903. University of Kansas Science Bulletin, 2: 323-350.

1906. Some results of the University of Kansas entomological expeditions to Galveston and Brownsville, Tex., in 1904 and 1905. Transactions of the Kansas Academy of Science, 20: 136-154.

1906. Some results of the University of Kansas entomological expeditions to Arizona in 1904 and 1905. Transactions of the Kansas Academy of Science, 20: 155-181.

1907. Results of the entomological collecting expedition of the University of Kansas to Pima County, Arizona, in June and July, 1906. Transactions of the Kansas Academy of Science, 20: 140-164.

Solbreck, C.

1978. Migration, diapause, and direct development as alternative life histories in a seed bug, *Neacoryphus bicrucis*. Pages 195-217. *In* Dingle, H. (editor). Evolution of Insect Migration and Diapause. New York: Springer-Verlag.

1979. Induction of diapause in a migratory seed bug, *Neacoryphus bicrucis* (Say) (Heteroptera: Lygaeidae). Oecologia, 43: 41-49.

Solbreck, C. and I. Pehrson

1979. Relations between environment, migration and reproduction in a seed bug *Neacoryphus bicrucis* (Say) (Heteroptera: Lygaeidae). Oecologia, 43: 51-62.

Somes, M. P.

1916. Some insects of *Solanum carolinense* L. and their economic relations. Journal of Economic Entomology, 9: 39-44.

Southwood, T. R. E.

1953. The morphology and taxonomy of the genus *Orthotylus* Fieber (Hem., Miridae), with special reference to the British species. Transactions of the Royal Entomological Society of London, 104: 415-449.

Southwood, T. R. E. and D. Leston

1959. Land and Water Bugs of the British Isles. Fredrick Warne and Co. Ltd., London and New York. xi + 436 pages, 63 plates.

Spalding, J. B.

1979. The aeolian ecology of White Mountain Peak California: windblown insect fauna. Arctic and Alpine Research, 11: 83-94.

Spangler, P. J., R. C. Froeschner, and J. T. Polhemus

1985. Comments on a water strider, *Rheumatobates meinerti* from the Antilles and a checklist of the species of the genus (Hemiptera: Gerridae). Entomological News, 96: 196-200.

Spinola, M. M.

1837. Essai sur les genres d'Insects appartenants á l'ordre des Hémiptères Lin. ou Rhyngotes, Fab. et a la section Hétéroptères, Dufour. Chez Yves Gravier, Gènes. 383 pages, 15 fold-out tables [2nd edition, 1840].

1852. Orden VII. Hemipteros. *In* Gay, C. Historia Fisica y Politica de Chile. Maulde and Renou, Paris. Zoologia, 7: 113-320.

Sprague, I. B.

1956. The biology and morphology of *Hydrometra martini* Kirkaldy. University of Kansas Bulletin, 38: 579-693.

1967. Nymphs of the genus *Gerris* (Heteroptera: Gerridae) in New England. Annals of the Entomological Society of America, 60: 1038-1044.

Stahler, N.

1946. A new lacebug in Florida. Journal of Economic Entomology, 39: 545-546.

Stål, C.

1854a. Nya Hemiptera. Öfversigt af Kongliga Svenska Vetenskaps-Akademiens Förhandlingar, 11(8): 231-255.

1855. Nya Hemiptera. Öfversigt af Kongliga Svenska Vetenskaps-Akademiens Förhandlingar, 12(4): 181-192.

1858. Hemipterologiska bidrag. Öfversight af Kongliga Svenska Vetenskaps-Akademiens Förhandlingar, 15(9-10): 433-454.

1858. Beitrag zur Hemipteren-Fauna Sibiriens und des Russischen Nord-Amerika. Stettiner Entomologische Zeitung (Entomologische Zeitung Herausgegeben von dem Entomologischen Vereine zu Stettin), 19: 175-198, 1 plate.

1859. Hemiptera. Species novas descripsit. Konglika Svenska Fregattens Eugenies Resa Omkring Jorden. III. (Zoologi, Insekter). Pages 219-298, plates 3-4.

1859. Till kännedomen om Reduvini. Öfversight af Kongliga Svenska Vetenskaps-Akademiens Förhandlingar, 16(10): 175-204, 363-386.

1859. Till kännedomen om Coreida. Öfversigt af Kongliga Svenska Vetenskaps-Akademiens Förhandlingar, 16(10): 449-475.

1859. Monographie der Gattung *Conorhinus* und Verwandten. Berlinger Entomologische Zeitschrift, 3(1-3): 99-117.

1859. Synopsis specierum Spinigeri generis. Stettiner Entomologische Zeitung (Entomologische Zeitung Herausgegeben von dem Entomologischen Vereine zu Stettin), 20(10-12): 395-404.

1860-1862. Bidrag till Rio Janeiro-traktens Hemipter-fauna. Kongliga Svenska Vetenskaps-Akademiens Handlingar, 2(7): 1-84 (1860); 3(6): 1-75 (1862).

1861. Nova methodus familias quasdam Hemipterorum disponendi. Öfversight af Kongliga Svenska Vetenskaps-Akademiens Förhandlingar, 18(4): 195-212.

1861. Miscellanea hemipterologica. Stettiner Entomologische Zeitung (Entomologische Zeitung Herausgegeben von dem Entomologischen Vereine zu Stettin), 22(4-6): 129-153.

Stål, C. (continued)

1862. Hemiptera Mexicana enumeravit species-que novas descripsit. Stettin Entomologische Zeitung (Entomologische Zeitung Herausgegeben von dem Entomologischen Vereine zu Stettin), 23(1-3): 81-118; 23(4-6): 273-281; 23(7-9): 289-325; 23(10-12): 437-462.

1862. Synopsis Coreidum et Lygaeidum Suecia. Öfversigt af Kongliga Svenska Vetenskaps - Akademiens Förhandlingar, 19(3): 203-225.

1863. Beitrag zur Kenntniss der Fulgoriden. Stettiner Entomologische Zeitung (Entomologische Zeitung Herausgegeben von dem Entomologischen Vereine zu Stettin), 24: 230-251.

1863. Verzeichnifs der Mononychiden. Berliner Entomologische Zeitschrift, 7: 405-408.

1864. Hemiptera nonnulla nova vel minus cognita. Annales de la Société Entomologique de France, Series 4, 4: 47-68.

1865-1866. Hemiptera Africana. Norstedtiana, Stockholm. Volumes 1-4. [1865, 1: iv + 1-256; 1866, 2: 1-181; 1866, 3: 1-200; 1866, 4: i + 1-275, corrigenda, 1 plate]. [Dates for this work follow Mayr (1866, Verh. Zool.-Bot. Ges. Wien, 15: 429-430) and Bergroth (1919, Ent. Mitteil, 8: 190-191)].

1865. Hemiptera nova vel minus cognita. Annales de la Société Entomologique de France, Series 4, 5: 163-188.

1866. Bidrag till Reduviidernas kännedom. Öfversigt af Kongliga Vetenskaps-Akademiens Förhandlingar, 23(9): 235-302.

1867. Analecta hemipterologica. Berliner Entomologische Zeitschrift, 10: 151-172, 381-394.

1867. Bidrag till Hemipterernas Systematik. Öfversigt af Kongliga Svenska Vetenskaps-Akademiens Förhandlingar, 24(7): 491-560.

1868. Synopsis Hydrobatidum Sueciae. Öfversigt af Kongliga Svenska Vetenskaps-Akademiens Förhandlingar, 25(6): 395-398.

1868. Hemiptera Fabriciana. Fabricianska Hemipterarter, efter de i Köpenhamn och Kiel förvarade typexemplaren granskade och beskrifne. 1. Kongliga Svenska Vetenskaps-Akademien Handlingar, 7(11): 1-148.

1869. Analecta hemipterologica. Berliner Entomologische Zeitschrift, 13: 225-242.

1870-1876. Enumeratio Hemipterorum: Bidrag till en företeckning öfver alla hittills kända Hemiptera, jemte systematiska meddelanden. Parts 1-5. Kongliga Svenska Vetenskaps-Akademiens Handlingar, 1870, part 1, 9(1): 1-232; 1872, part 2, 10(4): 1-159; 1873, part 3, 11(2): 1-163; 1874, part 4, 12(1): 1-186; 1876, part 5, 14(4): 1-162.

1872. Genera Lygaeidarum Europae disposuit. Öfversigt af Kongliga Svenska Vetenskaps-Akademiens Förhandlingar, 29(7): 37-62.

1872. Genera Pentatomidarum Europae disposuit. Öfversigt af Kongliga Vetenskaps - Akademiens Förhandlingar, 29(3): 31-40.

1872. Genera Coreidorum Europae disposuit. Öfversigt af Kongliga Svenska Vetenskaps - Akademiens Förhandlingar, 29(6): 49-58.

1874. Genera Tingitidarum Europae disposuit. Öfversigt af Kongliga Svenska Vetenskaps - Akademiens Förhandlingar, 31(3): 43-60.

Stam, P. A., D. F. Clower, J. B. Graves, and P. E. Schilling

1978. Effects of certain herbicides on some insects and spiders found in Louisiana cotton fields. Journal of Economic Entomology, 71: 477-180.

Stanger, N. W.

1942. New species of *Lygus* from California (Hemiptera, Miridae). University of California Publications in Entomology, 7: 161-168.

Starks, K. J., A. J. Casady, O. G. Merkle, and D. Boozaya-Angoon

1982. Chinch bug resistance in pearl millet. Journal of Economic Entomology, 75: 337-339.

Stearns, L. A.

1958. Transient insects in Delaware's apple and peach plantings. Journal of Economic Entomology, 51: 81-82.

Stein, J. P. E. F.

1857. Die Gattung *Prostemma* Laporte. Berliner Entomologische Zeitschrift, 1: 81-96.

Steinhaus, E. A., M. M. Batey, and C. L. Boerke

1956. Bacterial symbiotes from the caeca of certain Heteroptera. Hilgardia, 24: 495-518.

Stephens, J. F.

1829. The nomenclature of British insects being a compendious list of such species as are contained in the systematic catalogue of British insects. London. 68 columns.

1829. A systematical catalogue of British insects, being an attempt to arrange all the hirtherto discovered indigenous insects in accordance with their natural affinities: containing also the references to every English writer on entomology, and to the principal foreign authors; with all the published British genera to the present time. London. XXXIV + 416 + 338 pages.

Sterling, W. L. and D. A. Dean

1977. A bibliography of the cotton fleahopper *Pseudatomoscelis seriatus* (Reuter). Texas Agricultural Experiment Station, MP-1342. 28 pages.

Stern, V. M., R. van den Bosch, and H. T. Reynolds

1960. Effects of Dylox and other insecticides on entomophagous insects attacking field crop pests in California. Journal of Economic Entomology, 53: 67-72.

Stevenson, C.

1903. A new capsid. Canadian Entomologist, 35: 214.

Stewart, M. A.

1930. The insect visitants and inhabitants of *Melilotus alba*. Journal of the New York Entomological Society, 38: 43-46.

Steyskal, G. C.

1973. The grammar of names in Slater's Catalogue of Lygaeidae of the World (Heteroptera). Proceedings of the Entomological Society of Washington, 75: 276-279.

1973. The grammar of names in the Catalogue of the Miridae (Heteroptera) of the World by Carvalho, 1957-1960. Studia Entomologica, 16: 203-208.

Stichel, W.

1925. Die systematische Stellung der Gattung *Myodocha* Latreille (Hem., Het.) (Versuch der Lösung einer schwierigen Nomenklaturfrage). Konowia, 4: 392-398.

1926. Die Gattung *Microtomus* Illiger (Hem., Het., Reduv.). Deutsche Entomologische Zeitschrift, pages 179-190, plate 1.

1925-1938. Illustrierte Bestimmungstabellen der Deutschen Wanzen (Hemiptera-Heteroptera). Verlag W. Stichel, Berlin-Hermsdorf. Parts 1-14. 1925, 1-3: 1-90; 1926, 4: 91-120; 1927, 5: 121-146; 1930, 6-7: 147-210; 1930, 8: 211-242; 1933, 9: 243-274; 1934, 10: 275-306; 1935, 11-12: 307-362; 1937, 13: 363-394; 1938, 14: 395-499.

1958. Illustrierte Bestimmungstabellen der Wanzen. II. Europa (Hemiptera-Heteroptera Europae). Martin-Luther, Berlin-Hermsdorf. 2(24) 737-768.

Stinner, B. R.

1975. Observations on predacious Miridae. Proceedings of the Pennsylvania Academy of Science, 49: 101-102.

Stock, M. W. and J. D. Lattin

1976. Biology of intertidal *Saldula palustris* (Douglas) on the Oregon coast (Heteroptera: Saldidae). Journal of the Kansas Entomological Society, 49: 313-326.

Stoltz, R. L. and V. M. Stern

1978. Cotton arthropod food chain disruptions by pesticides in the San Joaquin Valley California. Environmental Entomology, 7: 703-707.

Stone, J. D. and J. N. Fries

1986. Insect fauna of cultivated Guayule, *Parthenium argentatum* Gray (Campanulatae: Compositae). Journal of the Kansas Entomological Society, 59: 49-58.

Stone, T. B., H. N. Pitre, and A. C. Thompson

1984. Relationships of cotton phenology, leaf soluble protein, extrafloral nectar carbohydrate and fattyacid concentrations with populations of five predator species. Journal of the Georgia Entomological Society, 19: 204-212.

Stonedahl, G. M.

1983. New records for palearctic *Phytocoris* in western North America (Hemiptera: Miridae). Proceedings of the Entomological Society of Washington, 85: 463-471.

1984. Two new conifer-inhabiting *Phytocoris* from western North America (Hemiptera: Miridae). Pan-Pacific Entomologist, 60: 47-52.

1984. A new species of *Sigara* from western Oregon and Washington (Hemiptera: Corixidae). Journal of the New York Entomological Society, 92: 42-47.

1986. *Phytocoris adenostomae*, a new mirine plant bug (Heteroptera: Miridae) from southern California. Journal of the New York Entomological Society, 93: 1271-1274.

Stonedahl, G. M. and J. D. Lattin
1982. The Gerridae or water striders of Oregon and Washington (Hemiptera: Heteroptera). Oregon State University Agricultural Experiment Station Technical Bulletin, 144: 1-36.
1986. The Corixidae of Oregon and Washington (Hemiptera: Heteroptera). Oregon State University Agricultural Experiment Station Technical Bulletin, 150: ii-iv + 1-84.

Stonedahl, G. M. and R. T. Schuh
1986. *Squamocoris* Knight and *Ramentomiris*, new genus (Heteroptera: Miridae: Orthotylinae). A cladistic analysis and description of seven new species from Mexico and the Western United States. American Museum Novitates, 2852: 1-26.

Stonedahl, G. M. and M. D. Schwartz
1986. Revision of the plant bug genus *Pseudopsallus* Van Duzee (Heteroptera: Miridae). American Museum Novitates, 2842: 1-58.

Stoner, A.
1970. Plant feeding by a predaceous insect, *Geocoris punctipes.* Journal of Economic Entomology, 63: 1911-1915.
1972. Plant feeding by *Nabis*, a predaceous genus. Environmental Entomology, 1: 557-558.

Stoner, D.
1917. A new species of *Apateticus* from Louisiana (Hem., Het.). Entomological News, 28: 462-463.
1926. Pentatomoidea from western Canada. Canadian Entomologist, 58: 28-30.

Stout, R. J.
1978. Migration of the aquatic hemipteran *Limnocoris insularis* (Naucoridae) in a tropical lowland stream (Costa Rica, Central America). Brenesia, 14-15: 1-11.
1981. How abiotic factors affect the distribution of two species of tropical predaceous aquatic bugs (Family: Naucoridae). Ecology, 62: 1170-1178.

Streams, F. A. and S. Newfield
1972. Spatial and temporal overlap among breeding populations of New England *Notonecta.* University of Connecticut Occasional Papers, Biological Science Series, 2: 139-157.

Streu, H. T.
1973. The turfgrass ecosystem: Impact of pesticides. Bulletin of the Entomological Society of America, 19: 89-91.

Streu, H. T. and C. Cruz
1972. Control of the hairy chinch bug in turf grass in the northeast with Dursban insecticide. Down to Earth, 28: 1-4, cover photo.

Strickland, E. H.
1953. An annotated list of the Hemiptera (s.l.) of Alberta. Canadian Entomologist, 85: 193-214.

Stringfellow, T. L.
1967. Studies on turfgrass insect control in south Florida. Proceedings of the Florida State Horticultural Society, 80: 486-491.
1968. Studies on turfgrass insect control in south Florida. Proceedings of the Florida State Horticultural Society, 81: 447-454.
1969. Developments in Florida turfgrass insect control, 1969. Proceedings of the Florida Turf-Grass Management Conference, 17: 94-100.

Stroble, J.
1971. Turfgrass. Proceedings of the Florida Turf-Grass Management Conference, 19: 19-29.

Strong, F. E., J. A. Sheldahl, P. R. Hughes, and E. M. K. Hussein
1970. Reproductive biology of *Lygus hesperus* Knight. Hilgardia, 40: 105-147.

Stuart, J., G. Wilde, and J. H. Hatchett
1985. Chinch-bug (Heteroptera: Lygaeidae) reproduction, development, and feeding preference on various wheat cultivars and genetics sources. Environmental Entomology, 14: 539-543.

Štys, P.
1970. On the morphology and classification of the family Dipsocoridae s. lat., with particular reference to the genus *Hyp-sipteryx* Drake (Heteroptera). Acta Entomologica Bohemoslovaca, 67: 21-46.
1975. Suprageneric nomenclature of Anthocoridae (Heteroptera). Acta Universitatis Carolinae, Biologica, 4: 159-162.
1976. *Velohebria antennalis* gen. n., sp. n.-- a primitive terrestrial Microveliine from New Guinea, and a revised classification of the family Veliidae (Heteroptera). Acta Entomologica Bohemoslovaca, 73: 388-403.

1982. A new Oriental genus of Ceratocombidae and higher classification of the family (Heteroptera). Acta Entomologica Bohemoslovaca, 79: 354-376.

Štys, P. and I. Kerzhner
1975. The rank and nomenclature of higher taxa in recent Heteroptera. Acta Entomologica Bohemoslovaca, 72: 65-79.

Summers, H. E.
1891. The true bugs, or Heteroptera, of Tennessee. Bulletin of the Agricultural Experiment Station, University of Tennessee, 4: 75-94.

Swederus, N. S.
1787-1788. Et nytt genus, och femtio nya species af Insekter beskrifne. Vetenskaps Akademien nya Handlingar, 1787, 8: 181-201, 276-290; 1788, 8: 77-192, 266-279.

Sweet, M. H.
1960. The seed bugs: A contribution to the feeding habits of the Lygaeidae (Hemiptera: Heteroptera). Annals of the Entomological Society of America, 53: 317-321.
1963. A new species of *Ligyrocoris* Stål with a key to the northeastern species (Hemiptera: Lygaeidae). Psyche, 70: 17-21.
1964a. The biology and ecology of the Rhyparochrominae of New England (Heteroptera: Lygaeidae). Part I. Entomologica Americana, (New Series) 43: 1-124.
1964b. The biology and ecology of the Rhyparochrominae of New England (Heteroptera: Lygaeidae). Part II. Entomologica Americana, (New Series) 44: 1-201.
1967. The tribal classification of the Rhyparochrominae (Heteroptera: Lygae-idae). Annals of the Entomological Society of America, 60: 208-226.
1977a. Elevation of the seedbug *Eremocoris borealis* (Dallas) from synonymy with *Eremocoris ferus* (Say) (Hemiptera: Lygaeidae). Entomological News, 88: 169-176.
1977b. The systematic position of the seedbug genus *Neosuris* Barber, 1924 (Hemiptera: Lygaeidae) with a discussion of the zoogeographical significance of the genus and notes on the distribution and ecology of *N. castanea* (Barber, 1911) and *N. fulgida* (Barber, 1918). Journal of the Kansas Entomological Society, 50: 569-574.
1979. On the original feeding habits of the Hemiptera (Insecta). Annals of the Entomological Society of America, 72: 575-579.
1980. *Peggichisme* Kirkaldy, 1904 (Hemiptera Heteroptera: Lygaeidae): Proposal to designate a type species by use of the plenary powers. Z.N.(S.) 2197. Bulletin of Zoological Nomenclature, 37: 37-39.
1981. The external morphology of the pregenital abdomen and its evolutionary significance in the order Hemiptera (Insecta). Rostria, 33 (Supplement): 41-51.
1986. *Ligyrocoris barberi* (Heteroptera: Lygaeidae), a new seedbug from the southeastern United States with a discussion of its ecology, life cycle, and reproductive isolation. Journal of the New York Entomological Society, 94: 281-290.

Sweet, M. H. and J. A. Slater
1961. A generic key to the nymphs of North American Lygaeidae (Hemiptera-Heteroptera). Annals of the Entomological Society of America, 54: 333-340.
1974. The taxonomic status of *Exptochiomera nana* Barber (Hemiptera: Lygaeidae). Proceedings of the Entomological Society of Washington, 76: 82-85.

Swezey, O. H.
1945. Insects associated with orchids. Proceedings of the Hawaiian Entomological Society, 12: 343-403.

Taft, H. M., H. R. Agee, A. R. Hopkins, and W. James
1972. Field evaluation of artificial light for control of bollworms on cotton. Environmental Entomology, 1: 295-300.

Takahashi, R.
1923. Observations on the Ochteridae. Bulletin of the Brooklyn Entomological Society, 18: 67-68.

Tamaki, G.
1972. The biology of *Geocoris bullatus* inhabiting orchard floors and its impact on *Myzus persicae* on peaches. Environmental Entomology, 1: 559-565.
1981. Biological control of potato pests. Pages 178-192. *In* Lashomb, J. H. and R. Casagrande (editors). Advances in Potato Pest Management. Hutchinson Ross, Stroudsburg, Pennsylvania.

Tamaki, G., B. Annis, and M. Weiss
1981. Response of natural enemies to the green peach aphid in different plant cultures. Environmental Entomology, 10: 375-378.

Tamaki, G., L. Fox, and P. Featherston
1982. Laboratory biology of the dusky sap beetle and field interaction with the corn earworm in ears of sweet corn. Journal of the Entomological Society of British Columbia, 79: 3-8.

Tamaki, G. and G. E. Long
1978. Predator complex of the green peach aphid on sugarbeets: Expansion of the predator power and efficacy mode. Environmental Entomology, 7: 835-842.

Tamaki, G. and D. Olsen
1977. Feeding potential of predators of *Myzus persicae.* Journal of the Entomological Society of British Columbia, 74: 23-26.

Tamaki, G., D. P. Olsen, and R. K. Gupta
1978. Laboratory evaluation of *Geocoris bullatus* and *Nabis alternatus* as predators of *Lygus.* Journal of the Entomological Society of British Columbia, 75: 35-37.

Tamaki, G. and R. E. Weeks
1972a. Efficiency of three predators, *Geocoris bullatus, Nabis americoferus,* and *Coccinella transversogutata,* used alone or in combination against three insect prey species, *Myzus persicae, Ceramica picta,* and *Mamestra configurata,* in a greenhouse study. Environmental Entomology, 1: 258-263.
1972b. Biology and ecology of two predators, *Geocoris pallens* Stål and *G. bullatus* (Say). United States Department of Agriculture, Agricultural Research Service Technical Bulletin, 2(1466). i-ii + 1-46 pages.
1973. The impact of predators on populations of green peach aphids on field-grown sugarbeets. Environmental Entomology, 2: 345-349.

Tamaki, G., M. A. Weiss, and G. E. Long
1981. Evaluation of plant density and temperature in predator-prey interactions in field cages. Environmental Entomology, 10: 716-720.

Tamanini, L.
1958. Revisione del genere *Carpocoris* Klt. con speciale riguardo alle specie italiane (Hemiptera Heter., Pentatomidae). Memorie del Museo Civico di Storia Naturale, Verona, 6: 333-388.

Tanada, Y.
1959. Microbial control of insect pests. Annual Review of Entomology, 4: 277-302.

Tanada, Y. and F. G. Holdaway
1954. Feeding habits of the tomato bug, *Cyrtopeltis (Engytatus) modestus* (Distant) with special reference to the feeding lesion on tomato. University of Hawaii Agricultural Experiment Station Technical Bulletin, 24: 1-40.

Tappan, W. B.
1967. Chemical-control and physical autecology of insects attacking cigar-wrapper tobacco. Florida Agricultural Experiment Station Annual Report, 1967: 156.
1970. *Nysius raphanus* attacking tobacco in Florida and Georgia. Journal of Economic Entomology, 53: 658-660.

Tate, H. D.
1933. Notes on potato insects in Iowa. Journal of Economic Entomology, 26: 1173.

Tejada, L. O. and J. Garza Blanc
1976. Estudio de la dinamica de pablaciones de insectos entomófagos asociados con la palomilla del *Homoeosoma electellum* (Hulst), en Apodaca, N.L. Folia Entomológica Méxicana, 36: 26-27.

Thead, L. G., H. N. Pitre, and T. F. Kellog
1981. Radioactive 32P techniques used in determining predator effectiveness. Journal of the Mississippi Academy of Sciences, 26: 23.
1985. Feeding behavior of adult *Geocoris punctipes* (Say) (Hemiptera: Lygaeidae) on nectaried and nectariless cotton. Environmental Entomology, 14: 134-137.

Thomas, C.
1865. Insects injurious to vegetation in Illinois. Transactions of the Illinois State Agricultural Society, 5: 401-468.

Thomas, D. B., Jr.
1974. The genus *Codophila* Mulsant in North America (Hemiptera: Pentatomidae). Pan-Pacific Entomologist, 50: 441-442.
1983. Taxonomic status of the genera *Chlorochroa* Stål, *Rhytidilomia* Stål, *Liodermion* Kirkaldy, and *Pitedia* Reuter, and their included species (Hemiptera: Pentatomidae). Annals of the Entomological Society of America, 76: 215-224.
1985. Revision of the genus *Chlorocoris* Spinola (Hemiptera: Pentatomidae). Annals of the Entomological Society of America, 78: 674-690.

Thomas, D. B., Jr. and F. G. Werner
1981. Grass feeding insects of the western

ranges: An annotated checklist. University of Arizona Agricultural Experiment Station Technical Bulletin, 243: 1-50.

Thomas, D. B., Jr. and T. R. Yonke
1981. A review of the nearctic species of the genus *Banasa* Stål (Hemiptera: Pentatomidae). Journal of the Kansas Entomological Society, 54: 233-248.
1985. Cladistic analysis of zoogeography and polyploid evolution in the stinkbug genus *Banasa* Stal (Hemiptera: Pentatomidae). Annals of the Entomological Society of America, 78: 855-862.

Thompson, L. S.
1964. Insect survey of forage crops in Prince Edward Island. Journal of Economic Entomology, 57: 961-962.

Thomson, C. G.
1869. Öfversigt of Sveriges Coriser. Opuscula Entomologica, 1: 26-40.
1870. Öfversigt af de i Sverige funna arter af slägtet *Lygaeus* Fallén. Opuscula Entomologica, 1: 180-202.
1871. Öfversigt af Sveriges *Salda-arter.* Opuscula Entomologica, 4: 403-409.
1871. Öfversigt af de i Sverige fauna arter af gruppen Capsina. Opuscula Entomologica, 4: 410-452.

Thorpe, K. W. and B. J. Harrington.
1979. Observations of seed-bug (Hemiptera: Lygaeidae) parasitism by a species of *Catharosia* (Diptera: Tachinidae). Psyche, 86: 399-405.
1981. Sound production and courtship behavior in the seed bug *Ligyrocoris diffusus.* Annals of the Entomological Society of America, 74: 369-373.

Thunberg, C. P.
1781-1791. Dissertatio entomologica novas insectorum species, sistens, cujus partem secundum, cons. exper. facult. med. upsal., publice ventilandam exhibent. J. Edman, Upsaliae. 6 parts. [1781, 1: 1-28, plate 1; 1783, 2: 29-52, plate 2; 1784, 3: 53-68, plate 3; 1784, 4: 69-84, plate 4; 1789, 5: 84-106, plate 5; 1791, 6: 107-130, plate 6].

Tinker, M. E.
1952. The seasonal behavior and ecology of the boxelder bug *Leptocoris trivittatus* in Minnesota. Ecology, 33: 407-414.

Todd, E. L.
1954. New species of *Nerthra* from California (Hemiptera: Gelastocoridae). Pan-Pacific Entomologist, 30: 113-117.
1955. A taxonomic revision of the family Gelastocoridae (Hemiptera). University of Kansas Science Bulletin, 37: 277-475.
1959. The Gelastocoridae of Melanesia (Hemiptera). Nova Guinea, New Series, 10: 61-94.

Tonks, N. V.
1953. Annotated list of insects and mites collected on brambles in the lower Fraser Valley, British Columbia, 1951. Proceedings of the Entomological Society of British Columbia, 49: 27-28.

Torre-Bueno, J. R. de la
1903. A preliminary list of the Pentatomidae within fifty miles of New York. Journal of the New York Entomological Society, 11: 128-129.
1903. Brief notes toward the life history of *Pelocoris femorata* Pal. B. with a few remarks on habits. Journal of the New York Entomological Society, 11: 166-173.
1904. A palearctic *Notonecta.* Entomological News, 15: 220.
1904. [Notes in minutes of meeting]. Journal of the New York Entomological Society, 12: 62.
1904-1905. A list of certain families of Hemiptera occurring within seventy miles of New York. Journal of the New York Entomological Society, 12: 251-253 (1904); 13: 29-47 (1905).
1905. The genus *Notonecta* in America north of Mexico. Journal of the New York Entomological Society, 13: 143-167, 1 plate.
1905. The tonal apparatus of *Ranatra quadridentata,* Stal. Canadian Entomologist, 37: 85-87.
1905. *Hydrometra australis,* Say. Canadian Entomologist, 37: 264.
1906. Life histories of North-American waterbugs. II. Life history of *Ranatra quadridentata* Stal. Canadian Entomologist, 38: 242-252.
1906. On some aquatic Hemiptera from Costa Rica, Central America. Entomological News, 17: 54-57.
1907. Two undescribed water bugs from the United States. Canadian Entomologist, 39: 225-228.
1907. On some heteropterous Hemiptera from N. Carolina. Entomological News, 18: 433-443.
1908. Hemiptera Heteroptera of Westchester County, N.Y. Journal of the New York

Torre-Bueno, J. R. de la (continued)
Entomological Society, 16: 223-238.
1910. Westchester Heteroptera.--II. Additions, corrections and new records. Journal of the New York Entomological Society, 18: 22-33.
1910. Life-histories of North American water-bugs.--III. *Microvelia americanus* Uhler. Canadian Entomologist, 42-176-186.
1911. On *Halobatopsis beginii* Ashm. Canadian Entomologist, 43: 226-228.
1911. The gerrids of the Atlantic states (subfamily Gerrinae). Transactions of the American Entomological Society, 37: 243-252.
1912. Three days in the pines of Yaphank. Records of captures of Hemiptera Heteroptera. Canadian Entomologist, 44: 209-213.
1912. A new *Corizus* from northeastern United States (Hemip., Coreidae). Entomological News, 23: 217-219.
1912. Records of Heteroptera from Brownsville, Texas (Hemip.). Entomological News, 23: 120-122.
1913. Some heteropterous Hemiptera from Southern Pines, N.C. Canadian Entomologist, 45: 57-60.
1913. Some new and little-known Heteroptera from the western United States. Entomological News, 24: 20-23.
1914. European Heteroptera alleged to occur in the United States. Entomological News, 25: 230.
1915. Heteroptera in beach drift. Entomological News, 26: 274-279.
1916. [Notes in] Weiss, H. B. Additions to insects of New Jersey, No. 3. Entomological News, 27: 9-13.
1916. A new tingid from New York State. Bulletin of the Brooklyn Entomological Society, 11: 39-40.
1916. The Veliinae of the Atlantic states. Bulletin of the Brooklyn Entomological Society, 11: 52-61.
1916. Aquatic Hemiptera. A study in the relation of structure to environment. Annals of the Entomological Society of America, 9: 353-365.
1917. Life history of the northern *Microvelia*--*Microvelia borealis* Bueno (Hem., Het.). Entomological News, 28: 354-359, 1 plate.
1917. A new species of tingid from New York. Bulletin of the Brooklyn Entomological Society, 12: 19-20.
1918. Some Heteroptera from the Parry Sound District, Ont. Canadian Entomologist, 50: 24-25.
1919. Virginia Heteroptera. Bulletin of the Brooklyn Entomological Society, 14: 124-125.
1920. On *Rhamphocorixa balanodis* Abbott. Bulletin of the Brooklyn Entomological Society, 15: 88.
1921. New records of aquatic Hemiptera for the United States, with description of new species. Entomological News, 32: 273-276.
1922. Distributional records of aquatic Hemiptera. Bulletin of the Brooklyn Entomological Society, 17: 120-121.
1923. Family Nerthridae. Pages 392-396. *In* Britton, W. E. (editor). Guide to the Insects of Connecticut. Part IV. The Hemiptera or sucking insects of Connecticut. Connecticut State Geological and Natural History Survey Bulletin, 34: 1-807, 20 plates.
1923. Family Saldidae (Acanthiidae). Pages 408-416. *In* Britton, W. E. (editor). Guide to the insects of Connecticut. Part IV. The Hemiptera of sucking insects of Connecticut. Connecticut Geological and Natural History Survey Bulletin, 34: 1-807, 20 plates. [See also: Addenda et Corrigenda, 4 unnumbered pages issued separately by the Connecticut Geological Survey, apparently in 1924; and Britton, W. E., 1924, Additions & corrections to the Hemiptera of Connecticut, Entomological News, 35: 367].
1923. A saldid genus new to the United States and a new species, with notes on other water bugs from the Adirondacks. Bulletin of the Brooklyn Entomological Society, 18: 149-154.
1924. A preliminary survey of the species of *Microvelia* Westwood (Veliidae, Heteroptera) of the Western World, with descriptions of a new species from the southern United States. Bulletin of the Brooklyn Entomological Society 19: 186-194.
1924. The nearctic Rhagoveliae (Heteroptera; Veliidae). Transactions of the American Entomological Society, 50: 243-252.
1924. Three Canadian Acanthiidae (Saldidae, Heteroptera). Canadian Entomologist, 56: 296-300.

1925. Methods of collecting, mounting and preserving Hemiptera. Canadian Entomologist, 57: 6-10, 27-32, 53-57.

1926. Some remarks, Al Vuelo, on tingitid names. Bulletin of the Brooklyn Entomological Society, 21: 116-117.

1926. The family Hydrometridae in the Western Hemisphere. Entomologica Americana, (New Series) 7: 83-128.

1929. [Note in minutes of meeting on distribution of 3 gerrid species]. Bulletin of the Brooklyn Entomological Society, 24: 193.

1929. On some New England Heteroptera. Bulletin of the Brooklyn Entomological Society, 24: 310-313.

1929. [Review of] Flowers and Insects by Charles Robertson. Bulletin of the Brooklyn Entomological Society, 24: 335-337.

1930. Records of Anthocoridae, particularly from New York. Bulletin of the Brooklyn Entomological Society, 25: 11-20.

1931. Heteroptera collected by G. P. Engelhardt in the South and West--II. Bulletin of the Brooklyn Entomological Society, 26: 135-139.

1933. On Floridian Heteroptera, with new state records from the Keys. Bulletin of the Brooklyn Entomological Society, 28: 28-31.

1939. Two new United States records of Heteroptera. Bulletin of the Brooklyn Entomological Society, 34: 214.

1939. Remarks on the subgenus *Tivarbus* Stål of the genus *Hyalymenus* A. & S. with descriptions of five new species (Hemiptera, Alydidae). Bulletin of the Brooklyn Entomological Society, 34: 177-197.

1939. A synopsis of the Hemiptera-Heteroptera of America north of Mexico. Part I. Families Scutelleridae, Cydnidae, Pentatomidae, Aradidae, Dysodiidae and Termitaphididae. Entomologica Americana (New Series), 19: 141-304.

1939. *Andrallus spinidens* Fabricius in the U.S. Bulletin of the Brooklyn Entomological Society, 34: 118.

1940. Biological notes on Arizona Heteroptera. Bulletin of the Brooklyn Entomological Society, 35: 157.

1940. Food-plant of a coreid bug. Bulletin of the Brooklyn Entomological Society, 35: 45.

1940. *Thasus gigas* Burmeister, a correction. Bulletin of the Brooklyn Entomological Society, 35: 102.

1940. *Tollius vanduzeei* n. sp., with notes on the genera *Tollius* Stål and *Stachyocnemis* Stål (Heteroptera, Alydidae). Bulletin of the Brooklyn Entomological Society, 35: 159-161.

1940. Synonymic notes on *Dysdercus* A. & S. (Hemip.). Bulletin of the Brooklyn Entomological Society, 35: 12.

1941. Help notes toward a revision of the genus *Harmostes* Burm. 1835. Bulletin of the Brooklyn Entomological Society, 36: 82-92.

1941. Remarks on the genus *Corizus* of authors, not of Fallen. Annals of the Entomological Society of America, 34: 284-288.

1941. A synopsis of the Hemiptera-Heteroptera of America north of Mexico. Part II. Families Coreidae, Alydidae, Corizidae, Neididae, Pyrrhocoridae and Thaumastotheriidae. Entomologica Americana (New Series), 21: 41-122.

1942. Notes on Coreidae. Bulletin of the Brooklyn Entomological Society, 37: 180.

1942. Notes on distribution of Heteroptera. Bulletin of the Brooklyn Entomological Society, 37: 183-185.

1942. Notes on *Arhaphe cicindeloides* Walker and *Japetus mimeticus* Barber. Bulletin of the Brooklyn Entomological Society, 37: 68-69.

1944. *Plinthisus martini* Van Duzee 1921 (Hemiptera, Lygaeidae), a synonym. Bulletin of the Brooklyn Entomological Society, 39: 170.

1944. New records of *Oncopeltus* (Hemiptera, Lygaeidae) and a new species from the U.S. Bulletin of the Brooklyn Entomological Society, 39: 135-136.

1946a. A synopsis of the Hemiptera-Heteroptera of America north of Mexico. Part III. Family XI - Lygaeidae. Entomologica Americana (New Series), 26: 1-141.

1946b. A North American *Dieuches* (Heteroptera, Lygaeidae, Beosini). Bulletin of the Brooklyn Entomological Society, 41: 126-128.

Torre-Bueno, J. R. de la and C. S. Brimley

1907. On some heteropterous Hemiptera from N. Carolina. Entomological News, 18: 433-443.

Torre-Bueno, J. R. de la, and G. P. Engelhardt

1910. Some Heteroptera from Virginia and North Carolina. Canadian Entomologist, 42: 147-151.

Torre-Bueno, J. R. de la and R. F. Hussey
1923. A new North American *Notonecta*. Bulletin of the Brooklyn Entomological Society, 18: 104-107.

Tostowaryk, W.
1971. Life history and behavior of *Podisus modestus* (Hemiptera: Pentatomidae) in boreal forest in Quebec. Canadian Entomologist, 103: 662-664.

Townsend, C. H. T.
1891. Hemiptera collected in southern Michigan. Proceedings of the Entomological Society of Washington, 2: 52-56.
1892. Biologic notes on New Mexico insects. Canadian Entomologist, 24: 193-200.
1894a. Notes on some south-western Hemiptera. Canadian Entomologist, 26: 312-316.

Trichilo, P. J. and T. F. Leigh
1986. The impact of cotton plant resistance on spider mites and their natural enemies. Hilgardia, 54: 1-20.

Trumbo, S. T. and N. J. Fashing
1980. Aggregation behavior in the large milkweed bug, *Oncopeltus fasciatus*, and small milkweed bug, *Lygaeus kalmii* (Hemiptera: Lygaeidae). Virginia Journal of Science, 31: 103.

Truxal, F. S.
1949. A study of the genus *Martarega* (Hemiptera, Notonectidae). Journal of the Kansas Entomological Society, 22: 1-24.
1953. A revision of the genus *Buenoa* (Hemiptera, Notonectidae). University of Kansas Science Bulletin, 35: 1351-1523.
1979. Family Notonectidae/backswimmers. Pages 139-147. *In* Menke, A. S. (editor). The Semiaquatic and Aquatic Hemiptera of California (Heteroptera: Hemiptera). Bulletin of the California Insect Survey, 21: 1-166.

Tucker, E. S.
1907. Some results of desultory collecting of insects in Kansas and Colorado. Kansas University Science Bulletin, 4: 51-112. [List of Hemiptera-Heteroptera, pages 52-62].
1907. Contributions towards a catalogue of the insects of Kansas. Results of personal collecting. Transactions of the Kansas Academy of Science, 20: 190-193 (1906).
1911. Random notes on entomological field work. Canadian Entomologist, 43: 22-32.

Tugwell, P., E. P. Rouse, and R. G. Thompson
1973. Insects in soybeans and a weed host (*Desmodium* sp.). Arkansas Agricultural Experiment Station Report, 214: 3-18.

Turnipseed, S. G.
1973. Management of insect pests of soybeans. Proceedings Tall Timbers Conference on Ecological Animal Control by Habitat Management, 4: 189-203.

Turnipseed, S. G., J. W. Todd, and W. V. Campbell
1975. Field activity of selected foliar insecticides against geocorids, nabids, and spiders on soybeans. Journal of the Georgia Entomological Society, 10: 272-277.

Udine, E. J.
1962. A lace bug (*Dictyla echii* Schrank). United States Department of Agriculture, Cooperative Economic Insect Report, 12: 778.

Ueshima, N.
1963. Chromosome study of *Thyanta pallidovirens* (Stål) in relation to taxonomy (Hemiptera: Pentatomidae). Pan-Pacific Entomologist, 39: 149-154.
1963. Chromosome behavior of the *Cimex pilosellus* complex (Cimicidae: Hemiptera). Chromosoma, 14: 511-521.
1979. Hemiptera II. Heteroptera, in Animal Cytogenetics 3, Insecta 6. Gebrüder Bornträger, Berlin. ix + 118 pages.

Ueshima, N. and P. D. Ashlock
1980. Cytotaxonomy of the Lygaeidae (Hemiptera-Heteroptera). University of Kansas Science Bulletin, 51: 717-801.

Uhler, P. R.
1860. Hemiptera of the North Pacific Exploring Expedition under Com'rs Rodgers and Ringgold. Proceedings of the Academy of Natural Science of Philadelphia, 12: 221-231.
1861. Descriptions of a few new species of Hemiptera and observations upon some already described. Proceedings of the Entomological Society of Philadelphia, 1: 21-24.
1861. Descriptions of four species of Hemiptera collected by the North-Western Boundary Survey. Proceedings of the Academy of Natural Sciences of Philadelphia, 13: 284-286.
1862. Hemiptera. Pages 11-12, 192-256, plate I. *In* Harris, T. W. (editor). A treatise on

some of the insects injurious to vegetation. 3rd edition. William White, Boston.

1863. Hemipterological contributions.--No. 1. Proceedings of the Entomological Society of Philadelphia, 2: 155-162.

1863. Hemipterological contributions.--No. 2. Proceedings of the Entomological Society of Philadelphia, 2: 361-366.

1868. Notices of the Hemiptera Heteroptera in the collection of the late T. W. Harris, M. D. Proceedings of the Boston Society of Natural History, 11: 365-446.

1870. [Description of *Podisus placidus*, page 203]. *In* Saunders, W. Notes and experiments on currant worms. American Entomologist and Botanist, 2: 200-203.

1871. [Descriptions of *Hygrotrechus robustus*, *Corixa decolor* and *Salda interstitialis* Say]. *In* Packard, A. S., Jr. On insects inhabiting salt water. No. 2. American Journal of Sciences and Arts, Series 3, 1: 100-110.

1871. A list of Hemiptera collected in eastern Colorado and northeastern New Mexico, by C. Thomas, during the expedition of 1869. Pages 471-472. *In* Hayden, F. V. Preliminary Report of the United States Geological Survey of Wyoming and Portions of Contiguous Territories, pages 471-472 (1870).

1871. Notices of some Heteroptera in the collection of Dr. T. W. Harris. Proceedings of the Boston Society of Natural History, 14: 93-109.

1872. Notices of the Hemiptera of the western territories of the United States, chiefly from the surveys of Dr. F. V. Hayden. *In* Hayden, F. V. Preliminary Report of the United States Geological Survey of Montana and Portions of Adjacent Territores, 5: 392-423 (1871).

1875. Report on the collections of Hemiptera made in portions of Nevada, Utah, California, Colorado, New Mexico, and Arizona. During the years 1871, 1873, and 1874. Report upon the United States Geological and Geographical Explorations and Surveys West of the One Hundreth Meridian, in Charge of First Lieut. Geo. M. Wheeler, Corps of Engineers, U.S. Army, 5: 829-842, plate XLII.

1876. List of Hemiptera of the region west of the Mississippi River, including those collected during the Hayden explorations of 1873. Bulletin of the United States Geological and Geographical Survey of the Territories, 1: 267-361, plates 1-21.

1877. Report on the insects collected by P. R. Uhler during the explorations of 1875, including monographs of the families Cydnidae and Saldae, and the Hemiptera collected by A. S. Packard, Jr., M. D. Bulletin of the United States Geological and Geographic Survey of the Territories, 3: 355-475, 765-801, plates 27-28.

1877. Report upon the Hemiptera collected during the years 1874 and 1875, by Mr. P. R. Uhler. *In* Wheeler, G. M. Annual Report upon the Geographical Surveys (West of the One-hundredth Meridian) of the Chief Engineer for 1877. Appendix NN: 1322-1334.

1878. Notices of the Hemiptera Heteroptera in the collection of the late T. W. Harris, M.D. Proceedings of the Boston Society of Natural History, 19: 365-446.

1878. On the Hemiptera collected by Dr. Elliot Coues, U. S. A., in Dakota and Montana, during 1873-74. Bulletin of the United States Geological and Georgraphical Survey of the Territories, 4: 503-512.

1884. Order VI.--Hemiptera. Volume II. Pages 204-296. *In* Kingsley, J. S. (editor). The Standard Natural History. S. E. Cassino & Co., Boston, Massachusetts.

1886. Check-list of the Hemiptera Heteroptera of North America. Brooklyn Entomological Society, Brooklyn, New York. iv + 32 pages.

1886. A new noxious capsid. Canadian Entomologist, 18: 208.

1887. Observations on some North American Capsidae. Entomologica Americana, 2: 229-231 (1886).

1887. Observations on some Capsidae with descriptions of a few new species. Entomologica Americana, 3: 29-35.

1887. [*Lygus monachus* n.sp.]. Pages 63-64. *In* Murtfeldt, M. E. Notes from Missouri for the season of 1886. United States Department of Agriculture, Division of Entomology Bulletin, 13: 59-65.

1887. Partial list of Capsidae taken at Buffalo, N. Y. Canadian Entomologist, 19: 69-73.

1887. Observations on North American Capsidae with descriptions of new species. Entomologica Americana, 3: 67-72.

1887. Observations on Capsidae with descriptions of new species. Entomologica Americana, 3: 149-151.

Uhler, P. R. (continued)

1889. Observations upon the Heteroptera collected in southern Florida by Mr. E. A. Schwarz. Proceedings of the Entomological Society of Washington, 1: 142-144.

1890. Observations on North American Capsidae, with descriptions of new species. No. 5. Transactions of the Maryland Academy of Sciences, 1: 73-88.

1891. Observations on some remarkable forms of Capsidae. Proceedings of the Entomological Society of Washington, 2: 119-123.

1892. Observations on some remarkable Heteroptera of North America. Transactions of the Maryland Academy of Sciences, 1: 179-184.

1893. Hemiptera, Heteroptera of the Death Valley Expedition. Pages 260-265. *In* North American Fauna. No. 7. United States Department of Agriculture, Division of Ornithology and Mammalogy Publication.

1893. A list of the Hemiptera-Heteroptera collected in the island of St. Vincent by Mr. Herbert H. Smith; with descriptions of new genera and species. Proceedings of the Zoological Society of London, pages 705-719.

1893. Summary of the collection of Hemiptera secured by Mr. E. A. Schwarz in Utah. Proceedings of the Entomological Society of Washington, 2: 366-385.

1894. Observations upon the heteropterous Hemiptera of Lower California, with descriptions of new species. Proceedings of the California Academy of Science, Series 2, 4: 223-295.

1894. On the Hemiptera-Heteroptera of the island of Grenada, West Indies. Proceedings of the Zoological Society of London, pages 167-224.

1895. [Descriptions of new genera and species of Hemiptera]. *In* Gillette, G. P. and C. F. Baker. A preliminary list of the Hemiptera of Colorado. Colorado Agricultural Experiment Station Bulletin, 31 (Technical Series No. 1). 137 pages.

1897. Contributions towards a knowledge of the Hemiptera-Heteroptera of N. America.--No. 1. Transactions of the Maryland Academy of Sciences, pages 383-394.

1897. [Two new species]. *In* Kirkland, A. H. Notes on predaceous Heteroptera, with Prof. Uhler's description of two species. Canadian Entomologist, 29: 115-118.

1899. A new destructive capsid. Entomological News, 10: 59.

1901. Some new genera and species of North American Hemiptera. Proceedings of the Entomological Society of Washington, 4: 507-515.

1904. List of Hemiptera-Heteroptera of Las Vegas Hot Springs, New Mexico, collected by Messrs. E. A. Schwarz and Herbert S. Barber. Proceedings of the United States National Museum, 27: 349-364.

Usinger, R. L.

1930. Two new species of *Vanduzeeina* from California (Scutelleridae, Hemipt.). Pan-Pacific Entomologist, 6: 131-133.

1931. A new species of *Platylygus* (Miridae, Hemiptera). Pan-Pacific Entomologist, 7: 129-130.

1932. Miscellaneous studies in the Henicocephalidae. Pan-Pacific Entomologist, 8: 145-156.

1933. A new species of *Gastrodes* from California (Lygaeidae-Hemiptera). Pan-Pacific Entomologist, 9: 127-128.

1933. New distributional and host plant records of Heteroptera for California. I. Pan-Pacific Entomologist, 9: 171-172.

1936. New distributional records of Hawaiian Heteroptera. Proceedings of the Hawaiian Entomological Society, 9: 209-210.

1936. The genus *Geocoris* in the Hawaiian Islands (Lygaeidae, Hemiptera). Proceedings of the Hawaiian Entomological Society, 9: 212-215.

1936. Studies in the American Aradidae with descriptions of new species (Hemiptera). Annals of the Entomological Society of America, 29: 490-516.

1938. Review of the genus *Gastrodes* (Lygaeidae, Hemiptera). Proceedings of the California Academy of Sciences, Fourth Series, 23: 289-301.

1938. Dorsal abdominal scent glands in nymphs of Lygaeidae. Pan-Pacific Entomologist, 14: 83.

1939. Descriptions of new Triatominae with a key to genera (Hemiptera, Reduviidae). University of California Publications in Entomology, 7: 33-51, plate 1.

1940. Fossil Lygaeidae (Hemiptera) from Florissant. Journal of Paleontology, 14: 79-80.

1941. The present status and synonymy of some orsilline species (Hemiptera, Lygaeidae). Bulletin of the Brooklyn Entomological Society, 36: 129-132.

1941. Rediscovery of *Emesaya brevicoxa* and its occurrence in the webs of spiders (Hemiptera, Reduviidae). Bulletin of the Brooklyn Entomological Society, 36: 206-208.

1941. A remarkable immigrant leptopodid in California. Bulletin of the Brooklyn Entomological Society, 36: 164-165.

1941. Key to the subfamilies of Naucoridae with a generic synopsis of the new subfamily Ambrysinae (Hemiptera). Annals of the Entomological Society of America, 34: 5-16.

1942. The genus *Nysius* and its allies in the Hawaiian Islands (Hemiptera, Lygaeidae, Orsillini). B. P. Bishop Museum Bulletin, 173: 1-167, 12 plates.

1942. A brachypterous *Reduvius* from Lower California (Heteroptera: Reduviidae). Entomological News, 53: 198-200.

1943. A taxonomic note on *Aradus depictus* Van Duzee. Pan-Pacific Entomologist, 19: 138.

1944. The Triatominae of North and Central America and the West Indies and their public health significance. United States Public Health Bulletin, 288: i-iv, 1-83.

1945. Review of the genus *Saldoida* with new records for Georgia and Virginia (Hemiptera, Saldidae). Bulletin of the Brooklyn Entomological Society, 40: 116-118.

1945. Biology and control of ash plant bugs in California. Journal of Economic Entomology, 38: 585-591.

1945. Classification of the Enicocephalidae (Hemiptera, Reduvioidea). Annals of the Entomological Society of America, 38: 321-342.

1945. Notes on the genus *Cryptostemma* with a new record for Georgia and a new species from Puerto Rico (Hemiptera: Cryptostemmatidae). Entomological News, 56: 238-241.

1946. Poyctenidae. Fascicle V. *In* China, W. E. and H. M. Parshley (editors). General Catalogue of the Hemiptera. Smith College, Northampton, Massachusetts. 18 pages.

1946. Notes and descriptions of *Ambrysus* Stal with an account of the life history of *Ambrysus mormon* Montd. (Hemiptera, Naucoridae). University of Kansas Science Bulletin, 31: 185-210.

1947. Classification of the Cryphocricinae (Hemiptera: Naucoridae). Annals of the Entomological Society of America, 40: 329-343.

1948. [Book review of] A synopsis of the Hemiptera-Heteroptera of America north of Mexico by J. R. de la Torre-Bueno. Bulletin of the Brooklyn Entomological Society, 42: 173-175.

1952. Two new Orsillini from Kilimanjaro (Hemiptera: Lygaeidae). Proceedings of the Royal Society of London, (B)21: 140-146.

1953. Notes on the genus *Metrobates* in California with description of a new subspecies (Hemiptera: Gerridae). Pan-Pacific Entomologist, 29: 178-179.

1954. Obituary. Howard Madison Parshley (1884-1953). Pan-Pacific Entomologist, 39: 1-4.

1955. Hemiptera. Pages 533-535. *In* A Century of Progress in the Natural Sciences--1853-1953. California Academy of Sciences, San Francisco. vii + 807 pages.

1956. Aquatic Hemiptera. Chapter 7. Pages 182-228. *In* Usinger, R. L. (editor). Aquatic Insects of California with keys to North American Genera and California Species. University of California Press, Los Angeles, Berkeley. 508 pages.

1959. New Species of Cimicidae (Hemiptera). Entomologist, 92: 218-222.

1960. Observations on the biology of *Chiloxanthus stellatus* (Curtis) and *C. arcticus* (Sahlberg)(Hemiptera: Saldidae). Pan-Pacific Entomologist, 36: 189-190.

1963. Animal distribution patterns in the tropical Pacific. Pages 255-261. *In* Gressitt, J. L. (editor). Pacific Basin Biogeography: A symposium. Bishop Museum Press, Honolulu.

1966. Monograph of Cimicidae (Hemiptera-Heteroptera). Thomas Say Foundation. Entomological Society of America, College Park, Maryland, 7: i-xi, 1-585.

1968. R. L. Usinger Hemiptera collection. Pan-Pacific Entomologist, 44: 257.

1972. Robert Leslie Usinger: Autobiography of an Entomologist. Pacific Coast Entomological Society, San Francisco. xiii + 330 pages.

Usinger, R. L. and P. D. Ashlock

1959. Revision of the Metrargini (Hemiptera, Lygaeidae). Proceedings of the Hawaiian Entomological Society, 17: 93-116.

Usinger, R. L. and R. Matsuda

1959. Classification of the Aradidae. British Musuem, London. 410 pages.

Usinger, R. L. and R. I. Sailer

1944. Nomenclature of the genus *Nysius* and its allies (Lygaeidae: Heteroptera). Proceedings of the Entomological Society of Washington, 46: 260-262.

1951. Proposed use of the plenary powers to designate type species of the genera "*Nysius*" Dallas, 1852 and "*Artheneis*" Spinola, 1837. Bulletin of Zoological Nomenclature, 2: 313-314.

Usinger, R. L. and N. Ueshima

1965. New species of bat bugs of the *Cimex pilosellus* complex (Hemiptera: Cimicidae). Pan-Pacific Entomologist, 41: 114-117.

van den Bosch, R. and K. Hagen

1966. Predaceous and parasitic arthropods in California cotton fields. California Agricultural Experiment Station Bulletin, 820. 32 pages.

van den Bosch, R., T. F. Leigh, L. A. Falcon, V. M. Stern, D. Gonzales, and K. S. Hagen

1971. The developing program of integrated control of cotton pests in California. Pages 377-394. *In* C. B. Huffaker (editor). Biological Control. Plenum Press, New York and London.

van den Bosch, R., T. F. Leigh, D. Gonzales, and R. E. Stinner

1969. Cage studies on predators of bollworm in cotton. Journal of Economic Entomology, 62: 1486-1489.

van den Bosch, R. and V. M. Stern

1969. The effect of harvesting practices on insect populations in alfalfa. Proceedings Tall Timbers Conference on Ecological Animal Control by Habitat Management, 1: 47-54.

van Doesburg, P. H., Jr. [see Doesburg, P. H., Jr., van]

Van Duzee, E. P.

1887. Partial list of Capsidae taken at Buffalo, N. Y. CanadianEntomologist, 19: 69-72.

1889. List of Hemiptera from Muskoka Lake District. Canadian Entomologist, 21: 1-11.

1894. Note on *Scolopostethus*. Entomological News, 5: 108.

1894. A list of the Hemiptera of Buffalo and vicinity. Bulletin of the Buffalo Society of Natural Sciences, 5: 167-204.

1903. Hemiptera of Beulah, New Mexico. Transactions of the American Entomological Society, 29: 107-116.

1904. Annotated list of the Pentatomidae recorded from America north of Mexico, with descriptions of some new species. Transactions of the American Entomological Society 30: 1-80.

1905. List of Hemiptera taken in the Adirondack Mountains. *In* Felt, E. P. 20th Report of the [New York] State Entomologist for 1904. New York State Museum Bulletin, 97: 546-556.

1906. New North American Heteroptera. Entomological News, 17: 384-391.

1907. Notes on Jamaican Hemiptera: A report on a collection of Hemiptera made on the island of Jamaica in the spring of 1906. Bulletin of the Buffalo Society of Natural Sciences, 8: 1-79.

1908. List of Hemiptera taken by W.J. Palmer about Quinze Lake, P. Que., in 1907. Canadian Entomologist, 40: 109-116, 157-160.

1909. Observations on some Hemiptera taken in Florida in the spring of 1908. Bulletin of the Buffalo Society of Natural Sciences, 9: 149-230. [Heteroptera, pages 153-184].

1909. North American Heteroptera. Entomological News, 20: 231-234.

1909. Synonymical and descriptive notes on North American Heteroptera. Canadian Entomologist, 41: 369-375.

1910. Monograph of the genus *Crophius* Stal. Bulletin of the Buffalo Society of Natural Sciences, 9: 389-398.

1910. Descriptions of some new or unfamiliar North American Hemiptera. Transactions of the American Entomological Society, 36: 73-88.

1912. Synonymy of the Provancher Collection of Hemiptera. Canadian Entomologist, 44: 317-329.

1912. A few days' work and play in Canada. Ottawa Naturalist, 26: 68-70.

1912. Hemipterological gleanings. Bulletin of the Buffalo Society of Natural Sciences, 10: 477-512.

1914. A preliminary list of the Hemiptera of San Diego County, California. Transactions of the San Diego Society of Natural History, 2: 1-57.

1914. Nomenclatorial and critical notes on

Hemiptera. Canadian Entomologist, 46: 377-389.

1915. New genera and species of North American Hemiptera. Pomona Journal of Entomology and Zoology, 7: 109-121.

1916. Check List of the Hemiptera (Excepting the Aphididae, Aleurodidae and Coccidae) of America, North of Mexico. New York Entomological Society, New York. xi + 111 pages.

1916. Synoptical keys to the genera of the North American Miridae. University of California Publications, Technical Bulletins, Entomology, 1: 199-216.

1916. New or little known genera and species of Orthotylini (Hemiptera). University of California Publications, Technical Bulletins, Entomology, 1: 217-227.

1916. Notes on some Hemiptera taken near Lake Taho, California. University of California Publications, Technical Bulletins, Entomology, 1: 229-249.

1916. Review of the genus *Macrotylus* Fieb. (Hemiptera). Pomona Journal of Entomology and Zoology, 8: 5-11.

1916. Monograph of the North America species of *Orthotylus* (Hemiptera). Proceedings of the California Academy Sciences, Fourth Series, 6: 87-128.

1917. Catalogue of the Hemiptera of America north of Mexico excepting the Aphididae, Coccidae and Aleurodidae. University of California Publications, Technical Bulletins, Entomology, 2: i-xvi + 1-902.

1917. Report on a collection of Hemiptera made by Walter M. Giffard in 1916 and 1917, chiefly in California. Proceedings of the California Academy of Sciences, Fourth Series, 7: 249-318.

1918. New species of Hemiptera chiefly from California. Proceedings of the California Academy of Sciences, Fourth Series, 8: 271-308.

1919. Report of the Canadian Arctic Expedition 1913-18. Vol. III: Insects. Part F: Hemiptera. Southern Party 1912-16. Ottawa. 5 pages.

1920. New hemipterous insects of the genera *Aradus*, *Phytocoris* and *Camptobrochys*. Proceedings of the California Academy of Sciences, Fourth Series, 9: 331-356.

1921. Characters of some new species of North American hemipterous insects, with one new genus. Proceedings of the California Academy of Science, Fourth Series, 11: 111-134.

1921. Characters of eight new species of North American Anthocoridae or flower bugs. Proceedings of the Calfornia Academy of Sciences, Fourth Series, 11: 137-144.

1921. A study of the North American grass-bugs of the genus *Irbisia*. Proceedings of the California Academy of Sciences, Fourth Series, 11: 145-152.

1922. A new North American genus of Cydnidae (Hem.). Entomological News, 33: 270-271.

1923. Expedition of the California Academy of Sciences to the Gulf of California in 1921. The Hemiptera (true bugs, etc.). Proceedings of the California Academy of Science, Fourth Series, 12: 123-200.

1923. Notes on *Lygaeus kalmii* Stal and allies (Hemip.). Canadian Entomologist, 55: 214.

1923. A new subspecies of *Euryophthalmus cinctus* (Hemiptera). Canadian Entomologist, 55: 270.

1923. A rearrangement of our North American Thyreocorinae (Hemip.). Entomological News, 34: 302-305.

1925. Notes on a few Hemiptera from the San Bernardino Mountains, California. Bulletin of the Brooklyn Entomological Society, 20: 89-90.

1925. A new mirid from Arizona (Hemiptera). Pan-Pacific Entomologist, 2: 35.

1925. New Hemiptera from western North America. Proceedings of the California Academy of Sciences, Fourth Series, 14: 391-425.

1927. Notes on western Aradidae. Pan-Pacific Entomologist, 3: 139-142.

1928. Two interesting additions to the hemipterous fauna of California. Pan-Pacific Entomologist, 4: 190-191.

1928. A misidentified *Hadronema* (Hemiptera). Pan-Pacific Entomologist, 4: 182.

1929. Our first *Rhyparochromus* (Hemip. Lygaeidae). Pan-Pacific Entomologist, 5: 47.

1929. A new *Corimelaena* (Hemiptera). Pan-Pacific Entomologist, 6: 10.

1929. Notes on two Berytidae. Pan-Pacific Entomologist, 5: 166.

1929. Some new western Hemiptera. Pan-Pacific Entomologist, 5: 186-191.

1931. A new *Ischnorrhynchus* (Hemiptera: Lygaeidae). Pan-Pacific Entomologist, 7: 110.

Van Duzee, E. P. (continued)

1933. A new *Lopidea* from California. Pan-Pacific Entomologist, 9: 96.

1934. A new *Brochymena*. Pan-Pacific Entomologist, 10: 22.

1934. An apparently new pentatomid. Pan-Pacific Entomologist, 10: 96.

1935. Four hitherto undescribed Hemiptera. Pan-Pacific Entomologist, 11: 25-29.

1937. A few new Hemiptera. Pan-Pacific Entomologist, 13: 25-31.

Vangeison, K. W. and J. E. McPherson

1975. Life history and laboratory rearing of *Proxys punctulatus* (Hemiptera: Pentatomidae) with descriptions of immature stages. Annals of the Entomological Society of America, 68: 25-30.

Van Steenwyk, R. A., N. C. Toscano, G. R. Ballmer, K. Kido, and H. T. Reynolds

1975. Increases of *Heliothis* spp. in cotton under various insecticide treatment regimes. Environmental Entomology, 4: 993-996.

Viana, M. J. and d. J. Carpintero

1981. Una nueva especie de "*Discocoris*" Kormilev, 1955 (Hemiptera, Xylastodoridae). Comunicaciones del Instituto Nacional de Investigacion de las Ciencias Naturales "Bernardino Rivadavia" Entomologica, 1: 63-74.

Villers, C. J.

1789. Caroli Linnaei entomologia faunae Suecicae descriptionibus aucta: D. D. Scopoli, Geoffroy, De Geer, Fabricii, Schrank, etc., speciebus vel in systemate non enumeratis, vel nuperime detectis, vel speciebus Galliae australis locupletata, generum specierumque rariorum iconibus ornata, curante et augente Carolo de Villers. Lugduni (Lyon). 1: 1-765.

Vinokurov, N. N.

1977. On the systematics and intraspecific variability of capsid bugs of the genus *Capsus* (Heteroptera, Miridae). Entomologischeskoye Obozreniye, 56: 103-115 [English translation in Entomological Review, 56: 76-85].

Voigt, W. G. and R. Garcia

1976. Keys to the *Notonecta* nymphs of the West Coast United States (Hemiptera: Notonectidae). Pan-Pacific Entomologist, 52: 172-176.

Vojdani, S.

1961. The nearctic species of the genus *Eurygaster* (Hemiptera: Pentatomidae: Scutellerinae). Pan-Pacific Entomologist, 37: 97-107.

1961. Bio-ecology of some *Eurygaster* species in central California (Pentatomidae-Scutellerinae). Annals of the Entomological Society of America, 54: 567-578.

Vollenhoven, S. C. S.

1868. Diagnosen van eenige nieuwe soorten van Hemiptera Heteroptera. Verslagenen Mededeelingen der Koniklijik Akademie van Wetenschappen, Letterkunde en Schoone Kunsten te Amsterdam, Series 2, 2: 172-188.

Vosler, E. J.

1913. A new fruit and truck crop pest (*Irbisia brachycera* Uhler). Monthly Bulletin of the California State Commission of Horticulture, 2: 551-553.

Waddell, D. B.

1952. A preliminary list of the Hemiptera of the Kootenay Valley. Proceedings of the Entomological Society of British Columbia, 48: 93-96.

Waddill, V. and M. Shepard

1974. Potential of *Geocoris punctipes* (Hemiptera: Lygaeidae) and *Nabis* spp. (Hemiptera: Nabidae) as predators of *Epilachna varivestis* (Coleoptera: Coccinellidae). Entomophaga, 19: 421-426.

Waddill, V. H., B. M. Shepard, S. G. Turnipseed, and G. R. Carner

1974. Sequential sampling plans for *Nabis* spp. and *Geocoris* spp. on soybeans. Environmental Entomology, 3: 415-419.

Wadley, J.

1962. Studies of two species of milkweed bugs. Proceedings of the Utah Academy of Sciences, Arts and Letters, 39: 212-215.

Wagner, E.

1951. Zur Systematik der Gattung *Dicyphus* (Hem. Het., Miridae). Societas Scientiarum Fennica Commentationes Biologicae, 12: 1-36.

1952. Die europäischen Arten der Gattung *Orius* Wff. (Hem. Het. Anthocoridae). Notulae Entomologicae, 32: 22-59.

1952. Blindwanzen oder Miriden. *In* Dahl, F. Die Tierwelt Deutschlands. G. Fischer, Jena. 41: 1-218.

1956. *Trigonotylus coelestialium* Kirk. in Nordwestdeutschland (Heter. Miridae). Schriften Naturwissenschaftlichen Vereins für Schleswig-Holstein, 28: 69-71.
1956. On the genus *Strongylocoris* Blanchard, 1840 (Hemiptera, Heteroptera, Miridae). Proceedings of the Entomological Society of Washington, 58: 277-281.
1960. Über einige neue Miriden-Arten aus dem Zoologischen Museum Helsingfors (Hem. Heteropt.). Notulae Entomologicae, 40: 112-122.
1962. Comment on the proposed designation of a type-species under the plenary powers of *Blissus* Burmeister, 1835. Z.N.(S.) 1471. Bulletin of Zoological Nomenclature, 19: 172.
1970-1978 Die Mirdae Hahn, 1831, des Mittelmeerraumes und der Makaronesischen Inseln (Hemiptera, Heteroptera). Entomologische Abhandlungen, 37(supplement): 1-273 (1970), 274-484 (1971); 39(supplement): 1-421 (1973); 40(supplement): 1-483 (1975); 42(supplement): 1-96 (1978).

Wagner, E. and J. A. Slater
1952. Concerning some holarctic Miridae (Hemiptera, Heteroptera). Proceedings of the Entomological Society of Washington, 54: 273-281.
1964. Zur Systematik der Blissinae Stal in der Paläarktis (Hem. Het. Lygaeidae). Entomologische Berichten, 24: 66-76.

Waive, C. M. and D. F. Clower
1976. Natural enemies of the bandedwing whitefly in Louisiana. Environmental Entomology, 5: 1075-1078.

Walker, F.
1867-1873. Catalogue of the specimens of Heteropterous-Hemiptera in the collection of the British Museum. 8 parts. British Museum, London. [1867, 1: 1-240, 2: 241-417; 1868, 3: 419-599; 1871, 4: 1-211; 1872, 5: 1-202; 1873, 6: 1-210, 7: 1-213, 8: 1-220].
1873. Catalogue of Hemiptera Heteroptera in the British Museum. Supplement. E. W. Janson, London. 63 pages.

Walker, J. T. and S. G. Turnipseed
1976. Predatory activity, reproductive potential and longevity of *Geocoris* spp. treated with insecticides. Journal of the Georgia Entomological Society, 11: 266-271.

Walker, J. T., S. G. Turnipseed, and M. Shepard
1974. Nymphal development and fecundity of *Geocoris* spp. surviving insecticide treatments to soybeans. Environmental Entomology, 3: 1036-1037.

Wallengren, H. D. J.
1894. Revision af släktet *Corisa* Latr. beträffande dess skandinaviska arter. Entomologiske Tidskrift, 15: 129-164.

Walley, G. S.
1929. Two new species of *Eremocoris* with notes and a key to the species of *Peritrechus* (Hemip., Lygaeidae). Canadian Entomologist, 61: 41-44.
1929. Notes on *Homaemus* with a key to the species (Hemip., Scutelleridae). Canadian Entomologist, 61: 253-256.
1930. Heteroptera from the north shore of the Gulf of St. Lawrence. Canadian Entomologist, 62: 75-81.
1930. A review of the genus *Palmacorixa* Abbott (Hemip., Corixidae). Canadian Entomologist, 62: 99-106.
1930. Notes and descriptions of species of *Arctocorixa* from Ontario and Quebec (Hemip., Corixidae). Canadian Entomologist, 62: 280-286.
1930. A new *Arctocorixa* with a note on synonymy (Hemipt., Corixidae). Bulletin of the Brooklyn Entomological Society, 25: 203-206.
1931. Corixidae from the environs of Hudson Bay (Hemip., Corixidae). Canadian Entomologist, 63: 238-239.
1932. A second report on Hemiptera from the north shore of the Gulf of St. Lawrence. Canadian Entomologist, 64: 152-155.
1934. Pages 143-144. *In* Brown, W. J. The entomological record 1931, 1932, 1933. Annual Report of the Quebec Society for the Protection of Plants, 25-26: 140-162.
1935. Three rare Canadian Hemiptera. Canadian Entomologist, 67: 159-160.
1936. New North American Corixidae with notes (Hemiptera). Canadian Entomologist, 68: 55-63.

Wallis, R. L. and J. E. Turner
1972. Insects overwintering in the warm microenvironment of drainage ditches in central Washington. Environmental Entomology, 1: 107-109.

Waloff, N.
1966. Scotch broom (*Sarothamnus scoparius* (L.) Wimmer) and its insect fauna introduced into the Pacific Northwest of America. Journal of Applied Ecology, 3: 293-311.

Walsh, B. D.

1864. On certain remarkable or exceptional larvae, coleopterous, lepidopterous and dipterous, with descriptions of several new genera and species and of several species injurious to vegetables which have been already published in agricultural journals. Proceedings of the Boston Society of Natural History, 9: 286-318. [Heteroptera, page 313].

1864. On phytophagic varieties and phytophagic species. Proceedings of the Entomological Society of Philadelphia, 3: 403-430.

1869. The imported currant worm fly (*Nematus ventricosus*, Klug.) and its parasite (*Hemiteles nemativorus*, Walsh). Canadian Entomologist, 2: 31-33.

Walt, J. F. and J. E. McPherson

1973. Descriptions of immature stages of *Stethaulax marmorata* (Hemiptera: Scutelleridae) with notes on its life history. Annals of the Entomological Society of America, 66: 1103-1107.

Walton, G. A.

1940. Classification of the family. [Corixidae]. Pages 343-345. *In* Hutchison, G. E. A revision of the Corixidae of India and adjacent regions. Transactions of the Connecticut Academy of Sciences, 33: 339-476.

1943. The natural classification of the British Corixidae (Hemip.). Transactions for the Society for British Entomology, 8: 155-168.

Ward, C. R., C. W. O'Brien, L. B. O'Brien, D. E. Foster, and E. W. Huddleston

1977. Annotated checklist of New World insects associated with *Prosopis* (Mesquite). United States Department of Agriculture, Agricultural Research Service Technical Bulletin, 1557. (i) + 115 pages.

Watson, J. R.

1917. [No title]. Florida Bulletin, 14: 72.

1919. An outbreak of the cotton stainer on citrus. Florida Buggist, 2: 88-90.

Watson, S. A.

1928. The Miridae of Ohio. Ohio State University Bulletin 33. Ohio Biological Survey Bulletin, 16: 1-44.

Watson, T. F., D. Langston, D. Fullerton, R. Rakickas, B. Engroff, R. Rokey, and L. Bricker

1972. Environmental improvement through biological control and pest management. University of Arizona Cooperative Extension Service, (P)24: 86-87.

Watts, J. G.

1963. Insects associated with black grama grass, *Bouteloua eriopoda*. Annals of the Entomological Society of America, 56: 374-379.

1973. Adaptation of indigenous insects to *Tamarix pentandra*. Folia Entomológica Méxicana, 25/26: 51-52.

Webster, R. L.

1886. Insects affecting fall wheat. Pages 311-319. *In* Annual Report of the United States Department of Agriculture for 1885, Report for Entomology.

Webster, R. L. and D. Stoner

1914. The eggs and nymphal stages of the dusky leaf bug *Calocoris rapidus* Say. Journal of the New York Entomological Society, 22: 229-234.

Weiss, H. B.

1916. Additions to insects of New Jersey, No. 3. Entomological News, 27: 9-13.

1919. Notes on *Corythuca bulbosa* O. & D. Ohio Journal of Science, 20: 17-20.

Weiss, H. B. and E. West

1925. The insects and plants of a strip of New Jersey coast. Psyche, 32:231-241.

Wellhouse, W. H.

1920. Wild hawthorns as hosts of apple, pear and quince pests. Journal of Economic Entomology, 13: 388-391.

Wene, G. P.

1953. The false chinch bug (*Nysius raphanus*). Proceedings of the Rio Grande Valley Horticulture Institute, 7: 74-76.

Wene, G. P., L. A. Carruth, A. D. Telford, and L. Hopkins

1965. Descriptions and habits of Arizona cotton insects. Arizona Agricultural Experiment Station Bulletin, A-23. 61 pages.

Wene, G. P. and L. W. Sheets

1962. Relationship of predators and injurious insects in cotton fields in the Salt River Valley of Arizona. Journal of Economic Entomology, 55: 395-398.

Werner, F. G.

1969. Terrestrial insects of the Rita Blanca Lake deposits. Pages 123-130. *In* Anderson, R. Y. and D. W. Kirkland (editors). Paleoecology of an Early Pleistocene Lake on the High Plains of Texas. Memoirs of the Geological Society of America, 113. 215 pages.

West, A. S. and B. DeLong

1955. Notes on the biology and laboratory rearing of a predatory insect, *Zelus exsanguis* (Stal) (Hemiptera: Reduviidae). Proceedings of the Entomological Society of Ontario, 86: 97-101.

Westigard, P. H.

1973. The biology of and effect of pesticides on *Deraeocoris brevis piceatus* (Heteroptera: Miridae). Canadian Entomologist, 105: 1105-1111.

Westwood, J. O.

1837-1842. A Catalogue of Hemiptera in the Collection of the Rev. F. W. Hope, with short Latin descriptions of new species. J. Bridgewater, London. 1837, 1: 1-46; 1842, 2: 1-26.

1840. An Introduction to the Modern Classification of Insects; Founded on the Natural Habits and Corresponding Organization of the Different Families. Longman, Orme, Brown, Green, and Longmans, London. Volume 2. 587 pages + Synopsis of the Genera of British Insects, 158 pages. [Order Heteroptera. 2: 450-488; synopsis 119-158].

1841. Observations upon the hemipterous insects composing the genus *Syrtis* of Fabricius, or the family Phymatides of Laporte, with a monograph of the genus *Macrocephalus*. Transactions of the Entomological Society of London, 3: 18-28, plate 2.

1874. Thesarus entomologicus Oxoniensis; or, Illustrations of new, rare, and interesting insects, for the most part contained in the collections presented to the University of Oxford by the Rev. F. W. Hope. Clarendon Press, Oxford. xxiv + 205 pages, 40 plates.

Whalon, M. and B. L. Parker

1978. Immunological identification of tarnished plant bug predators. Annals of the Entomological Society of America, 71: 453-456.

Wheeler, A. G., Jr.

1970. *Berytinus minor* (Hemiptera: Berytidae) in North America. Canadian Entomologist, 102: 876-886.

1971. An additional note on *Berytinus minor* (Hemiptera: Berytidae) in North America. Canadian Entomologist, 103: 497.

1976a. Life history of *Kleidocerys resedae* on European white birch and ericaceous shrubs. Annals of the Entomological Society of America, 69: 459-463.

1976. *Anaptus major* established in eastern North America (Hemiptera: Nabidae). Proceedings of the Entomological Society of Washington, 78: 382.

1976. Yucca plant bug, *Halticotoma valida*: Author-ship, distribution, host plants, and notes on biology. Florida Entomologist, 59: 71-76.

1977. Life history of *Niesthrea louisianica* (Hemiptera: Rhopalidae) on rose of Sharon in North Carolina. Annals of the Entomological Society of Washington, 70: 631-634.

1977. Spicebush and sassafras as new North American hosts of andromeda lace bug, *Stephanitis takeyai* (Hemiptera: Tingidae). Proceedings of the Entomological Society of Washington, 79: 168-171.

1977. A new name and restoration of an old name in the genus *Fulvius* Stål (Hemiptera: Miridae). Proceedings of the Entomological Society of Washington, 79: 588-592.

1978. *Neides muticus* (Hemiptera: Berytidae): Life history and description of fifth instar. Annals of the Entomological Society of America, 71: 733-736.

1980. First United States records of *Lygocoris knighti* (Hemiptera: Miridae). Entomological News, 91: 25-26.

1980. Life history of *Plagiognathus albatus* (Hemiptera: Miridae), with a description of the fifth instar. Annals of the Entomological Society of America, 73: 354-356.

1981. Hawthorn lace bug (Hemiptera: Tingidae), first record of injury to roses, with a review of the host plants. Great Lakes Entomologist, 14: 37-43.

1981. The distribution and seasonal history of *Slaterocoris pallipes* (Knight) (Hemiptera: Miridae). Proceedings of the Entomological Society of Washington, 83: 520-523.

1982. Bed Bugs and Other Bugs. Pages 319-351. *In* Mallis, A. Handbook of Pest Control. Sixth edition. Franzak & Foster Company, Cleveland. 1101 pages.

1982. *Coccobaphes sanguinarius* and *Lygocoris*

Wheeler, A. G., Jr. (continued)
vitticollis (Hemiptera: Miridae): Seasonal history and description of fifth-instar, with notes on other mirids associated with maple. Proceedings of the Entomological Society of Washington, 84: 177-183.
1983a. The small milkweed bug, *Lygaeus kalmii* (Hemiptera: Lygaeidae): Milkweed specialist or opportunist? Journal of the New York Entomological Society, 91: 57-62.
1983b. *Stygnocoris rusticus*: New records in eastern North America, with a review of its distribution (Hemiptera-Heteroptera: Lygaeidae). Entomological News, 94: 131-135.
1983c. Outbreaks of the apple red bug: Difficulties in identifying a new pest and emergence of a mirid specialist. Journal of the Washington Academy of Sciences, 73: 60-64.
1984a. *Aufeius impressicollis* (Hemiptera: Rhopalidae): Easternmost U.S. record, host plant relationships, and laboratory rearing. Journal of the New York Entomological Society, 92: 174-178.
1984b. Seasonal history, habits, and immature stages of *Belonochilus numenius* (Hemiptera: Lygaeidae). Proceedings of the Entomological Society of Washington, 86: 790-796.
1985. Seasonal history, host plants, and nymphal descriptions of *Orthocephalus coriaceus*, a plant bug pest of herb garden composites (Hemiptera: Miridae). Proceedings of the Entomological Society of Washington, 87: 85-93.
1986. A new host association for the stilt bug *Jalysus spinosus* (Heteroptera: Berytidae). Entomological News, 97: 63-65.

Wheeler, A. G., Jr. and T. J. Henry
1973. *Camptozygum aequale* (Villers), a pine-feeding mirid new to North America (Hemiptera: Miridae). Proceedings of the Entomological Society of Washington, 75: 240-246.
1974. *Tropidosteptes pacificus*, a western ash plant bug introduced into Pennsylvania with nursery stock (Hemiptera: Miridae). United States Department of Agriculture Cooperative Economic Insect Report, 24: 588-589.
1975. Recognition of seven Uhler manuscript names, with notes on thirteen other species used by Heidemann (1892) (Hemiptera: Miridae). Transactions of the American Entomological Society, 101: 355-369.
1976. First records of the predaceous mirid *Phytocoris tiliae* (F.) from the United States. Entomological News, 87: 25-28.
1976. Biology of the honeylocust plant bug, *Diaphnocoris chlorionis*, and other mirids associated with ornamental honeylocust. Annals of the Entomological Society of America, 69: 1095-1104.
1977. Rev. Modestus Wirtner: Biographical sketch and additions and corrections to the Miridae in his 1904 list of western Pennsylvania Hemiptera. Great Lakes Entomologist, 10: 145-157.
1977. Miridae associated with Pennsylvania conifers: 1. Species on arborvitae, false cypress, and juniper. Transactions of the American Entomological Society, 103: 623-656.
1978. *Ceratocapsus modestus* (Hemiptera: Miridae), a predator of grape phylloxera: Seasonal history and description of fifth instar. Melsheimer Entomological Series, 25: 6-10.
1978. Isometopinae (Hemiptera: Miridae) in Pennsylvania: Biology and descriptions of the fifth instars, with observations of predation on obscure scale. Annals of the Entomological Society of America, 71: 607-614.
1980. Seasonal history and host plants of the ant mimic *Barberiella formicoides* Poppius, with description of the fifth-instar (Hemiptera: Miridae). Proceedings of the Entomological Society of Washington, 82: 269-275.
1980. *Brachynotocoris heidemanni* (Knight), a junior synonym of the palearctic *B. puncticornis* Reuter and pest of European ash. Proceedings of the Entomological Society of Washington, 82: 568-575.
1981. *Jalysus spinosus* and *J. wickhami*: Taxonomic clarification, review of host plants and distribution, and keys to adults and 5th instars. Annals of the Entomological Society of America, 74: 606-615.
1983. Seasonal history and host plants of the plant bug *Lygocoris atrinotatus*, with description of the fifth-instar nymph (Hemiptera: Miridae). Proceedings of the Entomological Society of Washington, 85: 26-31.

1984. Host plants, distribution, and description of fifth-instar nymphs of two little-known Heteroptera, *Arhyssus hirtus* (Rhopalidae) and *Esperanza texana* (Alydidae). Florida Entomologist, 67: 521-529.

1985. *Trigonotylus coelestialium* (Heteroptera: Miridae), A pest of small grains: Seasonal history, host plants, damage, and descriptions of adult and nymphal stages. Proceedings of the Entomological Society of Washington, 87: 699-713.

Wheeler, A. G., Jr., T. J. Henry, and T. L. Mason, Jr.

1983. An annotated list of the Miridae of West Virginia (Hemiptera - Heteroptera). Transactions of the American Entomological Society, 109: 127-159.

Wheeler, A. G., Jr. and J. L. Herring

1979. A potential insect pest of azaleas. Quarterly Bulletin of the American Rhododendron Society, 33: 12-14, 1 color figure.

Wheeler, A. G., Jr. and E. R. Hoebeke

1982. *Psallus variabilis* (Fallén) and *P. albipennis* (Fallén), two European plant bugs established in North America, with notes on taxonomic changes (Hemiptera: Heteroptera: Miridae). Proceedings of the Entomological Society of Washington, 84: 690-703.

1985. The insect fauna of ninebark, *Physocarpus opulifolius* (Rosaceae). Proceedings of the Entomological Society of Washington, 87: 356-370.

1985. *Dictyla echii*: Seasonal history and North American records of an immigrant lace bug (Hemiptera: Tingidae). Journal of the New York Entomological Society, 93: 1057-1063.

Wheeler, A. G., Jr. and J. P. McCaffrey

1984. *Ranzovius contubernalis*: Seasonal history, habits, and description of fifth instar, with speculations on the origin of spider commensalism in the genus *Ranzovius* (Hemiptera: Miridae). Proceedings of the Entomological Society of Washington, 86: 68-81.

Wheeler, A. G., Jr. and G. L. Miller

1981. Fourlined plant bug (Hemiptera: Miridae), a reappraisal: Life history, host plants, and plant response to feeding. Great Lakes Entomologist, 14: 23-35.

1983. *Harmostes fraterculus* (Hemiptera: Rhopalidae): Field history, laboratory rearing, and descriptions of immature stages. Proceedings of the Entomological Society of Washington, 85: 426-434.

Wheeler, A. G., Jr., G. L. Miller, and T. J. Henry

1979. Biology and habits of *Macrolophus tenuicornis* (Hemiptera: Miridae) on hayscentedfern (Pteridophyta: Polypodiaceae). Melsheimer Entomological Series, 27: 11-17.

Wheeler, A. G., Jr. and C. W. Schaefer

1982. Review of stilt bug (Hemiptera: Berytidae) host plants. Annals of the Entomological Society of America, 75: 498-506.

Wheeler, A. G., Jr., B. R. Stinner, and T. J. Henry

1975. Biology and nymphal stages of *Deraeocoris nebulosus* (Hemiptera: Miridae), a predator of arthropod pests on ornamentals. Annals of the Entomological Society of America, 68: 1063-1068.

Whitaker, J. O., Jr., D. Rubin and J. R. Munsee

1977. Observations of food habits of four species of spadefoot toads, genus *Scaphiopus*. Herpetologica, 33: 468-475.

Whitcomb, W. D.

1953. The biology and control of *Lygus campestris* L. on celery. University of Massachusetts Agricultural Experiment Station Bulletin, 473: 1-15.

Whitcomb, W. H.

1967. Field studies on predators of the second-instar bollworm, *Heliothus zea* (Boddie) (Lepidoptera: Noctuidae). Journal of the Georgia Entomological Society, 2: 113-118.

1974. Natural populations of entomophagous arthropods and their effect on the agroecosystem. Pages 150-169. *In* Maxwell, F. G. and F. A. Harris (editors). Proceedings of the summer institute on biological control of plant insects and diseases. University Press of Mississippi, Jackson. 647 pages.

Whitcomb, W. H. and K. Bell

1964. Predaceous insects, spiders, and mites of Arkansas cotton fields. Arkansas Agricultural Experiment Station Bulletin 690. 84 pages.

White, A.

1839. Description of two hemipterous insects. Annals and Magazine of Natural History, New Series, 3: 537-543.

White, F. B.

1873. Notes on *Corixa*. Entomologist's Monthly Magazine, 10: 60-63, 75-80.

White, F. B. (continued)
1877. Description of new species of heteropterous Hemiptera collected in the Hawaiian Islands by the Rev. T. Blackburn. Annals and Magazine of Natural History, Series 4, 10: 110-114.
1878. Descriptions of heteropterous Hemiptera collected in the Hawaiian Islands by the Rev. T. Blackburn.--No. 2. Annals and Magazine of Natural History, 1: 365-374.
1879. List of the Hemiptera collected in the Amazons by Prof. J. W. H. Trail, M.A., M.D., in the years 1873-1875, with descriptions of the new species. Transactions of the Entomological Society of London, 4: 267-276.
1879. Descriptions of new Anthocoridae. Entomologist's Monthly Magazine, 16: 142-148.
1883. Report on the pelagic Hemiptera procured during the voyage of H. M. S. Challenger in the years 1873-76. *In* Report on the Scientific Results of the Voyage of H. M. S. Challenger during the years 1873-76. Zoology. Volume VII. London. 82 pages, 3 plates.

Whitehead, D. R.
1974. Variation and synonymy in *Hypselonotus* (Hemiptera: Coreidae). Journal of the Washington Academy of Sciences, 64: 223-233.

Wickham, H. F.
1894. On the habits of some oceanic Hemiptera. Entomological News, 5: 33-36.

Wilcox, R. S.
1975. Sound-producing mechanisms of *Buenoa macrotibialis* Hungerford. International Journal of Insect Morphology and Embryology, 4: 169-182.

Wilde, G., A. Kadoum, and T. Mize
1984. Absence of synergism with insecticide combinations used on chinch bugs (Heteroptera: Lygaeidae). Journal of Economic Entomology, 77: 1297-1298.

Wilde, G. and T. Mize
1984. Enhanced microbial degradation of systemic pesticides in soil and its effect on chinch bug *Blissus leucopterus leucopterus* (Say) (Heteroptera: Lygaeidae) and green-bug *Schizaphis graminum* Rondani (Homoptera: Aphididae) control in seedling sorghum. Environmental Entomology, 13: 1079-1082.

Wilde, G., T. Mize, J. Stuart, J. Whitworth, and R. Kinsinger
1984. Comparison of planting-time applications of granular or liquid insecticides and liquid fertilizer plus insecticide combinations for control of chinch bugs (Heteroptera: Lygaeidae) and greenbugs (Homoptera: Aphididae) on seedling sorghum. Journal of Economic Entomology, 77: 706-708.

Wilde, G. and J. Morgan
1978. Chinch bug on sorghum: Chemical control, economic injury levels, plant resistance. Journal of Economic Entomology, 71: 908-910.

Wilde, G., O. Russ, and T. Mize
1986. Tillage, cropping, and insecticide use practice: Effects on efficacy of planting time treatments for controlling greenbug (Homoptera: Aphididae) and chinch bug (Heteroptera: Lygaeidae) in seedling sorghum. Journal of Economic Entomology, 79: 1364-1365.

Wilde, G., J. Stuart, and H. H. Hatchett
1986. Chinch bug (Heteroptera: Lygaeidae) reproduction on selected small grains and genetic sources. Journal of Kansas Entomological Society, 59: 550-551.

Wiley, G. O.
1922. Life history notes on two species of Saldidae (Hemiptera) found in Kansas. Kansas University Science Bulletin, 14: 301-311.
1923. A new species of *Rheumatobates* from Texas (Heteroptera, Gerridae). Canadian Entomologist, 55: 202-205.
1924. On the biology of *Curicta drakei* Hungerford (Heteroptera, Nepidae). Entomological News, 35: 324-331.

Wilkinson, J. D., K. D. Biever, and C. M. Ignoffo
1979. Synthetic pyrethroid and organophosphate insecticides against the parasitoid *Apanteles marginiventris* and the predators *Geocoris punctipes*, *Hippodamia convergens*, and *Podisus maculiventris*. Journal of Economic Entomology, 72: 473-475.

Wilkinson, J. D., K. D. Biever, C. M. Ignoffo, W. J. Pons, R. K. Morrison, and R. S. Seay
1978. Evaluation of diflubenzuron formulations on selected insect parasitoids and predators. Journal of the Georgia Entomological Society, 13: 227-236.

Williams, F. X.
1944. Biological studies of Hawaiian water-loving insects. 5. Proceedings of the Hawaiian Entomological Society, 12: 149-197.
1946. Two new species of Astatinae, with notes on the habits of the group (Hymenoptera: Sphecidae). Proceedings of the Hawaiian Entomological Society, 12: 641-650.

Wilson, C. A.
1953. A new *Velia* from Mississippi (Hemiptera - Veliidae). Florida Entomologist, 36: 27-29
1953. A new corixid from Mississippi (Hemiptera: Corixidae). Florida Entomologist, 36: 67-69.
1958. Aquatic and semiaquatic Hemiptera of Mississippi. Tulane Studies in Entomology, 6: 115-170.

Wilson, L. T. and A. P. Gutierrez
1980. Within plant distribution of predators on cotton, *Gossypium hirsutum*: Comments on sampling and predator efficiencies. Hilgardia, 48: 3-11.

Wilson, M. F. and P. W. Price
1977. The evolution of inflorescence size in *Asclepias* (Asclepiadaceae). Evolution, 31: 495-511.

Wilson, R. L. and R. L. Burton
1980. Feeding and oviposition of selected insect pests on proso millet cultivars. Journal of Economic Entomology, 73: 817-819.

Wing, J. F.
1982. The predictability of reversals in solar-terrestrial correlations. Ohio Journal of Science, 82: 100.

Wirtner, M.
1904. A preliminary list of the Hemiptera of western Pennsylvania. Annals of the Carnegie Museum, 3: 183-232.
1917. A new genus of Bothynotinae, Miridae (Heter.). Entomological News, 28: 33-34.

Wolfenbarger, D. O.
1963. Insect pests of the avocado and their control. Bulletin of the University of Florida Agricultural Experiment Station, 605A: 1-52.

Wolff, J. F.
1800-1811. Icones cimicum descriptionibus illustratae. J. J. Palm, Erlangan. [1800, 1:i-vii, 1-40, plates 1-4; 1801, 2: 43-84, plates 5-8; 1802, 3: 85-126, plates 9-12; 1804, 4: 127-166, plates 13-16; 1811, 5: 167-208, plates 17-20].

Wood, E. A., Jr. and K. J. Starks
1972. Damage to sorghum by a lygaeid bug *Nysius raphanus* Journal of Economic Entomology, 65: 1507-1508.

Woodroffe, G. E.
1967. The natural occurrence of apparent hybrids between two species of *Scolopostethus* Fieber (Hem., Lygaeidae). Entomologist's Monthly Magazine, 103: 174.

Wray, D. L.
1950. Insects of North Carolina. Second Supplement. North Carolina Department of Agriculture, Division of Entomology, Raleigh. 59 pages.
1967. Insects of North Carolina. Third Supplement. North Carolina Department of Agriculture, Division of Entomology, Raleigh. 181 pages.

Wroblewski, A.
1968. Klucze do oznacania owadów Polski. Czesc XVIII. Pluskwiaki róznoskrzydle--Heteroptera. Zeszyc 3. Leptopodidae, Nabrzezkowate--Saldidae. Polskie Towarzystwo Entomologiczne, 58: 1-35.

Wu, C. F.
1935. Catalogus insectorum Sinensium (Catalog of Chinese Insects). Fan Memorial Institute of Biology, Peiping. 2: 1-634.

Wyman, J. A., E. R. Oatman and V. Voth
1979. Effects of varying two spotted spider mite (*Tetranychus urticae*) infestation levels on strawberry yield. Journal of Economic Entomology, 72: 747-753.

Wygodzinsky, P.
1947. Sôbre duas novas espécies de "Emesinae" do Brasil, com notas sôbre *Stenolemoides arizonensis* (Banks) (Reduviidae, Hemiptera). Revista Brasileira de Biologia, 6: 509-519 (1946).
1948. On some Reduviidae belonging to the Naturhistorisches Museum at Vienna (Hemiptera). Revista Brasileira de Biologia, 8: 209-224.
1949. Elenco sistematico de los Reduviiformes americanos. Instituto de Medicina Regional, Tucumán, Monográfia, 1: 1-102.
1955. Description of a new *Cryptostemma* from North America (Hemiptera: Cryptostemmatidae). Pan-Pacific Entomologist, 31: 199-202.
1966. A monograph of the Emesinae (Reduviidae, Hemiptera). Bulletin of the American Museum of Natural History, 133: 1-614.

Wygodzinsky, P. and P. Štys
1970. A new genus of aenictopecheine bugs from the Holarctic (Enicocephalidae, Hemiptera). American Museum Novitates, 2411: 1-17.

Wygodzinsky, P. and R. L. Usinger
1964. The genus *Reduvius* Fabricius in western North America (Reduviidae, Hemiptera, Insecta). American Museum Novitates, 2175: 1-15.

Yokoyama, V. Y.
1978. Relation of seasonal changes in extrafloral nectar and foliar protein and arthropod populations in cotton. Environmental Entomology, 7: 799-802.
1980. Method for rearing *Geocoris pallens* (Hemiptera: Lygaeidae) a predator in California cotton. Canadian Entomologist, 112: 1-3.

Yokoyama, V. Y. and J. Pritchard
1984. Effect of pesticides on mortality, fecundity and egg viability of *Geocoris pallens* (Hemiptera: Lygaeidae). Journal of Economic Entomology, 77: 876-879.

Yokoyama, V. Y., J. Pritchard, and R. V. Dowell
1984. Laboratory toxicity of pesticides to *Geocoris pallens* (Hemiptera: Lygaeidae), a predator in California cotton. Journal of Economic Entomology, 77: 10-15.

Yonke, T. R.
1972. A new genus and two new species of neotropical Chariesterini (Hemiptera: Coreidae). Proceedings of the Entomological Society of Washington, 74: 283-287.

Yonke, T. R. and J. T. Medler
1967. Observations on some Rhopalidae (Hemiptera). Proceedings of the North Central Branch, Entomological Society of America, 22: 74-75.
1969. Description of immature stages of Coreidae. 1. *Euthochtha galeator*. Annals of the Entomological Society of America, 62: 469-473.
1969. Description of immature stages of Coreidae. 2. *Acanthocephala terminalis*. Annals of the Entomological Society of America, 62: 474-476.
1969. Description of immature stages of Coreidae. 3. *Archimerus alternatus*. Annals of the Entomological Society of America, 62: 477-480.
1970. New records of parasites from *Zelus exsanguis* and *Pselliopus cinctus*. Journal of the Kansas Entomological Society, 43: 441-443.

Yonke, T. R. and D. L. Walker
1970. Field history, parasites and biology of *Harmostes reflexulus* (Say) (Hemiptera: Rhopalidae). Journal of the Kansas Entomological Society, 43: 444-450.
1970. Description of the egg and nymphs of *Harmostes reflexulus* (Hemiptera: Rhopalidae). Annals of the Entomological Society of America, 63: 1749-1754).

York, G. T.
1944. Food studies of *Geocoris* spp., predators of the beet leafhopper. Journal of Economic Entomology, 37: 25-29.

Young, E. C.
1966. Observations on migration in Corixidae (Hemiptera: Heteroptera) in southern England. Entomologist's Monthly Magazine, 101: 217-229 (1965).

Young, W. R.
1970. Sorghum insects. *In* Wall, J. S. and W. M. Ross (editors). Sorghum production and utilization. Avi Publishing Co., Westport. 702 pages.

Yuasa, H.
1929. An ecological note on *Speovelia maritima* Esaki. Annals and Magazine of Natural History, 10: 346-349.

Zetterstedt, J. W.
1818-1824. Några nya Svenska Insect-arter fundne och beskrifne. Kongliga Svenska Vetenskaps-Akademiens Handlingar. 1818: 249-262; 1819: 69-86; 1824: 149-159.
1838-1840. Insecta Lapponica descripta. Sumtibus Leopoldi Voss, Lipsiae. 1139 pages. [Appeared in 6 parts; Heteroptera, 1838: 253-314].

Zheng, L.-Y. and J. A. Slater
1984. A revision of the lygaeid genus *Pseudopachybrachius* (Hemiptera). Systematic Entomology, 9: 95-115.

Zimmer, J. T.
1910. Two new species of Pentatomidae from Nebraska. Canadian Entomologist, 42: 166-167.
1912. The Pentatomidae of Nebraska. University of Nebraska, Contributions from the Department of Entomology, 4: 1-33 (1911).

Zimmerman, E. C.
1948. Insects of Hawaii. Volume 3. Heteroptera. University of Hawaii Press, Honolulu. 255 pages.

COMPREHENSIVE INDEX

Note: Currently used names in Roman; synonyms, misidentifications, and obsolete combinations in italics; currently used names above species in bold. Page numbers for principal entries of currently used names bold when multiple pages occur.